PLANETS, STARS AND NEBULAE

studied with photopolarimetry

(Courtesy J. S. Hall, Lowell Observatory)

Merope Nebula and directions of electric vector maximum; largest polarizations are about 12 percent. North is at top, east is left.

PLANETS, STARS AND NEBULAE

studied with photopolarimetry

T. GEHRELS, *editor*

The University of Arizona Press

Tucson, Arizona

THE UNIVERSITY OF ARIZONA PRESS

Copyright © 1974
The Arizona Board of Regents
All Rights Reserved
Manufactured in the U.S.A.

I.S.B.N. 8165-0428-8
L. C. No. 73-86446

TO JEAN R. STREETER

CONTENTS

FOREWORD   xiii
John S. Hall
PREFACE   xv
T. Gehrels

Part I — GENERAL THEORY AND TECHNIQUES

INTRODUCTION AND OVERVIEW   3
T. Gehrels

POLARIMETRIC DEFINITIONS   45
D. Clarke

MECHANISMS THAT PRODUCE LINEAR AND CIRCULAR POLARIZATION   54
J. R. P. Angel

ELECTROMAGNETIC RADIATION AND DISSYMMETRY   64
Dennis J. Caldwell and Henry Eyring

COHERENCE AND ITS APPLICATION IN THE BEAM-FOIL LIGHT SOURCE   88
C. H. Liu and S. Bashkin

SOME EXAMPLES OF EXACT AND APPROXIMATE SOLUTIONS IN SMALL PARTICLE SCATTERING: A PROGRESS REPORT   107
J. Mayo Greenberg

POLARIMETERS FOR OPTICAL ASTRONOMY   135
K. Serkowski

APPLICATION OF MUELLER CALCULUS IN ASTRONOMICAL POLARIMETRY: ACHROMATIC MODULATORS AND POLARIZATION CONVERTERS, AND DEPOLARIZERS   175
J. Tinbergen

OPTICAL POLARIMETERS IN SPACE   189
David L. Coffeen

PHOTOGRAPHIC POLARIZATION MEASUREMENTS OF VENUS   218
Reta Beebe

SPATIAL DISTRIBUTION OF POLARIZATION OVER THE
 DISKS OF VENUS, JUPITER, SATURN, AND THE MOON 223
 John W. Fountain

DIFFRACTION GRATING POLARIZATION 232
 James B. Breckinridge

ATTEMPT TO MEASURE STELLAR MAGNETIC FIELDS
 USING A LOW-LIGHT LEVEL TELEVISION CAMERA 237
 *G. G. Fahlman, J. W. Glaspey, O. Jensen, G. A. H. Walker,
 and J. R. Auman*

THE HIGH ALTITUDE OBSERVATORY STOKES
 POLARIMETER 246
 T. Baur, G. W. Curtis, H. Hull, and J. Rush

THE HIGH ALTITUDE OBSERVATORY CORONAL-
 EMISSION-LINE POLARIMETER 254
 C. W. Querfeld

STELLAR AND SOLAR X-RAY POLARIMETRY 262
 R. Novick

THE BERKELEY INFRARED POLARIZATION SURVEY 318
 Robert Landau

ASTRONOMICAL POLARIMETRY IN THE FAR INFRARED
 (10–1000 μm) 322
 *G. Dall'Oglio, B. Melchiorri, F. Melchiorri, V. Natale,
 S. Aiello, and F. Mencaraglia*

RADIO MEASUREMENTS OF POLARIZATION 352
 R. G. Conway

RADAR POLARIMETRY 359
 G. Leonard Tyler

Part II — SURFACES AND MOLECULES

POLARIZATION EFFECTS IN THE OBSERVATION
 OF ARTIFICIAL SATELLITES 371
 Kenneth E. Kissell

POLARIZATIONS OF ASTEROIDS AND SATELLITES 381
 Edward Bowell and Ben Zellner

POLARIZATION IN A MINERAL ABSORPTION BAND 405
 Carle E. Pieters

MIE SCATTERING OF THE INTERPLANETARY MAGNETIC
 FIELD BY THE WHOLE MOON 419
 C. P. Sonett and D. S. Colburn

CONTENTS

HEURISTIC ARGUMENTS FOR THE PATTERN OF
   POLARIZATION IN DEEP OCEAN WATER   434
John E. Tyler

THE POLARIZATION OF LIGHT IN THE ENVIRONMENT   444
K. L. Coulson

POLARIMETERS IN ANIMALS   472
Talbot H. Waterman

THE CIRCULAR POLARIZATION OF LIGHT REFLECTED
   FROM CERTAIN OPTICALLY ACTIVE SURFACES   495
Ramon D. Wolstencroft

POLARIMETRIC INVESTIGATIONS OF THE TURBIDITY
   OF THE ATMOSPHERE OVER LOS ANGELES   500
T. Takashima, H. S. Chen, and C. R. Nagaraja Rao

MEASUREMENTS OF THE ELLIPTICAL POLARIZATION
   OF SKY RADIATION: PRELIMINARY RESULTS   510
Dieter Hannemann and Ehrhard Raschke

NOCTILUCENT CLOUDS   514
G. Witt

POLARIZATION STUDIES OF PLANETARY ATMOSPHERES   518
D. L. Coffeen and J. E. Hansen

INVARIANT IMBEDDING AND CHANDRASEKHAR'S
   PLANETARY PROBLEM OF POLARIZED LIGHT   582
S. Ueno, S. Mukai, and A. P. Wang

POLARIZATION MEASUREMENTS OF JUPITER AND
   THE GREAT RED SPOT   593
John S. Hall and Louise A. Riley

POLARIZATION INVESTIGATIONS OF THE PLANETS
   CARRIED OUT AT THE MAIN ASTRONOMICAL
   OBSERVATORY OF THE UKRAINIAN ACADEMY
   OF SCIENCES   599
O. I. Bugaenko, A. V. Morozhenko, and E. G. Yanovitskii

CIRCULAR POLARIZATION OF PLANETS   607
James C. Kemp

POLARIZATION IN ASTRONOMICAL SPECTRA:
   THEORETICAL EVIDENCE   617
A. L. Fymat

ASTRONOMICAL FOURIER SPECTROPOLARIMETRY   637
F. F. Forbes and A. L. Fymat

Part III — STARS AND NEBULAE

POLARIZATION FROM ILLUMINATED NONGRAY STELLAR ATMOSPHERES *George W. Collins, II, and Paul F. Buerger*	663
OBSERVATIONAL ASPECTS OF COHERENCE IN RADIO POLARIZATION MEASUREMENTS OF AREA SOURCES *G. Feix*	676
POLARIZATION MEASUREMENTS ON THE SUN'S DISK *Donald L. Mickey and Frank Q. Orrall*	686
THE FRENCH SOLAR PHOTOELECTRIC POLARIMETER AND ITS APPLICATIONS FOR SOLAR OBSERVATIONS *Audouin Dollfus*	695
POLARIZATION OF SOLAR EMISSION LINES *E. Tandberg-Hanssen*	730
POLARIZATION MEASUREMENTS WITHIN STELLAR LINE PROFILES *D. Clarke and I. S. McLean*	752
K–CORONA AND SKYLIGHT INFRARED POLARIMETRY *J.-L. Leroy and G. Ratier*	762
EXTRATERRESTRIAL POLARIZATION OF THE ZODIACAL LIGHT: ROCKET MEASUREMENTS AND THE HELIOS PROJECT *C. Leinert, H. Link, and E. Pitz*	766
MULTICOLOR POLARIMETRY OF THE NIGHT SKY *Ramon D. Wolstencroft and John C. Brandt*	768
POLARIZATION OF THE ZODIACAL LIGHT *J. L. Weinberg*	781
THE LINEAR POLARIZATION OF THE COUNTERGLOW REGION *F. E. Roach, B. Carroll, L. H. Aller, and J. R. Roach*	794
ZODIACAL LIGHT MODELS BASED ON NONSPHERICAL PARTICLES *Richard H. Giese and Reiner Zerull*	804
COMETARY POLARIZATIONS *D. Clarke*	814
COMET BENNETT 1970 II *L. R. Doose and D. L. Coffeen*	818
POLARIMETRY OF LATE-TYPE STARS *Stephen J. Shawl*	821

CONTENTS xi

CIRCUMSTELLAR POLARIZATIONS OF EARLY-TYPE STARS
   AND ECLIPSING BINARY SYSTEMS   845
Andrzej Kruszewski

ASTRONOMICAL POLARIMETRY FROM 1 TO 10 μm   858
H. M. Dyck

POLARIZATION STUDIES OF REFLECTION NEBULAE   867
Ben Zellner

ORION NEBULA POLARIZATION   881
Richard Hall

POLARIZATION BY INTERSTELLAR GRAINS   888
George V. Coyne

MICROWAVE ANALOGUE STUDIES   901
Reiner Zerull and Richard H. Giese

EFFECTS OF PARTICLE SHAPE ON THE SHAPE OF
   EXTINCTION AND POLARIZATION BANDS IN GRAINS   916
J. Mayo Greenberg and Seung Soo Hong

INTERSTELLAR CIRCULAR POLARIZATION: A NEW
   APPROACH TO THE STUDY OF INTERSTELLAR GRAINS   926
P. G. Martin

INTERSTELLAR CIRCULAR POLARIZATION OF
   EARLY-TYPE STARS   939
James C. Kemp and Ramon D. Wolstencroft

POLARIZATION OF STARS IN ORION AND OTHER
   YOUNG REGIONS   946
Michel Breger

INTERSTELLAR DUST IN DARK CLOUDS   954
L. Carrasco, K. M. Strom, and S. E. Strom

INTERSTELLAR POLARIZATION AND THE GALACTIC
   MAGNETIC FIELD   960
G. L. Verschuur

THE ORIENTATION OF THE LOCAL INTERSTELLAR DARK
   CLOUDS WITH RESPECT TO THE GALACTIC
   MAGNETIC FIELD   972
W. Schlosser and Th. Schmidt-Kaler

STARLIGHT POLARIZATION BETWEEN BOTH
   MAGELLANIC CLOUDS   976
Thomas Schmidt

OBSERVATIONS OF MAGNETIC CIRCULAR POLARIZATION
   OUTSIDE THE SOLAR SYSTEM   981
J. D. Landstreet

INTRINSIC POLARIZATION AND THE TRANSVERSE ZEEMAN
    EFFECT IN MAGNETIC Ap STARS   988
    James C. Kemp and Ramon D. Wolstencroft

COMPUTATION OF STRONG MAGNETIC FIELDS IN
    WHITE DWARFS   992
    R. F. O'Connell

POLARIZATION OF PULSAR RADIATION   997
    W. J. Cocke

POLARIZATION AND STRUCTURE OF THE CRAB NEBULA   1014
    James E. Felten

RADIO POLARIMETRIC OBSERVATIONS OF SUPERNOVA
    REMNANTS   1029
    D. K. Milne and John R. Dickel

POLARIZATION OF EXTRAGALACTIC RADIO SOURCES   1030
    A. G. Pacholczyk

EXTRAGALACTIC OPTICAL POLARIMETRY   1059
    Natarajan Visvanathan

GLOSSARY   1084

LIST OF CONTRIBUTORS   1087

INDEX   1091
    Mildred Shapley Matthews

FOREWORD

During my lifetime the number of astronomers engaged in photoelectric photometry has risen from half a dozen to well over a hundred.

The early observations were limited to eclipsing variables and to the measurement of the colors of stars. Now it is difficult to think of any class of celestial object, other than meteors, that has not been measured photoelectrically. Wavelengths used extend from the far ultraviolet to the near infrared. Since about 1950, radio astronomers have made rapid and sensational progress by measuring energy at the radio wavelengths.

Within the past ten years many astronomers, both optical and radio, have devoted more and more attention to the measurement of the polarization of a wide variety of objects ranging from the moon to extremely faint galaxies. They have used many techniques, an enormous spectral range, and have achieved results the importance of which is emphasized by the contributions in this book.

JOHN S. HALL

PREFACE

This book comes at a time of greatly increasing interest in polarimetry spurred by discoveries of elliptical polarization, by the combination of photometry and polarimetry, by the introduction of new computers and instruments, including those on spacecraft, and by the growing realization of the usefulness of these techniques when supported by well-developed interpretive theories. The purpose of this volume is to cover all these developments and to be a source and reference book where none existed before.

In order to bring about such a book, we developed a new scheme (tried previously on "Physical Studies of Minor Planets," NASA SP-267, 1971). About a dozen reviewers were invited, long in advance, so that major topics in this wide field would be properly covered. Next, a colloquium was arranged in order to complete the coverage and to ensure that everyone involved in polarimetry could have a chance to contribute. After the colloquium, several papers and discussions were added to the book, as were the Introduction and Overview and, of course, the Index. Much editing and refereeing were needed to improve some of the papers and to make a cohesive whole of all of them.

The colloquium was useful not only for this book, but also for bringing together some 120 workers in the various fields of polarimetry for lively interchanges in an informal atmosphere. As Colloquium 23 of the International Astronomical Union, it was held in Tucson, 15–17 November 1972, in the Optical Sciences Building of the University of Arizona. The Organizing Committee consisted of Audouin Dollfus, Aina Elvius, Tom Gehrels, Mayo Greenberg, Andrzej Kruszewski, and Nicolai Shakhovskoj, while the Local Committee included David Coffeen, George Coyne, Mel Dyck, Fred Forbes, Tom Gehrels, and Krzysztof Serkowski. There was an opening word by Executive Vice President A. B. Weaver, while the chairmen were Roger Angel, Alex Goetz, Don Rea, Charles Sonett, Stephen Strom, and Ray Weymann. The sessions were held in the mornings and evenings, with afternoons left free for informal discussion after the luncheons, at which the speakers were Vice President A. Richard Kassander on the origin of the Salton Sea, Don Rea on Mariner 9, and Ray Weymann on the multi-mirror telescope. On the day before the colloquium, there was a hike to Seven Falls in the Santa Catalina foothills, and in the afternoon and evening we made an excursion to the telescopes in the Catalinas. On the day following the colloquium we visited the Kitt Peak National Observatory.

Several acknowledgments may be made for this cooperative effort. Much of the research described in this book was supported by private, industrial, or governmental funding agencies, and we dedicate this book to our colleagues who serve in agencies as monitoring officers. We thank them for their contribution to the research which requires hard work, especially in times of tight budgets. To represent these officers for the book's dedication, we singled out Jean R. Streeter (ONR, 1950–1970) because of the friendly professional manner in which she conducted the Astronomy Program at the Office of Naval Research, a pioneering program that helped the start of photoelectric polarimetry.

We thank the University of Arizona Press for their pleasant cooperation, particularly through the editorial efforts of Mary A. Davis. We thank the National Science Foundation and the International Astronomical Union for their support of the colloquium. The Polaroid Corporation and the National Science Foundation made it possible for the price of this book to be reduced to a minimum. We thank the referees and the contributors listed on pp. 1087–1089. I am especially grateful for the dedicated assistance of Mildred Shapley Matthews.

TOM GEHRELS

PART I

General Theory and Techniques

INTRODUCTION AND OVERVIEW

T. GEHRELS
University of Arizona

> *This overview is written for readers who already are familiar with polarimetry, and its associated photometry, in order to provide cross references among the 74 papers in this book. Special attention is paid to suggestions for future work. For readers who are new in the field, some introductory remarks are made particularly in the subsection Basic Concepts.*

The book is divided into three parts: Part I gives general theories and techniques that are applicable to more than one specific problem. This part also has some papers concerning the basic nature of matter and light. Part II contains the papers that deal primarily with the earth and other planets. Part III is for the stars and for the particles near and between the stars. There might have been a Part IV for pulsars, quasars, nova remnants, and extragalactic nebulae, but this would have been such a small section that it is placed at the end of Part III.

We treat the sun as a star, the zodiacal cloud as a reflection nebula, and the sky as a planetary atmosphere. In the future we will, no doubt, move more toward that direction, away from the trend in the past when stellar astronomers would largely ignore papers on the sun. The techniques in stellar polarimetry have improved so drastically in the past few years that spectral polarimetry will soon be done with high resolution and the comparison with solar work will become more fruitful.

Now may be the last chance to produce a single volume containing all polarimetry; the field is expanding so rapidly. It seemed an impossible task for one man to write a suitable book because of the great variety of objects and the wide range of applications; even if one author could have produced a good book, it would certainly have removed him from active research for several years. We therefore put into effect the scheme of production described in the Preface.[1] Clearly, this book is not the proceedings of a conference. The effective date of the material presented here is March 1973. Our scheme for producing a source book has the

---

[1] See p. xv.

disadvantage that the volume is more disjointed than if it had been written by one author. More than one topic may be discussed in a certain paper. These disadvantages I shall try to minimize with this introduction. For example, the paper by Novick[2] is a general review and it belongs in Part I, but I shall remind you of its discussion of the sun for Part III.

I shall point out a few gaps that have not been treated in this volume. I make some introductory remarks, especially in the section Basic Concepts,[3] for the reader new to the field. This introduction, however, has gaps of its own and it does not give due credit especially to the longer review papers; these are so complete that we need only to list them in the Table of Contents and mention them by page number in this introduction. The Index should also be useful in tying together the various parts of this book.

PREVIOUS LITERATURE

A general text and source book does not exist. In the early 1950s, Öhman compiled a set of classnotes resulting from his lectures at the University of Colorado, but this has been available only in limited circulation. The Lyot (1929) thesis still stands as an authoritative summary of polarization observed in visible light on the planets and on laboratory samples. There was a conference in 1960 (Hall 1960) on interstellar polarization, but other topics were also discussed and the suggestions for future work now are of historical interest. The paperback by Shurcliff and Ballard (1964) and the book of Shurcliff (1966) treat the detection of polarization and the usage of Mueller and Jones calculus. The books of Feofilov (1959) and of Clarke and Grainger (1971) deal with the physics of polarized light. A brief semipopular overview of astronomical polarimetry was made by Gehrels (1973).

Historical notes and references are given at the start of many of the papers in this book. It is surprising to note that the wavelength dependence of circular polarization was studied as early as 1911, although this was in a very special case, namely of the effects observed on insects.[4]

Now follows a chronology, compiled with the help of some of our authors, for the principal events in polarimetry. The last entry is for 1970, after which followed an outburst of papers; that activity, and its future, is what the present book is all about. References can be found in Shurcliff (1966), from which much of this chronology was taken, while especially for the modern references various papers in this book may be consulted.

---

[2] See p. 262.   [3] See p. 7.   [4] See p. 495.

1669 Erasmus Bartholinus, in Denmark, discovered double refraction on Iceland spar, although he was not aware of the polarization.

1690 Christiaan Huygens made a descriptive theory of double refraction on Iceland spar and discovered the intrinsic difference between the doubly refractive beams.

1808 Etienne-Louis Malus discovered the polarizations of light reflected by glass and metallic surfaces, and the proportionality of the intensity to the square of the cosine of the angle (Malus' law). He was looking through a calcite crystal — because of a prize offered by the French Academy for a mathematical theory of double refraction — at the light reflected obliquely from a window of the Luxembourg Palace, in Paris, and observed that the two images produced by the calcite were extinguished alternately when he rotated the crystal.

1809 D. F. J. Arago found that the light from the sunlit sky is partially polarized. He established the fact that the polarization maximum in the sky is located at about 90° from the sun and discovered the existence of a point of zero polarization (the neutral point of Arago) at a position 20°–25° above the antisolar direction. Arago also discovered the rotary polarization of quartz, observed polarization of two comets and of the moon, and found larger polarization on the maria than on the highlands.

1812 Sir David Brewster discovered the relationship between refractive index and the polarizing angle (Brewster's law). Later, he found the neutral point of skylight polarization located 20°–25° below the sun (the Brewster point).

1815 Jean B. Biot found the rotary polarization in liquids and described dichroism of tourmaline.

1816 Augustin Fresnel, taking up Thomas Young's notion that light vibrations are transverse, gave the first theoretical interpretation of Malus' observations. He also derived, using a mechanical model, the formulae for the reflection coefficients of dielectric materials according to the angle of incidence and the direction of vibration of the waves (Fresnel's laws).

1844 Wilhelm K. Haidinger discovered a phenomenon by which the eye can recognize strong polarization (Haidinger's brush). Later he also discovered circular dichroism in crystals of amethyst quartz.

1845 Michael Faraday discovered the rotation of the plane in a beam of linearly polarized light traversing certain media parallel to magnetic field lines.

1848 Louis Pasteur described hemihedral crystals that are optically active.
1852 W. B. Herapath, an English physician, discovered that a drug combination of iodine and quinine gave crystals with optical properties similar to those of tourmaline.
1852 Sir George G. Stokes described the four Stokes parameters.
1858 E. Liais, a French astronomer observing a solar eclipse in Brazil, discovered that the light of the corona is partially linearly polarized.
1869 D. Tyndall established the fact that the character of the polarization of light scattered by particles changes strongly with the dimensions of the particles, explaining earlier observations by G. Govi on smoke particles.
1871 Lord Rayleigh began his studies of the polarization of the sunlit blue sky.
1872 Lord Rosse found some polarization on Venus.
1875 John Kerr, a Scottish physicist, discovered the birefringence of electrified media (Kerr effect). Later he also discovered changes in metallic reflection of polarized light in the presence of magnetic fields (Kerr magneto-optic effect).
1896 Pieter Zeeman found that spectral lines can be broadened when the radiating atoms are in the presence of an intense magnetic field. Observation of the splitting of spectral lines by magnetic fields and their associated polarization (Zeeman effect) was observed a few years later.
1908 George Ellery Hale showed the existence of strong magnetic fields in sunspots by means of polarization measurements of the Zeeman effect.
1908 Gustav Mie, and Pieter Debye independently, developed the theory of light scattering by spherical particles of arbitrary size.
1928 Edwin H. Land, then a 19-year-old student at Harvard, invented the first successful sheet-type dichroic polarizer.
1929 Bernard Lyot published his thesis containing major discoveries on the linear polarization of light from planets and terrestrial substances. This work, also for the sun, is continued by Audouin Dollfus.
1942 Yngve Öhman found polarization in galaxy M 31.
1943 Hans Mueller developed a phenomenological approach to problems involving wide-band partially polarized light, using 4×4 matrices.
1947 H. W. Babcock discovered that peculiar A-type stars have strong magnetic fields.

INTRODUCTION

1949   J. S. Hall and W. A. Hiltner discovered linear interstellar polarization, and they subsequently published extensive catalogs of polarized starlight.

1950   S. Chandrasekhar published the solution of radiative transfer in a sunlit plane-parallel planetary atmosphere with Rayleigh scattering. The solutions were later tabulated in convenient form by K. L. Coulson, J. V. Dave, and Z. Sekera.

1954   V. A. Dombrovskij discovered large polarization in the Crab Nebula predicted with synchrotron radiation a year before by I. S. Shklovskij.

1957   H. C. van de Hulst published his fundamental text on light scattering by small particles.

1957   C. H. Mayer, T. P. McCullough, and R. M. Sloanaker detected the radio polarization of the Crab Nebula. They also were the first to measure the polarization of an extragalactic radio source, in 1962.

1959   Strong wavelength dependence was found for the linear polarization of planets, stars, and nebulae. The results are published by several authors in a series of papers in the *Astronomical Journal*.

1962   N. M. Shakhovskoj found variable polarization in early-type eclipsing binary β Lyr.

1966   Krzysztof Serkowski found strong intrinsic polarization for Mira stars.

1970   James C. Kemp discovered circular polarization on a white-dwarf star. The discovery stimulated a variety of findings of circular polarizations on planets, stars, and nebulae.

CONCEPTS AND TERMINOLOGY

Basic Concepts

A critical discussion of our most basic term *polarimetry* is made by Clarke[5] immediately following this introduction. The word *photopolarimetry,* in the title and used throughout the book, is to convey the conviction that the observations and analysis of photometry and polarimetry should be combined. Polarimetry evolved from photometry as explained in the Foreword.[6] The techniques of photometry and polarimetry are similar and so are the methods of interpretation.

In most astronomical problems, a large number of parameters must be solved for. The geometry and aspect, of a reflecting object with respect to the light source and the observer, may be known, or they may have to be assumed as a part of the model of interpretation. As a further com-

---

[5] See p. 43.   [6] See p. xiii.

plexity, the apparent geometry may be dependent on wavelength, as we see in the example of a planetary atmosphere where the penetration into the atmosphere depends on the wavelength of the incident and emerging light. Next there are the refractive index, with its real and imaginary parts, and the number of scatterers or the optical depth of an atmosphere, and these also are a function of the wavelength. Distributions of the sizes and refractive indices must be taken into account. The particles may have various shapes and they may be preferentially aligned.

In order to solve for the several parameters that are unknown, many independent parameters or phenomena must be observed, and here exist great possibilities not yet fully explored. Because the objects are so remote, a solution of an astronomical problem requires that observation of the light emitted or reflected by the body be made in the fullest detail. The results of radio astronomy, spectroscopy, astrometry, radial velocity studies, direct photography, and visual observations should be added to those of photometry and polarimetry. In order to increase the number of parameters of the observed data, the observations should be made over the largest possible range of phases and wavelengths. New possibilities are opening up for the extension of the phase- and wavelength-ranges through the use of spacecraft. Some of the techniques used in radio astronomy should be of interest to optical observers as well.

In this book, we have tried to include the results of photometry where it relates to polarimetry. This has increased the size of the book, but it is noted that, although a fair degree of completeness has been reached in polarimetry, this certainly was not attempted for photometry.

The wave model of electromagnetic radiation, whereby the stellar radiation consists of an ensemble of wave trains, is applicable for photopolarimetry. Each of the emitted wave trains lasts about 10^{-8} secs, is about 1 meter long, and is polarized linearly or circularly. When the time average shows no polarization it is called *natural light*. The concept of the microscopic lightsources leads to the concept of the superposition, and decomposition, of streams of light of various polarization states. They are analyzed with *Stokes parameters* and *Mueller matrices*. Apparently, the great paper of Stokes (1852) had been overlooked until it was reintroduced by Mueller and by Chandrasekhar in the 1940s. There are several treatments of the Stokes parameters and their matrix presentation in the book;[7] special mention should be made of the case where the analyzer is used at relative position 0° and 45°, as in the paper by Hannemann and Raschke.[8] The two most useful matrices, and perhaps the only ones needed for the design and description of astronomical

---

[7] See pp. 175, 360, 531, and 732.   [8] See p. 510.

polarimeters, are given by Serkowski.[9] The graphical presentation with the Poincaré sphere is carefully described by Shurcliff (1966), but it seems to be used very little in astronomy and in this book only by Dollfus.[10]

When there is a predominance of a certain orientation in which the vibrations occur, we speak of *partial linear polarization*. Similarly, when either the right-hand or left-hand circular polarization predominates, we speak of *partial circular polarization* of the emerging light. These polarizations usually are expressed in fractions, percentages, or in astronomical magnitudes. The astronomical magnitude of polarization is approximately 2.17 times the degree of polarization (see Behr 1959b) but the nonlinearity is more than 10^{-4} for 0.07 polarization, and the magnitudes should therefore not be used in modern work.

Generally, the light emerging from a stellar photosphere is natural light to within a high degree of precision. On isolated regions of the sun linear polarizations as large as 4×10^{-4} are observed, but regions have various position angles and on the average the resulting polarization of sunlight may be less than 10^{-8}. Frequently, the sun and stars are merely used as lightsources. In the measurements of interstellar polarization it is not necessary to observe only one star. A double star, a group of stars, or even a cluster may be enclosed in the focal-plane diaphragm.

Observable polarizations occur after scattering, and we generally consider only the cases in which the inhomogeneities scatter *independently*. Van de Hulst (1957) has stated that a mutual distance of three times the radius of the scattering particles is sufficient; this condition should be verified. If the positions of the individual inhomogeneities vary sufficiently, so that the vibrations are not in phase, the assembly is said to scatter *incoherently*. If there is only one particle, this is called single scattering. The general term is *extinction*, which is the sum of the absorption and of the scattering without loss of energy.

The *plane of scattering* is the plane through the lightsource, the scatterer, and the observer. In French publications it is sometimes called the plane of vision, while some authors speak of the principal plane. In that plane, subtended at the scatterer, we have the *phase angle, α*, and its supplement, the *scattering angle* Θ. For solar-system objects viewed from the earth, the phase angle usually has a minus sign before opposition, that is 0° phase, and a plus sign afterwards. This usage of the word "phase" is not to be confused with the *position angle* of polarization, which is the orientation of the electric-vector maximum with respect to the north direction, increasing through east, etc. The *plane of polarization* is

---

[9] See p. 145.   [10] See p. 695.

understood to be the plane that contains the directions of propagation and of the maximum of the electric vector. This definition is, however, not yet followed by everyone (see Clarke[11]).

When a perfect analyzer is rotated in the plane perpendicular to the direction of propagation of the light, the maximum intensity, I_{max}, is observed when the axis of the analyzer is in the plane of the predominant vibration, that is, parallel to the electric-vector maximum. When the analyzer is rotated over an angle χ with respect to the electric-vector maximum, the intensity is found to follow the relation $I(\chi) = I_{max} \cos^2 \chi$. This is *Malus' law,* and it is a direct consequence of the proportionality of the intensity of the lightwave to the square of its amplitude. Malus' law allows us to recognize two physical concepts. Natural light has no directional vector; it is seen through the axis of an analyzer as the time average of Malus' law, and $<\cos^2\chi> = \frac{1}{2}$. Secondly, $\cos^2 \chi = (\cos 2\chi + 1)/2$, i.e., the polarized component of the light shows a double cosine curve.

A systematic discussion of scattering domains is made by Coffeen and Hansen.[12] When a scattering element is large compared to the wavelength of light, *geometrical optics* can be used to fit parameters of the theory to those of the observations. In the other extreme, where the particle is appreciably smaller than the wavelength of the incident light, the electromagnetic fields are uniform within the particle and it will oscillate as a simple dipole; in this case we speak of *Rayleigh scattering* which has a strong wavelength dependence. Particle sizes intermediate to those described by geometrical optics and Rayleigh scattering are treated with the *Mie theory* for spherical grains, the van de Hulst theory for elongated grains, Greenberg's program[13] for ellipsoids, etc. The patterns of intensities and polarization as a function of phase angle usually are quite irregular. In other words, when the particle size is near that of the wavelength of the incident light, the dependencies on α and λ are complex, and strong effects occur when α and λ are changed. These dependencies are a function of the size of the particle, the shape, the composition, and the texture. In analysis the inverse problem is posed, namely from the observations as a function of α and λ, usually giving startling effects, the nature of the scattering material is studied.

Remarks on the Theoretical Papers

Three basic methods are used in order to gather information from the polarization and brightness measurements.

---

[11] See p. 45.   [12] See p. 518.   [13] See p. 916.

1. The most fundamental is to make comparisons with the scattering theories like the Rayleigh-Chandrasekhar theory of multiple molecular scattering and the Mie theory of scattering by small particles. However, a decreasing order of applicability is seen. A rigorous solution may be made in the case of multiple scattering in a clear molecular atmosphere overlying a ground that reflects in a prescribed manner. Separately, also the problem of single scattering on a spherical particle of known size and refractive index has a rigorous solution. The study of interstellar grains is already further complicated because the sizes and refractive indices have distributions, and there are effects of aspect and particle shape to be considered. A molecular atmosphere with aerosols and "ground" effects is highly complex; the cases of the Venus and Jupiter atmospheres may be the ultimate challenge in our work, especially because of nonuniformities in the cloud deck.[14]
2. A second method that is applicable even when one cannot apply rigorous theory is to compare one observed object with another. For instance, if it is true that circular polarization observed on the planets has limited usefulness[15] in theoretical analysis, there still will be great merit in comparing various planets and observed hemispheres.
3. Laboratory results on appropriate samples can help in the understanding of phenomena observed at the telescope. Examples of such work for the study of surfaces of the moon and the asteroids are given by Pieters[16] and by Bowell and Zellner.[17]

A first introduction to some of the polarization mechanisms is made in the paper by Dall'Oglio et al.[18] The theoretical foundations for all the work in this book are laid in the review paper of Angel.[19] The year 1972 was one of exciting discoveries of circular polarizations, and this is reflected in the emphasis that Angel makes on circular polarization,[20] somewhat out of proportion to the larger effects of linear polarization encountered everywhere in the universe. The papers by Caldwell and Eyring,[21] Liu and Bashkin,[22] and Feix[23] also deal with the fundamental nature of light and coherence. The topic of angular momentum of circularly polarized light is referred to a few times.[24] The paper of Pacholczyk[25] appears at the end of this book, but it could have been put in the introductory chapters that give basic theory, because of its discussion of the propagation of polarized light within a plasma. Optical astronomers have

---

[14] See p. 560.
[15] See p. 616.
[16] See p. 405.
[17] See p. 381.
[18] See p. 322; also see p. 1059.
[19] See p. 54.
[20] Also see p. 984.
[21] See p. 64.
[22] See p. 88.
[23] See p. 676.
[24] See pp. 53 and 433.
[25] See p. 1030.

much to learn from chemists Caldwell and Eyring[26] who review the fundamentals of optical activity. The book of Feofilov (1959) deals with polarizations due to resonance, luminescence and fluorescence.

For the scattering by molecules in planetary atmospheres, including the effects of aerosols and surface reflections, the reader is referred to the phenomenological treatment by J. E. Tyler,[27] the introductory theoretical sections of Coulson[28] and the discussion of the theory and applications for planetary atmospheres by Coffeen and Hansen.[29] A specialized paper on the Rayleigh-Chandrasekhar theory of multiple molecular scattering was written by Ueno et al.[30]

Much of the theory on scattering by small particles is in the paper by Greenberg.[31] A curious application of the Mie theory is given for the case of the "particle" being the moon (see p. 430)! For the scattering by the surface of the moon the *Umov effect* is sometimes quoted,[32] which states that there is a reciprocity between the brightness and the amount of polarization.

While the Mie theory is for spherical particles and the theory of van de Hulst (1957, Ch. 15) for elongated infinite cylinders, the intermediate shapes of spheroids and truncated cylinders are treated by Greenberg[33] and by Greenberg and Hong.[34] Such intermediate cases of particle shapes are usually studied in the laboratory by a microwave analogue technique which is described by Zerull and Giese[35] and also by Greenberg,[36] one of the pioneers in this field. With the success of that work, Greenberg became an astronomer, much to our gain.

Terminology Adopted for This Book

One editorial task we struggled with conscientiously was to bring at least some unification in the conventions. It was a difficult problem because so many people, from a variety of disciplines, contributed to the book. The problem of positive/negative and right-handed versus left-handed circular polarizations is presented by Clarke,[37] and what was adopted is stated in the discussion following Clarke's paper. Here I believe an advance has been made and a documented recommendation to adopt that convention will be made to the International Astronomical Union and to all workers in this field. Note, however, that a definition of ellipticity has not been uniformly adopted in this book.

The adopted symbols are listed in the Glossary[38] and here again there

---

[26] See p. 64.
[27] See p. 434.
[28] See p. 444.
[29] See p. 518.
[30] See p. 582.
[31] See p. 107.
[32] See pp. 381, 405, and 518.
[33] See p. 107.
[34] See p. 916.
[35] See p. 901.
[36] See p. 126.
[37] See p. 45.
[38] See p. 1084.

were some successes and some failures. It seems rather elegant to recommend for further usage what we have used in the book: θ for position angle, Θ for scattering angle, and ϑ as used in spherical coordinate systems or, more specifically, the angle between the direction of observation of a point on a planet and the local vertical.

It may not be the best solution, but at least some consistency is reached to use p for the amount of linear polarization in an arbitrary frame (P is with respect to the plane of scattering) while the symbol q then follows for the amount of circular polarization. The usage of q was started by Kemp and he mentions that origin on page 608. However, the problem of what symbol to use for the total amount of elliptical polarization is still before us, although I have tentatively suggested the usage of s, for the *sum*.[39]

We switched to the metric system rigorously, following the adoption, also by NASA, in the previous book (Gehrels 1971). Astronomers still sentimentally refer to telescopes in inches, as for instance the 200-inch, but we will get used to calling it the 5-meter telescope; the European Southern Observatory, for instance, has it this way. On the other hand, we probably were too dogmatic in eliminating the word *micron;* the contribution of Meeus[40] arrived too late to restore the word in this book. Ångstroms were kept in certain cases because their elimination would have interfered too much with ingrained usage. For the future, one might prefer the simplification of using only nm and μm (calling it *micron*). Coffeen (personal communication) has suggested that we consider the word *microm*.

We tried to eliminate the term *percentage polarization,* in favor of fractions, especially now that the precisions generally are improving; 10^{-5}, for instance, makes more sense than $10^{-3}\%$. This in the future may indeed be the way to go, but for this book it turned out to be impractical because the percentage concept is still used so much. One can see the new usage for instance in the paper by Bowell and Zellner[41] who carefully avoided the use of the word percentage.

We would enrich our vocabulary by distinguishing the concept of *accuracy,* a measure of absolute certainty, from that of *precision,* a measure of observational consistency; this distinction has, however, not been enforced in this book.[42]

Clarke[43] objects to the term *negative polarization,* but I believe it is here to stay because there is a concept attached to it. The negative sign indicates that the electric-vector maximum lies in the plane of scattering

[39] See p. 52.   [41] See p. 381.   [43] See p. 47.
[40] See p. 53.   [42] See p. 211.

and it is indicative of (internal) scattering by particles, while for molecular scattering, and for external reflection, the sign is reversed. Similarly, Kemp[44] refers to a *solid sign* for the circular polarization when it is indicative of scattering by particles, while for molecular scattering the sign is reversed.

Our editorial efforts to reach unification in the conventions came to a halt for the papers by Pacholczyk[45] and Greenberg.[46] Their own set of symbols they clearly defined, and changing these would have caused major interference with their own books and disciplines. I hope the other authors can forgive us when symbols were changed in the manuscripts in order to reach at least some uniformity. We indeed tried to find the most acceptable set of conventions in the hope that this book may set a pattern for usage by all workers in the various fields of photopolarimetry.

OPTICAL TECHNIQUES

Techniques of Polarization Observations

If the human eye had been as sensitive to polarization as it is to contrast and color, the history of polarimetry would have developed in an entirely different manner and this first general source book would not have been published in 1973! On the other hand, it is remarkable how our eyes have evolved to just that region of the spectrum where the sun gives maximum emission and where the earth's atmosphere is transparent over a relatively large range of wavelengths, bounded by ozone absorption in the ultraviolet and by water-vapor absorptions in the infrared. The human eye has a slight sensitivity to polarization as it can perceive strong effects, for instance on the blue sky about 90° away from the sun, as a vague ellipsoidal figure called after the discoverer Haidinger's brush (described, for instance, by Minnaert 1954). What is not generally known is that there also is a sensitivity to elliptical polarization (Shurcliff 1955), and that the sensitivity of the eye to colors and polarization can be refreshed by turning the head over 90°. Mechanisms of eyes are discussed by Waterman,[47] and we find, in fact, that some animals have much more developed polarization-sensing eyes.

Only photoelectric polarimetry is good enough to give a precision of $\pm 10^{-3}$ or better. The exception occurs in visual polarimetry in the work of Lyot and Dollfus which has a precision on the order of $\pm 10^{-3}$. Visual polarimeters have not been mentioned in this book, other than in passing by Coffeen.[48] They have been adequately described elsewhere:

[44] See p. 610. [46] See p. 107. [48] See p. 189; also see p. 456.
[45] See p. 1030. [47] See p. 472.

Lyot's visual polarimeter in his thesis (Lyot 1929), Dollfus' in his thesis (Dollfus 1955), and the work has been summarized by Dollfus (1961).

For photographic polarimetry, two papers were submitted for inclusion in this book. There was some debate with referees because the precision in photographic work does not seem good enough to warrant consideration for a modern book. The possible argument that photographically one can cover extended areas efficiently does not seem to have much validity any longer because of the photoelectric area scanning, pioneered by Hall,[49] which, in the near future, may be done even better with multi-element sensors.[50] Nevertheless, there seemed to be some merit in showing just how difficult it is to overcome the disadvantages of photographic techniques, as evidenced in the paper by Beebe.[51] The Fountain[52] paper was accepted because the superposition technique, of positive and negative photographs, is interesting by itself. One of our referees urged that these papers be published because not everyone can afford to build sophisticated (electronic) gadgetry, but this does not apply to the Lyot polarimeter, which is commercially available for about $300, nor to Hiltner's system described below.

The instruments with the highest precision are those developed for the study of the sun as described by Dollfus,[53] Leroy and Ratier,[54] Mickey and Orrall,[55] Tandberg-Hanssen,[56] Baur et al.,[57] and Querfeld.[58] The heavy reliance on computers is noted in the case of the American instruments.[59] Clarke and McLean[60] also describe instrumentation, while the new polarimeters of Kemp and of Angel are described in some of the references that they give in their papers in this book. Serkowski[61] reviews the techniques in a systematic fashion, gives references to detailed descriptions, and describes his own new polarimeter, which also relies completely on the usage of a computer at the telescope. The high precision, on the order of 10^{-5}, that used to be attained only in solar work, now is reached also on stars and planets, and it is noted that this increase in precision has come about rather suddenly; the precision during the years 1922–1972 was typically 10^{-3}.

I see three stages in the evolution of polarimetric techniques, namely the Lyot and Polaroid, the Wollaston, and a hybrid stage. (These stages supply a generalization that may be forgiven, I hope, by owners of other instruments.)

1. During the years 1922–1957, there were few people in polarimetry, they worked alone, their instruments were simple and/or these men

---

[49] See p. 593.
[50] See p. 164.
[51] See p. 218.
[52] See p. 223.
[53] See p. 695.
[54] See p. 762.
[55] See p. 686.
[56] See p. 730.
[57] See p. 246.
[58] See p. 254.
[59] Also see p. 154.
[60] See p. 752.
[61] See p. 135.

were experts in obtaining the best performance from their instruments. Lyot's success is enthusiastically described by Coffeen and Hansen.[62] Hiltner tried a Wollaston prism, and he pointed out the basic merit of a two-beam polarimeter,[63] but his great catalogue (Hiltner 1956) was made with a Polaroid turned over 90° (for photometry it was simply removed), the photometer was rotated at the base of the telescope, and Barbara Perkins did all the reductions by hand.
2. The intermediate period, 1958–1972, was that of the Wollaston polarimeter, using both beams simultaneously and calibrating closely with a Lyot depolarizer. The great advantage of the Wollaston polarimeter was simple reliability, provided the wavelength band was wide enough so that the Lyot depolarizer could not introduce systematic errors. The polarimetry was straightforward, even for astronomers with little technical talent, and a precision on the order of $\pm 10^{-3}$ was achieved consistently on a wide variety of objects and wavelength dependencies. The instrumentation was, however, not efficient, and the precision as well as the wavelength resolution are now being surpassed.
3. Starting about now, hybrid instruments combine observations of brightness, linear and circular polarizations, and spectroscopy. Arrays of detectors such as in the Digicon are mentioned by Serkowski,[64] while a paper by Fahlman et al.[65] shows the possibilities of television imaging. The instrumentation has become complicated and teamwork of engineers and astronomers is required, especially in the interface with the computer.

Hiltner's method, mentioned above, still is quite applicable if the precision of $\pm 10^{-3}$ is sufficient. High precision is not always needed.[66] Sheet polarizers, commonly called Polaroids, are commercially available for the wavelength range 0.25–2.2 μm, although no single one will work over the entire range.

A few prisms are sketched in Fig. 1. Serkowski[67] makes a comparison of various polarimeters, with a display of their optical principles, and he discusses the Lyot depolarizer in detail together with other retarders (also see Tinbergen).[68] The reliability of the Lyot depolarizer has been doubted especially during the early stages of wavelength-dependence studies. Until 1959, observers paid little attention to the wavelength dependence of polarization, which is surprising because the Rayleigh-

---

[62] See p. 518, also see p. 381.
[63] See p. 18.
[64] See p. 136.
[65] See p. 237.
[66] See p. 209.
[67] See p. 149.
[68] See p. 175.

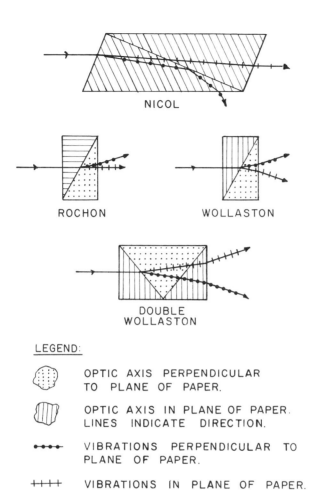

Fig. 1 Optical diagrams of a few prisms.

Chandrasekhar and Mie theories predict strong effects. The sunlit blue sky had been observed with filters ranging between about 0.4 and 0.8 μm and so had Mars (Dollfus 1955), but these were not considered striking results; they were not plotted as a function of wavelength. For the interstellar polarization, some contradictory results had been published. Hiltner (1950) emphasized that the interstellar polarization was in first approximation independent of wavelength; he had actually observed two stars at 0.42, 0.55 and 0.80 μm, one showing no wavelength dependence and the other somewhat smaller polarization at 0.80 μm than at shorter wavelengths. Hiltner also had an unpublished observation at 2μm, namely of one-fifth the amount of polarization observed in visible light (see Strömgren 1956). Behr (1959a) found between 0.37 and 0.52 μm a small decrease towards shorter wavelengths for 4 stars and no such effect for 2 other stars. Teska and I assembled a Wollaston polarimeter rather by trial and error; we were not even aware of Behr's results nor of his polarimeter.[69] The suggestion to use a Wollaston prism and perhaps a depolarizer had been made by Hiltner (1951), but there was the problem of how to execute the calibration. In despair, after trials with only the Wollaston, I mounted a Lyot depolarizer for close calibration, by sliding the depolarizer in and out of the beam at each orientation of the Wollaston, on 16 April 1959. That same night was unforgettable as the start of an observing run discovering strong wavelength dependences on interstellar grains, Mars, Venus, the blue sky and lunar regions (p. 119 of Gehrels and Teska, 1960). However, delay occurred in publication and in further observation of interstellar polarization because of criticism that the Lyot depolarizer was unreliable (see, for instance, pp. 294, 301, and 309 of the proceedings edited by Hall, 1960).

Two fundamental tests must be made in polarimetry, namely in measurement of zero polarization and of 100% polarization; the lack of perfection is corrected with an *instrumental polarization correction* and a *depolarization correction*. Serkowski[70] gives a systematic survey of sources of error in polarimetry, he lists stars of known near-zero polarizations, and he indicates the need for further work to determine more precisely these small polarizations. The essential observations that are presently needed could be made with the rotating telescopes at the Yerkes and the Siding Spring observatories. Rotating telescopes are expensive because of their peculiar mounting design, allowing the whole tube of the telescope to be turned, and Serkowski[71] points out that there is no need for additional rotating telescopes. The idea of having a large telescope for polarimetry on alt-azimuth mounting has occasionally been discussed.[72] With an alt-az mounting, the field rotates and one could, during the night, observe the same object with say 90° field rotation, thereby calibrating for the instrumental effects, provided the optical performance is constant during the time that it takes to rotate the field 90°. An older method, that may not as yet be obsolete, is to select stars of low polarization from Serkowski's table or from the catalogue of Behr (1959b); they must be within $\sim$ 50 parsecs of the sun; they must have normal spectral type (no red variables or supergiants[73]), and must occur in various parts of the sky. One may trust then that the *average* polarization of a sufficiently large selection of stars is $\leq 10^{-4}$.

Cassegrain and prime focus reflectors are the only ones considered suitable for polarization work, as any inclined mirror will introduce

---

[69] See p. 154. [71] See p. 140. [73] See p. 32.
[70] See p. 135. [72] See, for instance, pp. 140 and 358.

INTRODUCTION

linear polarization if the incident light is unpolarized; furthermore, it will cause, by the metallic reflection on the aluminum coating, a transformation of incident linear polarization into partial elliptical polarization. The transformation is dependent on position angle so that the resultant effect is a complicated function of the amount and orientation of the incident polarization. Problems due to instrumental polarizations are discussed in various papers. There is the striking case of strong instrumental polarization after reflection on gratings.[74] Serkowski[75] mentions the possible polarization effects caused by diaphragms in the focal plane and how to avoid these by using dielectric materials rather than metallic ones. Strong instrumental effects are reported by Landau[76] when 90° reflections *before* the analyzer occur.

Coudé systems should be avoided for polarimetry because of their inclined mirrors. Serkowski and I once measured instrumental linear polarizations at the Steward 229-cm reflector, of which the Coudé has 6 mirrors, and found $p = 0.059$ at 0.36 μm, 0.090 at 0.52 μm, and 0.099 at 0.94 μm, while the circular instrumental polarizations, q, were respectively, -0.022, -0.022, and -0.013. Furthermore, these amounts showed some dependence on the declination of the star.[77]

Appreciable effects of instrumental polarization may be caused by improper aluminization of telescope mirrors. If in the vacuum tank for aluminization there is asymmetry in the configuration of the glow-discharge electrodes, the sweeping effect of the glow discharge apparently sets up an electromagnetic field on the front surface of the mirror. Possibly this occurs within a residual oil film on the optics, where the oil originates from the vacuum pump. It follows that the instrumental polarization effects can be minimized by having a filtered pump line and by using glow discharge for cleaning of the mirrors as little as possible. The stripping, cleaning, and drying of the mirror surface is done symmetrically as a complete unit; any local work is repeated over the entire optical surface. The electrode of the glow discharge should preferably be a wire grid, parallel to the surface of the mirror so that the discharge occurs everywhere perpendicular to the mirror. The sources of the aluminum evaporation should also be suspended symmetrically with respect to the surface of the mirror. They are preflashed filaments, loaded with high-purity aluminum rinsed previously in reagent alcohol. The aluminum evaporation is a rapid one at low pressure, below 5×10^{-5} torr. If possible, the secondary mirror of the Cassegrain system is aluminized at the same time with the primary, suspended in the

---

[74] See p. 232.
[75] See p. 141.
[76] See p. 318; see, however, also p. 234.
[77] See p. 232.

aluminization tank in the place of the central hole of the primary. The relative orientations are marked, and in the telescope the secondary is mounted at 90° to the orientation in which the two mirrors were aluminized. These procedures we developed together with J. H. Richardson of the Kitt Peak National Observatory.

For convenient checks of calibration of polarization amounts and position angles, Serkowski[78] gives a list of standard stars. I have, however, some misgivings about all subsequent papers in the literature using the same calibration values of such a list of standard stars. Independent fundamental checks appear essential, and it is not difficult to make the absolute calibrations.

The calibration of the amount of polarization is made by inserting a Polaroid in the beam. (The Polaroid itself may not be 100% effective, but this defect is easily measured by observing the transmission of two pieces, of the same sheet, when their axes are at 90° to each other.) A simple depolarization correction factor, for instance $100/99.50 = 1.0050$, is applied before publication of the results. One could further check the performance of the polarimeter by making observations at various position angles with respect to the axis of the Polaroid and comparing the results with Malus' law.

A method for calibration of position angles is described by Serkowski.[79] I have used a method of pointing the telescope, with drive turned off, toward the horizon at zero hour angle and observing a lamp through a Polaroid sheet in front of the polarimeter. This Polaroid is hung from a plumb line, and the analyzer in the polarimeter is turned until crossed with the Polaroid. In this position, very little light is transmitted, and the orientation of the analyzer that gives minimum light can be read with precision. The Polaroid is next turned 180° about the plumb line, and again a reading of the analyzer orientation of minimum light is obtained. The average of the two readings gives the orientation at which the analyzer is crossed with the plumb line; this is the position of the celestial meridian if the telescope is at exactly zero hour angle. If the position of zero hour angle is not precisely known, the telescope is next turned towards the other horizon, leaving the hour angle unchanged, and the measurement is repeated; the average of the four orientations is position angle 0° in the equatorial frame of reference.

We encountered in this method a problem, namely that the Polaroid appeared to be slightly inhomogeneous. This has not been precisely verified, but there is an indication, at least for the Polaroid we used, that the average — over the measurements with the Polaroid in one position

---

[78] See p. 168.   [79] See p. 144.

and after turning 180° about the plumb line — may have been rotated by about 0°.2.

Another check on the position angle is by making observations of the integrated disks of Venus and Mars, at phase angles and with filters such that there are appreciable amounts of polarization. A central region of the moon can be used too, but the computation of the scattering plane must be made with great care.[80] This method assumes that the position angle of a scattering sphere is precisely at right angles to the plane of scattering; this seems reasonable on first principle, and has been verified by previous observation.

For the measurements of circular polarization one needs to know the handedness. One could simply buy a circular Polaroid sheet with handedness stated by the manufacturer. It is more interesting, however, to make that fundamental check independently, and for this the principles have been described by Swindell (1971).

The precise determination of effective wavelengths has often been ignored in astronomical photometry and polarimetry. In many papers in the literature, the filter transmission is obtained from the manufacturer's figure or from actual tracing, the effect of the earth's atmosphere is computed, and allowance is made for the response of the detector, but this is usually based on an average curve supplied by the manufacturer. Instead, Serkowski[81] obtains the effective wavelength for each object observed at the telescope.

SPECIAL TECHNIQUES

X-ray and γ-ray polarimetry are in their beginning stages; the subject is reviewed by Novick,[82] who shows how much progress will soon be made. He finds spatial resolution important for the future and speaks of an imaging polarimeter. It is stressed in the Novick paper that polarimetry is essential and that the many puzzling phenomena cannot be resolved by photometry and spectroscopy alone. Novick[83] mentions a computer program to design the parameters of a polarimeter. The work in the X-ray domain is done with space polarimeters, and these possibilities and instruments on board spacecraft are reviewed.

Coffeen[84] tabulates the optical polarimeters in space, while those for zodiacal light measurements are mentioned by Weinberg,[85] Leinert et al.[86] and Roach et al.[87] Pieters[88] and Caldwell and Eyring[89] discuss

---

[80] See p. 25.
[81] See p. 143.
[82] See p. 262; also see pp. 58, 1005, and 1021.
[83] See p. 305.
[84] See p. 190.
[85] See p. 781.
[86] See p. 766.
[87] See p. 794.
[88] See p. 405.
[89] See p. 84.

possibilities for space research. For faint extended sources, the principal gain in the future will be made with measurements from spacecraft. From ground-based observatories the difficulties due to sky emissions and scattered skylight are a hurdle in addition to the already difficult problem of the faintness of the zodiacal light and the added contribution of the galactic background.[90]

However, one should consider that extensive effort and partial success with ground-based telescopes may still be more economical in costs and time. I once sketched (Gehrels 1972) a set of steps rising from ground-based facilities, to aircraft, to balloons and rockets on the third step, and to spacecraft at the top, in order to illustrate a *Stairway Rule*: "Whatever can be learned with ground-based techniques should not be considered for the more expensive modes. What can be done with aircraft should not be done with balloons and rockets. Spacecraft should be employed only when the others are entirely insufficient."

For some problems in the infrared part of the spectrum, there are intermediate solutions offered on high altitude aircraft, for instance on NASA's C-141. Ballooning is much more difficult than working with aircraft because of the severe restrictions on weather conditions for launch, and the uncertainty in the predictions of winds at altitudes greater than 30 km, making it impossible to plan the duration or the monitoring of the flight. Wigand and Dollfus may be the only persons in *manned* ballooning who have made polarization observations. From an open basket they observed the polarization of the sky and of the environment, respectively.[91]

For wavelengths shorter than 310 nm, high-altitude balloons may be used; we flew a 71-cm telescope at an altitude of 36 km for measurements near 285 and 225 nm. Photomultipliers for these and shorter wavelengths can be supplied in the so-called "solar blind" form, that is, with low sensitivity at wavelengths longer than 330 nm. Even with such tubes, however, red leakage problems are severe for solar-system objects because the sun emits only about 1% of its brightness at 225 nm compared to that near 350 nm. References to this "Polariscope" ballooning are given by Coffeen,[92] and a photograph of our telescope at launch has been published (Gehrels 1973).

For the infrared part of the spectrum a general review of techniques and theory is made by Dall'Oglio et al.[93] Their own work in ballooning has not been successful as yet, but their description indicates some of the difficulties. Quoting Dyck (personal communication 1973) on far-

---

[90] See pp. 768 and 780.   [92] See p. 189.
[91] See p. 189.   [93] See p. 322.

INTRODUCTION 23

infrared polarimetry: "All currently used techniques are exploratory but they will probably lead to the correct solution if sufficiently interesting results can be found. One is trying to make observations of relatively faint objects against the background which is some 10^5 times brighter. The observations of a 0-magnitude star at 10 μm, which is about midway between the brightest and faintest objects observable, would be equivalent to trying to observe in the visible part of the spectrum a star of visual magnitude 6 during the daytime!" It is quite obvious from Dyck's own paper[94] that highly interesting results are being found and great activity in this field is clearly seen for the future.

Fundamental reviews[95] of spectropolarimetry, with enthusiasm for its possibilities, are presented by Fymat[96] and by Forbes and Fymat,[97] and these papers may be useful to all workers in Fourier interferometry[98] (called "Fourierists" by the authors!). The instrument of Serkowski[99] also allows high wavelength resolution, Tandberg-Hanssen[100] gives a thorough discussion of the polarization in emission lines and Clarke and McLean[101] review that in broadened line profiles. R. Hall[102] finds low polarization in emission lines. Coyne[103] describes the beginning stages of this type of work on stars. Weinberg[104] mentions that some airglow line emissions may show intrinsic polarization. Pieters[105] reports appreciable linear polarization found in a mineral electronic-transition absorption band.

The microwave analogue studies of Greenberg and of Zerull and Giese have already been mentioned.[106] These studies now are for linear polarization and one wonders[107] if the extension to circular polarization would not be of great interest, in view of the results of Martin.[108]

In principle, the microwave analogue work has techniques similar to that in radar work. In practice, however, the subject is very different. With radar work, much larger bodies such as the moon and Mercury are considered, and the equipment with the large radio dishes is highly specialized. The basic types of radio polarimetry are reviewed by Conway,[109] who makes a separation between the work using a single telescope and that using interferometry; he also reviews the reception of total power and differential work using a cross-correlation detector. Feix[110] discusses coherence and the polarimetry of the sun at millimeter

---

[94] See p. 858.
[95] Also see p. 12, and Feofilov (1959).
[96] See p. 617.
[97] See p. 637.
[98] Also see pp. 365 and 526.
[99] See p. 136.
[100] See p. 730.
[101] See p. 752.
[102] See p. 881.
[103] See p. 888.
[104] See p. 781.
[105] See p. 405.
[106] See p. 12.
[107] See p. 915.
[108] See p. 926.
[109] See p. 352.
[110] See p. 676.

wavelengths. A fundamental theoretical review of radio polarimetry is made by Pacholczyk.[111] G. L. Tyler[112] reviews radar work especially for optical astronomers, and the areas where further work is needed. With the improvements to the dish at Arecibo, great advances are to be expected in this field toward new targets like the asteroids, the satellites of Jupiter, and comets, in addition to improvements on such difficult targets as, for instance, the rings of Saturn. Goldstein et al. (1973) are publishing a paper on the faint radar returns they obtained on asteroid Toro during the flyby in August 1972, and Goldstein (1973) appears to have found boulders in Saturn's rings, but polarization has not as yet been studied.

SURFACES

The Fresnel equations are given in the paper by Kissell.[113] In his Fig. 5 one sees the general shape in geometrical optics of the polarization-phase function, that is, with strong polarizations that have electric-vector maximum perpendicular to the plane of scattering and with the maximum occurring near twice the Brewster angle (a phase angle near 100°). The moon and the asteroids essentially show that profile,[114] but with a dilution of polarization proportional to the fraction of the light that is reflected. In addition there are second-order effects near small phase angles due to multiple reflections on the rough micro-texture of the surface, resulting in some negative polarization (electric-vector maximum *in* the plane of scattering).

The brightness of the moon increases from quarter-phase to near-full by about a factor of 9; the magnitude-phase relation is strikingly linear at phases larger than 7°. (The whole moon has been observed by Lane and Irvine [1973]; calibration was a special problem since the full moon is near -13 mag, and a further difficulty is the problem of observing the whole moon, an object one-half a degree in diameter.) Already a century ago it was recognized that only rough objects could give such a steep brightness-phase relation. The problem became more complicated when the *opposition effect* was found, which causes an additional brightness increase of a factor of about 2 between 7° and 0° phase angle. The general behavior of the brightness-phase relation has been fitted to numerical theory for a porous micro-structure by Hapke (1963), but a physical explanation of the opposition effect has not as yet been completed. The moon is redder than sunlight and shows further reddening with increasing phase angle. The outstanding problem of the

---

[111] See p. 1030. [113] See p. 371; also see pp. 381, 466, and 526.
[112] See p. 359. [114] See p. 381.

photometry of the lunar surface is to explain why the highlands are brighter than the maria.

There are various reports of luminescence on the lunar surface, but so far these effects have not been definitely established. The study of magnetic and electrical fields of the lunar surface, and the interaction with the interplanetary fields, is an important discipline (see Sonett and Colburn; p. 419). One might further speculate that this study should be brought into contact with the problems of optical photometry and polarimetry of the lunar surface, as follows. As the frequency increases, the skin depth for electromagnetic induction decreases, and the resulting fields may be strong enough to sustain some suspension of charged particles (Lal 1972, p. 58) and to provide a mechanism for luminescence.

The polarization of the moon and other objects without atmospheres was observed by Lyot in visible light, and the general behavior is shown in Fig. 1 of Bowell and Zellner.[115] The characteristic features of these curves are a maximum near 100° phase angle and the largest amount of negative polarization near 11°, with an *inversion angle* near 20° phase. Lyot already found a difference in maximum polarization between waxing and waning moon, caused by the apparent difference in the proportion of dark and bright regions. Deviations from the Umov effect,[116] the inverse proportionality of polarization and reflectivity, are indicated in the paper by Fountain.[117]

The orientation of the electric-vector maximum for all regions on the moon, at phase angles larger than the inversion angle, is precisely perpendicular to the plane of scattering. Would regions on a smooth sphere show this phenomenon? There have been some reports in the literature of rotational effects of the plane of polarization, but they presumably are due to systematic errors in the instrumental polarization or in the computation of the scattering plane. The plane of scattering goes through the center of the sun, the region on the moon, and the position of the observer on the earth at the time of observation. Drastic errors can result if this is not precisely determined, especially near opposition. The proper method to compute the geometry near full moon has been given by Lyot (1929, p. 35).

Work on the asteroids is relatively new, as can be seen in the review paper by Bowell and Zellner.[118] Some work done by Lyot with a photographic method may not be reliable. Polarimetry on the asteroids is an important tool for determining the reflectivity of the surface, especially since it is now apparent that the direct methods, with micrometers and disk meters, have consistently given diameters that are too small. Precise

---

[115] See p. 383.   [116] See p. 12.   [117] See p. 223.   [118] See p. 381.

diameters are needed together with mass determinations (Schubart 1971) for the derivation of the density, which presently appears to be about 2 gm/cm^3 for Ceres and Vesta while only a few years ago we believed it to be 6 gm/cm^3.

The inverse relationship between reflectivity and polarization can be employed to determine whether the lightcurves of asteroids are caused by the rotational variation in the projected area or by reflectivity variation over the surface. In a recent paper by Dunlap et al. (1973), this method has been applied to asteroid Toro for which it was found that, although the lightcurve is mostly determined by a time variation in the projected area, i.e., by an elongated shape of a rotating body, there are some second-order effects due to reflectivity differences over the surface.

Whereas the Mie theory is applicable to the study of light scattered by small spherical particles, the theoretical treatment of grains packed together on the surface of the moon becomes very difficult because they are no longer independent scatterers. The present studies are therefore based on laboratory comparison, and this topic is extensively reviewed by Bowell and Zellner.[119] The paper by Pieters[120] gives a discussion of laboratory techniques and the equipment used for obtaining brightness and polarization measurements on laboratory samples. The study of lunar samples, and of effects observed on the whole moon, remains astronomically important for the comparison with objects that still are too remote for direct sampling, like the asteroids and Mercury. Incidentally, the *de*polarization properties of the lunar surface, for optical frequencies, have been studied only once, namely with observations of the earthshine by Dollfus.[121]

The planet Mercury[122] is mentioned in this book and so are Mars[123] and Pluto.[124] For Mercury there is a paper by Coyne and Gehrels (1973) and Dollfus also is about to publish additional measurements. In the future, observations of Mercury are needed because of appreciable differences found in the polarimetry with that on the moon and also because there is some indication of time variations on Mercury. These effects, however, could be proven only with an extensive series of measurements with a single telescope on an excellent mountain site having free horizons so that Mercury can be observed with the same equipment on both sides of the sun. The discovery by Landau[125] of strong negative polarization in the infrared opens a new field of investigation.

Little polarimetry has been done on the rings of Saturn. Lyot (1929) reported anomalous orientations of the position angles in certain cases.[126]

[119] See p. 381. [121] See p. 566. [123] See p. 606. [125] See p. 321.
[120] See p. 405. [122] See p. 387. [124] See p. 573. [126] Also see p. 24.

INTRODUCTION

The detection of tenuous atmospheres has been attempted for Mars with polarimetry,[127] but this is a difficult topic because of the complexity of the problem; not only the optical thickness of the atmosphere is uncertain but also the effect of the underlying surface. For Mars, the detection of the atmosphere is an obsolete problem now that it has been completely resolved with the use of spacecraft. Coffeen and I have unpublished polarizations of Mars observed over several years that no longer appear to be useful.[128] On the other hand, continued patrolling of Mars polarimetrically from the earth may provide a tool for detecting dust storms and for indicating the nature of the dust particles. Measurements of circular polarization of Mars and other surfaces are made by Wolstencroft.[129]

OUR ENVIRONMENT

We measure with accuracy and precision the color, the brightness, and the polarization of celestial objects, but we hardly know what they are for objects in our immediate environment. For example, great effort is spent on observing the B-V colors for G2 V stars, that have spectral types similar to that of the sun, but the B-V of the Arizona desert is not at all known. A knowledge of the polarizations in our natural environment and their variations with change in angles of incidence and emergence would form a practical background for evaluation of similar data obtained in the laboratory and on astronomical objects.

The paper by Coulson[130] shows how much linear polarization there is in our environment. The mechanisms of animals for using these polarizations for navigation are reviewed by Waterman.[131] Wolstencroft[132] surprised us by showing how much circular polarization there is. He gives ample suggestions for further study of these effects — for example, laboratory work to study the circular polarization of green leaves. Generally, the wavelength dependence of polarization should be studied much more for our environment.

It is a great delight to use Polaroid glasses and to inspect the polarization in our environment, where one finds a large variety of effects on various substances. The reduction of glare, using Polaroids, is described by Shurcliff (1966), Shurcliff and Ballard (1964), and mentioned by Waterman.[133] Also without Polaroid glasses, one can observe polarization (other than through the usage of the Haidinger brush, see p. 14). For instance, when one observes the reflection of the blue sky upon a still

---

[127] See p. 566.
[128] See, however, p. 606.
[129] See p. 499; also see p. 610.
[130] See p. 444.
[131] See p. 472.
[132] See p. 495.
[133] See p. 472.

surface of water, the polarization effects in certain directions are enhanced; the clouds, that polarize little, and the blue sky may be seen in greater contrast. To the delightful book by Minnaert (1954) on light and color in our environment, more data can always be added. As I write this, early in the morning, there are several hundred grasshoppers outside on the concrete sidewalk, all facing away from the building, facing due south to within 10°. It would have been understandable if they had been alternating between pointing north and south, warming themselves after a cool night. But why do they all face south, and so exactly?

The optical aspects of oceanography are reviewed by J. E. Tyler,[134] and it is surprising to see how little polarimetry has been done under water. There seems to be a rich field for theoretical studies of oceanographical polarimetry, as apparently little if any application of the Rayleigh-Chandrasekhar and Mie theories has yet been made. How inviting it would be to take one of our advanced instruments under water! Tyler gives a set of predictions of polarization effects at various depths in the oceans. Surprising also is his prediction of an appreciable amount of polarization at large depths because it would appear, in the case of conservative molecular scattering, that the polarization should vanish for large optical thickness. Should one then also expect appreciable amounts of polarization observable during penetration experiments into the atmospheres of Venus and Jupiter? This seems to be a question with some urgency, as penetration into the atmosphere of Venus is being executed in the Soviet space program and planned for the American one of the Pioneer/Venus missions.

PLANETARY ATMOSPHERES

I see a progression of increasing difficulty in astronomical photopolarimetry, namely from that on single scatterers to the complicated case of a nonuniform planetary atmosphere. The polarization due to a single scatterer is illustrated in the paper by J. E. Tyler.[135] Coulson[136] describes the polarization phenomena of the earth's atmosphere, which is an introduction to the more complicated general case of planetary atmospheres. A nice summary of the phenomena that can be observed on the blue sky is in the abstract of the Coulson paper.

Coulson mentions especially the *neutral points* where the polarization is zero, called *inversion points* by Bowell and Zellner.[137] A simple

---

[134] See p. 434 and also p. 472.
[135] See p. 434.
[136] See p. 444.

[137] See p. 25; "neutral point" seems a preferable name; neutral points are also found for zodiacal light (see p. 789).

physical explanation of such neutral points I have not seen in previous literature and I will therefore try one here (similar physical explanations, of the strong polarizations at the poles of Jupiter, are given on p. 194 of *Astron. J.* 74, 1969). Consider the Babinet point: let the sun be near the horizon and we face that direction. In the column of air above us, in the direction 90° away from the sun, the transverse vibrations are seen edge-on with the result of strong polarization having electric-vector maximum perpendicular to the plane of scattering. But a beam of light close to the horizon, just above the sun, will have received multiple scattering much more from the left and right than from below (where the earth is) and from above (because the atmosphere in that direction is thinner than to the left and right), and the resultant vibrations are predominantly in the plane of scattering. Somewhere between the two directions of our observations there has to be a neutral point.

Typically, 75% polarization is found on a clear sky 90° away from the sun in visible light. Comparison of the Rayleigh-Chandrasekhar theory with observations explains the difference from 100% (expected from a single isotropic scattering) as follows. Multiple scattering dilutes 6%, molecular anisotropy 6%, and reflection by the ground 5%, while a residual 8% is presumably due to aerosols suspended in the atmosphere even when the sky is seen as hard blue.

The effects of aerosols in the terrestrial atmosphere is studied by Takashima et al.[138] While several people in the past have looked for elliptical polarization of the earth's atmosphere and not found it, there is a discovery reported in this book by Hannemann and Raschke[139] who find appreciable amounts of circular polarization on aerosols in the polluted air above industrial West Germany. Witt[140] is convinced that the noctilucent clouds, at about 82 km altitude, are caused by terrestrial particles. Because of that conclusion, his paper is in this part of the book rather than with the particles of the zodiacal light in Part III. Following the Witt paper, however, Martha Hanner at our request has given references to other papers on the possible origin of the noctilucent particles; the paper that differs most extremely with Witt's views is the one by Hemenway et al. (1972) suggesting that the particles come from the sun.

The photometry of lunar eclipses has been reviewed by Barbier (1961), and we should not forget the curious phenomenon of having brighter and darker eclipses. Peculiar effects of polarization have been reported by Coyne and Pellicori (1970).

The "planetary problem" is that where the sun shines onto a planet

---

[138] See p. 500. [140] See p. 514.
[139] See p. 510.

consisting of a surface, a molecular atmosphere, and aerosols suspended in that atmosphere.[141] The surface reflection may or may not be polarized, and that problem is further discussed by Coulson,[142] while Ueno et al.[143] assume that the radiation from that surface is scattered isotropically or by specular reflection. Molecular scattering was studied in a classical investigation by Lord Rayleigh and is therefore called Rayleigh scattering. Chandrasekhar provided the mathematical analysis, and we therefore refer to the molecular multiple-scattering theory as the Rayleigh-Chandrasekhar theory. For practical applications there are the extensive tables computed by Coulson et al. (1960), while several people now have computer routines to make their own calculations. The effect of the aerosols is computed with the theory of Mie, and there are a few computer routines available now that will take into account the complete scattering by molecules and by various size distributions and compositions of suspended aerosols and will include the effects of a surface, although again the question comes up of what type of scattering law for the surface should be assumed. The application of such calculations to observations of planets is reviewed by Coffeen and Hansen.[144] The Mie theory is for spherical particles, but we have the paper by Giese and Zerull[145] who study particles of various shapes. Their study is for the zodiacal light specifically, but for aerosols in planetary atmospheres this type of study also is of a fundamental nature, and the results may be useful in several other areas. The theoretical paper by Ueno et al.[146] gives several references to the pertinent literature. The theory of the production of spectral features in planetary atmospheres is reviewed by Fymat,[147] while the Fourier analysis for observing polarization in spectral features is pioneered in the paper by Forbes and Fymat.[148]

Coffeen and Hansen[149] report on the secular variation of the amount of polarization observed on Venus, which indicates that the gas pressure at the top of the clouds may vary between about 30 and 60 mb, and they point out that a continuing patrol of the global polarization of Venus is therefore useful. Bugaenko et al.[150] present their observations on Venus and Jupiter and also report on a first attempt to derive the refractive index and particle size for Jupiter, as Coffeen and Hansen did so successfully for the clouds of Venus. Fountain[151] concludes that the Saturn atmosphere shows appreciable differences from that of Jupiter, a conclusion that has been recently confirmed by Hall and Riley (1973).

Hall developed the first polarimeter to scan planetary disks, and results obtained with that instrument are found in the paper by Hall and Riley[152]

---

[141] See p. 11.
[142] See p. 444.
[143] See p. 582.
[144] See p. 518.
[145] See p. 804.
[146] See p. 582.
[147] See p. 617.
[148] See p. 637.
[149] See p. 552.
[150] See p. 599.
[151] See p. 223.
[152] See p. 593.

INTRODUCTION 31

for Jupiter, and in the paper by Coffeen and Hansen[153] for Venus. Jupiter, because of its large distance from the earth, is never observed at phase angles larger than 12°, and the polarization effects therefore are rather limited. However, near the limb of the planet, especially at the north and the south poles, the polarizations are strong because the clouds are apparently at low level and the overlying molecular atmosphere is thick so that by multiple molecular scattering the polarizations obtain appreciable values. It is not generally known that the cloud heights drop sharply with the discontinuity occurring near 45° latitude (Gehrels 1969).

Hall and Riley[154] do not find different amounts of polarization over the Great Red Spot of Jupiter, whereas direct photography, spectroscopy, and photometry indicate strong differences over the Red Spot with respect to the surrounding cloud deck. The polarimetry thus far has had a precision on the order of $\pm 2 \times 10^{-3}$. With a scanning polarimeter that has an array of detectors in the focal plane and with careful work under excellent seeing conditions, it may be possible to improve that precision by an order of magnitude. Appreciable progress may then be expected for the study of cloud heights in and near the Red Spot. This should next be combined with the study of cloud dynamics using high-resolution photography for a series of plates taken on the same night and on consecutive nights in order to observe the cloud motion near the Red Spot. Fymat[155] makes suggestions for spectropolarimetric scanning over planetary disks.

Kemp[156] is pioneering studies of circular polarization of planetary atmospheres. The significance of circular polarization for deducing the microstructure of cloud particles has been questioned, but differential effects may be of great interest.[157]

STELLAR EFFECTS

Most often the term *intrinsic polarization* is used to differentiate from interstellar polarization; the intrinsic effect can be due to circumstellar shells (see next section) or it can be inherent in the stars. A historical event is critically discussed in the paper by Collins and Buerger,[158] namely Chandrasekhar's prediction of intrinsic polarization in a binary system where a hot early-type star is partially eclipsed by a cooler companion. The flux of the observed part of the hot star has been generated deeper down and is subsequently scattered over nearly 90° by electrons. The electron scattering yields strong polarization that is,

---

[153] See p. 552. [155] See p. 617. [157] See p. 11.
[154] See p. 593. [156] See p. 607. [158] See p. 663.

of course, diluted by the light of the cooler companion. Hall (1949) and Hiltner (1949) looked for the effect predicted by Chandrasekhar but, instead, they made the more important discovery of interstellar polarization. There followed a period of great activity of observing interstellar polarization for the purpose mostly of delineating the lines of force of the galactic magnetic field.[159] Not until some 15 years later were intrinsic effects found on certain stars; the most convincing evidence that these polarizations were intrinsic was their time variability, usually correlated with variability in the amount of light. Now it appears that we have made almost a full cycle, with intrinsic polarizations known for many stars. It becomes, in fact, difficult to study the interstellar polarization entirely free from intrinsic effects; practically all stars that show emission lines in their spectrum, including those of early spectral types, have some intrinsic polarization.

Mickey and Orrall[160] give a general review of present and future work in polarization measurements on the sun's disk. In the papers by Mickey and Orrall and by Tandberg-Hanssen[161] it is seen that the polarization at the solar limb has the electric-vector maximum in a tangential direction; this is reminiscent of the Chandrasekhar prediction: the energy generation occurs at greater depths and the scattering causes some tangential polarization, which is, of course, strongly diluted by the light emitted all around the observed region. These papers, as well as the ones by Dollfus[162] and Clarke and McLean,[163] contain references to and/or description of equipment, while other specialized papers on solar equipment are by Baur et al.,[164] and by Querfeld for coronal mapping.[165] Feix[166] discusses observations of solar regions at millimeter wavelengths. Additional papers for the sun have recently been published in the proceedings of IAU Symposium No. 43 (Howard 1971).

The paper by Dollfus deals with photospheric phenomena, but the detection of the solar corona is also discussed. Tandberg-Hanssen reviews the polarization of prominences and of the inner corona. The polarization of solar flares in the X-ray domain is discussed by Novick.[167] Leroy and Ratier[168] use the polarization-wavelength dependence to separate out the observations of the F-corona, that shows the solar Fraunhofer spectrum diffracted by interplanetary particles, from the observations of the K-corona, that shows a continuous spectrum scattered by electrons (the L-corona is due to coronal line emission).

Magnetic fields are studied with polarimetry of fine structures in the

---

[159] See p. 964.　[162] See p. 695.　[165] See p. 254.　[168] See p. 762.
[160] See p. 686.　[163] See p. 752.　[166] See p. 676.
[161] See p. 730.　[164] See p. 246.　[167] See p. 268.

INTRODUCTION

wings of spectral lines (effects discovered by Zeeman and named after him) and for the sun they are discussed in the above-mentioned papers. For linear polarization of magnetic stars there is a contribution by Kemp and Wolstencroft.[169] When strong magnetic fields are present, Kemp predicted one should find circular polarization with broadband filters as well; he has proven his prediction in the laboratory and has found the effect in a white dwarf; that star, incidentally, had exceptionally high circular polarization even for white dwarfs. The white dwarf discoveries and studies are reviewed by Angel[170] and by Landstreet,[171] while O'Connell[172] has a theoretical paper on this topic.

As for the intrinsic polarization of pulsars, we are fortunate enough to have one of the discoverers of the optical pulsar in the Crab Nebula to review this new field.[173]

CIRCUMSTELLAR PARTICLES

The cloud of particles that gives rise to the faint glow along the zodiac is considered in this book as a circumstellar shell. Even the comets are considered a part of this circumstellar nebula, for which earthling observers have the great advantage of being able to observe from within. Just what connection there may be between the sun's nebula and the interstellar grains in the galaxy is not established at this time, although it appears that the sun is not in a dense galactic dust cloud (Behr 1959b, p. 224). Alfvén was interviewed for *Physics Today* (May 1972), and he suggested that there may be a magnetic field screening the solar system from the galaxy. Such a magnetic field would align the interstellar grains; the effect would be observable on nearby stars if the field is strong enough and the particle density high enough. Some alignment, with the electric vector perpendicular to the ecliptic, is indeed seen in the map of Behr (1959b). This could be verified with observations of the highest possible precision. Definitive proof would be to find a 90° variation per quarter year in the orientation of the electric vector observed on a nearby star near the ecliptic pole.

The history and the present status of linear polarization measurements of the zodiacal light is reviewed by Weinberg.[174] Wolstencroft and Brandt[175] present new results of their ground-based observations made at Chacaltaya, and they review the circular polarization of the night sky. Giese and Zerull[176] discuss various models of the zodiacal light and they

---

[169] See p. 988 and also p. 985.
[170] See p. 54.
[171] See p. 981.
[172] See p. 992.
[173] See p. 997 and also p. 39.
[174] See p. 781.
[175] See p. 768.
[176] See p. 804.

find that the particles are not spherical; the paper of Zerull and Giese[177] may also be considered in this connection. There is a recent paper with a model of a zodiacal cloud by Roach (1972). Progress with zodiacal light studies may be expected especially with observations made from spacecraft, as shown by Leinert et al.[178] and by Weinberg. New results from the Pioneer missions have been published by Hanner (1973) and by Hanner and Weinberg (1973), and from the OSO-5 spacecraft by Sparrow and Ney (1972a). The Pioneer missions should soon resolve whether or not particles of sizes 10 μm and smaller originate from the asteroids or from a source outside the asteroid belt, such as the comets or the interstellar grains. Weinberg notes that to date there have been no observations of the zodiacal light in the ultraviolet nor in the infrared parts of the spectrum and that very little work has been done on the wavelength dependence of polarization. It looks promising, in comparison with the theoretical work of Martin,[179] that the characteristics of the grains can be further studied from observations of circular polarization.

The counterglow is a faint lightsource opposite the sun that is rarely seen and then only by people with good eyes for such phenomena, at locations where the sky is dark and at times when the counterglow is away from the Milky Way. Roosen (1970) has abstracted the literature on the counterglow. Weinberg[180] reports a new observational result from the Pioneer mission, namely that the counterglow cannot be due to a gaseous tail of the earth, as had been proposed occasionally in the past, but that it is due to scattering by particles at great distances from the earth. This result had been derived before from careful ground-based work, but now the classical observation has been made on the counterglow when a spacecraft was sufficiently far (10^7 km) from the earth, thereby removing any further doubt. However, there is the surprising result of Roach et al.,[181] and their Figs. 4 and 5 indicate drastic changes in the amount and orientation of the polarization with a time scale of hours. These changes must be due to some phenomenon not very far from the earth, in apparent contradiction to the report by Weinberg. One should consider at the same time the controversy on the nature of the noctilucent clouds.[182]

In organizing this book we had not invited a review article on the polarization of comets. The reason for this is not a lack of observations but a lack of conclusive interpretations. Clarke[183] presents interesting observational results and so do Doose and Coffeen,[184] but the latter argue that the interpretations of polarimetry, when combined with photometry

---

[177] See p. 901.
[178] See p. 766.
[179] See p. 926.
[180] See p. 781.
[181] See p. 794; also see p. 780.
[182] See p. 29.
[183] See p. 814.
[184] See p. 818.

— and this turns out to be crucial — are inconclusive. Following the paper by Doose and Coffeen, I made some comments about the possibility that the particles in the coma are hydrate clathrates and that further work should concentrate on making comparisons of such icy conglomerates with the results of photometry and polarimetry. This work would be quite a challenge and it may be necessary to attempt a comparison in the laboratory as well as with theoretical work using an adaptation of the Mie theory to clusters of particles.

After some reports of intrinsic polarizations on certain variable stars, especially in Russian literature, Serkowski in 1965 found as much as 0.07 polarization in V CVn. This discovery immediately brought doubt as to a possible interstellar origin for this polarization, because the star is located near the north galactic pole and little interstellar matter occurs in that direction. The fact that we are dealing here with intrinsic effects became clear beyond argument when the polarization was observed to be time dependent: from 0.072 to 0.027 polarization was observed with a blue filter.

The general conclusions on circumstellar polarizations can be summarized as follows:

a. red variables may show strong intrinsic polarizations;
b. the wavelength dependence differs from star to star, but a common feature is a fast increase with decreasing wavelength;
c. most stars show a pronounced time variability in the amount and position angle of polarization;
d. the stars that show intrinsic polarization may have asymmetric shells, and they may be ejecting matter into the general interstellar medium; here might be a source of condensation nuclei for interstellar grains.

The observations in the infrared on circumstellar shells around various stars are described by Dall'Oglio et al.,[185] Dyck,[186] and Landau.[187] Dyck makes specific suggestions for future observation of effects in emission and absorption, and he mentions the very likely possibility of unexpected discoveries. Shawl[188] and Kruszewski[189] show some astonishing wavelength dependencies of the position angle as well as the amount of polarization, and various observers are finding drastic variations with time. A vast amount of work has to be done in the case of intrinsic polarizations. Some of these variable stars could by themselves merit a Ph.D. dissertation for the study of spectroscopy, photometry, and polarimetry combined. This indeed is a field of the future, and the application of circular polarization observations is only just starting! Angel and Martin (1973) report on four stars showing circular polarization; these

---

[185] See p. 322.   [186] See p. 858.   [187] See p. 318.   [188] See p. 821.   [189] See p. 845.

stars have M-type spectra, they are giants or supergiants, and they are in highly obscured regions. Early-type stars that show emission lines in their spectrum, Be stars, are discussed in various papers.[190]

A general reference to intrinsic polarizations of variable stars, including novae and R Mon, may be found in the paper of Shawl[191] and in a paper by Zellner and Serkowski (1972). BL Lac type objects are reviewed by Visvanathan.[192]

INTERSTELLAR POLARIZATIONS

Coyne[193] shows a characteristic curve of the interstellar polarization. It has a maximum in visible light, a gradual decline toward shorter wavelengths, and a steep drop in the infrared part of the spectrum. It is curious that the interstellar grains have effective sizes that give such interesting phenomena observable in our optical window where our best photoelectric detectors, as well as our eyes, are effective. The sizes of the interstellar grains have an upper limit, presumably due to radiative sputtering, at just the right dimension. The largest particles are the most effective scatterers, and the result is an apparent size distribution that is narrow; had it been wide, the wavelength dependence of the polarization would have been washed out. But for these strokes of luck, we would not have been able to observe interstellar polarization at all, let alone the characteristic curve of the wavelength dependence.

Many attempts have been made to fit the van de Hulst theory for aligned infinite cylinders ("picket fence model") to the measurements of interstellar polarization, and the theory of Mie to the observations of the interstellar reddening and also to those of reflection nebulae. (The terminology is clearly defined. One speaks of interstellar polarization effects when the scattering angle is zero, and of reflection nebulae when all scattering angles are considered.) It is frustrating that no perfect fits have been obtained with these theories. Further trials in the future will no doubt be made, including attempts with particles of irregular shapes using the laboratory comparisons of Greenberg and of Zerull and Giese.[194]

It does not seem realistic to try to fit the characteristic curve of interstellar polarization with a single type of particle or even with a composite particle precisely defined as, for instance, having a graphite core with an icy mantle. It also seems an oversimplification to state whether or not the interstellar grains are dielectric.[195] The cosmic abundances indicate

---

[190] See pp. 663, 845, and 858.
[191] See p. 821.
[192] See p. 1076; also see p. 986.
[193] See p. 888.
[194] See p. 12.
[195] See pp. 926, 942, and 1059.

a variety of composition in interstellar space, and a complex mixture within each particle is to be expected. It seems likely that the grains have an absorptive core, because no particles seem to form without a condensation nucleus. But these cores should have a mixture of substances, while on the mantles, mostly dielectric, there should also be materials of various composition. The scattering phenomena on the skin must be studied in detail, because in the theoretical approach at present only smooth particles are considered. (I once wrote a brief note on this [in *Astron. J.* 71: 62–63, 1966] which might be useful.)

With observations at wavelengths shorter than 0.2 μm, some of the molecular scattering phenomena may be observable (see Bless and Savage 1972). It is noted that polarimetry at these wavelengths has not as yet been done nor is it being considered for future space missions. On the other end of the spectrum, Dyck[196] points out that to date only very few stars have been observed for interstellar polarization at wavelengths longer than 1 μm.

The interstellar grains can be further studied by polarization measurements within spectral absorption features.[197] The new techniques for using circular polarizations are reviewed by Angel.[198] While Kemp and Wolstencroft[199] add the observations of circular polarization to the usual ones of linear interstellar polarization, Martin[200] uses the rare case of having a source, the Crab Nebula, that illuminates the grains with strong linear polarization. The curious phenomenon of wavelength dependence of the position angle of linear polarization is discussed by Coyne[201] as well as by Martin. Martin also encourages an extensive campaign for observations of the wavelength dependence of linear and circular polarization.

Breger[202] and Carrasco et al.[203] give summaries of their recent work published elsewhere. They use the interstellar polarization observations, that is, again, those of scattering angle zero, carefully sorting out the contributions from: circumstellar shells, reflection nebulae in the immediate vicinity of the star, dense clouds that are in between the members of a star cluster, and the general interstellar medium. There is a general conclusion that when the dust is dense, the particles are large and the magnetic fields are strong ($\sim 10^{-4}$ gauss compared to $\sim 10^{-5}$ gauss for the general interstellar medium).

The proceedings of a conference on interstellar grains are edited by Greenberg and van de Hulst (1973).

---

[196] See p. 858. [199] See p. 939. [202] See p. 946.
[197] See p. 916. [200] See p. 926. [203] See p. 954.
[198] See p. 54.   [201] See p. 888.

NEBULAE

An interesting application of polarimetry was made by Herbig (1972)[204] who wanted to establish whether there are stellar companions near VY CMa or if the extension to the star that is visually seen is a reflection nebula. He photographically observed that the polarization is very strong (0.70) and concluded that this is due to scattering by dust particles.

Reflection nebulae that are caused by starlight scattered over a wide range of scattering angles have already been mentioned in the previous section. The review paper of Zellner[205] shows that only a beginning has been made with the study of the reflection nebulae. The first-order theory works quite well, and some tentative knowledge of the grain characteristics can be gained from the application of that theory, but the nebulae appear to have optical depths so great that multiple scattering effects have to be taken into account. This combination of theory and observations has not yet been developed. There is also the problem of unknown geometries in these nebulae, which often appear to be quite irregular on direct photographs. Observationally, it is difficult work that requires the best telescopes and a large time investment on them; this situation can be seen also in the paper by Visvanathan.[206] Nevertheless, these are problems where future work is challenging as they carry promise of new gains in a relatively unexplored territory. Also little explored is the polarization of the diffuse galactic light (see Sparrow and Ney 1972b).[207]

The paper by Dall'Oglio et al.[208] presents a survey of various mechanisms that have been proposed for the alignment of interstellar grains. There still is some question as to how they are aligned. The prevailing model at present has the galactic magnetic field responsible for the alignment, and the *Davis-Greenstein mechanism* is generally invoked. Davis and Greenstein (1951) considered the interstellar grains as having an elongated shape and a predominantly dielectric composition. They have, therefore, paramagnetic properties rather than ferromagnetic ones as metallic particles would have had. Collisions with photons cause a fast spin, of the order of 10^5 cycles per second. The grains spin with their long axis perpendicular to the lines of force of the galactic magnetic field; any deviation from 90° causes induced magnetization, which in turn causes a torque on the spinning particles. The result is a near-equilibrium configuration with a fast spin in the plane nearly perpen-

---

[204] See p. 834.
[205] See p. 867; also see p. 881.
[206] See p. 1059.
[207] See pp. 768, 779; also see p. 932.
[208] See p. 322.

dicular to the galactic magnetic field. In the Perseus direction we see the particles spinning edge on; the lesser absorption of the light is in the direction of the short axis of the particles and that is therefore the direction of the electric-vector maximum, i.e., parallel with the plane of the galaxy. In the Cygnus direction we apparently are looking along a spiral arm (i.e., along the lines of force), the particles are seen spinning nearly pole-on, and little polarization is observed on the average.

Carrasco et al.[209] also comment on the alignment by the galactic magnetic field, and they apparently rule out alignments by radiation pressure and by exchange of photon angular momentum. Angel[210] brings up the question whether the galactic magnetic field is strong enough to align by the Davis-Greenstein mechanism. The possibility of observational verification that the grains are indeed spinning is discussed.[211]

Interstellar polarization was first used to learn the patterns of the galactic magnetic field, rather than for the study of the physical parameters of interstellar grains; the alignments are beautifully shown in the second figure of the paper by Verschuur.[212] Such alignment supposedly is due to the interstellar magnetic field, but the shape of the general field is argued. Verschuur shows that the magnetic field is longitudinal, with local disturbances, rather than one having a helical structure as proposed by others in the past. This point of view is also taken by Schmidt[213] and by Schlosser and Schmidt-Kaler;[214] the arguments appear quite conclusive. The degree of local disturbance is observable.[215] Faraday rotation is discussed by Caldwell and Eyring[216] as well as by Verschuur.

Schmidt studied the large Magellanic cloud (LMC) and the small Magellanic Cloud (SMC), which are at declinations of about $-70°$ and the observations are therefore made at the European Southern Observatory in Chile. The southern Milky Way is studied by Schlosser and Schmidt-Kaler, and a beautiful photograph of the Milky Way taken in ultraviolet light is shown on page 973. The papers are progress reports and the work is continuing with the new facilities in Chile.

One of the most dramatic displays of polarization patterns is seen in the Crab Nebula.[217] The literature of the nebula itself is reviewed by Felten,[218] and that of the pulsar in the Crab Nebula by Cocke.[219] Novick[220] reports on the X-ray results, while the radio observations of supernovae remnants are mentioned by Milne and Dickel.[221] Felten expresses surprise that not more high-resolution polarimetry has been

---

[209] See p. 954; also see p. 774.
[210] See p. 60.
[211] See p. 942.
[212] See p. 964.
[213] See p. 976.
[214] See p. 972.
[215] See p. 936.
[216] See p. 77.
[217] See p. 1016.
[218] See p. 1014; also see p. 54.
[219] See p. 997; also see p. 1078.
[220] See p. 262.
[221] See p. 1029.

done on the Crab Nebula, but I believe this is a case where the job had originally been done so well that it could hardly be improved. Felten does, however, indicate special detailed studies that should be made in the future. With a small telescope, it may be possible to actually observe the net optical polarization of the entire Crab Nebula, in order to compare with results at longer (radio) wavelengths. The precise value of the interstellar polarization, which has to be applied as a correction in order to find the real values for the Crab and the pulsar, is still unknown, but a new precise determination may soon be available (Angel 1973, personal communication). The problem has been that the interstellar polarization was derived from the stars surrounding the Crab Nebula, while these stars may be at different distances than that of the Crab and pulsar.

Our book ends with the fundamental review of techniques in radio astronomy by Pacholczyk[222] with application to extragalactic radio sources, and with the review by Visvanathan[223] of the frontier of astronomy, the extragalactic nebulae. In the future one might observe the wavelength dependence in the dust lane of NGC 4594, for instance, for comparison with that of our own galaxy. Novick[224] reviews the X-ray work on galaxies. The possible existence of antimatter is suggested by Caldwell and Eyring.[225] Circular polarization, at optical frequencies, has not as yet been found.[226] That the frontier is a lively one is seen in the new results of strong polarization observed on galactic nuclei by Heeschen (1973) and Kinman (1973).

Acknowledgments. I thank D. L. Coffeen, M. S. Matthews, K. Serkowski, W. Swindell, and B. H. Zellner for their helpful comments on this paper.

REFERENCES

Angel, J. R. P., and Martin, P. G. 1973. Observations of circumstellar circular polarization in 4 more infrared stars. *Astrophys. J.* 180: L39–L41.

Barbier, D. 1961. Photometry of lunar eclipses. *Planets and satellites.* (G. P. Kuiper and B. M. Middlehurst, eds.) Chicago, Illinois: Univ. of Chicago Press, pp. 249–271.

Behr, A. 1959*a*. Beobachtugen zur Wellenlängenabhängigkeit der interstellaren Polarisation. *Zs. für Astrophys.* 47: 54–58.

―――. 1959*b*. Die interstellare Polarisation des Sternlichts in Sonnenumgebung. Nachr. Akad. Wiss. Göttingen. *Math.-Phys. Kl.,* pp. 185–240.

[222] See p. 1030.   [224] See p. 262.   [226] See p. 986; however, also see pp. 941 and 1055.
[223] See p. 1059.   [225] See p. 86.

Bless, R. C., and Savage, B. D. 1972. Interstellar extinction in the ultraviolet. *The scientific results from the Orbiting Astronomical Observatory (OAO-2)*. (A. D. Code, ed.) Washington, D.C.: National Aeronautics and Space Administration, SP-310. pp. 175–198.

Clarke, D., and Grainger, J. F. 1971. *Polarized light and optical measurement*. Oxford: Pergamon Press.

Coulson, K. L.; Dave J. V.; and Sekera, Z. 1960. Tables related to radiation emerging from a planetary atmosphere with Rayleigh scattering. Berkeley & Los Angeles: Univ. Calif. Press.

Coyne, G. V., and Gehrels, T. 1973. Wavelength dependence of polarization. The planet Mercury. *Astron. J*. Vol. 78. In preparation.

Coyne, G. V., and Pellicori, S. F. 1970. Wavelength dependence of polarization. XX. The integrated disk of the moon. *Astron. J*. 75: 54–60.

Davis, L., and Greenstein, J. L. 1951. The polarization of starlight by aligned dust grains. *Astrophys J*. 114: 206–240.

Dollfus, A. 1955. Étude des planètes par la polarisation de leur lumière. Dissertation, Univ. Paris (Trans. Study of the planets by means of the polarization of their light. TTF-188, National Aeronautics and Space Administration, Washington, D.C., 1964.)

———. 1961. Polarization studies of planets. *Planets and Satellites*. (G. P. Kuiper and B. M. Middlehurst, eds.) Chicago, Illinois: Univ. of Chicago Press.

Dunlap, J. L.; Gehrels, T.; and Howes, M. L. 1973. Minor planets and related objects. IX. Photometry and polarimetry of (1685) Toro. *Astron. J*. Vol. 78. In press.

Feofilov, P. P. 1959. *Polyarizovannaya lyuminestsentsiya atomov, molekul i kristallov*. Moscow: State Physico-Mathematical Press. (In English: *The physical basis of polarized emission*. Consultants Bureau, New York, 1961.)

Gehrels, T. 1969. The transparency of the Jovian polar zones. *Icarus* 10: 410–411.

———. 1972. Remarks on techniques for future studies. *From Plasma to Planet*. (A. Elvius, ed.) Stockholm: Almqvist and Wiksell.

———. 1973. Photopolarimetry of planets and stars. *Vistas in Astronomy*. (A. Beer, ed.) Vol. 15. Oxford: Pergamon Press.

Gehrels, T., ed. 1971. *Physical Studies of Minor Planets*. SP-267. Washington, D.C.: National Aeronautics and Space Administration.

Gehrels, T., and Teska, T. M. 1960. A Wollaston photometer. *Publ. Astron. Soc. Pac*. 72: 115–122.

Goldstein, R. M. 1973. Radar observations of the rings of Saturn. *Bull. Am. Astron. Soc*. In press.

Goldstein, R. M.; Holdridge, D.; and Lieske, J. H. 1973. Minor planets and related objects. XII. Radar observations of (1685) Toro. *Astron. J*. Vol. 78. In press.

Greenberg, J. M., and Hulst, H. C. van de 1973. *Interstellar Grains*. Dordrecht, Holland: D. Reidel Publ. Co.

Hall, J. S. 1949. Observations of the polarized light from stars. *Science* 109: 166–167.

Hall, J. S., ed. 1960. Polarization by starlight of the interstellar medium. *Lowell Obs. Bull*. 4: 264–321.

Hall, J. S., and Riley, L. A. 1973. Polarization observations of Saturn made in ultraviolet and visual light since 1968. *Bull. Am. Astron. Soc*. In press.

Hanner, M. S. 1973. Interplanetary results from Pioneer 10. *Bull. Am. Astron. Soc*. In press.

Hanner, M. S., and Weinberg, J. L. 1973. Gegenschein observations from Pioneer 10. *Sky and Tel.* 45: 217–218.
Hapke, B. W. 1963. A theoretical photometric function for the lunar surface. *J. Geophys. Res.* 68: 4571–4586.
Heeschen, D. S. 1973. Optical polarization in the nuclei of galaxies. *Astrophys. J.* 179: L93–L96.
Hemenway, C. L.; Hallgren, D. S.; and Schmalberger, D. C. 1972. Stardust. *Nature* 238: 256–260.
Herbig, G. H. 1972. VY Canis Majoris. III. Polarization and structure of the nebulosity. *Astrophys. J.* 172: 375–381.
Hiltner, W. A. 1949. Polarization of light from distant stars by interstellar medium. *Science* 109: 165.
———. 1950. On polarization of radiation by interstellar medium. *Phys. Rev.* 78: 170–171.
———. 1951. Compensation of seeing in photoelectric photometry. *Observatory* 71: 234–237.
———. 1956. Photometric, polarization, and spectrographic observations of O and B stars. *Astrophys. J. Suppl.* 2: 389–462.
Howard, R., ed. 1971. *Solar magnetic fields.* Dordrecht, Holland: D. Reidel Publ. Co.
Hulst, H. C. van de 1957. *Light scattering by small particles.* New York: Wiley.
Kinman, T. D. 1973. Optical polarization in the nucleus of M87. *Astrophys. J.* 179: L97–L99.
Lal, D. 1972. Accretion processes leading to formation of meteorite parent bodies. *From plasma to planet.* (A. Elvius, ed.) Stockholm: Almqvist & Wiksell.
Lane, A. P., and Irvine, W. M. 1973. Monochromatic phase curves and albedos for the lunar disk. *Astron. J.* Vol. 78. In press.
Lyot, B. 1929. Recherches sur la polarisation de la lumière des planètes et de quelques substances terrestres. *Ann. Obs. Paris* (Meudon) Vol. 8. (Trans. Research on the polarization of light from planets and from some terrestrial substances. TTF-187. Washington, D.C.: National Aeronautics and Space Administration, 1964.)
Minnaert, M. G. J. 1954. *The nature of light & colour in the open air.* Dover Publications.
Roach, F. E. 1972. A photometric model of the zodiacal light. *Astron. J.* 77: 887–891.
Roosen, R. G. 1970. An annotated bibliography on the Gegenschein. *Icarus* 13: 523–539.
Schubart, J. 1971. Asteroid masses and densities. *Physical studies of minor planets.* (T. Gehrels, ed.) SP-267. Washington, D.C.: National Aeronautics and Space Administration.
Shurcliff, W. A. 1955. Haidinger's brushes and circularly polarized light. *J. Opt. Soc. Am.* 45: 399.
———. 1966. *Polarized light, production and use.* Second printing. Cambridge, Mass.: Harvard Univ. Press.
Shurcliff, W. A., and Ballard, S. S. 1964. *Polarized light.* Princeton, N.J.: Van Nostrand Co.
Sparrow, J. G., and Ney, E. P. 1972a. Observations of the zodiacal light from the ecliptic to the pole. *Astrophys. J.* 174: 705–716.
———. 1972b. Polarization of the diffused galactic light. *Astrophys. J.* 174: 717–720.

Stokes, G. G. 1852. On the composition and resolution of streams of polarized light from different sources. *Cambridge Philosoph. Soc.* 9: 399–416.

Strömgren, B. 1956. Report of the Yerkes and McDonald observatories. *Astron. J.* 61: 45.

Swindell, W. 1971. Handedness of polarization at the metallic reflection of linearly polarized light. *J. Opt. Soc. Am.* 61: 212–215.

Zellner, B. H., and Serkowski, K. 1972. Polarization of light by circumstellar material. *Publ. Astron. Soc. Pac.* 84: 619–626.

DISCUSSION

HARWIT: Now you should consider writing a textbook on photopolarimetry. It would have the advantages of consecutive presentation and of uniform style, avoiding multiplicity of definitions.

GEHRELS: The merits of a good textbook are seen in the one you just finished writing (Harwit 1973). During the work for the present book and the above paper, however, I learned that this scheme, if rigorously followed (see pages xv and 3), can produce the best in texts. Its principal advantage is in objectivity, without predilection of a single author.

As for "Proceedings of Meetings," they should not be published anymore. The above scheme is more work for the editorial staff, but the improvement is worthwhile (the reader be the judge). The meeting — having the original goals of assembly, presentation, and debate — is built into the above scheme of book production.

CLARKE: The term *polarization* is most inapt for describing what we now know to be statistical fluctuations of the electric vector in a light beam. In his Baltimore lectures of 1884, Lord Kelvin notes the following:

> ... 'polar' and 'polarization' were, as is now generally admitted, in the very beginning unhappily chosen words for the differences of action in different directions around a ray of light.

Its use seems to have arisen out of a remark by Sir Isaac Newton in his *Opticks,* published just after the turn of the eighteenth century. He had realized that the phenomenon of double refraction (discovered by Bartholinus in 1669) implied that a ray of light, obtained by double refraction, had special properties related to special directions at right angles to its own direction. In 'Question 29' of Book III of the *Opticks* we read:

> ... the unusual Refraction of Island-Crystal looks very much as if it were perform'd by some kind of attractive Virtue lodged in certain Sides both of the Rays, and of the Particles of the Crystal ... since the Crystal by this disposition or Virtue does not act upon the Rays unless when one of their Sides of unusual Refraction look towards that coast, this argues a Virtue or Disposition in those Sides

of the Rays which answers to, and sympathizes with that Virtue or Disposition of the Crystal, as the poles of two Magnets answer to one another. . . .

Newton's reference to *poles* was clearly an analogy to describe the observed behavior. Newton appreciated that the significant parameter in describing double refraction is the relative orientation of two directions. But we are stuck with the word 'polarization' and now there is little point in trying to displace it from the context of the description of the vibrations of light. In any case, there is no obvious single word that could be used in its place.

DISCUSSION REFERENCE

Harwit, M. 1973. *Astrophysical Concepts.* New York: Wiley.

POLARIMETRIC DEFINITIONS

D. CLARKE
The University, Glasgow

Comment is made on some polarimetric definitions. For the fourth Stokes parameter, it is apparent from the current literature that different definitions and sign conventions are being used. Various alternatives are presented for consideration.

In the opening chapter of his book on polarized light, Shurcliff (1962) comments, "In preparing this book the author faced a major problem as to conventions." His book aimed at covering a wide range of disciplines throughout science, and he noted that researchers in different branches of science employ various sets of conventions for signs, handedness, etc. Current literature on the recent advances in astronomical polarimetric measurements reveals that there is no accepted system of conventions even within this subject alone. I do not know if this presents us with a "major problem," but it does cause confusion. My purpose here is to comment on some polarimetric definitions and to point out differences of convention used in current astronomy literature.

The most convenient way to express polarizational information is to use the four Stokes parameters symbolized by I, Q, U, V. (A variety of symbols occurs in the literature. For example, some authors prefer I, M, C, S, and in Stokes' [1852] original presentation, A, B, C, D were used.) Stokes parameters are phenomenological in that they are directly related to the signals produced by polarimeters (e.g., see Clarke and Grainger [1971] in optical astronomy, and Cohen [1958] in radio astronomy). They are also readily describable in terms of the parameters used for the classical description of electromagnetic waves (e.g., see Chandrasekhar [1950]).

There now appears to be universal agreement on the use of the *electric vector* to describe the orientation of the wave vibrations. A quasi-monochromatic electromagnetic disturbance may be written in terms of the components of this vector in orthogonal planes xz, yz as:

$$E_x = E_{x_0} e^{i(\omega t - \frac{2\pi z}{\lambda} + \delta_x)}$$

and

$$E_y = E_{y_0} e^{i(\omega t - \frac{2\pi z}{\lambda} + \delta_y)},$$

where E_x, E_y are the values of the electric field in the x- and y- directions at the position z and at time t; E_{x_0}, E_{y_0} are the amplitudes of the x- and y- vibrations; δ_x, δ_y are the phases (advancements) of the x- and y- vibrations at $z = 0$; ω is the angular frequency, λ is the wavelength and i is the square root of minus 1. The system x, y, z is usually defined to be a right-hand frame.

For such a beam as described by the two wave equations above, the Stokes parameters may be expressed as:[1]

$$I = <E_{x_0}^2> + <E_{y_0}^2>$$
$$Q = <E_{x_0}^2> - <E_{y_0}^2>$$
$$U = <2 E_{x_0} E_{y_0} \cos(\delta_y - \delta_x)>$$
$$V = <2 E_{x_0} E_{y_0} \sin(\delta_y - \delta_x)>.$$

The degree of polarization, p, may be defined as:

$$p = \frac{(Q^2 + U^2 + V^2)^{1/2}}{I}.$$

For any beam of radiation, the parameters Q and U take on particular values according to the chosen reference axes, i.e., if the Stokes parameters are known in terms of axes x, y, then only the values of Q and U are affected if the axes of reference are rotated to new directions x', y'. It is these two middle parameters that define an angle ζ relating to the direction of vibration of any linear polarization or the azimuth of the major axis of any elliptical polarization. Thus

$$\tan 2\zeta = U/Q.$$

So that the correct value of ζ is taken from this identity, note must be taken of the individual signs of U and Q.

---

[1] Also see p. 359 and p. 730.

For a celestial source, an obvious choice of reference axes are those of declination (equivalent to x-axis) and right ascension (equivalent to y-axis), with the direction of propagation completing a right-hand frame. Hence a direction of vibration might be expressed as an angle running from 0° to 180°, increasing in a sense going from north, through east (e.g., see Kristian et al. 1970).[2]

For some observations, alternative reference frames are used; for example, in lunar and planetary studies, the perpendicular to the *scattering plane* formed by the sun, scattering body, and observer may act as the reference axis.[3] Related to these studies, comment should perhaps be made on the terms "positive" and "negative" polarization. In many situations where polarization is produced by the scattering of light, the direction of vibration occurs at either one of two orthogonal positions. Generally, it is found at right angles to the scattering plane. This type of polarization has been termed "positive." Thus, when unpolarized light is reflected by a dielectric surface, Fresnel's laws show that, over the whole range of angles of incidence, the electric vector resolved perpendicularly to the plane of incidence is always reflected more strongly than the parallel component, with the reflected light exhibiting "positive" polarization. Occasionally scattering processes give rise to polarization with the direction of vibration lying in the scattering plane. This type of polarization has been called "negative." However, usage of the terms "positive" and "negative" in this context should perhaps not be encouraged as the degree of polarization is essentially always a positive quantity, and in some studies, the observed direction of vibration is neither "positive" nor "negative." The values of Q and U or ζ are sufficient to describe the direction of vibration. It will be recalled here also that the terms "positive" and "negative" are sometimes used in relation to the fourth Stokes parameter, V.

Perhaps the worst confusion over definitions is in connection with the fourth parameter, V. This parameter describes the shape of the polarization ellipse and the sense of description, or handedness of it. In describing the shape of the polarization ellipse, it appears that, especially in the field of optics, the term *ellipticity* is sometimes used synonymously with *axial-ratio;* this point has been made previously by Kraus (1950). Usually, ellipticity is defined[4] as the ratio of the difference of the major and minor axes to the major axis, i.e., $(a - b)/a$. With this definition, the value of ellipticity increases as the ellipse departs more from a circle

---

[2] In that frame θ is used in this book. — Ed.
[3] See the definition of θ_r in the Glossary. — Ed.
[4] An alternate definition of ellipticity is given on p. 733.

towards a straight line. If η is defined as the axial ratio, the ellipticity is given by $(1-\eta)$. The value of η may be obtained from:

$$\frac{|V|}{I} = \frac{2\eta}{1+\eta^2}$$

Current literature shows that different definitions and conventions for V are used by different authors. This results chiefly from the arbitrary nature of these definitions and conventions, although it must be said that on aesthetic grounds, perhaps one particular set has merit and should be preferred.

If one considers circularly polarized light as it passes through a particular plane, the electric vector in that plane rotates with a constant frequency. The sense of rotation depends on the viewpoint of the plane. A clockwise rotation viewed by an observer receiving the radiation would be declared as counterclockwise if the observer were to move his position to that of the source. Thus, terms such as *right-handed* or *left-handed* are arbitrary. For example, Gehrels (1972) writes "... right-handed if the electric vector when approaching the observer is seen rotating clockwise; ..." and Martin, Illing, and Angel (1972) write, "..... (right-handed) when, to an observer facing the Nebula, the electric vector in a fixed plane rotates counterclockwise," i.e., clockwise when viewed from the source.

The above dilemma need not exist if, instead of considering the behavior of the electric vector in a particular plane, it is explored along the z-axis at a particular instant. An instantaneous snapshot picture of the tip of the electric vector as it is distributed along the z-axis would show a helix. As a helix has a handedness that is defined independently of the observer's viewpoint, the polarization handedness may be labeled by that helix (left- or right-handed).

In considering the link between the snapshot picture and what happens to the electric vector in a particular plane, say the xy-plane at $z = 0$, confusion again unfortunately exists in the literature. For example, Born and Wolf (1970) give the impression that with the passage of time the helix "screws" its way through the plane; with this erroneous concept, a right-handed helix would provide a vector tip that would rotate counterclockwise in a fixed plane when viewed by an observer receiving the radiation. They write:

> It seems natural to call the polarization right-handed or left-handed according as to whether the rotation of **E** and the direction of propagation form a right-handed or left-handed screw. But the traditional terminology is just opposite — being based on the apparent behaviour of E when "viewed" face-on by the observer. ... Thus we

say that the polarization is *right-handed* when to an observer looking in the direction from which the light is coming, the endpoint of the electric vector would appear to describe the ellipse in the clockwise sense.

However, it is more correct to say that the helix moves along the direction of propagation *without rotation* and that its point of intersection with the plane transverse to this direction executes the polarization ellipse; the sense of execution, as seen by the observer receiving the radiation, is clockwise for a right-handed helix. This is illustrated in Fig. 1 where a beam of circularly polarized light has been resolved into linear components, the y-component being $\pi/2$ in advance of the x-component. At a particular instant, the distribution in space of the tip of the electric vector is in the form of a right-handed helix. As the two components advance, keeping the same phase relationship, the electric vector would appear to rotate clockwise in the xy-plane at $z = 0$ to an observer receiving the radiation. Thus, the traditional terminology is *not* just opposite, as Born and Wolf suggest, but is in exact accord. If the snapshot concept of the helical distribution of the tip of the electric vector is interpreted properly, it provides an unequivocal means of defining hand-

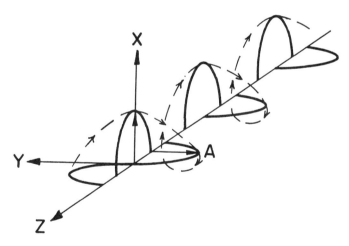

Fig. 1. The resolution of a circularly polarized beam into two linear components with a phase difference of $\pi/2$ (y-component in advance of x-component). For clarity, along negative z, only positive values of the x-component and negative values of the y-component have been drawn. This snapshot picture illustrates the right-handed helical distribution of the tip of the electric vector (the dashed line). As the two components advance keeping the same phase relationship, the next wave crest to arrive in the xy-plane at $z = 0$ corresponds to the point A, and the observer sees a clockwise rotation of the electric vector in this plane.

edness; this in turn may be related to the time-dependent behavior of the electric vector in a fixed plane according to the viewpoint of the observer.

In addition to the alternative definitions of handedness that are in use, the choice of the sign V to relate a particular handedness is also arbitrary, although perhaps there is an innate preference to associate a *positive* value to *right* hand.

From the above equations defining the Stokes parameters, it can be seen that a positive value of V shows that the y-component is in advance of the x-component, and, to an observer receiving the radiation, the electric vector would appear to rotate clockwise. However, it is equally valid to allow the y-component to act as reference when the phase difference between the components is considered and to write the fourth Stokes parameter as:

$$V = <2\, E_{x_0} E_{y_0} \sin(\delta_x - \delta_y)>.$$

With this definition, it is obvious that an alternative sign/handedness convention results. (This alternative also results if the terms δ_x, δ_y are given negative signs in the equations of the wave disturbance, thus signifying retardations rather than advancements, and if the phase differences are formed as before as $[\delta_y - \delta_x]$). It is quite understandable, therefore, to find Gehrels (1972), for example, writing "... the plus sign is called right-handed if the electric vector when approaching the observer is seen rotating clockwise;..." and Pospergelis (1969) writing, "... the quantity $S(\equiv V)$ is negative for positive γ (the electric vector rotates clockwise looking toward the direction of propagation of the light beam)." With this situation, it goes without saying that references to V as being either positive or negative are meaningless, unless they are accompanied by explicit definitions.

If, as in the case of defining the direction of vibration of linear polarization, the north direction is taken as reference axis, and if by a positive phase an advancement is implied, then a *positive* value of the phase difference $(\delta_{RA} - \delta_\delta)$ implies a *right-hand* helix.

Before promoting any particular definition for handedness and the sign of V in optical astronomy, comparison should perhaps be made with the well-established practices of radio astronomy. Their handedness follows the Institute of Radio Engineers standard (IRE 1942) which states that a right-handed elliptically polarized wave is one in which the rotation of the direction of displacement is clockwise for an observer looking in the direction in which the wave is traveling, i.e., the snapshot picture would be that of a left-handed helix. Kraus (1950) has pointed out the convenience of this definition in relation to helical antennas.

In early solar polarimetric studies, a negative value of V was used to imply the radio right-hand definition (Cohen 1958; Kraus 1966); in the more recent galactic studies the opposite convention is preferred (Gardner and Whiteoak 1966; Conway,[5] personal communication, 1972). Table I shows alternative definitions of handedness and their sign conventions used to describe circular polarization. Two questions come to mind: (1) Should one combination be selected for use in optical astronomy, and if so, (2) which combination is to be preferred?

TABLE I

Examples of Handedness Definitions and Sign Conventions Associated With Stokes Parameter, V

Helix in Space	Preferred Sign	Optical			Radio	
		Gehrels (1972)	Pospergelis (1969)	Martin, Illing, and Angel (1972)	Gardner and Whiteoak (1966) (Galactic)	Cohen (1958) (Solar)
Right	+	Right +	Right −	Left −	Left −	Left +
Left	−	Left −	Left +	Right +	Right +	Right −

REFERENCES

Born, M., and Wolf, E. 1970. *Principles of optics.* 4th ed. Oxford: Pergamon Press.
Chandrasekhar, S. 1950. *Radiative transfer.* London: Oxford Univ. Press.
Clarke, D., and Grainger, J. F. 1971. *Polarized light and optical measurement.* Oxford: Pergamon Press.
Cohen, M. H. 1958. Radio astronomy polarization measurements. *Proc. IRE* 46: 172–183.
Gardner, F. F., and Whiteoak, J. B. 1966. The polarization of cosmic radio waves. *Ann. Rev. Astron. and Astrophys.* 4: 245–292.
Gehrels, T. 1972. On the circular polarization of HD 226868, NGC 1068, NGC 4151, 3C273 and VY Can Maj. *Astrophys. J.* 173: L23–L25.
Institute of Radio Engineers 1942. Definition of terms. *Proc. IRE* 30: no. 7 pt. III suppl. IW 47.
Kraus, J. D. 1950. Wave polarization. *Antennas.* New York: McGraw Hill.
——— 1966. *Radio astronomy,* ch. 4. New York: McGraw Hill.
Kristian, J.; Visvanathan, N.; Westphal, J. A.; and Snellen, C. H. 1970. Optical polarization and intensity of the pulsar in the Crab Nebula. *Astrophys. J.* 162: 475–483.
Martin, P. G., Illing, R., and Angel, J. R. P. 1972. Discovery of interstellar circular polarization in the direction of the Crab Nebula. *Mon. Not. R. Astr. Soc.* 159: 191–201.

---

[5] Also see p. 352.

Pospergelis, M. M. 1969. Measuring the Stokes parameters for scattered light. *Sov. Astr. A. J.* 12: 973–977.

Shurcliff, W. A. 1962. *Polarized light*. Cambridge, Mass: Harvard Univ. Press.

Stokes, G. G. 1852. *Trans. Cambridge Phil. Soc.* 9:399; see also, ———, 1901. *Mathematical and physical papers*, vol. 3. London: Cambridge Univ. Press.

DISCUSSION

DICKEL: The radio astronomers have defined planetary phase angle as the planetocentric angle between the earth and the sun measured westward from the sun to the earth, increasing from 0° through 180° to 360?

GEHRELS: After several interchanges with various authors, I have for this book adopted the following:

1. Stokes parameters: I, Q, U, V.

2. Degree of linear polarization (lower-case p):

$$p = \frac{(Q^2 + U^2)^{1/2}}{I}.$$

Degree of circular polarization $q = V/I$.

The degree of polarization to be expressed in power notation; percentages are, however, acceptable especially when appreciable amounts of polarization are discussed.

It remains to be seen if we should use (lower-case s):

$$s = \frac{(Q^2 + U^2 + V^2)^{1/2}}{I}.$$

3. With reference to a scattering plane, the degree of linear polarization may be defined as (upper-case P):

$$P = \frac{I\perp - I\|}{I\perp + I\|}.$$

See the Glossary for the definition of the position angle θ_r. The expression negative polarization is used so often in the discussions of solid surfaces (see p. 381) that we have made no attempt to change it.

4. Define axial ratio as $\eta = b/a$;

ellipticity is then $(1 - \eta) = \dfrac{a-b}{a}$.

A different definition of ellipticity is, however, given by Tandberg-Hanssen (p. 733).

5. The term "direction of vibration" is used in preference to "plane of polarization," and the electric vector is used to define the direction of vibration. We have used the position angle, θ, measured eastward from north, that is, in a coordinate frame with the x-axis in the direction of positive declination, y in the direction of increasing right ascension, and z in the direction of propagation of the radiation.

6. Positive circular polarization is when the electric vector has increasing θ with time. The term *hand*edness can be defined with a hand: 1) stick out a thumb to indicate the direction of propagation of the radiation; 2) the curved fingers indicate the direction in which the electric vector moves with time; 3) which hand described the situation? One finds that right-handed circular polarization is positive. (An observer of right-handed circular polarization sees a counter-clockwise rotation of E. Care must be taken so that data are presented after correction for handedness reversals by telescope mirrors.)

Note that the above definition is independent of whether one observes along the direction of propagation or against the direction. This definition is also consistent with the general definition of momentum vectors; the angular momentum of circular polarization is positive for "positive circular polarization." Finally, people generally prefer to associate *right* with *positive*. This definition is the same as that of Martin et al. (1972) and Gardner and Whiteoak (1966) in Clarke's Table I.

MEEUS: I regret that in this book the term "micrometer" and its symbol μm are used, instead of the classical "micron." I know that in 1967 the Bureau International des Poids et Mesures had officially adopted the micrometer instead of the micron, but it is a stupid resolution. The word micron has been used for many years, it is still being used by most people, and old terms should not be changed without serious reasons. But there is no serious reason to pass from the micron to the micrometer; it is not more difficult to work with microns than with micrometers. In fact, now there is confusion because the word micrometer already has another meaning: it is an instrument to measure small lengths. Now we may write texts such as: "The length of 1 micrometer has been measured with a micrometer." If the Bureau has made the change just to have a completely uniform system (km, m, mm, μm, nm), why did it not also change the ton to the megagram?

Similarly, I feel the Ångstrom should be kept. All our spectroscopic lists and catalogs are using Ångstroms. For an astrophysicist, the number 1216, for example, means the Lyman alpha line.

MECHANISMS THAT PRODUCE LINEAR AND CIRCULAR POLARIZATION

J. R. P. ANGEL
Columbia University

A compilation of different polarizing mechanisms is given, with emphasis on those of current interest.

Polarization arises in many of the radiation generation and transfer processes of astrophysics. In this review we hope to cover at least some of the more interesting polarization mechanisms that have been proposed or discovered in the past few years. In particular, we will emphasize mechanisms causing circular polarization, which until recently was scarcely known in optical astronomy except in the Zeeman effect.

Magnetic fields play an important role generally in polarization mechanisms, either directly in the emission process itself or indirectly in the alignment of scattering grains. Essentially, the field provides the ordering over sizable regions of emission or scattering that is necessary for polarization. As well as determining the direction of linear polarization, the magnetic field, which is a pseudovector with a definite sense of rotation, can give rise to circular polarization. The other important source of polarization in optical astronomy is scattering in a nonspherically symmetric geometry, which is well known as a source of linear polarization. In addition, an asymmetric scattering geometry with a definite handedness can give rise to circular polarization.

In the following discussion we will consider first ordered emission and absorption mechanisms, nearly all magnetic, that give polarization at the source of the radiation, and then scattering processes that introduce or modify polarization in radiative transfer. We first examine the simple emission process of radiation by free electrons in a magnetic field.

Nonthermal Emission in a Magnetic Field

The most familiar astronomical example of ordered emission in a magnetic field is synchrotron radiation by highly relativistic electrons. Essentially pure linear polarization is produced in regions of uniform magnetic field (see, for example, Jackson 1962). This mechanism is thought to explain the linear polarization of many nonthermal sources in radio and optical astronomy as, for example, in the strong polarization of the Crab Nebula observed from radio to X-ray wavelengths (Novick et al. 1972).[1] Synchrotron radiation can also show circular polarization, and recent results for an optically thin synchrotron emitter are given by Legg and Westfold (1968). In a region of uniform magnetic field in which the total emission is averaged over electrons with a uniform distribution of pitch angle, a component of circular polarization of order $V \simeq (\omega_B/\omega)^{1/2}$ is predicted. Here ω is the radiation frequency and ω_B the gyration frequency. At fields of order or greater than 0.1 gauss this becomes detectable at radio frequencies and is thought to be responsible for the circularly polarized radio emission from some quasars (see, for example, the recent results by Roberts et al. 1972). In fact, these objects are not optically thin, and a development of the theory to cover optically thick sources is given by Pacholczyk.[2]

A way in which synchrotron radiation can be strongly circularly polarized even at optical frequencies has been pointed out by Rees (1971) and has been further developed by Arons (1972) and Blandford (1972). If the relativistic electrons are moving in the strong wave field of a rotating neutron star, then provided the gyration frequency ω_B is greater than the wave frequency ω_{rot}, the analysis is similar to that for a DC field. However, away from the equatorial plane of the neutron star where the low frequency wave is circularly polarized, the electron radiation will also be circularly polarized with $V \simeq \omega_{rot}/\omega_B$. If this process were responsible for the light from the Crab Nebula, then circular polarization of a few percent would be expected. However, the observational upper limit of an effect with the correct spatial dependence in the Crab is 0.1% (Landstreet and Angel 1971a; Martin, Illing, and Angel 1972).

At nonrelativistic velocities, electrons spiraling in a magnetic field emit circularly polarized cyclotron radiation with frequency $\omega_B = 2.8$ MHz gauss $^{-1}$. This type of emission is observed at radio frequencies, for example, in the decametric bursts from Jupiter (Burke and Franklin 1955).

Thermal Processes in a Magnetic Field

This is a more complicated situation in which atomic or molecular processes — bound-bound, bound-free, and free-free transitions — have

---

[1] See pp. 262 and 1014.   [2] See p. 1030.

to be considered in the presence of a magnetic field. The effect of a magnetic field on bound-bound transitions (the Zeeman effect) is the splitting of spectral lines into components of different polarization. There are many astronomical examples of ordered magnetic fields that result in detectable polarization effects, and magnetic fields have been measured over the enormous range from 10^{-5} gauss in the interstellar medium (Verschuur 1970)[3] to 10^7 gauss in the magnetic white dwarfs (Angel and Landstreet 1972).[4] The theory of the atomic Zeeman effect is well known, and we will not discuss it here. However, there is an important qualitative difference in the molecular Zeeman effect seen in the magnetic white dwarfs. Circular polarization of molecular bands is not caused predominantly by the splitting of oppositely polarized Zeeman components, which is generally very small for molecules. Instead, it is the difference in intensity of the left- and right-handed transitions that results in a net polarization of entire branches and band heads (Angel 1972). This second-order effect due to mixing of states with different J values becomes important in molecules in which closely spaced rotational levels can be coupled by the magnetic interaction. An accurate calculation for the G band of CH, seen in the magnetic white dwarf G99-37, shows that the P and R branches exhibit opposite polarization, the net circular polarization for the entire P branch being 0.17% per 10 kilogauss. The effect has been measured in the laboratory using fields up to 19 kilogauss and is in good agreement with the theoretical prediction. In G99-37 a strong polarization feature is seen at the G band, values of circular polarization of -3.8% and $+2.8\%$ being measured in 160-Å-wide bands centered at 4220 and 4380 Å. These are interpreted as due to opposite circular polarization of the P and R branches and indicate an effective field of 2.5×10^6 gauss.

The polarizing effect of a magnetic field on thermal continuum absorption and emission is detectable at radio wavelengths for fields $\gtrsim 10$ gauss and in the optical spectrum for fields $\gtrsim 10^6$ gauss. The latter effect, pointed out by Kemp (1970a), has proved to be valuable in detecting strong fields in white dwarfs. A magnetic field acting along the direction of propagation introduces in the absorption coefficient a dependence on circular polarization. The differential effect for free-free absorption given by magnetoionic theory (see, for example, Ratcliffe 1959) is

$$\frac{\kappa_L}{\kappa_R} = \left(\frac{\omega + \omega_B}{\omega - \omega_B}\right)^2 ,$$

where κ_L and κ_R are the absorption coefficient for left- and right-handed

---

[3] See p. 960. [4] See p. 981.

circular polarization and propagation is in the magnetic field direction. For emission along the field lines from an optically thin plasma, this yields a circular polarization

$$V = \frac{2\omega_B}{\omega} \quad (\omega_B \ll \omega).$$

Circular polarization of a few percent produced by this mechanism is observed in the enhanced millimeter emission associated with sunspots, found by Khangil'din (1964) and recently discussed by Edelson, Mayfield, and Shimabukuro (1971), Kundu and McCullough (1972), and Feix.[5] The magnetic fields deduced for the above formula are in the range 25–500 gauss, consistent with the emission coming from the chromosphere above the sunspot. In the optical spectrum, free-free interactions play a role in producing the continuum circular polarization observed in magnetic white dwarfs (see review by Landstreet[6]). Bound-bound and bound-free mechanisms must also be involved though, and a simple interpretation is not possible because of the effect of radiative transfer. Lamb, Pethick, and Pines (1973) and Rees[7] have pointed out that X-rays from the pulsed sources may also be circularly polarized by the same mechanism. This would be the case if the emission is thermal bremsstrahlung from matter accreted at the poles of a magnetic neutron star with the field strength of the order 10^{12} gauss, a model suggested by several authors for the X-ray pulsars for Cen X-3 and Her X-1.

A recent analysis of bound-free optical polarization effects in strong magnetic fields has been given by Lamb and Sutherland (1972). These authors give expressions for the circular and linear polarization introduced by passage through an optically thin atmosphere in terms of the zero field opacity. Applied to hydrogen, away from the absorption edges they find the circular polarization is given by $V = -2\bar{\tau} \frac{\omega_B}{\omega} \cos \psi$, where $\bar{\tau}$ is the mean optical depth of the absorbing layer and ψ is the angle between the propagation and field directions. Linear polarization of maximum value $p = \frac{5}{4} \bar{\tau} (\omega_B/\omega)^2$ occurs in the bound-free case. At wavelengths near the absorption edges strong circular polarization of opposite sign is found.

Optical cyclotron radiation will be generated in a hot plasma in strong magnetic fields, $B \gtrsim 10^8$ gauss. This has been suggested as a possible contributor to the circular polarization of magnetic white dwarfs by Kemp (1970b) and Lamb and Sutherland (1972).

---

[5] See p. 676. [6] See p. 981. [7] Personal communication.

Other Polarized Emission Processes

A directed stream of electrons interacting with matter can give rise to polarized bremsstrahlung and line emission. An analysis of the expected effect in solar X-ray flares by this mechanism was made by Korchak (1967), and X-ray polarization has recently been reported by Tindo et al. (1972).

A discussion of the infrared thermal emission from aligned dust grains, which is linearly polarized, has been given recently by Capps and Dyck (1972).[8] Measurements by these authors at 10 μm do not show the strong polarization that could be produced if there was good alignment.

Finally, a unique emission mechanism, giving pure polarized radiation, is the generation of low-frequency electromagnetic waves by a rotating magnetic neutron star.

POLARIZATION EFFECTS OF SCATTERING

Scattering is well known as the cause of linear polarization in many astronomical objects. The property of some scatterers of introducing a circular component when linearly polarized light is incident has recently been demonstrated in several examples. We must, therefore, consider the effect of scattering on both unpolarized and polarized incident light. A detailed treatment of scattering is given by van de Hulst (1957). In the following, we summarize the processes of astrophysical importance.

Electron and Compton Scattering

Unpolarized or linearly polarized light incident on free electrons is, in general, partially linearly polarized after scattering, complete polarization with the electric vector perpendicular to the plane of scattering occurring for 90° scattering. The cross section for electron scattering is the same for all wavelengths $\lambda \ll \hbar/m_ec$, the Compton wavelength of the electron. An example of polarization by electron scattering in an optically thin situation is that seen in the continuum light of the sun's corona. Polarization arising in radiative transfer by electron scattering was calculated by Chandrasekhar (1950)[9] and results in a small polarization of rotationally distorted or eclipsed stars. The linear polarization of Be stars is very likely caused by electron scattering, thought to take place in an ionized disk around the star.

---

[8] See p. 858.   [9] See p. 663.

If circularly polarized light is scattered by electrons, then the scattered light (except at 90°) is also circularly polarized. This is true even when the electrons are relativistic and photon energies are greatly increased by Compton scattering. Because of this property, Sciama and Rees (1967) have suggested that circular polarization of the optical emission from quasars would constitute strong proof of Compton scattering, since direct optical synchrotron radiation must have very small circular polarization (see above). As yet there is no certain detection of optical circular polarization in quasars or Seyfert nuclei, even those showing radio circular polarization (Landstreet and Angel 1972). Another situation in which Compton scattering of circularly polarized radiation may take place is in X-ray stars if there is accretion onto a magnetic white dwarf or neutron star. This process, which could result in circularly polarized light and X-rays, is discussed by Gnedin and Sunyaev (1972).

Atomic and Molecular Scattering

The angular-dependence polarization of light scattered by atoms and molecules is represented ideally by Rayleigh scattering. In practice the polarization is not complete in any direction (see, for example, Chandrasekhar 1950). Free atoms show very strong scattering of their resonance lines, the peak polarization depending on the Zeeman sublevels of the states involved and their lifetimes. If an atom in a magnetic field precesses significantly while in the excited state, then the polarization on decaying will be rotated and in a strong field lost altogether (the Hanle effect). A general discussion of polarization effects in resonance radiation is given by Mitchell and Zemansky (1961), and astrophysical examples are given by the emission lines in solar prominences and the solar corona, discussed by Hyder (1965) and Warwick and Hyder (1965).

Polarization in molecular scattering is very familiar in being the origin of the blue day sky polarization[10] and also of polarization in some of the planets.[11] In this case, incomplete polarization is understood in terms of aerosols, "ground" reflections, multiple molecular scattering, and anisotropy of the molecules, averaged over random orientation.

Scattering by Dust Particles

Scattering by dust particles has long been recognized as an important source of linear polarization in optical astronomy. The result of grain

---

[10] See p. 444.   [11] See p. 518.

alignment causing preferential extinction of one sense of linear polarization and, hence, interstellar linear polarization does not need much discussion here. The optical effects are well understood (see, for example, van de Hulst 1957, or Greenberg 1968), and recent observations of the wavelength dependence are in good agreement with a model of imperfectly aligned cylindrical grains (Serkowski 1973a).[12] However, the mechanism by which the grains are aligned is not clear. Recent calculations indicate that the magnetic fields required for grain alignment are higher than deduced from other observations (Purcell and Spitzer 1971). Nevertheless, the model of aligned grains causing linear polarization of transmitted light is generally accepted as valid, and recently the other polarization effects of radiative transfer through a medium with aligned grains have been studied and detected. It was pointed out by van de Hulst (1957) that such a medium is birefringent and will thus introduce ellipticity to a linearly polarized wave. Martin (1972) has recently calculated these effects in detail. The phase shift introduced between perpendicular components of linearly polarized light is of the same order as the fractional linear polarization that would be introduced in unpolarized light, the actual relation depending strongly on the assumed complex refractive index of the grains. Ellipticity caused by this mechanism has been discovered in the polarized light from the Crab Nebula (Martin, Illing, and Angel 1972), where the sign and wavelength dependence of the circular component are characteristic of dielectric grains.[13] The effect in the Crab Nebula can be clearly identified as interstellar rather than the intrinsic strong wave effect discussed above, from the observed dependence of circular polarization on strength and position angle of linear polarization in different parts of the Nebula. This is exactly as expected if a weak wave plate were held up in front of the Nebula with its axis along the independently known direction of grain alignment.

If unpolarized light travels through a medium in which the alignment direction is not constant, then linear polarization introduced early can later be made elliptical. With a favorable geometry this results in ellipticity of the same order as in the Crab Nebula (~ 0.01). This effect was searched for first unsuccessfully by Serkowski and Chojnacki (1969) but has recently been found in the bright stars σ Sco and o Sco (Kemp and Wolstencroft 1972).

Grain alignment also has an effect on the polarization of scattered light. It has recently been pointed out by Schmidt (1973) that unpolarized light, scattered at an angle by a nonspherical grain, can be ellip-

---

[12] See p. 888. [13] See p. 926.

tically polarized. He finds a maximum ellipticity of 0.15 under the most favorable conditions. This mechanism is probably responsible for the circular polarization of the diffuse galactic light reported by Staude, Wolf, and Schmidt (1973).[14]

When dust scattering takes place in an optically thick cloud, circular polarization can arise by multiple scattering and without the need of alignment, provided the grains are of finite size. Linearly polarized light found on the first scattering and then scattered again will in general become elliptically polarized, provided that the incident electric vector is not parallel or perpendicular to the plane of scattering. This mechanism probably accounts for the circular polarization discovered first in VY CMa by Gehrels (1972) and apparently common in infrared stars with circumstellar shells (Serkowski 1973b; Angel and Martin 1973). It also explains the circular polarization seen in planets with scattering atmospheres (Hansen 1971; Kemp, Wolstencroft, and Swedlund 1971),[15] including the earth (Angel, Illing, and Martin 1972).

Acknowledgments. I wish to thank Peter Martin for useful discussions on polarization by grains. This work was supported by the Research Corporation and the National Science Foundation under grant NSF GP-31356X.

REFERENCES

Angel, J. R. P. 1972. Interpretation of the Minkowski bands in Grw+70°8247. *Astrophys. J.* 171: L17–L21.
Angel, J. R. P.; Illing, R.; and Martin, P. G. 1972. Circular polarization of twilight. *Nature* 238: 389–390.
Angel, J. R. P., and Landstreet, J. D. 1972. Observation of the Zeeman effect in the magnetic white dwarf G99-37. *Bull. Am. Astron. Soc.* 4: 409.
Angel, J. R. P., and Martin, P. G. 1973. Observations of circumstellar circular polarization in four more infrared stars. *Astrophys. J.* 180: L39–L41.
Arons, J. 1972. Nonlinear inverse Compton radiation and circular polarization of diffuse radiation from the Crab Nebula. *Astrophys. J.* 177: 395–410.
Blandford, R. D. 1972. The polarization of synchro-Compton radiation. *Astron. Astrophys.* 20: 135–144.
Burke, B. F., and Franklin, K. L. 1955. Observations of a variable radio source associated with the planet Jupiter. *J. Geophys. Res.* 60: 213–217.
Capps, R. W., and Dyck, H. M. 1972. The measurement of polarized 10-micron radiation from cool stars with circumstellar shells. *Astrophys. J.* 175: 693–697.
Chandrasekhar, S. 1950. *Radiative transfer.* Oxford, England: Clarendon Press.
Edelson, S.; Mayfield, E. B.; and Shimabukuro, F. I. 1971. Polarized solar radio emission at mm wavelengths associated with sunspot magnetic fields. *Nature Phys. Sci.* 232: 82–84.

---

[14] See p. 779. [15] See p. 607.

Gehrels, T. 1972. On the circular polarization of HD 22868, NGC 1068, NGC 4151, 3C 273, and VY Canis Majoris. *Astrophys. J.* 173: L23–L25.

Gnedin, Y. N., and Sunyaev, R. A. 1972. Luminosity of thermal X-ray sources with a strong magnetic field. Submitted for publication.

Greenberg, J. M. 1968. Interstellar grains. *Stars and stellar systems*. (B. M. Middlehurst and L. H. Aller, eds.) Vol. 7, pp. 221–364. Chicago: Univ. Chicago Press.

Hansen, J. E. 1971. Circular polarization of sunlight reflected by clouds. *J. Atmos. Sci.* 28: 1515–1516.

Hulst, H. C. van de 1957. *Light scattering by small particles*. New York: Wiley.

Hyder, C. L. 1965. The polarization of emission lines in astronomy. III. The polarization of coronal emission lines. *Astrophys. J.* 141: 1382–1389.

Jackson, J. D. 1962. *Classical electrodynamics*. New York: Wiley.

Kemp, J. C. 1970a. Circular polarization of thermal radiation in a magnetic field. *Astrophys. J.* 162: 169–179.

———. 1970b. Quantum magnetic features in the polarized light of Grw+70°8247. *Astrophys. J.* 162: L69–L72.

Kemp, J. C., and Wolstencroft, R. D. 1972. Interstellar circular polarization: data for six stars and the wavelength dependence. *Astrophys. J.* 176: L115–L118.

Kemp, J. C.; Wolstencroft, R. D.; and Swedlund, J. B. 1971. Circular polarization: Jupiter and other planets. *Nature* 232: 165–168.

Khangil'din, U. V. 1964. Characteristics of solar active regions obtained from observations in millimeter wavelengths. *Sov. Astron. A. J.* 8: 234–242.

Korchak, A. A. 1967. Possible polarization of bremsstrahlung from solar flares. *Sov. Phys.-Dokl.* 12: 192–194.

Kundu, M. R., and McCullough, T. P. 1972. Polarization of solar active regions at 9.5-mm wavelength. *Solar Phys.* 24: 133–141.

Lamb, F. K.; Pethick, C. J.; and Pines, D. 1973. A model for compact X-ray sources: accretion by rotating magnetic stars. *Astrophys. J.* In press.

Lamb, F. K., and Sutherland, P. G. 1972. Continuum polarization in magnetic white dwarfs. *Physics of dense matter,* I.A.U. Symp. No. 53, Boulder, Colorado.

Landstreet, J. D., and Angel, J. R. P. 1971a. Search for optical polarization in the Crab Nebula. *Nature* 230: 103.

Landstreet, J. D., and Angel, J. R. P. 1972. Search for optical polarization in quasars and Seyfert nuclei. *Astrophys. J.* 174: L127–L129.

Legg, M. P. C., and Westfold, K. C. 1968. Elliptic polarization of synchrotron radiation. *Astrophys. J.* 154: 499–514.

Martin, P. G. 1972. Interstellar circular polarization. *Mon. Not. R. Astron. Soc.* 159: 179–190.

Martin, P. G.; Illing, R.; and Angel, J. R. P. 1972. Discovery of interstellar circular polarization in the direction of the Crab Nebula. *Mon. Not. R. Astron. Soc.* 159: 191–201.

Mitchell, A. C. G., and Zemansky, M. W. 1961. *Resonance radiation and excited atoms*. Cambridge, England: Cambridge Univ. Press.

Novick, R.; Weisskopf, M. C.; Berthelsdorf, R.; Linke, R.; and Wolff, R. S. 1972. Detection of X-ray polarization of the Crab Nebula. *Astrophys. J.* 174: L1–L8.

Purcell, E. M., and Spitzer, L. Jr. 1971. Orientation of rotating grains. *Astrophys. J.* 167: 31–62.

Ratcliffe, J. A. 1959. *The magneto-ionic theory*. Cambridge, England: Cambridge Univ. Press.

Rees, M. J. 1971. Implications of the 'wave field' theory of the continuum from the Crab Nebula. *Nature Phys. Sci.* 230: 55–57.

Roberts, J. A.; Ribes, J. C.; Murray, J. D.; and Cooke, D. J. 1972. Measurements of the circular polarization of radio sources at 1.4 and 5 GHz. *Nature Phys. Sci.* 236: 3–4.

Sciama, D. W., and Rees, M. J. 1967. Possible circular polarization of compact quasars. *Nature* 216: 147.

Schmidt, T. 1973. Elliptical polarization by light scattering by submicron ellipsoids. *Interstellar dust and related topics*. (H. C. van de Hulst and J. M. Greenberg, eds.) Dordrecht, Holland: D. Reidel Publ. Co.

Serkowski, K. 1973a. Interstellar polarization (review). *Interstellar dust and related topics*. (H. C. van de Hulst and J. M. Greenberg, eds.) Dordrecht, Holland: D. Reidel Publ. Co.

———. 1973b. Infrared circular polarization of NML Cygni and VY Canis Majoris. *Astrophys. J.* 179: L101–L106.

Serkowski, K., and Chojnacki, W. 1969. Polarimetric observations of magnetic stars with two-channel polarimeter. *Astron. Astrophys.* 1: 442–448.

Staude, J.; Wolf, K.; and Schmidt, T. 1973. A surface polarization survey of the Milky Way and the zodiacal light. *Interstellar dust and related topics*. (H. C. van de Hulst and J. M. Greenberg, eds.) Dordrecht, Holland: D. Reidel Publ. Co.

Tindo, I. P.; Ivanov, V. D.; Mandel'stam, S. L.; and Shuryghin, A. I. 1972. New measurements of the polarization of X-ray solar flares. *Solar Phys.* 24: 429–433.

Verschuur, G. L. 1970. Observational aspects of the galactic magnetic fields. *Interstellar gas dynamics*. (H. J. Habing, ed.) pp. 150–167. Dordrecht, Holland: D. Reidel Publ. Co.

Warwick, J. W., and Hyder, C. L. 1965. The polarization of emission lines in astronomy. I. Resonance polarization effects in the field-free cases. *Astrophys. J.* 141: 1362–1373.

ELECTROMAGNETIC RADIATION AND DISSYMMETRY

DENNIS J. CALDWELL and HENRY EYRING
University of Utah

> The basic principles of dispersion and absorption are discussed with the aid of the classical electromagnetic theory. From these, the phenomenological equation of natural and magnetically induced optical activity is developed. The presentation is illustrated by specific models including the complex oscillator. The general behavior of dispersion and circular dichroism curves for simple systems are given. The basic features of these experimental methods are discussed relative to astronomical measurements.

The simplest light-wave experiments that provide an insight into the structure of a material medium generally involve a randomly polarized monochromatic source, an intervening sample, and an analyzer, which compares the intensity or direction of the incident beam with that of the emerging one. Although such experiments require considerable modification for astronomical purposes, it is worthwhile to have a firm foundation for their execution and interpretation.

Natural optical activity is manifested when electromagnetic radiation traverses a dissymmetric medium, which may be either microscopic or macroscopic. In order to obtain intelligible data, it is necessary for the light to be linearly or circularly polarized. Making an early advance in the study of optical activity, Pasteur (1848) discovered that, on close examination, crystals of tartaric acid existed in two forms that were mirror images of one another. It was possible to separate the two forms by a rudimentary mechanical process, and it was determined that the plane of polarization of a light wave was rotated in opposite directions on passage through solutions of the two mirror image forms.

There are two types of optically active crystals: one type is composed of molecules that are themselves dissymmetric and optically active in solution, and the other type is composed of molecules or ions that are identical with their mirror images and do not rotate the plane of polarized light in solution. This same distinction may be made for any optically active medium, whether it is composed of individual molecules or large aggregates (Lowry 1935).

The optimum wavelength for the investigation of a medium depends on its resonant frequencies and on the mass of the moving charges. The most interesting results are obtained in the region of one of the resonant frequencies; however for optical activity, instrumentation has not yet been developed to give definable results at all such frequencies. Particularly lacking, for example, are data on IR vibrations. It has been demonstrated that optical activity can be manifested by an array of macroscopic conductors such as springs, which play the role of dissymmetric molecules. At radar frequencies, a random array of such objects would behave in much the same way as a solution of dissolved molecules with visible and UV frequencies.

Optically active substances can manifest their unique properties in ways other than the rotation of plane polarized light. For example, there are certain X-ray techniques (Bijvoet 1955) that, by the introduction of heavy atoms into a crystal, can determine even the absolute configuration of a molecule. Not only will diffracted light have a unique behavior in an optically active medium, but so will reflected light, although relatively little has been done of practical significance to date.

The Faraday effect is generally associated with natural optical activity, but in reality the two are greatly different. Strictly speaking, an induced optical activity would comprise the distortion of molecules into a dissymmetric form, whereas the Faraday effect arises from the modification of electronic motion in the presence of a magnetic field. Under the circumstances in which both experiments are carried out, the end result—the rotation of plane polarized light or the differential absorption of left and right circularly polarized light—is the same, but the mechanisms are vastly different.

In the classic Faraday effect experiment, plane polarized light travels from a source to an observer parallel to a constant, uniform magnetic field. The direction of rotation depends both on the direction of the magnetic field and on the medium. Unlike natural optical activity, the Faraday effect is exhibited by all atoms and molecules, which can be either an advantage or a disadvantage.

One particular difference between the two phenomena might be of significance in terrestrial applications. When a beam of light traverses a naturally optically active medium and is reflected back toward the source, the angle of polarization returns to its initial value and the effect is completely canceled. When reflection occurs after passage through a magnetic field in a Faraday effect experiment, reinforcement occurs and the angle of rotation is doubled. Thus, in principle, one can send a probing beam through space in the presence of the natural magnetic fields there and obtain information about the intervening space from the reflected beam.

Before discussing the theory of natural and magnetically induced optical rotations, it will be expedient to summarize the two methods for obtaining data: dispersion and absorption. The intensity of a light wave can be written in the complex form

$$\mathbf{E} = Re\left[\mathbf{E}_0 \exp -i\omega\left(t - \frac{nz}{c}\right)\right], \quad (1)$$

where ω is the frequency of a wave traveling along the z-direction through a medium with index of refraction n (Caldwell and Eyring 1971). Mathematically speaking, those phenomena that depend on the real part of n are called *dispersive*, whereas those depending on the imaginary part are called *absorptive*.

If one writes $n = n_1 + in_2$, a linearly polarized wave would be given by

$$\mathbf{E} = \mathbf{E}_0 \cos \omega\left(t - \frac{n_1 z}{c}\right) e^{-n_2 z \omega/c}. \quad (2)$$

A graph of this function at $t = 0$ is shown in Fig. 1. A measure of the exponential decrease in amplitude is a measure of the absorption of the medium (cf. dotted lines), while a measure of the effective wavelength in the medium by any of the classic methods for obtaining the index of refraction over a range of frequencies gives the dispersion.

The medium may either be regarded as a continuum with a given charge density or as a matrix of charges whose motion is specified by atomic parameters. In the latter instance, the component of charge motion along the direction of the impressed field is governed by the equation

$$\frac{d^2 x}{dt^2} + \gamma \frac{dx}{dt} + \omega_0^2 x = \frac{e}{m} F_0 e^{-i\omega t}, \quad (3)$$

where m and e are the mass and charge of the particle, ω_0 is the natural frequency of the charges, and γ is a dissipative parameter. This equation is satisfied by $x = x_0 e^{-i\omega t}$, where

$$x_0 = \frac{eF_0}{m} \frac{1}{\omega_0^2 - \omega^2 - i\omega\gamma}, \quad (4)$$

from which it follows that

$$x = \frac{eF_0}{m} \left\{ \left[\frac{\omega_0^2 - \omega^2}{(\omega_0^2 - \omega^2)^2 + \omega^2 \gamma^2}\right] \cos \omega t \right.$$
$$\left. + \left[\frac{\omega \gamma}{(\omega_0^2 - \omega^2)^2 + \omega^2 \gamma^2}\right] \sin \omega t \right\}. \quad (5)$$

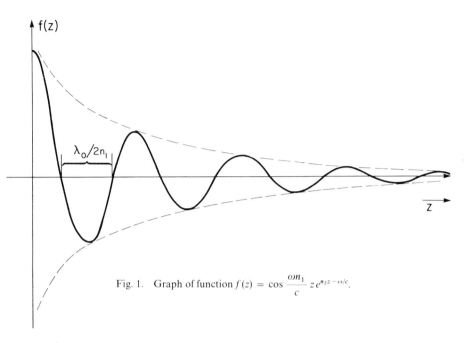

Fig. 1. Graph of function $f(z) = \cos \dfrac{\omega n_1}{c} z\, e^{n_2 z - \omega/c}$.

The first term, or in-phase component, determines the dispersive properties of the medium, in particular the refractive index. The second term, or out-of-phase component, indicates a continual loss in energy by the system which must be restored by the driving force, F; hence, this describes absorption. The graph of these two functions is shown in Fig. 2. This form of behavior is common to ordinary dispersion and absorption as well as to natural optical activity and the Faraday effect. As will be seen, the latter two phenomena depend on the difference between the complex index of refraction (cf. Equations [1] and [2]) for left and right circularly polarized light. The in-phase behavior is termed optical rotatory dispersion (ORD) and magnetic optical rotatory dispersion (MORD), while the out-of-phase or differential absorptive behavior is termed circular dichroism (CD) and magnetic circular dichroism (MCD).

The behavior depicted in Fig. 2 is general for any mechanical or electromagnetic phenomenon and its corresponding absorption of energy. In principle, both absorption and dispersion give identical information about a system. There are, however, important practical differences between the two, which can be strongly dependent on the sensitivity of the instrumentation.

By means of the Kronig-Kramer theorem (Kronig 1926; Kramer 1929), it can be shown that the mathematical form of the dispersion curve can be

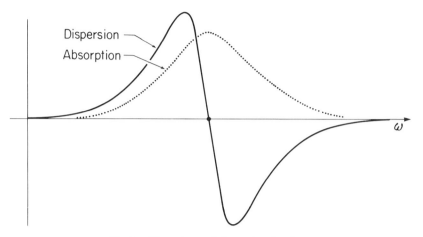

Fig. 2. Dispersion and absorption functions.

deduced from that of the absorption curve, and vice versa. Due to the practical differences in measuring these phenomena, one curve may exhibit more illuminating fine structure than the other. This can be determined not only by the different physical principles by which the data are obtained but also by the details of apparatus construction.

If it were possible to construct an experiment with absolutely monochromatic light, one could in principle obtain dispersion and absorption curves that completely satisfy the mathematical transformation properties. Since electromagnetic experiments always deal with a narrow band of frequencies, a simple mathematical equivalence of the two will give an approximation.

The simple model expressed in Equations (4) and (5), as well as the more general theories, indicates a decisive practical difference in data analysis. Consider the dispersion and absorption functions,

$$\frac{\omega_0^2 - \omega^2}{(\omega_0^2 - \omega^2)^2 + \omega^2\gamma^2} \quad \text{and} \quad \frac{\omega\gamma}{(\omega_0^2 - \omega^2)^2 + \omega^2\gamma^2},$$

depicted in Fig. 2. By recognizing that the term $\omega^2\gamma^2$ is only important near the center of the band and that the first expression is approximately an odd function of $(\omega_0 - \omega)$, one can see that the total area of the dispersion curve is essentially zero. At frequencies far removed from the absorption, the dispersion function has the form $\frac{1}{\omega_0^2 - \omega^2}$; this is independent of the parameter γ, which is sensitive to the conditions of the experiment such as slit width.

On the other hand, the area under the absorption curve is approximated by

$$\int_0^\infty \frac{\omega\gamma}{(\omega_0^2 - \omega^2)^2 + \omega^2\gamma^2} d\omega \cong \int_{-\infty}^\infty \frac{\omega_0\gamma \, d\omega}{(2\omega_0)^2(\omega_0 - \omega)^2 + \omega_0^2\gamma^2}$$

$$= \frac{\pi}{2\omega_0}, \quad (6)$$

which is independent of γ; while the asymptotic behavior of the integrand is given by $\frac{\omega\gamma}{(\omega_0^2 - \omega^2)^2}$, which is not independent of γ.

A historical note is in order on this subject. At first, optical activity was measured by the classical polarimeter experiment, in which the angle of polarization was compared before entering and after leaving an optically active liquid. Furthermore, the experiment was made at only one frequency, generally far removed from the absorption band. For this reason, it was logical to attempt to represent the data, in those few cases where several frequencies far from absorption were investigated, by an expression of the form $A/(\omega_0^2 - \omega^2)$.

As techniques were refined, more data were obtained within the absorption band, and the familiar form of dispersion behavior was observed. Attempts to interpret such irreproducible data as half maximum widths and differences between peak and trough values proved to be of limited value. The situation was remedied when workers in this field realized the practical advantages of the absorption or circular dichroism curve, whereby it is possible to relate the area under this curve to molecular parameters.

A second great advantage of circular dichroism over rotatory dispersion arises from the fact that absorption curves for different resonance frequencies tend to overlap to a much lesser extent than dispersion curves, which invariably have the asymptotic dependence of $1/(\omega_0^2 - \omega^2)$, a relatively slowly varying quantity.

NATURAL OPTICAL ACTIVITY

The electromagnetic properties of a medium are determined by the equations

$$\nabla \times \mathbf{H} = \frac{1}{c} \frac{\partial \mathbf{D}}{\partial t} \quad (7a)$$

$$\nabla \times \mathbf{E} = -\frac{1}{c} \frac{\partial \mathbf{B}}{\partial t} \quad (7b)$$

In free space $\mathbf{B} = \mathbf{H}$ and $\mathbf{E} = \mathbf{D}$. For a dielectric, the two vectors associated with the magnetic field are approximately equal; $\mathbf{E}$ is the suitably averaged macroscopic electric field in the medium, and $\mathbf{D}$ has been suitably defined to preserve the simple symmetrical form of Maxwell's equations for free space. If $\mathbf{P}$ is the dipole moment per unit volume, the familiar relation $\mathbf{D} = \mathbf{E} + 4\pi\mathbf{P}$ follows.

For most experimental work, a linear relation between $\mathbf{D}$ and $\mathbf{E}$ has been adequate:

$$\mathbf{D} = \boldsymbol{\varepsilon} \cdot \mathbf{E}, \qquad (8)$$

where $\boldsymbol{\varepsilon}$ is the tensor dielectric constant. Since both in- and out-of-phase motion must be considered, $\boldsymbol{\varepsilon}$ will be complex. If it is first assumed that the medium is nonabsorbing, the electric energy term $\mathbf{E} \cdot \mathbf{D}$ must be real; otherwise there is a net gain or loss of energy through the appearance of a $\cos \omega t \sin \omega t$ term. If ε_1, ε_2, and ε_3 are the principal dielectric constants, $\mathbf{D}$ may be expressed in the form

$$\mathbf{D} = (\mathbf{ii}\varepsilon_1 + \mathbf{jj}\varepsilon_2 + \mathbf{kk}\varepsilon_3) \cdot \mathbf{E} + i\boldsymbol{\gamma} \times \mathbf{E}. \qquad (9)$$

If now the general phenomenon of absorbing media is considered, ε_1, ε_2, ε_3, and γ will be complex, but the form of Equation (9) is still preserved.

If $\gamma = 0$ the medium is an ordinary anisotropic dielectric. In general, waves with an arbitrary direction and polarization cannot be transmitted. When such a plane wave enters a surface of the medium whose normal makes an arbitrary angle with each of the principal dielectric axes, it is split into two waves with different directions and polarization determined by the initial conditions of entry. This makes the measurement of γ for an optically active medium rather difficult.

Of particular interest are axially symmetric media for which $\varepsilon_1 = \varepsilon_2$. In this case, the direction of γ, which in general bears no definite relation to the principal dielectric axes, is prescribed by symmetry to be along the axis $\mathbf{k}$. It next remains to calculate from Equation (7) the properties of waves transmitted along this axis of symmetry. One may write

$$\mathbf{D} = [\varepsilon_1(\mathbf{ii} + \mathbf{jj}) + \varepsilon_3 \mathbf{kk}] \cdot \mathbf{E} + i\gamma\mathbf{k} \times \mathbf{E}. \qquad (10)$$

By taking the curl of Equation (7b) and the time derivative of Equation (7a) and assuming that $\mathbf{B} = \mathbf{H}$, one obtains

$$\nabla \times \nabla \times \mathbf{E} = -\frac{1}{c^2}\frac{\partial^2 \mathbf{D}}{\partial t^2}. \qquad (11)$$

If one assumes a plane wave of the form

$$\mathbf{D} = \mathbf{D}^{(0)} \exp i\omega\left[t - \frac{n\mathbf{k}\cdot\mathbf{r}}{c}\right] \quad (12a)$$

$$\mathbf{E} = \mathbf{E}^{(0)} \exp i\omega\left[t - \frac{n\mathbf{k}\cdot\mathbf{r}}{c}\right], \quad (12b)$$

the use of the above two equations gives

$$-n^2[\mathbf{k}(\mathbf{k}\cdot\mathbf{E}^{(0)}) - \mathbf{E}^{(0)}] = \mathbf{D}^{(0)}.$$

From the auxiliary electromagnetic requirement, $\mathbf{V}\cdot\mathbf{D} = 0$, it follows that $\mathbf{D}$ is perpendicular to k; Equation (10) requires $\mathbf{k}\cdot\mathbf{E}$ also to vanish. One may then write $(\varepsilon_1 - n^2)\mathbf{E}^{(0)} + i\gamma\mathbf{k}\times\mathbf{E}^{(0)} = 0$, which leads to the individual equations

$$(\varepsilon_1 - n^2)E_1^{(0)} - i\gamma E_2^{(0)} = 0 \quad (13a)$$

$$+i\gamma E_1^{(0)} + (\varepsilon_1 - n^2)E_2^{(0)} = 0. \quad (13b)$$

The allowable solutions for wave polarization are found by solving the equation

$$(\varepsilon_1 - n^2) - \gamma^2 = 0,$$

which has roots $n^2 = \varepsilon_1 \pm \gamma$. The two solutions for polarization vectors have the form

$$\mathbf{E}^{(0)} = E^{(0)}(\mathbf{i} \pm i\mathbf{j}), \quad (14)$$

which represents right, or clockwise, polarization as viewed by the observer for the upper sign, and the opposite polarization for the lower sign.

Since $\varepsilon_1 \gg \gamma$, as experiment will verify, one may write

$$n_\pm = \sqrt{\varepsilon_1} \pm \frac{1}{2}\frac{\gamma}{\sqrt{\varepsilon_1}}. \quad (15)$$

The parameter γ is called the gyration vector, and its evaluation provides a particular insight into the nature of a medium's dissymmetry. Since $\gamma \ll \bar{n}_1 = \sqrt{\varepsilon_1}$, it is most convenient to obtain this quantity by means of an experiment that measures the difference, $n_+ - n_- = \gamma/\sqrt{\varepsilon_1}$. Since plane polarized light may be resolved as the difference of two circularly polarized components, the simplest methods of experimentation first concentrated on the changes in the polarization of a plane wave brought about by an optically active medium.

Accordingly, a plane polarized wave in free space with **E** along the **i** axis may be written

$$\mathbf{E}_{vac} = \frac{1}{2} E^{(0)}[(\mathbf{i} - i\mathbf{j}) + (\mathbf{i} + i\mathbf{j})] \exp i\omega \left[t - \frac{z}{c} \right]. \quad (16)$$

In the medium, plane polarized waves as such are not transmitted, and Equation (16) becomes

$$\mathbf{E}_{med} = \frac{1}{2} E^{(0)} \left\{ (\mathbf{i} - i\mathbf{j}) \exp i\omega \left[t - \frac{n_- z}{c} \right] + (\mathbf{i} + i\mathbf{j}) \exp i\omega \left[t - \frac{n_+ z}{c} \right] \right\}$$

$$= \frac{1}{2} E^{(0)} \exp i\omega \left[t - \frac{\bar{n} z}{c} \right] \{ (\mathbf{i} - i\mathbf{j}) e^{+i\delta z} + (\mathbf{i} + i\mathbf{j}) e^{-i\delta z} \} \quad (17)$$

$$= E^{(0)} \exp i\omega \left[t - \frac{\bar{n} z}{c} \right] [\mathbf{i} \cos \delta z + \mathbf{j} \sin \delta z],$$

where

$$\delta = \frac{1}{2} \frac{\omega}{c} \frac{\gamma}{\bar{n}}.$$

This represents a plane polarized wave whose electric vector has been rotated through an angle δz:

$$\phi = -\frac{1}{2} \frac{\omega}{c} \frac{\gamma}{\bar{n}} z. \quad (18)$$

When γ is negative, the rotation is clockwise and the medium is said to be dextrorotatory; when γ is positive, the medium is levorotatory. If the region near a resonant frequency is considered, the complex form of δ must be used. In this case, one must write $\delta = \delta_1 + i\delta_2$; then by using such relations as $\cos(\delta_1 + i\delta_2) = \cos\delta_1 \cos h\delta_2 - i\sin\delta_1 \sin h\delta_2$, Equation (17) becomes

$$\mathbf{E}_{med} = E^{(0)} \exp i\omega \left[t - \frac{\bar{n} z}{c} \right] \{ (\mathbf{i} \cos \delta_1 z + \mathbf{j} \sin \delta_1 z) \cos h\delta_2 z$$

$$+ i(-\mathbf{i} \sin \delta_1 z + \mathbf{j} \cos \delta_1 z) \sin h\delta_2 z \}. \quad (19)$$

This represents an elliptically polarized wave with major and minor axes along $\mathbf{i} \cos \delta_1 z + \mathbf{j} \sin \delta_1 z$ and $-\mathbf{i} \sin \delta_1 z + \mathbf{j} \cos \delta_1 z$, respectively. If δ_2 as well as δ_1 are positive, the sense of the polarization is clockwise.

The ratio of the minor to the major axis is called the ellipticity and is a direct measure of the differential absorption of left and right circularly polarized light. Thus, one may write for the differential absorption parameter of the system

$$\theta = -\tanh\frac{1}{2}\frac{\omega\gamma 2}{c\bar{n}}z. \qquad (20)$$

For laboratory path lengths, this is a small quantity and may be approximated by taking $\tanh\chi = \chi$.

It should be noted from Equation (19) that the dispersion parameter δ_1 is measured by the angle that the principal axis of the ellipse makes with the initial direction of polarization. The absorption parameter δ_2 is measured by the eccentricity of the ellipse, and its sign is determined by the sense of circulation (clockwise or counterclockwise) of the electric vector in relation to the direction of rotation of the principal axis, as prescribed by Equation (19).

For long path lengths, the angle of rotation ϕ exceeds 360° and the ellipticity θ asymptotically approaches unity. Thus, the practical difficulties for this method of measurement render an interpretation difficult. Even the small-path-length experiments in the laboratory on optically active materials are more readily conducted by means of an electronically governed production of left and right circularly polarized light, with the difference in absorption measured directly. Here, the only restriction on path length for unambiguous interpretation of data is determined by the sensitivity of the instrument in determining the final intensity of transmitted light.

The crucial phase of any such investigation is the interpretation of the gyration vector γ in terms of the medium's structure. The dielectric properties are determined by the phases of motion parallel and perpendicular to the force vector **E**. In the absence of absorption, optical activity has been shown to arise from out-of-phase motion perpendicular to **E**, as expressed in Equation (9).

If at every point of a continuum, a linear anisotropic restoring force were to operate on a point charge, then only in-phase motion could occur and γ would vanish. In order to account for optical activity, a microscopic substratum must be postulated in which the force on an element of charge at one point is affected by the motion of its neighbors.

Consider the interaction of the two oscillating dipoles shown in Fig. 3. If the motion of the two dipoles is in phase, then so will be their interaction; if the two are slightly out of phase and, in addition, are oriented in such a way that when **E** is perpendicular to one, it is not so to the other, then, provided the force between the two dipoles does not vanish, there will arise an out-of-phase motion that is perpendicular to **E**.

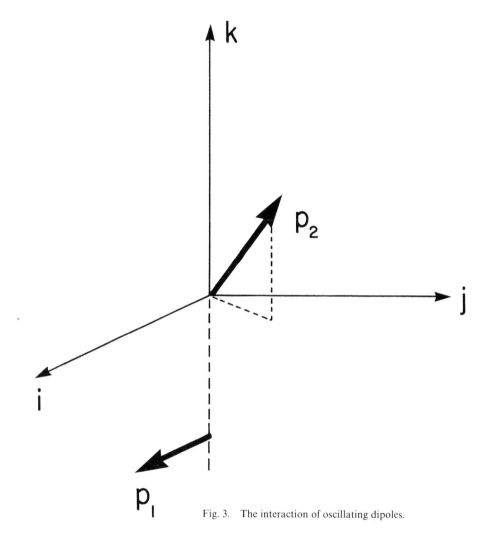

Fig. 3. The interaction of oscillating dipoles.

The desired out-of-phase component may be computed with reference to the figure by observing that a wave momentarily polarized along the i-axis is represented by the expression, $\mathbf{E} = E_0 \mathbf{i} \exp i\omega \left[t - \dfrac{\bar{n}z}{c} \right]$. At the position of the first oscillator, the induced dipole moment is given by

$$\boldsymbol{\mu}_1 = \mathbf{i}\alpha_1 E_0 \exp i\omega \left[t + \dfrac{\bar{n}a}{c} \right] \cong \mathbf{i}\alpha_1 E_0 e^{i\omega t} \left[1 + \dfrac{i\omega \bar{n}a}{c} \right], \quad (21)$$

where α_1 is the polarizability.

This induced dipole produces a field at the origin given by $-\boldsymbol{\mu}_1/a^3$, which in turn induces a dipole $\boldsymbol{\mu}_2$:

$$\boldsymbol{\mu}_2 = \alpha_2(-\boldsymbol{\mu}_1/a^3) \cdot \mathbf{b}\mathbf{b}, \tag{22}$$

where $\mathbf{b} = b_1\mathbf{i} + b_2\mathbf{j} + b_3\mathbf{k}$ is a unit vector along the direction of the second dipole. Then, the perpendicular out-of-phase component of $\boldsymbol{\mu}_2$ is given by

$$\boldsymbol{\mu}_2^{\ddagger} = -\frac{i\omega\bar{n}}{ca^2}\alpha_1\alpha_2\, b_1 b_2\, \mathbf{j}\, E_0 e^{i\omega t}, \tag{23}$$

from which it follows that

$$\gamma = -4\pi N \frac{\omega\bar{n}}{ca^2}\alpha_1\alpha_2\, b_1 b_2\, \mathbf{k}. \tag{24}$$

The factor of $4\pi N$ arises through the relation, $\mathbf{D} = \mathbf{E} + 4\pi\mathbf{P} = \mathbf{E} + 4\pi N\boldsymbol{\mu}$, where $\mathbf{P}$ is the dipole moment per unit volume, $\boldsymbol{\mu}$ is the individual oscillator moment, and N is the number of oscillators per unit volume.

The maximum effect occurs when the vector $\mathbf{b}$ bisects the angle between $\mathbf{i}$ and $\mathbf{j}$; i.e., $\mathbf{b} = \dfrac{\sqrt{2}}{2}(\mathbf{i} + \mathbf{j})$. In this instance, a compromise is reached between the maximum dipole interaction, which occurs for $\mathbf{b} = \mathbf{i}$, and the maximum perpendicular out-of-phase dipole moment occurring at $\mathbf{b} = \mathbf{j}$.

In a fluid of such oscillators with random orientation, an average over all of the coupled system with respect to $\mathbf{E}$ and $\mathbf{k}$ is performed, which merely has the effect of introducing an additional numerical factor in Equation (24). Strictly speaking, a correct formulation is only possible through an average of the orientations in a fluid or in an ordered lattice structure with a well-defined unit cell; otherwise, misleading results can be obtained. The above example is meant for the instructive purpose of demonstrating the origin of perpendicular out-of-phase motion in a conceptually simple manner.

The idea of coupled oscillators (Kuhn 1961) is particularly useful for understanding optical activity that is based on a microscopic substratum. Such occasions may arise through the interaction of large molecules, clusters of molecules, or even objects visible to the naked eye. Again, the most appropriate frequencies for the investigation of the phenomenon will be in regions near the resonant frequencies of the system.

It is possible to make certain glass-like substances optically active by the purely mechanical process of giving them a torque. If a thin layer of the material is placed between two glass plates to which it adheres, a rotation of the two plates creates a spiral effect. The individual molecules may be considered to interact along the lines discussed in conjunction with Fig. 3. In this manner, very large rotations may be obtained, which are dependent on the sign and magnitude of the angle of torque between the plates.

If visible and UV frequencies are used, information is obtained on molecular structure. A detailed quantum mechanical summary would be beyond the scope of this synopsis; it will be sufficient to say that γ is related to matrix elements of the molecule under investigation. In the simple case of nonoverlapping absorption bands that are not too broad, the area under the circular dichroism curve for a given nondegenerate absorption band is proportional to the quantity $\mathbf{\mu}_{on} \cdot \mathbf{m}_{no}$, which is called the rotational strength of the band. The quantities $\mathbf{\mu}_{on}$ and $\mathbf{m}_{no}$ are defined by

$$\mathbf{\mu}_{on} = \int \psi_o^* \mathbf{\mu} \psi_n \, d\tau \tag{25a}$$

$$\mathbf{m}_{no} = \int \psi_n^* \mathbf{m} \psi_o \, d\tau, \tag{25b}$$

where $\mathbf{\mu}$ and $\mathbf{m}$ are the electric and magnetic moment operators of the molecular system. (It should be noted that the area under the curve for ordinary absorption is measured by the quantity $\mathbf{\mu}_{on} \cdot \mathbf{\mu}_{no}$.)

Typical electric and magnetic dipole transitions are shown in Fig. 4 in terms of the familiar hydrogen-like wave functions. When molecules are superimposable on their mirror images, if $\mathbf{\mu}_{on} \neq 0$, then either $\mathbf{m}_{no} = 0$ or $\mathbf{m}_{no}$ is perpendicular to $\mathbf{\mu}_{on}$; and for $\mathbf{m}_{no} \neq 0$, a similar consideration applies to $\mathbf{\mu}_{on}$. When these molecules are asymmetrically substituted, $\mathbf{\mu}_{on} \cdot \mathbf{m}_{no}$ no longer vanishes. The optically active transitions are called electric dipole, a typical example being $\pi - \pi^*$ in an optically active olefin (double bond), or magnetic dipole, such as the $n \to \pi^*$ of a carbonyl group. Since both $\mathbf{\mu}_{on}$ and $\mathbf{m}_{no}$ enter on an equal footing, it follows that both strong and weak transitions can have large rotational strengths.

The behavior of degenerate and vibrationally controlled transitions is more complex, and there are often observed sign changes within the absorption band of the electronic transition. It is not possible to give an exhaustive account of the matter, but a typical and important example is the coupled oscillator system described above. This type of degeneracy is exhibited in long-chain polymeric molecules such as peptides, which consist of a coil of identical monomeric units bonded together in a highly ordered conformation. The individual monomers have identical resonant frequencies; the aggregate polymer will have a band that results from the existence of identical groups in the same molecule. The circular dichroism of this band is sigmoid in shape, consisting of approximately equal and

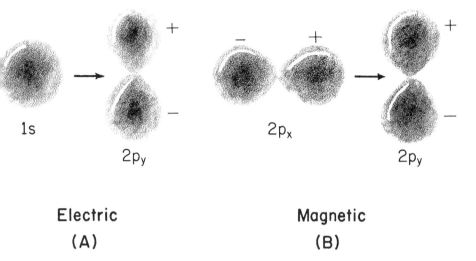

Fig. 4. Transition prototypes.

opposite signed contributions. When the coiled form of the polymer is broken down by denaturation and the molecule assumes a random chain conformation, this characteristic behavior disappears in the CD spectrum. When properly interpreted, optical activity data can provide important information on molecular structure, particularly conformation (Djerassi 1960; Velluz, LeGrand, and Grossjean 1960).

The fact that optical activity disappears when a system assumes a net nondissymmetrical conformation will be of central importance in devising and interpreting experiments on circular dichroism. An initially optically active system can become inactive either by assuming a macroscopic symmetry or by achieving a microscopic symmetry in which the individual components are symmetrical or consist of equal numbers of mirror image antipodes. It should be emphasized that the individual building units need not be dissymmetric but only the net aggregate; for example, there are crystals such as $NaClO_3$ that are formed by molecules or ions which themselves would not exhibit optical activity in solution.

THE FARADAY EFFECT

Closely related to natural optical activity is the phenomenon of magnetically induced optical rotation, which is exhibited by any substance in the presence of a magnetic field (Buckingham and Stephens 1966; Caldwell and Eyring 1971; Shieh, Lin, and Eyring 1972). Consider first a continuum for which the electrical restoring force is given by the tensor, $\kappa = \mathbf{ii}\kappa_1 + \mathbf{jj}\kappa_2 + \mathbf{kk}\kappa_3$. If this medium is subjected to a constant magnetic field $\mathbf{H}_0$ and an electromagnetic field with $\mathbf{E} = \mathbf{E}_0 e^{i\omega t}$ and $|\mathbf{H}_0| \gg |\mathbf{E}_0|$, the equation of motion for a representative charge e is given by

$$m\ddot{\mathbf{r}} = -\kappa \cdot \mathbf{r} + e\left[\mathbf{E} + \frac{\dot{\mathbf{r}} \times \mathbf{H}_0}{c}\right]. \tag{26}$$

If a solution of the form $\mathbf{r} = \mathbf{r}_0 e^{i\omega t}$ is assumed, the three components of the resulting vector equation are

$$\Gamma_1 x_0 - i\sigma H_3 y_0 + i\sigma H_2 z_0 = \frac{e}{m} E_1 \tag{27a}$$

$$i\sigma H_3 x_0 + \Gamma_2 y_0 - i\sigma H_1 z_0 = \frac{e}{m} E_2 \tag{27b}$$

$$-i\sigma H_2 x_0 + i\sigma H_1 y_0 + \Gamma_3 z_0 = \frac{e}{m} E_3, \tag{27c}$$

where $\Gamma_i = \omega_i^2 - \omega^2$, $\omega_i = \sqrt{\frac{\kappa_i}{m}}$, $\sigma = \frac{e\omega}{mc}$.

The solution to this system of linear equations in the variables x_0, y_0, and z_0 can, by means of standard methods, be expressed in the form

$$\mathbf{r}_0 = \frac{\frac{e}{m}\Gamma^{(2)} \cdot \mathbf{E} - \frac{e}{m}\sigma^2 \mathbf{H}_0 \mathbf{H}_0 \cdot \mathbf{E} - i\frac{e\sigma}{m}(\Gamma^{(1)} \cdot \mathbf{H}_0) \times \mathbf{E}}{\Gamma_1 \Gamma_2 \Gamma_3 - \sigma^2 \mathbf{H}_0 \cdot \Gamma^{(1)} \cdot \mathbf{H}_0}, \tag{28}$$

where

$$\Gamma^{(1)} = \mathbf{ii}\Gamma_1 + \mathbf{jj}\Gamma_2 + \mathbf{kk}\Gamma_3$$

$$\Gamma^{(2)} = \mathbf{ii}\Gamma_2\Gamma_3 + \mathbf{jj}\Gamma_1\Gamma_3 + \mathbf{kk}\Gamma_1\Gamma_2.$$

The five terms in Equation (28) may be analyzed in the following manner:

1. $\frac{e}{m}\Gamma^{(2)} \cdot \mathbf{E}$ represents the effect of the electromagnetic field alone.

   When $\mathbf{H}_0 = 0$, Equation (28) becomes

   $$e\mathbf{r}_0 = \frac{e^2}{m}\left(\mathbf{ii}\frac{1}{\Gamma_1} + \mathbf{jj}\frac{1}{\Gamma_2} + \mathbf{kk}\frac{1}{\Gamma_3}\right) \cdot \mathbf{E} = \boldsymbol{\alpha} \cdot \mathbf{E},$$

   where $\boldsymbol{\alpha}$ is the familiar expression for the polarizability tensor.

2. $-\dfrac{e}{m}\sigma^2\mathbf{H}_0\mathbf{H}_0\cdot\mathbf{E}$ is seen to be in phase with $\mathbf{E}$ and represents a corrective term to the polarizability tensor brought about by the constant magnetic field. Even in an isotropic medium, this term will lead to electrical birefringence unless $\mathbf{H}_0$ and $\mathbf{E}$ are perpendicular.

3. $-\dfrac{ie\sigma}{m}(\mathbf{\Gamma}^{(1)}\cdot\mathbf{H}_0)\times\mathbf{E}$ describes an out-of-phase motion perpendicular to $\mathbf{E}$, which leads to optical activity (cf. Equation [9] et seq.).

4. $\Gamma_1\Gamma_2\Gamma_3$ is simply the product of frequency factors that leads to the correct expression for the field-free polarizability as expressed in number 1 above.

5. $-\sigma^2\mathbf{H}_0\cdot\mathbf{\Gamma}^{(1)}\cdot\mathbf{H}_0$ represents a corrective term to the frequency factor product $\Gamma_1\Gamma_2\Gamma_3$. By a suitable choice of ω_i', one may write

$$\alpha' = \dfrac{\dfrac{e^2}{m}\Gamma^{(2)}}{\Gamma_1\Gamma_2\Gamma_3 - \sigma^2\mathbf{H}_0\cdot\mathbf{\Gamma}^{(1)}\cdot\mathbf{H}_0} = \dfrac{e^2}{m}\left(\mathbf{ii}\dfrac{1}{\Gamma_1'} + \mathbf{jj}\dfrac{1}{\Gamma_2'} + \mathbf{kk}\dfrac{1}{\Gamma_3'}\right),$$

where $\Gamma_i' = \omega_i'^2 - \omega^2$. Generally under the prevailing experimental conditions, these frequency corrections are small and may be neglected.

If N is the effective density of vibrating charge e with mass m, the displacement vector may be written

$$\mathbf{D} = \mathbf{E} + 4\pi N e\mathbf{r}$$

$$= \left[\varepsilon - 4\pi N\dfrac{e^2}{m}\dfrac{\sigma^2}{\Gamma_1\Gamma_2\Gamma_3}\mathbf{H}_0\mathbf{H}_0\right]\cdot\mathbf{E} - i4\pi N\dfrac{\dfrac{e^2\sigma}{m}(\mathbf{\Gamma}^{(1)}\cdot\mathbf{H}_0)\times\mathbf{E}}{\Gamma_1\Gamma_2\Gamma_3}, \quad (29)$$

where ε is the dielectric tensor of the medium,

$$\varepsilon = \mathbf{I} + 4\pi\dfrac{Ne^2}{m}\left(\mathbf{ii}\dfrac{1}{\Gamma_1} + \mathbf{jj}\dfrac{1}{\Gamma_2} + \mathbf{kk}\dfrac{1}{\Gamma_3}\right).$$

This equation is in the form of Equation (9) discussed in conjunction with natural optical activity, and the results deduced for the transmission of waves with $\mathbf{D}$ expressed in that form apply equally well here.

In a structurally ordered medium, such as a crystal, it is not possible to transmit plane or circularly polarized waves in the presence of a magnetic field unless ε has axial symmetry and $\mathbf{H}_0$ is along the preferred axis. In this case, one may take $\Gamma_1 = \Gamma_2$ and $\mathbf{\Gamma}^{(1)}\cdot\mathbf{H}_0 = \Gamma_3\mathbf{H}_0$. For light transmitted along this axis of symmetry, one may use Equation (10) with

$$\varepsilon_1 = 1 + 4\pi \frac{Ne^2}{m} \frac{1}{\Gamma_1} \quad (30\text{a})$$

$$\varepsilon_3 = 1 + 4\pi \frac{Ne^2}{m} \frac{1}{\Gamma_3} \quad (30\text{b})$$

$$\gamma = -4\pi \frac{Ne^2\sigma}{m} \frac{1}{\Gamma_1^2} H_0. \quad (30\text{c})$$

From Equation (18), the angle of rotation of plane polarized light is given by

$$\phi = \frac{1}{2} \frac{\omega^2}{c^2 \bar{n}_1} 4\pi \frac{Ne^3}{m^2 \Gamma_1^2} H_0 z. \quad (31)$$

Since $\bar{n}_1^2 - 1 = \dfrac{4\pi Ne^2}{m\Gamma_1}$, this may be written in the form

$$\phi = \frac{1}{8\pi Ne} \frac{\omega^2}{c^2} \frac{(\bar{n}_1^2 - 1)^2}{\bar{n}_1} H_0 z. \quad (32)$$

Although this result has been obtained by a model that ignored damping, the above result obtains equally well when a frictional term, $-\gamma \dot{r}$, is included in the equation of motion. The index of refraction has the dispersive behavior shown in Fig. 2. The Faraday rotation is displayed in Fig. 5. The corresponding dichroism curve is interesting and is best inferred from the following discussion.

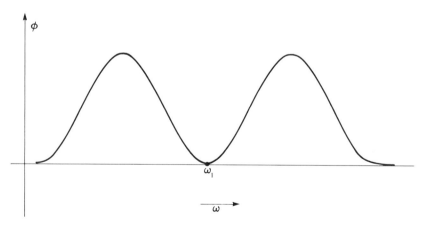

Fig. 5. Faraday rotation for axial symmetry.

In nonisotropic media, the magnetic birefringence is generally masked by that of the medium. In order to display the phenomenon of a medium that displays only magnetic birefringence and magnetic optical activity, it is worthwhile to examine the case of a medium of anisotropic particles with random orientation. For this purpose, Equation (29) may be averaged over all orientations of the $\mathbf{i}$, $\mathbf{j}$, $\mathbf{k}$ coordinate system with respect to the fixed axes, $\mathbf{E}$ and $\mathbf{H}_0$.

By using the relations $\langle \mathbf{ii} \cdot \mathbf{a} \rangle_{av} = \frac{1}{3}\mathbf{a}$ and $\langle \mathbf{i}(\mathbf{j} \cdot \mathbf{a})(\mathbf{k} \cdot \mathbf{b}) \rangle_{av} = \frac{1}{6}\mathbf{a} \times \mathbf{b}$, where $\mathbf{a}$ and $\mathbf{b}$ are fixed vectors, it is found that

$$\mathbf{b} = (\varepsilon \mathbf{I} - f_1 \mathbf{H}_0 \mathbf{H}_0) \cdot \mathbf{E} - i f_2 \mathbf{H}_0 \times \mathbf{E}, \tag{33}$$

where
$$\varepsilon = 1 + \frac{4\pi N e^2}{3m}\left(\frac{1}{\Gamma_1} + \frac{1}{\Gamma_2} + \frac{1}{\Gamma_3}\right),$$

$$f_1 = \frac{4\pi N e^2}{m} \frac{\sigma^2}{\Gamma_1 \Gamma_2 \Gamma_3},$$

$$f_2 = \frac{4\pi N e^2}{m} \frac{1}{3}\sigma\left(\frac{1}{\Gamma_2 \Gamma_3} + \frac{1}{\Gamma_1 \Gamma_3} + \frac{1}{\Gamma_1 \Gamma_2}\right).$$

Again from Equation (18), the rotation for $\mathbf{H}_0 \cdot \mathbf{E} = 0$ (no birefringence) is given by

$$\phi = \frac{1}{2}\frac{\omega}{c\bar{n}}f_2 H_0 z. \tag{34}$$

In Equation (34) the magnetic field is positive when the north pole is nearest the observer. It is emphasized that birefringence is eliminated only when $\mathbf{H}_0$ is parallel to the direction of wave propagation. When $\mathbf{E}$ is parallel to $\mathbf{H}_0$ the optical activity vanishes, but the birefringence does also, since $\mathbf{D}$ and $\mathbf{E}$ have the same direction. When $\mathbf{H}_0$ and $\mathbf{E}$ make an oblique angle, the effects of birefringence and optical activity are superimposed. Generally speaking, magnetic birefringence tends to be regarded as a necessary evil, and most experiments are set up in such a way as to eliminate it; however, it is not always possible to control all the conditions of an experiment, and further investigation of the phenomenon may be necessary.

By the decomposition into fractions,

$$\frac{1}{\Gamma_1 \Gamma_2} = \frac{1}{(\omega_1^2 - \omega^2)(\omega_2^2 - \omega^2)}$$

$$= \frac{1}{\omega_2^2 - \omega_1^2}\left[\frac{1}{\omega_1^2 - \omega^2} - \frac{1}{\omega_2^2 - \omega^2}\right],$$

etc., f_2 may be written

$$f_2 = \frac{4\pi Ne^2}{m} \frac{1}{3} \sigma \left\{ \left[\frac{1}{\omega_2^2 - \omega_1^2} + \frac{1}{\omega_3^2 - \omega_1^2} \right] \frac{1}{\omega_1^2 - \omega^2} \right.$$
$$+ \left[\frac{1}{\omega_3^2 - \omega_2^2} + \frac{1}{\omega_1^2 - \omega_2^2} \right] \frac{1}{\omega_2^2 - \omega^2}$$
$$\left. + \left[\frac{1}{\omega_1^2 - \omega_3^2} + \frac{1}{\omega_2^2 - \omega_3^2} \right] \frac{1}{\omega_3^2 - \omega^2} \right\}. \quad (35)$$

If $\omega_1 > \omega_2 > \omega_3$, then the first and third dispersion terms will have the opposite signs with the middle indeterminate, depending on the actual magnitudes of the three frequencies.

By reference to Fig. 2 and the preceding development, it is evident that the magnetic optical rotatory dispersion (MORD) and the magnetic circular dichroism (MCD), depicted in the approximate expression Equation (35) (representing the limiting behavior outside absorption bands), are shown in Fig. 6.

The sign of the Faraday effect depends on the signs of both the charge e and magnetic field H_0. Since e is negative for an electron, the preceding development indicates that the sign of the lowest transition of an anisotropic system is negative. We may now return to the degenerate situation described in Fig. 5. By letting $\omega_1 \to \omega_2$ in Equation (35), one obtains

$$\frac{1}{\omega_2^2 - \omega_1^2} \left[\frac{1}{\omega_1^2 - \omega^2} - \frac{1}{\omega_2^2 - \omega^2} \right] \to \frac{1}{(\omega_1^2 - \omega^2)^2} = \frac{1}{\Gamma_1^2}, \quad (36)$$

which has the same frequency dependence as Equation (31). If, guided by the above considerations, we let $\omega_1 \to \omega_2$ in Fig. 6, the resulting behavior for a degenerate resonant frequency is obtained (Fig. 7).

In the literature on this subject, two separate sign conventions have been used: the physicist's, from the point of view of an observer looking along the direction of wave propagation, and the chemist's, which envisions the problem from the point of view of an observer looking toward the oncoming wave. These two conventions lead to the opposite sign; the chemist's has been used in most of the recent literature and is employed here.

A more exhaustive analysis of data on chemical compounds indicates that not only the sign and magnitude of the Faraday effect but also the detailed fine structure of the curves can provide information on molecules. A detailed discussion on the subject is beyond the scope of this review, but it can be said that the effect is capable of verifying certain hypotheses

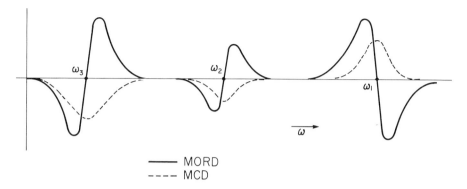

Fig. 6. The magnetic optical activity of a random anisotropic oscillator system.

on a medium's structure that go somewhat beyond a mere enumeration of the resonant frequencies. The above example, which is the simplest imaginable, clearly indicates the relationship to anisotropy.

The quantum mechanical expression is an interesting example:

$$f_{on} = Im\left[\sum_s \frac{\boldsymbol{\mu}_{on} \cdot \mathbf{m}_{ns} \times \boldsymbol{\mu}_{so}}{E_s - E_n} - \sum_s \frac{\boldsymbol{\mu}_{on} \cdot \boldsymbol{\mu}_{ns} \times \mathbf{m}_{so}}{E_s - E_o}\right]. \quad (37)$$

This measures the MCD of a transition $o \to n$ in terms of magnetic and electric transition moments of all the system's transitions. It shows that in general all the modes of a system interact, and the observed intensity of a given transition is discussed in terms of its interaction with those transitions having the symmetry necessary to prevent the above vector products from vanishing.

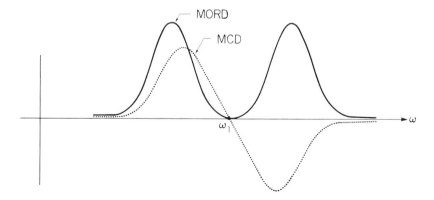

Fig. 7. Behavior of a degenerate resonant frequency.

Consider the interaction of the two transitions shown in Fig. 8. The appropriate matrix elements are

$$\mu_{1s-2p_x} = \mu \mathbf{i} \tag{38a}$$

$$\mu_{1s-2p_y} = \mu \mathbf{j} \tag{38b}$$

$$\mathbf{m}_{2p_x-2p_y} = -i \frac{eh}{2mc} \mathbf{k}. \tag{38c}$$

Equation (37) gives

$$f_{1s-2p_x} = -f_{1s-2p_y} = \frac{1}{\omega_2 - \omega_1} \mu^2 \frac{eh}{2mc}, \tag{39}$$

which is analogous to the preceding discussion on the resonant frequencies of an anisotropic oscillator.

The situation is particularly interesting if the two transitions actually consist of an interacting band system of many closely spaced transitions with intricate symmetry properties. In this case, the MCD curve can consist of alternating signed subbands within the given absorption band.

EXTRATERRESTRIAL APPLICATIONS

Some general concluding words are in order on the devising and interpretation of extraterrestrial polarized light experiments. Such investigations can be performed either by analyzing light originating from space or by transmitting a beam from earth that is reflected from a celestial body and then analyzed like a radar pulse.

It should be borne in mind that all laboratory experiments, as indicated in the preceding sections, maintain a strict control over the polarization of the incident light so that the comparison of the polarization of incident and transmitted light so necessary for the conventional experiments can be made. Thus, the analysis of light from outer space, which has traveled through a natural or magnetically induced optically active medium, is complicated by a lack of knowledge of the initial intensity and polarization.

On the other hand, the transmission of a beam of light from earth limits the experimenter to objects in our own solar system. In addition, if a conventional optical activity experiment is considered, a second complication arises, for polarized light transmitted from a source through a naturally optically active medium and reflected back toward the source undergoes a 180° phase change which nullifies the rotation or the dichroism. This is not true, however, for the Faraday effect, since the 180° phase change of the electric vector is compensated for by the reversal of the

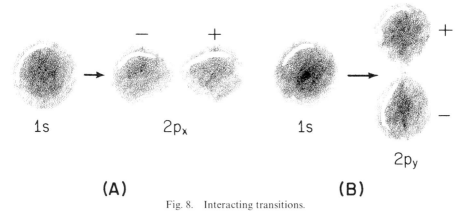

Fig. 8. Interacting transitions.

beam's direction with respect to the fixed magnetic field H_0; the effect is doubled rather than nullified.

While it is true that magnetic fields exist in space, it is unlikely that the necessary uniform field parallel to the transmitted beam will exist. A modification of the above theory is required to take into account the field inhomogeneity. In principle, if independent information on the magnetic field around the celestial body is available, information on its atmosphere can be obtained, or vice versa.

Whatever direction this research takes, it is more than likely that considerable modification of existing practices will be necessary; nevertheless a firm foundation in these laboratory experiments will be invaluable for those seeking to explore the unknown.

The principles discussed here are general and can be applied to either molecular media or a plasma. In any case, the most interesting results will be obtained with frequencies near the appropriate resonant frequencies of the system. An experiment sophisticated enough to analyze a planetary atmosphere for the existence of natural optical activity might be of assistance in establishing the possibility of other life forms.

CHIRALITY IN BIOLOGICAL SYSTEMS

Chirality[1] plays a decisive role in many biological reactions (Eyring and Johnson 1957). The proteins in all living things on earth, for example, are composed exclusively of the L amino acids. The DNA of the chromo-

---

[1] Chirality is a topological description of the corkscrew patterns which molecules exhibit either by virtue of their conformation or configuration of atoms about a central one, such as an asymmetric carbon center.

somes and ribonucleic acid, on the other hand, contain only the D-sugars. If life migrated from planet to planet as spores or some other living entity, the same chiralities should persist. This migration seems extremely unlikely; one would expect either biological chirality as we find it or the mirror image of our chiral compounds to occur with equal likelihood on another planet.

The occurrence of molecules of one chirality to the exclusion of their mirror image poses an interesting problem. When a chemist makes such molecules from inactive starting material, he gets equal amounts of a compound and its mirror image. In a restricted locality one sometimes finds quartz crystals of one chirality, all having originated as a result of a single nucleation. However, in a second location the mirror image is equally likely to be found.

In the evolving of living things, our L-amino-acid–D-sugar world would have had equal a priori probability with its mirror image. Whether our particular biological world suddenly got the ascendency over the competing mirror-image world or whether it slowly won out by a series of small steps needs definitive clarification by studying the amino acids in the fossil record. It seems most likely that the two systems were in equal competition until a single extremely improbable step took place that effectively shut the door on the competition. A likely candidate for this step is the first effective replication of enzyme-like and chromosome-like material. If it took, say, a billion years for this very unlikely step to be taken, the competition would in all probability not have arrived until about half as long again, even if conditions stayed equally favorable; in the meantime, the rapidly replicating first arrival would use up the available material and shut the door on the competition. If this view is correct, as we predict, the fossil record will only show L-amino-acid optical activity or a racemic mixture in the proteins.

NUCLEAR CHIRALITY

Chirality at the nuclear level, as exemplified by Cobalt 60 emitting its beta particles out of its south magnetic pole, raises the same kind of questions as to the origin of nuclei and where the mirror images of our kind of nuclei are to be found (Eyring 1967). There is insufficient reason to give up the idea of symmetry. Thus, the existence somewhere of antimatter worlds follows, in which, for example, Cobalt 60 with a negative nucleus emits positive electrons from its north pole. The question then arises whether the conversion of energy into material can be so catalyzed that the energy is drained off preferentially into matter or antimatter, or whether there is an effective separation into vast islands of matter and antimatter after their twin birth. Chirality thus poses unanswered questions on the cosmic scale paralleling those still perplexing the chemists.

Acknowledgment. The authors wish to thank the National Science Foundation, GP 28631, the National Institute of Health, GM 12862-08, and the U. S. Army, Contract DA-ARO-D-31-124-72-G15, for support of this research.

REFERENCES

Bijvoet, J. M. 1955. Determination of absolute configuration of optical antipodes. *Endeavour* 14: 71–77.

Buckingham, A. D., and Stephens, P. J. 1966. Magnetic optical activity. *Ann. Rev. Phys. Chem.* 17: 399–432.

Caldwell, D. J., and Eyring, H. 1971. *The theory of optical activity.* New York: Wiley.

———. 1972. Theory of optical rotation. *Adv. Quant. Chem.* 6: 143–158.

Djerassi, C. 1960. *Optical rotary dispersion.* New York: McGraw-Hill.

Eyring, H. 1967. Near symmetry and model structures. *Proceedings of the Robert A. Welch Foundation Conferences on Chemical Research.* XI: 7–28. Houston, Texas.

Eyring, H., and Johnson, F. H. 1957. The critical complex theory of biogenesis. *The influence of temperature on biological systems.* Am. Physiological Soc. pp. 1–8.

Kramer, H. A. 1929. Paramagnetic rotation of the plane of polarization in uniaxial crystal of the rare earth. *K. Acad. Amsterdam, Proc.* 32 (9): 1176–1189.

Kronig, R. de L. 1926. On the theory of dispersion of x-rays. *J. Opt. Soc. Am.* 12: 547–558.

Kuhn, W. 1961. Rotary dispersion and the vibrating momentum of optically active absorption bands. *Tetrahedron* 13: 1–12.

Lowry, T. M. 1935. *Optical rotary power.* New York: Longmans Green and Co.

Pasteur, L. 1848. *Oeuvres.* p. 78, A.C.P. (iii) 24, 457.

Shieh, D. J., Lin, S. H., and Eyring, H. 1972. Magnetic circular dichroism of molecules in dense media. I. Theory. *J. Phys. Chem.* 76: 1844–1848.

Velluz, L.; LeGrand, M.; and Grossjean, M. 1960. *Optical circular dichroism.* New York: Academic Press.

COHERENCE AND ITS APPLICATION IN THE BEAM-FOIL LIGHT SOURCE

C. H. LIU and S. BASHKIN
University of Arizona

Alignment and coherence are created among the excited states of atoms or ions during the impulsive collision between a fast-moving ion beam and a thin carbon foil. The subsequent spatially extended optical emission associated with the decay of the non-degenerate excited states is partially polarized and exhibits periodic intensity variations with or without an external electrostatic or magnetic field. These periodic intensity variations are called quantum beats. Though the nature of the creation of coherence and alignment by the beam-foil collision is not completely clear, the frequencies of quantum beats have been utilized to measure the separations of the multiplets and the Landé g-factors of excited states in various stages of ionization to which conventional techniques are not easily applicable. The measurement of mean lives of excited states is made with negligible effect of cascades by taking advantage of the coherence and alignment features.

Beam-foil spectroscopy (Bashkin 1964) is well established as a versatile technique for the observation and study of the spectra of atoms in various stages of ionization. Previously unobserved transitions have been recorded and analyzed in a number of elements, the mean lives of many excited states have been measured, and the stages of ionization from which the spectral lines emerge have been ascertained (Bashkin 1968). What is of interest in the present instance is that the beam-foil interaction also creates coherent excitation and alignment among the levels and sublevels of the excited states (Bashkin et al. 1965; Martinson, Bromander, and Berry 1970; Andrä 1970a,b; Liu et al. 1971). Although the coherent excitation and alignment mechanism is still a subject for investigation, the brevity of the collisions between the fast-moving ion beam and the thin foil, and the weakness of a direct interaction with the electron moment, indicate that spin-dependence is negligible compared to the actual beam-foil interaction (Macek 1970).

With the help of Fig. 1, we review how the beam-foil light source is created. One uses particle accelerators, which, depending on the element and the stage of ionization of interest, yield ions with energies between

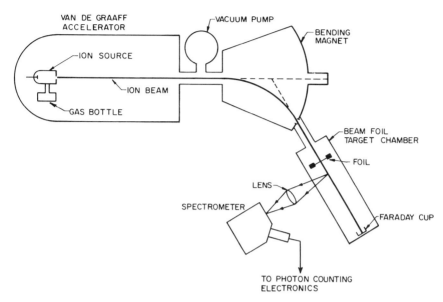

Fig. 1. Schematic arrangement of a typical beam-foil experiment.

10 keV and 400 MeV. After the beam leaves the accelerator, a magnet selects the particular beam component that is to be admitted into the target chamber. The target chamber includes a thin foil of carbon, perhaps 0.05 μm thick, which serves as the target, and a Faraday cup for measuring beam current.

The ions penetrate through the carbon foil; when they emerge, they are in various stages of ionization and excitation. De-excitation is accompanied by the emission of light that contains information on the states from which it originated. This is the beam-foil light source (Bashkin 1964).

This source is very useful because it has the remarkable properties that (1) all excited states are created at the same place ($x = 0$); (2) the motion of the ions, the interaction with the thin foil, and subsequently, the spatial-extended optical decays all occur in a vacuum in which collisions are negligible; and (3) compared with the conventional methods of excitation and ionization, such as electron bombardment, *rf* discharge, optical pumping, etc., the beam-foil source has a far wider range of excitation and ionization, especially as it is not restricted to resonance transitions. Almost all the excited and ionized states of ions can be created in the beam-foil source by adjusting the energy of the incident beam. Many of the excited states are of astrophysical interest, and measurement of the mean lives of those states has an important bearing on the assessment of the relative abundances of the elements and the theory of stellar behavior.

To the eye, the ion beam emerging from the foil looks like a glowing streak which fades in intensity downstream from the foil, decreasing approximately according to the well-known law of exponential decay. The optical decay of interest is picked out and detected by an optical arrangement that generally consists of a spectrometer and photon-detecting device, as shown in Fig. 1.

Like other drastic methods of excitation and ionization, excited levels higher than the level of interest are also produced in the beam-foil source. Since some of the higher excited levels can themselves decay optically to the one of interest, the population of the level of interest can have a time-dependence different from that of a simple exponential decay. In some cases, especially in the mean-life measurements made by simply looking at the decay of intensity, the cascade effects are appreciable and can affect the reliability of the lifetimes unless elaborate analyses are made.

Because of the relatively high speed of the ions ($\gtrsim 10^8$ cm/sec) and the nonzero acceptance angle of the spectrometer, the Doppler broadening of spectral lines imposes serious limits on the information one can abstract from the beam-foil source. By refocusing the spectrometer (Stoner and Leavitt 1971), this broadening has recently been considerably reduced.

INTENSITY BEATS, QUANTUM BEATS, AND THEIR APPLICATION

Coherent Excitation

Breit (1933) showed that coherent excitation of nondegenerate levels was possible if the exciting resonance radiation were pulsed in a time, Δt, much shorter than the relaxation time of the excited state. A reason for this is that the shorter the light pulse, the broader the range of spectral frequencies it contains. Therefore, if the width, Δv, of the line in the pulse is larger than the natural line width and the separation of the levels, these levels can be excited simultaneously. This happens when

$$\Delta t \simeq \frac{1}{\Delta v} < \tau, \tag{1}$$

where τ is the natural lifetime of the excited states. This means that all the nondegenerate states are excited by one wave packet, provided the excitations are allowed by the optical selection rule. Under such excitation conditions, the phase differences among the levels are perfectly definite. Such excited states are called coherently excited states. The interference of resonance radiations emitted or absorbed by coherently produced states creates beat phenomena.

That coherent excitation of ions can be achieved by beam-foil collision is obvious. Because of the uncertainty principle and the fact that the ions pass through the foil in roughly 10^{-14} sec, the necessary condition for coherent excitation—that the spread of energy of excitation be larger than the separations of the excited states—is often well satisfied. In order to observe the subsequent beats in the optical emission from the coherently excited states, there must be preferential or selective excitation of some of the nondegenerate states. In the beam-foil light source, the selective excitation is such that the levels are aligned, i.e., the excitation probability is a function of $|m_J|$. We define the polarization, P, of the emitted light as

$$P = \frac{I_\parallel - I_\perp}{I_\parallel + I_\perp}, \qquad (2)$$

where $I_\parallel$ and $I_\perp$ are the optical emission intensities measured in a direction perpendicular to the ion beam, with the electric vectors parallel and perpendicular respectively to the ion beam. This polarization is used to measure the degree of preferential excitation of the states. From the definition, alignment cannot occur unless $J > 1/2$.

An externally applied electric field or magnetic field can also cause coherence among the beam-foil-excited nondegenerate states. For simplicity, assume that of all the eigenstates ϕ_α of the field-free Hamiltonian $\mathscr{H}_0$, only one is created in the beam-foil interaction. With the application of a weak external field, say an electric field, ε, a new set of eigenstates, u_ρ, of the new Hamiltonian, $\mathscr{H}$, is found:

$$\mathscr{H} = \mathscr{H}_0 + e\boldsymbol{\varepsilon} \cdot \mathbf{r}. \qquad (3)$$

The new eigenfunctions corresponding to the energy eigenvalues, E_ρ, are then linear combinations of the field-free ones, i.e.,

$$u_\rho = \sum_\alpha a_{\rho\alpha} \phi_\alpha \exp\left[-\left(\frac{1}{2}\Gamma_\rho + i\frac{E_\rho}{\hbar}\right)t\right], \qquad (4)$$

where $\Gamma_\rho = \dfrac{1}{\tau_\rho}$ is the reciprocal of the lifetime of the state ρ. To insure the creation of coherence by a weak electric field, the ion should enter into the electric field from the field-free region in a short time $\Delta t < \dfrac{\hbar}{\Delta E}$, with ΔE being the separation of the states of interest. This is possible, due to the relatively high speed of the ions and the localization of the external field. A sudden rise of the electric field forces the ion, with an initial $\psi(t=0) = \phi_\alpha$, into a linear combination of the new eigenfunctions that are coherently produced:

$$\psi(0) = \phi_\alpha = \sum_\rho a^*_{\rho\alpha} u_\rho. \qquad (5)$$

Therefore,

$$\psi(t) = \sum_{\rho,\mu} a_{\rho\alpha}^* a_{\rho\alpha} \phi_\mu \exp\left[-\left(\frac{1}{2}\Gamma_\rho + i\frac{E_\rho}{\hbar}\right)t\right]. \quad (6)$$

The subsequent optical emission can thus be obtained as the square of the matrix element:

$$I(t) \propto \sum_f \sum_\rho a_{\rho\alpha}^* \exp\left[-\left(\frac{1}{2}\Gamma_\rho + i\frac{E_\rho}{\hbar}\right)t\right] |\langle\psi_f|V|\sum_\mu a_{\rho\mu}\phi_\mu\rangle|^2, \quad (7)$$

where V is the electric-dipole operator. The nonvanishing cross-terms oscillate according to $\cos\left(\frac{\Delta E_{\rho\rho'}}{\hbar}t\right)$, where the $\Delta E_{\rho\rho'}$ are the separations of the states of interest and are field dependent. The intensity patterns are called quantum beats.

Electric-Field-Induced Intensity Beats

S. Bashkin et al. were the first ones to observe beat phenomena (Martinson, Bromander, and Berry 1970; Bashkin et al. 1965; Andrä 1970a, b; Liu et al. 1971). The observed beats were induced by an electric field that was applied to the radiating beam. Their experimental arrangement was similar to that in Fig. 2. A beam of HHH^+ ions, with an energy of 200 keV,

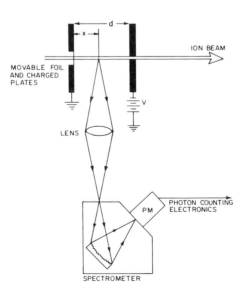

Fig. 2. Experimental arrangement of the electric-field-induced beats.

passed successively through two thin carbon foils. The collision between the first foil and the beam dissociated the molecules into atoms; the second collision excited the atoms. The excited atoms then entered a uniform electric field that was either a static one produced by a pair of charged parallel metal plates or a motional one created by the motion of the atoms in an applied magnetic field. Optical radiations from a segment of the beam were spectroscopically analyzed and photographically recorded. Detailed investigations have since been performed on H atoms and He$^+$ ions (Bickel and Bashkin 1967; Andrä 1970a, b; Sellin et al. 1969a, b), and photon-counting devices have been employed as well as photographic plates. Figure 3 shows in (a) the beat patterns of some spectral lines from He$^+$ on photographic plates, and in (b) the beat patterns in hydrogen Lyman α displayed on a multiscaler from an experiment that used photon counting.

This technique has been used to measure fine-structure separations in hydrogenic species such as H and He$^+$. The procedure for obtaining the separations $\Delta E_{\rho\rho'}$ can be described with the help of Fig. 3b. The experiment consists of measuring the beat spacing $\bar{x}$ on spatial decay curves for various electric fields $\varepsilon = V/d$. The energy separation ΔE and speed of the ion v are related to $\bar{x}$ as

$$\Delta E = \frac{hv}{\bar{x}}, \qquad (8)$$

derived by rewriting the $\cos\left(\frac{\Delta E_{\rho\rho'}}{\hbar} t\right)$, from the previous section, as $\cos\left(\frac{\Delta E}{\hbar} \frac{x}{v}\right)$ and then equating $\left(\frac{\Delta E}{\hbar} \frac{\bar{x}}{v}\right)$ to 2π. We here assume that only two states, with energies E_ρ and $E_{\rho'}$ respectively, are involved, or that only one of all the beats corresponding to all the different fine-structure separations of interest is predominant. Otherwise one must analyze the fine structure of interest, beginning with a nonequilibrium excitation. Without more information concerning the nonequilibrium excitation, the accuracy of the calculated contributions (amplitudes) of each of the beats to the resultant experimental beat pattern is not reliable. One plots ΔE as a function of ε and extrapolates to $\varepsilon = 0$ to obtain the field-free fine-structure separation. In using the beam-foil source to measure the fine-structure as well as the hyperfine-structure separations where the beat patterns are not sinusoidal and where lifetimes are in the range of 10^{-8} to 10^{-9} sec, the level-crossing technique is much more accurate and promising, and this effort has been made in several laboratories (Hadeishi et al. 1973).

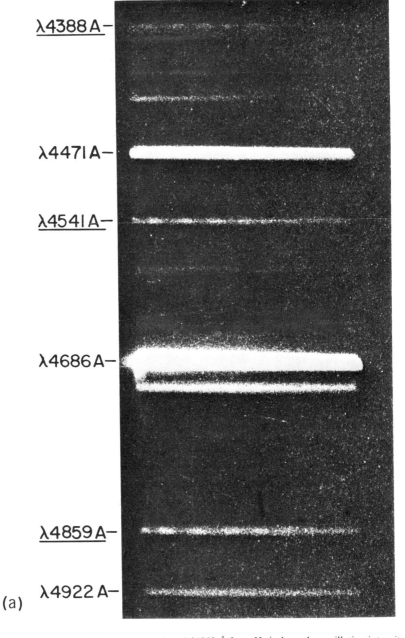

Fig. 3. (a) Spectral lines of $\lambda 4541$ A and $\lambda 4859$ Å from He$^+$ show the oscillating intensity patterns.

(b) Beat patterns in H L$_\alpha$ emission as a function of electric field $\varepsilon = \dfrac{V}{d}$.

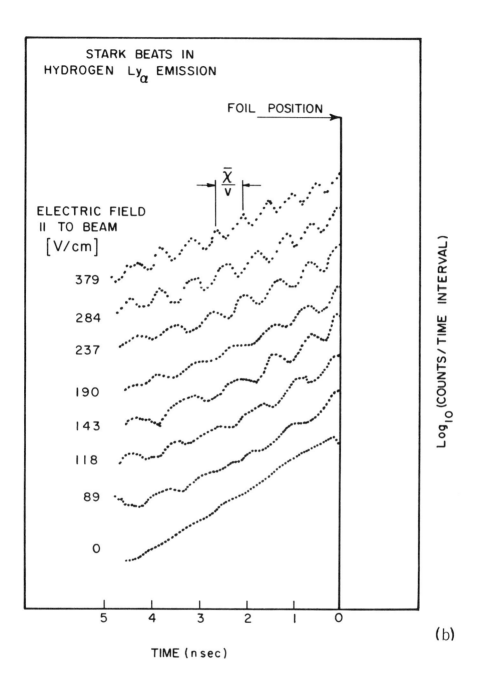

Magnetic-Field-Induced Quantum Beats

Let a beam of foil-excited multielectron atoms or ions with nuclear spin $I = 0$ enter a magnetic field that is perpendicular to the beam, the particles being suddenly exposed to a motional electric field $\varepsilon = \dfrac{\mathbf{v}}{c} \times \mathbf{B}$. Since the fine-structure levels are widely separated, the electric-field-induced intensity beats are too weak in amplitude and too fast in frequency to be seen with our present detection techniques. However, with the presentation of the magnetic field of intensity H, the degeneracy of the Zeeman magnetic sublevels is removed. This can give state separations of the right order for beats to be seen.

The formation of the quantum beats can be easily visualized by a classical picture. An oscillating electric dipole in the atom is created by the foil excitation and the subsequent optical emission. Because of the Larmor precession of the atom from the magnetic-dipole interaction with the external magnetic field, H, the electric-dipole oscillator or its component perpendicular to H precesses around the magnetic field H. From the

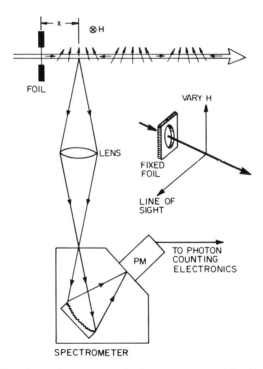

Fig. 4. Experimental arrangement for the measurement of Landé g-factors.

geometry of the experimental arrangement as shown in Fig. 4 and from the formula for the energy radiated from an electric dipole, it is easy to see that the detected optical emission is modulated at twice the Larmor frequency ω_L. Because the states of interest are excited coherently, they precess together in the same phase.

Since alignment of the excited levels is created in the beam-foil collision, the observed subsequent optical emissions are partially polarized. With the aid of a linear polarizer, the degree of alignment of the excited states can be determined by measuring the polarization (see Eq. [2]). Whether a linear polarizer is needed or not in the quantum-beat experiments depends on the individual case and the geometry of the experimental arrangement. A simple example is shown in Fig. 5. A more complicated example involves the measurements of Landé g-factors by the quantum-beat technique, as described below. In Fig. 6a, assume that the electric dipole oscillator associated with the optical decay of the aligned excited states has only two components: $E_\parallel$ is parallel and $E_\perp$ is perpendicular

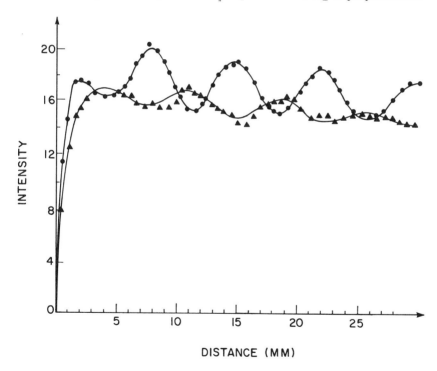

Fig. 5. Zero-field quantum beats pattern from 2s ^{3}S–4p ^{3}P, He I transition. The circles and triangles are the measurements in light polarized parallel and perpendicular to the beam, respectively.

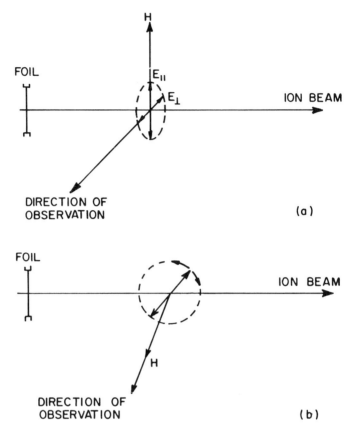

Fig. 6. (a) An experimental arrangement in which the linear polarizer is not necessary for detecting the quantum beats.
(b) A linear polarizer is needed in this slightly different geometrical experimental arrangement.

to the external magnetic field H. The beat pattern described in the classical model results from the change of the intensity radiated by a precessing electric dipole around the external H field. Therefore, a linear polarizer with axis of polarization parallel to the ion beam is perhaps desirable but not necessary. Since the $E_\parallel$ component cannot precess around H, it only contributes to the background of the observation. Because the polarizer is not needed, this geometrical arrangement can be used to observe the oscillations in the vacuum ultraviolet region, thereby letting one measure the g-factors of highly excited atoms. On the other hand, if H is applied parallel to the direction of observation, as shown in Fig. 6b, a linear polarizer is definitely needed, since in this experimental arrangement, only the two circularly polarized emissions from the excited levels will be seen. Furthermore, due to the coherent excitation, the emissions are combined with a certain phase difference, resulting in a linearly polarized emission, that is, a linear oscillating dipole, as shown in Fig. 6b. The precession of the emitting dipole around the magnetic field, H, can be detected only by looking through a linear polarizer.

The quantum beats, namely the modulation of the intensity of light emitted by the excited atoms, may be written (Church and Liu 1972) as

$$I(t) = A(1 + B \cos 2\omega_L t) e^{-\Gamma t}. \tag{9}$$

Here Γ is the reciprocal mean life of the excited level. The Larmor frequency is $\omega_L = g_J \frac{\mu_B}{\hbar} H$, where g_J is the Landé g-factor of the excited level, μ_B is the Bohr magneton, and $\hbar$ is Planck's constant over 2π. $A(1 + B)$ is the total intensity at the foil, while B is the relative quantum-beat amplitude. A density matrix analysis of an analogous experiment involving a coherently excited and aligned level (Hadeishi 1967) shows that nonzero B implies both coherence and alignment. If there were no coherence, B would be zero whether alignment existed or not. Equation (9) can also be written as

$$I(H) = A\left(1 + B\cos\left[2g_J \frac{\mu_B}{\hbar} \frac{x}{v} H\right]\right) e^{-\Gamma \frac{x}{v}}. \tag{10}$$

Equation (10) shows that, for a fixed value of t or, equivalently, for a fixed point of observation, x, downstream from the foil, $I(H)$ should vary sinusoidally with H as shown in Fig. 7. Then

$$g_J = \frac{\pi \hbar v}{\mu_B x} \frac{1}{\bar{H}}, \tag{11}$$

where $\bar{H}$ is the change in magnetic field strength required to produce a single period of the intensity oscillation. The values of v, x, and $\bar{H}$ must be measured to about 0.1% in order to produce an accuracy of half a percent in an absolute measurement of the Landé g-factor, g_J. Of the three, perhaps measuring the velocity, v, presents the greatest difficulty, since one needs the velocity of the ions after transmission of the ions through the foil. In the absence of a high-precision post-foil velocity analyzer, one is reduced to calculations that are not always satisfactory. An alternative approach is to obtain the g-values, g_J, in terms of a known one, g_s, of the same element and in terms of the relative periods of the light intensity oscillations in the same linear magnetic field sweep. Then

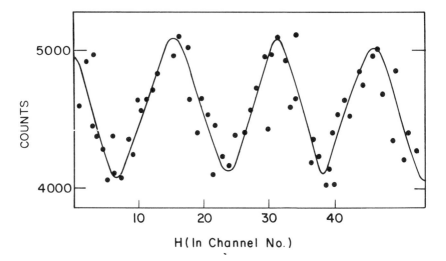

Fig. 7. A typical oscillating quantum beats pattern.

$$g_J = g_s \frac{\overline{H}_s}{\overline{H}}. \tag{12}$$

In this case the velocity drops out, even for different charge states of the same element. If the standard is taken from a different element accelerated to the same energy, E, then

$$g_J = g_s \frac{\overline{H}_s}{\overline{H}} \left(\frac{m_s}{m}\right)^{\frac{1}{2}} (1 + \varepsilon), \tag{13}$$

where m_s and m are the respective ionic masses and $\varepsilon - \frac{\Delta_s - \Delta}{2E}$. Δ_s and Δ refer to the absolute mean-energy losses experienced by the different ions in passing through the foil (Fastrup, Hvelplund, and Sautter 1966); ε is generally less than 1%.

Classical optical Zeeman spectroscopy is a powerful method for measuring g-factors and also for determining the J-value of many levels. Unfortunately, the light sources generally used for such work give only the arc and first spark spectra. Magnetic-resonance experiments are also interesting, but the frequency and power ranges of rf oscillators are seriously limiting factors. If we turn to the beam-foil light source, many levels of ions are accessible, but we find that the Doppler broadening interferes with the Zeeman patterns, although one such experiment has been reported (Stoner and Radziemski 1972). The application of rf to the beam is also difficult, unless one turns to the "artificial" *rf*-generators that have been used (Martinson, Bromander, and Berry 1970; Bashkin et al. 1965; Andrä 1970*a,b*. Liu et al. 1971). Hence, each method has its handicap.

Measurements of *g*-factors by the beam-foil quantum-beat technique should aid in identifying the atomic levels from which the observed spectral lines arise and in finding the coupling scheme of the atomic angular momenta. Knowing the *g*-factor of a level also permits a lifetime measurement by using the Hanle effect (Church, Druetta, and Liu 1971; Liu, Gardiner, and Church 1973).

Lifetime Measurements and the Problem of Cascades

Optical cascades to the level of interest, as mentioned earlier, introduce difficulties in analyzing the lifetimes of the levels of interest from the spatial optical-decay curves. This problem has been attacked in terms of the coherence effects. First, efforts were made to measure the degree of transfer of the coherence through optical-cascade effects. The results showed that the coherence transfer is too small to be detected in the process of measuring the *g*-factors using quantum beats. In other words, the effects of cascades on the *g*-values are negligible within the measurement precision of $\simeq 0.5\%$ (Liu, Druetta, and Church, 1972; Dufay, Gaillard, and Carre 1973). Second, because of the negligible transfer of coherence through optical cascades, a beam-foil technique was developed, using the spatial decay of quantum beats, to measure the lifetime of a level that is subject to strong optical cascades (Liu and Church 1972). This method as described below is quite straightforward and requires little analysis.

As in Fig. 8*a*, light emitted from the fast-moving, foil-excited ions is collected from a narrow region ($\simeq 0.1$ mm wide) about a fixed point x_0 in space. This light, after transmission through a linear polarizer, is spectroscopically analyzed and detected. The foil is advanced at a uniform rate until it passes x_0; it is then retracted at the same rate until it has returned to its initial position. The detected light intensity increases as the foil approaches x_0, drops to a background level as the foil passes x_0, and follows a mirror-image curve when the foil is retracted. The usual beam-foil-source mean life is obtained from such a measurement.

Now let a fixed, uniform, external magnetic field H be applied perpendicular to both the beam and the direction of observation. The observed light intensity is then described by

$$I(t,H) = (A + B\cos 2\omega_L t)e^{-\Gamma t} + C(t,H) + D, \qquad (14)$$

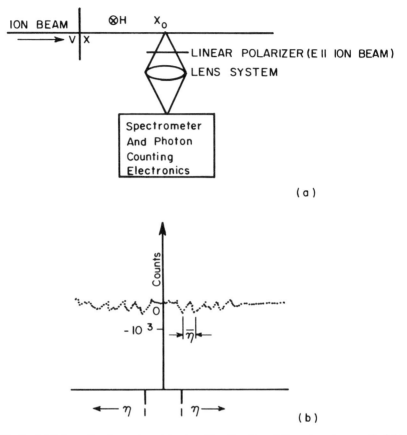

Fig. 8. (a) Schematic of experimental setup. The direction of the external magnetic field H is into the paper.
(b) The quantum-beats signals shown are accumulated over a time interval of alternating modes of data collection. Each cycle, with magnetic field H on and the photons counted in an additive mode is followed by another cycle with H off and photons counted in a subtractive mode.

where $C(t,H)$ describes light that emanates from the level of interest by virtue of cascade population from higher excited levels, and D is the background. The difference between $I(t)$ without and with applied magnetic field H is

$$\Delta I \equiv I(t,0) - I(t,H) = B(1 - \cos 2\omega_L t)e^{-\Gamma t} + [C(t,0) - C(t,H)]. \quad (15)$$

Thanks to the negligible transfer of coherence through cascade, $C(t,H) \simeq C(t,0)$. Now set $2\omega_L t = a\eta$, where η is the channel number in a multiscaler display corresponding to the distance, $x - x_0$. Then ΔI can be written as

$$\Delta I(\eta) = B(1 - \cos a\eta) \exp\left(-\frac{a\Gamma}{2\omega_L}\eta\right), \tag{16}$$

where $a = \dfrac{2\pi}{\bar{\eta}}$ (see Fig. 8b) in terms of $\bar{\eta}$, the spatial period of the beat. With the help of Equation (16), mean lives $\tau = \dfrac{1}{\Gamma}$ are obtained in terms of the Landé g-factors and the external magnetic field H. No measurements on the ion velocity v or the decay length are needed. Most importantly, the cascade effect has been virtually eliminated. An example of the utility of this method is shown in Table I, where column 5 gives the lifetimes of certain O^+ levels as measured in a standard beam-foil experiment. Column 7 gives the effective cascade lifetime as found in that work. Column 4 shows the lifetimes as determined by the g-value technique (Liu and Church 1972), which, as described above, is cascade-free. The difference between the methods is striking.

Zero-Field Quantum Beats

As discussed in the first section, quantum beats result from the impulsive collision between the beam and the foil even in the absence of any external electric or magnetic field. Work concerned with this aspect has been reported recently (Bashkin et al. 1965; Martinson, Bromander, and Berry 1970; Andrä 1970a,b; Liu et al. 1971; Lynch et al. 1971; Berry and Subtil 1971). The zero-field quantum beats were used to measure the fine-structure or hyperfine-structure separations in hydrogen-like and helium-like atoms. The early observations of the zero-field quantum beats on He I and II (Bashkin and Beauchemin 1966) were interpreted as the kind of beat discussed in the second section, namely, a coherent mixing of fine-structure levels of the excited states due to an intrinsic field associated with the beam itself. (The motional electric field due to running the experiment in the earth's magnetic field is too small to induce beats.) This interpretation has since been discarded.

The modulated light intensity behind the foil as a function of time is of the form

$$I(t) = \sum_i (A_i + B_i \cos \omega_i t) e^{-\Gamma t}, \tag{17}$$

the summation being over all possible energy separations ω_i within the upper spectroscopic term, and Γ is the common decay constant. The

TABLE I
Comparison of Lifetime Measurements

Charge state	Lower and upper level	Wave length (Å)	Mean life (nsec)		Calculated transition probability[a] (10^8 sec^{-1})	Other observed decay constants (nsec)		Wavelengths of cascades into levels from which transitions were observed (Å)
			This method	Beam foil source		Beam foil source[b]	This method	
O II	$3s'\ ^2D_{5/2} - 3p'\ ^2F^\circ_{7/2}$	4590.97	13.7 ± 0.4	14.0[b], 12.8[b], 9.5[c], 10.7[c], 12.0[c], 14[c], 8.7[d]	1.11	4.0	None	4448, 4190, 4114
O II	$3p'\ ^2F^\circ_{7/2} - 3d'\ ^2G_{9/2}$	4189.79	6.8 ± 0.2	6.1[b], 5.05[b]	2.51	20.0	None	4253
O II	$3s\ ^2P_{3/2} - 3p\ ^2P^\circ_{3/2}$	3973.26	6.8 ± 0.1	7.1[b], 6.0[c]	1.27		None	—

[a] Wiese, Smith, and Glennon 1966.
[b] Druetta, Poulizac, and Dufay 1971; Druetta and Poulizac 1969.
[c] Kernahan, Lin, and Pinnington 1970.
[d] Pinnington and Lin 1969.

coefficients B_i are zero if there is no alignment in the upper term; the relative values of A_i and B_i depend on the initial coherence and populations of the angular momentum sublevels. A typical zero-field quantum-beat pattern is shown in Fig. 5. The difficulty in analyzing the data is the same as discussed in the section on electric-field-induced intensity beats.

CONCLUSION

The beam-foil light source has been shown to be very useful in spectroscopic work. Not only the lifetimes of highly excited, multiply charged atoms can be measured in a straightforward way, but also the fine-structure and hyperfine-structure separations and the Landé g-factors can be obtained, due to the fact that coherent excitations are created in the impulsive beam-foil collision. The beam-foil technique for measuring the g-factors has been improved to a degree of accuracy (Church, Druetta, and Liu 1971; Liu, Gardiner, and Church 1973) that is comparable to that of the conventional optical Zeeman effect measurement, but the accessible range of the former is much greater than that of the latter. The theories suggested to explain the origin of coherence are presently incomplete. This field, therefore, is still open to investigation.

Acknowledgments. This research is supported in part by the Office of Naval Research, the National Aeronautics and Space Administration, and the National Science Foundation.

REFERENCES

Andrä, H. J. 1970a. Zero-field quantum beats subsequent to beam-foil excitation. *Phys. Rev. Letters* 25: 325–327.

———. 1970b. Stark-induced quantum beats in H Ly α emission. *Phys. Rev. A* 2: 2200–2207.

Bashkin, S. 1964. Optical spectroscopy with Van de Graaff accelerators. *Nucl. Instr. Methods* 28: 88–96.

———. (ed.) 1968. *First International Conference on Beam-Foil Spectroscopy, Tucson, Arizona, November, 1967.* New York: Gordon & Breach.

Bashkin, S., and Beauchemin, G. 1966. Production de raies spectrales de He I et He II par l'excitation d'atomes He$^+$ accélérés au moyen d'un Van de Graaff. *Can. J. Phys.* 44: 1603–1607.

Bashkin, S.; Bickel, W. S.; Fink, D.; and Wangsness, R. K. 1965. Interference of fine structure levels in hydrogen. *Phys. Rev. Letters* 15: 284–285.

Berry, H. G., and Subtil, J. L. 1971. Fine-structure measurements by the beam-foil technique. *Phys. Rev. Letters* 27: 1103–1104.

Bickel, W. S. 1968. Electric-field and multiple-foil-excitation experiments on beam-foil-excited hydrogen atoms. *J. Opt. Soc. Am.* 58: 213–221.

Bickel, W. S., and Bashkin, S., 1967. Interference of fine-structure levels of He II. *Phys. Rev.* 162: 12–15.

Breit, G. 1933. Quantum theory of dispersion, Parts VI & VII. *Rev. Mod. Phys.* 5: 91–137.

Church, D. A.; Druetta, M.; and Liu, C. H. 1971. Hanle-effect mean-life measurements on aligned fast ions. *Phys. Rev. Letters* 27: 1763–1765.

Church, D. A., and Liu, C. H. 1972. Measurement of g factors by a beam-foil quantum-beat technique. *Phys. Rev.* A 5: 1031–1036.

Druetta, M., and Poulizac, M. C. 1969. Radiative lifetimes for some excited electronic states of oxygen. *Phys. Letters* 29A: 651–652.

Druetta, M.; Poulizac, M. C.; and Dufay, M. 1971. Assignments and mean-life measurements for O II and O III ions. *J. Opt. Soc. Am.* 61: 515–518.

Dufay, M.; Gaillard, M. L.; and Carré, M. 1973. In draft.

Fastrup, B.; Hvelplund, P.; and Sautter, C. A. 1966. *Kgl. Danske, Videnskab, Selskab. Mat. Fys. Medd.* Vol. 35, No. 10.

Hadeishi, T. 1967. Phenomenon of the metastable state coherently excited by electron impact. *Phys. Rev.* 162: 16–22.

Hadeishi, T.; Michael, M.; Yellin, J.; and Geneux, F. 1973. In draft.

Kernahan, J. A.; Lin, C. C.; and Pinnington, E. H. 1970. Radiative lifetimes of some energy levels of O II and O III ions. *J. Opt. Soc. Am.* 60: 986–987.

Liu, C. H.; Bashkin, S.; Bickel, W. S.; and Hadeishi, T. 1971. Alignment of atoms and ions with the beam-foil light source. *Phys. Rev. Letters* 26: 222–224.

Liu, C. H., and Church, D. A. 1972. Suppression of cascade effects in beam-foil mean-life measurements. *Phys. Rev. Letters* 29: 1208–1211.

Liu, C. H.; Druetta, M.; and Church, D. A. 1972. Quantum beats in transitions from levels subject to optical cascades. *Phys. Letters* 39A: 49–50.

Liu, C. H.; Gardiner, R. B.; and Church, D. A. 1973. Improved Hanle effect measurement technique for fast ions. *Phys. Letters* 43A: 165–166.

Lynch, D. J.; Drake, C. W.; Alguard, M. J.; and Fairchild, C. E. 1971. Zero-field quantum beats in Lyman-β radiation from beam-foil-excited hydrogen atoms. *Phys. Rev. Letters* 26: 1211–1213.

Macek, J. 1970. Theory of atomic lifetime measurements. *Phys. Rev.* A 1: 618–627.

Martinson, I.; Bromander, J.; and Berry, H. G., eds. 1970. Second International Conference on Beam-Foil Spectroscopy, Lysekil, Sweden, June 1970. *Nucl. Instr. Methods.* Vol. 90.

Pinnington, E. H., and Lin, C. C. 1969. Radiative lifetimes for some carbon and oxygen energy levels. *J. Opt. Soc. Am.* 59: 780–781.

Sellin, I. A.; Moak, C. D.; Griffin, P. M.; and Biggerstaff, J. A. 1969a. Periodic intensity fluctuations of Balmer lines from single-foil excited fast hydrogen atoms. *Phys. Rev.* 184: 56–57.

———. 1969b. Stark-perturbed Ly α decay in flight. *Phys. Rev.* 188: 217–221.

Stoner, J. O., Jr., and Leavitt, J. A. 1971. Reduction of Doppler broadening of spectral lines in fast-beam spectroscopy. *Appl. Phys. Letters* 18: 477–479.

Stoner, J. O., Jr., and Radziemski, L. J. Jr. 1972. Observations of Zeeman effect in the optical spectra of fast ion beams. *Appl. Phys. Letters* 21: 165–166.

Wiese, W. L.; Smith, M. W.; and Glennon, B. M. 1966. *Atomic transition probabilities.* U.S. National Bureau of Standards.

SOME EXAMPLES OF EXACT AND APPROXIMATE SOLUTIONS IN SMALL PARTICLE SCATTERING: A PROGRESS REPORT

J. MAYO GREENBERG
State University of New York at Albany
and
Dudley Observatory

Approximate methods are derived for obtaining the scattering of electromagnetic waves from spheroids and truncated cylinders. Some microwave experimental results for spheroids are shown.

In the field of astrophysics one of the principal processes leading to the polarization of light is that due to the scattering and/or extinction of light by small solid particles. Based on the observations, there is overwhelming evidence that the particles span a size range from very small to those that are comparable to the wavelength or larger.

For a variety of reasons we can show that we are required to deal with nonspherical inhomogeneous particles possessing a considerable variety of optical properties characterized by their complex indices of refraction.

The most generally applied method of solving the scattering problem is to solve the vector wave equation in some appropriate coordinate system in which both the solutions and the boundary conditions are separable. We are considering here the cases represented by bodies whose indices of refraction are constant within their boundaries, although not necessarily homogeneous or isotropic. The only bodies for which such solutions may be obtained are the sphere and the infinite cylinder if the material is not perfectly reflecting. This fact never ceases to surprise those who are not familiar with the scattering problem. It may be stated that no general solution exists—even as an infinite series—for such a simple finite nonspherical body as a prolate spheroid.

In this paper those scattering problems for which computable solutions have already been adequately developed and are readily available elsewhere are only presented to give a proper basis for new methods of approximation and extensions of well-known methods. It turns out that the approximate results make explicit use of the exact results.

FUNDAMENTAL SCATTERING FORMULATION

This section is concerned with the formulation of basic equations from which the scattering of radiation by a localized variation in a medium can be discussed. These equations are developed in both the differential and the integral form.[1]

We are primarily interested in the scattering of electromagnetic waves for which we must consider the solution of the vector wave equation with appropriate boundary conditions. However, in practice, the first step is generally to find solutions of the scalar wave equation from which the solution of the electromagnetic equation may be generated. Furthermore, the solutions of the scalar wave equation contain a considerable portion of the fundamentals of electromagnetic scattering theory without the complications introduced by the vector forms. Finally, the solutions of the scalar wave equation are of considerable interest in direct application to many physical problems, including the electromagnetic ones in certain limiting cases. For these reasons, we consider in detail the solution and basic scattering results as they apply to the scalar Helmholtz equation

$$\nabla^2 \psi + k^2(r)\psi = 0, \qquad (1)$$

where $k^2(r)$ deviates from the free space value k^2 in a limited volume or approaches the free space value sufficiently rapidly (this will be specified more completely, later) as one departs from some localized volume element.

The Scattering Problem—Basic Definitions

We consider the situation in which a plane wave is incident on a region whose propagation properties differ from those of the surrounding medium. We specify the exterior medium by the propagation function k and the scatterer by the propagation function $k(r)$. The incident wave is then given by $\exp[-i\mathbf{k}_1 \cdot \mathbf{r}]$, and the scattering problem consists in finding solutions of Equation (1) satisfying

$$\psi = \psi_i(\mathbf{r}) + \psi_s(\mathbf{r}) = \exp\{-i\mathbf{r} \cdot \mathbf{k}_1\} + \psi_s(\mathbf{r}) \qquad (2a)$$

$$\psi_s(\mathbf{r}) \sim r^{-1} f(\mathbf{k}_1, \mathbf{k}_2) e^{-ikr} \quad \text{as } r \to \infty, \qquad (2b)$$

where $\mathbf{k}_1/k$ and $\mathbf{k}_2/k$ are unit vectors along incident and outgoing directions so that

$$|\mathbf{k}_2| = |\mathbf{k}_1|. \qquad (2c)$$

---

[1] *Editorial note*: We have tried to unify symbols throughout this book but in the case of this paper we let the symbols stand because they are consistent with the usage in the field of scattering theory and that defined in Greenberg's Fig. 1.—T. Gehrels.

We call $f(\mathbf{k}_1, \mathbf{k}_2)$ the scattering amplitude. Asymptotically, the scattered wave is proportional to r^{-1} so that an appropriate density satisfies an inverse square law. Note that the definition of f given here differs by a factor of (ik) from that in van de Hulst (1957).

Regardless of the nature of the waves whose propagation is represented by a time-dependent wave equation which leads to Equation (1), we may justify expressing an appropriate density and a current density, respectively, by

$$\text{density} \equiv |\psi|^2$$
$$\text{current density} \equiv (A)(\psi^*\nabla\psi - \psi\nabla\psi^*). \tag{3}$$

Consequently, the incident wave intensity as obtained by substituting $\psi_i = \exp[-i r \cdot \mathbf{k}_1]$ into Equation (3) is

$$I_i = 2ikA|\psi_i|^2 = v|\psi_i|^2, \tag{4}$$

where the constant A is normalized to give the current density as the product of the propagation velocity and the density. The scattered current in direction $\mathbf{k}_2$ as $r \to \infty$ is given by

$$I_s(\mathbf{k}_1, \mathbf{k}_2) \sim r^{-2} v |f(\mathbf{k}_1, \mathbf{k}_2)|^2. \tag{5}$$

The total scattered current is obtained by integration over a very large spherical surface

$$I_s = v \int_0^\pi \int_0^{2\pi} |f(\mathbf{k}_1, \mathbf{k}_2)|^2 \, d\Omega. \tag{6}$$

We see that only when the scattered wave has the form as given in Equation (2) does the integral on the right of Equation (6) yield a finite (neither zero nor infinite) result.

We define the total scattering cross section by

$$\sigma_s = I_s/I_i, \tag{7}$$

and denote the differential scattering cross section by

$$\sigma(\mathbf{k}_1, \mathbf{k}_2) = |f(\mathbf{k}_1, \mathbf{k}_2)|^2. \tag{8}$$

The differential scattering cross section is then (for incident current density $I_i = v$, $|\psi_i|^2 = 1$) the energy (for light or sound or particles—for Schrödinger wave equation) per unit solid angle emerging in the direction given by $\mathbf{k}_1$. The total scattering cross section is then the total energy (or number of particles) scattered per unit time per unit incident current density. Note that it is convenient and fairly common to use an incident current density equal to v rather than to unity.

In general, energy is absorbed as well as scattered. In this event we define the absorption cross section as

$$\sigma_a = \frac{\text{energy absorbed per unit time}}{\text{incident current density}} \quad (9)$$

and the total cross section by

$$\sigma_t = \sigma_s + \sigma_a. \quad (10)$$

The analogous equation that defines the electromagnetic cross section makes use of the Poynting definition of current density. However, the solution of the vector equation must be precisely of the same form as that given in Equation (2) except for the vector form electric field as given by

$$\mathbf{E}(r) \sim \mathbf{E}_i(r) + r^{-1} e^{-ikr} \mathbf{f}(\mathbf{k}_1, \mathbf{k}_2). \quad (11)$$

The relationship between $\mathbf{E}$ and $\mathbf{H}$ is, of course, established via the basic Maxwell equations, and subsequently one may calculate the incident and scattered current densities using the Poynting vector. We let $\mathbf{E}_i$ be plane polarized radiation given by

$$\mathbf{E}_i = \mathbf{n} E_0 e^{-i\mathbf{k}_1 \cdot \mathbf{r}} \quad (12)$$

where $\mathbf{n} \cdot \mathbf{k}_1 = 0$. Then the incident current density is

$$\mathbf{I}_i = \frac{\mathbf{k}_1}{k} \frac{c}{8\pi} \left(\frac{\varepsilon}{\mu}\right)^{1/2} |E_0|^2. \quad (13)$$

Similarly, the outgoing current is given by

$$\mathbf{I}_s = \frac{c}{8\pi} \frac{\mathbf{k}_2}{k} \left(\frac{\varepsilon}{\mu}\right)^{1/2} \frac{1}{r^2} [\mathbf{f}^*(\mathbf{k}_1, \mathbf{k}_2) \cdot \mathbf{f}(\mathbf{k}_1, \mathbf{k}_2)]. \quad (14)$$

Thus the differential scattering and total cross sections are given by

$$\sigma_s(\mathbf{k}_1, \mathbf{k}_2) = |f(\mathbf{k}_1, \mathbf{k}_2)|^2 \quad (15)$$

$$\sigma_s = \int_0^\pi \int_0^{2\pi} |f(\mathbf{k}_1, \mathbf{k}_2)|^2 \, d\Omega. \quad (16)$$

Integral Formulation

An integral formulation and some basic reciprocity relationships are shown here in some detail for the scalar and vector wave equation. The use of the integral formulation is not primarily for computational purposes. Its main contributions are in demonstrating clearly some of the physical aspects of scattering, and also, as we shall see later, it points the way to extremely valuable approximation techniques.

SMALL PARTICLE SCATTERING

Let us rewrite Equation (1) in the form

$$\nabla^2 \psi(r) + k^2 \psi(r) = [k^2 - k^2(r)] \psi(r). \tag{17}$$

When no scatterer is present, the right side is identically zero, and we may regard this as the source free equation, one solution of which is the undisturbed plane wave $e^{-i\mathbf{k}\cdot\mathbf{r}}$. We may, from the complete set of plane wave solutions construct a Green's function for the inhomogeneous problem. The result is quite simple to understand physically as the point radiating source given by

$$G(\mathbf{r},\mathbf{r}') = -\frac{1}{4\pi} \frac{e^{-ik|\mathbf{r}-\mathbf{r}'|}}{|\mathbf{r}-\mathbf{r}'|}, \tag{18}$$

where $G(\mathbf{r},\mathbf{r}')$ is a symmetric function in $\mathbf{r}$ and $\mathbf{r}'$ and represents a spherical outgoing wave. It is easy to verify that

$$\nabla_r^2 G(\mathbf{r},\mathbf{r}') + k^2 G(\mathbf{r},\mathbf{r}') = \delta(\mathbf{r} - \mathbf{r}') \tag{19}$$

by integrating the equation.

Let $\psi(\mathbf{r})$ in Equation (17) be written as

$$\psi(\mathbf{r}) = \psi_i(\mathbf{r}) + \psi_s(\mathbf{r})$$
$$= e^{-i\mathbf{k}_1\cdot\mathbf{r}} + \psi_s(\mathbf{r}). \tag{20}$$

Then substitution into Equation (17) yields

$$\nabla^2 \psi_s(\mathbf{r}) + k^2 \psi_s(\mathbf{r}) = [k^2 - k^2(\mathbf{r})](\psi_i + \psi_s). \tag{21}$$

Multiply Equation (21) by G and Equation (19) by ψ_s and subtract. Then, integrating over a large spherical volume V and applying Green's theorem, we have

$$\int_S [G\nabla\psi_s - \psi_s \nabla G] \cdot d\mathbf{S} = \int G[k^2 - k^2(\mathbf{r})] \psi(\mathbf{r}) d^3\mathbf{r}$$
$$- \int \psi_s \delta(\mathbf{r} - \mathbf{r}') d^3\mathbf{r}. \tag{22}$$

It is readily seen that the surface integral on the left vanishes identically when one employs the asymptotic form $\psi_s \to r^{-1} f(\mathbf{k}_1, \mathbf{k}_2) e^{-ikr}$ and lets $r \to \infty$. Reconstituting $\psi = \psi_i + \psi_s$, Equation (22) implies that (after interchanging $\mathbf{r}$ and $\mathbf{r}'$),

$$\psi(\mathbf{r}) = \psi_i(\mathbf{r}) - \frac{1}{4\pi} \int \frac{e^{-ik|\mathbf{r}-\mathbf{r}'|}}{|\mathbf{r}-\mathbf{r}'|} [k^2 - k^2(\mathbf{r}')] \psi(\mathbf{r}') d^3\mathbf{r}'. \tag{23}$$

The physical picture evoked by this integral equation is that each point within a scatterer is excited by the incident wave to an amplitude proportional to the value of the product of the wave function at that point and the strength of the scatterer at that point. Each of these elemental point scatterers then radiates so that the total wave is the sum of the spherical waves emitted by the elemental point scatterers. In order to bring the basic scattering definitions into analytical form, we consider the asymptotic limit of Equation (23). As $|r| \to \infty$ we get

$$\psi(\mathbf{r}) \sim e^{-i\mathbf{r}\cdot\mathbf{k}_1} - \frac{e^{-ikr}}{4\pi r} \int e^{i\mathbf{k}_2\cdot\mathbf{r}'} [k^2 - k^2(\mathbf{r})] \psi(\mathbf{r}') d^3\mathbf{r}', \qquad (24)$$

which when combined with the definition of ψ_s [Eq. (2)] gives the scattering amplitude as

$$f(\mathbf{k}_1, \mathbf{k}_2) = -\frac{1}{4\pi} \int e^{i\mathbf{k}_2\cdot\mathbf{r}'} [k^2 - k^2(\mathbf{r}')] \psi(\mathbf{r}') d^3\mathbf{r}'. \qquad (25)$$

At this point it is useful to present what is variously known as the optical theorem or forward dispersion relation. It has a curious history, having been rediscovered numerous times in the last 70 years or so, with the frequency of rediscovery being proportional to the interest in scattering theory at a given time.

The optical theorem for the scalar wave is given by

$$-(4\pi/k) Im\, f(\mathbf{k}_1, \mathbf{k}_1) = \int |f(\mathbf{k}_1, \mathbf{k}_2)|^2 d^3\Omega_{\mathbf{k}_2}$$
$$+ k^{-1} \int |\psi|^2\, Im[k^2(\mathbf{r})]\, d^3\mathbf{r}. \qquad (26)$$

The first term on the right is the scattering cross section; the second is the absorption cross section. If the second term on the right-hand side is not zero, it represents a sinc (absorption) or source (radiation) depending on its sign. In general, the sign of the imaginary part of k^2 is such that the right-hand side gives the absorption cross section. The quantity on the left of Equation (26) is, therefore, the total cross section. Our final result is then that the total cross section is given by the forward scattering amplitude

$$\sigma_t = -(4\pi/k) Im\, f(\mathbf{k}_1, \mathbf{k}_1) \qquad (27)$$

and this is quite general.

SMALL PARTICLE SCATTERING

It is possible to demonstrate that there always exists a reciprocity between incoming and outgoing waves in the sense that

$$f(\mathbf{k}_1, \mathbf{k}_2) = f(-\mathbf{k}_2, -\mathbf{k}_1). \tag{28}$$

If, furthermore, the scatterer is sufficiently symmetric, $k^2(\mathbf{r}) = k^2(-\mathbf{r})$; i.e., if the interaction is inversion invariant, we may write

$$\psi_{\mathbf{k}_1}(\mathbf{r}) = \psi_{-\mathbf{k}_1}(-\mathbf{r}),$$

which immediately implies that

$$f(\mathbf{k}_1, \mathbf{k}_2) = f(-\mathbf{k}_1, -\mathbf{k}_2). \tag{29}$$

Comparing Equations (28) and (29), we have

$$f(\mathbf{k}_1, \mathbf{k}_2) = f(\mathbf{k}_2, \mathbf{k}_1). \tag{30}$$

Although Equation (30) is subject to the restriction of inversion symmetry for the scatterer, there are so many situations in which this condition is satisfied that we shall hereafter refer to it as a basic reciprocity relation.

The integral form of the solution of the electromagnetic scattering problem is clearly similar in broad outline to that for the scalar wave (Schrödinger) equation but is considerably more complicated.

We define the dyadic Green's function by

$$\Gamma(\mathbf{r}, \mathbf{r}') = -\left\{ I + \frac{1}{k^2} \nabla_{r'} \nabla_r \right\} G(\mathbf{r}, \mathbf{r}'), \tag{31}$$

where $G(\mathbf{r}, \mathbf{r}')$ is the scalar Green's function defined in Equation (18). The solution may be given in a form directly related to that of Equation (23)

$$\mathbf{E}(\mathbf{r}) = \mathbf{E}_i(\mathbf{r}) + \int d^3r' [k^2 - k^2(\mathbf{r}')] \mathbf{E}(\mathbf{r}') \cdot \Gamma(\mathbf{r}, \mathbf{r}'). \tag{32}$$

However, this form is not particularly suitable for further computation. After some manipulation we find

$$\mathbf{E}(\mathbf{r}) = \mathbf{E}_i(\mathbf{r}) - \int d^3\mathbf{r} [k^2 - k^2(\mathbf{r}')] \mathbf{E}(\mathbf{r}') \cdot (I + k^{-2} \nabla_r \nabla_{r'}) G(\mathbf{r}, \mathbf{r}')$$

$$= \mathbf{E}_i(\mathbf{r}) - \mathbf{I}_1 + k^{-2} \mathbf{I}_2, \tag{33}$$

where

$$\mathbf{I}_1 = \int [k^2 - k^2(\mathbf{r}')] \mathbf{E}(\mathbf{r}') G(\mathbf{r}, \mathbf{r}') d^3r' \tag{34}$$

$$\mathbf{I}_2 = \nabla_r \int d^3r' [\nabla_{r'} G(\mathbf{r}, \mathbf{r}')] \cdot \mathbf{E}(\mathbf{r}') [k^2(\mathbf{r}') - k^2]. \tag{35}$$

Asymptotically, as $r \to \infty$

$$\mathbf{I}_1 \sim -\frac{e^{-ikr}}{4\pi r} \mathbf{D}(\mathbf{n}_1, \mathbf{n}_2), \tag{36}$$

where

$$\mathbf{D}(\mathbf{n}_1, \mathbf{n}_2) = \int \mathbf{E}(\mathbf{r}')[k^2 - k^2(\mathbf{r}')] \exp(i\mathbf{k}_2 \cdot \mathbf{r}') \, d^3 r' \tag{37}$$

and

$$\mathbf{n}_1 = \mathbf{k}_1/k \quad \text{and} \quad \mathbf{n}_2 = \mathbf{k}_2/k,$$

which we see is (except for the vector form) essentially the same as Equation (25). However, there is now the additional term I_2 which has the form as $r \to \infty$

$$\mathbf{I}_2 \sim \nabla_r \left\{ \frac{\exp(-ikr)}{4\pi r} \int d^3 r' (\nabla_{r'} \exp i\mathbf{k}_2 \cdot \mathbf{r}') \cdot \mathbf{E}(\mathbf{r}')[k^2(\mathbf{r}') - k^2] \right\}$$

$$= -ik \left[\nabla_r \frac{\exp(-ikr)}{4\pi r} \right] [\mathbf{n}_2 \cdot \mathbf{D}(\mathbf{n}_1, \mathbf{n}_2)]$$

$$\sim -\frac{k^2 \exp(-ikr)}{4\pi r} \mathbf{n}_2 [\mathbf{n}_2 \cdot \mathbf{D}(\mathbf{n}_1, \mathbf{n}_2)]. \tag{38}$$

By combining Equations (36), (37), and (38), we find the asymptotic form of the electric field, Equation (33), as $r \to \infty$ to be

$$\mathbf{E}(\mathbf{r}) \sim \mathbf{E}_i(\mathbf{r}) + r^{-1}(\exp - ikr) \mathbf{f}(\mathbf{n}_1, \mathbf{n}_2), \tag{39}$$

where

$$\mathbf{f}(\mathbf{n}_1, \mathbf{n}_2) = \frac{1}{4\pi} \{ \mathbf{D}(\mathbf{n}_1, \mathbf{n}_2) - \mathbf{n}_2 [\mathbf{n}_2 \cdot \mathbf{D}(\mathbf{n}_1, \mathbf{n}_2)] \}. \tag{40}$$

We have thus finally cast the amplitude of the scattered electromagnetic wave in a form very closely resembling that for the scalar wave. Note that for scattering along the forward direction of propagation of the incident wave, the I_2 term vanishes, and one has only the form given by I_1 which, analytically, is exactly like the scalar wave result.

Thus, the optical theorem for a plane electromagnetic wave is a rather simple generalization of that for the scalar wave

$$\sigma_t = -(4\pi/k) Im \{ \mathbf{q} \cdot f_\mathbf{q}(\mathbf{n}_1, \mathbf{n}_1) \} \tag{41}$$

where $\mathbf{q}$ is a unit vector along the direction of polarization of the incident wave.

Similarly, we may specify the appropriate reciprocity relation for the electromagnetic wave scattering amplitudes.

SCALAR SCATTERING BY AN INFINITE HOMOGENEOUS ISOTROPIC CIRCULAR CYLINDER

As in the case of the scattering by a sphere, we start our discussion with the scalar problem. Besides being interesting in its own right, it serves as a basis for the vector electromagnetic scattering.

Consider a plane wave propagated at an angle χ with respect to the axis of an infinite circular cylinder. Let the propagation vector lie in the x-z plane. We require solutions of the wave equations

$$\nabla^2 \psi + m^2 k^2 \psi = 0 \quad \text{if } r > a \qquad (42a)$$

$$\nabla^2 \psi + k^2 \psi = 0 \quad \text{if } r < a, \qquad (42b)$$

r being the cylindrical coordinate and a the cylinder radius.

The general solution of the wave equation is

$$\psi_n = Z_n(r[k^2 - h^2]^{1/2}) \exp -i(n\theta \pm hz), \qquad (43)$$

where Z_n denotes the nth cylindrical Bessel function J_n or the nth cylindrical Hankel function $H_n^{(2)}$. The incident wave is

$$\psi_i = e^{-i\mathbf{r}\cdot\mathbf{k}_1} = e^{-ikz \cos \chi} e^{-ikr \sin \chi \cos \theta}$$

$$= e^{-ikz \cos \chi} \sum_{n=-\infty}^{\infty} (-i)^n e^{-in\theta} J_n(rk \sin \chi). \qquad (44)$$

The transmitted and scattered waves are represented by

$$\psi_t = \sum_{n=-\infty}^{\infty} a_n(-i)^n J_n(rmk \sin \chi_1) e^{-in\theta - ikmz \cos \chi_1} \qquad (45a)$$

$$\psi_s = \sum_{n=-\infty}^{\infty} b_n(-i)^n H_n^{(2)}(rk \sin \chi) e^{-in\theta - ikz \cos \chi}. \qquad (45b)$$

In order for $\psi_i + \psi_s = \psi_t$ at $r = a$, for all z, χ and χ_1 must be related by Snell's law

$$k \cos \chi = mk \cos \chi_1. \qquad (46)$$

The coefficients a_n and b_n may then be found by simultaneously imposing continuity of ψ and its radial derivative at $r = a$. The solutions one obtains for the coefficients are

$$-a_n = (2/i\pi l)/[(l_1/l)H_n^{(2)}(al) - J_n'(al_1)H_n'^{(2)}(al)] \qquad (47a)$$

$$-b_n = \frac{-(l_1/l)J_n'(al_1)J_n(al) + J_n(al_1)J_n'(al)}{(l_1/l)H_n^{(2)}(al)J_n'(al_1) - J_n(al_1)H_n'^{(2)}(al)}, \qquad (47b)$$

where $l = k \cos \alpha = k \sin \chi$

$$l_1 = mk \cos \alpha_1 = k(\mu\varepsilon - \sin^2\alpha)^{1/2}. \qquad (48)$$

The asymptotic form of the scattered wave may be obtained by using the relation (as $r \to \infty$)

$$H_n^{(2)}(lr) \sim (2/\pi lr)^{1/2} \exp\{-i[lr - (2n + 1)\pi/4]\}$$

in Equation (46). We obtain (as $r \to \infty$)

$$\psi_s \sim e^{-ikz \cos \chi} e^{-ilr} (2/\pi lr)^{1/2} \sum_{n=-\infty}^{\infty} b_n e^{+i\pi/4} e^{-in\theta}. \qquad (49)$$

From the above expression for the scattered wave, we note that the scattered amplitude must be taken as the coefficient of $e^{-ilr}/r^{1/2}$. This not-unexpected result is due to the fact that, in essence, we are dealing with a two-dimensional problem even when the radiation is incident at an angle other than 90° with respect to the cylinder. We may write the total wave function in the form

$$\psi \sim e^{-ikz \cos \chi}[e^{-ilr \cos \theta} + r^{-1/2}f(\theta)e^{-ilr}]. \qquad (50)$$

In this form we see that the case of oblique incidence is obtained from that for normal incidence by a simple adjustment of the internal and external propagation constants; i.e., the external k goes into l and the internal mk is replaced by l_1.

The scattered flux per unit length along z is finite only when the scattered wave is of the form

$$\psi_s \sim r^{-1/2}f(\theta)e^{-ilr}, \qquad (51)$$

this being the modified radiation condition for cylindrical waves.

The scattering cross section is still given by the squared magnitude of the scattering amplitude

$$\sigma(\theta) = |f(\theta)|^2$$

when

$$f(\theta) = (2/\pi)^{1/2} \sum_{n=-\infty}^{\infty} b_n e^{i\pi/4} e^{-in\theta}.$$

The total scattering cross section per unit length is, therefore,

$$\sigma_s = \int_0^{2\pi} \sigma(\theta)\, d\theta$$

$$= (4/l) \sum_{n=-\infty}^{\infty} |b_n|^2. \qquad (52)$$

The scattering efficiency for cylindrical targets is defined as the ratio of the cross section to the normally projected geometrical area of the cylinder. For a unit length of circular cylinder, the projected area is $2a \sin \chi$, and the scattering efficiency is then

$$ZQ_s = \sigma_s/2a \sin \chi = (2/ka) \sum_{n=-\infty}^{\infty} |b_n|^2$$

$$= (2/ka)\left\{[|b_0|^2 + 2 \sum_{n=1}^{\infty} |b_n|^2]\right\}. \qquad (53)$$

SCATTERING OF ELECTROMAGNETIC WAVES BY INFINITE CIRCULAR CYLINDERS

We again require solutions of the scalar equation

$$\nabla^2 \psi + k^2 m^2 \psi = 0 \qquad (54a)$$

with

$$m^2 = \begin{cases} 1 & \text{if } r > a \\ \mu(\varepsilon - 4\pi i \sigma \omega^{-1}) & \text{if } r < a \end{cases} \qquad (54b)$$

and whose general solution is given by Equation (43).

Generally we set the magnetic inductive capacity inside the cylinder equal to μ and outside equal to the free space value μ_0. In most applications, however, $\mu = \mu_0 = 1$. In the equations that follow, the parameter ε will be given the value ε_0 when the equations refer to empty space, and $\varepsilon - 4\pi i \sigma \omega^{-1}$ when they refer to a region in the interior of the cylinder. In principle, the role of the inside and outside could be reversed. However, if the external medium is conducting, the radiating field might be absorbed and the asymptotic boundary conditions would have to be modified.

We now require two solutions of the scalar wave equation, U and V (see van de Hulst 1957).

The components of the electric and magnetic field vectors are expressed in terms of U and V by

$$E_\theta = -\mu^{1/2}\frac{\partial V}{\partial r} - \frac{h}{kr\varepsilon^{1/2}}\frac{\partial U}{\partial \theta}; \quad H_\theta = \varepsilon^{1/2}\frac{\partial U}{\partial r} - \frac{h}{kr\mu^{1/2}}\frac{\partial V}{\partial \theta} \quad (55a)$$

$$E_r = -\frac{\mu^{1/2}}{r}\frac{\partial V}{\partial \theta} + \frac{h}{k\varepsilon^{1/2}}\frac{\partial U}{\partial r}; \quad H_r = \frac{\varepsilon^{1/2}}{r}\frac{\partial U}{\partial \theta} + \frac{h}{k\mu^{1/2}}\frac{\partial V}{\partial r} \quad (55b)$$

$$E_z = U\varepsilon^{-1/2}(m^2k^2 - h^2); \quad H_z = V\mu^{-1/2}(m^2k^2 - h^2)/k. \quad (55c)$$

The choice of ε and μ follows the rule given above which depends on whether the fields are measured inside or outside the scattering cylinder.

$$m^2 = \begin{cases} (\varepsilon_0\mu_0) & \text{outside} \\ [\mu(\varepsilon - 4\pi i\sigma/\omega)] & \text{inside} \end{cases} \quad (56a)$$

$$h = k\cos\chi. \quad (56b)$$

Consider two forms of polarization of the incident wave; first that with the electric field vector parallel to the cylinder axis, and second with the magnetic field vector parallel to the cylinder axis. The solutions of the scalar wave equation U and V, denoted by a superscript G, take on the values E and H respectively.

The external solutions ($r > a$) are

$$U_{\text{out}}^G = \sum_{n=-\infty}^{\infty} F_n[\delta_E^G J_n(rl) - b_n^G H_n^{(2)}(rl)] \quad (57a)$$

$$V_{\text{out}}^G = \sum_{n=-\infty}^{\infty} F_n[\delta_H^G J_n(rl) - a_n^G H_n^{(2)}(rl)]. \quad (57b)$$

By imposing Snell's law (Equation 46) so that the boundary conditions may be satisfied for all z, we may write the internal solutions as ($r < a$)

$$U_{\text{in}}^G = \Sigma F_n d_n^G J_n(rl_1) \quad (58a)$$

$$V_{\text{in}}^G = \Sigma F_n c_n^G J_n(rl_1), \quad (58b)$$

where

$$\delta_E^G = \begin{cases} 0 & \text{if } G \neq E \\ 1 & \text{if } G = E \end{cases} \begin{matrix} 1 \\ 0 \end{matrix} = \delta_H^G \quad (59)$$

and

$$F_n = i^n \exp\{-i\omega t + ikz\cos\chi + in\theta\}. \quad (60)$$

As previously defined in the scalar wave case, the effective propagation constants inside and outside are $l = k \cos \alpha$ and $l_1 = mk \cos \alpha_1$ [see Eq. (48)].

In order to determine the coefficients a_n, b_n, c_n, and d_n, we must match the tangential components of the electric (E) and magnetic (H) fields as given by Equation (55) across the cylinder boundary $r = a$.

The resulting values are given by (Lind and Greenberg 1966):

$$a_n^E = +inSR_n \sin \alpha [B_n(\mu) - A_n(\mu)]/\Delta_n \tag{61}$$

$$b_n^E = R_n[A_n(\mu) B_n(\varepsilon') - n^2 S^2 \sin^2 \alpha]/\Delta_n \tag{62}$$

$$a_n^H = R_n[A_n(\varepsilon') B_n(\mu) - n^2 S^2 \sin^2 \alpha]/\Delta_n \tag{63}$$

$$b_n^H = -inSR_n \sin \alpha [B_n(\mu) - A_n(\mu)]/\Delta_n \tag{64}$$

$$c_n^G = \mu^{1/2} \frac{(al)^2 H_n(al)}{(al_1)^2 J_n(al_1)} [-a_n^G + R_n \delta_H^G] \tag{65}$$

$$d_n^G = (\varepsilon')^{1/2} \frac{(al)^2 H_n(al)}{(al_1)^2 J_n(al_1)} [-b_n^G + R_n \delta_E^G], \tag{66}$$

where

$$\Delta_n = A_n(\varepsilon') A_n(\mu) - n^2 S^2 \sin^2 \alpha \tag{67a}$$

$$S = (al)^{-2} - (al_1)^{-2} \tag{67b}$$

$$R_n = J_n(al)/H_n(al), \tag{67c}$$

and the functions $A_n(\xi)$ and $B_n(\xi)$ are defined by

$$A_n(\xi) = \frac{H_n'(al)}{al\, H_n(al)} - \xi \frac{J_n'(al_1)}{al_1\, J_n(al_1)} \tag{67d}$$

$$B_n(\xi) = \frac{J_n'(al)}{al\, J_n(al)} - \xi \frac{J_n'(al_1)}{al_1\, J_n(al_1)} \tag{67e}$$

$$\varepsilon' = m^2/\mu = \varepsilon - 4\pi i \sigma \omega^{-1}. \tag{68}$$

For a perfectly conducting cylinder these coefficients are considerably simplified, being

$$a_n^E = 0 = b_n^H \tag{69}$$

$$a_n^H = J_n'(al)/H_n'(al) \tag{70}$$

$$b_n^E = J_n(al)/H_n(al). \tag{71}$$

We see that the coefficients for the tilted conducting cylinder of radius a are the same as those for a normally oriented conducting cylinder, but with the radius reduced by the factor $\cos \alpha$. The similarity between the scalar and electromagnetic case is seen by comparing the values of b_n^E in Equation (71) with the value of b_n in Equation (47b), which are identical in the limiting case $l_1/l \to \infty$.

We shall now explicitly evaluate the fields far from the cylinder in order to calculate the scattered intensity and, subsequently, the scattering cross section.

In the case in which the E field is polarized parallel to the cylinder axis, the scattered far fields follow from applying the condition $kr \cos \alpha \gg 1$ to the above equations when the superscript G is given the value E. Then

$$E_r^E = (ithl/k) T_b^E(\theta); \quad H_r^E = -(thl/k) T_a^E(\theta) \qquad (72a)$$

$$E_\theta^E = tl T_a^E(\theta); \quad H_\theta^E = itl T_b^E(\theta) \qquad (72b)$$

$$E_z^E = (-itl^2/k) T_b^E(\theta); \quad H_z^E = (tl^2/k) T_a^E(\theta), \qquad (72c)$$

where

$$t = (2/\pi lR)^{1/2} \exp\{-i[hz + lR - \pi/4]\} \qquad (73a)$$

$$T_a^E(\theta) = 2 \sum_{n=1}^{\infty} a_n^E \sin n\theta \qquad (73b)$$

$$T_b^E(\theta) = b_0^E + 2 \sum_{n=1}^{\infty} b_n^E \cos n\theta. \qquad (73c)$$

The sums $T(\theta)$ are analogous to the scattering amplitude functions of the sphere. The associated intensities of scattered radiation as obtained by evaluating the Poynting vector are given by

$$I_r^E = (c/8\pi)(2/\pi kR)\{|T_a^E(\theta)|^2 + |T_b^E(\theta)|^2\} \qquad (74a)$$

$$I_z^E = (h/l) I_r^E; \quad I_\theta^E = 0. \qquad (74b)$$

The scattered radiation lies on cones of half angle χ and the phase fronts lie on oppositely directed cones of half angle α_0.

The total scattered energy per unit time crossing a cylindrical region of radius R and length L concentric with our scatterer is

$$W = \int_0^L \int_0^{2\pi} R I_r \, d\theta \, dz. \qquad (75)$$

We finally obtain the cross section per unit cylinder length in the case in which the incident E field is polarized in the direction of the axis of the scatterer:

$$\sigma_{sc}^E = (4/k)\left\{|b_0^E|^2 + 2 \sum_{n=1}^{\infty} [|a_n^E|^2 + |b_n^E|^2]\right\}. \qquad (76)$$

When the incident magnetic field is polarized in the direction of the axis of the cylindrical scatterer, we find

$$\sigma_{sc}^H = (4/k)\left\{|a_0^H|^2 + 2\sum_{n=1}^{\infty}[|a_n^H|^2 + |b_n^H|^2]\right\}, \qquad (77)$$

which is obtained from the scattered intensities

$$I_r^H = (c/8\pi)(2/\pi kR)\{|T_a^H(\theta)|^2 + |T_b^H(\theta)|^2\} \qquad (78a)$$

$$I_z^H = (h/l)I_r^H; \quad I_\theta^H = 0, \qquad (78b)$$

where

$$T_a^H(\theta) = a_0^H + 2\sum_{1}^{\infty} a_n^H \cos n\theta \qquad (79a)$$

$$T_b^H(\theta) = 2\sum_{1}^{\infty} b_n^H \sin n\theta. \qquad (79b)$$

Application of the optical theorem may be made to give the extinction efficiencies as

$$Q_{ext}^E = \sigma_{ext}^E/2a = (2/ka)Re\left\{b_0^E + 2\sum_{n=1}^{\infty} b_n^E\right\} \qquad (80a)$$

$$Q_{ext}^H = \sigma_{ext}^E/2a = (2/ka)Re\left\{a_0^H + 2\sum_{n=1}^{\infty} a_n^H\right\}. \qquad (80b)$$

The replacing of Im by Re is due here to a $\pi/2$ phase difference in the definition of the scattering amplitude from that given in Equation (41).

FINITE CIRCULAR CYLINDER

The extension of the electromagnetic infinite cylinder results to the finite cylinder is made by an approximation that neglects the effect of the truncated ends on the field distribution elsewhere on the cylinder. This is analogous to the method first used by Montroll and Hart (1951) in calculating the scalar wave scattering by a finite cylinder. The principal difference here (Lind 1966) is that instead of integrating the fields over the volume of the particle, we use the vector Green's theorem in surface form rather than volume form. Only the final formulae are presented so that direct comparison may be made with the infinite cylinder and the differences may be explicitly seen. Computer programs are presently being written and the numerical results will appear in a later publication.

Reference is made to Fig. 1 for definition of the various angles. We define a new parameter

$$p = k \sin \Theta$$

analogous to the quantity l in Equation (48).

The far field components for the electric vector of the incident radiation along the cylinder axis are given by

$$E_\Theta{}^E = t'D\left\{\xi_0 b_0{}^E + 2\sum_{n=1}^{\infty} \xi_n b_n{}^E \cos n\Phi \right.$$

$$\left. + (2i/ka)\,\Omega \sum_{n=1}^{\infty} n\zeta_n a_n{}^E \cos n\Phi \right\} \quad (81a)$$

$$E_\Phi{}^E = -t'D\left\{2i \sum_{n=1}^{\infty} \xi_n a_n{}^E \sin n\Phi + (2/ka)\,\Omega \sum_{n=1}^{\infty} n\zeta_n b_n{}^E \sin n\Phi \right\} \quad (81b)$$

$$E_\Theta{}^H = t'D\left\{2i \sum_{n=1}^{\infty} \xi_n b_n{}^H \sin n\Phi - (2/ka)\,\Omega \sum_{n=1}^{\infty} n\zeta_n a_n{}^H \sin n\Phi \right\} \quad (81c)$$

$$E_\Phi{}^H = t'D\left\{\xi_0 a_0{}^H + 2\sum_{n=1}^{\infty} \xi_n a_n{}^H \cos n\Phi \right.$$

$$\left. - (2i/ka)\,\Omega \sum_{n=1}^{\infty} n\zeta_n b_n{}^H \cos n\Phi \right\}, \quad (81d)$$

where

$$t' = kaLe^{-ikR}/2R$$

$$D = \frac{\sin\left[\dfrac{kL}{2}(\cos\Theta - \sin\alpha)\right]}{\dfrac{kL}{2}(\cos\Theta - \sin\alpha)} \xrightarrow[L\to\infty]{} \frac{\pi}{kL}\delta(\cos\Theta - \sin\alpha) \quad (82)$$

$$\xi_n = \cos\alpha\, H_n(la) J_n'(pa) - \sin\Theta\, H_n'(la) J_n(pa)$$

$$\zeta_n = H_n(la) J_n(pa)$$

$$\Omega = \frac{\cos\Theta\cos^2\alpha - \sin\alpha\sin^2\Theta}{\sin\Theta\cos\alpha}.$$

The differences from the infinite cylinder results [Eqs. (72) and (73)] are explicitly shown through the introduction of the new terms: t', D, ξ_n, ζ_n, Ω. The factor t' demonstrates the spherical wave as distinct from the cylindrical wave.

SMALL PARTICLE SCATTERING

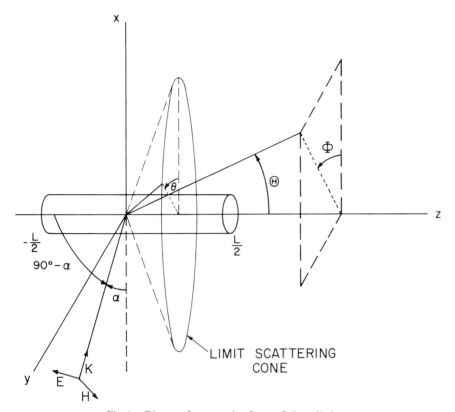

Fig. 1. Diagram for scattering from a finite cylinder.

The factor D exhibits a diffraction effect introduced by the finite cylinder length. In the limit as $L \to \infty$, D approaches the character of a delta function and again the scattering is limited to a cone as has been seen for the infinite cylinder.

The finite cylinder approximation given by van de Hulst contains only an equivalent diffraction coefficient which was limited to the case of normal incidence on the cylinder. However, the new approximation includes additional terms produced by $\xi_n \neq 1$ and $\Omega \neq 0$ in Equation (81). Until exact numerical computations for the finite cylinder are performed, it is not clear whether these additional correction terms produce a significant improvement. A priori, it would seem that they should do so.

In the limit as $L \to \infty$ the only contribution to the scattering occurs where $\cos \Theta = \sin \alpha$. Therefore $\Omega \to 0$ as $L \to \infty$.

The factors ξ_n become, as $L \to \infty$ and $\cos \Theta \to \sin \alpha$, $l \to p$, simply the Wronskian times $\cos \alpha$,

$$\xi_n \to \cos \alpha [H_n(la) J'_n(la) - H'_n(la) J_n(la)] = \cos \alpha \left(\frac{2i}{\pi la} \right) = \frac{2i}{\pi ka}. \quad (83)$$

Incorporating all the above limits for D, ξ_n, Ω in Equation (81) gives

$$E_\Theta^E = \frac{iL}{\pi R} e^{-ikR} T_b^E(\Phi)$$

$$E_\Phi^E = \frac{L}{\pi R} e^{-ikR} T_a^E(\Phi)$$

$$\qquad\qquad\qquad\qquad L \to \infty. \quad (84)$$

$$E_\Theta^H = -\frac{L}{\pi R} e^{-ikR} T_b^H(\Phi)$$

$$E_\Phi^H = \frac{-iL}{\pi R} e^{-ikR} T_a^H(\Phi)$$

A PROCEDURE FOR OBTAINING ANGULAR SCATTERING DISTRIBUTIONS FROM SPHEROIDS

The method used by Greenberg (1960) to obtain the extinction properties of soft spheroids gives the small angle scattering properties as well. The precise limitation of the applicability of the scattering results is presently under investigation by comparison with "exact" results.

We first consider the scalar wave case for a spheroid of semiminor axis a and semimajor axis b whose symmetry axis makes an angle χ with respect to the propagation vector. We note that the scattering amplitude as given in Equation (25) reduces—with some minor phase changes—to that of Greenberg (1960),

$$f(\mathbf{k}_1, \mathbf{k}_2) = -ikAB \int_0^\infty \rho \, d\rho \, J_0(q'\rho)$$

$$\times \left\{ 1 - \exp\left[\frac{-iC}{2k} \int_{-\infty}^\infty (k^2(\rho, \zeta) - k^2) d\zeta \right] \right\}, \quad (85)$$

where

$$A^2 = b^2 \sin^2 \chi + a^2 \cos^2 \chi$$
$$B^2 = a^2$$
$$C^2 = (ab)^2/A^2 \qquad (86)$$
$$K' = [(AK_x)^2 + (BK_y)^2]$$
$$\mathbf{K} = \mathbf{k}_1 - \mathbf{k}_2.$$

We may integrate Equation (85) to obtain

$$f(\mathbf{k}_1,\mathbf{k}_2) = kAB\left[\sum_{n=1}^{\infty}\left(\frac{C\mu}{a}\right)^{2n}\frac{J_{n+1}(K')}{(K')^{n+1}}\frac{2^n n!}{(2n-1)!} + (C\mu/a)(\pi y/2)^{1/2}J_{3/2}(y)y^{-2}\right], \quad (87)$$

where

$$\mu = m^2 - 1$$
$$y^2 = (C\mu/a)^2 + (K')^2.$$

Looking directly at Equation (85) and comparing it with that for an "equivalent" sphere, we may infer that the spheroid scattering amplitude at an angle Θ, Φ would be given by (AB/C^2) times the scattering amplitude at an angle Θ_{sph} from a sphere of radius C where

$$\sin \Theta_{sph} = \sin[\Theta(A^2 \cos^2 \Phi + B^2 \sin^2\Phi)^{1/2}/C]. \qquad (88)$$

By this procedure we reduce the spheroid scattering problem to that of a set of appropriate equivalent spheres.

We may expect that the angular distributions obtained from such a modification of exact sphere solutions will be superior to those obtained from the completely consistent application of the ray approximation which is assumed valid only in the region of small angles.

The extension to the electromagnetic case within the framework of this same approximation is contained in the fact that the integral defined in Equation (37) is, except for its vector character, exactly like that for the scalar wave case. It should, however, be remembered that the vector scattering amplitude includes an extra term as given in Equation (40). This is a very important addition. Explicitly, then, we may obtain the electromagnetic scattering distribution from spheroids by appropriate combinations of the very well-known Mie theory results for spheres. In other words, the appropriate quantity $\mathbf{D}(\mathbf{n}_1,\mathbf{n}_2)$ appearing in Equation

(40) is obtained by: (1) Inserting the internal field obtained from Mie theory for a sphere of radius C and, for each Θ, Φ, choose a direction for k_2 corresponding to the appropriate Θ_{sph} defined in Equation (88). This then reduces Equation (40) to a series of terms involving coefficients (c_n and d_n in van de Hulst) related to the Mie theory coefficients.

MICROWAVE ANALOGUE RESULTS

The microwave analogue method has been used to obtain some angular distributions for scattering by spheroids. We present here in Figs. 4–7 some selected results for lucite ($m = 1.61$, $b/a = 2$) spheroids measured by Wang (1970). A measure of the accuracy of the experiment is indicated in the comparison between the results for spheres and the Mie theory in Figs. 2 and 3. The oscillations in amplitude for the horizontal polarization were produced by second-order scattering effects. A new laboratory now under construction will make it possible to eliminate such spurious terms. The vertical polarization compares rather well with the exact calculations.

In Figs. 8 and 9 are shown angular scattering distributions for several configurations of a stacked sphere model of a rough sphere (Greenberg, Wang, and Bangs 1971).

Acknowledgment. This work has been supported in part by National Aeronautics and Space Administration Grant No. NGR-33-011-043 and National Science Foundation Grant No. GP-27085.

REFERENCES

Greenberg, J. M. 1960. Scattering by nonspherical particles. *J. Applied Phys.* 31: 82–84.
Greenberg, J. M.; Wang, R. T.; and Bangs, L. 1971. Extinction by rough particles and the use of the Mie theory. *Nature Phys. Sci.* 230: 110–112.
Hulst, H. C. van de 1957. *Light scattering by small particles.* New York: Wiley.
Lind, A. C. 1966. Ph.D. Thesis, Rensselaer Polytechnic Institute, Troy, New York.
Lind, A. C., and Greenberg, J. M. 1966. Electromagnetic scattering by obliquely oriented cylinders. *J. Applied Phys.* 37: 3195–3203.
Montroll, E. W., and Hart, R. W. 1951. Scattering of plane waves by soft obstacles. II. Scattering by cylinders, spheroids, and disks. *J. Applied Phys.* 22: 1278–1289.
Wang, R. T. 1970. Unpublished experimental results.

DISCUSSION

SERKOWSKI: Polarization of a star and the central part of an envelope, within the radius of about 0.05 arcsec, can be measured separately from polarization of the outer parts of an envelope. A stellar interferometer (e.g., the one proposed by KenKnight [1972]) forms an image of the telescope mirror intersected by parallel fringes. A reticle transmits to the polarimeter first the bright fringes, where stellar light predominates, then the dark fringes, in which the light originates mostly in the outer envelope. Such measurements may provide a crucial test for different hypotheses explaining intrinsic polarization of starlight.

DISCUSSION REFERENCE

KenKnight, C. E. 1972. Autocorrelation methods to obtain diffraction-limited resolution with large telescopes. *Astrophys. J.* 176: L43–L45.

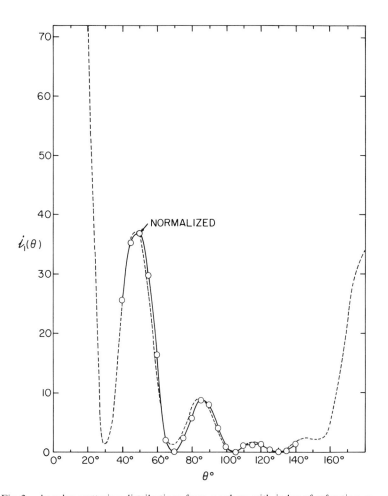

Fig. 2. Angular scattering distributions from a sphere with index of refraction $m = 1.61$ (Lucite) and radius $a = 2.522$ cm. The wavelength is $\lambda = 3.183$ cm. The dashed curve is calculated by Mie theory. The experimental points are obtained by the microwave analogue method. The case shown is for polarization perpendicular to the scattering plane.

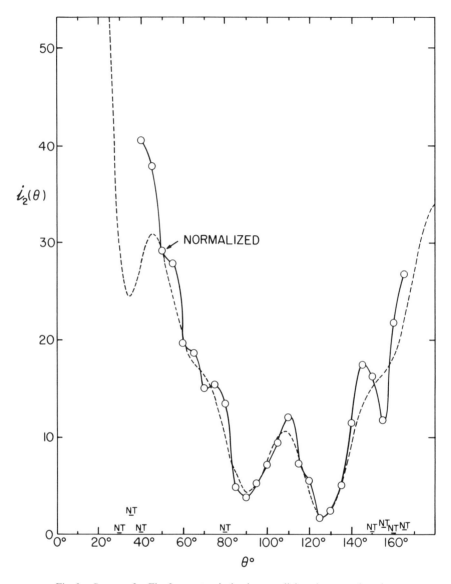

Fig. 3. Same as for Fig. 2 except polarization parallel to the scattering plane.

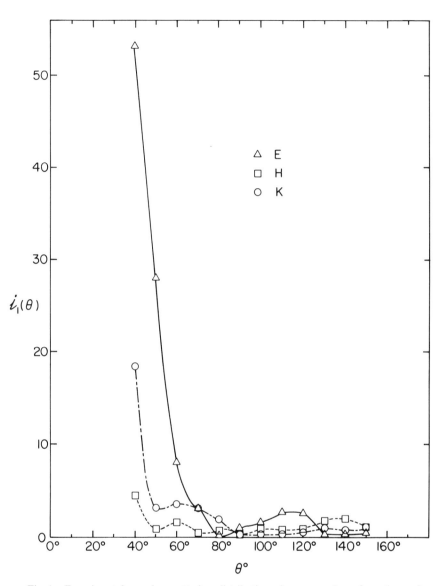

Fig. 4. Experimental angular scattering distributions from several configurations of a spheroid with $m = 1.61$, $a = 1.27$, $x = 2.507$, $b/a = 2$. Polarization perpendicular to the scattering plane. E = spheroid axis parallel electric vector of incident radiation; H = spheroid axis parallel magnetic vector of incident radiation; k = spheroid axis parallel propagation direction of incident radiation.

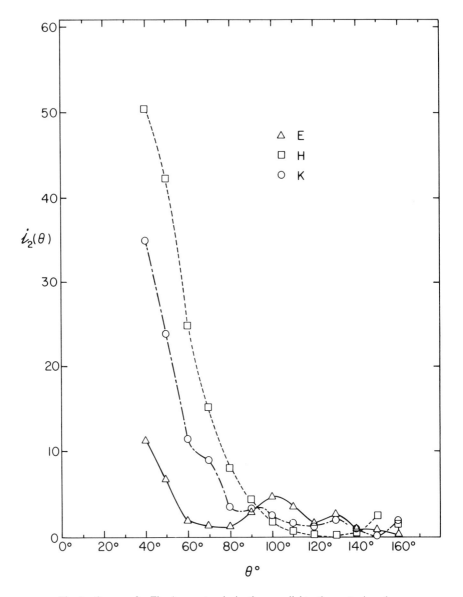

Fig. 5. Same as for Fig. 4, except polarization parallel to the scattering plane.

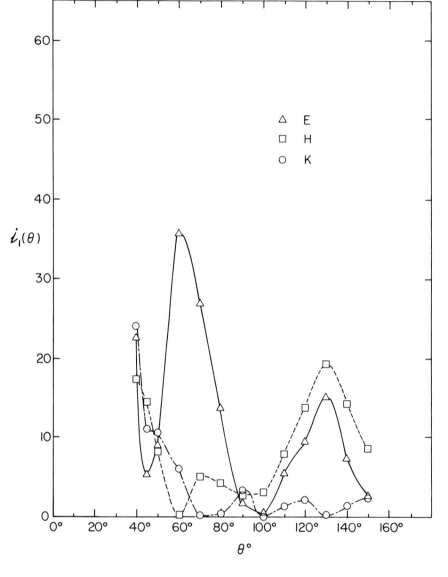

Fig. 6. Same as for Fig. 4, except $a = 1.902$ cm, $x = 3.754$.

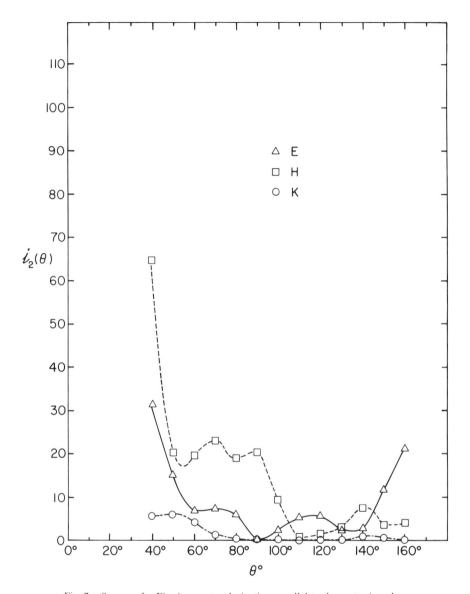

Fig. 7. Same as for Fig. 6, except polarization parallel to the scattering plane.

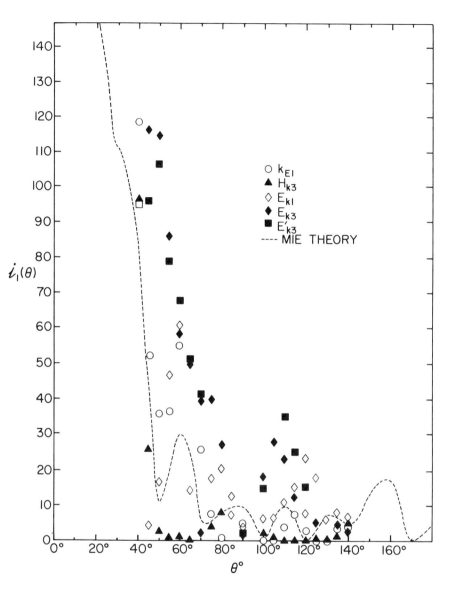

Fig. 8. Experimental angular scattering distributions for several configurations of a stacked cylinder model of a roughened sphere with $m = 1.352$, $a = 3.903$ cm. For a description see Greenberg, Wang, and Bangs (1971). Polarization perpendicular to the scattering plane. The symbols are: k_{E1} = propagation along main cylinder axis with E plane passing through only the central cylinder; $H_{k3} = H$ parallel main axis with principal ray passing through center plus two outer cylinders; $E_{k1} = E$ parallel main axis with principal ray passing through central cylinder; $E_{k3} = E$ parallel main axis with principal ray passing through three cylinders; E'_{k3} = same basic configuration as E_{k3} but particle rotated through 60°.

133

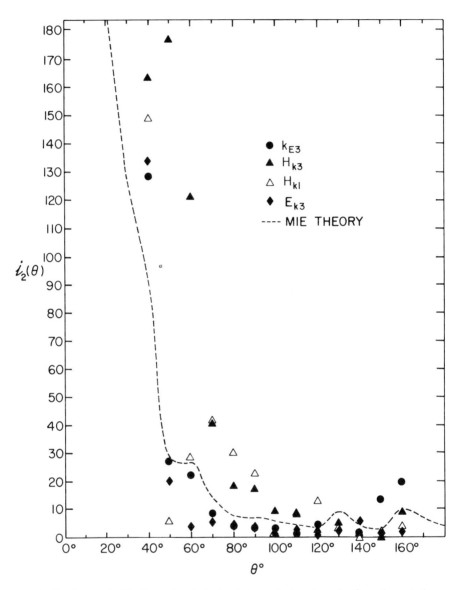

Fig. 9. Same as for Fig. 8, except polarization in the scattering plane. Configuration notation analogous to Fig. 8.

POLARIMETERS FOR OPTICAL ASTRONOMY

K. SERKOWSKI
University of Arizona

Sources of error in polarimetry and different designs of polarimeters are discussed using a new concept of efficiency for a polarimeter. A method for calibrating the zero point of position angles and a method for avoiding wavelength dependence for the orientation of the optical axis for Pancharatnam achromatic half-wave plates are proposed. A theory of the Lyot depolarizer is given, and its optimum orientation relative to an analyzer is specified. Elimination of polarization introduced by a tilted plane mirror and of sky background polarization is described. Measurements of linear polarization produced by a metallic focal plane diaphragm are reported. A design for a scanning polarimeter is presented.

For many astronomical objects the observed polarization is very small, making high polarimetric accuracy essential. Polarimetric precision can be orders of magnitude higher than photometric precision because the effects of atmospheric scintillation, seeing, and extinction can be eliminated. Nevertheless, in proportion to the number of published papers, more false announcements of discoveries were made in the field of stellar polarimetry than in any other field of astronomy. Discrepancies between observers using the best existing polarimeters are so large that even for the well-studied bright standard stars with large interstellar polarization, linear polarization is not known to a better accuracy than $\pm 0.1\%$ and the position angle to $\pm 1°$ (cf. Appendix). Theoretically, precision on the order of $\pm 0.003\%$ should be easy to achieve for these stars. This indicates that improvement in the observing techniques is badly needed.

The present review complements the discussion of polarimetric techniques in optical astronomy published by Serkowski (1973a), where a more complete bibliography may be found. Another review of this subject has been published by Shakhovskoj (1971). Different types of analyzers and retarders are discussed by Clarke and Grainger (1971).

SOURCES OF ERROR IN POLARIMETRY

Various sources of error are listed in Table I. Some of them are discussed in more detail below.

Photon Noise

The principal limitation of precision in astronomical polarimetry results from photon statistics. If the polarimeter is not exceptionally bad, we may expect that whenever the error of percentage polarization exceeds 0.2%, this error results from photon statistics. Other sources of error become important only when better accuracy is sought.

Ideally, the mean error of each of the simultaneously determined normalized Stokes parameters Q/I and U/I, describing linear polarization (cf. Clarke and Grainger 1971; Serkowski 1962) is

$$\varepsilon(Q/I) = \varepsilon(U/I) = \sqrt{2/N}, \tag{1}$$

where N is the total number of photons counted. Of course, the only method to reduce the error resulting from photon statistics is to count more photons. To make the most efficient use of the light available, we should observe in a wide range of wavelengths simultaneously, using many detectors. For each spectral range, the detectors of highest quantum efficiency in this range can be chosen. Different wavelengths are separated either with *dichroic filters*, as was done successfully in a 10-channel *UBVRI* polarimeter built in Australia (Serkowski, Mathewson, and Ford 1973), or with a spectrometer coupled to a photon-counting image tube.

In the latter case, high polarimetric accuracy is obviously possible only for very high photon-counting rates. The only presently available photon-counting image tube capable of counting at rates in excess of a few thousand photons per second per resolution element is a semiconductor diode image tube, called *Digicon* (Electronic Vision Corporation, San Diego, Calif. 92121; cf. Beaver et al. 1972*a, b*). The photoelectrons emitted by a conventional photocathode are accelerated to high energy and form an image of the cathode on a long array of silicon diodes. Each diode is connected by a separate wire to an external amplifier. Every photoelectron produces a strong pulse in a diode; these pulses can be counted easily at rates as high as several hundred thousand per second per diode.

Any device producing spectral dispersion changes the state of the polarization of light (cf. Breckinridge 1971; Poulsen 1972).[1] Therefore an

---

[1] See p. 232.

TABLE I
Sources of Error in Polarimetry

Source of error	Remedy
Photon noise	Subdividing light among the largest feasible number of high quantum efficiency detectors by using a beam-splitting analyzer (e.g., a Wollaston prism), dichroic filters, and/or a spectrometer.
Atmospheric scintillation	Using at each spectral region two detectors for orthogonal polarizations and/or rapid modulation of signal.
Instrumental polarization or depolarization and linear-to-circular conversion	Avoiding filters, lenses, and tilted mirrors in front of an analyzer or, in case of a tilted mirror, placing a rotatable retarder in front of it. Rotating the entire telescope or using an altazimuth mounting. Using dielectric focal-plane diaphragms. Measuring circular polarization alternately with and without a half-wave plate inserted between the rotating half-wave plate and the remaining parts of the polarimeter.
Motion of light beam on the photocathode caused by:	
a. moving polarizing optics	Moving optical components should be accurately plane parallel and/or with an image of the telescope mirror formed on them. Modulation frequency differing by a large factor from the frequency of rotation of the modulating optical components.
b. inaccurate telescope guiding or bad seeing	Achromatic Fabry lens accurately focusing an image of the telescope mirror on the photocathode of the detector. Rapid modulation of the signal.
Unnecessary reflections from optical components	Tilting the Wollaston prism and other stationary optical components that have flat surfaces.
Variable sky background	Rapid switching between object and sky. Cancelling polarization of sky background by superimposing perpendicularly polarized images of the sky.
Interference from neighboring objects	Avoiding optical components in front of a focal-plane diaphragm. Observing a starlike object on the nebular background while alternating two focal-plane diaphragms of different size.
Inaccurately known effective wavelength at which polarization is measured	Determining effective wavelength for each object and each spectral region by measuring polarization with a polarizer and a special retarder inserted in front of the polarimeter.

analyzer should be placed in a *fixed orientation* in front of the spectrometer or filters. This orientation should be such that the light emerging from an analyzer is polarized in a plane making 45° with the plane of incidence on a spectrometer grating or on dichroic filters; this minimizes undesirable polarization effects in the instrument. As will be discussed further, the most efficient way of measuring polarization is by modulating light with a retarder, e.g., a half-wave or quarter-wave plate, placed in front of an analyzer. Constructing achromatic retarders capable of working in a very wide spectral range is therefore essential if the light is to be used most efficiently. Such retarders are discussed in another section of this review.

Atmospheric Scintillation and Seeing

Since air is not birefringent, scintillation is the same for both perpendicularly polarized components of light from an astronomical object. The *ratio of intensities* of two such beams, emerging, e.g., from a Wollaston prism, is free of the effects of atmospheric scintillation and is not affected by the presence of thin clouds. Extinction by clouds is nearly neutral in the visible region (Serkowski 1970), and the accuracy of polarimetry through clouds is reduced only because of fluctuations in sky background and the smaller number of photons received.

On the other hand, the atmospheric seeing, i.e., the fluctuations and spread in the *direction* from which we receive stellar light, affects the ratio of signals from two beams emerging from the Wollaston prism. Because of the inhomogeneous sensitivity of detectors, atmospheric seeing and imperfections in telescope guiding would spoil any hope of achieving high polarimetric accuracy in a system where an image of an astronomical object is formed on photosensitive surfaces, unless the signal were modulated with high frequency. Without rapid modulation, an essential condition for high accuracy is the formation of an image of the telescope mirror, rather than that of the object, on a photosensitive surface by means of a lens, called a *Fabry lens*. The focal length of the Fabry lens should be as long as possible to diminish the shifts of the image of the telescope mirror on the photosensitive surface caused by the motion of the image of the observed object in the diaphragm. If the Fabry lens is not perfectly focused, the ratio of such a shift to the diameter of the image of the telescope mirror is inversely proportional to the *square* of the focal length of the Fabry lens (Serkowski and Chojnacki 1969). The Fabry lens should be achromatic, unless different detectors are used for various spectral regions separated by dichroic filters.

An image of the telescope mirror formed by stellar light is not, however, free of the effects of atmospheric seeing; the distribution of illumination in such an image (*shadow pattern*) changes rapidly with time. This causes fluctuations in the signal if the image of the telescope mirror is formed on a photosensitive surface of inhomogeneous sensitivity. For typical photomultipliers, the seeing errors resulting from this effect for the ratio of signals from two beams of starlight emerging from a Wollaston prism amount to 10% to 20% of the relative errors caused by atmospheric scintillation for signals from each of these beams separately.

The harmful effects of both atmospheric scintillation and seeing can be suppressed by a rapid modulation of the signal. According to unpublished calculations by A. T. Young, the sinusoidal modulation with frequency f diminishes the error of atmospheric origin in the amplitude of this modulation by a factor $(f/f_c)^{5/6}$, where f_c is a "cutoff" frequency equal to

$$f_c = V_\perp/(\pi D). \qquad (2)$$

Here D is the telescope diameter and $V_\perp$ is the speed at which the wind drags the shadow pattern past the telescope aperture; a typical value for $V_\perp$ is 3000 cm/sec. Assuming this value, we find that the critical frequency of modulation, below which the photometric errors caused by atmospheric scintillation and seeing are not diminished, equals 20 Hz for a telescope of 50-cm diameter, and 2 Hz for 500-cm diameter.

If modulation is obtained by rotating a half-wave plate in front of an analyzer, the modulation frequency f equals four times the number of revolutions of a half-wave plate per second. Polarimetric errors for this type of polarimeter are listed in Table II which gives the error ε_{phot} of degree of polarization p resulting from photon statistics and errors ε_{scint} resulting from atmospheric scintillation (cf. Young 1967) for different modes of operation of the polarimeter. All errors ε_{scint}, except those for the ratios of two channels without modulation, refer to results obtained by considering each of two channels separately and averaging the results. The errors listed in Table II for the cases of rapid modulation would be diminished by a factor of about 6 if polarization were calculated from the ratios of simultaneous signals from two channels. The errors in Table II were calculated on the assumption that the total integration time is 60 secs, the quantum efficiency of detectors is 20%, and the total transmittance of the telescope and polarimeter (both channels together) for the blue spectral region of UBV photometric system is 50%.

Instrumental Polarization

Linear polarization by the mirrors of a Cassegrain telescope usually does not exceed 0.1% if special precautions, described by Thiessen and

TABLE II

Polarimetric Errors Caused by Photon Noise and Atmospheric Scintillation for a Polarimeter with a Half-Wave Plate and a Wollaston Prism at Integration Time 60 Seconds in B Spectral Region ($T = 50\%$, $Q. E. = 20\%$)

Telescope diameter		50 cm		155 cm		500 cm	
Air mass		1.0	2.0	1.0	2.0	1.0	2.0
$\varepsilon_{phot}(p)$	$B = 4.5$ mag	±.010%		±.003%		±.001%	
	$B = 7.0$ mag	±.032%		±.010%		±.003%	
	$B = 9.5$ mag	±.10%		±.032%		±.010%	
	$B = 12.0$ mag	±.32%		±.10%		±.032%	
	$B = 14.5$ mag	±1.0%		±.32%		±.10%	
$\varepsilon_{scint}(p)$	scintillation not eliminated	±.035%	±.10%	±.017%	±.048%	±.008%	±.022%
	ratio of two channels	±.006%	±.015%	±.003%	±.008%	±.001%	±.004%
	modulation 30 sec^{-1}	±.025%	±.072%	±.005%	±.014%	±.001%	±.002%
	modulation 80 sec^{-1}	±.011%	±.031%	±.002%	±.006%	±.000%	±.001%

All errors ε_{scint}, except those for ratios of two channels, refer to measurements in each of two channels reduced separately and the results averaged.

Broglia (1959), are taken when aluminizing the mirrors. This instrumental polarization can be determined by observing nearby standard stars (listed in the Appendix) which are known to be unpolarized from observations made with one of the telescopes with a rotatable tube. Two such 61-cm telescopes, designed by Hiltner (Hiltner and Schild 1965), are now in operation, one at Yerkes and the other at Siding Spring Observatory. These two telescopes seem to be sufficient for establishing the unpolarized standards, and there is no obvious need for building more telescopes with rotatable tubes; a telescope on *alt-azimuth mounting* can serve the same purpose.[2]

More difficult to eliminate than instrumental linear polarization is the conversion of linear to circular polarization by the telescope optics. Such *linear-to-circular conversion* diminishes the actual degree of linear polarization for the object observed, rotates the plane of polarization, and causes instrumental circular polarization. Considerable linear-to-circular

---

[2] See p. 352.

conversion, caused by stress birefringence or metallic reflection, is expected to occur in lens objectives, correcting plates, glass filters, and tilted plane mirrors. Such optical components should be avoided in telescopes used for polarimetric measurements, unless they follow an analyzer.

Introducing linear polarization and linear-to-circular conversion by a tilted plane mirror (*Newtonian flat* or *heliostat mirror*) can be avoided if a rotatable retarder is placed in front of such a mirror. A plastic retarder, which can be obtained in practically any size, can be used for that purpose. Its rotation modulates all three Stokes parameters Q, U, and V describing the polarization of an astronomical object. If the position angles θ, used for defining the Stokes parameters [cf. Eqs. (3) of the next section] are counted from the plane of incidence of light on a tilted mirror, reflection from this mirror changes only the Stokes parameters U and V, while the modulated component of parameter Q is only multiplied by a constant factor close to unity. The light reflected from a tilted mirror goes through a Wollaston prism remaining in a fixed orientation such that the intensities of transmitted light beams depend only on parameter Q. This scheme gives, with maximum efficiency, a measurement of linear and/or circular polarization of astronomical objects free of the adverse effects of a tilted plane mirror.

A rotating achromatic half-wave plate which eliminates the linear polarization (Billings 1951) should always be used in front of a *circular polarimeter* whenever the highest precision of circular polarization is sought. Since the half-wave plate changes the sign of circular polarization, the measurements can be done alternately with and without another half-wave plate, following the first one and either stationary, or better, rotating with frequency incommensurable with that of the first half-wave plate. Such a pair of half-wave plates, placed in front of all other optical components, eliminates the troubles with linear-to-circular conversion and instrumental circular polarization.

The *focal plane diaphragms* in a polarimeter should be made of a dielectric material, e.g., ebonite (Pospergelis 1965), because metallic diaphragms polarize light. I have made measurements of linear polarization of an unpolarized standard star, the image of which, about 0.15 mm in diameter, was on the edge of a metallic diaphragm 0.5 mm in diameter so that less than half of the light entered the polarimeter. Linear polarization of over 0.2% was found, with the electric vector parallel to the edge of the diaphragm.

Motion of Light Beam on Photocathode

As the distribution of sensitivity on photocathodes is usually very nonuniform, accurate polarimetry with photomultipliers is possible only if the image of the telescope mirror on a photocathode does not shift during the measurement by more than about 0.01% of its diameter. An optical element that is either rotated or inserted into the light beam during the measurement should be plane parallel with an accuracy of a few seconds of arc to avoid shifting an image on the photocathode. The requirements for plane parallelism are relaxed if an image of the telescope mirror is formed on the rotating optical element which is then re-imaged on the photocathode (cf. Serkowski 1973a).

The problems caused by the inhomogeneous sensitivity of a photocathode become particularly serious when, instead of an image of telescope mirror, an image of an astronomical object or of its spectrum is formed on the photocathode of a photon-counting image tube. Such an image is subject to shifts caused by inaccurate telescope guiding or bad seeing. Achieving high polarimetric accuracy is then possible only with a rapid modulation of the signal by a rotating retarder in front of a stationary analyzer. This retarder should be placed as close as possible to the telescope focal plane to relax the requirements for its plane parallelism. A rotating half-wave plate has a convenient feature of modulating the polarization at a frequency four times higher than that of a mechanical rotation. This again relaxes considerably the requirements for plane parallelism of the retarder; nevertheless, for precise polarimetry in most cases a half-wave plate should be plane parallel to an accuracy of at least 1 arcmin.

Unnecessary Reflections from Optical Components

In a polarimeter, care should be taken to eliminate the unnecessarily reflected light from optical components. Particularly harmful is the light doubly reflected from surfaces of a Wollaston prism. Admixture of such reflected light makes the light beams emerging from a Wollaston prism less than 100% polarized. The amount of reflected light that reaches the detectors usually depends strongly on the position of the image of the observed object in the focal plane diaphragm. This makes the resulting systematic errors particularly difficult to eliminate.

One way to prevent the doubly reflected light from reaching the detectors is to tilt the Wollaston prism, and all other stationary optical components with flat surfaces, with respect to the axis of the polarimeter.

Variable Sky Background

Polarization of sky background can be eliminated by observing a star centered in the middle one of three identical focal plane diaphragms. The light from the diaphragms, after going through a Wollaston prism, should form images of the star close[3] to two Fabry lenses placed in front of two photomultipliers. The centers of three focal plane diaphragms should lie on a straight line spaced so that an ordinary image of the central diaphragm on one of the Fabry lenses is superimposed upon an extraordinary image of the left diaphragm; in such a pair of superimposed images, the light of the sky background becomes unpolarized. Similarly on the other Fabry lens an extraordinary image of the central diaphragm is superimposed upon an ordinary image of the right diaphragm.

An advantage of this method of eliminating the polarization of the sky background is that this background needs to be measured much less frequently than would otherwise be necessary. We need now only to know the brightness of sky background, not the polarization. For faint objects, for which the signal is not more than twice as strong as the signal from sky background, we should be able to obtain the desired polarimetric accuracy in half the time by using three diaphragms rather than by using the conventional single diaphragm. An invisible faint star that may happen to be present in one of the outer diaphragms will be noticed easily as it will cause a discrepancy between the polarimetric results from two photomultipliers. With conventional methods of measuring the sky background, such a star may go unnoticed, causing a systematic error in the results.

Effective Wavelengths

An important source of discrepancies between the observations made with different polarimeters is the inaccurate knowledge of the effective wavelengths of the spectral regions used. Such discrepancies could be easily avoided because every polarimeter has an inherent ability of measuring the effective wavelengths with high accuracy. All that is needed is to measure the polarization of the objects studied with a polarizer and a suitable retarder inserted in front of a polarimeter.

For wide-band spectral regions between 0.3 and 1.1 μm, a quartz retarder, which is a quarter-wave plate at 0.45 μm, is the most suitable. If an optical axis of this retarder makes 45° with the principal plane of the polarizer, having good ultraviolet transmittance, the degree of linear polarization for light emerging from the retarder is approximately proportional to the inverse wavelength, with the position angle flipping by 90° at 0.45 μm. Measuring polarization with a precision of $\pm 0.1\%$ gives an effective wavelength accurate to ± 3 Å in the blue spectral region. Similarly, a thick wide-angle retarder can be used for calibrating a

---

[3]Not exactly *on* Fabry lenses to avoid the possibility of forming a sharp stellar image on a dust particle on the lens surface.

spectrum scanner with an accuracy of ± 0.01 Å or better (Serkowski 1972), which makes possible the accurate measurements of radial velocities with wide open (~ 1 Å) entrance and exit slits of the scanner.

Zero Point of Position Angles

A very accurate calibration of position angles in an equatorial coordinate system can be obtained by replacing the diagonal mirror, which reflects the light to the viewing eyepiece in the polarimeter, by a plane-parallel stress-free glass plate. The telescope, with clock drive stopped, is pointed in such a direction that a spirit level put on this glass plate indicates its exact horizontal orientation. The position angle of the plane of incidence of the telescope axis on a glass plate can now be calculated from the readings of the declination and hour angle circles. This is compared with the position angle of polarization measured for any unpolarized standard star through the glass plate remaining tilted to the telescope's optical axis at the same angle of about 45°. Since the linear polarization introduced by such a tilted plate amounts to about 9%, the position angles can be easily measured with an accuracy on the order of a minute of arc. To eliminate the effects of the deviations of the glass plate from the plane parallelism and its strain birefringence, the calibration should be repeated at different orientations of the glass plate.

RETARDERS

Transformation Equations for Stokes Parameters

The most general form of light is *partially elliptically polarized light*; all possible states of polarization are its special cases. The partially elliptically polarized light can be described either by its intensity I, *degree of linear polarization* p, *position angle* θ in the equatorial coordinate system, and *degree of circular polarization* q, or by the Stokes parameters:

$$\begin{aligned} & I, \\ & Q = Ip \cos 2\theta, \\ & U = Ip \sin 2\theta, \\ & V = Iq. \end{aligned} \quad (3)$$

The Stokes parameters I', Q', U', and V' of the light transmitted through a perfect *analyzer* with the principal plane at position angle φ are connected with Stokes parameters I, Q, U, and V, describing the incident light, by a matrix transformation equation

$$\begin{bmatrix} I' \\ Q' \\ U' \\ V' \end{bmatrix} = \tfrac{1}{2} \begin{bmatrix} 1 & \cos 2\varphi & \sin 2\varphi & 0 \\ \cos 2\varphi & \cos^2 2\varphi & \tfrac{1}{2}\sin 4\varphi & 0 \\ \sin 2\varphi & \tfrac{1}{2}\sin 4\varphi & \sin^2 2\varphi & 0 \\ 0 & 0 & 0 & 0 \end{bmatrix} \cdot \begin{bmatrix} I \\ Q \\ U \\ V \end{bmatrix}. \quad (4)$$

From this equation we obtain the intensity of the light transmitted through a perfect analyzer

$$I' = \tfrac{1}{2}(I + Q\cos 2\varphi + U\sin 2\varphi). \quad (5)$$

The transformation equation for a perfect retarder of retardance τ and optical axis at position angle ψ is[4]

$$\begin{bmatrix} I' \\ Q' \\ U' \\ V' \end{bmatrix} = \begin{bmatrix} 1 & 0 & 0 & 0 \\ 0 & G + H\cos 4\psi & H\sin 4\psi & -\sin\tau\sin 2\psi \\ 0 & H\sin 4\psi & G - H\cos 4\psi & \sin\tau\cos 2\psi \\ 0 & \sin\tau\sin 2\psi & -\sin\tau\cos 2\psi & \cos\tau \end{bmatrix} \cdot \begin{bmatrix} I \\ Q \\ U \\ V \end{bmatrix}$$

(6)

where

$$G = \tfrac{1}{2}(1 + \cos\tau), \quad H = \tfrac{1}{2}(1 - \cos\tau). \quad (7)$$

If the direction of incident light makes a small angle i with the normal to the surface of the retarder, and the plane of incidence makes an angle ω with the optical axis of crystal, the retardance at wavelength λ equals

$$\tau \cong 2\pi(n_e - n_o)(s/\lambda)\left[1 - \frac{i^2}{2n_o}\left(\frac{\cos^2\omega}{n_o} - \frac{\sin^2\omega}{n_e}\right)\right], \quad (8)$$

where s is the thickness of the retarder, while n_e and n_o are the refractive indices of its material for the *extraordinary* and *ordinary* rays, i.e., for the vibrations of the electric vector of the light wave which are parallel and perpendicular to the optical axis of the retarder, respectively.

From Equations (5) and (6) we obtain the intensity of light transmitted through a retarder with the optical axis at position angle ψ followed by an analyzer with the principal plane at position angle $\varphi = 0°$ (upper signs) or $\varphi = 90°$ (lower signs):

$$I' = \tfrac{1}{2}[I \pm Q(G + H\cos 4\psi) \pm UH\sin 4\psi \mp V\sin\tau\sin 2\psi]. \quad (9)$$

---

[4] See p. 175.

For a *quarter-wave plate* $\tau = 90°$, $G = H = \frac{1}{2}$, and

$$I' = \tfrac{1}{2}(I \pm \tfrac{1}{2}Q \pm \tfrac{1}{2}Q\cos 4\psi \pm \tfrac{1}{2}U\sin 4\psi \mp V\sin 2\psi). \quad (10)$$

For a *half-wave plate* $\tau = 180°$, $G = 0$, $H = 1$, and

$$I' = \tfrac{1}{2}(I \pm Q\cos 4\psi \pm U\sin 4\psi). \quad (11)$$

The transformation equation for two retarders in series is obtained by replacing the square matrix in Equation (6) with a product of two such matrices for the two retarders. The intensity of light transmitted by two retarders of retardances τ_1 and τ_2 and optical axes at position angles ψ_1 and ψ_2, followed by an analyzer with the principal plane at position angle $\varphi = 0°$ (upper signs) or $\varphi = 90°$ (lower signs) is (cf. Ramachandran and Ramaseshan 1961)

$$\begin{aligned}I' = \tfrac{1}{2}\{I &\pm Q[G_1G_2 + H_1H_2\cos 4(\psi_1 - \psi_2) + H_1G_2\cos 4\psi_1 \\ &+ G_1H_2\cos 4\psi_2 - \sin\tau_1\sin\tau_2\sin 2\psi_1\sin 2\psi_2] \\ &\pm U[H_1H_2\sin 4(\psi_1 - \psi_2) + H_1G_2\sin 4\psi_1 + G_1H_2\sin 4\psi_2 \\ &+ \sin\tau_1\sin\tau_2\cos 2\psi_1\sin 2\psi_2] \mp V[H_2\sin\tau_1\sin(2\psi_1 - 4\psi_2) \\ &+ G_2\sin\tau_1\sin 2\psi_1 + \cos\tau_1\sin\tau_2\sin 2\psi_2]\}. \quad (12)\end{aligned}$$

In a special case of a quarter-wave plate followed by a half-wave plate and an analyzer at $\varphi = 0°$ or $90°$, we have

$$\begin{aligned}I' = \tfrac{1}{2}\{I &\pm \tfrac{1}{2}Q[\cos 4(\psi_1 - \psi_2) + \cos 4\psi_2] \pm \tfrac{1}{2}U[\sin 4(\psi_1 - \psi_2) \\ &+ \sin 4\psi_2] \mp V\sin(2\psi_1 - 4\psi_2)\}; \quad (13)\end{aligned}$$

this combination offers interesting possibilities for the simultaneous measurement of all Stokes parameters.

For two half-wave plates followed by an analyzer, Equation (12) takes a simple form

$$I' = \tfrac{1}{2}[I \pm Q\cos 4(\psi_1 - \psi_2) \pm U\sin 4(\psi_1 - \psi_2)]; \quad (14)$$

the intensity of the transmitted light beam depends now on the angle between the optical axes of two half-wave plates.

Polarization by Refraction on Surfaces of Retarder

While deriving Equations (6) through (14), the effects of reflection of light by the surfaces of the retarder have been neglected. Since for a birefringent material reflection coefficients for normal incidence depend on the position angle of the plane of vibrations for incident light, unpolarized light may become polarized after passing through a retarder;

in the case of a calcite retarder, polarization by refraction may amount to $p_R \cong 2.4\%$. When a rotating retarder is followed by an analyzer, a term $Ip_R \cos 2\psi$ describing the polarization by refraction should be added to Equations (9), (10), and (11). This term may cause systematic errors in the measurements of circular polarization if the orientation of the optical axis of the retarder is not known accurately.

Polarization by refraction can be diminished considerably if the retarder is cemented or optically contacted between two plates of transparent isotropic material with a refractive index similar to indices of the retarder, or even if the retarder is given an evaporated coating of such a material. Antireflection coating on the surfaces of these cover plates is desirable, not only to save light but also to diminish the effects of interference of reflected rays (Holmes 1964).

Polarization by refraction cancels out completely for a retarder made of *two optically contacted plates* of the same birefringent material with optical axes crossed. The resultant retardance equals the difference of retardances of the two plates. Usually quartz retarders are made this way because otherwise they would be too thin to manufacture. This type of retarder is better than a single retarder whenever circular polarization is measured with the retarder rotated in front of an analyzer.

Using retarders made of mica is not recommended. Not only do such retarders produce linear polarization by refraction and absorb strongly in the near ultraviolet, but also they often have considerable *linear dichroism*, producing similar effects as polarization by refraction.

Lyot Depolarizer

Two retarders with retardances $\tau_1 = \tau \gg 1$ and $\tau_2 = 2\tau$ having optical axes at $\psi_1 = 0°$ and $\psi_2 = 45°$ form a *Lyot depolarizer*. The intensity of light transmitted through such a depolarizer followed by an analyzer at $\varphi = 0°$ (upper signs) or $90°$ (lower signs) is related to Stokes parameters I, Q, U, and V of the incident light by an equation

$$I' = \tfrac{1}{2}[I \pm Q \cos 2\tau \pm \tfrac{1}{2}U(\cos \tau - \cos 3\tau) \mp \tfrac{1}{2}V(\sin \tau + \sin 3\tau)], \quad (15)$$

which is a special case of Equation (12).

For different wavelengths occuring in the light of a wide spectral band, the retardance τ may take values ranging over many multiples of π. Therefore, for such wide bands the mean values of sines and cosines of τ and its multiples occurring in Equation (15) are close to zero; the degree of polarization averaged over the wavelength becomes very small for the light transmitted through a Lyot depolarizer.

Because of the polarization by refraction, Lyot depolarizers may polarize linearly the incident unpolarized light; for a Lyot depolarizer made of *calcite* this polarization amounts to about 1.2%, while for crystalline quartz it is 0.07%. This effect is eliminated if the depolarizer is followed by an analyzer with the principal plane making 45° with the second plate of the depolarizer. Such an optimum orientation of the depolarizer was assumed when deriving Equation (15). Another advantage of such orientation is that the depolarizing action is not impaired if the angle between the axes of the depolarizer plates deviates somewhat from 45° (Serkowski 1973a).

The intensity of light transmitted through a *rapidly rotating* Lyot depolarizer, with the optical axes of two plates making an angle $\pi/4 + \varepsilon$ radians, followed by an analyzer at $\varphi = 0°$ or $90°$, equals

$$I' \cong \tfrac{1}{2}[I \pm \tfrac{1}{2}Q(\cos \tau + \cos 2\tau) \mp U(\varepsilon - \tfrac{1}{4}\cos \tau + \tfrac{1}{4}\cos 3\tau)]; \quad (16)$$

the depolarizing action is somewhat improved as compared to a stationary depolarizer, but only if the angle between the optical axes of the depolarizer plates equals 45° exactly. Rotating a Lyot depolarizer eliminates polarization produced by refraction on its back surface and averages out the shift of the light beam on the photocathode caused by inserting a depolarizer which is not perfectly plane parallel. A *calcite* Lyot depolarizer, which cannot be made accurately plane parallel and cannot be optically contacted because of the softness of the material, should always be rotating while in use.

Achromatic Retarders

Retarders for which the retardance τ is approximately independent of wavelength can be obtained by combining two or more retarders made of materials with a different wavelength dependence of birefringence (Beckers 1971a). Theoretically, the best pair is magnesium fluoride and KDP (potassium dihydrogen phosphate) giving almost perfect achromatism from 0.3 to 1.2 μm. Obtaining such a good achromatism in practice requires, however, that the surfaces of KDP are flat to $\lambda/10$ and plane parallel to about 1 arcsec, which seems impossible to achieve because of the softness of this crystal. The next best pairs are magnesium fluoride with quartz or with colorless sapphire. Magnesium fluoride 0.262 mm thick and quartz 0.304 mm thick form an achromatic half-wave plate for which retardance does not deviate by more than 45° from 180° over the spectral range 0.3–1.0 μm (Fig. 1).[5]

If a *wide-field* achromatic retarder is needed, for which the retardance depends little on the angle of incidence, Equation (8) indicates that the

---

[5] Achromatic retarders of this type are manufactured, e.g., by Continental Optical Corp., Farmingdale, N. Y. 11735.

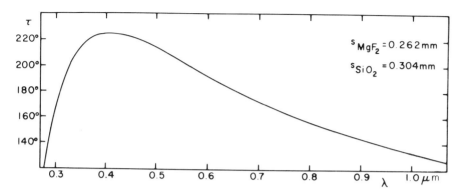

Fig. 1. Wavelength dependence of retardance for a half-wave plate of magnesium fluoride and quartz.

two materials used should have birefringence $n_e - n_o$ of the opposite sign; two plates of each material should be used in a subtractive mode so that a suitable ratio of the thickness of two materials is achieved (Beckers 1972) such that the dependence on the angle of incidence cancels out. Magnesium fluoride and colorless sapphire (Jeppesen 1958; I. H. Malitson[6]) form a suitable pair, which has the advantage of being highly transparent over the spectral range from 0.20 to 6 μm. *Infrared transmittance* may be shifted to 8 μm by using lanthanum fluoride (Wirick 1966) instead of sapphire. The birefringent materials transparent in the farther infrared are cadmium sulphide (Walsh 1972), transparent to 16 μm, and quartz, transparent between 60 and 400 μm (Palik 1965).

Another method of constructing achromatic retarders was proposed by Pancharatnam (1955). A combination of three retarders, the first and the last having their optical axes parallel and with an identical retardance τ_1, behaves as a single retarder with retardance τ and with an equivalent optical axis making an angle ψ with the optical axis of the first retarder, where τ and ψ are given by

$$\cos(\tau/2) = \cos\tau_1 \cos(\tau_2/2) - \sin\tau_1 \sin(\tau_2/2) \cos 2c, \qquad (17)$$

$$\cot 2\psi = [\sin\tau_1 \cot(\tau_2/2) + \cos\tau_1 \cos 2c]/\sin 2c; \qquad (18)$$

here τ_2 is the retardance of the central retarder and c is the angle between its optical axis and that of the other two retarders. Equations (17) or (18) written for different wavelengths give conditions for achromatism. For a

---
[6] Personal communication.

Pancharatnam quarter-wave plate, the achromatism is not as good as for his half-wave plate (cf. Tinbergen 1973). The latter consists of three half-wave plates; the optical axis of the central one makes an angle of about 57° with the axes of the outer two half-wave plates.

A very high degree of achromatism can be achieved by using a Pancharatnam retarder in which each of three components is already an achromatic pair of two birefringent materials. Properties of such a *superachromatic half-wave plate* designed by Serkowski (1971)[7] are shown in Fig. 2. Similar designs were described by Tinbergen (1972).[8] Deviations of retardance from 180° do not exceed 3° over the 0.28 to 1.05 μm spectral range; hence, they are completely negligible because only the cosine of the retardance enters Equation (9) when linear polarization is measured with a half-wave plate.

The main disadvantage of Pancharatnam retarders is the wavelength dependence of the position angle of their effective optical axis (Fig. 2), described by Equation (18). When wide spectral bands are used, corrections that depend on the spectral energy distribution for the astronomical object observed must be applied to the observed position angles of polarization. Fortunately, this inconvenience can be avoided when measuring linear polarization with a rotating Pancharatnam half-wave plate. What is needed is to place another, stationary, Pancharatnam half-wave plate between the rotating plate and an analyzer (Serkowski 1971). As results from Equation (14), the intensity of the transmitted light beam now becomes independent of the orientation of the optical axes of the individual Pancharatnam half-wave plates and depends on an angle $\psi_1 - \psi_2$ between the axes of two plates. This angle is *independent of wavelength* if the two Pancharatnam half-wave plates are *exactly* identical.

The stationary half-wave plate should have an equivalent optical axis approximately parallel (or perpendicular) to the principal plane of the analyzer, because then the terms proportional to $\sin 2\psi_2$ and $\sin 4\psi_2$ in Equation (12) become negligibly small. This stationary half-wave plate can be cemented or optically contacted to an analyzer to diminish the light losses by reflection. The Pancharatnam retarders should always be cemented or optically contacted between the two plates of isotropic

---

[7] In the paper quoted here several figure legends are incorrect. In Fig. 18, $\Delta\theta$ denotes an error in the position angle of polarization resulting from neglecting the wavelength dependence of the effective optical axis. The polarimeter shown in Fig. 19 can be used as a birefringent interferometer, described in another paper (Serkowski 1973a), when prisms P_1, P_2, and Rochon prisms are inserted into the light beam. In Figs. 6 and 7 the observations by Gehrels mentioned in the legend are not included.

[8] See p. 175.

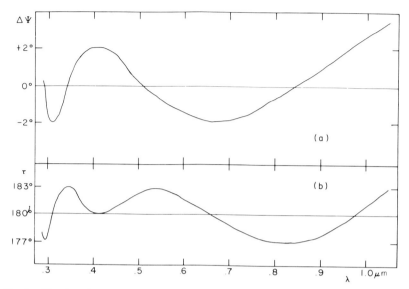

Fig. 2. Wavelength dependence of (a) the position angle of the equivalent optical axis, and (b) the retardance for the Pancharatnam achromatic half-wave plate of magnesium fluoride and quartz.

material to diminish the effects of polarization by refraction. These effects cause serious systematic errors in the case of circular polarization measured with a Pancharatnam quarter-wave plate.

COMPARISON OF VARIOUS POLARIMETERS

Efficiency of the Polarimeter

Let the light of an astronomical object incident on a polarimeter be described by the Stokes parameters I, Q, U, and V. Denoting the signals (photon counts) from two beams emerging from a beam-splitting analyzer (e.g., a Wollaston prism) by I_1 and I_2, all the information on the state of polarization of incident light should be contained in the difference $I_1 - I_2$. This difference can be represented in the form

$$I_1 - I_2 = Q f_Q(t) + U f_U(t) + V f_V(t) + c(I, Q, U, V), \qquad (19)$$

where the mean values of the functions f_Q, f_U, and f_V, averaged over time t during the measurement, are equal to zero, and the function c is independent of the time. The values of these four functions can be found from Equations (5) and (9) through (14). Equation (19) holds also for a one-channel polarimeter, in which case $I_2 = 0$.

The efficiencies of a polarimeter for linear and circular polarization are defined as

$$E_{\text{lin}} = \langle f_Q^2(t) + f_U^2(t) \rangle, \tag{20}$$

$$E_{\text{cir}} = \langle f_V^2(t) \rangle, \tag{21}$$

where angular brackets denote the averaging over the duration of measurement lasting a unit of time. The efficiencies E_{lin} and E_{cir} are inversely proportional to the amount of time needed for obtaining a given polarimetric precision for the incident light of intensity I; they equal 1 for a perfect polarimeter. The meaning of Equation (19) may be more easily visualized when both sides of this equation are divided by the intensity I of incident light and the equation takes the form

$$\frac{(I_1/I_2) - 1}{(I_1/I_2) + 1} T = \frac{Q}{I} f_Q(t) + \frac{U}{I} f_U(t) + \frac{V}{I} f_V(t) + c'\left(\frac{Q}{I}, \frac{U}{I}, \frac{V}{I}\right), \tag{22}$$

where T is a transmittance of the polarimeter for unpolarized light.

The efficiencies calculated from Equations (20) and (21) for the maximum possible value of transmittance T are listed in Table III, which contains a comparison of different types of polarimeters. Among other entries in this table the ratio of modulation frequency to frequency of mechanical motion is listed. If the modulating optical element is alternately in and out of the beam, this ratio was assumed as 1. The higher this ratio, the better is the separation of polarization modulation from the modulation caused by a wedge-shaped or tilted moving optical component, dust specks on this component, etc. The dichroism of a modulating optical component or the polarization produced by refraction on its surfaces produces a modulation of the signal with a frequency twice as high as the frequency of the rotation. The ratio of the frequencies of modulation and rotation should therefore be larger than 2 to eliminate the systematic errors arising from these sources. High values of this ratio are also desirable for better elimination of atmospheric scintillation and seeing.

Polarimeters without Rapid Modulation of the Signal

The simplest type of polarimeter is a Polaroid rotated in discrete steps in front of a detector. Since only the light linearly polarized in the principal plane of a Polaroid is transmitted, the efficiency E_{lin}, calculated from Equations (5) and (20), cannot exceed $\frac{1}{2}$. A depolarizer must be placed between the Polaroid and the detector to eliminate the dependence of sensitivity on the plane of polarization, occurring for most detectors. This limits the applications of such a polarimeter to wide spectral regions because constructing monochromatic depolarizers (Billings 1951) is difficult.

The most widely used type of polarimeter without rapid modulation of the signal is a so-called Wollaston polarimeter (No. 2 in Table III). Since the entire polarimeter is rotated, the ratio of the sensitivities of two photomultipliers is different at each orientation of the instrument. This is caused by shifts of the images of the telescope mirror on the photocathodes caused by the mechanical flexure and by changing orientation of the photomultipliers in the earth's magnetic field. Therefore, the ratio of the sensitivities of the photomultipliers is determined at each orientation of the instrument from the measurement with a Lyot depolarizer inserted in front of the Wollaston prism, preferably at the optimum orientation discussed previously. The measurement with the depolarizer fulfils its aim only if the depolarizer is either plane parallel to within a few seconds of arc or is continuously rotating so that the shifts of the images of the telescope mirror on the photocathodes due to inserting the depolarizer cause negligible change in the sensitivity of the photomultipliers.

The efficiency of the Wollaston polarimeter, as results from Equations (5), (20), and (22) for $T = 1$, would equal $E_{\text{lin}} = 1$ if there were no need to use the depolarizer. Actually, instead of Equation (22), we have for the Wollaston polarimeter

$$\left(\frac{I_1 I_{2d}}{I_2 I_{1d}} - 1\right) \Big/ \left(\frac{I_1 I_{2d}}{I_2 I_{1d}} + 1\right) = \frac{Q}{I}\cos 2\varphi + \frac{U}{I}\sin 2\varphi, \quad (23)$$

where subscript d denotes the measurement with depolarizer and φ is the position angle of polarimeter. If each of the ratios I_1/I_2 and I_{1d}/I_{2d} is measured with a mean error ε, the mean error of $I_1 I_{2d}/I_2 I_{1d}$ equals $2^{\frac{1}{2}}\varepsilon$. If a depolarizer were not used, twice as much of the observing time would be spent on observing I_1/I_2, and its mean error would decrease to $2^{-\frac{1}{2}}\varepsilon$. Therefore, in unit observing time, the left side of Equation (23) is measured with a mean error twice as large as that for the left side of Equation (22); hence, when a depolarizer is used, the efficiency of the Wollaston polarimeter equals $E_{\text{lin}} = \frac{1}{4}$. Obtaining any desired precision with this Wollaston polarimeter takes four times as much observing time as with an ideal polarimeter.

The simplest method of increasing the efficiency of a Wollaston polarimeter to $E_{\text{lin}} \cong 1$ is to replace the measurements with and without a depolarizer by the measurements at two orientations of a Wollaston prism relative to the polarimeter, differing by 180° (polarimeter No. 3 in Table III). The Wollaston prism must be, in this case, relatively thin and must consist of three components (Soref and McMahon 1966) so that the shift of the images of the telescope mirror on the photocathodes caused by rotation of the Wollaston prism is negligibly small. An image of the telescope mirror should be formed on the Wollaston prism to diminish this shift and to make it independent of small deviations of the angle of

TABLE III

Comparison of Various Types of Optical Polarimeters

No.	First astronomical use	Scheme	Moving components	Efficiency linear	Efficiency circular
A. POLARIMETERS WITHOUT RAPID MODULATION OF SIGNAL					
a. Linear Polarimeters					
1	Hiltner 1956	Glan, Depol.	Glan prism or Polaroid rotated in discrete steps	$\frac{1}{2}$	—
2	Behr 1956	Depol., Wollaston	Depolarizer in and out, entire polarimeter rotated in discrete steps	$\frac{1}{4}$	—
3	Serkowski 1973b	Wollaston, Depol.	3-component Wollaston at 0° and 180°, entire polarimeter rotated in 45° steps	1	—
4	Appenzeller 1967	$\lambda/2$, Wollaston, 22.5° steps	Half-wave plate rotated in 22°.5 steps	1	—
b. Circular Polarimeters [b]					
5	Serkowski and Chojnacki 1969	$\lambda/4$, Wollaston, 90° steps	Quarter-wave plate rotated in 90° steps	—	1

[a] Bibliography of polarimeters in space is given by Coffeen, p. 189.

[b] Placing a rotating achromatic half-wave plate in front of a circular polarimeter is always recommended as it eliminates the linear-to-circular conversion.

POLARIMETERS FOR OPTICAL ASTRONOMY 155

Ratio of frequencies of modulation to mechanical motion		Availability of achromatic components for 0.3–1.0 μm	Critical requirements for high accuracy	Descriptions of ground-based[a] polarimeters currently in use	Observatory
linear	circular				
2	—	+	Analyzer plane parallel, followed by a good depolarizer	Dombrovskij et al. 1965	Leningrad
1	—	+	Depolarizer very accurately plane parallel	Gehrels and Teska 1960 Clarke 1965 Elvius and Engberg 1967 Behr 1968 Hall 1968 Serkowski and Chojnacki 1969 Serkowski, Mathewson, and Ford 1973 (10-channel)	Univ. of Ariz. Glasgow Stockholm ESO, Chile Lowell Obs. Warsaw Siding Spring
2	—	+	Wollaston prism thin, plane parallel, not tilted, with telescope mirror imaged on its lower surface	Serkowski 1973a,b (infrared)	Univ. of Ariz.
4	—	+	Half-wave plate plane parallel	Appenzeller 1967 Behr 1968 Serkowski, Mathewson, and Ford 1973 (10-channel)	Cerro Tololo and Yerkes Obs. ESO, Chile Siding Spring
—	2	+	Quarter-wave plate plane parallel	Gehrels 1972	Univ. of Ariz.

Continued on next pages

TABLE III — continued
Comparison of Various Types of Optical Polarimeters

No.	First astronomical use	Scheme	Moving components	Efficiency linear	Efficiency circular
	B. POLARIMETERS WITH RAPID MODULATION OF SIGNAL				
	a. Linear Polarimeters				
6	Öhman 1943, Hall and Mikesell 1950	Glan, Depol.	Glan prism, Polaroid, or wire grid polarizer rapidly rotating	½	—
7	Dollfus 1963[c]	λ/4 λ/2 λ/4 Wollaston (45° steps) (90° steps)	Half-wave plate rapidly in and out, 1st quarter-wave plate rotated in 45° steps	1	—
8	Lyot 1948	λ/2 Wollaston	Half-wave plate continuously rapidly rotating	1	—
9	Livingston and Harvey 1971	±λ/2 or zero, ±λ/2 Wollaston, 22°.5, 45°	No moving components; two Pockels cells or piezooptical modulators	1[f]	—
	b. Circular Polarimeters[b]				
10	Babcock 1953	±λ/4 Wollaston	No moving components; Pockels cell or piezooptical modulator	—	1[f]

[a] Bibliography of polarimeters in space is given by Coffeen, p. 189.
[b] Placing a rotating achromatic half-wave plate in front of a circular polarimeter is always recommended as it eliminates the linear-to-circular conversion.
[c] See p. 695.
[d] See p. 752.

POLARIMETERS FOR OPTICAL ASTRONOMY

Ratio of frequencies of modulation to mechanical motion		Availability of achromatic components for 0.3–1.0 μm	Critical requirements for high accuracy	Descriptions of ground-based[a] polarimeters currently in use	Observatory
linear	circular				
2	—	+	Analyzer plane parallel, followed by a good depolarizer	Oskanjan et al. 1969	Belgrade
				Hashimoto et al. 1970 (infrared)	Kyoto
				Bugaenko et al. 1971	Kiev
				Dyck and Sandford 1971	Kitt Peak
				Dyck et al. 1971 (infrared)	Kitt Peak
				Visvanathan 1972	Hale Obs.
				Shakhovskoj and Efimov 1973	Crimean Obs.
				Ksanfomaliti et al. 1973	Abastumani
1	—	—	Half-wave plate very accurately plane parallel, the first quarter-wave plate cannot be of Pancharatnam type	Marin 1965	Meudon
				Tinbergen 1972, 1973	Leiden
4	—	+	Half-wave plate plane parallel	Hämeen-Anttila 1972 and p. 161	Univ. of Ariz.
				Clarke and McLean[d]	Glasgow
				Querfeld[e] (solar corona)	High Altitude Obs., Colorado
∞	—	—	——	Livingston and Harvey 1971 (solar)	Kitt Peak
—	∞	—	——	Angel and Landstreet 1970	Columbia Univ., N. Y.
				Kemp et al. 1972	Hawaii
				Swedlund (unpublished)	Batelle Obs.
				Beckers 1971b lists solar instruments	

[e] See p. 254.

[f] Such efficiency is achieved only for square-wave modulation of Pockels cell or piezooptical modulator.

Continued on next pages

TABLE III — continued
Comparison of Various Types of Optical Polarimeters

No.	First astronomical use	Scheme	Moving components	Efficiency linear	Efficiency circular
		b. Circular Polarimeters[b] *(continued)*			
11	Dollfus 1963	λ/2 λ/4 Wollaston, 90° steps	Half-wave plate in and out, quarter-wave plate in 90° steps[g]	—	1
		c. Stokes-meters (simultaneous linear and circular polarization)			
12	Orrall 1971	λ/4 Wollaston	Quarter-wave plate continuously rapidly rotating	¼	½
13	—	λ/4 λ/2 Wollaston, 45° steps	Half-wave plate continuously rapidly rotating, quarter-wave plate in 45° steps	½	½
14	—	λ/4 λ/2 Wollaston, same speed, opposite direction	Quarter-wave plate and half-wave plate continuously rapidly rotating in opposite directions	½	½
15	Ikhsanov and Platonov 1967	λ/2 ±λ/4 or zero Wollaston, 45°	Continuously rapidly rotating half-wave plate modulates linear polarization; variable retarder modulates circular polarization	½	½ [f]
16	Beckers 1971*b*	±λ/2.6 ±λ/6.8 λ/4 Wollaston, 0° 45° 45°	No moving components; two Pockels cells or piezooptical modulators with sinusoidal modulation	0.87	¼

[a] Bibliography of polarimeters in space is given by Coffeen, p. 189.
[b] Placing a rotating achromatic half-wave plate in front of a circular polarimeter is always recommended as it eliminates the linear-to-circular conversion.
[f] Such efficiency is achieved only for square-wave modulation of Pockels cell or piezooptical modulator.

POLARIMETERS FOR OPTICAL ASTRONOMY 159

Ratio of frequencies of modulation to mechanical motion		Availability of achromatic components for 0.3–1.0 μm	Critical requirements for high accuracy	Descriptions of ground-based[a] polarimeters currently in use	Observatory
linear	circular				
—	1 or 2	+	Modulating retarder plane parallel	Marin 1965 Tinbergen 1972, 1973. p. 161	Meudon Leiden Univ. of Ariz.
4	2	+	Quarter-wave plate plane parallel	Orrall 1971 Nordsieck 1972	Hawaii Lick Obs.
4	4(2)	+	Half-wave plate plane parallel	p. 161	Univ. of Ariz.
4	6	+	Both retarders plane parallel	p. 161	Univ. of Ariz.
4	∞	—	Half-wave plate plane parallel	Ikhsanov and Platonov 1967 (solar)	Pulkovo
∞[h]	∞[h]	—	—	Beckers 1971b (solar)[i]	High Altitude Obs., Colorado

[g] Rapid modulation is obtained either by a half-wave plate rapidly inserted in and out (while rotations of a quarter-wave plate are made less often) or by rapid rotation of a quarter-wave plate by a stepper motor, in 90° steps.

[h] Different modulation frequency for each of Stokes parameters Q, U, V.

[i] See p. 246.

rotation of the Wollaston prism from 180°. The Wollaston prism must be followed by a thick retarder, with an optical axis at 45° to the principal plane of the Wollaston prism, to act as a depolarizer. The need for this depolarizer and the necessity for rotating the entire instrument are the main disadvantages of this type of polarimeter.

In some of the early Wollaston polarimeters the difference between the signals from two photomultipliers was recorded. This is clearly inferior to the accurate independent recording of signals from both photomultipliers, either by photon counting or by readout of integrating capacitors with a digital voltmeter. The polarimetric results obtained by any of the latter methods are not affected by the presence of a thin layer of clouds causing rapid fluctuations in the intensity of the signals. When using photon counting, the usual corrections for overlapping pulses (Serkowski 1973a) must be applied. The measurements with integrating capacitors are affected by fatigue effects in the photomultipliers (Gex and Tassart 1971), by fluctuations in photocathode voltage, and by dielectric hysteresis in the capacitors; this last effect is least troublesome with polystyrene capacitors.

Polarimeters with Rapid Modulation of Signal

Rapid modulation of the signal is the only way to eliminate the polarimetric errors caused by atmospheric seeing and by inaccurate telescope guiding. These are the main sources of error for bright stars observed without rapid modulation, and they make it virtually impossible to reach a precision better than $\pm 0.01\%$ in the degree of polarization when using such methods.

A modulation at very high frequency can be obtained by using a Pockels cell or piezooptical modulator. In the Pockels cell (Billings 1952) a crystal, KDP for example, changes its birefringence in phase with a rapidly changing high voltage applied to its surface. In the piezooptical modulator (Kemp 1969) the stress birefringence is produced in a transparent isotropic material by acoustic vibrations.

Achromatic variable retarders, operating on either of these principles, have not yet been constructed. This seriously limits the attractiveness of variable retarders for astronomical applications. Many different types of polarimeters using variable retarders have been constructed, mainly for solar applications. Only a few examples are listed in Table III; a more complete listing is given by Beckers (1971b). Most of these polarimeters require a square-wave modulation of retardance for high polarimetric efficiency. This square-wave modulation may be difficult to achieve at high frequencies; consequently, a sinusoidal modulation is often used.

For a variable half-wave plate with a sinusoidal modulation of angular frequency ω, we have (Beckers 1968; Ikhsanov and Platonov 1967)

$$\sin \tau = \sin[(\pi/2) \sin \omega t]$$
$$= 2 J_1(\pi/2) \sin \omega t \simeq 1.132 \sin \omega t, \quad (24)$$
$$\cos \tau = \cos[(\pi/2) \sin \omega t]$$
$$= J_0(\pi/2) + 2 J_2(\pi/2) \cos 2\omega t$$
$$\simeq 0.472 + 0.498 \cos 2\omega t, \quad (25)$$

where J_n stands for the n-th order Bessel function. We notice that the terms proportional to cos τ in Equations (9) or (12) now produce a modulation at a frequency twice as high as the terms proportional to sin τ. In some polarimeters based on variable retarders (e.g., No. 16 in Table III) this is utilized for producing a different frequency of modulation for each of the Stokes parameters, which is clearly advantageous for reducing systematic errors.

As long as achromatism is not possible for variable retarders, Table III indicates that the best linear polarimeter, No. 8 in Table III, is an achromatic half-wave plate rotating in front of a Wollaston prism, [cf. Eq. (11)], as proposed by Lyot (1948). The best circular polarimeter is one of Dollfus' (1963) design (No. 11 in Table III), and the best instrument for simultaneous measurements of linear and circular polarization consists of achromatic quarter-wave and half-wave plates [cf. Eq. (13)] rotating with the same speed but in opposite directions in front of a Wollaston prism (No. 14 in Table III). This last instrument remains to be built.

DESCRIPTION OF A SCANNING POLARIMETER

In the studies of changes in polarization with time, observed for a wide variety of astronomical objects, simultaneous polarimetry and spectrophotometry would greatly increase the value of observational results. A scanning polarimeter that will make these observations possible is now under construction at our Laboratory under sponsorship of the National Science Foundation. The mechanical design is made by E. H. Roland and R. H. Toubhans, while J. E. Frecker is building the electronic system (described in general outline by Hämeen-Anttila 1972) and writing the computer programs. The polarimeter can work in three modes briefly described below.

Wide-Band Polarimetry

An image of a star or starlike object is formed on the middle one of three focal plane diaphragms, the centers of which are separated by 8 mm. The two outer diaphragms are used for eliminating the polarization of the sky background as described earlier in this paper. A fused silica stress-free lens T (Fig. 3) forms an image of the telescope mirror on a continuously rotating superachromatic half-wave plate, the properties of which are illustrated in Fig. 2. The same lens forms the images of the diaphragms about 2 cm in front of the Fabry lenses. Lens T can be rotated in $22°.5$ steps to determine any depolarization that it may produce.

After passing through a rotating half-wave plate, the light goes through a stationary achromatic half-wave plate and through a three-component calcite Wollaston prism which can take two orientations differing by $180°$. The two beams of stellar light emerging from the Wollaston prism are reflected by a mirror M toward three dichroic filters similar to those used by Serkowski, Mathewson, and Ford (1973). The first dichroic filter reflects the light of wavelengths $0.30\,\mu\text{m} < \lambda < 0.395\,\mu\text{m}$ toward two bialkali photomultipliers and transmits the remaining wavelengths. The second filter reflects the wavelengths $\lambda > 0.77\,\mu\text{m}$ toward a pair of infrared-sensitive photomultipliers. The third dichroic filter reflects $0.59\,\mu\text{m} < \lambda < 0.77\,\mu\text{m}$ toward two S20 photomultipliers, and transmits $0.395\,\mu\text{m} < \lambda < 0.59\,\mu\text{m}$ toward a pair of S11 photomultipliers. Each of these four spectral regions can be, if desired, subdivided into two narrower spectral regions by glass or interference filters placed in front of the Fabry lenses; each of the eight photomultipliers could then measure polarization in a different spectral region. The light loss on each dichroic filter does not exceed 15%. All transmitting optical components have a single-layer antireflection coating with a minimum reflectance at $0.46\,\mu\text{m}$; the light loss by reflection from each surface is less than 1.5% for 0.37 to $0.60\,\mu\text{m}$ and less than 3% outside of this spectral region.

The spectral regions separated by the dichroic filters depend on the ratio of the intensities of light waves vibrating parallel and perpendicular to the plane of incidence on the filter. To minimize this effect, the Wollaston prism used in the polarimeter has a somewhat unusual design. The optical axis of each of three components of the prism makes a $45°$ angle with the refracting edge of the prism; therefore, the two light beams emerging from the Wollaston prism are polarized at $45°$ to the plane defined by the axes of these beams and at $45°$ to the plane of incidence on the dichroic filters.

A Glan-Taylor prism can be inserted in front of the focal plane diaphragms. Since the polarization of the sky background is eliminated, the deviation from 100% of stellar polarization measured with the Glan prism indicates the brightness of the sky background, which therefore does not need to be measured separately. Some other corrections, in particular the correction for overlapping pulses, are also determined from the measurements with the Glan prism (Serkowski 1973a).

The polarimeter has a provision for using simultaneously up to three retarders rapidly (20 rev/sec) rotated by stepper motors. Two of them

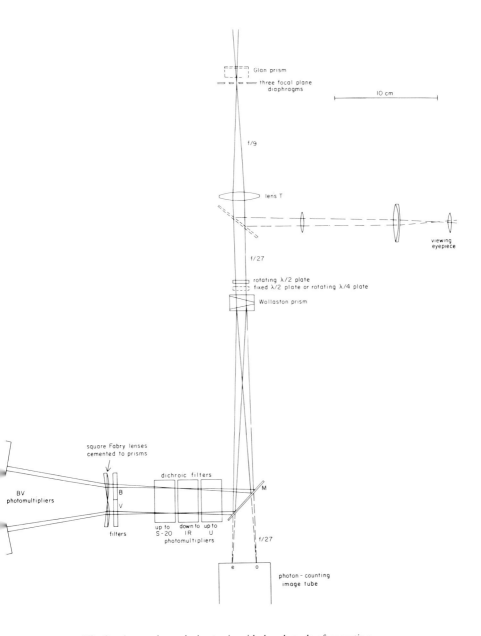

Fig. 3. A scanning polarimeter in wide-band mode of operation.

can be inserted immediately below the focal plane diaphragm, the third one in front of the Wollaston prism. Measurements of circular polarization or simultaneous measurements of linear and circular polarization according to schemes Nos. 11, 13, or 14 of Table III will therefore be possible with the new polarimeter. The stepper motor operating the uppermost retarder can also be used for rapid interchanging of two or more focal plane diaphragms of different size or differently located in the focal plane; this device will be used for polarimetry of the satellites of planets or for observing stars embedded in nebulae.

The electronic system will be basically the same as the one that is presently used to operate a continuously rotating half-wave plate which has been recently installed in front of the Wollaston prism in the polarimeter described by Coyne and Gehrels (1967); the polarimetric precision obtained with an instrument thus modified is similar to that indicated by Table II. Twelve locations in the memory of a Nova computer (Data General Corp., Southboro, Mass. 01772) are reserved for each photomultiplier. Every millisecond the photon counts accumulated during the last millisecond are added to the next one of these memory locations. A half-wave plate rotates during this millisecond by $7°.5$; a full sine wave of modulation is covered in 12 msecs. The synchronization of the rotation of the half-wave plate with the computer is checked after every integration using a light-emitting diode sensed by a silicon fotofet. At the end of the stellar measurement, usually lasting 1 min, the Nova computer calculates and types on a teletypewriter the degree of polarization, its mean error, position angle of polarization, time, and stellar magnitude of an object. The degrees of polarization are multiplied by a factor $(\pi/12)/\sin(\pi/12) \cong 1.0115$ to correct for the effects of averaging the photon counts over 12 sections of the sine curve.

Area Scanning

For polarimetric area scanning of extended objects (planets, nebulae, galaxies) the focal plane diaphragms are replaced by a slit, and the mirror M (Fig. 3) is removed from the light beam. Another three-component Wollaston prism is inserted about 2 cm below the first one, making the ordinary and extraordinary beams parallel to each other. The two perpendicularly polarized images of the focal plane slit, magnified about 5 times, are formed by an achromatic lens on the S20 photocathode of a Digicon photon-counting image tube having 200 silicon diodes in two parallel rows (Beaver et al. 1972*a*, *b*). The polarization of a hundred areas of an extended object, imaged along the slit, is measured simultaneously using a rotating superachromatic half-wave plate placed immediately behind the focal plane slit. When observing a planet, two narrow strips of the planetary surface, separated by about 1/3 of planet's diameter, are measured alternately, using the rapid switching provided by a deflection coil of the Digicon. The ratio of the lengths of the corresponding chords determines the position of the observed strips on the disk of a planet.

Spectrum Scanning and High Precision Radial Velocity Measurements

A negative lens is inserted in front of two Wollaston prisms (Fig. 4) to change a converging light beam into a divergent one of focal ratio about $f/5$. This negative lens of fused silica forms with the positive lens T an achromatic telephoto system. Mirror m reflects the light toward a grating spectrometer in the Ebert mounting. A grating with 600 grooves/mm, blazed at 0.75 μm in the 1st order, is interchangeable with an echelle of 79 grooves/mm with 102 mm × 206 mm ruled area. The tilt of any of them can be changed by a stepper motor over a range of 9°, to obtain spectrum scanning.

Two images of the spectrum, ordinary and extraordinary, are formed on two rows of diodes in a photon-counting image tube. Different orders of the spectrum are separated by glass or interference filters.

An echelle gives on the image tube a reciprocal dispersion of 2 Å/mm and a free spectral range of 85 Å at Hγ; a focal plane slit 0.2 mm wide is imaged along the width of a single Digicon diode corresponding to 0.3 Å of the spectrum. The echelle will be used for measuring the changes of linear and circular polarization across the profiles of spectral lines and for determining stellar *magnetic fields* from these measurements. In an hour of observing a 7-mag star with a 225-cm telescope, the polarization will be measured with $\pm 0.1\%$ accuracy for every resolution element 0.3 Å wide along the profile of the spectral line or band. If these are the circular polarization measurements in the blue spectral region for a strong line which is rotationally broadened to a half-width of about 4 Å and susceptible to Zeeman splitting ($z = 1.5$), this observation will give the magnetic field with a precision of about ± 50 gauss (cf. Beckers 1971b).[9]

The echelle will also be used with a *radial velocity meter* (Serkowski 1972), inserted in a thermostat (a Dewar filled with a mixture of ice and water) immediately in front of a rotating half-wave plate. This radial velocity meter consists of a Polaroid, a wide-angle retarder (Beckers 1972) of sapphire and magnesium fluoride, about 3 cm thick, and a quarter-wave plate with an optical axis at 45° to that of the wide-angle retarder. The light emerging from this radial velocity meter is fully linearly polarized with the position angle changing by 180° over every $\Delta\lambda = 6.5$ Å at Hγ. To avoid the shifts of the light beam across the wide-angle retarder, an

---

[9] See p. 237.

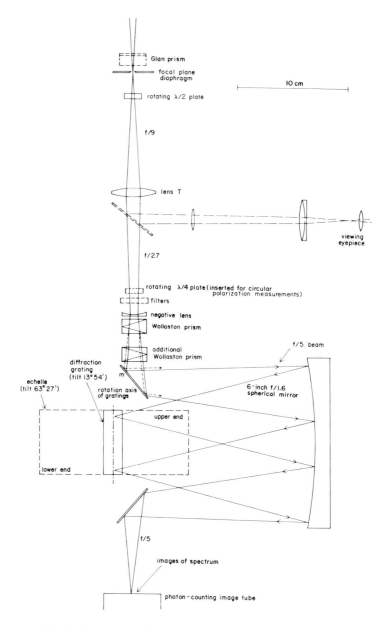

Fig. 4. A scanning polarimeter in narrow-band mode of operation.

image of the telescope mirror, about 2 cm in diameter, is formed inside the retarder.

For determining stellar radial velocities, the polarization measurements with the Digicon will be made with a star image centered in a large focal plane diaphragm. A seeing disk 2 arcsecs in diameter will correspond to a spectral purity of about 0.3 Å. The spectrophotometry of a spectral region about 30 Å wide will be made at several settings of the deflection coil in the Digicon, shifting the spectrum in steps corresponding to about 0.1 Å. With 5 mins of observing a 10-mag star with a 225-cm telescope, the radial velocity will be obtained with an accuracy of about ± 1 km/sec from a profile of a strong spectral line which is rotationally broadened to a half-width of about 4 Å. For an F-, G-, or K-type star of fifth magnitude or brighter, changes in radial velocity as small as a *few meters per second* may be detectable when several observations, each lasting about an hour, are intercompared. Numerous crowded absorption lines and bands in a spectral region about 30 Å wide around 0.43 μm will be used in this case for radial velocity determination.

The radial velocity meter can also be used without the spectrometer for studying the distribution of radial velocities over the surface of a *nebula* or *galaxy* in the light of a single emission line, Hα for example. The position angle of polarization in each point of a strip of the nebula imaged on the Digicon will indicate the radial velocity at that point. An error of radial velocity at each point equals

$$\varepsilon(V_r) = \frac{c}{\lambda} \frac{\Delta\lambda}{\pi(2N)^{\frac{1}{2}}}, \qquad (26)$$

where c is the velocity of light, $\Delta\lambda$ is a wavelength interval over which the position angle of light emerging from the radial velocity meter changes by 180°, and N is the number of emission-line photons counted. Since for our retarder $\Delta\lambda = 15$ Å at Hα, we need to detect only 1000 emission-line photons from a small area of the nebula to obtain its radial velocity with ± 5 km/sec accuracy.

With our scanning polarimeter, a photoelectric registration of the entire spectrum from 0.3 to 0.8 μm can be obtained from twelve integrations with the Digicon, made with different order-separating filters and at different tilts of the grating. Resolution of about 5 Å between 0.3 and 0.5 μm (2nd order) and 10 Å between 0.5 and 0.8 μm (1st order) will be obtained with a focal plane diaphragm 5 arcsecs in diameter. With 5 minutes of observing, the spectral scan with this resolution will have an accuracy of about $\pm 2\%$ per resolution element from 0.35 to 0.7 μm for a 12-mag early-type star observed with a 225-cm telescope. By obtaining such spectrophotometric data every time that the wavelength dependence of polarization is observed for a variable star, more meaningful conclusions can be drawn from the polarimetric measurements than is possible at the present time.

Acknowledgments. This work was supported by the National Science Foundation.

APPENDIXES: STARS THAT CAN BE USED AS POLARIMETRIC STANDARDS

Appendix 1

Unpolarized Nearby Standard Stars Observed with Rotatable Tube Telescopes

HD	Star	α_{1975}	δ_{1975}	m_v	Spectral type MK	Distance (pc)	Percentage polarization[a] $p(\%)$ m.e.	Position angle θ(degrees)	Reference number[b]
432	β Cas	0h 07.8m	+59°01′	2.2	F2 IV	14	0.009 ± .009	32	1
10476	107 Psc	1 41.2	+20 10	5.2	K1 V	8	.016 ± .006	175	2
20630	κ Cet	3 18.1	+3 17	4.8	G5 V	10	.006 ± .008	135	3
38393	γ Lep A	5 43.5	−22 27	3.6	F6 V	8	.005 ± .008	130	2, 3
39587	χ^1 Ori	5 52.9	+20 16	4.4	G0 V	10	.013 ± .007	20	2
43834	α Men	6 11.0	−74 45	5.1	G5 V	9	.009 ± .010	142	2, 3
61421	α CMi	7 38.1	+5 19	0.3	F5 IV	4	.005 ± .009	145	2, 3
100623	−32°8179	11 33.3	−32 42	6.0	K0 V	10	.016 ± .012	57	2
102870	β Vir	11 49.4	+1 55	3.6	F8 V	10	.017 ± .014	162	2
114710	β Com	13 10.8	+28 00	4.3	G0 V	8	.018 ± .014	116	2
115617	61 Vir	13 17.2	−18 08	4.8	G6 V	9	.010 ± .006	132	2, 3
142373	χ Her	15 51.9	+42 30	4.6	F9 V	18	.012 ± .009	31	4
155885/6	36 Oph AB	17 13.9	−26 32	4.3	K1 V	5	.005 ± .007	61	3
165908	99 Her AB	18 06.2	+30 33	5.0	F7 V	17	.002 ± .007	39	4
185395	θ Cyg	19 35.9	+50 09	4.5	F4 V	15	.003 ± .007	139	4
188512	β Aql	19 54.1	+6 20	3.7	G8 IV	14	.012 ± .005	154	4
198149	η Cep	20 44.8	+61 44	3.4	K0 IV	14	.006 ± .005	101	4
209100	ε Ind	22 01.4	−56 53	4.7	K5 V	4	.006 ± .008	88	2, 3
210027	ι Peg	22 05.9	+25 13	3.8	F5 V	14	.002 ± .006	45	4
216956	α PsA	22 56.0	−29 45	1.2	A3 V	7	0.006 ± .009	89	3

[a] Spectral region 0.4 to 0.6 μm.
[b] Reference numbers: 1. Appenzeller 1966; 2. Serkowski, Mathewson, and Ford 1973; 3. Serkowski 1968; 4. Walborn 1968.

Appendix 2

Nearby Stars of Spectral Type G Fainter than 6.5 Mag

These stars, which may be useful for determining the instrumental polarization using the photon-counting equipment, were selected from the catalogue by Gliese (1969). Polarization of these stars has not as yet been observed with rotatable tube telescopes.

HD	BD CoD	α_{1975}	δ_{1975}	m_v	Spectral type MK	Distance (pc)
9540	$-24°$ 658	$1^h\ 32.1^m$	$-24°18'$	7.0	G8 V	16
9407	$+68°$ 113	1 32.7	$+68\ 37$	6.5	G6 V	20
18803	$+26°$ 503	3 01.0	$+26\ 31$	6.7	G6 V	18
42807	$+10°1050$	6 11.8	$+10\ 39$	6.5	G6 V	18
65583	$+29°1664$	7 59.0	$+29\ 18$	7.0	G8 V	17
90508	$+49°1961$	10 26.5	$+48\ 56$	6.5	G1 V	19
98281	$-4°3049$	11 17.0	$-4\ 55$	7.3	G8 V	20
102438	$-29°9337$	11 46.0	$-30\ 09$	6.5	G5 V	18
103095[a]	$+38°2285$	11 51.4	$+37\ 56$	6.5	G8 VI	9
125184	$-6°3964$	14 16.7	$-7\ 25$	6.5	G8 V	16
144287	$+25°3020$	16 03.0	$+25\ 19$	7.1	G8 V	19
144579	$+39°2947$	16 04.1	$+39\ 16$	6.7	G8 V	12
154345	$+47°2420$	17 01.9	$+47\ 06$	6.8	G8 V	16
202573	$+24°4357$	21 14.9	$+25\ 20$	7.0	G5 V	18
202940	$-26°15541$	21 18.4	$-26\ 29$	6.6	G5 V	19

[a] $p = 0.005\% \pm 0.037\%$ (m.e.), $\theta = 137°$ according to Appenzeller (1968).

Appendix 3

Standard Stars with Large Interstellar Polarization

In this table, λ_{max} denotes the wavelength of maximum linear polarization taken from the paper by Serkowski, Mathewson, and Ford (1973). Errors of $p(\lambda_{max})$ and $\theta(\lambda_{max})$ are likely to be less than $\pm 0.1\%$ and $\pm 1°$, respectively. The degrees of polarization $p(\lambda)$ at wavelengths λ other than the wavelength of maximum polarization, λ_{max}, given in this table can be calculated from the formula describing the wavelength dependence (Fig. 5) of interstellar polarization (cf. Serkowski 1971)[a]:

$$p(\lambda)/p(\lambda_{max}) = \exp[-1.15\, ln^2(\lambda_{max}/\lambda)]. \qquad (27)$$

HD	Star	α_{1975}	δ_{1975}	m_v	Spectral type MK	λ_{max} (μm)	$p(\lambda_{max})$ (%)	$\theta(\lambda_{max})$ (degrees)
7927	φ Cas	1ʰ 18.5ᵐ	+58°05′	5.0	F0 Ia	0.51	3.4	94
14433	+56°568	2 20.1	+57 07	6.4	A1 Ia	0.51	3.9	112
21291	2H Cam	3 27.0	+59 51	4.2	B9 Ia	0.53	3.5	115
23512	+23°524	3 45.1	+23 33	8.1	A0 V	0.61	2.3	30
43384	9 Gem	6 15.4	+23 44	6.2	B3 Ia	0.53	3.0	170
80558	HR 3708	9 17.9	−51 27	5.9	B7 Iab	0.61	3.3	162
84810	ℓ Car	9 44.5	−62 23	3.3–4.0	F0–K0 Ib	0.57	1.6	100
111613	HR 4876	12 49.8	−60 12	5.7	A1 Ia	0.56	3.2	81
147084	o Sco	16 19.1	−24 07	4.6	A5 II–III	0.68	4.3	32
154445	HR 6553	17 04.3	−0 51	5.7	B1 V	0.55	3.7	90
160529	−33°12361	17 40.2	−33 29	6.7	A2 Ia+	0.54	7.3	20
183143	+18°4085	19 26.3	+18 13	6.9	B7 Ia	0.56	6.4	0
187929	η Aql	19 51.2	+0 56	3.5–4.3	F6–G2 Ib	0.56	1.8	93
198478	55 Cyg	20 48.1	+46 01	4.9	B3 Ia	0.53	2.8	3
204827	+58°2272	21 28.3	+58 37	7.9	B0 V	0.47	5.7	60

[a] Where the formula is incorrectly printed, with the exponent not superscripted.

Fig. 5. Normalized wavelength dependence of interstellar polarization. The solid line is calculated from Equation (27). Every open circle is an average of 60 observations, each of about 15 mins duration, made with a Wollaston polarimeter at the Lunar and Planetary Laboratory (Coyne and Gehrels 1967); an open circle at $\lambda_{max}/\lambda = 1.96$ is based on only 35 observations. The small filled circles are the observations of HD 183143 and star No. 12 in the association VI Cyg by Dyck*, small squares are the infrared and balloon ultraviolet observations of ζ Oph by T. Gehrels (1973), crosses are the observations of HD 38563 B, for which $\lambda_{max} = 1.04$ μm, by B. Zellner (unpublished).

*See p. 859.

REFERENCES

Angel, J. R. P., and Landstreet, J. D. 1970. Magnetic observations of white dwarfs. *Astrophys. J. (Letters)* 160: L147–L152.

Appenzeller, I. 1966. Polarimetrische, photometrische und spektroskopische Beobachtungen von Sternen im Cygnus und Orion. *Zeitschr. f. Astrophys.* 64: 269–295.

———. 1967. A new polarimeter for faint astronomical objects. *Publ. Astron. Soc. Pac.* 79: 136–139.

———. 1968. Polarimetric observations of nearby stars in the directions of the galactic poles and the galactic plane. *Astrophys. J.* 151: 907–918.

Babcock, H. W. 1953. The solar magnetograph. *Astrophys. J.* 118: 387–396.

Beaver, E. A.; McIlwain, C. E.; Choisser, J. P.; and Wysoczanski, W. 1972a. Counting image tube photoelectrons with semiconductor diodes. *Adv. Electronics & Electron Physics* 33B: 863–871.

Beaver, E. A.; Burbidge, E. M.; McIlwain, C. E.; Epps, H. W.; and Strittmatter, P. A. 1972b. Digicon spectrophotometry of the quasi-stellar object PHL 957. *Astrophys. J.* 178: 95–103.

Beckers, J. M. 1968. Principles of operation of solar magnetographs. *Solar Phys.* 5: 15–28.

———. 1971a. Achromatic linear retarders. *Applied Optics* 10: 973–975.

———. 1971b. The measurement of solar magnetic fields. *Solar magnetic fields.* (R. Howard, ed.), pp. 3–23. Dordrecht, Holland: D. Reidel Publ. Co.

———. 1972. Achromatic linear retarders with increased angular aperture. *Applied Optics* 11: 681–682.

Behr, A. 1956. Eine differentielle Methode der lichtelektrischen Polarisations messung des Sternlichts. *Veröff. Univ.-Sternw. Göttingen* 6: (No. 114) 401–435.

———. 1968. The polarimeter of the 1 m photometric telescope. *ESO Bull.* No. 5: pp. 9–13.

Billings, B. H. 1951. A monochromatic depolarizer. *J. Opt. Soc. Amer.* 41: 966–975.

———. 1952. The electro-optic effect in uniaxial crystals of the dihydrogen phosphate type. *J. Opt. Soc. Amer.* 42: 12–20.

Breckinridge, J. B. 1971. Polarization properties of a grating spectrograph. *Applied Optics* 10: 286–294.

Bugaenko, O. I.; Galkin, L. S.; and Morozhenko, A. V. 1971. Polarimetric observations of the major planets. *Astron. Zh.* 48: 373–379. (Transl. *Sov. Astr. A.J.* 15: 290–295).

Clarke, D. 1965. Studies in astronomical polarimetry. *Mon. Not. R. Astr. Soc.* 130: 75–81.

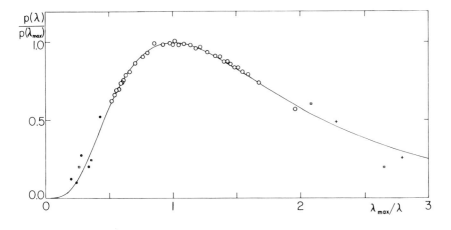

Clarke, D., and Grainger, J. F. 1971. *Polarized light and optical measurement*. Oxford: Pergamon.

Coyne, G. V., and Gehrels, T. 1967. Wavelength dependence of polarization. X. Interstellar polarization. *Astron. J.* 72: 887–898.

Dollfus, A., 1963. Un modulateur de lumière polarisée pour polarimètre photoélectrique. *C. R. Acad. Sci. Paris* 256: 1920–1922.

Dombrovskij, V. A.; Gagen-Torn, V. A.; Gutkevich, S. M.; Poljakova, T. A.; Svechnikov, M. A.; and Shulov, O. S. 1965. A 20-inch telescope with astrophotometer for photometric, colorimetric, and polarimetric studies. *Trudy Astron. Observ. Leningrad* 22: 83–94.

Dyck, H. M.; Forbes, F. F.; and Shawl, S. J. 1971. Polarimetry of red and infrared stars at 1 to 4 microns. *Astron. J.* 76: 901–915.

Dyck, H. M., and Sandford, M. T. 1971. Multicolor polarimetry of some Mira variables. *Astron. J.* 76: 43–49.

Elvius, A., and Engberg, M. 1967. A two-cell photoelectric polarimeter. *Arkiv f. Astron.* 4: 387–403.

Gehrels, T. 1972. On the circular polarization of HD 226868, NGC 1068, NGC 4151, 3C273, and VY Canis Majoris. *Astrophys. J. (Letters)* 173: L23–L25.

———. 1973. Wavelength dependence of polarization. Interstellar polarization from 0.22 to 2.2 μm. *Astron. J.* In preparation.

Gehrels, T., and Teska, T. M. 1960. A Wollaston polarimeter. *Publ. Astron. Soc. Pac.* 72: 115–122.

Gex, F., and Tassart, J. 1971. Influence de la température sur la réponse des photomultiplicateurs à couche photosensible S 11. *Nouv. Rev. d'Optique appliquée* 2: 175–178.

Gliese, W. 1969. Catalogue of nearby stars. *Veröff. Astr. Rechen-Inst. Heidelberg* Nr. 22: 1–117.

Hall, J. S. 1968. A scanning polarimeter. *Lowell Obs. Bull.* 7: (No. 143) 61–66.

Hall, J. S., and Mikesell, A. H. 1950. Polarization of light in the galaxy. *Publ. U. S. Naval Obs.* 17/1: 7–62.

Hämeen-Anttila, J. 1972. An inexpensive astronomical control computer system. *Publ. Astron. Soc. Pac.* 84: 185–189.

Hashimoto, J.; Maihara, T.; Okuda, H.; and Sato, S. 1970. Infrared polarization of the peculiar M-type variable VY Canis Majoris. *Publ. Astron. Soc. Japan* 22: 335–340.

Hiltner, W. A. 1956. Photometric, polarization, and spectrographic observations of O and B stars. *Astrophys. J. Suppl.* 2: 389–462.

Hiltner, W. A., and Schild, R. 1965. A rotatable telescope for polarization studies. *Sky and Telescope* 30: 144–147.

Holmes, D. A. 1964. Exact theory of retardation plates. *J. Opt. Soc. Amer.* 54: 1115–1120.

Ikhsanov, R. N., and Platonov, Y. P. 1967. The new magnetograph of the Pulkovo Observatory. *Solnechnye dannye* 1967, No. 11, pp. 78–89.

Jeppesen, M. A. 1958. Some optical, thermo-optical and piezo-optical properties of synthetic sapphire. *J. Opt. Soc. Amer.* 48: 629–632.

Kemp, J. C. 1969. Piezo-optical birefringence modulators: new use for a long known effect. *J. Opt. Soc. Amer.* 59: 950–954.

Kemp, J. C.; Wolstencroft, R. D.; and Swedlund, J. B. 1972. Circular polarimetry of fifteen interesting objects. *Astrophys. J.* 177: 177–189.

Ksanfomaliti, L. V., and Dzhapiashvili, V. P. 1973. Polarovisor. *Astr. Zh.* 50: 357–361.

Livingston, W., and Harvey, J. 1971. The Kitt Peak magnetograph. *Kitt Peak Contr.* No. 558, pp. 1–22.

Lyot, B. 1948. Un polarimètre photoélectrique. *C. R. Acad. Sci. Paris* 226: 25–28.

Marin, M. 1965. Mesures photoélectriques de polarisation à l'aide d'un télescope coudé de 1 m. *Rev. Optique* 44: 115–144.

Nordsieck, K. H. 1972. A search for circular polarization in extragalactic objects. *Astrophys. Letters* 12: 69–74.
Öhman, Y. 1943. On some applications of the flicker method in photoelectric measurements. *Stockholms Obs. Meddelande* No. 54. pp. 1–7.
Orrall, F. Q. 1971. A complete Stokes vector polarimeter. *Solar magnetic fields*. (R. Howard, ed.), pp. 30–36. Dordrecht, Holland: D. Reidel Publ. Co.
Oskanjan, V.; Kubičela, A.; and Arsenijević, J., 1969. Polarimètre photoélectrique de l'Observatoire Astronomique de Belgrade. *Bull. Obs. Astron. Belgrade* 27 (No. 2): 1–11.
Palik, E. D. 1965. A Soleil compensator for the far infrared. *Applied Optics* 4: 1017–1021.
Pancharatnam, S. 1955. Achromatic combinations of birefringent plates. II. *Proc. Indian Acad. Sci.* A41: 137–144.
Pospergelis, M. M. 1965. The "Taimyr" electronic polarimeter. *Astron. Zh.* 42: 398–408. (Transl. *Sov. Astr. A.J.* 9: 313–321).
Poulsen, O. 1972. Polarization effects in grating spectrometers. *Applied Optics* 11: 1876–1878.
Ramachandran, G. N., and Ramaseshan, S. 1961. Crystal optics. *Handbuch der Physik*. (S. Flügge, ed.) vol. 25/1, pp. 1–217. Berlin: Springer Verlag.
Serkowski, K. 1962. Polarization of starlight. *Adv. Astron. Astroph.* 1: 289–352.
———. 1968. Correlation between the regional variations in wavelength dependence of interstellar extinction and polarization. *Astrophys. J.* 154: 115–134.
———. 1970. Neutrality of extinction by atmospheric clouds in UBVR spectral regions. *Publ. Astron. Soc. Pac.* 82: 908–910.
———. 1971. Polarization of variable stars. Proc. I. A. U. Colloquium No. 15, *Veröff. Remeis Sternw. Bamberg,* 9 (No. 100): 11–31.
———. 1972. A polarimetric method of measuring radial velocities. *Publ. Astr. Soc. Pac.* 84: 649–651.
———. 1973a. Polarization techniques. *Methods of experimental physics, Vol. 12: Astrophysics, Part A.* (M. L. Meeks and N. P. Carleton, eds.) New York: Academic Press. In press.
———. 1973b. Infrared circular polarization of NML Cygni and VY Canis Majoris. *Astroph. J. (Letters)* 179: L101–L106.
Serkowski, K., and Chojnacki, W. 1969. Polarimetric observations of magnetic stars with two-channel polarimeter. *Astron. Astrophys.* 1: 442–448.
Serkowski, K.; Mathewson, D. S.; and Ford, V. L. 1973. Observations of the wavelength dependence of interstellar polarization with a multi-channel polarimeter. *Astrophys. J.* In preparation.
Shakhovskoj, N. M. 1971. Methods of studying polarization of variable stars. (In Russian) *Methods of studying variable stars.* (V. B. Nikonov, ed.), pp. 199–224. Moscow.
Shakhovskoj, N. M., and Efimov, Yu. S. 1973. *Izv. Crimean Astrophys. Obs.* 45: In press.
Soref, R. A., and McMahon, D. H. 1966. Optical design of Wollaston-prism light deflectors. *Applied Optics* 5: 425–434.
Thiessen, G., and Broglia, P. 1959. Über einen Polarisationseffekt an aufgedampften Aluminiumschichten bei senkrechter Lichtinzidenz. *Zeit. f. Astrophys.* 48: 81–87.
Tinbergen, J. 1972. Achromatic polarization modulators for multichannel polarimeters. Proc. of ESO/CERN Conference on "Auxiliary Instrumentation for Large Telescopes." (S. Laustsen and A. Reiz, eds.), pp. 463–471. Geneva.
———. 1973. Precision spectropolarimetry of starlight. *Astron. Astrophys.* 23: 25–48.
Young, A. T. 1967. Photometric error analysis. VI. *Astron. J.* 72: 747–753.
Visvanathan, N. 1972. An automatic fast digital-photoelectric photometer with polarimeter. *Publ. Astron. Soc. Pac.* 84: 248–253.
Walborn, N. R. 1968. Twenty-five polarimetric standard stars within twenty parsecs of the sun. *Publ. Astron. Soc. Pac.* 80: 162–164.

Walsh, T. E. 1972. Birefringence of cadmium sulphide single crystals. *J. Opt. Soc. Am.* 62: 81–83.

Wirick, M. P. 1966. The near ultraviolet optical constants of lanthanum fluoride. *Applied Optics* 5: 1966–1967.

DISCUSSION

SCHMIDT: Using a rotating phase plate with a retardation angle τ defined by $\sin \tau = \sin^2 \tau/2 \equiv 0.8$, or $\tau = 126°9 = 0.352\lambda$, followed by a fixed polarizer, it is possible to measure the linear and circular polarization simultaneously. The polarization amplitudes are equal to 80% of the possible maximum, analyzing the 4ω and 2ω oscillations, respectively, where ω is the frequency of rotation.

SERKOWSKI: The polarimeter that you are proposing gives the same precision for all three Stokes parameters describing polarization: Q, U, and V. According to my definitions of efficiency, it will have $E_{\text{lin}} = 0.64$ and $E_{\text{cir}} = 0.32$.

CLARKE: I would like to point out the danger of using rotating wave plates before an angular-dispersive spectrometer. If the plate is not exactly normal to the rotational axis, small displacements of the light rays can occur during the polarimetric analysis; this produces a "spectral jitter" and may invalidate the polarimetry.

SERKOWSKI: To prevent such a jitter, the rotating retarder should be very close to the slit of the spectrometer; the amount of jitter is proportional to the distance between them. When my new polarimeter is used for spectropolarimetry, the rotating half-wave plate will be located only a few millimeters below the focal plane diaphragm.

WOLSTENCROFT: Astronomers and physicists using rotating elements in polarimeters used to worry about the "flicker effect," i.e., the spurious polarization introduced when there is a speck of dust or blemish on the rotating element. How serious a problem is this in your instrument?

SERKOWSKI: Such a speck of dust on a rotating half-wave plate will produce a modulation of signal that can be expressed as a Fourier series. Only the term corresponding to a frequency four times higher than the frequency of revolution of the half-wave plate may produce a systematic error in the measured linear polarization. This systematic error will be eliminated by averaging the results obtained with a three-component Wollaston prism at two orientations differing by 180°.

APPLICATION OF MUELLER CALCULUS IN ASTRONOMICAL POLARIMETRY: ACHROMATIC MODULATORS AND POLARIZATION CONVERTERS, AND DEPOLARIZERS

J. TINBERGEN
Sterrewacht te Leiden
Netherlands

The Mueller matrix calculus is the most important tool in the design of astronomical polarimeters. This paper discusses achromatic modulators and polarization converters for multichannel polarimetry of high precision. The Mueller calculus is discussed in relation to depolarizer design.

ACHROMATIC MODULATORS

For reliable results in precision polarimetry, modulation methods are the obvious choice. Many modulators have been devised (Hall and Mikesell 1950; Takazaki, Okazaki, and Kida 1964; Pospergelis 1965; Kemp 1969; Angel and Landstreet 1970). The more successful of these employ time-varying "retarders," such as on-off or reversing quarter-wave or half-wave plates. Because of this, polarization modulators are generally more-or-less monochromatic, though some of them can be tuned.

Astronomical polarization measurements need large numbers of photons therefore long observing times; there is thus a strong case for multichannel instruments, as I have stressed before (Tinbergen 1972*a,b*, 1973). The wavelength-splitting components generally affect the state of polarization; hence, the modulator must precede them and it must be achromatic, i.e., effective over a wide range of wavelengths. We wish to know how one constructs achromatic modulators and how well they work.

The basic tool for computations is the Mueller calculus. (Shurcliff [1962] is the best introduction.) The four Stokes parameters of the output beam of an optical component are related to those of the input beam by a 4 × 4 "Mueller matrix." The effect of a series of components is represented by multiplication of the matrices of the components. Rotation of a component about the optical axis is represented by multiplication by suitable rotation matrices.

The last component of a polarimeter is always a linear polarizer; this constrains the polarimeter matrix to the form

$$\tfrac{1}{2} \cdot \begin{bmatrix} A & B & C & D \\ A & B & C & D \\ 0 & 0 & 0 & 0 \\ 0 & 0 & 0 & 0 \end{bmatrix}.$$

The top row of the matrix determines the intensity striking the detector; for a modulation polarimeter,

	(a)	(b)	(c)
linear	$(B,C)/A = \mathrm{f}(t)$	$\mathrm{Alt}(D)/A \approx \epsilon$	$\mathrm{Alt}(A)/A \approx \epsilon^2$
circular	$D/A = \mathrm{f}(t)$	$\mathrm{Alt}(B,C)/A \approx \epsilon$	$\mathrm{Alt}(A)/A \approx \epsilon^2$,

where $\mathrm{f}(t)$ is an alternating function of time of amplitude of order unity, A is the transmission for unpolarized light, Alt (x) is short for "the part of x that alternates with the same frequency as $\mathrm{f}(t)$," and ϵ is a small quantity (hopefully ≈ 0.01). Condition (a) expresses the *modulation efficiency;* conditions (b) and (c) are *spurious sensitivities* to polarized and unpolarized radiation, respectively. Which condition is the most important will depend on the particular application one has in mind.

For the basic modulating component, the most likely choice is a half-wave retarder, since this is simplest to construct in time-varying achromatic form. One may use it in light that is linearly polarized (Fig. 1) or circularly polarized (Fig. 2); component 1 in each figure is not part of the basic modulator but serves to convert from linear to circular polarization or vice versa. The main differences between the two types are:

1. The circular modulator is square-wave in its action, whereas the linear one is sinusoidal. For circular polarization or for linear polarization of known orientation, the circular modulator is therefore the more efficient. The linear modulator requires two *de*modulators per channel, the circular only one.

2. The linear modulator depends on mechanical rotation, making the modulation frequency a few hundred Hz at most. The circular modulator could be made in electro-optic form and could modulate at MHz rate, if this is desired.

3. The linear modulator is the simpler to construct in very compact form while still retaining its high quality of performance.

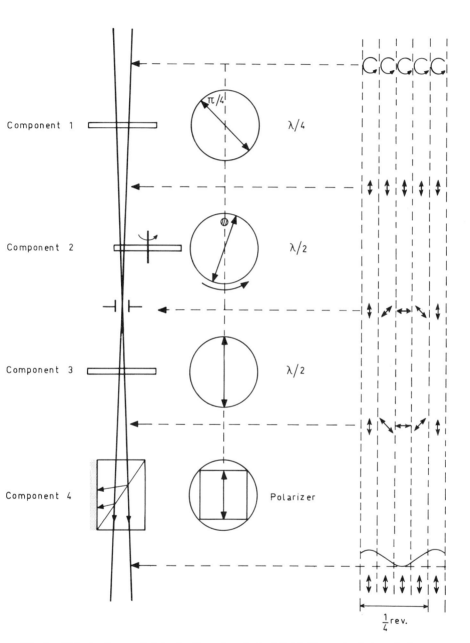

Fig. 1. Polarization modulator for linearly polarized light. The state of the polarization information is sketched on the right, for 100% polarized radiation. (Figure from *Proc. ESO/CERN Conf.*, 1972.)

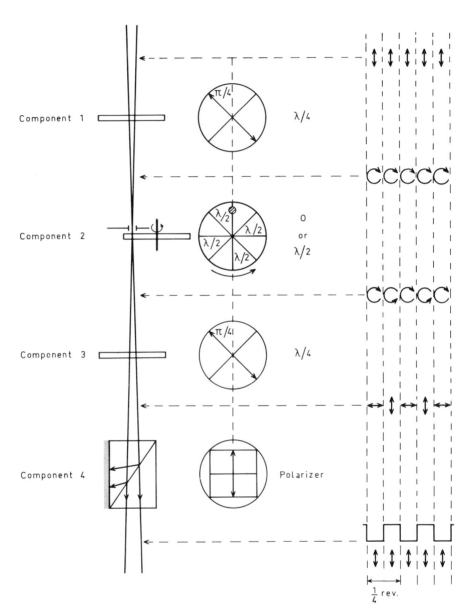

Fig. 2. As for Fig. 1, but for circularly polarized light.

The usual procedure has been to start with an arrangement that works well with ideal components, then to put all imperfections into the matrices and obtain the polarimeter matrix by tedious multiplying out (e.g., Tinbergen 1973). Priebe (1969) uses an operational form of the Mueller calculus for finding equivalents (to first order) to the starting arrangement, which may then be investigated in detail by full multiplication. McIntyre and Harris (1968, and in a series of earlier papers cited there) describe analysis of the properties of complicated systems by the Jones calculus, but van de Hulst (1957) gives the transformation from Jones to Mueller matrices. Many of these computations can be programmed very easily for numerical work when the analytical expressions become too complicated. Mueller matrices in graphical or pictorial form are a concise way of showing the complete polarization behavior of systems or components (Tinbergen 1973).

It seems feasible (Serkowski 1972; Tinbergen 1973) to construct a modulator for the range 0.3–1 μm, which has a modulation efficiency of more than 0.7 (in fact, for most of the range, > 0.95), spurious sensitivity to polarized radiation of less than 3×10^{-3}, and spurious sensitivity to unpolarized radiation of less than 10^{-4} or 10^{-5} (in other words, $\epsilon < 0.01$). Such a modulator allows one to convert an existing multichannel photometer or spectrometer into a precision polarimeter.

ACHROMATIC POLARIZATION CONVERTERS

The achromatic components one may use are of three types, each of which has its own uses; a polarimeter I am using in fact incorporates all three. They have been discussed more fully elsewhere (Tinbergen 1973).

1. *Total-internal-reflection retarders.* The Fresnel rhomb is the oldest and best known, but better types have been invented (Clapham, Downs, and King 1969; Bennett 1970), some of which are actually too good for use in a telescope instrument, where they are often exposed to outside air and collect surface films, spoiling their performance. Figure 3 shows the spectral properties of fused-silica devices using various angles of incidence.

2. *Combinations of two birefringent materials,* such as quartz and magnesium fluoride (Clarke 1967; Chandrasekharan and Damany 1968; Beckers 1971, 1972). Figure 4 shows how one such design performs, but within wide limits the quality of the achromatism may be exchanged for a spectral range to suit a particular application.

3. *Combinations of several layers of the same material.* Pancharatnam (1955) uses three layers, not necessarily of the same thickness; McIntyre and Harris (1968) use any number of layers, but all of the same thickness (the general case has so far defied analysis). Some of the resultant assemblies are analogous to Solc filters of one kind or another (Tinbergen 1973). Some of the assemblies are equivalent to a single wave plate with retardation and orientation that may both depend slightly on wavelength, but others have completely different properties and can only be called "polarization converters" (Fig. 5; see also Table 2 in Tinbergen 1973).

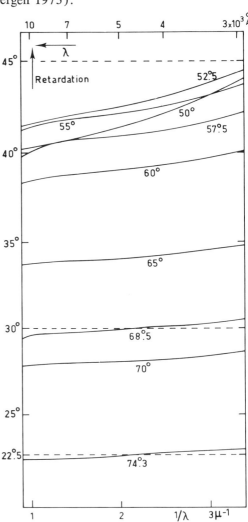

Fig. 3. Retardation for one total internal reflection on an interface from fused silica to air. The curves are labeled with their angle of incidence.

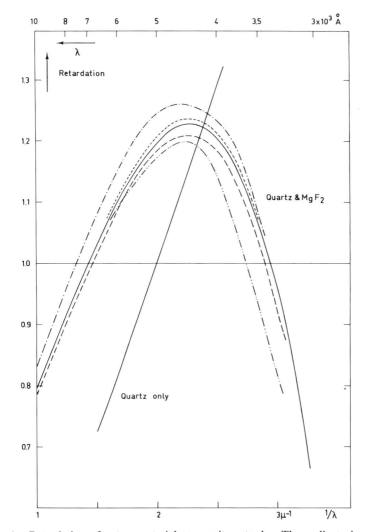

Fig. 4. Retardation of a two-material composite retarder. The ordinate is actual retardation divided by nominal retardation. (Figure from *Astron. Astrophys.*, Tinbergen, 1973)

——————: quartz only (nominal retardation at 5000 Å)
——————: MgF_2-and-quartz composite (nominal retardation at 3400 Å and 7000 Å, computed for 20 °C)
— — — — — —: the same retarder, now at 0 °C
- - - - - - - -: the same retarder, but for a ray 2° from the normal (at an azimuth 90° different, the effect will be in the other direction)
—·—·—·—·—: the composite retarder with construction errors; the thickness of MgF_2 increased by 10% that of quartz by 10.5%.
—··—··—··—: the same, but the thickness of the quartz increased by 11%.

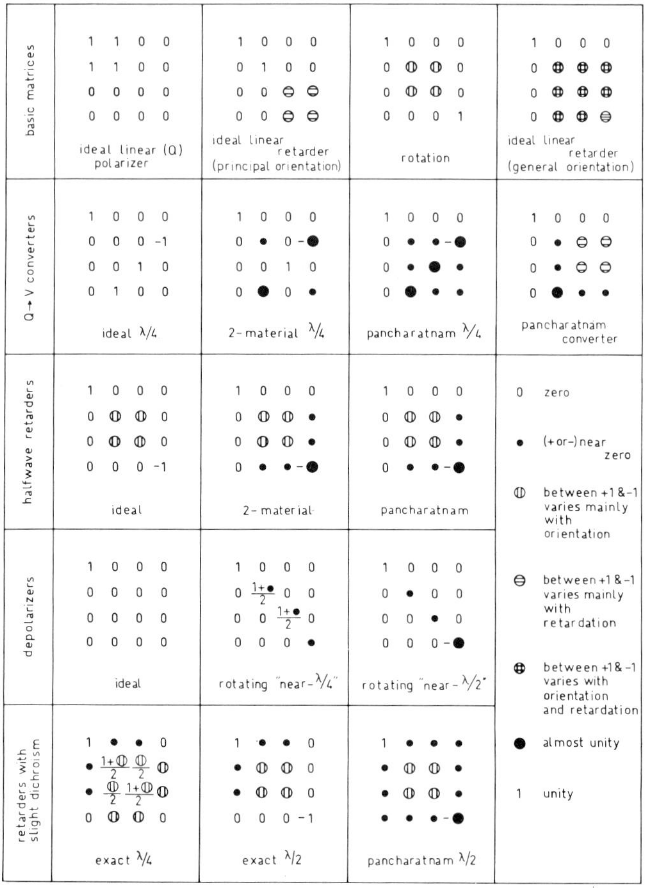

Fig. 5. General form of selected Mueller matrices. Constant factors outside the matrix have been neglected. Dichroism has been neglected, except where stated. (Figure from *Astron. Astrophys.,* Tinbergen, 1973)

Types 2 and 3 may be combined to form very achromatic components (Fig. 6 and Serkowski 1972); Serkowski and I have such components on order from two different firms. Components of type 1 are bulky and often displace the beam, so they are not suitable for applications where they would have to be rotated. Components of types 1 and 2 have principal planes that are constant with wavelength; for those of type 3 the orientation of the principal planes is a function of wavelength. Components of types 2 and 3 can be very thin and compact. Type 3 is the most suitable for the time-varying component in the circular modulator (though this could change in favor of type 2 as new materials become

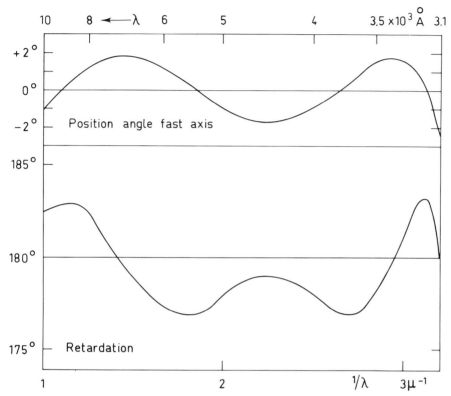

Fig. 6. Wavelength dependence of retardation and orientation of fast axis of a 6-layer achromatic half-wave retarder. The retarder is a Pancharatnam combination of three identical half-wave retarders, each of which is a combination of magnesium fluoride and quartz. (Figure from *Proc. ESO/CERN Conf.*, 1972.)

available); Mueller matrix analysis and physical layout of the modulator determine which type is best for each of the other components.

DEPOLARIZERS

In many older polarimeters, measurements with a depolarizer in the beam were used for calibration. This practice is discouraged, since it is unreliable and, for most types of polarimeter, unnecessary. However, there may be an occasion sometimes to put a depolarizer just in front of the photomultiplier, or to use a linear or circular depolarizer to improve performance of a modulator. If one does use a depolarizer, its action must be thoroughly understood. Mueller calculus is suitable for this, provided the conditions are such that intensities (Stokes parameters) may be added. The discussion here will be brief; a more detailed one is being prepared for publication.

The Mueller matrix definition of a depolarizer (generalized from Billings 1951) is:

$$[D] \equiv \int\int\int\int [M] \cdot d\lambda\, dt\, dA\, d\theta = \begin{bmatrix} d_{11} & 0 & 0 & 0 \\ 0 & 0 & 0 & 0 \\ 0 & 0 & 0 & 0 \\ 0 & 0 & 0 & 0 \end{bmatrix},$$

where $[D]$ is the average Mueller matrix for the depolarizer
$[M(\lambda,t,A,\theta)]$ is the Mueller matrix of the component to be used as a depolarizer.

$\left.\begin{array}{l} d\lambda \\ dt \\ dA \\ d\theta \end{array}\right\}$ are elements of $\left\{\begin{array}{l} \text{wavelength} \\ \text{time} \\ \text{cross-sectional area of the beam} \\ \text{position angle of the component.} \end{array}\right.$

The limits of integration are determined by the experimental conditions, and $[D]$ is normalized in such a way that d_{11} is unity when absorption is absent. $[M]$ may be a function of even more variables (such as, for instance, the direction of a ray through the element of area), and the definition may be extended accordingly. Not all the variables are necessarily independent (for instance, one might take $\theta = \theta(t)$, as will be mentioned). Which variable(s) of integration one chooses to use depends on the light beam to be depolarized: normally, the elements of the matrix $[M]$ will be much faster functions of the variable of integration than the Stokes parameters of the beam that is to be depolarized.

Single retarders can never be complete depolarizers since some of the elements of their matrices contain constant terms and therefore will not integrate to zero. One may construct two-component depolarizers whose action is complete. Those of Lyot (1948) for wavelength integration and Billings (1951) for time integration are of this kind. Other two-component depolarizers are possible.

In the modulator of Fig. 2, the spurious sensitivities can be much reduced if linear polarization of the beam is largely eliminated just before component 2. A suitable linear depolarizer for this is a slowly rotating retarder, close to half wave, for which the Mueller matrix, averaged over an integral number of half-revolutions, is

$$\begin{bmatrix} 1 & 0 & 0 & 0 \\ 0 & l & 0 & 0 \\ 0 & 0 & l & 0 \\ 0 & 0 & 0 & -\cos b \end{bmatrix},$$

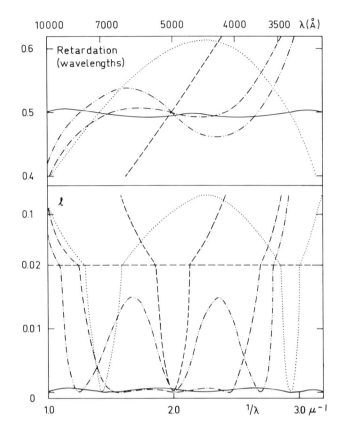

Fig. 7. Performance of various achromatic half-wave plates as linear depolarizers:
– – – – – –: single quartz plate cut for 5000 Å.
. . . .: quartz and magnesium fluoride, designed for exact half-wave retardation at 3400 Å and 7000 Å.
—·—·—·—: Pancharatnam combination, using only quartz, for central wavelength 5000 Å.
—··—··—··—: the same, but for a wider "achromatic range."
—————: the combination of Fig. 6.

where $l = (1 - \cos b)/2 \approx b^2/4$ and $\pi - b$ is the (wavelength-dependent) retardation. Figure 7 shows the performance for different designs of a half-wave plate. It should be noted that this type of depolarizer is effective separately at each wavelength within a wide range; if it has sufficient optical quality, it is suitable for reducing the linear polarization of a Coudé telescope at the entrance to a spectrometer (Breckinridge 1971).[1]

---

[1] See p. 232.

Polarimetric design need no longer be an art practiced by the few; it can be a normal part of engineering, thanks mainly to the power of the Jones and Mueller calculus. It seems likely that we have made merely a beginning, particularly as regards components and systems that would only be of interest to astronomers.

REFERENCES

Angel, J. R. P., and Landstreet, J. D. 1970. Magnetic observations of white dwarfs. *Astrophys. J.* 160: L147–L152.

Beckers, J. M. 1971. Achromatic linear retarders. *Applied Optics* 10: 973–975.

———. 1972. Achromatic linear retarders with increased angular aperture. *Applied Optics* 11: 681–682.

Bennett, J. M. 1970. A critical evaluation of rhomb-type quarterwave retarders. *Applied Optics* 9: 2123–2129.

Billings, B. H. 1951. A monochromatic depolarizer. *J. Opt. Soc. Am.* 41: 966–975.

Breckinridge, J. B. 1971. Polarization properties of a grating spectrograph. *Applied Optics* 10: 286–294.

Chandrasekharan, V., and Damany, H. 1968. Birefringence of sapphire, magnesium fluoride and quartz in the ultraviolet, and retardation plates. *Applied Optics* 7: 939–941.

Clapham, P. B.; Downs, M. J.; and King, R. J. 1969. Some applications of thin films to polarization devices. *Applied Optics* 8: 1965–1974.

Clarke, D. 1967. Achromatic halfwave plates and linear polarization rotators. *Optica Acta* 14: 343–350.

Hall, J. S., and Mikesell, A. H. 1950. Polarization of light in the galaxy, as determined from observations of 551 early-type stars. *Pub. U.S. Naval Obs.* 17: Part I.

Hulst, H. C. van de. 1957. *Light scattering by small particles.* p. 44. New York: Wiley.

Kemp, J. C. 1969. Piezo-optical birefringence modulators: new use for a long-known effect. *J. Opt. Soc. Am.* 59: 950–954.

Lyot, B. 1929. Recherches sur la polarisation de la lumièrè des planètes et de quelques substances terrestres. *Ann. Obs. Paris* (Meudon) 8: Fasc. I, 102–103.

———. 1948. Un polarimètre photoélectrique intégrateur utilisant des cellules à multiplicateurs d'electrons. *C. R. Acad. Sci. Paris* 226: 137–140.

McIntyre, C. M., and Harris, S. E. 1968. Achromatic wave plates for the visible spectrum. *J. Opt. Soc. Am.* 58: 1575–1580.

Pancharatnam, S. 1955. Achromatic combinations of birefringent plates. *Proc. Indian Acad. Sci.* A41: 130–144.

Pospergelis, M. M. 1965 .The "Taimyr" electronic polarimeter. *Astron. Zhurnal* 42: 398–408 (*Sov. Astr. A. J.* 9: 313–321).

Priebe, J. R. 1969. Operational form of the Mueller matrices. *J. Opt. Soc. Am.* 59: 176–180.

Serkowski, K. 1972. Polarization techniques. Preprint of a chapter for *Methods of experimental physics, vol. 12: Astrophysics, Part A.* (M. L. Meeks and N. P. Carleton, eds.). New York: Academic Press.

Shurcliff, W. A. 1962. *Polarized light.* Cambridge, Mass.: Harvard University Press.

Takazaki, H.; Okazaki, N.; and Kida, K. 1964. An automatic polarimeter. II: Automatic polarimetry by means of an ADP polarization modulator. *Applied Optics* 3: 833–837.

Tinbergen, J. 1972*a*. Applications of the Dollfus polarization modulator in stellar polarimetry. *Veröff. Remeis Sternwarte Bamberg* IX, No. 100 (IAU Coll. No. 15), pp. 36–40.

———. 1972*b*. Achromatic polarization modulators for multichannel polarimeters. *Auxiliary instrumentation for large telescopes* (S. Laustsen and A. Reiz, eds.), pp. 463–471, ESO-Cern, Geneva.

———. 1973. Precision spectropolarimetry of starlight: Development of a wide-band version of the Dollfus polarization modulator. *Astron. Astrophys.* 23: 25–48.

DISCUSSION

WOLSTENCROFT: The exact theory of retardation plates that allows for internal reflections in the plates, which was carried out by Holmes (1964), shows that the retardation as a function of plate thickness is not a linear function but rather a sinusoidal function that has large amplitudes at short wavelengths superposed on a sloping line. The result is that the retardation is very sensitive to the slight variations of thickness over the surface of the plate. If your plate is not plane parallel to within a very small fraction of a wavelength, your wave plate will not behave as the classical theory predicts. Can you comment on this, and can you say whether the curves for achromatic retarders are theoretical or experimental?

SERKOWSKI: The theory by Holmes describes the effects of interference on the interface between the air and the birefringent material of the retarder. These effects become negligible if the retarder is enclosed between cover plates or thin film coatings with an index of refraction intermediate between the indices of the birefringent material for ordinary and extraordinary rays.

TINBERGEN: The curves are theoretical, without allowing for the effect described by Holmes. I have used Pancharatnam combinations of Polaroid plastic retarders which were uniformly achromatic over 10 cm or more; the combinations were airspaced so the Holmes effect should have been operative, unless the cover material meets the condition mentioned by Serkowski.

J. S. HALL: Why are the two Lyot depolarizers in series no better than one?

COFFEEN: In the laboratory, with 100% linear polarization incident, if one Lyot depolarizer would reduce this to 1%, for example, then with two in series the emergent polarization could be as high as 30% or more, depending on relative azimuth of the plates. Thus, the four retarder plates in series were no longer a good depolarizer, and one should treat this with the appropriate matrices for four separate components.

TINBERGEN: If the two depolarizers are identical, I am not surprised. In any case, the Lyot depolarizer is just a combination of two retarders (high retardation). Adding a second depolarizer disturbs the 1:2 relation between them, and one might expect to spoil the performance. Analogous to this is the case of two half-wave plates rotating together; they are not two linear depolarizers in series, but they merely rotate the plane of polarization by a constant angle.

DISCUSSION REFERENCE

Holmes, D. A. 1964. Exact theory of retardation plates. *J. Opt. Soc. Am.* 54: 1115–1120.

OPTICAL POLARIMETERS IN SPACE

DAVID L. COFFEEN
University of Arizona and Columbia University

> *Several dozen instruments have been carried aloft on aircraft, balloons, rockets, satellites, and spacecraft in order to measure linear and circular polarization of astronomical and terrestrial targets. These range from a hand-held visual polarimeter carried on a free hydrogen balloon, to an automatic instrument sent on a flyby of Jupiter. Several exemplary devices are discussed and a comprehensive tabulation of optical space polarimeters is presented.*

This is a review of instrumentation used for measurement of optical polarization from aircraft, balloon, rocket, satellite, and spacecraft platforms. The purpose is to survey the instrument techniques and to provide a source of references.

Only four parameters can be measured at a given wavelength, time, and direction in an incoherent beam of radiation: intensity, linear polarization, its position angle, and circular polarization. These are fundamental characteristics of the light beam. An instrument that by design is made sensitive to any of the three polarization parameters (i.e., Q, U, and/or V, in terms of Stokes parameters) is here considered a polarimeter. This review is restricted to passive polarimeters operating in the range 0.1–1000 μm. X-ray, radio, and radar polarimeters, as well as active optical instruments, are excluded.

Table I is an itemization of space polarimeters and their essential characteristics, with relevant literature references.[1] All known instruments are included, even though some failed during launch and some are still in the design stage.

---

[1] Throughout this paper, the position of polarimeters in Table I is indicated by the table part (A, B, C, D) and entry number, e.g. (A-1).

TABLE I
Optical Space Polarimeters A. BALLOON

Instrument	Altitude	Pointing	Analyzer	Calibration	Detector	Wavelength Coverage
1. Halle Univ. polarisc.	3–6 km	by hand	Savart polarisc.	—	A. Wigand	Visual
2. Obs. de Paris vis. fringe polarisc.	0–6 km	by hand	Lyot polarim. Savart polarisc.	Polariz. comp. thru incl. plate	A. Dollfus	Broadband red, yellow green, blue
3. Univ. of Minn. photom.	30 km	Az. scan by rotation	Rotating Polaroid	Standard source	S-13 PMT	Broadband blue and visual
4. Univ. of Ariz. auto. polarim.	35 km	2-axis rotation/ moon tracker	2 Woll. ADP prisms, 45° apart	Lyot depol., ground tests	4 S-13 PMT's	Intermed. band filts., 0.20–0.30 μm
5. Univ. of N. Mex. photom. polarim.	30 km	Balloon rotation	Rotating Polaroid	Ground tests	S-4 PMT	Broadband 0.40, 0.435 0.55 μm
6. U.C.L.A. gain-comp. polarim.	28 km	2-axis solar control	Rotating Glan prism (2.3 rps)	Ground tests	S-11 PMT	Intermed. band 0.36, 0.40, 0.50, 0.60 μm
7. Tübingen Univ. radiosonde	0–30 km	Fixed near zenith	Rotating Polaroid (1.5 rpm)	Neut. dens. filters	3 CdSe photocells	Broadband 0.43, 0.53, 0.72 μm simultaneo
8. Georgia Tech. spectrophotom.	19–30 km	Elev. scan in solar vert.; sun tracker	Rotating retard. plate (20 rps), & partial polarizer	Ground tests	2 PMT's in Ebert-Fastie spectrom.	20 Å resol. over 0.20– 0.40 μm range
9. Univ. of Ariz. POLARISCOPE	37 km	3-axis gyroservo control (±5 arcsec rms)/ star tracker	Rotatable ADP Woll. prism	Lyot depol., stars, ground tests	2 solar blind PMT's	Intermed. band filts., 0.20–0.30 μm

OPTICAL POLARIMETERS IN SPACE 191

Telescope aperture	Field of View	Mass of Polarimeter	Goals	Status	Reference
—	—	—	Terr. atmos. neut. points	3 May 1914 ascent; sun elev. +7° to −6°	Wigand (1917)
.1 cm	0°025 in center of 0°25 image	0.2 kg	Earth surf., atmos., clouds	Various ascents	Dollfus (1956, 1957)
.0 cm	10°	2.3 kg	Zod. light	—	Gillett (1966)
two .5 cm	1°3	17 kg	Moon in ultrav.; space prototype	3 ascents; observed moon once	Pellicori & Gray (1967)
) cm	1°8	2.7 kg	Zod. light	7 ascents 1965, 1966; polariz. at 25°–50° elong.	Regener & Vande Noord (1967); Vande Noord (1970)
—	3°0	7 kg	Terr. atmos.	1965, 1968 ascents; polar.-phase angle curves obtained	Rao (1968); Rao & Sekera (1967)
three cm	12°	—	Terr. aerosols vs. height	9 ascents, 1965–1966; deduced part. concentration vs. altitude	Unz (1969)
—	12°	—	Terr. atmos.	1964, 1965, 1966 ascents; sky spectra obtained	Hodgdon (1965); Hodgdon & Edwards (1966)
cm	2 arc-min	—	Planets & stars in ultrav.	4 ascents, 1966–1969; observed Venus, Mars, ζ Oph, Sirius, etc.	Gehrels (1967); Coffeen & Gehrels (1970)

TABLE I — continued

Optical Space Polarimeters A. BALLOON

Instrument	Altitude	Pointing	Analyzer	Calibration	Detector	Wavelength Coverage
10. Tokyo Astr. Obs. photoelect. polarim.	26 km	Gond. az. contr. by mag. sensor	Cont. rotating $\lambda/4$ plate, plus 45°-step rotating analyzer	Lab. standard lamp; intern. lamp	PMT	Intermed. band, 0.48, 0.52, 0.60 μm; narrowb. 0.56 μm
11. High Alt. Obs. coronagraph	~30 km	Solar pointer	Polaroids	In flight	Film	Broadband 0.65–0.85 μm
12. C.N.R.-Firenze polarim.	30 km	Mag. az. contr., elev. motor dr.	Rotating wire grid polarizer (15 cps)	Intern. thermal source	Ge bolom. cooled by liquid helium	Broadband filters 100–2000 μm
13. MPI-Astr., Heidelberg THISBE	32 km	Az. & elev. scanning	Polaroid 0°, 45°, 90°	Ground test	PMT	Intermed. band 0.50 μm
14. THISBE	43 km	Az. scanning	Polarizer	Ground test	PMT	Intermed. band 0.205 μm
15. Obs. de Paris coronagraph	32 km	3-axis stab. gond. "Astrolabe," sun sensor, ±15 arcsec	Polaroid sections $\perp$ and $\parallel$ to solar limb	Solar disk	Photographic plate	Near infrar.

[a] Personal communication, 1972.
[b] Personal communication, 1973.
[c] See p. 322.
[d] Personal communication, 1972.
[e] Personal communication, 1973.

Telescope aperture	Field of View	Mass of Polarimeter	Goals	Status	Reference
cm	3°	20 kg	Circ. polariz. of zod. light	——	H. Tanabe[a]
cm	3°.2	150 kg	Solar corona	7 June 1970 ascent	L. L. House[b]
cm	0°.5	——	Galactic ctr., sun, stars in far infrar.	16 Sept 1971 ascent	G. Dall'Oglio et al.[c]
cm	~2°	~2 kg	Zod. light mapping	1970, 1971, 1972 ascents	Gabsdil (1971)
cm	~2°	~2 kg	Zod. light	Oct 1973 ascent planned	D. Lemke[d]
cm	3°	——	Solar corona streamers	7 Nov 1970, 13 Sept 1971 ascents	A. Dollfus[e]

TABLE I — continued
Optical Space Polarimeters B. AIRCRAFT

Platform	Instrument	Altitude	Pointing	Analyzer	Calibration	Detector	Wavelength Coverage
1. Airplane	NRL Macbeth illuminometer	0–6 km	Alt. & az. adj.	Rotatable polarizing plate	Standard lamp	—	Broadband visual
2. B-29	Nav. Res. Lab. sky photom.	5–12 km	Rotatable viewing prism	Rotatable Glan-Thompson prism	—	PMT	Broadb. vis narrowband 0.35, 0.43, 0.56, 0.69 μ
3. B-29	Univ. of Colo. Schmidt camera, F/0.9	10 km	by hand	Polaroid wedge array	Step-wedge exposures	Kodak Super Orthopress film	Broadband visual
4. Lincoln	Cambridge 8-lens cam.	9 km	by hand	Polaroids	Standard lamp and step-wedge	Ilford HP3 plate	Broadband visual
5. Sunderland	Cambridge 3-lens cam, F/1.8	2.7 km	by hand	HN-22 Polaroids	Standard lamp and step-wedge	Ilford HP3 plate	Broadband visual
6. NASA CV-990	Univ. of Minn. camera, F/6.8	12 km	by hand	3-section mosaic of HN-32 Polaroid, man. rot.	Standard source	Tri-x pan film; SO-166 Kodak exp. film	Broadband red
7. KC-135	Sacramento Peak Obs. (AFCRL) polarim.	10 km	—	Rotatable Savart plate	Solar disk after eclipse; tiltable plate	103aG film	Narrowbar 0.5303 μm
8. Airplane	U.S.S.R. polarim.	—	Fixed at nadir	Polarizer 0°, 60°, 120°	—	—	0.3–2.5 μm
9. USAF C-135	UCLA/ NOTS polarim.	12 km	Fixed	Rotating KN-36 Polaroid (2 rps)	Ground tests	S-11 PMT	Intermed. band 0.48 and 0.60 μ simultaneo
10. NASA CV-990	High Alt. Obs. telesc.	—	Gyro. stab.	Rotatable Polarizer	Mean solar disk	Image tube with film	Narrowbar 1.0747 and 1.06 μm
11. NCAR Airplane	Univ. of Ariz. (IAP)	0.5 km	Manual about 1 axis	Rotatable Polaroid	Brewster angle ground tests	PMT	Broadband 0.43 μm
12. USAF NC-135	Los Alamos cam., F/6	12 km	Gyro-stab. ($\pm$5 arcsec)	Three insertable HN-38 Polaroids	—	Kodak 2475 film	Broadb. re (0.55–0.70 and 0.61–0 μm)

lescope erture	Field of View	Mass of Polarimeter	Goals	Status	Reference
—	4°	—	Terrestrial atmos.	11 flights; brightness & polarization of sky-light implies aerosol	Tousey & Hulburt (1947)
—	2°.5	—	Terrestrial atmos.	15 flights, March–May 1948; skylight obs.	Packer & Lock (1951)
—	40° image	—	Inner zod. light during solar eclipse	25 Feb 1952 flight; observed at 5°.5–13° solar elongation	Rense, Jackson, & Todd (1953)
—	14° image	—	Outer corona of sun during total eclipse	30 June 1954; polariz. at 1°–5° elongation	Blackwell (1955)
ree m	60° image	—	Inner zod. light	2 flts., May–June 1955; polariz. at 21°–31° elongation	Blackwell (1956)
cm	10° image	14 kg	Outer corona of sun during total eclipse	30 May 1965; 12 Nov 1966; polarization at 1°–3° elongation	Pepin et al. (1965); Pepin (1970a,b)
—	—	—	Inner corona of sun during total eclipse	30 May 1965; radial polarization 2–25% at 0°.3 elongation	Hyder, Mauter, & Shutt (1968)
—	—	—	Terrestrial clouds, atmos., and surfaces	Instrument difficulties; cloud polariz. ~20%	Chapurskii (1966)
—	4°.0	—	Terrestrial atmos. during solar eclipse	30 May 1965, 12 Nov 1966; polarization of multiply scattered light	Moore & Rao (1966); Rao, Takashima, & Moore (1972)
:m	1°.5	—	Solar corona during eclipse, line emission and continuum	12 Nov 1966	Newkirk (1970); Eddy, Lee & Emerson (1973)
—	1°.5, 3°	~4.5 kg	Desert terrain, skylight polarization	4 flights Jan 1968	Fernald, Herman, & Curran (1969)
:m	~6° image	—	Outer corona of sun during total eclipse	March 1970, July 1972 flights; polarization at 0°.5–3° elongation	Keller et al. (1970); Keller (1971, 1973)

TABLE I — continued

Optical Space Polarimeters B. AIRCRAFT

Platform	Instrument	Altitude	Pointing	Analyzer	Calibration	Detector	Wavelength Coverage
13. Ilyushin 18	Cent. Aerological Obs. (Moscow) photom./polarim.	—	Scanning in vert. plane	Rotating Polaroid (40.5 rps)	—	Bolometer	Broadband (0.3–0.8 μ)
14. Cessna 401	NASA Ames Res. Ctr. diff. radiom.	0.15–12 km	Fixed at nadir	Two fixed polarizers, signals altern. chopped	Channels balanced over oil-free ocean	Si photo-diodes	Narrowband 0.38 μm
15. NASA CV-990	Univ. of Ariz. (LPL) AEROPOL	11 km	Manual rot. about horiz. axis	Rotating (2 rps) polarizer, deposited wire grid	Insertable polarizer, standard sources	PbS chip cooled by Freon-13	Intermed. band filter 1.2–3.5 μm

TABLE I — continued

Optical Space Polarimeters C. ROCKET

Platform	Instrument	Altitude	Pointing	Analyzer	Calibration	Detector	Wavelength Coverage
1. Aerobee 150	NRL scanning coronagr.	130–200 km	Biaxial sun pointer ($\pm$ 1 arc-min) with spiral scanning	Radial and tangential polarization scans	—	Photoel.	Broadband visual
2. Aerobee 150	Polarim. photom.	150 km	Scanning with attitude control system	Rotating quarter-wave plate (0.5 rps)	Grnd. tests, stars, int. lamp	4 PMT's	Intermed. band 0.70 and 0.45 μ
3. Aerobee 150	GSFC polariz. photom.	90–200 km	Inertial control system	Rotating LiF stack at Brewster angle for Lyman-α	Laboratory	Solar-blind PMT	4 passes: from 0.120 0.123, 0.13 and 0.143 to 0.200 μ

Telescope aperture	Field of View	Mass of Polarimeter	Goals	Status	Reference
—	—	—	Earth clouds and surfaces	5 flights, July 1970; polariz. vs. phase angle	Bazilevskii et al. (1972)
—	~10°	1 kg	Ocean oil-spill detection	Polarization different. demonstrated	Millard & Arvesen (1972)
2 cm	1°5	~9 kg	Terrestrial cloud part. micro-structure, cloud heights	10 flights, Jan 1972; obtained polarization-phase angle curves for clouds	Coffeen, Hämeen-Anttila & Toubhans (1973)

Telescope aperture	Field of View	Mass of Polarimeter	Goals	Status	Reference
5 cm	—	—	Outer corona of sun	28 June 1963 launch; polarization at 1°–3° elongation	Tousey (1965)
m	4°7	—	Linear and circular polarization of zodiacal light	15 Sept 1964 launch; polarization at 35°–175° elongation at $\lambda = 0.7 \mu m$	Wolstencroft & Rose (1967)
—	4°	—	Polarization and intensity of hydrogen Lyman-α day airglow emission	29 Aug 1966 launch; Lyman-α unpolarized ($<1.5\%$)	Heath (1967)

TABLE I — continued
Optical Space Polarimeters   C.   ROCKET

Platform	Instrument	Altitude	Pointing	Analyzer	Calibration	Detector	Wavelength Coverage
4. Nike-Apache and Centaur	Stockholm Univ. Meteorological Inst. polarim.	130 km	Spin-stabilized about optic axis	105 UVR, HNP'B, and KS-Pro fixed polariz.	Laboratory	PMT's	Narrowb. 0.215, 0.25 0.309, 0.36 0.453, 0.53 0.589, 0.762 µm
5. Aerobee 150, 170	GSFC scanning spectrom.	175–225 km	Gyro-stabilized, rolled || and ⊥ to polariz.	Polacoat, LiF stack at Brewster angle	Laboratory	3 PMT's	50 Å resolution over 0.12–0.40 range
6. ESRO Rocket S73	Heidelberg Obs. dbl. Polarim. Exper. R214	220 km	Spin-scan about sun line	3 polarizing filters	Int. lamp; stars	2 PMT's	Broadband 0.47 µm
7. Terrier-Sandhawk	Los Alamos 6-channel polarimeter	320 km	3-axis ACS	Rotating HNP'B Polaroid (10 rps)	Stand. lamp	6 S-20 (ext. red) PMT's	Intermedia band 0.35, 0.45, 0.55, 0.65 µm
8. Rocket	Tokyo Astro. Obs. photoelectric polarimeter	270 km	Scanning; star sensor	Three parallel telescopes with fixed analyzers at 0°, 60°, 120°	Lab.; grnd.-based star measurement; int. lamp	——	Intermedia band 0.52 0.60 µm; narrow band 0.63 µm
9. Rocket	Nagoya U. photoelectric polarimeter	250 km	Gyro-controlled	3 polarizers at 0°, 60°, 120°	Laboratory; int. lamp	PbS photo-conductor	2 passes: 0.5–1.2 µm 1.2–1.8 µm
10. Rocket	Dudley Obs. polarimeter	>60 km	Fixed by direction of launch	Fixed HN-32 polaroid using rocket spin	Int. phosphor source	2 S-11 PMT's	Intermedi band 0.41 and 0.54 µ
11. Skylark	Heidelberg Observatory polarimeters	~250 km	3-axis stabilization	3 polarizing filters	Int. lamp; stars	3 PMT's	Intermedi band 0.26 0.48, 0.59 0.90 µm

f See also p. 514.
g See p. 766.
h Personal communication, 1972.
i Personal communication, 1972.
j Personal communication, 1972.

Telescope aperture	Field of View	Mass of Polarimeter	Goals	Status	Reference
5–3.5 cm	6°	—	Noctilucent cloud scattering, twilight airglow emissions	Launches in July-Aug 1967–1971; vertical profiles through NLC's	Witt et. al. (1971)[f] Witt, Dye, & Wilhelm (1973)
cm	0°.2	—	ζ Ophiuchi ultraviolet polarization	Oscillatory polarization detected	Stecher (1970, 1972)
5 cm and 0 cm	1°.2, 1°.5	3 kg	Zod. light at 15° and 21° solar elongation, simultaneous	2 July 1971 launch	C. Leinert, H. Link, & E. Pitz[g]
cm	Six 2°.8 fields	154 kg	Inner zodiacal light during total eclipse	10 July 1972 launch	Sandford, Theobald & Horak (1973)
ree 6 cm	3°	30 kg	Inner zodiacal light	—	H. Tanabe[h]
cm	3°	15 kg	Infrared polarization of zod. light	Sept 1972 launch; ACS failure	S. Hayakawa[i]
cm	~2°	4 kg	Noctilucent cloud particle location and optical properties	—	Tozer & Beeson (1973)
4, 9 cm	2°	~14 kg total	Zod. light at 30° elongation; Milky Way regions	1975 launch planned	E. Pitz, H. Link, & C. Leinert[j]

TABLE I — continued

Optical Space Polarimeters D. SATELLITE AND SPACECRAFT [k]

Platform	Instrument	Pointing	Analyzer	Calibration	Detector	Wavelength Coverage
1. OSO-2, 600 km earth orb. 33° inc.	U. of Minn. photom./ polarim.	Spin-scan	Fixed polaroids, spinning vehicle (30 rpm)	Standard sources	4 PMT's	Broadband blue and visual
2. OSO-5, 550 km earth orb. 33° inc.	U. of Minn. Dim Light Monitor	Spin-scan	Fixed polaroids, spinning vehicle (30 rpm)	Standard sources	2 PMT's	Broadband 0.42 and 0.68μm
3. Surveyors 6 and 7 on lunar surface	JPL TV camera	Pointable mirror in front of F/4 objective lens	Three KN-36 polaroids in filterwheel, at 0°, 45°, 90°	Lab. photometric targets on spacecraft	Vidicon tube	Broadband green
4. Apollos 12–17 on lunar surface	Hasselblad camera	Hand-held	Rotatable polaroid in detent wheel in front of camera; 0°, 45°, 90°	Gnomon photography on lunar surface	SO-267 film	—
5. Mariner IX Mars orbiter	JPL TV wide-angle camera	3-axis stab. spacecraft; scan platform	Polaroids in filter wheel	—	Vidicon tube	—
6. OSO-6, 520 km earth orb. 33° inc.	Rutgers U. zod. light polarim./ photom.	Spin-scan in a plane, all solar elongations 10°(5°)180°	Rouy Prism, insertable retarder plates	Selected stars	PMT	Intermediate band 0.40, 0.50, 0.61μ
7. OGO-1, ecc. earth orb., 282/ 14,900 km 31° inc.	G S F C photom.	3-axis stab. spacecraft	Polaroids	—	S-13 and S-1 PMT's	Broadband 0.5 μm and 1.0 μm
8. Diamant B-4 (D2A polar)	C.N.E.S. Service d' Aéronomie polarimeter	Spin-stabilized satellite; views in plane perpendicular to sunline	2 LiF plates at Brewster angle	—	—	Narrow-band 0.1216μm
9. Meteor 8, 620 km earth orb. 81° inc.	Central Aerological Obs. (Moscow) radiom. polarim.	Scanner	—	Laboratory; ground tests on sea surface	—	Broadband (0.3–0.8μr

[k] Some information taken from *TRW Space Log,* Vol. 10, 1972.
[l] Personal communication, 1972.
[m] See also p. 794.

OPTICAL POLARIMETERS IN SPACE 201

Telescope Aperture	Field of View	Mass of Polarimeter	Goals	Status	Reference
.5 cm	10°	11 kg total	Zodiacal light	3 Feb 1965 launch; polarization measured at 90° solar elongations	Gillett (1967); Sparrow & Ney (1968)
.5 cm	10°	13.5 kg total	Zodiacal light; terrestrial lightning; galactic light	22 Jan 1969 launch	Burnett, Sparrow, & Ney (1972); Sparrow & Ney (1972)
———	6°.5 or 25° image	7.3 kg camera	Lunar soil and rocks, solar corona, earth	TV image Nov/Dec 1967, Jan/Feb 1968	Holt (1970); Shoemaker et al. (1968)
———	———	———	Lunar material differentiation	Photography performed on Apollos 12 and 16	Holt & Rennilson (1970); H. E. Holt[1]
.2 cm	11°×14° image	———	Mars surface and atmosphere	In orbit Nov 1971	Masursky et al. (1970)
cm	2°	———	Zodiacal light: linear & elliptical polarization	9 Aug 1969 launch; Gegenschein variations found	Roach & Lillie (1972)[m]
———	0°.5	3.8 kg	Counterglow photometry in 10° × 20° field	4 Sept 1964 launch; spacecraft failed to orient	Richter (1966); Corliss (1967)
———	———	2.8 kg	Geocoronal Lyman-α emission	Launch failure of second stage, 5 Dec 1971	Huriet (1970)
———	4°×5°	———	Earth clouds	17 April 1971 launch; polarization maps obtained	Bazilevskii & Vinnichenko (1972)

TABLE I — *continued*

Optical Space Polarimeters D. SATELLITE AND SPACECRAFT[k]

Platform	Instrument	Pointing	Analyzer	Calibration	Detector	Wavelength Coverage
10. Pioneer 10/11: Jupiter flybys 1973, '75	U. of Ariz. Imaging Photopolarimeter	Spin-scan and one axis of telescope articulation	Fixed Wollaston prism, insertable half-wave plates	Lyot depolarizer, standard sources	Dual S-11 and dual S-20 channeltrons	Broadband 0.45 and 0.65μm simultaneous
11. Skylab ATM, 435 km earth orb.	High Alt. Obs. white light coronagraph	Solar pointed, stabilized CMG platform	Polaroids	In flight	Film	0.35–0.70 μm
12. Mariner Venus/Mercury	JPL TV cameras	3-axis stab. vehicle; scan platform	One polaroid in each camera, parallel	Laboratory	Vidicon tube	Broadband 0.36μm
13. Skylab, 435 km earth orb.	Dudley Obs. polarim., Exper. S073	Alt-az. mount. on 18′ ext. from vehicle, scans sky to within 15° of sun	Rotating HN32 Polaroid	Internal phosphor source	PMT	Intermediate band 0.40, 0.48, 0.51, 0.53, 0.56, 0.61, 0.63, 0.64, 0.71, 0.82μm
14. Helios, solar orb.; within 0.3 AU of sun	Heidelberg polarimeter, Experiment E9	Spin-stabilized spacecraft	Polarizing filters; rotating Polaroid	Internal lamp; stars	3 PMT's	Broadband 0.36, 0.42, 0.52μm
15. Mariner Jupiter flybys 1979, Saturn 1981	Univ. of Colorado photom./polarim.	3-axis stab. spacecraft; scan platform	3 polarizers in aperture wheel, at 0°, 60°, and 120°	Stars	PMT	Intermediate band filters, 0.22–0.73 μm

[k] Some information taken from *TRW Space Log*, Vol. 10, 1972.
[n] See p. 766.
[o] Personal communication, 1973.

Telescope Aperture	Field of View	Mass of Polarimeter	Goals	Status	Reference
5 cm	0°.5, 2°.3	4.2 kg	Jupiter cloud particle microstructure, cloud heights; zodiacal light beyond earth	3 March 1972, 5 April 1973 launches	Pellicori, Russell, & Watts (1973), KenKnight (1971)
2 cm	3°.2	150 kg	Coronal brightness and polarization between 1.5 and 6.0 solar radii	14 May 1973 launch	Watts (1973)
cm	0°.4×0°.5 image	——	Venus clouds and Mercury surface	Nov 1973 launch planned	Murray et al. (1972)
cm	1°, 3°, 6°	——	Zodiacal light, starlight, airglow, spacecraft corona	14 May 1973 launch	Weinberg (1973)
cm, 5 cm	1°, 2°, 3°	8.9 kg total	Zodiacal light on circles of ecliptic latitude −16°, −31°, −90°	1974–75 launches planned	C. Leinert, H. Link, & E. Pitz[n]
cm	0°.25, 4°	1.6 kg	Zodiacal light, planets, satellites	1977 launches planned	NASA News Press Release (1972); C. F. Lillie[o]

INSTRUMENT OPERATION IN SPACE

Accurate polarimetry has been accomplished from every type of space platform, using a variety of rather straightforward techniques. The design of a space polarimeter can be the same as that of a ground-based instrument,[2] except for considerations of the special environment and limitations on weight, power, volume, and bit rate. Several aspects of space operation need special attention. High voltages must be very carefully shielded to avoid glow discharge and arcing at moderately low pressures. Pulses due to cosmic rays are unavoidable in many cases; a frequently observed mechanism is Cherenkov radiation in the fore optics detected by photomultiplier tubes. Finally, an observing window or port is often required. It should have low strain birefringence (note that birefringence may be induced if there is a pressure differential across the port), and should be exactly perpendicular to the incident beam.

A man-operated instrument can have minimum weight. Figure 1 shows Dollfus' hand-held polarimeter (A-2) weighing 0.2 kg. A completely automatic instrument can be miniaturized if the cost is justified; Fig. 2 shows the Imaging Photopolarimeter (D-10) now enroute to Jupiter (weight 4.2 kg., cost $\sim \$10^6$). Similar considerations apply to power and volume. The bit rate, however, tends to be set by the available recording and telemetry systems and by vehicle power. A Skylab instrument (D-13) uses a rotating analyzer and will transmit *all* measured points to the ground for later analysis of the modulation; this requires 7000 bits per second. The airborne AEROPOL instrument (B-15), shown in Fig. 3, is tied to a minicomputer. As the analyzer rotates, the computer receives 970 bits per second, makes a least-squares solution each 2.5 sec in real time, and outputs 15 bits per second; the individual measurements are not saved.

POLARIZATION ANALYSIS

Of principal importance is the optical means of polarization analysis, requiring either a rotary modulation (of polarizer, wave plate, or entire instrument) or multiple detectors (with attendant difficulties of calibration).

Dollfus (A-2) uses a null technique whereby a plane-parallel plate in the incident beam is intentionally tilted, introducing sufficient polarization to just cancel the incident polarization, as indicated by a Savart

---

[2] See page 135 and Serkowski (1973) for a review of ground-based polarimeters and their design.

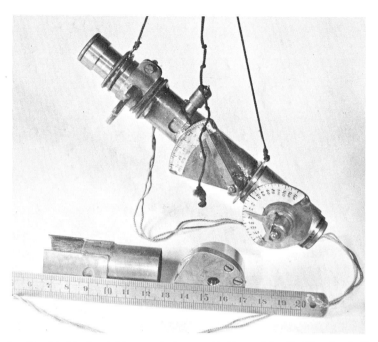

Fig. 1. Hand-held visual fringe polarimeter (A-2) used by Dollfus for balloon flight measurements of the polarization of sunlight scattered by clouds and by ground surfaces, through the earth's atmosphere. (Photo courtesy of A. Dollfus.)

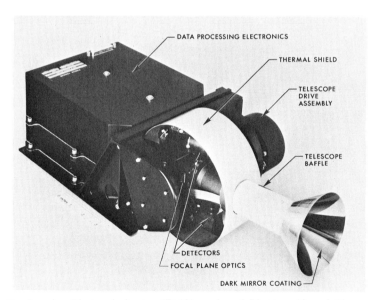

Fig. 2. Imaging Photopolarimeter (D-10) on board Pioneers 10 and 11 enroute to Jupiter. The clouds of Jupiter will be observed over a wide range of phase angles during flyby with this two-color spin-scan instrument. (Photo courtesy of Santa Barbara Research Center.)

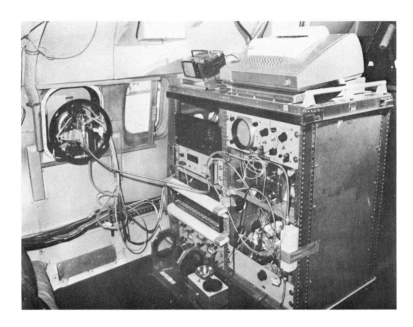

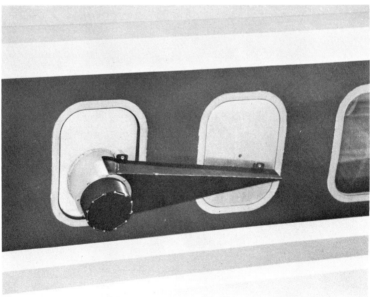

Fig. 3. Two views of the AEROPOL infrared polarimeter (B-15) mounted in the NASA CV-990 jet aircraft. The instrument can be rotated in a pressure seal in the airplane window plate. A minicomputer provides instrument control and real-time solutions for the degree of polarization. (Photo courtesy of Airborne Science Office, NASA Ames Research Center.)

fringe polariscope. The angle of tilt is then a measure of the incident polarization. This tilt is set manually by trial and error, but the settings quickly converge for a bright object.

Many of the space polarimeters use rotating Polaroids, prisms, wire grid analyzers, or wave plates. Figure 4 is a schematic of the Imaging Photopolarimeter (D-10) which has a fixed Wollaston prism and insertable retarders on an aperture wheel. In some cases the entire instrument rotates, viewing along the axis of a spin-stabilized rocket (e.g., Witt, C-4 and Fig. 5) or spacecraft (e.g., OSO, D-1 and D-2).

All rotation can be avoided by using multiple detectors. Figure 6 shows a four-channel balloon instrument (A-4). Each telescope has a Wollaston prism and two photomultipliers; the Wollaston axes differ by 45° and are permanently fixed. With no moving parts, one can measure the degree of linear polarization and its direction of vibration. But the relative sensitivity of the different photomultipliers must be carefully determined.

ACCURACY AND CALIBRATION

Polarization is measured differentially by comparing the intensity of an object as viewed through different polarization optics (different

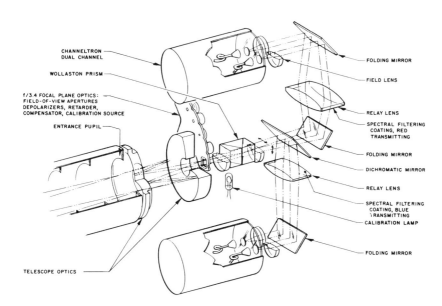

Fig. 4. Imaging Photopolarimeter (D-10) schematic. A single Wollaston prism and an insertable half-wave plate permit measurement of linear polarization. A dichromatic filter separates the red and blue passbands.

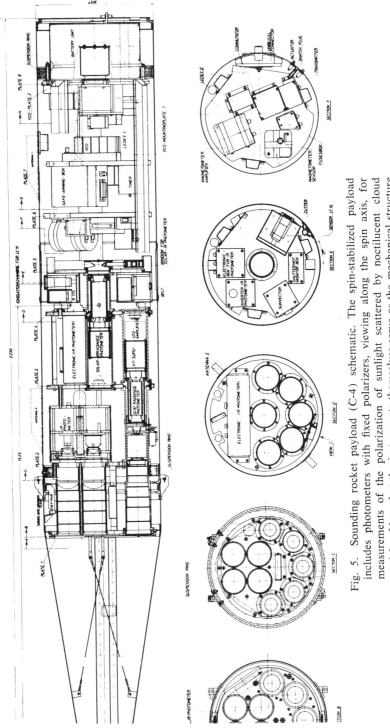

Fig. 5. Sounding rocket payload (C-4) schematic. The spin-stabilized payload includes photometers with fixed polarizers, viewing along the spin axis, for measurements of the polarization of sunlight scattered by noctilucent cloud particles. Note that the instruments themselves serve as the mechanical structure of the scientific section. (After Witt et al. 1971.)

Fig. 6. Four-channel automatic balloon polarimeter (A-4). The two Wollaston prisms are mounted at 45° to each other, for instantaneous measurement of linear polarization of astronomical objects. (After Pellicori and Gray 1967.)

azimuths of Polaroid, etc.). Polarization can therefore be measured with rather high precision, simply by minimizing the sources of noise, maximizing the photon collection, increasing the recorder resolution, etc. The degree of *precision* is then indicated by the repeatability of measurements on a given object.

The absolute *accuracy*, however, is much harder to estimate and to control because of systematic errors in the instrument. The required accuracy depends on the scientific objectives and on the degree of polarization of the object. Many objects introduce only a few percent linear polarization by scattering and, therefore, require a measurement accuracy of a few tenths of a percent. To reach this accuracy, the instrument must be carefully designed and constructed and will need *in-flight* calibration.

Calibration requires measuring a controlled source of polarization. In principle, one must measure all possible amounts of polarization at all possible position angles. In practice, if the instrument response is

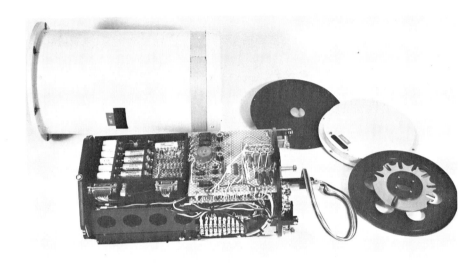

Fig. 7. Two views of the ultraviolet polarimeter (A-9) used on the POLARISCOPE balloon system. The entire inner assembly, containing a Wollaston prism and two solar-blind photomultiplier tubes, rotates to six different position angles.

rather linear, it is sufficient to observe an unpolarized source, and a source of 100% linear polarization at a series of position angles. If the instrument measures circular polarization, one should observe sources of 100% right-handed and left-handed polarization, as well as the source of linear polarization to determine the degree of linear-to-circular conversion in the instrument.

The creation of controlled laboratory sources of polarization is not simple, especially in the case of small polarizations. The light from an incandescent bulb typically shows several percent linear polarization. Symmetry properties are useful; e.g., a source can be built for minimum polarization and then can be measured once, turned by 90°, measured again, and the results averaged.

Ground tests are usually insufficient for the highest accuracy with space polarimeters. In-flight calibrations should be made, assuring the same optical ray paths, temperature and pressure environment, etc. "Unpolarized" stars are often suitable targets.

The AEROPOL instrument (Fig. 3, B-15) is equipped with calibration slides so that a polarizer and/or an unpolarized light source may be inserted in front of the entrance pupil at any time during flight.

The highest accuracy ($\pm 0.2\%$ absolute) obtained with a polarimeter in space was probably that of the POLARISCOPE balloon instrument (Fig. 7; A-9), which included a depolarizer for frequent in-flight calibration of the instrument response to unpolarized light. The polarimeter went through a very thorough laboratory calibration, and as a final check, unpolarized stars were included in the observing program at altitude.

High accuracy is not a noble goal in itself; the accuracy should be governed by the needs of science. An excellent example of a simple polarimeter well suited to its task is an airborne instrument used to evaluate the feasibility of detecting ocean oil slicks (Fig. 8; B-14). The absolute degree of polarization was of no particular concern. The instrument succeeded in detecting polarization *variations* over oil slicks.

In actual practice, the *absolute accuracy* obtained with a particular instrument is typically *10 times* the *instrumental precision*. This guideline applies only if careful calibration corrections are made. Improving the precision does not automatically improve the accuracy.

SCIENCE GOALS

The principal use of optical polarimeters in space has been in the study of the solar system. For the instruments in Table I, the breakdown by objects of primary interest is as follows:

zodiacal light	20 instruments
earth atmosphere and surface	18
sun and corona	9
planets	4
moon	3
stars	2
	56 instruments

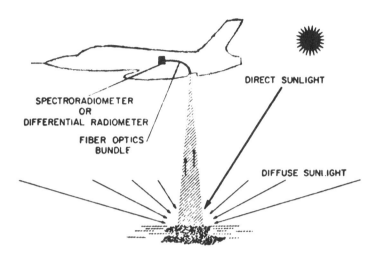

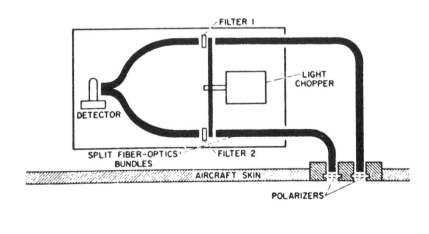

Fig. 8. Airborne differential polarimeter (B-14) schematic. The instrument is capable of detecting oil slicks on the ocean, by the change of polarization. (After Millard and Arvesen, 1972.)

The instruments are divided almost equally among balloons, aircraft, rockets, and satellites/spacecraft.

There are several reasons for sending instruments aloft. For zodiacal light studies, it is the removal of local light sources (atmospheric scattering, airglow, etc.). For earth studies, it is the new vantage point, the synoptic view, as well as the opportunity to probe the vertical structure. For the solar corona, it is the removal of atmospheric scattering. For planets, moon, and stars, it is the extension in wavelength coverage, especially into the ultraviolet, as well as the opportunity for detailed mapping and unique coverage in phase angle, using space probes to the planets.

CONCLUSION

We can expect polarimeters to flourish in space as long as there is a need for remote sensing of scattered light. For some studies, the measurement of polarization is the key to understanding intrinsic properties of the object, as indicated throughout this book. The classification of instruments as "polarimeters," "spectrometers," "vidicons," "radiometers," etc., may become less well defined as more versatile instruments are created. We already have "photopolarimeters" and "spectropolarimeters." All of these various instruments simply measure different combinations of the four basic parameters of light at different rates and resolutions.

Acknowledgments. I am indebted to many investigators for assistance in the preparation of Table I. Unfortunately the list of instruments is probably incomplete; I apologize for any omissions. In particular, no classified instruments are included here. The preparation of this paper was supported by the National Aeronautics and Space Administration.

REFERENCES

Bazilevskii, K. K.; Pakhomov, L. A.; Tsitovich, T. A.; and Skliarevskii, V. G. 1972. Polarization characteristics of shortwave radiation. *Space Research XII.* 1:511–516. Berlin: Akademie-Verlag.

Bazilevskii, K. K., and Vinnichenko, N. K. 1972. Personal communication; satellite results will be published in *Space Research XIII* (1973).

Blackwell, D. E. 1955. A study of the outer corona from a high altitude aircraft at the eclipse of 1954 June 30. *Mon. Not. R. Astron. Soc.* 115:629–649.

———. 1956. Observations from an aircraft of the zodiacal light at small elongations. *Mon. Not. R. Astron. Soc.* 116:365–379.

Burnett, G. B.; Sparrow, J. G.; and Ney, E. P. 1972. OSO-5 dim light monitor. *Applied Optics* 11:2075–2081.

Chapurskii, L. I. 1966. Experimental studies of the spectral brightness characteristics of clouds, the atmosphere, and the underlying surface in the wavelength range from 0.3–2.5 μm. *Trudy Glavnaia Geofizicheskaia Observatoriia*, Leningrad 196:110–119 (in Russian).

Coffeen, D. L., and Gehrels, T. 1970. Ultraviolet polarimetry of planets. *Space Research X*:1036–1042.

Coffeen, D. L.; Hämeen-Anttila, J.; and Toubhans, R. H. 1973. Airborne infrared polarimeter. *Space Science Instrumentation* 1. In press.

Corliss, W. R. 1967. *Scientific satellites*. NASA SP–133. Washington: U.S. Government Printing Office.

Dollfus, A. 1956. Polarisation de la lumière renvoyée par les corps solides et les nuages naturels. *Ann. d'Astrophys.* 19:83–113.

———. 1957. Étude des planètes par la polarisation de leur lumière. *Suppl. Ann. Astrophys.* No. 4. (In English as NASA TT F-188.)

Eddy, J. A.; Lee, R. H.; and Emerson, J. P. 1973. The λ 10747 coronal line at the 1966 eclipse I. Emission line polarization. *Solar Physics*. In press.

Fernald, F. G.; Herman, B. M.; and Curran, R. J. 1969. Some polarization measurements of the reflected sunlight from desert terrain near Tucson, Arizona. *J. Appl. Met.* 8:604–609.

Gabsdil, W. 1971. *Measurement of the zodiacal light with the balloon telescope THISBE*. Thesis, Univ. of Heidelberg.

Gehrels, T. 1967. Ultraviolet polarimetry using high altitude balloons. *Applied Optics* 6:231–233.

Gillett, F. C. 1966. *Zodiacal light and interplanetary dust*. Astrophysics report #2, School of Physics and Astronomy, Univ. of Minnesota.

———. 1967. Measurement of the brightness and polarization of zodiacal light from balloons and satellites. *The zodiacal light and the interplanetary medium.* (J. Weinberg, ed.) NASA SP-150, 9–15. Washington, D.C.: U. S. Government Printing Office.

Heath, D. F. 1967. Polarization and intensity measurements of hydrogen Lyman-alpha in the day airglow. *Astrophys. J.* 148:L97–L100.

Hodgdon, E. B. 1965. Theory, design, and calibration of a UV spectrophotopolarimeter. *Applied Optics* 4:1479–1483.

Hodgdon, E. B., and Edwards, H. D. 1966. A UV spectrophotopolarimeter for the study of natural sky backgrounds. Report #AFCRL-66-397.

Holt, H. E. 1970. Photometry and polarimetry of the lunar regolith as measured by Surveyor. *Radio Science* 5:157–170.

Holt, H. E., and Rennilson, J. J. 1970. Photometric and polarimetric properties of the lunar regolith. *Apollo 12 Preliminary Science Report*. NASA SP-235, 157–161. Washington, D.C.: U.S. Government Printing Office.

Huriet, J. R. 1970. Les expériences optiques du satellite D2A. *Nouv. Rev. d'Optique appliquée* 1:213–219.

Hyder, C. L.; Mauter, H. A.; and Shutt, R. L. 1968. Polarization of emission lines in astronomy. VI. Observations and interpretations of polarization in green and red coronal lines during 1965 and 1966 eclipses of the sun. *Astrophys. J.* 154:1039–1058.

Keller, C. F. 1971. Results of polarization observations of the outer corona from a jet aircraft. *Solar Physics* 21:425–429.

———. 1973. Airborne white light polarimetry of the outer corona during July 1972 solar eclipse. *Bull. Amer. Astron. Soc.* 5:19–20.

Keller, C. F.; Strait, B.; Winslow, O.; and Rice, L. 1970. Airborne photography of the solar corona during the March 1970 eclipse. *Bull. Amer. Astron. Soc.* 2:325.

KenKnight, C. E. 1971. Observations in the asteroid belt with the imaging photopolarimeter of Pioneers F and G. *Physical studies of minor planets.* (T. Gehrels, ed.) NASA SP-267, pp. 633–637. Washington, D.C.: U.S. Government Printing Office.

Masursky, H., and 25 co-authors. 1970. Television experiment for Mariner Mars 1971. *Icarus* 12:10–45.

Millard, J. P., and Arvesen, J. C. 1972. Airborne optical detection of oil on water. *Applied Optics* 11:102–107.

Moore, J. G., and Rao, C. R. N. 1966. Polarization of the daytime sky during the total solar eclipse of 30 May 1965. *Ann. de Géophys.* 22:147–150.

Murray, B., and 14 members of the MVM '73 Imaging Science Team. 1972. *Mariner Venus/Mercury 1973 imaging sequence document* (preliminary).

NASA News Press Release. 1972. Scientists selected for Jupiter/Saturn missions. #72-239, by H. Allaway and F. Bristow, for release 10 Dec 1972.

Newkirk, G. 1970. Reports on astronomy, Commission 9: Instruments and techniques. *Transactions of the International Astronomical Union Vol. XIV A.* (C. de Jager, ed.) p. 61.

Packer, D. M., and Lock, C. 1951. The brightness and polarization of the daylight sky at altitudes of 18,000 to 38,000 feet above sea level. *J. Opt. Soc. Am.* 41:473–478.

Pellicori, S. F., and Gray, P. R. 1967. An automatic polarimeter for space applications. *Applied Optics* 6:1121–1127.

Pellicori, S. F.; Russell, E. E.; and Watts, L. A. 1973. Pioneer imaging photopolarimeter optical system. *Applied Optics* 12:1246–1258.

Pepin, T. J. 1970a. *Observation of the brightness and polarization of the outer corona during the November 12, 1966 total eclipse of the sun.* Ph.D. Thesis, Univ. of Minnesota.

———. 1970b. Observations of the brightness and polarization of the outer corona during the 1966 November 12 total eclipse of the sun. *Astrophys. J.* 159:1067–1075.

Pepin, T. J.; Gillett, F. C.; and Mantis, H. T. 1965. Measurements of the brightness and polarization of the outer corona. *Proceedings of the AAS-NASA solar eclipse symposium.* Ames Research Center. pp. 199–203.

Rao, C. R. N. 1968. Electronic-servo governed photoelectric polarimeter for airborne measurements of skylight polarization. *Proc. XIX International Astronautical Congress.* (M. Lunc, ed.) 1:151–159. New York: Pergamon.

Rao, C. R. N., and Sekera, Z. 1967. A research program aimed at high altitude balloon-borne measurements of radiation emerging from the earth's atmosphere. *Applied Optics* 6:221–225.

Rao, C. R. N.; Takashima, T.; and Moore, J. G. 1972. Polarimetry of the daytime sky during solar eclipses. *J. Atmos. Terres. Physics* 34:573–576.

Regener, V. H., and Vande Noord, E. L. 1967. Observations of the zodiacal light by means of telemetry from balloons. *The zodiacal light and the interplanetary medium.* (J. Weinberg, ed.) NASA SP-150, 45–47. Washington, D.C.: U. S. Government Printing Office.

Rense, W. A.; Jackson, J. M.; and Todd, B. 1953. Measurements of the inner zodiacal light during the total solar eclipse of February 25, 1952. *J. Geophys. Res.* 58:369–376.

Richter, H. L., Jr. (ed.). 1966. *Instruments and Spacecraft*. NASA SP-3028. Washington: U.S. Government Printing Office.

Roach, F. E., and Lillie, C. F. 1972. A comparison of surface brightness measurements from OAO-2 and OSO-6. *The scientific results from the Orbiting Astronomical Observatory (OAO-2)*. NASA SP-310, 71–79. Washington, D.C.: U. S. Government Printing Office.

Sandford, M. T., II; Theobald, J. K.; and Horak, H. G. 1973. Observations of the F corona and inner zodiacal light during the 1972 July 10 total solar eclipse. *Astrophys. J.* 181:L15–L17.

Serkowski, K. 1973. Polarization techniques. *Methods of experimental physics Vol. 12: Astronomical and astrophysical methods.* (M. L. Meeks and N. P. Carleton, eds.) Academic Press. In press.

Shoemaker, E. M., and 7 co-authors. 1968. Television observations from Surveyor. *Surveyor Project final report. Part II. Science results.* (JPL TR 32-1265), pp. 115–119, 122–124.

Sparrow, J. G., and Ney, E. P. 1968. OSO-B2 satellite observations of zodiacal light. *Astrophys. J.* 154:783–787.

―――. 1972. Polarization of the diffuse galactic light. *Astrophys. J.* 174:717–720.

Stecher, T. P. 1970. Stellar spectrophotometry from a pointed rocket. *Astrophys. J.* 159:543–550.

―――. 1972. The wavelength dependence of interstellar polarization in the direction of ζ Ophiuchi. *Bull. Amer. Astron. Soc.* 4:270.

Tousey, R. 1965. Observations of the white-light corona by rocket. *Ann. d'Astrophys.* 28:600–604.

Tousey, R., and Hulburt, E. O. 1947. Brightness and polarization of the daylight sky at various altitudes above sea level. *J. Opt. Soc. Amer.* 37:78–92.

Tozer, W., and Beeson, D. 1973. In preparation.

Unz, F. 1969. Die Konzentration des Aerosols in Troposphäre und Stratosphäre aus Messungen der Polarisation der Himmelsstrahlung im Zenit. *Beiträge zur Physik der Atmosphäre* 42:1–35.

Vande Noord, E. L. 1970. Observations of the zodiacal light with a balloon-borne telescope. *Astrophys. J.* 161:309–316.

Watts, R. N., Jr. 1973. Progress report on Skylab. *Sky and Telescope* 45:24–26.

Weinberg, J. 1973. A coordinated program of satellite and ground-based observations of the light of the night-sky. In preparation.

Wigand, A. 1917. Beobachtungen der neutralen Polarisationspunkte aus grösserer Höhe. *Phys. Zeit.* 18:237–240.

Witt, G.; Dye, J. E.; and Wilhelm, N. 1973. *Measurements of scattered sunlight in the mesosphere during twilight.* Univ. of Stockholm Inst. of Meteorology Report AP-8.

Witt, G.; Wilhelm, N.; Stegman, J.; Williams, A. P.; Holback, B.; Llewellyn, E.; and Pedersen, A. 1971. *Sounding rocket experiment for the investigation of high-latitude summer atmospheric conditions between 60 and 110 km (ESRO Experiment R-175, Payload C-59).* Univ. of Stockholm Inst. of Meteorology Report AP-7.

Wolstencroft, R. D., and Rose, L. J. 1967. Observations of the zodiacal light from a sounding rocket. *Astrophys. J.* 147:271–292.

THE FOLLOWING INSTRUMENT REFERENCES WERE ADDED IN PROOF

Arnquist, W. N., and Waddell, J. H. 1965. Polarization photography of the 1965 solar eclipse corona. *Proceedings of the AAS-NASA solar eclipse symposium.* Ames Research Center. pp. 187–198.

Ingham, M. F., and Jameson, R. F. 1968. Observations of the polarization of the night sky and a model of the zodiacal cloud normal to the ecliptic plane. *Mon. Not. R. Astr. Soc.* 140:473–482.

Smith, S. M.; Henderson, M. E.; and Torrey, R. A. 1965. Coronal photographs, isophotes, and a flash spectrum from the solar eclipse of May 30, 1965. *Proceedings of the AAS-NASA solar eclipse symposium.* Ames Research Center. pp. 234–272.

Stone, S. N. 1965. Photographic photometry of the solar corona at the 1965 eclipse from 40,000 feet altitude. *Proceedings of the AAS-NASA solar eclipse symposium.* Ames Research Center. pp. 205–226.

PHOTOGRAPHIC POLARIZATION MEASUREMENTS OF VENUS

RETA BEEBE
New Mexico State University

A program for measuring polarization of Venus photographically to obtain high spatial resolution has been carried out. Average values of polarization of the disk are compared with those of Dollfus and Coffeen. Trends and errors in spatial mapping are discussed.

A program for measuring polarization of Venus photographically is being carried out. Although this is a difficult measurement to make with photographic techniques, it is of interest because a spatial mapping of the polarization can be obtained and compared directly with relative intensities across the disk. For instance, such an approach would allow a direct correlation of polarization with ultraviolet markings in Venus' upper cloud deck.

In this preliminary study, plates obtained on 26 July 1972 were selected for analysis. The phase angle of Venus was 115°, which should yield near-maximum values of polarization according to Dollfus and Coffeen (1970). Both red and blue plates were analyzed in order to compare wavelength dependence with values obtained by Coffeen and Gehrels (1969). Methods of reduction and sources and magnitudes of errors are discussed.

PHOTOGRAPHIC SYSTEM

The f/75 system of the 61-cm planetary telescope at New Mexico State University (NMSU) was used. The plate scale of 4"53/mm yields images on the selected plates with diameters of 8 mm. A polarizing module was incorporated into the NMSU planetary camera replacing the standard filter box. This module contained the filter and polarizing unit, which consisted of a sheet of HN-38 polarizer that could be rotated through four positions at 45° intervals.

Plates were obtained in two colors with central wavelengths of 0.42 μm and 0.66 μm and band passes of approximately 0.045 μm. These were obtained with Kodak III-O/Wratten-35 and III-F/Schott RG-2 plate and filter combinations. Exposures were 0.25 and 0.20 seconds, respectively. Twelve exposures were obtained on each plate, with three sequences containing each orientation of the Polaroid. A calibration strip was placed on each plate.

REDUCTION OF THE DATA

The plates were scanned on the Sacramento Peak Observatory microdensitometer. Using a 42 × 42 μm slit, the images were scanned perpendicular to the intensity equator and sampled every 20 μm in both the X and Y directions, where X is perpendicular and Y is parallel to the intensity equator. This yields a data array for each image of 250 × 500 data points, covering an area 0.5 × 1.0 cm. The two best images of each orientation of the Polaroid were scanned along with the calibration strip, plate background, dark current, and a sample of machine noise measured by scanning the same scan five times without stepping in the Y-direction.

The logarithm of transmission obtained from the eleven-step calibration wedge was fit to standard values of the logarithm of exposure with a five-term expression

$$\log E = C_1 + C_2 \log T + C_3 (\log T)^2 + C_4 \log(1-T) + C_5/T, \quad (1)$$

where T is the transmission normalized to the plate background. The second-order term in $\log T$ was added to the expression proposed by Honeycutt and Chaldu (1970) to improve the fit near the sky background. Fitting errors of less than 1.9% and 1.2% were obtained for the blue and red plates, respectively.

SELECTION OF IMAGES

Filter systems with band passes of 0.045 μm were chosen for this study to utilize good transient seeing conditions and to obtain high resolution of the images. Inspection of intensities averaged over the disk indicates that uncertainties in exposures and atmospheric transmission cause differences of up to 8% in the mean intensities for images taken with the polarizer in a given position. This requires normalization of the images before compositing them.

Since positions 1 and 3 and positions 2 and 4 of the polarizer form two systems in which the second axis is oriented 90° to the first and the second system is rotated 45° relative to the first, constraints can be placed on the data as follows. For average intensities I_i on all acceptable images, where i denotes the position of the polarizer, the following constraints are imposed on all possible combinations of the images:

$$I_1 + I_3 = I_{\text{TOTAL}} = I_2 + I_4. \tag{2}$$

The angle the polarization vector forms with the polarizer's principal axis is determined in system A (positions 1, 2, and 3) and in system B (positions 2, 3, and 4) where

$$\tan 2\psi_A = \frac{I_1 + I_3 - 2I_2}{I_1 + I_3} \quad \text{or} \quad \tan 2\psi_B = \frac{I_2 + I_4 - 2I_3}{I_2 + I_4}, \tag{3}$$

and

$$\psi_A - \psi_B = 45°, \tag{4}$$

since coordinate system B is rotated $45°$ relative to coordinate system A.

The four images that yielded the most consistent values for these two constraints were chosen for normalizing the intensities before compositing images of the same orientation of the polarizer.

IMAGE REGISTRATION

Two methods of registration were used. A detailed cross-correlation of all other images relative to one image was carried out. Since the sampling array (250×500) was large, a scheme was developed that selected every second data point of every second scan of image II to be correlated and compared with a preselected grid from the comparison image, image I. By correlating image II to image I four times using X_0,Y_0; X_0+1,Y_0; X_0,Y_0+1; and X_0+1,Y_0+1 as initial starting values in image II, all data points in II could be correlated to a grid from I with a mesh size twice that of II. This allowed spatial discrimination with the limited core storage of the computer.

A second method of registering the images was investigated. This method consisted of calculating the first moments of the intensity of the illuminated disk about axes perpendicular and parallel to the intensity equator where

$$X_0 = \frac{\sum\limits_{i=1}^{N} I_i X_i}{\sum\limits_{i=1}^{N} I_i} \quad \text{and} \quad Y_0 = \frac{\sum\limits_{i=1}^{N} X_i Y_i}{\sum\limits_{i=1}^{N} I_i}. \tag{5}$$

I_i is the measured intensity minus the average intensity of the sky. If the value of I_i is not greater than one-tenth the maximum intensity of the disk, it is excluded from the calculation.

Results from the two methods were consistent. The moment method was chosen for further investigation because it is more efficient in terms of computer time and space. Since the limb-darkening on Venus defines the limb very sharply, either method is more strongly influenced by the center-to-limb behavior of the intensity than by local variations in polarization.

CALCULATIONS

Average intensities after correcting for sky background were obtained for the four orientations of the polarizer. The magnitude of the average polarization was calculated as

$$P = \frac{I_1 + I_3}{(I_1 + I_3)\cos 2\psi}, \qquad (6)$$

where ψ is the angle between the first position of the polarizing axis and the polarization vector and is determined by Equation 3.

Intensity values that were corrected for sky background and registered with respect to each other were averaged over 120×120 μm regions for the four composited images. This yields approximately 70 samples across the diameter of the disk.

The magnitude and direction of polarization were obtained for each point in the arrays in both systems A and B. Differences between local position angles of the polarization vector in the two systems were calculated and compared to 45°.

RESULTS

Average polarization at 0.42 μm and 0.66 μm were 6.0% and 3.5% respectively. The direction of the polarization vector was rotated counterclockwise 18° at 0.42 μm and clockwise 10° at 0.66 μm with respect to the intensity equator. The red value agrees well with Dollfus and Coffeen[1] (1970); however, the blue value is larger than their values by a factor of two. Close examination of intensity profiles across the disk of the planet indicates that there is some scattering in the system in blue light. These trends are not present in red light.

High-resolution analysis at 0.66 μm yields polarization values of less than 10% polarization to within 240 μm of the limb of the image. Values

---

[1] See p. 551.

of polarization vary locally ±2%, having values of 3% near the center of the illuminated disk, and increasing outward toward the limb and terminator. There is no indication that the direction of polarization varies significantly over the disk. The position of the polarization vector relative to the intensity equator was calculated in systems A and B and was parallel to the intensity equator to within ±10° as indicated by Coffeen and Gehrels (1969).

The 0.42 μm data show a rotation of the polarization vector near the limb; however, results are not consistent in systems A and B, and hence, no conclusions can be drawn without further analysis of scattering within the system.

DISCUSSION AND CONCLUSIONS

The values of polarization with high spatial resolution found here are strongly influenced by the nonuniformity of exposures obtained in each orientation of the polarizer and by errors in registration of the images. Values near the limb are also influenced by variations in the physical size of the image. In order to improve the system, a method of monitoring the unpolarized light must be utilized. An obvious method is to remove a portion of the Polaroid which would allow a sample of the sky well away from the image to be used as a normalizing value for the intensity.

Excellent registration of the images and rigorous control of the plate scale are required in order to obtain consistent results. A system for obtaining the exposures more rapidly is desirable to reduce differential seeing conditions over the plate.

Acknowledgments. The author is indebted to S. Murrell who obtained the photographs and to A. Herzog for assistance in scanning the plates. This work is supported by NASA Grant No. NGL 32-003-001.

REFERENCES

Coffeen, D. L., and Gehrels, T. 1969. Wavelength dependence of polarization. XV. Observations of Venus. *Astron. J.* 74: 433–445.

Dollfus, A., and Coffeen, D. L. 1970. Polarization of Venus. I. Disk observations. *Astron. Astrophys.* 8: 251–266.

Honeycutt, R. K., and Chaldu, R. S. 1970. *The Wavelength dependence of the gradient for three Kodak spectroscopic emulsions.* Am. Astron. Soc. Photobulletin No. 2, 14.

SPATIAL DISTRIBUTION OF POLARIZATION OVER THE DISKS OF VENUS, JUPITER, SATURN, AND THE MOON

JOHN W. FOUNTAIN
University of Arizona

> The method of photographic subtraction, which superposes positive and negative photographs taken with the analyzer rotated through 90°, is used to analyze polarization photographs of Venus, Jupiter, Saturn, and the moon. For Venus, near 90° phase angle, variation in polarization in ultraviolet light appears to correspond generally with the position of the cloud markings. The northern hemisphere of Saturn shows higher polarization in blue light than does the rest of the planet. The polarization of the moon is shown to deviate significantly from Umov's law for reciprocity of polarization and reflectivity in certain regions.

Photographic subtraction involves placing a negative contact copy in perfect registration with the original plate. The result is a change in the image contrast. Thus, if the gamma of the copy is low, then the contrast of the two in register is reduced. As the gamma of the copy approaches unity, the contrast of the two in register goes to zero, and the image disappears. If the planet has been imaged through Polaroid filters that are orthogonally oriented, and a negative copy of the first is made, processed to a gamma of one and placed in register with the second, and if all densities on the copy lie on the straight-line portion of its characteristic curve, the remaining differences in density are due to differences in polarization.

This method has the advantage of displaying in a continuous way the variation in polarization over the disk and thus allows convenient qualitative interpretation. Polarization anomalies of small extent can easily be seen if present, and the photographic image provides an unambiguous means of comparing the relative brightness and polarization. The method can be especially useful in detecting changes in polarization related to unusual features or activity, such as after a South Equatorial Belt disturbance on Jupiter or during the rare periods when markings are seen on Venus in yellow light. The extent to which deviations from the albedo-polarization relationship occur on the moon may also be checked in this manner. If seeing, filter defects, or other problems make a pair of images unsuitable for analysis, it is usually quite obvious. The method does not provide values for the amount of polarization, but rather shows polarization differences. The accuracy is low compared to that of photoelectric work, and valid results depend critically upon uniform processing, good contrast control, and proper registration of the images.

The method of photographic subtraction has been employed by Zwicky (1953) to show color differences in stars and nebulae and by Whitaker (1972) to show color differences on the moon. The multiple detector array described by Serkowski (p. 164) retains many of the advantages of this method while greatly increasing the photometric accuracy.

A calcite crystal and an HNPB Polaroid filter were used as analyzers in this study. The high transmission at short wavelengths, the efficient polarization, and the possibility of recording two orthogonally polarized images simultaneously make the calcite crystal useful for this purpose. However, the crystal available was not thick enough to prevent overlapping of the images of the larger planets and the moon, and the focal planes of the two images are not precisely coplanar due to differences in path length through the crystal. The Polaroid filter had objectionable defects, and it was necessary to expose the images separately; thus rotation of the planet (in the case of Jupiter) between exposures, causing a relative displacement of the features, and the effect of differential image motion due to atmospheric seeing prevented proper registration over the entire image in some cases.

In Fig. 1, craters near the edge of the polarization image are clearly seen because a good fit was not possible over the entire image. This should not be attributed to polarization differences associated with the craters. The effect of differential image motion was minimized by selecting images of somewhat inferior resolution. With any polarizer, the method is limited by the photometric accuracy of the photographic emulsion. Good results may be obtained only if all densities lie on the straight-line portion of the characteristic curve, if the development of the film is extremely uni-

Fig. 1. *Upper photograph:* the full moon near Triesnecker. (Photograph from U.S. Naval Observatory.) *Lower photograph:* the polarization image. Brighter areas have higher polarization than darker areas.

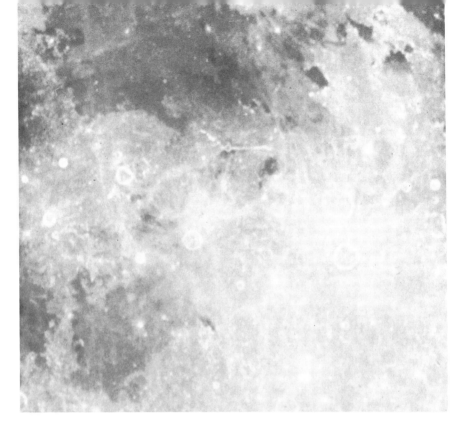

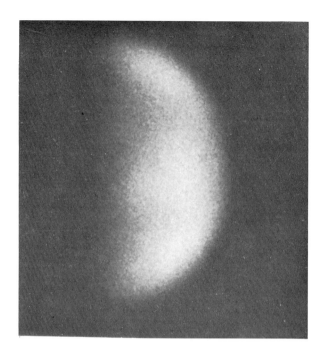

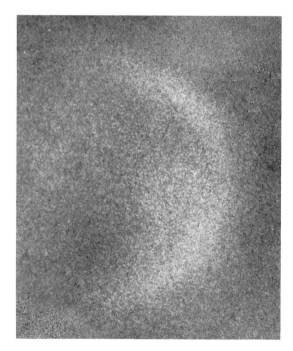

Fig. 2. *Top:* Venus, 1968 December 30 23:44:15 UT at 0.37 μm. Note the relatively high contrast ultraviolet markings. *Bottom:* polarization image. Note general agreement with upper photograph.

form, and if a gamma of one is achieved. Through the use of an automatic film processor, the problems in meeting these conditions were largely overcome.

The photography was done at the Catalina 154-cm telescope. The photographs are reproduced here north up, and the polarization is represented in the sense that lighter parts of the image have higher polarization than darker parts. In Figs. 2 and 3, the normal image shown for comparison is one of the images used in making the polarization image. In Fig. 1 a full-moon image is reproduced for albedo comparison. The polarization images are all reproduced to approximately the same contrast, and the components of the electric vector are parallel and perpendicular to the plane of scattering.

VENUS

The higher polarization of the poles of Venus in ultraviolet light, reported by Coffeen and Gehrels (1969), was observed, but compared with the poles, the polarization of the bright limb was variable with time. It was found that the variation in polarization in ultraviolet light over the disk was higher when the contrast of the ultraviolet markings was also high, and the distribution was generally the same as that of the markings, the brighter areas having higher polarization than the darker areas. In Fig. 2, the image of Venus taken at 0.37 μm shows rather high contrast markings. There is fairly close agreement between markings bright in ultraviolet light and large amounts of polarization.

JUPITER

The electric vector maxima are approximately radial on Jupiter (Hall and Riley 1968).[1] Since photographic subtraction shows the polarization resolved into two perpendicular components according to the orientation of the analyzer, it displays the polarization variations appropriately only at one part of the limb for one pair of plates. I have seen no polarization variations associated with any of the prominent belts or spots. Strong polarization is seen in blue light within 15°–20° of the limb, with the highest polarization being at the poles. The polarization is nearly constant over the disk at 0.8 μm.

---

[1] Also see p. 593.

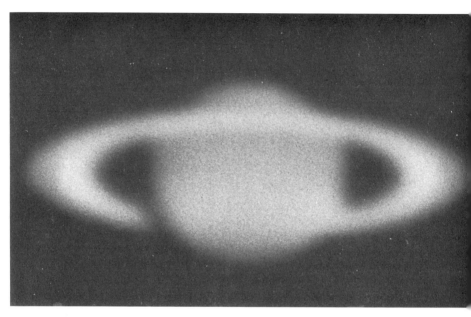

Fig. 3. *Left:* Saturn, 1970 February 6 03:05:38 UT at 0.44 μm. *Right:* polarization image. The northern part of the planet is more highly polarized than the rest of the disk. Variations in the rings are due to distortion by seeing.

SATURN

In blue light, the northern hemisphere of Saturn has higher polarization than does the rest of the planet. Small variations seen in the rings (Fig. 3) are not real but are due to imperfect alignment because of differential image motion. Like Jupiter, Saturn shows no correspondence between features in the comparison image and polarization differences. Unlike Jupiter, there does not appear to be any strong radial polarization, suggesting a rather different condition in the upper atmosphere.

MOON

Figure 1 shows the moon near the crater Triesnecker. Most of the darkest mare regions have high polarization, and some highland areas have lower polarization than the mare. The lowest polarization appears

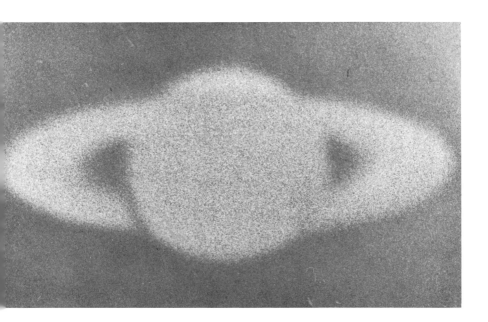

to be in the region of mixed highland and mare terrain in the vicinity of Boscovich. The polarization boundaries are not sharp compared with the clearly defined albedo boundaries, and they do not appear correlated with color or terrain boundaries. Polarization boundaries may cross varied terrain and even craters (Agrippa and d'Arrest). The wedge-shaped southern edge of Mare Vaporum south of Ukert has a similar polarization boundary, but the apex of the wedge lies within the mare and the wedge makes a smaller angle. Ray material is sometimes seen as dark diffuse lanes in the polarization image, like that to the right of Godin, but some prominent rays have no counterpart in the polarization image. Crater rims and rilles that appear very bright at full moon show no change in polarization from their floors or surroundings (crater rims visible on the polarization image are due to imperfect fit because of atmospheric seeing; when adjusted for a *local* fit, no polarization variation is seen). These data support the finding of Pellicori (1969) that a strict adherence to the inverse relationship between brightness and polarization (Umov's law) does not pertain to the moon.

Acknowledgments. I thank S. F. Pellicori and D. L. Coffeen for their helpful comments and G. P. Kuiper for his support of this work through the planetary photography program. S. M. Larson, C. Campbell, J. W. Barrett, and R. B. Minton participated in some of the observations. The planetary research program is supported by NASA Grant NGL 03-002-002.

REFERENCES

Coffeen, D. L., and Gehrels, T. 1969. Wavelength dependence of polarization. XV. Observations of Venus. *Astron. J.* 74: 433–445.
Hall, J. S., and Riley, L. A. 1968. Photoelectric observations of Mars and Jupiter with a scanning polarimeter. *Lowell Obs. Bull.* 7: 83–92.
Pellicori, S. F. 1969. Polarization-albedo relationship for selected lunar regions. *Nature* 221: 162.
Whitaker, E. A. 1972. Lunar color boundaries and their relationship to topographic features: A preliminary survey. *The Moon* 4: 348–355.
Zwicky, F. 1953. Species of cosmic matter. *Astron. J.* 58: 237–238.

DISCUSSION

REA: Have you considered using a vidicon system with a computer analysis of the output to get polarization maps?

FOUNTAIN: No, the straightforward and inexpensive method seemed most appropriate for a preliminary survey.

SERKOWSKI: At the present state of the art only the photon-counting television and photon-counting image tubes may give high photometric and polarimetric accuracy, on the order of 0.1% per resolution element. The television camera and the two-dimensional arrays of silicon diodes, relying on processing signals in analogue form, presently do not seem capable of achieving accuracy better than 1% or 2% per pixel. The accuracy is limited by the difficulty of gain calibration across the image, by constraints on linearity caused by limitations of dynamic range, and by imperfect erasing of the previous television frame. These limitations are absent for photon counting.

ANONYMOUS: Why is the sky density not the same as the image density when there is no polarization difference?

FOUNTAIN: Well-exposed original images have densities much greater than the sky density. In order to insure that all of the densities of the planet image lay on the straight-line portion of the characteristic curve of the copy film, it was often necessary to make exposures that led to sky densities somewhat off the straight-line portion.

ANONYMOUS: What is the accuracy of the method?

FOUNTAIN: The accuracy is not only dependent on the limitations of the photographic process but also on the person doing the work. Under best conditions, one certainly would not feel comfortable about differences corresponding to a polarization of 0.5%.

DIFFRACTION GRATING POLARIZATION

JAMES B. BRECKINRIDGE
Kitt Peak National Observatory

The purpose of these remarks is to warn astronomers of pitfalls in measurements made for photoelectric photometry and polarimetry with grating instruments. No new scientific information on diffraction grating polarization is provided, but a brief review of current work is given. Polarization introduced by narrow spectrograph entrance slits is discussed.

Study of polarization of light by diffraction gratings is an active field (Jovicevic and Sesnic 1972; McPhedran and Waterworth 1972). Analytic work is mainly concerned with computer solutions. Wave equations with a variety of boundary values for different groove profiles have been examined to obtain intensity as a function of angle and polarization state (Pavageau and Bousquet 1970; Kalhor and Neureuther 1971). Experimental work has involved measurements of the percentage polarization or of absolute power in polarized light as a function of grating angle (Breckinridge 1971; Stewart and Gallaway 1962). It appears impossible to manufacture or derive analytically a groove profile that does not introduce some polarization over a band width of several hundred angstroms.

One should view a grating spectrograph as an optical system containing a partial linear polarizer and a dispersive element (the grating). This partial linear polarizer has some interesting properties. The degree of polarization is wavelength and diffraction-order dependent. Breckinridge (1971) has shown that the percentage instrumental polarization can be as high as 90%.

At least two phenomena appear to contribute to grating polarization. These are: (1) the anomalies of Wood (1902), which appear at the Rayleigh (1907) points; and (2) a wave interaction introduced when groove dimension is nearly the same as the wavelength. Wood's anomalies appear as instrumentally induced depressions in continuum spectra. This appears at a wavelength and in a diffraction order such that radiation of the same wavelength but in a different order is leaving the grating at a grazing angle. Users of diffraction-grating instruments for studies of polarized light (e.g., solar magnetographs) are aware of the necessity of designing experiments in which grating polarization has little effect on data (Livingston and Harvey 1971).

However, unexpectedly large radiometric and photometric errors may arise in measurements made at the spectrograph focal plane if the radiation on the entrance slit is partially polarized. Let $T_\parallel$ stand for spectrograph transmissivity for incident light linearly polarized parallel to the entrance slit and $T_\perp$ be the transmissivity for light polarized perpendicular to the entrance slit. Breckinridge (1971) has measured the ratio $T_\parallel/T_\perp$ photoelectrically as a function of wavelength for five diffraction grating orders, with the 13.7-meter spectrograph of the McMath Solar Telescope at Kitt Peak. Figure 1 shows this ratio as a function of wavelength. Values of $T_\parallel/T_\perp$ vary from 0.3 to 20.

If light incident on the entrance slit is partially polarized by the feed optics or because the apparent source radiates partially polarized light, the transmissivity of the spectrograph changes its wavelength dependence compared to that measured for unpolarized light. From the image plane of the spectrograph, it is as though one looks through wavelength-

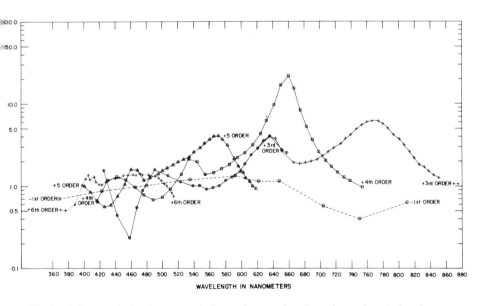

Fig. 1. Linear polarization transmission ratio as a function of wavelength for the 254 × 158 mm, 600-groove-per-millimeter grating in the 13.7-meter spectrograph of the McMath Solar Telescope at Kitt Peak. Reproduced here with permission from *Applied Optics* 10: 286–294, 1971.

dependent crossed partial polarizers. One polarizer represents the polarization of the light at the slit; the other, that caused by the grating.

Instrumental polarization in astronomical telescope-spectrograph feed mirrors arises primarily from non-normal angle of incidence of the light. Using the Fresnel equations, unpolarized light incident at a 45° angle (90° total deviation) from a fresh aluminum film in air gives 6% linear polarization in the reflected light. In many astronomical telescope imaging geometries, the amount of instrumental polarization at a spectrograph entrance slit depends on hour angle and declination of the telescope. This is particularly the case at the Coudé focus. The Coudé arrangement of the 223-cm reflector at Kitt Peak is given as an example. The degree of linear polarization is dependent on declination. Orientation of the preferred electric vector relative to the entrance slit depends on the hour angle of the telescope. At a wavelength of 5500 Å, the Fresnel equations give 9% polarization at the equator and 14% when the telescope is pointed poleward. If one were to design a grating spectrograph scanner for use at this focus or to perform broadband photometry using spectrograph-image plane masks with photomultipliers, large colorimetric errors would be found, requiring a polarization analysis of the system to correct the data.

Let us consider an example. Assume we wish to measure the color of a star with filter band passes 200 Å wide, one at 5500 Å, the other at 7500 Å. We shall use the third order of the grating, whose properties are given in Fig. 1, and assume the light on the entrance slit is 10% polarized. Inspection of the figure shows that $T_\parallel/T_\perp$ at 7500 Å is about five times its value at 5500 Å. The maximum percentage photometric error was given by Breckinridge (1970) as

$$E_{max} = 100 \times p \left(1 - \frac{T_{\lambda_1}}{T_{\lambda_2}}\right),$$

where p is the degree of linear polarization incident on the slit; T_{λ_1} the linear polarization transmissivity or $T_\parallel/T_\perp$ at wavelength λ_1; and T_{λ_2} the linear polarization transmissivity or $T_\parallel/T_\perp$ at wavelength λ_2. The percentage error for the above example is then

$$E_{max} = (100) \times (0.10) \times (1.0 - \frac{1}{5}) = 8\%.$$

This should be readily detectable, even on a photographic plate. A star may be observed to have an unusual color for its spectral type, when, in fact, the polarization transmissivity of the grating spectrograph may be responsible for altering the intensity distribution.

An example of current interest is the case of observing a star immersed in a polarizing nebula, such as the Crab Nebula. The apparent radiation from the object consists of two parts: that directly from the star (unpolarized), and that from the nebula (polarized). If a grating spectrograph were to be used, it is possible that, without care, the polarization vector of the nebula could be crossed with that associated with the grating. The spectrogram would then show a bias toward the light of the illuminating star and not the nebula. Of course, this technique can be used to artificially enhance the spectrum of the nebula.

Many small grating spectrographs are used with a narrow entrance slit. An interesting consequence of the measures obtained by Jones and Richards (1954) and the measures given here on the polarization transmissivity of a grating is revealed. Unpolarized light of 1 μm wavelength becomes 30% polarized on passing through a steel slit 20 μm wide. Hence, for grating spectrographs using narrow entrance slits, the transmissivity with wavelength will be dependent on slit width. Particular caution is required for infrared spectroscopy. Light having a wavelength of 5 μm may be passed through a 200-μm slit. Unintentionally, the observer may illuminate the grating in polarized light and alter a calibration.

Photometric errors due to irradiating a grating with partially polarized light will, in general, affect only broadband data. Profiles of individual absorption features will remain relatively unaltered. The exception are those lines near the anomalies described by Wood (1902), where an instrumentally induced steep slope in the continuum line may occur.

In conclusion, we have shown that observers should be most careful when recording spectra-photometric data with grating instruments. Astronomers attempting to obtain precision photometric data, without concern for the polarization, may find that polarization inadvertently reduces the precision of their work.

REFERENCES

Breckinridge, J. B. 1970. Polarization and image-forming properties of a large grating spectrograph. M.S. Thesis Univ. of Arizona. Issued as *Optical Sciences Center Technical Report 56,* Univ. of Arizona, Tucson.

———. 1971. Polarization properties of a grating spectrograph. *Applied Optics* 10: 286–294.

Jones, R. V., and Richards, J. C. S. 1954. The polarization of light by narrow slits. *Proc. Roy. Soc. Lond.* A225: 122–135.

Jovicevic, S., and Sesnic, S. 1972. Diffraction of a parallel- and perpendicular-polarized wave from an Echelette grating. *J. Opt. Soc. Am.* 62: 865–877.

Kalhor, H. A., and Neureuther, A. R. 1971. Numerical method for the analysis of diffraction gratings. *J. Opt. Soc. Am.* 61: 43–48.

Livingston, W., and Harvey, J. W. 1971. The Kitt Peak magnetograph I. Principles of the instrument. *Kitt Peak Contribution No. 558.*

McPhedran, M. C., and Waterworth, M. D. 1972. A theoretical demonstration of properties of grating anomalies (S-polarization). *Optica Acta* 19: 877–892.

Pavageau, J., and Bousquet, J. 1970. Diffraction par un réseau conducteur nouvelle méthode de résolution. *Optica Acta* 17: 469–478.

Rayleigh (Strutt, J. W., 3rd Baron Rayleigh) 1907. Note on the remarkable case of diffraction spectra described by Prof. Wood. *Phil. Mag.* 14: 60–65.

Stewart, J. E., and Gallaway, W. S. 1962. Diffraction anomalies in grating spectrophotometers. *Applied Optics* 1: 421–429.

Wood, R. W. 1902. On a remarkable case of uneven distribution of light in a diffraction grating spectrum. *Phil. Mag.* 4: 396–402.

ATTEMPT TO MEASURE STELLAR MAGNETIC FIELDS USING A LOW-LIGHT LEVEL TELEVISION CAMERA

G. G. FAHLMAN, J. W. GLASPEY, O. JENSEN,
G. A. H. WALKER and J. R. AUMAN
Institute of Astronomy and Space Science
University of British Columbia, Vancouver

A refrigerated Isocon television camera has been used as the detector in a magnetograph arrangement of the vacuum spectrograph of the McMath telescope at Kitt Peak in attempts to measure weak longitudinal magnetic fields in the brighter stars. Displacement of selected absorption lines observed alternately in circularly polarized light of opposite handedness was measured from the difference spectrum. Signal-to-noise was improved by appropriate filtering of the data. Jitter of the order of 1/10 to 1/20 of a data point in the scanning raster appears to impose the main limitation on precision at this time.

Spectral lines are distorted by the presence of a magnetic field in a stellar atmosphere. For a uniform field parallel to the line of sight, the wavelength of a line in levogyrate polarized light is displaced relative to the line in the dextrogyrate polarization[1] by an amount

$$\triangle \lambda_H = 0.93\, Hz\lambda^2 \times 10^{-12}, \qquad (1)$$

where H is the field strength in gauss, λ is the mean wavelength of the line, and z is calculated from the term values of the transition to account for the effects of spin-orbit coupling on the classical Zeeman pattern.

Resolved Zeeman patterns have been observed in only three Ap stars, thereby permitting a calculation of the mean surface magnetic field (Preston 1971). More typically, the Zeeman components are blended, and it is possible to calculate only the integrated longitudinal component of the magnetic field from the displacements between line profiles in the levogyrate and dextrogyrate spectra. In photographic spectroscopy, these are recorded simultaneously and immediately adjacent to each other (see, e.g., Babcock 1960). The precision of the technique is

---

[1] The terms levogyrate and dextrogyrate are used throughout this paper to refer respectively to the left-hand and right-hand modes of circular polarization.

limited by the information-storage capacity of the plate and the number of lines recorded, which can be on the order of 100. Preston (1969) has noted that probable errors of ±150 gauss are to be expected in measuring a single plate.

If a photomultiplier is illuminated by the light within one wing of an absorption line alternately in levogyrate and dextrogyrate polarization, the modulation of the output signal is proportional to $\triangle \lambda_H$ and to the central depth of the line. In phase-sensitive detection, the signal-to-noise ratio that can be achieved in measuring the modulation is limited only by the exposure time and the magnitude of the star. This technique has been used very successfully in measurements of solar fields and has been applied to stellar observations by Landstreet.[2]

Severny (1970) has used a single-channel photoelectric scanner in a phase-sensitive detection between levogyrate and dextrogyrate spectra of a number of bright stars. From modulation in the difference spectrum, he has estimated the Zeeman shift for the line λ4254 Cr I and calculated the associated fields.

In this paper, we shall discuss the use of an Image Isocon television camera as a multichannel detector in a magnetograph arrangement on the vacuum spectrograph of the McMath telescope at Kitt Peak to measure longitudinal magnetic fields of bright stars. Of the methods mentioned above, this resembles most closely the photographic technique.

EXPERIMENTAL ARRANGEMENT

For the observations to be discussed here, the analyzing optical train consisted of a fixed quarter-wave plate 30 cm within the telescope focus and an image slicer followed by a polarizing beam splitter which gave two spectra in orthogonally plane polarized light (referred to the beam splitter) at the focus of the spectrograph camera. These two spectra were reduced to 0.5 mm height by two cylindrical lenses 44 mm below the camera focus. No compensation was made for circular polarization introduced by the telescope mirrors, as this effect was expected to be small in view of measurements made by solar astronomers.

The image slicer gave an effective entrance-slit width of 250 μm, which was close to the spectrograph resolution and matched the spatial resolution of the Image Isocon English Electric P850 tube that was used. The tube was cooled to about −10°C in order to eliminate significant deterioration of the Modulation Transfer Function through charge spread on the glass target of the camera. During an exposure, the camera reading beam was left blanked which allowed analogue integration at the target.

---
[2] See p. 981.

On reading out, the target was scanned in 30 ms by an 840-line raster normal to the two spectra, and the camera video output was integrated and digitized on line to a 12-bit word at each crossing of a spectrum. The data were transferred to magnetic tape after checking in a small control computer. On completion of read-out, the target was wiped for a further 9 frames to eliminate any residual charge pattern. Simultaneously, the fixed quarter-wave plate was rotated 90° before the next integration. As a result of this procedure, each spectrum was of alternate polarization. A full description of the Image Isocon detection and data acquisition system has been given by Walker et al. (1971).

Spectra of the brightest stars were obtained at 7 mm/Å and the fainter ones at 0.9 mm/Å. This corresponds to resolutions of 0.03 Å and 0.25 Å, respectively.

Under these circumstances, small guiding errors would be expected to show up in differences between single frames. However, the techniques of taking polarizations alternately and accumulating several hundred integrations should have largely eliminated the guiding errors. Small collimation errors, possibly introduced by switching the quarter-wave plate, together with a number of other small systematic effects should have been eliminated by recording both spectra simultaneously in addition to being interchanged on alternate integrations.

ANALYSIS

The principle of the measurement technique may be outlined as follows. We have two orthogonally polarized line profiles, one dextrogyrate $R(\lambda)$ and one levogyrate $L(\lambda)$. We assume that both profiles may be described by the same intrinsic function $P(\lambda)$, but that one — say $L(\lambda)$ — is shifted, or delayed, relative to the other by a small amount $\triangle \lambda_H$ due to the Zeeman effect in a magnetic field of strength H; i.e.,

$$R(\lambda) = P(\lambda), L(\lambda) = P(\lambda + \triangle \lambda_H). \qquad (2)$$

The difference spectrum, defined as

$$D(\lambda, \triangle \lambda_H) = R(\lambda) - L(\lambda),$$

becomes

$$D(\lambda, \triangle \lambda_H) = - \triangle \lambda_H P'(\lambda) + O(\triangle \lambda_H)^2. \qquad (3)$$

For small shifts, the amplitude of the difference spectrum at a fixed wavelength is proportional to $\triangle \lambda_H$, the quantity we wish to measure.

A null technique for measuring $\triangle \lambda_H$ was developed and used to analyze observations made in April 1971 (Auman, Jensen, and Walker

1971). This involves applying an artificial delay, $\triangle\lambda_s$, to $L(\lambda)$ and constructing the difference spectrum

$$D(\lambda, \triangle\lambda_H, \triangle\lambda_s) = R(\lambda) - L(\lambda - \triangle\lambda_s)$$

which, to first order, becomes

$$\begin{aligned}D(\lambda, \triangle\lambda_H, \triangle\lambda_s) &= -(\triangle\lambda_H - \triangle\lambda_s) P'(\lambda) \\ &= \triangle\lambda_s P'(\lambda) - \triangle\lambda_H P'(\lambda).\end{aligned} \quad (4)$$

To exploit the linear relationship between the amplitude and the delay, a series of delays, $\triangle\lambda_s$, is applied and a plot of $D(\lambda, \triangle\lambda_H, \triangle\lambda_s)$ versus $\triangle\lambda_s$ is constructed. This plot is a straight line, and the intercept on the $\triangle\lambda_s$-axis gives the value of $\triangle\lambda_H$. This method is, however, sensitive to scaling and noise within the data.

The technique adopted for the present observations is a modification of the above and represents an attempt to lessen the sensitivity of the measurement to the scaling and the noise. Consider the function

$$Y(\triangle\lambda_s) = \sum_\lambda D^2(\lambda, \triangle\lambda_H, \triangle\lambda_s), \quad (5)$$

which, to terms of order $(\triangle\lambda_s - \triangle\lambda_H)^2$, may be written

$$Y(\triangle\lambda_s) = [(\triangle\lambda_s)^2 - 2\triangle\lambda_H\triangle\lambda_s + (\triangle\lambda_H)^2] \sum_\lambda [P'(\lambda)]^2. \quad (6)$$

It is apparent that $Y(\triangle\lambda_s)$ is a parabola that has a minimum when $\triangle\lambda_H = \triangle\lambda_s$. The advantages of using this function to determine $\triangle\lambda_H$ are: (1) the entire profile is used to determine the Zeeman shift; (2) $Y(\triangle\lambda_H)$ is a measure of the variance between the two profiles, and in principle, gaussian noise will not affect the value of the measured shift; (3) if the line profile is symmetrical, scaling differences between $R(\lambda)$ and $L(\lambda)$ do not affect the measurement. Of course, in reality, the noise within a line profile is not strictly gaussian nor is it always possible to use symmetric profiles. Nevertheless, this method seems to offer the best chance of minimizing spurious results and was the one finally adopted.

A possible source of error, which has not been mentioned, is the crossover effect. This refers to the observation that in most magnetic stars the dextrogyrate and levogyrate profiles show systematic differences, particularly, but not always, when the polarity of the magnetic field changes sign. This invalidates our basic assumption expressed by Equation (2). We did not, however, discover any convincing evidence for crossover effects in our observational data.

A single observation made with the Isocon system consists of a series of exposure frames in which each frame includes both a dextrogyrate and a levogyrate spectrum in separate positions on the tube face. These positions are interchanged on alternate frames that are subsequently added together to form two sets of mean spectra, one set corresponding to the upper position, the other, to the lower position. These spectra are analyzed separately because of the different response of the tube in the two positions.

The digital stellar spectra so obtained are inaccurate, to some degree, due to noise within the system. In fact, as illustrated by Fig. 1, the noise dominates the mean-difference spectrum. Clearly, if the data are to be reliably analyzed, the noise must be suppressed. The noise characteristics of the Isocon system are not known a priori, and it is necessary to construct a noise model with the observational data at hand. Such a model was constructed by differencing two of the individual frames within a single observation. Under the assumption that the signal should increase linearly with the number of frames added, whereas the noise should

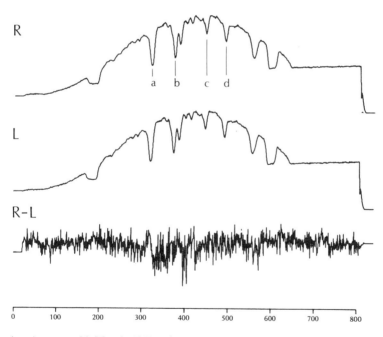

Fig. 1. Arcturus, 29 March 1972. The two upper tracings show the mean dextrogyrate and levogyrate spectra respectively, taken from the same position on the Isocon camera. The lower tracing shows the difference between the two. The lines indicated are: (a) λ5348 Cr I, (b) λ5349 Ca I, (c) λ5351 Ti I, and (d) λ5352 Co I. The scale on the abscissa indicates sample points.

increase only as the square root of the number, it is possible to extract a noise model by comparing the mean difference and the individual difference spectra. Unfortunately, the Fourier characteristics of the noise model were found to be very similar to those of the mean-difference spectrum, and so optimum gathering techniques will not work (see, e.g., Brault and White 1971).

The signal we are searching for results from slight differences between the positions of the line profiles in the two polarized spectra. It is easy to see that such a signal should be sinusoidal with a period equal to the width of the line profile. For the high-dispersion data, the lines are sufficiently well sampled, so that the expected signals have a fairly low Fourier frequency. Therefore, we can apply bandpass filters to the difference spectra and remove all the high-frequency components that are expected to be mostly noise. A typical filter is shown in Fig. 2. This filter and all others used in this study were truncated with a Lancoz spectral window in order to minimize the height of the side lobes. The high-frequency cutoff is determined by the width of the narrowest spectral line to be analyzed, while the low-frequency cutoff should accommodate at least the broadest line. In this context, the width of a spectral line

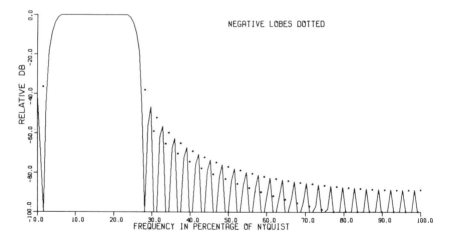

Fig. 2. An example of the bandpass filters used to smooth the data. This particular filter is designed to pass all Fourier frequencies between 5% and 25% of the Nyquist frequency. The Nyquist frequency is defined as $f_N = \frac{1}{2}\Delta$ and is the highest frequency that can be detected with data sampled at intervals Δ. Here $\Delta = 1/N$ where $N = 840$, the number of sample points per spectrum. The decibel (DB) scale is such that a difference of 20 DB corresponds to an amplitude ratio of 10.

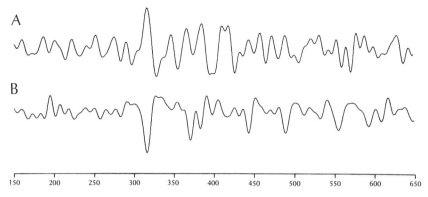

Fig. 3. The mean-difference spectra of Arcturus after applying the filter shown in Fig. 2. The upper tracing, A, corresponds to the mean difference shown in Fig. 1. Tracing B is the mean difference for the other position on the camera. Note that only the central region of the spectrum in Fig. 1 is shown on an expanded scale.

refers only to the number of sample points involved rather than a physical width in, say, angstroms.

The result of applying the filter of Fig. 2 to the mean-difference spectrum of Fig. 1 is shown in Fig. 3. Clearly, the application of the filter has considerably improved the visual appearance by suppressing the high-frequency noise. Appropriate bandpass filters were applied to all the high-dispersion observations discussed here.

The lower-dispersion observations of fainter stars, including the standard magnetic stars β CrB and 78 Vir, were analyzed in much the same way but without success. The principal difficulty is that the line widths, particularly in the case of the sharp-line standard stars, are so narrow that any signal in the difference spectra is seriously contaminated by the dominant high-frequency noise in the data. A really satisfactory method of separating the signal from the noise in such cases has not been found. So far, we have been unable to extract reliable results from the low-dispersion data, and consequently, we shall only present results for the high-dispersion data. We emphasize that the noise problem is much less severe for the high-dispersion data and we feel that separating the data in this way is justifiable.

RESULTS

Up to thirty values of $\triangle \lambda_s$ in the range ($-1.5 \leq \triangle \lambda_s \leq 1.5$ sample points) were used to establish the function $Y(\triangle \lambda_s)$ for each line to be analyzed. The lines typically are sampled at 15 points or more so that the first-order approximation to the difference spectrum should be valid.

A separate parabola was then fitted to each function $Y(\triangle \lambda_s)$ by a least-squares method. The minimum of the fitted parabola was taken to be the value of $\triangle \lambda_H$ (in sample points). A dispersion curve, angstroms versus sample points, was constructed for each night of observations and used to convert $\triangle \lambda_H$ to angstroms. The effective magnetic field H, was then calculated from Equation (1) for each line. The spectra obtained in the two positions on the Isocon tube face were analyzed separately and, in most cases, a systematic difference between the measured values of $\triangle \lambda_H$ was found. The origin of this effect, which differed from observation to observation, is not certain. The Zeeman shift finally adopted was the simple mean of the two measurements, as this seemed to give the most satisfactory results for the lunar spectra.

The results for all the high-dispersion observations are summarized in Table I. The mean field quoted is the unweighted average for all the

TABLE I

High Dispersion Observations of Zeeman Effect

Object	Date 1972	Wavelength Region (Å)	No. Lines Used	Mean Field ± St. Dev. (gauss)
α Ori	March 28	5348	2	(16.5 − 11.3)
α Ori	March 29	"	2	(106 − 82.4)
Moon	"	"	3	16.1 ± 29.6
α Boo	"	"	4	9.3 ± 23.8
Moon	"	"	3	17.8 ± 47.9
α Sco	"	"	2	(45.0 − 16.4)
Moon	"	5170	4	1.8 ± 23.4
α CMa	"	"	3	7.1 ± 38.0
α CMa	"	"	3	− 10.8 ± 22.0
α CMa	March 30	"	3	− 4.2 ± 30.3
α CMi	"	"	5	6.3 ± 37.7
Sunspot	March 31	"	5	68.5 ± 12.8

lines used, and the standard deviation refers only to this mean. The errors introduced by fitting a parabola to $Y(\triangle \lambda_s)$ are entirely negligible by comparison. In general all the identifiable, unblended atomic lines in the spectra were measured for the Zeeman effect, but due to the limited spectral range covered — some 6 Å — the number of lines available was always small, and the few measures that seemed obviously discrepant were not used to calculate the mean field.

An inspection of Table I shows that we did not positively detect a magnetic field in any of the five bright stars observed. The results for these stars do not differ from those obtained by analyzing linear spectra in precisely the same way. A rather weak field was detected in a sunspot observed on 31 March 1972. The small field may be explained by noting that the spot was small, and high winds during the observation made guiding extremely difficult. It is perhaps worth noting that the quoted mean field for the sunspot corresponds to a shift of approximately $\triangle \lambda_H = 0.09$ sample points for a line with $z = 1$.

The only published measurements of comparable accuracy are those of Severny (1970). For α CMa, he found a positive field of 38 gauss $\pm$ 12 (p.e.) using the single line $\lambda 4254$ Cr I. Our experience with the Isocon indicates that measurements of stellar fields based on a single line can be misleading. Our results, however, can not be considered contradictory to his. For α CMi, Severny reports that no effect was detected within slightly lower limits than what we report.

In summary, it appears that the Isocon camera can be used, at least at very high dispersion, to detect stellar magnetic fields with an accuracy superior to the standard photographic technique and comparable to the other photometric techniques now being used.

Acknowledgment. This work was supported by the National Research Council of Canada.

REFERENCES

Auman, J. R.; Jensen, O. G.; and Walker, G. A. H. 1971. A search for magnetic fields in four bright stars. *Bull. Am. Astr. Soc.* 3: 442.

Babcock, H. W. 1960. Measurement of stellar magnetic fields. *Stars and stellar systems* (W. A. Hiltner, ed.), Vol. 2, pp. 107–125. Chicago: Univ. of Chicago Press.

Brault, J. W., and White, O. R. 1971. The analysis and restoration of astronomical data via the fast Fourier transform. *Astron. Astrophys.* 13: 169–189.

Preston, G. W. 1969. The periodic variability of 78 Virginis. *Astrophys. J.* 158: 243–249.

———. 1971. The mean surface fields of magnetic stars. *Astrophys. J.* 164: 309–316.

Severny, A. 1970. The weak magnetic fields of some bright stars. *Astrophys. J.* 159: L73–L76.

Walker, G. A. H.; Auman, J. R.; Buchholz, V.; Goldberg, B.; and Isherwood, B. 1971. Direct analysis of spectra by Image Isocon and computer. *Publ. Roy. Obs. Edinburgh* 8: 86–89.

THE HIGH ALTITUDE OBSERVATORY STOKES POLARIMETER

T. BAUR, G. W. CURTIS, H. HULL and J. RUSH
High Altitude Observatory
National Center for Atmospheric Research

A photoelectric polarimeter or Stokesmeter spectrograph intended for observing the sun has been constructed at the High Altitude Observatory. It measures all four Stokes parameters simultaneously as a function of wavelength. The instrument currently uses the existing 40-cm coronagraph and 13-m Littrow spectrograph at Sacramento Peak Observatory. Encoding of the polarization information into intensity modulations of the light beam is achieved by two electro-optic crystals, an achromatic quarter-wave plate, and a linear polarizer at the prime focus of the coronagraph.

The High Altitude Observatory Stokes polarimeter or Stokesmeter spectrograph operates as a polarization-measuring solar spectrograph. By measuring all four Stokes parameters, it provides a complete description of the polarization of the radiation field as a function of wavelength inside a line profile.

The heart of the instrument is the modulation and calibration package, which is mounted at the prime focus of the telescope (Fig. 1). It modulates the intensity of the incoming light beam in a manner that is dependent on the polarization of this beam. The light leaving the modulator has the polarization description encoded into these intensity modulations. The intensity of the incoming light is proportional to the D.C. level of the exit beam, and the values for the other three Stokes parameters are proportional to the strength of the light modulation at three specific frequencies.

A limiting aperture at the prime focus of the telescope defines the region of the sun observed, and an exit slit on the spectrograph defines the wavelength of the observation. A pair of photomultipliers at the exit slit detect the intensity modulations, and phase-locked amplifiers measure the power at the three frequencies. Taken together with a D.C. measurement of the light level, these measurements provide, after calibra-

THE H.A.O. STOKES POLARIMETER

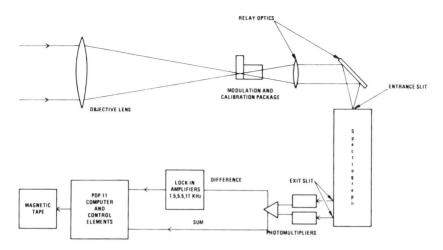

Fig. 1. A schematic diagram of the polarimeter.

tion, the values of the four Stokes parameters at a particular wavelength. Scanning the spectrum past the exit slit gives a wavelength scan in all four Stokes parameters.

THE MODULATOR

A combination of two electro-optic crystals (KD*P), an achromatic quarter-wave plate, and a calcite rhomb arranged according to a scheme suggested by J. Beckers (1969) modulate the light intensity (Fig. 2). The time variations of the retardations delivered by the two electro-optic crystals are

and
$$\delta_1 = A_1 \sin \Omega_1 t \qquad (1)$$
$$\delta_2 = A_2 \Omega_2 t \qquad (2)$$

for the first and second crystals, respectively. The retardation frequencies, $\Omega_1/2\pi$ and $\Omega_2/2\pi$, are 5.5 kilohertz and 1.0 kilohertz. The fast axis of the first crystal is either horizontal or vertical, and the fast axes of the second crystal and the quarter-wave plate are at 45° to the horizontal. The calcite rhomb splits the light into two beams, one that is linearly polarized horizontally and another that is linearly polarized vertically.

From the Mueller matrices for this system (Shurcliff 1962),[1] the intensities of the two beams are

$$I_H = \tfrac{1}{2}I - \tfrac{1}{2}Q \sin \delta_2 - \tfrac{1}{2}U \sin \delta_1 \cos \delta_2 - \tfrac{1}{2}V \cos \delta_1 \cos \delta_2 \qquad (3)$$

and

$$I_V = \tfrac{1}{2}I + \tfrac{1}{2}Q \sin \delta_2 + \tfrac{1}{2}U \sin \delta_1 \cos \delta_2 + \tfrac{1}{2}V \cos \delta_1 \cos \delta_2, \qquad (4)$$

---
[1] See p. 175.

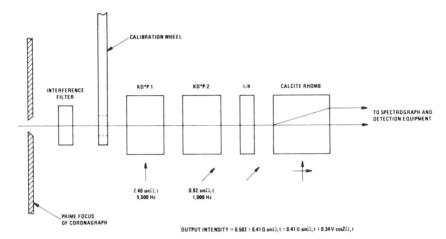

Fig. 2. Analyzer package containing calibration and modulation optics. The arrows below the optical elements indicate directions of retarder fast axes and polarizer transmission directions as seen along the optical axis.

where I_H and I_V are the intensities in the horizontally and vertically polarized beams, respectively. The horizontal direction is chosen so that it corresponds to the geocentric east-west direction in the plane of the sky. Using the Bessel function expansions of sin $(A \sin \Omega t)$ and cos $(A \sin \Omega t)$ and keeping only first-order terms, Equations (3) and (4) for the intensity in the two beams become

$$I_{H,V} = \tfrac{1}{2} I \mp Q J_1(A_2) \sin \Omega_2 t \mp U J_1(A_1) \sin \Omega_1 t$$

$$\left[J_0(A_2) + 2 J_2(A_2) \cos 2 \Omega_2 t \right] \mp \tfrac{1}{2} V \left[J_0(A_1) + 2 J_2(A_1) \cos 2 \Omega_1 t \right]$$

$$\left[J_0(A_2) + J_2(A_2) \cos 2 \Omega_2 t \right], \quad (5)$$

where the upper signs are for the horizontally polarized beam.

To eliminate the D.C. term from the V term we require that

$$J_0(A_1) = 0. \quad (6)$$

To minimize the retardation required from the first crystal, we choose the smallest value of A_1 that will satisfy Equation (6), $A_1 = 2.405$. This value also gives the maximum modulation, at the frequencies to be monitored, for a given polarization of the incoming sunlight. The low-order terms in Equation (5) give signals at four frequencies; 0, Ω_2, Ω_1, and $2\Omega_1$:

$$\text{D.C.} : \tfrac{1}{2} I \quad (7)$$

$$\Omega_2 : \tfrac{1}{2} Q J_1(A_2) \sin \Omega_2 t \quad (8)$$

$$\Omega_1 : \tfrac{1}{2} U J_0(A_2) J_1(A_1) \sin \Omega_1 t \quad (9)$$

$$2\Omega_1 : V J_0(A_2) J_2(A_1) \cos 2\Omega_1 t. \quad (10)$$

Signals at all other frequencies are ignored. The higher-order terms neglected in writing Equation (5) can be shown to introduce no crosstalk terms at these frequencies that are larger than 10^{-9} of the main signals produced by a given polarization.

To keep all four signals in Expressions (7)–(10) about the same size for like magnitudes of the Stokes parameters, we set

$$J_0(A_2) J_1(A_1) = J_1(A_2) \quad (11)$$

from Expressions (8) and (9). Solving this gives $A_2 = 0.923$. Then Equation (5) can be written as

$$I_{H,V} = 0.500\, I \mp 0.414\, Q \sin \Omega_2 t$$
$$\mp 0.414\, U \sin \Omega_1 t \mp 0.345\, V \cos 2\Omega_1 t. \quad (12)$$

Each beam is detected by a separate photomultiplier at the exit slit of the spectrograph, and the output signals from these two photomultipliers are summed to give the D.C. signal

$$I' = a_1 I. \quad (13)$$

The signals are differenced to give signals at frequencies Ω_2, Ω_1, and $2\Omega_1$,

$$Q' = a_2\, 0.928\, Q \sin \Omega_2 t \quad (14)$$
$$U' = a_3\, 0.928\, U \sin \Omega_1 t \quad (15)$$
$$V' = a_4\, 0.690\, V \cos 2\Omega_1 t, \quad (16)$$

where I', Q', U', and V' are the signals recorded from the detection equipment that measures power at the four frequencies. The a values are gain factors in the electronics.

CALIBRATION

In principle, all that is necessary to compute I, Q, U, and V from Equations (13)–(16) is to introduce light of known polarization and intensity into the polarimeter to determine the a values. In fact the four output signals have more complicated dependencies on the Stokes parameters due to crosstalk. The main sources of this crosstalk are small misalignments of the retarder axes and off-axis rays through the retarders. In addition to crosstalk, the signals depend on the polarizing properties of the telescope objective lens and an interference filter (Fig. 2), since these optical elements act on the sunlight before the calibration optics.

We find that the following linear equations adequately describe the four signal dependencies for the purposes of calibration, provided the values of the Stokes parameters introduced for calibration are of the same order of magnitude as the observed values.

$$I' = \alpha_1 I + C_1 \tag{17}$$

$$Q' = \alpha_2 I + \beta_2 Q + \gamma_2 U + \delta_2 V + C_2 \tag{18}$$

$$U' = \alpha_3 I + \beta_3 Q + \gamma_3 U + \delta_3 V + C_3 \tag{19}$$

$$V' = \alpha_4 I + \beta_4 Q + \gamma_4 U + \delta_4 V + C_4. \tag{20}$$

The calibration technique must determine all 17 of the constants in Equations (17)–(20). The crosstalk puts rather stringent requirements on the accuracies to which we must know the polarizations introduced by the calibration optics. For example, if a calibration light beam supposedly containing only U polarization is fed through the modulator and if a small V' signal as well as the U' signal is observed, we need to know if the V' signal is crosstalk or if there is a bit of V polarization in the calibration beam. The extent to which the crosstalk can be calibrated out is limited by the accuracy to which this contamination of the calibration beam is known.

We calibrate the polarimeter by introducing a series of linear polarizers and a retarder of known optical constants into a beam of unpolarized light. The source of unpolarized light is a 50-Å band of continuum from the center of the solar disk, centered at the wavelength used for the observation. Dollfus (1965) finds that the amount of polarization in such broad spectral bands of sunlight is less than 0.01%. The linear polarizers are Polaroids or tilted glass plates, depending on the desired magnitude of Q or U, and the retarder is an achromatic quarter-wave plate. Varying the tilt of the glass plate varies the linear polarization. The quarter-wave plate following the glass plate supplies the circular polarizations. In all there are ten calibration points, five positive and five negative, for each of Q, U, and V. The polarizations vary from 100% to 0.01% in steps of approximately a decade.

The 50-Å band of sunlight from an interference filter on the analyzer is fed through the exit slit of the spectrograph by the grating in zero order. Neutral density filters at the entrance slit to the spectrograph reduce the intensity of the calibration beam to something near that of the observation. The four signals I', Q', U', and V' are sufficiently linear functions of intensity that it is not necessary to calibrate at more than one intensity. The polarimeter performance, and in particular the electro-optic crystal performance, is sufficiently stable that it is not necessary to recalibrate more than about once every hour. Also the instrumental polarization from the telescope objective lens and the interference filter is small, less than 0.1%, and relatively constant in spite of the stresses induced in these elements by solar heating.

Deviations of the calibration beam from ideal pure polarization forms are caused by:
1. Rotational misalignments about the optical axis of the calibrating optics,
2. Stress-induced retardations from the glass plates,
3. Departure of the wave plate from exactly quarter-wave retardation,
4. Dichroism of the quarter-wave plate.

Rotational misalignments of calibration optical elements never exceed 6 arcmins. We measure an average stress-induced retardation from the glass plates of 0.0012 waves, and the accuracy of this value as well as our measurement of the wave-plate retardation is $\pm$ 0.005 waves. Our measurement of the dichroism in the wave plate gives a ratio of principal transmittances of $K_2/K_1 = 0.999 \pm 0.003$. In most cases these uncertainties prevent measuring values for Q, U, or V less than 1% of the largest Stokes component in the measurement.

TEST RESULTS

We have successfully operated a test version of the modulator and calibration package on the 40-cm coronagraph and the 13-m Littrow spectrograph in the Big Dome at Sacramento Peak Observatory. This instrument differed from the version described above in that it operated as a single-beam instrument, and some of the details of the calibration scheme were different. Also, the instrument was not under computer control, and the data were recorded on chart paper rather than magnetic tape. Figure 3 shows one of the uncalibrated chart recordings made during our tests. The line scanned here is H α from a part of a prominence.

For this test version of the instrument as well as for the completed instrument, a typical field-of-view size for an observation is 3 $\times$ 80 seconds of arc. For the Sacramento Peak Littrow spectrograph, this implies a spectral resolution of about 0.02 Å. This spectral resolution can be improved to about 0.005 Å by reducing the narrow dimension of the field of view, but this increases the necessary integration time. Typical integration times to obtain a Stokes spectral scan of a line with a 3 $\times$ 80 arc second field are one minute for a solar disk observation and five to ten minutes for a prominence observation in Hα or D_3. At these integration times, polarizations of the order of 0.05% are easily detectable provided the seeing is good. By using both beams from the modulator we should be able to improve this sensitivity. This not only doubles the available light but also eliminates both the effect of spurious high-frequency intensity fluctuations from poor seeing and scattered light

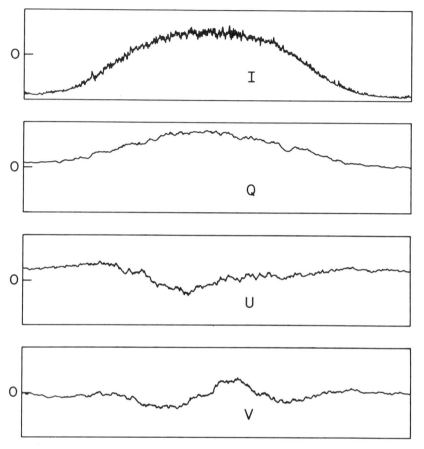

Fig. 3. An uncalibrated wavelength scan in all four Stokes parameters of the Hα emission line from part of a prominence. A full-scale deflection on the Q, U, and V traces represents a polarization change of about 0.1%.

from airborne dust particles. If these spurious intensity fluctuations are not eliminated, the measured polarizations will be incorrect because the high-frequency noise contributes to the power at the four frequencies, 0, Ω_1, Ω_2, and $2\Omega_1$.

COMPUTER CONTROL

Early in 1973 the instrument will be permanently installed at Sacramento Peak and will begin operating on a regular basis as a double-beam instrument. A Digital Equipment Corporation PDP-11 computer[2] will

---

[2] See p. 258.

handle the data recording on magnetic tape as well as the control and monitoring of the modulation, calibration, spectral scan, and positional scan functions. This same computer can convert the raw data into a smoothed and calibrated form once the observation is completed. We plan to measure each of the four Stokes parameters at several hundred points through the spectral line scan and to integrate over many such scans.

Acknowledgments. E. Tandberg-Hanssen and J. Beckers initiated the program to build this instrument. I. Lee and L. Lacey have assisted with parts of the mechanical design, and S. Rogers helped design the data gathering and control electronics. E. Harper assisted with the PDP-11 computer programming. We would like to thank J. Beckers, L. House, and E. Tandberg-Hanssen for their helpful suggestions regarding the design of this instrument.

REFERENCES

Beckers, J. 1969. Unpublished.
Dollfus, A. 1965. *Stellar and solar magnetic fields* I.A.U. Symposium No. 22, R. Lüst, ed., pp. 176–183. North-Holland, Amsterdam.
Shurcliff, W. A. 1962. *Polarized light.* Cambridge, Mass: Harvard University Press.

THE HIGH ALTITUDE OBSERVATORY CORONAL-EMISSION-LINE POLARIMETER

C. W. QUERFELD
High Altitude Observatory
National Center for Atmospheric Research

> *A scanning solar polarimeter is being constructed by the High Altitude Observatory in collaboration with the Sacramento Peak Observatory to measure the linear polarization of light emitted at 10747Å and 10798 Å by FeXIII ions and of light from the continuum corona at 10690 Å. The polarimeter is to be mounted as an on-axis coronagraph of 40-cm aperture on the large spar at the Sacramento Peak Observatory. Both the emission line and continuum channels are modulated by a rotating half-wave plate. The signals are synchronously demodulated and integrated in a PDP-11 computer which controls and points the polarimeter. The resulting data are subsequently processed to infer the direction but not the magnitude of the magnetic field in the solar corona.*

In recent years it has become increasingly apparent that the density, temperature, and velocity structure of the solar corona is largely determined by the coronal magnetic field (Newkirk 1971). The theoretical understanding of the coronal magnetic field has improved to the point that it is highly desirable to have a quantitative knowledge of the magnetic field. Until now this need has been met by the availability of current-free potential fields mapped into the corona from photospheric fields measured with a conventional magnetograph (Altschuler and Newkirk 1969). Although conventional magnetographs are unsuitable for measurements of coronal fields, the polarization of coronal emission lines may yield the direction but not the magnitude of the coronal magnetic field (Charvin 1965; Hyder 1965; House 1972).[1]

Few attempts have been made to measure coronal-emission-line polarization and to infer magnetic-field configurations from the observed polarization. Measurements at eclipse (Hyder 1966; Eddy and Malville 1967; Eddy, Firor, and Lee 1967; Hyder, Mauter, and Shutt 1968; Beckers and

---

[1] Also see p. 762.

Wagner 1971; Eddy, Lee, and Emerson 1973) have found polarizations that are reasonably consistent with field configurations inferred from coronal density structures, but the eclipse data do not themselves admit good inferences of magnetic-field configuration. Charvin (1971) has observed the polarization of the FeXIV 5303 Å green line outside of eclipse to make the first systematic inferences of magnetic fields from emission-line measurements. The success of these measurements has encouraged Charvin to refine his experiment[2] and the High Altitude Observatory (HAO) to build a coronal-emission-line polarimeter.

The HAO emission-line polarimeter is a major collaborative effort by HAO and the Sacramento Peak Observatory (SPO) to obtain synoptic maps of the coronal-magnetic-field configuration between 1 $R_\odot$ and 2–2.5 $R_\odot$ outside of eclipse. The 10747 Å and 10798 Å lines of FeXIII were selected as the best compromise of emission-line equivalent width, line polarization, sky brightness, and detector responsivity. The polarimeter will measure the first three Stokes parameters, I, Q, and U ($V = 0$ in integrated coronal emission lines), of the 10747 Å line, the 10798 Å line, and the continuum (K) corona at 10690 Å. The K-corona data are needed to subtract sky and K-corona Stokes vectors from the emission-line measurements and also to obtain electron-density maps (Altschuler and Perry 1972) for use in the inversion of the emission-line data. The polarimeter is now under construction at HAO and will be installed and operated at SPO early in the summer of 1973.

THE POLARIMETER

The polarimeter is composed of a telescope and the polarimeter proper. The design of these elements is intended to minimize signal-integration time so that coronal maps may be completed in one or two hours. The telescope, shown in Fig. 1, is an on-axis coronagraph with a 40-cm aperture and a 615-cm focal length. The telescope will occupy the east side of the large spar in the Big Dome at SPO. The forward end of the telescope tube is gimbal mounted just forward of the Coudé optics, and the after end is held in a double eccentric at the lower end of the spar. The rotation of the inner with respect to the outer eccentric determines the scan radius about the center of the disk (the spar is centered on the disk), and the rotation of the outer eccentric provides the scan about the disk in the plane of the sky. The scan radii may vary from 0 to 2.6 $R_\odot$. The telescope tube rotates independently. The objective is focused by moving it in its supporting cage.

---

[2]Personal communication, 1971.

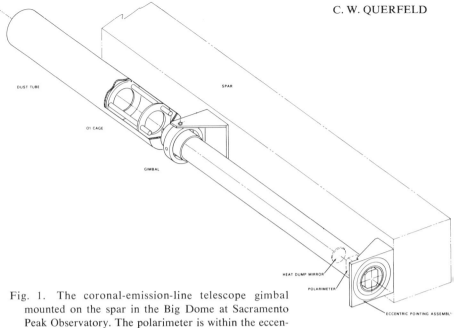

Fig. 1. The coronal-emission-line telescope gimbal mounted on the spar in the Big Dome at Sacramento Peak Observatory. The polarimeter is within the eccentrics at the lower end.

A metal mirror (Fig. 2) at the prime focus occults the disk when scanning off the limb by imaging the objective into a Rayleigh horn. The central aperture in the mirror admits light to a field stop in a turret behind the mirror. The turret contains a shutter and field apertures of ¼, ½, 1, 1½, 2, 2½, and 3 arc minutes. The field of view is varied under manual and computer control to obtain an appropriate balance of spatial resolution and integration time. The turret also holds the first-field lens ($F1$) which images the objective ($O1$) onto the second objective ($O2$).

The polarimeter itself is a separate package mounted within and extending from the telescope tube. The $O2$ with the usual Lyot diaphragm and spot is the entrance window of the polarimeter. The $O2$ collimates the light passing through the field stop and sends it to the rotating half-wave plate modulator. A tilt-plate polarization calibrator may be inserted between the $O2$ and the half-wave plate to provide polarization of known amount and azimuth. The quartz half-wave plate is mounted in the hollow shaft of an optical encoder which emits an index pulse once per rotation, a quadrant pulse once per modulation cycle, and sector pulses 32 times per modulation cycle. The latter pulses transmit data to and control synchronous data demodulation in the PDP-11 computer, which controls the polarimeter. The encoder is rotated at 25 revolutions per second by a hysteresis synchronous motor to give a modulation frequency of 100 Hz. The half-wave plate is followed by a compensated Wollaston prism whose mutually diverging, orthogonally polarized beams are made

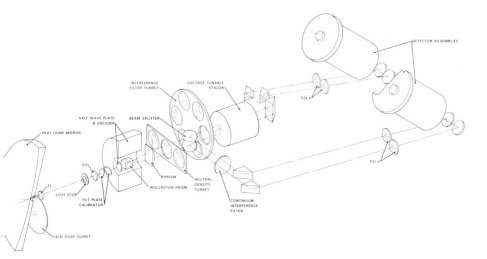

Fig. 2. The optical train of polarimeter and the field-lens–heat-dump assembly.

parallel by a biprism. The light then passes to a beam splitter which sends 55% of each orthogonally polarized beam to detectors that see line, continuum, and sky and the remainder to detectors that see only continuum and sky. Neutral-density filters ($D = 3.4$ and 4.5) in a turret between the biprism and the beam splitter are used to calibrate the instrument against the center of the disk.

The emission-line corona (L) channel has interference prefilters in a turret just behind the beam splitter which are selectable under manual or computer control and isolate the 10747 Å and 10798 Å lines of FeXIII, the 10830 Å helium line, and the 10938 Å hydrogen Pγ line. Narrow-band spectral filtration is done with a voltage tunable Fabry-Pérot etalon under computer and manual control. The etalon has a passband of 2 Å (the approximate width of the emission lines), a free spectral range of about 60 Å, and a transmission of 80%. Light from the etalon is routed to the detectors via steering prisms, second-field lenses ($F2$), and mirrors. The $F2$ lenses image the $O2$ into the detector apertures. The continuum corona (K) channel is similarly constructed with a single interference filter centered at 10690 Å and no etalon. The detectors are cooled germanium photodiodes mounted in pairs with their preamplifiers in liquid nitrogen dewars. The detectors have good quantum efficiency (55%), low noise ($< 6 \times 10^{-14}$ W/H$_z^{1/2}$) and high responsivity (10^8 V/W). Each detector sees only one of the two orthogonally polarized beams. The use of detector pairs helps to suppress the effects of seeing and sky-transparency fluctuations.

SIGNAL PROCESSING AND CONTROL

Signals from the detector-preamplifier go to chopper-stabilized sum and difference amplifiers (see Fig. 3), which provide sum signals Σ_L and Σ_K and difference signals $\triangle_L$ and $\triangle_K$ for the L and K channels respectively. Additional difference amplifiers produce signals $\Sigma_L - \Sigma_K$ and $\triangle_L - \triangle_K$ which with Σ_K and $\triangle_K$ go through low-pass filters, sample and hold amplifiers, and a multiplexed analog-to-digital converter to the PDP-11 computer. When gains are properly balanced, the $\Sigma_L - \Sigma_K$ signal is a D.C. voltage proportional to emission-line intensity I_L, and the $\triangle_L - \triangle_K$ signal is a 100-Hz voltage proportional to Q_L and U_L in phase quadrature. The Σ_K gives a D.C. voltage proportional to the line plus sky intensity I_K, and $\triangle_K$ gives Q_K and U_K at 100 Hz. The four analog signals are simultaneously clamped at the sample and hold amplifiers by each of the sector pulses from the half-wave plate encoder. The four clamped analog voltages are digitized sequentially during the 312-μs interval between sector pulses and are accumulated as double-word (32 bit) sums in the computer memory. The 100-Hz signals are synchronously demodulated by multiplication with tabulated sines and cosines prior to addition to the Q_L, U_L, Q_K, and U_K accumulators. The data integration is done in blocks of ten modulation cycles for which signal variances are computed to estimate signal-to-noise values and to set the total integration time per scan point. Integration times per scan point will vary between 0.1 and 10 seconds to maintain a signal-to-noise ratio of about 100. The signal-processing algorithm comprises a digital filter with pass bands at D.C. and 100 Hz, and zeros at other multiples of 10 Hz. Aliasing of noise beyond the Nyquist frequency of 1600 Hz is suppressed by the 500-Hz bandwidth of the detector preamplifiers and by the low-pass filters in the signal lines. Apodization is not used so that filter zeros remain at power-line harmonics. Data for each scan point are moved to an output buffer and are periodically dumped onto magnetic tape for final processing.

The computer, which is shared with the HAO Stokes polarimeter,[3] performs most of the control functions of the polarimeter in addition to doing the signal processing. The scan program, for example, will ordinarily consist of scans at constant radius about the disk, but the closure of the eccentric and tube azimuth control loops in the computer pointing interface provides great scanning flexibility. Radial, sector, and spiral scans will be included as standard scan programs. The azimuth loops may be operated either in a position-control mode or in a constant-rate mode. Similarly, the usual control program will actively keep the tunable etalon centered on the emission line, but it is also

---
[3] See paper by Baur et al. on p. 252.

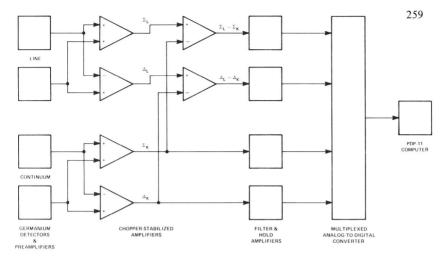

Fig. 3. The data flow from detectors to the PDP-11.

possible to modulate the line signal by tuning the etalon through the line. The absolute separation of K-corona intensity and polarization from the sky contribution can be made by scanning through the Pγ line, which is present in the sky light but is absent from the K corona. One particularly important control function is the ability to control integration time by changes in the field of view. Charvin notes, however, that his green-line measurements show that emission-line signals are independent of the field of view.[4] If this is true for the infrared lines, then the effectiveness of the field-of-view, integration-time trade off will be compromised.

The final processing of the emission-line data will be done at the computer center in another division of the National Center for Atmospheric Research (NCAR). The immediate product will be maps of the emission-line Stokes vector and three-dimensional maps of the electron density. Initially, magnetic field maps obtained from the potential field construction will be used to predict the emission-line Stokes vector, and a comparison of the predicted with the observed maps may be used to modify the potential field construction. The most efficient way of achieving this is unclear. A better procedure is to invert the three line-of-sight integrals for the first three Stokes parameters which have the form (House 1972)

$$(I, Q, U) = \int_{-\infty}^{+\infty} f_{I,Q,U}(a, A, B, N_e, T_e) \, dx,$$

where a is the angle between the observed polarization and the magnetic

---

[4] Personal communication, 1971.

field projected on the plane of the sky, A is the polar angle of the field line relative to the local radius, B is the azimuth angle of the field line about the local radius, N_e is the electron density (known from the K-channel data inversion), T_e is the temperature (known only crudely, but excitation conditions do not depend strongly on it [Chevalier and Lambert 1969]), and x is the line-of-sight coordinate. The inversion algorithm, now being derived by Altschuler,[5] will be similar to that used to invert the K-coronameter data (Altschuler and Perry 1972). Further refinement of the data inversion will incorporate new non-LTE calculations of the excitation equilibrium among magnetic sublevels. This calculation, now being completed by House, will permit the proper inclusion of both radiative and collisional processes.

Acknowledgments. The instrument is being constructed at the suggestion of G. Newkirk and is part of the broad effort in coronal physics at HAO. R. Dunn provided the basic design of the telescope, and the design has been completed by R. Wendler. The polarimeter mechanical design has been done by L. Laramore. J. Rush designed the optical train. V. Borgogno designed the electronics and the interfaces with the PDP-11. E. Harper and M. Gay have provided much of the software. Supervision and many helpful suggestions have come from H. Hull, L. Lacey, and R. Lee. E. Tandberg-Hanssen and L. House have also given many useful suggestions. This research was supported by the National Center for Atmospheric Research through sponsorship of the National Science Foundation.

REFERENCES

Altschuler, M. D., and Newkirk, G. 1969. Magnetic fields and the structure of the solar corona, I: Methods of calculating coronal fields. *Solar Phys.* 9: 131–149.
Altschuler, M. D., and Perry, R. M. 1972. On determining the electron density distribution of the solar corona from K-coronameter data. *Solar Phys.* 23: 410–428.
Beckers, J. M., and Wagner, W. J. 1971. The polarization of coronal emission lines. *Solar Phys.* 21: 439–447.
Charvin, P. 1965. Etude de la Polarisation des Raies Interdites de la Couronne Solaire. Application au Cas de la Raie Verte λ5303. *Ann. d'Astroph.* 28: 877–934.
———. 1971. Experimental study of the orientation of magnetic fields in the corona. *Solar magnetic fields.* R. Howard, ed., pp. 580–587. Dordrecht, Holland: D. Reidel.
Chevalier, R. A., and Lambert, D. L. 1969. The excitation of the forbidden coronal lines. *Solar Phys.* 10: 115–134.
Eddy, J. A., Firor, J. W., and Lee, R. H. 1967. Near infrared measurements of coronal polarization in line and continuum. *Astron. J.* 72: 793.
Eddy, J. A., Lee, R. H., and Emerson, J. P. 1973. The λ10747 coronal line at the 1966 eclipse: I. Emission line polarization. *Solar Phys.* In press.

---

[5] Personal communication, 1972.

Eddy, J. A., and Malville, J. M. 1967. Observations of the emission lines of FeXIII during the solar eclipse of May 30, 1965. *Astrophys. J.* 150: 289–297.

House, L. L. 1972. Coronal emission line polarization. *Solar Phys.* 23: 103–119.

Hyder, C. L. 1965. The polarization of emission lines in astronomy III. The polarization of coronal emission lines. *Astrophys. J.* 141: 1382–1389.

———. 1966. Plane polarization and solar magnetic fields. *Atti del Convegno Sui Campi Magnetici Solari,* (Rome 1964), pp. 110–119.

Hyder, C. L., Mauter, H. A., and Shutt, R. L. 1968. Polarization of emission lines in astronomy. VI. Observations and interpretations of polarization in green and red coronal lines during 1965 and 1966 eclipses of the sun. *Astrophys. J.* 154: 1039–1058.

Newkirk, G. 1971. Coronal magnetic fields. *Physics of the solar corona.* C. J. Macris, ed., pp. 66–87. Dordrecht, Holland: D. Reidel.

STELLAR AND SOLAR X-RAY POLARIMETRY

R. NOVICK
Columbia University

Stellar and solar X-ray polarimetry hold great promise for elucidating the emission mechanism and structure of X-ray objects and events. Polarization is expected whenever nonthermal processes are important or when there is strong scattering in a nonspherical thermal source. The recent detection of X-ray polarization in the Crab Nebula and in solar flares is taken as evidence of synchrotron emission and linear bremsstrahlung, respectively. X-ray polarization is expected in one or more knots in the jet of M 87, in NGC 1275, in 3C 273, and in a number of other sources that exhibit polarization in the radio and optical bands. Phase-dependent polarization is expected in pulsars such as NP 0532 and in Cen X-3, Her X-1, and other pulsating X-ray binary stars. In the case of the X-ray binary stars, it may be possible to infer the inclination of the orbit from the position angle of the polarization. Polarization of a few percent may be present in Sco X-1 and other compact sources if the accretion model is correct. Flare events on Sco X-1, Cyg X-1, and the short-lived X-ray flare stars may be polarized if linear bremsstrahlung contributes significantly to the emission process. It will be possible to interpret the X-ray observations in terms of intrinsic source properties since interstellar grains are not expected to produce significant X-ray polarization.

In this paper we review the various experimental approaches that can be used for X-ray polarimetry, with particular emphasis on Bragg-crystal and Thomson-scattering devices. We indicate the properties, statistical limitations, predicted performance, and the space flight schedules for a number of approved and proposed instruments. We emphasize the need for imaging polarimeters for studying individual areas in extended objects such as M 87 and the Crab Nebula, individual stars in external galaxies such as M 31, and for isolating individual sources near the center of the galaxy. These objectives can only be met by an instrument placed at the focus of a large-area high-resolution grazing incidence X-ray telescope. In the case of solar flares, a modulation collimator can be used to isolate individual flare events and active regions on the solar disk. We emphasize the need for improved solar-flare polarimeters with high sensitivity over a broad energy band, time resolution of one second or less, and with energy resolution comparable to that achieved with gas proportional counters. Finally, we briefly mention the problems of high-energy X-ray and gamma-ray polarimeters and of X-ray circular polarization analyzers.

INTRODUCTION

Stellar X-Ray Polarimetry

While X rays were first detected from the sun in 1948 (Burnight 1949), the first stellar source was discovered in 1962 by Giacconi and co-workers (1962), and the first satellite devoted exclusively to stellar X-ray astronomy was launched in the fall of 1970. In spite of this short history, almost two hundred X-ray sources have been found. These have been associated with almost every known class of astronomical object such as quasi stellar objects (QSO), Seyfert galaxies, radio galaxies, galactic clusters, supernova remnants, and ordinary galaxies, and a number of sources have been revealed for which no presently known radio or optical counterpart exists. The distribution of X-ray sources from the 2 Uhuru catalog (Giacconi et al. 1972) is plotted in galactic coordinates in Fig. 1.

Noteworthy among the discoveries of this period was the detection of the diffuse X-ray background which exhibits very complex structure at low energies (0.25 keV) and whose origin is not fully understood (Gorenstein et al. 1969; Bowyer et al. 1968; Henry et al. 1968; Yentis et al. 1972a, b). There is growing evidence that at least part of the low-energy background arises from a large number of unresolved galactic sources (Gorenstein and Tucker 1972; Culhane and Fabian 1972).

A further important discovery has been that of eclipsing, pulsating X-ray binary sources (Schreier et al. 1972; Tananbaum et al. 1972). The pulse period of these objects is on the order of a few seconds and the

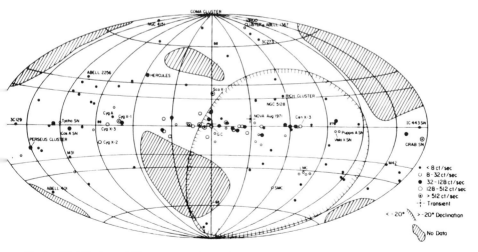

Fig. 1. Distribution of X-ray sources from the 2 Uhuru catalog, plotted in galactic coordinates (Giacconi et al. 1972).

eclipse period, a few days. The most interesting binary object is Her X-1, which has a pulsation period of 1.24 sec and an eclipse period of 1.7 days. Recently an optical candidate has been identified that exhibits smooth photometric variation of about one magnitude at the eclipse period (Bahcall and Bahcall 1972; Liller 1972). Two groups claim to have occasionally detected very low-level, 1.24-sec optical pulsations at certain phases of the eclipse (Lamb and Sorvari 1972; Davidson et al. 1972). If this claim is confirmed, this object will become the second known optical pulsar (NP 0532 is the first).[1] The X-ray emission ceases for about 23 days every 35 days, and very recently a survey of the Harvard patrol plates revealed that the 1.7-day optical variations were present from 1945 to 1948, were not visible from 1949 to 1956, and then recommenced. The data are not complete enough to determine whether this long-term behavior is periodic or not (Liller[2]).

Her X-1 is certainly one of the most unusual objects in the sky. Many authors have suggested that it is a contact binary, one member of which is a rotating magnetic (10^{12} gauss) neutron star (Pringle and Rees 1972; Davidson and Ostriker 1973; Lamb et al. 1972; Gnedin and Sunyaev 1972). The X-ray emission is presumed to arise from the accretion of matter onto the poles of the neutron star. Brecher (1972) suggested that the 35-day period may result from precession of the axis of the neutron star. If this model is correct, then the X-ray emission will exhibit both linear and circular polarization that varies with the pulse phase. The average position angle of the linear polarization may provide a clue to the inclination of the binary orbital plane. This information, of course, is essential for determining the masses of the two stars from the mass function

$$\frac{M^3 \sin^3 i}{(M + m)^2}. \qquad (1)$$

At present the mass function has not been determined in a convincing way for any of the eclipsing X-ray sources since it has not been possible to obtain unambiguous spectroscopic data on both members of the binary. The data that do exist are probably strongly contaminated with spectral features arising in a gas stream.

Another binary source, Cyg X-1, exhibits large random variations in the X-ray emission on all time scales from a fraction of a second to months. This object has been identified with a highly variable radio source and with a 9th magnitude B0 I supergiant (HDE 226868) spectroscopic binary with a 5.6-day period. Bolton (1972) has found a He II 4686 Å emission line in this object that appears to exhibit a velocity curve that is in anti-

---

[1] See p. 997.
[2] Personal communication, 1972.

phase to the absorption-line velocity curves of the B0 I primary star. Assuming that the 4686 Å line arises on the secondary, Bolton finds that the mass ratio is 1.5. The distance and spectral data suggest that the primary is a normal B0 I star with a mass in the range from 15 to 20 $M_\odot$. This means that the secondary has a minimum mass of 10 $M_\odot$, and the rapid X-ray fluctuations require that it be compact. When these facts are considered together, it would seem probable that the secondary is a black hole orbiting the B0 I primary. In this model X rays are assumed to result from the heating of gas that is accreting from the supergiant and falling into the black hole. Spectroscopic studies by Smith et al. (1973) show that the time dependence of the 4686 Å line is quite complex. These authors conclude that the line profiles at He II 4686 Å for HDE 226868 are not consistent with a model where the emission component arises from an unseen secondary and that the system cannot be solved as a double-lined spectroscopic binary. The determination of the mass of the secondary is further complicated by the fact that the spectral type and luminosity class of the primary are seen to change as a function of phase, due to the close presence of the compact secondary. The 4686 Å profiles suggest that the emission component originates in a gas stream falling toward the secondary, thus providing observational evidence that accretion powers this X-ray source. Their best estimate of the distance to HDE 226868 is 2 kpc. At the present time there is no conclusive evidence regarding the structure of the Cyg X-1 X-ray object. Clearly, substantial additional work must be done to clarify these important questions.

In the case of Cyg X-3, the flux from its radio counterpart suddenly increased by a factor of 1000, briefly making it one of the strongest radio sources (Gregory et al. 1972). The X-ray emission showed no such change (Parsignault et al. 1972).

It is obvious that nonthermal processes dominate the emission from these and many other X-ray objects and that we can expect to find X-ray polarization in a significant number of the sources. Clearly, we cannot hope to distinguish between the various theoretical models with only photometric data. Many of the more interesting sources are so dense (10^{16} particles cm^{-3}) that there is little hope of detecting sharp spectral features (Angel 1969a; Kestenbaum et al. 1971a; Felten et al. 1972). Thus, the powerful spectroscopic tools that are so important to visible-light astronomy may not be available to X-ray astronomy, and it appears that polarimetry will be essential for a complete understanding of these objects. It is worth noting that the radio galaxies exhibit smooth power-law-type spectra and appreciable linear polarization. This is taken to indicate synchrotron emission. Thus, much of the physical insight that we have for many radio sources has come from polarimetric observations rather than high-resolution spectroscopic studies.

Identifications have been made for X-ray sources associated with the extragalactic sources NGC 1275, NGC 4151, M 87, NGC 5128, and 3C 273. The nature of the X-ray emission is still obscure, and thermal and non-thermal models have been proposed. In NGC 5128 the X-ray emission has been shown to come from the central region rather than from the extended radio lobes. Polarized optical radiation has been observed in the jet of M 87. Individual knots have polarizations of about 20% (Hiltner 1959), indicating synchrotron emission. Depending upon the relative intensities of these knots, integrated polarization of perhaps 5%–10% should be expected for X rays emitted in the jet (Woltjer[3]). Radio polarization observations of small-diameter variable sources show that, although sources like NGC 1275 and 3C 273 have small net polarizations (2%), the variable components have rather large polarizations (10%–20%) (Aller 1970). In some of these objects strong infrared emission may deplete the high-energy electrons by Compton-type processes, in which case not much X-ray emission by the synchrotron mechanism would occur, but it is by no means clear that this happens in all or most of the sources.

X-ray polarization observations of these objects would provide information about the radiation mechanism and, if positive results are obtained, could give much information on the small-diameter sources that probably are responsible for the energetics of radio galaxies and QSO's. The OSO-I and other polarimeters described below will have sufficient sensitivity to answer essentially all of the above questions.

A number of radio and optical sources exhibit both circular and linear polarization, and in some cases this polarization is time dependent. For example, in the case of pulsars, it is found that the polarization vector rotates through an angle of roughly 180° during the pulse (Kristian et al. 1970; Wampler et al. 1969).[4] This has been taken as strong evidence for the magnetic rotator pulsar model. The recent observation of circular polarization in a number of white dwarf stars (Landstreet and Angel 1971a) shows that some highly evolved stars have very large magnetic fields, supporting the soundness of the basic view of the magnetic neutron star model of pulsars (Gold 1968, 1969). The observation of linear polarization in the Crab Nebula provided the first evidence for the synchrotron emission mechanism in celestial objects.[5] Interstellar optical polarization provides a measure of the distribution of the interstellar magnetic field and the properties of interstellar matter (van de Hulst 1957; Martin 1972; Martin et al. 1972).[6,7] Interstellar grains are not expected to produce X-ray polarization. Finally, the radio polarization maps of radio galaxies give the magnetic-field pattern in these objects. Such polarization

---

[3]Personal communication, 1971.
[4]See p. 997.   [5]See p. 1014.   [6]See p. 888.   [7]See p. 926.

studies have proved to be invaluable in the determination of the structure of sources and their emission mechanisms. Without such data, in many cases there would be considerable ambiguity as to the radiation process.

So far, stellar X-ray polarization studies have been made of only Sco X-1 and the Crab Nebula. The instruments of the type used in this work are discussed in detail below, and the results have been published elsewhere. In the case of sources such as Sco X-1 which are thought to be thermal emitters, we would not expect to observe polarization. However, if these sources are not spherically symmetric and if they exhibit a large amount of internal electron scattering, then a few percent polarization is expected (Angel 1969b). In the case of Sco X-1, the emitting plasma is thought to arise from the accretion of matter onto a white-dwarf member of a contact binary. The view is not universally accepted since the optical spectrum of Sco X-1 reveals broad, variable lines that do not exhibit the line shifts expected for binary orbital motion, but the changes in the spectrum are so complex that this view cannot be excluded. If the binary model is correct, then we might expect asymmetry in the emitting region. Furthermore, many independent observations indicate an optical depth for electron scattering in the range 5–17 (Felten et al. 1972). Thus, there are good reasons to suspect that this source is slightly polarized. With 99% confidence, the X-ray polarization in Sco X-1 has been shown to be less than 20% (Angel et al. 1969). The OSO-I instrument described below will allow us to improve the accuracy of this result by at least a factor of 20. At this level, we would expect to observe polarization if the binary model is correct. If a positive result is obtained, the position angle will provide a clue to the orbital inclination. Thus, polarimetry may reveal the true nature of this source.

In the case of the Crab Nebula, three experiments have been performed (Wolff et al. 1970; Novick et al. 1972). By combining the data from these, it was shown with 99.7% confidence that the X-ray emission is polarized with $15.4 \pm 5.2\%$ polarization at a position angle of $156 \pm 10°$. The agreement of this result with the observed radio and optical polarization and the fact that the entire Nebula emission can be fitted with power-law-type spectra are taken as proof that the X-ray emission arises by the synchrotron process. If the thermal mechanism of Sartori and Morrison (1967) obtained, then no polarization would have been detected. Thus, the Crab Nebula is the first X-ray object for which we have a confirmed emission mechanism. This observation poses new problems for pulsar theories since it implies that the synchrotron electron spectrum extends to 6×10^4 GeV if the Nebula magnetic field is taken as 10^{-4} gauss. Since the radiative lifetime for such electrons is a fraction of a year, they must either be continuously generated by the pulsar, or they must be continuously accelerated within the Nebula (Gunn and Ostriker 1971; Rees, 1971). Continuous injection of 6×10^4-GeV electrons appears to be impossible since in the high magnetic field (10^{12} gauss) of the pulsar, the electrons will lose their energy by either synchrotron or curvature radiation before they escape from the neighborhood of the pulsar (Ruderman[8]). Continuous acceleration by the process proposed by Gunn and Ostriker and by Rees seems to be ruled out by the fact that Landstreet and Angel (1971b) failed to find the circular polarization in the optical band that has been predicted for this process. Thus, we must look elsewhere for either an injection or acceleration mechanism. The OSO-I polarimeter described below will allow us to improve the accuracy of the present Crab Nebula result by an order of magnitude in one day of observing time. In addition, we will be able to determine the average polarization in the main pulse of the Crab X-ray pulsar with a 3 σ precision comparable to the average observed optical polarization of 7% at a position angle of 98° (Kristian et al. 1970). If the X-ray pulsar polarization is comparable to the known optical polarization, then we will confirm the presently held view that the optical and X-ray pulsar emission mechanisms are identical. Below we give the predicted performance of the OSO-I and other polarimeters for a number of other important sources.

Solar X-ray Polarimetry

Solar X-ray phenomena are extremely complex, and it is inappropriate to discuss the entire field here. An excellent review has been given by Neupert (1969). Here we will emphasize the high-energy events that require polarimetry for their elucidation (see also Wolff 1972).

There is an abundance of evidence showing that the early phase of flares involves nonthermal electron distributions (McKenzie 1972). In this early phase, high-energy X rays are often emitted in short impulsive bursts that show strong temporal correlation with similar burst phenomena in the microwave region. Typical examples of these correlated X-ray and radio bursts are shown in Fig. 2. The X-ray spectra in these bursts are characterized by a power law rather than a thermal distribution. As the flare develops, these bursts decrease in intensity, and the X-ray emission gradually takes on a thermal character. It has been suggested that the X-ray and microwave bursts are caused by a nonthermal high-energy electron flux that is guided by the local magnetic field in the active region. In this model the microwave emission occurs by the synchrotron emission resulting from the motion of the electrons about the field lines. Observationally it is found that the microwave bursts exhibit the strong circular and linear polarization expected for this process (Takakura 1967). In the

---

[8] Personal communication, 1972.

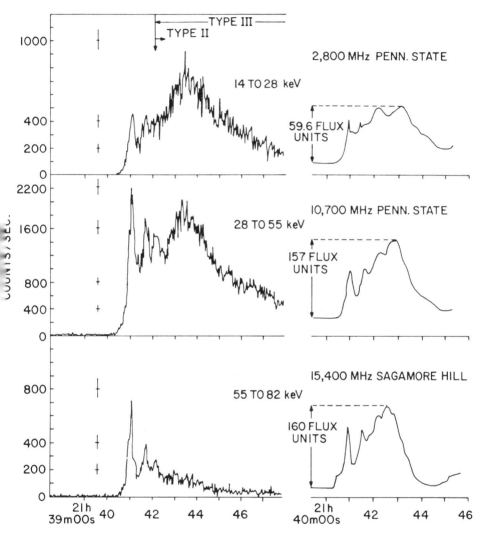

Fig. 2. Comparison of hard X-ray and radio bursts of 1 March 1969 measured by OSO-5 (Frost 1969).

same model, the X rays result from the impact of the energetic electrons on the ambient gas. The direction of this motion is parallel to the field lines. Under these conditions one expects strong linear X-ray polarization. The sign of the polarization changes with photon energy; at low energies, the electric vector of the emitted photons is transverse to the direction of the electron motion, while at high energies it is parallel to the electron velocity vector. Detailed predictions of the X-ray polarization have been made by Elwert (1968) and by Korchak (1967) for a variety of viewing angles and electron energy distribution functions. Two examples of these are shown in Fig. 3. This process is known as linear bremsstrahlung since it results from the impact of monodirectional electrons. Observationally it can be distinguished from synchrotron radiation by the energy dependence of the polarization; in synchrotron emission the polarization is independent of photon energy, and the electric vector is perpendicular to the magnetic field lines, while in linear bremsstrahlung the polarization reverses sign as discussed above. In synchrotron emission we expect circular polarization, while in linear bremsstrahlung we expect pure linear polarization. Unfortunately, it appears to be extremely difficult to construct an X-ray circular polarization analyzer. Thus, we must depend on the energy dependence of the linear polarization to distinguish the two processes. The solar physics group in the P. N. Lebedev Physical Institute has flown Thomson-scattering X-ray polarimeters on Intercosmos-1 and -4, and they report the observation of large (20%) X-ray polarization in the early phases of three flares (Tindo et al. 1970, 1971). In a fourth flare, there were two maxima in the intensity and two corresponding maxima in the polarization (Tindo et al. 1972). These data strongly support the model discussed above, but as yet no data have been published on the energy dependence of the polarization. Neupert,[9] of the NASA Goddard Space Flight Center, has a small scattering polarimeter on OSO-7, but he is still analyzing the data and has not reported any evidence for polarization. The solar physics group at the U.S. Naval Research Laboratory is planning on including a small scattering polarimeter on Solrad 11 (Doschek[10]). It appears that more sensitive polarimeters of the type discussed below will be required to verify the existence of polarization and to determine the energy dependence of the polarization. All of the present instruments view the entire solar disk, and in the future it will be necessary to improve the spatial resolution of the instruments.

---

[9] Personal communication, 1972.
[10] Personal communication, 1972.

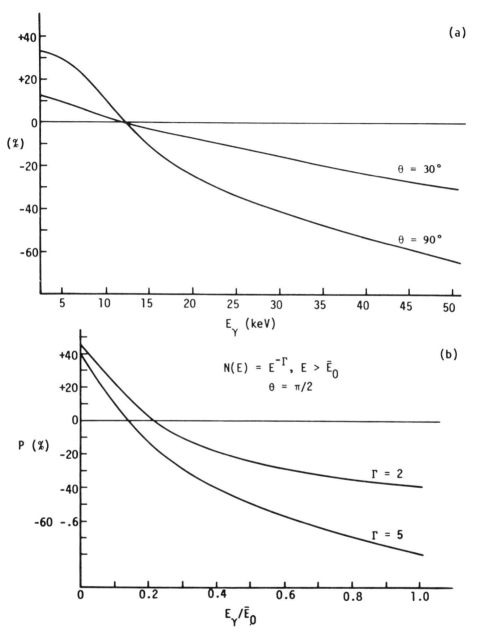

Fig. 3. (a) Polarization of bremsstrahlung as a function of the photon energy E for two viewing angles relative to the direction of the electron flux. Electron energy $E_0 = 100$ keV (Korchak 1967). (b) X-ray polarization as a function of energy for bremsstrahlung from a power-law distribution of electrons $N(E)$. The viewing angle is $90°$ (Korchak 1967).

X-RAY POLARIMETERS

In this section we will discuss the statistical limitations on all polarimeters and present a brief description of photoelectric, secondary fluorescence, and Borrmann-effect polarimeters. Thomson-scattering and Bragg-crystal polarimeters are described on p. 291 and p. 280, respectively.

Statistical Limitations

When an ideal polarimeter is illuminated with 100% linearly polarized radiation, the signal is unattenuated when the polarimeter pass axis is parallel to the polarization vector, and it is zero in the orthogonal direction. In any real polarimeter, the transmission in the pass direction is always less than 100%, and the transmission in the orthogonal direction is finite. We define the modulation factor F as the apparent polarization for fully polarized radiation. If $N_\parallel(100)$ and $N_\perp(100)$ are the signal counts obtained with 100% polarized radiation when the pass axis of the polarimeter is parallel and perpendicular to the polarization vector, respectively, then we define the polarimeter modulation factor F as

$$F = \frac{N_\parallel(100) - N_\perp(100)}{N_\parallel(100) + N_\perp(100)}. \tag{2}$$

When such a polarimeter is used to measure a source with polarization P and the parallel and perpendicular signal counts are $N_\parallel(P)$ and $N_\perp(P)$, respectively, we can readily show that the polarization of the source is given by

$$P = \frac{1}{F} \frac{N_\parallel(P) - N_\perp(P)}{N_\parallel(P) + N_\perp(P)}. \tag{3}$$

Since polarization is a positive, definite quantity, we find, when we take account of the statistics of the photon-counting process, that all observations must always yield a finite result even if the source is unpolarized. The magnitude of this spurious polarization depends on the number of photons counted as well as the properties of the polarimeter.

In the ideal case of a polarimeter with no background signal and having a 100% modulation factor F, we can readily show for Poisson counting statistics that the probability of detecting a fractional polarization P in an unpolarized source is

$$W(P)dP = \frac{1}{2} PN \exp(-P^2 N/4)dP, \tag{4}$$

where N is the total number of photon counts in the observation. The $1\,\sigma$ variance of this distribution is given by

$$P(1\,\sigma) = (2/N)^{1/2}. \tag{5}$$

In Equations (4) and (5) we have assumed the number of counts is large enough so that

$$P(1\,\sigma) \ll 1. \tag{6}$$

That is, the $1\,\sigma$ variance corresponds to a small polarization. In the case of stellar X-ray polarimeters, this statistical limitation is often quite severe, and long observing times are needed to obtain enough counts to reduce $P(1\,\sigma)$ to a physically interesting level.

For a real polarimeter, the modulation factor F for 100% polarized X rays is always less than unity, and cosmic-ray and other background signals are present. In such case, it can be readily shown from Equation (5) that the apparent polarization arising from the statistical fluctuations in the data obtained on an unpolarized source is given by

$$P(3\,\sigma) = 3[2(S+B)/T]^{1/2}/FS, \tag{7}$$

where T is the observing time, S is the signal or source counting rate, B is the background counting rate, and F is the modulation factor for the polarimeter for a 100% polarized source. Here we have taken a $3\,\sigma$ criterion for the spurious polarization so that we can have greater than 99.7% confidence that any indicated polarization greater than that given by Equation (7) is real and not spurious. In discussing the various instruments below, we will use Equation (7) to determine the minimum detectable polarization in any given source.

In the following sections we will discuss different types of X-ray polarimeters, their physical basis, their limitations, and, where appropriate, specific designs with the predicted performance for a number of important objects.

Photoelectric Polarimeters

Photoelectric polarimeters exploit the fact that in the photoionization of a K shell the electron is ejected in the direction of the electric vector of the incident photon (Heitler 1954). In the nonrelativistic limit where the photon energy is small compared to the rest mass of the electron, the angular distribution of the photoelectron is given by

$$P(\psi)d\psi = \cos^2\psi\,d\psi, \tag{8}$$

where ψ is the angle between the electric vector of the photon and the momentum vector of the ejected photoelectron. If we could readily detect the direction of the electron, then we would know the polarization state of the photon. This technique has been successfully used in the laboratory, but there are serious difficulties in applying it in X-ray astronomy. Briefly, the problem has to do with the fact that, even near an absorption edge, the range of the ejected electron is only about 1% of the photon absorption depth in the gas (Kanter and Sternglass 1962; Plechaty and Terrall 1968). Since the cosmic X-ray sources are very weak, it is essential to absorb all of the photons, and the depth of a photoelectric polarimeter must be at least equal to the photon absorption depth. In order to detect the direction of the electron track, the transverse dimensions of the counter must be comparable to the electron range or about one hundredfold less than the depth. Thus, we are constrained to considering long and very narrow counters with an unfavorable ratio of sensitive area to total area for cosmic-ray backgrounds, and we expect an unfavorable signal-to-background ratio. In addition, it is necessary to point this type of polarimeter toward the source with an angular accuracy very much less than 0.01 radian (0.5°) to avoid spurious polarization effects.

A polarimeter of this general type has been constructed by the X-ray astronomy group at the California Institute of Technology (Riegler et al. 1970). This polarimeter used a three-anode proportional counter chamber operating with an argon-methane mixture at 1 atm. The X rays are incident along the axis of the central counter, and the side counters are held in anticoincidence with the central signal counter. In this scheme the net counting rate is least when the photon polarization is parallel to the plane of the three counters and greatest when it is perpendicular to the plane. In the former case, the photoelectrons trigger both the central counter and one of the side veto counters, while in the latter case only the signal counter is triggered. This polarimeter was tested with 18-keV Mo Kα radiation, and it was found that the modulation factor F was about 10%. This low modulation presumably results from the multiple scattering of the photoelectron as well as from the natural width of the primary photoelectron angular distribution function [Eq. (8)]. As expected, the authors found that the incident radiation had to be carefully aligned with the axis of the counter to avoid spurious polarization effects. They also found that the pulse-height threshold setting in the side anticoincidence was quite critical; the modulation factor increased from about 5% to 10% as the lower-level threshold was decreased from 7.5 to 2.5 keV. This dependence on threshold level results from range straggling as well as multiple scattering. If this technique can be applied at energies less than a few kiloelectron volts, then it may be possible to use it at the focus of a high-resolution grazing-incidence telescope. This will probably require the use of a low-pressure, low-atomic number counting gas to insure a reasonable electron range (Kanter and Sternglass 1962) as well as a very efficient anticoincidence shield counter to reduce the non-X-ray background to a manageable level.

An interesting variation of the photoelectric polarimeter has been reported by Sanford et al. (1970). In this polarimeter the rise time of the pulse in a proportional counter is found to depend on the direction of the electric vector in an incident polarized X-ray beam. The rise time is greatest when the electric vector (electron track) is parallel to the anode wire and least when the polarization vector and anode wire are orthogonal. Unfortunately, the effect is small so that this does not appear to be a promising approach for either a stellar or a solar X-ray polarimeter.

Secondary Fluorescence X-Ray Polarimeters

Recently Kifune (1972) has reported on a new type of X-ray polarimeter that may be useful in the low-energy domain. This polarimeter is based on the fact that the secondary fluorescence accompanying photoionization of a P shell is not spherically symmetric if the initial photon is polarized. In the nonrelativistic limit with hydrogen wave functions, it can be shown that the angular distribution of the fluorescent X-ray resulting from a 3S to 2P transition following photoionization of the 2P shell by polarized X radiation is given by

$$f(\beta,\phi) \propto 2 - \sin^2\beta \cos^2\phi, \qquad (9)$$

where β is the angle formed by the direction of the initial photon and the ejected electron and ϕ is the angle between the polarization vector and the plane formed by the photon and electron momentum vectors.

In an actual polarimeter the asymmetry would be reduced by the symmetrical contribution of the 3P to 2S transition. Also, in light atoms most of the excited atoms decay by Auger emission rather than fluorescence, and finally, this effect only occurs with L-shell photoionization; K-shell photoionization necessarily yields symmetrical fluorescence, and therefore the X-ray energy should lie between the K-shell and L-shell excitation energy. Accordingly, at energies below 10 keV we are restricted to argon or krypton gas, in spite of their small fluorescence yield.

To test this type of polarimeter, Kifune constructed a special counter consisting of a central counter surrounded by four side counters (see Fig. 4). The wall separating the central and side counters is constructed of high-transmission mesh. The primary X rays are incident along the axis of the central counter, and the side counter detects the secondary fluorescence. Using argon gas with 10% methane at a total pressure of 6 cm Hg, this polarimeter indicated a modulation factor of 7 ± 2% when the polarized radiation was restricted to energies less than the K edge

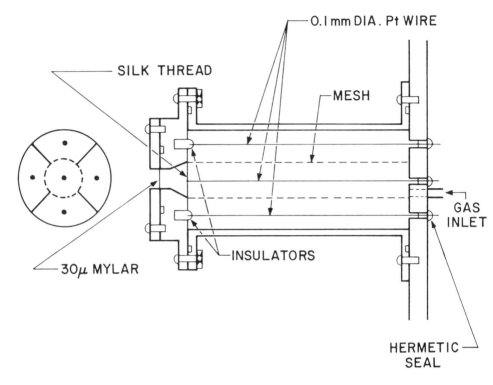

Fig. 4. Structure of the detection system. Platinum wire has a diameter of 0.1 mm (Kifune 1972).

and $0.5 \pm 2\%$ when the energy was greater than the K edge in argon. These results are in reasonable agreement with the prediction obtained by integrating Equation (9) over the aperture of the side counters, i.e., that the modulation should be 10% if the K shell is not ionized and essentially zero if it is ionized.

It appears that this polarimeter has merit, particularly in the low-energy domain, but further development is clearly required.

Borrmann-Effect Polarimeters[11]

The kinematic theory of X-ray diffraction by a crystal lattice considers each scattering center as independent of every other center and neglects any effects that the gradual extinction of the primary beam by the scattering of photons out of it may cause. For most purposes it correctly de-

---

[11] This section is based on material provided by Dolan (1966).

scribes the phenomenon of diffraction, giving the Bragg angle θ_B at which all scattering centers in a given set of atomic planes will scatter in phase as

$$\sin \theta_B = n\lambda/2d, \qquad (10)$$

where d is the perpendicular distance between the planes, λ is the wavelength of the scattered radiation, and n is any positive integer. θ is the angle between the atomic plane and the incident beam in the plane of scattering.

With the availability of perfect cyrstals, it was to be expected that certain phenomena might occur which the kinematic theory would fail to predict. To explain such effects, the interaction between the scattered beams from other scattering centers and the scattering centers themselves must be taken into account. This dynamical theory of X-ray diffraction (Ewald 1965) studies the coupling between the scattering centers, treated as dipole radiators, and the electromagnetic wave field in the crystal, composed of the primary (or incident) beam and the scattered waves from every scattering center in the crystal. In the dynamic theory of scattering, a diffracted beam will occur when the dipole radiators scatter coherently in some direction; in this case, the interaction between wave field and radiators is such that a steady state is set up inside the crystal, with the scattering dipoles ideally neither receiving nor emitting energy into the electromagnetic wave field. Under such conditions, a standing wave pattern is set up between the atomic planes containing the scattering centers by the interference of the primary beam and the coherent diffracted beams from each of the dipoles. Because a standing wave can transport no energy, no energy can flow perpendicular to the atomic planes; rather, it must flow parallel to the atomic planes until it emerges from the other side of the crystal.

Ordinarily, with no diffracted beam present, the scattering is incoherent, and thus the primary beam is attenuated as it passes through the crystal. In the case of diffraction, however, the dipole radiators that act as coherent scatterers absorb no energy from the electromagnetic field. Thus, we must have an anomalous transmission of X rays through perfect crystals set in the diffraction position given by the Bragg equation, a phenomenon known experimentally as the Borrmann effect.

In reality, we are dealing not with point scatterers or radiating dipoles, but with scattering from electrons bound to the atoms in the crystal lattice sites. When a diffracted beam is produced, two standing wave patterns exist inside the crystal lattice: one with nodes at the atomic scattering planes, the other with antinodes there. Energy is absorbed from the transmitted beam in the X-ray region almost exclusively by the photoelectric effect, which is proportional to the square of the electric field strength at the atomic shell location. We may thus expect that the standing wave with nodes at the absorbing sites, the atomic planes, would undergo no absorption. This is very nearly true in the Borrmann effect, with the small absorption that does occur being caused by the spatial extension of the electrons away from the lattice planes of the crystal.

Since the primary and diffracted beam define a plane (the plane of scattering), we may expect that plane-polarization effects may occur in the transmitted beam. This is, in fact, the case. The component of dipole oscillation projected into this plane of scattering is less strongly coupled to the wave in the crystal (and thus to the other oscillators) than that component perpendicular to this plane. Thus, the transmitted wave with polarization in the plane of scattering is more strongly absorbed than the wave with orthogonal polarization, and for real crystals, with absorption effects, the transmitted beam (or more properly, the forward diffracted beam) is polarized orthogonal to the plane of scattering. The Borrmann effect may thus be used as a polarizing device for an initially unpolarized beam or, conversely, as an analyzing filter for a polarized beam.

For a crystal of thickness t, let the optical depth for the wave with nodes at the atomic planes which is polarized perpendicular to the plane of scattering be τ_n, and let that for the wave polarized in the plane of scattering be τ'_n. Then (Cole et al. 1961)

$$\tau_n = \mu_e t(1 - \varepsilon), \tag{11a}$$

$$\tau'_n = \mu_e t(1 - |\cos 2\theta_B|\varepsilon), \tag{11b}$$

where μ_e is the ordinary linear absorption coefficient of the crystal for non-Bragg orientations. For the waves with antinodes at the atomic planes in the crystal,

$$\tau_a = \mu_e t(1 + \varepsilon), \tag{12a}$$

$$\tau'_a = \mu_e t(1 + |\cos 2\theta_B|\varepsilon). \tag{12b}$$

ε is a parameter lying between 0 and 1 (generally being close to 1) which varies inversely as the spatial extent of the photoelectric absorption of the atoms composing the crystal lattice. It depends to a certain extent upon the temperature, as higher temperatures lead to larger amplitude atomic vibrations within the lattice (Batterman 1962).

For the (220) reflection of germanium, $\theta_B = 22°\ 41'$, and ε has been measured at room temperature as 0.95. Thus, $\tau'_n = 6.6\ \tau_n$, and τ_a and τ'_a are 39 and 33 times the optical depth τ_n. Thus, for a thick enough crystal, the forward diffracted beam is nearly 100% plane polarized perpendicular to the plane of scattering.

Because of the diffraction process involved, only a very small bandwidth of energies centered about the Bragg energy corresponding to θ_B

fulfill the conditions necessary for diffraction and, thus, anomalous transmission. If E is the photon energy in keV, d the interplanar spacing in Å, and $d\theta$ the width of the diffracted beam in arc seconds, the bandwidth dE is given by

$$\frac{dE}{E} = 4.85 \times 10^{-6}(2.59 \times 10^{-2} E^2 d^2 - 1)^{1/2} d\theta. \qquad (13)$$

At 3 Å, dE/E is equal to 1.8×10^{-5} for the (220) reflection in silicon, with $d\theta$ a typical 10″. dE/E is 8.80×10^{-5} for the (111) reflection (Table I).

TABLE I
Bandwidth of Borrmann Effect
(Dolan 1966). dE/E for Silicon

λ (Å)	$d\theta$ (arc sec)	Crystal Plane (111)	(220)
1.54	10	1.91×10^{-4}	1.10×10^{-4}
3	10	8.80×10^{-5}	1.80×10^{-5}

Since the transmitted beam is not undergoing diffraction, its optical depth τ is equal to $\mu_e t$. Thus, $\tau = 20\ \tau_n$, and the problem in practice becomes one of choosing a crystal thick enough to filter out the transmitted beam, yet thin enough not to seriously attenuate the forward diffracted one. For the low fluxes coming from celestial sources, thin crystals may be necessary, so that some of the transmitted beam may pass through the crystal. If such thin crystals are necessary, a calibration can be made with the aid of a beam of known partial polarization. Further details are given by Cole et al. (1961).

Because of the small bandwidth of the Borrmann effect, satellite observations rather than rocket-borne experiments are necessary to determine the polarization of celestial X-ray sources. Since we may assume most X-ray spectra are continuous, although they may be peaked, we can estimate the fluxes passed by an experimental Borrmann arrangement as 5×10^{-5} the total flux in the 1–15 Å bandwidth. This is about 10^{-4} photons cm^{-2} sec^{-1} for a typical source. If the detector is of 100 cm^2 cross-sectional area, i.e., composed of several individual crystals, the fluxes passing through it are about 10^{-2} sec^{-1}. Thus, we need about one hour of observation of any one source to obtain enough counts to determine its polarization. In view of this very low efficiency, it appears that the Borrmann effect may be better suited to solar than to stellar X-ray polarimetry. In any case, this effect is only useful at certain discrete energies corresponding to certain Bragg planes in germanium and silicon.

These are the only materials that are available with sufficient purity to exhibit the Borrmann effect.

Circular Polarization Analyzers, Gamma-Ray Polarimeters and Multilayer Reflection Polarimeters

At energies appreciably above 1 MeV we can take advantage of the fact that electron-positron pair production becomes dominate. Since the plane determined by the paths of the charged pair is parallel to the electric vector of the incident photon, the polarization can be determined by means of a spark chamber or other suitable charged-particle track detectors. The photon converter plates must of course be thin enough to prevent appreciable scattering of the electron-positron pair. As far as we know, no effort has been made either to evaluate this effect fully or to exploit it in gamma-ray astronomy.

In the case of circular polarimetry, the only work that appears to have been done is that of Goldhaber et al. (1958) who detected the sense of the circular polarization of the 960-keV gamma-ray line in Sm^{152}. This was accomplished by taking advantage of the polarization dependence on Compton scattering from polarized electrons. The d-shell electrons in magnetized iron were used for this purpose. While the experiment was successful, the effect was very small, being on the order of 1.7%. It does not appear to be useful in astronomy.

Finally, we remark that Spiller (1972) has performed an analysis of multilayer X-ray reflectors in which alternate layers have different absorption coefficients. He finds that such multilayer mirrors may have large reflection coefficients even at normal incidence. He also points out that such mirrors at 45° incidence with proper materials and thicknesses can be used as nearly perfect polarization analyzers. As yet no experimental work has been done with these devices.

BRAGG CRYSTAL POLARIMETERS

The coherent scattering from a Bragg crystal at 45° provides a very effective technique for determining X-ray polarization within a narrow energy range. In this section we will primarily discuss an instrument of this type that is scheduled to fly on the OSO-I satellite in 1974. A similar instrument was flown in an Aerobee-350 rocket in February 1971 by the Columbia Astrophysics Laboratory. The flight resulted in the detection of X-ray polarization from the Crab Nebula. The rocket instrument has been described elsewhere (Weisskopf et al. 1972).

Theory of the Polarimeter Operation

The operation of crystal polarimeters depends upon the fact that at 45° incidence a Bragg crystal reflects only those X rays that are polarized perpendicular to the plane of incidence. Each bound electron in the crystal lattice acts as a Hertzian dipole excited along the electric vector of the incident photon, and such dipoles do not radiate along their axes. The interference between the individual scattering centers produces the strong reflection at the Bragg angle but does not modify the polarization dependence of the individual centers. If a beam of 100% polarized radiation is scattered through 90° by a Bragg crystal, the intensity of the diffracted beam varies by 100% as the crystal is rotated about the line of sight to the source. Thus, a Bragg crystal is a perfect polarization analyzer for radiation that satisfies the Bragg condition at 45° incidence. With such a polarimeter, the crystal reflects only those X rays whose energy E_n satisfies the Bragg condition for 45° reflection

$$E_n = nhc/d\sqrt{2}, \qquad (14)$$

where d is the crystal lattice spacing, n is the diffraction order, h is Planck's constant, and c is the velocity of light.

If nearly perfect crystals are used, the reflection coefficient may be 50% or higher but only over an energy bandwidth of a small fraction of an electron volt. The bandwidth can be greatly increased without excessively reducing the reflection efficiency for monochromatic radiation if we use mosaic instead of perfect crystals (Angel and Weisskopf 1970). This results in a large increase in the reflection of continuum radiation. Mosaic crystals consist of a randomly ordered matrix of very small crystal domains. Each domain has a regularly ordered atomic structure and strongly reflects those X rays that satisfy the Bragg condition for its particular orientation. Each domain is small enough so that X rays that do not satisfy the Bragg condition pass through it with negligible absorption. Thus, if we consider a polychromatic X-ray beam to be incident on the mosaic crystal, each wavelength will pass through the crystal until it intercepts a domain that satisfies the Bragg condition corresponding to its wavelength. It is then strongly reflected at 90° to its original direction, passes out of the crystal, and is detected with a proportional counter.

The effective bandwidth and reflection efficiency of mosaic crystals depend upon the absorption coefficient of the material, the atomic scattering factor, and the domain distribution; the relevant theory is given by Angel and Weisskopf (1970). These authors show that if the Bragg angle of a crystal is varied from θ_1 to θ_2 at a constant angular velocity ω, then the total reflected energy at wavelength λ, $W(\lambda)d\lambda$, is given by

$$W(\lambda)\mathrm{d}\lambda = \int_{\theta_1}^{\theta_2} I(\lambda)\mathrm{d}\lambda\, R(\theta,\lambda)\frac{\mathrm{d}\theta}{\omega}$$

$$= \frac{1}{\omega} I(\lambda)\mathrm{d}\lambda \cdot \Delta\theta(\lambda), \tag{15}$$

where

$$\Delta\theta(\lambda) = \int_{\theta_1}^{\theta_2} R(\theta,\lambda)\mathrm{d}\theta. \tag{16}$$

Here, $I(\lambda)\mathrm{d}\lambda$ is the power at wavelength λ in the spectral range $\mathrm{d}\lambda$ incident on the crystal, $R(\theta,\lambda)$ is the coefficient of reflection of X rays of wavelength λ incident at angle θ, and $\Delta\theta$ is the integrated reflectivity, a measure of the reflection efficiency for continuum radiation. The full width at half maximum of the reflection function $R(\theta,\lambda)$ provides a measure of the degree of misorientation of the crystal domains. This quantity is known as the mosaic spread. The maximum value of $R(\theta,\lambda)$ is called the peak crystal reflectivity. In the case of graphite with a mosaic spread of $0.8°$, the peak reflectivity at 2.6 keV is about 8%.

Mosaic graphite crystals were selected for the OSO-I polarimeter because they possess a very high integrated reflectivity at the energy where many X-ray objects are most intense, and they are available in a form such that large areas of uniform quality can easily be fabricated. The integrated reflectivity, peak reflectivity, and mosaic spread of a large number of graphite crystals have been measured. We find that the integrated reflectivity of crystals with $0.°8$ mosaic spread is about 9×10^{-4} radians in first order (2.6 keV) and 5×10^{-4} radians in second order (5.2 keV).

Instrument Design

The minimum detectable polarization in any source depends on the signal strength, the background counting rate, the observing time, and systematic effects (see p. 300). If we employ a flat crystal panel, then the detector area must be comparable to the crystal area, and the cosmic-ray-induced background will be excessively large even for the strongest nonsolar X-ray sources. For continuum sources, we may greatly reduce the size of the detector by mounting the crystals on a sector of a parabolic surface (see Fig. 5). Using this technique on the OSO-I polarimeter, we are able to employ a small, 2.54-cm diameter detector and thereby reduce the background counting rate by a factor of about 20:1. The curvature of the panel increases the bandwidth of the detected 2.6-keV photons from 0.04 to 0.4 keV (see Table II), but this is of no concern since we are dealing with continuum sources and the bandwidth is still smaller than the 0.7-keV resolution of the proportional counter. However, since the angle of

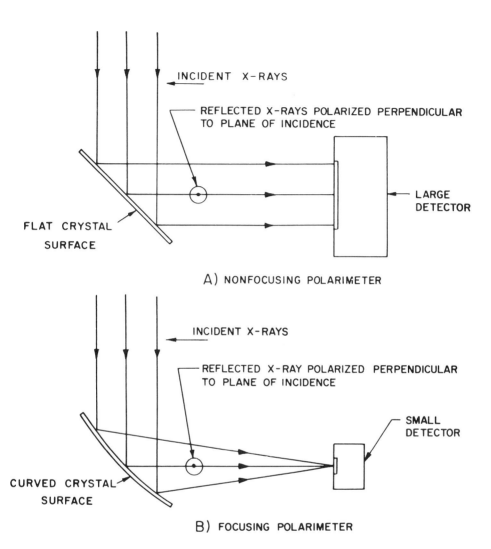

Fig. 5. Conceptual diagram of nonfocusing and focusing polarimeters.

incidence for some of the crystals departs slightly from 45°, the flux with the polarization component in the plane of scattering is not completely absorbed and produces a small unmodulated signal. The theoretical modulation of the counting rate as the polarimeter is rotated about the line of sight to a 100% polarized source is 100% for a flat crystal panel. It is about 96% for the focusing crystal panels that will be used in the OSO-I polarimeter.

TABLE II
Characteristics of Crystals to Be Used in OSO-I Polarimeter

Material:	Grade ZYC Pyrolytic Graphite
Manufacturer:	Union Carbide Corporation
Lattice Spacing:	3.35 Å
Mosaic Spread:	0.8°
Range of Bragg Angles:	40° to 50°
	First Order *Second Order*
Photon-Energy Range:	2.4–2.8 keV 4.8–5.6 keV
	At a Bragg Angle of 45° (*First Order*)
Energy:	2.62 keV
Energy Resolution:	37 eV
Wavelength:	4.74 Å
Wavelength Resolution:	0.066 Å
Resolution:	72

The instrument is shown in Fig. 6. Proportional counters with thin beryllium windows are used as photon detectors. X rays satisfying the Bragg condition incident on the crystals scatter to the proportional counter and stop in the signal counter gas volume. The combined projected area of the two polarimeter panels is 280 cm^2.

The use of the curved graphite crystal panel and the focusing of the incident X rays onto a small detector limit the field of view to within about 2° of the spacecraft spin axis (see Fig. 7). Under normal operation of the OSO spacecraft it should be possible to observe a cosmic source for at least six days twice a year. If the source is polarized, then as the spacecraft rotates with angular frequency ω, the signal counting rate in each detector is modulated sinusoidally at twice the spacecraft rotation frequency (the polarization signal is proportional to $\sin 2\omega t$). The depth of the modulation yields the magnitude of the source polarization, and the phase gives the position angle of the polarization vector. As an additional check on the data, there will be a predictable phase relationship in the counting rates of the two detectors if the source is polarized. This phase relationship makes it possible to differentiate between polarization modulation and time variability of the source.

For each valid event, 12 bits of azimuth data, 5 bits of pulse-height data, and 1 bit of counter-location information will be recorded. Timing bins of 2.4 msec are derived from the azimuth data for use in pulsar data analysis. The maximum average event counting rate, 25 events sec^{-1}, is determined by the telemetry rate of 500 bits sec^{-1}. An experiment memory unit acts as a queuing buffer.

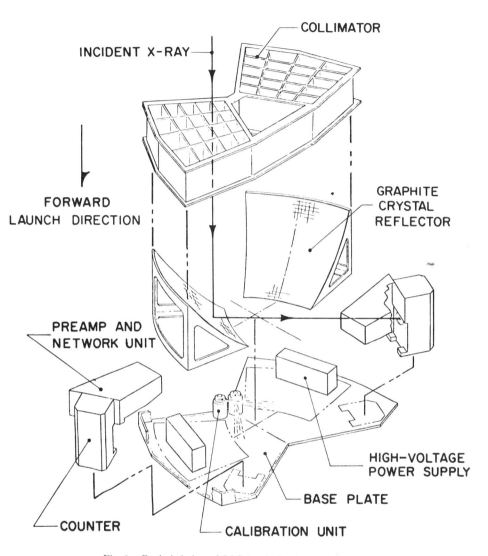

Fig. 6. Exploded view of OSO-I polarimeter assembly.

Pulse-height and rise-time discrimination will be used to reduce the cosmic-ray-induced background counting rate in the polarimeter detectors. In addition, the main counter volume will be surrounded by a three-sided anticoincidence-shield counter in the same gas volume. Pulse-shape discrimination is effective because the characteristic rise times of charged particles and X rays are easily distinguished. Charged cosmic-ray particles generally will produce long ionization paths in the counter gas, and

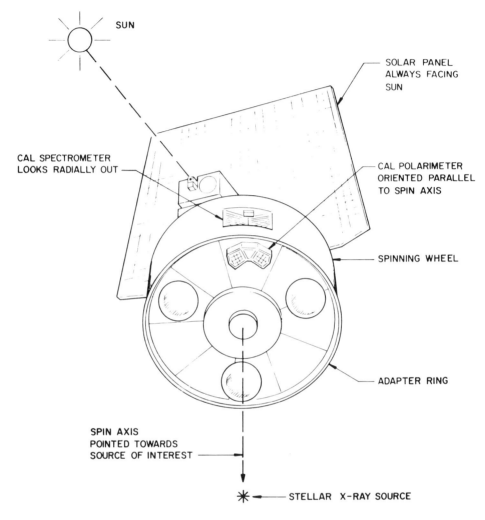

Fig. 7. Polarimeter in OSO-I wheel compartment.

since the charges along the track must drift different distances to the anode, the pulse will have a relatively long rise time. On the other hand, an X ray will be absorbed in a single photoelectric event in the gas and will produce a quickly rising pulse.

Two independent and redundant polarimeters will be used as a safeguard against single-point failures; if a serious detector leak or electronics malfunction develops in one polarimeter, that channel will be shut off by ground command, and the other would still function. In order to insure an adequately long counter lifetime in orbit, a CO_2 quenching gas will be used, and the programed proportional-counter high-voltage power supply will be automatically shut off when the spacecraft passes through the South Atlantic Anomaly where the high-energy charged-particle density is very high.

The polarimeter detector will be calibrated in flight by the use of two small, high-purity ^{55}Fe (5.9 keV) radioactive sources. The sources will be mounted on the shafts of permanent-magnet stepping motors and will be used to calibrate the two polarimeter detectors. The position of each calibration rod and the counter high voltage will be commanded from the ground.

A preengineering model polarimeter has been constructed and calibrated at the Columbia Astrophysics Laboratory. Bremsstrahlung X rays from a Machlett platinum-anode X-ray tube are incoherently scattered by a lithium-hydride crystal through 90° to produce a highly polarized X-ray continuum which is collimated with an evacuated tube, 20 m long. The detector assembly, located at the end of the tube, consists of a collimator, a paraboloidal crystal surface, and a proportional counter. Some of the results of these modulation tests are shown in Fig. 8. Here we see

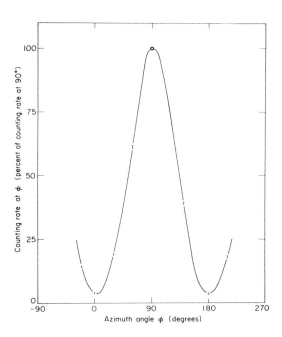

Fig. 8. Modulation response of the preengineering model OSO-I polarimeter measured with a collimated beam of 5.2-keV (second-order) polarized X rays.

that the measured modulation in second-order (5.2 keV) reflection is 93%.

In Table III we estimate the minimum detectable polarization for several sources, derived from the experimental values for the integrated reflectivities, the known spectra for a number of X-ray sources, and the geometry of the crystal panel. These estimates are based on the minimum detectable polarization as given in Equation (7) above. From this table, we see, for example, that it should be possible in a 24-hr observation time to measure with 99% confidence (3 σ) a minimum detectable polarization of 0.5% and 4% for Sco X-1 and Cyg X-1, respectively. In the same observing time, the polarization of the Crab Nebula can be measured to 0.6% (1 σ). This threshold decreases inversely with the square root of the observing time.

Measurement of the polarization within the pulse of the highly variable pulsating X-ray source Cen X-3 is feasible. This X-ray pulsar has a

TABLE III

Minimum Detectable Polarization (MDP) of the OSO-I Graphite Polarimeter at the 99% (3 σ) Confidence Level

Object	Observation Time (days)[a]	First Order (2.6 keV)		Second Order (5.2 keV)		Combined MDP (percent)
		Assumed Flux (keV cm^{-2} sec^{-1} keV^{-1})	MDP (percent)	Assumed Flux (keV cm^{-2} sec^{-1} keV^{-1})	MDP (percent)	
Crab Nebula	1	3.4	1.8	1.6	3.9	1.7
Sco X-1	1	32	0.58	19	1.0	0.51
Sco X-2	1	1.8	2.6	1.7	3.8	2.2
Cyg X-1	1	0.9	4	0.5	8	4
Cyg X-2	1	0.60	5.0	0.34	11	4.7
Cyg X-3	1	0.37	6.9	0.43	9.5	5.6
Nor X-1	1	1.2	3.3	0.7	6.7	3
Nor X-2	1	0.4	7	0.3	13	6
Nor X-3	1	2.1	2.4	1.2	4.7	2.1
Cir X-1	1	2.3	2.3	1.1	5.0	2.1
Lup X-1	1	1.2	3.3	1	5.3	2.8
Cen X-3	1	0.5	6	1	5	4
Cen X-3 Pulsar	6	0.1	8	0.2	7	5
Crab Primary Pulse	6	0.14	5.8	0.09	14	5.7
NGC 1275 (Per A)	6	0.27	3.5	0.03	40	4.0
M 87 (Vir X-1)	6	0.09	8	0.03	40	9
LMC	6	0.23	3.9	0.06	21	4.3

[a] One observation day requires about 1.8 orbit days because of earth occultation. The assumed background is 0.023 count (sec·counter)$^{-1}$.

period of 4.8 sec and a pulse width of about 1 sec with about 70% of the X-ray flux in the pulse (Giacconi et al. 1971). In each of four 0.25-sec intervals within the pulse, the minimum detectable polarization at the 99% confidence level is 5% for an integration time of six days. Similarly, in the case of the Crab pulsar, we can measure the average polarization in the main pulse with a 3 σ limit of about 6%.

Systematic Errors

In order to be able to measure polarization of strong sources at the 1% level, it is essential to eliminate large sources of spurious modulation and to understand and calibrate the remaining small systematic effects. Of greatest concern are spurious modulation effects that have a large component of twice the rotation rate of the satellite, since these would be interpreted as source polarization. Almost all spurious modulation effects are associated with the possible misalignment of the polarimeter axis and the X-ray source direction. To minimize the effects of off-axis spurious polarization, the spacecraft will be commanded to orient its spin axis to within 0.5° of an X-ray source of interest. This pointing error will be known to about 0.1°, after the fact, at the time of data analysis. Appropriate small corrections derived from preflight laboratory calibrations will be made at that time.

A number of steps have been taken to minimize off-axis effects. The efficiency of the polarimeter detector should be constant over the entrance window since the X-ray focal spot for an off-axis source moves around the center of the counter due to the rotation of the spacecraft. A circular counter window will be used, and the image of an on-axis source will be centered on the window. A window or window support structure with a twofold symmetry will not be used because either one might produce a spurious polarization signal at 2ω for an unpolarized off-axis source.

The focal spot size will be made as small as possible so that a small low-background counter may be used. If the spot size is comparable to the window diameter and the polarimeter alignment is not perfect, then the signal will be spuriously modulated as part of the focused beam alternately falls on and off the window. The diameter of the visible-light focal spot of the preengineering-model polarimeter is 0.51 cm, which is significantly smaller than the 2.54-cm diameter of the counter window. As shown in Fig. 9, 99% and 95% of the flux fall within circles of diameters 1.78 and 1.02 cm, respectively.[12] The counter window was made large

---

[12] The validity of these optical tests is based on a previously determined correlation between optical and X-ray imaging to within a few minutes of arc (Kestenbaum et al. 1971b).

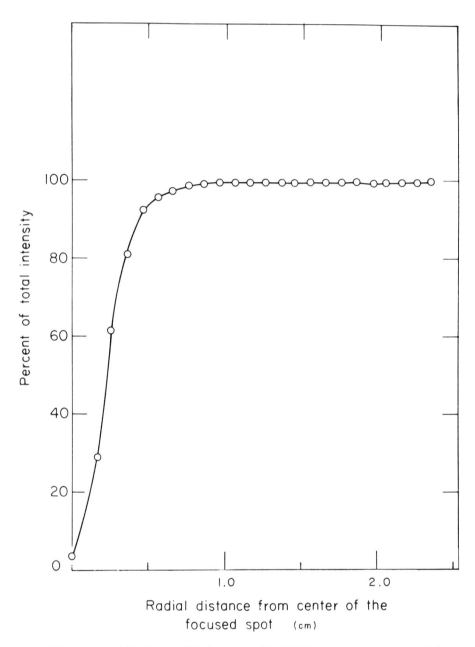

Fig. 9. Integral distribution of the focal spot of the OSO-I preengineering model polarimeter measured with visible light; fraction of total intensity falling within a circle as a function of circle radius.

enough so that the entire image of a source 2° off axis falls onto the detector window.

A counter with an anode wire oriented perpendicular to the direction of the reflected beam and centered behind the window could conceivably produce a 2 ω modulation from an off-axis source as the focused rays trace a circular path about the center of the window. Such a modulation could occur because the detector efficiency may change due to variations in the signal pulse height or rise time. A multisignal anode counter design was tried in an attempt to alleviate these problems. However, it was not possible to fabricate a counter with good resolution and rise-time properties that also had a cell size (anode spacing) small compared with the focal spot radius (see Fig. 9). Variations in geometry in a 9-signal anode counter produced large modulation effects. End-window counters with cylindrical symmetry were considered but were also rejected because of poor resolution and rise-time properties. A laboratory measurement with a preengineering model of a single-signal anode counter indicated a $\pm 2\%$ modulation at 1 ω signal modulation and less than a $\pm 1\%$ effect at 2 ω for a source equivalent to 1°5 off axis. It was therefore decided to use this type of polarimeter counter.

An internal misalignment of the polarimeter could result in a spurious modulation. In order to minimize this effect, the polarimeter will be mounted on an optical bench, and the counter will be carefully aligned with respect to the crystal panel. The collimator used is so coarse that its alignment will not be critical. A laser will be used to align the polarimeter-alignment mirrors with a corresponding mirror mounted to the spacecraft. No moving parts will be used that could change the polarimeter alignment. The rotation of the spacecraft results in the suppression of spurious systematic modulation effects associated with differences in detector-channel sensitivities.

INCOHERENT- OR THOMSON-SCATTERING POLARIMETERS

Historically, X rays were shown to be transverse waves with a double scattering experiment (Barkla 1906). The success of this experiment was based on the fact that electromagnetic waves scatter incoherently at right angles to the direction of the electric vector of the incident photon. That is, the scattering cross section is given by

$$\sigma = \sigma_T \sin^2\phi, \qquad (17)$$

where σ_T is the Thomson cross section and ϕ is the angle between the polarization vector of the incident photon and the propagation vector of the scattered photon. In using the Thomson cross section, we neglect the effect of relativity and of atomic binding. The photon energy is assumed to be large compared to the binding energy but small compared to the rest energy of the electron. Since most X-ray sources are most intense at photon energies of a few kiloelectron volts or less, the use of the nonrelativistic cross section is fully justified, but the effects of binding are quite severe. The most important binding effect is that photoelectric absorption competes appreciably with scattering. Since the absorption of a photon of given energy increases rapidly with the atomic number Z of the scattering atom, we are required to use scattering blocks made from low-Z material. In practice, Li, LiH, and Be, have been used. Hydrogen and helium would be preferable, but these would necessitate the use of cryogenic techniques to obtain short enough scattering lengths. In Fig. 10 we show how the ratio of scattering cross section to total cross section (scattering plus photoelectric absorption) varies with photon energy for

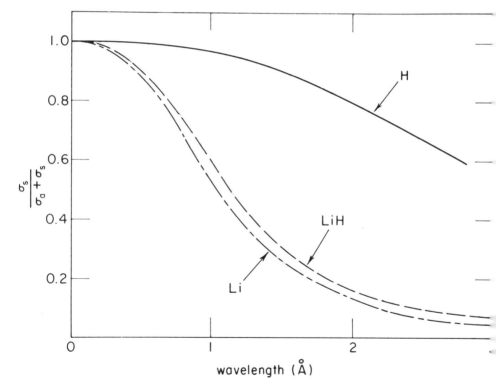

Fig. 10. The scattering efficiency of Li, LiH, and H as a function of wavelength; σ_s is the scattering cross section, and σ_a is the absorption cross section.

hydrogen, lithium hydride, and lithium. In Fig. 11 we show the scattering efficiency for hydrogen and lithium as a function of photon energy, and we also show the spectra of two of the brightest stellar X-ray sources, Sco X-1 and the Crab Nebula. If we restrict ourselves to noncryogenic scattering materials, then it is clear that the effective low-energy cutoff of

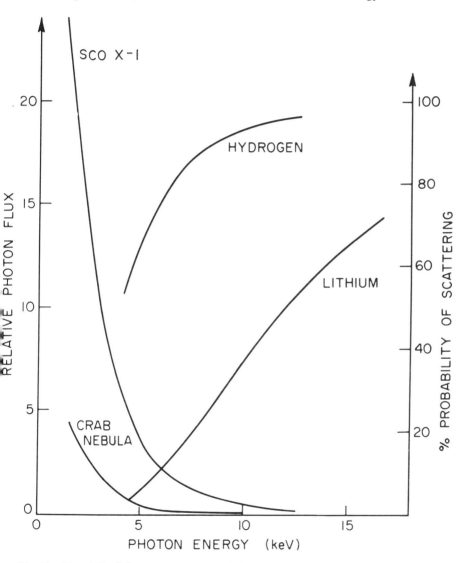

Fig. 11. Li and H efficiency curves compared with the spectra of Sco X-1 and the Crab Nebula.

the polarimeter will be about 6 keV if we use lithium and that this cutoff will increase rapidly with increasing atomic number of the scattering material. In the case of beryllium, the scattering efficiency drops to 32% at 12.3 keV, while in the case of lithium the efficiency is 52% at the same energy. The beryllium efficiency drops very rapidly with decreasing photon energy.

From the above discussion, it is clear that Thomson-scattering polarimeters are useful at energies above several kiloelectron volts. In spite of this limitation, they have had extensive use in both stellar and solar X-ray astronomy, and it is clear that they will be of even greater value in the future. Scattering polarimeters are distinguished by the fact that they provide a measurement of polarization over a large energy bandwidth. At low energy, they are limited by photoelectric absorption in the scattering blocks, while at high energy, the cutoff is determined by the source spectrum, the observing time, and the non-X-ray background. In the case of the Crab Nebula, a lithium-scattering polarimeter provides a measure of polarization from about 6–25 keV with the largest signal occurring at about 10 keV.

The Columbia X-ray astronomy group has flown metallic lithium scattering polarimeters in three sounding rockets (Angel et al. 1969; Wolff et al. 1970; Novick et al. 1972). The solar physics group of the P. N. Lebedev Physical Institute has flown beryllium-plate scattering polarimeters on Intercosmos-1 and -4 (Tindo et al. 1970, 1972), and Neupert (1972), of the Goddard Space Flight Center, has a beryllium-cylinder scattering polarimeter on OSO-7.

In the next section the Thomson-scattering polarimeter that was used to detect polarization in the Crab Nebula (Novick et al. 1972) is described in detail [prepared by Linke (1972)]. It is followed by a summary of Landecker's (1972) detailed study of a stellar X-ray scattering polarimeter suitable for use in a large satellite.

A ROCKET THOMSON-SCATTERING POLARIMETER

The angular dependence of Thomson scattering is given by (Heitler 1954)

$$\frac{d\sigma}{d\Omega} = (e^2/mc^2)^2 (1 - \sin^2\theta \cos^2\phi), \tag{18}$$

where θ is the scattering angle and ϕ is the angle between the initial electric vector and a projection of the scattered momentum in a plane perpendicular to the initial momentum. For the case of right-angle scattering, $\theta = 90°$, and the scattering cross section is proportional to $\sin^2\phi$ $[= \frac{1}{2}(1 - \cos 2\phi)]$. Thus the linear polarization state of an incident X-ray flux can be determined from the azimuthal distribution of the scattered photons. If one scatters photons in some medium and rotates a detector of small solid angle about the scatterer in a plane perpendicular to the incident photon beam, linear polarization of the beam will be exhibited as a cos 2ϕ modulation of the detected flux. The amplitude of the modulation will be proportional to the degree of polarization, 100% modulation corresponding to 100% polarization only for the idealized case of a point detector.

Design and Construction

Choice of geometry. The polarimeter was designed to fit into a 91-cm long cylindrical section of the 56-cm diameter Aerobee-350 sounding rocket. The geometry of scattering material and detectors was chosen to optimize the polarization sensitivity. The design employs rectangular "scattering blocks" of metallic lithium surrounded on four sides by proportional counters in the configuration of Fig. 12. Factors relevant to this choice of geometry are discussed by Wolff (1969). Briefly one must consider that: (1) increasing the length of the scattering blocks increases the fraction of incident X rays that will be scattered but also increases the size of the detector required and therefore increases the cosmic-ray background signal; (2) decreasing the width of the blocks increases the chance that a scattered X ray will reach the detector without being photoabsorbed but also decreases the area presented by the top of the block to the source; and (3) increasing the proportional counter depth improves its efficiency but subtracts from the total cross-sectional area available for scattering material. The design finally chosen employs 24 right-square cylinders, $5.08 \times 5.08 \times 14.45$ cm, and 8 right-triangular cylinders of lithium set in an array. The array presents 723 cm^2 of scattering material to the Nebula. The rectangular sides of the blocks are viewed by 52 proportional counters. In use, the entire polarimeter is rotated about the line of sight to the source. Polarization is then detected as a modulation in the X-ray counting rates at twice the rocket rotation frequency. This geometry yields a modulation of 30% for completely polarized X rays in the absence of a background signal, as demonstrated experimentally by Wolff (1969).

Selection of lithium as scattering material. Since the anticipated flux from the Crab Nebula is small, it is desirable to have as high an instrumental efficiency as possible. The X-ray spectrum of the Crab above 2.5 keV has been fairly well determined and is given by (Henry et al. 1972):

$$I(E) = 9.2 \times 0.9 \ E^{-2.0 \pm 0.2} \text{ photons cm}^{-2} \text{ sec}^{-1} \text{ keV}^{-1}. \quad (19)$$

When integrated over a range of 1 to 20 keV, this expression yields ~ 8.8 photons cm^{-2} sec^{-1}, while integration from 5 to 20 keV yields only

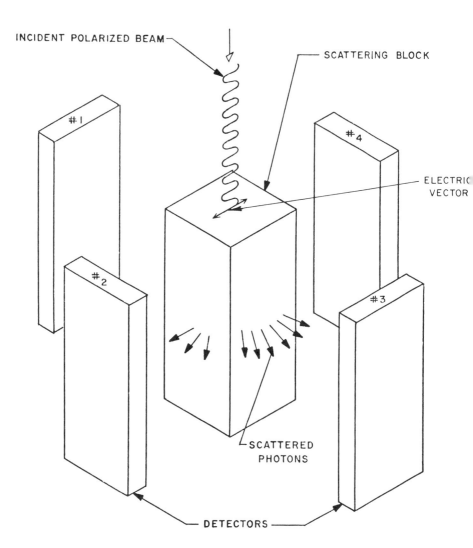

Fig. 12. Schematic representation of Thomson-scattering polarimeter. With the electric vector as shown, detectors #1 and #3 will receive more X rays than detectors #2 and #4. When the instrument is rotated, counting rates will be modulated at twice the roll frequency, and modulation in #1 and #3 will be out of phase with that of #2 and #4.

~ 1.4 photons cm^{-2} sec^{-1}. This sharp decrease in flux with increased energy makes low-energy sensitivity of the instrument extremely desirable. Unfortunately, both coherent scattering and photoelectric absorption tend to reduce the fraction of photons that are scattered into the detectors.

The photoelectric absorption cross section is proportional to (Heitler 1954) $\sigma_T Z^5 E^{-7/2}$, where σ_T is the Thomson-scattering cross section, Z is the nuclear charge, and E is the photon energy. Thus, we expect high scattering efficiencies only for low-Z scattering materials and high photon energies. For lithium, $Z = 3$, the photoelectric and Thomson cross sections are equal at 4.6 keV. Hydrogen and helium are more efficient scatterers than lithium but are useful only in their dense liquid or solid states. Considerations of cryogenic engineering made these choices of scattering materials untenable for a rocket flight. Lithium hydride is another possible scattering material and, partially as a result of the presence of hydrogen, is approximately 20% more efficient than lithium metal as an X-ray scatterer in the relevant energy range. Unfortunately, contamination by oxidation and difficulties in fabrication prevented the use of lithium hydride as a scattering medium. Since lithium metal of relatively high purity (99.9+%) is commercially available and can be shaped easily, lithium was ultimately chosen as the scattering material.

Fabrication of the scattering blocks. Lithium metal ingots, obtained from Lithium Corporation of America, were compression molded into blocks 5.08 × 5.08 × 14.45 cm under pressures exceeding 70 kg cm^{-2}. A mounting base of high-density polyethylene was compressed into each block during the molding process. Each block was covered with a three-layered plastic: 0.0013 cm of Mylar for abrasion resistance, 0.0003 cm of Saran for protection from water vapor, and 0.005 cm of polyethylene for heat sealing. A single thickness of this film was found to absorb 13% of the incident X rays at 5.9 keV but a negligible fraction of X rays at 10.0 keV.

The detectors. Proportional counters were used to detect the scattered X-ray flux. The detectors incorporated magnesium cathodes and 0.038-cm beryllium windows on two sides and were filled with 3 atm of 90% xenon +10% carbon dioxide. Because of the 180° symmetry in the Thomson-scattered flux, X rays from adjacent scattering-block faces were detected in a single proportional-counter gas volume with no loss of polarization information. The calculated efficiency of these proportional counters to X rays in the range of 5 to 25 keV is shown in Fig. 13.

The total array of scattering blocks and proportional counters was constructed in four 90° pie sections. Each section consisted of 13 proportional counters, 6 "full"-sized (5.08 × 5.08 × 14.45 cm) scattering blocks, and 2 "half"-sized blocks mounted on a machined magnesium base plate. Also mounted on the aft side of the plate were 13 signal preamplifiers and 1 miniature dc-to-dc converter which provided a bias voltage of 2500 V for the proportional-counter anodes.

Efficiency of the instrument. The efficiency of this polarimeter geometry was calculated by means of a stochastic photon-scattering computer pro-

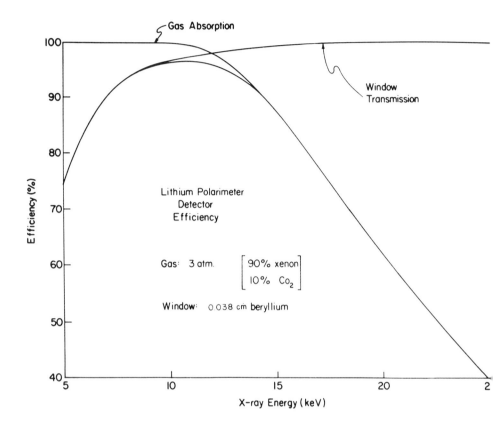

Fig. 13. X-ray detection efficiency of proportional counter.

gram (Wolff 1969), and the results were confirmed experimentally for certain incident photon energies (Wolff 1969; Landecker[13]). The polarimeter efficiency as a function of energy is shown in Fig. 14a together with an observed Crab Nebula spectrum for reference. The drop in efficiency at low photon energies is due to photoelectric absorption in the lithium and in the counter window, as well as to coherent scattering effects. The turnover in efficiency above ~ 18 keV is a result of the increased transparency of both the lithium blocks and the proportional-counter gas to the harder X rays. The response of the instrument to the Crab Nebula's spectrum was obtained by weighting the total (anticipated) flux incident on the lithium blocks by the polarimeter efficiency. The result (Fig. 14b) indicates a broad peak in the detected flux at ~ 8.5 keV and an absence of detected flux below ~ 5.5 keV. The integrated rate for a bandwidth of 5 to 25 keV is 137.7 counts sec^{-1} (or 2.65 counts sec^{-1} counter^{-1}).

---

[13] Personal communication.

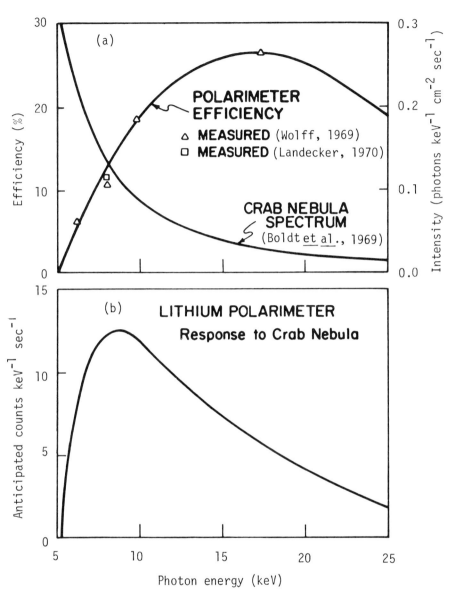

Fig. 14. (a) Crab Nebula X-ray spectrum and polarimeter efficiency defined as the ratio of the detected flux to the flux incident on the scattering block. (b) Response to Crab Nebula, i.e., product of two functions given in (a) for an area of 723 cm^2 (28 scattering blocks).

Background Reduction

The counting rates recorded by the polarimeter detectors were expected to contain a background contribution from the isotropic diffuse X-ray flux as well as from the cosmic-ray (i.e., charged-particle and gamma-ray) flux which penetrates the instrument during the flight. Since the minimum polarization measurable by this instrument is limited by the background, it is highly desirable to reduce this rate wherever possible.

Diffuse X-ray background. The diffuse X-ray background contribution can be reduced by an X-ray collimator placed in front of the scattering blocks that limits the angular field of view of the instrument. The desired collimator was constructed of 0.013-cm thick stainless steel in the form of 104 boxes 4.29 cm^2 × 6.98 cm deep. The collimator alone has a field of view of 0.38 sterad; however, the detector cathodes become part of the collimation system in the geometry employed (see Fig. 15).

In spite of the collimator, background X rays may enter the detectors either by scattering within the lithium metal or, for certain incident angles, simply by penetrating a lithium block (Fig. 15). An upper estimate of the X-ray background rate is provided by the flux reaching the top of the lithium blocks. For the geometry employed and a background spectrum given by $I(E) = 6.3 \pm 0.5 \, E^{-1.3 \pm 0.1}$ photons cm^{-2} sec^{-1} keV^{-1} sterad^{-1} (Boldt et al. 1969), we obtain a flux of 1.1 ± 0.1 photons cm^{-2} sec^{-1} (14.2 ± 1.2 photons sec^{-1} counter^{-1}) for the energy range of 5–25 keV. The detected X-ray background was expected to be considerably lower than this value since photoelectric absorption in the lithium will necessarily reduce the flux reaching the counters. For an estimated integrated detection efficiency of 10%, we find a detected X-ray background rate of 1.4 ± 0.1 photons sec^{-1} counter^{-1}.

Cosmic-ray-induced background. From previous measurements (Wing 1968; Wolff 1969) of the background counting rate induced in xenon-filled proportional counters by cosmic rays, we can estimate that this background component for the lithium polarimeter would exceed the diffuse X-ray background rate by a factor of ~30 in the absence of background suppression. Consequently, a considerable effort was expended to reduce the non-X-ray background rate.

1. Anticoincidence detector. In order to reduce the background signal resulting from high-energy charged particles penetrating the instrument, the polarimeter was surrounded by an anticoincidence detector consisting of a cylindrical shell of scintillation plastic. The plastic was viewed by three 12.7-cm diameter photomultiplier tubes (RCA C31027), and signals from these tubes were summed and used to veto coincident signals from the polarimeter proportional counters. The efficiency of the plastic shield in the detection of charged cosmic-ray

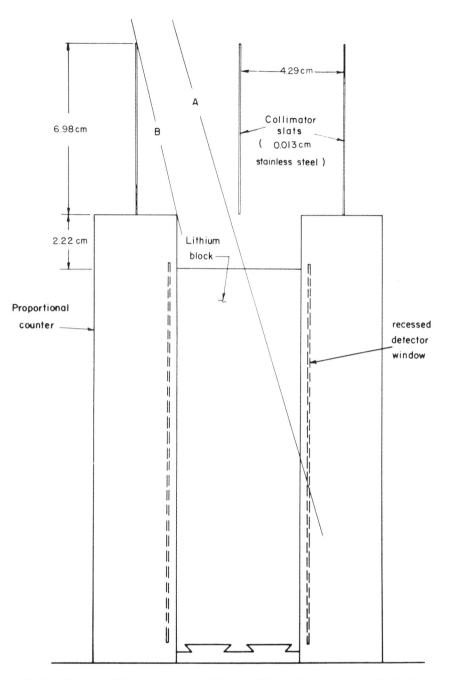

Fig. 15. Side view of detectors, scattering block, and X-ray collimator showing that background X rays may enter detector by penetrating lithium block (ray *A*) and that the detectors as well as collimator limit the X-ray field of view of the instrument (ray *B*). Dimensions are given in cm.

particles was measured by means of a simple cosmic-ray telescope. The efficiency was found to vary from ~90% to ~95%, depending on the point at which the particles penetrated the plastic; the efficiency decreased with increased distance from the phototubes. The anticoincidence shield was not, however, expected to be effective in eliminating counts initiated by cosmic gamma rays. For this reason electronic discrimination based on pulse amplitude and rise time was also included as a means of background reduction.

2. Pulse amplitude discrimination. Since the instrument is not sensitive to Crab Nebula X rays below ~5 keV or above ~25 keV, proportional counter pulses corresponding to X rays outside of this range were rejected by the on-board signal processing electronics. In addition, pulse-height information was telemetered to earth for each count in this range. This allowed for bandwidth optimization during postflight analysis.

3. Pulse rise-time discrimination. It is well known (e.g., Gorenstein and Mickiewicz 1968) that the pulses produced in proportional counters by X rays have a steeper rise, on the average, than those produced by cosmic rays. This is due in part to the longer path length traveled by a highly energetic particle in depositing the equivalent of an X ray's energy in the counter gas. This difference in rise time was exploited to help reduce the cosmic-ray-induced background signal. All pulses with rise times longer than a preset time were rejected by the signal processing electronics. Unfortunately, xenon's high electronegativity leads to a slow electron drift velocity. Consequently X-ray pulse rise times in xenon are on the order of 0.5 to 0.8 μsec, compared with ~0.1 μsec in argon-filled proportional counters, and rise-time discrimination is not as effective when used with xenon-filled counters as with argon- or neon-filled counters since the X rays themselves tend to be rejected by the circuitry.

The rise-time circuits were preset before launch to an X-ray acceptance of ~80% to ~90% for X rays from 5.9 to 22 keV and a simultaneous rejection of 38% (55%) of the ^{137}Cs (^{60}Co) gamma-ray-induced events with X-ray pulse amplitudes. This performance is quite poor compared to that attainable with argon.

Sensitivity Estimates

Through a simple statistical argument, it has been demonstrated that the standard deviation of a component of polarization measured by an instrument similar to that described above can be approximated by Equation (7). The modulation factor F for this instrument is 0.299. From the results of previous experiments, we expected a background rate of 0.0035 counts $\sec^{-1}$ cm^{-2} keV^{-1} after background suppression. This rate implies a background of 90,000 counts for 300 sec of observation time over a bandwidth of 5 to 25 keV. Assuming the spectrum of the Crab Nebula to be that given in Equation (19), the efficiency of the instrument to be that shown in Fig. 14, and an X-ray rise-time acceptance of 85%, we predicted a total signal of 35,000 counts. These estimates imply that a polarization of 14% is measurable to a 3 σ (99%) confidence level.

The polarimeter described above was flown on 22 February 1971 from the Wallops Island, Virginia, test range. This flight resulted in the detection of X-ray polarization from the Crab Nebula. The results and their analysis have been published by Novick et al. (1972).

A SATELLITE THOMSON-SCATTERING POLARIMETER

Landecker (1972) has analyzed the performance to be expected with a Thomson-scattering polarimeter suitable for use in an OSO or larger type of satellite such as HEAO. The polarimeter is designed to measure the polarization of stellar X-ray sources in the energy range 4–24 keV. It consists of compacted lithium-hydride scattering blocks, packaged in thin beryllium cases and surrounded on the sides by sealed, gas proportional counters (Fig. 16). The instrument would be located in the wheel compartment of the OSO-I satellite (see Fig. 7).

Many X rays incident on the tops of the scattering blocks would be scattered out of the sides and stop in the proportional counter. A two-dimensional tubular collimator is used to limit the field of view of the scattering blocks within a few degrees of the spacecraft spin axis so that much of the diffuse cosmic X-ray background and nearby cosmic sources would be masked. Orientation of the spacecraft spin axis within 0.5° of an X-ray source would avoid off-axis spurious polarization effects, and a small 1-mm lip placed around the entrance perimeter of the scattering blocks would prevent unscattered X rays from entering the proportional counter region as long as the spin axis is pointed within 0.5° of the source.

Since the normal to the OSO-I spacecraft solar panel is constrained to point within 4° of the sun-satellite line and the galactic sphere rotates at about 1° per day with respect to this line, under ideal conditions it should be possible to observe a cosmic source for about eight days twice a year with this configuration. If the source were polarized, the signal counting rate in each detector wire would be modulated sinusoidally at twice the spacecraft rotation frequency. The magnitude of the polarization of the source is determined by the depth of the modulation, and the position angle of the polarization vector by its phase. There should also be an

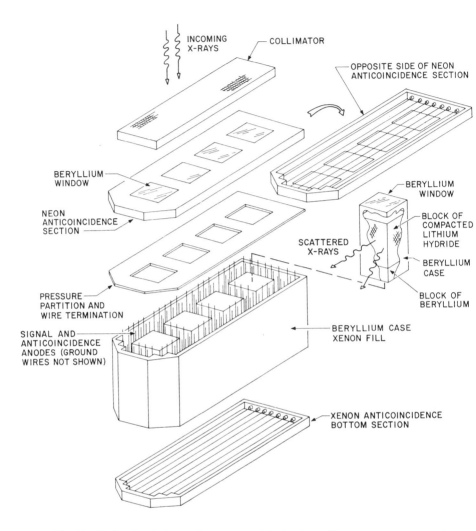

Fig. 16. Designed polarimeter detector assembly showing collimator, gas-counter anodes, and scattering blocks.

antiphase relation in the counting rates of the orthogonal anodes in each of the two detectors if the source is polarized, so that polarization modulation can be differentiated from the time variability of the source. Spurious modulation effects associated with differences in detector-channel sensitivities would be suppressed by the spacecraft rotation.

To reduce the cosmic-ray-induced background counting rate in the polarimeter, pulse-height and rise-time discrimination would be used.

Adjacent counter cells would be wired in anticoincidence, and many cells would register the signal counts from scattering blocks located on either side of the cell. Further, the main counter volume would be completely surrounded by a six-sided 4π-sterad anticoincidence shield (see Fig. 16) consisting of (a) the perimeter of the outer anode wires in the main counter, (b) a 1.27-cm, 1-atm neon anticoincidence section located at the top between the collimator and the main chamber, and (c) a 1.5-atm xenon anticoincidence section located at the bottom of the main chamber.

Monte Carlo Computer Program

The optimum design parameters of the polarimeter were chosen by a Monte Carlo computer program. The parameters specified in the program were:

a. Dimensions, position, number, and composition (including the fraction by weight of impurities and density of the Li or LiH scattering blocks). The option to insert a beryllium block at the base of the scattering blocks in order to scatter higher-energy X rays into the proportional-counter region was also included.
b. The composition, pressure, dimensions, and location of the proportional-counter gas.
c. The dimensions, pressure, and location of the anticoincidence counters.
d. The thickness of the beryllium windows surrounding the anticoincidence and signal counters.
e. The energy band of each energy bin.
f. The assumed polarization of the incident X-ray beam.
g. The number of Monte Carlo tries at each energy.
h. An initial nine-digit random number.
i. The background rate, observation time, and the source signal spectra.

Values for the photoelectric coherent and incoherent scattering mass-attenuation coefficients as a function of energy for all the elements in question (Storm and Israel 1967) were entered together with values for the densities of Li, LiH, and Be, and for the STP densities of the gases.

In order to determine the optimum configuration, only one or two parameters were varied at a given time. By a process of iteration, the optimum configuration was obtained. During the final iteration, one or two parameters were varied, while the others were fixed at the following values: a 1.27-cm thick, 1-atm neon top anticoincidence counter with 0.013-cm beryllium windows; a 1.5-atm xenon signal proportional-counter gas at 20°C; two perimeter signal cells surrounded by one perimeter anticoincidence cell; and eight scattering blocks of compacted lithium hydride (100% pure), each 4.57 × 4.57 × 10.16 cm, with eight beryllium

TABLE IV
Minimum Detectable Polarization (MDP) at 99% Confidence Level in a 24-Hour Observation Time as a Function of Assumed Background Rate

Stellar Source	Background ($cm^{-2} sec^{-1} keV^{-1}$)		
	4×10^{-4}	8×10^{-4}	1.6×10^{-3}
	MDP (%)		
Tau X-1	1.1	1.4	1.6
Sco X-1	0.36	0.38	0.40
Cyg X-1	2.4	3.2	4.0
Vir X-1	5.3	7.6	9.9
Cen X-3	1.4	1.6	2.1

blocks, each 4.57 × 4.57 × 1.27 cm, encased in a 0.051-cm beryllium container, and recessed 0.64 cm below the neon anticoincidence counter.

Results of Calculations

Cosmic-ray background. The greatest uncertainty in the sensitivity of the polarimeter derives from the uncertainty in the prediction of the background counting rate. The variation in the capability of the polarimeter for three values of the assumed background rate is presented in Table IV.

The best estimate of the expected background with this configuration in orbit would be $\left(8 \begin{array}{c} +8 \\ -4 \end{array}\right) \times 10^{-4}$ counts $cm^{-2} sec^{-1} keV^{-1}$ at a pressure of 1 atm of xenon in the main chamber of the proportional counter. The largest sectional area of each signal volume was used to estimate the background rate which is assumed to result from interactions in the counter gas and, consequently, is proportional to the gas density. This estimate would not apply in a region of high charged-particle density such as the South Atlantic Anomaly. The programmed proportional-counter high-voltage power supply would be automatically turned off when the spacecraft passes through this area.

The background counting rate would be measured when the spacecraft axis is not pointing at an X-ray source. During this time, it would be possible to check for spurious instrumental polarization induced by anisotropy in the primary cosmic-ray flux and by albedo gamma rays from the earth's atmosphere. In a balloon flight in November 1967 no significant background-induced asymmetry effects were recorded in a payload similar to the one proposed here (Wing 1968).

Scattering blocks. Lithium hydride was the primary scattering-block material selected for this satellite package. It has a greater density than metallic lithium (0.78 versus 0.53 g cm^{-3}) and a greater ratio of scattered to absorbed X rays. When the hydride is used instead of lithium metal, the average minimum detectable polarization for Tau X-1, Sco X-1, Vir X-1, and Cyg X-1 is 22% higher.

Tests have shown that it is possible to compact $-8+35$ mesh lithium hydride (2.4–0.42 mm diameter) into a single block of more than 99% of the theoretical density of solid lithium hydride when a pressure of 6300 kg cm^{-2} is applied. Lithium hydride is extremely hydroscopic, fairly active chemically, relatively easy to contaminate, and, of course, quite dangerous to handle. It has been produced at 99.9% purity. The effect of the presence of impurities in the lithium hydride on the sensitivity of the polarimeter is given in Table V.

The number and dimensions of the scattering blocks were varied with the constraint that the package must fit within the area of the lower half of the OSO-I 40°-wide wheel compartment. More than 20 possible configurations were tried from four 5.08 × 5.08 cm blocks to sixteen 3.18 × 3.18 cm blocks in order to find the optimum size of the block for the overall sensitivity of the proportional counter. Multiple scattering occurs if the width is too great, and the effective area is reduced if it is too small. Our results showed only a relatively small variation in the sensitivity for

TABLE V

Effect of Impurities in Lithium Hydride on Minimum Detectable Polarization (MDP) of X-Ray Sources for 24-Hour Observation Time

Type of Lithium Hydride	Assumed Weight (%) of Impurities			MDP (%) at 99% Confidence Level			
	Oxygen	Calcium	Lead	Tau X-1	Sco X-1	Vir X-1	Cyg X-1
100% pure	0	0	0	1.36	0.384	7.54	3.17
Atomergic Co.	0.05	0.005	0	1.41	0.395	7.85	3.30
LRL	0.3	0.025	0.0035	1.53	0.427	8.61	3.59
Lithcoa and Foote Co.	1.0	0.025	0.0035	1.76	0.486	10.0	4.14
2% oxygen impurity	2.0	0	0	1.83	0.511	10.4	4.30

the different geometrical arrangements. However, there was a 36% difference between the best case (seven 5.08 × 5.08 cm blocks) and the worst case (six 4.06 × 4.06 cm blocks). The configuration selected was eight 4.57 × 4.57 cm blocks because two identical, redundant assemblies could be incorporated in the design parameters with a sensitivity only 7% lower than the best case.

In Fig. 17 is shown the sensitivity of the polarimeter as a function of the height of the scattering block, with and without a 1.27-cm high beryllium plug at the base of the block. With the beryllium plug, when the height of the lithium-hydride block was varied, a broad maximum in the polarimeter sensitivity appeared at a composite height of 11.43 cm. When the beryllium plug was replaced with lithium hydride, the sensitivity was slightly decreased. Only 3% of the 10-keV incident photons passed through the bottom of the composite block without an interaction.

Proportional-counter gas. The polarimeter sensitivity as a function of type and pressure of the signal proportional-counter gas is given in Fig. 18. A broad minimum in detectable polarization occurred in the pressure range 1.0–1.5 atm of xenon. This conclusion is based on our assumption of the source of the background counting rate and its dependence on the counter gas pressure.

Other variables. If the number of signal perimeter cells were reduced from two to one, the average increase of the threshold would be 4% of the values given above. If the top anticoincidence cell were eliminated, the average signal counting rate would increase by 5%, but the background rate would increase by an amount difficult to determine. If the beryllium side thickness were reduced from 0.051 to 0.025 cm, the minimum detectable polarization would be decreased only about 1% of the computed values.

The program used to design the polarimeter could also be used to find a distribution of modulation arising from random polarization data. Since a small positive result could be consistent with a null result, such a program would authenticate any postlaunch results. For example, a signal modulation amplitude of only 0.8% was obtained on a run with randomly polarized photons compared with 27% for 100% linearly polarized photons.

Pulsar Capability

Meaningful observations of the polarization could be made at different phases within the pulse of the highly variable pulsating X-ray source Cen X-3. This X-ray pulsar has a period of 4.8 sec and a pulse width of

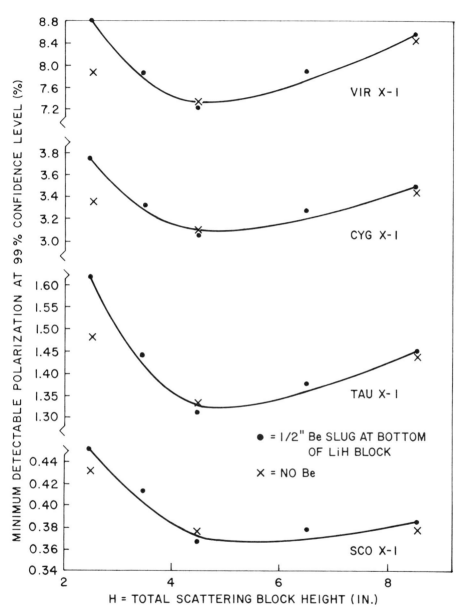

Fig. 17. Polarimeter sensitivity versus scattering-block height and composition (with and without 1.27-cm beryllium at base of lithium-hydride block); 1.5-atm xenon, eight 4.57 cm × 4.57 cm × H (cm) blocks, 0.051-cm beryllium on side, 0.013 cm × 5.08 cm beryllium on top, 4–24 keV, 24-hour exposure.

about 1 sec with about 70% of the X-ray flux in the pulse (Giacconi et al. 1971). If the X-ray spectrum is assumed to be

$$\frac{dN}{dE} = 1.3 \ E^{-1} \ \exp^{-[(E/15.7)+(3.8/E)^{8/3}]} \ \text{photons cm}^{-2} \ \text{sec}^{-1} \ \text{keV}^{-1}$$

(E in keV), then, in each of four 0.25-sec intervals within the pulse, the minimum detectable polarization at the 99% confidence level is 3% for an integration time of 24 hr. It should therefore be possible to evaluate the likelihood that this object is a magnetic rotator.

It would be possible to measure the X-ray polarization of the Crab pulsar with an OSO-I-type polarimeter if the polarization is similar to that measured in the optical range. For example, the minimum detectable average polarization of the main pulse at the 3 σ level is about 6% in an observation (actual counting) time of four days (about seven orbit days because of earth eclipse). This is to be compared with the measured average optical polarization of 7% at a position angle of 98° (Kristian et al. 1970).[14] In addition, for the 1-μsec bin centered near the primary pulse and an observation time of nine days, a lithium-hydride polarimeter would have a capability of measuring about 10% polarization with 99% confidence versus about the same amount already measured in the optical region for this time bin (Kristian et al. 1970; Wampler et al. 1969). If the X-ray pulsar polarization is comparable to the known optical polarization, then the present belief that the optical and X-ray pulsar emission mechanisms are identical would be confirmed.

Larger Satellite Experiments

A polarimeter for use in a satellite much larger than OSO-I could be easily designed by using the computer program described in this paper; it would be correspondingly more sensitive. For example, in a 24-hr observation time, a polarimeter for HEAO of dimensions 1 m × 1 m, containing 196 scattering blocks would have a minimum detectable polarization (at the 99% confidence level) of 0.27% for Tau X-1; 0.078% for Sco X-1; 0.64% for Cyg X-1; 1.5% for Vir X-1; and 0.33% for Cen X-3. At the same statistical confidence level, it would be possible with an integration time of four days to measure a polarization of (a) each quarter segment within the pulse greater than 0.3% and 3% for the pulsars Cen X-3 and Tau X-1, respectively, or (b) nonflaring Sco X-1 in the 32–50 keV energy range greater than about 5%. Since Sco X-1 characteristically triples in intensity during a flare, it would be possible in a typical flare-time duration

---

[14] See p. 997.

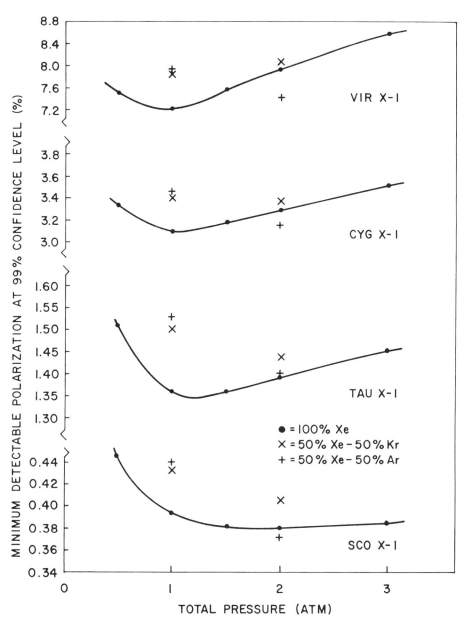

Fig. 18. Polarimeter sensitivity versus proportional-counter gas pressure and composition; eight 4.57 cm × 4.57 cm × 10.16 cm 100% lithium-hydride blocks plus eight 4.57 cm × 4.57 cm × 1.27 cm 100% beryllium blocks, 0.051-cm beryllium on side, 0.013 cm × 5.08 cm beryllium on top, 4–24 keV, 24-hour exposure.

of 20 min to measure the polarization of the flaring component of the flux in the 5–10 keV region with 99% confidence, provided that it is greater than 0.6%.

CONCLUSION

We have discussed the status of X-ray polarimeters in terms of their scientific objectives, in terms of the measurements that have been obtained thus far, and in terms of a number of different types of instruments. From the predictions made above and on the basis of flight data that presently exist, it is clear that these instruments are capable of elucidating a number of important questions relating to the emission mechanisms and other phenomena in solar and stellar X-ray astronomy. All of the previously described instruments have relatively large fields of view and would provide a measurement of the average polarization over the entire area of most of the presently known sources. It is clear from radio and optical data on such important objects as M 87, 3C 273, and the Crab Nebula that the polarization is strongly spatially dependent. Large time-variable polarization is observed in particular knots in M 87, in small areas in 3C 273, and in other objects.

Clearly, instruments with large fields of view cannot be used to study the spatial dependence of the polarization. High-energy processes which are not presently understood certainly play an important role in the compact region. Possibly they contain very large, massive, rotating magnetic objects. It is of critical importance that we have the means to measure the X-ray polarization not only over the entire area of these objects but also within these compact regions with large optical and radio polarization. For this purpose, we require an instrument with good spatial resolution and reasonable polarization sensitivity. It appears that our only hope for accomplishing this is with a paraboloid-hyperboloid type of X-ray telescope, combined with a focal plane polarimeter. If we use a large-area telescope with four nested surfaces, then it is possible to obtain a collecting area of 200 cm^2 at 2.6 keV. If we place a mosaic graphite crystal near the focus of the telescope and if the crystal is oriented 45° with respect to the optical axis, then the diffracted X rays will consist of only one state of polarization and the entire instrument can be used as an imaging polarimeter. Kestenbaum[15] has estimated the performance that one might expect with such an instrument, and the results are given in Table VI. It is clear that this instrument can, in fact, be used to study many of the questions of importance to X-ray astronomy, and through

---

[15] Personal communication, 1970.

TABLE VI
Estimated Sensitivity of Imaging Focal Plane Polarimeter[a]

Source	Type	$I(E)$ (keV keV^{-1} cm^{-2} sec^{-1})	Minimum Detectable Polarization (%) at 3 σ Level Observation Time 1 day	6 days
Sco X-1	Variable blue star	33	0.61	
Crab Nebula	SNR	3.5	1.9	0.78
Crab Pulse	SNR pulsar	0.158	16	6.5
Crab Primary Pulse	SNR pulsar	0.079	21	8.6
Crab Secondary Pulse	SNR pulsar	0.079	25	10.2
Cen X-3	Pulsing binary	0.47	5.2	2.1
Cen X-3 (pulse)	Pulsing binary	0.33	6.8	2.8
Her X-1	Pulsing binary	0.24	7.5	3.1
Her X-1 (pulse)	Pulsing binary	0.19	8.7	3.6
Cyg X-1	Binary, black hole (?)	1.0	3.6	1.5
Cyg X-2		0.79	4.0	
Cyg X-3	Variable radio	0.37	5.9	2.4
Sco X-2		1.8	2.6	
Cir X-1		2.3	2.3	
Ser X-1		0.61	4.6	
Ser X-2		1.4	3.0	
Lup X-1		1.2	3.2	
Nor X-1		1.2	3.2	
Nor X-2		0.42	5.5	
GX 5-1		1.3	3.1	
GX 9+9		0.34	6.2	
GX 9+1		0.65	4.4	
GX 13+1		0.26	7.1	
GX 340+0		1.0	3.6	
Vela X-2		0.72	4.2	
Cep X-1	Tycho's SNR	0.13	10	
Cas A	SNR	0.41	5.6	

[a] Assumed values: $A(E) = 200$ cm^2 $\varepsilon(E) = 0.70$
$\Delta\theta(E) = 12 \times 10^{-4}$ $T = 86{,}400$ sec (1-day observation)

its use, we can hope to understand the physical processes occurring in compact energetic sources.

Acknowledgments. I wish to thank J. F. Dolan, P. B. Landecker, R. A. Linke, H. L. Kestenbaum, T. Kifune, M. C. Weisskopf, and R. S. Wolff for their contributions to this paper. This work was supported by the National Aeronautics and Space Administration under grant NGR 33-008-102. It is Columbia Astrophysics Laboratory Contribution No. 78.

REFERENCES

Aller, H. D. 1970. The polarization of variable radio sources at 8 GHz. I. The observations. II. Interpretation. *Astrophys. J.* 161: 1–18; 19–31.

Angel, J. R. P. 1969a. X-ray line emission from Sco X-1. *Nature* 224: 160–161.

———. 1969b. Polarization of thermal X-ray sources. *Astrophys. J.* 158: 219–224.

Angel, J. R. P.; Novick, R.; Vanden Bout, P.; and Wolff, R. 1969. Search for X-ray polarization in Sco X-1. *Phys. Rev. Letters* 22: 861–865.

Angel, J. R. P., and Weisskopf, M. C. 1970. Use of highly reflecting crystals for spectroscopy and polarimetry in X-ray astronomy. *Astron. J.* 75: 231–236.

Bahcall, J. N., and Bahcall, N. A. 1972. HZ Herculis. I.A.U. Central Bureau for Astronomical Telegrams. (B. G. Marsden, ed.). Circular No. 2427.

Barkla, C. G. 1906. Polarisation in secondary Röntgen radiation. *Proc. Roy. Soc.* A 77: 247–255.

Batterman, B. W. 1962. Effect of thermal vibrations on diffraction from perfect crystals. I. The case of anomalous transmission. *Phys. Rev.* 126: 1461–1469.

Boldt, E. A.; Desai, U. D.; and Holt, S. S. 1969. 2–20 keV spectrum of X-rays from the Crab Nebula and the diffuse background near galactic anticenter. *Astrophys. J.* 156: 427–436.

Bolton, C. T. 1972. Dimensions of the binary system HDE 226868 = Cygnus X-1. *Nature Phys. Sci.* 240: 124–127.

Bowyer, C. S.; Field, G. B.; and Mack, J. E. 1968. Detection of an anisotropic soft X-ray background flux. *Nature* 217: 32–34.

Brecher, K. 1972. Her X-1: A precessing binary pulsar. *Nature* 239: 325–326.

Burnight, T. R. 1949. Soft X-radiation in the upper atmosphere. *Phys. Rev.* 76: 165.

Cole, H.; Chambers, F. W.; and Wood, C. G. 1961. X-ray polarizer. *J. Appl. Phys.* 32: 1942–1945.

Culhane, J. L., and Fabian, A. C. 1972. Origin of the low-energy diffuse cosmic X-ray flux. *Nature* 237: 379–381.

Davidson, A.; Henry, J. P.; Middleditch, J.; and Smith, H. E. 1972. HZ Herculis. I.A.U. Central Bureau for Astronomical Telegrams. (B. G. Marsden, ed.). Circular No. 2433.

Davidson, K., and Ostriker, J. P. 1973. Neutron-star accretion in a stellar wind: Model for a pulsed X-ray source. *Astrophys. J.* 179: 585–598.

Dolan, J. F. 1966. *The polarization of celestial X rays*, Smithsonian Researches in Space, Special Report No. 212.

Elwert, G. 1968. The significance of the polarization of solar short-wavelength X-rays. *Structure and development of solar active regions*, I.A.U. Symp. No. 35. (K. O. Kiepenheuer, ed.) pp. 444–448. Dordrecht, Holland: D. Reidel Publ. Co.

Ewald, P. P. 1965. Crystal optics for visible light and X rays. *Rev. Mod. Phys.* 37: 46–56.

Felten, J. E.; Rees, M. J.; and Adams, T. F. 1972. Transfer effects on X-ray lines in optically thick celestial sources. *Astron. Astrophys.* 21: 139–150.

Frost, K. J. 1969. Rapid fine structure in a burst of hard solar X-rays observed by OSO-5. *Astrophys. J.* 158: L159–L163.

Giacconi, R.; Gursky, H.; Kellogg, E.; Schreier, E.; and Tananbaum, H. 1971. Discovery of periodic X-ray pulsations in Centaurus X-3 from *Uhuru*. *Astrophys. J.* 167: L67–L73.

Giacconi, R.; Gursky, H.; Paolini, F. R.; and Rossi, B. B. 1962. Evidence for X rays from sources outside the solar system. *Phys. Rev. Letters* 9: 439–443.

Giacconi, R.; Murray, S.; Gursky, H.; Kellogg, E.; Schreier, E.; and Tananbaum, H. 1972. The *Uhuru* catalog of X-ray sources. *Astrophys. J.* 178: 281–308.

Gnedin, Y. N., and Sunyaev, R. A. 1972. The beaming of radiation from an accreting magnetic neutron star and the X-ray pulsars. In draft.

Gold, T. 1968. Rotating neutron stars as the origin of the pulsating radio sources. *Nature* 218: 731–732.

———. 1969. Rotating neutron stars and the nature of pulsars. *Nature* 221: 25–27.

Goldhaber, M.; Grodzins, L.; and Sunyar, A. W. 1958. Helicity of neutrinos. *Phys. Rev.* 109: 1015–1017.

Gorenstein, P.; Kellogg, E. M.; and Gursky, H. 1969. The spectrum of diffuse cosmic X-rays, 1–13 keV. *Astrophys. J.* 156: 315–324.

Gorenstein, P., and Mickiewicz, S. 1968. Reduction of cosmic background in an X-ray proportional counter through rise-time discrimination. *Rev. Sci. Instr.* 39: 816–820.

Gorenstein, P., and Tucker, W. H. 1972. On a galactic origin for the soft X-ray background. *Astrophys. J.* 176: 333–344.

Gregory, P. C.; Kronberg, P. P.; Seaquist, E. R.; Hughes, V. A.; Woodworth, A.; Viner, M. R.; Retallack, D.; Hjellming, R. M.; and Balick, B. 1972. The nature of the first Cygnus X-3 radio outburst. *Nature Phys. Sci.* 239: 114–117.

Gunn, J. E., and Ostriker, J. P. 1971. On the motion and radiation of charged particles in strong electromagnetic waves. I. Motion in plane and spherical waves. *Astrophys. J.* 165: 523–541.

Heitler, W. 1954. *The quantum theory of radiation*. London: Clarendon Press, Oxford University, 3rd edition.

Henry, R. C.; Fritz, G.; Meekins, J. F.; Chubb, T. A.; and Friedman, H. 1972. Absorption of Crab Nebula X-rays. *Astrophys. J.* 174: 389–397.

Henry, R. C.; Fritz, G.; Meekins, J. F.; Friedman, H.; and Byram, E. T. 1968. Possible detection of a dense intergalactic plasma. *Astrophys. J.* 153: L11–L18.

Hiltner, W. A. 1959. Photoelectric polarization observations of the jet in M 87. *Astrophys. J.* 130: 340–343.

Hulst, H. C. van de 1957. *Light scattering by small particles*. New York: John Wiley and Sons.

Kanter, H., and Sternglass, E. J. 1962. Interpretation of range measurements for kilovolt electrons in solids. *Phys. Rev.* 126: 620–626.

Kestenbaum, H.; Angel, J. R. P.; and Novick, R. 1971a. X-ray spectrum of Scorpius X-1 obtained with a Bragg crystal spectrometer. *Astrophys. J.* 164: L87–L93.

———. 1971b. A Bragg spectrometer for stellar X-ray astronomy. *New techniques in space astronomy*, I.A.U. Symp. No. 41. (F. Labuhn and R. Lüst, ed.) pp. 137–144. Dordrecht, Holland: D. Reidel Publ. Co.

Kifune, T. 1972. Development of X-ray polarimeter by detecting secondary fluorescent X-rays. Preprint.

Korchak, A. A. 1967. Possible polarization of bremsstrahlung from solar flares. *Soviet Phys.—Doklady* 12: 192–194.

Kristian, J.; Visvanathan, N.; Westphal, J. A.; and Snellen, G. H. 1970. Optical polarization and intensity of the pulsar in the Crab Nebula. *Astrophys. J.* 162: 475–483.

Lamb, D. Q., and Sorvari, J. M. 1972. HZ Her as a possible optical pulsar. I.A.U. Central Bureau for Astronomical Telegrams. (B. G. Marsden, ed.). Circular No. 2422.

Lamb, F. K.; Pethick, C. J.; and Pines, D. 1972. A model for compact X-ray sources: Accretion by rotating magnetic stars. Preprint.

Landecker, P. B. 1972. Design of a celestial Thomson-scattering X-ray polarimeter. *IEEE Trans. Nucl. Sci.* NS-19: 463–475.

Landstreet, J. D., and Angel, J. R. P. 1971a. Discovery of circular polarization in the white dwarf G99-37. *Astrophys. J.* 165: L67–L70.

―――. 1971*b*. Search for optical circular polarization in the Crab Nebula. *Nature* 230: 103.
Liller, W. 1972. HZ Herculis. I.A.U. Central Bureau for Astronomical Telegrams (B. G. Marsden, ed.) Circular No. 2427.
Linke, R. A. 1972. Measurement of the X-ray polarization of the Crab Nebula by means of a Thomson-scattering polarimeter. Ph.D. Dissertation, pp. 23–51. Columbia University, New York.
Martin, P. G. 1972. Interstellar circular polarization. *Mon. Not. R. Astron. Soc.* 159: 179–190.
Martin, P. G.; Illing, R.; and Angel, J. R. P. 1972. Discovery of interstellar circular polarization in the direction of the Crab Nebula. *Mon. Not. R. Astron. Soc.* 159: 191–201.
McKenzie, D. L. 1972. Correlation studies of solar X-ray and radio bursts. *Astrophys. J.* 175: 481–492.
Neupert, W. M. 1969. X rays from the sun. *Annual review of astronomy and astrophysics* Vol. 7, pp. 121–148. Palo Alto, California: Annual Reviews, Inc.
Novick, R.; Weisskopf, M. C.; Berthelsdorf, R.; Linke, R.; and Wolff, R. S. 1972. Detection of X-ray polarization of the Crab Nebula. *Astrophys. J.* 174: L1–L8.
Parsignault, D. R.; Gursky, H.; Kellogg, E. M.; Matilsky, T.; Murray, S.; Schreier, E.; Tananbaum, H.; Giacconi, R.; and Brinkman, A. C. 1972. Observations of Cygnus X-3 by *Uhuru. Nature Phys. Sci.* 239: 123–125.
Plechaty, E. F., and Terrall, J. R. 1968. *Photon cross sections, 1 keV to 100 MeV*. Lawrence Radiation Laboratory, University of California, UCRL-50440, Vol. VI, TID-4500, UC-34 (Physics).
Pringle, J. E., and Rees, M. J. 1972. Accretion disc models for compact X-ray sources. *Astron. Astrophys.* 21: 1–9.
Rees, M. J. 1971. Implications of the "wave field" theory of the continuum from the Crab Nebula. *Nature Phys. Sci.* 230: 55–57.
Riegler, G. R.; Garmire, G. P.; Moore, W. E.; and Stevens, J. C. 1970. A low-energy X-ray polarimeter. *Bull. Am. Phys. Soc.* 15: 635.
Sanford, P. W.; Cruise, A. M.; and Culhane, J. L. 1970. Techniques for improving the sensitivity of proportional counters used in X-ray astronomy. *Non-solar X- and gamma-ray astronomy*. (L. Gratton, ed.) pp. 35–40. I.A.U. Symposium No. 37 Dordrecht, Holland: D. Reidel Publishing Co.
Sartori, L., and Morrison, P. 1967. Thermal X-rays from non-thermal radio sources. *Astrophys. J.* 150: 385–403.
Schreier, E.; Levinson, R.; Gursky, H.; Kellogg, E.; Tananbaum, H.; and Giacconi, R. 1972. Evidence for the binary nature of Centaurus X-3 from *Uhuru* X-ray observations. *Astrophys. J.* 172: L79–L89.
Smith, H. E.; Margon, B.; and Conti, P. S. 1973. Spectroscopic observations of the Cygnus X-1 optical candidate. *Astrophys. J.* 179: L125–L128.
Spiller, E. 1972. Low-loss reflection coatings of absorbing materials. Spring Meeting, Optical Society of America, New York, N.Y., April 11–13, 1971.
Storm, E., and Israel, H. I. 1967. *Photon cross sections from 0.001 to 100 meV for elements 1 through 100*, Los Alamos Scientific Laboratory, Bull. No. LA 3753.
Takakura, T. 1967. Theory of solar bursts. *Solar Phys.* 1: 304–353.
Tananbaum, H.; Gursky, H.; Kellogg, E. M.; Levinson, R.; Schreier, E.; and Giacconi, R. 1972. Discovery of a periodic pulsating binary X-ray source in Hercules from *Uhuru. Astrophys. J.* 174: L143–L149.
Tindo, I. P.; Ivanov, V. D.; Mandel'stam, S. L.; and Shuryghin, A. I. 1970. On the polarization of the emission of X-ray solar flares. *Solar Phys.* 14: 204–207.

———. 1972. New measurements of the polarization of X-ray solar flares. *Solar Phys.* 24: 429–433.

Tindo, I. P.; Valnicek, B.; Livshitz, M. A.; and Ivanov, V. D. 1971. Preliminary interpretation of the polarization measurements performed on 'Intercosmos-4' during three X-ray solar flares. Presented at International Meeting on Solar Activity, Izmiran, November 15–19.

Wampler, E. J.; Scargle, J. D.; and Miller, J. S. 1969. Optical observations of the Crab Nebula pulsar. *Astrophys. J.* 157: L1–L10.

Weisskopf, M. C.; Berthelsdorf, R.; Epstein, G.; Linke, R.; Mitchell, D.; Novick, R.; and Wolff, R. S. 1972. A graphite crystal polarimeter for stellar X-ray astronomy. *Rev. Sci. Instr.* 43: 967–976.

Wing, T. E. 1968. A polarimeter suitable for making polarization measurements of cosmic X-ray sources. Ph.D. Dissertation, Columbia University, New York.

Wolff, R. S. 1969. Measurement of the polarization of the X-ray emission of the Crab Nebula. Ph.D. Dissertation, Columbia University, New York.

———. 1972. High-energy solar X-ray polarimetry, presented at NASA Symposium on High-Energy Phenomena on the Sun, Greenbelt, Maryland, September 28–30.

Wolff, R. S.; Angel, J. R. P.; Novick, R.; and Vanden Bout, P. 1970. Search for polarization in the X-ray emission of the Crab Nebula. *Astrophys. J.* 160: L21–L25.

Yentis, D. J.; Novick, R.; and Vanden Bout, P. 1972a. Positive detection of an excess of low-energy diffuse X-rays at high galactic latitude. *Astrophys. J.* 177: 365–373.

———. 1972b. Galactic-latitude dependence of low-energy diffuse X-rays. *Astrophys. J.* 177: 375–386.

DISCUSSION

DICKEL: The polarization of the integrated radio and optical pulsar fluxes from the Crab Nebula is not 15% (the radio is more like 8%–10%).[1] Exactly what size is your X-ray source and exactly where is it positioned on the source?

NOVICK: The lunar occultation experiment at the Naval Research Laboratory and the modulation collimator experiment carried on jointly by the American Science and Engineering Inc. and MIT show that the X-ray emission is confined to a region of about 1 arcmin radius centered on the pulsar. The weighted average optical polarization in this region is about 14% at a position angle of 154°.

---

[1] See p. 997.

THE BERKELEY INFRARED POLARIZATION SURVEY

ROBERT LANDAU
University of California, Berkeley

We describe some of the problems of polarimetry in the infrared. Large instrumental polarizations may be encountered, and careful sky subtraction is necessary. The instrument used in the Berkeley polarization survey of stars with circumstellar shells is described. Real-time computation of the polarization is provided by a computer that also controls much of the equipment. Results are presently sparse.

A survey is being conducted at Berkeley to measure the polarization between 2 and 11 μm of all the sources bright enough for our system to detect in ten minutes integration time. (If detected, an object may be measured for much longer than ten minutes.) This criterion requires an object to be brighter than -1 mag at 11 μm. There are two or three dozen such objects, mostly late-type stars, but also some Be stars and some F and G supergiants, all with presumed circumstellar shells that may radiate in the infrared. We have also measured Venus and Mercury.

Since the sky and the telescope radiate strongly in the infrared, photometric measurements there are the difference between two large numbers: the sky-plus-the-source and the sky alone. To do infrared polarimetry, this small difference must be analyzed for another small difference, namely, that between the intensities in two orthogonal directions.

This work is still in progress, and our conclusions are tentative. The first version of our system had an instrumental polarization of 25% at 2 μm, decreasing to about 10% at 11 μm. For such large values, it is improper to derive the true polarization of a source by vector subtraction of the instrumental polarization from the observed polarization; such procedure would introduce an error of some percent. Furthermore, the discordance between repeated measurements of the same source on different days (occasionally on the same day) is about an order of magnitude larger than the formal statistical error. (We understand that M. Dyck,[1]

---
[1] See p. 858.

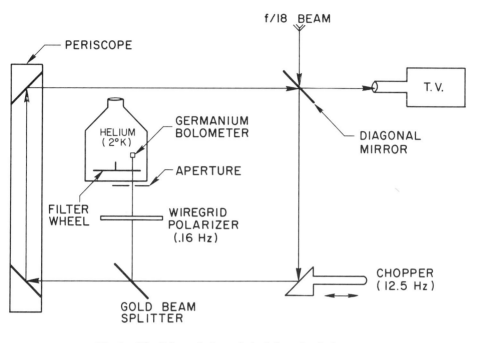

Fig. 1. The light path through the infrared polarimeter.

who is engaged in a similar program of infrared polarimetry, has had the same difficulty.)

Figure 1 shows the light path through the instrument which is used at the Cassegrain focus of the 3-m reflector of the Lick Observatory. With the diagonal mirror in place, the television camera sees the field of the object; the limit is 20 mag, although with enough neutral density filters, the camera can be used during the day. After the object is found, the diagonal mirror is removed and the beam is reflected at 45° from the chopper and then again at 45° from the beam splitter, which passes the visible light back through the periscope to the television camera for guiding. The infrared beam passes through the wire-grid polarizer which rotates at about 10 rpm; each revolution is divided into 20 integration periods, covering 18° each and lasting about 300 millisecs.

The signal is detected by the bolometer through cooled filters, and the portion of the signal that appears at the chopping frequency is amplified and digitized and stored in successive locations of a PDP8/I computer. The polarizer rotates at a rate that exactly maintains four chops of the chopper per integration period.

To subtract sky variations at the end of three revolutions of the polarizer, the computer puts the star in the other beam of the chopper (by moving the telescope) and, after six revolutions there, puts it back in the first beam for three more revolutions. This constitutes one observation, and six or so of these observations in succession are made for each filter.

The major trouble with the system is the beam splitter which causes the large instrumental polarization. In laboratory tests of different beam splitters, we have found that the ones coated most thinly (which make the best beam splitters) introduce the largest amount of polarization. The lowest polarization we can get from a coating, and still have a beam splitter, still gives about 10% polarization at 2 μm and about 4% at 11 μm. We have replaced the beam splitter with a mirror but have not had an opportunity to test this new system yet. Since we can no longer use the periscope, it requires offset guiding and a diagonal mirror with a central opening.

When the observations are being made, the computer makes a least-squares fit to the observed polarization, and the result is displayed together with the original data and the errors so that we can judge when to stop integrating. We currently fit a function that allows a 360° period of variation, presumably arising from imperfections in the polarizer, as well as the 180° variation expected for the polarization itself.

Our observing procedure is to measure, as standards, stars bright in the infrared like α Her, α Tau, and β Peg that we have no reason to believe are polarized. The measured polarizations of different standards do not differ from each other by much more than the internal errors of observation of any standard by itself.

TABLE I

Infrared Observations of Linear Polarization

Star	Spectral Type	Wavelength							
		2.2 μm		3.5 μm		4.8 μm		10.6 μm	
		% Pol	θ	% Pol	θ	% Pol	θ	% Pol	θ
NML Cyg	M I	2.2±0.5	65°	1.0±0.5	80°	1			
α Sco	M I	6 ±4	50°	5 ±2	50°	4.5±1	60°	5±1	45°
μ Cep	M I	<3		<3		3 ±3		6±4	25°
VX Sgr	M I	<5		<5		<10		<5	
W Cep	K I	too faint		too faint		too faint		3±2	30°
AX Sgr	G I	too faint		too faint		too faint		8±6	70°
IRC+10216	Carbon	12 ±8	50°	<10		7 ±3	20°	4±2	40°
MR-80	Wolf-Rayet	6 ±2	60°	4 ±3	70°	3 ±1.5	25°	5±4	

The star for which we have the best data is NML Cygnus; the position angles shown in Table I agree with the observations of Dyck, Forbes and Shawl.[2] At 5 µm the polarization continues to fall and is less than 1%.

We have also measured Venus at 110° phase at both eastern and western elongations. At 2.2 µm our observations of 2% positive polarization agree with the measurements of Forbes.[3] (At this phase there is a neutral point at about 2 µm.) At 3.5 µm we find about 30% positive polarization, indicating the approach of the Rayleigh-Jeans portion of the scattering curve. At 5 µm we are no longer seeing the sun's light scattered from the Venus atmosphere but, instead, its thermal emission, which is not polarized.

We have also measured polarization on Mercury at 3.5 µm. We get 15% at 93° phase and 20% at 120° phase; in both cases, the polarization is negative. However, the measurement is difficult and the uncertainty large.

Notes added in proof:

1. Now that the beam splitter is replaced with a mirror, the instrumental polarization is reduced to a few percent.

2. On Mercury, the diaphragm, having a width of 4 arcsecs, was centered not equatorially but near the southern cusp.

3. Further measurements of Mercury at 3.5 µm obtained with the 3-m reflector on 12 April 1973 at 89° phase (with instrumental polarization ~ 1%) showed a linear polarization of ~ 10%, again negative. This time, although the beam was still only about half the diameter of Mercury, it was centered approximately on the subsolar point.

---

[2] See p. 862.
[3] See p. 653.

ASTRONOMICAL POLARIMETRY IN THE FAR INFRARED (10–1000 µm)

G. DALL'OGLIO, B. MELCHIORRI, F. MELCHIORRI, V. NATALE
Laboratorio TESRE-CNR, Firenze
and
S. AIELLO, F. MENCARAGLIA
Cattedra di Fisica dello Spazio, Firenze

The infrared excess of many celestial sources could be polarized, and a measurement of the degree of polarization together with its spectral dependence might allow us to discriminate between the various mechanisms of emission. The expected percentage polarization in the presence of synchrotron emission or thermal emission by magnetically oriented grains has been evaluated. Experimental details are given for the materials suitable for IR polarimetry. Some experimental results obtained during a balloon flight with a far-infrared polarimeter at the focus of a 25-cm balloon-borne telescope are given.

The aim of the present work is to describe the state of art of astronomical polarimetry in the far infrared (FIR), i.e., beyond 10 µm. Until recently, research in this field of investigation has suffered from a complete lack of knowledge about the properties of FIR celestial sources. The discovery of intense FIR radiations both in galactic and extragalactic objects has led various authors to discuss mechanisms of emission responsible for the observed excess; many papers have been devoted to the calculation of possible models in terms of synchrotron or thermal emission. In both cases the produced radiation could be polarized, although the degree of polarization and its spectral features have not been investigated in papers available to date.

In any case, the possibility of detecting polarized radiations in the far infrared is receiving increased attention. In the present paper we will briefly discuss the degree of polarization that one could expect from an FIR source; we will also describe the performance of available FIR polarizers and the problems arising from spurious instrumental polarization. Finally, some preliminary results will be given relative to a balloon-borne polarimeter operating in the 100–2000 µm range of wavelengths.

CELESTIAL INFRARED SOURCES AND MECHANISMS OF EMISSION

A number of sources have been discovered having an IR excess. Detailed discussion of the nature of these sources and the possible criteria of classification is beyond the scope of this work. We should point out that any attempt to classify the sources is destined to be unsuccessful until more measurements are available. Among the various classifications we recall that of Burbidge and Stein (1970) where the sources are divided into galactic (grain emission) and extragalactic (synchrotron emission), and that of Pecker (1971) where the sources are studied with respect to the dimensions of the surrounding gaseous envelope.

In view of the present state of knowledge, we think that a less ambitious program should be undertaken—that of dividing the sources following phenomenological criteria, on the basis of their spectral characteristics. We have roughly three types of spectra: (1) spectra showing a near-IR excess (between 1 and 30 μm); (2) spectra showing a far-IR excess without a radio enhancement; and (3) spectra showing a far-IR excess with a radio enhancement.

In the first type falls a large number of hot stars studied by Pecker (as shown in Fig. 1); however, η Car and Z C Ma seem to be members of the

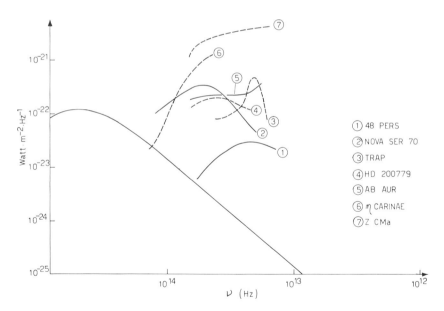

Fig. 1. Infrared emission from galactic near-infrared sources (hot stars). The continuous line corresponds to a $10^{4\circ}$K blackbody emission.

second type. Also in the second type fall the largest part of the galactic objects as shown by Burbidge (Fig. 2); however, the Orion Nebula has a radio enhancement. In the last type we have many extragalactic objects (Fig. 3); again we should note that BL Lac seems to be a member of the second type.

Various mechanisms of emission have been proposed to date: (a) coherent photon scattering by electrons in plasma oscillations; (b) thermal bremsstrahlung; (c) molecular or atomic line emission; (d) synchrotron emission; and (e) thermal emission by grains.

The first two mechanisms can easily be rejected on the basis of simple considerations developed by Burbidge and Stein (1970). The third mechanism has been proposed by Goldberg (1968) to explain the emission of NGC 7027; however the theoretical considerations suggested by Pecker (1971) together with the observations of Gillett and Stein (1971) seem to eliminate this interpretation completely. On the other hand, until recently suggestions of possible contributions from gaseous lines have been proposed, as for instance the 112-μm deuterium line (Dalgarno and Wright 1972). Thermal emission by grains and synchrotron emission appear to be preferred in explaining the observed spectra. Since synchrotron emission should be polarized, measurements of the state of polarization have been suggested to discriminate between the two mechanisms (Marsh 1971); in spite of this we should point out that a certain amount of polarization is expected by magnetically aligned grains, so that we may conclude that spectral measurements of the degree of polarization are required to distinguish between the possible mechanisms of emission.

POLARIZATION IN THE FAR-INFRARED SYNCHROTRON EMISSION

Theoretical Considerations

The general theory of synchrotron emission together with its application to astrophysics is given by Ginzburg and Sirovantzkii (1965).

The relevant formulae are (1)–(4):

$$\phi(v) = 3 \times 10^{-36} N_0 \frac{l^3}{r^2} \frac{v_m^2}{v_1} \left(\frac{v}{v_1}\right)^{\frac{\gamma+2}{2}} \left[1 - \exp\left(-\frac{v_1}{v}\right)^{\frac{\gamma+4}{2}}\right], \quad (1)$$

where $\phi(v)$ is the power flux in watt m^{-2}Hz^{-1}; N_0 is the density of relativistic electrons in m^{-3} (typical value 10^5 m^{-3}); γ is the spectral index of relativistic electrons ($\gamma \simeq 3$); l is the linear dimension of the source in meters; r is the distance of the source from the center in meters; v_m is the cyclotron frequency $= \dfrac{eB}{2\pi m} = 3.10^6 \, B$, with B in gauss; and v is the frequency of the observed radiation in Hz.

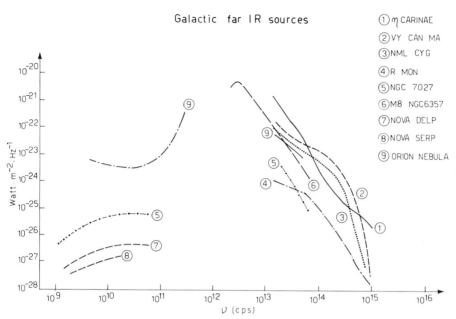

Fig. 2. Infrared emission from galactic sources. Although no measurements are available between 10^{12} and 10^{13} Hz, the behavior of near IR and radio emission suggests that the largest amount of radiation falls in the far infrared around 100 μm.

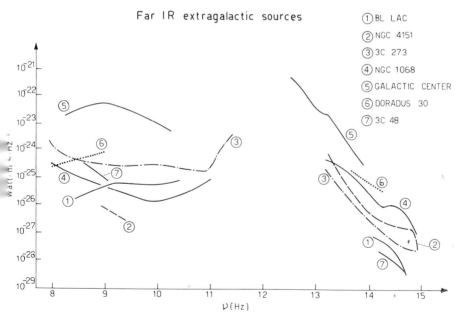

Fig. 3. Infrared emission from extragalactic sources. The emission from the galactic center is added for comparison (Dalgarno and Wright 1972). The same considerations of Fig. 2 suggest a far infrared excess.

$$v_1 = \left[3.5 \times 10^{-6} N_0 v_m^{\frac{\gamma+2}{2}} l \right]^{\frac{2}{\gamma+4}} \tag{2}$$

where v_1 is the frequency corresponding to the maximum value of $\phi(v)$; for $v > v_1$ the source is optically thin, while for $v \leq v_1$ the source is optically thick and synchrotron self-absorption occurs.

Assuming $\gamma = 3$ and substituting Equation (2) in Equation (1), we obtain

$$\phi(v) = 10^{-36} \frac{l^2}{r^2} B^{-1/2} v^{5/2} \left[1 - \exp\left(-\frac{v_1}{v} \right)^{\frac{\gamma+4}{2}} \right]. \tag{3}$$

If we assume that the magnetic field and the pitch-angle distribution are uniform, we have a linear polarization p independent of the wavelength, given by

$$p(\gamma) = \frac{3\gamma + 3}{3\gamma + 7}, \tag{4}$$

and with $\gamma = 3$ we have $p \simeq 75\%$.

We must point out that Equation (4) is indicative of the maximum percent of polarization that one should expect from a synchrotron source; in practice the polarization may be much smaller due to the various depolarization effects. Before discussing these effects, we should point out that a small circular polarization may rise from synchrotron emission as a second-order effect (Roberts and Komesaroff 1965), or as a transformation of linear polarization due to Faraday pulsation in an appropriate layer (Rosenberg 1972).

However, at present the observations of circular polarization are questionable, with the maximum detected polarization being on the order of 0.15% (Gilbert and Conway 1970; Conway et al. 1971); moreover, the percentage polarization seems to decrease at shorter wavelengths as shown by Biraud (1972), and no polarization has been detected at $\lambda = 9$ mm.

Depolarization Mechanisms

Depolarization mechanisms may be distinguished in the two cases $v > v_1$ and $v \leq v_1$. We first discuss $v > v_1$, i.e., an optically thin source in which the depolarization may be attributed to the following processes:

1. Spatial depolarization, which occurs due to the various possible orientations of the polarization vector in the field of view of the optical system in use. Assuming that the field of view is divided into n cells not resolved by the optical system, having the same percent

of polarization p but random orientation, the resulting total polarization p_t may be shown to be $p_t = pn^{-1/2}$.

2. Depolarization due to the inhomogeneity of the magnetic field in the source. Let us suppose that we have a dominant magnetic field B_0 together with a number of random fields B_r. In this case the polarization is reduced to the ratio

$$p = p_0 \frac{h_0^2}{h_0^2 + 1},$$

with h_0^2 being the energy in the uniform magnetic field divided by the energy in the random magnetic field (Gardner and Whiteoak 1966).

3. Faraday dispersion, which occurs both in the source and in the interstellar medium. Faraday rotation has been discussed by Woltjer (1962)[1] and has received more attention recently as a possible method for investigating both the galactic and extragalactic magnetic fields (Kronberg, Conway, and Gilbert 1972; Reinhardt 1972). Briefly the mechanism is the following: when an extragalactic polarized radiation traverses a magnetoionic medium of electronic density N_i and a magnetic field B_i, it suffers a Faraday rotation given by

$$\psi = 87 N_i B_i l \lambda^2, \tag{5}$$

where B_i is measured in gauss, l is the dimension of the medium in parsec, and λ is the wavelength measured in cm. In a variable magnetoionic medium the radiation will be randomly rotated and consequently depolarized. Although various possible source models have been investigated to evaluate the Faraday dispersion, in most cases the polarization seems to decrease as $\dfrac{\sin \psi}{\psi}$ so that depolarization becomes important when ψ exceeds a few radians (Cohen 1958).

Next we discuss $v \leq v_1$, i.e., an optically thick source. In this case the polarizations are differentially absorbed, and depolarization occurs as studied by Pacholczyk and Swihart (1967); the percent of resulting polarization depends on the ratio between the absorption coefficient K and the Faraday rotation per unit length, where K is given by

$$K = 3.5 \times 10^{-6} N_0 v_m^{\frac{\gamma+2}{2}} v^{-\frac{\gamma+4}{2}} m^{-1}$$

and equals $\dfrac{1}{l}$ when $v = v_1$.

---
[1] Also see p. 960.

If $\dfrac{Kl}{\psi} \gg 1$, the percentage polarization tends to its maximum value which is about 10% (Fig. 4).

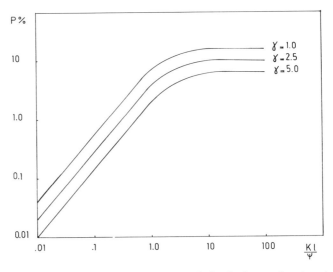

Fig. 4. Percentage polarization of synchrotron radiation in the wavelength region where the source is optically thick (synchrotron self-absorption). Kl is the absorption and ψ the Faraday dispersion.

Evaluation of the Degree of Polarization in Far-Infrared Sources

Since the Faraday dispersion depends on the wavelength as shown by Equation (5), one should expect an increase of the percentage polarization when the wavelength decreases from radio to IR regions. This fact is true for a number of radio sources, as shown in Fig. 5. In these cases we are able to study the source without the effect of any intermediate magnetoionic medium, so that the degree of polarization in the far infrared gives us information about the internal structure and homogeneity of the source. Moreover, we should point out that the IR excess observed in many celestial sources cannot be explained without assuming that a self-absorption occurs for $v_1 = 4 \times 10^{12}$ Hz (Burbidge and Stein 1970) so that the percentage polarization p must show a strong decrease when v varies from optically thin to optically thick conditions.

These facts should allow us to discriminate the synchrotron process among the various proposed mechanisms if a polarimeter of suitable sensitivity could be realized. As an example, let us try to calculate the polarization of the galactic center assuming synchrotron emission by irtrons (Low 1970).[2] As a model of the galactic center, we assume a dense cloud of ionized gas having a density of thermal electrons $N_i = 5 \times 10^2$ cm^{-3} and a diameter $l = 10$ pc. Let us suppose that the magnetic field in the cloud is randomly oriented with an intensity of $B_i = 10^{-4}$ gauss. These data are taken from radio observations by Downes and Maxwell 1966; Lequex 1967; Maxwell and Taylor 1968; Thompson, Riddle, and Lang 1969; and Lipovka 1971. Finally, let us suppose that about 10^3 irtrons having a diameter of about 10^{12} cm and a uniform magnetic field of 10^2 gauss are present in this medium. The radiation from every irtron will be strongly polarized ($p \simeq 75\%$) for $v > v_1$, with $v_1 = 4 \times 10^{12}$ Hz,

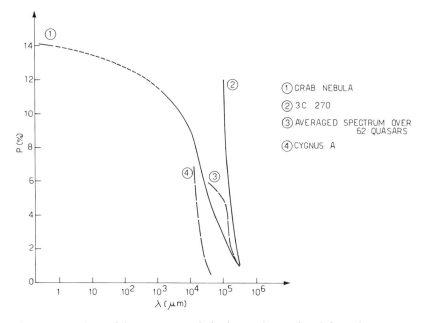

Fig. 5. Dependence of the percentage polarization on the wavelength for various sources: in the case of Crab Nebula the solid line is referred to the averaged polarization over the dimension of the object. Polarization as high as 50% has been observed in the radio waves by using high resolution instruments to observe regions of small dimension.

---

[2]"Irtrons" were suggested (Low 1970) as the fundamental unit in the nuclei of galaxies which emits the enormous far-infrared continuum. It had been observed that the IR spectrum remained remarkably the same from one galaxy to another over the full seven orders of magnitude range in infrared luminosity. Limits on the size of the emitting regions come from variability and from restrictions imposed by the steep slope of the far-infrared spectrum. Thus, a given nucleus would contain large numbers of small, essentially identical sources, called "Irtrons."

and slightly polarized ($p \simeq 10\%$) for $v \leq v_1$. Crossing the nuclear region the radiation will suffer a rotation

$$\psi = 87 \times 10^{-4} \times 5 \times 10^2 \times 10 \times 10^{-4} \text{ rads}$$
$$\simeq 5 \times 10^{-3} \text{ rads} (v = 4 \times 10^{12} \text{ Hz}).$$

So, in this case, the Faraday dispersion is negligible, and this result is quite insensitive to the choice of the parameters B_i and N_i.

However, a spatial depolarization is expected in the source itself due to random orientation of the polarization vectors of the 10^3 unresolved irtrons, leading to

$$p = \frac{75}{10^{3/2}} \simeq 2\% \quad \text{for} \quad v > 4 \times 10^{12} \text{ Hz}$$

$$p = \frac{10}{10^{3/2}} \simeq 0.3\% \quad \text{for} \quad v \leq 4 \times 10^{12} \text{ Hz}.$$

In this situation the polarization measurements performed by Low et al. (1969) at 10 μm wavelength appear to be inadequate due to their uncertainty on the order of $\pm 2\%$ in the minimum detectable polarization.

It may be concluded that the spurious instrumental polarization must be reduced to about 0.1% in order to obtain significant results. Obviously, the idea of irtrons is not proven and is questionable. In any case, if the number of the unresolved sources is lower or if they are similarly oriented, a larger polarization may be expected.

INFRARED POLARIZATION BY INTERSTELLAR GRAINS

Physical and Chemical Properties of the Grains

The properties of interstellar grains could be in principle deduced from the observations in the optical range, i.e., from the interstellar extinction, albedo, and polarization of the starlight in the ultraviolet, visible, and near-infrared wavelengths.[3] Following this theory many papers have been devoted to the identification of the chemical composition of interstellar grains. Among the various proposed components we may recall here ice grains, graphite grains, iron grains, silicate grains and silicon carbide grains (Hoyle and Wickramasinghe 1969; Wickramasinghe and Nandy 1971a,b). These refractory materials could be formed through processes involving mass ejection from cool stars, explosions of massive objects, and supernovae. Each material shows characteristics able to explain a few experimental data. Ice grains, for instance, give a good

---
[3] See p. 888.

agreement with the extinction curve in the visible; graphite grains also fit the extinction curve in the visible and, moreover, explain a broad peak in extinction curve over the waveband 0.2–0.25 μm; silicate grains are able to explain the extinction in the far ultraviolet ($\lambda < 0.2\,\mu$m) and the polarization, while iron grains are able to explain the albedo.

Nevertheless, at present no material seems able to explain the whole of the experimental data. It is beyond the purpose of the present work to discuss in detail the solutions suggested by various authors. We only want to point out that the present trend is to reject the hypothesis of ice grains both because grains have been found in hot H-II regions (O'Dell and Hubbard 1965) and because the 3-μm absorption band of ice has not been observed despite several attempts (Knacke, Cudaback, and Gaustad 1969).

Various mixtures have been proposed to explain the optical data, for instance, one consisting of iron, graphite, and silicates, while the dimensions of the grains may be distributed accordingly to

$$n(r) = Ar^2 \exp(-\beta r^3)$$
$$\beta = 2/3 r_0^{-3},$$

with
$$r_0(\text{graphite}) = 0.045\text{--}0.07\ \mu\text{m}$$
$$r_0(\text{iron}) = 0.02\ \mu\text{m}$$
$$r_0(\text{silicates}) = 0.15\text{--}0.18\ \mu\text{m}.$$

The average density of the grains along the galactic plane may then be 5×10^{-13} cm^{-3} for graphite, 2×10^{-11} cm^{-3} for iron, and 2×10^{-14} cm^{-3} for silicates. In another mixture (Gilra 1971) the iron component is substituted by silicon carbide, and similar parameters are used to describe the properties of the grains.

Little attention has been devoted up to now to the shape of the grains and to the structure of their surface, probably due to the fact that no relevant changes are to be expected depending on the shape (Greenberg 1971a,b).

INFRARED EMISSION BY INTERSTELLAR GRAINS

Interstellar grains are associated with gas clouds in space and are concentrated in small regions along the galactic plane that form dark nebulae with linear dimensions ranging between 0.1 and several tens of parsecs. The density of the grains in these globules may be up to 10^2–10^3 times higher than the mean densities given in the previous section.

The dark nebulae should emit infrared radiation depending on the emissivity and the temperature of the grains. The emissivity of the various types of grains has been studied by a number of authors (Aiello and Borghesi 1970; Krishna Swamy 1970, 1971). Roughly we have

graphite: $1.9 \times 10^{-21} T^6$ erg sec^{-1}, or

iron: $8.8 \times 10^{-25} T^6$ erg sec^{-1}, or

silicates: $1.1 \times 10^{-16} T^5$ erg sec^{-1} for every grain.

More detailed data may be obtained from Fig. 6 (Greenberg 1971a,b).

The temperature of the grains depends on their absorption coefficients and on the heating electromagnetic field. Two very different cases may occur:

1. The dark nebula is located in the "normal" interstellar medium, and the only heat source is the starlight and also, for silicate grains, the 3°K blackbody radiation.
2. The dark nebula is located around an H-II region or, generally speaking, it contains a very hot object such as a star with $T \geq 10^{4°}$K or a galactic nucleus.

The behavior in the two cases is different mainly due to the various absorption coefficients of the grains in the visible and ultraviolet band of the spectrum, which, in turn, give rise to a different temperature of the grains in the two cases. Let us discuss this argument briefly.

Emission from a dark cloud heated by the starlight: The starlight spectrum may be treated as the sum of three dilute blackbodies having temperatures T_i and dilution factors W_i given by Werner and Salpeter (1969)

T_i	14,500°K	7,500°K	4,000°K
W_i	4×10^{-16}	1.5×10^{-14}	1.5×10^{-13}

From Fig. 6 one may observe that silicate grains cannot efficiently absorb the starlight in the 0.3-8 μm region (Wickramasinghe and Nandy 1971a, b) so that their temperature at the thermal equilibrium will be lower than that of graphite or iron grains. The calculation for a single grain exposed to starlight gives an equilibrium temperature (Greenberg 1971a, b; Aiello and Borghesi 1970; Werner and Salpeter 1969; Stein 1967) of 30°K for graphite, 70°K for iron, and 10°K for silicates. Moreover, in a dark cloud the temperature of the grain depends on the distance of the grain itself from the center of the cloud, because a large amount of starlight radiation is absorbed by external grain layers.

For instance, in Table I we calculated the temperature in a cloud 7 psc in diameter at various distances from the center following the model proposed by Wickramasinghe and Nandy (1970, 1971a, b).

The most relevant features with respect to the emission of the dark nebula are:

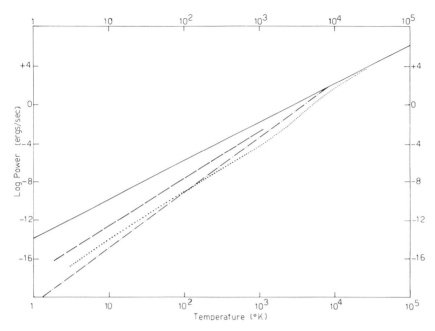

Fig. 6. Emission from various types of grains versus the grain temperature. Solid line is the blackbody emission. The dashed line is the maximum emission expected from a grain 0.05 μm in radius. The dotted line is for silicate grains. The dashed-dotted line is for graphite grains.

TABLE I

Equilibrium Temperature Reached by Various Types of Grains at Various Depths of a Dark Cloud Heated by Starlight

Distance (parsecs)	Temperature in °K		
	Graphite	Iron	Silicates
0	19.5	39.6	3.5
1	19.7	40.0	3.5
2	19.9	40.5	3.5
3	20.5	42.1	3.6
4	21.3	44.3	3.9
5	22.4	47.4	4.2
6	24.0	52.4	5.0
7	27.1	62.0	6.4

a. The contribution of silicate grains becomes significant only beyond $\lambda \sim 400$ μm; in any case this contribution could be detected on the earth from ground-based infrared observatories.
b. The maximum emission occurs around 90 μm, but in the presence of iron the emission remains practically flat down to 20–30 μm.

Until now no object showing these spectral features has been detected, although the flatness in the spectrum of NGC 1061 and other extragalactic objects cannot be excluded on the basis of the most recent observations by Low.[4]

Emission of a dark cloud heated by an ultraviolet excess: In the presence of an ultraviolet excess, the temperature of the silicate grains will be increased up to 20–30°K due to the large absorption of the silicates for wavelengths shorter than 0.3 µm. Under these conditions silicate grains become the most important source of infrared emission due to their large emissivity. In this case the emission spectrum could perhaps fit the emission observed from the galactic center, although the attempts made by Krishna Swamy (1970, 1971) do not appear to be satisfactory (Harwit et al. 1972), as shown in Fig. 7.

Mechanisms of Orientation and Percentage of Linear Polarization Expected in the Infrared

Observations performed in the optical band from the near ultraviolet to the near infrared have proved that the starlight is polarized by grains due to their anisotropies in shape and/or in physical parameters (Coyne and Gehrels 1967). Various mechanisms have been proposed to explain the alignment of the grains necessary to give rise to the observed polarization. We may briefly review these mechanisms:

1. Alignment produced by the interaction between dust and gas at supersonic velocity (Gold 1953).
2. Magnetic orientation: for a general review see Jones and Spitzer (1967); for the case of paramagnetic resonance see Davis and Greenstein (1951), and for the case of ferromagnetic resonance see Henry (1958).
3. Alignment due to the absorption of polarized photons (Harwit 1970).
4. Alignment due to the interaction with soft X-rays (Wickramasinghe 1969b) from the galactic center.

It has been suggested that graphite grains cannot be aligned efficiently (Purcell 1969), while silicate grains could explain the observed polarization when conveniently aligned (Wickramasinghe 1969a). This idea is in agreement with recent observations of circularly polarized radiation, which may only be attributed to dielectric grains, although more measurements in different sky regions are required before any conclusion can be drawn (Martin 1972; Martin, Illing, and Angel 1972).[5] It has been suggested by Stein (1967) that the grains should emit polarized radiation in the infrared. Stein quoted values of about 70% and 30% of linear polarization respec-

---

[4]Personal communication.   [5]See p. 926.

THE FAR INFRARED (10–1000 μm)

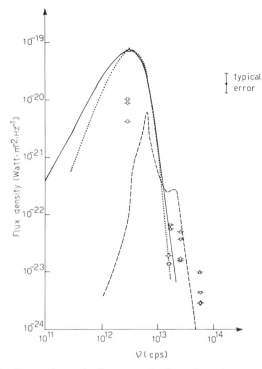

Fig. 7. Emission from various galactic sources: the dots refer to the sources observed by the Cornell University Group (see the text). The solid line is for 50°K blackbody radiation. The dotted line is a dilute 40°K blackbody having an emissivity proportional to $1/\lambda$. The dashed line is for silicate-grain emission.

tively for metallic and dielectric grains perfectly aligned. However, the alignment of the grains is a very questionable matter. We will attempt to give a brief description as follows:

a. In the case of a dark cloud heated by starlight as previously discussed, we have three components of which iron could be strongly oriented despite its higher temperature: so, a significant percentage of polarization is to be expected in the 10–50 μm range. Beyond 50 μm the contribution of graphite grains becomes predominant, but the percentage of polarization should decrease if the hypothesis of a poor graphite orientation (Purcell 1969) is valid. Finally, beyond 400 μm the percentage of polarization should increase due to the contribution of silicate grains (Fig. 8).

b. In the case of a dark nebula surrounding a hot core, the contribution of silicates should be predominant all over the spectrum, and a polarization of the order of 10–20% should be expected if the align-

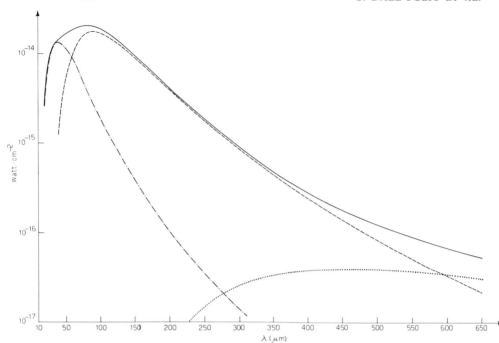

Fig. 8. Calculated emission from a dark nebula about 7 parsecs in diameter, heated by starlight. The computation is relative to a grain mixture suggested by Wickramasinghe (see the text). The solid line is the total emission. The dashed line shows the contribution by graphite grains. The dashed-dotted line refers to iron grains while the dotted line is for silicates.

ment of silicate grains were of the order of 60% as suggested by Wickramasinghe (1969b); in this case the percentage of polarization should be practically independent of the wavelength.

It is obvious that the above quoted results could be valid only in regions in which a homogeneous magnetic field is present, giving rise to a significant orientation of the grains. In any case detailed measurements are needed to confirm or reject the previously stated results.

EXPERIMENTAL PROBLEMS RELATED TO FAR-INFRARED POLARIMETRY

Far Infrared Polarizers

Various types of polarizers suitable for infrared polarimetry have been described in the literature. We will briefly survey the situation:

Dichroic and birefringent polarizers: This type of polarizer may be considered the extension in the infrared of the polarizers commonly used in the ultraviolet and visible region of the spectrum. HR-type Polaroid sheets may be used up to 2.7 μm (Blake, Makas, and West 1949). Other materials such as calcite (Bridges and Klüver 1965), CdS, rutile (Vergnoux 1957) have been used up to 16 μm. Pyrolitic graphite has been employed both in reflection and transmission, and good results have been obtained in the 5–500 μm range despite its brittleness (Rupprecht, Ginzberg, and Leslie 1962). More recently, proustite ($Ag_3 AsS_3$) has been employed as a polarizer at the focus of a telescope operating at 5 μm (Marsh 1971).

Brewster angle polarizers: Since the reflectivity of a dielectric surface tends to zero at the Brewster angle for radiation linearly polarized in such a way that the electric vector vibrates in the plane of incidence, a polarizer may be made with a pile of plates of a suitable dielectric operating at the Brewster angle. Due to the usually large values of the Brewster angle, these polarizers are quite large in dimension, and moreover, they displace the beam by refraction into the plates. Nevertheless, Greenler, Adolph, and Emmons (1966) obtained a compact polarizer by folding the plates many times. The percentage of polarization p obtained by using N sheets having refractive index m is given by

$$p = \frac{1 - \left[\frac{2m}{m^2+1}\right]^{4N}}{1 + \left[\frac{2m}{m^2+1}\right]^{4N}}.$$

In the far infrared, mylar or polyethylene films have been used (Mitsuishi et al. 1960), although interference effects between the foils may give rise to unwanted features in the transmission curve. Other solutions have been described by various authors, which are intended to compensate for the lateral displacements of the beam (Makas and Shurcliff 1955; Harrick 1959).

Grid polarizers: They consist of a sequence of parallel, equally spaced metallic strips. Beginning from the radio waves, these polarizers have been used for smaller and smaller wavelengths as the photoetching technique progressed. In the beginning, plastic replicas of optical gratings were used as polarizers by evaporating a metal onto the tips of the groove (Bird and Parrish 1960). Although a wide acceptance angle is obtained for this type of polarizer, the absorption bands of the plastic substrate appear to be a limiting factor when the polarizer has to be used in the near infrared. So the grooves have been obtained by directly ruling a suitable substrate such as Irtran 2 or Irtran 4 (Young, Graham, and Paterson 1965). Photolithographic techniques have been used on various substrates, and commercial grid polarizers are now available for wavelengths as short as 2 μm (Cambridge Physical Science). A new technique is listed in the literature, which consists of exposing *photoresist* to the interference pattern obtained with a laser source. A sinusoidal grating is obtained in this case, which may be converted into a polarizer by using the previously described method of evaporation (Bird and Parrish 1960). By using this technique, grid polarizers may be obtained that cover the whole infrared spectrum (Auton and Hutley 1972). In the very far infrared (say beyond 100 μm) a simple and inexpensive method for preparing grid polarizers has been described by Martin and Puplett (1970), which consists of scratching a metallic film evaporated on a mylar substrate.

The general theory of grid polarizers has been discussed by various authors; a review may be found by Auton (1967). Let us say that d is the spacing between the strips of a grid polarizer; then the transmission $T_\perp$ for an electric vector vibrating in a plane perpendicular to the metallic strips increases by increasing the ratio λ/d, while the transmission $T_\parallel$ for an electric vector vibrating in a plane parallel to the metallic strips decreases by the same factor when increasing λ/d so that for $\lambda \gg d$, the system acts as a polarizer with a high transmittance for an electric vector perpendicular to the strips. The transmittance depends on the ratio a/d between the width a of a single metallic strip and the spacing d between the strips. Figure 9 gives the dependence of $T_\perp$ versus the ratio a/d for an ideal polarizer.

The properties of two identical polarizers may be evaluated by means of four measurements performed in monochromatic light. The four measurements are of:

1. The total incident light I_0,
2. The power transmitted by a single polarizer having horizontal strips $I_\parallel$,
3. The power transmitted by a single polarizer having vertical strips $I_\perp$,
4. The power transmitted by a couple of crossed polarizers I_+.

It may be shown that

$$T_\perp = \frac{I_\parallel + I_\perp}{2I_0} \left\{ 1 + \left[1 - \frac{4I_+ I_0}{(I_\parallel + I_\perp)^2} \right]^{1/2} \right\}$$

$$T_\parallel = \frac{I_\parallel + I_\perp}{2I_0} \left\{ 1 - \left[1 - \frac{4I_+ I_0}{(I_\parallel + I_\perp)^2} \right]^{1/2} \right\}$$

$$P = \frac{T_\perp - T_\parallel}{T_\perp + T_\parallel}$$

where P is defined as the polarizing efficiency of the polarizer (remembering that for an ideal polarizer $T_\perp = 1$ while $T_\parallel = 0$). Figure 10 shows the behavior of various polarizers.

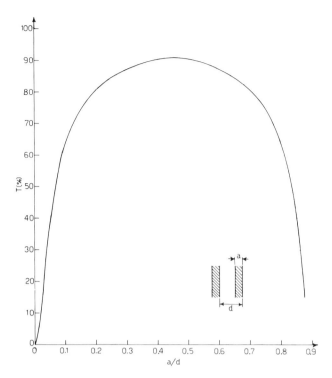

Fig. 9. Transmittance of a grid polarizer, for electric vector perpendicular to the strip, as a function of the ratio a/d between the width a of a single metallic strip and the spacing d between the strips.

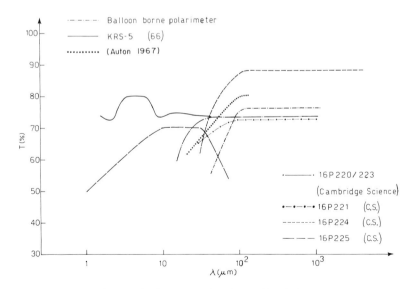

Fig. 10. Performances of various grid polarizers.

Spurious Instrumental Polarization

It is beyond the purpose of this work to study in detail the spurious polarization introduced by various spectroscopic instruments such as grating monochromators or interferometers (Hadni 1967). We only want to discuss here the spurious polarization effects introduced by some optical components that are commonly used in astronomical equipment, namely flat mirrors, spherical mirrors, and filters.

Polarization introduced by a flat mirror: Let us suppose that a flat mirror is used in the optical system, forming an angle of incidence i with respect to the optical axis, for instance as in the case of the bending Cassegrain focus of the NASA infrared observatory on the C-141 plane or as in the case of some commercially available germanium bolometers (Infrared Laboratories, Tucson, Arizona) using Pfund condensing optics. In these cases we may expect a polarization introduced by reflection on the mirror given by

$$P = \sqrt{\frac{v}{\sigma}} \left(\frac{1}{\cos i} - \cos i \right), \quad (6)$$

where v is the optical frequency of the radiation and σ is the electrical conductivity. Equation (6) is plotted in Fig. 11 for various materials. It must be pointed out that Equation (6) is valid only when assuming that the optical constant may be evaluated from the static conductivity. This is not true, at least up to 30 μm, where reflection measurements are available for various materials (Lonke and Ron 1967). In the case of astronomical applications, an important source of spurious polarization must be taken into account, i.e., the thermal emission of the mirror which is polarized. Let us say R_s and R_p are the reflectivities for s and p polarized waves. Then the polarization P_r introduced by a single reflection over the mirror is

$$P_r = \frac{R_s - R_p}{R_s + R_p},$$

while for the polarized emission P_e of the same mirror we have

$$P_e = \frac{\varepsilon_s - \varepsilon_p}{\varepsilon_s + \varepsilon_p} = P_r \frac{2}{\varepsilon_s + \varepsilon_p},$$

where $\varepsilon_s, \varepsilon_p$ are the emission coefficients for s and p polarized waves respectively. This emission may be modulated by the rotating analyzer of a polarimeter, giving rise to an intense spurious signal (Worthing 1926; Vincent 1972; Conn and Eaton 1954).

Depolarization introduced by spherical mirrors: A small depolarization effect is introduced when using a spherical mirror (Jäger and Oetken

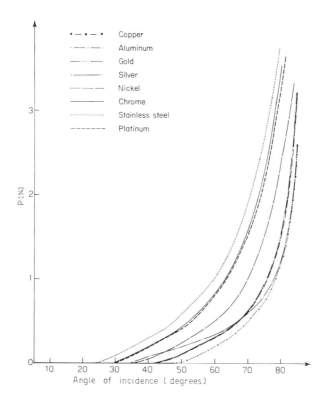

Fig. 11. Polarization introduced by a single reflection on various flat metallic mirrors for various beam incidence angles, computed at $\lambda = 100 \, \mu$m.

1963). The depolarization increases when increasing the field of view of the optical system, and it may be shown that the observed polarization P^* is given by

$$P^* = P_0 [\tfrac{1}{2} + (t + t^{-1})^{-1} \cos \tau]^2,$$

where P_0 is the actual polarization of the source, τ is the phase lag between the s and p polarized waves, and t is the ratio between the reflection coefficients relative to the s and p polarized waves. Since $t \simeq 1$ and $\tau \simeq 0$, P^* cannot differ from P_0 much more than 1%.

Polarization introduced by the filters: There are several types of filters commonly used in the infrared, namely single crystal filters, echelette filters, Yoshinaga-type filters, and metallic mesh filters. Of these, echelette filters have been proposed as high-pass far-infrared filters since 1963 (Möller and McKnight 1963; Ginzerl et al. 1970). Recently the polarization introduced by echelette filters has been studied (Carli and Mencaraglia 1972). The result is that a polarization on the order of 20% may be introduced by an echelette having a groove angle of 45°, while the polarization ranges around 5% for a groove angle of 30°. This means that crossed echelette filters must be used to eliminate the unwanted polarization; on the other hand, the transmission curve of two crossed echelettes seems to show a sharper cutoff at short wavelengths, which is surely desirable. Metallic mesh filters will also introduce spurious polarization when used at wide incidence angles (Mitsuishi et al. 1963), while no polarization is to be expected when using Yoshinaga-type filters.

EXPERIMENTAL RESULTS

Only a few measurements have been performed until now in infrared polarimetry. Low et al. (1969) have tried to measure the polarization of the galactic center at 10 μm with a polarimeter having a spurious polarization on the order of 2%. More recently, Capps and Dyck (1972)[6] have observed polarized radiation on the order of 1% from ten cool stars despite their instrumental spurious polarization, which ranges between 5% and 7%.

In the following we will describe the experimental results obtained during our balloon flight. The general layout of the polarimeter has been described elsewhere (Dall'Oglio et al. 1973). It consists of (Fig. 12):

1. A germanium bolometer cooled at liquid helium temperature; the helium is pumped at ambient pressure.
2. A three-filter system selecting radiation in the spectral range 100–2000 μm.
3. A rotating analyzer (grid polarizer) whose efficiency is higher than 76% beyond 50 μm. The linearly polarized radiation is modulated at a frequency of 30 Hz.
4. A metallic mirror telescope $f/4$, 25 cm in diameter.
5. An on-board polarized calibration source heated to $64 \pm 2°$C.
6. A lock-in amplifier with two channels $\approx 90°$ out of phase and two corresponding derivative channels.

The minimum detectable flux for a 100% linearly polarized diffuse radiation is about 5×10^{-12} W cm^{-2} ster^{-1}, with an integration time of 1 sec and a signal-to-noise ratio of 1.

The balloon was at ceiling altitude (pressures less than 12 mb) from 18:00 to 20:07 UT on 16 September 1971; the maximum altitude (pressure 8 mb) was reached at 18:10 UT. The axis of the telescope while at ceiling altitude was kept in the N-S direction within 1°. The scanning of the sky was done by rotating the axis of the telescope in a plane normal to the horizon between 5° and 27° in elevation with a period of 3 mins. The minimum percentage of detectable polarization was found in the labora-

---

[6]See p. 858; also see p. 318.

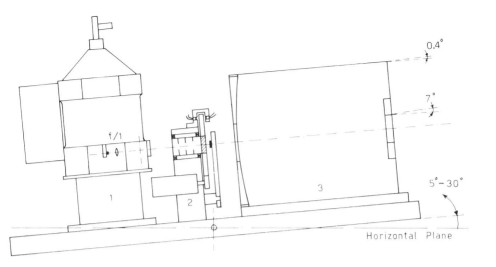

Fig. 12a. General view of the balloon-borne polarimeter:
1—Detector assembly 2—Rotating analyzer assembly 3—$f/4$, 25-cm telescope

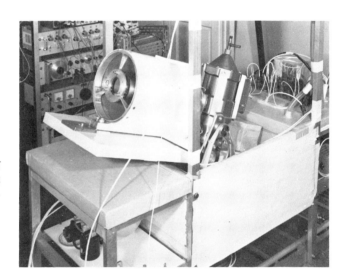

Fig. 12b. Picture of the gondola during laboratory calibrations.

tory to be of about 0.1%. The sensitivity of the instrument would allow us to detect the galactic center if its emission were polarized more than 0.3%. An atmospheric effect has been observed which falls within the limits of spurious polarization of the instrument (Dall'Oglio et al. 1973b).

Various sources were detected during the flight. A polarized component from the sun was observed (Fig. 13) having polarization perpendicular to the sun's equator; the electric vector maximum is in a plane perpendicular to the sun's equator within $10°$ (Dall'Oglio et al. 1973a). The observed polarization is 6.4%, in the 100–400 μm range. Although the mechanism

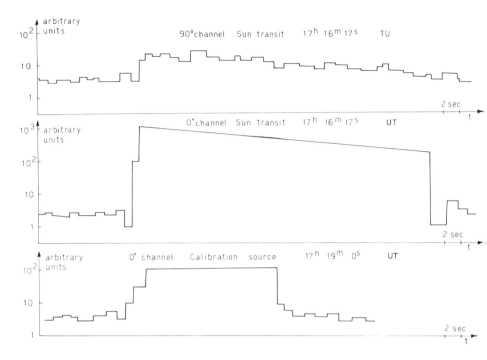

Fig. 13. Signal detected from the sun in the two channels, compared with the on-board calibration source.

responsible for this polarization is unknown, we may note that an infrared excess has been detected in this same region of wavelength by Müller, Kneubühl, and Stettler (1972).

The galactic center was scanned as shown in Fig. 14, but no significant signal was detected, as shown in Fig. 15; i.e., an upper limit of about 0.3% may be assumed for the polarized component from the galactic center in the 100–2000 μm region.

Some signals have been recorded during the flight, and the possible identification of the observed sources is listed in Table II. Since our instrument is sensitive only to polarized radiation, we are not able to give the values of the percentage of polarized radiation, because the total infrared fluxes from the sources listed in Table II are unknown. Figure 16 shows the sky zone that has been explored by the telescope, and in Fig. 17 the corresponding signals are shown. If the identifications are valid (we must recall the quite large field of view of our instrument: $\pm 0.°50$), it seems that polarized radiation comes from dark nebulae heated by stars, although the exact nature of these stars is unknown, and this fact could be in agreement with the considerations developed above.[7]

---

[7] See p. 334.

THE FAR INFRARED (10–1000 μm) 345

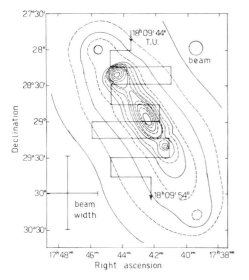

Fig. 14. Scanning of the galactic center. The movement of the beam is due to random gondola oscillations.

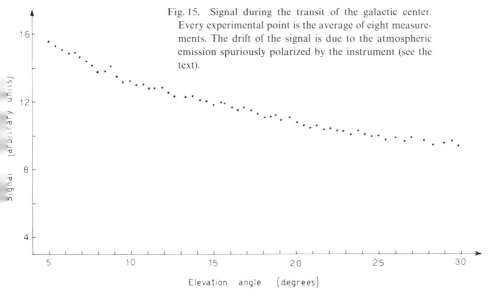

Fig. 15. Signal during the transit of the galactic center. Every experimental point is the average of eight measurements. The drift of the signal is due to the atmospheric emission spuriously polarized by the instrument (see the text).

In the case of n.28 Hoffmann source, the polarized intensity in the 100–2000 μm region seems to be much higher than the total intensity observed by Hoffmann in the 80–120 μm region, which suggests that the mechanism of emission could be a radio continuum and perhaps the source could be detected from ground-based observatories in the 400 μm atmospheric window.

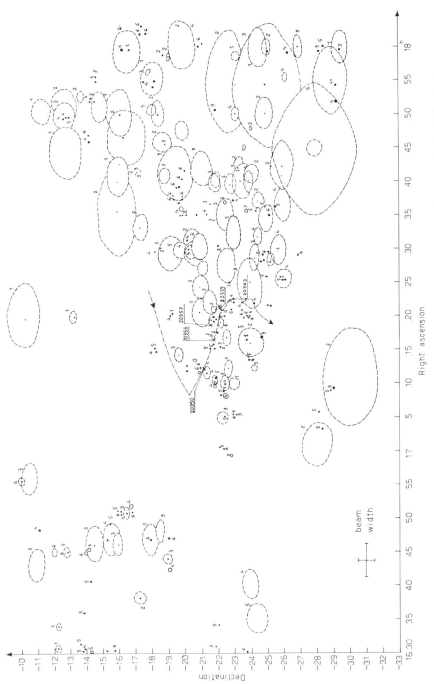

Fig. 16. Part of the sky surveyed by the balloon-borne telescope. The zones indicated in the figure are dark nebulae having absorption coefficients ranging between 1 and 6. The map of the dark nebulae has been obtained on the basis of data given by Lynds (1962). The numbers correspond to IR stars observed by Leighton and Neugebauer during their 2.2 μm survey.

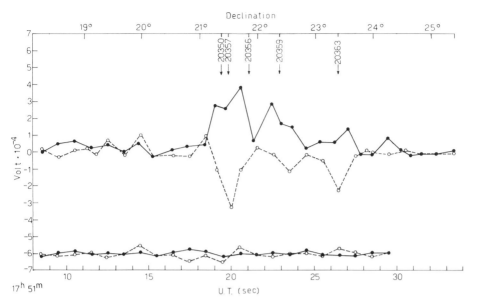

Fig. 17. Signal recorded during the scan indicated in Fig. 16. The solid and dashed lines refer to the two channels. The two upper lines represent the actual signal during the transit of the sources, while the other two lines represent the typical noise level in absence of sources. The latter two lines have been shifted, away from the 0 level, to facilitate the comparison.

TABLE II
Tentative Identification of the Sources Observed During the Balloon Flight

Sources	Observed R. A. (h m s)	Decl.	Tentative Ident. R. A. (h m s)	Decl.	Type	Flux[a] (watt·cm^{-2})	No. of measur.
1	17 12 26	$-21°22'$	17 12 26	$-21°26'$	20350	$3 \cdot 10^{-15}$	8
2	17 16 50	$-21°35'$	17 16 55	$-21°41'$	20356	$4 \cdot 10^{-15}$	8
3	17 18 20	$-21°38'$	17 18 20	$-21°31'$	20357	$3 \cdot 10^{-15}$	8
4	17 21 22	$-22°10'$	17 21 23	$-22°20'$	20359	$3 \cdot 10^{-15}$	16
5	17 25 20	$-23°10'$	17 25 52	$-23°24'$	20363	$1.5 \cdot 10^{-15}$	8
6	17 30 00	$-20°$	17 28 13	$-20°13'$	20367	$3 \cdot 10^{-14}$	4
7	17 25 32	$-34°40'$	17 25 34	$-34°30'$	n. 28 Hoffmann	$1.2 \cdot 10^{-13}$	4

[a] The indicated fluxes are not corrected for the filter system and atmospheric absorption.

CONCLUSIONS

Far-infrared polarimetry could give important information about the mechanisms of emission responsible for the infrared excesses observed in some galactic and extragalactic objects, provided that measurements of their spectra can be accomplished by means of low resolution filters.

A strong dependence of the polarization on the wavelength may be expected for synchrotron radiation around the wavelength beyond which the self-absorption occurs. A flat spectrum is expected in the case of grain emission when the grains are heated by a central hot core, so that the contribution of silicates is dominant; otherwise polarization could occur only before ~ 60 μm.

With the present state of knowledge, the first step in infrared polarimetry should be that of observing whether polarized emission does exist in the infrared. A balloon-borne experiment carried out with low angular resolution and large optical bandwidth has shown that polarized emissions are coming to the earth from highly obscured dark nebulae, perhaps heated by stars, and from other sources that could be of the synchrotron type. More investigation is, therefore, justified to obtain measurements both of the percentage of polarization and of the spectral features.

REFERENCES

Aiello, S., and Borghesi, A. 1970. The dark nebula near rho-Ophiuchi as a far infrared source. *Quaderni della Ricerca Scientifica CNR* 64: 29–40.

Auton, J. P. 1967. Infrared transmission polarizers by photolithography. *Applied Optics* 6: 1023–1027.

Auton, J. P., and Hutley, M. C. 1972. Grid polarizers for the near infrared. *Infr. Phys.* 12: 95–100.

Biraud, F. 1972. A search for circular polarization in compact sources at 9 mm wavelength. *Astron. Astrophys.* 19: 310–311.

Bird, G. R., and Parrish, M. 1960. The wire grid as a near infrared polarizer. *J. Opt. Soc. Am.* 50: 886–891.

Blake, R. P.; Makas, A. S.; and West, C. D. 1949. Molecular type dichroic film polarizer for the 0.75–2.8 micron radiations. *J. Opt. Soc. Am.* 39: 1054.

Bridges, T. S., and Klüver, J. W. 1965. Dichroic calcite polarizer for the infrared. *Applied Optics* 4: 1121–1125.

Burbidge, G. R., and Stein, W. A. 1970. Cosmic sources of infrared radiation. *Astrophys. J.* 160: 573–593.

Cambridge Physical Science. Bar Hill Cambridge, England.

Capps, R. W., and Dyck, H. M. 1972. The measurement of polarized 10 micron radiation from cool stars with circumstellar shells. *Astrophys. J.* 175: 693–697.

Carli, B., and Mencaraglia, F. 1972. Optical properties of echelette transmission high pass filters. *Infr. Phys.* 12: 187–195.

Cohen, M. H. 1958. Radio astronomy polarization measurements. *Proc. Inst. Radio Engrs.* 46: 172–183.

Conn, G. K. T., and Eaton, G. K. 1954. On the analysis of elliptically polarized radiation in the infrared region. *J. Opt. Soc. Am.* 44: 546–552.
Conway, R. G.; Gilbert, S. A.; Raymond, E.; and Weiler, K. W. 1971. Circular polarization of quasars at 21 cm. *Mon. Not. R. Astr. Soc.* 152: 1P–4P.
Coyne, G. V., and Gehrels, T. 1967. Wavelength dependence of polarization. X. Interstellar polarization. *Astron. J.* 72: 887–898.
Dalgarno, A., and Wright, E. L. 1972. Infrared emissivities of H_2 and HD. *Astrophys. J.* 174: L49–L51.
Dall'Oglio, G.; Melchiorri, B.; Melchiorri, F.; Natale, V., and Gandolfi, E. 1973a. Evidence for polarized emission from the sun in the far infrared. *Infra. Phys.* 13: 1–6.
Dall'Oglio, G.; Melchiorri, F.; Mencaraglia, F.; and Moreno, G. 1973b. Observation of atmospheric emission in the far infrared. *Lettere al Nuovo Cimento.* 5: 1019–1023.
Davis, L., and Greenstein, J. L. 1951. The polarization of starlight by aligned dust grains. *Astrophys. J.* 114: 206–240.
Downes, D., and Maxwell, A. 1966. Radio observations of the galactic center region. *Astrophys. J.* 146: 653–665.
Gardner, F. F., and Whiteoak, J. B. 1966. The polarization of cosmic radio waves. *Ann. Rev. Astron. Astrophys.* 4: 245–292.
Gilbert, J. A., and Conway, R. G. 1970. Circular polarization of quasars at $\lambda = 19$ cm. *Nature* 227: 585–586.
Gillett, F. C., and Stein, W. A. 1971. Infrared studies of galactic nebulae II, B stars associated with nebulosity. *Astrophys. J.* 164: 77–82.
Gilra, D. P. 1971. Composition of interstellar grains. *Nature* 229: 237–241.
Ginzburg, V. L., and Sirovantzkii, S. I. 1965. Cosmic magnetobremsstrahlung (synchrotron radiation). *Ann. Rev. Astron. Astrophys.* 3: 297–350.
Ginzerl, M.; Moreno, G.; Melchiorri, B.; Melchiorri, F. 1970. Transmittance and reflectance of an echelette high pass filter. *Quaderni della Ricerca Scientifica CNR* 64: 41–54.
Gold, T. 1953. The alignment of galactic dust. *Mon. Not. R. Astr. Soc.* 112: 215–218.
Goldberg, L. G. 1968. The infrared emission of NGC 7027. *Astrophys. Letters* 2: 101.
Greenberg, J. M. 1971a. Interstellar grain temperature; effect of shape. *Astron. Astrophys.* 12: 250–257.
——— . 1971b. Interstellar grain temperature; effect of grain materials and radiation fields. *Astron. Astrophys.* 12: 240–249.
Greenler, R. G.; Adolph, K. W.; and Emmons, G. H. 1966. A compact polarizer for the infrared. *Applied Optics* 5: 1468–1469.
Hadni, A. 1967. *Essentials of modern physics applied to the study of far infrared.* London: Pergamon Press.
Harrick, N. J. 1959. Infrared polarizer. *J. Opt. Soc. Am.* 49: 379.
Harwit, M. 1970. Is magnetic alignment of interstellar dust really necessary? *Nature*, 226: 61–62.
Harwit, M.; Soifer, B. T.; Houck, J. R.; and Pipher, J. L. 1972. Why many infrared astronomical sources emit at 100 microns. *Nature Phys. Sci.* 236: 103.
Henry, J. 1958. Polarization of starlight by ferromagnetic particles. *Astrophys. J.* 128: 206–240.
Hoyle, F., and Wickramasinghe, N. C. 1969. Interstellar grains. *Nature* 223: 459–462.
Jäger, F. W., and Oetken, L. 1963. Zur theorie und Praxis der instrumentellen Polarization. *Pub. Astrophys. Obs. Potsdam* Vol. 31: No. 107 pp. E1–E41.
Jones, R. V., and Spitzer, L. 1967. Magnetic alignment of interstellar grains. *Astrophys. J.* 47: 943–964.
Knacke, R. F.; Cudaback, D. D., and Gaustad, J. E. 1969. Infrared spectra of highly reddened stars: A search for interstellar ice grains. *Astrophys. J.* 158: 151–160.

Krishna Swamy, K. S. 1970. Intensity of infrared radiation from interstellar grains in the solar neighborhood. *Astrophys. Space Sci.* 6: 474–480.

———. 1971. Thermal emission from grains in the galactic center. *Astrophys. J.* 167: 63–69.

Kronberg, P. P.; Conway, R. G., and Gilbert, S. A. 1972. The polarization of radio sources with appreciable red-shift. *Mon. Not. R. Astr. Soc.* 156: 275–282.

Lequex, J. 1967. The galactic nucleus: Sagittarius A. *IAU Symp. No. 31*, 393–404.

Lipkova, N. M. 1971. The structure of radio source Sagittarius A. *Soviet Astron.-AJ.* (English translation) Vol. 15: 203–209.

Lonke, A., and Ron, A. 1967. Infrared reflection from metals. *Phys. Rev.* 160: 577–584.

Low, F. J. 1970. The infrared-galaxy phenomenon. *Astrophys. J.* 159: L173–L179.

Low, F. J.; Kleinmann, D. E.; Forbes, F. F.; and Aumann, H. H. 1969. The infrared spectrum, diameter, and polarization of the galactic nucleus. *Astrophys. J.* 157: L97–L101.

Lynds, B. T. 1962. Catalogue of dark nebulae. *Astrophys. J. Suppl.* 64: 1–52.

Makas, A. S., and Shurcliff, W. A. 1955. A new arrangement of silver chloride polarizer for the infrared. *J. Opt. Soc. Am.* 45: 998.

Marsh, J. C. D. 1971. Conference on infrared techniques: Proceeding No. 22 p. 295. Univ. of Reading.

Martin, D. H., and Puplett, E. 1970. Polarized interferometric spectrometry for the millimeter and submillimeter spectrum. *Infr. Phys.* 10: 105–109.

Martin, P. G. 1972. Interstellar polarization. *Mon. Not. R. Astr. Soc.* 159: 179–190.

Martin, P. G.; Illing, R.; and Angel, J. R. P. 1972. Discovery of interstellar circular polarization in the direction of the Crab Nebula. *Mon. Not. R. Astr. Soc.* 159: 191–201.

Maxwell, A., and Taylor, J. H. 1968. Lunar occultation of the radio source Sagittarius A. *Astrophys. Letters* 2: 191–194.

Mitsuishi, A.; Otsuka, Y.; Fujita, S.; and Yoshinaga, H. 1963. Metal mesh in the far infrared region. *Japan J. Appl. Phys.* 2: 574–577.

Mitsuishi, A.; Yamada, Y.; Fujita, S.; and Yoshinaga, H. 1960. Polarizer for the far infrared region. *J. Opt. Soc. Am.* 50: 433–445.

Möller, K. D., and McKnight, R. V. 1963. Far infrared filter gratings. *J. Opt. Soc. Am.* 53: 760–761.

Müller, E. A.; Kneubühl, F. K.; and Stettler, P. 1972. Absolute measurement of the solar brightness in the spectral region between 100 and 500 microns. *Astron. Astrophys.* 20: 309–312.

O'Dell, C. R., and Hubbard, W. B. 1965. Photoelectric spectrophotometry of gaseous nebulae I: The Orion Nebula. *Astrophys. J.* 142: 591–603.

Pacholczyk, A. G., and Swihart, T. L. 1967. Polarization of radio sources I: Homogeneous sources of arbitrary optical thickness. *Astrophys. J.* 150: 647–650.

Pecker, J. C. 1971. Colloque n. 17, Section IV Symp. de Liège. (Unpublished.)

Purcell, E. M. 1969. On the alignment for interstellar dust. *Physica* 41: 100–127.

Reinhardt, M. 1972. Interpretation of rotation measurements of radio sources. *Astron. Astrophys.* 19: 104.

Roberts, J. A., and Komesaroff, M. M. 1965. Observations of Jupiter's radio spectrum in the range from 6 cm to 100 cm. *Icarus* 4: 127–156.

Rosenberg, H. 1972. On circularly polarized radiation from extragalactic radiosources. *Astron. Astrophys.* 19: 66–70.

Rupprecht, G.; Ginzberg, D. M.; and Leslie, J. D. 1962. Pyrolitic graphite transmission polarizer for the infrared. *J. Opt. Soc. Am.* 52: 665–669.

Stein, W. 1967. Infrared radiation from interstellar grains. *Astrophys. J.* 144: 318–325.

Thompson, A. R.; Riddle, A. C.; and Lang, K. R. 1969. Observation of the occultation of Sagittarius A by the moon. *Astrophys. Letters* 3: 49–50.

Vergnoux, A. M., and Duverney, R. 1957. Polarimétrie dans l'infra-rouge. *J. Phys. Radium* 18: 527–536.
Vincent, R. K. 1972. Emission polarization study on quartz and calcite. *Applied Optics* 11: 1942–1945.
Werner, M. W., and Salpeter, E. E. 1969. Grain temperature in interstellar clouds. *Mon. Not. R. Astr. Soc.* 145: 249–269.
Wickramasinghe, N. C. 1969*a*. Interstellar polarization by graphite silicate grain mixtures. *Nature* 224: 656–658.
―――. 1969*b*. Galactic soft X-rays and the alignment of interstellar grains. *Nature* 228: 540–542.
Wickramasinghe, N. C., and Nandy, K. 1970. Interstellar extinction by graphite iron silicate grains. *Nature* 227: 51–53.
―――. 1971*a*. Optical properties of graphite iron silicate grain mixtures. *Mon. Not. R. Astr. Soc.* 153: 205–227.
―――. 1971*b*. Interstellar dust: Graphite, iron and silicate. *Nature* 229: 81–82.
Woltjer, L. 1962. The polarization of radio sources. *Astrophys. J.* 136: 1152–1154.
Worthing, A. G. 1926. Deviation from Lambert's law and polarization of light emitted by incandescent tungsten, tantalum and molybdenum and changes in the optical constants of tungsten with temperature. *J. Opt. Soc. Am.* 13: 635.
Young, J. B.; Graham, H. A.; and Paterson, E. W. 1965. Wire grid infrared polarizer. *Applied Optics* 5: 1023–1026.

RADIO MEASUREMENTS OF POLARIZATION

R. G. CONWAY
Nuffield Radio Astronomy Laboratories
Jodrell Bank

A review is given of different methods of measuring the polarization of radio sources. They are classified according to whether a single telescope or an interferometer is used, and whether the receiving system involves a total power or a cross-correlation detector.

The majority of radio sources emit by nonthermal mechanisms and, hence, exhibit some degree of polarization. Synchrotron radiation from quasars and galaxies is linearly polarized, usually by not more than 5%–10% (a figure that reflects the degree of randomization of the magnetic field). At long wavelengths the degree of polarization is reduced and the position angle rotated by the Faraday effect. Circular polarization in quasars has been detected, but at very low levels, usually less than 0.1%. Pulsars are also believed to emit by the synchrotron process and exhibit very high degrees of polarization, often over 50% of either linear or circular polarization. Measurements of such sources require special techniques because the signals vary within the length of the pulse, i.e., on the time scale of milliseconds. Similar problems arise in the measurement of radiation from active solar regions, though here the time scales are usually longer, on the order of seconds or minutes.

Strong polarization is also detected in the line radiation at λ 18 cm of the OH radical. The process is believed to involve maser action, with characteristic modes that are circularly polarized. Up to 100% circular polarization has been detected in narrow bandwidths, much narrower than the total bandwidth within which the OH radiation can be detected.

In every case polarimetric measurements are of great interest for the evidence they provide about the physical processes occurring within the source. The technical demands are invariably high because the polarized signals are strongly dependent upon frequency or because they vary rapidly with time or because they are simply very faint. This paper describes briefly some of the radio astronomers' responses to these demands.

APPARATUS

The polarization accepted by a reflecting telescope is determined by the feed element at the focus. The most commonly used elements are linearly polarized devices — the dipole for use at frequencies lower than 1–2 GHz, the horn plus rectangular wave-guide at higher frequencies. A single linear feed receives only half the total power incident on the telescope but achieves rather good rejection of the unwanted (orthogonal) mode; values of 0.1 or 0.2% by voltage are typical. This figure is important in determining the instrumental polarization for which correction must be made.

If a pair of crossed linear detectors is used to receive all the incident power, greater crosstalk may be expected, on the order of 0.3% to 1% by voltage. At low frequencies, crossed linear feeds consist of perpendicular dipoles; at high frequencies, a common arrangement is for the feed horn to be attached to an oversized circular wave-guide (which will pass both polarizations) with exit ports of rectangular wave-guides to separate the two polarizations.

Circularly polarized feeds may be constructed by combining the orthogonal linear outputs in phase quadrature[1]

$$R = (X + jY)/\sqrt{2}$$
$$L = (X - jY)/\sqrt{2}.$$

The combination in coaxial systems may be made by means of either a coupler or a hybrid ring. There are several different possible arrangements in wave-guide that fulfill this function (which is analogous to that of the optical quarter-wave plate). The quarter-wave phase shift may be achieved by inserting screws or a dielectric or metallic fin, or even by mechanically squashing the circular guide into an elliptical shape. These devices are all alike in providing both circular modes simultaneously. It appears to be difficult to reduce the crosstalk for such devices to better than 1%–3% by voltage. Circularly polarized feeds can also be

---

[1] The deplorable confusion that at present surrounds the conventions of sign and terminology is admirably documented by Clarke (see p. 50). Radio astronomers, though guilty in the past of wavering as to the sign of V, are constrained by an Institute of Radio Engineers' definition to consider right-handed radiation as that in which the **E** vector rotates clockwise with time at a fixed point in space, looking from source to observer. Positive V corresponds to right-handed radiation as so defined. The X direction is in zero position angle, the Y direction is in position angle 90°, i.e., in the directions of increasing declination and right ascension, respectively. (This convention is adopted and used throughout this book. — ED.)

constructed from helical antennas, but an arrangement to provide right- and left-handedness simultaneously is not possible.

MODE OF USE

How may the apparatus described above be used to determine the complete state of polarization of the radiation under study, i.e., to find all four Stokes parameters, I, Q, U, V? It will be convenient to consider separately the ways of processing signals from a single telescope and those appropriate to an interferometer, in which there are two elements whose polarization can be adjusted at will.

Single Telescope

1. *Single Linear Feed*. Linear polarization can be detected by rotating the feed either continuously or in steps over a range of at least 180° in position angle. The instrumental correction is very low, but the method is not suitable if the source itself varies with time during the rotation cycle. The radio polarization of the Crab Nebula was detected in this way by Mayer, McCullough, and Sloanaker (1957).

2. *Crossed Linear Feed*. The outputs from the two feeds can solve the polarization completely, without feed rotation, if the power in each mode and the cross-correlation between them is determined. Let the feeds be labeled X and Y, and let the power channels have outputs labeled X^2 and Y^2. The correlated component consists of a phase and quadrature part, which may be labeled $Re\ (XY^*)$ and $Im\ (XY^*)$. This terminology follows the concept that each voltage (Ex and Ey) is complex, varying as e^{iwt}, and that the correlator measures the mean value of $Ex\ Ey\ \exp\ (i\phi x - i\phi y)$ where ϕ signifies a phase advance. The Stokes parameters are then

$$I = X^2 + Y^2$$
$$Q = X^2 - Y^2$$
$$U = 2\ Re(XY^*)$$
$$V = 2\ Im(XY^*).$$

A receiver to study solar radiation by this method has been used by Hatanaka, Suzuki, and Tsuchiya (1955).

If the whole system is rotated, then both the power difference and the correlator output vary sinusoidally with rotation and either may be used to deduce the linear polarization. For an example of the use of the correlated output see the Dutch measurements of the galactic background by Westerhout et al. (1962).

3. *Opposite Circular Feeds.* In an exactly similar way, if circularly polarized feeds are constructed, outputs may be obtained proportional to R^2, L^2, and the real and imaginary parts of RL^*. The Stokes parameters are given by

$$I = R^2 + L^2$$
$$V = R^2 - L^2$$
$$Q = 2\,Re(RL^*)$$
$$U = 2\,Im(RL^*).$$

Although the initial construction of the feeds presents more problems than the dual linear feed, the dual circular system has a practical advantage in that the two parameters (Q and U) that together denote the linear polarization of the source come from similar outputs, whose calibration and systematic effects are therefore more nearly equal. This system has been used at Jodrell Bank for measuring the polarization of pulsars by Lyne, Smith, and Graham (1971).

Interferometer

An interferometer is of great value in the study of extra-galactic sources because of its property of canceling partially or completely the polarized background radiation from the galaxy.

If, as before, we let letters such as X or R denote the polarization of the feed of one element of the interferometer, then the amplitude and phase of the interferometer fringes are obtained from the (complex) output of a correlator such as XX^*. Let X, Y, S, and T refer to linear polarization at phase angle $0°$, $90°$, $-45°$, and $+45°$, respectively. In terms of the Stokes parameters, the outputs are

$XX^* = \frac{1}{2}(I + Q)$		$I = XX^* + YY^*$
$YY^* = \frac{1}{2}(I - Q)$	so that	$Q = XX^* - YY^*$
$XY^* = \frac{1}{2}(U + iV)$		$U = XY^* + YX^*$
$YX^* = \frac{1}{2}(U - iV)$		$iV = XY^* - YX^*$

or, if rotated by $45°$,

$SS^* = \frac{1}{2}(I - U)$		$I = SS^* + TT^*$
$TT^* = \frac{1}{2}(I + U)$	so that	$U = TT^* - SS^*$
$ST^* = \frac{1}{2}(Q + iV)$		$Q = ST^* + TS^*$
$TS^* = \frac{1}{2}(Q - iV)$		$iV = ST^* - TS^*$

while circular feeds give

$$RR^* = \tfrac{1}{2}(I+V)$$
$$LL^* = \tfrac{1}{2}(I-V)$$
$$RL^* = \tfrac{1}{2}(Q+iU)$$
$$LR^* = \tfrac{1}{2}(Q-iU)$$

so that

$$I = RR^* + LL^*$$
$$V = RR^* - LL^*$$
$$Q = RL^* + LR^*$$
$$iU = RL^* - LR^*.$$

We may make the following comments on these equations. Four measurements are needed to determine the four unknowns $I, Q, U,$ and V. Since each interferometer observation measures two quantities, an amplitude and phase, each set of four observations listed above provides some redundancy. In practice this redundancy is valuable for discriminating against instrumental effects.

At least one observation with "parallel" feeds is needed (i.e., with both feed elements having the same polarization) in order to measure the total intensity, I. The linear polarization may be found either by observing with feeds that are rotated, remaining always parallel to each other, or by observing with orthogonal feeds. With orthogonal circular feeds, no mechanical rotation is necessary.

Circular polarization (V) can be measured with circularly polarized feeds, but usually it consists of the small difference between two large quantities. Alternatively, it may be measured using crossed linear feeds, in the form of a phase shift of the output.

Clearly, many different combinations are possible in principle. We quote four examples, the first of which uses only "parallel" combinations, i.e., with the feed elements always alike in the two elements of the interferometer. Two further examples determine polarization from "crossed" or orthogonal combinations, while the fourth example uses an intermediate arrangement, with the feeds neither parallel nor orthogonal.

1. Mitton (1971) measured the polarization of Cygnus A with the 1.6-km telescope at Cambridge with linearly polarized feeds, always parallel, rotated in steps of $45°$.

2. Morris, Radhakrishnan, and Seielstad (1964) at Owens Valley, California, measured the distribution of polarization of eight sources using linearly polarized feeds in configurations $XX, XY,$ and ST.

3. Conway et al. (1972) at Jodrell Bank have measured the linear polarization of extra-galactic sources using circularly polarized feeds, in configurations $RR, RL,$ and LR; and Gilbert and Conway (1970) have measured values of circular polarization using linearly polarized feeds in the XY and YX configurations.

4. A novel "mixed" method has been developed by Weiler at Westerbork, the Netherlands (unpublished), using combinations of linearly polarized feeds at 45° to each other.

$$XS^* = (I + Q - U - iV)/2\sqrt{2}$$
$$XT^* = (I + Q + U + iV)/2\sqrt{2}$$
$$YS^* = (-I + Q + U - iV)/2\sqrt{2}$$
$$YT^* = (I - Q + U - iV)/2\sqrt{2}.$$

The polarization parameters may be recovered by adding and subtracting these four outputs. No feed rotation is required during observations.

CRITIQUE

The variety of methods currently in use follows the idiosyncrasy and expertise of individual workers but shows that no one method is demonstrably superior to all others. An important distinction may be drawn between those methods that require mechanical rotation at the focus and those that do not. For example, rotation is essential with the "parallel linear" combination, optional with the "crossed linear" method, and not required with crossed circular feeds. Because of the size and weight of radio frequency equipment, mechanical rotation entails the provision of motors, which will be heavy, and read-outs, which must be accurate, at a remote focus, where space and weight will certainly be at a premium. In any case, a method requiring rotation is not possible for sources that vary rapidly, like pulsars.

A further important criterion is the magnitude of the spurious instrumental polarization (I.P.) to which all methods are liable and which must be calibrated and removed. A contribution to the I.P. may be expected from the very act of rotating the feed system (see Gardner and Davies 1966). A second contribution to the I.P. is due to the crosstalk in the feed from unwanted modes. From the figures given above, the lowest crosstalk occurs with single linear feeds, followed by dual orthogonal linear feeds, followed by dual circular feeds. Weiler's 45° method is noteworthy in that a crosstalk of 100% is deliberately maintained. An additional contribution to the I.P. arises in the case of large reflectors, especially if working near their shortest usable wavelength, from distortions of the reflecting surface. In particular, gravitational distortions can introduce astigmatism and, hence, an I.P. that varies with zenith angle.

These contributions must be removed by suitable calibration. The true figure of merit is not so much the magnitude of the I.P. as its repeatability and the degree to which it may be eliminated. Numerical estimates for this parameter are not always given, but it would appear that most of the systems discussed here are comparable in this respect. The most stringent requirement to date has been for the measurement of circular polarization of quasars, where the signals do not exceed 0.1% of the total intensity. In this experiment the use of crossed linear feeds was undoubtedly the optimum mode (Gilbert and Conway 1970).

For the measurement of polarization, there is clearly an extra advantage in the use of an alt-azimuth mounting over an equatorial mounting, as the change of parallactic angle with hour angle itself acts as a useful tool in the calibration of instrumental effects (Conway and Kronberg 1969).

REFERENCES

Conway, R. G.; Gilbert, J. A.; Kronberg, P. P.; and Strom, R. G. 1972. Polarization of radio sources at λ 49-cm and λ 73-cm. *Mon. Not. R. Astr. Soc.* 157: 443–459.

Conway, R. G., and Kronberg, P. P. 1969. Interferometric measurement of polarization distribution in radio sources. *Mon. Not. R. Astr. Soc.* 142: 11–32.

Gardner, F. F., and Davies, R. D. 1966. The polarization of radio sources. *Aust. J. Phys.* 19: 441–459.

Gilbert, J. A., and Conway, R. G. 1970. Circular polarization of quasars at λ 49 cm. *Nature* 227: 585–586.

Hatanaka, T.; Suzuki, S.; and Tsuchiya, A. 1955. Polarization of solar radio bursts at 200 Mc/s. I. A time sharing radio polarimeter. *Publ. Astro. Soc. Japan* 7: 114–120.

Lyne, A. G.; Smith, F. G.; and Graham, D. 1971. Characteristics of the radio pulses from the pulsars. *Mon. Not. R. Astr. Soc.* 153: 337–382.

Mayer, C. H.; McCullough, T. P.; and Sloanaker, R. M. 1957. Evidence for polarized radio radiation from the Crab Nebula. *Astrophys. J.* 126: 468–470.

Mitton, S. 1971. Observations of the distribution of polarized emission of Cygnus A at 6-cm wavelength. *Mon. Not. R. Astr. Soc.* 153: 133–143.

Morris, D.; Radhakrishnan, V.; and Seielstad, G. A. 1964. Preliminary measurements on the distribution of linear polarization over eight radio sources. *Astrophys. J.* 139: 560–569.

Westerhout, G.; Seeger, C. L.; Brouw, W. N.; and Tinbergen, J. 1962. Polarization of the galactic 75-cm radiation. *Bull. Astron. Netherlands* 16: 187–224.

RADAR POLARIMETRY

G. LEONARD TYLER
Center for Radar Astronomy
Stanford University

Radar astronomy experiments can completely define the scattering properties of a planetary surface in terms of a matrix relationship joining the incident and reflected Stokes vectors. Experimentally, it is necessary to determine all of the Stokes parameters at the receiver while transmitting two or more polarizations. Typical radar observations leave certain properties of the scattered waves undetermined. Two groups have determined the complete polarization states for limited observations. Further theoretical and experimental work is required before such observations can realize their full potential.

Radar astronomy observations of solar-system objects now include the moon (Evans and Hagfors 1968), all the terrestrial planets (Evans 1969; Pettengill 1968), and the asteroids Icarus (Goldstein 1969) and Toro (Goldstein[1]). Radar echoes from Jupiter may have been detected on one occasion (Goldstein 1964). Whatever advantages radar observations enjoy over studies of natural emissions in the same frequency range accrue from the ability of the experimenter to precisely control and measure the properties of the transmitted and echo waveforms. Frequency, time origin, modulation, and polarization may be chosen to optimize experimental sensitivity to particular attributes of the radar target. Here I shall only describe the use of polarization in radar astronomy experiments; general radar astronomy results will not be discussed. It is assumed that the reader is generally familiar with the principles of range, Doppler, and range-Doppler mapping (Green 1968). I will first review the two principal representations of polarization used in radar astronomy and then describe three main types of experiments that have been carried out.

Signals employed in radar astronomy are highly coherent in comparison with radio astronomy and optical counterparts. In optical terms, radar

---

[1] Personal communication, 1972.

astronomy signals would be classified as quasi-monochromatic. But, in absolute terms, the signal bandwidths are readily accommodated by current electronic signal processing and high-speed sampling technology. Thus, in contrast with optical experiments, the time variation of electric fields is observable. Consequently, radar data are often analyzed and manipulated directly in terms of their complex signal representations rather than as signal intensities.

Given such measurements of the electric fields, it is a simple matter to characterize the electromagnetic wave in terms of its second-order statistics. Let E_1 and E_2 represent time-varying quasi-monochromatic fields received on identical linear antennas oriented along orthogonal $\overline{a_1}$ and $\overline{a_2}$ directions. Then, the covariance matrix (the coherency matrix of Born and Wolf [1959]) is

$$\mathbf{J} = \begin{bmatrix} J_{11} & J_{12} \\ J_{21} & J_{22} \end{bmatrix} \text{ where } J_{ij} = <E_i E_j^*>, i, j = 1, 2 \qquad (1)$$

and $<>$ denote time or ensemble averaging, as appropriate.

The representation **J** is extremely useful for data processing because of the ease with which it can be manipulated. Once **J** has been formed, a matrix multiplication can be used to provide coordinate transformations or corrections for imperfections in the receiving antenna system (Tyler 1970). It can be shown that two important parameters of the matrix, total received power

$$J_{11} + J_{22} = \text{Trace } (\mathbf{J}) \qquad (2)$$

and fractional polarization

$$p = \sqrt{1 - \frac{4 \text{ Det } \mathbf{J}}{[\text{Trace } (\mathbf{J})]^2}}, \qquad (3)$$

are invariant under all sets of orthogonal polarizations. That is, if E_1 and E_2 represent signals received on any two mutually orthogonal polarizations from which **J** is formed, then the quantities Trace (**J**) and p as calculated above will not change. Further, if E'_1 and E'_2 represent signals from antennas that are not mutually orthogonal (but not the same) and **J**′ is formed, then a correction matrix **C** can be found in which

$$\mathbf{J} = \mathbf{C} \mathbf{J}' \qquad (4)$$

corresponds to the coherency matrix for orthogonal antennas. To determine **C**, the primed polarizations must be known. An example of the use of these techniques is given in the Appendix, where the formation of polarization spectra is discussed. Techniques similar to those given there can be derived using the range-Doppler mapping principle, but thus far these have not been carried out.

While the coherency matrix formulation is useful for data manipulation, the Stokes vector is better adopted for understanding scattering from planetary surfaces. The relationship between the two formulations is direct. For example, letting E_x and E_y represent linearly polarized field components along the x and y directions, respectively,

$$
\begin{aligned}
I &= (J_{xx} + J_{yy}) & J_{xx} &= (I - Q)/2 \\
Q &= (J_{xx} - J_{yy}) & J_{yy} &= (I + Q)/2 \\
U &= (J_{xy} - J_{yx}) & J_{xy} &= (U + iV)/2 \\
V &= i(J_{yx} - J_{xy}) & J_{yx} &= (U - iV)/2
\end{aligned}
\tag{5}
$$

where I, Q, U, and V are the components of the Stokes vector.

The effects of scattering upon an arbitrary quasi-monochromatic electromagnetic wave are described by a linear transformation as

$$\mathbf{S}' = \mathbf{M}\,\mathbf{S}, \tag{6}$$

where $\mathbf{S}$ and $\mathbf{S}'$ are Stokes vectors before and after scattering, and $\mathbf{M}$ is the Mueller matrix. For backscattering from a statistically isotropic surface with the x-axis oriented in the plane of incidence, it has been argued on the basis of symmetry that $\mathbf{M}$ reduces to

$$
\mathbf{M}^0 = \begin{bmatrix}
M^0_{11} & M^0_{12} & 0 & 0 \\
M^0_{12} & M^0_{22} & 0 & 0 \\
0 & 0 & M^0_{33} & 0 \\
0 & 0 & 0 & M^0_{44}
\end{bmatrix}. \tag{7}
$$

Thus, five numbers completely describe backscattering from a rough surface; similar results hold for forward scattering. Given the ability to control $\mathbf{S}$ (at the radar transmitter) and measure $\mathbf{S}'$ (at the receiver), the five non-zero M^0_{ij} can be determined by a sequence of experimental observations. The interested reader is referred to Hagfors (1967) for details of the arguments.

However, in spite of the simplicity of the observational relationships, the lack of a firm theoretical foundation connecting the scattering matrix $\mathbf{M}$ with specific surface models has caused most experimenters to ignore them. To our knowledge, only two groups (see below) have attempted to measure the complete polarization characteristics of waves scattered from the moon. Of these two, only one has been able to determine the five non-zero elements of $\mathbf{M}^0$. No such measurements have been carried out for the planets.

Typical monostatic (transmitter and receiver on the earth) radar astronomy experiments have employed circular polarization on transmission ($S = 1,0,0, \pm 1$) and reception. Normally, the receiver is always equipped for the circular polarization opposite to that transmitted. Since the sense of rotation of a circularly polarized wave is reversed upon backscatter from a smooth spherical surface, the signal received in this mode is denoted polarized. Following historical usage, the circularly polarized component with the same sense as that of the transmitted wave is denoted *de*polarized (Evans and Pettengill 1963). In cases in which linear polarization is employed, the polarized and depolarized components are those associated with the transmitted and orthogonal linear components, respectively. The basis of the terminology is the polarization state expected from a smooth planet and that which is orthogonal to it. Clearly this usage of the term *polarized* is distinct from its normal meaning in optics.

There is, however, considerable empirical justification for this procedure. Using the terminology just introduced, the polarized echo from radar astronomy targets is always of considerably greater strength than the depolarized one. Per unit area of planet surface at the subradar point, the power in the polarized return from terrestrial planets exceeds that of the depolarized by factors on the order of 100. Thus, particularly in early experiments, polarized echoes were used to maximize experimental signal-to-noise ratios. Polarized echoes are quite specular, however, and when mapped into planetary coordinates, give rise to a strong highlight at near-zero angles of incidence. Depolarized echoes follow scattering laws that vary as low-order powers of cos i, where i is the angle of incidence, and they do not display the specular highlight. Further, circular transmission and reception is technically simple to implement and is not affected by Faraday rotation. Consequently, this type of experiment has been used extensively. It is known that for low angles of incidence, i.e., in the region of the specular highlight, the polarized signal can be explained through the use of physical optics analyses of well-behaved, random rough surfaces. It is argued heuristically, albeit on firm physical grounds, that the depolarized signals arise from small-scale structure or from impedance discontinuities on or within the surface (Barrick and Peake 1967). Initially, some authors evidently equated the polarized echo with a quasi-specular scattering mechanism, while the depolarized echo was believed to be associated with a more diffuse process. However, some of the polarized echo, especially in regions remote from the specular highlight, must also be due to random small-scale surface structure. Also, it is now known that even very well-behaved dielectric interfaces produce some depolarization (Barrick 1965). Results from experiments of this type have been used in model-fitting programs in attempts to determine lunar and planetary *rms* slopes, slope distributions, and small-scale structure. For the lunar case, several authors (e.g., see Pollack and Whitehill 1972) have concluded that subsurface structure must play an important role in the production of the depolarized component. In radar mapping, a comparison of images constructed from polarized and depolarized return signals are of importance in assessing the relative abundance of small-scale structure in adjacent regions (Thompson et al. 1970*a,b*). Still, the relative apportionment of polarized and depolarized echoes among large- and small-scale surface structure has been and remains one of the principal problems of data interpretation in radar astronomy.

Two groups have reported more extensive measurements, but only for the moon. The first (Hagfors et al. 1965; Hagfors 1967), in a monostatic experiment conducted from the earth, utilized a controlled illumination and direct measurement of individual components in the Stokes vector to determine the non-zero elements of $\mathbf{M}^0$. For example, by transmitting either a right or left circularly polarized wave ($S = 1,0,0, \pm 1$) and measuring received power on both right and left circular polarization, one can determine M^0_{11} and M^0_{44} directly. Similarly, transmitting either circular polarization and receiving on orthogonal, properly oriented linear antennas yields M^0_{12}. The quantities M^0_{22} and M^0_{33} may be determined also. In the experiments carried out, each of the elements of $\mathbf{M}^0$ has been determined as a function of angle of incidence, i, either for a radar range ring (an annulus on the surface perpendicular to the radar line of sight), or for a range-Doppler area resolution element. Thus, the behavior with i represents an average over a considerable fraction of the lunar disk. Further, not all the M_{ij} are determined for the same area. These measurements are uncorrected for variations in topography. Again, a somewhat heuristic approach to data interpretation is required. One clear result, however, is the failure of the physical optics depolarization models for the lunar surface at angles of incidence greater than about 20°. Also, the data provide the basis for strong arguments that subsurface scattering plays an important role in lunar radar returns.

The second set of complete observations was conducted using entirely different methods. Transmissions from lunar-orbiting spacecraft were scattered by the lunar surface and observed on the earth (Tyler and Simpson 1970; Howard and Tyler 1971). The area of the lunar surface from which echoes are received is limited by the spacecraft horizon. In general, the scattering is strongly specular so that most of the signal energy comes from a small region about the point of specular reflection on the mean surface. In this case, the incident polarization is determined by the antenna characteristics of the space vehicle and its orientation with respect to the lunar surface. Thus, it was not possible to modify the incident polarization. However, by coherently receiving the echo signals on orthogonal polarizations, the complete coherency matrix of the echo signal was determined. Such observations are sufficient to determine four of the elements of M^0 in terms of the remaining one. This technique permits the determination of the polarized and *un*polarized parts in the optical sense. From Equations (2) and (3), the polarized part is $p(J_{11} + J_{22})$ and the unpolarized part is $(1-p)(J_{11} + J_{22})$. (In contrast, the degree of *de*polarization is simply J_{22}/J_{11}, where J_{11} is the expected component.) Since it represents the random component of the scattered wave, the unpolarized echo is associated with randomly oriented small-scale structure. Evidently, based on tests of quasi-specular and diffuse-scattering laws, this separation by polarization does correspond to these two basic types of scattering mechanisms. The polarized portion of the echo can be related quantitatively to large-scale lunar surface structures. Unfortunately, further theoretical guidance is required for a quantitative understanding of the mechanisms responsible for the unpolarized component.

At present, analysis of radar data can yield reliable information, depending on signal-to-noise ratio, regarding certain aspects of lunar and planetary surfaces. Root-mean-square slopes, slope-frequency distribution, surface reflectivity, and usually, surface density can be obtained quantitatively with some confidence. In general, such information is limited to a small region about either the subearth or specular point, depending on whether the experiment is monostatic or bistatic. In some cases average values over the disk may be obtained (Rea, Hetherington, and Mifflin 1964). Qualitative assessments of localized variations in small-scale structure are also possible from mapping experiments. The broad aspects of scattering from planetary surfaces are understood, especially in terms of large-scale surface structure. While certain refinements can be expected, it seems unlikely that further large advances will occur in this area. However, there is a broad range of models for scattering from small-scale structure. At present, we are unable to choose among such models on the basis of simple observations of polarized and depolarized echoes or the polarized and unpolarized parts. It seems likely that further progress will depend on an increased awareness among radar astronomers of the physics governing the generation and evolution of planetary surfaces, on more detailed calculations of the scattering proper-

APPENDIX

Computation of Polarization Spectra

Consider the block diagram in Fig. 1 which shows signals from orthogonal antennas as synchronously sampled. If d_p represents one of a long series of data samples, then those samples are grouped according to

$$d_j^n = d_p, \text{ where } p = nN + j, j \leq N$$

and n, N, j are all positive integers. The weighted complex Fourier coefficients are

$$f_k^n = \sum_{j=0}^{N-1} W(\frac{2\pi}{N} j) \; d_j^n e^{i(2\pi/N)jk}, \quad (8)$$

where $0 \leq j \leq N - 1$ and $i = \sqrt{-1}$.

Normally such coefficients are formed into spectral estimates such as

$$F_k' = \sum_{m=1}^{M} \sum_{l=1}^{L} | f_{Mk+m}^l |^2 \quad (9)$$

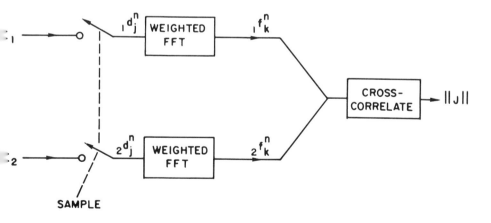

Fig. 1. Block diagram of method for forming polarization spectra.

where the indices are all integers. However, the complete description of the received signal may be obtained by forming the matrix

$$\mathbf{J}_k = \begin{bmatrix} \sum_{n=1}^{N} |{_1f_k^n}|^2 & \sum_{n=1}^{N} {_1f_k^n}{_2f_k^{n*}} \\ \sum_{n=1}^{N} {_1f_k^{n*}}{_2f_k^n} & \sum_{n=1}^{N} |{_2f_k^n}|^2 \end{bmatrix}. \quad (10)$$

The components of $\mathbf{J}_k$ provide four independent numbers ($J_{12} = J^*_{21}$) that completely describe the received signal at a frequency determined by the index k, the sampling rate, and the spectral window associated with the data weighting function $W(\cdot)$. Obviously many variations of the J matrix manipulations are possible. This example is one method used to obtain spectral decomposition of polarization of continuous wave radar data.

REFERENCES

Barrick, D. E. 1965. A more exact theory of backscattering from statistically rough surfaces. Ph.D. Thesis, Ohio State Univ., Columbus, Ohio.

Barrick, D. E., and Peake, W. H. 1967. Scattering from surfaces with different roughness scales. *Research Report BAT-197A-10-3*, Battelle Memorial Institute.

Born, M., and Wolf, E. 1959. *Principles of optics.* New York: Pergamon Press.

Evans, J. V. 1969. Radar studies of planetary surfaces. *Ann. Rev. Astron. and Astrophys.* 7: 201–248.

Evans, J. V., and Hagfors, T. 1968. Radar studies of the moon. *Radar astronomy* (J. V. Evans and T. Hagfors, eds.) Ch. 5. New York: McGraw-Hill.

Evans, J. V., and Pettengill, G. H. 1963. The scattering behavior of the moon at wavelengths of 3.6, 68 and 784 centimeters. *J. Geophys. Res.* 68: 423–447.

Goldstein, R. M. 1964. Radar observations of Jupiter. *Science* 144: 842–843.

———. 1969. Radar observations of Icarus. *Icarus* 10: 430–431.

Green, P. E. 1968. Radar measurements of target scattering properties. *Radar astronomy* (J. V. Evans and T. Hagfors, eds.) Ch. 1. New York: McGraw-Hill.

Hagfors, T. 1967. A study of the depolarization of lunar radar echoes. *Radio Sci.* 2: 445–465.

Hagfors, T.; Brockelman, R. A.; Danforth, H. H.; Hanson, L. B.; and Hyde, G. M. 1965. Tenuous surface layer on the moon: Evidence derived from radar observations. *Science* 150: 1153–1156.

Howard, H. T., and Tyler, G. L. 1971. Bistatic-radar investigations. *Apollo 14 preliminary science report.* NASA SP-272. Washington, D.C.: U.S. Government Printing Office.

Pettengill, G. H. 1968. Radar studies of the planets. *Radar astronomy* (J. V. Evans and T. Hagfors, eds.) Ch. 6. New York: McGraw-Hill.

Pollack, J. B., and Whitehill, L. 1972. A multiple scattering model of the diffuse component of lunar radar echoes. *J. Geophys. Res.* 77: 4289–4303.

Rea, D.; Hetherington, N.; and Mifflin, R. 1964. The analysis of radar echoes from the moon. *J. Geophys. Res.* 69: 5217–5223.

Thompson, T. W.; Masursky, H.; Shorthill, R. W.; Zisk, S. H.; and Tyler, G. L. 1970. A comparison of infrared, radar, and geologic mapping of lunar craters. Contrib. No. 16, Lunar Sciences Institute, Clear Lake, Texas.

Thompson, T. W.; Pollack, J. B.; Campbell, M. J.; and O'Leary, B. T. 1970. Radar maps of the moon at 70 cm and their interpretation. *Radio Sci.* 5: 253–262.

Tyler, G. L. 1970. Estimation of polarization with arbitrary antennas. *Stanford Electronics Laboratories Report SR 3610-1, Stanford University,* Stanford, California.

Tyler, G. L., and Simpson, R. A. 1970. Bistatic radar measurements of topographic variations in lunar surface slopes with Explorer 35. *Radio Sci.* 5: 263–271.

PART II
Surfaces and Molecules

POLARIZATION EFFECTS IN THE OBSERVATION OF ARTIFICIAL SATELLITES

KENNETH E. KISSELL
Wright-Patterson Air Force Base

Photometric observations of artificial satellites are being used to determine rotational periods and the dynamical motion of these objects. Frequently the spacecraft lightcurves exhibit specular glints that are produced by reflections from dielectric surfaces. It is shown that such reflections will be highly polarized. Measurements of such polarization by space targets are discussed, and a theoretical treatment of the polarization state of the optical glints resulting from a mixed diffuse and specular surface are compared with the observations. Agreement of theory and experiment is sufficient to suggest usage of photopolarimetry to study space environmental effects on materials.

The polarization of sunlight scattered by orbiting spacecraft can reach nearly 100% in some cases. This is due to the regular geometrical nature of these man-made objects and to the relatively smooth surfaces of coatings applied for thermal control. In some cases the scattering surfaces are nearly of optical quality, e.g., the cover slips of solar cell arrays. These latter surfaces produce brief glints of light that must be highly polarized, but because these events occur for only 10–50 msec, the measurement of such polarization is difficult. The observation of linearly polarized light from dielectric coatings on curved spacecraft surfaces is somewhat easier and is the subject of this paper. The agreement of these observations with the predicted polarization properties of the scattered sunlight is sufficient to suggest its use in determining changes in the physical properties of spacecraft surfaces brought about by exposure to the space environment. It was the possibility of inferring surface properties and dynamics of motion of the spacecraft that lead us to begin space-object photometry (Kissell 1969).

REFLECTION THEORY

The reflection of sunlight from spacecraft can be of several types—specular or diffuse from dielectric coatings, specular or diffuse from smooth or roughened metallic surfaces, or even multiple reflections from a combination of these processes. One can expect, at any given moment, that several of these processes from different regions of the object contribute to the light received by a distant observer. The reflection by metallic surfaces will result in an elliptically polarized component as a result of the absorption. Only linear polarization produced by dielectric surfaces and the effects of simultaneous diffuse scattering will be treated here.

A typical surface coating used for thermal control of a spacecraft consists of a pigment, generally a powdered oxide such as TiO_2 or ZnO, suspended in a binder of silicon resin. The surface layer normally has a luster, although it is not highly polished. This is due to the incomplete filling of the surface area by the pigment grains so that some fraction of surface is formed by transparent resin. In the scattering process, the incident sunlight first passes a vacuum/resin interface in these areas and then reaches the pigment where the diffuse reflection occurs (Fig. 1). If the spacecraft is observed from the direction of regular reflection, two scattering coefficients determine the brightness of the target: the Fresnel reflectivity, α_{pol}, and the diffuse albedo, α_{diff}. The incident unpolarized beam can be represented as two equal but orthogonal components of totally plane-polarized light, $E_{o\perp} + E_{o\|}$; similarly we can express the Fresnel reflectivity as

$$\alpha_{pol} = \alpha_\perp + \alpha_\| = \frac{\sin^2(i-r)}{\sin^2(i+r)} + \frac{\tan^2(i-r)}{\tan^2(i+r)}, \qquad (1)$$

where the two terms are recognized as Fresnel's equations. The argument r is determined from the refraction equation

$$r = \arcsin\left(\frac{\sin i}{m}\right). \qquad (2)$$

With values for the Fresnel reflectivity and the diffuse albedo, we can predict the polarization state of light received from a distant spacecraft, taking into account the percentage of the surface that is occupied by interstitial resin between the pigment grains. This percentage will range from a fraction of 1% for a nearly diffuse coating to nearly 100% for a polished, glossy surface.

SCATTERING BY A CYLINDRICAL SPACECRAFT

The brightness of a diffuse cylindrical spacecraft has been studied by several authors (Stiles 1956; Sussman 1958). We will add a thin layer of smooth dielectric material over the diffuse pigment, of index of refraction m, and compute the percentage polarization when a specular glint occurs. Kissell (1969) expresses the diffuse brightness in terms of δ_s, the angle of incidence of the sun onto the figure axis of the cylinder; δ_0, the angle of

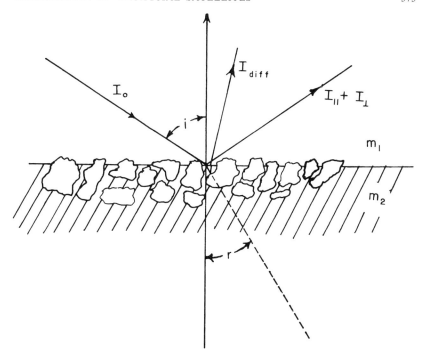

Fig. 1. Regular reflection from a semigloss resin-bonded pigment.

emergence toward the observer; and the dihedral angle ε between the planes through figure axis and the sun, and through figure axis and the observer. This yields an expression for the flux, E_{diff}, scattered to an observer at a distance R

$$E_{\text{diff}} = \frac{E_s h r \alpha_{\text{diff}}}{2\pi R^2} [(\pi - \varepsilon) \cos \varepsilon + \sin \varepsilon] \cos \delta_s \cos \delta_0, \qquad (3)$$

where h and r are the length and radius of the cylinder and E_s is the solar flux incident on the space object. This diffuse flux will consist of components of parallel and perpendicular polarization that should not differ by more than 10%.

The specularly scattered light will be observed to originate from a narrow strip along the cylinder, a strip which subtends $d\phi = 16$ arc-min at the cylinder axis (half the solar subtense). The flux at a distant observer can be calculated by representing this strip as a plane mirror of projected area $hrd\phi \cos i$ and brightness $E_s \alpha_{\text{pol}}$. The specular flux is then

$$E_{\text{pol}} = f \frac{E_s h r \alpha_{\text{pol}}}{\pi R^2 d\phi} \cos i, \qquad (4)$$

where f is the fraction of the surface that is specular. The flux scattered specularly is confined to the solid angle of the sun, $2\,d\phi$.

We can break Equations (3) and (4) into their component polarizations and obtain

$$E_\perp = \frac{1}{2} E_{\text{diff}} + E_{\perp\,\text{pol}} = \frac{E_s hr}{\pi R^2}\left(\frac{\alpha_{\text{diff}} G}{4} + \frac{\alpha_\perp f}{d\phi}\cos i\right) \tag{5}$$

$$E_\| = \frac{1}{2} E_{\text{diff}} + E_{\|\,\text{pol}} = \frac{E_s hr}{\pi R^2}\left(\frac{\alpha_{\text{diff}} G}{4} + \frac{\alpha_\| f}{d\phi}\cos i\right), \tag{6}$$

where $G = [(\pi - \varepsilon)\cos\varepsilon + \sin\varepsilon]\cos\delta_s \cos\delta_0$. The percentage polarization is the practical observable quantity, defined as

$$\frac{E_\perp - E_\|}{E_\perp + E_\|} = \frac{f(\alpha_\perp - \alpha_\|)\cos i}{0.5\alpha_{\text{diff}} G d\phi + f(\alpha_\perp + \alpha_\|)\cos i}. \tag{7}$$

From consideration of the geometry of single reflection from the side of the spacecraft, the perpendicular and parallel notation for the electric vector refer to the basic plane of incidence and reflection (the plane of scattering), which is also the plane of the phase angle.

APPARATUS

The basic equipment used to make photometric measurements at the Sulphur Grove Field Site of Wright-Patterson Air Force Base consists of a 61-cm Cassegrain telescope (Kissell 1969) that allows tracking of a satellite to within 1 arcmin. Light received from the target and the sky background in the field of view falls onto a photomultiplier after passing through a rotating analyzer. The output signal and a reference time from WWV and other essential data are recorded simultaneously on an FM tape recorder.

Photoelectric Photometer

The photomultiplier is an RCA 7029 tube selected for low dark current and high cathode sensitivity (S-17). This tube is especially suited for use under high ambient light levels. Its spectral response peaks at 0.5 μm. When used with a sheet of Polaroid HN-32 material, the spectral response is a band about 0.16 μm in width at the half-power points. The photomultiplier is energized by a Sweet-type (logarithmic) feedback power supply to maintain a constant output current from the photomultiplier by regulating the dynode voltage. This gives a nearly logarithmic response over a range of 10 to 15 stellar magnitudes and effectively prevents overloading of the photomultiplier by accidental overexposure.

Fig. 2. AbleStar Serial No. 5, preparatory to launch from Vandenberg Air Force Base, showing the conical flare at the base of the rocket.

Polarization Analyzer

The simple polarization analyzer consists of a plane sheet of HN-32 Polaroid rotated in the optical path between the Cassegrain secondary and the focal plane of the telescope. It produces a net DC-signal loss of approximately 2 stellar magnitudes, i.e., a factor of 6.3 reduction in available light at the multiplier. By rotating the analyzer at approximately 7 rps, a modulation frequency of 14 Hz was obtained, sufficient to modulate glint events of 1 to 4 secs duration.

EXPERIMENTAL DATA

The type of spacecraft selected for study of diffuse albedo and polarization effects was the AbleStar class of upper stage rocket, made by Aerojet-General, and coated almost uniformly with a slightly glossy white paint of TiO composition held in a silicon binder.

Figure 2 shows a typical AbleStar vehicle being prepared for launch. It consists of a cylinder-cone combination 2.9 m long and 0.7 m in radius. The cone half-angle is approximately 6°. These objects yield lightcurves as they tumble in orbit, of which Fig. 3 is typical. The ordinate is in units we define as "apparent visual stellar magnitude," i.e., the brightness of the spacecraft compared to a set of F5 to G8 standard stars measured with the same photocathode (but no filters of the *UBV* type are used)

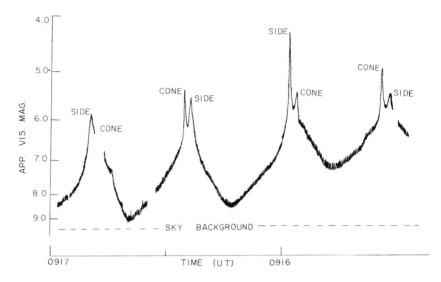

Fig. 3. Lightcurve of an AbleStar rocket, 1962 Beta Mu 2, on 25 April 1967. The distance to the rocket was about 1000 km.

through the same air mass. The AbleStar observed by photopolarimetry was 1964-63A, placed into a near-circular polar orbit (inclination of 89°, height of about 1050 km) on 28 July 1964. On 21 November 1966, a photometric polarization record was obtained which contained three sets of double or triple peaks, each peak of each set exhibiting measurable polarization. Figure 4 shows the modulation produced by the polarizer on two of the sets in comparison with observations, 36 hours later, without the polarizer.

PERCENTAGE POLARIZATION

The maximum and minimum brightness for each peak was measured and their corresponding stellar magnitudes obtained from standard-star calibration data for that evening. These values were then converted to obtain values for the maximum and minimum intensities. The percentage polarizations given in Table I were compiled from the data of Fig. 4 where the peaks are designated in accordance with the concept that they originate alternately with the cylindrical and the conical portions of the body. Corrections for 1% to 2% instrumental polarization have been applied to these data. Also given is the phase angle at the instant of observation. Some polarized light from the twilight crescent of the earth

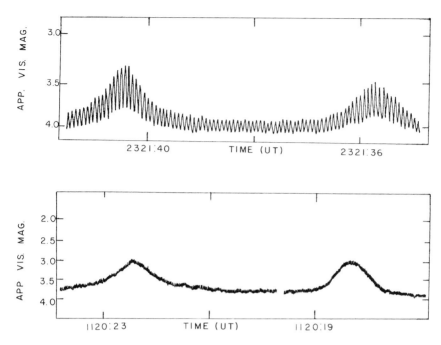

Fig. 4. Comparison of analyzer-modulated signal from AbleStar 1964-63A on 21 November 1966 with the unmodulated signal on 23 November 1966.

was incident on the object, but it should be negligible in these late-twilight observations.

DEGREE OF POLARIZATION TO BE EXPECTED FROM THEORY

If the simple model of a dielectric scattering surface is employed, one can use Equations (2), (3), and (7) to compute the percentage polarization for a body exhibiting no diffuse scattering, i.e., an idealized body where all of the refracted ray is absorbed without diffuse scattering. Such a model sets an upper limit for the polarization. This leads to the data of Fig. 5 where the degree of polarization (for bodies of different indices of refraction) is predicted to be high over a wide range of phase angles (phase angle = 2 × angle of incidence). Also shown are the data of Table I for the AbleStar rocket. Diffuse unpolarized light scattered from the target and partially depolarized refracted light caused a reduction in the observed polarization.

To predict the polarization to be expected from a semiglossy surface,

TABLE I

Peak Polarization Percentages

Set	Time (U.T.)	Peak	Percentage Polarization	Phase Angle (degree)
1	2321:36	A	28.5	88.2
	2321:40	B	39	87.7
2	2323:08	B	17	82.7
	2323:14	A	19	82.0
3	2323:35	A	18	80.2
	2323:40	B	13	78.7

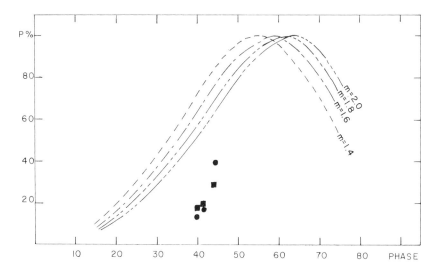

Fig. 5. Percentage polarization versus angle of incidence for space object 1964-63A compared with polarization to be expected from a simple dielectric surface of refractive index, m. Squares denote reflections from the conical skirt, circles from the body cylinder.

we must know not only the complex index of refraction but also the fraction, f, of the surface where pure resin is exposed. The latter was unfortunately not available at this writing. Taking the diffuse reflectivity as 30%, after 37 months in orbit (Kissell 1969; p. 65) and examining a range of values $0.05 < f < 0.005$ with an index of refraction $m = 1.5$, we

TABLE II

Percent Polarizations for a
Semiglossy Cylinder with
$\alpha_{\text{diff}} = 0.60, m = 1.50, i = 45°$

f	Percentage Polarization
0.005	28
0.010	36
0.020	50
0.050	66

arrive at the polarizations in Table II, computed for an angle of incidence of 45° and again ignoring the refracted light.

CONCLUSIONS

The agreement between the observed high polarizations and that to be expected on the basis of a simple theory of specular reflections is encouraging. This type of measurement can be carried out for other types of remote objects where the scattering geometry can be defined. Changes in the optical properties of spacecraft protective coatings should be measurable by choice of the orbital and observational parameters to maximize the sensitivity to the parameter under study.

Acknowledgments. The polarization analyzer used in these measurements was constructed by Richard P. Stead, USAF, who also performed the observations when he was a graduate student at the Air Force Institute of Technology (Stead 1967). I thank C. E. KenKnight for stimulating discussions and R. J. Rebillot for assistance with the figures.

REFERENCES

Kissell, K. E. 1969. Diagnosis of spacecraft surface properties and dynamical motions by optical photometry. *Space Research IX*, North Holland Publ. Co., pp. 53–75.

Stead, R. P. 1967. An investigation of polarization produced by space objects. Thesis GSP/PH/67-7, Air Force Institute of Technology, Wright-Patterson Air Force Base, Ohio.

Stiles, G. J. 1956. Prediction of apparent magnitude of distant missiles in sunlight. Memo. Rept. No. 1008, Aberdeen Proving Ground, Maryland, Ballistic Research Laboratories.

Sussman, M. 1958. Note on perfect Lambertian diffusers. *J. Opt. Soc. Am.* 48: 275–276.

DISCUSSION

COFFEEN: Are there any measurements for Echo I, the large balloon satellite that showed such large fluctuations in brightness as early as a few days after launch? It would be interesting if you could determine the balloon shape by a deconvolution of the intensity data, as in the asteroid problem.

KISSELL: There are no such measurements for Echo I. There are measures for Echo II and for Pageos which was very similar in surface properties and in brightness fluctuations to Echo I. We have attempted deconvolution of the radius of curvature (Vanderburgh and Kissell 1971), but with limited success. These vehicles show only small polarizations.

DISCUSSION REFERENCE

Vanderburgh, R. C., and Kissell, K. E. 1971. Measurements of deformation and spin dynamics of the Pageos balloon-satellite by photometry. *Planet. Space Sci.* 19: 223–231.

POLARIZATIONS OF ASTEROIDS AND SATELLITES

EDWARD BOWELL
Observatoire de Paris
and
BEN ZELLNER
University of Arizona

Polarization data for Mercury, the moon, asteroids, and satellites of Mars, Jupiter, and Saturn are compiled along with the relevant laboratory results. It is shown that the slope of the ascending branch of the polarization-phase curve is a discriminant of the bulk albedo only, while the maximum polarization and the depth of the negative branch are sensitive also to the microscale roughness and opacity. Recently published data are used for new calibrations of the laws relating these polarimetric parameters to geometric albedo. It is demonstrated that the albedos of asteroids and small satellites derived from their polarization properties are in good agreement with infrared diameters, and that a microscopically intricate surface is characteristic of atmosphereless solar-system bodies.

One of the triumphs of astronomical polarimetry was the demonstration that the moon's surface is covered with small highly absorbing particles having physical characteristics similar to terrestrial volcanic ashes. This conclusion was arrived at by Lyot (1929) after extensive lunar observations and comparative laboratory measurements on a wide variety of terrestrial rocks and minerals. Lyot's work has set the pattern for subsequent research: ground-based telescopic polarization measurements on solar-system objects are interpreted in the light of comparable laboratory measurements on terrestrial, meteoritic, and lunar samples. The recent acquisition of Apollo and Luna samples from the moon has, as well as fully upheld Lyot's qualitative conclusions, greatly sharpened the physical interpretation of the surfaces of other atmosphereless bodies in the solar system by providing a catalogue of polarimetric data for samples that have been exposed to a space environment over very long periods of time.

We have three aims in this paper: (1) to assemble accurate polarization data on atmosphereless bodies in the solar system; (2) to analyze these data in what we consider to be the most useful way, making liberal use of laboratory results; and (3) to indicate what future lines of research might be most profitable. Our compilation of astronomical data has generally been limited to polarization measurements made with instruments having a precision of ± 0.001 or better. We include Mercury, the moon, asteroids, and the satellites of Mars, Jupiter, and Saturn, but not Titan or Mars itself. The surface of Mars is often obscured by atmospheric veils with diverse polarimetric properties (e.g., Dollfus 1961), and its optical characteristics may be controlled by weathering effects (Dollfus, Focas, and Bowell 1969). The polarimetric properties of Titan are evidently the result of an optically thick atmosphere (Veverka 1970, 1973a; Zellner 1973).

PHOTOPOLARIMETRIC QUANTITIES

Figure 1 defines the parameters associated with a polarization-phase curve. The phase angle α is measured at the scattering surface, between the source and the observer. The sign of polarization is taken to be positive or negative according as the predominant electric-vector is perpendicular to or contained in the plane of scattering. Since the time of Lyot, it has been abundantly demonstrated that other azimuths do not occur for asteroids and satellites. P_{min} is treated as an algebraically positive quantity. In principle the slope h should be measured near the inversion angle $\alpha(0)$ (Widorn 1967); all well-observed astronomical objects, however, have shown very nearly linear ascending branches. Throughout, polarizations will be expressed as the fraction of polarized light, and phase angles will be given in degrees.

The readily observable brightness parameter for a reflecting object is its *geometric albedo A*, given by the ratio of its integrated apparent brightness to that of a flat, normally illuminated Lambert screen at the same position and having the same projected area. The quantity A is thus uniquely defined for a body or surface element of any shape, orientation, scattering properties, or condition of illumination. It is determined by the refractive power, opacity, and microscale structure of the surface in a way not yet completely clear. For solar-system objects at zero phase angle, the definition reduces to

$$\log A = -0.4 [M(1,0) - M_\odot] - 2 \log r,$$

where r is the mean projected radius of the object, $M(1,0)$ is its apparent magnitude at unit heliocentric and geocentric distances and at zero phase

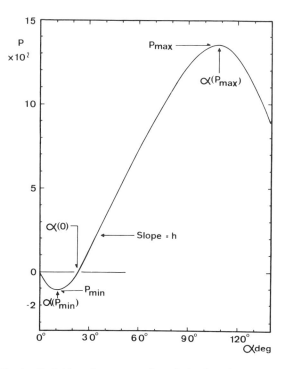

Fig. 1. Definition of parameters for polarization-phase curves.

angle, and $M_\odot$ is the apparent magnitude of the sun at unit distance. With the values $V = -26.77$ and $B - V = +0.63$ for the sun (Gehrels et al. 1964), we obtain

$$\log A_V = 5.642 - 0.4\, V(1,0) - 2 \log r,$$
$$\log A_B = 5.894 - 0.4\, B(1,0) - 2 \log r,$$

where r is the radius in kilometers. It has been recommended (Gehrels 1970) that the photometric phase function $m(\alpha)$ be linearly extrapolated to zero phase without taking into account the *opposition effect* or surge in brightness at small phase angles. The absolute magnitude $M(1,0)$ may then be determined from only two brightness measurements. Also, for practical purposes the astronomical albedos are then directly comparable to laboratory measurements, which are usually made by comparison with an MgO-smoked plate at some fixed phase angle near $3°$ and may include part but not all of the opposition surge (KenKnight, Rosenberg, and Wehner 1967; Dollfus, Bowell, and Titulaer 1971). With this definition, however, the commonly expressed equivalence between the geometric albedo and the ratio of the Bond albedo to the phase integral must be used with caution.

PHYSICAL FACTORS CONTROLLING POLARIZATION

An inverse correlation between albedo and degree of polarization is commonly found in polarimetry. For surface scattering, this so-called "Umov effect" may be understood phenomenologically by an analysis similar to those of Gehrels (1970) and of Dollfus and Titulaer (1971). We may write

$$P = \frac{I_{\text{pol}}}{I_{\text{pol}} + I_{\text{unpol}}} = \frac{I_{\text{pol}}}{kA}, \qquad (1)$$

where k is a constant defining the intensity scale. Now for surfaces of similar structure, the albedo is determined by *internal* processes, whereas the polarization is primarily produced by *surface* effects which should, to the first order, be independent of such internal processes. Thus, we may take the *quantity of polarized light* I_{pol} to be constant, with the result

$$\frac{dP}{P} = -\frac{dA}{A},$$

or

$$\log A = -\log P + \text{const.} \qquad (2)$$

More generally we may write

$$\log A = -c_1 \log P + c_2; \qquad (3)$$

insofar as increased internal transmission for high-albedo materials tends to depolarize, we may expect values of c_1 somewhat less than unity. Dollfus and Bowell (1971) found $\log A = -0.724 \log P_{\max} - 1.81$ for lunar regions measured in orange light. Widorn (1967) and KenKnight, Rosenberg, and Wehner (1967) independently noted that the polarization slope h follows a law of this form for astronomical objects and particulate laboratory samples, with $c_1 \approx 1.3$.

Thus, a straight line in a log P, log A plane for materials that have the same surface structure is a natural consequence of the way in which the degree of polarization is defined. *Deviations* from such a law are likely to be the prime indicators of surface structure. For a qualitative physical explanation of the effects of opacity and microstructure on the polarization-phase curve, we refer to Dollfus (1956). The polarization of smooth surfaces is completely described by Fresnel's laws. It is always positive and total at the Brewster angle for opaque dielectric materials. For translucent smooth surfaces, the polarization is diluted by internal scattering and reduced by the negative polarization component due to the refraction of penetrating and emergent beams. In the case of a microscopically intricate or particulate surface, however, the strong polarization peak tends to disappear owing to dilution by surface scattering.

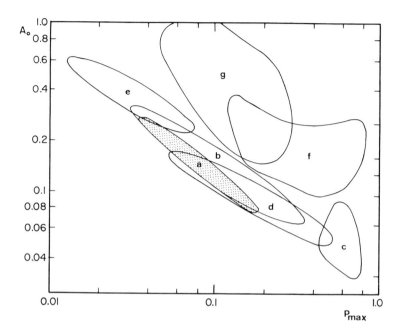

Fig. 2. Geometric albedo versus P_{max} (adapted from Dollfus and Titulaer 1971). Domains characterize telescopically observed lunar regions and various pulverized terrestrial substances: (a) lunar regions; (b) basaltic lavas; (c) vitric basalts; (d) volcanic ashes; (e) ignimbrites; (f) granites and gabbros; (g) sands, clays, and chalks.

This effect is demonstrated for laboratory samples in Fig. 2; a lower limit of P_{max}, which correlates with albedo according to Equation (3), is reached (presumably) when the roughness scale approaches the wavelength of light. Since the slope h is measured at small phase angles away from the Fresnel peak, it should be relatively insensitive to surface structure.

Surface roughness also tends to remove the azimuthal dependence of brightness and polarization. For a rough, relatively opaque surface such as that of the moon, the polarization is almost independent of the macro-surface orientation, varying only with phase angle. This property does not extend to vitreous or icy surfaces; however, the symmetry of spherical astronomical bodies still requires that the polarization position angle be either perpendicular or parallel to the plane of scattering.

The negative branch of polarization at small phase angles is not predicted by Fresnel's equations, and no theoretical understanding of its origin as yet exists. The conditions necessary for its formation, however, can be demonstrated by two experiments (Dollfus 1956; Bowell, Dollfus, and Geake 1972): (1) Carbon particles ascending from a smoky flame

TABLE I
Polarization Parameters for Mercury and the Moon

Object	λ (μm)	A	Negative Branch $P_{min} \times 10^3$	$\alpha(0)$	Positive Branch Slope $h \times 10^3$	$P_{max} \times 10^2$	Source[a] Albedo	Polarization
Mercury	0.36	0.07:			2.08 ± 0.05	12.3	1	3,4,5
	0.44	0.085			1.72 0.05	9.9		3,4,5
	0.55	0.104	14 ± 1	$25° \pm 2°$	1.47 0.02	8.2		3,4,5,6
	0.85	0.14:				6.4		4
Moon	0.36	0.072	10 ± 0.5	21.3 ± 0.5	2.30 0.07	15.4	2	7,8,9
	0.44	0.094	11.2 0.5	22.5 0.5	1.75 0.05	11.7		7,8,9
	0.55	0.117	12 0.5	23.6 0.5	1.48 0.05	8.6		6,7,8,9
	0.85	0.17:	12 0.5	25.0 0.5	1.22 0.05	5.8		7,8

[a] 1. The visual albedo of Mercury is from $V(1,0) = -0.36$ according to Harris (1961), diameter 4850 km from Dollfus (1970), and $V_\odot = -26.77$. The spectral dependence is taken from Irvine et al. (1968) and from McCord and Adams (1972), but corrected for reddening with phase in the blue and ultraviolet via coefficients given by Gehrels, Coffeen, and Owings (1964) for the moon. 2. The listed albedos for the whole moon are interpolated from compilations by Harris (1961) by Gehrels, Coffeen, and Owings (1964), corrected for 0ᵐ56 of opposition effect, and by Lane and Irvine (1973). 3. From observations reported by Dollfus at COSPAR, Madrid, 1972. 4. Coyne and Gehrels (in draft). 5. P_{max} derived from Ingersoll (1971). 6. Lyot (1929). 7. Gehrels, Coffeen, and Owings (1964). 8. Dollfus and Bowell (1971). 9. Coyne and Pellicori (1970).

polarize approximately in accordance with the molecular scattering law $P = \sin^2\alpha/(1 + \cos^2\alpha)$. No negative branch occurs. Deposited on a plate, such particles exhibit a very pronounced negative branch with $P_{\min} = 0.019$. Experiments using other materials confirm this finding provided that the particles are small (< 1 mm). (2) If blue $CuSO_4$ crystals are observed in red light, a well-developed negative branch is measured, but when illuminated with white light and observed through a cell containing $CuSO_4$ solution, the negative branch is absent. $CuSO_4$ is highly opaque in the red and partially transparent in the blue. Thus, the negative branch results from scattering by small opaque particles in mutual contact. As the particle opacity is decreased, the albedo increases and the negative branch becomes weaker.

OBSERVATIONAL RESULTS

Mercury and the Moon

For Mercury we have examined four polarization data sources. These are listed in Table I together with smoothed and averaged values of A, h, and $P_{\max}$ at the same four wavelengths used for the moon. All of the polarimetric quantities vary less than lunar ones over the wavelength range considered, in accordance with the more nearly neutral colors of Mercury.

Whole-disk polarization-phase curves for the moon have been presented by Lyot (1929) and by Coyne and Pellicori (1970). Both showed that waxing and waning phases have different polarizations due to the asymmetric distribution of maria and highlands. Polarization-phase curves for regions as small as 5 arcsecs diameter have been obtained by Gehrels, Coffeen, and Owings (1964), Wilhelms and Trask (1965), Dollfus and Bowell (1971), and Pellicori (1971). These data comprise measurements on 72 diverse regions in colors spanning the wavelength range 0.327 μm through 1.05 μm. The results may be summarized as follows:
1. Regions of low albedo have high h and $P_{\max}$, and vice versa.
2. $P_{\max}$ decreases monotonically from the ultraviolet to the infrared.
3. There is little dependence of P on the orientation of the macrosurface.
4. The phase angle of maximum polarization $\alpha(P_{\max})$ increases with $P_{\max}$.
5. The negative branch of polarization is almost independent of morphology and wavelength.
6. The inversion angle $\alpha(0)$ increases slightly with wavelength.

Specific experiments and tests have been devised to verify and enlarge some of these results: Lyot (1929) observed the moon at inversion angles as found for highland regions and found the maria to possess slightly positive polarization; Dollfus (1955) inferred the depolarizing ability of the lunar surface from earthshine polarization; P_{max} was measured for a large number of regions by Dollfus and Bowell (1971); Bowell, Dollfus, and Geake (1972) observed many lunar regions near $\alpha(P_{min})$ and detected no significant variations; polarization and brightness have been determined simultaneously by Riley and Hall (1972) using a scanning photopolarimeter.

Table I contains a compilation of A, P_{min}, $\alpha(0)$, h, and P_{max} for the whole earth-side disk of the moon. Figure 3 is a composite plot of P_{max}, $\alpha(0)$, and P_{min} versus wavelength, and shows the empirically fitted curves used to derive some of the data in Table I. Agreement between the results of the four independent sets of polarization measurements is quite good; a variation of P_{min} with wavelength of amplitude only 0.002 is indicated.

Laboratory polarization measurements on terrestrial and returned lunar samples have been used to draw the following conclusions:

1. On scales greater than about 100 km² the moon's entire surface is polarimetrically particulate (Lyot 1929; Dollfus 1956). Lunar rocks and breccias and consolidated terrestrial rocks polarize much more strongly than the telescopically observed moon, and it is clear that *exposed* consolidated surfaces are nowhere very abundant on the lunar surface (Bowell, Dollfus, and Geake 1972). However, a thin coating of fines material suffices to reduce the polarization of consolidated surfaces to observed lunar values (Wilson 1968). Returned lunar fines exhibit polarizations identical in all respects to those of same regions on the telescopically observed moon (Geake et al. 1970).

2. A mean lunar surface particle size of about 20 µm has been derived by KenKnight, Rosenberg, and Wehner (1967), Pellicori (1969), and Bowell (1971). Note, however, that polarization is not a good size discriminator for particles smaller than about 50 µm.

3. Among terrestrial samples, lunar polarization is best matched by basalts and volcanic ashes (see Fig. 2; also Lyot 1929; Dollfus 1961). However, polarization cannot readily be applied to the determination of lunar surface composition.

Asteroids and Planetary Satellites

Tables II and III give the available polarimetric parameters for asteroids and satellites.[1] Asteroids cannot ordinarily be observed at phase

---

[1] The tabular data have been updated to reflect improved results through April 1973; no corresponding changes, however, have been made in the figures.

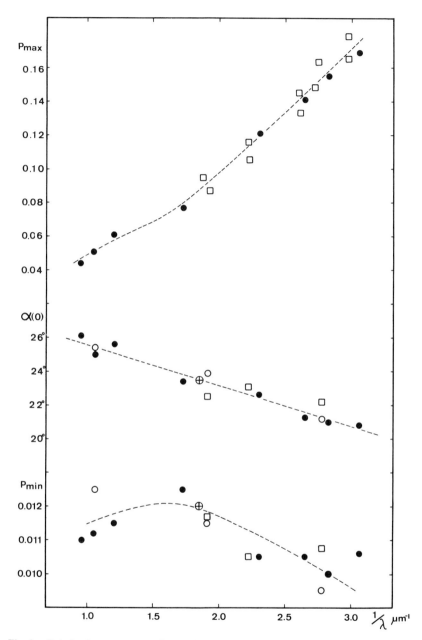

Fig. 3. Polarization parameters P_{max}, $\alpha(0)$, and P_{min} versus wavenumber per micron for the whole moon. (◯) data by Gehrels, Coffeen, and Owings (1964); (□) Coyne and Pellicori (1970), average of leading and trailing hemispheres for two lunations; (●) Dollfus and Bowell (1971); (⊕) Lyot (1929).

TABLE II
Polarization Properties of Asteroids

Object		λ (μm)	Negative Branch		Positive Branch[a]		Source[b]
			$P_{min} \times 10^3$	α(0)	Slope $h \times 10^3$	N	
1	Ceres	0.44	17.0 ± 1.0	18°.0 ± 0°.5	2.55 ± 0.08	9	1
2	Pallas	0.44	13.5 ± 1.0	19.0 ± 0.5	2.25 ± 0.12	4	1
3	Juno	0.44	7.8 ± 0.5	20.0 ± 0.5	0.97 ± 0.04	8	1
4	Vesta	0.55	5.5 ± 0.5	21.5 ± 0.5	0.76 ± 0.07	13	1,2
6	Hebe	0.55	9				1
8	Flora	0.5	6	19	1.20 ± 0.05	5	1,3
9	Metis	0.5	8	21	1.0		4
15	Eunomia	0.5	7	21	1.3		4
20	Massalia	0.5	7	19			4
89	Julia	0.5	10	22	1.5		4
324	Bamberga	0.44	14.0 ± 1.0				1
433	Eros	0.44	7.2 ± 0.5	24:	1.05 ± 0.18	4	1
1566	Icarus	0.44			1.1	4	5
1620	Geographos	0.44			0.66	4	1
		0.55					1
1685	Toro	0.44		23:	1.35 ± 0.06	6	1

[a] Generally the slope h and the inversion angle α(0) were found by linear least-squares fitting to the measured polarizations at phase angles near and above α(0). N is the number of data points used.

[b] 1. Unpublished observations by Dunlap, Gehrels, and Zellner. 2. Veverka (1971b). 3. Veverka (1971c). 4. Veverka (1970, 1973b). 5. Gehrels et al. (1970).

TABLE III
Polarization Properties of Satellites

Object	λ (μm)	A	$P_{\min} \times 10^3$	$\alpha(0)$	Slope $h \times 10^3$	Source[a] Albedo	Source[a] Polarization
Deimos	0.44	0.07	14 ± 2	19°.5	2.55 ± 0.12	1	5
Io	0.55	0.82	1.8 ± 0.5			2,3	6,7,8
Europa	0.55	0.75	1.0 ± 0.5	6.5:	0.27:	2,3	6,7,8
Ganymede	0.55	0.42	2.8 ± 0.5	10:	0.44:	2,3	6,7,8
Callisto W	0.55	0.22	6 ± 1			2,3	6,7,8
Callisto E	0.55	0.21	9 ± 1			2,3	6,7,8
Dione	0.52		4 ± 1				8
Rhea	0.52	0.57	4 ± 1			4	8
Iapetus W	0.52	0.28	2 ± 1			4	9
Iapetus E	0.52	0.04	13 ± 1			4	9

[a] 1. Pollack et al. (1972); Pascu (1973). 2. From diameters by Dollfus (1970). 3. From V magnitudes by Harris (1961). 4. From infrared diameters by Murphy, Cruikshank, and Morrison (1972). 5. Zellner (1972a). 6. Dollfus (1971a). 7. Veverka (1971a). 8. Zellner (unpublished). 9. Zellner (1972b).

angles larger than about 25°. For the satellites of Jupiter and Saturn, this limit shrinks to 12° and 6° respectively, so that the positive branch can be observed only from spacecraft. In many cases the listed parameters are provisional, being based on incomplete data; measurements from Harvard (Veverka), Meudon (Dollfus), and the University of Arizona (Gehrels and Zellner), however, were generally found to agree within ±0.001. The photographic polarimetry by Lyot (1934) and the early photoelectric work of Provin (1955) first demonstrated the presence of a negative branch for asteroids, but these measurements were not of a precision compatible with more recent results.

Harris (1961, p. 292) pointed out that the leading hemisphere of Callisto, seen at eastern elongations, is photometrically rougher than the trailing hemisphere, with little difference in albedo. Analogous variations in the depth of the negative branch have been found by Veverka (1971a) and Zellner (unpublished) but are not confirmed in data by Dollfus (1971a, and unpublished). For Iapetus, a much deeper negative branch for the dark leading hemisphere is explained primarily as an albedo effect (Zellner 1972b; see also Murphy, Cruikshank, and Morrison 1972; and Fig. 8). Further examples of *rotational* variations of polarization will doubtless be found when data of sufficient precision and completeness have been acquired (see Gehrels 1970, p. 359). For well-observed asteroids, however, polarization curves tend to be remarkably smooth and reproducible (±0.0005) with no identified rotational effects (Veverka 1971d). *Wavelength dependence* of polarization has been noted for a few Mars-crossing asteroids which can occasionally be observed at large phase angles (e.g., Gehrels et al. 1970); the polarization and albedo as functions of wavelength are found to vary inversely.

The apparent opposition diameters of the largest asteroids and satellites of Saturn fall in the range 0.1 to 0.5 arcsec and, hence, are very difficult to measure directly. Visual diameter measurements, principally by Barnard (1902) and by Kuiper (1955; see compilations by Dollfus 1970, 1971b), have recently been complemented by the method of infrared diameters, in which the observed thermal radiation in the spectral region 8–25 μm is modeled as a function of the diameter, the Bond albedo, and the infrared emissivity. As can be seen from Table IV, the infrared results have given substantially larger diameters, and therefore lower albedos, than the visual work (see, for a debate of the discrepancies, Dollfus 1971b, p. 30). The polarization slope h as an indicator of albedo has been exploited for indirect diameter determinations (e.g., Gehrels et al. 1970; Veverka 1971b; Zellner 1972a) and is considered at length in the following section.

TABLE IV
Geometric Albedos of Asteroids

Object		λ (μm)	Visual Results		Infrared Results	
			Albedo	Source[a]	Albedo	Source[b]
1	Ceres	0.44	0.12	1	0.07	1,2,3
2	Pallas	0.44	0.11	1	0.09	3
		0.44	0.03	2		
3	Juno	0.44	0.22:	1	0.15	1,3
4	Vesta	0.55	0.41	1,3	0.21	1,2,3
15	Eunomia				0.15	3
324	Bamberga	0.55			0.02	2

[a] 1. Barnard (1902). 2. Dollfus (1971b). 3. Compilation by Dollfus (1970).
[b] 1. Allen (1970,1971). 2. Matson (1971a, b). 3. Cruikshank and Morrison (1973).

RECALIBRATION OF POLARIZATION-ALBEDO RELATIONSHIPS

With the abundance of new laboratory and astronomical data made available in the past two years, new calibrations of the relationships between geometric albedo and the polarization parameters P_{max}, h, and P_{min} are in order. Figure 4 shows A and P_{max} values for 146 lunar

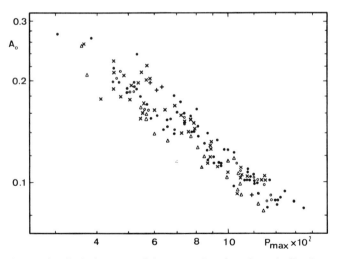

Fig. 4. Geometric albedo in orange light versus P_{max} for telescopically observed lunar regions (adapted from Dollfus and Bowell 1971). (+) Gehrels, Coffeen, and Owings (1964), region diameter < 10 arcsec; (Δ) Wilhelms and Trask (1965), 15 arcsec; (○) Dollfus, 65 arcsec; (×) Dollfus, <5 arcsec; (●) Bowell, <5 arcsec.

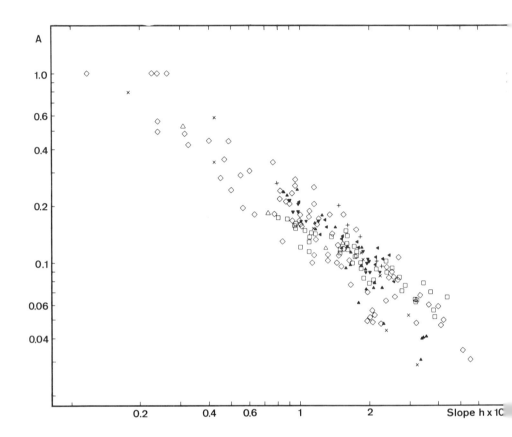

Fig. 5. Geometric albedo versus slope h of the ascending branch of the polarization-phase curve, for lunar regions and laboratory samples: (▲) lunar regions in three wavelengths, from observations by Gehrels, Coffeen, and Owings (1964), analyzed by Ken-Knight, Rosenberg, and Wehner (1967); (▼) lunar regions in orange light, from observations by Dollfus and Bowell (1971), with albedo scale corrected according to Dollfus, Geake, and Titulaer (1971); (▶) lunar regions in orange light, from Wilhelms and Trask (1965), with similarly corrected albedos from Dollfus and Bowell (1971); (□) lunar fines, (+) lunar rocks, and (Δ) lunar breccias from Apollos 11, 12, 14, and Luna 16; (×) terrestrial rocks; and (◇) terrestrial powders, including dry snows, and crushed meteorites. Sources for the laboratory data include Lyot (1929); Ken Knight, Rosenberg, and Wehner (1967); Geake et al. (1970); Dollfus, Bowell, and Titulaer (1971); Dollfus and Titulaer (1971); Dollfus, Geake, and Titulaer (1971); Bowell, Dollfus, and Geake (1972); and unpublished measurements at Meudon.

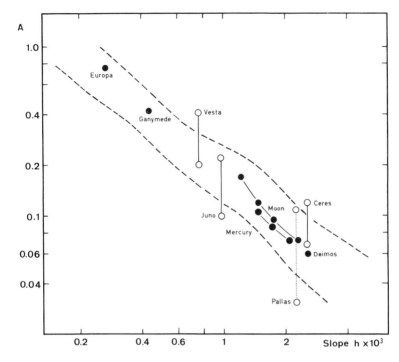

Fig. 6. Geometric albedo versus slope h of the ascending polarization branch for solar-system bodies. All data points in Fig. 5 lie between the dashed lines. Filled circles represent Europa and Ganymede in visual light; the whole moon at 0.85 μm; the moon and Mercury at 0.55 μm, 0.44 μm, and 0.36 μm; and Deimos in blue light. Crosses represent Mercury in the same four wavelengths as the moon. Solid vertical lines connect results from visual (top) and infrared (bottom) diameter determinations for the asteroids Vesta, Juno, and Ceres, and a dotted line connects the two visual albedo determinations for Pallas. Sources for the data are given in the tables.

regions of diverse stratigraphy. An analysis of the probable errors in A and P_{max} for 67 of these regions by Dollfus and Bowell (1971) showed that the finite width of the domain defined by these observations may be attributed to observational error. However, although points for all returned lunar fines so far measured fall within this domain (Bowell, Dollfus, and Geake 1972), inherent scatter about a straight line is unquestionably present. From the combined data of Fig. 4 along with measurements on 10 samples of lunar fines, we find $\log A = -0.71 \log P_{max} - 1.65$.

Figure 5 displays the slope-albedo law for all surfaces for which we found appropriate data: telescopically observed lunar regions, returned lunar rocks and fines, powdered meteorites, and a wide variety of terrestrial samples. Measurements at wavelengths from the ultraviolet to the infrared are included. Solar-system bodies are added in Fig. 6. The

TABLE V
Recalibration of the Slope-Albedo Law $\log A = -c_1 \log h + c_2$

Material	Least-Squares Regression		c_2 if $c_1 = 1$ assumed
	c_1	c_2	
88 terrestrial samples	1.04	−3.90	−3.772
90 lunar rocks and powders [a]	0.96	−3.66	−3.773
52 lunar regions	1.23	−4.44	−3.777
5 astronomical objects [b]	1.10	−4.07	−3.768
Straight average	1.08	−4.02	−3.772
	±0.10		±0.003

[a] Including 35 new data points, not plotted in Fig. 5, from unpublished measurements of Apollo 15 fines.
[b] Europa, Ganymede, Deimos, the whole moon in U, B, and V wavelengths, and Mercury in V.

agreement between laboratory and astronomical data is impressive, both with regard to the overall linearity and to the lack of wavelength sensitivity. In view of the mixed absolute precision of the data and the difficulties of reading slopes from published curves, it seems unlikely that the mean intrinsic scatter from a straight line exceeds ±0.1 in the logarithm. In the controlled experiments of KenKnight, Rosenberg, and Wehner (1967), powders of different grain sizes were found to move along parallel but distinct loci in the $\log A$, $\log h$ plane when artificially darkened by proton irradiation. Such a particle size effect is not apparent from the more heterogeneous data of Fig. 5; in particular, no systematic difference between powders and solid rocks is evident. Thus for practical purposes *we take the polarization slope h to be a discriminant of albedo only.* The dimensionless parameters c_1 and c_2 describing this law are rederived in Table V, with the assumption that roughly equal errors occur in the albedo and polarization variables. Veverka and Noland (1973) have carried out a similar analysis using an almost identical body of data but differently weighted, with the result $c_1 = 1.12$ and $c_2 = -4.15$ in our notation. Since our computed values of c_1 are consistent with the unit gradient of Equation (2), we adopt

$$\log A = -\log h - 3.77. \tag{4}$$

For Europa, Ganymede, Deimos, Mercury, and the moon in three wavelengths, the rms deviation from this line is only 0.04 in $\log A$.

Slopes and albedos for the four major asteroids are plotted in Fig. 6 using albedos from both the visual and the infrared diameter determinations. The visual diameter of Vesta and the lower of the visually derived albedos for Pallas are rejected as being inconsistent with all laboratory data. With somewhat less confidence we may similarly reject the visual diameter of Ceres. Asteroid diameters calculated from Equation (4) are listed in Table VI. The polarimetric diameters of Ceres, Pallas, and Vesta are in good agreement with the radiometric results, and we would be surprised if they prove to be in error by more than 15%.

The relationship between albedo and the depth P_{min} of the negative branch for laboratory samples is illustrated in Fig. 7. Since polarizations on the order of 0.01 are involved, the relative precision here is inferior to that in Figs. 4 and 5. Moreover, high-albedo materials are poorly represented. It is nevertheless clear that, for a given albedo, P_{min} has a sharp upper limit which increases with decreasing albedo, and that consolidated rocks tend to exhibit shallower negative branches than powders and breccias. The laboratory domains of rocks and powders in the log A, log h plane are compared with astronomical results in Fig. 8. All of the solar-system bodies, including Mercury and Mars, apparently have powder-covered surfaces; for Vesta and the photometrically smooth trailing hemispheres of Callisto and Iapetus, the possibility of exposed consolidated rock is not excluded. The extremely dark asteroid 324 Bamberga falls outside the range of available laboratory data, and we are not yet able to say anything about it polarimetrically.

PROSPECTS FOR THE FUTURE

We have as yet no mathematical model that predicts the polarization phenomena exhibited by solid surfaces, and it does not seem likely that an analytic description comparable to the Mie theory for interstellar

TABLE VI
Asteroid Diameters in Kilometers

Object	$V(1,0)$[a]	A_V	Polarimetric Diameter	Visual Diameter	Infrared Diameter
1 Ceres	3.40	0.069	1060 ± 130	770	1045 ± 75
2 Pallas	4.53	0.074	600 ± 75	490:	550 ± 40
3 Juno	5.62	0.207	220 ± 30	195	260 ± 20
4 Vesta	3.54	0.227	550 ± 70	410	565 ± 30
8 Flora	6.62	0.14	165		
9 Metis	6.42	0.17	165		
15 Eunomia	5.49	0.13	295		275 ± 15
89 Julia	7.40	0.11	130		
1566 Icarus	16.82	0.19	1.3[b]		
1620 Geographos	15.15	0.22:	2.6:		
1681 Toro	13.8:	0.15:	6:		

[a] From Gehrels (1970, and unpublished).
[b] Radar observations (Goldstein 1971) set a lower limit of 1 km to the diameter of Icarus.

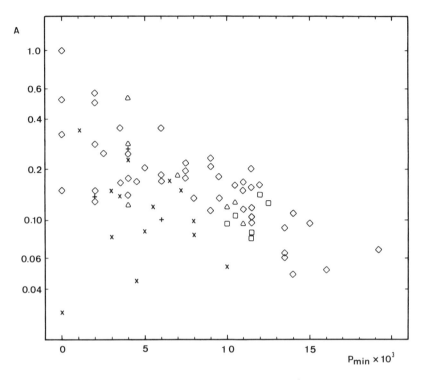

Fig. 7. Geometric albedo versus depth P_{min} of the negative polarization branch for laboratory samples. Symbols and data sources are the same as in Fig. 5.

particles or the Rayleigh-Chandrasekhar theory for molecular atmospheres is imminent. Thus, we use laboratory results and observations of astronomical objects with known properties for the calibration of empirical relationships. Substantial progress has been made for the two factors dominating the polarization, namely the albedo and the surface structure. Given suitable polarization data for a body of otherwise unknown properties, we are now able to (1) identify the polarization-phase curve as being characteristic of a solid surface rather than an atmosphere; (2) obtain the geometric albedo within fairly narrow limits from the slope of the ascending polarization branch; (3) check this albedo against the lower limit implied by the maximum polarization and the upper limit consistent with the depth of the negative branch; and (4) distinguish between microscopically rough and smooth surfaces by the position in the log A, log P_{max} and log A, log P_{min} planes.

Two principal deficiencies in this first-order approach are felt at present. First, most laboratory data have been collected with a view toward the lunar problem, so that high albedos comparable to those of the satellites of Jupiter and Saturn are under-represented. More importantly, we have no quantitative parametrization of the factor loosely described as surface roughness. Experiments are needed in which the size, texture, and opacity of individual particles as well as their state of compaction are under careful control.

As yet little compositional information has been obtained from polarization-phase curves except that carried by the bulk albedo. For example, we are not yet able to distinguish, from integrated-disk measurements, between a snowy satellite surface and one covered with a bright rock powder.

We have fairly complete and homogeneous polarization data with state-of-the-art precision for only a handful of asteroids; with present technology and telescopes of moderate size, this number could be extended to 50 main-belt objects and half-a-dozen Mars-crossers. Such a statistically adequate sample is needed to determine whether the small objects

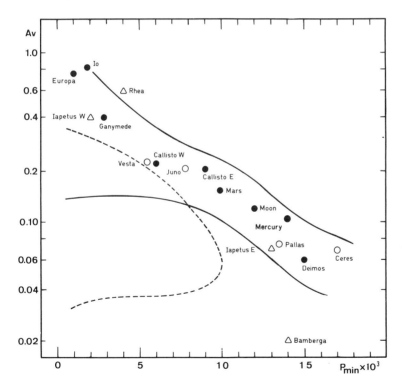

Fig. 8. Geometric albedo at 0.55 μm versus P_{min} for astronomical bodies. For the asteroids, open circles are plotted with polarimetric albedos from Table VI. Triangles represent albedos determined from infrared diameters. Laboratory rocks fall inside the dashed domain; laboratory powders and breccias generally fall between the solid lines.

have systematically different albedos or less surface dust than large bodies like Ceres and Vesta, and in particular to find out if some of the Mars-crossers may be extinct cometary nuclei. Rotational variations of the satellites of Jupiter are poorly understood, and the satellite system of Saturn is still almost virgin territory.

Great stimulus to polarimetry is provided by the battery of other astrophysical techniques that are only recently being turned toward asteroids and satellites. Without the new diameters provided by infrared radiometry, for example, our understanding of the slope-albedo law for asteroids would still be in a most confused state. We expect similar assistance from radar work and developing interferometric techniques capable of accurate diameter measurements at 0.1 arcsec and smaller. Space missions into the satellite systems of Jupiter and Saturn will open a new domain of high-albedo surfaces that may be very different from those of the asteroids.

Acknowledgments. We are indebted to A. Dollfus, T. Gehrels, D. Morrison, and J. Veverka for stimulating discussions and the use of unpublished data. Our work was supported by the European Space Research Organization (E.B.) and by the National Aeronautics and Space Administration (B.Z.).

Note added in proof. We would like to call attention to an unpublished manuscript by Milo Wolff of the Aerospace Corporation, in which polarization-phase curves complete with a negative branch are modeled by multiple scattering between randomly oriented crystal surfaces of various refractive indices. This theory looks very promising for interpretation of the polarizations exhibited by airless solar-system bodies.

REFERENCES

Allen, D. A.1970. The infrared diameter of Vesta. *Nature* 227: 158–159.

———. 1971. The method of determining infrared diameters. *Physical studies of minor planets.* (T. Gehrels, ed.), pp. 41–44. NASA SP-267. Washington, D.C.: U.S. Government Printing Off.

Barnard, E. 1902. On the dimensions of the planets and satellites. *Astron. Nachr.* 157: 260.

Bowell, E. 1971. Polarimetric studies. *Geology and physics of the moon* (G. Fielder. ed.), ch. 9. New York: American Elsevier Publ. Co.

Bowell, E.; Dollfus, A.; and Geake, J. E. 1972. Polarimetric properties of the lunar surface and its interpretation, Part 5. Apollo 14 and Luna 16 lunar samples. *Proc. Third Lunar Sci. Conf., Geochim. Cosmochim. Acta Suppl. 3,* Vol. 3. Cambridge, Mass.: MIT Press.

Coyne, G. V., and Pellicori, S. F. 1970. Wavelength dependence of polarization, XX. The integrated disk of the moon. *Astron. J.* 75: 54.
Cruikshank, D. P., and Morrison, D. 1973. Radii and albedos of nine asteroids. *Bull. Am. Astron. Soc.* In press.
Dollfus, A. 1955. Étude des planètes par la polarisation de leur lumière. Thesis, University of Paris; Study of the planets by means of the polarization of their light. *NASA Tech. Transl. F-188,* 1964.
———. 1956. Polarisation de la lumière renvoyée par les corps solides et les nuages naturels. *Ann. Astrophys.* 19: 83–113.
———. 1961. Polarization studies of planets. *Planets and Satellites.* (G. P. Kuiper and B. M. Middlehurst, eds.) Chicago: Univ. of Chicago Press.
———. 1970. Diamètres des planètes et satellites. *Surfaces and interiors of planets and satellites* (A. Dollfus, ed.), ch. 2. New York: Academic Press.
———. 1971*a*. Physical studies of asteroids by polarization of the light. *Physical studies of minor planets* (T. Gehrels, ed.), pp. 95–116. NASA SP-267. Washington, D.C.: U.S. Government Printing Off.
———. 1971*b*. Diameter measurements of asteroids. *Physical studies of minor planets* (T. Gehrels, ed.), pp. 25–31. NASA SP-267. Washington, D.C.: U.S. Government Printing Off.
Dollfus, A., and Bowell, E. 1971. Polarimetric properties of the lunar surface and its interpretation, Part 1. Telescopic observations. *Astron. Astrophys.* 10: 29–53.
Dollfus, A.; Bowell, E.; and Titulaer, C. 1971. Polarimetric properties of the lunar surface and its interpretation, Part 2. Terrestrial samples in orange light. *Astron. Astrophys.* 10: 450–466.
Dollfus, A.; Focas, J.; and Bowell, E. 1969. La planète Mars: la nature de sa surface et les propriétés de son atmosphère d'apres la polarisation de sa lumière, II. La nature du sol de la planète Mars. *Astron. Astrophys.* 2: 105–121.
Dollfus, A.; Geake, J. E.; and Titulaer, C. 1971. Polarimetric properties of the lunar surface and its interpretation, Part 4. Apollo 11 and Apollo 12 lunar samples. *Proc. Second Lunar Sci. Conf., Geochim. Cosmochim. Acta Suppl. 2,* Vol. 3, pp. 2285–2300. Cambridge, Mass.: MIT Press.
Dollfus, A., and Titulaer, C. 1971. Polarimetric properties of the lunar surface and its interpretation, Part 3. Volcanic samples in several wavelengths. *Astron. Astrophys.* 12: 199–209.
Geake, J. E.; Dollfus, A.; Garlick, G.; Lamb, W.; Walker, G.; Steigmann, G.; and Titulaer, C. 1970. Luminescence, electron paramagnetic resonance and optical properties of lunar material from Apollo 11. *Proc. Apollo 11 Lunar Sci. Conf., Geochim. Cosmochim. Acta Suppl. 1,* Vol. 3, pp. 2127–2147. Oxford: Pergamon.
Gehrels, T. 1970. Photometry of asteroids. *Surfaces and interiors of planets and satellites* (A. Dollfus, ed.), ch. 6. New York: Academic Press.
Gehrels, T.; Coffeen, D.; and Owings, D. 1964. The wavelength dependence of polarization, III. The lunar surface. *Astron. J.* 69: 826–852.
Gehrels, T.; Roemer, E.; Taylor, R.; and Zellner, B. 1970. Minor planets and related objects, IV. Asteroid (1566) Icarus. *Astron. J.* 75: 186–195.
Goldstein, R. M. 1971. Asteroid characteristics by radar. *Physical studies of minor planets* (T. Gehrels, ed.), pp. 165–171. NASA SP 267. Washington, D.C.: U.S. Government Printing Off.
Harris, D. L. 1961. Photometry and colorimetry of planets and satellites. *Planets and satellites* (G. P. Kuiper and B. M. Middlehurst, eds.). Chicago: Univ. of Chicago Press.

Ingersoll, A. P. 1971. Polarization measurements of Mars and Mercury: Rayleigh scattering in the Martian atmosphere. *Astron. J.* 163: 121–129.

Irvine, W. M.; Simon, T.; Menzel, D. H.; Pikoos, C.; and Young, A. T. 1968. Multicolor photoelectric photometry of the brighter planets. III. Observations from Boyden Observatory. *Astron. J.* 73: 807–828.

KenKnight, C. E.; Rosenberg, D. L.; and Wehner, G. K. 1967. Parameters of the optical properties of the lunar surface powder in relation to solar wind bombardment. *J. Geophys. Res.* 72: 3105–3129.

Kuiper, G. P. 1955. Progress of research: Diameters *Trans. IAU* 9: 250–251.

Lane, A. P., and Irvine, W. M. 1973. Monochromatic phase curves and albedos for the lunar disk. *Astron. J.* 78: 267–277.

Lyot, B. 1929. Recherches sur la polarisation de la lumière des planètes et de quelques substances terrestres. *Ann. Obs. Paris.* Vol. 8, No. 1; Research on the polarization of light from planets and from some terrestrial substances. *NASA Tech. Transl. TT F-187*, 1964.

———. 1934. Polarisation des petites planètes, *C. R. Acad. Sci. Paris* 199: 774.

Matson, D. L. 1971a. Infrared emission from asteroids. Ph.D. Thesis, California Institute of Technology, Pasadena, California.

———. 1971b. Infrared observations of asteroids. *Physical studies of minor planets* (T. Gehrels, ed.), pp. 45–50. NASA SP-267. Washington, D.C.: U.S. Government Printing Off.

McCord, T. B., and Adams, J. B. 1972. Mercury: surface composition from the reflection spectrum. *Science* 178: 745–747.

Murphy, R. E.; Cruikshank, D. P.; and Morrison, D. 1972. Radii, albedos, and 20-micron brightness temperatures of Iapetus and Rhea. *Astrophys. J.* 177: L93–L95.

Pascu, D. 1973. Photographic photometry of the Martian satellites. *Astron. J.* In press.

Pellicori, S. F. 1969. Polarizing properties of pulverized materials: application to the lunar surface. *Tech. Rep. 42.* Optical Sciences Center, Univ. of Arizona.

———. 1971. Polarizing properties of pulverized materials with special reference to the lunar surface. *Applied Optics* 10: 270–285.

Pollack, J. B.; Veverka, J.; Noland, M.; Sagan, C.; Hartmann, W. K.; Duxbury, T. C.; Born, G. H.; Milton, D. J.; and Smith, B. A. 1972. Mariner 9 television observations of Phobos and Deimos. *Icarus* 17: 394–407.

Provin, S. 1955. Preliminary observations of the polarization of asteroids. *Publ. Astron. Soc. Pac.* 67: 115; also see Dollfus (1961).

Riley, L. A., and Hall, J. S. 1972. High resolution measures of polarization and color of selected lunar areas. *Lowell Obs. Bull.* 7: 255–271.

Veverka, J. 1970. Photometric and polarimetric studies of minor planets and satellites. Ph.D. Thesis, Harvard Univ. Cambridge, Mass.

———. 1971a. Polarization measurements of the Galilean satellites of Jupiter. *Icarus* 14: 355–359.

———. 1971b. The polarization curve and the absolute diameter of Vesta. *Icarus* 15: 11–17.

———. 1971c. Photopolarimetric observations of the minor planet Flora. *Icarus* 15: 454–458.

———. 1971d. Asteroid polarimetry: a progress report. *Physical studies of minor planets* (T. Gehrels, ed.), pp. 91–94. NASA SP-267. Washington, D.C.: U.S. Government Printing Off.

ASTEROIDS AND SATELLITES 403

———1973a. Titan: polarimetric evidence for an optically thick atmosphere? *Icarus.* In press.
———. 1973b. Polarimetric observations of 9 Metis, 15 Eunomia, 89 Julia, and other asteroids. *Icarus.* In press.
Veverka, J., and Noland, M. 1973. Asteroid reflectivities from polarization curves: calibration of the "slope-albedo" relationship. *Icarus.* In press.
Widorn, T. 1967. A photometric method of estimating the diameters of minor planets. *Ann. Univ. Sternw. Wien* 27: 112–119.
Wilhelms, D. E. and Trask, N. J. 1965. Polarization properties of some lunar geologic units. *USGS Astrogeol. Studies.* Part A, pp. 63–80.
Wilson, L. 1968. The interpretation of lunar photometry. Ph.D. Thesis, Univ. of London, London.
Zellner, B. 1972a. Minor planets and related objects, VIII. Deimos. *Astron. J.* 77: 183–185.
———. 1972b. On the nature of Iapetus. *Astrophys. J.* 174: L107–L109.
———. 1973. The polarization of Titan. *Icarus.* In press.

DISCUSSION

CHAPMAN: The errors you give in your application to asteroids, of the albedo versus polarization-slope relation, imply that you believe your laboratory data define a perfect relationship between the two variables. Nevertheless, the laboratory data show a spread in the albedos of various samples with same h of at least a factor of two. Have you any reason for believing this variability to be due to measuring errors? It seems more plausible to me that the variations are mostly real, due to differences between samples. In that case one cannot positively apply the relationship to asteroids to better than a factor of two.

ZELLNER: The total scatter in Fig. 5 is on the order of a factor of two in *albedo,* i.e., plus or minus a factor of $\sqrt{2}$ deviation from the mean line. Thus, the maximum uncertainty in a derived asteroid *diameter* is on the order of 20%. With the heterogeneous data currently available, it seems impossible to say how much of the observed scatter may be genuine. The albedos listed by Lyot, for example, were little more than good guesses. The problems of an absolute albedo scale for telescopically observed lunar regions are well known. Also, the photopolarimetric properties of laboratory samples are preparation-dependent, but rarely have albedos and polarizations been measured simultaneously.

The essential point, however, is that the slope h does not seem to depend in any systematic way on parameters (wavelength, surface texture, mineralogy, etc.) other than the bulk albedo. We may find second-order effects when we have better data.

FINK: Does it not seem rather strange that you should get such a "universal curve" of albedo versus the slope of the polarization curve? Is there any theoretical reason why you should expect such a curve?

ZELLNER: As noted in the text, a law of the form $\log A = -c_1 \log P + c_2$ with c_1 near unity means that roughly the same amount of polarized light is scattered from all surfaces. For the P_{max} of relatively opaque powders, this does not seem unreasonable. The more general applicability of the slope-albedo law is presumably due to the fact that h is measured at phase angles in the range $10°-30°$, far away from the Fresnel peak of intensity and polarization for smooth surfaces.

SERKOWSKI: There is a possibility of using polarimetry for very accurate measurements of the *oblateness* of asteroid disks. After the light of an asteroid goes through a polarizer, an enlarged image of the seeing disk of an asteroid, about 1 mm in diameter, is formed by a zoom lens on the wedge-shaped retarder. The retardance of this retarder, having optical axis at $45°$ to that of the polarizer, changes along its diameter by $360°$ for every millimeter. The retarder is followed by a quarter-wave plate with optical axis parallel to that of the polarizer. The light emerging from this entire polarizing device is fully linearly polarized with the plane of polarization changing by $180°$ for every millimeter along its diameter. This means that the plane of polarization at the center of the seeing disk is perpendicular to that at half-width points, and the polarization averaged over the seeing disk approximately cancels out. Now let the entire polarizing unit rotate rapidly. If the seeing disk is slightly elongated, the degree of polarization will change with a frequency twice as high as that of the mechanical rotation. These changes in polarization, rapidly measured, e.g., with a Pockels cell, will be completely independent of centering an asteroid in the focal plane diaphragm. Since polarization can be easily measured with an accuracy of $\pm 0.1\%$, oblateness can be also measured with 0.1% precision. This method besides being useful for measuring oblateness of asteroids may find application for detecting very close *visual binaries,* particularly those that cannot be detected with an interferometer either because the angular separation is too small or because the magnitude difference between the components is too large.

POLARIZATION IN A MINERAL ABSORPTION BAND

CARLE E. PIETERS
Massachusetts Institute of Technology

A theoretical and experimental study was undertaken to examine the polarimetric properties of light reflected from a particulate surface in the spectral region (0.7 to 1.1 μm) of a mineral electronic-transition absorption band. The purpose of the investigation was to show that spectral polarimetry is an alternative diagnostic tool to absolute reflectivity measurements for some applications, notably the determination of absorption-band positions for the lunar surface.

The major results are: (1) Polarization increases significantly in an absorption band at large phase angles. (2) The wavelength of the maximum of the polarization variation corresponds directly with the center of the absorption band. (3) The amount of increase of polarization for minerals with an absorption band is dependent on particle size. (4) The fractional change of polarization with wavelength is greater for mixtures with transparent minerals than with absorbing minerals. The magnitude of change, however, is greater for mixtures with absorbing minerals.

These polarization variations are interpreted as the result of the changing ratio of specular to diffuse components of light reflected from a particulate surface.

Reflectivity measurements of planetary objects and related laboratory measurements of rocks have demonstrated and defined absorption bands in the visible and near infrared spectral region (McCord, Adams, and Johnson 1970; McCord et al. 1972; McCord and Adams 1973). With careful work, these absorption bands can be used as diagnostic tools to determine the composition of the dominant absorbing (but not opaque) mineral on a planetary surface.

The mineral absorption bands of the spectrum in this region are generally d-d orbital electron transitions of transition metal ions (Fe, Cr, Ti, Ni, etc.) in a crystal structure. The d orbital energy levels of these ions in a crystal environment are no longer degenerate but have been split due to the effect of the surrounding negative ions, called ligands. The magnitude of this energy level split, and thus the wavelength of the energy absorbed to make the transition, is strongly dependent on the type of ion, the crystal environment, and the metal-ligand distance. Briefly, the wavelength of the absorption band center is very dependent on the composition and crystal symmetry of the mineral. (See Burns 1970 for an extensive discussion of mineral absorption.) The strength of an absorption band in a reflection spectrum is dependent on the ratio of the crystalline to glassy content of the surface material (Adams and McCord 1971a,b).

To obtain mineral compositional information, a critical measurement is the wavelength position of the absorption band center (Adams and McCord 1972). Normally, this information is obtained by measuring the spectral reflectivity of the substance using a filter photometer or spectrometer. The technique requires high spectral resolution (at least 0.01 μm), good calibration of equipment, and absolute reflectivity standards. The calibration is usually the greatest source of error in planetary measurements. An alternative technique to determine the wavelength of an absorption band involves spectral polarimetry rather than absolute reflectivity measurements.

Light reflected from a particulate surface contains both a specular and a diffuse component. The relative proportion of the components is dependent on particle size, compaction, and opacity of the surface particles as well as on the geometry of reflection. The specular component is governed by Fresnel's laws of reflection and accounts for the positive polarization observed at large phase angles. The diffuse component contains light that has been transmitted through one or more particles and eventually scattered out of the soil. Thus, the spectrum of the diffuse component contains absorption features of the surface particles. The amount of Fresnel polarization is dependent on the magnitude of the diffuse component (relative to the specular), which in turn is dependent largely on the opacity of the particles. At large phase angles, bright particulate surfaces (relatively transparent particles with a large diffuse component) have lower polarization than dark surfaces (opaque or absorbing particles and small diffuse components). Spectral polarimetry at large phase angles should reveal an absorption band in the form of an increase in polarization since the absorption itself will decrease the diffuse component.

To examine this hypothesis a three part program was initiated: (1) A computer analysis of reflectivity and polarization using a complex refractive index tailored to describe a crystal field absorption band, (2) laboratory measurements of reflectivity and polarization (at large phase angle) for mineral mixtures in the spectral region of a mineral absorption band, and (3) telescopic observations of small areas on the moon to measure the wavelength of absorption bands using polarimetric techniques. The first two parts have been completed, and the results are briefly discussed. The third part is in progress.

RESULTS OF COMPUTER MODEL ANALYSIS

Fresnel's equations of reflection and transmission involve the complex refractive index $m = n - in'$, where n is the real refractive index, $-n'$ is the extinction coefficient, and i is $\sqrt{-1}$. The real refractive index n of a dielectric changes in the spectral region of an absorption band (Fowles 1968). Specular reflection is expected to be affected by any variation of either part of the index. It was found, however, that the variation of the refractive index with wavelength in the region of a crystal-field absorption band has essentially no detectable effect on the specular reflection of the material. This is due primarily to the tiny magnitude of the variation ($\pm 10^{-4}$). Stronger absorption bands farther in the infrared have greater index variations which may have an observable effect on specular reflection (Garbuny 1965). Diffuse reflection, since it contains transmitted light, is strongly affected by the extinction coefficient $-n'$.

To model the reflectivity of a particulate surface, the specular and diffuse components were calculated. The two planes of polarization of the specular component were produced directly from Fresnel's equations using the complex refractive index with a real part $n = 1.5$. The diffuse component was produced by assuming the transmitted light had traveled a mean optical path length ($MOPL$) of 1 mm through the particles, had been scattered, and had emerged as unpolarized light. This diffuse component can be calculated from the combined transmitted components of Fresnel's equations and assumes no absorption other than that which causes an absorption band centered at 0.95 μm.

The major difficulty of the modeling process is choosing the correct relative amounts of the specular and diffuse components. Two models were dealt with, neither of which correctly models a particulate surface for all phases, but they do bring to light some of the properties of reflection from a particulate surface. *Model I* used the specular and diffuse components in the relative amounts that appear directly from Fresnel's equations. *Model II* ensured the specular component was 10% of the total for all phases.

Figures 1 and 2 show the results for these two simple models. In Fig. 1, polarization is plotted versus phase for two wavelengths, one in the absorption band (0.95 μm) and one outside the band (0.80 μm). Even though the same refractive index was used, the position of maximum polarization is very dependent on the mixing of the two components. Figure 2 shows the polarization versus wavelength for two large phase angles, 90° and 120°. In all cases polarization increases directly as a function of absorption band strength.

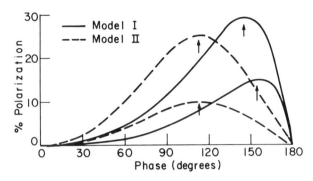

Fig. 1. Calculated polarization versus phase for two models of reflection. Solid lines are for Model I; dashed lines are for Model II. Lower lines are for a wavelength of 0.80 μm; upper lines are for a wavelength of 0.95 μm.

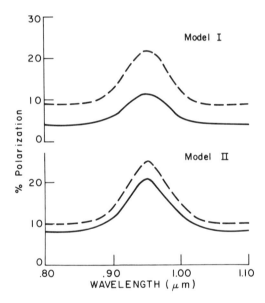

Fig. 2. Calculated polarization versus wavelength for two models of reflection. Solid lines are for a phase angle of 90°; dashed lines are for a phase angle of 120°.

LABORATORY MEASUREMENTS

Description of Samples

The primary mineral chosen for the study is an orthopyroxene which is the dominant mineral of a Websterite from North Carolina. The mineral contains 6.04% iron of which 98.5% is Fe^{+2}. The sample is En_{89} in the enstatite-ferrosilite series $(Mg,Fe)_2Si_2O_6$.

A diffuse reflection spectrum of this sample made on a Cary 17 spectrometer is shown in Fig. 3. The Fe^{+2} spin-allowed absorption band is centered at 0.91 μm. There is a second absorption band out of range in the infrared caused by further splitting of the d orbital energy levels of the Fe atom in a distorted octahedral site of the pyroxene. The small bands and spike around 0.4 and 0.5 μm are spin-forbidden bands of Fe^{+2} and are therefore much less intense.

Table I lists the samples prepared for this study and the particle size. Samples designated E are 100% enstatite$_{89}$. Evf is a very fine sample prepared by grinding a sample of E < 63 for about 10 minutes. E100 and E < 125 are the same sample. The samples with particle size < 125 μm contain roughly equal amounts of [125, 63] μm and < 63 μm. Those listed as Pxx contain xx% by volume plagioclase; the remainder is enstatite. The plagioclase is a hand-picked transparent sample of oligoclase. No traces of contaminants could be seen visually. Those designated Mxx contain xx% magnetite by volume, with the rest enstatite. The magnetite contains small amounts (less than 1%) of a red translucent regularly shaped mineral.

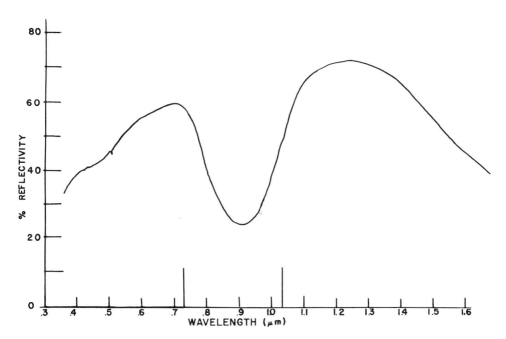

Fig. 3. Diffuse reflection spectrum of En$_{89}$. The 0.91 μm band is due to Fe^{+2} in the crystal structure.

TABLE I
Band Depth of Samples Measured

Sample	Particle size (μm)	Rp(0.91)/Rp(0.73) (Diffuse)
E [250, 125]	250 > ps > 125	0.29 [a]
E [125, 63]	125 > ps > 63	0.29
E < 125	less than 125	0.37
E < 63	less than 63	0.47 [b]
Evf	very fine	0.62 [b]
P100	less than 125	1.01
P90	less than 125	0.74
P50	less than 125	0.50
P10	less than 125	0.37
E100	less than 125	0.37
M10	less than 125	0.49
M50	less than 125	0.74
M90	less than 125	0.87
M100	less than 125	0.87

[a] Saturated band
[b] Partially packed

Diffuse Reflection Measurements

Diffuse reflection measurements were made on a Cary 17 spectrometer with a Type II diffuse reflection attachment for all samples listed in Table I. A smoked MgO surface was used as a standard. The measured reflectivity is the ratio of the reflectivity of the sample to that of the standard and is recorded on a chart recorder intricately calibrated with the Cary rates. The results are shown in Figs. 4 and 5.

The values of reflectivity in Fig. 4 for different particle sizes demonstrate the principle "the smaller the brighter" described by Adams and Filice (1967). The apparent saturation that occurs in the band for sample E[250,125] also occurs for other samples of larger particle sizes (not shown here) and indicates that the *MOPL* of the diffuse component for the large particles is so long that most of the light is absorbed.

The effect of particle size on the total *MOPL* for these samples is best seen by measuring the band depth. The ratio of the intensity in the center of the band [Rp(0.91)] to that near the edge [Rp(0.73)] is listed in column 3 of Table I. Except for the case of saturation, as the particle size decreases the band depth also decreases, indicating a decrease in the total *MOPL*. Preparation of the smallest size samples for a vertical position measurement involved pouring into a sample dish and covering with a glass cover. This procedure causes the sample to be partially packed. Thus, for Evf and E < 63 the *MOPL* is shorter than the other

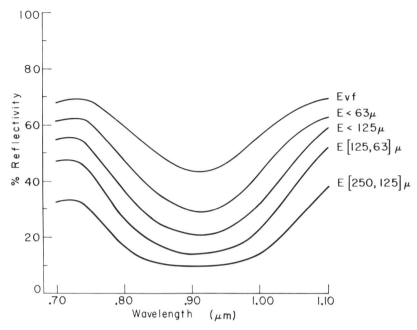

Fig. 4. Diffuse reflection spectra of the absorption band of En_{89} for various particle sizes.

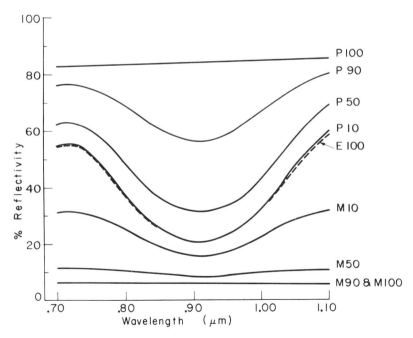

Fig. 5. Diffuse reflection spectra of the absorption band of En_{89} for mixtures with plagioclase and magnetite. The particle size for all mixtures is < 125 μm.

samples not only because the particles are smaller, but also because they are packed. E < 125 was chosen as the particle size for mixing with other minerals since it is similar to particle sizes for lunar soil and has an *MOPL* that is fairly large but without saturation.

The magnetite and plagioclase of Fig. 5 have an opposite effect on the mixtures. The opaque material absorbs light, while the transparent mineral transmits it essentially without absorption, thus allowing a greater number of reflections and a greater probability of return to the surface. Concerning the band-depth ratios for the mixtures, the results indicate, as expected, that the *MOPL* through the enstatite decreases with dilution by another mineral. What is most striking is the effect of the *type* of mixture. The results indicate that only 10% opaques reduce the *MOPL* by an amount comparable to 50% dilution by a transparent mineral, and 50% opaques are comparable to 90% transparent mineral.

Polarization Measurements at 90° Phase Angle: Laboratory Techniques

Figure 6 shows schematically the laboratory arrangement for the measurements at 90° phase angle. A Heath 701 series monochromoter and tungsten lamp were used to provide monochromatic light to 1.03 μm with a bandwidth of 40Å The detector used for the measurements was an ITT FW-118 (S-1) photomultiplier cooled to dry-ice temperatures. A Fabritek instrument computer was used in the pulse-counting mode to measure the data. The results were recorded with a Cipher magnetic tape recorder. By combining the scan rate of the monochromoter, the integration time of the Fabritek, and the data process averaging, the effective sampling was every 20 Å.

Prior to sample measurement, two Polaroid HR filters were calibrated. The filters were found to act as reliable polarization analyzers from 0.73 μm into the near infrared. The two Polaroid filters were aligned so they could be fixed easily in either of two orthogonal positions: L position parallel to the plane of scattering (in the plane of the page for Fig. 6A), and R position perpendicular to the plane of scattering. Each filter was also oriented so that its surface was perpendicular to the direction of propagation of the light passing through it.

The source was found to be approximately 20% polarized in the L direction as measured from position B in Fig. 6. Six measurements are required for a valid polarization spectrum: two measurements of the source and four of the sample. Table II gives the equation sequence for accomplishing the correction, as follows. The light from the source is analyzed with a Polaroid filter (P2) into light of two planes of polarization, S_L and S_R. For both S_L and S_R, the intensity of the light reflected from the sample contains a specular component, C, and a diffuse component D. Thus, if one analyzes the reflections of S_L and S_R with

TABLE II

Sequence of Equations to Correct For a Polarized Light Source

P2	Source	Reflection from sample	P1	Components measured
L	S_L	$(C_L + D_L)S_L$	L	$(C_L + \tfrac{1}{2}D_L) S_L$
			R	$\tfrac{1}{2}D_L\, S_L$
R	S_R	$(C_R + D_R)S_R$	L	$\tfrac{1}{2}D_R\, S_R$
			R	$(C_R + \tfrac{1}{2}D_R) S_R$

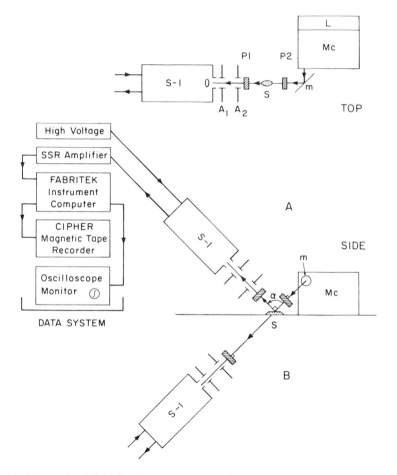

Fig. 6. Schematic of the laboratory arrangement for measuring polarization. Position A is used to measure the four reflection components of the sample. Position B of the same equipment with the sample removed measures the two components of the source. $\alpha = 90°$, m = front surfaced mirror, P1, P2 = HR Polaroids, S = sample, $A_1\, A_2$ = two apertures, S-1 = cooled photomultiplier detector.

another Polaroid filter (P1), one obtains the four measurable components listed on the right of Table II. The total intensity of the light reflected in the two planes of polarization, I_L and I_R, is the sum of the L components and R components from P1 respectively:

$$I_L = (C_L + \tfrac{1}{2}D_L)S_L + \tfrac{1}{2}D_R S_R$$

$$I_R = (C_R + \tfrac{1}{2}D_R)S_R + \tfrac{1}{2}D_L S_L$$

To correct for the unequal source intensities, one forms the correction factor $N = \dfrac{S_L}{S_R}$ and applies it to obtain corrected values

$$I'_L = (C_L + \tfrac{1}{2}D_L)S_L + \tfrac{1}{2}D_R S_R \cdot N$$

$$I'_R = (C_R + \tfrac{1}{2}D_R)S_R \cdot N + \tfrac{1}{2}D_L S_L$$

From these intensities one can calculate the true percentage polarization

$$P = \frac{I'_R - I'_L}{I'_R + I'_L} \times 100$$

This technique was used to obtain the polarization measurements reported here.

An alternative and simpler technique for eliminating the effects of a polarized source would involve inserting a good depolarizer before the sample. No acceptable depolarizer was available. The six-measurement technique described above does provide a direct measurement of the diffuse and specular components, which cannot be obtained otherwise. The reflectivity of smoked MgO was found to be polarized, and absolute reflectivity measurements at 90° phase angle were prevented for the two planes of polarization.

RESULTS

For all samples, the polarization of reflected light increased in the spectral region of an absorption band. The wavelength dependence of polarization is shown in Figs. 7, 8, and 9. The error bars on the left of the figures represent the maximum extent of the data variation due to statistics of photons detected. The greatest error occurs below 0.8 μm where the light level was the lowest. Some critical values are listed in Table III for comparison. The only dubious distinction is between M90 and M100 where the random error is larger than the difference between samples. The polarization effect is strongest for larger particles where the strongest absorption band occurs. With decreasing particle size, the diffuse component of the reflection increases while the *MOPL* decreases.

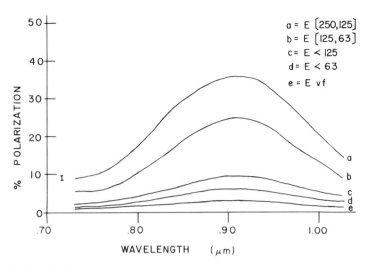

Fig. 7. Polarization spectra of the absorption band of En_{89} for various particle sizes.

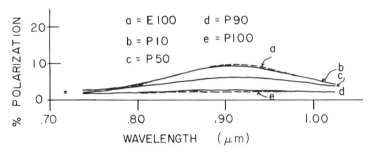

Fig. 8. Polarization spectra of the absorption band of En_{89} for mixture with plagioclase.

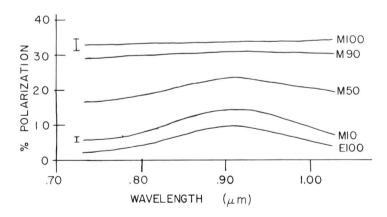

Fig. 9. Polarization spectra of the absorption band of En_{89} for mixtures with magnetite.

TABLE III

Values of Percentage Polarization at 90° Phase Angle

Sample	Percentage of Polarization		
	0.73 μm	0.91 μm	△P
E [250, 125]	9	36	27
E [125, 63]	5.5	24.5	19
E < 125	2	9.5	7.5
E < 63	1.5	6	4.5
Evf	1	3	2
P100	1.75	1.75	—
P90	2	3	1
P50	2.5	6	3.5
P10	2.5	9.5	7
E100	2	9.5	7.5
M10	5.75	14.2	8.45
M50	17	23.5	6.5
M90	29	31	2
M100	32.5	34	1.5

All samples were poured onto a flat surface. For samples < 125 μm this created a porous, hummocky (on the scale of millimeters) texture, which caused reflection to be relatively diffuse.

Absorbing mixtures enhance the wavelength dependence of polarization caused by a mineral with an absorption band. Mixtures of 50% magnetite have a polarization change of 6.5 percentage points in an absorption band, while mixtures of 50% plagioclase have only a 3.5 percentage point change. The reflectivity measurements showed that it is relatively difficult to detect an absorption band in dark mixtures.

The polarization of a constant specular component (Fresnel reflection) causes the polarization for all samples at this large phase angle. For the bright plagioclase mixtures, the diffuse component is large. A ratio of the intensity of the specular component to the intensity of the diffuse component is small, as indicated by the low polarization. For magnetite, however, the ratio of specular/diffuse components is larger, and changes in the intensity of the diffuse component in the spectral region of an absorption band cause a larger magnitude of change in polarization. Thus, it is easier to detect polarization differences in material that is already strongly polarized than in a more diffuse substance. The magnitude of polarization change is greater for dark material even though the *MOPL* has been severely reduced.

CONCLUSIONS

The major conclusion supported by both theoretical and experimental evidence is that if there is significant positive polarization of the light reflected from a particulate surface, the magnitude of the polarization will increase in the spectral region of an absorption band. The amount of change of polarization is a function of particle size, as well as type and proportion of minerals in a mixture. For pure enstatite$_{89}$ samples, the polarization increases three- to fourfold in an absorption band, with the greatest change of magnitude for the larger particles (lower overall reflectivity). The absorption that occurs in the diffuse component changes the ratio of the specular to diffuse components and, thus, causes the polarization to increase. Measurements of both reflectivity and polarization can detect an absorption band. For low-albedo surfaces especially, polarimetry may provide a new, sensitive tool.

Other conclusions from the theoretical discussion are: (1) The angle of maximum polarization of light reflected from a particulate surface is not an accurate indication of the real refractive index of the surface material. It is a function of the relative portions of diffuse and specular components of reflection. (2) The variation of the real refractive index in the region of a transition-metal absorption band of a mineral has essentially no effect on the specular reflection properties of the surface.

APPLICATIONS

For detecting absorption bands, polarimetry can be applied to only a few planetary objects. The moon and Mercury are the only objects that can be observed at large phase angles, unless space probes are used. Mercury has suggestions of absorption bands in its spectrum (McCord and Adams 1972). The most promising application is for the moon, which not only has absorption bands in its spectrum (McCord and Johnson 1969) but has a well-defined variation of the band center resulting from different surface composition of various areas of the moon (Adams and McCord 1972; McCord et al. 1972).

With the conclusion of the Apollo program, earth-based measurements can be a powerful tool in extending the knowledge gained to other locations on the front face of the moon. In particular, to determine the composition of the average pyroxene at any location, the methods previously used to detect the absorption bands must be refined to determine more accurately the wavelength of the band center. For reflectivity measurements, this requires very good spectral resolution and still involves relative lunar measurements, stellar measurements, and stellar calibrations. Unfortunately, the major lunar absorption band occurs in a small telluric water-vapor band which increases observational difficulties. For polarimetric measurements, the same spectral resolution is required, but each lunar area can be measured independently from all others. The only calibration required is that of the polarization of the equipment, which can be made by observing known unpolarized sources.

Acknowledgments. I am indebted to Thomas McCord for many helpful and enjoyable discussions and to Roger Burns for his encouraging advice and for the use of his spectrometer. This work was made possible by NASA grants NGL 22-009-350 and NGR 22-009-583.

REFERENCES

Adams, J. B., and Filice, A. L. 1967. Special reflectance 0.4 to 2.0 microns of silicate rock powders. *J. Geophys. Res.* 72: 5705–5715.

Adams, J. B., and McCord, T. B. 1971a. Alternation of lunar optical properties: age and composition effects. *Science* 171: 567–571.

———. 1971b. Optical properties of mineral separates, glass, and anorthositic fragments from Apollo mare samples. *Proceedings of the Second Lunar Science Conference* (A. A. Levinson, ed.), Vol. 3, pp. 2183–2195. Cambridge, Mass.: MIT Press.

———. 1972. Electronic spectra of pyroxenes and interpretation of telescopic spectral reflectivity curves of the moon. *Proceedings of the Third Lunar Science Conference 3*. (D. R. Criswell, ed.) Cambridge: MIT Press.

Burns, R. G. 1970. Mineralogical applications of crystal field theory. London: Cambridge Univ. Press.

Fowles, G. R. 1968. *Introduction to modern optics.* New York: Holt, Reinhart and Winston, Inc.

Garbuny, M. 1965. Optical physics. New York: Academic Press.

McCord, T. B., and Adams, J. B. 1972. Mercury: surface composition from the reflection spectrum. *Science* 178: 745–746.

———. 1973. Progress in the remote optical analysis of lunar surface composition. *The moon.* In press.

McCord, T. B.; Adams, J. B.; and Johnson, T. V. 1970. Asteroid Vesta: Spectral reflectivity and compositional implications. *Science* 168: 1445–1447.

McCord, T. B.; Charette, M. P.; Johnson, T. V.; Lebofsky, L. A.; Pieters, C.; and Adams, J. B. 1972. Lunar spectral types. *J. Geophys. Res.* 77: 1349–1359.

McCord, T. B., and Johnson, T. V. 1969. Relative spectral reflectivity 0.4-1 μ of selected areas of the lunar surface. *J. Geophys. Res.* 74: 4395–4401.

MIE SCATTERING OF THE INTERPLANETARY MAGNETIC FIELD BY THE WHOLE MOON

C. P. SONETT and D. S. COLBURN
NASA Ames Research Center

It is known from the Apollo magnetometer experiments that significant electromagnetic induction takes place in the lunar interior. This induction is excited by fluctuations of the interplanetary magnetic field and is detected by the induced fields on the surface of the moon. This paper reviews these results briefly and discusses the formal properties of the theory. It is shown that the mathematical treatment parallels that for classical electromagnetic scattering. Further, the wavelength spectrum of the fluctuations of the interplanetary magnetic field includes scales consistent with the radius of the moon. The result is that the moon is excited in several modes. Quadrupole and possibly octupole magnetic multipoles are found in the data. The electric-type radiation corresponding to transverse magnetic excitation appears suppressed and far below the detection threshold of the magnetometers, but electric multipoles should be present as well. If so, they would tend to be restricted to the outer layers of the moon.

The purpose of this paper is to demonstrate the formal similarity between classical Mie scattering as known in optical phenomena and electromagnetic induction in the moon caused by fluctuations of the interplanetary magnetic field. It is now well known that the interplanetary magnetic field contains a continuum of disturbances that are seen by the moon as time-dependent variations in the direction and strength of the magnetic field (Sonett et al. 1971). In turn, these fluctuations result in the flow of eddy currents in the lunar interior, whose magnetic field is detected on the surface of the moon by comparison of records taken there using the Apollo surface magnetometers and the free-stream solar-wind field simultaneously monitored by lunar satellite Explorer 35 (Sonett et al. 1972). From the concurrent Fourier spectra of the surface and solar-wind magnetic fields, the internal bulk electrical conductivity profile of the moon can be determined using the classical equations of electromagnetic scattering.

The actual theory of scattering as it is known for optical wavelengths and for vacuum conditions must be appropriately modified for the boundary conditions that exist at the contact between the moon and solar wind. However, the general formalism remains intact, and we shall show later that the wavelength regime found in the solar wind that is responsible for the induction in the moon is identifiable with magnetic multipole radiation, including at least quadrupole and octupole terms. Proof of the effect of the solar wind in modifying the boundary conditions comes from the observation that the lunar induction signal rises to values of three to four at the peak frequency, some six to eight times that for vacuum induction. The increase is due to the dynamic pressure of the solar wind, which forces the induced magnetic-field lines back under the surface of the moon. Skin depth arguments show that the confinement of the field varies with depth in the moon; the confinement is frequency dependent, and the shell within which the field is confined shrinks as the frequency is increased.

The topology of the idealized interplanetary-magnetic-field lines is given by Parker (1963), whose equation for the lines is that of an Archimedean spiral. This follows from a combination of the solar rotation period, the high electrical conductivity of the solar gas, and the assumption of constant outward velocity for the solar wind. The latter is substantially correct for most of the flow field because of the strongly supersonic flow. Components of the Parker field (in the ecliptic plane) are $B_r = B_{r0}(r_0/r)^2$ and $B_t = (\omega r/V_s\, B_r$, where B_r is the component of the field radially outward from the sun at r, B_{r0} is the radial component at the surface of the sun, r_0 is the solar radius, r is the radius vector to the position of observation, B_t is the tangential component, ω is the angular velocity of the sun, and V_s is the bulk-flow speed of the solar wind. The angle of the field measured between the outward-pointing radius vector and the tangent to the field line at the position, r, is $\eta = \tan^{-1}(\omega r/v_s)$. Near the earth during quiet times, the ideal angle based upon $V_s = 400$ km/sec is $45°$. Thus, the k vector field at the position of the moon during quiet times is pointed at about $45°$ away from the radius from the sun.[1] During times of solar disturbance the field can take on any direction, and the spiral geometry is destroyed (Colburn and Sonett 1966). Observation of induction by the solar-wind field is limited to times when the moon is in the solar wind and free of the magnetic tail of the earth, though observation can be continued when the moon dwells in the post-shocked plasma of the transition region behind the bow shock

---

[1] Based upon Alfvén waves propagating along the field lines.

wave. The reason for the latter is that at lunar distance the post-shocked plasma has returned nearly to the parameters found in the solar wind, and therefore the two are not easily distinguished.

Our aim in the following sections is to show the formal similarity between lunar electromagnetic induction and Mie scattering. For this demonstration we shall draw heavily upon earlier work. The key to the similarity lies in the applicable wave equations given in the next section which, as will be seen, are just equations of classical vacuum scattering of light from spherical objects. In the lunar case, complications arise from the radial stratification of the conductivity function for the moon, from modification of the boundary conditions because of the pressure of the solar wind which results in the formation of a surface current layer (Blank and Sill 1969; Schubert and Schwartz 1969), and from the existence, in principle, of steady-state excitation corresponding to ohmic dissipation for the transverse magnetic (TM) mode, which has no analogue in vacuum scattering (Sonett and Colburn 1967). Further, for the moon there is a vanishingly small radiation field, at least for frequencies considered here; the induced field is primarily due to magnetic multipole excitation and includes only a near field. Electric multipoles are by no means ruled out but have so far not been detected. The dominance of magnetic excitation arises naturally from the large bulk electrical conductivity.

The basic equations below have general applicability even for the more exact asymmetric induction problem which takes into account the existence of the lunar plasma cavity; this paper, however, is restricted to spherically symmetric induction. The two approaches are not basically different, but asymmetric induction has been solved so far only for the low-frequency limit (Schubert et al. 1973). Modification of the mode spectrum by the introduction of higher frequencies is presently being worked on.

FIELD EQUATIONS

The transverse magnetic (TM) wave equation is

$$\nabla^2 \Omega^m + k^2 \Omega^m = 0, \qquad (1)$$

and the transverse electric (TE) mode equation is

$$\nabla^2 \Omega^e - \frac{1}{k_1 r} \frac{dk_1}{dr} \frac{\partial}{\partial r}(r\Omega^e) + k^2 \Omega^e = 0, \qquad (2)$$

where $k^2 = \omega^2 \mu \varepsilon + i\sigma\mu\omega$, $k_1 = i\omega\varepsilon - \sigma$, and μ and ε are the magnetic permeability and electrical permittivity, respectively. The field vectors

$\mathbf{E}$ and $\mathbf{H}$ are determined from the potentials Ω^m and Ω^e. The potentials Ω^m and Ω^e are given by

$$\Omega^m = \mu V_s H_0 \sin\phi \sum_{l=1}^{\infty} \beta_l A_l^m G_l^m(r) P_l^1 (\cos\Theta) \qquad (3)$$

and

$$\Omega^e = H_0 \frac{a}{r} \cos\phi \sum_{l=1}^{\infty} \beta_l A_1^e G_1^e(r) P_l^1 (\cos\Theta), \qquad (4)$$

where V_s is the speed of the solar wind; A, the lunar radius; r, Θ, and ϕ, spherical polar coordinates; $P_l^1 (\cos\Theta)$, the associated Legendre polynomials; and

$$\beta_l = i^l \frac{(2l+1)}{l(l+1)}.$$

The radial functions are solutions of the differential equations

$$\frac{d^2 G_l^m}{dr^2} + \left\{ k^2 - \frac{l(l+1)}{r^2} \right\} G_l^m(r) = 0 \qquad (5)$$

and

$$\frac{d^2 G_l^e}{dr^2} - \frac{1}{k^2} \frac{d(k^2)}{dr} \frac{dG_l^e}{dr} + \left\{ k^2 - \frac{l(l+1)}{r^2} \right\} G_l^e(r) = 0. \qquad (6)$$

A more complete theory of the induction has recently been given by Schubert et al. (1973), which includes the asymmetry introduced by the flow field of the solar wind. However, the fundamental properties of the induction can, in most part, be seen from the behavior of the spherically symmetric equations; departures are important, but the theory is still under development for the time-dependent case including the higher-order time multipoles.

Induction in the moon is expressed most conveniently by the formalism of Schubert and Schwartz (1969), where

$$\frac{\mathbf{H}_s}{H_0} = \mathbf{a}_y + (\sigma_s \mu V_s R_m) \sin\gamma \cdot \chi \frac{\left[\frac{1}{2} + \frac{R_c^3}{R_m^3} + \frac{\sigma_s}{\sigma_c} \frac{p_1(k_c R_c)}{2j_1(k_c R_c)} \left(1 - \frac{R_c^3}{R_m^3}\right) \right]}{\left[1 - \frac{R_c^3}{R_m^3} + \frac{\sigma_s}{\sigma_c} \frac{p_1(k_c R_c)}{2j_1(k_c R_c)} \left(2 + \frac{R_c^3}{R_m^3}\right) \right]}$$

$$- \frac{3}{2} \sin\alpha \cdot \alpha \frac{\left(\frac{R_c}{R_m}\right)^3 \cdot B_l(k_c R_c)}{\left[1 - \left(\frac{R_c}{R_m}\right)^3 \cdot B_l(k_c R_c) \right]} \qquad (7)$$

for the magnetic field on the surface of the moon. Here the first term represents the unit forcing field; the second, the induction due to the TM mode; and the last term, that due to the TE mode. H_0 is the forcing field; $\mathbf{a}_y$, a unit vector in the direction of H_0; χ and α, unit vectors in the directions along the axis of toroidal symmetry and poloidal symmetry, respectively; σ_s and σ_c, the shell and core conductivities for our model of a two layer moon; R_c and R_m, the core and lunar radius, respectively; μ, the magnetic permittivity; V_s, the solar-wind speed; p_1 and j_1, spherical Bessel functions in the notation of Stratton (1941); and B_1, an induction number given by

$$B_1(k_c R_c) = 1 - \frac{3}{k_c^2 R_c^2} + \frac{3 \cot(k_c R_c)}{k_c R_c}.$$

Lastly, γ and α are, respectively, the polar angles between the toroidal or poloidal axes to the position of observation on the surface of the moon, and k_c is the wave number in the core. Equation (7) represents the lunar response for the lowest-order induction, but higher orders are obtainable from the general formulae of Schubert and Schwartz (1969). For illustrative purposes, the dipole order suffices, until later when higher-order TE interaction is specifically addressed.

In Equation (7) the second (TM) term is linearly dependent upon the speed of the solar wind, V_s, which determines the interplanetary electric field $\mathbf{E}_m = \mathbf{V}_s \times \mathbf{B}$, which drives the TM mode. However, experimental findings show that this mode is suppressed and the lunar response is dominated by the TE mode, as expected for a partially conducting material. We note that $B_1 < 1$ for all $k_c R_c$, and since $R_c < R_m$, the denominator can never vanish; however, large values for the ratio can be attained as $(R_c/R_m)B_1 \to 1$, confirming that a strong current layer exists on the sunward side of the moon consistent with the dynamic pressure of the solar wind forcing into the moon the induced magnetic field, and yielding amplifications well in excess of those found for vacuum induction.

Although the customary treatment of scattering concentrates primarily on the far field, little radiation field is found in the present case. The near field dominates the response, and most attention is concentrated upon the interior of the scatterer, the moon. We shall return to the properties of Equation (7) in a later section.

THE GEOMETRY OF LUNAR INDUCTION AND THE RESPONSE (TRANSFER) FUNCTION

Viewed as a scattering problem, the basic geometry of the excitation of lunar response by the solar-wind electromagnetic field is shown in

Fig. 1. E, B, and k are taken mutually orthogonally, as in the case for Alfvén waves propagating along the mean interplanetary magnetic field, which is taken along k (not shown). For the case of Apollo 12 (from which the data shown in this paper are taken), the magnetometer lies nearly on the lunar equator; the diagrammatic representation of k in Fig. 1 is indicative of the mean geometry, but time-dependent deviations take place that are significant and must be accounted for. Such details would not materially modify the present discussion. The solar-wind vector is shown as V_s, making an angle δ with k and V_p (phase velocity). The position of the lunar surface magnetometer (LSM) is also shown, and Θ is the scattering angle measured from the direction of k to the outward pointing radius vector from the center of the moon to LSM. The response of the moon for modal order higher than dipole is asymmetric and qualitatively indicated by the two functions A_Θ and A_ϕ.

Data taken in the experiment discussed here are from the lunar satellite Explorer 35, which monitors the free-stream solar-wind magnetic field, free of perturbing influences of the moon, and also from the LSM, which determined the surface field on the moon.

Data taken in orbit are free of influence of the moon because of the supersonic streaming of the solar wind, which prevents upstream propagation of waves from the moon. We define an operational transfer function by

$$h_{2i}(f) + h_{1i}(f) = A(f)_i \cdot h_{1i}(f),$$

where $h_{2i}(f)$, $h_{1i}(f)$ are the Fourier-transformed i^{th} components of the forcing field and response field, f is the frequency, and $A_i(f)$ is the ith component of the transfer function, all defined in a rectangular coordinate system where x is vertical locally at the LSM site, y locally eastward, and z northward. LSM measures the sum of the forcing field and of the induction, while Explorer 35 in lunar orbit measures only the former.

To give the reader some idea of the properties of the data obtained, Fig. 2 shows the transfer function as defined above. These data have been corrected for plasma noise on the lunar surface and are representative of the response in the frequency interval shown. As we shall see shortly, the rollover at high frequency and of the scale is indicative of excitation of higher-order magnetic multipoles. The transfer function attains values higher than three, verifying that containment of field lines is taking place. The model moons shown in the insert are best-fit monotonic models. They are used to make the forward calculated curves that are shown superimposed upon the data and correspond to different values of phase velocity, which is explained by Sonett et al. (1972).

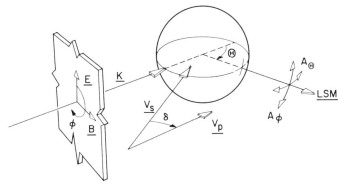

Fig. 1. Geometry of the scattering process. The plane containing **E** and **B** is the plane of the disturbance wave in the solar wind. The propagation is shown along **k** which is also the direction of the interplanetary magnetic field for Alfvén waves. V_s is the solar-wind bulk velocity and V_p the phase velocity, including the convection of the disturbance that dominates the actual propagation of waves in the solar wind. The position of the surface is indicated by LSM, and the scattering angle, Θ, is defined as the angle between **k** and the position vector of LSM referred to the center of the moon. A_Θ and A_ϕ are the response functions in the east-west and north-south directions; they are different for higher-order modes.

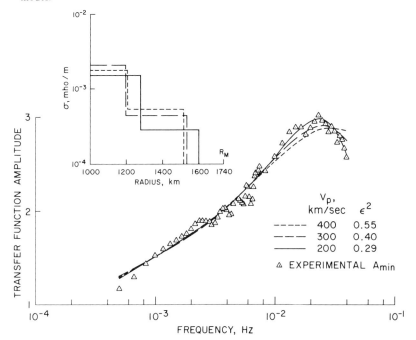

Fig. 2. Transfer function amplitude for lunar response corrected for plasma-generated noise at the lunar surface. The insert shows best-fit model conductivity profiles obtained by inversion of the transfer function together with an assumed value for phase velocity, V_p. Forward calculations of the lunar response are shown as the lines superimposed upon the data.

PARTIAL WAVES

Equations (1) and (2) show that the partial wave excitation is practically restricted to the lowest-order mode when the wavelength of the exciting radiation field is large compared to the scale size of the moon, i.e., $\lambda >> R_m$. This corresponds in the optical case to the Rayleigh limit. Actually, for the moon the problem is considerably more complex as higher order modes have been inferred in the iteration of the conductivity profile.

Assuming the idealized case of Alfvén waves propagating along the direction of the mean solar-wind magnetic field B_0, since B_0 is spiraled because of the high conductivity of the plasma and the rotation of the sun, the angle with respect to the radius vector from the sun is about 45° at 1 AU. Therefore, the wave normals for the Alfvén fields are slanted at about this angle. Since the angle of the mean field $\eta = \tan^{-1}(r/V_s)$, η changes according to the regional value of V_s. Other effects can also change the direction of B_0 over more-or-less short periods of time. Furthermore, the wave field does not necessarily consist of Alfvén waves all the time, and the direction of the wave normals, k, can vary. In general, however, a reasonable representation is obtained by the assumption of a mean field direction of about 45°.

Figures 3 and 4 show the partial fields for two cases calculated using the best three-layer moon conductivity profile. Figure 3 is for a scattering angle $\Theta = 150°$ and Fig. 4 for $\Theta = 120°$. Both use a phase velocity, $V_p = 200$ km/sec, somewhat lower than the usual value, but not seriously in error. The total induced field is given by Σ, while the partial contributions are shown separated into real and quadrature parts. Both figures show significant increases in the higher-order contributions as $2\pi r_m/\lambda$ increases, and the general conclusion that the Mie scattering range is attained, is supported.

Although not a central issue in this paper, we show in Fig. 5 all the model conductivity functions determined by iterative techniques (Sonett et al. 1972). These indicate generally the magnitude of the internal conductivity and justify the conclusion that displacement currents are usually ignorable in the moon, except perhaps near the surface. The various models give a concise statement of the limitation of uniqueness, permitting various combinations of monotonic and current layers. All are electrically equivalent when the amplitude of the induced field is considered, but differences in phase are likely. For the demonstrative purposes of this paper, we use the best-fit three-layer (3L) model.

Variations in V_p and in scattering angle lead to different scattering amplitudes when they are calculated using the 3L model. Several cases are shown in Fig. 6 for comparative purposes. The use of a small value

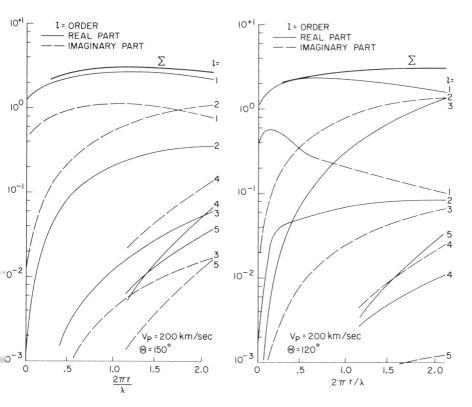

Fig. 3. Partial wave contributions to the scattered (near) wave field showing the sum (Σ), and the real and imaginary parts of the partial waves for $1 \leq l \leq 5$, assuming $V_p = 200$ km/sec and scattering angle $\Theta = 150°$.

Fig. 4. Similar to Fig. 3, except scattering angle $\Theta = 120°$.

of V_p shifts the wavelength spectrum to smaller values and, therefore, increases the contributions from the higher-order modes, as is seen by the steeper rollover. Cases calculated are for a scattering angle of $\Theta = 150°$, which shows the rollover strongly. That the rollover is a strong function of scattering angle is seen by inspection of the upper part of the figure, which uses a fixed value for V_p, namely at 200 km/sec. This clearly is an effect of mutual interference between the different multipoles of the scattered field.

WAVELENGTHS

So far we have shown that the wave equations for lunar scattering are identical to those for the generalized case of induction in a spherical object; it is necessary now to show that the wavelength regime fits that

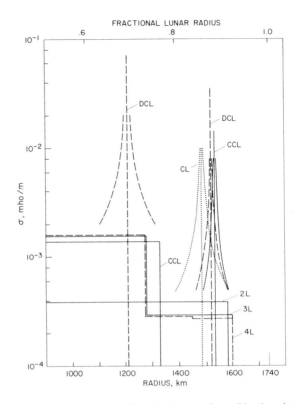

Fig. 5. Lunar internal conductivity profiles showing two-layer (2L), three-layer (3L), four-layer (4L), core-plus-layer (CCL), current-layer (CL), and double-current-layer (DCL) cases; $V_p = 200$ km/sec, $\Theta = 150°$. All are approximately equivalent electrically, but significant differences do exist in the residual errors. Also, no account has been taken of phase up to this time. For details see Sonett et al. (1972).

required for Mie scattering. The wavelength of disturbances in the solar wind is inferred from the Fourier-transformed time series record for each component of the magnetic field. This forcing field is measured on Explorer 35 lunar orbiter. Without knowledge of the solar-wind speed, it is impossible to infer the wavelength spectrum. Furthermore, the wavelength is that taken normal to the surfaces of constant phase, which generally are not oriented with the k vector along the solar-wind direction but along the spiral direction.

We assume a mean velocity for the solar wind, V_s, of 400 km/sec, which is the approximate observable mean value. Next, the direction of the mean interplanetary field is determined using minimum variance techniques. This is based upon determining the total spectral variance

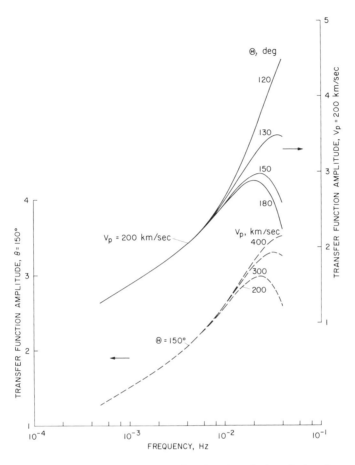

Fig. 6. Forward calculation of transfer functions based on the best-fit three-layer model of Fig. 5. The effects of varying V_p while holding Θ fixed are shown in the lower part of the figure (dashed curves, left ordinate). All cases show some rollover. The upper part of the figure demonstrates the effect of changing scattering angle while holding V_p fixed.

for each of the three components of the field in an arbitrary coordinate system and then rotating until the direction of minimum variance is found. The mean field direction is also found using an averaging technique. Analysis shows the direction of minimum variance and mean field to be colinear; discussion is restricted to these cases, which form the basis for the comments made here.

Assumption of a reasonable solar-wind speed and knowledge of the direction of the interplanetary magnetic and k fields permit a simple geometrical construction to be made to find the phase velocity and, from this, the wavelength spectrum. Figure 7 shows a diagram of the geometry. The figure is a view down upon the plane of the ecliptic. The line OA is the radius vector from the sun along which V_s is measured. (It is assumed that the solar wind is free of aberration, unimportant in the present context.) The line WW is the intersection of the ecliptic with the wave front; $W'W'$ is the wave front transported through the distance PP' by solar-wind convection. (The propagation speed is small compared to this velocity.) The wavelength is defined by QP', so that a wave front is moved normal to itself in the time it takes for the wave front WW to convect through PP'. Therefore the phase velocity, $V_p = V_s \cos \psi$, and the wavelength are given by $\lambda = V_s \cos \psi / f$, where f is the measured frequency. The wavelength regime for the frequencies encountered in the Apollo magnetometer experiment range from $7.7 \times 10^{-3} \leq S \leq 1.4$, measured in units of the scattering parameter $S = 2\pi R_m / \lambda$, where R_m is the lunar radius. Thus, it is seen that the scattering range involved encompasses the Rayleigh limit at low frequency but extends well into the Mie regime at the high-frequency end of the spectrum. The important consequence of this is the addition of higher order modes to the induction spectrum; the dominant magnetic multipole radiation of the excited field includes quadrupole, octupole, and perhaps higher orders, depending upon the forcing field configuration at the time of observation. Waves in the solar wind that are thought to be responsible for the bulk of the induction are consistent with Alfvén waves propagating along the lines of force of the mean field. Such waves have been demonstrated by Belcher and Davis (1971) to be outward propagating and therefore "blue" shifted by the solar wind. Their real frequencies are, therefore, probably still lower than those found by direct measurement in the frame co-moving with the moon. Such waves are known to be nondispersive and have propagation speeds given by $V_a = B/(4\pi\rho)$, where B is the field intensity, and ρ the mass density of plasma. For typical solar-wind conditions, $V_a \cong 50$ km/sec, and therefore, the propagation speed can be ignored in a first-order analysis when compared to the convection speed, V_s, of the solar wind.

SPECIAL PROPERTIES, POLARIZATION, ETC., OF THE MOON

This paper is concerned primarily with the electromagnetic excitation of the moon by the solar wind and with the formal similarity to Mie scattering. We have discussed the excitation of higher-order modes and shown this to be consistent with magnetic-type radiation, although a radiation field is likely to be vanishingly small. Thus, the far field in

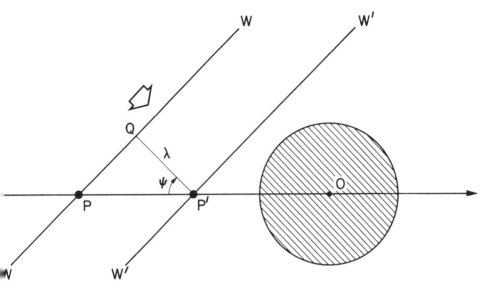

Fig. 7. Geometry of plane waves propagating and convecting with the solar wind. The wave front WW moves one wavelength normal to itself in the time taken for the distance PP' to be covered by solar-wind convection. Thus, wavelengths must be corrected downward by the factor $\cos \psi$.

the sense of radiation theory is important, and the principal response of the moon that is detected by magnetometers consists of the near field, which is nonradiative. The conditions are analogous to an antenna at very low frequency where radiation is trivially small and only the close-in nonradiating electromagnetic field is seen.

The moon is a somewhat special object. Its high internal conductivity is consistent with the suppression of electric-type excitation. As is well known, such media are anomalously dispersive because the phase velocity increases with frequency. Displacement currents are ignored, since $\sigma/\omega\epsilon \gg 1$, where σ is the bulk electrical conductivity, ω the angular frequency, and ϵ the inductive capacity.

The TM mode can probably be shown to be unimportant by other arguments, at least when analysis of existing data undertaken up to now can be used as a guide. As pointed out earlier in this paper, the TM mode has anomalous behavior because of the presence of the conducting solar wind, which permits ohmic currents to flow in the steady state. This has no analogue in vacuum scattering. It can be shown that the ohmic currents must decrease monotonically as the frequency is increased because skin-depth effects come into play and force the stream lines of the lunar interior current nearer to the surface. Therefore, the total resistance increases with an attendant decrease in the integrated current.

However, the above argument cannot be correct for the reactive currents that flow in the TM mode. Using a simple two-layer model for illustration, the capacity of the moon can be shown to be the same as that for a parallel plate capacitor of the same surface area. However, the effective plate separation depends upon the assumed skin depth, which depends on the frequency. The plate separation will decrease as the frequency is increased, so that the capacitive current will increase with frequency. Therefore, a peak in the TM response can take place at some frequency that depends upon various factors such as the conductivity profile in the lunar interior.

Calculation suggests that these effects must be small; the absence of a detectable bow shock wave about the moon shows that the steady-state interaction is virtually suppressed. For these reasons, present consideration of the moon usually ignores the TM mode.

Polarization of signals from the moon has not so far been detected, but theory indicates that the induced field should be elliptically polarized. The reason for the difficulty is that the incoming Alfvén wave field carried in the solar wind appears, when tested, to have random polarization. Until sufficiently "clean" signals — i.e., swaths of data with linearly polarized waves having fixed polarization — are found in the overall data set, it is difficult if not impossible to make an adequate polarization test of the induction. Difficulties also exist with the coherence between Explorer 35 and the lunar surface because of decoherence by the intervening solar wind. Nevertheless, theoretical calculations show that the expected lunar echo at the high-frequency end of the data should be elliptically polarized.

REFERENCES

Belcher, J. W., and Davis, L. 1971. Large amplitude Alfvén waves in the interplanetary medium, 2. *J. Geophys. Res.* 76: 3534–3563.

Blank, J. L., and Sill, W. R. 1969. Response of the moon to the time varying interplanetary magnetic field. *J. Geophys. Res.* 74: 736–743.

Colburn, D. S., and Sonett, C. P. 1966. Discontinuities in the solar wind. *Sp. Sci. Rev.* 5: 439–506.

Parker, E. N. 1963. *Interplanetary dynamical processes.* New York: Wiley.

Schubert, G., and Schwartz, K. 1969. A theory for the interpretation of lunar surface magnetometer data. *The Moon* 1: 106–117.

Schubert, G.; Sonett, C. P.; Schwartz, K.; and Lee, H. J. 1973. The induced magnetosphere of the moon. *J. Geophys. Res.* In preparation.

Schwartz, K. 1971. *Interim final report.* American Nucleonics Corp.

Sonett, C. P., and Colburn, D. S. 1967. Establishment of a lunar unipolar generator and associated shock and wake by the solar wind. *Nature* 216: 340–343.

Sonett, C. P.; Dyal, P.; Parkin, C. W.; Colburn, D. S.; Mihalov, J. D.; and Smith, B. F. 1971. Whole body response of the moon to electromagnetic induction by the solar wind. *Science* 172: 256–258.

Sonett, C. P.; Smith, B. F.; Colburn, D. S.; Schubert, G.; and Schwartz, K. 1972. The induced magnetic field of the moon: Conductivity profiles and inferred temperature. *Proc. 3rd Lunar Science Conf.* Geochim. et Cosmochim. Acta, Suppl. 2, Vol. 3, pp. 2309–2336. Cambridge, Mass.: M.I.T. Press.

Stratton, J. A. 1941. *Electromagnetic theory*. New York: McGraw Hill.

DISCUSSION

SERKOWSKI: If the signals could be interpreted as electromagnetic radiation at a frequency of a small fraction of hertz, and if circular polarization were present, a tremendous angular momentum would be carried by photons of circularly polarized radiation, the angular momentum of each photon being equal to $h/2\pi$.

SONETT: I think that the effect that you propose is far too small to be measurable. The excitation of the moon cannot exceed the total input from the solar wind since some would anyway be lost by joule heating and scattering. Assume a solar wind speed of 400 km/sec and for the disturbance magnetic field of the solar wind, integrated over frequency, $<B^2> = 1$ gamma^2. If we naively assume rectilinear propagation, the Poynting flux would be no greater than (0) 10^{-4} ergs/ (cm^{2}sec). Let this flux be intercepted by the moon. The total cross section is very nearly the geometric value for the moon, i.e. (0) 10^{17} cm^2, so the intercepted energy rate is (0) 10^{13} ergs/sec. The sea of low-frequency photons induced in the moon cannot have a production at a rate exceeding this. The induced magnetospheric spectrum of the moon begins to roll over at the low end at about 10^{-4} Hz. This would represent the case of greatest photon angular momentum M_z which would be about 10^{17} gm cm^2/sec^2. The moment of inertia of the moon is (0) 10^{41} gm cm^2. The angular acceleration of the moon due to the photon field would be $\dot{\omega} = M_z/I$ where I is the moment of inertia. This has a value of about 10^{-24} rad/sec^2. For one year this would be 3×10^{-17} rad/(sec yr). I should caution that in a plasma of finite conductivity it is not clear that the wave field is entirely transverse.

HEURISTIC ARGUMENTS FOR THE PATTERN OF POLARIZATION IN DEEP OCEAN WATER

JOHN E. TYLER
Scripps Institution of Oceanography, La Jolla

Visual and photographic observations have confirmed that the natural light under water is partially polarized and that the plane of vibration exhibits variable tilt.[1] *It is argued that the basic pattern of the amount as well as the tilt of the polarization can be predicted from the classic picture of unpolarized light scattered by a single charged particle. Heuristic arguments are put forth to show that the pattern of polarization is related to the distribution of radiance and, hence, will be a function of wavelength. It is also argued that an asymptotic pattern of polarization will be associated with the asymptotic radiance distribution at great depth. Recently published observations on underwater polarization are reviewed in support of these arguments.*

From simple considerations it can be predicted that the radiant energy from the sun and sky that penetrates into the upper layers of the ocean (or into any large body of water) will exhibit partial polarization. At least three causative mechanisms can be isolated that contribute to this partial polarization. The first arises from the fact that the radiant energy from certain regions of the sky dome is itself partially polarized. The transmission of this partially polarized energy through a roughened air-water interface at various angles of incidence can result in a complex and rapidly changing pattern of polarized light below the surface. The second mechanism arises from the angle of incidence of the rays from the sun (or from some point in the sky) on any small area of the air-water interface. The degree of polarization generated by this mechanism is a function of the angle of incidence and becomes maximum at Brewster's angle. Since the air-water interface controls, to some extent, the angle of incidence, the polarization generated in this way will vary with surface roughness as well as with the altitude of the sun. The third

---

[1] The *tilt* is defined as the angle of the observed direction of electric-vector maximum with respect to the horizontal.

mechanism is a consequence of the scattering of electromagnetic radiation by small particles and molecules within the water body, a mechanism that generates partial polarization.

Of these three mechanisms, only the third is independent of the condition of the air-water interface and is, therefore, the only one that persists with depth in the ocean. In this paper we will assume that the impinging electromagnetic radiation from the sun and sky dome is unpolarized and that polarization due to reflection at the air-water interface does not occur. The discussion will thus be limited to polarization generated by the scattering of electromagnetic radiation by the small particles within a homogeneous body of water.

GENERAL DESCRIPTION

The existence of partially polarized light under water has been known for some time and was mentioned, for example, by Le Grand (1939), who simply stated that the scattered light under water is partially polarized and that his submersible photographic photometer could be equipped with a birefringent component for measuring the extent of the polarization. Kalle (1939) also mentioned the polarization of underwater daylight in his paper "Die Farbe Des Meeres," which is best known for his discussion of Gelbstoff.[2] However, prior to about 1950 the existence of polarized light under water does not seem to have excited the curiosity of oceanographers.

New activity began with Waterman (1954)[3] who made visual observations of the pattern of polarization in clear water in the sea around Bermuda to depths of about 15 meters. Waterman (1955) published detailed information based on photographs of interference patterns taken near Barbados, through a polarization analyzer. Together with Westell (Waterman and Westell 1956) he published a very complete description of the partial polarization of the near-surface light field under water. This latter paper is based on more than 1000 observations at depths ranging from 3 to 41 meters, obtained for the most part in clear water under sunny conditions. Figure 1 shows the polarization pattern they observed; it fits neatly into the classical picture of a beam of unpolarized light (along the z axis) impinging on a single charged particle at 0. If an observer could station himself in Fig. 1 so as to observe the hori-

---

[2] "Yellow substance." In his investigations, Kalle found a variable amount of dissolved, blue-absorbing material in ocean water and argued that this yellow substance, which he could not identify, was the cause of the ocean being green instead of blue.

[3] See also p. 472.

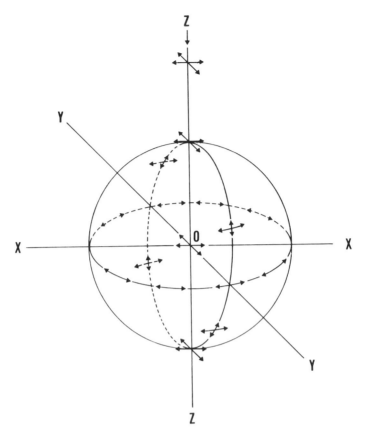

Fig. 1. Unpolarized light is indicated traveling down the z axis and impinging on a single charged particle at 0. The theoretical pattern of polarization is indicated by the arrows. 100% linear polarization occurs in the x-y plane; 0% polarization is found in the z direction. Intermediate between the x-y plane and the z axis, the amount of polarization decreases as indicated by the arrows in the z-y plane. In the text it is imagined that the entire pattern is rotated away from the horizontal around the y axis. An observer looking in the y direction will therefore observe tilting of the plane of vibration with no change in the amount of polarization. An observer looking horizontally in the x-z plane will observe a change in percent polarization with no change in the tilt of the plane of polarization.

zontal pattern of polarization from the charged particle as the impinging beam of unpolarized light changed angle relative to the vertical, he would observe, in the direction at right angles to the vertical plane containing the beam, changes in the angle of the plane of vibration with no accompanying changes in the amount of polarization, which in the single-particle case would remain 100%. If, on the other hand, he made observations in the direction within the vertical plane containing the beam of light, he would observe changes in the percent polarization with no accompanying changes in the angle of the plane of vibration, which would remain horizontal for observations thus confined. In the ocean, 100% linearly polarized light is not observed because of the existence of multiple scattering and also because the impinging radiant energy is multidirectional. In their work, Waterman and Westell (1956) found the amount of polarization in the real ocean to range from 5 to 50% depending on the direction of observation. Ivanoff (1956b), who published similar observations from the French Mediterranean, found the amount of polarization to range from 6 to 34%.

The classic picture of single-particle scattering also helps explain these observed variations in percent polarization with the direction of observation, since for a single particle, maximum polarization occurs in the plane normal to the beam and diminishes to zero along the axis of the beam.

To explain partial polarization due to the penetration of natural radiant energy into deep ocean water, it is necessary to depart from the single-particle single-beam-of-radiant-energy concept. In Fig. 1, if a second beam having the same radiant-energy content as the first were directed along the x axis, a pattern of polarization identical to the first pattern, but oriented relative to the second beam, would be superimposed on the pattern generated by the first beam. This would result in the complete depolarization of the scattered energy along the direction through the particle and at right angles to the plane containing the two beams of radiant energy (i.e., along the y axis in Fig. 1). If, however, two such beams of energy were to impinge upon the particle at some acute angle, partial polarization would be observed along this same direction (i.e., along the y axis as above).

If this reasoning is extended to the more complex situation of multiple beams of light of different energy content coming from all directions, which is the real situation under water, it can be seen that partial polarization must result from every conceivable radiance distribution except isotropic radiance distribution.

In ocean water, every elemental volume is surrounded by a natural radiance distribution, initiated by the natural light from the sun and sky and altered under water by scattering and absorption. The shape of the radiance distribution found in any specific water body will be directly responsible for the pattern of polarization and its angular orientation. In the near-surface layers, the refracted rays of the sun will have the greatest influence on the magnitude and also on any observed tilt of the polarization.

It has been shown by Whitney (1941), Jerlov (1960), Tyler (1960), and others that the natural radiance distribution found near the surface of the ocean will undergo a change in shape as the radiant energy penetrates to greater depths. As a result of this change in shape, the direction of maximum radiance for the distribution will occur nearer and nearer the zenith direction as depth in the ocean increases. Since the pattern of polarization is closely aligned with respect to the maximum radiance vector, it would be expected that for a fixed sun altitude, the tilted planes of vibration that have been observed in the surface waters would exhibit less and less tilt at greater and greater depths.

It has been further demonstrated (Preisendorfer 1959) that at very great depths, the distribution of radiant energy around a point tends asymptotically toward a fixed shape that depends upon the scattering and absorbing properties of the water. In all cases this fixed shape will be symmetrical around the vertical and erect, with the maximum radiance in the zenith direction. At this asymptotic depth, the plane of vibration will consequently be horizontal in all azimuths, and since no further changes in the shape of the distribution would occur at still greater depths, it would be expected that the pattern of percentage polarization and its orientation would also remain unchanged at greater depths. It is axiomatic, therefore, that asymptotic radiance distribution will be associated with an asymptotic pattern of polarization. Because of the intimate relationship between the pattern of polarization and the distribution of radiance around a point under water, it is of special interest to determine the shape of the radiance distribution and identify the factors that influence its shape. A major factor influencing the near-surface radiance distribution is the distribution of energy from the sky dome, which can range, of course, from complete overcast to bright sun with the sun at any altitude. A second and overriding factor governing the underwater radiance distribution is the ratio of the scattering to the absorbing properties of the water itself (Preisendorfer 1959). If the water were a strongly scattering medium having zero absorption, the asymptotic distribution of radiant energy at great depth would be equal in all directions and there would be no detectable polarization.[4] If, on the other hand, the water were an absorbing medium with zero scattering, the distribution of energy at great depth would be confined to a single vector in the vertical direction. Under these latter circumstances, a maximum amount of polarization would be generated and the plane of vibration would be horizontal in all directions.

When these arguments are applied to the different spectral components of the natural light under water, it can be seen that in clear water and

---

[4] This situation cannot occur under water but can be experienced above water in the form of the Arctic "white out."

at great optical depth the amount of polarization in the horizontal direction would be smallest for monochromatic wavelengths near 0.45 μm where water absorption is minimum and scattering is maximum. The percentage polarization for monochromatic red light at great depth would, on the other hand, be relatively large since this is a spectral region of strong water absorption, resulting in an asymptotic radiance distribution that is more nearly collimated. From this it follows that under a completely overcast sky, the amount of polarization for red light will increase with depth because, near the surface, the radiance field is multidirectional whereas at great depth it will become nearly collimated.

OBSERVATIONS OF LINEAR POLARIZATION

These heuristic arguments are, of course, based on simple theory, and it is of considerable interest, therefore, to examine existing experimental data in the real ocean to determine their validity.

From Snell's law and Fig. 1 it can be seen that the angle from the vertical of an underwater beam of light is equal to the angle of refraction of the beam at the ocean surface and also to the tilt, relative to the horizontal, of the plane of electric-vector maximum observed horizontally at 90° from the sun's azimuth bearing. If the heuristic concepts of polarization due to scattering are applicable to complex radiance distributions in the real ocean, a high correlation should be found between measurements of these angles. The paper by Waterman and Westell (1956) provides data (in their Fig. 3) that exhibit this high correlation, and their Fig. 5 offers additional confirmation of the predicted relationship between the angle of refraction of sunlight and the orientation of the plane of vibration observed at right angles to the vertical plane containing the sun.

Lundgren (1971) has recently made measurements in certain spectral regions as a function of depth. In his Table 2 he gives data at several depths for wavelengths near 0.5 μm on both a clear sunny day and an overcast day. His data for the clear day show that the maximum polarization decreased with depth, which is consistent with the observed fact that for sunny conditions the radiance distribution is less collimated at depth than it is near the surface. For a sun elevation of 85° and a depth of 1 meter, Lundgren found the plane of vibration to be horizontal. At 50 meters' depth and for a decreased sun elevation of 75°, he reports the maximum tilt angle of the plane of vibration to be 12°. These angles are consistent with the calculated angles of refraction of the sun's rays at the water surface.

For the overcast day the data of Lundgren (1971, p. 24 and Table 2) seem to report a nonuniform overcast with the sun at an altitude of about 63°. On this basis, the data given in his Table 2 for the tilt of the plane of vibration and the decrease of tilt with depth are consistent with heuristic arguments, since under overcast conditions the radiance distribution will rapidly become more erect with depth. Lundgren's data also show the predicted increase of maximum percentage polarization with depth under overcast-sky lighting conditions. His measured ratio, E_0/E, which is scalar irradiance divided by the down-welling irradiance on a horizontal surface, is a rough measure of the shape of the radiance distribution. Experimental determinations of this ratio have shown that in natural waters, its value does not vary greatly from 1.3. The data given by Lundgren for this ratio range from 1.26 to 1.40, including both sunny and overcast measurements, and although there should be a correlation between this ratio and the maximum amount of polarization, the ratio is probably not a very sensitive indicator, as Lundgren points out.

The earlier work of Ivanoff and Waterman (1958a) and Ivanoff (1955, 1956a,b) also supports the heuristic arguments given here for the pattern of polarization under water. Their measurements near Bermuda of polarization versus depth (to over 100 m) under clear sunny skies show asymptotically decreasing amounts of polarization as depth increases. They also found larger polarizations in the red region of the spectrum, where absorption is high, than in the blue region where absorption is low. In discussing their data, Ivanoff and Waterman (1958a) have themselves suggested the influence of the shape of the underwater radiance distribution on the pattern of polarization.

Sasaki et al. (1959) have made observation of the pattern of polarization from the submersible vessel "Kuroshio" and have reported results similar to those of Waterman and Ivanoff (1958a). In their summary, Sasaki et al. conclude, "... the directionality of penetrating light is one of the essential factors governing the polarization patterns." Sasaki and his co-workers have also expressed the idea that the pattern of polarization is a function of radiance distribution and have deduced that in a fixed water type the tilt of the plane of polarization relative to the horizontal will become zero at a depth closer to the surface on a cloudy day than on a sunny day, because asymptotic radiance distribution will be achieved at a shallower depth.

Timofeeva (1961, 1962) has made a very detailed study of underwater polarization and of the related property of radiance distribution. Her results, obtained from on board the research vessel "Mikhail Lomonosov," confirm the results obtained by Ivanoff and Waterman (1958a) and confirm, as well, the close relationship between radiance distribution and the pattern of polarization with respect to both the tilt and the degree of polarization. In her 1962 paper she makes special reference to the effect of wavelength on the degree of polarization and gives data which again confirm the heuristic arguments. Timofeeva's 1962 paper is especially interesting for its polar presentation of her measurements. These show that some degree of polarization is found in all directions in the ocean and illustrate the limitation of heuristic arguments.

Although the gross features of underwater polarization appear to be correctly explained by heuristic arguments, it remains for the experimentalist to obtain a detailed quantitative description and for the theoretician to devise practical equations.

ELLIPTICAL POLARIZATION

It is known that elliptical polarization exists in the surface layers of the oceans. This phenomenon has been studied by Ivanoff and Waterman (1958b) who concluded that, "... the hypothesis that elliptical light in the sea originates by total reflection of linearly polarized light produced by scattering of directional underwater illumination seems clearly consonant with the facts." It is not yet known if elliptical polarization is also generated at great depths in the ocean, and this possibility has not been considered in the heuristic arguments.

OBSERVATIONAL DIFFICULTIES

Accurate and detailed measurements of the pattern of polarization under water are extremely difficult to obtain. The number of controllable parameters is large. For example, it is necessary to preset the spectral and spatial resolution of the instrumentation and to incorporate means for scanning the spherical field in known increments of azimuth relative to the sun and of angle relative to the zenith. It is also necessary to provide the instrument with a watertight housing capable of withstanding the pressures encountered. For a complete analysis of the state of polarization, it is further necessary to plan a minimum of four measurements for every angular orientation, every depth, and every wavelength.

The amount of natural light available for these measurements will vary by at least a factor of 50 at any single depth and, near the surface, can vary by a factor of something like 10,000 (Tyler 1960). As a function of depth, the natural light in clear water will be exponentially attenuated by the factor 0.04 for each meter in the blue region of the spectrum and by the factor 0.60 for each meter in the red region of the spectrum (Tyler and Smith 1970). In less clear water these factors will, of course, be larger. For a useful package of data, it is consequently necessary to plan circuitry capable of linear response over a range of about 10^6.

The operational difficulties involved in obtaining a useful package of data are even more formidable. The natural light is subject to systematic as well as nonsystematic variations with time. The experimental measurements must be made from a floating platform in water having unknown optical inhomogeneity. Due to surface or submarine currents, there may be appreciable and unknown errors in depth determination as well as in the angular orientation of the instrument. Added to all of these problems are persistent annoyances such as: water surface irregularities that cause excessive fluctuations of the light field and changes in the instrument's depth location; ship shadow; and the curiosity of local fish.

In the ocean under real conditions the pattern of polarization will be more complex near the ocean surface than has been pictured above. The investigator is, therefore, faced with the most complex and variable pattern of polarization at a depth where the operational problems are minimized. At greater depths, the pattern of polarization is simple, but the radiant energy available for measurement is marginal and the operational problems are vastly more difficult.

THEORETICAL WORK

In 1968 Beardsley initiated an ambitious program with the object of determining the pattern of polarization under water, as well as the volume scattering function, from measurements of the Stokes parameters. His doctoral thesis (Beardsley 1966) and two papers (Beardsley 1968a,b) give his theoretical preparation. Continuation of this program has suffered as a result of his untimely death.

The theoretical literature on particle scattering and the resulting polarization is immense and somewhat inadequate for the complex problem represented by inhomogeneous particles in an ocean under natural lighting. An excellent review of this literature and an excellent reference list are given by Jerlov (1968, Ch. 2).

REFERENCES

Beardsley, G. F., Jr. 1966. The polarization of the near asymptotic light field in sea water. Ph.D. Thesis, Mass. Institute of Technology, Cambridge, Mass.
———. 1968a. Mueller scattering matrix of sea water. J. Opt. Soc. Am. 58: 52–57.
———. 1968b. The polarization of submarine daylight at near-asymptotic depth. J. Geophys. Res. 73: 6449–6457.
Ivanoff, A. 1955. Au sujet du facteur de polarisation de la lumière solaire dans la mer. C. R. Acad. Sc. Paris 241: 1809–1811.
———. 1956a. Facteur de polarisation du résidu sous-marin de lumière du jour. Ann. Géophys. 12: 45–46.

———. 1956b. Degree of polarization of submarine illumination. *J. Opt. Soc. Am.* 46: 362.
Ivanoff, A., and Waterman, T. H. 1958a. Factors, mainly depth and wavelength, affecting the degree of underwater light polarization. *J. of Mar. Res.* 16: 283–307.
———. 1958b. Elliptical polarization of submarine illumination. *J. of Mar. Res.* 16: 255–282.
Jerlov, N. G. 1960. Radiance distribution in the upper layers of the sea. *Tellus* 12: 348–355.
———. 1968. *Optical oceanography.* London: Elsevier Pub. Co.
Kalle, K. 1939. Die Farbe des Meeres. *Rapports et procès-verbaux des réunions. Conseil permanent international pour l'exploration de la mer* 109: Pt. 3, 98–105.
Le Grand, Y. 1939. La pénétration de la lumière dans la mer. *Ann. l'Institut Océanographique, Monaco* 19: 393–436.
Lundgren, B. 1971. On the polarization of the daylight in the sea. *Københavns Universitet Institut for Fysisk Oceanografi,* Report No. 17.
Preisendorfer, R. W. 1959. Theoretical proof of the existence of characteristic diffuse light in natural waters. *J. of Mar. Res.* 18: 1–9.
Sasaki, T.; Okami, N.; Watanabe, S.; and Oshiba, G. 1959. Measurements of submarine polarization. *Records of oceanographic works in Japan* (New Series) 5: 91–99.
Timofeeva, V. A. 1961. On problem of polarization of light in turbid water. *Izv. Geophysics Ser.,* pp. 766–774.
———. 1962. Spatial distribution of the degree of polarization of natural light in the sea. *Izv. Geophysics Ser.,* pp. 1843–1851.
Tyler, J. E. 1960. Radiance distribution as a function of depth in an underwater environment. *Univ. of California Bulletin of the Scripps Inst. of Ocean.,* Vol. 7, pp. 363–412.
Tyler, J. E., and Smith, R. C. 1970. *Measurements of spectral irradiance underwater.* New York: Gordon and Breach Science Publishers.
Waterman, T. H. 1954. Polarization patterns in submarine illumination. *Science* 120: 927–932.
———. 1955. Polarization of scattered sunlight in deep water. *Deep-Sea Research* 3 (Suppl.): 426–434.
Waterman, T. H., and Westell, W. E. 1956. Quantitative effect of the sun's position on submarine light polarization. *J. of Mar. Res.* 15: 149–169.
Whitney, L. V. 1941. A general law of diminution of light intensity in natural waters and the percent of diffuse light at different depths. *J. Opt. Soc. Am.* 31: 714–722.

THE POLARIZATION OF LIGHT IN THE ENVIRONMENT

K. L. COULSON
University of California, Davis

Light to which we are exposed in the environment is in general partially polarized, the degree of polarization varying from as much as 0.70 or more in the light from the cloudless sunlit sky, to very small values for light transmitted by a thick overcast. The main mechanisms for producing polarization in the natural environment are Rayleigh-type scattering by the gaseous molecules and other particles of sizes much smaller than the wavelength of light, Mie-type scattering by particles of sizes comparable to or larger than the wavelength, and reflection from surfaces of soils, sands, rocks, water, vegetation, and various man-made materials constituting the underlying surface. Computations and measurements of the polarization in a clear sunlit atmosphere show a maximum in a direction at approximately 90° from the direction of the sun, with points of zero polarization (neutral points) in the plane of the sun's vertical at roughly 25° above (Babinet) and below (Brewster) the sun for cases of high sun elevations. For low sun elevations the Brewster point is replaced by another neutral point (Arago) which appears in the opposite portion of the sky about 25° above the antisolar point. The magnitude of polarization and positions of the neutral points are functions of both wavelength of the radiation and reflection properties of the underlying surface. Dust, haze, and other particulates in the atmosphere tend to decrease the degree of polarization and cause a shift of the Babinet and Brewster points toward the sun, and of the Arago point toward the antisolar point. The maximum degree of polarization of light reflected from natural surfaces varies from 1.0 for reflection from a still water surface at the Brewster angle to less than 0.05 for light reflected from very white sands. Dark soils and vegetation produce maxima of 0.20–0.40, while light-colored soils and desert sands produce maxima of 0.10–0.20. It is observed that highly reflecting surfaces generally produce low polarization maxima while dark surfaces polarize strongly. Neutral points in the light reflected from natural surfaces appear in directions above and below the antisource direction in a manner somewhat analogous to those of skylight. The polarization field of light reflected from natural surfaces is dependent on wavelength, angle of incidence, and properties of the surface itself.

The fact that the light in the environment in which we live is partially polarized has been known since Arago observed in 1809 that the diffuse light from the sunlit sky showed effects of polarization. The polarization of light in the atmosphere has received much less emphasis, however, than it has in the classical laboratory of physics, and atmospheric effects in polarization is still a very active field of research. More progress has been made in both theory and measurements of the polarization of light in the outdoor natural environment starting in the 1940s than was made in the previous eighty years since Lord Rayleigh's explanation (Strutt 1871) of the scattering of electromagnetic waves by molecular size particles.

Of the many possible reasons for renewed interest in polarization of light in the atmosphere, probably the most important are, first, the development of powerful new theoretical techniques, by Chandrasekhar (1950), Sobolev (1963), and others for the calculation of radiative transfer in relatively realistic models of planetary atmospheres, and secondly, advances in instrumentation permitting objective and high-speed measurements of polarization. Also here, as in almost every field, the development of electronic computers opened new possibilities for numerical studies of radiative transfer in the atmospheric medium.

Following his discovery of skylight polarization, Arago established the position of the polarization maximum at approximately 90° from the direction of the sun and discovered the existence of a point in the sky, about 25° above the antisolar direction, at which the polarization vanishes (the neutral point of Arago). The other two neutral points that normally exist in the sunlit sky, the Babinet and Brewster points, were first observed by Jacques Babinet and Sir David Brewster in 1840 and 1842, respectively.

The mechanisms that produce polarization, such as scattering by molecules and aerosol particles and reflection from surfaces, are discussed by Angel (p. 54) and by Ueno (p. 582) in this book. Consequently, the discussion here will be confined mainly to the results of observations and calculations that apply to light in the atmosphere-surface system of the planet earth, although their application is more general than this. Likewise, the effect of atmospheric aerosols is discussed by Rao (p. 500) and Raschke (p. 510), so that subject will be touched on only lightly in this paper.

Rayleigh's elementary theory explained several of the features observed to exist in the field of polarization in the clear blue sky (e.g., position of the polarization maximum and existence of neutral points), but it did not explain the lack of complete polarization at the maximum, the existence of three neutral points in the skylight polarization, or the fact that neutral points do not occur at scattering angles of exactly 0° and 180°. We now know that the deviations of the actual polarization field of skylight from that given by the elementary scattering theory is due to a combination of molecular anisotropy, multiple scattering, atmospheric aerosols, and reflection of light from the underlying surface. These effects will be further discussed below.

In addition to the radiation fields of direct solar radiation and diffuse radiation from the sunlit sky, the natural environment includes the radiation that is reflected from the underlying surface, be it soil, sand, vegetation, water, or other type of material. The physical laws for the reflection of radiation from surfaces are known for only a few idealized cases, such as Fresnel reflection from polished surfaces, still water, etc., or Lambert reflection from a perfectly diffuse surface. Several investigations have been oriented to a parameterization of the reflectance of surfaces with specified roughness properties (Torrance and Sparrow 1965, 1967; Bennett 1963; Porteus 1963). In most such studies, only the scalar field has been included, thereby neglecting the polarization characteristics of the reflected radiation. In a few cases (e.g., Beckmann and Spizzichino 1963) the radiation has been assumed completely polarized, a situation that is also not representative of real reflection in nature. Rao and Chen (1969) used a twelve-term harmonic series to fit the azimuthal pattern of polarization measured for selected samples of soils, but no attempt was made at parameterization of the soil itself. A complete solution has yet to be found, but a step in that direction is being taken by Coulson and Walraven (1973) in a study presently under way, in which a representation for the complete reflection matrix, including polarization, for real surfaces is being obtained in terms of a few generalized parameters based on the physics of the problem. It is anticipated that the results of this work will appear in the near future.

POLARIZATION IN THE PURE MOLECULAR (RAYLEIGH) ATMOSPHERE

For application of scattering theory to the case of the sunlit sky, we assume the incident solar radiation at the top of the atmosphere to be parallel, with a net flux of energy in unit time and unit frequency interval across a unit surface oriented normal to the incident beam to be $\pi \mathbf{F}_0$, where $\mathbf{F}_0$ is a one-column matrix of the Stokes parameters written as

$$\mathbf{F}_0 = \begin{bmatrix} F_{0\|} \\ F_{0\perp} \\ U_0 \\ V_0 \end{bmatrix}. \tag{1}$$

Measurements show the extra-atmosphere incident radiation to be essentially unpolarized, for which case we have the relations

$$\begin{array}{l} F_{0\|} = F_{0\perp} = \tfrac{1}{2} F_0 \\ U_0 = V_0 = 0 \end{array}. \tag{2}$$

A portion of this incident radiation that is scattered by gaseous molecules and other particles in the atmosphere eventually shows up as diffuse light from the sunlit sky. The vector intensity **I** of the radiation scattered by a spherical particle is given in its most general form by the relation

$$\mathbf{I} = \begin{bmatrix} I_\parallel \\ I_\perp \\ U \\ V \end{bmatrix} = \mathbf{PF}_0, \quad (3)$$

where **P** is a sixteen-element scattering matrix with elements P_{ij}, i, $j = 1, 2, 3, 4$. For Rayleigh scattering, in which case the particles are small with respect to the wavelength, the vector intensity of the scattered radiation is given, with respect to scattering angle Θ, by the simplified relation

$$\mathbf{I}(\Theta) = \left(\frac{2\pi}{\lambda}\right)^4 \Delta^2 \mathbf{P}(\Theta) \cdot \mathbf{F}_0 = \left(\frac{2\pi}{\lambda}\right)^4 \Delta^2 \begin{bmatrix} \cos^2\Theta & 0 & 0 & 0 \\ 0 & 1 & 0 & 0 \\ 0 & 0 & \cos\Theta & 0 \\ 0 & 0 & 0 & \cos\Theta \end{bmatrix} \cdot \mathbf{F}_0 \quad (4)$$

where Δ is the depolarization factor.

The geometry of the scattering problem can be seen from Fig. 1, in which the scattering particle is assumed to be at 0, the unit vectors parallel and normal to the scattering plane (plane defined by directions of incident and scattered light) are indicated by $\parallel$ and $\perp$, respectively, and the scattering angle is Θ.

For monochromatic radiation of net flux $\pi \mathbf{F}_0(-\mu_0, \varphi_0)$ on a surface normal to the direction of propagation at the top of a plane-parallel atmosphere, the equation of radiative transfer may be written (Chandrasekhar 1950) as

$$\mu \frac{d\mathbf{I}(\tau; \mu, \varphi)}{d\tau} = \mathbf{I}(\tau; \mu, \varphi) - \tfrac{1}{4} e^{-\tau/\mu_0} \mathbf{P}(\mu, \varphi; -\mu_0, \varphi_0) \mathbf{F}_0(-\mu_0, \varphi_0)$$

$$- \frac{1}{4\pi} \int_0^{2\pi} \int_{-1}^{+1} \mathbf{P}(\mu, \varphi; \mu', \varphi') \mathbf{I}(\tau; \mu', \varphi') d\mu' d\varphi'. \quad (5)$$

Here τ is normal optical thickness of the atmosphere and $\mathbf{P}(\mu, \varphi; \mu', \varphi')$ is the matrix of scattering applicable to the atmospheric medium. The first term on the right represents the directly transmitted part of the original solar beam, the second term is for primary scattering, and the third is for multiple scattering of the radiation within the atmosphere.

The solution of Equation (5) has been obtained by a number of authors,

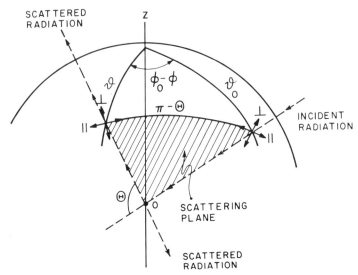

Fig. 1. Geometry of the scattering and reflection problems. The scattering or reflecting element is assumed to be at point 0, the incident radiation is propagated in direction (ϑ_0, φ_0), and the scattered and reflected radiation are propagated downward and upward, respectively, at angles (ϑ, φ). Symbols $\perp$ and $\parallel$ refer, respectively, to directions normal and parallel to the scattering plane (shaded); Θ is scattering angle.

the most widely used being that of Chandrasekhar (1950). By reducing the original integro-differential equation to a system of simultaneous integral equations and developing the principles of invariance for the atmospheric model, Chandrasekhar was able to derive the transmission matrix $\mathbf{T}(\tau_1; \mu, \varphi; \mu_0, \varphi_0)$ and reflection matrix $\mathbf{S}(\tau_1; \mu, \varphi; \mu_0, \varphi_0)$ to give the intensity of the diffuse light transmitted and reflected by an atmosphere of optical thickness τ_1 by means of the relations

$$\mathbf{I}(\tau_1; -\mu, \varphi) = \frac{1}{4\mu} \mathbf{T}(\tau_1; \mu, \varphi; \mu_0, \varphi_0) \cdot \mathbf{F}(\mu_0, \varphi_0) \tag{6a}$$

$$\mathbf{I}(0; \mu, \varphi) = \frac{1}{4\mu} \mathbf{S}(\tau_1; \mu, \varphi; \mu_0, \varphi_0) \cdot \mathbf{F}(\mu_0, \varphi_0). \tag{6b}$$

Computations of the field of radiation transmitted and reflected by a plane-parallel Rayleigh atmosphere have been carried out by a number of authors, including Chandrasekhar and Elbert (1951, 1954), Coulson (1952, 1959), Coulson and Sekera (1955), Sekera (1956, 1957), Dave (1964, 1965), Dave and Furukawa (1966), and de Bary and Bullrich (1964). An extensive set of tables of the intensity and state of polarization of diffuse light for a number of different Rayleigh optical thicknesses has been developed by Coulson, Dave, and Sekera (1960).

The degree of polarization in a Rayleigh atmosphere is shown as a function of angle in the plane of the sun's vertical for three different wavelengths (at sea level) in Fig. 2. The depolarization factor has been neglected in these and subsequent curves. The effects of multiple scattering are clearly evident in the decrease of positive polarization and increase of negative polarization with decreasing wavelength. The physical reason for multiple scattering changing the degree of polarization is that in a statistical sense the photons which are incident on a given scatterer, after having been scattered one or more times, arrive from all possible directions, so those scattered into a given direction by the scatterer encompass all possible scattering angles. This means that all magnitudes of polarization ($0 \leq p \leq 1$) are produced in the composite stream. Since, statistically, $p < 1$ in the multiply scattered component, multiple scat-

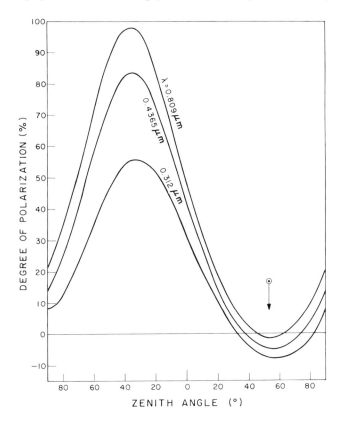

Fig. 2. Degree of polarization of skylight of three different wavelengths in a Rayleigh atmosphere as a function of direction in the sun's vertical for a solar zenith angle of 53° and no reflection from the underlying surface (data from Coulson, Dave, and Sekera 1960).

tering can only reduce the degree of polarization in the direction at right angles to the source, for which $p = 1$ for primary scattering. Conversely, in those directions in which primary scattering gives $p = 0$, multiple scattering introduces radiation characterized by the intensity component $I_\parallel$ parallel to the scattering plane, which is greater than the intensity $I_\perp$ normal to the scattering plane. By convention, this is called negative polarization. Thus, the composite stream at scattering angles $\Theta = 0°$ or $180°$ is changed from neutral ($p = 0$) for primary scattering to partially polarized for the composite. The neutral points of the field occur in those directions at which the positive polarization resulting from primary scattering is just balanced by the negative polarization of the multiple scattered component, each being weighted, of course, by the relative intensity of its contribution to the total. Symbolically, we can write for the neutral points

$$(pI)_{\text{primary}} + (pI)_{\text{multiple}} = 0. \tag{7}$$

On this basis, the increasing shift of the neutral points away from $\Theta = 0$ with decreasing wavelength, as shown in Fig. 2, can be understood to result from a relative increase of I_{multiple} and relative decrease of I_{primary} with decreasing wavelength, and there is no requirement for a wavelength dependence of p for the individual streams.

The polarization pattern moves across the sky as the sun changes position, as shown in Fig. 3. This is a manifestation of the fact that the scattering angle Θ for primary scattering is the dominant parameter for this case of a Rayleigh atmosphere of moderate optical thickness. A number of other changes of the polarization field with position of the sun are evident in the diagram. First, the degree of polarization at the maximum, p_{max}, reaches higher values at both very low and very high solar zenith angles ϑ_0 than at moderate values of ϑ_0. This effect has been studied by Coulson (1952), who found that p_{max} for this case shows a broad minimum at $\vartheta_0 = 50°$ to $60°$, with its value increasing by about 1% at $\vartheta_0 > 80°$ and by as much as 4% to 5% at small values of ϑ_0. This variation must be due to a complex of the polarizing effects of primary and multiple scattering in combination with the relative incident intensity and transmission of the scattered light through varying optical paths as the viewed direction moves between the zenith and horizontal directions.

While the effect is not immediately evident in the curves of Fig. 3, it has been known since the time of Arago that the position of p_{max} is generally at a position somewhat less than $90°$ from the sun. Fraser (1955), who has studied the problem from both theoretical and observational standpoints, found that the deviation of p_{max} from $90°$ as given by theory increases with both decreasing wavelength of the radiation and the amount of light reflected from the underlying surface. The deviation is shown to be

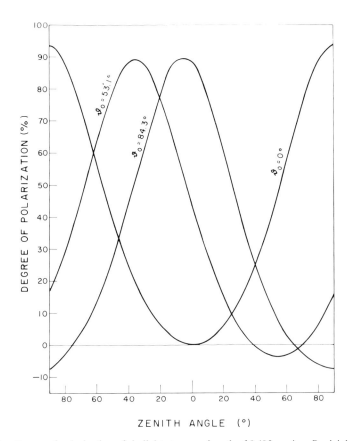

Fig. 3. Degree of polarization of skylight at a wavelength of 0.495 μm in a Rayleigh atmosphere as a function of direction in the sun's vertical for three different solar zenith angles and no reflection from the underlying surface (data from Coulson, Dave, and Sekera 1960).

as much as 5° in the ultraviolet. At longer wavelengths, the deviation is less but still places p_{max} less than 90° from the sun. Observations in the actual atmosphere tend to confirm the theoretical expectations in the ultraviolet, but observations at longer wavelengths, e.g., $\lambda = 0.625$ μm, show p_{max} to be greater than 90° from the sun. It is likely, however, that non-Rayleigh scatterers in the atmosphere are responsible for the discrepancy observed at the longer wavelengths.

Ample evidence shows that radiation that is reflected from the surface and scattered back down toward the surface has a significant effect on the polarization field of skylight. A theoretical method for accounting for this component of skylight was developed by Chandrasekhar (1950); many authors, including Chandrasekhar and Elbert (1954), and Coulson (1968), have studied the problem. The computations are particularly simple

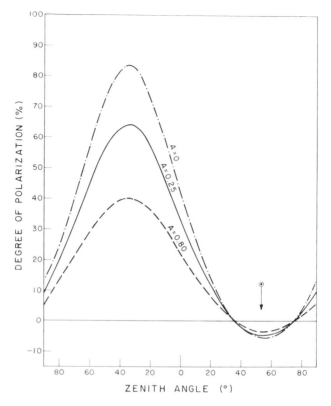

Fig. 4. Degree of polarization of skylight at a wavelength of 0.436 μm in a Rayleigh atmosphere as a function of direction in the sun's vertical for a solar zenith angle of 53° and three different values of (Lambert) surface reflection (data from Coulson, Dave, and Sekera 1960).

for the case of a Lambert surface (reflected radiation unpolarized and isotropic in the outward hemisphere), although Coulson (1968) has discussed the effects of reflection from a real desert surface, and the case of reflection from a Fresnel (mirror-type) surface has been discussed by Fraser (1964, 1966).

Curves of skylight polarization in the principal plane for a Rayleigh atmosphere overlying a Lambert surface are shown in Fig. 4. Results for albedos typical of a completely black surface ($A = 0$), a desert surface ($A = 0.25$), and a new snow surface ($A = 0.80$) are given. One may interpret the effects introduced by surface reflection by considering the stream of skylight radiation to be composed of that which would exist in the absence of a ground or for a black ground ($A = 0$) and that which results from surface reflection. The degree of polarization P of the composite is given for the principal plane by the relation

$$P = \frac{(I_\parallel - I_\perp) + (I_\parallel^* - I_\perp^*)}{I_\parallel + I_\perp + I_\parallel^* + I_\perp^*}, \tag{8}$$

where the asterisks indicate the contribution by surface reflection. Since the outward-directed light at the surface is assumed unpolarized and is of a diffuse character, scattering by the atmosphere introduces little polarization into this component. Thus, to a first approximation, $I_\parallel^* - I_\perp^* = 0$, in which case reflection does not affect the numerator of Equation (8). But $I_\parallel^* + I_\perp^* \neq 0$ in general, and the denominator of Equation (8) may be significantly increased by surface reflection, with the result shown in Fig. 4. Since, by definition, the addition of the unpolarized light contributed by reflection in the direction of the neutral points of the original field would produce no change of polarization, the positions of the neutral points would be very insensitive to Lambert surface reflection. This expected insensitivity is seen in the curves.

Coulson (1968) has shown that for a Rayleigh atmosphere overlying a desert surface, for which the measured reflection characteristics are taken into account, the degree of polarization of the backscattered radiation is not greater than 1% to 3%, for which case the Lambert surface assumption is a good approximation for the polarization of skylight. It is not a good approximation, however, for the light emerging to space from the top of the atmosphere, as surface effects are much more important for the outward radiation than for skylight.

The pattern of polarization over the entire hemisphere of the sky is shown in Fig. 5 for a Rayleigh atmosphere in the absence of surface reflection. The position of the polarization maximum is maintained at a total scattering angle of approximately 90° from the sun, and its magnitude is practically independent of position in the sky. There is a considerable region of negative polarization surrounding the position of the sun, and the positions of the Babinet and Brewster points, above and below the sun, respectively, are well defined in the pattern. At other solar positions (not shown), the polarization field is shifted corresponding to the shift of the sun, but it is qualitatively similar to that of Fig. 5. Similarly, the general hemispheric pattern is largely independent of wavelength, although the degree of polarization is higher at longer wavelengths and lower at shorter wavelengths than that shown in the diagram.

The positions of the neutral points, particularly the Babinet and Arago points, have been studied more extensively than any other feature of the polarization field, the main reasons being that they are quite easy to observe visually and they have been found to be sensitive indicators of atmospheric turbidity (Neuberger 1950). The Brewster point is difficult to observe by eye, so most of the measurements of Brewster-point positions have been accumulated since the advent of electronic polarimeters.

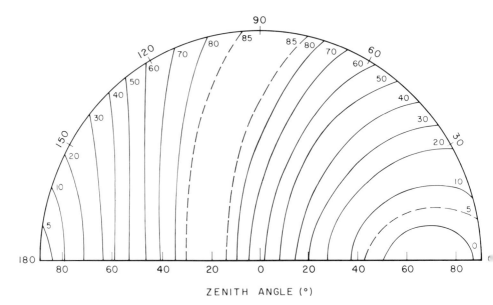

Fig. 5. Hemispheric pattern of the degree of polarization of skylight at a wavelength of 0.495 μm in a Rayleigh atmosphere for a solar zenith angle of 66.4° and no surface reflection. Only half of the hemisphere is shown; the other half is a mirror image of this one (data from Coulson, Dave, and Sekera 1960).

The positions of the neutral points for a Rayleigh atmosphere have been discussed by Chandrasekhar and Elbert (1954), Sekera (1956), Coulson (1954) and others. Their general behavior can be seen from Fig. 6, in which the angular distances of the Babinet and Brewster points from the sun, and of the Arago point from the antisun, are plotted as a function of zenith angle of the sun for two different values of Lambert surface reflectance ($A = 0$ and $A = 0.80$). Since the existence of neutral points at positions other than in the exactly forward and backward directions is due to multiple scattering, the wavelength dependence shown by the curves is caused by increasing multiple scattering with decreasing wavelength. From symmetry considerations, the Babinet and Brewster points must both coincide with the sun at $\vartheta_0 = 0$. The curves for Arago and Brewster points join at the cusps, at which solar zenith angle both points are exactly on the horizon. Chandrasekhar and Elbert (1954) pointed out that for a Rayleigh atmosphere, the Brewster point rises as the Arago point sets, and vice versa. An interesting indication in the diagram is the possibility of a double Arago point occurring in short wavelengths at $\vartheta_0 \approx 65°$ for the case of very low surface reflectance.

The dotted curves of Fig. 6 show a slight modification of neutral-point positions for the case of a highly reflecting (Lambert) surface, such as a new snow surface, in the sense of a shift of the Babinet and Brewster points away from the sun and the Arago point away from the antisun. The effect is minor, however, particularly at the longer wavelengths, and it is likely that a non-Lambert character of any natural surface and non-Rayleigh-type scattering would make the effect very difficult to observe in the real atmosphere. For instance, Fraser's (1964) calculations for the case of a specularly reflecting (mirror) surface indicate some very significant changes of neutral-point behavior from that shown in Fig. 6. Under conditions of small optical thickness (long wavelengths) and small values of ϑ_0, the Babinet and Brewster points merge and disappear, while at the same time neutral points appear at some distance to the right and left of the principal plane. Further, the doubling property of the Arago point appears to be enhanced by specular reflection. Both of these features have been observed over bodies of water, the appearance of neutral points on each side of the principal plane by Soret (1888) and a double Arago point by Jensen (1942).

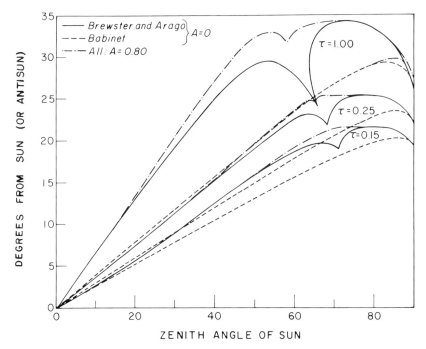

Fig. 6. Angular distance of the Babinet and Brewster points from the sun, and of the Arago point from the antisun, at three different optical thicknesses in a Rayleigh atmosphere for two different values of (Lambert) surface reflectance. Optical thicknesses of 1.0, 0.25, and 0.15 correspond, respectively, to wavelengths of 0.312, 0.436, and 0.495 μm at sea level.

SKYLIGHT POLARIZATION IN THE REAL ATMOSPHERE

Before about the time of World War II, measurements of skylight polarization were made primarily by visual means. Brewster (1864) made a large number of measurements of neutral-point positions, as did Cornu (1890), Dorno (1919), and Jensen (1932), all by visual means. The degree of polarization of skylight also received considerable attention in the late 19th and early 20th centuries, the most notable authors perhaps being Soret (1888), Cornu (1890), Dorno (1919), Gockel (1920), Tichanowski (1926, 1927), Kalitin (1926), and Jensen (1932, 1942).

Although several types of electronic polarimeters were developed after about 1945, some workers continued to obtain significant results by visual means. Pyaskovskaya-Fesenkova (1958, 1960), in particular, reported many measurements of skylight polarization for different atmospheric and surface conditions, all taken with visual instruments, and found empirical expressions for describing the distribution of polarization in the sun's vertical and in the almucantar of the sun. Some of her results for the solar almucantar are shown by the plot of degree of polarization versus azimuth from the sun in Fig. 7. The measurements (curve and points) were taken in the Libyan desert at a height of 200 m above sea level by means of a visual polarimeter with a yellow Schott filter on a cloudless day when the solar zenith angle was between $73°$ and $80°$. The atmospheric transparency at the time of the measurements was 0.88, indicating a relatively clean atmosphere. The crosses show values of polarization p given by the simple relation

$$p(\Theta) = p_{\max} p_R, \qquad (9)$$

where p_R is the polarization for Rayleigh scattering at angle Θ, and

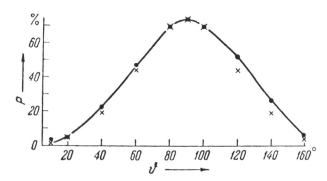

Fig 7. Degree of polarization in the solar almucantar (curve and points) as observed by means of a visual polarimeter at a height of 200 m above sea level in the Libyan desert. The crosses are values computed from Equation (9) (after Pyaskovskaya-Fesenkova 1960).

p_{max} is the maximum polarization observed at the time. A better fit to the data was obtained at a mountain observatory (1450 m). There appeared to be little relation between atmospheric transparency and quality of the fit of Equation (9), although of course p_{max} was a strong function of turbidity. Stamov (1955, 1963, 1970) also carried out empirical studies based on visual observations of skylight polarization, and Neuberger (1950) used visual means for studying the behavior of the Arago point in a turbid atmosphere.

In spite of their extensive use, visual types of instruments suffered from being slow in operation, from restricting measurements to the visible spectrum, and from the fact that considerable skill is required to obtain reliable results. These disadvantages are all overcome in electronic types of polarimeters.

Electronic polarimeters have been used for skylight measurements beginning in the 1950s by many workers, including Sekera et al. (1955), Holzworth and Rao (1965), Gehrels (1962), Gehrels and Teska (1963), Coulson (1971), Bullrich et al. (1966a, b), Ivanov (1971), Ivanov et al. (1971), and Isaev (1968). Typical results of measurements in the sun's vertical for a cloudless atmosphere are shown for two different wavelengths and two different sun elevations in Fig. 8. The data were taken under very clear atmospheric conditions in a rural location near Davis, California. The approximation to a Rayleigh atmosphere must have been very good in this case, as indicated by the horizontal visibility being in excess of 130 km and the absence of discernible whitening of the sky near the horizons. The general configuration of the curves is similar to that for a Rayleigh atmosphere, the maxima being at about 90° from the sun and the neutral-point positions being at roughly their expected positions. The major discrepancies between these results and those expected in the Rayleigh case are a shift of the Babinet-point position toward the sun and a lower maximum of polarization for a wavelength of 0.652 μm.

The shift of the Babinet point toward the sun at the longer wavelengths has been observed many times in the past (e.g., Sekera 1956) and is caused by aerosol effects, which are always present in even the clearest cases of the real atmosphere. The reason that the maximum polarization is not as high as might be expected for such a clear atmosphere is more difficult to explain. The computations of Coulson, Dave, and Sekera (1960) indicate that for a wavelength of 0.652 μm and no surface reflection, the maximum for the Rayleigh atmosphere should be approximately 96%. If a realistic value of surface albedo for Davis were considered, the maximum would be decreased to approximately 93%. The depolarization factor for scattering by anisotropic molecules, which was not included in the computations, would decrease the maximum by approximately another 4%, making the maximum polarization for an aerosol-free atmo-

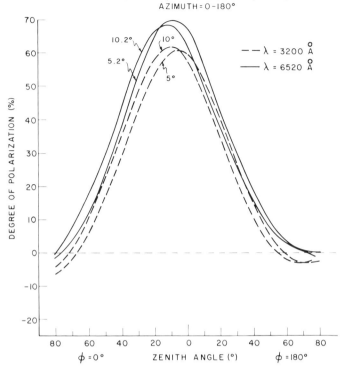

Fig. 8. Degree of polarization as a function of direction in the sun's vertical for two different wavelengths and at various solar elevations as observed in a very clear atmosphere at Davis, California.

sphere over a surface such as that in the Davis vicinity approximately 89%. The observed decrease of polarization from that value to about 70% must be ascribed to the depolarizing effects of atmospheric aerosols.

The depolarizing effects of the aerosols in the atmosphere can be seen more explicitly by the measurements in clear and polluted atmospheric conditions in Los Angeles, shown in Figs. 9 and 10. Interpolations among the data were necessary in some cases to obtain curves at these particular positions of the sun. Curves for the Rayleigh atmosphere are included for comparison. The most apparent modifications of the polarization field by the air pollution of Los Angeles are (a) a pronounced decrease of polarization over most of the dome of the sky, including both positive and negative areas of the field, (b) a small but consistent shift of the positions of the Babinet and Brewster points toward the sun and the Arago point toward the antisun, and (c) a shift of the position of the polarization maximum from about 90° from the sun at a sun elevation of 10° to less than 90° at the higher sun elevations.

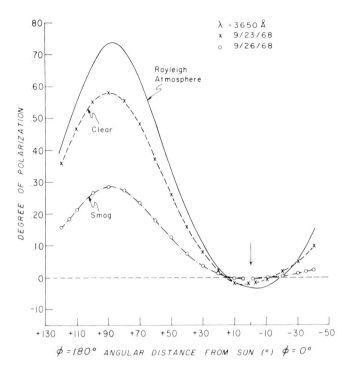

Fig. 9. Degree of polarization of skylight as a function of direction in the sun's vertical for a wavelength of 0.365 μm as computed for a Rayleigh atmosphere and as measured in clear and polluted atmospheric conditions in Los Angeles, California (after Coulson 1971).

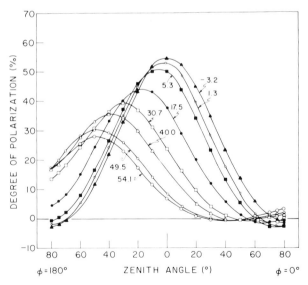

Fig. 10. Degree of polarization of skylight at a wavelength of 0.32 μm as a function of direction in the principal plane as measured at seven different solar elevations in a polluted atmospheric condition in Los Angeles (after Coulson 1971).

459

Observations of the neutral-point positions have spanned a century and a half, and they are still a subject of considerable interest because of their sensitivity to atmospheric particulates. The existence of volcanic ash in the atmosphere produces large changes of neutral-point behavior, which appear to be unique to volcanic ash. Cornu (1890) observed neutral points to be situated symmetrically on each side of the sun and antisun after the eruption of Krakatoa, and the positions of the Babinet and Arago points were strongly shifted after the eruptions of both Krakatoa (in 1883) and Mount Katmai (in 1912). Sekera (1950) has summarized the observations for normal atmospheric conditions and for post-volcanic conditions, the results of which are shown in Fig. 11. The observations were made visually in "white" light. The Babinet point appears to be particularly sensitive to volcanic ash, undergoing at least a $+16°$ shift of position at a sun elevation of $4°$, which changes to a $-2°$ shift when the sun is $4°$ below the horizon. The Arago point is less sensitive, but it also shows some anomalous effects produced by volcanic ash.

Series of measurements of neutral-point positions at different wavelengths have been made with electronic polarimeters by Sekera et al. (1955), Holzworth and Rao (1965), Coulson (1971), and others. Typical locations of the points, as observed at three visible wavelengths by Holzworth and Rao, are plotted as a function of sun elevation in Fig. 12.

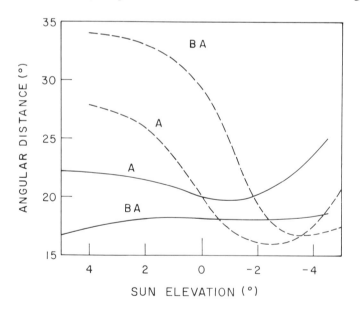

Fig. 11. Angular distance of the Babinet point (BA) from the sun and the Arago (A) point from the antisun for normal atmospheric conditions (solid curves) and as observed after the volcanic eruption of Krakatao (dashed curves) (adapted from Sekera 1950).

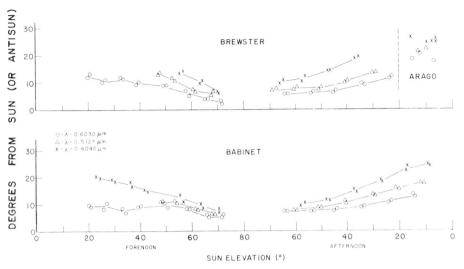

Fig. 12. Observed positions of the neutral points of skylight polarization as a function of sun elevation for three different wavelengths in relatively clear atmospheric conditions in Los Angeles (after Holzworth and Rao 1965).

The measurements were taken under good visibilities and clear skies in Los Angeles. The shift of neutral-point positions toward the sun (or antisun for the Arago point) with increasing wavelength is clearly seen in the data. This shift is a result of decreasing multiple scattering with increasing wavelength, and corresponds roughly to that predicted by theory as shown previously in Fig. 6.

Similar measurements of the location of the Babinet-point position in the ultraviolet by Coulson (1971) taken in a very clear atmosphere at a rural location near Davis, California, are compared in Fig. 13 with those for moderately clear and for heavily polluted atmospheric conditions in Los Angeles. The theoretical curve is for a Rayleigh atmosphere at a wavelength of 0.365 μm with no reflection from the underlying surface. It is seen that the Babinet point in the real atmosphere, in which both aerosol scattering and surface reflection play a part, is considerably closer to the sun than it would be in a theoretically pure atmosphere. The shift of the Babinet point toward the sun is particularly pronounced in the heavily polluted atmosphere. Multiple scattering is very strong in this case, as evidenced by the high intensity of skylight in polluted conditions. Since multiple-scattering effects are responsible both for the existence of neutral points and for the fact that neutral points in a Rayleigh atmosphere are farther from the sun in short than in long wavelengths, on the basis of multiple scattering alone one might expect the pollution to shift the

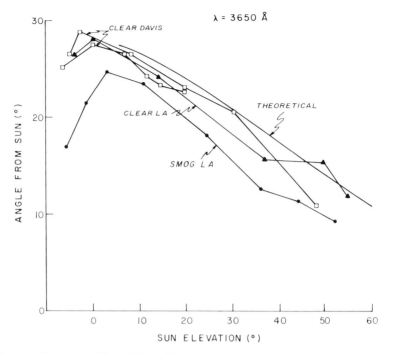

Fig. 13. Observed positions of the Babinet point at a wavelength of 0.365 μm as a function of sun elevation for very clear atmospheric conditions at Davis, California, and for relatively clear and heavily polluted conditions in Los Angeles. The theoretical curve is for a Rayleigh atmosphere.

Babinet point away from the sun. The fact that the observed shift is in the opposite direction is explained by the non-Rayleigh type of scattering produced by aerosol particles. Available measurements indicate that even in the clearest of atmospheres found in nature, there are sufficient aerosol particles to produce a significant deviation of neutral-point behavior from that predicted for the pure Rayleigh atmosphere. In addition, it is likely that the non-Lambert-type reflection of light from real ground surfaces will have a significant effect on the positions of the neutral points, but this aspect has received relatively little attention so far.

Scattering theory indicates the possibility of elliptical polarization in the atmosphere, as contrasted with plane polarization discussed above. Elliptic polarization may be introduced by the scattering of plane-polarized light by aerosol particles. Since aerosol scattering is necessary to produce elliptic polarization, it would appear that measurements of elliptic polarization, should be a unique tool for studying the aerosols themselves. In a long series of studies, Rozenberg and his colleagues (e.g., Rozenberg and Gorchakov 1967; Gorchakov 1966; Rozenberg 1960) have measured the ellipticity introduced by scattering of a plane-polarized light beam by atmospheric aerosols to determine the elements of the scattering matrix, and Eiden (1966) has used a similar method with artificial lighting. Unfortunately, the full potential of the method has not been realized under natural lighting conditions because of the very small magnitude of elliptic polarization that exists in the sunlit atmosphere. However, Dave (1970) has obtained theoretical indications that the ellipticity may be as much as 4% in some cases, and states that, "... one would expect the sky radiation to show strong elliptical polarization under such circumstances where the primary scattering contributes very little to the emergent radiation." Preliminary indications in this direction are reported in this book by Hanneman and Raschke (p. 510). New instrumentation of high precision (Coulson and Walraven 1973) is becoming available for application to skylight measurements, so it seems likely that elliptic polarization will receive increased attention in the future.

POLARIZATION BY REFLECTION FROM NATURAL SURFACES

As pointed out by Angel on p. 54, the reflection of light is an efficient polarizing mechanism, and as is well known, complete polarization may be produced by reflection from a Fresnel-type reflector. Natural surfaces such as soils, sands, and vegetation, although not generally Fresnel-type reflectors, also introduce greater or lesser amounts of polarization into the reflected radiation, the degree of polarization depending on the type and condition of the surface, wavelength, and angles of incidence and reflection of the radiation. The effect of the polarization of skylight by the introduction of this partially polarized radiation into the atmosphere is outside the scope of this paper; it has been discussed in some detail by Fraser and Sekera (1955), Sekera (1961), and Fraser (1964, 1966) for the case of a Fresnel reflector, such as a still-water surface, and by Coulson (1966, 1968) for surfaces more typical of continental locations. It has been well established that the polarization of light directed downward from the sky is much less strongly affected by surface reflection than is that for light directed outward from the top of the atmosphere.

Measurements of the polarizing properties of materials extend back to the pioneering work of Arago and to the set of measurements performed by Brewster (1865) on painted surfaces, paper, cloth, snow, and white powders of various kinds. Lyot and Dollfus (1949), by use of a visual polarimeter, determined the degree of polarization of light reflected by the moon and by Mars.

The polarizing properties of a number of types of natural and artificially produced materials have been measured by Dollfus (1957, 1961), Pellicori (1971), Coffeen (1965), Fernald, Herman, and Curran (1969), Coulson,

Bouricius, and Gray (1965), and others for interpreting the composition of the lunar and Martian surfaces. Since such extraterrestrial use is covered elsewhere in this book (see Bowell and Zellner, p. 381), the present discussion will be confined to selected surfaces that occur on earth.

Curves showing the degree of polarization of radiation of 0.492 μm reflected in the principal plane from various types of natural surfaces are shown for angle of incidence $\vartheta_0 = 53°$ in Fig. 14 and for $\vartheta_0 = 0°$ in Fig. 15. The incident light was unpolarized for all of these measurements. Features of the polarization field of reflected light that have been found most interesting are the magnitude and position of the polarization maximum, the positions of the neutral points (which occur above and below the direction of the source for small angles of incidence and, for

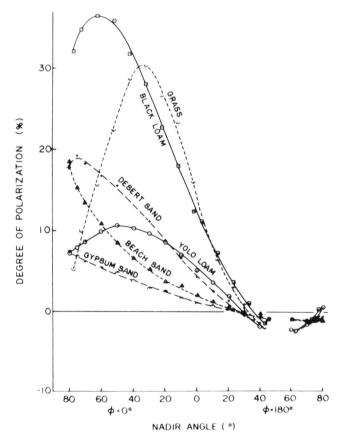

Fig. 14. Laboratory measurements of the degree of polarization of light at 0.492 μm reflected in the principal plane from various types of natural surfaces for an angle of incidence of 53°.

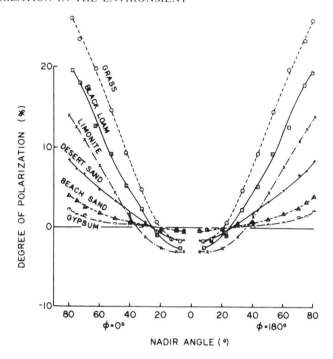

Fig. 15. Laboratory measurements of the degree of polarization of light at 0.492 μm reflected in the principal plane from various types of natural surfaces for an angle of incidence of 0°.

some surfaces, above the antisource direction for large incidence angles), and the slope of the positive branch of the polarization curve near the neutral point. On p. 392 of this book, Bowell and Zellner have discussed determination of sizes of asteroids and satellites by measurements of the slope of the polarization curve for light reflected from those bodies.

A general characteristic of all of the surfaces investigated is that highly reflecting surfaces show relatively low polarization of the reflected radiation, whereas surfaces that have low polarization polarize strongly. Similarly, the wavelengths that are strongly reflected have low polarization, and vice versa. This inverse proportionality between reflectance and polarization is known as the "Umov effect," after the Russian physicist who discovered the characteristic.[1] The effect is well illustrated by Fig. 14, the highly reflecting gypsum sand showing less than 10% polarization and the dark surfaces of black loam and green-grass turf polarizing as much as 30% to 40%. The positions of the neutral points are also sensitive to total reflectance, as are the slopes of the polarization curves.

---

[1] See, however, p. 223.

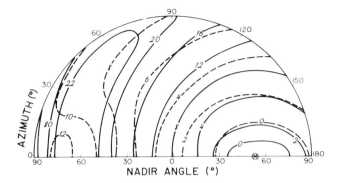

Fig. 16. Hemispheric patterns of the degree of polarization of light of wavelength 0.492 μm reflected from black loam soil (solid isopleths) and desert sand (dashed isopleths) for an angle of incidence of 53° (after Coulson 1966).

The hemispheric patterns of the degree of polarization of light of wavelength 0.643 μm reflected from desert sand and black loam soil are shown in Fig. 16. The patterns for the two surfaces are quite similar, the most obvious feature being the curvature of the isopleths around the antisource direction. This means that the angle between the antisource direction and the direction of view is the dominant geometric parameter of the polarization field. Although the details vary, the fields shown by Fig. 16 are typical of other wavelengths and incident angles, as well as for other natural surfaces for which measurements have been made.

For smooth surfaces for which the index of refraction of the material is known, it is a simple matter to compute the degree of polarization of the reflected light by the well-known Fresnel law of reflection. Laboratory measurements of the degree of polarization of light reflected from a still-water surface, as given by Chen and Rao (1968), for three different wavelengths are shown in Fig. 17. These authors found a considerable deviation from the polarization predicted by the Fresnel laws, in the sense of scattering in nonspecular directions in the principal plane. It is thought possible that the effect may be an artifact of the measurements, and it needs further verification.

Although the effects of moisture in porous surfaces such as soils has received relatively little attention, soils in the moist state are generally darker than in the dry state, and in accord with the Umov effect, they polarize more strongly. This is shown by the measurements for a red clay soil at a wavelength of 0.52 μm in Fig. 18. Although no pronounced reflectance maximum was found in the measurements, it is likely that a Fresnel-type reflection from the water films surrounding the grains of soil are responsible for the pronounced increase of polarization produced by surface moisture.

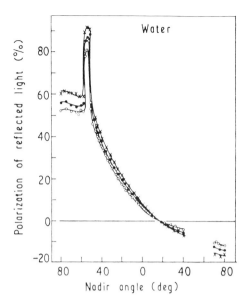

Fig. 17. Degree of polarization of light of three different wavelengths (circles: 0.3975 μm; dots: 0.50 μm; crosses: 0.605 μm) reflected in the principal plane from a smooth water surface (after Chen and Rao 1968).

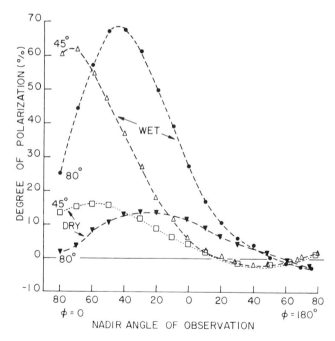

Fig. 18. Degree of polarization of light of wavelength 0.52 μm reflected in the principal plane from red clay soil, for two different angles of incidence and two conditions of surface moisture.

REFERENCES

Bary, E. de 1964. Influence of multiple scattering on the intensity and polarization of diffuse sky radiation. *Applied Optics* 3: 1293–1303.

Bary, E. de, and Bullrich, K. 1964. Effects of higher-order scattering in a molecular atmosphere. *J. Opt. Soc. Am.* 54: 1413–1416.

Beckmann, P., and Spizzichino, A. 1963. *The scattering of electromagnetic waves from rough surfaces.* New York: Macmillan.

Bennett, H. E. 1963. Specular reflectance of aluminized ground glass and the height distribution of surface irregularities. *J. Opt. Soc. Am.* 53: 1389–1394.

Brewster, D. 1864. Observations of the polarisation of the atmosphere, made at St. Andrews in 1841, 1842, 1843, 1844, and 1845. *Trans. Roy. Soc. Edin.* 23: 211.

―――. 1865. On the polarization of light by rough and white surfaces. *Trans. Roy. Soc. Edin.* 23: 205–210.

Bullrich, K.; Eiden, R.; Jaenicke, R.; and Nowak, W. 1966a. *Optical transmission in the atmosphere in Hawaii.* Final Tech. Rept., DA-91-591-EVC-3458, Gutenburg Univ., Mainz, Ger.

Bullrich, K.; Eiden, R.; and Nowak, W. 1966b. Sky radiation, polarization and twilight radiation in Greenland. *Pure App. Geophys.* 64: 220–242.

Chandrasekhar, S. 1950. *Radiative transfer.* Oxford: Clarendon Press.

Chandrasekhar, S., and Elbert, D. D. 1951. Polarization of the sunlit sky. *Nature* 167: 51–54.

―――. 1954. Illumination and polarization of the sunlit sky on Rayleigh scattering. *Trans. Am. Phil. Soc.* 44: 643–728.

Chen, H. S., and Rao, C. R. N. 1968. Polarization of light reflected by some natural surfaces. *Brit. J. Appl. Phys.* Ser. 2, 1: 1191–1200.

Coffeen, D. L. 1965. Wavelength dependence of polarization. IV. Volcanic cinders and particles. *Astron. J.* 70: 403–413.

Cornu, A. 1890. Sur l'application du photopolarimetrie à la meteorologie. C. R. Ass. franc. Av. Sci., Session a Limoges, pp. 267–290.

Coulson, K. L. 1952. *Polarization of light in the sun's vertical.* Sci. Rept. No. 4, Contr. AF 19(122)-239, Univ. of California, Los Angeles.

―――. 1954. *Neutral points of skylight polarization in a Rayleigh atmosphere.* Sci. Rept. No. 7, Contr. AF 19(122)-239, Univ. of California, Los Angeles.

―――. 1959. Characteristics of solar radiation emerging from the top of a Rayleigh atmosphere. *Plan. Space Sci.* 1: 265–276.

―――. 1966. Effects of reflection properties of natural surfaces in aerial reconnaissance. *Applied Optics* 5: 905–917.

―――. 1968. Effect of surface reflection on the angular and spectral distribution of skylight. *J. Atmos. Sci.* 25: 759–770.

―――. 1971. On the solar radiation field in a polluted atmosphere. *J. Quant. Spectrosc. Radiat. Transfer* 11: 739–755.

Coulson, K. L.; Bouricius, G. M.; and Gray, E. L. 1965. Optical reflection properties of natural surfaces. *J. Geophys. Res.* 70: 4601–4611.

Coulson, K. L.; Dave, J. V.; and Sekera, Z. 1960. *Tables related to radiation emerging from a planetary atmosphere with Rayleigh scattering.* Berkeley: Univ. of California Press.

Coulson, K. L., and Sekera, Z. 1955. *Distribution of polarization and the orientation of the plane of polarization of sky radiation over the entire sky in a Rayleigh atmosphere.* Final Rept., App. B., Contr. AF 19(122)-239, Univ. of California, Los Angeles.

Coulson, K. L., and Walraven, R. L. 1973. A photon-counting polarizing radiometer. *Applied Optics.* In press.
Dave, J. V. 1964. Importance of higher order scattering in a molecular atmosphere. *J. Opt. Soc. Am.* 54: 307–315.
———. 1965. Multiple scattering in a non-homogeneous Rayleigh atmosphere. *J. Atmos. Sci.* 22: 273–279.
———. 1970. Intensity and polarization of the radiation emerging from a plane parallel atmosphere containing monodispersed aerosols. *Applied Optics* 9: 2673–2684.
Dave, J. V. and Furukawa, P. M. 1966. Intensity and polarization of the radiation emerging from an optically thick Rayleigh atmosphere. *J. Opt. Soc. Am.* 56: 394–400.
Dollfus, A. 1957. Étude de planètes par la polarisation de leur lumière. *Ann. d'Astrophys.,* Supp. 4 (English Translation NASA TT F-188, 1964).
———. 1961. Polarization studies of the planets. *Planets and satellites.* (G. P. Kuiper and B. M. Middlehurst, eds.), Ch. 9. Chicago: Univ. of Chicago Press.
Dorno, C. 1919. Himmelshelligkeit, Himmelspolarisation, und Sonnenintensität in Davos (1911 bis 1918). *Met. Zeit.* 36: 109–124, 181–192.
Eiden, R. 1966. The elliptical polarization of light scattered by a volume of atmospheric air. *Applied Optics* 5: 569–575.
Fernald, F. G.; Herman, B. M.; and Curran, R. J. 1969. Some polarization measurements of the reflected sunlight from desert terrain near Tucson, Ariz. *J. App. Met.* 8: 604–609.
Fraser, R. S. 1955. *Theoretical positions of maximum degree of polarization.* Final Rept. App. C, Contr. AF 19(122)-239, Univ. of California, Los Angeles.
———. 1964. *Theoretical investigation: the scattering of light by a planetary atmosphere.* Final Rept. Contr. NAS 5-3891, TRW Space Technology Labs., Redondo Beach, Calif.
———. 1966. *Theoretical investigation: the scattering of light by a planetary atmosphere.* Final Rept. Contr. NAS 5-9678, TRW Space Technology Labs., Redondo Beach, Calif.
Fraser, R. S., and Sekera, Z. 1955. *The effect of specular reflection in a Rayleigh atmosphere.* Final Rept. App. E. Contr. AF 19(122)-239, Univ. of California, Los Angeles.
Gehrels, T. 1962. Wavelength dependence of the polarization of the sunlit sky. *J. Opt. Soc. Am.* 52: 1164–1173.
Gehrels, T., and Teska, T. M. 1963. The wavelength dependence of polarization. *Applied Optics* 2: 67–77.
Gockel, A. 1920. Beitrage zur Kenntnis von Farbe und Polarisation des Himmelslichtes. *Ann. Phys. Leipzich* (4), 62: 283–292.
Gorchakov, G. I. 1966. Light scattering matrices in the atmospheric surface layer. *Izv. Atmos. & Oceanic Phys.* 2: 359–366.
Holzworth, G. C., and Rao, C. R. N. 1965. *Investigations of the polarization of the sunlit sky.* Sci. Rept. No. 1, Contr. AF 19(628)-3850, AFCRL-65-167, Univ. of California, Los Angeles.
Isaev, G. S. 1968. Some results of photoelectric measurements of atmospheric polarization in the city of Shachtii. *Atmospheric Optics,* Akad. Nauk., USSR, pp. 92–96 (in Russian).
Ivanov, A. I. 1971. On the possibility of polarimetric control of the optical stability of the atmosphere. *Works of the Astrophysical Institute,* Akad. Nauk., Kaz. SSR, 18: 59–62 (in Russian).
Ivanov, A. I.; Livshitz, G. Sh.; Tashenov, B. T.; and Fedulin, A. V. 1971. Brightness and polarization of the sky in the solar alucantar in the near infrared region of the spectrum. *Scattering and absorption of light in the atmosphere.* Vol. 18, Science Publishers, Kaz. SSR, Alma Ata (in Russian).
Jensen, C. 1932. Normale, gestorte und pseudonormale Polarisations-erscheinungen der Atmosphäre. *Met. Zeit.* 49: 419–430.

Jensen, C. 1942. Die Polarisation des Himmelslichtes. *Handbuch der Geophysik* 8: 527–620, Berlin.

Kalitin, N. N. 1926. Zum Studium spektraler Polarisation des Himmelslichtes. *Met. Zeit.* 43: 132–140.

Lyot, B., and Dollfus, A. 1949. Polarisation de la lumière cendreé de la lune. *C. R. Acad. Sci. Paris* 228: 1773–1775.

Neuberger, H. 1950. Arago's neutral point: a neglected tool in meteorological research. *Bull. Am. Met. Soc.* 31: 119–125.

Pellicori, S. F. 1971. Polarizing properties of pulverized materials with special reference to the lunar surface. *Applied Optics* 10: 270–285.

Porteus, J. O. 1963. Relation between the height distribution of a rough surface and the reflectance at normal incidence. *J. Opt. Soc. Am.* 53: 1394–1402.

Pyaskovskaya-Fesenkova, E. V. 1958. On scattering and polarization of light in desert conditions. *Doklady, Akad. Nauk.,* USSR, 123 (6): 1006–1009 (in Russian).

———. 1960. Some data on the polarization of atmospheric light. *Doklady, Akad. Nauk.,* USSR, 131(2): 297–299 (in Russian).

Rao, C. R. N., and Chen, H. S. 1969. *An atlas of polarization features of light reflected by desert sand, white sand, and soil.* Sci. Rept. No. 3, Contr. F19628-67-C-0196, Univ. of California, Los Angeles.

Rozenberg, G. V. 1960. Light scattering in the earth's atmosphere. *Soviet Phys.* USPEKHI, 3(3): 346–371 (English Trans.).

Rozenberg, G. V. and Gorchakov, G. I. 1967. The degree of ellipticity of the polarization of light scattered by atmospheric air as a tool in the investigation of aerosol microstructures. *Izvestia Atmos. & Oceanic Phys.* 3: 400–407.

Sekera, Z. 1950. Polarization of skylight. *Compendium of Meteorology.* pp. 79–90, Am. Met. Soc., Boston.

———. 1956. Recent developments in the study of the polarization of skylight. *Advances in geophysics.* Vol. III, pp. 43–104.

———. 1957. Polarization of skylight. *Encyclopedia of physics.* Vol. 48, pp. 288–328. New York: Springer Publ. Co.

———. 1961. *The effect of sea surface reflection on the sky radiation.* Monograph No. 10, Union Géodesique et Géophysique Internationale.

Sekera, Z.; Coulson, K. L.; Deirmendjian, D.; Fraser, R. S.; and Seaman, C. 1955. *Investigation of the polarization of skylight.* Final Rept., Contr. AF 19(122)-239, Univ. of California, Los Angeles.

Sobolev, V. V. 1963. *A treatise on radiative transfer.* (S. I. Gaposchkin, trans.) Princeton: D. Van Nostrand.

Soret, J. L. 1888. Influence des surfaces d'eau sur la polarisation atmosphérique et observation de deux points neutres à droite et à gauche du soleil. *C. R. Acad. Sci. Paris* 107: 867–870.

Stamov, D. G. 1955. On the question of the dependence of the polarization of skylight on meteorological conditions in the earth's atmosphere. *Izvestia Crimean Pedagogical Inst.* 21: 301–312 (in Russian).

———. 1963. Influence of the optical inhomogeneities of the atmosphere on the diurnal course of skylight polarization. *Actinometry and atmospheric optics.* pp. 176–181 (in Russian).

———. 1970. The necessity of accurate quasi-empirical relations for summarizing observations of skylight polarization. *Atmospheric optics.* pp. 138–144 (in Russian).

Strutt, J. W. (Lord Rayleigh). 1871. On the light from the sky, its polarization and colour. *Phil. Mag.* 41: 107–120, 274–279.

Tichanowski, J. J. 1926. Resultate der Messungen der Himmelspolarisation in verschiedenen Spektrumabschnitten. *Met. Zeit.* 43: 288–292.

———. 1927. Die Bestimmung des Optischen Anisotropiekoeffizienten der Luftmolekulen durch Messungen der Himmelspolarisation. *Phys. Zeit.* 28: 252–260.

Torrance, K. E., and Sparrow, E. M. 1965. Biangular reflectance of an electric nonconductor as a function of wavelength and surface roughness. *J. Heat Trans.* C87: 283–292.

———. 1967. Theory of off-specular reflection from roughened surfaces. *J. Opt. Soc. Am.* 57: 1105–1114.

DISCUSSION

WITT: From rocket observations in the downward direction near 90° scattering angle, ∼94.5% polarization has been found, in accordance with recent determinations of the anisotropy of air molecules. This was from altitudes above 55 km and at wavelengths 366 and 536 nm; when noctilucent clouds are present, the polarization is increased.

POLARIMETERS IN ANIMALS

TALBOT H. WATERMAN
Yale University

*In apparent response to the considerable degree of linear polarization in the Rayleigh scattered sunlight of the atmosphere and hydrosphere, many animals have evolved a capacity to perceive **E**-vector direction. Because natural polarization patterns depend mainly on the sun's position in the sky, they provide an extension of the sun-compass by which an animal may steer required azimuth directions. Most known cases of **E**-vector sensitivity occur in arthropods and cephalopods whose eyes have rhabdoms as their photoreceptor organelles. Rhabdoms have been shown to constitute intraretinal dichroic analyzers. Their selective absorptance depends on the fine structure of their extensive membrane system, comprised of regularly arrayed microvilli, as well as on the molecular dichroism and orientation of the visual pigment which they contain. Some insect rhabdoms comprise three, six or more apparent **E**-vector analyzing components. In other insects, most decapod crustaceans, and in cephalopods, there are two orthogonal polarization-sensitive channels per rhabdom. For a fixed analyzer, Stokes parameters I and Q could be readily specified by such a system with two channels. Several ways of determining U are feasible: e.g., successive measurements with different rhabdom orientations or fixed occurrence of more than one rhabdom orientation in local areas of the retina. But the means actually used to measure the third parameter if it is required has yet to be experimentally demonstrated.*

Since much natural irradiation is partially plane polarized, it is not surprising that many animals can perceive the E-vector of linearly polarized light (Waterman 1966a). Obviously, the polarimeters involved were not designed by astronomers, physicists, or engineers, like most of those discussed in this book. Instead they were evolved in nature because of the adaptive advantage of the organism's being able to "read" just one more potentially informative parameter in the visual environment. In general, the more perfectly its sense organs can track all significant stimuli, the more likely an organism is to survive and propagate.

Much of the polarized electromagnetic energy discussed elsewhere in this volume is either invisible to earthbound animals or is presumably too faint or too weakly polarized to be adaptively significant. Nevertheless, there are three particular sources of polarized light that compose a substantial if not ubiquitous part of animals' visual environment: (1) Rayleigh scattered light in the atmosphere (Sekera 1957), (2) Rayleigh scattered light underwater (Waterman 1954; Ivanoff 1973; Tyler[1]), and (3) differentially reflected light from surfaces, particularly air-water interfaces (Coulson[2]).

The polarization of such natural light is almost entirely linear, and no biological significance can yet be assigned to the known quantitatively very minor exceptions to this rule.[3] While any directional light ray is subject to polarization by scattering and reflection, all of the biological significance so far known for natural polarized light is restricted to sunlight.

Consequently, this review will deal primarily with the specific mechanism evolved in certain animal groups to analyze the position angle of linearly polarized sunlight in the atmosphere and hydrosphere. The polarimeter concerned is located within the retina of the rhabdom-[4] bearing eyes of arthropods (such as insects and crustaceans) as well as cephalopods (such as squids and octopuses). E-vector analysis has been shown in these eyes to depend on the fine structure of the photoreceptor cells and on the orientation of visual pigment chromophores.[5] Together these form a characteristic array of dichroic analyzers.

BIOLOGICALLY SIGNIFICANT POLARIZATION

Before introducing the evidence for this polarimetric system, it may be well to describe briefly the natural polarization patterns that must be analyzed and to discuss the known or probable significance of perceiving them in the life of the animals concerned. The degree of polarization of the clear sky ranges from $P = 0$ at the neutral points in lines of sight near the sun and antisun, to $P = 90\%$ or more in sight lines $90°$ to the sun. Hence, under optimal conditions, strong linear polarization

---

[1] See p. 434.   [2] See p. 444.

[3] For example, there is some elliptical polarization present underwater in lines of sight directed toward the surface just beyond the critical angle (Waterman 1954; Ivanoff and Waterman 1958a). Also, the light reflected from the wing covers of certain beetles is strongly circularly polarized (Wolstencroft, see p. 495). Circular dichroism is a well-known property of many large organic molecules (e.g. Beychok 1966; Mommaerts 1969), but no case is specifically known where this becomes visually significant in nature.

[4] A rodlike photoreceptive element containing visual pigment and lying axially in each structural unit of the compound eye.

[5] The photon-absorbing part of a pigment molecule, such as visual purple.

will be present in some parts of the blue sky at all times of day. Both P_{max} and P for the sky around the zenith will be least at noon and greatest at sunrise and sunset (Sekera 1957). Clouds and fog generally depolarize the sky, but complex effects may result from the nature and shape of certain aerosol particles.

Underwater, the maximum degree of polarization is less than in clear air, but the whole spherical irradiance distribution is partially polarized, not just the hemisphere centered about the zenith as in air. Near the water surface, the sky polarization pattern is visible within the critical angle but will gradually be replaced in this sector by polarization originating in the water as depth increases (Waterman 1954, 1972). Maximum P measured in clear water near the surface is about 60% (Ivanoff and Waterman 1958b). Polarization decreases with depth but, nevertheless, is estimated to be 20%–30% at equilibrium depths and below (Tyler[6]).

The third extensive source of polarized light in nature is reflected irradiance. Its E-vector is perpendicular to the plane of incidence, and $P = 100\%$ at the critical angle. As a result, partial to fully linearly polarized light is ubiquitous, being especially prominent on wet surfaces in the air and at air-water interfaces. The value of P required for satisfactory operation of biological polarimeters has not been studied extensively. However, both in the honeybee (von Frisch 1965) and in the small aquatic crustacean *Daphnia* (Waterman 1966b), the behavioral threshold for E-vector discrimination is somewhere about 10% linear polarization.

USEFULNESS OF E-VECTOR ANALYSIS

A final important consideration governing the design of an animal's polarization analyzer is the use to which the device is to be put. This will establish the parameters that must be measured as well as the accuracy and frequency with which they are monitored. There are two apparent applications for the three major sources of polarization identified above.

One is the visual advantage that may accrue from the elimination of glare and surface reflection from natural bodies of water. Since these are predominantly polarized horizontally, a photoreceptor system sensitive only to vertically oscillating photons could be of considerable advantage, as in the familiar sunglasses operating on this principle. Although there is no direct evidence yet, there are two insect eyes whose organization suggests that this particular possibility has been exploited (Schneider

---

[6] See p. 434.

and Langer 1969; Bohn and Täuber 1971; Trujillo-Cenóz and Bernard 1972).

The other application is undoubtedly a more important one in terms of general biological usefulness. This depends on the fact that polarization patterns both in the atmosphere and in the hydrosphere arise primarily from the Rayleigh scattering of directional rays of the sun in the medium (Fig. 1). As a result, the E-vector orientation in various lines of sight depends on the sun's position in the sky. The resulting

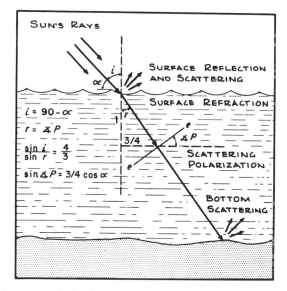

Fig. 1. Underwater polarization originating by Rayleigh scattering of refracted sun's rays. Overall E-vector pattern depends on the sun's position in the sky and lines of sight. (Waterman and Westell 1956).

pattern provides a potential extension of the sun compass to any organism that can perceive it (Fig. 2).

This is an orienting mechanism used by a wide range of animals for determining azimuth directions through observing the bearing and elevation of the sun.[7] These, of course, change characteristically as a function of time of day and the observer's location. Hence, the effective use of a sun compass for more than a local direction finder over brief time

---

[7] Orientation to the moon and stars is also well documented for certain animals but will not be considered here. For recent general reviews of animal orientation and navigation, see Adler (1971) and Galler, Schmidt-Koenig, Jacobs, and Belleville (1972).

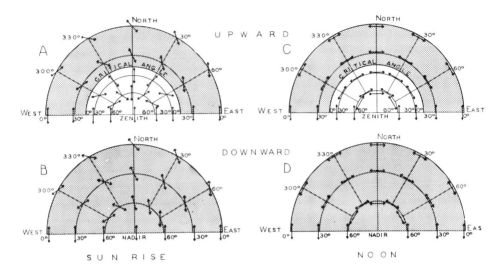

Fig. 2. Underwater polarization patterns in shallow depths at sunrise (*A*, *B*) and noon (*C*, *D*). Polar plots in upward (*A*, *C*) and downward (*B*, *D*) sight lines. E-vector orientation indicated by double arrows for various azimuths and elevation angles of the line of sight. Note that at shallow depths sky polarization is visible above the critical angle (*A*, *C*). (Waterman 1954).

intervals requires information on the solar path as a function of time of day plus a chronometer to measure the latter variable.

The above summary makes evident the basic characteristics required for an animal polarimeter that can perceive the E-vector of natural light in biologically useful ways. It must measure the position angle for a given line of sight but need not measure degree of polarization or ellipticity. Only substantial values of P ($>10\%$) are known to be analyzed, so the operational threshold is rather high.

THE ANALYZER

From an abstract point of view, the required analyzer could be located in one of three sites. The simplest site would be extraocular and depend on differential reflection from surrounding objects. Such reflection is, of course, related by Fresnel's equations to the position angle and angle of incidence of the incoming beam. While such a mechanism could clearly be used under certain circumstances, it would not in general discriminate Stokes parameters I from U and would impose strong limitations on direction of view versus E-vector analysis.

Alternatively, a dichroic component could be located somewhere in the dioptric system of the eye. The cornea or any other element before the retina could be involved. However, if it were a perfect analyzer, it would subtract at least half of all light entering the eye and superimpose an E-vector biased signal on all other visual input channels, unless the dichroic area was characteristically restricted in extent.

Finally, the analyzer could be an inherent part of the light-detecting system and so organized that specific receptor channels are specialized for handling this particular irradiation parameter. While more than one of these three mechanisms is no doubt used by animals, only the retinal analyzer has been studied in some detail. In fact, its mechanism of E-vector analysis is basically understood in several particular cases.

Furthermore, this is functionally the most interesting of the possibilities. It is known to occur in the rhabdom-bearing eyes of arthropods and cephalopods. Their retinal fine structure and restricted orientation of visual pigment chromophores provide the necessary dichroic analyzers. In the retinas of such eyes, the photoreceptor cells collaborate in groups of two to eight or more to form the rhabdom, which lies parallel to the direction of illumination and contains the visual pigment, rhodopsin. To sharpen the present review, attention will be focused on the rhabdom of the crayfish which has been extensively studied from several points of view, including a direct spectrophotometric study (Waterman, Fernández, and Goldsmith 1969).

The crayfish compound eye comprises between 1000 and 2000 structural elements, the ommatidia (Fig. 3). Each is marked externally by a square corneal facet. Beneath the cornea lie the other dioptric components and the photoreceptor units. The latter are arrayed around the optic axis of each ommatidium in a cluster of seven regular retinular cells which contribute microvilli to the composite rhabdom. This ellipsoidal structure, about 20×120 μm, is made up of around 25 transverse layers. Electron microscopy has demonstrated that these layers are composed of closely packed microvilli about 0.08 μm in diameter. They are proximally continuous with their parent cell and closed at their tips. Within each layer, microvilli are perpendicular to the optic axis and all parallel to one another. Alternate bands have orthogonal microvilli which, at least in the central retina, approximate respectively vertical and horizontal directions relative to the body axes of the animal (Fig. 4).

On a cellular basis, three of the seven retinular cells give rise to alternate rhabdom layers with horizontal microvilli, and the other four to those with the vertical subunits. For any given retinular cell, all its microvilli are approximately parallel. The axial organelle thus formed is the light-absorbing part of the receptor; excitation triggered by photons is transmitted from the microvilli containing the photosensitive pigment via the corresponding retinular cell body and its proximal axon to the optic ganglia.

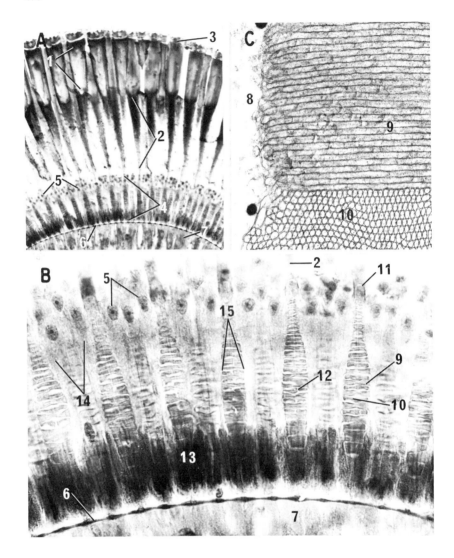

Fig. 3. Structural organization of the crayfish eye. *A*. Radial section showing ommatidia comprising crystalline cone (*1*), cone stalk (*2*), cone cells (*3*), rhabdom (*4*), nuclei of retinular cells (*5*), basement membrane (*6*) and primary optic fibers (*7*). ($\times$ 90). *B*. Rhabdom layer seen at greater magnification. Additional features are orthogonal layers of microvilli (*9, 10,* see also *C*), distal cap of rhabdom (*11*), break showing point where closed ends of microvilli meet in midline (*12*), screening pigment (mostly bleached away) (*13*), cytoplasm of retinular cells (*14*), width of rhabdom (*15*). ($\times$ 400). *C*. Electron micrograph of parts of two layers of rhabdom microvilli parallel (*9*) and perpendicular (*10*) to the plane of section. Cytoplasm (*8*) of the retinular cell from which microvilli in layer *9* arise. ($\times$ 24,400). (Waterman, Fernández, and Goldsmith 1969).

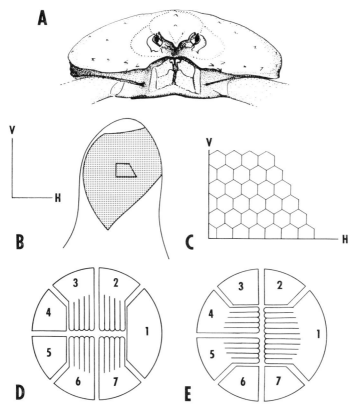

Fig. 4. Relation between dichroic channel orientation in the central retina and a spider crab's overall symmetry. (A). Location of stalks bearing the compound eyes (frontal view of female, 80 mm carapace width). (B). Anterolateral aspect of the stalked eye. Faceted cornea stippled. Trapezoid indicates site of sections cut. (C). Enlargement of corneal surface showing hexagonal facets (20 μm in diameter). (D) and (E). Cross sections of the seven regular retinular cells present in each ommatidium. The two dichroic analyzer channels are seen to have their microvilli vertical in one layer (R_2, R_3, R_6, R_7) and horizontal in the other (R_1, R_4, R_5). The visual pigment that absorbs photons and is itself dichroic is contained within the microvilli (0.09 μm in diameter) of the axial rhabdom. (Eguchi and Waterman 1968).

THE DICHROIC ELEMENTS

To one seeking a clue to the mechanism of E-vector discrimination, this elaborate structure immediately suggested a two-channel analyzer with orthogonal dichroic sensitivities (Waterman 1966c,d). Subsequently, several lines of evidence culminating in direct microspectrophotometry were found to support this notion. First, selective adaptation experiments on the electroretinogram showed that there were two independent orthogonal E-vector sensitive systems in the decapod crustacean retina (Waterman and Horch 1966). Then further adaptation studies coupled with electron microscopy proved that maximum absorption must occur parallel to the cylindrical axis of the microvilli (Eguchi and Waterman 1968).

Finally, fresh crayfish rhabdoms were isolated from dark-adapted eyes and studied microspectrophotometrically (Waterman, Fernández, and Goldsmith 1969). The banding of these structures could readily be seen using light microscopy; with some searching, it was possible to find an example so oriented that one set of microvilli was parallel to the plane of observation and the other perpendicular to that.

Using a linearly polarized test beam measuring 4×8 µm at the preparation, it was possible to determine the absorptance cf a single layer of microvilli. This was done at wavelengths from 400 to 675 nm with the E-vector parallel and perpendicular to the optic axis of the rhabdom. In favorable preparations, optical properties of several layers could be measured in succession (Fig. 5).

The results confirmed the conclusion from selective adaptation experiments (Eguchi and Waterman 1968) that the microvilli are dichroic, with the absorptance parallel to their long axis at λ_{max} being two times that in the perpendicular direction (Waterman, Fernández, and Goldsmith 1969). This dichroism was observed when the microvilli lay parallel to the plane of observation. When they were perpendicular to it (i.e., parallel to the test beam), there was no significant anisotropy. Due to the symmetry and orientation of the cylindrical microvilli, the dichroism measured transversely is also valid for light parallel to the rhabdom's normal optic axis.

Therefore, the rhabdomeres of three cells (Nos. 1, 4, and 5) are dichroic in vivo, absorbing twice as much light horizontally polarized as vertically. Similarly, the microvilli of the vertical cells (Nos. 2, 3, 6, and 7) absorb twice as much vertically polarized light as horizontally. This demonstrates directly that there is a pair of dichroic orthogonal input channels in each ommatidial unit of the polarized-light analyzing system in the crayfish (Fig. 6). However, there are two important but as yet experimentally unsettled matters relating to this conclusion.

One is that intracellularly measured electrical responses of retinular cells indicate apparent E-vector sensitivity ratios up to ten or more (Shaw 1969; Waterman and Fernández 1970), far in excess of the dichroism found optically (Langer and Thorell 1966; Hays and Goldsmith 1969; Kirschfeld 1969; Waterman, Fernández, and Goldsmith 1969). However, stimulus-induced current flow in light receptors is generally proportional to photon absorption (Penn and Hagins 1972) and theoretical analysis indicates that in a fused layered rhabdom like that of the crayfish, the ratio of the parallel and perpendicular absorption coefficients should be exactly equal to the ratio of the total power absorbed by two rhabdomeres with orthogonal microvilli (Snyder 1973).

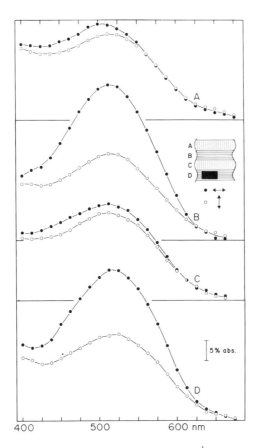

Fig. 5. Absorptance as a function of wavelength (abscissa) measured microspectrophotometrically in four contiguous microvillus layers of a crayfish rhabdom (λ_{max} about 525 nm). See insert for locations. Layers A and C are essentially isotropic, but B and D have dichroic ratios of nearly two. The greater absorption occurs parallel to the axes of the microvilli. (Waterman, Fernández, and Goldsmith 1969).

VISUAL INFORMATION PROCESSING

The other unsettled matter more central to the present context relates to how the sensory input is processed to evoke an appropriate response in the animal. We know, of course, that the polarization of any beam of electromagnetic energy is uniquely characterized by its four Stokes parameters (e.g., Chandrasekhar 1960; Clarke and Grainger 1971). Since no significant circular component is present in animals' normal visual experience, $V = 0$. Hence, I, Q, and U need to be determined to characterize the intensity and polarization fully. If the light were 100% linearly polarized, one of these variables would be redundant, since $I^2 = Q^2 + U^2$. However, in mixtures of unpolarized light with

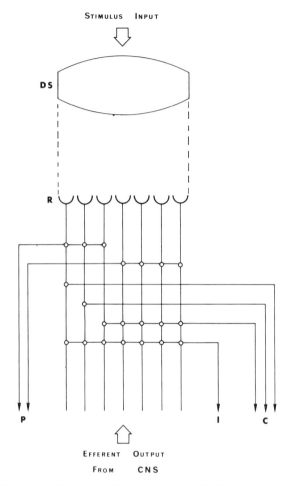

Fig. 6. Information channeling diagramed for a single ommatidium of a compound eye. The dioptric system (*DS*) conducts light to the photoreceptor layer (*R*) comprising seven regular retinular cells. In many forms, these have to analyze **E**-vector orientation and wavelength as well as register intensity and some component of form. *I* is proportional to the sum of the seven cellular channels; **E**-vector is analyzed by two orthogonal channels (*P*), and in various cases, color is separated by two, three, or more channels (*C*). (Waterman 1966*b*).

various amounts of linearly polarized light, $I^2 > Q^2 + U^2$, and all three variables might have to be specified by the animals' polarimeters.

Clearly *I*, which is just overall intensity, is a basic visual parameter for all photoreceptors; it might be determined from the input of **E**-vector sensitive channels or from other elements unaffected by the direction of the polarization plane. Q and U, defining as they do the rectangular and oblique components of the position angle, are obviously central to our problem. However, before proceding further, the kinds of response animals show to polarized light need to be briefly reviewed so that operational requirements of their polarimeters will be more explicit.

POLAROTAXIS

Two types of oriented E-vector response (polarotaxis) are known. In the so-called basitactic response to the plane of polarization (Fig. 7), the orientation is related to E-vector direction by one or more of the following fixed angles: 0°, 45°, 90°, 135° (Jander and Waterman 1960). In the second type of polarotactic behavior, which is a menotactic response to E-vector orientation, any azimuth direction may be temporarily determined from the E-vector position angle (von Frisch 1948). In this case, the direction of vibration is being used like the magnetic north of a magnetic compass. The course steered from the E-vector may lie in any arbitrary direction within 180°. Obviously, some other cue must be used to select one or the other of the two semicircles symmetrical with the plane of polarization.

The analysis of such azimuth orientation data can be effected if there is only one peak, by computing vector sums and then using statistical methods appropriate for circular distributions to determine significance levels (e.g., Batschelet 1965). For multiple peaks, periodic regression using Fourier component analysis has been applied to establish the location and amplitude of significant preferences (e.g., Waterman 1963). A provisional and much simpler way to characterize response data of this sort is to estimate the degree of orientation O by assuming that the peaks and troughs in the distribution approximate a sinusoid or are at least symmetrical about the mean.

Then by analogy with the usual way of defining the degree of linear polarization P let

$$O = \frac{d_{\max} - d_{\min}}{d_{\max} + d_{\min}}$$

where d is the data entry for the directions specified. For a polarotaxis where the orientation maximum $d_{\max}$ is parallel to the E-vector, $O_{\parallel}$ can range from nearly 0 to 1.0. Thus the example in Fig. 10 has $O_{\parallel} = 0.28$ indicating a 28% preference for orientation parallel to the overhead E-vector.

To relate basitactic and menotactic responses to the biological polarimeter, we must consider the following questions: (1) Where do the fixed angles of a basitaxis originate in the input processing system?

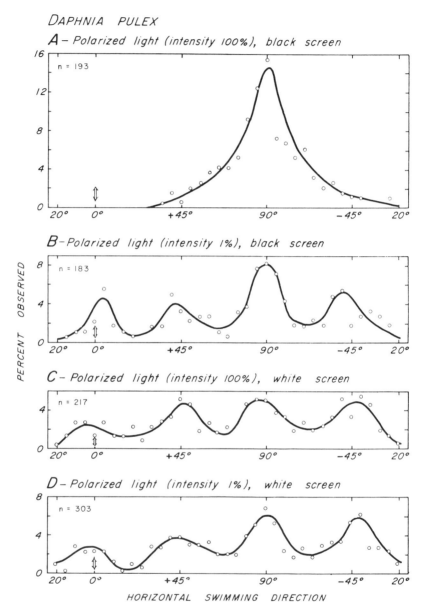

Fig. 7. Spontaneous orientation of *Daphnia* swimming in a vertical beam of linearly polarized light. Polarotaxis results in basic orientation at 0°, 45°, 90°, 135° relative to the polarization plane (0°). Note that very strong preference for 90° under one condition (*A*) is supplanted by milder preference for all four basitactic directions under three others (*B, C, D*). (Jander and Waterman 1960).

(2) How is the course "command" of menotaxis (the light-compass reaction) imposed on the functional organization of the sensory system? The latter question implies that the E-vector sensing mechanism must produce a signal whose value can be unambiguously recognized for all directions through 180°. Then this measure of the animal's spatial orientation must be compared with the command value and corrections effected if there is a discrepancy.

THE ANALYZER CHANNELS

Directly related to these questions are the number and nature of the sensory channels involved (Moody and Parriss 1961; Waterman 1966c,d; Kirschfeld 1972a,b). The models previously proposed have been mainly concerned with two-channel systems that correspond with the commonest type of receptor organization (Fig. 6). Nevertheless, some consideration has been given to single- and multiple-channel input. Hence, we need briefly to review these three possibilities.

A single-channel system, if it contains a perfect analyzer, will waste at least half of the incident light. In addition, it clearly cannot discriminate I from polarization (Q and U) if it is fixed in position. However, if the analyzer is rotated through 180°, I_{max} and I_{min} could be determined provided that I and P remain constant. Hence, a freely swimming organism with one polarizer channel could no doubt steer basitactic courses at 0° and 90° by maximizing or minimizing the channel input for a constant linearly polarized stimulus. If it could average I_{max} and I_{min}, this would of course be a measure of overall intensity. Also, if degree of polarization were estimated from $P = (I_{max} - I_{min})/(I_{max} + I_{min})$, Q and U would be defined by $I_P{}^2 = Q^2 + U^2$, where I_P is the fraction of the light that is polarized.

However, a single-channel system would not be able directly to steer unambiguous menotactic courses from the E-vector. Except at I_{max} and I_{min}, there would always be a pair of angles equidistant from either of these two directions that would give the same sensory input for $\cos^2 \theta$ (where θ is the angle between the E-vector and the reference coordinate [Fig. 8B]). In other words, the animal could not discriminate trigonometric quadrants I from IV or II from III. This ambiguity was pointed out by Moody and Parriss (1961), but they did not suggest a way around it. Furthermore, in a one-channel model, the 45° and 135° points do not appear to have any characteristics distinguishing them from other possible angles between 0° and 90° or 90° and 180°. Hence, there would be no obvious peripheral source in such a model for these common basitactic preferences.

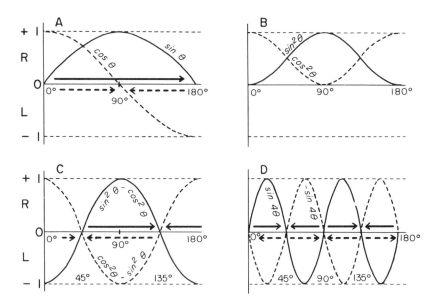

Fig. 8. Steering torques in animals responding to directional light parameters. In all cases, the model assumes that when the sinusoidal functions are positive, the animal tends to turn right with a force proportional to amplitude (see Fig. 9). Similarly, when they are negative, the comparable turning tendency is directed to the left. (A) Phototactic response to light-ray direction where turning force is a function of sin θ (positive or negative phototaxis depending on the sign) or of cos θ (transverse phototaxis like the dorsal light reflex; again of either sign). (B) The outputs of two orthogonal dichroic linear analyzers responding to linear polarization. Their sum is intensity and independent of θ (first Stokes parameter). (C) Their difference is either plus or minus cos 2 θ (the second Stokes parameter). (D) Result of multiplicative interaction between two orthogonal dichroic channels. As shown, these various possibilities (B, C, D) would result in preferential orientation at one or more of the four basitactic angles. (Waterman 1966d).

There is no evidence that such confusion occurs in the E-vector perception of animals at angles intermediate to the four fixed basitactic ones. Of course, with a moving animal variations in channel input for orientations around an intended menotactic direction might be differentiated. In that case, the sign of the slope would indicate whether the course held lay in the correct direction relative to the maximum or minimum. Indeed, Kirschfeld (1972a,b) has proposed, without specifying any mechanism, that a single channel is adequate for menotaxis provided successive inputs are analyzed from different eye positions. However, the ambiguity mentioned above was not considered and no experimental evidence was presented for any one-channel system mediating polarotaxis.

DETERMINING THE STOKES PARAMETERS

Whatever may be the potentialities and limitations of analyzing linearly polarized light with a single input channel, it is nevertheless true that rhabdoms with two orthogonal orientations of their microvilli appear to be the general rule in decapod and stomatopod crustaceans, many insects, and all cephalopods that have been studied. In fact, two fixed orthogonal dichroic channels can use all the entering light and are ideal for determining the first two Stokes parameters of the irradiance (integrated over the rhabdom's visual field). Thus, for perfect analyzers, the sum of the two channel signals would be proportional to the total light intensity: $I = I_{0°} + I_{90°}$. The second parameter would be the difference between these two inputs, since $Q = I_{0°} - I_{90°}$.

Since $Q = \pm \cos 2\theta$, with the sign depending on which channel is subtracted from the other, only the maxima and minima (0°, 90°) are unequivocally identifiable by the fixed double analyzer (Fig. 8C). This is generally similar to the result obtainable with successive determinations in different positions with a one-channel system. However, the angles 45° and 135°, while still indistinguishable from each other with the two perpendicular analyzers, would differ from all the other pairs of ambiguous angles in having a zero difference between the channel outputs.

When oblique polarotactic orientation is observed in animal behavior, equal preference for 45° and 135° is generally evident (Fig. 7B,C,D). This is consistent with some unique property shared by these directions. A more biologically explicit model has previously been proposed to account for basitactic orientation to E-vector (Waterman 1966c,d). In this, a reciprocally innervated steering system was hypothesized for the orienting animal. The model was so defined that positive values of Q gave rise to a motor command to turn to the right and negative values to turn to the left (Fig. 9). Both subtractive and multiplicative processing of the two orthogonal inputs were suggested as ways to account for the observed angular preferences (Waterman 1966d) (Fig. 8C,D). Note that with a simple directly coupled system of the sort hypothesized, only the sign of $\cos 2\theta$ or $\sin 4\theta$ needs to be determined for specific basitactic angles to be chosen.

Of course, intermediate angles pose the basic question for menotaxis. The fixed two-channel analyzer, like the rotated single-channel one, will confuse angles between the coordinate axes in quadrants I and IV as well as II and III. In analyzing their evidence that an octopus can discriminate vertical from horizontal E-vectors in addition to 45° and 135° ones, Moody and Parriss (1961) became quite concerned over this problem. However, despite considerable discussion, no solution to this dilemma was suggested. As stated above, there is no evidence that

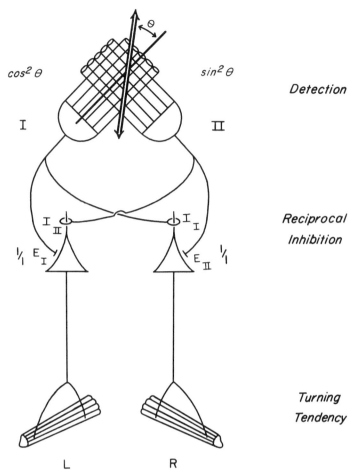

Fig. 9. Polarotactic steering model that would lead to orientation at $\theta = 45°$ as shown, but with reversal of the excitatory (E_I, E_{II}) and inhibitory (I_I, I_{II}) synapses, it would steer similarly at 135° (Fig. 8C). The sensory input of two orthogonal dichroic channels (I, II) reciprocally innervates motor neurons which cause right (R) or left (L) steering muscles to contract. (Waterman 1966d).

such bimodal confusion occurs generally in the E-vector perception of animals. Consequently, their polarimeters and information processing systems can avoid this difficulty.

With an orthogonal two-channel system, this might be done optimally by taking readings in two positions differing by 45°. Then only the sign of U (a sin 2θ function) would be required to identify the correct quadrant for the angle pairs given by Q. Alternatively, the effects of small continual displacements of the analyzer could be used as suggested above. Course wobble could well produce these, but there might be some special means of ensuring their occurrence such as the continuous jitter of the *Daphnia* compound eye and the small high-frequency oscillations of crustacean eyestalks (Horridge and Sandeman 1964). The little-studied small-scale movements of dipteran eyes may also be relevant (e.g., Franceschini and Kirschfeld 1971).

Still another possible mechanism would require more than one analyzer (rhabdom) fixed in different positions (Jander 1963). To be practical, this presumably would be limited to fairly close neighbors within the eye since they would have to share similar fields of view. We know, in fact, that rhabdom patterns differ in various areas of the retina in some compound eyes (e.g., in the water strider *Gerris* [Schneider and Langer 1969]) and that these differences can affect the E-vector sensitivity of the retinal areas involved (Bohn and Täuber 1971).

Comparisons between species also reflect similar relations. For example, in cephalopods and decapod crustaceans where at least in the central retina the orthogonal dichroic analyzers are vertically and horizontally oriented, better discrimination is shown between vertical and horizontal $\theta = 0°$ or $90°$ than for $\theta = 45°$ or $135°$ (Moody and Parriss 1961; Kirschfeld 1972*b*). On the other hand, maximal optomotor[8] responses induced by polarization patterns occur at $\theta = 45°$ and $135°$ in the house fly and honeybee (Kirschfeld and Reichardt 1970).

THREE OR MORE CHANNELS

A further alternative for analyzer design was discussed by Kirschfeld (1972*a,b*), who proposed that three channels are minimal for detecting position angle with a single fixed polarimeter. Consequently, each effective biological analyzer that functions in this way should have a minimum of three dichroic input channels. As pointed out above, current evidence is that this pattern is not present in a large number of E-vector detecting eyes (e.g., decapod crustaceans and cephalopods). It does, however, occur in a variety of insect and some other compound eyes, where two, three, five, six, or more rhabdomere directions may be present in each ommatidium.

However, an important caveat against premature generalization of such a model may be presented by dipteran flies. Here extensive study, especially by Reichardt's group in Tübingen, has shown that there are two functional systems of retinular cells in each ommatidium. Six cells

---

[8] The turning reflex of the whole body or a body part like the eye in response to movement of a large visual pattern, typically vertical black and white stripes moving horizontally.

(R_1 through R_6) make up one of these; two cells (R_7, R_8) make up the other. For peripheral retinular cells, R_1–R_6, the six rhabdomeres are aligned roughly in three different directions. Yet these particular receptor cells apparently do not function in E-vector perception by the flies but constitute a scotopic, long spatial- and spectral-wavelength selective system.

As their rhabdomere structure would imply, the receptor potentials of R_1–R_6 are systematically affected by E-vector direction (McCann and Arnett 1972; Smola and Gemperlein 1972). Interestingly enough, this differential information is not transmitted to their axons. As a consequence, higher-order fibers postsynaptic to R_1–R_6 do not show E-vector sensitivity. This lack would also be expected on the basis of the pattern of primary fiber convergence (neural superposition) in the lamina ganglionaris[9] (Trujillo-Cenóz 1965; Kirschfeld 1967).

In contrast to R_1–R_6, central cells 7 and 8, which have their microvilli in two orthogonal directions, do mediate E-vector sensitivity and function as a photopic, short spatial- and spectral-wavelength selective system. According to Kirschfeld (1969), R_7 in *Musca,* unlike all other retinular cells measured in this and other arthropods as well as cephalopods, has its dichroic maximum absorptance perpendicular to the axial direction of its microvilli. Hence, the E-vector sensing system of this house fly may have only one effective input direction per ommatidium, as suggested by the response of *Musca's* higher-order fibers to very short spatial wavelength polarized stimuli (McCann and Arnett 1972). However, further data are required to elucidate this.

These complex relations between fine structure and visual function in dipteran E-vector detection strongly reinforce the need for caution in drawing general conclusions before they are warranted. Thus, while electron microscopy often may suggest the definitive experiment, the latter must indeed be carried out to prove the point. Another sort of incomplete case is building up for a variety of vertebrates. Here the eyes patently do not contain rhabdoms to act as dichroic analyzers. Yet there is a growing body of evidence from behavioral experiments that E-vector direction can be perceived (e.g., in fishes see Fig. 10) (Forward, Horch, and Waterman 1972; Waterman and Forward 1972). Furthermore, many neurons in the goldfish tectum[10] are sensitive to stimulus E-vector direction (Hashimoto and Waterman, unpublished).

---

[9] The first relay and processing station between the retina and brain in the neural pathway of the compound eye.

[10] The primary visual projection area in the brain of lower vertebrate like fish.

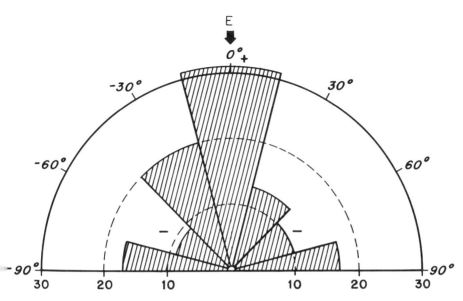

Fig. 10. Polarotactic responses of the West Pacific halfbeak fish *Zenarchopterus* to imposed overhead **E**-vector orientation. Six geographically different **E**-vectors were tested and the results for 18 fish superimposed ($N = 348$). Polar plot shows percentages of observations in 30° sectors centered in the polarization plane (0°). Significant preference ($p < 0.01$) for orientation parallel to the **E**-vector is evident. (Forward, Horch, and Waterman 1972).

Thus in the dipteran flies, dichroic receptor elements are present, but the differential output of the majority appears to be discarded; in fish, behavioral and central electrophysiological data demonstrate E-vector discrimination, but no appropriate analyzer has yet been found. Clearly, for a satisfactory conclusion both mechanism and functional relevance need to be known in some detail.

No doubt it is inappropriate here to discuss the biological uncertainties at greater length. Suffice it to say in conclusion that a large number of E-vector sensitive animals have rhabdoms with only two orthogonal orientations of their constituent microvilli and, hence, of their dichroic analyzers. Basitactic responses to polarized light probably can be accounted for relatively simply with such a system having two input channels. However, experimental proof of this still needs to be given. Furthermore, the mechanism mediating menotactic use of E-vector perception involves more untested hypotheses. Here, demonstration of the information processing that must occur between sensory input and motor output remains particularly challenging.

Acknowledgments. This research has been supported by United States Public Health Service grants as well as by additional funds from the Research Committee of the National Geographic Society and the National Aeronautics and Space Administration. Thanks are due George W. Kattawar and Gary D. Bernard for helpful discussions of the channeling problem.

REFERENCES

Adler, H. E. 1971. Orientation: Sensory basis. *Ann. N.Y. Acad. Sci.* Vol. 188. New York: N.Y. Acad. of Sci.

Batschelet, E. 1965. *Statistical methods for the analysis of problems in animal orientation and certain biological rhythms.* Washington, D.C.: Amer. Inst. Biol. Sciences.

Beychok, S. 1966. Circular dichroism of biological macromolecules. *Science* 154: 1288–1299.

Bohn, H., and Täuber, U. 1971. Beziehungen zwischen der Wirkung polarisierten Lichtes auf das Elektroretinogram und der Ultrastruktur des Auges von *Gerris lacustris* L. *Z. vergl. Physiol.* 72: 32–53.

Chandrasekhar, S. 1960. *Radiative transfer.* New York: Dover Publications.

Clarke, D., and Grainger, J. F. 1971. *Polarized light and optical measurement.* New York: Pergamon Press.

Eguchi, E., and Waterman, T. H. 1968. Cellular basis for polarized light perception in the spider crab, *Libinia. Z. Zellforsch.* 84: 87–101.

Forward, R. B., Jr., Horch, K. W., and Waterman, T. H. 1972. Visual orientation at the water surface by the teleost *Zenarchopterus. Biol. Bull.* 143: 112–126.

Franceschini, N., and Kirschfeld, K. 1971. Les phénomènes de pseudo-pupille dans l'oeil composé de *Drosophila. Kybernetik* 9: 159–182.

Frisch, K. von 1948. Gelöste und ungelöste Rätsel der Bienensprache. *Naturwiss.* 35: 38–43.

———. 1965. *Tanzsprache und Orientierung der Bienen.* Berlin: Springer-Verlag.

Galler, S. R., Schmidt-Koenig, K., Jacobs, G. J., and Belleville, R. E., eds. 1972. *Animal orientation and navigation.* Washington, D.C.: National Aeronautics and Space Administration.

Hays, D., and Goldsmith, T. H. 1969. Microspectrophotometry of the visual pigment of the spider crab *Libinia emarginata. Z. vergl. Physiol.* 65: 218–232.

Horridge, G. A., and Sandeman, D. C. 1964. Nervous control of optokinetic responses in the crab *Carcinus. Proc. Roy. Soc. Lond.* B 161: 216–246.

Ivanoff, A. 1973. Polarization measurements in the sea. *Optical aspects of oceanography.* (N. G. Jerlov, ed.) London: Academic Press.

Ivanoff, A., and Waterman, T. H. 1958a. Elliptical polarization of submarine illumination. *J. Mar. Res.* 16: 255–282.

———. 1958b. Factors, mainly depth and wavelength, affecting the degree of underwater light polarization. *J. Mar. Res.* 16: 283–307.

Jander, R. 1963. Grundleistungen der Licht- und Schwereorientierung von Insekten. *Z. vergl. Physiol.* 47: 381–430.

Jander, R., and Waterman, T. H. 1960. Sensory discrimination between polarized light and light intensity patterns by arthropods. *J. Cell. Comp. Physiol.* 56: 137–160.

Kirschfeld, K. 1967. Die Projektion der optischen Umwelt auf das Raster der Rhabdomere im Komplexauge von *Musca*. *Exp. Brain Res.* 3: 248–270.

———. 1969. Absorption properties of photopigments in single rods, cones and rhabdomeres. *Processing of optical data by organisms and by machines*. (W. Reichardt, ed.), pp. 116–136. New York: Academic Press.

———. 1972*a*. Die notwendige Anzahl von Rezeptoren zur Bestimmung der Richtung der elektrischen Vektors linear polarisierten Lichtes. *Z. Naturforsch.* 27b: 578–579.

———. 1972*b*. Vision of polarized light. *Symposia proceedings of the IV International Biophysics Congress, Moscow, Aug. 7-14, 1972*. In press.

Kirschfeld, K., and Reichardt, W. 1970. Optomotorische Versuche an *Musca* mit linear polarisiertem Licht. *Z. Naturforsch.* 25b: 228.

Langer, H., and Thorell, B. 1966. Microspectrophotometric assay of visual pigments in single rhabdomeres of the insect eye. *The functional organization of the compound eye*. (C. G. Bernhard, ed.), pp. 145–150. Oxford: Pergamon Press.

McCann, G. D., and Arnett, D. W. 1972. Spectral and polarization sensitivity of the dipteran visual system. *J. Gen. Physiol.* 59: 534–558.

Mommaerts, W. F. H. M. 1969. Circular dichroism and the conformational properties of visual pigments. *The retina*. (B. R. Straatsma, M. O. Hall, R. A. Allen, and F. Crescitelli, eds.), pp. 225–234. Berkeley: Univ. of California Press.

Moody, M. F., and Parriss, J. R. 1961. The discrimination of polarized light by *Octopus:* a behavioural-and morphological study. *Z. vergl. Physiol.* 44: 268–291.

Penn, R. D., and Hagins, W. A. 1972. Kinetics of the photocurrent of retinular rods. *Biophys. J.* 12: 1073–1094.

Schneider, L., and Langer, H. 1969. Die Struktur des Rhabdoms im "Doppelauge" des Wasserläufers *Gerris lacustris*. *Z. Zellforsch.* 99: 538–559.

Sekera, Z. 1957. Polarization of skylight. *Handbuch der Physik*. Vol. XLVII. (S. Flugge, ed.), pp. 288–328. Berlin: Springer-Verlag.

Shaw, S. R. 1969. Sense-cell structure and interspecies comparisons of polarized light absorption in arthropod compound eyes. *Vision Res.* 9: 1031–1040.

Smola, U., and Gemperlein, R. 1972. Übertragungseigenschaften der Sehzelle der Schmeissfliege *Calliphora erythrocephala*. 2. Die Abhängigkeit von Ableitort: Retina — Lamina ganglionaris. *J. Comp. Physiol.* 79: 363–393.

Snyder, A. W. 1973. Polarization sensitivity of individual retinula cells. *J. Comp. Physiol.* In press.

Trujillo-Cenóz, O. 1965. Some aspects of the structural organization of the intermediate retina of dipterans. *J. Ultrastruct. Res.* 13: 1–33.

Trujillo-Cenóz, O., and Bernard, G. D. 1972. Some aspects of the retinal organization of *Sympycnus lineatus* Lowe (Diptera, Dolichopodidae). *J. Ultrastruct. Res.* 38: 149–160.

Waterman, T. H. 1954. Polarization patterns in submarine illumination. *Science* 120: 927–932.

———. 1963. The analysis of spatial orientation. *Ergebn. Biol.* 26: 98–117.

———. 1966*a*. Specific effects of polarized light on organisms. *Environmental biology*. (P. L. Altman and D. S. Dittmer, eds.), pp. 155–165. Bethesda, Maryland: Fed. Amer. Soc. for Experimental Biology. (Second edition in press.)

——— 1966*b*. Polarotaxis and primary photoreceptor events in Crustacea. *The functional organization of the compound eye*. (C. G. Bernhard, ed.), pp. 493–511. Oxford: Pergamon Press.

Waterman, T. H. 1966c. Information channeling in the crustacean retina. *Proceedings of the symposium on information processing in sight sensory systems.* (P. W. Nye, ed.), pp. 48–56. Pasadena: National Institutes of Health and the California Institute of Technology.

———. 1966d. Systems analysis and the visual orientation of animals. *Amer. Sci.* 54: 15–45.

———. 1972. Visual direction finding by fishes. *Animal orientation and navigation.* (S. R. Galler, K. Schmidt-Koenig, G. J. Jacobs, and R. E. Belleville, eds.), pp. 437–456. Washington, D.C.: National Aeronautics and Space Administration.

Waterman, T. H., and Fernández, H. R. 1970. **E**-vector and wavelength discrimination by retinular cells of the crayfish *Procambarus*. *Z. vergl. Physiol.* 68: 154–174.

Waterman, T. H., Fernández, H. R., and Goldsmith, T. H. 1969. Dichroism of photosensitive pigment in rhabdoms of the crayfish *Orconectes*. *J. Gen. Physiol.* 54: 415–432.

Waterman, T. H., and Forward, R. B., Jr. 1972. Field demonstration of polarotaxis in the fish *Zenarchopterus*. *J. Exp. Zool.* 180: 33–54.

Waterman, T. H., and Horch, K. W. 1966. Mechanism of polarized light perception. *Science* 154: 467–475.

Waterman, T. H., and Westell, W. E. 1956. Quantitative effect of the sun's position on submarine light polarization. *J. Mar. Res.* 15: 149–169.

THE CIRCULAR POLARIZATION OF LIGHT REFLECTED FROM CERTAIN OPTICALLY ACTIVE SURFACES

RAMON D. WOLSTENCROFT
Institute for Astronomy
University of Hawaii

When unpolarized light is reflected from certain optically active surfaces, almost complete circular polarization may result at favored angles of incidence and wavelengths. Well-studied examples are the reflection from (1) the wing cases of certain classes of iridescent beetle, and (2) cholesteric liquid crystals of various compositions. The theory of de Vries relates this behavior to the helical and layered structure of these materials and, thence, to the selective reflection of circularly polarized light with a particular handedness.

Rather smaller circular polarization (1%) is produced when sunlight is reflected from a green leaf: this peak value occurs close to, though not precisely at, the center of the chlorophyll a *absorption. It is not known how rare or common this phenomenon is for optically active surfaces in general; if laboratory study proves this to be a common effect, some important applications to remote sensing may be possible.*

MATERIALS WITH A HELICAL LAYERED STRUCTURE

In 1911 A. A. Michelson, in a paper entitled "On the Metallic Colouring in Birds and Insects" (Michelson 1911), reported the discovery of strong circular polarization of the light reflected from the wing cases of *Plustiosis resplendens,* a beetle that shows strong metallic coloring (which we denote here by the term iridescence). The circular polarization was seen whether the incident light was polarized or unpolarized. He correctly attributed the effect to a screw structure of molecular dimensions in the wing cases and understood that the circular polarization arises because almost complete reflection occurs for one of the two circularly polarized beams (into which an incident unpolarized beam may be decomposed) and almost complete absorption of the other component. He understood that reflections between regularly spaced layers of the beetle exocuticle produced the iridescent colors. However, his observation of the wavelength dependence of the circular polarization and his interpretation in terms of the beetle's coloring have recently been refuted (Neville and Caveney 1969).

The study of this effect in other classes of iridescent beetles has been admirably reviewed by Robinson (1966); his principal conclusions are summarized as follows:

1. Not all iridescent beetles show this effect. Those that do show it all occur in the family Scarabaeidae. In all these cases the sense of the circular polarization is the same, as determined from visual examination through left-handed and right-handed circular polarizers.

2. The circular polarization of thin sections of wing cases, obtained by careful etching with nitric acid, has opposite handedness for the reflected and transmitted beams.

3. The magnitude of the circular polarization measured by Michelson was "nearly complete in the extreme red," which we take to mean that the modulus of the fractional circular polarization, $|q|$, was about 90% at $\lambda \approx 0.75$ μm. Quantitative measurements are quoted by Mathieu and Faraggi (1937). For example, for light reflected by *Potosia speciosissima*, $|q|$ has a maximum value of 100% at normal incidence, 75% at angle of incidence $i = 50°$, and zero at $i = 65°$ ($\lambda = 0.62$ μm; incident unpolarized light).

4. The similarity of the circular polarization properties exhibited by cholesteric liquid crystals and by beetle wing cases very probably means that they both have the same structural properties, namely a helical layered arrangement of aligned molecules. The axes of symmetry of the molecules in a given layer are all parallel, but there is a progressive twist to this alignment direction with distance, which is normal to the layers. The theory of de Vries (1951) relates the circular polarization of light reflected by the cholesteric liquid crystals to their structural and optical properties and is in qualitative agreement with the observations. More recent developments of the theory have been described by Chandrasekhar and Prasad (1971).

The wide occurrence of liquid crystals in biological systems has recently been emphasized by Fergason and Brown (1968); for example, structure in certain chromosomes is identical to that of cholesteric liquid crystals (Bouligand, Sayer, and Puiseun-Das 1968). This suggests the possibility that appreciable circular polarization may be found in the light reflected from a wide variety of biological materials.

GREEN PLANTS

Green plants are another class of materials that show circularly polarized reflection of sunlight. Pospergilis (1969) has measured the circular polarization spectrum of sunlight reflected from a green leaf of unspecified type between 0.4 μm and 0.8 μm; he found spectral structure that had a sharp increase shortward of 0.45 μm and a band of enhanced circular polarization close to 0.7 μm, with a peak value $q = -1.0\%$. This behavior is broadly consistent with the circular dichroism spectrum of *chlorophyll a* given by Houssier and Sauer (1970), and we may suppose that it is commonly found in all green plants with chlorophylls (mostly *chlorophyll a* and *b*) that exhibit appreciable circular dichroism. The mechanism of production of circular polarization in this case is presumably different from the case of liquid crystals and wing cases of beetles for which a layered, screw structure is required without the individual molecules being (necessarily) circularly dichroic.

The circular polarization of light transmitted through a green leaf may be calculated from the values of ϵ_R and ϵ_L, the absorption coefficients for right-handed and left-handed circularly polarized light, given in Table I of Houssier and Sauer (1970). The fractional circular polarization and the circularly polarized flux of the transmitted light are respectively

$$-q_T = \frac{e^{-\epsilon_R d} - e^{-\epsilon_L d}}{e^{-\epsilon_R d} + e^{-\epsilon_L d}} = -d \triangle \epsilon$$

and

$$q_T I_0 T = \frac{I_0}{2}(e^{-\epsilon_R d} - e^{-\epsilon_L d}) = -I_0 d \triangle \epsilon e^{-\epsilon d/2},$$

where I_0 is the incident intensity, T is the transmission, d is the thickness of the leaf, $\triangle \epsilon = \epsilon_R - \left(\frac{\epsilon_L + \epsilon_R}{2}\right)$, and it is assumed that $d\triangle\epsilon << 1$. As an example, consider a cottonwood leaf at the peak of the *chlorophyll a* band at 0.66 μm when measured in ether (Houssier and Sauer 1970) and 0.68 μm when measured out of solution:

$$T = 0.02 = e^{-(\epsilon_R + \epsilon_L)d},$$

$$\epsilon_R - \epsilon_L = 13.8$$

and

$$\epsilon_R + \epsilon_L = -86.3 \times 10^3$$

(units of ϵ are [mole/l]$^{-1}$ cm^{-1}), leading to $q_T = +0.031\%$. This is very small compared to the fractional circular polarization of the reflected light, $q_R = -1.0\%$, found for the leaf studied by Pospergilis (1969). However, this large ratio of $q_R : q_T$ is reasonable as we now show. A relation between q_R and q_T can be obtained using conservation arguments. Consider an unpolarized beam incident on the leaf. The fraction of the incident energy going into the absorbed, transmitted, and reflected beams is denoted by A, T, and R respectively, and the corresponding values of the fractional circular polarization are denoted by q_A, q_T, and q_R. Conservation of angular momentum and of energy require

$$q_A A + q_T T + q_R R = 0$$
$$A + T + R = 1$$

and thus
$$q_R = -\frac{q_A A - q_T T}{R}.$$

For the cottonwood leaf at 0.68 μm, $T = 0.02$, $R = 0.06$, $A = 0.92$, and $q_T = +3.1 \times 10^{-4}$; if we require that $q_R = -1 \times 10^{-2}$ (from Pospergilis), we obtain the value of the remaining variable $q_A = -6.7 \times 10^{-4}$. Being comparable in magnitude to q_T, this value of q_A is quite plausible. We note that in the above example $q_R >> |q_A| \sim |q_T|$, principally because $A >> R$.

APPLICATION TO EARTH RESOURCES AND ASTRONOMY

If the circular polarization spectra of many representative classes of optically active materials were known, some immediate and important applications to remote sensing of optically active materials could be made. Although the experimental data are extremely limited at present, some progress may be expected from theoretical studies of the relations between the reflection circular polarization and the transmission circular polarization, which is that commonly measured by chemists. Furthermore, the peak wavelengths in the absorption and circular polarization curves will coincide — although the latter curve will sometimes mimic the first derivative of the former — and thus give an important clue to the identity of the optically active constituent, as judged from circular dichroism studies. Another important difficulty associated with remote sensing will be the strong dilution of the polarization by unpolarized sources in the field of the polarimeter. As an illustration, a field of randomly aligned green leaves producing $q \approx 0.5\%$ near 0.7 μm when diluted by a factor of 10^3 would result in an observed polarization $q_0 = 5 \times 10^{-6}$, which is close to the limits of detection of current polarimeters set by instrumental polarization (see Kemp, Wolstencroft, and Swedlund 1972). An additional complication is that double scattering from optically inactive rough surfaces can lead to $q \sim 10^{-4}$ (see Bandermann, Kemp, and Wolstencroft 1972). Fortunately, because of symmetries in the behavior of the diffusely reflected component (Bandermann, Kemp, and Wolstencroft 1972), a separation of the optically active and diffuse components could be achieved in practice with a precision corresponding to a dilution factor of between 100 and 300 in the case of green leaves.

The application of these ideas to the observed circular polarization spectrum of Mars (Wolstencroft, Kemp, and Swedlund 1972) will be described in a forthcoming publication.

Acknowledgments. Helpful discussions with B. Malcolm, S. F. Mason, S. Siegel, and R. W. Woody are gratefully acknowledged.

REFERENCES

Bandermann, L. W.; Kemp, J. C.; and Wolstencroft, R. D. 1972. Circular polarization of light scattered from rough surfaces. *Mon. Not. R. Astr. Soc.* 158: 291–304.

Bouligand, Y.; Sayer, M.-O.; and Puiseun-Das, S. 1968. The fibril structure and orientation of the chromosomes of dinoflagellates. *Chromosoma* 24: 251–287.

Chandrasekhar, S., and Prasad, J. S. 1971. Theory of rotatory dispersion of cholesteric liquid crystals. *Molecular Crystals and Liquid Crystals* 14: 115–128.

Fergason, J. L., and Brown, G. H. 1968. Liquid crystals and living systems. *J. Amer. Oil Chem. Soc.* 45: 120–127.

Houssier, C., and Sauer, K. 1970. Circular dichroism and magnetic circular dichroism of the chlorophyll and protochlorophyll pigments. *J. Amer. Chem. Soc.* 92 (4): 779–791.

Kemp, J. C.; Wolstencroft, R. D.; and Swedlund, J. B. 1972. Circular polarimetry of fifteen interesting objects. *Astrophys. J.* 177: 177–190.

Mathieu, J., and Faraggi, N. 1937. Study of the circularly polarized light reflected from certain beetles. *C. R. Acad. Sci. Paris* 205: 1378–1380.

Michelson, A. A. 1911. On metallic colouring in birds and insects. *Phil. Mag.* 21: 554–567.

Neville, A. C., and Caveney, S. 1969. Scarabaeid beetle exocuticle as an optical analogue of cholesteric liquid crystals. *Biol. Rev.* 44: 531–562.

Pospergilis, M. M. 1969. Spectroscopic measurements of the four Stokes parameters for light scattered by natural objects. *Sov. Phys. Astron.* (transl) 12: 973–977.

Robinson, C. 1966. The cholesteric phase in polypeptide solutions and biological structures. *Molecular Crystals* 1: 467–494.

Vries, H. de 1951. Rotatory power and other optical properties of certain liquid crystals. *Acta. Cryst.* 4: 219–226.

Wolstencroft, R. D.; Kemp, J. D.; and Swedlund, J. B. 1972. Circular polarization spectrum of Mars. *Bull. Am. Astron. Soc.* 4: 372.

POLARIMETRIC INVESTIGATIONS OF THE TURBIDITY OF THE ATMOSPHERE OVER LOS ANGELES

T. TAKASHIMA, H. S. CHEN, AND C. R. NAGARAJA RAO
University of California, Los Angeles

> An elementary interpretation of ground-based measurements of skylight polarization made at Los Angeles has been attempted with the use of computed values of scattered radiation in a simple model of a turbid atmosphere. Measurements were made with a photoelectric polarimeter in four spectral intervals (bandwidth ~ 0.015 μm) centered on 0.362, 0.392, 0.500, and 0.590 μm. A multistream approximation was used to solve the radiative transfer problem in the three-layer model of a turbid atmosphere consisting of a concentrated aerosol layer between two molecular layers. Reasonable correspondence between theory and experiment is noticed. It is felt the "effective" turbidity of an "equivalent" atmosphere can be estimated if suitable constraints can be imposed on the range of variability of model parameters.

The polarization of scattered radiation in the earth's atmosphere in the visible and adjacent regions of the spectrum exhibits a very sensitive dependence on atmospheric turbidity caused by the presence of aerosols (dust and haze). It has been suggested that this dependence can be used as a means of monitoring atmospheric turbidity (Sekera 1956, 1957; Rozenberg 1960; Stamov 1963; Bullrich 1964). This entails, in practice, establishing correspondence between measurements of skylight polarization and numerical computations of the scattered radiation in models of turbid atmospheres. The validity of conclusions thus drawn about atmospheric turbidity would be decided by how truly representative the model is of the actual atmosphere and by the techniques used to solve the radiative transfer problem (de Bary 1963; Kano 1968; Ivanov et al. 1968; Herman, Browning, and Curran 1971).[1]

We report on the interpretation of a series of ground-based measurements of skylight polarization performed on the campus of the University of California, Los Angeles (34°04′N, 118°24′W, 128 m above mean sea level), during Fall 1969 and 1970, when the prevalent meteorological

---

[1] See p. 510.

conditions were favorable for the occurrence of strongly localized, low-level turbidity confined to within a few kilometers of the ground. Thus, the conditions under which the measurements were made were such that our layered model of a turbid atmosphere could be reasonably justified.

EXPERIMENTAL INVESTIGATION

Measurements were made with a simple photoelectric polarimeter shown schematically in Fig. 1. The field of view of the instrument is restricted to a cone of apex angle 3°00′. The incoming beam of skylight, after passing through the interference filter and the rotating Glan-Thompson prism, impinges upon the photocathode of the end window photomultiplier tube (Ascop Model 541A) with S-11 response. The sinusoidally varying photosignal is recorded after suitable amplification. Measurements are made in four spectral intervals isolated by interference filters (bandwidth ~ 0.015 μm) centered on 0.362, 0.392, 0.500, and 0.590 μm. The four colors are sampled in succession in one minute.

The degree of linear polarization of the incoming radiation is given by the ratio of the difference to the sum of the maximum and minimum values of the sinusoidally varying photosignal recorded over a 180°00′ rotation of the Glan-Thompson prism. The values thus obtained were periodically checked against the values obtained from the formula $p = (Q^2 + U^2)^{1/2}/I$ where I, Q, and U are the Stokes parameters of the incoming radiation. They were determined from conventional harmonic analysis of the photosignal recorded over one complete rotation of the Glan-Thompson prism (Chen and Rao 1968).

The instrument is linear in its response to variations of the intensity of the incoming radiation to within 2%. Its response to polarized light was determined with the use of a simple polarization calibrator (Rao

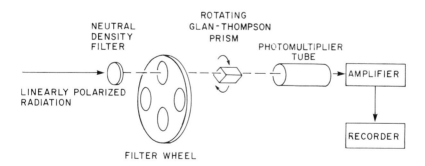

Fig. 1. Schematic diagram of the photoelectric polarimeter.

1966). It is estimated that the degree of polarization could be determined with an accuracy of $\pm 4\%$.

We shall be primarily concerned with measurements made in the principal plane of the sun, i.e., the vertical plane passing through the direction of illumination from the sun, in order to be able to use readily available theoretical computations for purposes of comparison.

THEORETICAL INVESTIGATION

The Radiative Transfer Problem

We have computed the degree of linear polarization of the diffusely transmitted skylight in the principal plane of the sun in a three-layer model of a purely scattering, plane-parallel cloudless turbid atmosphere illuminated at the top with unpolarized, parallel radiation from the sun (Fig. 2). The model consists of a highly concentrated aerosol layer located between two molecular layers. Scattering of radiation in the *non-mixing* aerosol and molecular layers is governed by the well-known Mie and

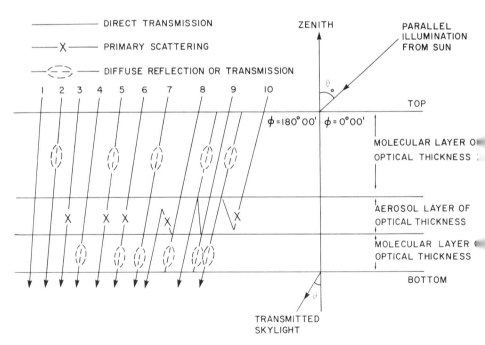

Fig. 2. The three-layer model of the turbid atmosphere; the principal plane of the sun is defined by the azimuth angles $\varphi = 0°00'$ and $180°00'$.

Rayleigh laws respectively. The ground reflects radiation according to the Lambert law; the reflected radiation is unpolarized and isotropically distributed in the outward hemisphere independently of the state of polarization and angle of incidence of the incident radiation.

We assume, insofar as scattering within the aerosol layer is involved, that such scattering is not accompanied by any change in the state of polarization of the incident radiation. Also, radiation that is incident on the aerosol layer is assumed either to be directly transmitted through the layer after being duly attenuated or to be scattered only *once*.

The radiation emerging at the bottom of the atmosphere in any given direction is interpreted as the weighted sum of various streams of radiation resulting from different modes of interaction of the incoming solar radiation with the composite turbid atmosphere. Referring to Fig. 2, where some of these modes have been illustrated, we see that all modes, other than mode 1 which gives the attenuated solar radiation directly transmitted through the entire atmosphere, involve at least one interaction between radiation scattered (transmitted) by the molecular (aerosol) layer with the aerosol (molecular) layer. Detailed expressions for the contributions of various modes have been obtained by Takashima (1971). For example, the contribution of a higher-order mode such as 10 may be written as

$$\mathbf{I}(-\Omega, -\Omega_0) = \frac{e^{-x/\mu_0}}{4\pi\mu}$$
$$\{e^{-y/\mu''} \mathbf{T}_R(z, -\Omega, -\Omega'')\{\mathbf{S}_R(x, -\Omega'', +\Omega')\mathbf{S}_A(y, +\Omega', -\Omega_0)\}\}\pi\mathbf{F}$$

where $\mathbf{I}(-\Omega, -\Omega_0)$ is the Stokes vector of the radiation emerging in the direction defined by $-\Omega$ and $\pi\mathbf{F}$ is the Stokes vector of the illuminating radiation from the sun in the direction defined by $-\Omega_0$; Ω stands for the conventional directional parameters (μ, φ); $\mu = |\cos\theta|$, and φ is the azimuth angle. $\mathbf{S}$ and $\mathbf{T}$ are the diffuse reflection and transmission matrices appropriate to the molecular and aerosol layers as indicated by the subscripts R and A respectively. The brackets denote hemispherical integration (Sekera 1966); thus

$$\{\mathbf{A}(\Omega')\mathbf{B}(\Omega')\} = \frac{1}{4\pi} \int_0^1 \int_0^{2\pi} \mathbf{A}(\Omega')\mathbf{B}(\Omega') \frac{d\mu'}{\mu'} d\varphi'.$$

The diffuse reflection and transmission matrices for the molecular layers are evaluated in the conventional manner using the well-known K and L matrices and the X and Y functions (Chandrasekhar 1950). For the aerosol layer, because of the assumptions we have made, it can be shown that

$$S_A(\tau_A^\lambda, +\Omega, -\Omega') = \frac{\mu\mu'}{\mu + \mu'}\left[1 - e^{-\tau_A^\lambda\left(\frac{1}{\mu} + \frac{1}{\mu'}\right)}\right]P_A(\Omega, -\Omega'),$$

where $P_A(\Omega, -\Omega')$ is the aerosol scattering matrix. Similarly,

$$T_A(\tau_A^\lambda, -\Omega, -\Omega') = \frac{\mu\mu'}{\mu' - \mu}\left[e^{-\frac{\tau_A^\lambda}{\mu'}} - e^{-\frac{\tau_A^\lambda}{\mu}}\right]P_A(-\Omega, -\Omega') \quad \text{for } \mu \neq \mu'$$

$$= \tau_A^\lambda e^{-\frac{\tau_A^\lambda}{\mu}} P_A(-\Omega, -\Omega') \quad \text{for } \mu = \mu'.$$

$\tau_A^\lambda(=y)$ is the normal optical thickness of the aerosol layer for radiation of wavelength λ.

Computational Work

The evaluation of the diffuse reflection and transmission matrices for the molecular layers was based on the earlier computations of radiation parameters for molecular atmospheres by Chandrasekhar and Elbert (1954), Sekera and Ashburn (1953), and Coulson, Dave, and Sekera (1960).

The aerosol layer is considered to be composed of either of two types of aerosols characterized by different size distributions and refractive indices. The first is the model L water haze described by Deirmendjian (1969). The particle size distribution is given by a modified gamma function of the form $n(r) = 4.9757 \times 10^6 \times r^2 \times \exp(-15.1186 r^{1/2})$, where $n(r)$ is the partial concentration of haze particles per unit volume per unit increment of radius r which lies between the limiting values of 0.005 and 2.9 μm. The second type corresponds to dust particles with a refractive index of 1.5 and a size distribution governed by the Junge power law $dN/d\log r = cr^{-4}$, where N is the number of particles of radius smaller than r and c is a constant. The particle radius r lies between the limiting value of 0.04 and 3.0 μm. We have used the phase function computations of de Bary, Braun, and Bullrich (1965) for this type of aerosol.

The model parameters used in the computations are listed in Table I. The values of τ_R^λ are based on the earlier computations of Elterman (1968). The albedo A is defined as the ratio of the outward normal flux of radiation to the inward normal flux of radiation incident on the ground. The turbidity factor T is given by $(\tau_A^\lambda + \tau_R^\lambda)/\tau_R^\lambda$, where $\tau_R^\lambda(= x + z)$ is the total normal molecular thickness of the composite atmosphere for radiation of wavelength λ. The concentrated aerosol layer was located 250 m above ground level. All the computations were performed on an IBM 360/91 computer.

TABLE I
Model Parameters

λ (μm)	τ_R^λ	τ_A^λ	T	A
0.362	0.560	1.232	3.2	0.075
0.392	0.376	1.203	4.2	0.125
0.500	0.144	0.576	5.0	0.150
0.590	0.070	0.315	5.5	0.175

RESULTS AND DISCUSSION

Since we did not include clouds in our model of the turbid atmosphere, we had to make measurements under cloudless skies so that comparison between experimental results and numerical computations would be meaningful. This requirement restricted the number of days on which measurements could be made to 28 during the months of September, October, and November of 1969 and 1970. We chose this part of the year for measurements because the prevalent meteorological conditions were favorable for the occurrence of strongly localized, low-level layered atmospheric turbidity—popularly called smog—over the Los Angeles basin (De Marrais, Holzworth, and Hosler 1965; Coulson 1969).

Measurements were made only in the principal plane of the sun for reasons mentioned earlier. The measured values of the degree of maximum polarization of skylight, generally observed in the region of sky approximately 90° removed from the direction of illumination from the sun, are shown in Figs. 3a and b along with the computed values in the model turbid atmosphere with the two different types of aerosols and in a molecular atmosphere of optical thickness τ_R^λ bounded by an identical Lambert ground. The computations for the molecular atmosphere were based on the earlier work of Coulson, Dave, and Sekera (1960). The values of the ground albedo used in the computations may be considered as characteristic of the terrain around the site of measurements (Middleton 1954).

The data shown in Figs. 3a and b are typical of days that could be described as having "moderate" (23 Sept. 1970) and "heavy" (24 Sept. 1970) smog. The depolarizing effects of atmospheric turbidity are obvious. The departure from what should prevail in a pure molecular atmosphere increases with increasing turbidity. The computational results also indicate, in so far as the two types of aerosols used in the present study are involved, that at any given value of the turbidity factor or, equivalently, of the aerosol optical thickness, the departure from what should prevail

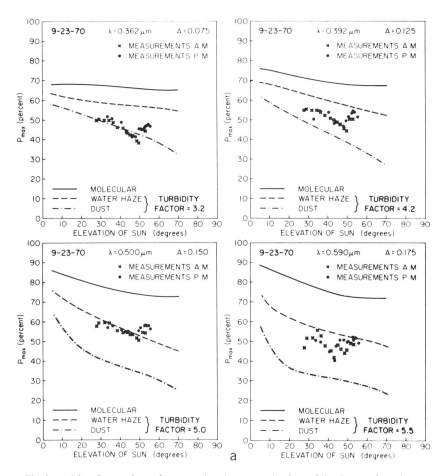

Fig. 3a and b. Comparison of measured and computed values of the degree of maximum polarization of skylight in the principal plane of the sun. A is the albedo of ground.

in a molecular atmosphere is greater for the dust particles which scatter more radiation in the lateral directions than does the water haze, thereby contributing more to multiple scattering and subsequent depolarization of radiation. These observations indicate that, in principle, systematic analysis of the correspondence between experiment and theory should yield information about "effective" values of atmospheric turbidity provided that suitable constraints can be imposed on the range of variability of model parameters such as the ground albedo, the aerosol type, and the aerosol optical thickness.

TURBIDITY OF THE ATMOSPHERE

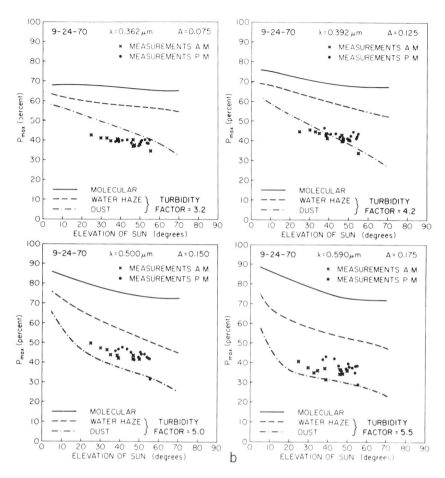

Since we are examining the diffusely transmitted skylight, slight departures of the ground from Lambert behavior or of the albedo values used in the computations from the optimum values—if indeed such optimum values can be determined for the vast variety of natural surfaces—will not appreciably affect the polarization features.[2] The choice of the remaining two parameters is subject to some ambiguity because of the limited amount of information available about the physico-chemical properties of the atmospheric aerosols (Rozenberg 1968).

It need hardly be stressed that the model of the turbid atmosphere employed and the techniques and assumptions underlying the solution

---

[2] See Coulson, p. 444.

of the radiative transfer problem in the model atmosphere affect the conclusions that can be drawn. The present model is a good approximation of the existing situation; similar models have been used by investigators elsewhere (Foitzik and Spankuch 1961; Kano 1968). One intriguing aspect of the computational results is that even with the simplifying assumptions of non-mixing aerosol and molecular layers and of single scattering in the aerosol layer, the computations and measurements are reasonably close to each other.

So far our discussion has been confined to the very basic and obvious aspects of retrieval of information about atmospheric turbidity from measurements of skylight polarization. The allied problem of interpretation of the polarization of the radiation diffusely reflected into space by the atmosphere-ground system has been discussed elsewhere (Rao, Takashima, and Toolin 1973). It is felt that in the present case inclusion in the discussion of details of local meteorological parameters responsible for the transport and localization of turbidity will enable us to judge the suitability of the model we have used and to improve upon it within the limits set by the feasibility of solution of the radiative transfer problem.

Acknowledgments. The authors wish to express their thankfulness to the late Zdenek Sekera for his interest and encouragement. Financial assistance from the Atmospheric Sciences Section, National Science Foundation, under Grant GA16617 and the Office of Air Programs, Environmental Protection Agency, under Grant AP-00768-02 is gratefully acknowledged.

REFERENCES

Bary, E. de 1963. Influence of multiple scattering of the intensity and polarization of diffuse sky radiation. *Applied Optics* 3: 1293–1303.

Bary, E. de; Braun, B.; and Bullrich, K. 1965. *Tables related to light scattering in a turbid atmosphere.* (Vol. 1), Air Force Cambridge Research Laboratories, Office of Aerospace Research, Bedford, Massachusetts, Special Reports No. 33.

Bullrich, K. 1964. Scattered radiation in the atmosphere and the natural aerosol. Advances in geophysics. (H. E. Landsberg, ed.) Vol. 10, pp. 99–260. New York: Academic Press, Inc.

Chandrasekhar, S. 1950. *Radiative transfer.* London: Oxford University Press.

Chandrasekhar, S., and Elbert, D. D. 1954. The illumination and polarization of the sunlit sky on Rayleigh scattering. *Trans. Am. Philos. Soc.* 44: 643–728.

Chen, H. S., and Rao, C. R. N. 1968. Polarization of light on reflection by some natural surfaces. *Brit. J. Appl. Phys.* (J. Phys.D) 1: 1191–1200.

Coulson, K. L. 1969. Measurements of ultraviolet radiation in a polluted atmosphere. Scientific Report No. 1 (Grant 5 R01 AP 00742-02), Department of Agricultural Engineering, University of California, Davis.

Coulson, K. L.; Dave, J. V.; and Sekera, Z. 1960. *Tables related to radiation emerging from a planetary atmosphere with Rayleigh scattering.* Berkeley and Los Angeles: University of California Press.

Deirmendjian, D. 1969. *Electromagnetic scattering on spherical polydispersions.* New York: American Elsevier Publishing Company, Inc.

De Marrais, G. A.; Holzworth, G. C.; and Hosler, R. C. 1965. Meteorological summaries pertinent to atmospheric transport and dispersion over southern California. Weather Bureau Technical Paper No. 54, U.S. Department of Commerce, Washington D.C.

Elterman, L. 1968. UV, visible and IR attenuation for altitudes to 50 km, 1968. Environmental Research Paper No. 285. Air Force Cambridge Research Laboratories, Office of Aerospace Research, Bedford, Massachusetts.

Foitzik, L., and Spankuch, D. 1961. Berechnung der Himmelslichtpolarisation bei dunstgeschichteter Atmosphäre. *Gerlands Beitr. Geophys.* 70: 79–89.

Herman, B. M.; Browning, S. R.; and Curran, R. J. 1971. The effect of atmospheric aerosols on scattered sunlight. *J. Atmos. Sci.* 28: 419–428.

Ivanov, A. I.; Livshits, G. Sh.; Pavlov, V. Ye.; and Teyfel, Ya. A. 1968. Polarization of daytime skylight and atmospheric aerosol, Izv. *Atmospheric and Oceanic Physics* (English translation), 4: 719–722.

Kano, M. 1968. Effects of a concentrated turbid layer on the polarization of skylight. *J. Opt. Soc. Am.* 58: 789–797.

Middleton, W. E. K. 1954. The color of the overcast sky. *J. Opt. Soc. Am.* 44: 793–798.

Rao, C. R. N. 1966. A simple polarization calibrator. *Applied Optics* 58: 1187–1189.

Rao, C. R. N.; Takashima, T.; and Toolin, R. B. 1973. Measurements and interpretation of the polarization of radiation emerging from the atmosphere at an altitude of 28 km over southwestern New Mexico (USA). *Quarterly Journal of the Royal Meteorological Society* 99: 294–302.

Rozenberg, G. V. 1960. Light scattering in the earth's atmosphere. *Soviet Physics Uspekhi* 3: 346–371.

———. 1968. Optical investigations of atmospheric aerosols. *Soviet Physics Uspekhi* 11: 353–380.

Sekera, Z. 1956. Recent developments in the study of the polarization of skylight. *Advances in geophysics*, Vol. 3. (H. E. Landsberg, ed.) pp. 43–104. New York: Academic Press.

———. 1957. Polarization of skylight. *Encyclopedia of physics.* (S. Flugge, ed.) Vol. 48, pp. 288–328. Berlin: Springer Verlag.

———. 1966. Recent developments in the theory of radiative transfer in planetary atmospheres. *Rev. Geophys.* 4: 101–111.

Sekera, Z., and Ashburn, E. V. 1953. Tables relating to Rayleigh scattering of light in the atmosphere. NAVORD Report 206, U.S. Naval Ordnance Test Station, China Lake, California.

Stamov, D. G. 1963. The advantages of the polarimetric method in observing and determining atmospheric turbidity as compared with actinometric and diaphonoscopic methods. *Proceedings of the conference on scattering and polarization of light in the atmosphere.* Alma Ata: Academy of Sciences of the Kazak SSR (1961). Translated into English by Israel Program for Scientific Translations, pp. 149–155.

Takashima, T. 1971. Dependence of the polarization features of the diffuse radiation field of the earth's atmosphere on the location of a concentrated aerosol layer. Doctoral Dissertation, University of California, Los Angeles.

MEASUREMENTS OF THE ELLIPTICAL POLARIZATION OF SKY RADIATION: PRELIMINARY RESULTS

DIETER HANNEMANN and EHRHARD RASCHKE
Ruhr-Universität, Bochum

A polarimeter has been built to measure quasi-monochromatically all four Stokes parameters of scattered incident solar radiation within the spectral range 0.5–1.1 μm. Ground-based measurements at 0.6 μm and 0.8 μm showed a considerable circular component in the sky radiation for polluted skies.

Multiple scattering in turbid planetary, as well as stellar, atmospheres causes linear and circular components of the polarization of scattered light. The degree and direction of polarization, the magnitude of the circular components, and the direction of rotation of the electrical vector of light, as described very elegantly by the four-dimensional Stokes vector, can be considered as signatures of various particles contained in these atmospheres. However, detailed theoretical studies with multiple scattering calculations of the relative importance of these four properties are still missing.

The investigations envisaged with the instrument described here are primarily concerned with water- and ice-clouds in the earth's atmosphere. They are in a preliminary state; thus, only ground-based measurements of sky radiation from cloudless and cloudy skies could be made.

PRINCIPLE OF THE MEASUREMENTS

The determinations of all four Stokes parameters I, Q, U, V are based in a straightforward manner on simultaneous measurements of six polarized radiances L as described by Born and Wolf (1965, Ch. 10.8) for quasi-monochromatic radiation. This is (first argument = direction of oscillation; second argument = retardation):

$$I = L(0°, 0) + L(90°, 0)$$
$$Q = L(0°, 0) - L(90°, 0)$$
$$U = L(45°, 0) - L(135°, 0)$$
$$V = L\left(45°, \frac{\pi}{2}\right) - L\left(135°, \frac{\pi}{2}\right).$$
(1)

The instrumental coordinates are chosen to coincide with the local vertical (0°) and horizontal (90°) directions at the instrument. The degree of polarization is obtained from

$$s = (Q^2 + U^2 + V^2)^{1/2}/I \qquad (0 \leq s \leq 1). \tag{2}$$

The direction of the major axis (= direction of polarization) is determined by

$$\tan 2\psi = \frac{U}{Q}, \qquad (0 \leq \psi \leq \pi), \tag{3}$$

and the ellipticity is described by

$$\sin 2\chi = V/(Q^2 + U^2 + V^2)^{1/2} \qquad \left(-\frac{\pi}{4} \leq \chi \leq +\frac{\pi}{4}\right) \tag{4}$$
$$= VI/s.$$

The sign of V determines the direction of rotation of the vector ($V > 0$: right-handed rotation).

INSTRUMENTAL CHARACTERISTICS

In brief, the instrument consists of an objective lens, an aperture for 7°.5 × 7°.5 field of view, and a field lens in front of a package with six analyzers and one retarder plate (for two analyzers). A filter is placed behind the analyzers to select the desired spectral range before radiation reaches 6 Si-avalanche-foto-diodes (TIXL 56). A multiplexer device enables a convenient storage for an A/D conversion before all six measurements are punched on tape for further automatic processing.

CALIBRATION

All optical components in front of the package of analyzers and the analyzers themselves change the polarization properties of incident radiation which can be described by an instrumental 4 by 4 vector matrix.

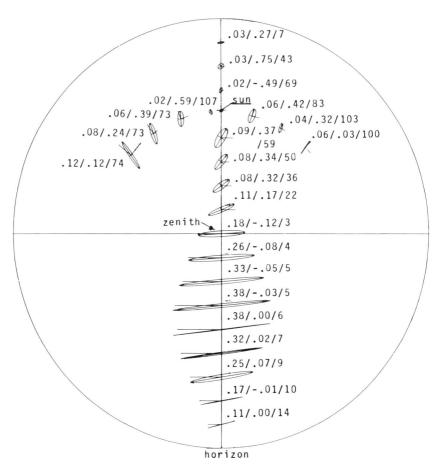

Fig. 1. Measurements obtained, for cloudless but very turbid skies at Bochum (19.9.72). Sun's elevation was 39°. Numbers at each polarization ellipse give the degree of polarization, the ellipticity (positive = right-handed), and the angle of the major axis.

The elements of this matrix (Jäger and Oetken 1963)[1] can be determined experimentally using, in four successive measurements: natural (unpolarized) light, linearly polarized light in 0° and in 45° directions, and elliptically polarized light. These four types of radiation can be produced in a simple device mounted in front of the instrument which enables easy calibrations during field measurements. Further calibration procedures include determinations of the characteristics of each of the six sensors; also the finite spectral width of the filter is taken into account in determining the average retardation.

---

[1] Also see p. 175.

PRELIMINARY RESULTS

Figure 1 shows polarization ellipses obtained on 19 Sept. 1972 in Bochum from skylight measurements at 0.6 μm. The sun's elevation was 39° while the sky was almost cloud free but very turbid and polluted. The fog was not uniformly distributed. In this figure the numbers beside each ellipse describe the degree of polarization, the ellipticity (positive numbers indicate right-handed rotation), and the angle of the major axis with respect to the instrument's coordinates.

In the sun's vertical, where measurements were obtained every 10°, the polarization was found to vary with the zenith angle of observation as is known from many other theoretical and observational studies. The circular component (second value) was found to be small where the polarization is large. The very large value of the circular component near the sun's position might also contain a relatively large error, since there the polarization is less than 10% (see Eq. 4).

More definite evaluation of our results will be possible once the error analysis of the entire instrument has been finished and theoretical results on the sky radiation have been obtained from multiple scattering calculations. Analyses of further measurements on completely clear and on uniformly covered skies are underway.

REFERENCES

Born, M., and Wolf, E. 1965. *Principles of optics* (3rd ed.). New York: Pergamon Press.

Jäger, F. W., and Oetken, L. 1963. Zur Theorie und Praxis der instrumentellen Polarisation. *Publik. d. Astrophys. Observ. Potsdam.*

NOCTILUCENT CLOUDS

G. WITT
University of Stockholm

The formulation of a theory that explains in a plausible way the mechanism of generation and maintenance of noctilucent clouds requires knowledge of both the seasonal, latitudinal, and altitudinal distribution of the phenomenon as well as the physical properties of the cloud particles. Also the structure of the ambient atmosphere and the relevant photochemical processes that determine its composition must be known. The appearance of the clouds is coincident with the season when the temperature at the arctic mesopause is exceedingly low ($130°K$ or less). Recent satellite observations (Donahue and Guenther 1972) indicate that noctilucent clouds are indeed permanently present in the polar cap regions during the appropriate summer season. It is therefore natural to look for a physical process that on the basis of a minimum number of assumptions explains as many as possible of the characteristic features of the clouds and to establish techniques that allow the measurement of these features.

On the basis of only one assumption — that the water vapor concentration at the mesopause is between 1 and 5 parts per million by number — and with the hypothesis that thermodynamically stable condensation nuclei exist or are generated near the mesopause, the set of equations describing the growth and decay of particles formed by water substance during free-fall conditions has been integrated by the authors. The cloud model is found to reproduce faithfully most of the observed features of noctilucent clouds. These include the altitude profile and wavelength dependence of the diffusely scattered downward radiance as well as the degree of polarization as a function of wavelength and scattering angle. The model calculations also support the explanation of the sharp lower boundary of the cloud layer in terms of sublimation as the cloud particles experience the increasing ambient temperature below the mesopause. The size distribution of particles produced by the proposed mechanism is peculiar to the mode of particle formation; it is a peaked distribution with the particle size sharply limited to below 0.10–0.15 μm. The cutoff in size is determined by the amount of available water vapor together with the amount of time the particle can spend in the altitude where the water deposition rate exceeds the rate of sublimation; the size depends, therefore, upon the temperature gradient and the actual position of the temperature minimum.

The optical parameters usually are observed by means of photoelectric filter photometers equipped with polarizing filters. The photometers are axially mounted in a forward-looking position in sounding rocket payloads. In addition to the photometers assigned purely to light-scattering studies, other instruments are included to measure airglow emissions relevant to the photochemistry of the arctic mesopause. The rocket roll causes the photometer output signal to be modulated at twice the roll frequency, while the precession of the vehicle permits the coverage of a range of different scattering angles. The photometer output is telemetered and recorded in the analog mode. The data are subsequently digitized and processed numerically in a digital computer. Using this technique the data can be adequately smoothed, and the Stokes parameters I and Q are obtained as continuous functions of altitude. From the data, the contributions from dark current, starlight, airglow emission and instrumentally scattered light are removed by subtraction. The necessary background data are obtained at apogee where the scattered light component is negligible. In spite of minute care in the design of the sun-protecting shade, the diffusely scattered and hence polarized light scattered by parts of the baffle system is a crucial limiting factor to the obtainable measuring accuracy.

Altogether eight rocket launchings with different photometric payloads have been executed to date within the research program of the Institute of Meteorology of the University of Stockholm. The data that were taken in different noctilucent cloud displays can be summarized as follows:

The degree of polarization decreases with decreasing wavelength. At scattering angles greater than about 70°, the polarization exceeds that of light scattered by anisotropic air molecules. The latter is limited to about 94%, which is the limiting value observed at lower altitudes where the scattered radiance is largely due to molecular scattering during twilight when the diffusely reflected light from the underlying atmosphere is negligible. At 90° scattering angle and at 536 nm wavelength, a value of 96% has been repeatedly observed in noctilucent clouds. Below about 70° scattering angle the observed polarization is less than that of molecular scattering, although the wavelength dependence remains unchanged. The observed color ratio of noctilucent clouds is systematically smaller than the value predicted for Rayleigh scattering. For the wavelengths

366 nm and 536 nm, the color ratio observed in different cloud displays was only about 70% of the Rayleigh value.

The optical information is interpreted on the basis of light-scattering properties of spherical particles. In view of the physical process assumed to be responsible for the particle formation, namely isotropic deposition of molecules onto the surface of the nucleus, this assumption is adequate. The interpretation supports the results of model calculations, i.e., that the size distribution of noctilucent cloud particles has a relatively narrow peak near 0.1 μm radius and that the proposed mechanism for particle generation is indeed plausible.

In summary, the high polarization near 90° scattering angle indicates that the particles have radii of 0.1 μm while the color implies slightly smaller radii; if the particles are ice spheres of equal size, a radius of about 0.08 μm is needed to give 70% of the Rayleigh color.

Further research by the Institute of Meteorology in this field will concentrate on the problems of clarifying the mechanism of nucleation and the establishment of the possible role of ion-water clustering in the nucleation process. Also a reliable method has still to be established for the direct measurement of the key parameter; the concentration of water vapor in the mesosphere.

REFERENCE

Donahue, T., and Guenther, B. T. 1972. Noctilucent clouds in daytime: Circumpolar particulate layers near the summer mesopause. *SRCC Report 169*, Univ. of Pittsburgh, Pittsburgh, Pennsylvania.

Editorial Note. At our request, Martha Hanner provided the following references on the question of whether the condensation nuclei are of atmospheric or extraterrestrial origin.

There are several articles in *Tellus,* 1964, by Hemenway, Soberman, Witt, and others.

In *Cospar Space Research IX* (K.S.W. Champion, P.A. Smith, and R. L. Smith-Rose, eds.) Amsterdam: North-Holland Publ. Co., 1969, are found the following:

G. Witt
The nature of noctilucent clouds. . . . . pp. 157–169

E. Hesstvedt
Nucleation and growth of noctilucent cloud particles. . . 170–174

A. D. Christie
A condensation model of noctilucent cloud formation. . . 175–182

R. K. Soberman
Extraterrestrial origin of noctilucent clouds. 183–189

B. A. Lindblad
Measurement of micrometeorite impacts from a soundnig
  rocket during a noctilucent cloud display. 190–197

F. Link
An optical model for the detection of cosmic dust in the upper
  atmosphere. 198–200

In *COSPAR Space Research XII* (S.A. Bowhill, L. D. Jaffe, and M. J. Rycroft, eds.) Berlin: Akademie-Verlag, 1972, are found the following:

N. H. Farlow and G. V. Ferry
Cosmic dust in the mesosphere 369–380

G. V. Ferry and N. H. Farlow
Upper atmospheric dust concentration in polar regions . . 381–390

P. Rauser and H. Fechtig
Combined dust collection and detection experiment during
  a noctilucent cloud display above Kiruna, Sweden . . . 391–402

A solar origin of the particles is proposed by: C. L. Hemenway, D. S. Hallgren, and D. C. Schmalberger 1972. Stardust. *Nature* 238: 256–260.

POLARIZATION STUDIES OF PLANETARY ATMOSPHERES

D. L. COFFEEN
University of Arizona
and
J. E. HANSEN
Goddard Institute for Space Studies

Remote measurements of scattered sunlight can be used for quantitative analysis of planetary atmospheres. For the purpose of determining the nature and distribution of atmospheric particles, the most useful measurements are of the degree of linear polarization as a function of wavelength and phase angle. Such measurements can be used to determine whether aerosols are spherical or nonspherical, and, in the case of spherical particles, to determine the particle refractive index. The linear polarization also yields specific information on the size of any scattering particles; for spherical particles it is often possible to obtain precise measures of the mean particle size and the dispersion of the size distribution. Regardless of the particle shape, it is possible to determine the optical thickness of Rayleigh scatterers above a cloud or haze layer.

Polarimetric observations of planetary atmospheres have increased considerably in the past decade with spectral, spatial, and temporal coverage. At the same time, theoretical techniques have been developed for the exact calculation of polarization for model atmospheres containing spherical particles. A comprehensive study of Venus permitted an accurate derivation of the shape and refractive index of the cloud particles, of the mean radius and dispersion of the particle size distribution, and of the cloud-top pressure. For terrestrial water clouds, polarimetric particle sizing has been demonstrated. For the outer planets, the observations are not as complete, but their number is increasing and will include measurement over a wide range of phase angle from flyby space probes.

This review paper deals primarily with the measurement and interpretation of polarization of sunlight reflected by planets having optically thick atmospheres. We confine ourselves to studies of Venus, Earth, Mars, Jupiter, Saturn, Uranus, and Neptune, although the techniques are directly applicable to other cases of independent scattering such as the zodiacal light and light scattered by comets and by the rings of Saturn.

The era of precise planetary polarimetry began with the work of Bernard Lyot (1929). After designing and building the necessary visual polarimeter, Lyot measured, with an absolute accuracy of $\sim 0.1\%$, the degree of linear polarization as a function of phase angle (source-object-observer angle) for Venus, Mars, Jupiter, and Saturn, including detailed regional studies for all of these. The thoroughness of his study is impressive. He demonstrated the low degree of polarization of direct sunlight, independent of solar zenith distance; he found the systematic nature of the polarization of Mars and its breakdown during a dust storm; he found the remarkable structure of the polarization-phase curve for Venus and its repeatability over several apparitions; he found the regional variation of polarization position angle for thick planetary atmospheres, in contrast to its invariance for solid mineral surfaces under most conditions; and he found the latitudinal variations of polarization on Jupiter and Saturn.

Equally important was his analysis based on his compendium of polarization measurements of laboratory samples. Mars was found to resemble certain mixtures of volcanic ashes, with a surface pressure less than 25 mb. His polarization curve for Venus was totally different from all the curves for solids but resembled those for clouds of spherical droplets, the best match being for a cloud of droplets ~ 2.3 μm in diameter. For Jupiter and Saturn, the regional polarizations were interpreted as arising from a mixture of single scattering from cloud particles and second scatterings within a gas. The latitudinal variations were interpreted as an increasing transparency of the atmosphere from the equator toward the poles.

Lyot truly founded the field of planetary polarimetry, entering upon it at a time when the planets were judged to have negligible polarization. He pursued this work throughout the 1920s, producing a 150-page document that answered many questions on the nature of the planetary atmospheres but raised even more. Despite this, the field lay almost dormant for a quarter of a century, perhaps due to its novelty, to the difficulty of making significant improvements on the comprehensive observations of Lyot, and to the absence of detailed theoretical work necessary for a full interpretation of polarization data. Some work was done by Öhman during this period, including in 1940 the first "spectropolarimetry" of a planetary atmosphere (see Dollfus 1961, Plate 7). The field was revived by A. Dollfus (1957), who added detailed studies of planetary regions and of the wavelength dependence of polarization, using colored filter glass with visual and photographic detection. Dollfus continued Lyot's high standard of observation, filled in many gaps in planetary data, and added to our understanding with laboratory measurements. A major contribution was his renewal of scientific interest in studies of planetary scattering.

A new breakthrough came with the availability of photoemissive detectors, which led to extensive measurements of the wavelength dependence of polarization of planets, now covering the range of 0.22 to 3.6 μm. This development was spurred by Gehrels (Gehrels and Teska 1963). The photoelectric measurement of planetary polarization is now widespread, with major activities in the United States, France and the Soviet Union.

Two very recent developments have been the inclusion of photoelectric polarimeters on planetary space probes[1] and the development by Kemp[2] of precise polarimeters for the measurement of elliptical polarization.

Simultaneous with progress in observations have been major advances in our capabilities for quantitative interpretation of polarization measurements. The advances in our understanding of how the polarization depends on the nature of scattering particles have followed from an improved ability to model theoretically the processes of single and multiple scattering. Part of this progress should be credited to the advent of digital computers, which raised the exact theory for single scattering by spheres to computational practicality and spurred the development of exact methods for multiple scattering by clouds.

The problem of multiple Rayleigh scattering was solved by Chandrasekhar (1950) and Sekera and his co-workers (Coulson, Dave, and Sekera 1960). These solutions were important for explaining the polarization of the earth's clear atmosphere, but the investigation of other planetary atmospheres and of the earth's atmosphere in cloudy and hazy conditions required the consideration of scattering by large particles. Coffeen (1968) used the Mie theory to investigate systematically the general dependence of the polarization on the scattering angle, size parameter, and refractive index for single scattering by spheres; from this investigation he could specify limits for the refractive index and size of the cloud particles on Venus. Several methods were developed that treat multiple scattering by either large or small particles, with perhaps the most successful being the "doubling" or "adding" method. This method was introduced apparently independently in gamma-ray transfer by Peebles and Plesset (1951), in transmission-line theory by Redheffer (1962), in astronomy by van de Hulst (1963), and in meteorology by Twomey, Jacobowitz, and Howell (1966). It was extended to azimuth-dependent problems by Hansen (1969) and to polarization by Hansen (1971a) and Hovenier (1971). This method was used to refine precisely our knowledge of several properties of the clouds of Venus.

Our theoretical capabilities are outlined in the next section, with a

---

[1] See p. 189.   [2] See p. 607.

review of relevant literature. This is followed by a review of polarimetric observations of specific planetary atmospheres and the state of their interpretations.

THEORY AND COMPUTATIONS

At the time of Lyot's (1929) investigation, the principal technique for the interpretation of planetary polarization data was comparison with measurements of light scattered by laboratory samples, such as rock, ice, and artificial clouds. Only for Rayleigh scattering were quantitative theoretical results available, and these only for single scattering. For large spherical particles a qualitative explanation based on ray optics was also useful. The comparative laboratory approach is still necessary for irregular shapes and surfaces. However, major advances have been made during the past several years in computations for single and multiple scattering by spherical particles. Future theoretical progress that can be anticipated includes atmospheric modeling with vertical and horizontal inhomogeneities, the calculation of polarization spectra across absorption and emission features, and the eventual development of computational procedures for nonspherical crystalline shapes and for rough surfaces. We anticipate expanded use of polarimetry in remote sensing for reliable determination of the microstructure of particles and surfaces.

In this section on theory and computations we briefly discuss the relative information content in the four Stokes parameters and then proceed to the theoretical models applicable to the interpretation of polarization. The major part of this section is based on an unpublished but available report (Hansen 1972). For more comprehensive reviews of single scattering theory, we refer to the works of Shifrin (1951), van de Hulst (1957), Volz (1961), Deirmendjian (1969), and Kerker (1969). For a review of techniques of multiple scattering, we refer to van de Hulst (1971) and Hunt (1971). For a wide variety of scattering topics, we mention the collections of the Interdisciplinary Conferences on Electromagnetic Scattering (Kerker 1963; Rowell and Stein 1967).

A variety of ways are used to represent radiation mathematically, but the most useful physical description is in terms of the *intensity,* the *linear polarization* (its degree and direction) and the *circular polarization.* All of these quantities can be measured as a function of wavelength, scattering geometry (the directions of the observer and of the sun), and location on the planet.

The intensity is the most commonly measured quantity. Observations of absorption lines at high spectral resolution yield information on the gaseous composition of an atmosphere and on the temperature and pressure distributions. Broadband measurements of the intensity of light scattered by planetary atmospheres and surfaces are useful for helping to determine the composition of the scatterers, though such measurements are seldom subject to unique interpretation. Imaging devices for the intensity can yield valuable information on the structure of clouds and planetary surfaces.

The linear polarization has the advantage of being obtainable from a relative measurement with a higher accuracy (typically $\sim 0.1\%$) than that with which the intensity can be measured. It has also been demonstrated with laboratory and theoretical work that errors of measurement can be kept much smaller than the characteristic features in the linear polarization, and that the features can be readily interpreted to yield information on the scattering material.

The circularly polarized component of sunlight reflected by planets has recently been measured and found to be very small.[3] Both the magnitude of the circular polarizations and the variations of its sense (or "handedness") are in good agreement with theoretical calculations (Hansen 1971c). Scattering of sunlight by molecules does not give rise to circular polarization. However, single scattering by oriented nonspherical particles and multiple scattering by particles with sizes $\gtrsim \lambda$ can produce circular polarization. More extensive observations and theoretical work on circular polarization can be anticipated in the near future; it will be interesting to see what new information on planetary atmospheres may result.

Small Particles: Rayleigh Scattering

Small particles provide the simplest scattering behavior. Rayleigh scattering requires that the particle size be much less than the wavelength both inside and outside of the particle. In this case, the electromagnetic fields are uniform throughout the particle, which thus oscillates as a simple dipole in response to the incident radiation (Fig. 1).

The strong polarization for Rayleigh scattering, its simple dependence on scattering angle, and its characteristic wavelength dependence (cross section or optical thickness $\propto \lambda^{-4}$) provide sure means for separating scattering due to Rayleigh particles from that due to large particles.

Various theoretical techniques for treating multiple scattering in a Rayleigh atmosphere are discussed by Ueno, Mukai, and Wang, who provide a useful bibliography.[4] For perfect Rayleigh scattering, a full multiple scattering theory for a plane-parallel layer over a Lambert ground was developed by Chandrasekhar (1950), who included numerical results

---

[3] See p. 607.   [4] See p. 582.

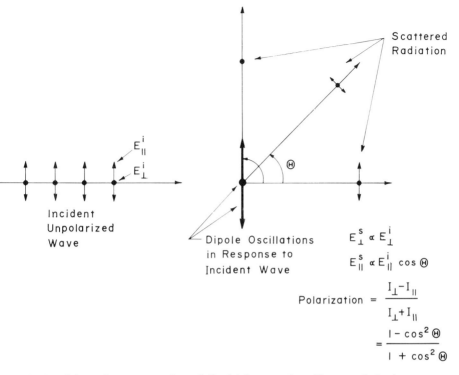

Fig. 1. Schematic representation of Rayleigh scattering. The unpolarized wave incident from the left can be represented by two linearly polarized waves vibrating at right angles to each other with equal electric field strengths ($E_\parallel = E_\perp$) and a random phase relationship. The electrons in a small particle will oscillate in response to the electric components of the incident wave, their motion being equivalent to two oscillating dipoles whose axes are represented by the heavy arrow and dot. The dipoles, which do not radiate in their direction of action, give rise to the indicated scattered fields and intensities. (After Hansen 1972.)

for the case of a semi-infinite atmosphere ($\tau = \infty$). Graphical results are given by Chandrasekhar and Elbert (1954). Tabular results for a wide range of geometric parameters, for optical depths up to $\tau = 1$, were published by Coulson, Dave, and Sekera (1960; see also Coulson 1959). Contours of intensity and polarization on a planetary disk are given by Sekera and Viezee (1961). The polarizations arising from single and from higher-order scattering are separated and mapped by de Bary and Bullrich (1964). The gap between small optical depths and the semi-infinite case was bridged by Sekera and Kahle (1966).

The above results are for the conservative case with no absorption by the Rayleigh scatterers. Real molecules and small particles can have various modes of absorption, represented by a single scattering albedo less than unity at a particular wavelength. Techniques and results for polarization by nonconservative Rayleigh scattering are given by Sekera (1963), Herman and Yarger (1965), Dave (1965), Abhyankar and Fymat (1970), and Bond and Siewert (1971).

In a real atmosphere the variation of temperature and pressure with altitude can have a substantial effect, requiring vertically inhomogeneous models. Using the auxiliary equation method, Molenkamp (1972) has computed the intensity and polarization across absorption lines in a stratified Rayleigh atmosphere. Figure 2 shows the variation of polarization across a line for atmospheres of various thicknesses. For pure single scattering, the polarization would be 100% in all cases, the phase

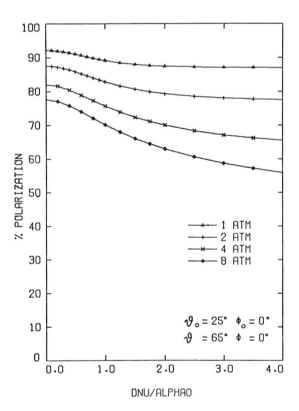

Fig. 2. Calculated variation of polarization across an absorption line for vertically inhomogeneous molecular atmospheres with different surface pressures and all with a completely absorbing ground. The abscissa is $(\nu - \nu_0)/\alpha_0$, where $(\nu - \nu_0)$ is the frequency separation from the line center, and α_0 is the half-width of the pressure-broadened Lorentz profile at a pressure of 1 atm. The phase angle is $90°$. (After Molenkamp 1972.)

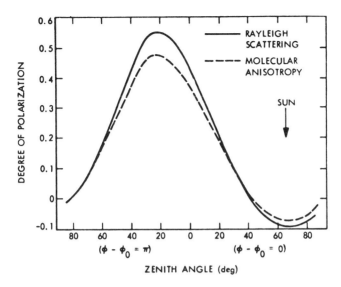

Fig. 3. Calculated degree of polarization for skylight radiation as a function of zenith angle, for $\mu_o = 0.4$, $A_g = 0$, and $\tau = 1$. Results are shown for Rayleigh scattering by isotropic molecules and for a pure CO_2 atmosphere with molecular anisotropy. (After Schiffer 1972.)

angle being 90°. As expected, the polarization is a maximum at the line center, where multiple scattering is a minimum. Similar results are found if a Lambert surface is included in the model.

Polarization line profiles have also been calculated by Lenoble (1970) and by Fymat, who includes the case of a gas absorption line in the presence of aerosols.[5] Such absorption band studies show the possibility of vertical probing of a planetary atmosphere with spectropolarimetry. A striking case is the 0.889 μm band of CH_4 on Jupiter and Saturn. Approximate calculations illustrate that, for a hypothetical two-cloud-layer model, measurements in the band could give the polarization-phase curve for particles in an upper thin haze layer, while out-of-band measurements at 0.92 μm give a curve dominated by the lower particles. The presence of the thin haze is revealed only by the absorption band data; the gas acts as a selective filter, damping the higher-order scatterings.

Real molecules do not all exhibit isotropic Rayleigh scattering; anisotropy reduces the degree of polarization, and for some molecules the effect is significant. Schiffer (1972) has included the effect of anisotropy in multiple scattering computations. Figure 3 shows an example of his

---

[5] See p. 617.

computations for the polarization of the diffuse radiation emerging at the bottom of a pure CO_2 atmosphere. For most other molecules, the effect of anisotropy is not as great as it is for CO_2 (van de Hulst 1952; Partington 1953; Fabelinskii 1968; Gucker et al. 1969).

In general, it is possible to separate the scattering by gas molecules from that due to large cloud particles in a thick planetary atmosphere regardless of the cloud particle type. The optical depth of molecules in a vertical column above the cloud can be determined in most cases by measuring the net brightness and polarization as a function of wavelength. For a known molecular composition, the cloudtop pressure can be derived; for measurements in the wavelength range 0.3–1.0 μm, reliable pressure determinations are possible throughout the range $\sim$ 10 mb to $\sim$ 1 atm. The derived gas amount refers to only gas above the clouds if the mean-free-path in the cloudtop is sufficiently small, as is the case for terrestrial clouds. In the case of diffuse haze, the derived amount refers also to gas mixed with the haze in the region where the first few scatterings occur, and thus, the derived pressure level does not refer to a sharp top of the particulate region of the atmosphere. The possibility of distinguishing between a diffuse haze, dense cloud, or some intermediate case can be easily understood because as the zenith angle of observation increases from 0° (i.e., normal to the planet's surface) toward 90°, the contribution to the polarization from Rayleigh scattering increases dramatically in the case of gas above a cloud deck, while the relative contribution of gas and particulates remains essentially constant in the case of a diffuse haze. If the cloud or haze particles are small compared to the wavelength, they will give Rayleigh scattering and will not be distinguishable from the gas; however, such small particles rarely occur with a large optical depth in nature.

Large Particles: Geometrical Optics

For large particles, the scattering can be understood in terms of the concepts of geometrical optics. In this case, the light incident on a particle may be thought of as consisting of separate rays of light that pursue independent paths. For a particle with a size at least several times the wavelength, it is possible to identify rays striking various local regions on the particle's surface.

Figure 4 illustrates the terminology used for the different contributions to the light scattered by a large particle. The division of rays into these different components is valid for all particle shapes, though the fraction of light that goes into the different components has some dependence on shape.

The light rays that miss the particle ($\ell = 0$) are partially diffracted into its geometrical shadow. The amount of diffracted light is equal to

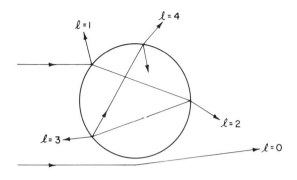

Fig. 4. Paths of light rays scattered by a sphere according to geometrical optics and diffraction. P is the phase function and Θ the scattering angle. The table on the right gives the fraction of the total scattered light contributed by each value of ℓ for nonabsorbing spheres with refractive indices 1.33 and 2.0. (After Hansen 1972.)

the amount striking the particle, and hence for nonabsorbing particles it constitutes exactly half of the scattered light. This fraction is independent of the particle shape and refractive index. The angular distribution of the diffracted light does depend on the particle shape; however, the dependence is slight if the particles are randomly oriented, and it can be easily computed for any particle shape. The polarization of diffracted light in the far field is zero in the limit of large particles, except for the fact that diffracted light can optically interfere with light which is reflected and refracted by the particle (see below).

The intensity and polarization of the light reflected from the outside of the particle ($\ell = 1$) may be computed from Fresnel's equations. For a transparent sphere, the externally reflected light makes up only a few percent of the total scattered light, though the fraction increases as the refractive index increases. For randomly oriented convex nonspherical particles, the external reflection makes up exactly the same fraction of the scattered light as it does for spheres, and the angular distribution and polarization of the reflected light are the same as for spherical particles.

The rays that are refracted twice without any internal reflections ($\ell = 2$) make up a large fraction of the scattered light for transparent or partially transparent spheres. This holds for nonspherical particles also, although the exact fraction does depend on the particle shape even if the particles are randomly oriented.

The light that is internally reflected in particles represents only a few percent of the scattered light. The angular distribution and polarization for these rays can be computed for any particle shape using Snell's law and the Fresnel reflection coefficients. For spheres, the $\ell = 3$ and $\ell = 4$ terms give rise to the primary and secondary "rainbows." The intensity of higher terms is negligible.

The intensity and polarization for single scattering by large spheres are given by Liou and Hansen (1971) for a variety of refractive indices, as computed by ray optics.

Intermediate Particle Sizes

The limiting cases of geometrical optics and Rayleigh scattering are sufficient to describe many of the characteristics of light scattered by particles of all sizes. This is illustrated in Fig. 5 which shows Mie calculations of the intensity and polarization for single scattering by a size distribution of spheres with a real refractive index $n = 1.33$. These are contour diagrams of the intensity and polarization as a function of the scattering angle, Θ (or the phase angle, $180° - \Theta$), and as a function of the mean size parameter for the distribution, $2\pi a/\lambda$, where a is equal to the mean effective particle radius. The angular scattering by a size distribution of particles of a given refractive index is primarily a function of only two parameters, a and the effective variance b, the higher moments of the distribution being negligible in most cases.[6]

For size parameters near zero, Rayleigh scattering occurs with its strong characteristic polarization. For the largest size parameters, the features are essentially those of geometrical optics: the primary rainbow at a scattering angle $\Theta \sim 145°$, the second rainbow for $\Theta \sim 120°$ (which is noticeable in the polarization for $2\pi a/\lambda < 50$, but not in the intensity) and the negatively polarized twice-refracted rays for $\Theta < 90°$.

As the size parameter decreases, the features of geometrical optics become increasingly blurred until they are essentially lost for size parameters less than 15. This is as expected since $2\pi a/\lambda = 15$ corresponds to a ratio of particle diameter to wavelength of ~ 5. Between here and the

---

[6] See p. 617.

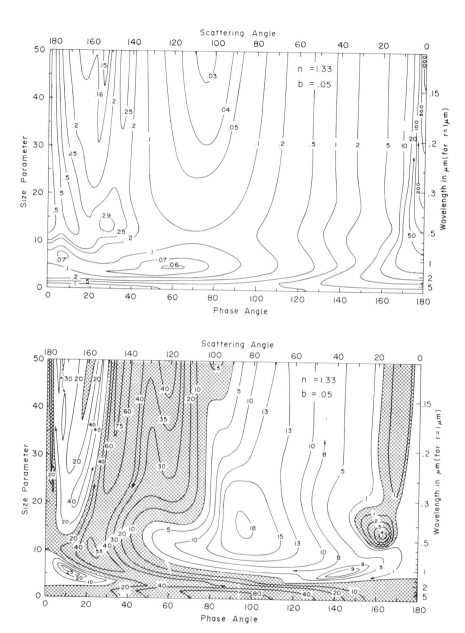

Fig. 5. *Below*, contour diagram of the percent polarization calculated for single scattering of unpolarized light by a size distribution of spheres with refractive index 1.33. The shaded areas indicate positive polarization ($I_\perp > I_\parallel$) and the unshaded indicate negative values. The size parameter is $2\pi a/\lambda$. a is equal to the effective mean radius and b the effective variance for the size distribution, for which the particular form was $N(r) \propto r^{(1-3b)/b} e^{-r/(ab)}$. *Above*, the same as below, except that the contours are for the phase function, or normalized intensity. (After Hansen 1972.)

domain of dipole (Rayleigh) scattering, the results can be understood in terms of radiation due to higher multipole moments and surface waves. The manners in which these features depend on the refractive index and size distribution are illustrated and discussed below.

Single Scattering by Spheres

The properties of radiation scattered by uncharged isotropic homogeneous spheres can be obtained exactly from classical electromagnetic theory. The results for a single sphere depend upon the index of refraction of the sphere relative to the surrounding medium, $m = n - in'$, where $i = (-1)^{1/2}$, and upon the ratio of the radius of the sphere to the wavelength of the incident radiation ($x = 2\pi r/\lambda$, where r is the particle radius and λ the wavelength). For these special particles, the properties of radiation scattered in an arbitrary direction depend additionally only on the scattering angle, Θ, and are completely defined for any polarization state of the incident light by four functions of m, x, and Θ. Using the notation of van de Hulst (1957), these four functions are:

$$M_1(\Theta) = |A_1(\Theta)|^2$$
$$M_2(\Theta) = |A_2(\Theta)|^2$$
$$S_{21}(\Theta) = |A_2(\Theta)A_1^*(\Theta) + A_1(\Theta)A_2^*(\Theta)|/2 \qquad (1)$$
$$D_{21}(\Theta) = |A_2(\Theta)A_1^*(\Theta) - A_1(\Theta)A_2^*(\Theta)|/2,$$

where A_1 and A_2 are the complex scattering amplitudes, respectively perpendicular and parallel to the scattering plane, and the asterisk indicates the complex conjugate. For a size distribution (polydispersion) of independent scatterers, the functions M_1, M_2, S_{21}, and D_{21} are obtained by simple addition of those functions for all particles in the distribution, e.g.,

$$M_1(\Theta) = \int_{r_1}^{r_2} M_1(r,\Theta) N(r) dr, \qquad (2)$$

where $N(r)$ is the size distribution of particles, and r_1 and r_2 correspond to the smallest and largest particles present.

For scattering problems, the most practical representation of a beam of radiation is in terms of its Stokes parameters, I, Q, U, and V. I is the intensity of the beam, Q and U specify the linear polarization, and V the circular polarization. The four functions in Equation (1) uniquely define the scattered radiation in terms of the Stokes parameters of the incident radiation:

$$\begin{pmatrix} I \\ Q \\ U \\ V \end{pmatrix} = \begin{pmatrix} \tfrac{1}{2}(M_1+M_2) & \tfrac{1}{2}(M_2-M_1) & 0 & 0 \\ \tfrac{1}{2}(M_2-M_1) & \tfrac{1}{2}(M_1+M_2) & 0 & 0 \\ 0 & 0 & S_{21} & -D_{21} \\ 0 & 0 & D_{21} & S_{21} \end{pmatrix} \times \begin{pmatrix} I_0 \\ Q_0 \\ U_0 \\ V_0 \end{pmatrix}. \quad (3)$$

It is convenient to further define a normalized phase matrix by means of

$$p^{11}(\Theta) = c[M_1(\Theta) + M_2(\Theta)]/2 = p^{22}(\Theta)$$
$$p^{12}(\Theta) = c[M_2(\Theta) - M_1(\Theta)]/2 = p^{21}(\Theta) \quad (4)$$
$$p^{33}(\Theta) = cS_{21}(\Theta) = p^{44}(\Theta)$$
$$p^{43}(\Theta) = cD_{21}(\Theta) = -p^{34}(\Theta) \quad ,$$

where the constant c is determined by the condition that

$$\int_{4\pi} p^{11}(\Theta)\, \frac{d\omega}{4\pi} = 1. \quad (5)$$

The phase function, or the probability for scattering of unpolarized light at the scattering angle Θ, is $p^{11}(\Theta)$. The degree of linear polarization for unpolarized incident light is $-p^{21}(\Theta)/p^{11}(\Theta)$; for simplicity we will sometimes refer to the latter as the "degree of polarization" or just the "polarization." The degree of circular polarization is zero for single scattering of unpolarized light by spheres.

Ray Optics Versus Mie Scattering

The ray tracing approach of geometrical optics provides a physical explanation for most of the features that occur in the exact theory for scattering by spheres (Mie scattering). In the computations for geometrical optics illustrated below, the contributions from rays undergoing diffraction ($\ell = 0$), reflection ($\ell = 1$), two refractions ($\ell = 2$), etc., are added without regard to phase. This is reasonable, since in nature there is generally a size distribution of particles present that tends to average out phase effects, assuming independent scattering. We will, however, also consider the exceptions, where features due to interference between different rays are noticeable.

Figure 6 illustrates the results of computations for both ray optics and Mie theory for two values of the refractive index, namely 1.33 and 1.50. These computations are for the particular size distribution

$$N(x) = x^6 e^{-9x/x_m} \quad (6)$$

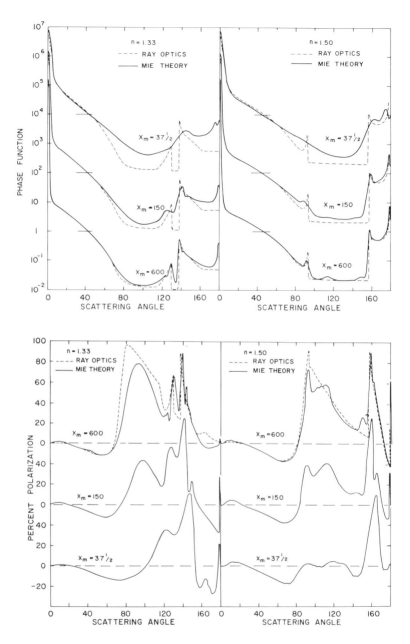

Fig. 6. Comparison of ray optics (geometrical optics and diffraction) and Mie calculations for the percent polarization *(below)* and phase function *(above)* for scattering by spheres. The size distribution is $N(x) = x^6 e^{-9x/x_m}$. Results are shown for two refractive indices and three values of the effective mean size parameter, x_m. For the phase function, the scale applies to the curves for $x_m = 600$, the other curves being successively displaced upward by factors of 100. (After Liou and Hansen 1971.)

for three values of x_m, which is the mean effective size parameter for the distribution. For the purpose of comparing geometrical optics and Mie theory, the shape of the size distribution is not critical.

In the curves for the phase function (intensity), the concentration of light near the scattering angle 0° represents the diffraction ($\ell = 0$), which is unpolarized. The contribution to the scattered light from $\ell = 1$ (external reflection) does not leave any apparent feature on the intensity, but it is strongly polarized and gives rise to the broad positive polarization feature for scattering angles in the range ~ 80°–120°. The energy contained in the $\ell = 2$ term (twice-refracted rays) is concentrated in the forward scattering hemisphere and is negatively polarized, as follows from Fresnel's equations. The components involving internal reflections ($\ell \geq 3$) contain only a few percent of the scattered light; however, they give rise to observable optical phenomena that are useful for cloud particle identification.

"Rainbows" occur when the scattering angle has an extremum as a function of the angle of incidence τ on the sphere. For example, for $n = 1.33$ as τ varies from 90° (central incidence) to 0° (grazing incidence), the scattering angle for rays internally reflected once, computed using Snell's law, decreases from 180° until it reaches 138° (the angle of "minimum deviation"), from which it then increases again. The resulting concentration of energy at 138° and slightly greater angles is the primary rainbow.

The minor feature on the large scattering angle side of the primary rainbow is the first "supernumerary bow." This is not rendered by the geometrical optics computations because it is an interference feature. At these scattering angles, there are $\ell = 3$ rays striking two different parts of the sphere but emerging with the same scattering angle. These rays optically interfere causing the supernumerary bows. The number and strength of the supernumerary bows depends on the shape of the size distribution, as discussed in the following section.

The enhanced scattering in the backward direction, $\Theta \sim 180°$, is the so-called "glory," caused specifically by the spherical shape of the scatterers.[7] More detailed discussions of the glory are given by van de Hulst (1957), Fahlen and Bryant (1968), and Liou and Hansen (1971).

Figure 6 illustrates that there is a close quantitative agreement between ray optics and Mie theory only if the value of the mean size parameter is at least several hundred. Dave (1969) has demonstrated rather detailed agreement for scattering by a single sphere, provided that the size parameter $2\pi r/\lambda \sim 800$ or larger. However, most of the features of ray optics remain visible to a much smaller particle size, and indeed, their variation

---

[7] See p. 550.

with decreasing size parameter can be qualitatively understood in terms of the reduced validity of the localization of rays. The decreasing size causes the light in the individual features to be blurred over a wider range of angles than predicted by ray optics, and it usually affects higher values of ℓ first because they have a more complex ray path within the sphere. Thus, the secondary rainbow ($\ell = 4$) is quite smooth in the intensity for $x_m = 150$ and is lost for $x_m = 37½$, while the primary rainbow ($\ell = 3$) is still easily visible in both cases. Because of the asymmetric shape of the rainbows, the smoothing due to a finite size parameter causes the peak of the rainbow to move slightly as the size parameter decreases.

Figure 6 also illustrates that the single scattering polarization, compared to the intensity, contains much stronger imprints of most of the features occurring in the scattered light. Furthermore, for the polarization these features remain visible to smaller size parameters. These conclusions hold for the rainbows, the supernumerary bows, the glory, and the external reflection. The strength of these features for single scattering is a major reason why there is a high information content in polarization observations. In addition, for multiple scattering the polarization features are much better preserved than are the intensity features.

Effect of Size Distribution

The characteristics of scattered light, its intensity and polarization in particular, depend upon the size distribution of particles as indicated by Equation (2). Generally, we do not know this distribution, and indeed, it is one of the quantities that we would like to obtain from measurements of the scattered light. Since there are an infinite number of possible size distributions, some authors have been uncertain as to how much unambiguous information can be extracted from scattered light measurements. Actually, the situation is much brighter than it may appear at first glance. The scattering properties of most physically plausible size distributions depend significantly on only a small number of characteristics of the distribution, e.g., on the average particle size and the width of the distribution. While it is usually not feasible to obtain the exact shape of the size distribution from scattered light, it is possible to extract the major characteristics of the size distribution, and as a consequence, we are able to solve for the other properties of the scatterers.

Clearly, the first parameter describing a size distribution should be some measure of the mean size. Since most particles scatter an amount of light in proportion to at least the square of the radius, it is logical to define a *mean effective radius* for a size distribution as

$$\langle r \rangle_{\text{eff}} = \frac{\int_0^\infty r\, \pi r^2 N(r)\, dr}{\int_0^\infty \pi r^2 N(r)\, dr}, \quad (7)$$

which differs from the simple mean radius only in the fact that the particle area is included as a weight factor multiplying $N(r)$. For some properties of the scattered light, $\langle r \rangle_{\text{eff}}$ is by itself an adequate specification of the size distribution. However, higher moments can be important, particularly for polarization. The *effective variance* is defined as

$$v_{\text{eff}} = \frac{\int_0^\infty (r - \langle r \rangle_{\text{eff}})^2 \pi r^2 N(r)\, dr}{\langle r \rangle_{\text{eff}}^2 \int_0^\infty \pi r^2 N(r)\, dr}, \quad (8)$$

where the factor $\langle r \rangle_{\text{eff}}^2$ in the denominator makes v_{eff} dimensionless and a relative measure. Additional moments of the size distribution may be defined analogously when useful.

Different particle size distributions having the same values of $\langle r \rangle_{\text{eff}}$ and v_{eff} have similar scattering properties. Thus, it is possible to use an analytic size distribution to represent a natural distribution. A particularly convenient analytic distribution, used by Hansen (1971b), is

$$N(r) = \text{constant} \times r^{(1-3b)/b} e^{-r/(ab)}, \quad (9)$$

which has the property that

$$\begin{aligned} a &= \langle r \rangle_{\text{eff}} \\ b &= v_{\text{eff}} \end{aligned}, \quad (10)$$

i.e., the two parameters in the distribution are equal to the physical quantities that characterize the scattering by the size distribution.

Figure 7 shows some measured size distributions for terrestrial water clouds compared to the standard distribution, Equation (9), with a and b specified by Equation (10). In Fig. 8 the phase function and percent polarization are plotted for both the observed and analytic distributions. The latter figure is a typical example of differences in the scattering that occur for distributions with the same $\langle r \rangle_{\text{eff}}$ and v_{eff}. The differences are largest for the stratus cloud in which the observed distribution

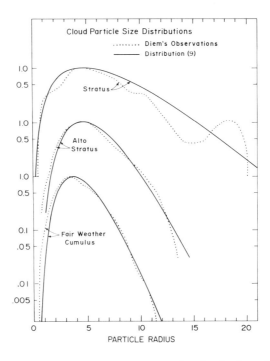

Fig. 7. Comparison of measured cloud particle size distributions (Diem 1948) to the analytical size distributions of Equation (9), $N(r) \propto r^{(1-3b)/b} e^{-r/(ab)}$. a and b are equal to the effective mean radius and the variance for the distributions. (After Hansen 1971b.)

is bimodal, where each mode contributes an approximately equal amount to the scattered light. However, even in this case, the only significant difference in the scattering is for $\Theta \sim 160°$; this is a region of optical interference, and hence, the results are particularly sensitive to the shape of the size distribution.

Since two parameters are sufficient to define adequately most size distributions, it is possible to illustrate in a small number of diagrams the effect of the size distribution on the scattered light. Figure 9 shows contour diagrams of the polarization for a size distribution of spheres, as a function of scattering angle on the horizontal axis and as a function of mean effective size parameter on the vertical axis. These results are for the standard size distribution, Equation (9), with the two parts of the figure corresponding to different values of b, but for the same refractive index, $n = 1.44$. The larger value of b, 0.20, is typical of terrestrial water clouds. Note that the results for a broader distribution can be anticipated from those for a narrower distribution because they are weighted averages along vertical lines.

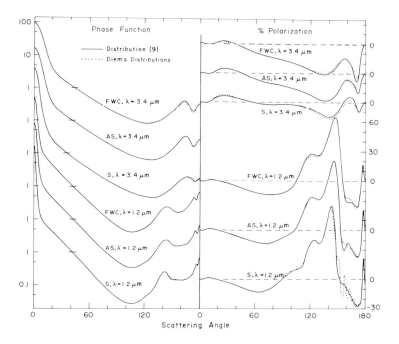

Fig. 8. Comparison of the calculated phase function and percent polarization for single scattering by spheres for the size distributions shown in Fig. 7. FWC, AS, and S represent fair weather cumulus, altostratus, and stratus, respectively. The phase functions are displaced with the dashed lines indicating zero polarization. The wavelengths λ are in μm. (After Hansen 1971b.)

It is useful to understand the features in these contour diagrams. The steep positive polarization ridge for scattering angles $\sim 140°–170°$ is, of course, the primary rainbow. As the size parameter decreases, the rainbow becomes smoother and sinks into the negative polarization at $2\pi a/\lambda \sim 10$, where the concepts of geometrical optics have lost their validity. Broadening the size distribution smooths the rainbow at its small size parameter extremity, because only in that region is the rainbow significantly size dependent.

The positive feature for scattering angles $\sim 90°–130°$ is the second rainbow, which, as expected, is broader than the primary rainbow and disappears at a somewhat larger size parameter. The second rainbow also has little dependence on the width of the size distribution.

For scattering angles smaller than those of the second rainbow, the polarization is essentially that of the twice-refracted rays ($\ell = 2$). However, for $\Theta \sim 5° -25°$, the external reflection ($\ell = 1$) wins out over $\ell = 2$ and gives rise to the positive polarization peninsula. This peninsula is connected to an "island" at $2\pi a/\lambda \sim 10$ for the case of the

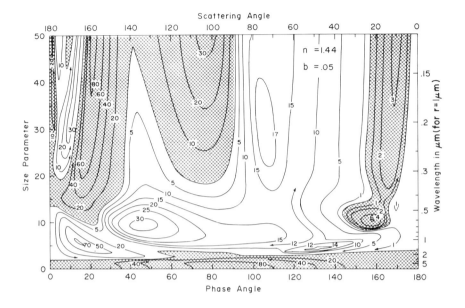
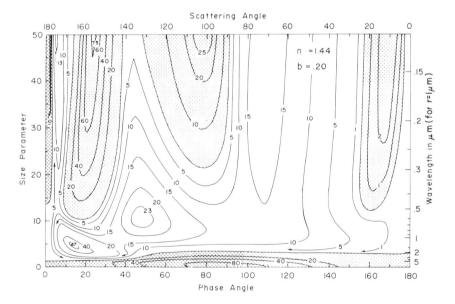

Fig. 9. Contour diagrams of the percent polarization calculated for single scattering of unpolarized light by a size distribution of spheres, similar to Fig. 5 but for refractive index 1.44, for two values of the effective variance b of the size distribution. (After Hansen 1972.)

narrow distribution ($b = 0.05$). Obviously this island must be washed away for a broad distribution as shown in Fig. 9. This particular feature is due to optical interference between rays passing through the particles and those passing outside of it. Van de Hulst (1957) refers to this interference as "anomalous diffraction." This feature can be observed in the polarization of sunlight reflected by Venus, for which it has been mapped in detail.[8]

For the smallest sizes, the positive polarization is Rayleigh scattering. The scattered radiation field can, in general, be expanded in a series in which the successive terms correspond to the radiation arising from the different multipole moments induced in the particle by the incident electromagnetic wave. The total number of terms required in this expansion is $\sim 2\pi a/\lambda$, with a minimum of one. Thus for $2\pi a/\lambda \ll 1$, the first term, representing the dipole radiation, is sufficient. The number of terms required can be understood with the help of the "localization principle" (van de Hulst 1957; Bryant and Cox 1966), which states that a term of order n in the multipole expansion of the scattered radiation corresponds to a ray passing the center of the sphere at a distance $n\lambda/2\pi$. (The basis for this principle is the same as in partial-wave analysis in quantum theory; the nth term is associated with an orbital angular momentum h/λ at an impact parameter $n\lambda/2\pi$.) Thus, the series converges shortly after n exceeds $2\pi a/\lambda$ because higher terms correspond to rays that do not pass through the sphere.

If the contributions of the successive terms in the multipole expansion are added one at a time to form contour diagrams of the type in Fig. 9, only the dipole term ($n = 1$) individually contributes a feature that is still readily apparent after the sum is taken over all terms. The quadrupole term moves the boundary between positive and negative polarization toward larger size parameters for small scattering angles; it is also primarily responsible for the positive feature at $x \sim 3$ and $\Theta \sim 110°$. The negative features for $2\pi a/\lambda = \sim 4\text{–}8$ arise as the combined result of a few terms with $n \sim 2\pi a/\lambda$ and just smaller values of n. For larger size parameters, the sharp feature in the polarization for $\Theta \sim 180°$, the glory, also arises from the few terms with $n \sim 2\pi a/\lambda$. This feature can thus be associated with edge (grazing) rays which set up surface waves on the sphere (van de Hulst 1957; Bryant and Cox 1966; Kerker 1969). The surface waves reradiate in all directions, but, because there are focal points at $\Theta = 0°$ and $180°$, the energy is concentrated in the forward and backward directions. For $\Theta = 0°$,

---

[8] See Fig. 16.

the effect is drowned in the much stronger diffracted light, but for $\Theta = 180°$, it is easily visible.

Effect of Refractive Index

The variation of the polarization with refractive index is partially illustrated by a comparison of Fig. 5 with Fig. 9. The primary changes that occur in the different features can be understood on the basis of the physical origin of the features. The location of the rainbows moves in accordance with Snell's law and geometrical optics. The relative contribution from Fresnel reflection increases with n. The island due to anomalous diffraction moves to a smaller size parameter as n increases, because of the reduced ratio of the wavelength of rays inside the particle to the wavelength of the rays outside the particle.

In the case of spherical particles, the variations of the polarization with refractive index can readily be used to establish the refractive index of the particles. An example of this for the clouds of Venus is given below.[9]

Multiple Scattering by Spheres

It has been demonstrated that singly scattered radiation carries a detailed signature of the scatterers. Thus, a large amount of information on planetary atmospheres can potentially be obtained from measurements of scattered radiation. Indeed, even for in situ atmospheric measurements of cloud and haze particles, it is difficult to find more reliable means for analyzing the nature of the particles than through measurements of singly scattered light (as is done with a nephelometer).

However, in most cases of interest, the radiation has been multiply scattered and the observable signal represents an average of photons scattered a different number of times at all possible scattering angles. This process tends to smooth out features that occur in the radiation as a function of scattering angle at a given wavelength. Spectral absorption features usually become stronger with multiple scattering, but these features then represent some average over the entire depth to which the photons have traveled. In either case, to interpret the observations it is necessary to be able to model the multiple scattering process. Multiple scattering computations are also important for finding the radiation characteristics that best survive multiple scattering, and are used as a guide for observations.

There are quite a number of ways to compute the multiple scattering of light, many of which include the effects of polarization. Several of these methods have been discussed by Hunt (1971), van de Hulst (1971) and Hansen (1971a). Here we will mention only a few techniques that

---

[9] See p. 550.

have been used extensively for computations of the polarization of sunlight reflected by model planetary atmospheres. The primary purpose of this section will then be to illustrate the effect of multiple scattering on the angular distribution of the intensity and polarization.

Chandrasekhar reduced the problem of multiple Rayleigh scattering to a set of integral equations for functions of a single angle. The solution represented a triumph of mathematical ingenuity and, to a large extent, was successful in explaining the polarization of the clear sky. Progress has been made toward extending Chandrasekhar's methods to more general phase matrices (e.g., Herman 1970; Sekera 1966; Fymat and Abhyankar 1969); however, numerical results for Mie scattering have not yet been obtained.

The first extensive multiple scattering computations with Mie phase matrices were obtained with the photon counting or Monte Carlo method (Kattawar and Plass 1968), which had been used previously for Rayleigh scattering by Collins and Wells (1965). This method can, in principle, yield accurate results if a sufficient number of photons are counted. Computations relevant to the clouds of Venus were made by Kattawar, Plass, and Adams (1971). One contribution made by using this method was the comparison of polarizations computed for a spherical model atmosphere with those computed for the more usual locally plane-parallel model. For these two models, the results integrated over a planetary disk are in close agreement both for multiple Rayleigh scattering and for certain aerosol scattering (Kattawar and Adams 1971; Collins et al. 1972).

In the last few years the "doubling" method (or "adding" method in the case of an inhomogeneous atmosphere) has become increasingly popular for multiple scattering computations with polarization. The basis for this method is the fact that the reflection and transmission functions for two layers can be combined to yield the corresponding functions for the composite layer, with the multiple reflections between the two layers taken into account. This method is also called the "matrix" method (Twomey, Jacobowitz, and Howell 1966), the "matrix operator" method (Plass, Kattawar, and Catchings 1973), the "star product" method (Redheffer 1962; Grant and Hunt 1969), and the "discrete ordinate" method (Kattawar, Plass, and Catchings 1972). The "transfer-matrix" method of Aronson and Yarmush (1966) is closely related to the doubling method. Herman, Browning, and Curran (1971), attacking the equation of transfer directly with a Gauss-Seidel iteration scheme, and Dave (1970), using both a successive-orders-of-scattering method and the Gauss-Seidel iterational approach, have also demonstrated methods that can yield accurate results for multiple scattering with Mie phase matrices.

Dave (1970) treated the case for multiple scattering by various plane-

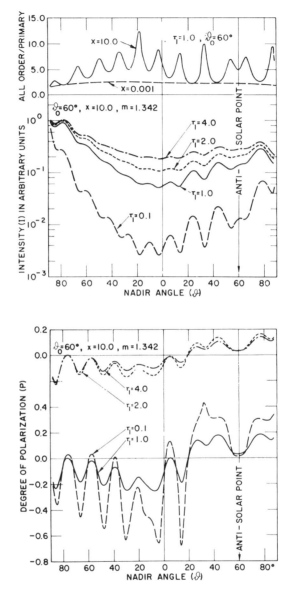

Fig. 10. *Below,* degree of polarization for light reflected by a plane-parallel cloud of spherical particles all of the same size ($x = 10.0$) and refractive index $n = 1.342$). The computations are all for the solar vertical plane, with the sun at 60° zenith distance. Note the dependence of polarization on the optical depth of the cloud. *Above,* the same, but for the reflected intensity, and showing the ratio of total intensity to the intensity coming from single scatterings, for unit optical depth of the cloud. (After Dave 1970.)

parallel layers of particles all having the same size and refractive index. Figure 10 shows his results for the reflected intensity and polarization in the solar vertical plane, with the sun at 60° zenith distance, for particles with $x = 10.0$ and $m = 1.342$. Note the dependence of polarization and intensity on the cloud optical depth. There is a general reduction of the polarization of reflected light for thicker clouds, but at some specific angles there is a reversal of the polarization sign. In this special situation, where all the scattering particles have an identical size, the polarization due to photons scattered two or three times has a significant smoothing effect on the net polarization, lowering peaks and filling in valleys.

Hansen (1971b) computed the intensity and polarization for the reflection of sunlight from a plane-parallel water cloud with a distribution of particle sizes. Figure 11 shows the results at $\lambda = 1.2$ μm. The size distribution is given by Equation (9), with $a = 6$ μm and $b = 1/9$, and is representative of terrestrial fair weather cumulus clouds. For this case the single scattering albedo is ~ 0.999, so there is much multiple scattering. The single scattering features in the angular distribution of the intensity are practically lost due to the multiple scattering. A "limb darkening" for $\Theta \sim 90°$ is introduced by the multiple scattering; this feature depends on the optical thickness, the single scattering albedo, and the asymmetry of the phase function, but it is not sufficiently sensitive to the nature of the scattering to be very useful for particle identification. The primary effect of multiple scattering on the degree of polarization is to reduce its value without changing its general form, as realized by Lyot (1929) and later Coffeen (1969). This generalization holds much better for size distribution (Fig. 11) than for a single particle size (Fig. 10), where some of the detailed interference features are distorted by multiple scattering.

The format of Fig. 11 is appropriate for comparison to observations, and thus for a practical test of how well single scattering features in the angular distribution of the reflected light survive the smoothing effect of multiple scattering. The intensity is plotted on a logarithmic scale, which is consistent both with the presentation of most observers and with the observational accuracies that have been obtained. The polarization is plotted on a linear scale that allows differences of a few tenths of a percent polarization to be resolved, because accuracies of this order are obtained in observations (Lyot 1929; Dollfus and Coffeen 1970).

Figure 12 shows the degree of polarization (I_p/I, where I is the total intensity and I_p is the intensity of polarized light) computed for water clouds. The wavelength and the particle size distribution are the same as for Fig. 11; thus the single scattering albedo is ~ 0.999. Included in Fig. 12 is an approximation for the polarization, I_p'/I, in which only

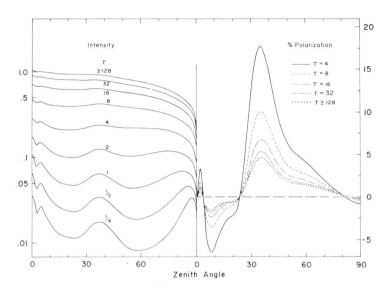

Fig. 11. Calculated intensity and percentage polarization ($-100\ Q/I$) of sunlight reflected by a plane-parallel water cloud with the sun overhead ($\theta_0 = 0°$). The wavelength is 1.2 μm and the results are shown for several optical thicknesses, τ. On the horizontal axis is the zenith angle of the reflected light, $\theta = \cos^{-1}\mu$. The calculations are for the size distribution Equation (9) with $a = 6$ μm and $b = 1/9$. (After Hansen 1971b.)

the contribution of single (first order) scattering is included in the numerator. Also illustrated is a modified approximation based on the assumption that diffracted photons may be counted as being unscattered; i.e., photons are allowed to contribute to I_p' according to the single scattering phase matrix on their first nondiffraction scattering (Hansen 1971b). This figure illustrates that the single scattering is primarily responsible for the polarization, although the second and higher orders of scattering do contribute (Hansen and Hovenier 1971).

The general effect of multiple scattering on the intensity and polarization can be readily understood. Photons emerging from the atmosphere after several scatterings are practically unpolarized, and they have an intensity that is more or less isotropic. Thus, in the degree of polarization, I_p/I, only the photons scattered once, or a small number of times, contribute significantly to I_p. The shape of the polarization versus scattering angle is, therefore, determined by the scatterers in the top of the atmosphere, where the optical thickness is $\tau \sim 1$. If variations of the properties of the scatterers occur at greater depths, the only significant effect on the degree of polarization is through the total intensity; this

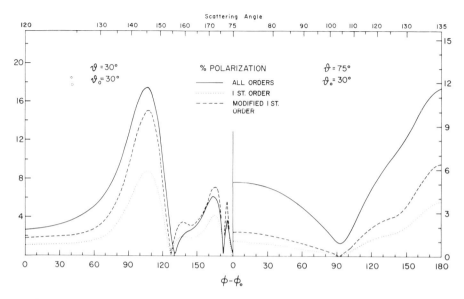

Fig. 12. Percentage polarization calculated as a function of azimuth angle, $\phi - \phi_0$, for sunlight reflected by a plane-parallel water cloud for $\lambda = 1.2$ μm, $\tau = 4$, and the indicated values of θ and θ_0. The curves for all orders are 100 $(Q^2 + U^2 + V^2)^{1/2}/I$; those for first order are computed including only the contribution of single scattering in the numerator, $(Q^2 + U^2 + V^2)^{1/2}$, and those for the modified first order are explained in the text. The calculations are for the size distribution of Equation (9) with $a = 6$ μm and $b = 1/9$. (After Hansen 1971b).

will not reduce the information that can be obtained on the top part of the atmosphere provided the emergent intensity is also measured.

The region of the atmosphere where $\tau \sim 1$ is one of great interest: this is approximately the depth to which we can see visually. Unless the particles are smaller than the wavelength, this region of the atmosphere will also be radiatively effective in the thermal part of the spectrum; i.e., the particles to which the polarization measurements refer will usually have a strong effect on the nature of the thermal infrared emission by the planet.

Nonspherical Particles

In the preceding sections we mentioned general differences and similarities in the theoretically predictable scattering behavior of spherical and nonspherical particles. Accurate numerical results for nonspherical particles of specific shapes have been obtained in only a few cases. However, laboratory and field measurements are available, and the prospects are good for obtaining improved laboratory data in the near future. Of particular interest are circular cylinders, hexagonal prisms, and irregular particles.

Anisotropic extinction by long circular cylinders has been studied in great detail for its application to the polarization of starlight.[10] However, the recent computation of the angular scattering by cylinders has special significance for atmospheric studies. Kerker (1969) and Liou (1972a) have made computations for scattering by single cylinders at different angles of incidence. Liou (1972b) integrated the results over a size distribution of particles and over a distribution of particle orientations. Figure 13 shows his results at four wavelengths for horizontal ice cylinders randomly oriented within a plane. Some of the features of spherical particles (e.g., the rainbow near 140° scattering angle at 0.7 μm wavelength) also exist for these cylinders because of the circular cross sections. Kerker (1969) and Liou (1972a) review other numerical studies of scattering by cylinders, as well as the history of the mathematical solutions for this problem.

For hexagonal cylinders, the polarization of scattered light has not been successfully calculated. However, for large hexagonal cylinders, two geometrical optics approaches have been used to obtain the scattered intensities, including the effect of orientation distributions and the effect of a finite particle length. Jacobowitz (1971) considered a parallel bundle of incident rays and, using geometrical optics, traced these rays for each desired orientation of infinitely long hexagonal cylinders. The computations for this case illustrate the absence of rainbows for particles that do not have a circular cross section. However, for hexagonal ice crystals, a new feature appears in the scattered light: at $\Theta \sim 22°$ there is a concentration of light due to the $\ell = 2$ (twice-refracted) rays. This "halo," like the rainbow for spheres, occurs at an angle of minimum deviation. Its location depends on the prism angle (in this case 60°) and on the refractive index. Greenler and Mallmann (1972) and Greenler et al. (1972) computed the emergent ray geometries for finite hexagonal prisms of different length-to-width ratio, and they reproduce in detail the circumscribed halos, sun pillars, and other features. Similar techniques could be used to compute the polarization of light scattered by large crystals.

Shape-specific features such as halos can potentially be used for particle identification by means of comparison with expected refractive-index/crystal-habit combinations. Any crystalline species can exhibit halo features corresponding to specific dihedral angles. If the particles are sufficiently large that geometric reflection and refraction laws apply, then an intensity maximum of characteristic polarization will appear at the scatter-

---

[10] See p. 60.

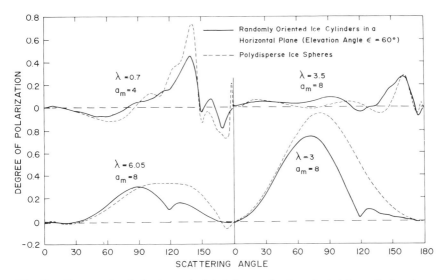

Fig. 13. Degree of polarization calculated for randomly oriented ice cylinders in a horizontal plane (solid lines) and polydisperse ice spheres (dotted lines). The horizontal dashed lines represent zero polarization. λ is the wavelength in μm; a_m, the mode radius, in μm, for the size distribution of cylinders; and ϵ, the angle between the direction of incidence and its projection on the horizontal plane. (After Liou 1972b.)

ing angle of minimum deviation. For a dihedral angle φ, formed by two faces not necessarily contiguous, this scattering angle is given by

$$2 \arcsin \left(n \sin \frac{\varphi}{2} \right) - \varphi, \qquad (11)$$

provided that φ is less than twice the critical angle of internal reflection (Tricker 1970, p. 75). This dihedral angle of 60° is the source of the familiar 22° halo of hexagonal ice cylinders. Additional features arise from reflected rays, from "skew" rays, and from nonrandom orientations of the shaped particles. The presence or absence of such specific halos provides a direct discrimination of particle shape and, therefore, a discrimination between liquid droplets and solid crystal forms.

Theoretical computations have been made for particles of a number of other shapes, including spheroids (Greenberg, Pederson, and Pederson 1961), elliptical cylinders (Yeh 1964), and several different shapes of totally reflecting particles. Furthermore, numerical methods are available (e.g., Kerker 1969) that can be applied to particles of almost any shape, and with modern computers these may prove to be useful. A wide variety of particle types are discussed by Beckmann (1968).

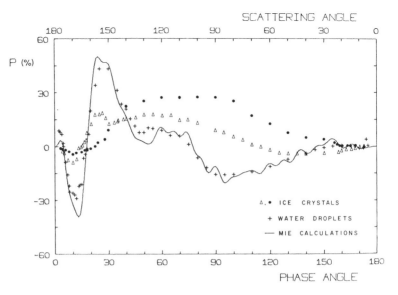

Fig. 14. Percentage polarization for clouds of ice crystals (dry •, partially melted △) and water droplets (+), measured in the laboratory by Lyot (1929), compared to the Mie calculations for a size distribution of spheres, $N(r) \propto r^{-2}$, with a mean radius $r_0 = 1.50$ μm and abrupt cutoffs at $0.75\ r_0$ and $1.25\ r_0$; $m = 1.335$, $2\pi r_0/\lambda = 17.0$.

Laboratory measurements have provided a powerful means of studying the scattering properties of nonspherical particles. Lyot (1929), using a visual polarimeter, was the first to obtain accurate measurements of the polarization properties as a function of scattering angle. Figure 14 shows his results for a dry fog of small ice crystals as well as measurements of similar particles about 5 μm in diameter that are partially melted. There was no indication of a halo at 22° scattering angle, probably because of the small size of the crystals.

Huffman (1970) measured the intensity and polarization for visible light scattered from a size distribution of hexagonal columns (mode length 25 μm) in a laboratory cloud chamber. He found negative polarization for the halo at 22° scattering angle and a positive branch between 30° and 150°.

Field measurements of terrestrial ice clouds provide another approach. Coffeen and Hansen (1973) measured the infrared polarization of cirrus clouds viewed from above. Figure 15 shows the results of measurements on a thick cirrus system in the intertropical convergence zone. The absence of any rainbow feature near 40° phase angle is significant; all liquid water clouds measured in the program showed a strong rainbow feature in the polarization.

Several authors have studied the scattering by irregular particles in the laboratory. Hodkinson (1963) measured the scattering by suspensions of irregular particles somewhat larger than the wavelength. In general, the polarization-phase curves showed little structure, being qualitatively similar for quartz, flint, diamond, and coal dusts several microns in diameter. Lyot (1929) found that a layer of crushed glass gives a smooth curve of positive polarization, with no indication of the strong rainbow present for a similar layer of 0.5 mm diameter spheres. Pritchard and Elliott (1960) and Holland and Gagne (1970) used photometric methods to measure all of the significant elements of the scattering matrix for a cloud of small irregular platelets of crystalline silica. Greenberg,[11] Pederson, and Pederson (1961) and Giese and Zerull[12] have employed microwave analogue techniques, with both the particle size and wavelength in the microwave region, to investigate the effects of nonspherical particles.

It has recently been demonstrated that laser illumination can be used to measure accurately the phase matrix for single scattering by spherical and nonspherical particles (Blau, McCleese, and Watson 1970), and it is likely that this technique will be widely used in the future for investigating nonspherical particles.

The above discussion was restricted to single scattering by nonspherical

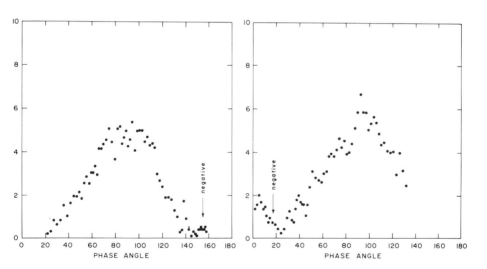

Fig. 15. Percent polarization of reflected sunlight observed at $\lambda = 2.25$ μm over thick cirrus clouds in the intertropical convergence zone. The two sets of observations were taken on the same day over clouds separated by several kilometers, using the AEROPOL polarimeter on board the NASA Convair-990 jet.

---

[11] See p. 916.   [12] See p. 901.

particles. However, once the phase matrix is known for a collection of randomly oriented nonspherical particles, multiple scattering calculations including polarization are not much more difficult than for spheres, provided that each particle has a plane of symmetry. The symmetries of the various matrices occurring in the equations have been derived by Hovenier (1969). The doubling method, for example, could be used to obtain results for a layer of any desired optical thickness. Calculations of this type can be expected in the near future.

Since there are so many different possible nonspherical particles, it is reasonable to ask what we can be confident of learning from measurements of scattered light for cases in which the particles may be of any shape. If observations are made over a wide range of scattering angles, it will certainly be possible to distinguish spherical particles from nonspherical ones. This is in itself a significant conclusion that is indicative of the phase (liquid or solid) of the particles. Of course, if the particles are spherical, more details on their microstructure can be extracted, and some possibilities for additional information exist for nonspherical particles of a regular shape. If the particles are irregular in shape, it will still be possible to determine whether the particle size is comparable to or much larger than the wavelength from the dependence or independence of the cloud particle polarization on wavelength.

In general, we observe more detailed structure in the scattering patterns from clouds of spheres than from clouds of nonspherical particles. And yet, for a single specific nonspherical shape, we should expect a highly structured scattering pattern. The difference must be due to the distribution of particle *orientations* in the cloud. Randomly oriented spheres retain their unique scattering characteristics, but for any other shape much information is lost in a distribution of orientations.

OBSERVATIONS AND INTERPRETATIONS

A major review of polarization observations of planetary atmospheres was given by Dollfus (1961), based largely on the theses of Lyot and of Dollfus. The state of the interpretations has been considered by van de Hulst in a review chapter (van de Hulst 1952) and in Chapters 20 and 21 of his book (van de Hulst 1957). We present here a review of the useful observations to date and of their interpretation in terms of particle characteristics and atmospheric structure.

Venus

The first measurement of the linear polarization of Venus was that of Lord Rosse (1878), who found -3.9% in the visual at intermediate phase angles. Landerer (1892) later concluded that the planet's light

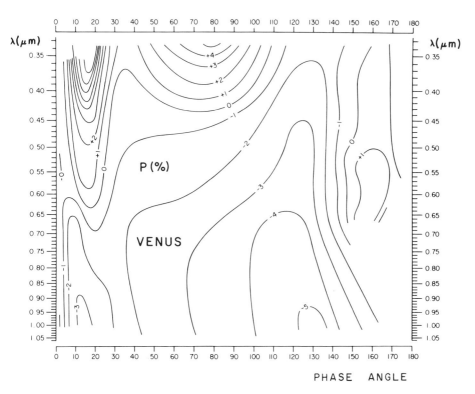

Fig. 16. Synthesis of observations showing the mean variation of the polarization of Venus as a function of phase angle and wavelength. The vertical scale is proportional to $\log_{10} 1/\lambda$. For phase angles larger than 140°, the polarizations are representative of the equatorial portion of the crescent; at all other phase angles the values refer to the complete disk, whose polarization variations differ little from those of the equatorial region. (After Dollfus and Coffeen 1970.)

was not polarized, but this difference was resolved by Lyot (1929), who obtained an accurate phase curve for Venus in the visual, over the range 2° to 175° phase angle. Modern techniques have not improved on Lyot's results at this wavelength.

The most important observational advance since Lyot's work has been the measurement of the wavelength dependence over all phase angles. Global (i.e., disk-integrated) coverage between 0.3 and 1.0 μm was presented in detail by Coffeen and Gehrels (1969) and by Dollfus and Coffeen (1970). A synthesis of these observations is given in Fig. 16. Extensions were made in the ultraviolet to 0.22 μm (Coffeen and Gehrels 1970) and in the infrared to 3.6 μm (Forbes 1971;[13] Landau[14]).

---

[13] See also p. 637.   [14] See p. 318.

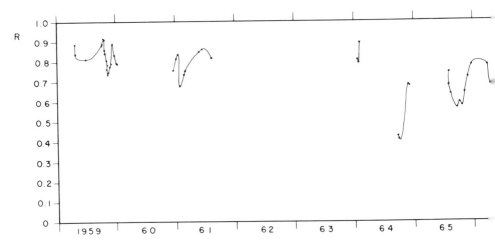

Fig. 17. Variation with time of the global polarization of Venus at 0.365 μm, observed by Coffeen, Gehrels, and Baker. The vertical scale is $R = (P - P_0)/(P_6 - P_0)$, where P is the observed polarization, P_0 is the calculated polarization for Hansen's cloud model with no Rayleigh scattering, and P_6 is calculated with a 6% admixture of Rayleigh scattering (see Fig. 20). The values of P_0 and P_6 are read off at the appropriate phase angle for each observation, restricted to the

Curves of polarization versus phase angle are very similar for each apparition, except in the ultraviolet. Based solely on the U.S. observations, the ultraviolet polarization is quite variable (Coffeen and Baker, unpublished) and was unusually low in 1964/65 and 1971/72, as shown in Fig. 17. The recent low polarizations have been confirmed by the French observations (Dollfus and Aurière, personal communication).

Regional studies have been made by Lyot (1929), Dollfus (1957, 1966, 1973), and Coffeen and Gehrels (1969). The principal variation over the disk is a geometrical one, symmetric about the light equator. In addition, local variations are found in the percent polarization and in its position angle. Figure 18 shows measurements made by Coffeen with Hall's scanning polarimeter; these suggest an inverse relation between ultraviolet intensity and polarization.

Kemp (personal communication) has recently measured the circular polarization of the northern and southern hemispheres of Venus.[15] At 53° phase angle, the polarization is appreciable ($\sim 10^{-4}$), is opposite in sign for the two hemispheres (same sense as for Jupiter; see section below), and increases from the red to the ultraviolet. However, the observations to date are few in number and not yet fully interpreted.

---

[15] See also p. 607.

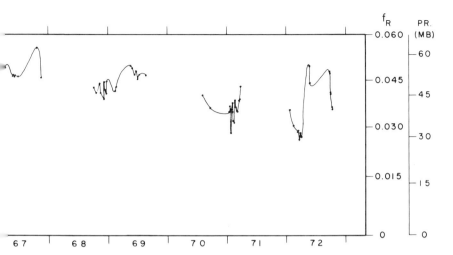

range 41°–127° phase angle. Thus R should be a measure of the contribution by molecules to the global scattering, free of any explicit dependence on phase angle. The right-hand scales show the corresponding fraction (f_R) of Rayleigh scattering, assuming uniformly mixed model, and the equivalent gas pressure in millibars at $\tau \sim 1$. Note the very appreciable variations, both short-term and long-term, in the equivalent cloud-top pressure.

The following analysis and interpretation for the clouds of Venus is based solely on the observations of linear polarization.

The fact that the mass of the Venus atmosphere is large (surface pressure ~ 100 atm) assures us that the polarization observations refer primarily to photons scattered within the atmosphere and not to reflections from a solid planetary surface. A complete theoretical interpretation of the observations must, therefore, be based on solutions of the radiative transfer equation. Exact solutions for a Rayleigh atmosphere have long been available, but multiple Rayleigh scattering produces a polarization at visible wavelengths that is opposite in sign to that observed on Venus.

It thus follows that in the atmosphere of Venus there must be cloud or haze particles that are primarily responsible for the polarization. Indeed, laboratory measurements, as well as theoretical calculations for single scattering by spherical particles, indicate that the polarization of Venus is characteristic of scattering by particles with $r \sim \lambda$. Using his laboratory measurements, Lyot (1929) found that clouds composed of water drops (refractive index ~ 1.33) with $r \sim 1.2$ μm were in reasonably good qualitative agreement with his observations of Venus. Coffeen (1969) compared observations as a function of wavelength with calculations for single scattering by spheres and concluded that $1.43 \leq n \leq 1.55$ and $\bar{r} = 1.25 \pm 0.25$ μm, ruling out the possibility of water

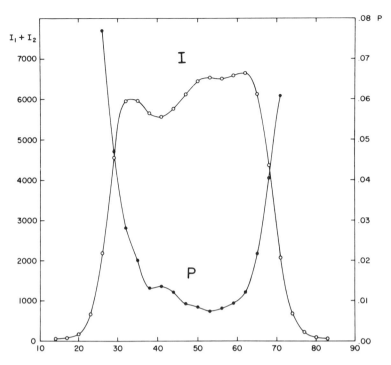

Fig. 18. Relative intensity and polarization observed on a north-south polar scan of Venus in the ultraviolet (3700 Å). The aperture size was 0.65 arcsec, and the planet diameter, 18.9 arcsecs. The two depressions in the intensity scan are apparent as dark markings on ultraviolet photographs taken at the same time. The positive polarization increases rapidly toward the poles and, in addition, shows a local maximum where the intensity has a minimum. The observations were made on 30 December 1968 by Coffeen at the Lowell Observatory using the scanning photometer/polarimeter of J. S. Hall, at 74° phase angle.

drops. Additional comparisons were made by Sobolev (1968), Loskutov (1971), Kattawar, Plass, and Adams (1971), Forbes (1971), and Morozhenko and Yanovitskii (1973), all of whom basically agree with the size and refractive index limits given by Coffeen.

A quantitative fitting with the observations requires accurate calculations of multiple scattering. The effect of multiple scattering on the polarization depends on the scattering geometry and on the particle characteristics, and cannot be reliably estimated a priori. The remedy is to make exact calculations for the polarization by multiple scattering and quantitative comparisons with the data.[16] This was done by Hansen and Arking (1971) and Hansen and Hovenier (1973), who

---

[16] In some cases of quantitative calculations for planetary scattering, it is important to integrate over the angular size of the sun, including limb-darkening. From Venus the sun subtends 0°.73, which has a negligible effect on the computations presented here.

obtained numerical solutions for the multiple scattering of light from a plane-parallel atmosphere consisting of a mixture of spherical particles and Rayleigh scatterers. The results were integrated over the visible part of the planetary disk to allow comparison with the observations of Venus. The polarizations obtained depend upon the real refractive index, the particle size distribution, and the admixture of Rayleigh scattering.

The sensitivity of the theoretical computations to changes in the refractive index is illustrated by Fig. 19, which includes Coffeen and Gehrels' observations at $\lambda = 0.99$ μm. For each refractive index, the particle size is shown that gives the best overall agreement for all wavelengths from 0.34 μm to 0.99 μm. The minimum near a phase angle of

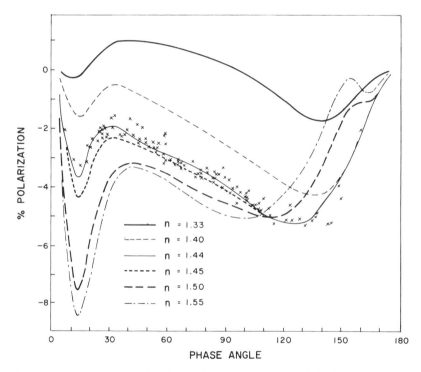

Fig. 19. The crosses show the observations of Coffeen and Gehrels (1969) of the polarization of Venus at a wavelength of 0.99 μm. For each refractive index n, the multiple-scattering calculations are for the particle size giving best agreement with the observations for all wavelengths: $\bar{r} =$ 0.7, 0.8, 1.1, 1.1, 1.2, and 1.2 μm, respectively, beginning with $n = 1.33$. The mean scattering radius $\bar{r}$ is approximately equal to a; $\bar{r}$ varies by a few percent with wavelength, but the size distribution is invariant. The curves for $n = 1.44$ and 1.45 are indistinguishable for phase angles greater than 110°. The albedo of Venus is assumed to be 90% at $\lambda = 0.99$ μm. (After Hansen and Arking 1971.)

15° is the glory, which arises from surface waves generated on spherical particles by edge rays; the glory is very broad at this wavelength because of the small size parameter $x = 2\pi\bar{r}/\lambda$. The broad maximum near a phase angle of 30° is the primary rainbow, which arises from rays internally reflected once in spherical particles; this feature becomes increasingly distinct toward shorter wavelengths.

The imaginary part of the refractive index must be very small for the particles in the upper cloud layer on Venus, and the single scattering albedo must be close to unity. To account for the spherical (Bond) albedo of Venus, which is less than 100%, it is possible to choose either a finite atmosphere with a partially absorbing ground or a single scattering albedo less than unity. Both of these alternatives were tested, and within the thickness of the curves, the results were identical.

As explained above, the effect of the size distribution can be accurately described by the mean effective radius and the effective variance. The calculations for the polarization in Figs. 19 and 20 were made for the size distribution of Equation (9) with the effective variance $b = 1/9 \sim 0.11$, which yields reasonably good agreement with the observations. As discussed below, a closer fit to the observations can be obtained with $b = 0.07$.

At visual and shorter wavelengths there is a nonnegligible contribution from Rayleigh scattering, which may be estimated best from the ultraviolet observations. Figure 20 shows the results of calculations at $\lambda = 0.365$ μm for a model with a uniform mixture of spherical cloud particles and Rayleigh scatterers. The derived amount of Rayleigh scattering may then be used to obtain the pressure at a level of significant optical depth ($\tau \sim 1$) in the clouds; the result is ~ 50 mb, which, for comparison, corresponds to the pressure in the earth's stratosphere at ~ 20 km. This result is somewhat model dependent, unlike the results for particle shape, size, and refractive index; however, the value 50 mb should be correct to ± 20 mb. Earlier deductions of the cloud-top pressure are in rather good agreement with this result. Based on polarization data, Lyot (1929) estimated 175 mb; Dollfus (1957), 90 mb; Sobolev (1968), 30 mb; Coffeen (1969), 30 mb (corrected value); and Sagan and Pollack (1969), 35 mb. The exact value depends on the definition of "cloud top" and on the assumed molecular composition. Note also that the ultraviolet polarization is rather variable (Fig. 17) and that the implied cloud-top pressure has been as low as 25 mb.

The long-term variations in the ultraviolet have not been explained satisfactorily. They suggest a global change in the mean cloud-top altitude (thereby changing the mean contribution of molecular scattering), a change in the planetary albedo (thereby changing the dilution of the

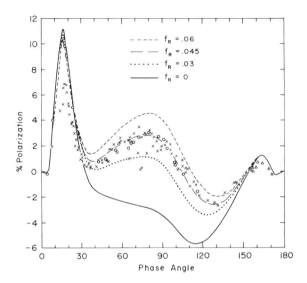

Fig. 20. Polarization of Venus at $\lambda = 0.365$ μm. The crosses show the disk observations of Coffeen and Gehrels (1969). The circles are more recent U.S. observations of the Venus disk, and the triangles are French observations of the equatorial region (Dollfus and Coffeen 1970). The theoretical curves are computed with a fraction (f_R) of the phase matrix being Rayleigh scattering and a fraction ($1 - f_R$) being the Mie phase matrix for $n = 1.46$ and $\bar{r} = 1.1$ μm. The spherical albedo of the planet is assumed to be 55% at this wavelength.

polarization by multiple scattering), or a change in the mean-free-path of photons in the clouds.

The visual observations of Lyot and the intermediate-bandwidth observations of Coffeen and Gehrels at $\lambda = 0.55$ μm can be used to determine the mean particle size. The computed polarization is quite sensitive to variations in the particle size and indicates that the mean particle radius is ~ 1.1 μm. In the results of Hansen and Arking (1971), there was a discrepancy with the observations at phase angles $\sim 160°$. However, there is better agreement if a somewhat narrower size distribution is used. Hansen and Hovenier (1973) have varied b and found that a better fit occurs for $b \sim 0.07$. Most significant is the fact that the feature would be entirely smoothed away if a broad distribution were used. It is clear that this feature for Venus is the polarization "island" of anomalous diffraction in the contour diagrams; for shorter wavelengths it becomes weaker, and for longer wavelengths it disappears.

The best fit to all of the observations occurs with a refractive index that decreases from ~ 1.45 in the ultraviolet region to ~ 1.43 at $\lambda =$

0.99 μm; the total uncertainty in n is ± 0.02 at each wavelength. The mean radius is 1.05 $\pm$ 0.1 μm. Most of the particles must be spherical; the glory, rainbow, and anomalous diffraction features are predicted for spheres, but they are not expected for irregular particles. The width of the size distribution is amazingly small. The value $v_{\text{eff}} \sim 0.07$ is smaller than for terrestrial water clouds (for which v_{eff} is usually in the range 0.10–0.25, the smaller values occurring for fair weather clouds and the larger ones for convective clouds) and terrestrial tropospheric hazes (for which $v_{\text{eff}} \sim 0.2$–0.4). However, two different measurements (Mossop 1965; Friend 1966) of the size distribution of particles in the earth's stratosphere (Junge layer) yielded $v_{\text{eff}} \sim 0.06$, and the particles appeared to be liquid; thus, perhaps for liquid particles in a convectively stable part of the atmosphere, it is reasonable to find a narrow size distribution. In any case, the derived particle shape and the width of the size distribution together strongly suggest that the cloud particles on Venus are liquid.

It should be emphasized that the cloud properties derived from the polarization refer to the top part of the clouds. Intensity measurements, for example, of weak absorption lines, include information on deeper layers where the cloud particle properties *may* differ. This must also be borne in mind when one considers the color and spectral reflectivity of Venus. Nevertheless, the "polarization clouds" must be equated with the "visible clouds" of Venus; the polarization clouds have a significant optical depth on a planet-wide basis.[17] The fact that these clouds have a substantial optical thickness high in the atmosphere, where $p \lesssim 50$ mb, makes them all the more interesting. In principle, it will be possible to explore the variation with height of cloud particle characteristics, using spectropolarimetry. Low concentrations of particles much higher than the 50 mb level will dominate the scattering in the centers of strong CO_2 absorption lines, while contributing little to the polarization in the broadband continuum measurements. Preliminary measurements in absorption bands have been reported by Forbes and Fymat.[18]

It should also be emphasized that the visible clouds are composed of particles with a single refractive index (note, for example, the sharp rainbow in the ultraviolet); they cannot be composed of a mixture such as dust and water. Indeed, there is only one type of particle contributing

---

[17] At an optical depth $\tau = 1$, the contrast for average observing conditions, $\mu = \mu_0 = \frac{1}{2}$, is already reduced to e^{-4}. Thus, $\tau = 1$ is approximately the depth to which we can "see."

[18] See p. 637.

significantly to the polarization. If there were two or more cloud layers with different particles (differing in refractive index or particle size or particle shape) contributing to the polarization, then the features due to each type of particle would be present; however, the only features in the observed polarization are those corresponding to the particles described above ($n \sim 1.44$, etc.), and moreover, all of the features predicted for such particles are present. However, the fact that one type of particle is responsible for the polarization does not mean that the atmosphere can necessarily be regarded as homogeneous for other purposes. Obviously, it would also be possible to match the polarization observations with computations for a layered model atmosphere (as has been considered by Kattawar, Plass, and Adams 1971) with two, three, or n layers, provided the top layer is sufficiently thick ($\tau \gtrsim 1$) and has the particle properties specified above.

A crystalline species in the upper clouds of Venus might be detected by a halo feature in the phase curves of intensity and polarization. A concentrated search has been made for a 22°-halo (at phase angle 158°) due to large hexagonal ice crystals. No substantial indication of a halo has been found in either the photometry (O'Leary 1970) or the polarimetry (Veverka 1971).

The above results stringently narrow the list of possible materials composing the visible clouds of Venus; indeed, most of the materials that have been proposed may be ruled out. The polarization data are incompatible with solid particulates such as SiO_2, $NaCl$, NH_4Cl, and $FeCl_2$ on several grounds, including their refractive indices. The refractive indices for pure water and ice are much too small, whereas the refractive indices for mercury and mercury compounds are much too large. Of course, the polarization results do not rule out the existence of these materials in some cloud deck beneath the visible clouds.

$HCl \cdot H_2O$ is a possibility for the cloud particle material. Lewis (1969, 1971), using observed abundances of the gases HCl and H_2O, predicted that cloud particles composed of an aqueous solution of hydrochloric acid ($HCl \cdot H_2O$), with $\sim 25\%$ HCl by weight, should exist at the ~ 50 mb level. The refractive index of HCl, ~ 1.42 (Arking and Rao 1971; Lewis 1971), may just barely be compatible with the polarization results.

Carbon suboxide (C_3O_2) has been proposed as a possible cloud particle material on Venus (Sinton 1953), and it has a refractive index of ~ 1.45. However, the spectroscopic upper limit on the abundance of its vapor (Jenkins, Morton, and Sweigart 1969; Kuiper 1969) is a few orders of magnitude less than the amount required in the stratosphere of Venus for equilibrium with the condensate of the monomer. The low polymers may provide an alternative, but there are so many doubts about C_3O_2

(e.g., see Hansen and Arking 1971) that it is not a likely candidate.

$H_2SO_4 \cdot H_2O$ has recently been suggested for the cloud particle composition on Venus by Young (1973). It appears that a solution of sulfuric acid $\sim 75\%$ H_2SO_4 by weight would have the observed refractive index and would be liquid at the cloud-top temperature. Sulfur is a major constituent in the particulate layer in the earth's stratosphere (the Junge layer) which exists at the 50 mb level. On photochemical grounds, Lewis argues against sulfuric acid in the clouds of Venus, but the possibility warrants a close examination. Based on Young's analysis, this material appears consistent with the observational data on the Venus clouds.

The distribution of polarization over the disk of Venus has not been studied in as great detail as has the global polarization. However, the regional polarization is a potentially rich source of information on the vertical and horizontal structure of the clouds.

Whitehill (1972) has recently computed the regional polarization for Venus at $\lambda = 0.99$ μm and compared the results to observations of Coffeen and Gehrels (1969). The computations were for a homogeneous cloud of particles, using preliminary values for the cloud properties deduced by Hansen from the global polarization (spheres with refractive index 1.45 and size distribution of Equation (9) with $a = 1.1$ μm, $b = 1/9$). Figure 21 shows Whitehill's results compared with the observations of Coffeen. The agreement is excellent, except at the poles. It is possible that the discrepancy at the poles is not real and may be attributed to numerical approximations in the integration scheme at that location (Whitehill 1972) or to difficulties in precise guiding during the observations (Coffeen and Gehrels 1969).

The primary conclusion that can be made from this example is that at the resolution employed (the aperture size being indicated in Fig. 21), the clouds are horizontally rather homogeneous, except possibly at the poles. This is not too surprising, considering the visual appearance of Venus and the fact that the global polarization also indicates only a single type of cloud particle.

Observations with high spatial resolution in the visual and ultraviolet, where both the cloud particles and gas molecules contribute significantly to the polarization, can be used to investigate both the vertical and horizontal structure of the atmosphere. For example, the gross cloud structure can be extracted in the sense that a diffuse haze (a mixture of molecules and particulates) can be distinguished from gas above a more optically dense cloud. Qualitatively, the basis for such a distinction is the fact that the contribution of Rayleigh scatterers above a particulate region increases with increasing zenith angle of either the sun or the observer, while the relative contribution of particles and Rayleigh

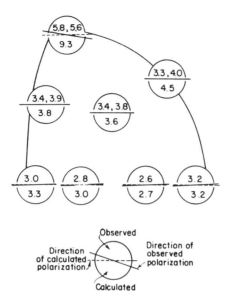

Fig. 21. Comparison of regional amounts of polarization on Venus observed by Coffeen (*upper* numbers; for the nonequatorial regions, the results for the northern hemisphere are on the left and those for the southern hemisphere on the right) with the calculations of Whitehill (*lower* numbers). The circles indicate the approximate size of the physical diaphragm used in the observations. The wavelength is 0.99 μm and the planetary phase angle 77°. (After Whitehill 1972.)

scatterers within a cloudy region remains essentially constant. The observations of Venus at $\lambda = 0.34$ and 0.52 μm by Coffeen and Gehrels (1969) qualitatively indicate that, at least above the poles, there is a significant clear gas region above the clouds. Their ultraviolet observations also show other, more random, regional variations that violate the symmetry principles of horizontally homogeneous atmospheres (Hovenier 1970). These are perhaps due to local variations of the height,[19] thickness, or density of the clouds.

A better knowledge of the regional cloud characteristics could be obtained with earth-based polarimetry of the highest practical spatial resolution in conjunction with precise numerical modeling. Photographic studies offer some hope of mapping the disk and correlating the polarization and intensity variations.[20]

---

[19] As suggested by Moroz (1971) on spectroscopic evidence.
[20] See pp. 218 and 223.

Space probes offer new possibilities for regional coverage. A detailed polarization mapping of Venus at 18° phase angle (the prominent rainbow angle, apparent in Figs. 16 and 20) will be attempted with the Mariner Venus/Mercury vidicon system, using an ultraviolet filter with and without a fixed polarizer.[21] However, a detailed analysis of local cloud structure could best be obtained from a polarimeter on a satellite orbiting Venus; this would allow measurements of small areas from different angles at short time intervals.

Earth

The earth has special importance for polarimetric studies of planetary atmospheres because we can examine the local structure of the atmosphere and sample the scattering particles directly. Thus, we can compare our theoretical expectations with in situ as well as remote observations, including cases in which the particles are spherical (e.g., cumulus clouds) and cases in which the particles are nonspherical (e.g., cirrus clouds). Our new ability to view the earth as a planet makes such comparisons all the more interesting. In this section we consider observations of three types: (1) ground and airborne observations of "skylight," the downward scattering from the cloud-free atmosphere, (2) airborne and space observations of light reflected upward from localized regions, and (3) space observations of the global earth.

Ground-based observations of skylight are useful for comparison with the Chandrasekhar theory of multiple Rayleigh scattering.[22] This is one of the most fundamental problems in planetary scattering, but it is complicated by molecular anisotropy, polarized reflection from the ground, and the variable presence of aerosols. Consequently, a complete theoretical treatment has been slow to evolve. The observational approach has been to isolate various scattering contributions; one example is the minimization of single scattering through observations of skylight during a total solar eclipse (Rao, Takashima, and Moore 1972; Dandekar and Turtle 1971). A related case occurs during twilight, with the earth used as a sunshade. Rozenberg (1966) treats the twilight case in detail. Gehrels (1962) and Bondarenko (1964) found striking changes in the wavelength dependence of zenith skylight polarization as the sun rises and sets, due to varying contributions of ground reflection and multiple scattering, and to the aerosol distribution with height. Large aerosol particles can be isolated by infrared measurements of skylight, but the technique is rather insensitive to the *natural* aerosol having small optical thickness. Coulson (1971), in the heavily polluted atmosphere of Los Angeles,

---

[21] See p. 189.   [22] See pp. 444 and 500.

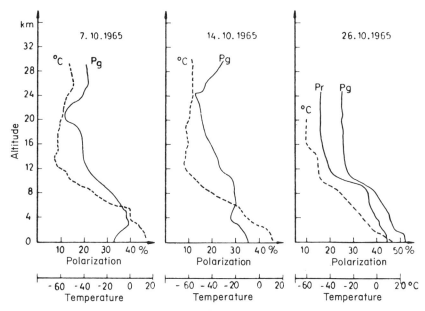

Fig. 22. Linear polarization of zenith skylight measured from an ascending balloon instrument in the green (0.535 μm) and red (0.725 μm). The solar zenith distance varied between 59° and 85°. The variations of polarization with altitude indicate different aerosol contributions. The dashed curves give the temperature profiles. (After Unz 1969.)

studied the characteristics of the linear polarization of skylight as a function of wavelength, solar elevation, and degree of man-made pollution.

Aerosols larger in size than the wavelength can introduce circular polarization for multiply scattered light; for molecules alone, the degree of circular polarization is zero. This gives a potential technique for recognizing the presence of aerosols (Rozenberg and Gorchakov 1967; Eiden 1971). Hannemann and Raschke find appreciable circular polarization in skylight measurements in the Ruhr region,[23] although the degree of circular polarization of skylight is usually rather small (Eiden 1970).

A novel approach is to measure the skylight polarization in situ from an ascending balloon platform, as carried out by Unz (1969). Figure 22 shows the measurements of polarization versus altitude on three different flights. The polarization ratio was measured near the zenith in the green (0.535 μm) and red (0.725 μm). The solar zenith angle was between 59° and 85°, depending on date and time. As the balloon rises, the

---

[23] See p. 510.

polarization can decrease and then increase depending on the relative scattering contributions from the molecules and aerosols in the vertical column and on the amounts of multiple scattering. The measurements can be inverted to yield the aerosol concentration as a function of height. High concentrations were usually found near the ground and again at altitudes between 10 and 25 km. For the highest concentration in the stratosphere, Unz found scattering by aerosols to be 10 to 20 times greater than scattering by molecules.

The principal studies of skylight have been done at the University of California at Los Angeles, in Germany, and in the Soviet Union. The following reviews and collections of papers are useful as an introduction to the extensive literature on the subject: Sekera (1957), Rozenberg (1960, 1968), Fesenkov (1962), Bullrich (1964), and Divari (1970, 1972).

The second stage in the study of the terrestrial planetary atmosphere has come with observations of light scattered upward from localized features, using balloon, aircraft, rocket, and satellite platforms. The emphasis of these observations is on cloud studies because the polarization varies strongly with cloud type and because clouds have a large optical thickness that allows the signal from the other parts of the atmosphere and from the earth's surface to be minimized. Measurements are feasible for other particulates (dust, smoke, pollution), though successful monitoring will require care in isolating the polarization due to the particulates. There is now little incentive for polarization studies of the molecular atmosphere itself, because its characteristics are reasonably well known and because of the difficulty in correcting for the contribution from ground reflection. The relative contribution of ground and atmosphere is shown in downward observations by Dollfus (1956) from a balloon at different altitudes.

Dollfus (1956) reported measurements of the polarization of sunlight reflected upward by clouds, as observed from a mountain. More recently, Coffeen and Hansen (1972) measured the linear polarization of water clouds from the NASA Convair-990 jet.[24] Similar measurements in the Soviet Union (Bazilevskii et al. 1972) led to the experimental use of a scanning polarimeter on the Meteor 8 earth satellite (Bazilevskii and Vinnichenko 1972).

Figure 23 shows the polarization measured by Coffeen and Hansen over two different cloud systems. Comparisons of these measurements with theoretical computations (Hansen and Coffeen 1972) indicate that the mean particle radius was ~ 5 μm for the case on the right, and

---

[24] See p. 189.

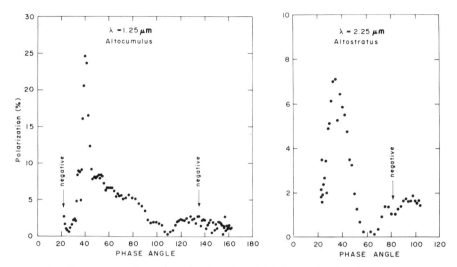

Fig. 23. Percentage polarization of reflected sunlight observed at $\lambda = 1.25$ μm (*left*) and $\lambda = 2.25$ μm (*right*) over two different cloud systems. "Negative" is used to indicate branches of the curves where the direction of polarization is predominately parallel to the plane of scattering. The deduced mean particle radius is $\gtrsim 25$ μm on the left, ~ 5 μm on the right. The measurements were made with the AEROPOL instrument on the NASA Convair-990 jet.

$\gtrsim 25$ μm for the case on the left. Although several features in the polarization of water clouds are correlated with particle size (Hansen 1971b; Coffeen and Hansen 1972), the most convenient indicator of particle size is the neutral point which occurs at phase angles $\sim 60°–110°$. Differentiation between ice and water clouds is also straightforward; compare Figs. 15 and 23.

Finally, to see the earth in planetary perspective, we must investigate its global scattering properties. Study of the global polarization and brightness could yield global averages of the mean terrestrial cloud-top particle size, the nonspherical fraction within the cloud tops, and the mean molecular optical depth above the reflecting surface(s), provided the measurements have sufficient coverage in wavelength and phase angle. Such measurements will require a polarimeter on a deep space probe. Spacecraft geometry precludes these measurements of the global polarization of the earth from Pioneers 10 and 11, but the Mariner Jupiter-Saturn missions could provide an excellent opportunity. Rough measurements already exist through the use of Polaroids in the filter wheel on Surveyor 7 (Holt 1970); about 15% polarization was found but with systematic errors due to the asymmetric camera mirror.

An indirect approach taken by Dollfus (1957), using the moon as a crude "mirror," gives the most reliable result to date. The earthshine on the moon is seen at a phase angle (earth-moon-earth angle) of essentially 0°, so that any polarization is a remnant of the polarization of light scattered by the earth. Dollfus measured the linear polarization of the earthshine over the range of earth phase angle (sun-earth-moon angle) 25°–145°. The curve is bell-shaped, with a peak polarization of 10% at 100° earth phase angle. Next, the appreciable depolarization by the lunar soil must be estimated; based on laboratory samples, all measured values should be multiplied by approximately 3.5, finally giving the curve in Fig. 24. The shape of the curve is more certain than is the absolute amount of polarization. Although more data would be needed for a meaningful analysis, the general shape is like that for a mixture of molecules, soil surfaces, snow, ocean, and/or cirrus clouds. The relatively large magnitude of the polarization suggests that there may be a large contribution from molecular and/or ocean scattering. We can speculate that the point at 36° phase angle is elevated due to contribution from water droplet rainbows, as in Fig. 23.

Mars

The thin atmosphere of Mars (mean surface pressure $\sim$ 5 mb) makes little contribution to the scattered polarization except in the ultraviolet or when the atmosphere is laden with aerosols. Lyot (1929) found the integrated disk polarization to be similar to that for the moon, until the

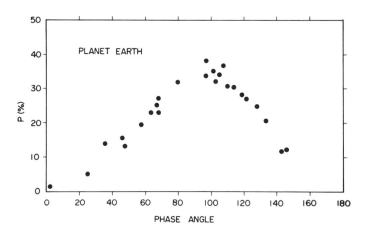

Fig. 24. Polarization of the global earth as deduced from Dollfus' earthshine measurements in visible light. The dark side of the moon serves as a depolarizing "mirror" in which we view ourselves, providing a complete phase curve in $\sim$14 days.

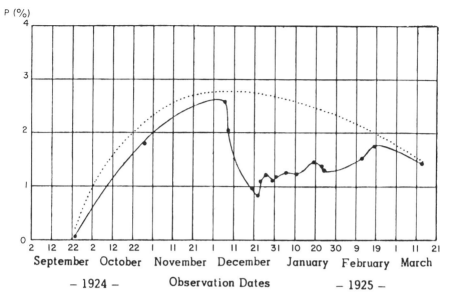

Fig. 25. Polarization of the disk of Mars observed in 1924–25 in the phase angle range 35°–42°, showing the onset of a thick haze. The dotted curve shows the polarization of the moon for the same phase angles. (After Lyot 1929.)

onset of a thick haze. Figure 25 shows his visual measurements of the disk polarization, with complete coverage through a planet-wide disturbance. With a haze present, the polarization is markedly reduced.

Lyot began the study of regional details; Dollfus (1957) extended this greatly, obtaining polarization-phase curves for characteristic atmospheric features, including white clouds, white polar veils, blue clouds at the limbs, and widespread yellow haze. Contamination by surface scattering, however, makes it difficult to solve for the aerosol particle characteristics.

The scattering by molecules and small particles can best be disentangled from the surface effects by considering the wavelength dependence. Extensive observations of Mars have been made by Dollfus and Focas (1969), Morozhenko (1964), and Coffeen (1973). The measurements were extended to 0.22 μm with the Polariscope balloon system (Coffeen and Gehrels 1970). A useful review is given by Coulson (1969).

Dollfus (1957; Dollfus and Focas 1966), Minin (1968), and Morozhenko (1970) have used wavelength-dependent polarization observations of Mars to infer several different estimates for the surface pressure and aerosol amount in the atmosphere of Mars. The derived surface pressures were generally higher than the estimates that were obtained from Mariner radio-occultation data. Ingersoll (1971a,b), however, observed the polarization of the bright area of Arabia near maximum elongation and in analysis of the wavelength dependence found Rayleigh optical depths consistent with the occultation data, without postulating a dust component.

These diverse interpretations could be due in part to real variations in aerosol content of the atmosphere of Mars. However, at least part of the problem lies in faulty interpretation of the data and is a consequence of the difficulty of separating the contributions to the polarization from gas, aerosol, and surface scattering. The major difficulty could be overcome if we had detailed knowledge of the scattering properties of the planetary surface. However, Fig. 25 illustrates that a simple assumption, such as unpolarized Lambert surface reflection, is entirely inadequate and would introduce large errors in the interpretation of atmospheric conditions. The effect of surface reflections can be reduced, but not eliminated, with measurements in the ultraviolet; the conclusions will, in general, be model dependent because of assumptions on the wavelength dependence of *surface* polarization and albedo. Detailed reviews of techniques for separating gas and surface scattering are given by Chamberlain and Hunten (1965), and by Cann et al. (1965).

A new approach is provided by measurement of circular polarizations. Kemp reports that the circular polarization on Mars showed a complex behavior during the recent dust storm.[25]

Jupiter

The outer planets present a limited range of phase angle as viewed from the earth; the range for Jupiter is $0°-12°$. The basis for polarimetric determination of particle microstructure is a full set of data as a function of phase angle and wavelength. Obviously, there is a severe handicap in earth-based studies of the outer planets; the eventual solution will come with space probes. A polarimeter is on board Pioneers 10 and 11 for the Jupiter flyby,[26] and as illustrated by Fig. 26 (Goetz et al. 1972), the proposed Mariner Jupiter-Saturn flybys would provide a full range of phase angle. We already have polarimetric data for the limited range of available phase angle, particularly for the variations of polarization over the planetary disk at several wavelengths.

The polarization-phase curve for the center of the disk has been measured by Lyot (1929), Morozhenko and Yanovitskii[27] (1973), and Doose (1973). The polarization becomes increasingly negative between $6°$ and $12°$. This may be a broad glory due to small spheres, corresponding to the similar feature in the polarization of Venus at $\lambda = 1.0$ μm. Such

---

[25] See. p. 610.   [26] See p. 189.   [27] See p. 599.

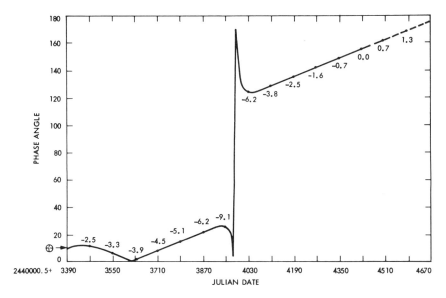

Fig. 26. Variation of phase angle versus time for viewing Jupiter from the Mariner Jupiter-Saturn spacecraft scheduled for launch in 1977. The maximum phase angle observable from earth is 12°. The magnitudes, V, of Jupiter are given along the curve.

an interpretation has been made by Loskutov (1972), who fit Lyot's Jupiter observations with particles of mean radius 0.3 μm and refractive index 1.38, and also by Morozhenko and Yanovitskii, who, with newer data, deduced a mean particle radius 0.2 μm and refractive index 1.36. However, these refractive-index/particle-size fits should not be considered unique, even if the cloud particles *are* spheres, because they are based on a feature in the Mie computations that persists from $n = 1.25$ to at least $n = 2.0$. Exact computations for multiple scattering over the complete wavelength range will be required to narrow the range of possibilities. Furthermore, there is no convincing evidence that the particles in Jupiter's atmosphere are spherical. This is in contrast to the case of the Venus clouds, for which there are several polarization features that unambiguously establish the spherical shape of the particles. The polarization-phase curve for Jupiter could have an origin similar to that of the negative branch observed for cirrus clouds (Fig. 15), in which case the particle size and refractive indices derived above have no meaning. An extensive program of laboratory measurements on ice clouds, as proposed long ago by Lyot, would be helpful; however, the essential thing is to extend the planetary observations to large phase angles.

With these reservations in mind, it is interesting to compare the above refractive indices with the values for methane and ammonia measured by Marcoux (1969) at $\lambda = 0.59$ μm. Methane (CH_4) has $n = 1.30$ in the liquid phase and 1.33 in the solid phase (both at the melting point, 91°K). Ammonia (NH_3) has $n = 1.36$ as a liquid (at 218°K) and 1.42 as a solid (at the melting point, 195°K). The effective (radiating) temperature of Jupiter is $\sim 135°$ (Aumann, Gillespie, and Low 1969), but the visible clouds may occur at considerably higher temperatures. It appears that spherical liquid droplets may comprise at least some of the Jupiter cloud tops.

A strong gradient of the polarization between the center of the disk and the poles of Jupiter was measured by Lyot (1929), Dollfus (1957), Hall and Riley[28] (1968), and Gehrels, Herman, and Owen (1969). The molecular optical depth above the equator is not large. Morozhenko and Yanovitskii (1973) find $\tau \sim 0.1$ at $\lambda = 0.37$ μm, based on numerical fits to the observed phase curve. The radial polarization, which is greatest at the edge of the disk, persists even at phase angle 0° and can be interpreted as second- and higher-order scatterings within an upper molecular layer. The vertical optical depth of this layer appears to be greatest at the poles; Gehrels, Herman, and Owen deduce an optical depth at $\lambda = 0.56$ μm of ~ 0.6 at the north pole, based on a comparison of calculations of multiple Rayleigh scattering with localized observations as a function of wavelength. Figure 27 shows intriguing measurements by Gehrels of the wavelength dependence of polarization, used in obtaining these molecular optical depths. The interpretation is, however, still incomplete and must be integrated with other observational data, including spectroscopic measurements of equivalent line widths for different regions (Teifel 1969; Moroz and Cruikshank 1969; McElroy 1969),[29] and with the simple fact that photographs in the strong methane band at 0.889 μm show bright regions at the poles (Owen 1969; Fountain and Larson 1973), suggesting a *smaller* molecular path length there. An important factor may be the variation of cloud particle composition (refractive index), size, and shape from region to region. Our understanding of the atmospheric structure will be much increased by accurate two-dimensional polarization mapping over the disk, combined with multiple scattering calculations for vertically and horizontally inhomogeneous models.

The discovery of circular polarization on Jupiter by Kemp et al. (1971a,b) is significant.[30] The handedness is opposite in the two hemispheres and inverts as Jupiter passes through opposition; the amplitude is not equal for the north and south poles and shows localized variations.

---

[28] See p. 593. [29] See p. 598. [30] See p. 607.

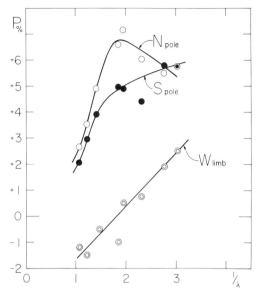

Fig. 27. Percentage linear polarization observed on Jupiter in November–December 1963, as a function of the reciprocal of the wavelength in microns. The phase angle was $\sim 9°$, and the focal plane aperture was 5 arcsecs in diameter. Positive polarization means that the electric vector maximum is in a radial direction; negative polarization means tangential. (After Gehrels, Herman, and Owen 1969.)

Hansen (1971c) has shown that multiple scattering in a cloudy atmosphere typically yields a circular polarization of the observed magnitude and also produces the observed symmetries of handedness. Observations of circular polarization for well-defined local regions over a range of phase angle may prove useful for defining cloud and cloud particle properties.

Saturn

Lyot (1929) observed the center of the disk of Saturn during three apparitions and found positive polarization between 2° and 6° phase angle, with the maximum polarization reaching 0.6% at 6°. This general trend is confirmed in measurements by Bugaenko, Morozhenko, and Yanovitskii[31] and by Coffeen (unpublished). The measured polarization curve resembles that for Rayleigh scattering, but the degree of polarization is even greater than that for single scattering by isotropic molecules. The polarization differs in sign from that observed by us for certain cirrus clouds (Fig. 15). However, it can be fit with Mie scattering calculations for size distributions of *spheres*. Figure 28 shows the locus of all possible

---

[31] See p. 599.

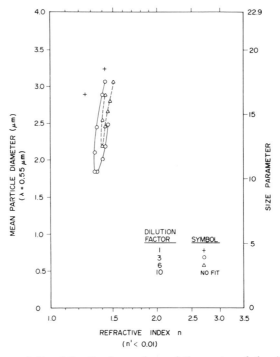

Fig. 28. Locus of fits of Lyot's observations of the center of the disk of Saturn with Mie calculations for size distributions of spheres, for various multiple scattering dilution factors. The single scattering calculations were made for the size distribution $N(r) \propto r^{-2}$, with abrupt cutoffs at 0.75 r_0 and 1.25 r_0, where r_0 is the mean particle radius. The isolated fits having dilution factor equal to one probably result from the narrowness of this size distribution, for which $\langle r \rangle_{\text{eff}} = r_0$ and $v_{\text{eff}} \simeq 0.02$.

fits over this specific domain of particle size and refractive index, for several assumed values of the multiple scattering dilution. These fits were found by comparing Lyot's visual measurements ($\lambda \sim 0.55$ μm) with Mie calculations and requiring that the theoretical results, after correction for multiple scattering, match within the total range of scatter of the data. The locus of possible fits could be further restricted by consideration of the wavelength dependence of polarization and by making exact calculations for multiple scattering, constrained by the measured albedo of Saturn. However, note that these fits have been made for spherical particles and that no shape-specific features have yet been observed in the polarization of Saturn for the small available range of phase angle. Thus, if the cloud particles are *not* spherical, the refractive index and particle size may be quite different from that indicated by Fig. 28.

Regional observations of linear polarization of Saturn have been made by Lyot (1929), Dollfus (unpublished), Hall and Riley (1969), Bugaenko, Galkin, and Morozhenko (1971), and Coffeen (unpublished). A latitudinal gradient is present, as for Jupiter. At the poles the polarization is a maximum more-or-less independent of phase angle, but with appreciable day-to-day variations. The position angle of polarization is much less systematic than for Jupiter and Venus. Bugaenko, Galkin, and Morozhenko find no detectable differences in the polarization for observations in and out of methane bands at 0.62 μm and 0.72 μm. However, Coffeen (unpublished) measured the pole of Saturn in the strong methane band at 0.889 μm and in the continuum at 0.92 μm and found the linear polarization to be twice as large in the absorption band as in the continuum.

Swedlund, Kemp, and Wolstencroft (1972) have measured the circular polarization for Saturn as a function of wavelength and phase angle.[32] Symmetries were found similar to those for Jupiter, but the handedness of circular polarization is consistently opposite for Saturn under the same scattering geometry. The circular polarization of the southern hemisphere is zero at $0°$ phase angle but rises to a sharp maximum at $\sim 2°.2$ phase angle. There is also considerable structure as a function of wavelength, the circular polarization being a minimum at ~ 0.50 μm and increasing toward the blue and red. These data are most intriguing, though more observations are needed. No detailed comparisons have yet been made with calculations for model atmospheres, but this would be a fruitful area of investigation.

For Saturn, as for Jupiter, we can expect to obtain a good description of the cloud heights (specifically the cloud-top pressures) and the gross structure of the visible clouds, as well as the curves of polarization versus phase angle for different particles in the atmosphere. This will require flyby or orbiter observations over a wide range of phase angle, which should be supplemented by further ground-based measurements. Our success at defining the refractive indices, sizes, and shapes of the cloud particles will depend in part on laboratory and theoretical progress for scattering by nonspherical particles.

Uranus and Neptune

Few polarization observations have been made for Uranus and Neptune, and apparently none have been published. Observations of the disks of these planets are not particularly difficult; indeed, such polarization measurements have been made even for Pluto (Kelsey and Fix 1972). For Uranus, regional measurements would also be feasible. Because of

---

[32] See also p. 607.

the very restricted range of available phase angles (Uranus 3°2 maximum, Neptune 1°9), the principal form of polarization will be radial or tangential linear polarization produced by multiple scattering, with the degree of polarization being maximum near the circumference of the planet and approaching zero at the center. These effects are also pronounced on Jupiter and Saturn, though latitudinal variations spoil the circular symmetry and give a nonzero net polarization for the disk. Disk measurements for Uranus and Neptune can be used to look for such asymmetry, with the potential for discrimination against the thick clear atmosphere model. The range of phase angle could be increased by space probes; from the vicinity of Saturn the proposed Mariner Jupiter-Saturn 1977 spacecraft would view Uranus at 6° and Neptune at 8°.

REFERENCES

Abhyankar, K. D., and Fymat, A. L. 1970. Imperfect Rayleigh scattering in a semi-infinite atmosphere. *Astron. Astrophys.* 4: 101–110.

Arking, A., and Rao, C. R. N. 1971. Refractive index of aqueous HCl and the clouds of Venus. *Nature Phys. Sci.* 229: 116–117.

Aronson, R., and Yarmush, D. L. 1966. Transfer-matrix method for gamma-rays and neutron penetration. *J. Math. Phys.* 7: 221–237.

Aumann, H. H.; Gillespie, C. M.; and Low, F. J. 1969. The internal powers and effective temperatures of Jupiter and Saturn. *Astrophys. J.* 157: L69–L72.

Bary, E. de, and Bullrich, K. 1964. Effects of higher-order scattering in a molecular atmosphere. *J. Opt. Soc. Am.* 54: 1413–1416.

Bazilevskii, K. K.; Pakhomov, L. A.; Tsitovich, T. A.; and Skliarevskii, V. G. 1972. Polarization characteristics of shortwave radiation. *Space Research XII, Vol. 1,* 511–516.

Bazilevskii, K., and Vinnichenko, N. K. 1972. Personal communication; satellite results will be published in *Space Research XIII* (1973).

Beckmann, P. 1968. *The depolarization of electromagnetic waves.* Boulder, Colorado: The Golem Press.

Blau, H. H.; McCleese, D. J.; and Watson, D. 1970. Scattering by individual transparent spheres. *Applied Optics* 9: 2522–2528.

Bond, G. R., and Siewert, C. E. 1971. On the nonconservative equation of transfer for a combination of Rayleigh and isotropic scattering. *Astrophys. J.* 164: 97–110.

Bondarenko, L. N. 1964. Spectral polarimetric measurements of the twilight sky polarization at the zenith. *Sov. Astr. A. J.* 8: 299–302.

Bryant, H. C., and Cox, A. J. 1966. Mie theory and the glory. *J. Opt. Soc. Am.* 56: 1527–1532.

Bugaenko, O. I.; Galkin, L. S.; and Morozhenko, A. V. 1971. Polarimetric observations of the major planets. I. Distribution of polarization over the disk of Saturn. *Sov. Astr. A. J.* 15: 290–295.

Bullrich, K. 1964. Scattered radiation in the atmosphere and the natural aerosol. *Adv. Geophys.* 10: 99–260.

Cann, M. W. P.; Davies, W. O.; Greenspan, J. A.; and Owen, T. C. 1965. *A review of recent determinations of the composition and surface pressure of the atmosphere of Mars.* NASA CR-298.

Chamberlain, J. W., and Hunten, D. M. 1965. Pressure and CO_2 content of the Martian atmosphere: A critical discussion. *Rev. Geophys.* 3: 299–317.

Chandrasekhar, S. 1950. *Radiative transfer.* Oxford Univ. Press. (Reprinted 1960, Dover.)

Chandrasekhar, S., and Elbert, D. D. 1954. The illumination and polarization of the sunlit sky on Rayleigh scattering. *Trans. Amer. Phil. Soc., new series* 44: 643–654.

Coffeen, D. L. 1968. *A polarimetric study of the atmosphere of Venus,* Ph.D. dissertation, Univ. of Arizona.

———. 1969. Wavelength dependence of polarization. XVI. Atmosphere of Venus. *Astron. J.* 74: 446–460.

———. 1973. Wavelength dependence of polarization. Mars. In draft.

Coffeen, D. L., and Gehrels, T. 1969. Wavelength dependence of polarization. XV. Observations of Venus. *Astron. J.* 74: 433–445.

———. 1970. Ultraviolet polarimetry of planets. *Space Research X,* 1036–1042.

Coffeen, D. L., and Hansen, J. E. 1972. Airborne infrared polarimetry. *Proceedings of Eighth International Symposium on Remote Sensing of Environment.* Ann Arbor, Mich.: Univ. of Michigan Press. In press.

———. 1973. Polarization of near-infrared sunlight reflected by terrestrial clouds. To be submitted to *J. Atmos. Sci.*

Collins, D. G.; Blättner, W. G.; Wells, M. B.; and Horak, H. G. 1972. Backward Monte Carlo calculations of the polarization characteristics of the radiation emerging from spherical-shell atmospheres. *Applied Optics* 11: 2684–2696.

Collins, D. G., and Wells, M. B. 1965. *Monte Carlo codes for study of light transport in the atmosphere.* Vols. I and II. Rept. RRA-T54. Radiation Research Associates, Inc., Fort Worth, Texas.

Coulson, K. L. 1959. Characteristics of the radiation emerging from the top of a Rayleigh atmosphere. I. Intensity and polarization. *Planet. Space Sci.* 1: 265–276.

———. 1969. Polarimetry of Mars. *Applied Optics* 8: 1287–1294.

———. 1971. On the solar radiation field in a polluted atmosphere. *J. Quant. Spectrosc. Radiat. Transfer* 11: 739–755.

Coulson, K. L.; Dave, J. V.; and Sekera, Z. 1960. *Tables related to radiation emerging from a planetary atmosphere with Rayleigh scattering.* Berkeley: Univ. of California Press.

Dandekar, B. S., and Turtle, J. P. 1971. Day sky brightness and polarization during the total solar eclipse of 7 March 1970. *Applied Optics* 10: 1220–1224.

Dave, J. V. 1965. Multiple scattering in a non-homogeneous, Rayleigh atmosphere. *J. Atmos. Sci.* 22: 273–279.

———. 1969. Scattering of visible light by large water spheres. *Applied Optics* 8: 155–164.

———. 1970. Intensity and polarization of the radiation emerging from a plane parallel atmosphere containing monodispersed aerosols. *Applied Optics* 9: 2673–2684.

Deirmendjian, D. 1969. *Electromagnetic scattering on spherical polydispersions.* New York: Elsevier.

Diem, M. 1948. Messungen der Grösse von Wolkenelementen II. *Meteor. Rund.* 1: 261–273.

Divari, N. B., ed. 1970. *Atmospheric optics.* Consultants Bureau. New York: Plenum Press.

———. 1972. *Atmospheric optics, Volume 2.* Consultants Bureau. New York: Plenum Press.

Dollfus, A. 1956. Polarisation de la lumière renvoyée par les corps solides et les nuages naturels. *Annales d' Astroph.* 19: 83–113.

———. 1957. Étude des planètes par la polarisation de leur lumière. *Suppl. Ann. Astrophys. No. 4.* (In English, NASA TT F-188.)

———. 1961. Polarization studies of planets. *Planets and satellites, Vol. III, The solar system.* (G. P. Kuiper and B. Middlehurst, eds.), pp. 343–399. Chicago: Univ. of Chicago Press.

———. 1966. Contribution au colloque Caltech-JPL sur la lune et les planètes: Vénus. *Proceedings of the Caltech-JPL Lunar and Planetary Conference.* JPL TM 33-266, 187–202.

———. 1973. Polarization of Venus. II. Regional observations. *Astron. Astrophys.* In draft.

Dollfus, A., and Coffeen, D. L. 1970. Polarization of Venus. I. Disk observations. *Astron. Astrophys.* 8: 251–266.

Dollfus, A., and Focas, J. 1966. Sur la pureté de l'Atmosphère de la planète Mars. *C. R. Acad. Sci. Paris* B262: 1024–1027.

———. 1969. La planète Mars: la nature de sa surface et les propriétés de son atmosphère, d'après la polarisation de sa lumière. I. Observations. *Astron. Astrophys.* 2: 63–74.

Doose, L. R. 1973. *A multicolor photometric and polarimetric study of the atmosphere of Jupiter.* Ph.D. dissertation, Univ. of Arizona. In draft.

Eiden, R. 1970. Influence of the atmospheric aerosol on the elliptical polarization of skylight. *Radiation including satellite techniques.* Geneva, WMO Tech. Note #104, 275–278.

———. 1971. Determination of the complex index of refraction of spherical aerosol particles. *Applied Optics* 10: 749–754.

Fabelinskii, I. L. 1968. *Molecular scattering of light.* (R. Beyer, trans.) New York: Plenum Press.

Fahlen, T. S., and Bryant, H. C. 1968. Optical backscattering from single water droplets. *J. Opt. Soc. Am.* 58: 304–310.

Fesenkov, V. G., ed. 1962. *Scattering and polarization of light in the atmosphere* (English title) 1965. Israel Program for Scientific Translations.

Forbes, F. 1971. Infrared polarization of Venus. *Astrophys. J.* 165: L21–L25.

Fountain, J. W., and Larson, S. M. 1973. Multicolor photography of Jupiter. *Comm. Lunar Planet. Lab.* No. 174. In press.

Friend, J. P. 1966. Properties of the stratospheric aerosol. *Tellus* 18: 465–473.

Fymat, A. L., and Abhyankar, K. D. 1969. Theory of radiative transfer in inhomogeneous atmospheres. I. Perturbation method. *Astrophys. J.* 158: 315–324.

Gehrels, T. 1962. Wavelength dependence of the polarization of the sunlit sky. *J. Opt. Soc. Am.* 52: 1164–1173.

Gehrels, T.; Herman, B. M.; and Owen, T. 1969. Wavelength dependence of polarization. XIV. Atmosphere of Jupiter. *Astron. J.* 74: 190–199.

Gehrels, T., and Teska, T. M. 1963. The wavelength dependence of polarization. *Applied Optics* 2: 67–77.

Goetz, A. F. H.; Coffeen, D.; Dollfus, A.; Gehrels, T.; Greene, T. F.; Hansen, J. E.; and Young, A. T. 1972. *A proposal for an MJS-77 photometer-polarimeter investigation.* Jet Propulsion Lab., Pasadena.

Grant, I. P., and Hunt, G. E. 1969. Discrete space theory of radiative transfer. I. Fundamentals. *Proc. Roy. Soc. Lond. A* 313: 183–197.

Greenberg, J. M.; Pederson, N. E.; and Pederson, J.C. 1961. Microwave analog to the scattering of light by nonspherical particles. *J. Appl. Phys.* 32: 233–242.

Greenler, R. G.; Drinkwine, M.; Mallmann, A. J.; and Blumenthal, G. 1972. The origin of sun pillars. *Amer. Sci.* 60: 292–302.

Greenler, R. G., and Mallmann, A. J. 1972. Circumscribed halos. *Science* 176: 128–131.

Gucker, F. T.; Basu, S.; Pulido, A. A.; and Chiu, G. 1969. Intensity and polarization of light scattered by some permanent gases and vapors. *J. Chem. Phys.* 50: 2526–2535.

Hall, J. S., and Riley, L. A. 1968. Photoelectric observations of Mars and Jupiter with a scanning polarimeter. *Lowell Obs. Bull.* 7: 83–92.

———. 1969. Polarization measures of Jupiter and Saturn. *J. Atmos. Sci.* 26: 920–923.

Hansen, J. E. 1969. Radiative transfer by doubling very thin layers. *Astrophys. J.* 155: 565–573.

———. 1971*a*. Multiple scattering of polarized light in planetary atmospheres. Part I. The doubling method. *J. Atmos. Sci.* 28: 120–125.

———. 1971*b*. Multiple scattering of polarized light in planetary atmospheres. Part II. Sunlight reflected by terrestrial water clouds. *J. Atmos. Sci.* 28: 1400–1426.

———. 1971*c*. Circular polarization of sunlight reflected by clouds. *J. Atmos. Sci.* 28: 1515–1516.

———. 1972. *Information contained in the intensity and polarization of scattered sunlight.* Rept., Goddard Institute for Space Studies, NASA, New York.

Hansen, J. E., and Arking, A. 1971. Clouds of Venus: Evidence for their nature. *Science* 171: 669–672.

Hansen, J. E., and Coffeen, D. L. 1972. Polarization of near-infrared sunlight reflected by terrestrial clouds. *Preprint volume of the conference on atmospheric radiation.* Amer. Meteor. Soc., 55–60.

Hansen, J. E., and Hovenier, J. W. 1971. The doubling method applied to multiple scattering of polarized light. *J. Quant. Spectrosc. Radiat. Transfer* 11: 809–812.

———. 1973. Interpretation of the polarization of Venus. To be submitted to *J. Atmos. Sci.*

Herman, B. M.; Browning, S. R.; and Curran, R. J. 1971. The effect of atmospheric aerosols on scattered sunlight. *J. Atmos. Sci.* 28: 419–428.

Herman, B. M., and Yarger, D. N. 1965. The effect of absorption on a Rayleigh atmosphere. *J. Atmos. Sci.* 22: 644–651.

Herman, M. 1970. Calcul théorique du régime asymptotique polarisé, dans un milieu absorbant, diffusant par des particules sphériques répondant à la théorie de Mie. *Nouv. Rev. d'Opt. Appl.* 1: 171–180.

Hodkinson, J. R. 1963. Light scattering and extinction by irregular particles larger than the wavelength. *Electromagnetic scattering (ICES I)* (M. Kerker, ed.) New York: Macmillan, pp. 87–100.

Holland, A. C., and Gagne, G. 1970. The scattering of polarized light by polydisperse systems of irregular particles. *Applied Optics* 9: 1113–1121.

Holt, H. E. 1970. Photometry and polarimetry of the lunar regolith as measured by Surveyor. *Radio Sci.* 5: 157–170.

Hovenier, J. W. 1969. Symmetry relationships for scattering of polarized light in a slab of randomly oriented particles. *J. Atmos. Sci.* 26: 488–499.

———. 1970. Principles of symmetry for polarization studies of planets. *Astron. Astrophys.* 7: 86–90.

———. 1971. Multiple scattering of polarized light in planetary atmospheres. *Astron. Astrophys.* 13: 7–29.

Huffman, P. 1970. Polarization of light scattered by ice crystals. *J. Atmos. Sci.* 27: 1207–1208.

Hulst, H. C. van de 1952. Scattering in the atmospheres of the earth and planets. *The atmospheres of the earth and planets.* 2nd ed. (G. P. Kuiper, ed.) pp. 49–111. Chicago: Univ. of Chicago Press.

———. 1957. *Light scattering by small particles.* New York: Wiley.

———. 1963. *A new look at multiple scattering.* Rept., Goddard Institute for Space Studies, NASA, New York.

———. 1971. Multiple scattering in planetary atmospheres. *J. Quant. Spectrosc. Radiat. Transfer* 11: 785–795.

Hunt, G. E. 1971. A review of computational techniques for analysing the transfer of radiation through a model cloudy atmosphere. *J. Quant. Spectrosc. Radiat. Transfer* 11: 655–690.

Ingersoll, A. P. 1971a. Polarization measurements of Mars and Mercury: Rayleigh scattering in the Martian atmosphere. *Astrophys. J.* 163: 121–129.

———. 1971b. Ultraviolet polarization measurements of Mars and the opacity of the Martian atmosphere. *Planetary atmospheres.* (C. Sagan et al., eds.) pp. 170–176. Dordrecht, Holland: D. Reidel Publ.

Jacobowitz, H. 1971. A method for computing the transfer of solar radiation through clouds of hexagonal ice crystals. *J. Quant. Spectrosc. Radiat. Transfer* 11: 691–695.

Jenkins, E. B.; Morton, D. C.; and Sweigart, A. V. 1969. Rocket spectra of Venus and Jupiter from 2000 to 3000 Å. *Astrophys. J.* 157: 913–924.

Kattawar, G. W., and Adams, C. N. 1971. Flux and polarization reflected from a Rayleigh-scattering planetary atmosphere. *Astrophys. J.* 167: 183–192.

Kattawar, G. W., and Plass, G. N. 1968. Radiance and polarization of multiple scattered light from haze and clouds. *Applied Optics* 7: 1519–1527.

Kattawar, G. W.; Plass, G. N.; and Adams, C. N. 1971. Flux and polarization calculations of the radiation reflected from the clouds of Venus. *Astrophys. J.* 170: 371–386.

Kattawar, G. W.; Plass, G. N.; and Catchings, F. E. 1972. *Discrete ordinate theory of radiative transfer. 2. Scattering from maritime haze.* NASA-CR-125817. Avail: NTIS CSCL 04A.

Kelsey, L. A., and Fix, J. D. 1972. Linear polarization measurements of Pluto. *Bull. Am. Astron. Soc.* 4: 321.

Kemp, J. C.; Swedlund, J. B.; Murphy, R. E.; and Wolstencroft, R. D. 1971a. Circularly polarized visible light from Jupiter. *Nature* 231: 169–170.

Kemp, J. C.; Wolstencroft, R. D.; and Swedlund, J. B. 1971b. Circular polarization: Jupiter and other planets. *Nature* 232: 165–168.

Kerker, M., ed. 1963. *Electromagnetic scattering (ICES I).* New York: Macmillan.

———, 1969. *The scattering of light and other electromagnetic radiation.* New York: Academic Press.
Kuiper, G. P. 1969. Identification of Venus cloud layers. *Comm. Lunar Planet. Lab.* Vol. 6, 229–250.
Landerer, J.-J. 1892. Sur la recherche de l'angle de polarisation de Vénus. *C. R. Acad. Sci. Paris* 114: 1524–1525.
Lenoble, J. 1970. Importance de la polarisation dans le rayonnement diffuse par une atmosphère planétaire. *J. Quant. Spectrosc. Radiat. Transfer* 10: 533–556.
Lewis, J. S. 1969. Geochemistry of the volatile elements on Venus. *Icarus* 11: 367–385.
———. 1971. Refractive index of aqueous HCl solutions and the composition of the Venus clouds. *Nature* 230: 295–296.
Liou, K.-N. 1972a. Electromagnetic scattering by arbitrarily oriented ice cylinders. *Applied Optics* 11: 667–674.
———. 1972b. Light scattering by ice clouds in the visible and infrared: A theoretical study. *J. Atmos. Sci.* 29: 524–536.
Liou, K.-N., and Hansen, J. E. 1971. Intensity and polarization for single scattering by polydisperse spheres. A comparison of ray optics and Mie theory. *J. Atmos. Sci.* 28: 995–1004.
Loskutov, V. M. 1971. Interpretation of polarimetric observations of the planets. *Sov. Astr. A. J.* 15: 129–133.
———. 1972. Interpretation of polarimetric observations of Jupiter. *Sov. Astr. A.J.* 15: 828–831.
Lyot, B. 1929. Recherches sur le polarisation de la lumière des planètes et de quelques substances terrestres. *Ann. Obs. Paris (Meudon)*, VIII. (In English, NASA TT F-187.)
Marcoux, J. E. 1969. Indices of refraction of some gases in the liquid and solid state. *J. Opt. Soc. Am.* 59: 998.
McElroy, M. B. 1969. Atmospheric composition of the Jovian planets. *J. Atmos. Sci.* 26: 798–812.
Minin, I. N. 1968. An optical model for the atmosphere of Mars. *Sov. Astr. A. J.* 11: 1024–1033.
Molenkamp, C. R. 1972. *Absorption lines in a stratified Rayleigh atmosphere.* Ph.D. dissertation, Univ. of Arizona.
Moroz, V. I. 1971. Height of the Venusian clouds at equatorial and polar latitudes. *Nature Phys. Sci.* 231: 36–37.
Moroz, V. I., and Cruikshank, D. P. 1969. Distribution of ammonia on Jupiter. *J. Atmos. Sci.* 26: 865–869.
Morozhenko, A. V. 1964. Results of polarimetric observations of Mars in 1962–1963. *Physics of the moon and planets.* Kiev: Naukova Dumka, 58–80. (In English, NASA TT F-382.)
———. 1970. The atmosphere of Mars according to polarization observations. *Sov. Astr. A. J.* 13: 852–857.
Morozhenko, A. V., and Yanovitskii, E. G. 1973. Atmosphere of Jupiter according to polarimetric observations. *Icarus*. In press.
Mossop, S. C. 1965. Stratospheric particles at 20 km altitude. *Geochim. Cosmochim. Acta* 29: 201–207.
O'Leary, B. 1970. Venus halo: Photometric evidence for ice in the Venus clouds. *Icarus* 13: 292–298.

Owen, T. 1969. The spectra of Jupiter and Saturn in the photographic infrared. *Icarus* 10: 355–364.

Partington, J. R., 1953. *An advanced treatise on physical chemistry, Vol. 4, Physicochemical optics.* London: Longmans.

Peebles, G. H., and Plesset, M. S. 1951. Transmission of gamma-rays through large thicknesses of heavy materials. *Phys. Rev.* 81: 430–439.

Plass, G. N.; Kattawar, G. W.; and Catchings, F. E. 1973. Matrix operator theory of radiative transfer. 1: Rayleigh scattering. *Applied Optics* 12: 314–329.

Pritchard, B. S., and Elliott, W. G. 1960. Two instruments for atmospheric optics measurements. *J. Opt. Soc. Am.* 50: 191–202.

Rao, C. R. N.; Takashima, T.; and Moore, J. G. 1972. Polarimetry of the daytime sky during solar eclipses. *J. Atmos. Terres. Phys.* 34: 573–576.

Redheffer, R. 1962. On the relation of transmission-line theory to scattering and transfer. *J. Math. Phys.* 41: 1–41.

Rosse, W. P. (Lord) 1878. Preliminary note on some measurements of the polarization of the light coming from the moon and from the planet Venus. *Sci. Proc. Roy. Dublin Soc.* 1: 19–20.

Rowell, R. L., and Stein, R. S., eds. 1967. *Electromagnetic scattering (ICES II).* New York: Gordon and Breach.

Rozenberg, G. V. 1960. Light scattering in the earth's atmosphere. *Sov. Phys. Uspekhi* 3: 346–371.

———. 1966. *Twilight.* New York: Plenum Press.

———. 1968. Optical investigations of atmospheric aerosol. *Sov. Phys. Uspekhi* 11: 353–380.

Rozenberg, G. V., and Gorchakov, G. I. 1967. The degree of ellipticity of the polarization of light scattered by atmospheric air as a tool in the investigation of aerosol microstructures. *Izv., Atmos. & Oceanic Phys.* 3: 400–407.

Sagan, C., and Pollack, J. B. 1969. On the structure of the Venus atmosphere. *Icarus* 10: 274–289.

Schiffer, R. A. 1972. A solution of the auxiliary equation of radiative transfer for a planetary atmosphere with molecular anisotropy. *Preprint volume of the Conference on Atmospheric Radiation.* Amer. Meteor. Soc.

Sekera, Z. 1957. Polarization of skylight. *Handbuch der Physik, XLVIII, Geophysik II,* 288–328.

———. 1963. *Radiative transfer in a planetary atmosphere with imperfect scattering.* RAND Rept. R-413-PR. Santa Monica, Calif.

———. 1966. *Reductions of the equations of radiative transfer for a plane-parallel planetary atmosphere. Parts I and II.* RAND Corp., RM-4951-PR and RM-5056-PR. Santa Monica, Calif.

Sekera, Z., and Kahle, A. B. 1966. *Scattering functions for Rayleigh atmospheres of arbitrary thickness.* RAND Report R-452-PR. Santa Monica, Calif.

Sekera, Z., and Viezee, W. 1961. *Distribution of the intensity and polarization of the diffusely reflected light over a planetary disk.* RAND Report R-389-PR. Santa Monica, Calif.

Shifrin, K. S. 1951. *Scattering of light in a turbid medium.* Published in Russian. Moscow. (In English, NASA TT F-477.)

Sinton, W. M., 1953. *Distribution of temperatures and spectra of Venus and other planets.* Ph.D. dissertation, Johns Hopkins Univ.

Sobolev, V. V. 1968. An investigation of the atmosphere of Venus. II. *Sov. Astr. A. J.* 12: 135–140.

Swedlund, J. B.; Kemp, J. C.; and Wolstencroft, R. D. 1972. Circular polarization of Saturn. *Astrophys. J.* 178: 257–265.
Teifel, V. G. 1969. Molecular absorption and the possible structure of the cloud layers of Jupiter and Saturn. *J. Atmos. Sci.* 26: 854–859.
Tricker, R. A. R. 1970. *Introduction to meteorological optics.* New York: American Elsevier.
Twomey, S.; Jacobowitz, H.; and Howell, H. B. 1966. Matrix methods for multiple scattering problems. *J. Atmos. Sci.* 23: 289–296.
Unz, F. 1969. Die Konzentration des Aerosols in Troposphëre und Stratosphäre aus Messungen dar Polarisation der Himmelsstrahlung im Zenit. *Beiträge zur Physik der Atmosphäre* 42: 1–35.
Veverka, J. 1971. A polarimetric search for a Venus halo during the 1969 inferior conjunction. *Icarus* 14: 282–283.
Volz, F. 1961. Optik der Tropfen. *Handbuch der Geophysik* 8: 943–1026.
Whitehill, L. 1972. *Solar system applications of Mie theory and of radiative transfer of polarized light.* Ph.D. dissertation, Cornell Univ.
Yeh, C. 1964. Scattering of obliquely incident light waves by elliptical fibers. *J. Opt. Soc. Am.* 54: 1227–1231.
Young, A. T. 1973. Are the clouds of Venus sulphuric acid? Submitted to *Icarus.*

INVARIANT IMBEDDING AND CHANDRASEKHAR'S PLANETARY PROBLEM OF POLARIZED LIGHT

S. UENO
University of Southern California
S. MUKAI
Kyoto University
and
A. P. WANG
Arizona State University

With the aid of the invariant imbedding technique, we show how to solve exactly the equation of transfer for Chandrasekhar's planetary problem of polarized light, i.e., the diffuse reflection and transmission problem of polarized parallel rays by finite atmospheres bounded by reflecting surfaces. The bottom surface is assumed to reflect radiation either isotropically or specularly. A Cauchy system for the scattering matrix of Chandrasekhar's planetary problem and the radiation field vector in the case of internal emitting source are presented, together with the initial conditions. Finally, the order-of-scattering theory of polarized light is shown. This theory facilitates the numerical computation, by successive iteration, of radiation emergent from the top of the planetary atmosphere. The characteristic of the invariant imbedding technique is to transform the two-point boundary-value problem into the initial value problem. The initial value problem of Riccati type is suitable for the numerical computation with the aid of high-speed digital computers.

In 1942, based on the superposition principle, Ambarzumian (1958) presented an elegant approach to the exact solution of an auxiliary equation of the Fredholm type. This first method was rigorously developed by several authors (Kourganoff 1952; Sobolev 1963, 1972; Busbridge 1960; Wing 1962; Sobolev, Gorbatskii, and Ivanov 1966; Hummer and Rybicki 1967). Introducing the principle of invariance for diffuse reflection, Ambarzumian (1958) showed that the solution reduces to a nonlinear integral equation for the scattering function given by his first method. Later, Ambarzumian's physical method was extensively developed as a particle-counting technique of invariant imbedding (Bellman, Kalaba, and Prestrud 1963; Bellman et al. 1964; Wing 1962). On the other hand, based on the invariance principles, Chandrasekhar (1950) ingeniously presented a complete set of integro-differential equations for the scattering and transmission functions in connection with the transfer equation and applied it to various kinds of transfer problems. Introducing the photon-emergence probability into radiative transfer, Sobolev (1963, 1972) has successfully developed the probabilistic method. From the stochastic aspect of Markovian-like processes, the probabilistic method has been further extended by a few authors (Ueno 1957, 1960, 1961; Uesugi 1963a,b). The Duhamel principle has also been applied to radiative transfer (Matsumoto 1966, 1968).

In recent years, the diffuse reflection and transmission problems of polarized light by finite atmospheres have been extensively discussed with the aid of an invariant-imbedding approach (Chandrasekhar 1950; Sobolev 1963, 1972; Busbridge 1960; Sekera 1963, 1966a; Sekera and Kahle 1966; Fymat 1967; Hovenier 1969; Abhyankar and Fymat 1970; Buell, Kalaba, and Ueno 1969; Buell et al. 1970; Sweigart 1970; Domke 1971, 1972; Takashima 1971, 1972). A Cauchy system for Chandrasekhar's planetary problem has been obtained by a few authors (Sobolev 1963; Kagiwada and Kalaba 1971; Bellman and Ueno 1972). The reduction of Chandrasekhar's planetary problem to the standard problem has been made (Chandrasekhar 1950; van de Hulst 1948; Sobolev 1963; Ueno 1963; Sekera and Kahle 1966; Kahle 1968; Kagiwada and Kalaba 1971; Bellman and Ueno 1972; Ueno and Wang 1972). Allowance for an emitting-source distribution in Chandrasekhar's planetary problem has been made by several authors (Sobolev 1963; Sekera 1963; Bellman et al. 1967).

Furthermore, a Cauchy system for the scattering and transmission matrices in a Rayleigh atmosphere bounded by a specular reflector is obtained by an initial value method (Ueno 1973a,b; see Appendix). The same specular problem for unpolarized light has been discussed (Kagiwada and Kalaba 1968; Casti, Kalaba, and Ueno 1969, 1970; Casti, Kagiwada, and Kalaba 1972; Ueno and Wang 1972; Matsumoto and Ueno 1972). The reduction of Chandrasekhar's planetary problem in the case of a specular reflection to the standard problem has been made by several authors (Ueno and Wang 1972; Mukai and Ueno 1972; Mukai 1973). Chandrasekhar's planetary problem for the case of a diffuse-and-specular reflector is discussed by Ueno and Wang (1973). In a manner similar to the multiple-scattering theory, the order-of-scattering theory of radiation has also been used successfully for polarized and unpolarized light (van de Hulst 1963; Dave 1964, 1965; Dave and Furukawa 1966; Abu-Shumays 1967; Uesugi and Irvine 1969, 1970; Uesugi, Irvine, and Kawata 1971; Hovenier 1971; Bellman, Ueno, and Vasudevan 1972a,b, 1973; Fymat and Ueno 1972; Mingle 1972).

Finally, it is interesting to note that in the study of radiative transfer in connection with invariant imbedding, the doubling method and the scattering-matrix method have also played an important role (van de Hulst 1963, 1971; van de Hulst and Grossman, 1968; Hansen 1969, 1971a,b; Hansen and Hovenier 1971; Redheffer 1962; Wang 1966, 1967, 1970; Twomey, Jacobowitz, and Howell 1966; Grant and Hunt 1969; Takashima 1972, 1973a,b; Poon and Ueno 1972).

MULTIPLE SCATTERING OF PARTIALLY POLARIZED LIGHT

Consider a plane-parallel, inhomogeneous atmosphere of optical thickness τ scattering partially polarized radiation according to a phase matrix. The bottom surface is assumed to reflect radiation either isotropically or specularly. Let parallel rays of net flux $\pi \mathbf{F} = \pi(F_l, F_r, F_u, F_v)$ per unit area normal to itself be incident on the top τ in a direction $-\mathbf{\Omega}_0 = (-\mu_0, \phi_0)$ $(0 < \mu_0 \leq 1, 0 \leq \phi_0 \leq 2\pi)$, where μ_0 represents the cosine of the angle of incidence with respect to the inwards normal, and ϕ_0 is the azimuth. The diffuse radiation field at level t in the atmosphere is characterized by an intensity vector $\mathbf{I}(t,\mathbf{\Omega})$, where $\mathbf{\Omega}$ stands for (μ, ϕ). Let the upwelling radiation field at level t in a direction $\mathbf{\Omega}$ be denoted by $\mathbf{I}(t, +\mathbf{\Omega})(0 < \mu \leq 1, 0 \leq \phi \leq 2\pi)$, and similarly, let the downwelling radiation field at level t in a direction $-\mathbf{\Omega}$ be denoted by $\mathbf{I}(t,-\mathbf{\Omega})$. In the modified Stokes representation of Chandrasekhar (1950), for example, $\mathbf{I}$ is the one-column matrix of the Stokes parameters (I_l, I_r, U, V) of the diffuse radiation in $\mathbf{\Omega}$ direction. Usually, $\mathbf{F}$ refers to the intensity vector of solar radiation with components $F_l = F_r = |\mathbf{F}|/2$, and $F_u = F_v = 0$.

Write the reflection integral operator for the atmosphere of optical thickness t as

$$\mathbf{R}(t, +\mathbf{\Omega})[\mathbf{f}(\mathbf{\Omega}')] = \frac{I}{4\pi\mu} \int_{2\pi} \mathbf{S}(t;\mathbf{\Omega},\mathbf{\Omega}')\mathbf{f}(\mathbf{\Omega}')\,d\mathbf{\Omega}', \quad (1)$$

where S-function is the four-by-four scattering matrix of Chandrasekhar's planetary problem. Starting with the equation of transfer for the Stokes vector $\mathbf{I}((t,\mathbf{\Omega})$ (Chandrasekhar 1950, Eq. [231], p. 44), and writing the upwelling radiation field in terms of the reflection operator

$$\mathbf{I}(t, +\mathbf{\Omega}) = \mathbf{R}(t, +\mathbf{\Omega})[\mathbf{I}^*(t, -\mathbf{\Omega}')], \quad (2)$$

where $\mathbf{I}^*$ represents the total downwelling radiation field, we have an invariant imbedding procedure that gives rise to a required Cauchy system for the scattering matrix (Fymat and Ueno 1972; Ueno and Wang 1972)

$$\left(\frac{1}{\mu} + \frac{1}{\mu_0} + \frac{\partial}{\partial \tau}\right) \mathbf{S}(\tau;\mathbf{\Omega},\mathbf{\Omega}_0) = \mathbf{P}(\tau;\mathbf{\Omega}, -\mathbf{\Omega}_0)$$

$$+ \frac{1}{4\pi} \int_{2\pi} \mathbf{P}(\tau;\mathbf{\Omega},\mathbf{\Omega}'') \, \mathbf{S}(\tau;\mathbf{\Omega}'',\mathbf{\Omega}) \, \frac{d\mathbf{\Omega}''}{\mu''}$$

$$+ \frac{1}{4\pi} \int_{2\pi} \mathbf{S}(\tau;\mathbf{\Omega},\mathbf{\Omega}') \, \mathbf{P}(\tau; -\mathbf{\Omega}', -\mathbf{\Omega}_0) \, \frac{d\mathbf{\Omega}'}{\mu'}$$

$$+ \frac{1}{16\pi^2} \int_{2\pi}\int_{2\pi} \mathbf{S}(\tau;\mathbf{\Omega},\mathbf{\Omega}') \, \mathbf{P}(\tau; -\mathbf{\Omega}',\mathbf{\Omega}'') \, \mathbf{S}(\tau;\mathbf{\Omega}'',\mathbf{\Omega}_0) \, \frac{d\mathbf{\Omega}'}{\mu'} \, \frac{d\mathbf{\Omega}''}{\mu''}, \qquad (3)$$

together with an initial condition

$$[\mathbf{S}(\tau;\mathbf{\Omega},\mathbf{\Omega}_0)]_{\tau=0} = 4\mu\mu_0 A, \qquad (4)$$

where $\mathbf{P}(\tau;\mathbf{\Omega},\mathbf{\Omega}_0)$ is a four-by-four phase matrix at the top and A is an albedo for Lambert's law of reflection at the bottom. When $A = 0$, Equation (3) reduces to that for the standard problem (Sekera 1963). A case of a specular reflector at the bottom has been solved by an initial value method (Ueno 1973a, b). A hybrid case of the diffuse-and-specular reflector by an integral operator method is discussed by Ueno and Wang (1973).

Furthermore, Chandrasekhar's planetary problem enclosing an internal emitting source has been solved by an invariant imbedding (Sekera 1963; Bellman et al. 1967). For the sake of simplicity, the thermal emission is considered in the state of hydrodynamical equilibrium. The elements of this vector assume the azimuth-independent form $\mathbf{B}(t) = [B(T)/2, B(T)/2, 0, 0]$, where $B(T)$ denotes the specific intensity of the blackbody radiation of the frequency v and the temperature T. Then, with the aid of the invariant imbedding technique, we get the required Cauchy system for the radiation field $\mathbf{I}$ emergent from the top as follows:

$$\left(\frac{1}{\mu} + \frac{\partial}{\partial \tau}\right) \mathbf{I}(\tau, +\mathbf{\Omega}) = \frac{1}{\mu}\left[\mathbf{B}(\tau) + \frac{1}{4} \mathbf{P}(\tau;\mathbf{\Omega}, -\mathbf{\Omega}_0)\mathbf{F}\right.$$

$$\left. + \frac{1}{4\pi} \int_{2\pi} \mathbf{P}(\tau;\mathbf{\Omega},\mathbf{\Omega}') \, \mathbf{I}(\tau,\mathbf{\Omega}') \, d\mathbf{\Omega}'\right] + \frac{1}{4\pi\mu} \int_{2\pi} \mathbf{S}(\tau;\mathbf{\Omega},\mathbf{\Omega}')$$

$$\cdot \left[\mathbf{B}(\tau) + \frac{1}{4\pi} \int_{2\pi} \mathbf{P}(\tau; -\mathbf{\Omega}',\mathbf{\Omega}'') \, \mathbf{I}(\tau,\mathbf{\Omega}'') \, d\mathbf{\Omega}''\right. \qquad (5)$$

$$\left. + \frac{1}{4} \mathbf{P}(\tau; -\mathbf{\Omega}', -\mathbf{\Omega}_0)\mathbf{F}\right] \frac{d\mathbf{\Omega}'}{\mu} - \frac{1}{4\mu\mu_0} \mathbf{S}(\tau;\mathbf{\Omega},\mathbf{\Omega}_0)\mathbf{F},$$

where the scattering matrix is given by Equation (3), together with an initial condition

$$[\mathbf{I}(\tau, +\mathbf{\Omega})]_{\tau=0} = \mu_0 A \mathbf{F}. \qquad (6)$$

ORDER-OF-SCATTERING OF PARTIALLY POLARIZED RADIATION

In this section, in place of a multiply scattered light, we shall consider the radiation field that has undergone scattering processes during the propagation exactly n times ($n \geq 1$). Let a finite-order scattering matrix for Chandrasekhar's planetary problem be denoted by $\mathbf{S}(n, \tau; \mathbf{\Omega}, \mathbf{\Omega}_0)$. Let a finite-order reflection operator be defined by

$$R(n, \tau, +\mathbf{\Omega})[\mathbf{f}(i, \mathbf{\Omega}')] = \frac{1}{4\pi\mu} \sum_{i=1}^{n-1} \int_{2\pi} \mathbf{S}(n - i, \tau; \mathbf{\Omega}, \mathbf{\Omega}') \mathbf{f}(i, -\mathbf{\Omega}') d\mathbf{\Omega}'. \qquad (7)$$

Writing the finite-order upwelling radiation field in terms of the reflection operator

$$\mathbf{I}(n, t, +\mathbf{\Omega}) = \mathbf{R}(n, t, +\mathbf{\Omega})[\mathbf{I}^*(i, t, -\mathbf{\Omega}')], \qquad (8)$$

after some manipulations, we have an invariant imbedding procedure that provides the required Cauchy system for the finite-order scattering matrix as follows (Fymat and Ueno 1972):

$$\left(\frac{1}{\mu} + \frac{1}{\mu_0} + \frac{\partial}{\partial \tau}\right) \mathbf{S}(n, \tau; \mathbf{\Omega}, \mathbf{\Omega}_0) = \mathbf{P}(\tau; \mathbf{\Omega}, -\mathbf{\Omega}_0) d(n, 1)$$

$$+ \frac{1}{4\pi} \int_{2\pi} \mathbf{P}(\tau; \mathbf{\Omega}, \mathbf{\Omega}'') \mathbf{S}(n - 1, \tau; \mathbf{\Omega}'', \mathbf{\Omega}_0) \frac{d\mathbf{\Omega}''}{\mu''}$$

$$+ \frac{1}{4\pi} \int_{2\pi} \mathbf{S}(n - 1, \tau; \mathbf{\Omega}, \mathbf{\Omega}') \mathbf{P}(\tau; -\mathbf{\Omega}', -\mathbf{\Omega}_0) \frac{d\mathbf{\Omega}'}{\mu'} \qquad (9)$$

$$+ \frac{1}{16\pi^2} \sum_{i=1}^{n-1} \int_{2\pi} \int_{2\pi} \mathbf{S}(n - i, \tau; \mathbf{\Omega}, \mathbf{\Omega}') \mathbf{P}(\tau; -\mathbf{\Omega}', \mathbf{\Omega}'')$$

$$\cdot \mathbf{S}(i - 1, \tau; \mathbf{\Omega}'', \mathbf{\Omega}_0) \frac{d\mathbf{\Omega}'}{\mu'} \frac{d\mathbf{\Omega}''}{\mu''},$$

where $d(n, 1)$ is a Kronecker delta function, together with a first-order scattering matrix

$$S(1,\tau;\mathbf{\Omega},\mathbf{\Omega}_0) = 4\mu\mu_0 \exp\left[-\tau\left(\frac{1}{\mu} + \frac{1}{\mu_0}\right)\right]A$$

$$+ \exp\left[-\tau\left(\frac{1}{\mu} + \frac{1}{\mu_0}\right)\right]\int_0^\tau \mathbf{P}(t;\mathbf{\Omega},-\mathbf{\Omega}_0)\exp\left[t\left(\frac{1}{\mu} + \frac{1}{\mu_0}\right)\right]dt. \quad (10)$$

Similarly, we can get a complete set of integro-differential recurrence relations of the scattering and transmission matrices for Chandrasekhar's planetary problems. Furthermore, in the case of a homogeneous atmosphere, the integro-differential recurrence relations reduce to the integral recurrence relations, which are more tractable from the computational aspect.

Finally, it is of interest that in the limit as n tends to infinity, the summation of the finite-order scattering matrix reduces to the total scattering matrix given by Equation (3).

APPENDIX

Consider a plane-parallel, inhomogeneous atmosphere of optical thickness τ scattering partially polarized radiation according to a Rayleigh phase matrix. It is also assumed that, while the upper surface at $t = \tau$ is illuminated by parallel rays of net flux $\pi\mathbf{F} = \pi(F_l, F_r, F_u, F_v)$ per unit area normal to itself in a direction $-\mathbf{\Omega}_0$, the bottom surface is bounded by a specular reflector. In what follows, our attention is focused on the azimuth-independent components of the partially polarized radiation, i.e., $I_l(t,\mu)$ and $I_r(t,\mu)(0 \le t \le \tau, -1 \le \mu \le 1)$ (Ueno 1973b).

Let an auxiliary matrix integral equation for the two-by-two source matrix $\mathbf{J}(t,\mu;\tau)$ be denoted by

$$\mathbf{J}(t,\mu;\tau) = \lambda(t)\left[e^{-(\tau-t)/\mu} + Ae^{-(\tau+t)/\mu}\right]\mathbf{E}$$

$$+ \lambda(t)\int_0^\tau \left[\mathbf{k}(|t-y|) + A\mathbf{k}(t+y)\right]\mathbf{J}(y,\mu;\tau)\,dy, \quad (11)$$

where $\mathbf{E}$ is an identity matrix

$$\mathbf{E} = \begin{pmatrix} 1 & 0 \\ 0 & 1 \end{pmatrix}, \quad (12)$$

$\lambda(t)(0 < \lambda \le 1)$ is an albedo for single scattering, and $A(0 < A \le 1)$ is a constant albedo for the specular reflection. In Equation (11) the kernel $\mathbf{k}$-matrix is given by

$$\mathbf{k}(|r|) = \int_0^1 e^{-|r|/\mu'}\mathbf{W}(\mu')\,d\mu', \quad (13)$$

where the characteristic matrix $\mathbf{W}$ is defined by

$$\mathbf{W}(\mu) = \frac{3}{8\mu} \begin{pmatrix} 1 + \mu^4 & \sqrt{2}(1 - \mu^2)\mu^2 \\ \sqrt{2}(1 - \mu^2)\mu^2 & 2(1 - \mu^2)^2 \end{pmatrix}. \tag{14}$$

In a manner similar to that given by Ueno (1973a) and with the aid of an integral equation approach, after some manipulations, we get the required Cauchy system for the scattering matrix

$$\left(\frac{1}{\mu} + \frac{1}{\mu_0} + \frac{\partial}{\partial \tau} \right) \mathbf{S}(\tau; \mu, \mu_0) = \lambda(\tau) [\mathbf{E} + Ae^{-2\tau/\mu} \mathbf{E}$$
$$+ \int_0^1 \mathbf{S}(\tau; \mu, \mu') \mathbf{W}(\mu') \, d\mu'] \, [\mathbf{E} + Ae^{-2\tau/\mu_0} \mathbf{E}$$
$$+ \int_0^1 \mathbf{W}(\mu'') \mathbf{S}(\tau; \mu'', \mu_0) \, d\mu''], \tag{15}$$

together with an initial condition

$$[\mathbf{S}(\tau; \mu, \mu_0)]_{\tau=0} = 0. \tag{16}$$

In the case of the standard diffuse-reflection and transmission problem, the Cauchy system for the scattering matrix is given by putting $A = 0$ in Equation (15), and the initial condition is given by Equation (16). The required intensity vector of azimuth-independent radiation, that is diffusely reflected by the finite atmosphere and bounded by a specular reflector, takes the form

$$\mathbf{I}(\tau, +\mathbf{\Omega}) = \frac{3}{16\mu} \mathbf{M}(\mu) \mathbf{S}(\tau; \mu, \mu_0) \mathbf{M}^*(\mu_0) \mathbf{F}, \tag{17}$$

where

$$\mathbf{M}(\mu) = \begin{pmatrix} \mu^2 & \sqrt{2}(1 - \mu^2) \\ 1 & 0 \end{pmatrix}, \tag{18}$$

and the superscript asterisk represents the transposition. In the paper by Ueno (1973b) the Cauchy system for the source matrix is provided; we can thus determine the internal radiation field of the azimuth-independent components with the aid of the scattering matrix given by Equation (15).

REFERENCES

Abhyankar, K. D., and Fymat, A. L. 1970. Imperfect Rayleigh scattering in a semi-infinite atmosphere. *Astron. Astrophys.* 4: 101–110.

Abu-Shumays, I. K. 1967. Generating functions and reflection and transmission functions. *J. Math. Analys. Appl.* 18: 453–471.

Ambarzumian, V. A., ed. 1958. *Theoretical astrophysics.* London: Pergamon Press.

Bellman, R.; Kagiwada, H.; Kalaba, R.; and Prestrud, M. C. 1964. *Invariant imbedding and time-dependent transport processes.* New York: American Elsevier Publ. Co. Inc.

Bellman, R.; Kagiwada, H.; Kalaba, R., and Ueno, S. 1967. Chandrasekhar's planetary problems with internal sources. *Icarus* 7: 365–371.

Bellman, R.; Kalaba, R.; and Prestrud, M. C. 1963. *Invariant imbedding and radiative transfer in slabs of finite thickness.* New York: American Elsevier Publ. Co. Inc.

Bellman, R., and Ueno, S. 1972. Invariant imbedding and Chandrasekhar's planetary problem of radiative transfer. *Astrophys. Space Sci.* 16: 241–248.

———. 1973. Invariant imbedding and the resolvent of Fredholm integral equation with composite symmetric kernel. *J. Math. Phys.* In press.

Bellman, R.; Ueno, S.; and Vasudevan, R. 1972a. Invariant imbedding and radiation dosimetry: I. Finite order scattering and transmission function. *Math. Biosci.* 14: 235–254.

———. 1972b. Invariant imbedding and radiation dosimetry: II. Integral recurrence relations for the finite order scattering and transmission functions. *Math. Biosci.* 15: 153–162.

———. 1973. Invariant imbedding and radiation dosimetry: IV. Finite order scattering of gamma radiation by a target slab. *Math. Biosci.* In press.

Buell, J.; Casti, J.; Kalaba, R.; and Ueno, S. 1970. Exact solution of a family of matrix integral equations for multiply scattered partially polarized radiation. II. *J. Math. Phys.* 11: 1673–1678.

Buell, J.; Kalaba, R.; and Ueno, S. 1969. Exact solution of a family of matrix integral equations for multiply scattered partially polarized radiation. I. Univ. of Southern California, Dept. Electrical Engineering. TR. No. 69-1; *J. Optimiz. Theory Appl.* 5: 170–177 (1970).

Busbridge, I. W. 1960. *The mathematics of radiative transfer.* Cambridge Tract. No. 50. London: Cambridge Univ. Press.

Casti, J.; Kagiwada, H., and Kalaba, R. 1972. Equivalence relationships between diffuse radiation fields for finite slabs bounded by a perfect specular reflector and a perfect absorber. *J. Quant. Spectrosc. Radiat. Transfer* 13: 267–272.

Casti, J.; Kalaba, R.; and Ueno, S. 1969. Reflection and transmission functions for finite isotropically scattering atmosphere with specular reflectors. *J. Quant. Spectrosc. Radiat. Transfer* 9: 532–552.

———. 1970. Source function for an isotropically scattering atmosphere bounded by a specular reflector. *J. Quant. Spectrosc. Radiat. Transfer* 10: 1119–1128.

Chandrasekhar, S. 1950. *Radiative transfer.* London: Oxford University Press.

Dave, J. V. 1964. Meaning of successive iteration of the auxilary equation in the theory of radiative transfer. *Astrophys. J.* 140: 1292–1303.

———. 1965. Multiple scattering in a non-homogeneous, Rayleigh atmosphere. *J. Atmos. Sci.* 22: 273–279.

Dave, J. V., and Furukawa, P. M. 1966. Intensity and polarization of the radiation emerging from an optically thick Rayleigh atmosphere. *J. Opt. Soc. Am.* 56: 394–400.

Domke, H. 1971. Radiation transport with Rayleigh scattering. I. Semi-infinite atmosphere. *Sov. Astr. A.J.* 15: 266–277.

———. 1972. Radiation transport with Rayleigh scattering. II. Finite atmosphere. *Sov. Astr. A.J.* 15: 616–623.

Fymat, A. L. 1967. Theory of radiative transfer in atmospheres exhibiting polarized resonance fluoresence. Ph.D. Thesis, Univ. of California, Los Angeles.

Fymat, A. L., and Ueno, S. 1972. Order-of-scattering of partially polarized radiation in inhomogeneous anisotropically scattering atmosphere. I. Fundamentals. Univ. of Southern California, Dept. Electrical Engineering. TR. No. 72–54. To be submitted to *Astrophys. J.*

Grant, I. P., and Hunt, G. E. 1969. Discrete space theory of radiative transfer. I. Fundamentals. *Proc. Roy. Soc., Lond. A.* 313: 183–197.

Hansen, J. E. 1969. Radiative transfer by doubling very thin layers. *Astrophys. J.* 155: 565–573.

———. 1971a. Multiple scattering of polarized light in planetary atmospheres. Part I. The doubling method. *J. Atmos. Sci.* 28: 120–125.

———. 1971b. Multiple scattering of polarized light in planetary atmospheres. Part II. Sunlight reflected by terrestrial water clouds. *J. Atmos. Sci.* 28: 1400–1426.

Hansen, J. E., and Hovenier, J. W. 1971. The doubling method applied to multiple scattering of polarized light. *J. Quant. Spectrosc. Radiat. Transfer* 11: 809–812.

Hovenier, J. W. 1969. Symmetry relationships for scattering of polarized light in a slab of randomly oriented particles. *J. Atmos. Sci.* 26: 488–499.

———. 1971. Multiple scattering of polarized light in planetary atmospheres. *Astron. Astrophys.* 13: 7–29.

Hulst, H. C. van de 1948. Scattering in a planetary atmosphere. *Astrophys. J.* 107: 220–246.

———. 1963. *A new look at multiple scattering.* Institute for Space Studies, New York.

———. 1971. Multiple scattering in planetary atmospheres. *J. Quant. Spectrosc. Radiat. Transfer* 11: 785–796.

Hulst, H. C. van de, and Grossman, K. 1968. Multiple scattering in planetary atmospheres. *Atmospheres of Venus and Mars.* (J. C. Brandt, and M. B. McElroy, eds.), pp. 35–54. New York: Gordon Breach Science Publishers.

Hummer, D. G., and Rybicki, G. 1967. Computational methods for NLTE line-transfer problems. *Methods in computational physics.* (B. Alder, S. Fernbach, M. Rotenberg, eds.) pp. 53–127. New York: Academic Press.

Kagiwada, H., and Kalaba, R. 1968. A Cauchy problem for Fredholm integral equation with kernels of the form $k(|t - y|) + k(t + y)$. *The Rand Corp.,* RM-5600-PR.

———. 1971. Invariant imbedding and radiation field in finite anisotropically scattering slab bounded by a Lambert's law reflector. *J. Quant. Spectrosc. Radiat. Transfer* 11: 1101–1109.

Kahle, A. B. 1968. Intensity of radiation from a Rayleigh scattering atmosphere. *The Rand Corp.,* RM-5620-PR.

Kourganoff, V. 1952. *Basic methods in transfer problems.* London: Oxford University Press.

Matsumoto, M. 1966. The functional equations in the radiation field. I. Monodirectional illumination of the upper surface. *Publ. Astr. Soc. Japan* 18: 445–455.

———. 1968. Duhamel's principle in the nonstationary radiation field. *J. Math. Analys. Appl.* 21: 445–457.

Matsumoto, M., and Ueno, S. 1972. Diffuse reflection and transmission of parallel rays by a finite anistropically scattering atmosphere with a specular reflector. Univ. of Southern California, Dept. Electrical Engineering. TR. In draft.

Mingle, J. O. 1972. Order-of-scattering theory for transport processes. *J. Math. Analys. Appl.* 38: 53–60.

Mukai, S. 1973. Reflection and transmission by planetary atmosphere with reflecting surface. To appear in *Publ. Astr. Soc. Japan.*

Mukai, S., and Ueno, S. 1972. Reduction of Chandrasekhar's planetary problem in the case of a specularly reflecting boundary to the standard problem. Univ. of Southern California, Dept. Electrical Engineering. TR. No. 72–56. *J. Quant. Spectrosc. Radiat. Transfer.* In press.

Poon, P. T. Y., and Ueno, S. 1972. Scattering matrix and doubling equations for the scattering and transmission functions. Univ. of Southern California, Dept. Electrical Engineering. TR. No. 72–51. To appear in *J. Atmos. Sci.*

Redheffer, R. 1962. On the relation of transmission line theory to scattering and transfer. *J. Math. Phys.* XLI: 1–41.

Sekera, Z. 1963. Radiative transfer in planetary atmosphere with imperfect scattering. *The Rand Corp.*, R-413-PR.

———. 1966a. Reduction of the equations of radiative transfer for a plane-parallel planetary atmosphere: Part I. *The Rand Corp.*, RM-4951-PR.

———. 1966b. Reduction of the equations of radiative transfer for a plane-parallel atmosphere: Part II. *The Rand Corp.*, RM-5056-PR.

Sekera, Z., and Kahle, A. B. 1966. Scattering functions for Rayleigh atmospheres of arbitrary thickness. *The Rand Corp.*, R-452-PR.

Sobolev, V. V. 1963. *A treatise of radiative transfer.* Princeton, New Jersey: D. Van Nostrand Co. Inc.

———. 1972. *Scattering of light in planetary atmosphere.* Moscow: Nauka Press.

Sobolev, V. V.; Gorbatskii, V. G.; and Ivanov, V. V. 1966. *Theory of stellar spectra.* Moscow: Nauka Press.

Sweigart, A. V. 1970. Radiative transfer in atmospheres scattering according to the Rayleigh phase function with absorption. *Astrophys. J. Suppl.* 182: 1–80.

Takashima, T. 1971. Dependence of the polarization features of the diffuse radiation field of the earth's atmosphere on the location of a concentrated aerosol layer. Ph.D. Thesis, Univ. of California, Los Angeles.

———. 1972. Computation on skylight polarization in a planetary atmosphere using a new approach of doubling method. To be submitted to *J. Atmos. Sci.*

———. 1973a. Emergent radiation from inhomogeneous plane-parallel planetary atmosphere by using adding method. To appear in *J. Quant. Spectrosc. Transfer.*

———. 1973b. Matrix method of evaluation of the internal radiation field in a plane-parallel atmosphere. To appear in *Astrophys. Space Sci.*

Twomey, S.; Jacobowitz, H. J.; and Howell, H. B. 1966. Matrix methods for multiple scattering problems. *J. Atmos. Sci.* 23: 289–296.

Ueno, S. 1957. The probabilistic method for problems of radiative transfer: II. Milne's problem. *Astrophys. J.* 126: 413–417.

———. 1960. The probabilistic method for problems of radiative transfer: X. Diffuse reflection and transmission in a finite inhomogeneous atmosphere. *Astrophys. J.* 132: 729–745.

———. 1961. On the diffusion matrix of radiative transfer. *Ann d'Astroph.* 24: 352–358.

———. 1963. Diffuse reflection of radiation by a planetary atmosphere. *Proc. Fourth Symp. on Space Technology and Science.* (Tokyo, 1962) pp. 451–458. Tokyo, Japan: Rocket Society of Japan, Publ.

———. 1973a. Scattering and transmission matrices of partially polarized light in a Rayleigh atmosphere bounded by a specular reflector. Univ. of Southern California, Dept. Electrical Engineering. TR. No. RB 73-9. To appear in *Astrophys. Space Sci.*

———. 1973b. Reflection and source matrices of partially polarized radiation in an inhomogeneous Rayleigh atmosphere bounded by a specular reflector. Univ. of Southern California, Dept. Electrical Engineering. TR. In draft.

Ueno, S., and Wang, A. P. 1972. Scattering and transmission functions of radiation by finite atmospheres with reflecting surfaces. Univ. of Southern California, Dept. Electrical Engineering, TR. No. 72-47. *Astrophys. Space Sci.* In press.

———. 1973. Scattering and transmission functions of radiation by a finite atmosphere bounded by a diffuse-and-specular reflector. In draft.

Uesugi, A. 1963a. An interpretation of photon diffusion process. *Ann d'Astroph.* 26: 263–278.

———. 1963b. On the generalized scattering and transmission functions in radiative transfer. *Publ. Astr. Soc. Japan* 15: 266–273.

Uesugi, A.; and Irvine, W. M. 1969. Computations of synthetic spectra for a semi-infinite atmosphere. *J. Atmos. Sci.* 26: 973–978.

———. 1970. Multiple scattering in a plane-parallel atmosphere. I. Successive scattering in a semi-infinite medium. *Astrophys. J.* 159: 127–135.

Uesugi, A.; Irvine, W. M.; and Kawata, Y. 1971. Formation of absorption spectra by diffuse reflection from a semi-infinite planetary atmosphere. *J. Quant. Spectrosc. Radiat. Transfer* 11: 797–808.

Wang, A. P. 1966. Scattering processes. Ph.D. Thesis, Univ. of California, Los Angeles.

———. 1967. On the relation of scattering matrix, Riccati equation and boundary conditions. *J. Math. Analys. Appl.* 17: 238–247.

———. 1970. Multiple scattering of anisotropic radiation through a layer of clouds. *App. Sci. Res.* 23: 221–236.

Wing, M. G. 1962. *An introduction to transport theory.* New York: Wiley.

POLARIZATION MEASUREMENTS OF JUPITER AND THE GREAT RED SPOT

JOHN S. HALL and LOUISE A. RILEY
Lowell Observatory

Polarization measurements of the whole disk of Jupiter were made in 1968 at effective wavelengths of 3760 Å and 5740 Å. Further measures of Jupiter with emphasis on the Great Red Spot were made in 1971 and 1972 in the ultraviolet region. Measures of the Red Spot showed no significant difference in polarization from those that would be expected at the same location had the Red Spot not been present; many scans across the Red Spot indicate that this difference is not greater than 0.002. Despite the low declination of Jupiter in the years 1971 and 1972 ($-19°$ and $-23°$, respectively), there was some evidence that the symmetrical radial pattern previously reported must be modified at some positions far from the limb by local conditions in the planet's atmosphere.

Observations of polarization of Jupiter (Hall and Riley 1968, 1969) were initiated at the Lowell Observatory in 1968 with a scanning polarimeter (Hall 1968). In this technique the resolution is limited by the seeing and, to some extent, by the guiding. The observations reported here were made near 3760 Å where the degree of polarization is much higher than in regions further to the red.

Lyot's early measurements (Dollfus 1961), made in the visual region, suggested day-to-day changes of polarization on Jupiter's disk. Recent photographic and infrared observations of intensity have shown some interesting anomalies and changes that signify substantial turmoil in the Jovian atmosphere. The polarization observations of Gehrels, Herman, and Owen (1969) indicate the presence of scattering by both aerosols and molecules. We believe that resolution and measuring techniques can be so improved that ground-based polarization measures might become very important indicators of atmospheric structure.

Figure 1 shows a polarization diagram obtained in the ultraviolet on a night of good seeing in 1968, when Jupiter was 12° north of the equator and the phase angle was 10°.

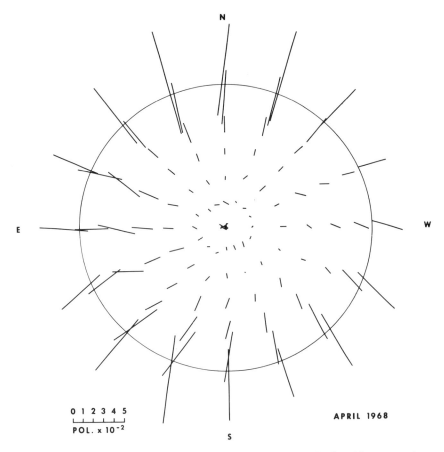

Fig. 1. Polarization measures of Jupiter obtained at 3760 Å with a scanning polarimeter on 24 April 1968. The directions of the lines indicate the planes of the electric-vector maxima, and their lengths indicate the amount of polarization. The focal plane aperture was 0.6 arcsec in diameter. (From Lowell Bull. No. 145, Vol. VII, 83–92, 1968.)

Because of improvements in equipment and technique, we have observed Jupiter again in the ultraviolet near opposition on several nights in 1971 and on a single night in 1972. Each vector shown in the diagrams represents more than half a million counts. The declination of Jupiter was $-19°$ and $-23°$ in 1971 and 1972, respectively; the seeing was never good, and therefore, high resolution could not be achieved, but the data do contain items of interest.

Dollfus has reported that his visual measures of the Red Spot made on

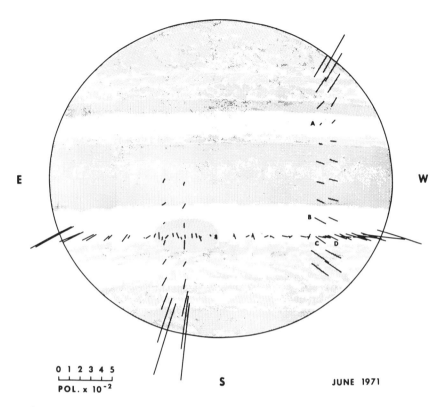

Fig. 2. Polarization measures of Jupiter obtained at 3760 Å on 8 June and 13 June 1971. Each scan passed through the Great Red Spot as it moved across the disk.

one night in 1952 showed no differences from those made of its surroundings. We have not seen any published quantitative data on this subject.

Figure 2 shows some 1971 observations in which each scan passed through the Red Spot as it moved westward with the rotation of the planet. Therefore, all polar scans covered the same longitude on the planetary disk. The polarization found at the two points north of the equatorial plane at A is substantially less than that found at the same location and latitude equally far south of the equator at B. The polarization found at points C and D is very close to what one would expect if the Red Spot were not present; the directions of the vectors are within three degrees of the expected angles, and the degree of polarization is within 0.002 of normal.

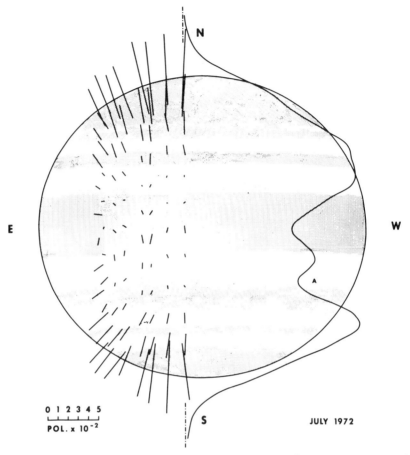

Fig. 3. Polarization measures of Jupiter obtained at 3760 Å on 5 July 1972. Each scan passed through the Great Red Spot as it moved across the disk. At the right side is the intensity curve observed along the scanline closest to the central meridian. The Great Red Spot is at A.

Figure 3 shows data obtained in 1972. The observed intensity curve shows the Red Spot at A and indicates rather poor resolution. The vector diagram shows mixed polarization in some zones and more nearly consistent values in others. Unfortunately, increasing clouds made it impossible to continue the observations to the western half of the disk on this night.

In Fig. 4 data obtained in the Great Red Spot are compared only with data obtained on the same scans and at the same longitude and latitude in the northern hemisphere. If we compute the mean polarization for the eight vectors in the northern quadrant and compare with the correspond-

JUPITER AND THE GREAT RED SPOT

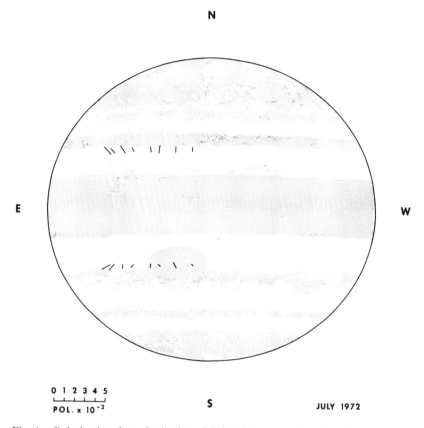

Fig. 4. Polarization data obtained on 5 July 1972 in the Great Red Spot only, are compared with data at the same latitude and similar longitudes in the northern hemisphere. In this figure the mean values are grouped differently than in Fig. 3 in order to give the mean polarization at the latitude of the Great Red Spot.

ing value for the eight vectors obtained at the Red Spot, we find that the mean direction of the two vectors corresponds very closely to what would normally be expected at each of these mean points on the disk, and that the mean polarization in the north is 0.003 greater than that found in the south. These and other observations of the Red Spot in the ultraviolet suggest that its polarization was within 0.002 of normal for the observed regions with the Red Spot absent.

The scanning technique requires accurate intensity curves for computing the polarization, which also provide a means of knowing precisely where each observation is obtained. The polarization near the north limb, for instance, can be found for different wavelength regions at any arbitrary fraction of the mean intensity. Our recent data indicate that the apparent polarization at the north limb at 3700 Å is about 0.005 greater than that at the south limb, in qualitative agreement with the observations of Gehrels, Herman, and Owen (1969).

The true value of the polarization at the limb, however, is found by using the intensity data and an accurate energy distribution function in order to eliminate loss of resolution because of scattering. We did this using a function for the accurate representation of the observed image profiles found by Franz[1] for a point at the north limb and discovered that the observed value would be increased by less than 0.010 when it is reduced to one square arc second and when the effects of scattered light are eliminated.

Although the observations obtained here indicate that polarization measures of the Red Spot may not provide new information, there is some evidence that systematic measures of other regions will yield interesting results. Several wavelength regions should be explored and several observers should participate. It is only in this way that marginal results can be either substantiated or denied.

REFERENCES

Dollfus, A. 1961. Polarization studies of planets. *Solar system Vol. III, Planets and satellites* (G. P. Kuiper and B. Middlehurst, eds.) Ch. 9, pp. 391–394. Chicago: Univ. of Chicago Press.

Gehrels, T.; Herman, B. M.; and Owen, T. 1969. Wavelength dependence of polarization. XIV. Atmosphere of Jupiter. *Astron. J.* 74: 190–199.

Hall, J. S. 1968. A scanning polarimeter. *Lowell Obs. Bull.* 7: 61–66.

Hall, J. S., and Riley, L. A. 1968. Photoelectric observations of Mars and Jupiter with a scanning polarimeter. *Lowell Obs. Bull.* 7: 83–92.

———. 1969. Polarization measures of Jupiter and Saturn. *J. Atmos. Sci.* 26: 920–923.

DISCUSSION

CRUIKSHANK: From spectroscopy and direct photography in the methane band at 8880 Å, one gains the impression that the Great Red Spot is higher than the surrounding cloud deck. It is, therefore, surprising that the ultraviolet polarization is not smaller over the Great Red Spot.

DOOSE: Photometric scans in this methane band show the Great Red Spot to be about 55% brighter than the surrounding clouds, while no brightening is seen in the continuum at 9215 Å. I should like to ask the author why polar scans are preferred over equatorial ones?

HALL: It is then easier to make the successive scans correspond to the same position on the disk because one can, by guiding better, compensate for the effects of planetary rotation.

---

[1] Personal communication, 1970.

POLARIZATION INVESTIGATIONS OF THE PLANETS CARRIED OUT AT THE MAIN ASTRONOMICAL OBSERVATORY OF THE UKRAINIAN ACADEMY OF SCIENCES

O. I. BUGAENKO, A. V. MOROZHENKO,
and E. G. YANOVITSKII
Main Astronomical Observatory of the
Ukrainian Academy of Sciences

A program of planetary polarimetry has been in progress at Kiev for the past ten years, with emphasis on the wavelength dependence of polarization. Extensive observations of Mars, Jupiter, and Saturn permit a deduction of aerosol particle characteristics by comparison with Mie calculations.

MARS

During 1962–1965 numerous measurements in the spectral range 0.355–0.600 μm were carried out to determine the degree of polarization of the entire Martian disk (Morozhenko 1964, 1966). The quantitative interpretation of these data (Morozhenko 1969) allowed us to determine the optical properties of the Martian atmosphere for the period in which the dust storm was absent. In particular it was found that the total optical thickness (τ_0) was equal to 0.064 for $\lambda = 0.355$ μm, and the Rayleigh component of the optical thickness was equal to 0.016 for $\lambda = 0.355$ μm. If we assume that the Martian atmosphere consists of pure CO_2, we find that this optical thickness of 0.016 corresponds to a surface pressure of about 6.2 mb. It is interesting to note that if we reduce τ_0 obtained above to the wavelength 0.305 μm, we get $\tau_0 = 0.1$, which is in good agreement with the results of the measurements of Mariner 6 and 7 (Barth and Hord 1971).

During 1971 (24 March–1 August) Morozhenko carried out the polarization measurements of the entire disk in the 0.37–0.80 μm region. The principal results of these observations are as follows:

During March 25–June 10, the phase angle of Mars changed from 41°.8 to 36°.8. The observational data for this period allowed us to investigate the dependence of the degree of polarization upon the longitude of the central meridian in different spectral intervals. It was found that this dependence was the same in the whole spectral range, though its amplitude $(P_{max}-P_{min})/P_{max}$ decreased from 0.22 for 0.800 μm to 0.08 for 0.373 μm. This fact seems to indicate that the dark and bright areas on Mars also have different polarizing properties in the ultraviolet (Fig. 1).

During May 13–June 16 a decrease of the degree of polarization was noted (especially in the ultraviolet) which was accompanied by an obscuration of the longitude effect. This is clear from Figs. 1 and 2. It is most probable that this decrease of polarization is caused by aerosol haze. Calculations for the model of the radial homogeneous haze with small optical thickness showed the best agreement with the observational data for particles having radii of about 0.1 μm and $\tau_0 = 0.063$ at 0.373 μm.

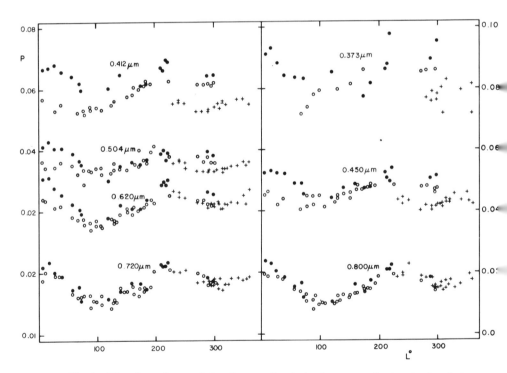

Fig. 1. The dependence of the degree of polarization upon the longitude of the Martian central meridian. Baselines for 0.412, 0.504 and for 0.373, 0.450 μm are the same. Dots: March 25–April 23 ($\alpha = 40°\!.5 - 41°\!.8$); Circles: April 30–May 28 ($\alpha = 41°\!.8 - 39°\!.5$); Crosses: May 28–June 10 ($\alpha = 39°\!.5 - 36°\!.8$).

JUPITER

In 1971, Morozhenko measured the polarization of the entire disk of Jupiter as well as of its central part (10″) in the spectral region 0.373–0.800 μm for the phase angle interval of $0°.4 \leq \alpha \leq 10°$. Figure 3 gives the results for the entire disk corrected for the effect of inhomogeneity of the disk. Figure 4 gives the results for the central part of the disk (circles and dots). The crosses in Fig. 4 represent the values of polarization of the disk center in visual light obtained by Lyot (1929).

For interpretation of these results, two models were considered according to Sobolev (1944, 1968): Model A: The aerosol particles and molecules are mixed in the atmosphere, their ratios being constant; Model B: A gas layer of small optical thickness is situated above the cloud layer that consists of aerosol particles.

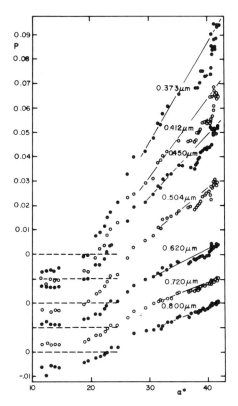

Fig. 2. The spectral phase curves of the Martian polarization reduced to L = 180°. The zero points for 0.373, 0.412, and 0.450 μm are the same.

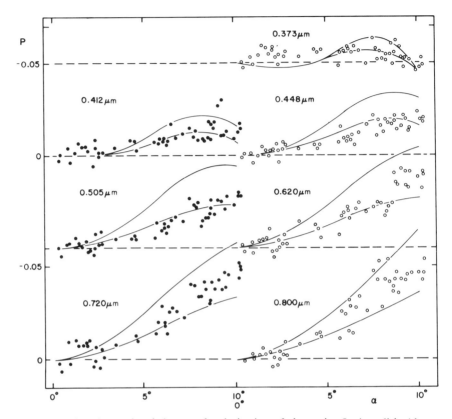

Fig. 3. The observational degree of polarization of the entire Jupiter disk (dots and circles). For explanation of solid curves see text.

The aerosol particles are considered to be nonabsorbing spheres, their size distribution being normal gaussian

$$\zeta(r)dr = \frac{1}{\sqrt{2\pi}\sigma} \exp\left[-\frac{ln^2 r/r_0}{2\sigma^2}\right] dln(r), \qquad (1)$$

where $\zeta(r)dr$ is the relative number of particles with radii between r and $r + dr$, σ is the dispersion of $ln(r)$, and r_0 is the geometrical mean of particle radii. Moreover, it was assumed that the refractive index of aerosol particles is constant in the wavelength range considered here.

An approximate method is used for the determination of parameters of Jupiter's atmosphere. It is assumed that for model A the observed polarization $P(\alpha)$ must satisfy the following inequality

$$|P_A^1(\alpha)| \leq |P(\alpha)| \leq 2|P_A^1(\alpha)|, \qquad (2)$$

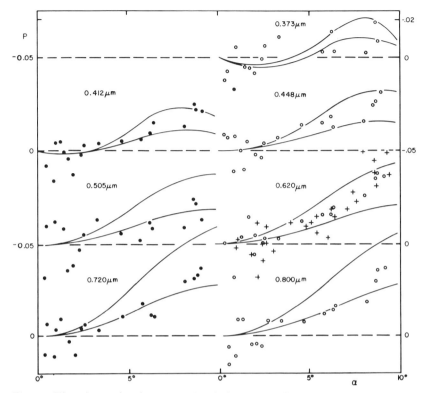

Fig. 4. The observational degree of polarization of the center of Jupiter's disk (dots and circles). Crosses indicate Lyot's (1929) observations in visual light. For explanation of solid curves see text.

where

$$P_A{}^1(\alpha) = \frac{f(\alpha)}{A(\alpha)} \left[\delta \frac{3}{4} \sin^2\alpha + (1-\delta)y(\alpha) \right], \quad (3)$$

$$f(\alpha) = \frac{1}{8}\left(1 - \sin\frac{\alpha}{2} \, tg\frac{\alpha}{2} \, lnctg\frac{\alpha}{4}\right),$$

$$y(\alpha) = X_r(\alpha) - X_l(\alpha).$$

Here $X_r(\alpha)$ and $X_l(\alpha)$ are the components of normalized aerosol scattering *indicatrix*, and $A(\alpha)$ is the observed value of visible albedo of the entire planetary disk.

The results of recent exact calculations by Hansen and Arking (1971), Hansen (1971), Hovenier (1971), and Kattawar, Plass, and Adams (1971) for polarization of the light multiply scattered by large particles confirm in general the inequality of Equation (2) at small phase angles ($\alpha \leq 30°$). An analogous inequality takes place also for model B.

This method of the determination of the optical properties of the atmosphere was tested by calculation of the parameters of the cloud layer of Venus. The refractive index was found to be $n = 1.435 \pm 0.015$, $\sigma^2 = 0.12$, $r_0 = 0.74$ μm, which agrees well with the exact values given by Hansen and Arking (1971).

For different values of the refractive index ($1.33 \leq n \leq 1.50$) such values of the parameters r_0 and σ^2 have been chosen so that the inequality of Equation (2) is satisfied. The best agreement with observations regarding the satisfaction of the inequality of Equation (2) was found for the parameter values $n = 1.36$, $\sigma^2 = 0.30$, $r_0 = 0.19$ μm, and $\delta = 0.3$ ($\lambda = 0.373$ μm) for model A. Figure 4 demonstrates the results of calculations of the function $P_A{}^1(\alpha)$ (lower solid curves) and $2P_A{}^1(\alpha)$ (upper solid curves) for the atmospheric parameters given above. Similar curves were obtained also for model B (see Fig. 3). For model B, $\tau_0 \approx 0.15$ was found at $\lambda = 0.373$ μm.

Thus, within the framework of the adopted model of the upper layers of Jupiter's atmosphere, the results of spectral polarization observations agree with calculations for the following atmospheric parameters: $n = 1.36 \pm 0.01$, $\sigma^2 \approx 0.3$, $r_0 \approx 0.2$ at $\lambda = 0.373$ μm, $\delta \approx 0.3$ for model A, and $\tau_0 \approx 0.1$ for model B. This refractive index for the particles agrees well with the ammonia cloud layer hypothesis. A more detailed account of the above results is given in a paper by Morozhenko and Yanovitskii (1973).

SATURN

The observations of the polarization distribution across Saturn's disk were started in 1969 (Bugaenko, Galkin, and Morozhenko 1971) and continued in 1970 by Bugaenko at the Main Astronomical Observatory of the Ukrainian Academy of Sciences and by Galkin at the Crimean Astrophysical Observatory of the Academy of Sciences of the USSR. The degree and orientation of the polarization were measured in 12 (1969) and 10 (1970) spectral intervals in the 0.36–0.75 μm range.

The observations were carried out with a round diaphragm in 1969 and 1970 and with a specially oriented segment diaphragm in 1969. The following regions of Saturn's disk were investigated: center, south, west, and east. Saturn's rings were also measured. Detailed tables of the observational results for 1969 have been published by Bugaenko, Galkin, and Morozhenko (1971), and the results for 1970 will be published in the same journal. As an example, Fig. 5 gives the phase dependence of the degree of polarization for six spectral intervals.

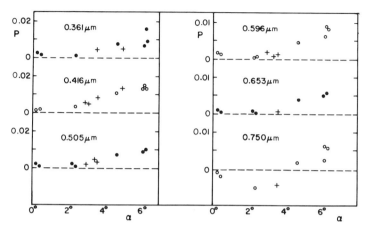

Fig. 5. The spectral phase curves of the center of Saturn's disk. Crosses are observations in 1969; dots and circles, 1970.

REFERENCES

Barth, C. A., and Hord, C. W. 1971. Mariner ultraviolet spectrometer: topography and polar cap. *Science* 173: 197–201.

Bugaenko, O. I.; Galkin, L. S.; and Morozhenko, A. V. 1971. Polarimetric observations of Jovian planets. I. Distribution of the polarization on the disk of Saturn. *Astron. Zh.* 48:373–379.

Hansen, J. E. 1971. Multiple scattering of polarized light in planetary atmosphere. Part III. Sunlight reflected by terrestrial water clouds. *J. Atmos. Sci.* 28: 1400–1426.

Hansen, J. E., and Arking, A. 1971. Clouds of Venus: evidence for their nature. *Science* 171: 669–672.

Hovenier, J. W. 1971. Multiple scattering of polarized light in planetary atmospheres. *Astron. Astrophys.* 13: 7–29.

Kattawar, G. M.; Plass, G. N.; and Adams, C. N. 1971. Flux and polarization calculations of the radiation reflected from the clouds of Venus. *Astrophys. J.* 170:371–386.

Lyot, B. 1929. Recherches sur la polarisation de la lumière de planètes et de quelques substances terrestre. *Ann. Obs. Meudon* 8: 1–161.

Morozhenko, A. V. 1964. Results of polarimetric observations of Mars in 1962–1963. *Naukova Dumka, Physika Luny i Planet* 58–80. Kiev.

———. 1966. Polarizable properties of atmosphere and surface of Mars. *Naukova Dumka, Physika Luny i Planet* 45–69. Kiev.

———. 1969. Mars atmosphere from polarized observations. *Astron. Zh.* 46: 1087–1094.

Morozhenko, A. V., and Yanovitskii, E. G. 1973. Atmosphere of Jupiter according to polarimetric observations. *Icarus.* In press.

Sobolev, V. V. 1944. On the optical properties of the atmosphere of Venus. *Astron. Zh.* 21:241–244.

———. 1968. Investigation of the atmosphere of Venus, II. *Astron. Zh.* 45: 169–176.

DISCUSSION

COFFEEN: I am pleased to see so much activity in the field of planetary polarimetry in the Soviet Union. The extensive measurements by Morozhenko and his colleagues are in rather good agreement with ours and those of the French.

Jaylee Mead, at Goddard Space Flight Center, has made a thorough comparison of the global data for Mars taken in the 1960s, and finds reasonable agreement among observers. For almost all combinations of the wavelength and phase angle, the data of Morozhenko and of Coffeen differ by less than 1% polarization (Mead 1973; personal communication). This level of agreement is similar to that for the French and U.S. data on Venus (Dollfus and Coffeen 1970).

For the integrated disk of Jupiter, the measured polarization is quite small ($<0.5\%$). The global polarization as given in Fig. 3 (above), is *negative* at all wavelengths and phase angles, approaching *zero* at $0°$ phase angle. Apparently these data have all been normalized to give zero polarization at $0°$ phase angle; in the words of the authors, these data are "corrected for the effect of inhomogeneity of the Jovian disk." The absolute amount of global polarization was measured in 1972 by Coffeen and Baker at 0.446 μm and 0.619μm, over the range $0°.1-11°.1$ phase angle. We found *positive* polarization, with a *maximum* of 0.3%–0.4% near $0°$ phase angle. Our data would agree with the Russian data if 0.35% were added to all of their points in Fig. 3. In other words, we agree on the relative variation of polarization with phase angle. The polarization does not have to be zero at $0°$ phase angle, for here the asymmetry of the cloudy regions can give a non-zero net polarization for the disk (see p. 594).

Only for Venus has a unique solution been made for the characteristics of the upper cloud particles (see p. 553). The results stated above for the Jupiter cloud particles are not really unique (see p. 569), although they demonstrate possible cloud species. In particular there is no direct evidence yet for the existence of spherical particles in the clouds of Jupiter.

DISCUSSION REFERENCE

Dollfus, A., and Coffeen, D. L. 1970. Polarization of Venus. I. Disk observations. *Astron. Astrophys.* 8: 251–266.

CIRCULAR POLARIZATION OF PLANETS

JAMES C. KEMP
Institute for Astronomy
University of Hawaii

The history of circular polarization studies on planets is reviewed, and the basic characteristics are outlined. Models are described that have been advanced to explain the polar scattering effect, in particular to account for the differences in sign (handedness of the polarization) for solid surfaces versus gas atmospheres. Finally, the role of aerosols is discussed.

This story begins with the discovery of a small circularly polarized component in the scattered light from Jupiter in early 1971 (Kemp et al. 1971a). Before that, there was no knowledge of any kind, either observational or theoretical, of circular or elliptical polarization in the light scattering from planets, with one exception. Lipskij and Pospergelis (1967) had found such an effect in the case of the moon. The possible contaminating effects in the instruments were not as well understood as they are now; and the measurements are extremely delicate, the main problem being that of false linear-circular conversion. Lipskij and Póspergelis seemed, in fact, to hint of some difficulties in reproducing their own data, particularly with regard to the phase dependence, and they did not follow the work up with a sequel. Our data for the moon (Kemp, Wolstencroft, and Swedlund 1971b) did indeed turn out to agree with theirs, at the phase reported. The sum total of our later results on five planets, the agreement on the data for the moon, and the many consistency checks that we carried out have erased any doubts that a real effect is being observed.

The magnitude of this effect varies among the planets, and of course it varies with the orbital phase angle, since the mechanisms involve the geometry of scattering, just as with the linear polarization of the planets. The largest circular polarization q detected so far is that on the south polar region of Saturn, in the red at critical phase angles near $2°$, where q approaches 0.1% (Swedlund, Kemp, and Wolstencroft 1972). More typically the q values on the planets fall in the range 10^{-5}–10^{-4}, i.e., 0.001% to 0.01%.

A broad question is whether this work is of use for the study of planets. In principle, knowledge of all four Stokes parameters — including the "circular" — is needed to characterize the light scattering and, in turn, the planetary surface. But the interactions producing the circular components are higher-order ones, usually double scattering, and they tend to be obscure. The circular polarization can give with ease only a qualitative check on structure models, for example, in distinguishing between solid versus gas surfaces (perhaps with aerosols). It may never be worth the time to work out the models to the point where circular-polarization data alone can be used to estimate particle sizes, the degree of surface roughness, or similar quantitative parameters. We have at best come up with only three or four surprising qualitative facts, e.g., the sign difference between the polar q values of Jupiter and Saturn, which indicates clearly a radical difference of some kind between the two atmospheres.

Our investigations were mostly impelled by pure curiosity, not so much about planetary surfaces but about light scattering. It was pleasing to find that: (a) the effects do exist, as expected from very fundamental symmetry considerations (see discussion below on models); and (b) models can be constructed that at least explain the signs. It is important also that upper limits have been set for the circular polarization due to two mechanisms other than polar scattering, viz. optical activity and magneto-reflection, of special interest for Mars and Jupiter respectively. Lastly, a spin-off from the work has been a perfection of the techniques of polarimetry, as told by Kemp, Wolstencroft, and Swedlund (1972).

OBSERVED FACTS

What is observed is the *polar scattering effect* (or briefly the *polar effect*). Light from the sun, scattered in the earth's direction from any restricted area on a planet, has in general a fractional circular polarization[1] q. The effect has two basic symmetries, portrayed in Fig. 1: at a given sun-planet-earth phase angle α, light from the north and south hemispheres have opposite senses of polarization; viewed at the opposite phase ($-\alpha$), both senses are reversed. At $\alpha = 0$ (opposition

---

[1] In our papers on the polar effect in planets we have used upper case Q, rather than q for the fractional circular polarization. This was an attempt to follow a usage established in laboratory magneto-optical work, particularly on metals, in which Q denotes a *reflective* polarization while q means the absorptive or emissive quantity. In astronomy the distinction is sometimes impossible, since we may not know which process is involved, or there is a mixture, say in light from a quasar or similar peculiar object of unknown structure. Henceforth q shall mean generally the observed polarization.

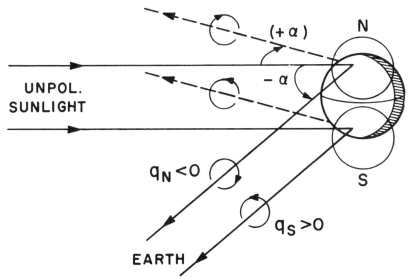

Fig. 1. The polar scattering effect on a typical planet. Light circles over the north and south polar regions or hemispheres indicate observing apertures defining the quantities q_N, q_S. Dashed scattered light paths indicate the situation at a later observation after opposition (for an outer planet).

or conjunction), $q = 0$ everywhere on the planet. Furthermore, at least approximately, $q_N(\alpha) = - q_S(\alpha)$, and $q_{N,S}(\alpha) = - q_{N,S}(-\alpha)$. By q_N and q_S here we refer to the crudest measure of the polar effect: a viewing aperture is placed over the planet image in the telescope, and we average over some representative large fraction of the visible northern hemisphere for q_N, then over a corresponding region for q_S.

All but a few of the observations to date have involved the simple averaging over N and S hemispheres, sometimes systematically emphasizing the polar regions (e.g., by using slightly smaller apertures than indicated in Fig. 1) where the effect is largest. This was mostly for the sake of economy in observing time; very small q values are measured, and in some cases the wavelength dependence was also desired, requiring long runs with a sequence of filters. Detailed study of q as a function of position on the disk would define the function q (ϕ,γ,ψ) in terms of a now-conventional set of angles (Pospergelis 1968; Bandermann, Kemp, and Wolstencroft 1972) (see Fig. 2).[2]

Without going into the symmetry question deeply, we can easily see

---

[2] Editorial Note: We have changed Kemp's ϕ into α (see p. 1084).

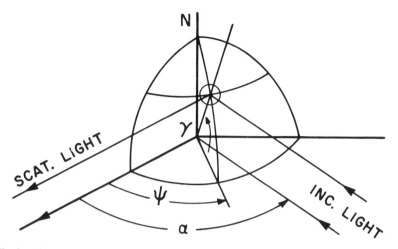

Fig. 2. An octant of a spherical planetary surface for defining a local surface element.

why the antisymmetries $q_N = -q_S$ and $q(\alpha) = -q(-\alpha)$ do not hold exactly. They would hold for perfectly homogeneous, spherical planetary surfaces (more generally surfaces with mirror symmetry through the scattering plane). Mars, as an extreme example, clearly has no such symmetry and its axis is tilted. Jupiter's axis is tilted and there is polar flattening. The effects of such asymmetries were immediately noticed in the observed polarization (Kemp, Wolstencroft, and Swedlund 1971b).

The moon, Mars, Venus, Mercury, Jupiter, and Saturn have been studied. The most striking early result was the discovery that the polar effect is opposite (i.e., all signs reversed) for solid as compared with atmospheric surfaces — although Saturn proved to be an exception. Jupiter and Venus showed, specifically, the signs of $q_{N,S}$ illustrated in Fig. 1; the moon and Mercury had the opposite sign. Mars was a mixed case, showing in Spring 1971 the solid signs in the mid-visible to near infrared, while in the blue-violet the northern hemisphere switched over to the atmospheric sign, the south polar region (dominated by its ice cap) retaining the solid-surface sign. Complex behavior occurred during the Martian dust storm of Fall 1971; occasionally the atmospheric sign was seen at certain wavelengths in the red. (The distinction between "atmospheric" and "solid-surface" signs here is essentially an observational one, although some progress has been made at associating these specific types of signs with models, as noted below.)

The phase dependence has been followed in detail only for Jupiter and Saturn. The wavelength variation has been systematically studied only for Mars and Saturn; the studies involved a series of interference filters of widths 0.01 to 0.02 μm, covering roughly the range 0.4–0.85 μm. In most of the other work, a far-red filter band was used. Brief checks on the moon and Jupiter with blue filters indicated the same respective signs as in the red, but smaller q values. Recent unpublished measurements of mine (November 1972) on Venus show $q_{N,S}(\lambda)$ *rising* toward the ultraviolet; the values at 0.38 μm were at least twice those at 0.74 μm, in all cases with the same "atmospheric" signs.

On Mars, we made an attempt to follow the circular polarization methodically through the opposition of August 1971, including the wavelength dependence. Within two weeks either side of the opposition, the q values were of course small, of the order 1×10^{-5}. A hope was that peaks in the spectral curve might then stand out, indicative of a component due to scattering from an optically active material; unfortunately we had problems with the apparatus at the time, and only modest upper limits could be set for such a component. Better records of $q_{N,S}(\lambda,\alpha)$ were obtained in September through November 1972, but the dust storm has made any interpretation very difficult, especially as regards the α dependence. Through all the data however, and only partially modified by the dust storm, we could see recurring spectral features in the $q(\lambda)$, notably in the range 0.6–0.75 μm. These may correlate with some extremely weak features in the absorption spectrum here (Wolstencroft, Kemp, and Swedlund 1972). Since some mechanisms for the polar effect require absorption, there was hope that the present technique would prove useful as a method for spectroscopy. We cannot claim to have demonstrated this for Mars, but further work may be worth the effort. (The advantage would be that interference from the solar spectrum and terrestrial absorption is automatically eliminated.)

The most methodical study of the polar effect has been on Saturn, over a six-month period through the opposition of November 1971. The key findings were: (a) over the range $0.4 < \lambda < 0.68$ μm and $-6° \leq \alpha \leq +6°$ (the observable range of α from earth), the polar q values on the planet disk or ball had the so-called "solid-surface" signs, i.e., opposite to those in Fig. 1 and opposite to Jupiter in the red; (b) at critical angles $\alpha \cong \pm 2.2°$ we saw sharp, enormous peaks in $q_S(\alpha)$, approaching $q = 0.1\%$ (− and + respectively on either side of opposition); and (c) the rings showed no circular polarization as large as 0.03%, except for a possible sharp rise in the ultraviolet ($\lambda \leq 0.39$ μm) which needs further investigation. The huge peaks in q_S are clearly some kind of opposition effect, such as penetration of light to a deep layer of aerosols (or even to a solid surface) at small incidence and scattering angles. These and other interesting aspects such as an asymmetry related to the spin-axis tilt are discussed by Swedlund, Kemp, and Wolstencroft (1972).

MODELS

Sazonov (1972) has elegantly treated the symmetry aspects of the polar scattering effect in terms of P,T (parity and time) invariance of the electromagnetic interaction. It is worthwhile to state a universal principle in physics that was in my mind during the first measurements on Jupiter: if symmetry alone does not forbid some effect, the effect exists (although sometimes at an undetectable level!).

The basic symmetries, really antisymmetries, $q(\alpha,\gamma,\psi) = -q(-\alpha,\gamma,-\psi)$ and $q(\alpha,\gamma,\psi) = -q(\alpha,-\gamma,\psi)$, obviously hold only if there is a certain degree of symmetry in the scattering body itself. If there is a north-south difference, i.e., for regions $+\gamma$ and $-\gamma$, the net q for the entire planet (what we call q_0) will not vanish. (This would be true for an odd-shaped asteroid.)

Even with an asymmetric body for which the above relations between the $q(\alpha,\gamma,\psi)$ do not hold, all the simple models that have been worked out have the feature $q(0,\gamma,\psi) = 0$, i.e., the polar effect vanishes identically at opposition. This has been established observationally with moderate accuracy in the cases of the moon, Jupiter, and Saturn. It is important to look into the generality of this feature.

All the models listed below are based on double scattering or on double specular reflection in the rough-surface models. At $\alpha = 0$, the incident, intermediate, and scattered ray vectors are coplanar — and as a consequence q vanishes identically for any possible double-scattering sequence, regardless of the nature of the individual single scatterings. (Materials that are optically active at the atomic level are of course excluded.) For *triple* scattering, however, this is not true, and one can construct hypothetical examples in which $q \neq 0$ at opposition. Consider a nonplanar sequence of four rays (incident, two intermediate, and exit rays) connected by three specular reflections, involving (a) a perfect dielectric, (b) a lossy dielectric or metal, and (c) a perfect mirror. The three surfaces are disposed at oblique angles but are oriented such that back scattering occurs, i.e., the incident and triply reflected rays are antiparallel ($\alpha = 0$). The perfect dielectric produces only linear polarization, which gets converted to elliptical by the lossy surface; the mirror serves only to bend the ray from the second surface back in the incident direction. A net q results from this sequence, and it is not cancelled out by the contribution from a "reversed" ray, since that undergoes the sequence mirror → lossy surface → dielectric, which is not equivalent to the first sequence. In any surface that is macroscopically homogeneous, it would appear that this, or other triple-scattering processes analogous to it, would average zero, i.e., would give zero net q at opposition — but that is not clear. In any case, there can exist a *large-*

scale geometry that permits such an effect. An example is Saturn, for which we must consider scattering of light between the rings and the planet. Either or both of those scatterings can be, separately, a double scattering, so that triple scattering is involved. (The rings could act as an efficient mirror.) It was concluded that any such contribution was in fact negligible; at any rate, the Saturn q values crossed through zero at or extremely close to opposition (Swedlund, Kemp, and Wolstencroft 1972).

Three explicit mechanisms for the polar effect have been discussed: (1) coherent, double Rayleigh scattering by absorbing atoms in a gas layer (Kemp and Wolstencroft 1971); (2) double Mie scattering by aersols (Hansen 1971); and (3) double reflections on an absorbing, rough solid surface (Bandermann, Kemp, and Wolstencroft 1972). The first was proposed to account for circular polarization from a pure gas atmosphere, without recourse to aerosols. Though it works in principle, it is evidently much too weak. The requirement is that a sufficiently sharp surface exist upon which the intermediate-ray directions of the double scattering can be correlated. The degree of sharpness needed to account for the observed $q_{N,S} \sim 10^{-4}$ (at $|\alpha| \sim 10°$) on Jupiter corresponds to a fractional, radial density gradient of at least 1 mm^{-1}, which seems far too sharp for a gas interface. Perhaps fortuitously though, this mechanism unambiguously and correctly gives the observed signs of $q_{N,S}$ on Jupiter and Venus.

Hansen suggests that the cloudiness or aerosol density on Jupiter (and presumably Venus) is enough to account for the measured values. The critical test of the absolute *signs* of $q_{N,S}$, however, has not been applied to the double Mie-scattering interpretation. Another test has to do with the role of absorption. The double Mie-scattering mechanism does not require absorption, although absorption can surely be added, modifying the resulting $q(\lambda)$ and introducing spectral features. While the $q(\lambda)$ have not been studied on Jupiter and Venus, it is interesting that the spectral behavior on Saturn (which admittedly may be a very different situation) shows some correlation with the known absorption — a general rise into the far red, with some sort of structure apparently related to the methane bands. The latter, of course, have nothing to do with aerosols. Perhaps some kind of mixed process, involving solid or liquid aerosols as well as refraction and absorption in the gas layers, may have to be called upon. Obviously, there are many unanswered questions in this field.

The situation with solid surfaces, while in a sense more complex than that in a gas atmosphere with areosols, seems not entirely mysterious after the model investigations by Bandermann, Wolstencroft, and myself. The key problem was to account for: (a) the absolute signs of the polarization at small phase angles, and (b) a sign change, not dictated by symmetry, that occurs at a characteristic phase angle $\alpha_c \sim 90°$. (The latter refers to hemispheric averages $q_{N,S}$, corresponding roughly to a mean $\psi \sim \alpha/2$; the sign change was found to appear at different critical phase angles for different ψ, when individual plane-surface elements are studied, as in Pospergelis' laboratory work. I have observed this sign change on the moon but without detailed study [see Bandermann, Kemp, and Wolstencroft 1972].) This second sign change is reminiscent of the well-known change in the moon's linear polarization (appearance of the "negative branch") at around $\alpha \sim 14°$, occurring however at a much larger angle. The approach we used was based on simple plane-wave or geometric optics, involving double specular reflection from lossy surfaces as described by Fresnel's laws. To simulate a rough surface we tried two surface-element distributions: one, that of cylindrical wells (termed the Puka model), and two, that of a planar assembly of spheres. The Puka model gave indeed the correct signs of $q_{N,S}$ at small α and predicted the sign change at an angle of the order $\alpha_c \sim 90°$; the model was inadequate for treating the detailed (α,γ,ψ) dependence and broke down at large phase angles — but that was expected for several reasons. At large phase, the "sphere-sphere" model seemed more successful. Whether more work is warranted on this very difficult problem remains to be seen. The machinery we developed, particularly in the sphere-sphere model, might also be of use for treating a planar (or semiplanar) layer of heavy aerosols, very large compared to a wavelength — a problem outside the realm of Mie theory.

REFERENCES

Bandermann, L. W.; Kemp, J. C.; and Wolstencroft, R. D. 1972. Circular polarization of light scattered from rough surfaces. *Mon. Not. R. Astr. Soc.* 158: 291–304.

Hansen, J. E. 1971. Circular polarization of sunlight reflected by clouds. *J. Atmos. Sci.* 28: 1515–1516.

Kemp, J. C.; Swedlund, J. B.; Murphy, R. E.; and Wolstencroft, R. D. 1971a. Circularly polarized visible light from Jupiter. *Nature* 231: 169–170.

Kemp, J. C., and Wolstencroft, R. D. 1971. Elliptical polarization by surface-layer scattering. *Nature* 231: 170–171.

Kemp, J. C.; Wolstencroft, R. D.; and Swedlund, J. B. 1971b. Circular polarization: Jupiter and other planets. *Nature* 232: 165–168.

———. 1972. Circular polarimetry of fifteen interesting objects. *Astrophys. J.* 177: 177–189.

Lipskij, Yu. N., and Pospergelis, M. M. 1967. Some results of measurements of the total Stokes vector for details of the lunar surface. *Astron. Zhur.* 44: 410–412. (Transl.: *Sov. Astron. A. J.* 11: 324–326.)

Pospergelis, M. M. 1968. Spectroscopic measurements of the four Stokes parameters for light scattered by natural objects. *Astron. Zhur.* 45: 1229–1234. (Transl.: *Sov. Astron. A. J.,* 12: 973–977 [1969].)

Sazonov, V. N. 1972. P,T invariance of electromagnetic interaction and circular polarization of the radiation of planets. *Astron. Zhur.* 49: 833–836 (in Russian).

Swedlund, J. B.; Kemp, J. C.; and Wolstencroft, R. D. 1972. Circular polarization of Saturn. *Astrophys. J.* 178: 257–265.

Wolstencroft, R. D.; Kemp, J. C.; and Swedlund, J. B. 1972. Circular polarization spectrum of Mars. *Bull. Am. Astron. Soc.* 4: 372.

DISCUSSION

GEHRELS: With the large diaphragms used (shown in Fig. 1) there is a large amount of light scattered by the earth's atmosphere. The diameters of your diaphragms must have been about 30 arcsecs; what were they exactly? Why did you not use smaller diaphragms, especially because of the excellent conditions you have at the Mauna Kea Observatory? Have you checked what effects such scattered light may have, for instance by putting the diaphragm partly off the edge of the moon?

KEMP: Naturally, we were always aware of the problem of background or night-sky light, and made the usual checks by integrating on the sky adjacent to the planet. This was not done for every single measurement, but was done sufficiently often. In our long series of observations on Saturn, for example, we found that circularly polarized flux from the adjacent night sky was normally negligible relative to that from the south polar region of the planet, even with a quarter moon in the sky. With respect to aperture size, for much of the work we used an aperture diameter of the order of the planet-image diameter; this was done in order to maximize the polarized flux (e.g., from a whole hemisphere), and also to provide a sort of averaging over an entire hemisphere (north or south). One reason why the hemispheric averaging was preferred, in this pioneering work, was to avoid getting into local variations over the planetary features. In some cases (see some of the measurements on Saturn in the paper by Swedlund et al. 1972) we used interference filters of widths 0.01–0.02 μm; and to extend the phase coverages, we often used the 61-cm rather than the 224-cm telescope at Mauna Kea. Photon collection was therefore sometimes a factor. Also, seeing is by no means always "excellent" at Mauna Kea. For overall consistency, therefore, we chose to use the large apertures indicated. Finally, regarding your worry about stray-light and sky contamination, note that the amount of such contaminating flux is proportional to the stop area; but so is the amount of circularly polarized flux, for apertures smaller than a hemisphere of the image. Thus, reducing the aperture diameter a factor of two or so might have reduced the contamination somewhat, but further reduction would be of little benefit in this connection.

COFFEEN: I would like to comment on the production of circular polarization in a thick planetary atmosphere. For pure Rayleigh scattering there is no way to get circular polarization, regardless of the optical depth. For neutral, homogeneous, isotropic spheres, the Mie theory gives the result: for unpolarized incident radiation there is still *no* circular polarization by independent single scattering, but it will be present for regions of a planetary disk in the case of multiple scattering, even if the particles are pure dielectrics. Typical for this multiple Mie scattering by clouds of particles is a circular polarization $|V|/I$ in the range 10^{-5}–10^{-3}. For a uniform planetary disk the net circular polarization will be zero, the circular polarization will be of opposite sign in the two hemispheres defined by the intensity equator, and the sign of each will reverse as the planet passes through $0°$ phase angle. I believe these conclusions agree with the observations for Venus, Jupiter, and Saturn. These comments are derived from a note by Hansen (1971).

Linear polarization appears to be the best tool for remotely determining particle microstructure (shape, size, refractive index) because it is sensitive to the single scattering. Circular polarization may be useful for disentangling scattering by molecules and aerosols and may provide a measure of the amount of multiple scattering in a cloud.

GEHRELS: The strong linear polarization at Jupiter's polar regions is present at all phases, including $\alpha = 0°$ (Gehrels 1969, and quoted references).

Regarding Saturn's upper atmosphere, a consistent picture seems to emerge (see p. 30). Saturn apparently differs greatly from Jupiter, with high aerosols, rather than molecules, causing most of the observed phenomena.

DISCUSSION REFERENCE

Gehrels, T. 1969. The transparency of the Jovian Polar Zones. *Icarus* 10: 410–411.

POLARIZATION IN ASTRONOMICAL SPECTRA: THEORETICAL EVIDENCE

A. L. FYMAT
Jet Propulsion Laboratory

Theoretical evidence for the existence and behavior of polarization in astronomical spectra is provided. The theory for the study of spectral multiple scattering of arbitrarily polarized light is first developed, and the detailed and integrated spectropolarimetry of a planetary atmosphere is then studied for cases in which the spectra are formed in the presence of either very small nonspherical particles (Rayleigh-Cabannes scattering) or large polydisperse spherical particles (Mie scattering). It is shown in both cases that polarization is indeed present; it increases with the line strength but decreases afterwards as the line becomes very strong and tends to saturation. A polarization reversal is also predicted during latitudinal (pole-to-equator) scan and possibly also during longitudinal (terminator-to-limb) scan of the planet. The reversal happened at all phase angles considered. Our companion article (Forbes and Fymat[1]) will provide observational substantiation to these theoretical predictions.

Astronomical absorption or emission spectra, as well as resonance-fluorescence spectra, are important sources of information regarding the composition, structure, thermal balance, and optical characteristics of the atmospheres in which they have been formed. In the case of planets, in particular, the spectra are observed in the diffuse reflection or transmission of solar radiation, or in the diffuse transmission of thermal radiation from the planet's atmosphere or from both its surface (if any) and atmosphere. They describe the interaction of the radiation field (whether of solar or thermal origin) with an atmosphere-surface system, and its variations with direction of propagation (scattering angle) and frequency of light. Spectral analysis and interpretation constitute, therefore, an indirect atmospheric sensor whose usefulness is only limited by the degree to which theory is able to represent physical reality and make valid predictions. Most spectral studies so far have disregarded the effects of scattering by both the gaseous and aerosol (dust, haze, clouds,

---
[1] See p. 637.

etc.) components of the atmosphere; they have also neglected the absorption or emission of radiation by the aerosols. This is the case, for example, of the CO_2 and H_2O infrared spectra of Earth and Mars for which, it has been claimed, all these effects are negligible (e.g., Gray 1966; and others). The basis for such a claim is not clear, however, since their systematic investigation, at least for aerosols, has not been carried out. In addition, the recent Mars data obtained with Mariner 9 (Barth et al. 1972; Hanel et al. 1972), both during and after the dust storm which spread over the planet, have conclusively shown, on the contrary, that dust has dramatic effects. On the other hand, it is now generally admitted that for Venus, Jupiter, and probably also for the other major planets, molecular and particulate scattering are highly relevant. A related consideration is the polarization induced by the scattering process in the radiation field even if, initially, the field was unpolarized. Polarization may modify its intensity in both spectral continuum and lines at least for a certain small-particle size range to be determined. In all situations, and even for those sizes for which such a modification may be extremely small, polarization is an additional source of information whose rejection cannot possibly be justified. Hence, spectral interpretation should not limit its considerations to the Stokes intensity parameter only, as is the case presently. The knowledge of the spectral distributions of the remaining three Stokes parameters (Fymat 1972; Fymat and Abhyankar 1970c,d; 1971; 1972) provides not only additional information but also three other constraints that will enable one to restrict the number of possible inverse solutions of the associated radiative transfer problem. Steps are therefore being taken toward the development of the appropriate radiative transfer theory for a realistic model atmosphere with complete consideration of the vector nature of the light field (Chandrasekhar 1950; van de Hulst 1957, 1971; Ueno 1960; Sekera 1963; Kano 1964; Herman and Browning 1965; Fymat 1967, 1970; Fymat and Abhyankar 1969a,b, 1970a,b; van de Hulst and Grossman 1968; Hovenier 1970; Hansen 1969a,b, 1971; Tanaka 1971; Takashima 1971; Fymat and Ueno 1973; Liou 1970; Hunt 1972).

Our aim in this paper is to present some results of computations that will illustrate the formation and behavior of spectral polarization. These computations form the origin of the experimental program to be described in a companion article (Forbes and Fymat[2]). Computations of this sort for the simple Rayleigh phase-matrix have been reported by Lenoble (1969). In the next section, the formalism required for the study

---

[2] See p. 637.

of the spectral multiple scattering of partially polarized light will be briefly set forth. In the remaining sections spectropolarimetry both detailed on and integrated over a planet's disk is studied for various laws of atmospheric scattering.

SPECTRAL MULTIPLE SCATTERING OF ARBITRARILY POLARIZED LIGHT

A detailed theoretical study on the subject of spectral multiple scattering of unpolarized light in homogeneous or inhomogeneous planetary atmospheres is available in a number of articles (see, e.g., Lenoble 1968; Belton 1968; Potter 1969; Hansen 1969a,b; Chamberlain 1970; Uesugi and Irvine 1969; Regas and Sagan 1970; Hunt 1972; and the references therein listed). As a preamble to the computations described later, this section provides a brief extension of the previous studies to the case of arbitrarily polarized light.

Observables

If $\mathbf{I}_v \equiv (I_l, I_r, U, V)_v = (I_i)_v$, $i = 1$ to 4, denotes the Stokes (field) vector of the diffuse radiation, the presently available observational data include spectral distributions of $I = I_l + I_r$ for a large variety of spectral resolutions, broadband measurements of $P = (I_r - I_l)/I = -Q/I$ at various wavelengths, and some similar isolated measurements of U. Spectra of Q, U, V, and of I_l and I_r, in particular, have never previously been recorded. The first spectra of these quantities, obtained for Venus in the 1–4 μm wavelength range (instrumental resolution of 0.5 cm^{-1}) will be reported on in our next paper.[3] Given that it is experimentally feasible to obtain such spectra, the spectral observations could therefore include some or all of the following quantities:

1. *Stokes intensity vector*, $\mathbf{I}_v$, if the planet's disk (or region of atmosphere) can be resolved spatially by the instrumentation used, or the *Stokes flux vector*, $\mathbf{F}_v \equiv (F_l, F_r, F_u, F_v)_v = (F_i)_v$, with

$$\mathbf{F}_v = \iint_{\substack{\text{illuminated}\\ \text{disk}}} \mathbf{I}_v \mu d\omega, \qquad (1)$$

otherwise. Here, μ is the direction cosine of the local zenith angle measured from the outward normal, and the integration is carried over the solid angle subtended by the unresolved disk area.

---

[3] See p. 637.

2. *Spectral variations of* I_ν *or* F_ν, i.e. $I_\nu(\nu)$ and $F_\nu(\nu)$, respectively. In the case of a spectral line, schematized in Fig. 1, these include:

a. *Line profiles* for all components of I_ν and F_ν when the spectral resolution achieved by the instrument is sufficient for resolving the lines. They are given by the expressions:

$$r_{l,i}(\theta, \Phi) = 1 - \left\{\frac{I_l(\theta, \Phi)}{I_c(\theta, \Phi)}\right\}_i, \quad R_{l,i} = 1 - \left\{\frac{F_l}{F_c}\right\}_i, \quad i = 1 \text{ to } 4, \qquad (2)$$

where the subscripts l and c apply to similar quantities taken on corresponding points in the line and in the continuum, and θ and Φ are the polar coordinates of the point of observation on the planet.[4]

If, however, the lines cannot be (spectrally) resolved, then, the quantities of interest are the:

b. *Equivalent widths* for either intensity or flux quantities:

$$w_i(\theta, \Phi) = \int_{\text{line}} r_{l,i}(\theta, \Phi)\, dl, \quad W_i = \int_{\text{line}} R_{l,i}\, dl, \, i = 1 \text{ to } 4; \qquad (3)$$

for any i, this is the width (expressed in frequency units) of the perfectly "black" line of rectangular profile (shown by the hatched area in Fig. 1) that would remove the same amount of energy from the spectrum of this Stokes component. In general, w (or W) is less seriously disturbed than r (or R) by the finite resolution of the spectrograph.

c. *Curves of growth* for intensities or fluxes; these are plots of w_i or W_i versus a quantity proportional[5] to Nf, where N is an "effective" number density of absorbing material, and f is the oscillator strength of the transition.

d. *Mean values per frequency interval* of these various quantities:

$$\langle X_{l,c}\rangle_i = \frac{1}{\Delta \nu}(X_{l,c})_i, \quad X \equiv I, F, w, W. \qquad (4)$$

3. *Directional variations of* I_ν *or* F_ν, i.e., $I_\nu(\mu)$ and $F_\nu(\mu)$, respectively. They include:

a. *Laws of darkening:*

$$[y(\tau;\mu)]_i, \quad y \equiv I_{l,c}, r_l, w; \quad 0 \leq \mu \leq 1; \qquad (5)$$

---

[4] Note the definition of θ for this paper. It is not the position angle mentioned on p. 1085.

[5] In a nonscattering atmosphere, N is the number density of absorbing gas. However, when continuum scattering is present along with absorption, this number is the so-called specific abundance (Belton et al. 1968; Chamberlain 1970) or abundance of absorbing gas in a column of unit cross-sectional area and unit optical thickness in the continuum. No such definition is available when line scattering is present such as in resonance lines.

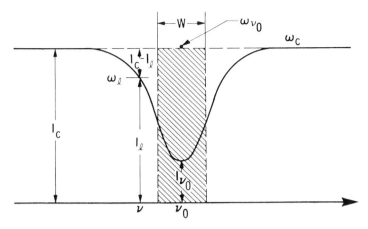

Fig. 1. Definitions of observational data for planetary polarization spectra.

b. *Center-to-limb variations:*

$$\left[\frac{y(\tau;\mu)}{y(\tau;0)}\right]_i, \quad 0 \leq \mu \leq 1, \tag{6}$$

where $\mu = 0$ corresponds to a direction of observation along the local zenith, and $\mu = 1$ along the horizon (grazing angle). In Equations (5) and (6), τ is the atmospheric thickness in either the line or the continuum. For a given direction of illumination of the atmosphere, the laws of darkening or the center-to-limb variations can also describe the *phase-angle variations* of intensities.

c. *Planetary albedo:*

$$(A_{l,c})_i = \frac{2}{(F_{l,c})_i} \int_0^1 \int_0^{2\pi} \int_0^1 [I_{l,c}(\mu, \Phi; \mu', \Phi')]_i \, \mu \, d\mu \, d\Phi' d\mu', \tag{7}$$

which is the ratio of the observed total Stokes flux component i to the corresponding incident total flux component. For the component sum $(i = 1) + (i = 2)$, and for the continuum, this is the well-known Bond-Russell albedo (see Harris 1961). Now, most of the measurements made in the case of a remote atmosphere refer to the whole illuminated disk; we can therefore define:

d. *Mean values* of flux quantities per unit area of illuminated disk by the relations

$$(\overline{Y})_i = \frac{1}{S}(Y)_i, \quad Y \equiv F_{l,c}, R_l, W, \tag{8}$$

where S is the total illuminated area.

4. *Heating (or cooling) rates*, or (differential) energy gain (or loss) per unit volume, unit frequency, and time intervals:

$$[h_{l,c}(z)]_i = -\frac{d}{dz}[F_{l,c}(z)]_i, \tag{9}$$

and their mean values:

$$\langle h_{l,c}\rangle_i = \frac{1}{\Delta v}(h_{l,c})_i, \quad \overline{(h_{l,c})}_i = \frac{1}{S}(h_{l,c})_i; \tag{10}$$

and the corresponding
5. *Temperature variations*

$$(dT_{l,c})_i = \frac{1}{\rho_a(z)C_p}[h_{l,c}(z)]_i, \tag{11}$$

and their mean values

$$\langle dT_{l,c}\rangle_i = \frac{1}{\Delta v}(dT_{l,c})_i, \quad \overline{(dT_{l,c})}_i = \frac{1}{S}(dT_{l,c})_i. \tag{12}$$

In Equation (11), $\rho_a(z)$ is the density of air at altitude z, and C_p is its specific heat capacity at constant pressure. It should be noted that the expression of the heating rate given in Equation (9) is the proper one to include in the hydrostatic system of equations used in dynamic circulation models of atmospheres.

Atmospheric Modeling

As already apparent, a basic quantity common to all previous observables is the intensity vector of the diffuse light field. Its theoretical evaluation is dependent on the model atmosphere selected. The underlying assumptions are the following:

1. *Geometry.* It is essentially as described by Horak (1950). If the light source (e.g., planet's disk) can be (spatially) resolved, then, at each k-point of the disk the atmosphere is taken to be plane-parallel stratified. Associating a local (xyz) coordinate system to this point [x: parallel to longitude circle, v_j, ($=\sin\zeta_j$, $\zeta=$ longitude) and directed, southwards; y: parallel to latitude circle, ψ_i, ($=\cos\eta_i$, $\eta=$ colatitude) and directed eastwards; z: along local normal], the only position variable of interest in the scattering computations is therefore z. Any light direction is referred to the local normal in terms of its spherical coordinates $\Omega \equiv (\mu = |\cos\theta|, \varphi)$, $0 \leq \mu \leq 1$, $0 \leq \varphi \leq 2\pi$. If $\Omega_0 \equiv (\mu_0, \phi_0)$ denotes the incident direction, the relations between the spherical and the disk coordinates are

$$\mu_0(k) = (1 - \psi_i^2)^{1/2}[(1 - v_j^2)^{1/2}\cos\alpha + v_j \sin\alpha]$$
$$\mu(k) = (1 - \psi_i^2)^{1/2}(1 - v_j^2)^{1/2} \quad \quad (13)$$
$$\cos\phi(k) = \frac{\mu(k)\mu_0(k) - \cos\alpha}{[1 - \mu^2(k)]^{1/2}[1 - \mu_0^2(k)]^{1/2}}, \quad \phi_0 = 0.$$

On the other hand, if the disk is unresolved, a disk integration becomes necessary and can be performed through usual cubature formulae for a grid of k-points on the disk.

2. *Optical Characteristics.* The optical properties of the atmosphere are completely defined by two functions: the single scattering albedo, ω, and the phase matrix of scattering, **P**. When the optical thickness of the atmosphere and these two functions are known, the emerging diffuse radiation field can be evaluated in all its detailed characteristics. Let σ, k, and $K = \sigma + k$ denote respectively the volume scattering (sink or source of radiation), absorption (or emission), and extinction (or corresponding emission quantity) coefficients; these quantities must be defined for both gases and particles active in the wavelength, ν, of interest and for both continuum and line frequencies (subscripts c and l, respectively). The optical thickness of the atmosphere above any level $z \geq 0$ is

$$\tau_\nu = \int_z^\infty [K_c(z) + K_l(z)]dz, \quad \quad (14)$$

and the albedos

$$\omega_\nu(\tau_\nu) = \left[1 + \frac{k_c + k_l}{\sigma_c + \sigma_l}\right]^{-1}, \quad k \equiv k(\tau_\nu), \quad \sigma \equiv \sigma(\tau_\nu). \quad (15)$$

The phase matrix is a weighted linear combination of the Mie matrix,[6] $\mathbf{P}_M$, for particles and the Rayleigh-Cabannes matrix, $\mathbf{P}_{RC}$, for nonspherical gas molecules or very small particles:

$$\mathbf{P}_\nu(\tau_\nu) = \frac{\sigma_{RC}\mathbf{P}_{RC} + \sigma_M \mathbf{P}_M}{\sigma_{RC} + \sigma_M}, \quad \quad (16)$$

where

$$\mathbf{P}_{RC} = q_1 \mathbf{P}_R + (1 - q_1)\mathbf{P}_I, \quad \quad (17)$$

$\mathbf{P}_R$ and $\mathbf{P}_I$ are respectively the Rayleigh phase matrix for spherical molecules and the phase matrix for isotropic scattering, $q_1 = 2(1 - \rho)/(2 + \rho)$,

---

[6]Strictly, one should use a modified form of the Mie matrix to include large nonspherical particles.

and ρ is the depolarization factor for incident natural light. In the case of resonance-fluorescence line spectra (H-Lyα; Na $-$ D$_1$, D$_2$; O$_2$ $-$ red doublet, He, etc.), q_1 is Hamilton's coefficient E_1 (Hamilton 1947) which depends on the total angular momentum quantum number, j, of the initial state and on the selection rule, Δj ($= \pm 1$ or 0) followed by the resonant transition in the Principal Series. For example, $q_1 = \frac{1}{2}$ for H-Lyα, $q_1 = 0$ for isotropic scattering, $q_1 = 1.0$ for Rayleigh scattering by spherical molecules, $q_1 = 0.4$ for rodlike scatterers, and $0.4 < q_1 < 1.0$ for all intermediary shapes between a rod and a sphere (Fymat 1967). Over the frequency range spanned by a line, $\mathbf{P}_\nu$ is usually assumed to remain invariant while the corresponding variations of ω_ν are of paramount importance in determining the line shape and its strength.

3. *Spectral line shapes.* Several (centrally symmetric) shapes are usually adopted for the line profile according to the atmospheric pressure at which the line is formed.

Doppler (low pressures)

$$k_l^D = \frac{1}{\sqrt{\pi}} k_{\nu_0}^D e^{-x_D^2} \tag{18}$$

Lorentz (high pressures)

$$k_l^L = k_{\nu_0}^L (1 + x_L^2)^{-1} \tag{19}$$

Voigt (intermediate pressures)

$$k_l^V = \frac{1}{S} \int_{-\infty}^{\infty} k_{\nu'}^D k_{\nu-\nu'}^L \, d\nu' = k_{\nu_0}^V V(x_D, x_L) \tag{20}$$

where $k_{\nu_0} = S/\pi\gamma$, $x = (\nu - \nu_0)/\gamma$, γ (half-width) $\equiv \gamma_L$, γ_D, S is the line intensity, and ν_0 refers to the line center.

Galatry (very high pressures, line narrowing)

$$k_l^G = \int_{-\infty}^{\infty} \cos[2\pi(\nu - \nu_0)t] \, e^{-Bt - C(\alpha t/c - 1 + e^{-\alpha t})} dt, \tag{21}$$

$$= k_{\nu_0}^G G(B, C, \alpha)$$

where $B =$ self-broadening coefficient, $C = 4\pi^2 mD^2/kT\lambda^2$, $D =$ self-diffusion coefficient, $k =$ Boltzmann constant, $T =$ temperature, $m =$ molecular mass, $\lambda =$ wavelength, and $\alpha = kT/mD$. It can be shown that Equation (21) reduces to Equations (18) and (20) at lower pressures.

4. *Characterization of lines.* When a model line shape has been adopted, two parameters are usually sufficient for entirely characterizing the line; these are the single scattering albedos at the line center, ω_{ν_0}, and in the continuum, ω_c. Two situations must then be considered: whether scatter-

ing is present in the line itself (resonance-fluorescence spectrum) or merely in the continuum (absorption or emission spectrum).

A. *Absorption or emission spectra* ($\sigma_l = 0$, $\sigma_c \equiv \sigma$).
   a. *absence of a continuum absorption or emission* ($k_c = 0$). The line center albedo which sets the magnitude (strength) of the line is then

$$\omega'_{v_0} = \left(1 + \frac{k_{v_0}}{\sigma}\right)^{-1}, \tag{22}$$

the continuum albedo $\omega'_c = 1$, and the albedo at any frequency in the line is written:

$$\omega'_l = \left[1 + \frac{1 - \omega'_{v_0}}{\omega'_{v_0}}\left(\frac{k_l}{k_{v_0}}\right)\right]^{-1}, \tag{23}$$

where k_l/k_{v_0} is provided by any one of Equations (18) to (21). In the present model, line absorption is only due to the gas, and the continuum results entirely from gaseous or particulate scattering or both.

   b. *presence of a continuum absorption or emission*, which may be due to gaseous or particulate absorption or both. The albedos in this case are:

$$\omega_{v_0} = \left(\frac{1}{\omega'_{v_0}} + \frac{k_c}{\sigma}\right)^{-1}, \tag{24}$$

$$\omega_c = \left(1 + \frac{k_c}{\sigma}\right)^{-1}, \tag{25}$$

and

$$\omega_l = \left(\frac{1}{\omega'_l} + \frac{1}{\omega_c} - 1\right)^{-1}. \tag{26}$$

B. *Resonance-fluorescence spectra.* The line scattering takes place according to the Rayleigh-Cabannes phase matrix, Equation (17). The albedos in this case can be written as previously with due consideration, however, of σ_l.

Computational Procedure

The diffusely reflected and transmitted light can be expressed with the help of the atmosphere's reflection and transmission matrices $\mathbf{S} \equiv (S_{ij})$, $\mathbf{T} \equiv (T_{ij})$, $i, j = 1$ to 4, respectively (Preisendorfer's linear interaction principle). These are:

$$\mathbf{I}(\tau_0; \Omega) = \frac{1}{4\pi\mu} \mathbf{S}(\tau_0, \tau_1; \Omega, \Omega_0) \mathbf{I}^{(0)} \tag{27}$$

and

$$\mathbf{I}(\tau_1; \Omega) = \frac{1}{4\pi\mu} \mathbf{T}(\tau_0, \tau_1; \Omega, \Omega_0) \mathbf{I}^{(0)}, \tag{28}$$

where τ_0 and τ_1 are the optical thicknesses of the top and bottom of the atmosphere, $\mathbf{I}^{(0)}$ is the incident light vector, and $\mathbf{S}$ and $\mathbf{T}$ obey the non-linear integro-differential equations (Sekera 1963):

$$\left(\frac{1}{\mu} + \frac{1}{\mu_0} - \frac{\partial}{\partial \tau}\right) \mathbf{S}(\tau, \tau_1; \Omega, \Omega_0) = \omega(\tau)\left[\mathbf{P}(\tau; \Omega, -\Omega_0)\right.$$
$$+ \frac{1}{4\pi} \int_{2\pi} \mathbf{P}(\tau; \Omega, \Omega'') \mathbf{S}(\tau, \tau_1; \Omega'', \Omega_0) \frac{d\Omega''}{\mu''}$$
$$+ \frac{1}{4\pi} \int_{2\pi} \mathbf{S}(\tau, \tau_1; \Omega, \Omega') \mathbf{P}(\tau; -\Omega', \Omega_0) \frac{d\Omega'}{\mu'}$$
$$+ \left(\frac{1}{4\pi}\right)^2 \int_{2\pi} \int_{2\pi} \mathbf{S}(\tau, \tau_1; \Omega, \Omega') \mathbf{P}(\tau; -\Omega', \Omega'')$$
$$\left. \mathbf{S}(\tau, \tau_1; \Omega'', \Omega_0) \frac{d\Omega'}{\mu'} \frac{d\Omega''}{\mu''}\right], \quad (29)$$

$$\left(\frac{1}{\mu_0} - \frac{\partial}{\partial \tau}\right) \mathbf{T}(\tau, \tau_1; \Omega, \Omega_0) = \omega(\tau)\left[\mathbf{P}(\tau; -\Omega, -\Omega_0) e^{-\tau/\mu}\right.$$
$$+ \frac{1}{4\pi} e^{-\tau/\mu} \int_{2\pi} \mathbf{P}(\tau; -\Omega, \Omega'') \mathbf{S}(\tau, \tau_1; \Omega'', \Omega_0) \frac{d\Omega''}{\mu''}$$
$$+ \frac{1}{4\pi} \int_{2\pi} \mathbf{T}(\tau, \tau_1; \Omega, \Omega') \mathbf{P}(\tau; -\Omega', \Omega_0) \frac{d\Omega'}{\mu'} + \left(\frac{1}{4\pi}\right)^2 \int_{2\pi} \int_{2\pi}$$
$$\left. \mathbf{T}(\tau, \tau_1; \Omega, \Omega') \mathbf{P}(\tau; -\Omega', \Omega'') \mathbf{S}(\tau, \tau_1; \Omega'', \Omega_0) \frac{d\Omega'}{\mu'} \frac{d\Omega''}{\mu''}\right], \quad (30)$$

where $-\Omega \equiv (-\mu, \phi)$, and $\mathbf{P}$ is given by Equations (16) and (17). The Stokes parameters of the reflected and transmitted light are then:

$$I_i(\tau_0) = \frac{1}{4\pi\mu} \sum_j S_{ij} I_j^{(0)} \qquad (31)$$

$$I_i(\tau_1) = \frac{1}{4\pi\mu} \sum_j T_{ij} I_j^{(0)}, \, i, j = 1 \text{ to } 4. \qquad (32)$$

With these last expressions, one can compute the various quantities listed earlier in this section. We have performed these computations for considerably more cases than it is practical to illustrate here. For convenience, therefore, we shall describe only the results pertaining to the diffuse reflection of the two orthogonal polarizations in the case of Lorentz absorption lines formed in the presence of scattering by Rayleigh-Cabannes and polydisperse Mie particles.

DETAILED SPECTROPOLARIMETRY

A detailed parametric study for several atmospheric and spectral models has been carried out. We shall now present some sample results for polarization spectra formed in the presence of very small and fairly large scatterers.

In the Presence of Very Small Nonspherical Particles

The relevant equations here are Equations (29) and (31), in which ω_{v_0}, ω_c, and ω_l are given by Equations (24), (25), and (26) in the presence of a continuum of absorption, and $\mathbf{P} \equiv \mathbf{P}_{RC}$ of Equation (17). The value of q_1 (anisotropy parameter) was here evaluated to correspond to an average value for the atmosphere of Venus, $q_1 = 0.8996$, and τ_1 was taken to be infinite. A closed analytical solution for this case has been obtained. It involves the evaluation of H- and $\mathbf{N}$-functions. These functions have been recently tabulated by the author (Fymat 1973) for all albedo and anisotropy parameter values. (For $q_1 = 0$ and 1.0, the corresponding tables have been given by Abhyankar and Fymat 1971.) For known phase angle and planetary coordinates of any k-point on Venus' disk, the corresponding spherical coordinates were obtained from Equation (13), and the light field vector was evaluated for these latter coordinates. A total of 28 points for a hemisphere was considered (the 18 usual grid points, 3 points on the central meridian, 6 on the equator, and the disk's center). For more details about this cubature scheme, the reader is referred to Horak (1950) (see also Fymat and Kalaba 1973).

The following quantities: I_l, I_r, $I = I_l + I_r$, $Q = I_l - I_r$, $-Q/I$, U, U/I and, when appropriate, V and V/I, were evaluated for both continuum and line frequencies. The corresponding line profiles were also obtained. In order to show that the observational results to be subsequently presented have been theoretically predicted, we shall only discuss the computations that refer to the line profiles $I_{l,v}/I_{l,c}$ and $I_{r,v}/I_{r,c}$ (l- and r-profiles). We shall further present only a few cases among the vast amount of computational data that we now have available. The results applicable to a Rayleigh-Cabannes atmosphere were obtained for four Lorentz absorption lines of different strengths: $\omega_{v_0} = 0.99$ (very weak line), 0.95 (weak line), 0.60 (strong line), and 0.20 (very strong line), all having a common continuum albedo $\omega_c = 0.975$. For incident solar illumination along the direction ($\mu_0 = 0.5$, $\phi_0 = 0$), these lines were to be observed in the

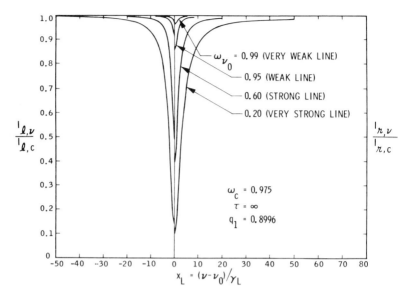

Fig. 2. Detailed spectropolarimetry of a semiinfinite homogeneous Rayleigh-Cabannes atmosphere for absorption lines of various strengths observed at phase angle $\alpha = 99°.2$. The coordinates of the observed point on the disk are $\eta = 25°.71$, $\zeta = 76°.32$ (in polar region close to illuminated limb). Note the dissymmetry between the profiles on the left and on the right of the diagram.

planetary diffuse radiation field at phase angle $\alpha = 99°.2$. Figure 2 gives the l- and r-profiles at a point on the disk in the polar region close to the illuminated limb ($\eta = 25°.71$, $\zeta = 76°.32$). For convenience, only half the profiles are illustrated; for better contrast, the l-profiles were drawn on the left half and the r-profiles on the right half of the diagram. If the diffuse light were spectrally unpolarized, the l- and r-profiles should be found identical (within computational errors). Departures from this situation are indicative of the existence of spectral polarization, larger differences between complementary r- and l-profiles corresponding to higher (algebraic) values of the degree of linear polarization. Under the conditions of Fig. 2, it is seen that *polarization is present in all the lines* studied (dissymmetry of corresponding profile halves), it is always *positive* (the equivalent width of any r-profile is always larger than that of the associated l-profile), and *it varies with the strength of the line* (variation of differences between the areas encompassed by the profiles; for a quick examination of this phenomenon, see the differences at the line center). It may be noted that the polarization first increases with the strength of the line and then decreases as the line saturates and becomes almost black at its center. This polarization increase is physically understandable, for as absorption in a line increases, scattering decreases—that is, we are concerned with the lower scattering orders which induce relatively higher polarizations. As the line becomes weaker, the scattering contribution increases; the corresponding added orders of scattering will dilute the polarization which will therefore decrease. The same situation applies when we move away from the line center and toward the continuum. The polarization decrease for very high line strengths can also be explained. At these strengths, absorption is overwhelmingly the dominant process; scattering becomes comparatively negligible and, hence, its polarization induction efficiency greatly reduced. To observe this phenomenon experimentally, it is important that the instrumental noise be inferior to the difference $|(I_{r,v}/I_{r,c}) - (I_{l,v}/I_{l,c})|_{x=0}$. Thus, strong unsaturated lines appear to be the best candidates. We must note, however, from a similar study conducted for various other points on the disk, that the polarization behavior just described is not universally valid across the illuminated disk and may in fact disappear entirely. Tendencies to a *reversal in polarization* at line strengths intermediary between $\omega_{v_0} = 0.95$ and 0.50 were also observed. This highly interesting and striking phenomenon will appear more dramatically in Fig. 4. The *longitudinal variations* of spectral polarization from terminator to limb are illustrated

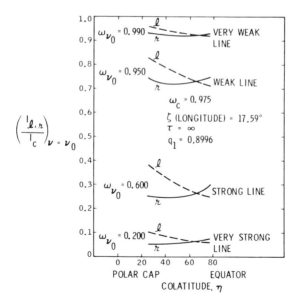

Fig. 3. Latitudinal variations of spectral polarization along the longitude circle $\zeta = 17°\!.59$ ($\alpha = 99°\!.2$).

in Fig. 5 along a latitudinal circle of colatitude $\eta = 77°.14$ for the same lines as before. For convenience in this graph, we have used only the line center values for both l and r components. It is seen that in all circumstances line polarization is present (r- and l-curves are not coincident), it varies with the line strength (the areas enclosed between corresponding l- and r-curves are different), and it is always positive (the r-curves are higher than the l-curves). The variation with line strength is not uniform; it increases first up to a certain maximum value and then decreases afterwards to become vanishingly small when the line becomes black at its center. There is an indication of polarization reversal for the weak line ($\omega_{v_0} = 0.95$) close to the limb. In Fig. 3, we have represented the *latitudinal variations* from polar to equatorial regions along the longitude circle $\zeta = 17°.59$. It should first be noted that, in opposition to the longitudinal variation case, the polarization is now mostly negative and slightly inferior in absolute value. Thus, a polarization reversal has taken place between the longitudinal and the latitudinal variations. A further reversal is happening in the latitudinal variation at $\eta \approx 70°$.

In the Presence of Large Polydisperse Spherical Particles

Absorption spectra formed in clear (Rayleigh) or cloudy (Mie) atmospheres composed of monodisperse H_2O particles of size $2\pi r/\lambda = 5$ were first presented by Fymat and Lenoble (1972). Four lines observed at the center of a planetary disk at phase angle $60°$ were considered. These lines had different strengths with albedo values: $\omega_c = 1.0$ (no continuum

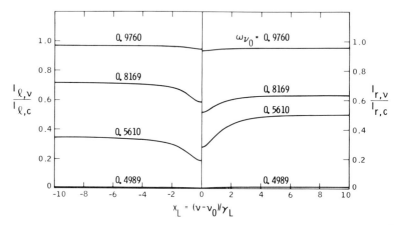

Fig. 4. Detailed spectropolarimetry of a polydisperse Mie atmosphere for absorption lines of various strengths observed at a planet's disk center ($\alpha = 30°$). Note the dissymmetry between the profiles on the left and on the right of the diagram and the polarization reversal at $0.5610 < \omega_v < 0.8169$.

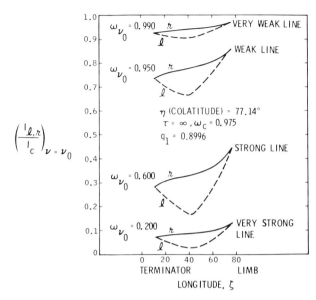

Fig. 5. Longitudinal variations of spectral polarization along the colatitude circle $\eta = 77°.14$ ($\alpha = 99°.2$).

absorption), 0.99, 0.983 and 0.975, and $\omega_{v_0} = 0.95$, 0.9406, 0.9347 and 0.9275, respectively. It was then found that, on transiting from a clear to a cloudy atmosphere, the Mie spectral lines were in every case stronger than the corresponding Rayleigh lines. This result was explained on the basis of a comparison between the Rayleigh and Mie scattering diagrams for the situation studied. The comparison showed the Rayleigh scattering to dominate the Mie scattering at the large scattering angle θ ($=\pi - 60°$) $= 120°$. Since prior to saturation the polarization increases with the strength of the line, we expected the Mie polarization to be stronger than in the previously studied Rayleigh-Cabannes case. We have, therefore, repeated the same procedure followed earlier, the phase matrix being now the one given by Equation (16). However, because $\sigma_{RC} \ll \sigma_M$ in our domain of interest ($\sigma_R = 0.0252$ and 0.0040 km^{-1} at $\lambda = 0.45$ and 0.70 μm, respectively, Deirmendjian 1969), we have limited the computations to the Mie matrix. A polydispersion of particles having the following characteristics was used: size parameter values ranging from $x_{min} = 0.1$ to $x_{max} = 57$ ($x = 2\pi r/\lambda$, r = radius, $\lambda = 0.55$ μm) and following Deirmendjian's cloud model C1 (distribution parameters $\alpha = 9.5$, $\gamma = 1.0$, $b = 12.5$), refractive index $m = m_r - im_i$ with $m_r = 1.44$ [values suggested by Hansen and Arking (1971) for the visible upper layers of Venus' clouds]. The continuum absorption corresponded to $m_i = 1 \times 10^{-4}$, and the line center absorptions were taken to correspond to $m_i = 1 \times 10^{-3}, 1 \times 10^{-2}$,

1×10^{-1} and 1.0, i.e., $\omega_c = 0.9975$, and $\omega_{v_0} = 0.9760$ (very weak line), 0.8169 (weak line), 0.5610 (strong line), and 0.4989 (very strong line), respectively. The corresponding efficiency factors, K, for extinction, scattering, and absorption are represented in Fig. 6 as functions of m_i. The lower part of Fig. 6 shows the associated albedo profiles. The continuum values are reached more rapidly in the case of weaker lines; for simplicity, in this example, we have truncated all the profiles at 10 Lorentz half-widths.

The profiles for the two orthogonal polarizations, $(I_v/I_c)_{l,r}$, corresponding to these four lines are drawn in Fig. 4 for phase angle $\alpha = 30°$. In this case also it is seen that the polarization is present and that, prior to saturation, it increases with the line strength. There is a polarization reversal between the two lines characterized by $\omega_{v_0} = 0.8169$ and 0.5610, the polarization values going from positive to negative. The same computations were repeated at various phase angles, and the reversal phenomenon appeared even more dramatically as α was increased. For the very weak ($\omega_{v_0} = 0.9760$) and the very strong ($\omega_{v_0} = 0.4989$) lines, the variation was weak. It was, however, very marked for the strong lines ($\omega_{v_0} = 0.8159, 0.5610$) and even more so for the stronger line. The polarization reversal mentioned earlier held at all the phase angles considered.

INTEGRATED SPECTROPOLARIMETRY

We have also studied the spectral polarization in the light diffusely reflected by the entire illuminated planetary disk (Fig. 7). The corresponding computations will be reported here only for the Rayleigh-Cabannes phase matrix of scattering. The disk integration was performed following the scheme of Horak (1950), all conditions being identical to those considered in the previous section. Most previously described phenomena are seen to be still present: existence of spectral polarization, its increase with line strength, and its subsequent decrease as the line becomes very strong and tends to saturation. In the present example, polarization was always positive; however, polarization reversal did not take place. Similar computations at different phase angles would almost certainly reveal a reversal.

SUMMARY AND CONCLUSIONS

We have provided the analytical theory for the study of the spectral multiple scattering of arbitrarily polarized light. The theory includes all polarization intensity parameters for absorption, emission, and resonance-fluorescence spectra. Although analytical expressions were used for the

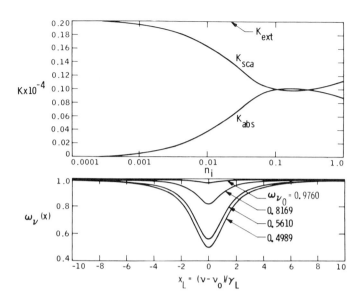

Fig. 6. *Above:* efficiency factors for absorption, scattering, and extinction, and *below:* single scattering albedos for four lines of different strengths formed in the presence of polydisperse Mie particles.

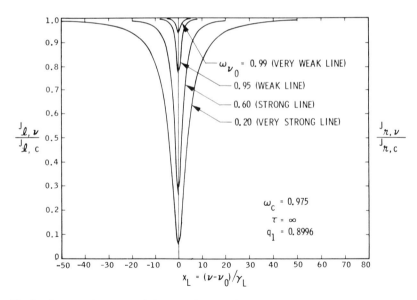

Fig. 7. Integrated spectropolarimetry of a Rayleigh-Cabannes atmosphere for the same lines as in Fig. 2.

spectral line shapes (Doppler, Lorentz, Voigt, Galatry), the treatment given is not necessarily restricted to these cases. An extensive study of the spectropolarimetry of an astronomical object, such as, for example, a planetary disk, was reported on with particular consideration to absorption lines. This was studied under the assumption that the spectra are formed in the presence of either very small nonspherical particles described by the Rayleigh-Cabannes phase matrix of scattering or large polydisperse spherical particles described by the Mie matrix. Lines of increasing strength were investigated and attention was focused mainly on the line profiles corresponding to the I_l and I_r components of radiation. It was thus determined that (a) spectral polarization is indeed present; (b) it varies in the same sense as the strength of the line, but as the line becomes very strong and tends to saturation, it decreases; and (c) a polarization reversal may take place at certain line strengths. The reversal may also happen during longitudinal or latitudinal variations or both. These effects were more marked in the case of the larger particles. These variations of spectral polarization between the Rayleigh and the Mie scattering cases are indicative of the change that must take place when lines are formed in clear or cloudy atmospheres or as we transit from a clear to a cloudy atmospheric region.

The integrated spectropolarimetry of the planetary disk was also studied. The same general conclusions have remained valid. It is clear however, that in the phase of research aimed at interpreting spectral polarization in terms of atmospheric physical parameters, the detailed spectropolarimetry should prove to be considerably more valuable than the integrated one whenever the object studied can be resolved telescopically.

Experimentally, differences in the line profiles for the l- and r-polarization components and their reversals from positive to negative values are of considerable value. It is thus of the utmost importance that the instrumental noise be kept at low values so as not to become larger than these differences. Strong lines must particularly be searched for; these lines, however, must not be saturated. Our following paper will provide observational substantiation to the theoretical conclusions here set forth.

Acknowledgments. This paper has received support from NASA Contract NAS7-100, JPL'S Director Discretionary Fund 730-00034-0-825.

REFERENCES

Abhyankar, K. D. and Fymat, A. L. 1971. Tables of auxiliary functions for nonconservative Rayleigh phase-matrix in semi-infinite atmospheres. *Astrophys. J.* 165: 673–674; *ibid.* Supplement 195: 35–101.

Barth, C. A.; Hord, C. W.; Stewart, A. I.; and Lane, A. L. 1972. Mariner Mars 1971 Project final report. Ultraviolet spectrometer experiment. *JPL Technical Report* 32–1550. 2: 39–46.

Belton, M. J. S. 1968. Theory of the curve of growth and phase effects in a cloudy atmosphere: Applications to Venus. *J. Atmos. Sci.* 25: 596–609.

Belton, M. J. S.; Hunten, D. M.; and Goody, R. M. 1968. Quantitative spectroscopy of Venus in the region 8,000–11,000 Å. *The atmospheres of Venus and Mars.* (J. C. Brandt and M. B. McElroy, eds.) pp. 69–97. New York: Gordon and Breach.

Chamberlain, J. W. 1970. Behavior of absorption lines in a hazy planetary atmosphere. *Astrophys. J.* 159: 137–158.

Chandrasekhar, S. 1950. *Radiative transfer*, Oxford: Clarendon Press.

Deirmendjian, D. 1969. *Electromagnetic scattering on spherical polydispersions*, New York: American Elsevier.

Fymat, A. L. 1967. Theory of radiative transfer in atmospheres exhibiting polarized resonance-fluorescence line scattering. UCLA Ph.D. Thesis.

———. 1970. Spectral multiple scattering of arbitrarily polarized light in inhomogeneous planetary atmospheres. JPL Technical Memorandum.

———. 1972. Interferometric spectropolarimetry: Alternate experimental methods. *Applied Optics* 11: 2255–2264.

———. 1973. Molecular anisotropic scattering of light. In draft.

Fymat, A. L., and Abhyankar, K. D. 1969a. Theory of radiative transfer in inhomogeneous atmospheres: I. *Astrophys. J.* 158: 315–324.

———. 1969b. Theory of radiative transfer in inhomogeneous atmospheres: II. *Astrophys. J.* 158: 325–335.

———. 1970a. Theory of radiative transfer in inhomogeneous atmospheres: III. *Astrophys. J.* 159: 1007–1018.

———. 1970b. Theory of radiative transfer in inhomogeneous atmospheres: IV. *Astrophys. J.* 159: 1019–1028.

———. 1970c. An interferometric approach to the measurement of optical polarization. *Applied Optics*: 9: 1075–1081.

———. 1970d. Interferometer for measurement of optical polarization. NASA Tech. Brief 70–10405.

———. 1971. A new field in Fourier spectroscopy: Interferometric polarimetry. *Proc. 1970 Aspen International Conference on Fourier Spectroscopy.* (G. A. Vanasse, A. T. Stair, and D. J. Baker, eds.) AFCRL 71-0019, Special Report, 114: 377–384.

———. 1972. Interferometer polarimeter, U.S. patent 3,700,334.

Fymat, A. L., and Kalaba, R. E. 1973. On Horak's and Sobolev's interpretations of the Venusian visual phase curve. *JPL Technical Report.*

Fymat, A. L., and Lenoble, J. 1972. Absorption profile of a planetary atmosphere: A proposal for a scattering independent determination. *Applied Optics* 11: 2249–2254.

Fymat, A. L., and Ueno, S. 1973. Order-of-scattering of partially polarized radiation in inhomogeneous, anisotropically scattering atmospheres: I. Fundamentals. *Astrophys. J.* Submitted for publication.

Gray, L. D. 1966. Transmission of the atmosphere of Mars in the region of $2\,\mu$. *Icarus* 5: 390–398.

Hamilton, D. R. 1947. The resonance radiation induced by elliptically polarized light. *Astrophys. J.* 106: 457–465.

Hanel, R. A.; Conrath, B. J.; Hovis, W. A.; Kunde, V. G.; Lowman, P. D.; Pearl, J. C.; Prabhakara, C.; Schlachman, B.; and Levin, G. V. 1972. Mariner Mars 1971 Project final report. Infrared spectroscopy experiment. *JPL Tech. Rept.* 32–1550. 2: 27–33.

Hansen, J. 1969a. Radiative transfer by doubling very thin layers. *Astrophys. J.* 155: 565–573.

Hansen, J. 1969b. Absorption line formation in a scattering planetary atmosphere: A test of van de Hulst's similarity relations. *Astrophys. J.* 158: 337–349.

———. 1971. Multiple scattering of polarized light in planetary atmospheres; I, II. *J. Atmos. Sci.* 28: 120–125; 28: 1400–1426.

Hansen, J., and Arking, A. 1971. Clouds of Venus: Evidence for their nature. *Science* 171: 669–672.

Harris, D. L. 1961. Photometry and colorimetry of planets and satellites. *Planets and satellites.* (G. P. Kuiper and B. M. Middlehurst, eds.) Chicago: Univ. of Chicago Press. pp. 272–342.

Herman, B. M., and Browning, S. R. 1965. A numerical solution to the equation of radiative transfer. *J. Atmos. Sci.* 22: 559–566.

Horak, H. G. 1950. Diffuse reflection by planetary atmospheres. *Astrophys. J.* 112: 445–463.

Hovenier, J. W. 1970. Polarized light in planetary atmospheres. Ph.D. thesis Univ. of Leiden, the Netherlands.

Hulst, H. C. van de 1957. *Light scattering by small particles.* New York: Wiley.

———. 1971. Multiple scattering in planetary atmospheres. *J. Quant. Spectros. Radiat. Transfer* 11: 785–796.

Hulst, H. C. van de, and Grossman, K. 1968. Multiple light scattering in planetary atmospheres. *The atmospheres of Venus and Mars.* (J. C. Brandt and M. B. McElroy, eds.), pp. 35–55. New York: Gordon and Breach.

Hunt, G. E. 1972. Formation of spectral lines in planetary atmospheres: I, II, III. *J. Quant. Spectros. Radiat. Transfer* 12: 387–404; 405–419; 1023–1028.

Kano, M. 1964. Effect of a turbid layer on radiation emerging from a planetary atmosphere. Ph.D. Thesis, Univ. of Calif., Los Angeles.

Lenoble, J. 1968. Absorption lines in the radiation scattered by a planetary atmosphere. *J. Quant. Spectros. Radiat. Transfer* 8: 641–654.

———. 1969. Remarques sur le rayonnement diffusé par une atmosphère planétaire épaisse suivant la loi de Rayleigh. *C. R. Acad. Sci. Paris.* Série B., 269: 232–234.

Liou, K. N. 1970. Calculations of multiple backscattered radiation and depolarization from water clouds for a collimated pulsed Lidar system, NYU Technical Report TR 70–8.

Potter, J. F. 1969. Effect of cloud scattering on line formation in the atmosphere of Venus. *J. Atmos. Sci.* 26: 511–517.

Regas, J., and Sagan, C. 1970. Line formation in planetary atmospheres: I, II, III. *Comments on Astrophysics and Space Physics* 11: 116–120; 138–143; 161–166.

Sekera, Z. 1963. Radiative transfer in a planetary atmosphere with imperfect scattering. Rand Report R-413, PR.

Ueno, S. 1960. The probabilistic method for problems of radiative transfer: X. *Astrophys. J.* 132: 729–745.

Uesugi, A., and Irvine, W. M. 1969. Computations of synthetic spectra for a semi-infinite atmosphere. *J. Atmos. Sci.* 26: 973–978.

Takashima, T. 1971. Dependence of the polarization features of the diffuse radiation field of the earth's atmosphere on the location of a concentrated aerosol layer. Ph.D. Thesis, Univ. of Calif., Los Angeles.

Tanaka, M. 1971. Radiative transfer in turbid atmospheres: I, II, III. *J. of Meteo. Soc. Japan.* 49: 296–311; 321–332; 333–342.

ASTRONOMICAL FOURIER SPECTROPOLARIMETRY

F. F. FORBES
University of Arizona
and
A. L. FYMAT
Jet Propulsion Laboratory

Spectra of the Stokes polarization parameters of Venus (resolution 0.5 cm^{-1}) are presented. They were obtained at the Cassegrain focus of the 154-cm telescope of the National Mexican Observatory, Baja California, Mexico, July 12 and 13, 1972, with the Fourier Interferometer Polarimeter (FIP). A preliminary, limited analysis of four spectral features and of the CO_2 rotational band structures at 6080 and 6200 cm^{-1} has demonstrated that spectral polarization is indeed present. These experimental results, confirmed by two series of observations, provide substantiation for this theoretically predicted phenomenon. They also tend to show that the FIP represents a novel astronomical tool for variable spectral resolution studies of both the intensity and the state of polarization of astronomical light sources.

In the presence of scattering, as exists for instance in planetary atmospheres and circumstellar shells envelopes, radiation fields are modified in their intensities (Chandrasekhar 1950; Lenoble 1969), but more importantly, they become polarized. The effect depends on two causes: (1) the relative importance of scattering and absorption in attenuating the intensity of the original beam (the single scattering albedo of the carrier medium), and (2) the scattering diagram of the particles appropriate for both intensity and state of polarization (the phase matrix of scattering of the medium). These two characteristics are wavelength dependent, and hence, the spectral variations of the observed radiation intensity and state of polarization, particularly the corresponding line profiles, can yield information about the medium composition and structure. The information contained in the intensity variations has been systematically exploited by use of the photometer and the monochromator, two instruments that have respectively enabled us to perform ultralow and moderate-to-high spectral resolution studies. An impetus to this activity has been given during the past twenty years by the introduction of a next generation multiplexing spectrometer, namely the Fourier interferometer, which is capable of effecting very high resolution studies over extended wavelength intervals.

Tandberg-Hanssen[1] and Clarke[2] report on the utilization of what we may term a polarization monochromator. Fymat and Abhyankar (1970a, b; 1971) and Fymat (1972a, b) have proposed that an interferometer polarimeter be developed to study moderate-to-high resolution spectra of polarization of absorption or emission bands spreading across large wavelength intervals. The origin of this proposal lies in certain theoretical computations (the article by Fymat[3], exhibits some of these; see also Lenoble 1969) that showed that polarization exhibits a marked variation across spectral lines and across the composite bands. It was clear that the photopolarimeter would not provide the necessary spectral resolution for recording these variations, and that a monochromator polarimeter would be restricted to single lines and to extremely bright sources if we were to obtain a workable signal-to-noise (S/N) ratio. On the other hand, the Fourier spectroscopic technique was known to possess both multiplex and luminosity advantages—hence, the concept of the Fourier Interferometer Polarimeter (FIP).

In this article, we wish to report on the first results of applying this new technique of polarimetry to astronomical polarization sources, particularly to Venus. A very brief review of the necessary aspects of Fourier spectroscopy is first given as an introduction. Next, the subject of Fourier spectropolarimetry is discussed within the context of its relevance to astronomy, and finally we present some of our observational results.

FOURIER TRANSFORM SPECTROSCOPY

Technique

During the past two decades, interferometric (Fourier) spectroscopy, particularly in the infrared region of the spectrum, has been made possible by the works of Fellgett (1958a, b), Jacquinot (1958), J. Connes (1961), J. and P. Connes (1966), and others. An excellent history and the current status of the subject can be found in Loewenstein (1966), Vanasse and Sakai (1967), P. Connes (1970), and in the Proceedings of the 1970 International Conference on Fourier Spectroscopy. Its relevance to astronomy is discussed by P. Connes (1970). Briefly, the technique consists of two distinct steps:

1. Recording of an interferogram, $I(\tau)$, with a two-beam amplitude division (Michelson) interferometer:

[1]See p. 730. [2]See p. 752. [3]See p. 617.

$$\mathrm{Var}\,[I(\tau)] = \int_0^\infty B(\sigma) \cos 2\pi\sigma\tau\, d\sigma, \tag{1}$$

where τ = variable path difference between the two beams; $\sigma = \lambda^{-1}$ = wave number; $2\pi\sigma\tau = \delta$, the phase difference between the beams; $B(\sigma)$ = spectrum (or spectral power distribution); and Var denotes variable part. Note that $\mathrm{Var}\,[I(\tau)]$ is an even function of τ unless τ is itself a function of σ, as may happen in the case of an improperly compensated instrument.

2. Computation of the corresponding spectrum, $B(\sigma)$, by the Fourier cosine transformation:

$$B(\sigma) = \int_0^\infty \mathrm{Var}\,[I(\tau)] \cos 2\pi\sigma\tau\, d\tau. \tag{2}$$

Thus, in Equation (1), each spectral element or value of $B(\sigma)$ is modulated sinusoidally at a frequency proportional to its σ. The modulation is performed at each variation of τ. The decoding is subsequently performed by the operation described by Equation (2). In this process, the transformations of δ (the phase of the modulating function relative to the carrier wave) as both σ and τ vary are considered.

Comparison With Michelson's Visibility Technique

In Michelson's original fringe visibility technique, the intensities at maximum and minimum brightness of the interference fringes, $I_{\max}$ and $I_{\min}$ respectively, were determined, and the visibility defined by

$$V = \frac{I_{\max} - I_{\min}}{I_{\max} + I_{\min}} \tag{3}$$

A thorough mathematical treatment of this technique (see, for example, Born and Wolf, 1964) has shown that a unique spectrum cannot be obtained from the visibility curve alone, except in the special case where the spectrum is symmetric about some wave number σ_0. The missing datum here is the phase δ. In other words, on preserving information on δ, the Fourier transform technique enables one to recover the spectrum uniquely, be it a narrow or wide-band, absorption or emission spectrum.

Fundamental Limitations

In practice, τ has an upper limit defined by the truncating rectangular function, $T(\tau)$, of width $\tau_{\max}$ so that the computed spectrum is truly $B(\sigma)$ convoluted with the sinc function of $T(\tau)$. [sinc $x = \sin \pi x / \pi x$.] The instrumental line shape, obtained by studying the computed spectrum when $B(\sigma)$ consists of a monochromatic line of wave number σ_0 and negligible width, is sinc $\pi(\sigma - \sigma_0)\tau_{\max}$. The effects of the intense secondary maxima of this function on the recovered spectrum are minimized by the procedure called apodization; this is the convolution of the computed spectrum with the Fourier cosine transform of a weighting function, $W(\tau)$. Unfortunately, the choice of $W(\tau)$ is not arbitrary because $\delta\sigma$, the half-width, of the central peak of the transformed function of W, must obey Heisenberg's uncertainty principle:

$$\delta\sigma \cdot \tau_{max} \simeq 1. \qquad (4)$$

The instrumental (spectral) resolution is then $\delta\sigma$, and Equation (4) shows that its value is determined by the achievable total path difference in the interferometer. Thus, the limitation $\tau \leq \tau_{max}$ is a fundamental one since $\delta\sigma$ has necessarily a finite nonzero value. One can also rewrite Equation (4) as

$$\delta\sigma \cdot c\delta t = \delta\sigma \cdot \delta l \approx 1, \qquad (5)$$

where δt = coherence time of the radiation, c = velocity of light, and δl = coherence length. The limitation is seen to result equivalently from these coherence parameters which cannot become unbounded. Also, from the practical side, there is only a finite path difference that one can achieve.

The other limitation is $\sigma \leq \sigma_{max}$. It has both instrumental (beam-splitter transmission, detector characteristics, etc.) and numerical (sampling of the interferogram for the Fourier transformation) causes.

Fourier spectroscopy, being admittedly an indirect, highly involved technique with two fundamental limitations, must present definite advantages over the usual spectroscopic methods to justify its use. These advantages, which have also proven to be invaluable for spectropolarimetry, will now be listed.

Advantages and Their Astronomical Relevance

There are essentially five fundamental advantages, some aspects of which are of a practical nature.

1. *Fellgett's multiplex advantage.* It has been shown that a Fourier interferometer using a single detector is equivalent to a spectrometer (monochromator) with an array of M detectors. Some relative gain in the S/N ratio, a factor $M^{1/2}$, where M (multiplexing factor) = $(\sigma_{max} - \sigma_{min})/2\delta\sigma$, results from this advantage. It is most notable in the infrared, but even there, it must be stressed, it depends on the magnitude of M, the number of resolution elements. Thus, if M is very small (realm of broadband photometry), no multiplex advantage is to be expected. This is also the situation for the opposite extreme case of a single (absorption or emission) line profile (realm of the

Fabry-Pérot). Likewise, multiplexing is without interest when the source is barely detectable. It is, on the contrary, most interesting for relatively bright sources and can be used in various alternate fashions: decrease of (telescope) integration time by $1/M$, or increase of S/N by $M^{1/2}$, or improvement of spectral resolution, or else increase of spectral range. In visible and near-infrared light, the gain from multiplexing is rather insignificant; however, in this case, the other advantages now to be described must also be considered.

2. *Jacquinot's luminosity (throughput or etendue) advantage.* If $\Omega = 2\pi/R$ and $\Omega' = l/2R$ (R = resolving power, l = slit angular length) are the solid angles of the light beams admitted by an interferometer and a grating spectrometer, respectively, the gain resulting from the use of the interferometer is $L = \Omega/\Omega' = 4\pi/l$. Thus, for a given telescope diameter and resolving power, the interferometer accepts considerably more incident flux than the spectrometer. For extended sources, however, l assumes a larger value, so that L is reduced.

3. *Connes' resolving power advantage.* The wave number resolution achieved by a grating spectrometer (Littrow mount, grazing incidence) is $\delta\sigma = 1/2d$, where d = length of the grating. The limitation of grating size reflects on the resolution. Likewise, for the Fabry-Pérot étalon, $\delta\sigma = 1/\lambda R$, $R = 2Nh/\lambda$, N = finesse coefficient (or number of spectral elements that can be studied), and h = separation of the two plates of the étalon. The limitations of both N and h also affect $\delta\sigma$. Thus, for monochromators, the wave number resolution is limited by optical surface considerations. Such is not the case, however, for interferometers with which a resolution of 6×10^{-3} has already been achieved (Pinard 1969). This advantage is obviously important for sharp lines such as can be found in the spectra of the sun, the major planets, and a number of bright stars. It is also of interest not only in infrared but equally in visible and ultraviolet light.

4. *Wave number measurement accuracy advantage.* It comes from the fact that only a single spectral line is necessary for the measurement of a Fourier spectrum, thus providing an accuracy that depends solely on that with which the line is known. An order of magnitude improvement in the measurement has been reported by Pinard (1969) for the interferometer. The interest of this advantage is similar to the one previously described.

5. *Size and weight reduction advantage.* The interferometer is far smaller and lighter than the spectrometer because it requires smaller diameters of corresponding optical elements and much shorter focal lengths of the lenses or mirrors used. Although this advantage is not of a crucial nature for ground-based observations, it is nevertheless important for measurements from satellites, aircraft, and balloons.

Examples of the enormous improvements thus brought on astronomical spectra by the use of Fourier spectroscopy are: (1) a comparison in the range (1.1 to 2.7 μm) of Venus spectra recorded with a Fourier and a grating spectrometer (P. Connes 1970); (2) the unexpected detection of comparatively minute amounts of HCl and HF in the atmosphere of Venus (P. Connes et al. 1967); (3) the conclusion that "CO_2 is probably the major constituent in the Venus atmosphere" about a year before the data of Venera 4 and Mariner 5 indicated that this is predominantly so (P. Connes et al.1967); (4) the discovery of CO (four isotopes) in the Martian atmosphere (Kaplan, Connes, and Connes 1969); (5) the satellite work on board Nimbus 3 for the earth (Conrath et al. 1970), and Mariner 9 for Mars (Hanel et al. 1972); (6) the determination of the composition of stellar atmospheres (Johnson and Mendez 1970) such as α Ori which was found to be "considerably different from a normal solar composition" (Beer et al. 1972); and (7) the spectral atlases recently published by Connes, Connes, and Maillard (1969) for Venus, Mars, Jupiter, and Saturn, and by Delbouille, Roland, and Neven (1973) for the sun.

FOURIER TRANSFORM SPECTROPOLARIMETRY

Since Fellgett's pioneering work, the efforts of all students of the Fourier system have concentrated on perfecting the technique, demonstrating its superiority over the conventional scanning technique in certain well-defined wavelength regions, and widening its field of application, although "applications to the visible and ultraviolet have not yet even been scratched" (Mertz 1971). Applications to refractometry and, most recently, to the detection of very weak lines over extended spectral intervals and in the presence of considerable total energy (see J. Connes et al. 1970, for complex atomic emission processes) deserve separate mention. Another potential use of Fourier spectroscopy, suggested by Beer,[4] is the determination of absolute intensities and oscillator strengths in a manner similar to that reported by d'Incan, Effantin, and Roux (1971) in the case of the OH radical. The use of the Fourier technique in astronomy, in particular the telescopic work of Connes, Connes, and Maillard (1969) and the satellite work of Conrath et al. (1970) and Hanel et al. (1972), obviously represents a milestone in the history of the technique. A last interesting application is the use of astronomical Fourier spectra for the inverse determination of spectroscopic constants (e.g., Beer 1973). However, the advantages of Fourier spectroscopy, described in the foregoing

---

[4]Personal communication, 1973.

section, were so far used only for measuring the brightness (or intensity) spectrum. A new avenue of research and experimentation was opened up by the work of Fymat and Abhyankar (1970a, b; 1971) and Fymat (1972a, b) when these authors demonstrated that the Fourier spectroscopic technique can be adapted for the measurement of all four Stokes parameters of a light source ($I = B$: brightness; Q: degree of polarization, U: orientation of plane of polarization; and V: ellipticity of polarization ellipse) with a comparable high degree of resolution and over as wide a spectral range.

Principle of the Instrument

A two-beam amplitude division (Michelson) interferometer is initially considered. Four methods (hereafter referred to as Methods 1 to 4) of converting this instrument into a polarimeter have been proposed (Fymat 1972a, b). These are illustrated in Fig. 1 in which A_i (θ_i) are added linear

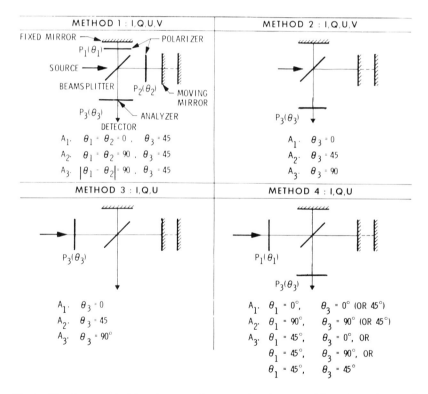

Fig. 1. Two-beam optical interferometer for measuring the spectral intensity and state of polarization of incident radiation. (Method 2 was used for obtaining the Venus data described in the text.)

polarizers (analyzers) with transmission axis azimuth θ_i. One could use the detector beam (DB, beam exiting from the end of the instrument) or the source beam (SB, beam returning in the direction of the source) or both. The interest of the latter possibility is only that, as in the usual Fourier spectroscopy, the difference between the complementary interferograms contained in the two beams may be used as a measure of the interference since it is more sensitive than that provided by either interferogram. The arrangement of Method 1 uses A_1 (θ_1) and A_2 (θ_2) in the two arms of the interferometer and A_3 (θ_3) in DB or SB, or both. Three interferograms recorded with ($\theta_1 = \theta_2 = 0°$, $\theta_3 = 45°$), ($\theta_1 = \theta_2 = 90°$, $\theta_3 = 45°$), and ($|\theta_1 - \theta_2| = 90°$, $\theta_3 = 45°$) were shown to provide the required polarization spectra of I, Q, U, and V. Three interferograms are also required in the remaining methods. In Method 2, only A_3 (θ_3) is used with $\theta_3 = 0°$, $45°$, and $90°$ in DB or SB, or both, and again I, Q, U, and V spectra are obtained. Methods 3 and 4 can only provide spectra of I, Q, and U. In Method 3, a single polarizer A_0 (θ_0) in the incident beam is used with $\theta_0 = 0°, 45°$ and $90°$. Finally, in Method 4, a hybrid of Methods 2 and 3, the polarizers' orientations used are ($\theta_0 = 0°$, $\theta_3 = 0°$ or $45°$), ($\theta_0 = 90°$, $\theta_3 = 90°$ or $45°$), and ($\theta_0 = 45°$, $\theta_3 = 0°$, or $45°$, or $90°$).

Analysis of the Interferogram

A thorough mathematical analysis of the interferogram has shown that Equation (1) now becomes:

$$\text{Var}\,[I(\tau)] = \int_0^\infty [a_2(\sigma)\cos\delta + a_3(\sigma)\sin\delta]\,d\sigma, \quad \delta = 2\pi\sigma\tau, \qquad (6)$$

where a_i ($i = 2, 3$) are linear combinations of the Stokes parameters:

$$a_i(\sigma) = i_i(\sigma)I(\sigma) + q_i(\sigma)Q(\sigma) + u_i(\sigma)U(\sigma) + v_i(\sigma)V(\sigma), \qquad (7)$$

and i, q, u, v are known coefficients for each of the above four Methods. Introducing the auxiliary quantities $\rho(\sigma)$ and $\psi(\sigma)$ by the equations:

$$\left.\begin{array}{l} a_2(\sigma) = \rho(\sigma)\cos\psi(\sigma) \\ a_3(\sigma) = \rho(\sigma)\sin\psi(\sigma) \end{array}\right\}, \qquad (8)$$

Equation (6) becomes

$$\text{Var}\,[I(\tau)] = \int_0^\infty \rho(\sigma)\cos[2\pi\sigma\tau - \psi(\sigma)]\,d\sigma. \qquad (9)$$

The polarizers have therefore introduced the phase term $\psi(\sigma)$, thus destroying the interferogram's symmetry as a function of τ. Hence, the spectral distribution of the incident radiation is no longer the Fourier cosine transform of the interferogram. We must resort to the full (exponential) transform for deriving $\rho(\sigma)$ and $\psi(\sigma)$ which, in turn, would give $a_2(\sigma)$ and $a_3(\sigma)$—that is, two relations in the four unknown Stokes parameters of the incident radiation. Under these circumstances—each interferogram containing both a cosine $[a_2(\sigma)]$ and a sine $[a_3(\sigma)]$ part—the three required interferograms together will in principle yield six relations, such as Equation (7) in the four unknowns. Actually, some of these relations are incomplete. For example, in Method 2, $u_2 = v_2 = u_3 = v_3 = 0$ for the first two interferograms so that only linear relations between I and Q can be obtained from them and, subsequently, I and Q separately. The third interferogram, on the other hand, will provide two complete relations which, with the previously acquired information on I and Q, will yield U and V. A similar situation exists for the other Methods [see Table I, where $I_l = \frac{1}{2}(I + Q)$ and $I_r = \frac{1}{2}(I - Q)$, I_l and I_r being the usual orthogonal states of polarization]. For a complete discussion of these questions, the reader is referred to Fymat (1972a, b).

Comparison with Photopolarimetry

Most of our current knowledge of the scattering components (gases, aerosols, lithosols) of astronomical bodies has been obtained from broad-band photopolarimetric observations, carried out in absorption or in emission, in the wavelength interval 0.33 to 3.4 μm. These ultra-low resolution measurements yield polarimetric data averaged over wide wavelength intervals preselected with filters of typical half-widths $\gtrsim 10^3$ cm^{-1}. Further, they are usually limited to the measurement of a particular case of a single polarization parameter, namely the degree of linear polarization. On the other hand, Fymat[5] has shown that in particular within absorption lines strong polarizations can be expected to be observed, the rates of polarization first increasing with the strength of the line and then decreasing as the line becomes saturated. If these spectral polarizations are astronomically relevant, it is clear that photopolarimetry neither possesses the required resolution even for a simple measurement of the difference between the integrated polarization within the line and the polarization of the surrounding continuum, nor yields a sufficient S/N ratio for the same resolution. By contrast, interferometric polarimetry possesses all the necessary advantages, as previously discussed. Indeed, the polarization spectra to be presented shortly were recorded with the moderate resolution of 0.5 cm^{-1}, i.e., with a resolution of 3.5 orders of magnitude better than the photopolarimetric one. An even better resolution could be obtained with an interferometer having a longer path difference than the one permitted by the instrument used.

---

[5] See p. 617.

TABLE I
Polarization Information Contained in Interferometric Arrangements A_1, A_2, and A_3 Using Linear Polarizers for the Four Methods of Spectropolarimetry

Spectra	Arrangements[a]		
	A_1	A_2	A_3
Method 1	$(x_1 \cos\delta + x_2 \sin\delta)I_t$	$(x_3 \cos\delta + x_4 \sin\delta)I_r$	$(x_5 U + x_6 V)\cos\delta + (-x_6 U + x_5 V)\sin\delta$
Method 2	$1/2(x_1 \cos\delta + x_2 \sin\delta)I_t$	$1/2(x_3 \cos\delta + x_4 \sin\delta)I_r$	$(x_7 I + x_8 Q + x_9 U + x_{10} V)\cos\delta$ $+ (x_{11} I + x_{12} Q + x_{13} U + x_{14} V)\sin\delta$
Method 3	$1/2(x_1 \cos\delta + x_2 \sin\delta)I_t$	$1/2(x_3 \cos\delta + x_4 \sin\delta)I_r$	$(x_{15} \cos\delta + x_{16} \sin\delta)(I - U)$
Method 4	$1/2(x_1 \cos\delta + x_2 \sin\delta)I_t$	$1/2(x_3 \cos\delta + x_4 \sin\delta)I_r$	$(x_1 \cos\delta + x_2 \sin\delta)(I - U)$

[a] x_i, $i = 1$ to 16, are known coefficients for any given apparatus.

Potential Advantages of High-Resolution Polarimetry

There are indeed several potential advantages of resorting to high-resolution polarimetry. We shall list the more prominent of them, not mentioning those associated with any Fourier spectroscopic techniques which have already been put forward. We may note, on general grounds, that an improvement over photopolarimetry comparable to that brought by Fourier spectroscopy on photometry or grating spectroscopy can reasonably be expected. In addition, we have the following advantages:

1. *Interpretation of spectra formed in the presence of scattering.* This interpretation is plagued by the crucial problem of separating the absorption (or emission) from the scattering. In the case of an absorption line, for example, the scattering takes place only in the continuum of the line and contributes along with the gaseous and particulate absorptions to the determination of its level. In order to interpret the line properly in terms of physical parameters such as gaseous abundances and their relative partial pressures, rotational temperatures, etc., one should like to quantify the scattering contribution. To this end, neither the spectroscopic nor the photopolarimetric techniques are helpful. It is known, however, that from the two physical processes, absorption (or emission) and scattering, only the latter induces polarization in the light field. The spectral polarization will, therefore, provide the required datum for effecting the separation.
2. *Probing the internal physical properties of a scattering medium.* Photopolarimetric measurements usually refer to the upper layers of the scattering medium, whereas spectra refer to these and deeper layers. In a scattering atmosphere, strong lines are formed in the upper regions, and progressively weaker lines in increasingly lower regions. On the other hand, Fymat[6] has shown that stronger polarizations happen in the cores of stronger unsaturated absorption lines with larger amplitudes of variation between this central position and the continuum. However, for moderately strong lines and even for weaker ones, the central polarization is still very high with a marked core-continuum variation that is well above instrumental sensitivity (see Fig. 2). It becomes, therefore, possible to probe different layers of an atmosphere by recording the polarization of spectral lines of varying strengths.
3. *Simultaneous spectral measurements of intensity and state of polarization.* This advantage is not entirely independent of the previous one. It is, of course, possible to operate simultaneously both a photometer

---

[6] See p. 617.

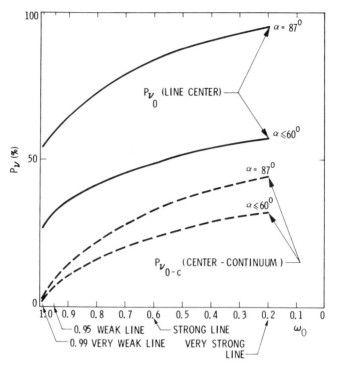

Fig. 2. Phase-angle variations of polarization in absorption lines of various strengths. Note the strong polarization amplitude (center-continuum) for moderate and strong lines (adapted from Lenoble 1969).

and a photopolarimeter in order to investigate the same upper atmospheric layers. These measurements, however, cannot be reconciled with the spectroscopic ones that refer to the deeper layers. This point has, incidentally, put to rest a controversy in the studies of the nature of Venus' clouds between apparently antagonistic spectral and polarization observations. A more positive use of polarization in the atmospheric regions covered by spectroscopy can only be provided by spectropolarimetry.

4. *Variable spectral resolution polarimetry.* Between the ultra-low- and the high-resolution techniques of polarimetry so far discussed, there is of course a broad range of interesting intermediate resolutions. These resolutions can in fact be easily achieved. One only has to shorten the total path difference in the interferometer [see Eq. (4)]. The same spectropolarimeter can be set to work at any resolution lower than its maximum resolution. A whole new area of study of variable spectral resolution polarimetry is thus afforded.

5. *Detailed wavelength dependence of polarization.* The works of Dollfus, Gehrels, Sekera, and their followers, and of Forbes (1971) have already demonstrated the added information contained in the wavelength variation of polarization. This new information was, however, recovered from the consideration of only a discrete set of bands (usually five) spread across the visible region of the spectrum. Along the same lines a wealth of additional information must be contained in the detailed wavelength variations that spectropolarimetry permits us to study.

OBSERVATIONAL RESULTS

We have started a program of observations of spectral polarization of planets and other astronomical bodies. In a first phase of this program, our objective was to verify some of the theoretical inferences of Fymat[7] concerning the existence of polarization within absorption lines formed in scattering atmospheres, more particularly the dissymmetrical aspect of the spectral line shapes corresponding to I_l and I_r. In the next phase of the program, we plan to bring to bear more specifically the advantages listed earlier of Fourier spectropolarimetry. For this purpose we have used the University of Arizona's interferometer, Block Engineering Model 1500 (maximum resolution $\delta\sigma = 0.5$ cm^{-1}), which has previously enabled us to obtain, in the usual fashion, intensity spectra of the sun, the moon, and a number of stars (α Her, μ Cep, α Sco, R Cas, χ Cyg, etc.). For a detailed description of this spectrometer see Johnson et al. (1973). The instrument was first converted into a polarimeter according to the suggested Method 2 by the addition of an HR-sheet analyzer in the detector beam (Fig. 3). The wavelength ranges 0.8–1.1 μm and 1.2–2.7 μm were successively examined using an Si- and a PbS-detector, respectively.

Laboratory Work

Both the above Methods 1 and 2 were explored in order to determine the most efficient observing procedure for our present instrumentation. Method 2 was found to be more satisfactory because the polarizers placed in the reflecting light paths of the instrument (Method 1) caused a significant decrease in modulation efficiency perhaps because of the nonuniform optical properties of the HR-sheet analyzers. In addition, the polarizer transmission of about 65% to 70% (Shurcliff 1962) further limited the instrumental sensitivity, indicating that the use of a single polarization analyzer was preferable to that of the three polarizers required by Method 1. In the next phase of our program, we plan to use wire-grid polarizers

---

[7]See p. 627.

which have a polarization efficiency greater than 95% up to a wavelength of 14 μm and a good transmission (Young, Graham, and Peterson 1965). Next, an optical scheme developed by Johnson et al. (1973), which allows the operation of the interferometer at a Cassegrain focus, was implemented in order to reduce the telescopic polarization. With this arrangement, the many reflections (causing polarization) of a Coudé optics were avoided. The polarization caused by the telescope primary and secondary optics was known from previous work to be relatively small (Forbes 1967).

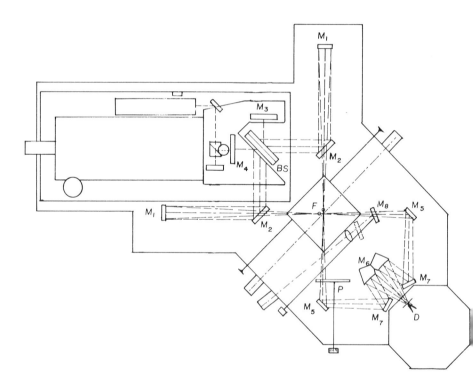

Fig. 3. Schematic diagram of the Fourier Interferometer Polarimeter (FIP). Light from the telescope enters the instrument at the two Cassegrain focal-plane flat field mirrors at point F. The light from these two 13-arcsec portions of field is directed into the interferometer by means of two collimator mirrors marked M_1 and flat mirror M_2. The interferometer mirrors M_3 (stationary) and M_4 (movable 1 cm) together with the beamsplitter BS comprise the interferometer. The beams emergent from the interferometer are directed into the liquid-nitrogen-cooled PbS detectors D by flat mirrors M_5 and transfer mirrors M_6 and M_7. The polarization analyzer P is positioned ahead of one of the M_5 flats. M_8 is inserted for guiding purposes and for whenever the instrument is used for spectropolarization.

Instrumental Polarization

The instrumental polarization was determined while operating the instrument in the polarimeter mode. Its value, obtained by measuring a bright star known to have low polarization such as α Ori, was found to be $8 \pm 2\%$. While this determination lacks the precision usually associated with photopolarimeters, it was considered adequate for the purpose of identifying spectral-line polarization. We should like to stress at this point the fundamental difference between photopolarization and spectropolarization. While the former typically observes values of a few percent in the case of Venus, the latter can be considerably larger, as may be seen from Fig. 2. This figure represents the phase-angle variations of polarization for Lorentz absorption lines of different strengths observed at a planet's disk center. The single scattering albedo in the continuum was assumed at the value 0.975. Decreasing values of the albedo in the line center corresponded to lines of increasing strength. It is seen in particular that, at the larger phase angles, the amplitude of polarization between the line center and the continuum is much greater than the instrumental polarization for moderate-to-strong lines. This situation was expected to hold approximately for the observing conditions under which we operated. In this first attempt at detecting spectral polarization, it was clear that the strong lines were the interesting ones. In the future, however, especially for the analysis of weak lines, the elimination of the instrumental polarization will be accomplished in the following manner. Consider that the interferometer is a perfectly linear system, i.e. $\mathbf{E} = \mathbf{K}\mathbf{E}^{(0)}$, where $\mathbf{E}^{(0)}$ and $\mathbf{E}$ are electric amplitude vectors for the incoming and outgoing light, respectively, and $\mathbf{K}$ is the Jones matrix of the polarimeter. From Fymat (1972a, b), it is easily seen that $\mathbf{K}$ is a diagonal matrix since the similar matrices characterizing the beamsplitter and the mirrors are also diagonal. In other words, only the l and r components of the incident radiation are influenced during their passage through the instrument. From Table I, we can write immediately

$$\frac{\text{Var}\,[I(0°)]}{\text{Var}\,[I^*(0°)]} = \frac{I_l}{I_l^*},\ \frac{\text{Var}\,[I(90°)]}{\text{Var}\,[I^*(90°)]} = \frac{I_r}{I_r^*}, \quad (10)$$

where quantities with an asterisk refer to a source of known polarization. These expressions provide a means for eliminating the instrumental polarization due to both the polarimeter and the telescope. Thus, the light source to be analyzed can be recorded using a depolarizer in front of the instrument. The essentially unpolarized light entering the instrument would then provide $I_l^* = I_r^* = \tfrac{1}{2}I$ whose spectrum is presumed to be known from independent experiments. Then I_l and I_r would follow immediately from Equations (10).

Calibration Spectra

The calibration of polarization spectra—that is, the elimination of the telluric spectral polarization—is a new problem for which we propose the following approximate solution. Assume that the earth's atmosphere is homogeneous, i.e., the 4 × 4 phase matrix of scattering, **P**, and the single scattering albedo, ω, are constant at all levels in the atmosphere, and assume only single scattering. Then, at any frequency, the transmitted vector at the surface, $\mathbf{I}(\tau_1; -\Omega) = \{I_l, I_r, U, V\}$, due to an incident monodirectional illumination at the top of the atmosphere, $\pi\mathbf{F} = \pi\{F_l, F_r, F_u, F_v\}$, is (see, for example, Deirmendjian 1969):

$$\mathbf{I}(\tau_1; -\Omega) = \frac{\omega}{4} \mathbf{P}(-\Omega, -\Omega_0) \mathbf{F}\left[\left(\frac{\mu_0}{\mu_0 - \mu}\right)(e^{-\tau_1/\mu_0} - e^{-\tau_1/\mu})\right], \quad (11)$$

where τ_1 = optical depth and $\Omega \equiv (\mu = \cos \vartheta, \phi)$ are the usual spherical coordinates, and the subscript zero refers to incident light. When looking directly at the source (forward-scattering situation in which the single-scattering approximation holds much better than in other directions), $\Omega \equiv \Omega_0$, and Equation (11) becomes:

$$\mathbf{I}(\tau_1; -\Omega_0) = \frac{\omega}{4} \mathbf{P}(-\Omega_0, -\Omega_0) \mathbf{F}\left[\frac{\tau_1}{\mu_0} e^{-\tau_1/\mu_0}\right]. \quad (12)$$

Assume also that a weighted linear combination of Rayleigh and Mie scattering describes adequately the atmospheric scattering. Then $\mathbf{P}(-\Omega, -\Omega_0)$ reduces to the diagonal matrix $a\mathbf{1}$, where $\mathbf{1}$ = unit matrix, $a = 1.5\sigma_R + \sigma_M P_1$, σ_R and σ_M are respectively the Rayleigh and the Mie scattering coefficients, and P_1 = Mie matrix element for the l or r intensity component. We can therefore write the components of Equation (12):

$$I_i(\tau_1; -\Omega) = cF_i, \quad i \equiv l, r, u, v, \quad (13)$$

where $c = (\omega a \tau_1 / 4\mu_0) e^{-\tau_1/\mu_0}$. Now, denoting by an asterisk all quantities referring to a calibration source (moon, sun, star), we have the following ratios:

$$\frac{I_i}{I_i^*} = \frac{F_i}{F_i^*} \quad (14)$$

which can be used for calibration purposes. This procedure is similar to the ratio technique of Johnson and Mendez (1970) for removing the telluric absorption. However, in the present case, scattering and the resulting polarization are also considered as well as the four Stokes components. Note from Equation (14) that, in order to determine the spectrum F_i in which all telluric effects have been removed from the telescopic observations of the spectra I_i of the source of interest and I_i^*

of the calibrating source, it is necessary to have the spectrum F_i^* of the calibrating source at the top of the atmosphere. This is not yet available, so the spectra in Figs. 4 and 5 have not as yet been calibrated. However, the spectra of these figures do provide positive evidence for the phenomenon of spectral polarization, as will be discussed later.

Spectropolarimetry can also yield the usual total intensity spectrum ($I = I_l + I_r$) given by the original Fourier spectrometer. This is accomplished simply by adding the spectra of the arrangements A_1 and A_2 of Table I. It is therefore possible to calibrate I since $F^*(= F_l^* + F_r^*)$ is already available from previous lunar observations on board the NASA Convair 990 flying observatory (altitude ≈ 12.5 km) according to the procedure described by Johnson and Mendez (1970).

Venus Spectra

We have made two series of observations of this planet in the wavelength range 0.8 to 2.7 μm at the full instrumental resolution of 0.5 cm^{-1}. This range was selected because the Fourier technique exhibits there its advantages at their best, and because Venus is known to possess strong continuum polarization over a large phase-angle range. Earlier work (Fig. 1 of Forbes 1971) has indeed shown that between phase angle $\alpha \cong 33°$ and 138° there is strong negative polarization reaching its absolute maximum at about $\alpha = 90°$ for both wavelengths 1.25 μm and 1.65 μm. At wavelengths longer than 2 μm, the polarization is still strong but has reversed its sign, now becoming positive with a maximum at about the same phase ($\alpha \simeq 90°$). The wavelength of transition from negative to positive polarization appears to be around 2 μm.

The first run of Venus observations (Forbes 1972) was performed during January, 1971 ($\alpha = 97°\!.6$) with the instrument mounted at the Coudé focus of the Steward Observatory's 230-cm telescope. The instrument was operated as a polarimeter (Method 2). Over twenty lines were found that appeared to exhibit changes in line strength for different polarization-analyzer orientations. The effects were clearly not due to noise; however only one set of observations could be made. Nevertheless, these results were sufficiently encouraging, and an observational program was devised in order to provide the necessary confirmation of the effects as well as to direct the study to a wavelength range known to have very strong Venus atmospheric lines.

The experiment was repeated using the Cassegrain focus of the 154-cm telescope of the new Observatorio Astronómico Nacional in the Sierra de San Pedro Martir, Baja California, Mexico, again under the favorable phase-angle value of $\alpha = 99°\!.2$. The instrument was operated in both the spectrometer and the polarimeter mode. An integration time of approximately 1.5 hours was used for recording the interferogram corre-

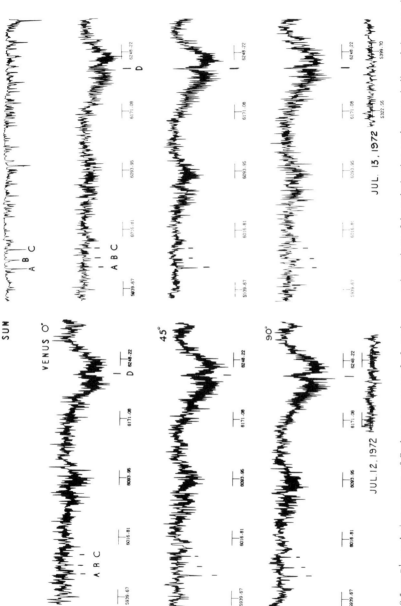

Fig. 4. 0.5 cm^{-1} resolution spectra of Stokes parameters I_l, I_r, and their combination with U and V, obtained with the Fourier Interferometer Polarimeter for Venus on July 12 and 13, 1972, at the National Mexican Observatory, Baja California, Mexico. (The three orientations of the polarization analyzer are indicated. A comparison solar spectrum, and instrumental noise tracings are also shown.) Note the polarization effects on the four features labeled A, B, C, and D, and on the CO_2 rotational band structures at 6080 and 6200 cm^{-1}.

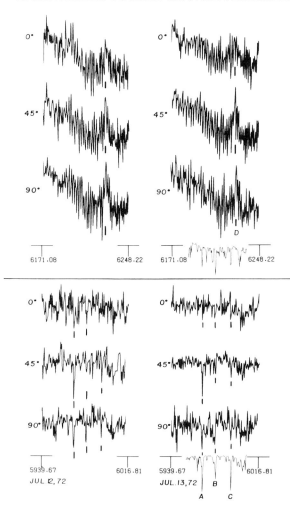

Fig. 5. Enlarged view of the polarization effects on the four spectral features A, B, C, and D of Fig. 4.

sponding to each of the three orientations of the polarization analyzer. A complete series of polarization measurements for the total number of orientations thus yielded about 4.5 hours of data. Since, as shown elsewhere,[8] polarization was expected to increase sharply with the spectral-line strength, we hoped in this next attempt to be able to record its effects particularly in the strong CO_2 bands. At 2.1 μm there is a region of major CO_2 absorption for Venus, and at this wavelength PbS is most sensitive. Unfortunately, this wavelength turns out to be, as mentioned earlier, that at which polarization is minimal regardless of phase. It was nevertheless

---

[8] See p. 628.

studied in order to investigate whether line polarization was, as theoretically predicted, more pronounced than the continuum one. We have not yet completed our analysis of the spectra obtained there.

In Fig. 4, we show a minute portion of the spectra in the interval 5900 to 6300 cm^{-1} (1.6 – 1.7 μm) obtained with the analyzer positioned in the exit beam along three orientations (0°, 45°, and 90°). The Stokes-parameters information provided by these spectra is as described in Table I (Method 2). In particular, the spectra associated with the orientations 0° and 90° correspond respectively to the quantities $I_{l,v}/I_{l,c}$ and $I_{r,v}/I_{r,c}$ (v = line frequency, c = continuum)—that is, the line profiles of the l and r components of intensity across the bands in the wavelength interval represented. The left and right parts of the diagram give data sets obtained on two successive days, July 12 and 13, 1972 (a total of nine hours of data). The lowest recordings represent typical noise tracings of the Venus data near the dates corresponding to the two sets. As seen, their values are small, and the S/N ratio is satisfactory. The noise in the data is attributed mainly, as discussed earlier, to the reduced sensitivity of the interferometer when it operates as a spectropolarimeter. Also, only the detector beam (DB) was utilized so that no benefit was drawn from the other beam (SB) in order to obtain a more sensitive measure of the interference. At present the system is background limited; therefore, an increase in observing time from 1.5 hours per analyzer position would be necessary to yield a higher S/N ratio. The right topmost curve is a comparison solar spectrum that was taken from a complete single tracing (1–6 μm) of the sun observed with the same interferometer placed on board the NASA Ames Lear Jet (cruise altitude 15.5 km) (Johnson et al. 1973). The high-altitude solar-line identifications were obtained from similar but lower-resolution work of Houghton et al. (1961).

An analysis in terms of polarization effects of the recorded data would go beyond the scope of this paper. We wish rather to concentrate our study here on the lines that we have for convenience labeled as A, B, C, and D. Fractions of the spectra encompassing these four lines are shown enlarged in Fig. 5. These lines have been selected for illustrative purposes only, and are by no means either typical or the best representatives of polarization effects on spectral-line shapes. The features A, B, and C have been identified as solar lines diffusely reflected by Venus' atmosphere; A and B are Al lines and C is an Fe and Si line. We have been unable to identify the feature D (6210 cm^{-1}) with any solar or telluric line, and have therefore tentatively associated it with Venus. We should mention, however, that we have not yet examined the sunlit sky at $\alpha = 99°$.

Looking first at line A, it is evident that its shape exhibits a marked dependence on the polarization-analyzer orientation. For the July 12 data, the line is more prominent at 45° than at the other two orientations which appear comparable, although the r-component line (90°) is slightly stronger than the l line (0°). On the other hand, for the following day, the 45° and l lines have both weakened to the point where the r line, now of comparable strength to the 45° line, is noticeably stronger than the l line. Thus, in particular, between the two series of observations, the difference between the r and l lines (related to the degree of polarization) has increased. Line polarization is therefore present; otherwise the various component lines (especially l and r) should have been identical during any given observation, which the observations strongly contradict. Similar remarks can be made during the other two lines B and C. We have not yet determined whether the lines were initially polarized and in that case what the effect was on this polarization caused by Venus, or whether the polarization must be attributed entirely to the diffuse reflection by this planet's atmosphere. Irrespective of that, the important phenomenon here is the existence of line polarization and the possibility of detecting it. Other lines (e.g., close to 5940 cm^{-1}, 6015 cm^{-1}, etc.) exhibit stronger l components than the associated r components, a phenomenon that is opposite to that found for A, B, and C. This is an $l - r$ reversal. Thus, there appears to be no universal behavior of the l and r components from line to line; according to the strength of the composite line $l + r$, the l line may be weaker than the complementary r line (positive polarization) or conversely (negative polarization). The preceding features provide observational substantiation to the theoretical predictions of Fymat.[9]

As for feature D, we first note that the rotational band structure (CO_2, $\lambda = 1.6$ μm) to which it belongs is itself globally polarization dependent (see both Figs. 4 and 5). On any given day of observation, the r-rotational structure is more pronounced than the associated l structure. This is even more dramatic when contrasting the 0°- and 45°-rotational structures at 6075 cm^{-1} on July 13. Here, the l structure is almost unsuspected, whereas at 45° it shows up much more clearly. Feature D exhibits likewise a strong polarization dependence. No $l - r$ reversal is apparently present in this case. Since too many features that display a polarization dependence are present, it is impractical to discuss them here.

In Fig. 6, we have displayed the portion (5940 cm^{-1}–6250 cm^{-1}) of the spectrum, $I(\sigma)$, of Venus with the interferometer operating as a spectrometer and the added spectrum $[I_l(\sigma) + I_r(\sigma)]$ obtained with the instrument used as a polarimeter. The comparison of these two spectra shows that, within experimental noise, all spectral features present in $I(\sigma)$ are also present in $[I_l(\sigma) + I_r(\sigma)]$. (In this connection, one must also

---

[9] See p. 617.

bear in mind that the two spectra were not obtained at the same time. Therefore, minor differences in certain line strengths must be attributed to differences in the air masses viewed.) This result provides a validation of the individual spectra $I_l(\sigma)$ and $I_r(\sigma)$ which, because they were obtained at orthogonal polarization angles, should indeed add to yield $I(\sigma)$. It shows, additionally, that the task of the spectrometer can also be accomplished by the spectropolarimeter. Portions of the spectrum obtained at the position angle 45° of the polarization analyzer are also represented in Fig. 6 so that their appearance may be contrasted conveniently with that of the corresponding portions in either of the foregoing two spectra. The contrast shows conclusively the existence of spectral polarization.

CONCLUSION

We have presented spectra of the Stokes polarization parameters I_l and I_r (and of their linear combination with U and V) in the case of Venus. These spectra were recorded with the Fourier Interferometer Polarimeter (FIP) at the full instrumental resolution of 0.5 cm^{-1}. They represent the first attempt at implementing the practically novel concept of moderate-to-high-resolution interferometric polarimetry. A limited sample only of the data gathered was exhibited, and a preliminary analysis of it was made. It was shown, in particular, that the rotational band structure of the planetary CO_2 component, the feature D of the band centered at 6210 cm^{-1}, and even reflected solar lines (features A, B, and C) do exhibit a marked polarization dependence. Both positive and negative polarizations were observed corresponding to a reversal in the relative strengths of the complementary l and r lines. It was also shown that the added spectrum $[I_l(\sigma) + I_r(\sigma)]$ compared very well with the spectrum $I(\sigma)$—obtained when the instrument is used as a spectrometer—in all its detailed features. Much remains to be done to refine the experiment (improvement of S/N, elimination of instrumental polarization, and recording of high-altitude calibration spectra). Nevertheless, we hope to have succeeded in showing that not only the concept is sound but that the FIP provides a new astronomical tool that can additionally accomplish the task of the Fourier spectrometer. During the second phase of our program we shall endeavor to demonstrate that this tool is a valuable one by showing that physical information not otherwise available can be extracted from our data. We would feel gratified if we succeeded equally in stimulating interest in this new technique of polarimetry.

Acknowledgments. This research was supported by NASA Contract NAS7-100, JPL's Director Discretionary Fund 730-00034-0-825, and NSF Grant GP-28057.

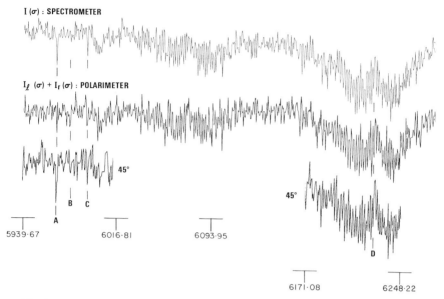

Fig. 6. Intensity spectra of Venus obtained with the instrument operating as a spectrometer and as a polarimeter. Portions of the polarization spectrum (analyzer orientation 45°) are also represented.

REFERENCES

Beer, R. 1973. Analysis of the $\Delta v = 2$ vibration-rotation bands of silicon monoxide (preprint).
Beer, R.; Hutchison, R. B.; Norton, R. H.; and Lambert, D. L. 1972. Astronomical infrared spectroscopy with a Connes-type interferometer. III. *Astrophys. J.* 172: 89–115.
Born, M., and Wolf, E. 1964. *Principles of optics.* 2nd ed., p. 320 ff. London: Pergamon.
Chandrasekhar, S. 1950. *Radiative transfer.* p. 262. New York: Dover Publications.
Connes, J. 1961. Spectroscopic studies using Fourier transformation. *Rev. d'Opt.* 40: 45–78, 116–140, 171–190, 231–265. (See also English transl. U.S. Naval Ordnance Test Station NAVWEPS Rept. 8099 NOTS TP3157, China Lake, Calif. 1963.)
Connes, J., and Connes, P. 1966. Near infrared planetary spectra of Fourier spectroscopy. I. Instruments and results. *J. Opt. Soc. Am.* 56: 896–910.
Connes, J.; Connes, P.; and Maillard, J.-P. 1969. Near infrared spectra of Venus, Mars, Jupiter and Saturn. Centre National de la Recherche Scientifique, Paris.
Connes, J.; Delouis, H.; Connes, P.; Guelachvili, G.; Maillard, J.-P.; and Michel, G. 1970. Spectroscopie de Fourier avec transformation d'un million de points. *Nouv. Rev. d'Opt.* 1: 3–22. (See also English transl. *NASA-JPL Tech. Mem.* 33–525, 1972, p. 53, R. Beer trans.)
Connes, P. 1970. Astronomical Fourier spectroscopy. *Rev. Astron. Astrophys.* 3: 209–230.
Connes, P.; Connes, J.; Benedict, W. S.; and Kaplan, L. D. 1967. Traces of HCl and HF in the atmosphere of Venus. *Astrophys. J.* 147: L1230–L1237.
Conrath, B. J.; Hanel, R. A.; Kunde, V. G.; and Prabhakara, C. 1970. The infrared interferometer experiment on Nimbus 3. *J. Geophys. Rev.* 75: 5831–5857.

Deirmendjian, D. 1969. *Electromagnetic scattering on spherical polydispersions.* p. 96. New York: American Elsevier.

Delbouille, L.; Roland, G.; and Neven, L. 1973. Photometric atlas of the solar spectrum from λ 3000 to λ 10000. Belgium: Université de Liège. In press.

Fellgett, P. B. 1958a. A propos de la théorie du spectromètre interférentiel multiplex. *J. Phys.* 19: 187–191.

———. 1958b. Spectromètre interférentiel multiplex pour mesures infrarouges sur les étoiles. *J. Phys.* 19: 237–240.

Forbes, F. F. 1967. The infrared polarization of the infrared star in Cygnus. *Astrophys. J.* 147: L1226–L1229.

———. 1971. Infrared polarization of Venus. *Astrophys. J.* 165: L21–L25.

———. 1972. Recent stellar spectra obtained with a 0.5 cm^{-1} interferometer. *Mém. Soc. Roy. Sci. Belgium* 3: 225–227.

Fymat, A. L. 1972a. Polarization effects in Fourier spectroscopy. *Applied Optics* 11: 160–173.

———. 1972b. Interferometric spectropolarimetry: Alternate experimental methods. *Applied Optics* 11: 2255–2264.

Fymat, A. L., and Abhyankar, K. D. 1970a. An interferometric approach to the measurement of optical polarization. *Applied Optics* 9: 1075–1081.

———. 1970b. Interferometer for measurement of optical polarization. *NASA Tech. Brief* 70–10405.

———. 1971. A new field in Fourier spectroscopy: Interferometric polarimetry. *Aspen International Conference on Fourier Spectroscopy 1970.* (G. A. Vanasse, A. T. Stair, Jr., and D. J. Baker, eds.) *AFCRL Rept.* 71–0019: 377–384.

Hanel, R. A.; Conrath, B. J.; Hovis, W. A.; Kunde, V. G.; Lowman, P. D.; Pearl, J. C.; Prabhakara, C.; Schlachman, B.; and Levin, G. V. 1972. Infrared spectroscopy experiment. *Mariner Mars 1971 project final report. JPL Tech. Rept.* 35-1550: 27–33.

Houghton, J. T.; Hughes, N. D. P.; Moss, T. S.; and Seeley, J. S. 1961. An atlas of the infrared solar spectrum from 1 to 6.5 μ observed from a high-altitude aircraft. *Roy. Soc. (London), Phil. Trans.* (Series A), 254: 47–123.

Incan, J. d'; Effantin, C.; and Roux, F. 1971. Intensités absolues et forces d'oscillateur de quelques raies des bandes de vibration-rotation 1-0 et 2-1 du radical OH. *J. Quant. Spectros. Radiat. Transfer.* 11: 1215–1224.

Jacquinot, P. 1958. Caractères communs aux nouvelles méthodes de spectroscopie interférentielle, facteur de mérite. *J. Phys.* 19: 223–229.

Johnson, H. L.; Forbes, F. F; Thompson, F. I.; Steimmetz, D.; and Harris, O. 1973. A new infrared Fourier-transform spectrometer. *P. Astron. Soc. Pacific.* In press.

Johnson, H. L., and Mendez, M. E. 1970. Infrared spectra for 32 stars. *Astron J.* 75: 785–817.

Kaplan, L. D.; Connes, J.; and Connes, P. 1969. Carbon monoxide in the Martian atmosphere. *Astrophys. J.* 157: L187–L192.

Lenoble, J. 1969. Remarques sur le rayonnement diffusé par une atmosphère planétaire épaisse suivant la loi de Rayleigh. *C. R. Acad. Sci. Paris* Ser. B, 269: 232–234.

Loewenstein, E. V. 1966. The history and current status of Fourier transform spectroscopy. *Applied Optics* 5: 845–854.

Mertz, L. 1971. Fourier spectroscopy: Past, present and future. *Applied Optics* 10: 368–389.

Pinard, J. 1969. Development of a very high-resolution Fourier transform spectrometer. *Ann. Phys.* 4: 147–196.

Shurcliff, W. A. 1962. *Polarized light.* p. 64. Cambridge, Mass.: Harvard Univ. Press.

Vanasse, G. A., and Sakai, H. 1967. Fourier spectroscopy. *Progress in Optics* 6: 259–330.

Young, J. B.; Graham, H. A.; and Peterson, E. W. 1965. Wire grid infrared polarizers. *Applied Optics* 4: 1023–1026.

PART III
Stars and Nebulae

POLARIZATION FROM ILLUMINATED NONGRAY STELLAR ATMOSPHERES

GEORGE W. COLLINS, II
Perkins Observatory
and
PAUL F. BUERGER
The Ohio State University

In this paper we shall outline the theory of the transport of polarized radiation in a nongray stellar atmosphere and, utilizing this theory, present results for model atmospheres with effective temperatures greater than $6000°K$. We find that the incorporation of the Rayleigh phase function in place of the isotropic phase function yields no significant changes in the atmosphere structure. However, under certain conditions large amounts of polarization can result in the emergent radiation field of a model atmosphere that is illuminated by an external source. This is particularly true for the far ultraviolet.

The dependence on wavelength of the polarization state of a radiation field can contain as much information about the physical nature of the source as does the variation of intensity. However, it is only recently that this potential wealth of information has begun to be exploited for astrophysical purposes. This results, primarily, from the small (and hence difficult to measure) amounts of polarization exhibited by most sources of astrophysical interest.

Also, the theory of transport of polarized radiation has been slow to develop. Indeed, after the formulation of the Stokes parameters (Stokes 1852) for describing the polarization state of a radiation field, very few attempts were made to apply them to specific problems of astrophysical interest until Chandrasekhar (1946a, b; 1947) formulated the equation of transfer allowing for anisotropic scattering. However, Chandrasekhar considered only problems involving pure-scattering gray atmospheres, i.e., $\kappa \neq f(\nu)$. Code (1950) extended the approach to include gray atmospheres having a pure absorption component. Collins (1970) and Harrington (1970) considered the transport problem from the standpoint of formulating Schwarzschild-Milne integral equations for functions related to the source functions of the Stokes parameters. Collins (1972) extended this formulation to include nongray atmospheres that are illuminated by an external source.

In this paper we shall present the results of using this formulation of the equations of radiative transfer to generate self-consistent nongray atmospheres. The first section will outline the formulation of the theory allowing for an arbitrarily polarized incident radiation field. We shall then present results dealing quantitatively with the extent to which polarized transport of radiation can alter the structure of the atmosphere and the degree to which the emergent radiation field will be polarized.

RADIATIVE TRANSFER THEORY

In developing the theoretical basis for the transport of radiation, one may adopt either the approach of solving the integro-differential equation for the specific intensity or utilize the integral equation for the source function together with the classical solution to the equation of transfer in order to determine the emergent radiation field. Normally, the choice is dictated by personal preference; however, in problems that do not exhibit axial symmetry one can make a strong case for solving the integral equation. Collins (1972) has described a method for separating the angle dependence from the depth dependence by developing Schwarzschild-Milne equations for various moments of the radiation field, which by definition are not functions of the angular variables in the problem. He shows that the source functions for the four Stokes parameters (I_l, I_r, U, V) can be expressed in terms of six linearly independent moments of the radiation field $[X(\tau), Y(\tau), Z(\tau), M(\tau), P(\tau), W(\tau)]$ as follows:

$$\left.\begin{aligned}S_l(\tau, \vartheta, \phi) &= \tfrac{1}{2}\varepsilon(\tau)B(\tau) + \tfrac{3}{4}\{2X(\tau) - \cos^2\vartheta\,[Y(\tau) \\ &\quad + Z(\tau)(\cos 2\phi - 2\xi \sin 2\phi)] \\ &\quad + \cos\vartheta \sin\vartheta\, M(\tau)(\cos\phi + \zeta \sin\phi)\} \\ S_r(\tau, \vartheta, \phi) &= \tfrac{1}{2}\varepsilon(\tau)B(\tau) + \tfrac{3}{4}[2X(\tau) - Y(\tau) \\ &\quad + Z(\tau)(\cos 2\phi - 2\xi \sin 2\phi)] \\ S_u(\tau, \vartheta, \phi) &= -\tfrac{3}{4}[2Z(\tau)\cos\vartheta(2\xi \cos 2\phi + \sin 2\phi) \\ &\quad - M(\tau)(\sin\vartheta - \zeta \cos\phi)\sin\vartheta] \\ S_v(\tau, \vartheta, \phi) &= \tfrac{3}{8}[P(\tau)\cos\vartheta + W(\tau)\sin\vartheta \cos\phi]\end{aligned}\right\}, \quad (1)$$

where $\varepsilon(\tau) = (\kappa/\kappa + \sigma)$, $B(\tau)$ is the Planck function, and ξ, ζ are constants that are specified by the surface boundary conditions (i.e., the incident radiation field). The remaining six functions are linear combinations of various moments in ϑ and ϕ of the radiation field, and satisfy the following integral equations:

$$
\left.\begin{aligned}
X(\tau) &= C_X(\tau) + \tfrac{1}{4}[\Lambda_1(\tau) - \Lambda_3(\tau)]|\varepsilon(t)B(t) + 3X(t)| \\
&\quad - \tfrac{3}{8}[\Lambda_3(\tau) - \Lambda_5(\tau)]|Y(t)| \\
Y(\tau) &= C_Y(\tau) + \tfrac{1}{4}[\Lambda_1(\tau) - 3\Lambda_3(\tau)]|\varepsilon(t)B(t) + 3X(t)| \\
&\quad + \tfrac{3}{8}[\Lambda_1(\tau) - 2\Lambda_3(\tau) + 3\Lambda_5(\tau)]|Y(t)| \\
Z(\tau) &= C_Z(\tau) + \tfrac{3}{16}[\Lambda_1(\tau) + 2\Lambda_3(\tau) + \Lambda_5(\tau)]|Z(t)| \\
M(\tau) &= C_M(\tau) + \tfrac{3}{8}[\Lambda_1(\tau) + \Lambda_3(\tau) - 2\Lambda_5(\tau)]|M(t)| \\
W(\tau) &= C_W(\tau) + \tfrac{3}{8}[\Lambda_1(\tau) - \Lambda_3(\tau)]|W(t)| \\
P(\tau) &= C_P(\tau) + \tfrac{3}{4}\Lambda_3(\tau)|P(t)|
\end{aligned}\right\}. \quad (2)
$$

Here, for the sake of brevity, the operator notation of Kurucz (1970) has been used, i.e.,

$$\Lambda_n(\tau)|G(t)| = \int_0^\infty \left(\frac{t-\tau}{|t-\tau|}\right)^{n+1} E_n|t-\tau|G(t)\,dt.$$

In addition, the terms $C_i(\tau)$ are determined by the boundary conditions (i.e., the incident radiation field). In the event that the incident radiation field is plane parallel, incident from an angle $\vartheta_0 = \cos^{-1}\mu_0$ and from a direction $\phi = 0$, and specified by Stokes parameters I_ℓ^0, I_r^0, U^0, V^0, then the inhomogeneous term, C_i, becomes

$$\begin{bmatrix} C_X(\tau) \\ C_Y(\tau) \\ C_Z(\tau) \\ C_M(\tau) \\ C_W(\tau) \\ C_P(\tau) \end{bmatrix} = \frac{[1-\varepsilon(\tau)]e^{-\tau/\mu_0}}{4\pi} \begin{bmatrix} (1-\mu_0^2)I_\ell^0 \\ (2-3\mu_0^2)I_\ell^0 - I_r^0 \\ I_r^0 - \mu_0^2 I_\ell^0 \\ -\mu_0(1-\mu_0^2)^{1/2}I_\ell^0 \\ (1-\mu_0^2)^{1/2}V^0 \\ -\mu_0 V^0 \end{bmatrix} \quad (\mu_0 > 0). \quad (3)$$

The boundary condition U^0 enters into the problem through the specification of ξ and ζ such that

$$\left.\begin{aligned} \xi &= +\mu_0 U^0/2(I_r^0 - \mu_0^2 I_\ell^0) \\ \zeta &= +U^0/2\mu_0 I_\ell^0 \end{aligned}\right\}. \quad (4)$$

The integral Equation (2) may be solved by any number of standard techniques, one of the simplest perhaps being the replacing of the Λ-operators by a gaussian sum and evaluating the resulting functional equation in τ at the points of the gaussian division. The resulting system of linear algebraic equations will then specify the solution at the points of the gaussian division. This technique can only be utilized if the singularity in the $\Lambda_1(\tau)$ operator is removed. This can be done by noting that

$$\Lambda_n(\tau)|G(t)| = \Lambda_n(\tau)|G(\tau) - G(t)| + G(\tau)\Lambda_n(\tau)|1|. \tag{5}$$

For the case $n = 1$, this yields

$$\Lambda_1(\tau)|G(t)| = \Lambda_1(\tau)|G(t) - G(\tau)| + G(\tau)[1 - \tfrac{1}{2}E_2(\tau)], \tag{6}$$

where the integrand of the first term on the right contains no singularities and is continuous if the solution to the equation is continuous. It is generally a good idea to evaluate all $\Lambda_n(\tau)$ operators in this manner even for $n \neq 1$ in order to assure continuity and differentiability of the integrand.

In order to construct the atmosphere, it is necessary to evaluate the net radiative flux as well as the flux derivative, as both of these parameters are required by the temperature correction scheme.

From the definition of the radiative flux, from the classical solution to the equation of transfer, and from the source function (Equations 1), one may immediately obtain an expression for the radiative flux carried vertically in the atmosphere, which is

$$F_v(\tau_v) = 4\Lambda_2(\tau_v)|\varepsilon_v(t)B_v(t) + 3X_v(t) - 3/4\,Y_v(t)| - 3\Lambda_4(\tau_v)|Y_v(t)|, \tag{7}$$

where v denotes the frequency of radiation. The value for the flux derivative is simply calculated in terms of the first moment of the equation of radiative transfer. Thus,

$$\frac{dF_v(\tau_v)}{d\tau_v} = J_v(\tau_v) - \frac{1}{4\pi}\int_\Omega [S_\ell(\tau_v, \vartheta, \sigma) + S_r(\tau_v, \vartheta, \phi)]\,d\Omega. \tag{8}$$

Calculating $J_v(\tau_v)$ from its definition and using Equations (1) for the source functions, we obtain

$$\frac{dF_v(\tau_v)}{d\tau_v} = \frac{\varepsilon_v(\tau_v)}{1 - \varepsilon_v(\tau_v)}\{[1 - \varepsilon_v(\tau_v)]B_v(\tau_v) - 3X_v(\tau_v) + Y_v(\tau_v)\}. \tag{9}$$

It is worth noting that only the moment functions $X(\tau)$ and $Y(\tau)$ play a role in determining the flux and flux gradient.

Both Equations (7) and (8) are required for the formulation of the Avrett-Krook temperature scheme. However, small modifications are made to account for the presence of incident flux. Following a suggestion by Karp (1972), Buerger (1972) has developed the formalism appropriate for the Avrett-Krook temperature correction scheme for illuminated atmospheres in which isotropic scattering is present. We have followed this procedure, modified by means of Equations (7) and (8), for illuminated anisotropically scattering atmospheres. The resulting procedure produces an iteration sequence that closely mirrors that of isotropic scattering atmospheres with identical effective temperature and surface gravity.

DISCUSSION OF RESULTS

In this section we shall describe the results of several calculations utilizing the theory described in the previous section. In order to isolate the effects of anisotropic scattering, a similar model with an isotropic scattering phase function was made for every anisotropic model. Care was taken to insure that all computational approximations were of the same accuracy for the pair of models and that in all respects the numerical methods were as identical as possible. Thus, we feel that the difference between the models can be attributed to the presence of anisotropic scattering to an accuracy of better than 0.1%.

In order to organize the results of this investigation, it is useful to divide our discussion into the effects directly related to the atmosphere structure itself and the modification of the emergent radiation field resulting both from structure changes and the anisotropic nature of the scattering functions.

Atmospheric Structure

Since the radiation field will depart from what is expected in an isotropically scattering atmosphere, one might expect that the somewhat different radiative flux would yield a slightly different temperature gradient for a model in radiative equilibrium. This effect should be amplified as one approaches early-type stars because, as the atmosphere becomes dominated by electron scattering (and thus nearly gray), the criteria of radiative equilibrium becomes much less sensitive in specifying the physical structure. This consideration was not taken into account by Harrington (1970) in reaching his conclusion, based on a gray atmosphere study, that the atmospheric structure would not be extensively modified by the presence of anisotropic scattering.

Since we would expect differences in the isotropic and anisotropic models to be maximized near the surface, let us investigate the behavior of the quantity

$$f_\nu(\tau) = \frac{F_\nu(\text{isotropic}) - F_\nu(\text{anisotropic})}{F_\nu(\text{anisotropic})} \qquad (10)$$

evaluated at optical depth $\tau = 0$ for models where we would expect large amounts of electron scattering. Figure 1 shows the variation of this quantity with frequency for models made with different values of effective temperature T_e and surface gravity g. From this it is clear that effects previously described are present but at a level much lower than anticipated. Indeed, for main-sequence B stars, the maximum redistribution of emergent flux attributable to the Rayleigh phase function is less than 0.5%. This rises to nearly 1% for supergiants suggesting that anisotropic scattering may become marginally important in extended atmospheres.

One can further see from Fig. 1 that the largest effects occur in the vicinity of the Balmer and Lyman absorption edges. This can be understood in terms of the very small redistribution in frequency space required to move the radiation from a regime where the opacity is dominated by scattering to one characterized by relatively large pure absorption. If one now considers the flux departures with depth, he will see that they rapidly diminish with optical depth. Since the radiation field near the surface plays little role in determining the physical structure, we conclude with Cassinelli and Hummer (1971) and Harrington (1970) that use of the Rayleigh phase function will have no effect on the physical structure of a normal stellar atmosphere. Indeed the largest change in any physical parameter was a depression of 1% in the surface temperature for atmospheres with $10,000°K < T_e < 30,000°K$ and $3 < \log g < 4$. Thus having verified the primary assumption of Collins (1970), we may proceed to compare the emergent radiation field with earlier work and consider the effect of scattering on the "reflected" incident radiation one might expect to be present in close binary systems.

The Emergent Radiation Field

Collins (1970), using atmospheres whose physical structure had been obtained with an isotropic phase function, noted that in the visible part of the spectrum only negligible (less than 1%) polarization would be found on the visible disk of an early-type star; similar results were obtained by Nagirner (1962) and Rucinski (1970). This was contrary to the expectation based on the work of Chandrasekhar (1946a, b) and Code (1950). This result can be most readily understood by noting that two conditions must be satisfied in order for a net polarization of light to occur in a situation where the scattering particles are not preferentially oriented. Firstly, there must be scattering particles present, and in our case that implies that the opacity must be largely due to electron scattering. Secondly, the radiation field illuminating the scattering particles must exhibit some anisotropy. Usually, the larger the anisotropy, the larger will be the degree of polarization in the scattered light. In stellar atmospheres the anisotropy of the

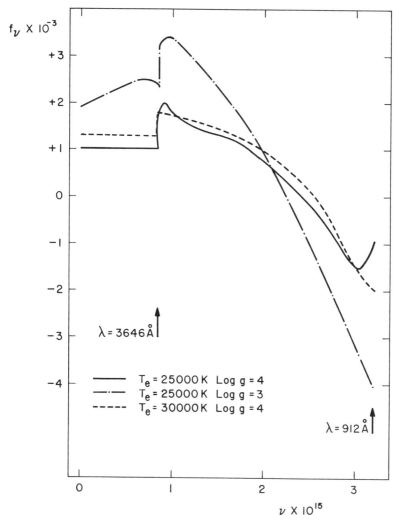

Fig. 1. The emergent flux for nonilluminated atmospheres that differ only in the nature of the assumed scattering function: $f_\nu = [F_\nu \text{(isotropic)} - F_\nu \text{(anisotropic)}]/[F_\nu \text{(anisotropic)}]$.

radiation field is directly related to the source function gradient. The source function gradient will in turn depend on the physical structure of the atmosphere and the wavelength of the radiation. The resulting dependence of the anisotropy on wavelength was not considered by Chandrasekhar (1946a, b), and thus, his calculations can only be expected to be applicable to atmospheres that are nearly gray and for wavelengths where the source function is roughly proportional to the fourth power of the temperature. For B stars the source function in the visible part of the spectrum is roughly proportional to the temperature to the first power or less, and the resultant anisotropy in the surface regions is greatly reduced and with it the net polarization.

For nonilluminated atmospheres, our results are in good quantitative agreement with those of Collins (1970) for $1 > \mu > 0.1$. Only at $\mu = 0$ is there any particular difference. It is possible to trace this difference to the nature of the basic atmospheres used in the two studies. Collins (1970) used atmospheres similar to those described by Underhill (1962, 1963). These atmospheres differ markedly from more recent calculations of Kurucz (1969) and Mihalas (1972) in the region $0 < \tau < 10^{-3}$. The rapid rise of the ionization in this region leads to a marked increase in $Y(\tau)$, which in turn will increase the value of the polarization at the limb. Even for atmospheres as cool as $T_e = 10,000°$K, $\log g = 4$, we have found limb values of the polarization as high as 2% in the visible.

With this one exception however, our results are in excellent quantitative agreement with those of Collins (1970), so let us now turn to the more interesting case of the polarization of the radiation field from an illuminated atmosphere. Buerger (1972) has shown that the effects of incident radiation on the structure of the atmosphere are not as simple as one would like. For instance, not only is the total flux important, but the angle of incidence for the same net incident flux will result in varying the atmospheric structure. This results from the fact that the depth at which energy is characteristically absorbed is governed by the angle of incidence and not by the amount of incident flux.

This problem seems to point out a second one, namely, a large number of linearly independent parameters are necessary to specify a model. Indeed, there are nine such parameters. In addition to the three parameters [i.e., T_e, $\log g$, and μ (chemical composition)] required to specify the normal atmosphere, six more are required to uniquely specify the incident radiation field: four Stokes parameters, a frequency distribution, and an incident angle. Clearly, we cannot investigate the properties of the emergent radiation field as a function of all nine parameters. Instead we shall limit ourselves to choices of three parameters that are most likely to provide insight into the problem of the reflection effect in close binaries. To this end we shall assume that the incident radiation is unpolarized for all our models (i.e., $U^0 = V^0 = 0$). In addition, let us consider an atmosphere that would have a nonilluminated effective temperature of 10^4°K and $\log g = 4$ illuminated by a radiation field appropriate to a 2.5×10^4°K, $\log g = 4$ star. Finally, we shall assume a dilution factor of 1/2 and an incident angle specified by $\mu_0 = 0.2$. Figure 2 indicates the

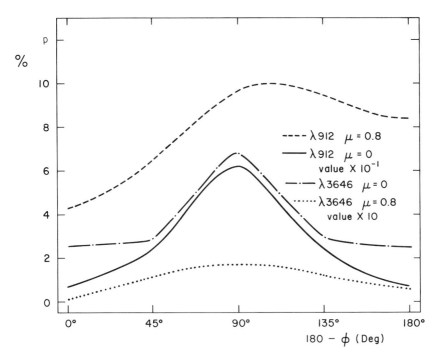

Fig. 2. Expected percentage polarization from an illuminated stellar atmosphere with a nonilluminated $T_e = 10^{4\circ}$K and log $g = 4$. The ordinate denotes the percentage of linear polarization while the abscissa indicates the azimuthal angle with respect to the incident beam plus π. The illuminating radiation field is characterized by a radiation temperature of $T_e = 2.5 \times 10^{4\circ}$K and log $g = 4$. It has been diluted so that $d = F_{\text{total}}(\text{incident})/T_e^4 = 0.5$. The angle of incidence is specified by $\mu_0 = 0.2$. The polarization values displaced for 912 Å, $\mu = 0$ have been decreased by a factor of 10, while the values for 3646 Å, $\mu = 0.8$ have been increased by a factor of 10.

degree of polarization at two wavelengths for emergent angles of $\mu = 0.8$ and 0.0 as a function of the azimuthal angle ϕ. Since the degree of polarization expected from the nonilluminated atmospheres is virtually zero for $\mu = 0.8$, we may conclude that the entire effect is due to the scattering of the incident radiation. The radiation field seen at $\mu = 0.8$ and $\phi = 0°$ corresponds to a scattering angle near $90°$, and therefore one might expect the polarization of the scattered radiation to be as high as 100%. However, the radiation is diluted both by the diffuse nearly nonpolarized radiation of the star and by the fact that the incident beam is scattered over 2π steradians. More important than those factors is the tendency for the incident beam to increase the isotropy of the radiation field in the surface regions. As we have already pointed out, it is the anisotropy of the radiation field, as determined by the source function gradient, that is the primary factor in determining the degree of polarization to be expected. Thus, if we are to encounter a physical situation where the back-scattered radiation exhibits significant polarization, we should attempt to satisfy the following conditions:

1. The incident flux should not be so large as to heat the surface layers sufficiently to significantly increase the isotropy of the surface radiation field.
2. The effective temperature of the illuminated atmosphere should be sufficiently low so as to insure that the wavelength maximum of the energy distribution is shifted to the red of the part of the spectrum where one expects to find polarization. For the visible, this requires $T_e < 8000°K$.
3. The surface gravity should be sufficiently low and/or the color temperature of the incident radiation sufficiently high so as to guarantee a fairly high degree of ionization in the surface region.

One might most likely encounter such conditions in binary systems consisting of a G giant and B dwarf which form a fairly close system, but not so close as to violate condition (1). Indeed, if one considers such an atmosphere (i.e., $T_e = 6000°K$, $\log g = 3$), illuminated by a B dwarf ($T_e = 25,000°K$, $\log g = 4$) with a $\mu_0 = 0.7$, we find that significant amounts of polarization will result in the visible, while in the far ultraviolet the degree of polarization may rise as high as 70%. Figure 3 shows the results of this calculation for wavelengths just to the long wavelength side of the Balmer and Lyman jumps. Clearly, the degree of polarization has been greatly enhanced over the previous case. This indicates that polarization in the reflection effect will only result in those cases where the incident field totally dominates the atmosphere structure and initial radiation field.

CONCLUSIONS

In this paper we have presented the results of self-consistent nongray model atmosphere calculations where the isotropic phase function has been replaced with the more physically correct Rayleigh phase function for electron scattering. We find that essentially no significant change in the physical structure results from the use of the anisotropic scattering law. It appears that the most plausible physical explanation for this is that the departures in J_v and F_v that arise from anisotropic scattering occur too high in the atmosphere to affect the physical structure. However, it is conceivable that these departures may very slightly affect the results of non–local-thermodynamic-equilibrium (LTE) model atmospheres since the maximum effects occur precisely in those regions where the departures from LTE become most dominant.

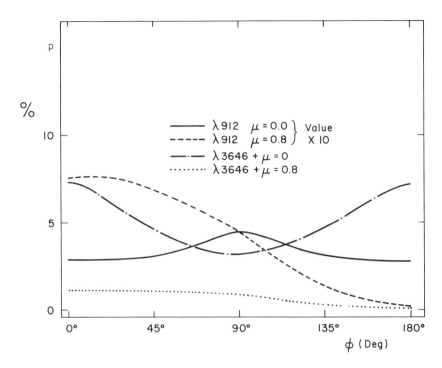

Fig. 3. As for Fig. 2, but the illuminated atmosphere is more nearly that of a G III star ($T_e = 6 \times 10^3$, log $g = 3$); $\mu_0 = 0.7$, $d = 1/6$; and the values for 3646 Å, $\mu = 0.8$ have not been multiplied by 10.

In addition to the atmospheric structure, we investigated the properties of the emergent radiation field for both nonilluminated and illuminated atmospheres. In the case of nonilluminated models, we found good agreement with earlier work except near the limb where the degree of polarization is somewhat larger than earlier predictions had indicated. This difference is attributed to improvement in the quality of the models, particularly in the surface regions. For models involving surface illumination, we found that significant polarization could result in visible light only for cases where the incident field contains sufficient ultraviolet radiation to ionize most of the hydrogen in the atmosphere, while the atmospheric temperature was low enough to yield a steep source function gradient in the visible part of the spectrum. In addition, we found many models for which large amounts of polarization would occur in the far ultraviolet part of the spectrum. Thus, we conclude that binary systems consisting of a late giant and early main-sequence dwarf may exhibit measurable polarization in the vicinity of the Balmer jump and most certainly would in the far ultraviolet. It appears that early-type atmospheres illuminated by a high temperature radiation field will not exhibit much polarization in visible light. This seems to result from the heating of the surface layers and the resultant increase in the isotropy of the radiation field. However, it is possible that if the calculation did not assume LTE, the heating of these layers would be reduced. This would most certainly increase the predicted polarization. Whether or not the increase would be sufficient to yield a measurable amount requires a detailed calculation. Lastly, it appears that any measured polarization in excess of 1% or 2% for either single or multiple star systems cannot arise in the atmospheres of these stars. The possible exception to this might be late-type supergiants that have been severely distorted either by rotation, pulsation, or large magnetic fields. Thus, we must look to circumstellar shells or vastly extended atmospheres to explain the observed polarization of Be stars.

REFERENCES

Buerger, P. F. 1972. Stellar atmospheres with radiation incident at the surface. *Astrophys.J.* 177: 657–664.

Cassinelli, J. P., and Hummer, D. G. 1971. Radiative transfer in spherically symmetric systems. II. The nonconservative case and linear polarized radiation. *Mon.Not.R.Astr.Soc.* 154: 9–21.

Chandrasekhar, S. 1946a. On the radiative equilibrium of a stellar atmosphere X. *Astrophys.J.* 103: 351–370.

———. 1946b. On the radiative equilibrium of a stellar atmosphere XI. *Astrophys.J.* 104: 110–132.

———. 1947. On the radiative equilibrium of a stellar atmosphere XV. *Astrophys.J.* 105: 424–460.

Code, A. D. 1950. Radiative equilibrium in an atmosphere in which pure scattering and pure absorption both play a role. *Astrophys.J.* 112: 22–47.

Collins, G. W. II. 1970. Intrinsic polarization in nongray atmospheres. *Astrophys.J.* 159: 583–591.

———. 1972. Transfer of polarized radiation in a stellar atmosphere. *Astrophys.J.* 175: 147–156.

Harrington, J. P. 1970. Polarization of radiation from stellar atmospheres. The gray case. *Astrophys.Space Sci.* 8: 227–242.

Karp, A. H. 1972. A modification of the Avrett-Krook temperature-correction procedure. *Astrophys.J.* 173: 649–652.

Kurucz, R. L. 1969. *Theory and observation of normal stellar atmospheres.* (O. Gingerich, ed.) pp. 401–405. Cambridge, Mass.: The MIT Press.

———. 1970. Atlas: A computer program for calculating model stellar atmospheres. Smithsonian Astrophys.Obs. Special Report 309.

Milhalas, D. 1972. Non-LTE model atmospheres for B and O stars. Nat.Center Atmospheric Research—TN/STR-76.

Nagirner, D. I. 1962. Polarization of light in stellar atmospheres. *Trudy Astron. Observ. Leningrad* 19: 79–87. (In Russian.)

Rucinski, S. M. 1970. An upper limit to Chandrasekhar's polarization in early-type stars. *Acta.Astr.* 20: 1–12.

Stokes, G. G. 1852. On the composition and resolution of streams of polarized light from different sources. *Trans.Camb.Philos.Soc.* 9: 399–416.

Underhill, A. 1962. A program for computing early-type model atmospheres and testing the flux integral. *Pub.Dom.Ap.Obs.* 11: 433–466.

———. 1963. Concerning the interpretation of line strengths in B-type spectra. *Bull. Astron. Inst. Netherlands* 17: 161–175.

DISCUSSION

SERKOWSKI: I have observed the polarization of HZ Her on two occasions. On September 29.19 UT, at minimum light I got $p = 0.4 \pm 0.5$ (m.e.)%, $\theta = 91°$, and right-handed circular $q = 0.2 \pm 0.2$(m.e.)%. On November 6.10, at phase 0.3, I got $p = 0.13 \pm 0.12$ (m.e.)%, $\theta = 90°$. Both observations were made at wavelengths around 0.5 μm.

LANDAU: Would your calculations be much different if you illuminated the B star with X rays as in Her X-1 or perhaps Cen χ-3?

COLLINS: No. I would not expect very much polarization as you are even further out on the tail of the energy distribution than that expected for early-type stars.

OBSERVATIONAL ASPECTS OF COHERENCE IN RADIO POLARIZATION MEASUREMENTS OF AREA SOURCES

G. FEIX
Astronomisches Institut
Bochum

The amount of polarization is discussed when independent elementary sources emit polarized radiation. It is assumed that the size of each source subtends a finite solid angle that is less than the half-power beamwidth of the antenna. As an example, the polarized emission at millimeter-radio wavelengths of a solar active region is discussed.

The emission from a physical source of finite extension may be expressed as the sum of components of a great many elementary sources. Since the emission received from such an incoherent source becomes quasi-coherent within a discrete area, the concept termed "area of coherence" (van Cittert 1934; Zernike 1938) has been introduced. For example, a source as large as the solar disk, i.e., having an angular extent of 32 arcmin, would give rise to an almost coherent illumination of an area on earth having a diameter on the order of only 18 wavelengths (Born and Wolf 1959). In general, we may express the relationship between the angular extent α of a uniform circular source and the distance of coherence d by

$$\alpha \leq 10° \, \lambda/d. \qquad (1)$$

On the other hand, the beamwidth of a two-element interferometer is given by

$$BWFN = 57°\!.3 \, \lambda/d, \qquad (2)$$

while for a uniformly illuminated parabolic dish, the half-power beamwidth is

$$BW_{3db} = 58°\!.4 \, \lambda/d; \qquad (3)$$

BWFN is the beamwidth between first nulls of the interferometer, and d is the spacing of the interferometer or in Equation (3), the diameter of the antenna. Comparing Equation (1) with Equations (2) and (3), one finds that a circular area source that subtends the beamwidth more than $10° \lambda/d$ will produce partially coherent emission. Thus, from the amount of the incoherent component, an estimate may be made for the decrease of the degree of polarization in an area source.

POLARIZATION AS A FUNCTION OF COHERENCE

We must take into account both the correlation that exists for received radiation, which is responsible for the described "area of coherence," and the correlation between orthogonal components of the electric vector, which implies the polarization of the resulting wave. In the first case, the degree of coherence may be obtained by measuring the sharpness of interference fringes that would result by combining the intensities from two arbitrarily chosen points at the receiver terminal. This observable quantity is called the visibility V_i and is related to the degree of coherence $|\mu_{12}|$ by

$$|\mu_{12}| = \frac{J_{12}}{\sqrt{J_{11}J_{22}}}, \tag{4}$$

$$V_i = \frac{\sqrt{J_{11}J_{22}}}{\frac{1}{2}(J_{11} + J_{22})} |\mu_{12}|, \tag{5}$$

where J_{mn} are the elements of the coherency matrix. Furthermore,

$$V_i \leq |\mu_{12}|. \tag{6}$$

In a similar way, the degree of coherence between orthogonal planes of the received wave is described by

$$|\mu_{xy}| = \frac{J_{xy}}{\sqrt{J_{xx}J_{yy}}}. \tag{7}$$

The absolute value of $|\mu_{xy}|$ may be measured by the degree of polarization p and the modified[1] Stokes parameter q,

$$q = Q/I_{\text{pol}}:$$

$$|\mu_{xy}| = p \frac{1 - q^2}{\sqrt{1 - p^2 q^2}}, \tag{8}$$

---

[1] This q should not be confused with the one in the Glossary (p. 1084) and elsewhere in the book — ED.

while the complex value of μ_{mn} is given by

$$\mu_{xy} = p \frac{u + iv}{\sqrt{1 - p^2 q^2}}$$

$$\mu_{yx} = p \frac{u - iv}{\sqrt{1 - p^2 q^2}}. \qquad (9)$$

It follows from Equation (8) that

$$\left| \mu_{xy} \right| \leq p; \qquad (10)$$

I_{pol} is defined by $\sqrt{Q^2 + U^2 + V^2}$. On the other hand, the parameters q, u, and v are dimensionless and related by

$$q^2 + u^2 + v^2 = 1. \qquad (11)$$

In dealing with the polarization of a source, it is evident that μ_{12} constitutes a primary limitation on polarization, while $|\mu_{xy}|$ always has a maximum when $q = 0$. Using Equation (8), we find that $|\mu_{xy}|$ is then equal to the degree of polarization. The xy-planes of reference can always be rotated so that this condition will be achieved.

POLARIZATION AS A FUNCTION OF THE BRIGHTNESS DISTRIBUTION OF THE SOURCE

Interferometric observations show that coherent emission is not necessarily polarized; on the other hand, polarization depends strictly on coherence. If we define the polarization in terms of intensities, we may write

$$p = \frac{I_{\text{pol}}}{I_{\text{tot}}}. \qquad (12)$$

I_{pol} is the polarized component of the total intensity I_{tot}. The presence of the coherent component in I_{pol} may be brought out if the degree of polarization is expressed by the ratio

$$p' = \frac{I_{\text{pol}}}{I_{\text{coh}}}, \qquad (13)$$

where I_{coh} is the coherent component of the total intensity received. From Equations (12) and (13) it follows that

$$p = \frac{I_{\text{pol}}}{I_{\text{coh}}} \cdot \frac{I_{\text{coh}}}{I_{\text{tot}}};$$

COHERENCE IN RADIO POLARIZATION

$$p = p' \cdot |\mu_{12}|, \qquad (14)$$

where the relation

$$|\mu_{12}| = I_{\text{coh}}/I_{\text{tot}}$$

has been used.

For a source of a definite angular extent, the polarization should be expressed by polarization per unit angle. Thus,

$$p(s,t) = p'(s,t) \cdot \mu_{12}(s,t). \qquad (15)$$

Equation (15) represents the angular distribution of the degree of polarization per unit angle, where s and t are the angular coordinates.

Since the coherency matrix of a combined wave is equal to the sum of the coherency matrices of all separate waves, the integral over the source yields the total polarization P as received by the antenna:

$$P = \iint_{\text{source}} p(s,t)\,\mathrm{d}s\,\mathrm{d}t \approx \int\int_{-\infty}^{+\infty} p(s,t)\,\mathrm{d}s\,\mathrm{d}t. \qquad (16)$$

When the source is not sufficiently small, the observed polarization will be less than that given by Equation (16) due to the weighting function of the antenna power pattern.

Taking the van Cittert-Zernike theorem that the degree of coherence $|\mu_{12}|$ is equal to the Fourier transform of the normalized intensity distribution of the source, and using the convolution theorem of the Fourier transforms, one obtains the distribution of polarization in terms of the spatial frequencies x/λ and y/λ; x and y are the coordinates of the source. Thus,

$$p\left(\frac{x}{\lambda},\frac{y}{\lambda}\right) = \int\int_{-\infty}^{+\infty} p'\left(\frac{x'}{\lambda},\frac{y'}{\lambda}\right) \cdot I_n\left(\frac{x'}{\lambda}-\frac{x}{\lambda},\frac{y'}{\lambda}-\frac{y}{\lambda}\right) \mathrm{d}\left(\frac{x'}{\lambda}\right)\mathrm{d}\left(\frac{y'}{\lambda}\right), \qquad (17)$$

where λ is the wavelength and x'/λ and y'/λ are the displacements of the convolution operation. This convolution integral implies the cross-correlation operation between two functions (Bracewell 1965). Equation (17) shows that the observed polarization is a weighted mean of the individual polarization distribution, i.e., weighted by the Fourier transform of the degree of coherence.

POLARIZED EMISSION FROM SOLAR PLAGES

Insight into the concepts described here may be obtained by using active areas of the sun as an example. In general, a region active at mm-radio wavelengths displays a strong emission center of about flare size, while the remaining radiation comes from a region of larger extent approximating that of the plage area. A photographic Hα representation is adopted here for the model of the mm-radio source. In Fig. 1 the bright spot of a flare embedded in a dark mottled plage area gives a typical illustration of the core and its surrounding solar microwave source. Although this region has not been completely resolved at radio wavelengths, the maximum of the total flux corresponds to the optically observable active core of the plage. However, the optical size and brightness of this flare-producing core are less than that shown in Fig. 1.

Fig. 1. Size and structure of a typical region active at mm-wavelengths. Note that a flare is observed in Hα + 0.5 Å. (Photograph courtesy of Sacramento Peak Observatory.)

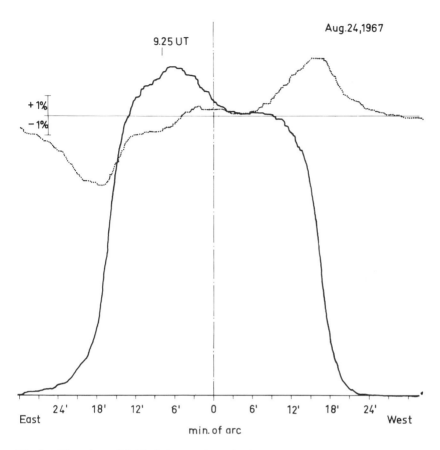

Fig. 2. The plage (McMath No. 8704) of Fig. 1 observed at mm-radio wavelengths (36 GHz). ——— total flux: ...... Stokes parameter V.

A mm-radio scan record of the same area is given in Fig. 2. Note the intrinsic polarization effects at the limb of the solar scan. For comparison, this region has also been recorded close to the solar limb, as shown in Figs. 3 and 4. These figures reveal linearly polarized regions at each side of the intensity maximum. The position angle of polarization is rotating almost 90°. On these records the source becomes smaller than the beamwidth of the antenna. Linear polarization has been obtained close to the limb, while the emission appears to be circularly polarized when the plage is approaching the solar center, i.e., facing the earth.

Apart from this effect there is also evidence for the decrease of the degree of polarization between the solar limb and the center. One possible explanation may be the "area of coherence." Let us assume a gaussian

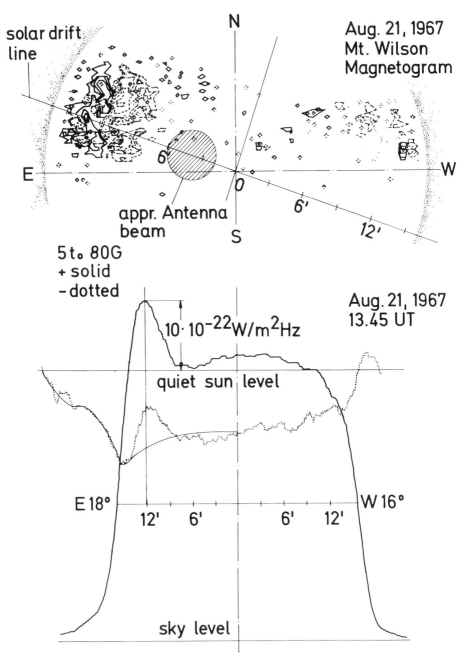

Fig. 3. The same plage close to the east limb observed at 36 GHz. Scale of polarization is twice that of Fig. 2. (Intrinsic polarization has been superimposed on the polarization profile.) ...... Stokes Parameter U.

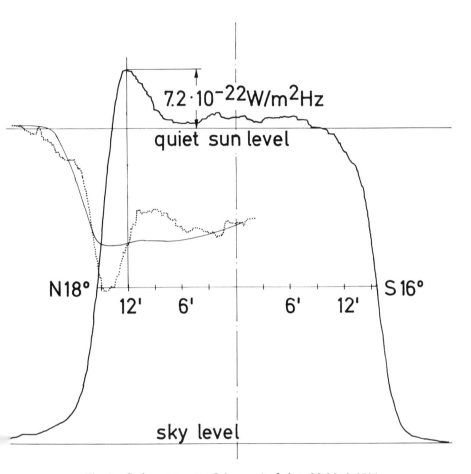

Fig. 4. Stokes parameter Q (......) of plage McMath 8704.

distribution for both the discrete polarized source as well as for the total flux source. In this case we may write

$$I(s,t) \propto e^{-\alpha \pi m^2}$$
$$p'(s,t) \propto e^{-\beta \pi m^2} \tag{18}$$

where

$$(x/\lambda)^2 + (y/\lambda)^2 = r^2$$
$$s^2 + t^2 = m^2.$$

Substituting the Fourier transforms of Equation (18) in Equation (17) and integrating, we have

$$p(x/\lambda, y/\lambda) = \frac{1}{\sqrt{\beta + \alpha}} \, e^{-\pi \frac{\alpha + \beta}{\alpha \beta} r^2}$$

Since

$$P = p(x/\lambda=0, y/\lambda=0) = \int\int_{-\infty}^{+\infty} p(s,t) \, ds \, dt,$$

the total polarization of the source becomes

$$P = \frac{1}{\sqrt{\beta + \alpha}}. \tag{19}$$

Equation (19) has been plotted in Fig. 5 for various source distributions α. The expression $\beta/\alpha = 1$ means that the polarized source extends over the whole main source. If the polarized source decreases, the polarization approaches unity, i.e., the polarized source appears to be a point source. This means that the degree of polarization will be equal to a maximum. On the other hand, when the source becomes larger with respect to the antenna main-lobe solid angle (see Figs. 2 and 3), the degree of polarization will decrease.

Thus, the degree of polarization depends on the size of the antenna beamwidth. This fact becomes important when data from different antennas are compiled in order to look for a polarization spectrum.

REFERENCES

Born, M., and Wolf, E. 1959. *Principles of optics.* P. 511. New York: Pergamon Press.

Bracewell, R. 1965. *The Fourier transform and its application.* P. 46. New York: McGraw-Hill.

Cittert, P. H. van 1934. Die Wahrscheinliche Schwingungsverteilung in einer von einer Lichtquelle Direckt oder Mittels einer Linse Beleuchteten Ebene. *Physica* 1: 201–210.

Zernike, F. 1938. The concept of degree of coherence and its application to optical problems. *Physica* 5: 785–795.

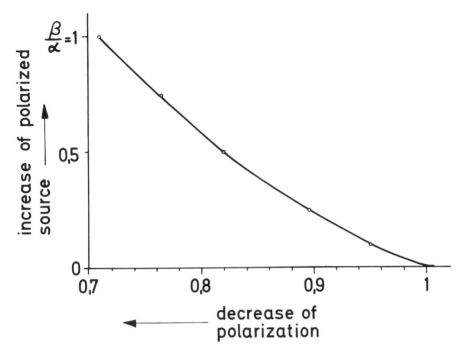

Fig. 5. Size of polarized source versus observed degree of polarization.

DISCUSSION

ANGEL: What kind of emission mechanism is applicable to the observed polarization?

FEIX: Probably a magneto-ionic process accounts for the polarization.

POLARIZATION MEASUREMENTS ON THE SUN'S DISK

DONALD L. MICKEY and *FRANK Q. ORRALL*
Institute for Astronomy
University of Hawaii

The nature and sources of polarization observed on the visible sun's disk in broad spectral bands are summarized. Observations from Mount Haleakala are used to illustrate the polarization near the limb on the quiet disk; two-dimensional raster scans of the complete Stokes vector in active regions show the distribution of broadband polarization within sunspots. A surprising result is that in sunspots there is a net circular polarization ($V/I \sim 10^{-4}$) as well as linear polarization. The importance of observing transient broadband polarization in solar flares is stressed.

Most polarization measurements of solar radiation have been made within spectral lines for the purpose of measuring photospheric magnetic fields by the Zeeman effect. In this paper we consider sources of broad-band polarization on the disk.

The measurements of disk polarization described here were made using a versatile scanning photoelectric polarimeter at Mount Haleakala (elevation 3054 m) on the island of Maui, Hawaii. The instrument has been described by Orrall (1971) and an analysis of the sources of noise and instrumental polarization was made by Mickey, Orrall, and Zane (1972). Briefly, the polarimeter consists of an $f/15$ coronagraph telescope of 2.5 m focal length. The scanning aperture in the primary focal plane is always on the optic axis of the telescope. Behind this aperture is a rotating quarter-wave plate followed by a Wollaston prism beam splitter, a collimating lens, and two photomultipliers. The system uses the principles of the Lyot coronagraph, even when used on the disk, in order to reduce scattered light and instrumental polarization. The two signals are processed by a PDP-8/I computer which computes the complete Stokes vector in real time from the Fourier components of the modulated signal. The entire instrument can perform a raster scan on the sun's disk or corona and build up two-dimensional scans of polarization.

The fundamental modulating frequency is 6.25 Hz, and the harmonic measuring linear polarization is 25 Hz. This is low, but by using the two orthogonal channels differentially, we can largely compensate for the effects of seeing and fluctuations in sky transparency. Nevertheless, such effects are the principal sources of noise in most applications. Systematic instrumental polarization seems to be introduced almost entirely by the scanning aperture. Although it may reach 10^{-4} when the smallest apertures are used, it can be reduced to $\sim 10^{-5}$ by calibration and real-time numerical compensation.

SOURCES OF BROADBAND POLARIZATION ON THE SUN'S DISK

We shall consider three sources of broadband disk polarization: (a) polarization near the sun's limb; (b) polarization in active regions within sunspot groups; and (c) polarization produced at the time of solar flares by nonthermal particles. The linear polarization near the limb and in sunspot groups is now well established, mainly because of the careful pioneering and modern work of the French observers, Lyot, Dollfus,[1] and Leroy,[2] and their associates at Meudon and Pic-du-Midi. To our knowledge, detailed observations of nonthermal, polarized, visible radiation from flares have not been successfully made, although Dollfus (1958) has observed a transient polarized event near a sunspot.

Polarization Near the Sun's Limb

The radiation field in the sun's atmosphere above the photosphere is not isotropic, and radiation scattered toward the observer from near the limb is linearly polarized with the electric vector tangent to the limb. Since the scattering is done by electrons (Thompson scattering), by atoms (Rayleigh scattering), and by scattering between the bound states of atoms (i.e., within Fraunhofer lines), the amount of polarization and its behavior near the limb will depend on wavelength. Center-to-limb observations in a number of wavelength bands have recently been made by Leroy (1972), who also gives a good review of earlier work. Recent theoretical studies of this broadband line and continuum polarization have been made by Pecker (1970), Dumont and Pecker (1971), and Débarbat, Dumont, and Pecker (1970*a,b*).

Figure 1 shows center-to-limb measurements of linear polarization p, made from Mount Haleakala in a wavelength band ~ 20 Å wide centered at 5834 Å — a region relatively free from Fraunhofer lines. The measurements are in good agreement with those of Leroy at 6034 Å.

---

[1] See p. 695. [2] See p. 762.

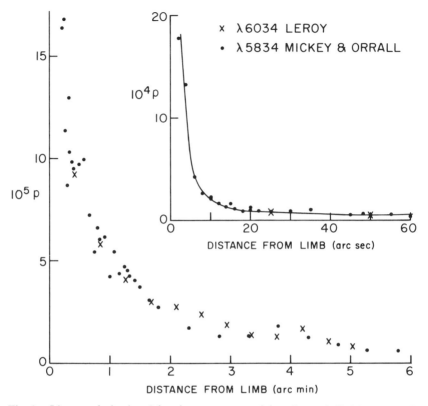

Fig. 1. Linear polarization (electric vector tangential to the sun's limb) measured along a radius of the quiet sun's disk toward the limb.

We are able to measure tangential linear polarization to within ~4 arcsec of the limb. The Stokes parameter measuring circular polarization and that measuring linear polarization oriented at 45° to the limb were measured simultaneously with the tangential component. Both are small and show no systematic behavior at the limb. We take this as evidence that the polarization measurements very close to the limb are real, and that the use of the two channels differentially is very effective in reducing spurious polarization signals caused by seeing noise.

Polarization in Sunspot Groups

Dollfus (1958) detected broadband linear polarization in sunspots, and detailed measurements of this effect have been made by Leroy (1962). They attribute this polarization to self-absorption in the transverse Zeeman effect. (In the simple Zeeman triplet, in an optically thin gas, the sum of the two σ components just equals the π component, and there is no net polarization. In a self-absorbing gas, however, the π component suffers a greater saturation effect, leaving a net linear polarization for the whole line.) Since a broadband measurement includes the polarization of many lines of different strengths and Landé factors, calibration of such measurements to determine the transverse field strength is difficult (Leroy 1962). Other sources of polarization (such as scattering) in both the lines and continuum are probably also important in sunspots. Dollfus (1965) has compared the polarization measured in wavelength bands containing few and many lines, and finds a factor of about 2 more polarization in the latter, but any Fraunhofer line formed in part by scattering can contribute polarization.

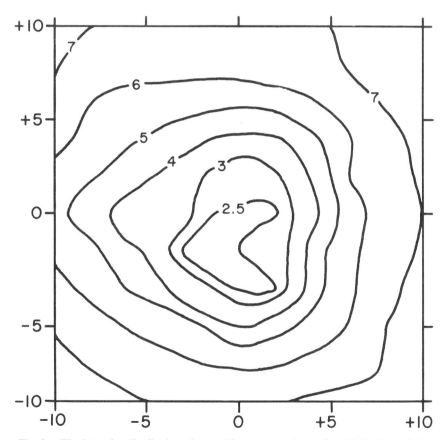

Fig. 2. The intensity distribution observed in a sunspot located $\sim 0.5\ R_\odot$ from disk center on 12 September 1972. Intensity units are arbitrary; x and y coordinates in this and following figures are shown in arcsec.

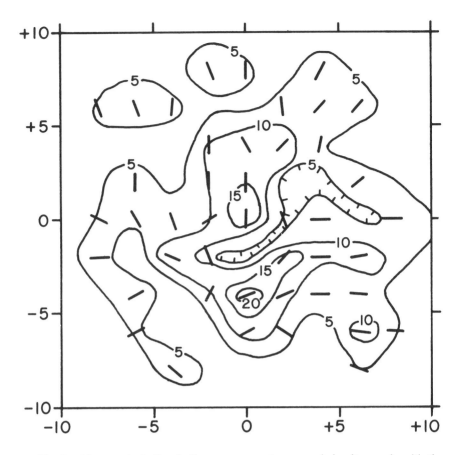

Fig. 3. Linear polarization in the same sunspot measured simultaneously with the intensity contours of Fig. 2. The contours give the polarization in units of 10^{-4} (0.01%), and the short lines indicate direction of electric vector maximum.

Figures 2, 3, and 4 are derived from a raster scan of a sunspot made at Mount Haleakala on 12 September 1972, in a wavelength band ~ 20 Å wide centered at 5834 Å. Measurements were made in 2 arcsec steps in the raster with a ~ 3.5 arcsec scanning aperture. The complete Stokes vector (I,Q,U,V) was measured at each point with a four-second integrating time. Figure 2 shows contours of the intensity, I, in the spot in arbitrary units. Figure 3 shows contours of the linear polarization (in units of 10^{-4}) and its direction, derived from Q/I and U/I.

Figure 4 shows contours of circular polarization, V/I, in units of 10^{-4}. The measurement is repeatable in the sense that a scan made an hour later shows the same general pattern, and similar measurements have been made in other sunspots. We consider the measurement tentative, but we cannot dismiss the observation as being simply due to seeing or to linear-to-circular conversion by the instrument. We have already pointed out that V/I is small at the limb and shows no strong systematic behavior there, even though linear polarization and seeing effects are strong. We have no physical explanation for broadband circular polarization in a magnetic field of a few kilogauss. We are acquiring more observations.

Polarization Effects in Flares

The generation of energetic particles is a very common phenomenon in the sun's atmosphere, but it occurs most dramatically in large solar

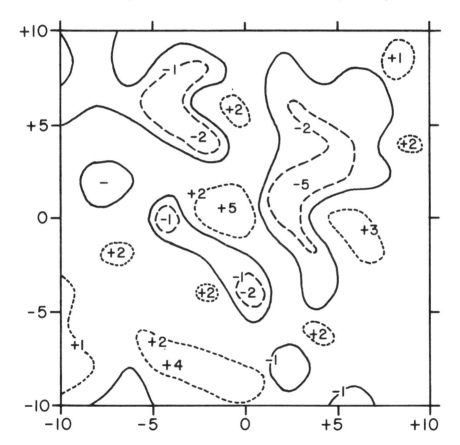

Fig. 4. Circular polarization measured simultaneously in the same sunspot with Figs. 2 and 3. The contours are V/I in units of 10^{-4} (0.01%). The sign convention is as adopted for this book (see p. 53).

flares of explosive onset. Although much is known about the flare process, little is known about the site of the particle acceleration or of the energy spectrum. The most direct observation of these particles is, of course, their in situ detection near the earth, but it is difficult to derive the initial spectrum at the sun and its development in time from its measured form near the earth because of the distortion caused by the intervening magnetic fields and plasma through which the particles must pass. Any observable manifestation of these particles in the immediate vicinity of the flare is therefore of great interest. Bursts of emission in the hard X-ray region,[3] in the extreme ultraviolet, and in the microwave region are observed in the early stages of some energetic flares and are believed to be due to streams of subrelativistic electrons impinging on the underlying chromosphere. These give good time resolution and set constraints on the lower end of the energy spectrum but, so far, have given little spatial resolution and give no direct information on the more energetic particles. Small patches of continuous visible emission are occasionally observed during the initial stages of flares, with easily attainable high spatial resolution. It has been variously suggested that these are due to (1) synchrotron radiation from highly relativistic electrons in the corona (e.g., Stein and Ney 1963); (2) radiation resulting from recombination in the chromosphere after ionization by beams of flare-produced Kev electrons (Hudson 1972); or (3) thermal emission from a small region of the photosphere underlying the flare which is heated locally by a beam of particles with sufficient energy to penetrate to the photosphere (>20 Mev) (Svestka 1970; Najita and Orrall 1970). (See for example the review by Svestka 1972.) Good photopolarimetric measurements of this emission are needed to interpret these white-light flares.

Unfortunately, these broadband events are rarely observed. No polarization measurements have been obtained, and only a few photographs, although a number have been observed visually. A patrol heliograph with polarizing optics to record the complete Stokes vector photographed the sun for months from Mount Haleakala, without catching a single white-light event (Najita and Orrall 1967). Yet it seems likely that white-light emission by the above processes and others (e.g., nonthermal bremsstrahlung or by purely thermal effects) is common. The difficulty is, of course, the overwhelming visible thermal radiation from the underlying quiet photosphere. Further, although the transient event may be strongly polarized, the polarization of the net (transient-plus-photospheric) radiation may be small. It is therefore perhaps not surprising that photographic patrols have recorded so few events.

---

[3] See p. 270.

SUMMARY AND FUTURE WORK

We have summarized the sources of broadband visible polarization on the sun's disk: linear polarization observed near the limb, polarization observed in active regions or sunspot groups, and transient polarization associated with flares. Linear polarization on the quiet sun's disk increases from $< 10^{-5}$ (0.001%) near the center to 10^{-3} (0.1%) or more close to the limb. (The electric vector is tangential to the limb.) Theory based on Thompson, Rayleigh, and resonance scattering successfully explains the general features of the observations, making polarization an important diagnostic tool for studying the atmosphere. More extensive observations, including measurements in the ultraviolet and infrared as well as in additional selected regions in the visible, are needed to make full use of existing theory.

Broadband linear polarization is observed in regions of strong magnetic field and may amount to 10^{-3} (0.1%) in sunspots. Broadband *circular* polarization with V/I as high as 5×10^{-4} has been measured at Mount Haleakala. The linear polarization has been interpreted as due to saturation effects in the transverse Zeeman effect; this would produce linear polarization in the direction of the field generally consistent with what is observed, although scattering, especially resonance scattering, may also contribute. We can suggest no mechanism to produce the large broadband circular polarization described here. The new mechanism required to explain this circular polarization may also be an important source of linear polarization.

Transient broadband visible radiation is expected from solar flares, but only a few events have been observed because of the intense thermal visible emission of the underlying photosphere. Polarization measurements of these white events would help to identify those caused by beams of energetic particles. Such observations would be of great value in studying the energy spectrum, time history, and location of the primary acceleration process.

REFERENCES

Débarbat, S.; Dumont, S.; and Pecker, J. C. 1970*a*. La polarisation du spectre continu au bord du disque solaire. *Astron. Astrophys.* 8: 231–242.

———. 1970*b*. Interpretation of optical solar polarization. *Astrophys. Lett.* 6: 251–256.

Dollfus, A. 1958. Premières observations avec le polarimètre solaire. *Comptes Rendus* 246: 3590.

———. 1965. Recherches sur la structure fine du champ magnetique photospherique. *Stellar and solar magnetic fields.* (R. Lust, ed.) p. 176, *IAU Symp. 22.* Amsterdam: North-Holland Publ. Co.

Dumont, S., and Pecker, J. C. 1971. Influence de l'absorption dans les rais sur la polarisation du spectre continu. *Astron. Astrophys.* 10: 118–127.

Hudson, H. S. 1972. Thick-target processes and white-light flares. *Solar Phys.* 24: 414–428.

Leroy, J.-L. 1962. Contributions à l'étude de la polarisation de la lumière solaire. *Ann. d'Astrophys.* 25: 127.

———. 1972. Nouvelles mesures de la polarisation de la lumière au bord du disque solaire. *Astron. Astrophys.* 19: 287.

Mickey, D. L.; Orrall, F. Q.; and Zane, R. 1972. A photon counting Stokes vector polarimeter. *Instrumentation in astronomy.* (L. Larmore, ed.) Proc. SPIE-AAS. Redondo Beach, Calif.: SPIE.

Najita, K., and Orrall F. Q. 1967. Unpublished.

———. 1970. White light events as photospheric flares. *Solar Phys.* 15: 176–194.

Orrall, F. Q. 1971. A complete Stokes vector polarimeter. *Solar magnetic fields* (R. Howard, ed.), p. 30. Dordrecht: Reidel Publ. Co.

Pecker, J. C. 1970. Some considerations from the direct comparison between the observations and the theory of solar disk polarization. *Solar Phys.* 15: 88–96.

Stein, W. A., and Ney, E. P. 1963. Continuum electromagnetic radiation from solar flares. *J. Geophys. Res.* 68: 65–81.

Svestka, Z. 1970 The phase of particle acceleration in the flare development. *Solar Phys.* 13: 471–489.

———. 1972. Spectra of solar flares. *Ann Rev. Astron. and Astrophys.* 10: 1–24.

DISCUSSION

GEHRELS: It would be interesting to try a dielectric aperture to see if your instrumental polarization would be reduced.[1]

NOVICK: The solar X-ray astronomy group of the Lebedev Institute in Moscow has detected linear X-ray polarization in solar flares.[2]

---

[1] See p. 141.   [2] See p. 270.

THE FRENCH SOLAR PHOTOELECTRIC POLARIMETER AND ITS APPLICATIONS FOR SOLAR OBSERVATIONS

AUDOUIN DOLLFUS
Observatoire de Paris, Meudon

A photoelectric polarimeter of very high accuracy was designed for solar work in 1953. The sensitivity threshold in the degree of polarization is 10^{-5}, on photospheric fields of 5 arcsecs diameter.

This instrument is used as a polarimeter for studying the scattering processes of polarizing light near the solar limb, as a magnetometer for mapping the azimuth and intensity of the photospheric magnetic field component perpendicular to the line of sight, and as a coronameter for studying and monitoring free electrons around the sun, both in enhancements and coronal streamers. The observing program initiated in 1956 continues at Meudon and Pic-du-Midi Observatories.

Early in the fifties photoelectric techniques reached a point whereby the design of a very sensitive polarimeter for solar observations seemed feasible. In France, Lyot (1948a, b) had already produced his polarimeter employing photomultipliers for stellar and planetary work and had published the results of his pioneering work on solar polarimetry undertaken in 1924. Öhman (1949), in Sweden, had also described several devices for astronomical photoelectric polarimetry. Photomultipliers began to be commercially available: Lallemand, in France, produced high-sensitivity photomultipliers specifically designed for astronomical purposes.

In 1951, we initiated studies for a photoelectric polarimeter of very high sensitivity, specially conceived for measurements on small areas of the sun's surface or around the disk. Our goal was to achieve an accuracy of 10^{-5} in polarization on photospheric fields of only 5 arcsecs diameter. These specifications were reached in the laboratory in January 1953 and have been described by Dollfus (1953); details of an application to the study of the solar corona were also given. This paper contained a curve, reproduced in Fig. 1, relating the sensitivity obtained to the light flux; the curve is also given by Dollfus (1958a). Coronal observations were begun in 1956 (Dollfus 1958c), and photospheric observations in 1958 (Dollfus 1958b). The prototype solar photoelectric polarimeter was con-

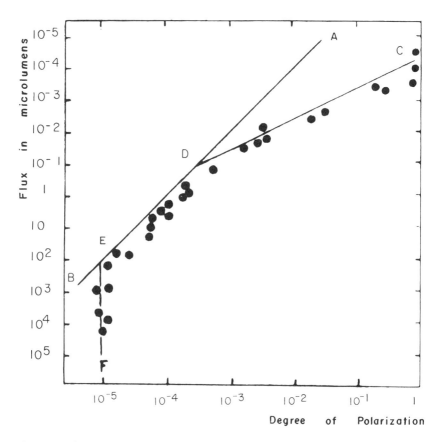

Fig. 1. Sensitivity diagram of the polarimeter: threshold polarization versus light flux (in microlumen) for an integration time of one second (after Dollfus 1953).

structed by Ets. R. Danger of Paris and was made available for the use of several observatories in 1963 (Figs. 2, 3 and 4; see also Dollfus 1968, Fig. 2).

POLARIMETER PRINCIPLES

The instrument is based on the zero deflection method. The degree of polarization $P = (I_\perp - I_\parallel)/(I_\perp + I_\parallel)$, or the intensity of polarized light $P \cdot I$ (see below) to be measured, is matched by an optical compensator. The various components are identifiable on the picture of the instrument in Fig. 2. A block diagram is given in Fig. 5. Light selected by the field stop passes through the polarization compensator and the modulator and is then split by a Wollaston prism into two orthogonal, totally polarized beams. These beams are collected by the cathodes of two photoelectric tubes. The action of modulation is to switch the azimuth of

Fig. 2. The solar photoelectric polarimeter (manufactured by R. Danger, Paris).

Fig. 3. The corona selector (manufactured by R. Danger).

polarization at a frequency of 20 Hz, thereby producing square-wave modulation of the polarized component only. In the absence of incident polarized light, the modulator does not modify the flux and the two cells receive constant equal D.C. signals. When the incident beam contains a

Fig. 4. View of the rear part of the polarization-free telescope with the photoelectric polarimeter attached, and the off-axis scanning device (manufactured by R. Danger).

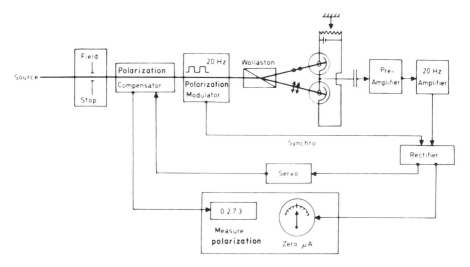

Fig. 5. Block diagram of the solar photoelectric polarimeter.

component of polarized light with the electric vector parallel to one of the polarizations introduced by the Wollaston prism, it passes, at the modulation frequency, into each of the two cells in turn and produces an alternating photoelectric current proportional to the intensity of the polarized component. This current is rectified synchronously with the modulation and supplies a microammeter whose gain and time constant are adjusted so that fluctuations in the photocurrent are easily observable.

The calibrated optical compensator can be adjusted with a precision that is noise limited. A servo-mechanism could perform the compensation necessary to match the signals, but in practice the operator makes the adjustment by hand. Calibration is stable with time and independent of the adjustment of other optical and mechanical parts. The zero reading obtained is, over a large range, insensitive to the gain in the electronics and also to its variation. The gain can be adjusted to the accuracy desired and to the amount of light available, without affecting measurement. Thus, this instrument permits high-sensitivity, calibrated polarimetry, limited only by the noise level of the photocurrent.

SENSITIVITY DIAGRAM

The sensitivity attained by an ideal polarimeter should be limited only by statistical fluctuations in the number of electrons delivered by the photoelectric conversion of photons. Let ϕ be the incident photon flux (expressed in lumens, or watts), S the quantum efficiency of the photocathode (or sensitivity expressed in microamperes/lumen or amperes/watt), and e the charge of the electron. The number of photoelectrons produced in one second is then $\phi S/e$, and over an integration time of t seconds this number is $\phi St/e$. Thus, the smallest observable signal is $K\sqrt{\phi St/e}$, K being the ratio of the threshold signal/noise necessary for safe detection of a departure from random fluctuations (we have used $K = 3$). The signal s generated is proportional to the intensity of the polarized component of the incident light: $s = kP\phi St/e$. The constant k is the overall modulation efficiency of the polarimeter, a number close to $8/\pi^2$ in an instrument that has square-wave modulation, two cells, and that uses sinusoidal amplification (see the discussion on modulation efficiency below). The threshold polarization detectable in t secs is then

$$P(\text{threshold}) = \frac{K}{k} \sqrt{e} \frac{1}{\sqrt{\phi St}}.$$

The ideal threshold polarization is plotted as a function of the light flux ϕ in Fig. 1 for $t = 1$ sec and for cathodes with sensitivity $S = 110\ \mu A/$lumen, such as were available in 1952 (selected tubes now give about $200\ \mu A/$lumen); it is represented by the straight line AB. The dots plotted in the diagram are measurements recorded with the instrument in the laboratory using an integration time of 1 sec. Within the domain DE (fluxes from 10^{-2} to 10^2 microlumens), the precision of the instrument is almost entirely limited by photoelectron noise. This covers a range of 10^4 in light flux. When the flux is higher than 10^2 microlumens, the smallest detectable polarization reaches 10^{-5} (segment *EF* in the diagram). The limit of 10^{-5} is imposed by spurious modulation due to residual defects in the modulator. This residual modulation can be split into two components, each of them proportional to the total flux. One of these components is in phase with the synchronous demodulator and is rectified like a polarized signal; it can be eliminated using the optical polarization compensator. The other is phase-shifted by $\pi/2$ and is not rectified as a signal but combines with the residual defects of the demodulator, giving spurious fluctuations resembling the noise limiting the threshold sensitivity.

For photon fluxes smaller than about 10^{-2} microlumens, the measurements depart significantly from the theoretical line *AD* and align along *CD*. This is because photons generate photoelectric noise smaller than the electronic noise of the preamplifier which follows the cells. This amplifier noise, independent of the photon flux, limits the detectable modulated flux to about 10^{-4} microlumens. It would be easy to greatly reduce the effect of electronic noise by replacing the two photoelectric cells with a photomultiplier, but this could not be done with the same signal/noise ratio, as discussed below. Since the instrument is intended to be used for solar work with fluxes larger than 10^{-2} microlumens and a sensitivity better than 10^{-3}, paired photocells provide higher accuracy than photomultipliers. Such is not the case for polarizations of planets and faint objects, and for these applications our instrument is used with a photomultiplier (Marin 1965).

The optical components are shown in Fig. 6.

POLARIZATION COMPENSATORS

Compensators are designed to measure either the degree of polarization $P = (I_\perp - I_\parallel)/(I_\perp + I_\parallel)$ of the incident beam, or the intensity of its polarized components $(I_\perp - I_\parallel)$ with respect to the intensity of a similar unpolarized beam at the center of the solar disk.

Compensators to Measure the Degree of Polarization

Measurement requires two stages of compensation. For degrees of polarization larger than 5×10^{-4}, the compensator is a thin plane-parallel plate, tiltable in the beam. The Fresnel laws, extended to light multiply reflected between the two faces of the plate, are used to compute polarization as a function of the tilt angle. Laboratory calibration is also

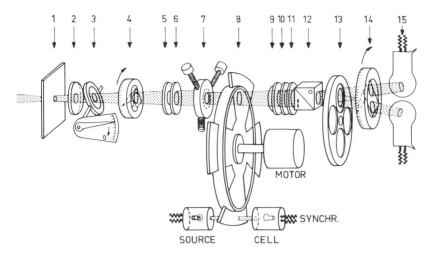

Fig. 6. Optical elements of the polarimeter: *(1)* field stop; *(2)* Fabry lens; *(3)* tilted plate compensator; *(4)* fine compensator; *(5)* and *(6)* half- and quarter-wave plates of the achromatized $\lambda/4$ retarder; *(7)* Lyot stop; *(8)* fan-shaped half-wave modulator; *(9)* relay lens; *(10)* and *(11)* components of the achromatized $\lambda/4$ retarder; *(12)* Wollaston prism; *(13)* color filter wheel; *(14)* atmospheric seeing compensator; *(15)* paired photocells.

undertaken. A full description of the tilted-plate compensator has been given by Lyot (1929). See also Dollfus (1955) and Fig. 3.

For degrees of polarization below 5×10^{-4}, the tilt angle required is less than 4°; the calibration takes the convergence of the beam into account and the compensator is increasingly sensitive to misalignment. In this case we use a fine compensator: the device is a dihedral polarization standard. Two thin plane-parallel glass plates are maintained at a fixed angle with respect to each other. The fine compensator is inserted into the incident beam with its plane of symmetry approximately perpendicular to the optic axis. It gives rise to a constant polarization with the electric vector perpendicular to the line of intersection of the two plates. This degree of polarization P_0, easy to compute by the Fresnel laws or to measure in the laboratory, is, to the second order, independent of any misalignment of the optic axis. It depends only on the dihedral angle being fixed. With plates of refractive index 1.52, a polarization of $P_0 = 1.0 \times 10^{-3}$ is given with a dihedral angle of about 8°. The device can be rotated about the optical axis of the polarimeter and produces a degree of polarization $P = P_0 \cos 2\beta$, β being the angle between the direction of polarization produced by the plates and the direction of polarization detected by the polarimeter (the azimuth of polarization transmitted by the Wollaston analyzer).

A description of the dihedral polarization standard has been given by Dollfus (1955). For $\beta = 45°$ the compensator introduces no signal. For $P_0 = 1.0 \times 10^{-3}$, a rotation of 1° gives a compensation of 3.0×10^{-5} and a rotation of 3° gives 1.0×10^{-4}. This device accordingly produces a very small polarization; calibration is almost insensitive to alignment and is stable with time.

Compensation of the Quantity of Polarization

Photometry of the polarized flux is a particularly valuable analytic tool in studying the properties of the corona. We must measure, in absolute intensity units, the quantity of polarized light $I_\perp - I_\parallel$, or $P \cdot I$, in the incident beam. For this purpose a flux of light is added, totally polarized orthogonally, and photometrically calibrated on an absolute scale. The intensity of this reference flux is adjusted to equal that of the polarized light to be measured. The center of the solar disk is used as a reference source.

The principle of the device is outlined in Fig. 7. The aperture itself is at A. The sun's image is centered on the metal disk B which is protected against heating by a reflecting cone C. The diameter of the disk B is slightly smaller than that of the solar image, so that a bright annulus is projected on the screen at D, facilitating accurate adjustment. The distance from the field stop to the edge of the solar image is controlled by the knob E. The position angle of the aperture with respect to the north point is adjusted by rotating the entire polarimeter about the axis joining the center of the objective to the center of the sun. The reference beam, collected at the center of the solar image by the prism F, passes to a scattering screen G (a piece of chalk) of geometric albedo A. A small fraction of the scattered light is intercepted by the stop H and polarized by the Polaroid K in a direction perpendicular to that of the incident beam (this latter is polarized in a direction tangential to the sun's limb). A second Polaroid L is rotated by the knob M to equalize the reference beam intensity, represented by the angle read at N.

The luminance of the screen G as seen from the stop H is

$$L_G = L_S \tau_1 \frac{A}{4} \cdot \left(\frac{1}{f}\right)^2 \cdot \cos b,$$

where L_S is the luminance of the center of the solar disk, τ_1 is the transmission of the prism F, f is the focal ratio of the telescope, A is the albedo of the scattering screen (usually close to 1.0), and b is the angle (usually 45°) between the plane surface of the scattering screen and the incident beam.

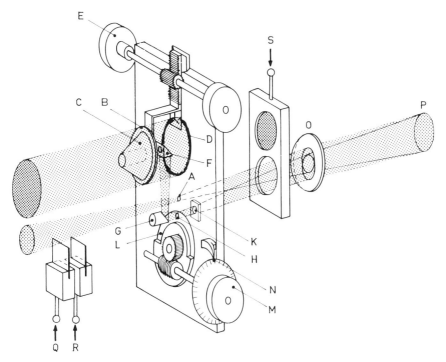

Fig. 7. Principle of the corona selector with polarized light photometer (see p. 702).

The reference flux is then

$$\phi_R = L_G \tau_2 \tau_3 \sin^2(\alpha - \alpha_0) \frac{\pi}{4} d_2^2 \left(\frac{1}{f}\right)^2 \frac{\pi}{4},$$

where τ_2 and τ_3 are the transmissions of the Polaroids K and L, α and α_0 are the angles pertaining to the measurement and to extinction with crossed Polaroids, and d_2 is the diameter of the stop H.

The flux to be measured, passing through an aperture of diameter d_1, is

$$\phi_X = \frac{\pi}{4} d_1^2 \frac{\pi}{4} \left(\frac{1}{f}\right)^2 L_X,$$

L_X being the unknown luminance of the polarized component, which is to be compared with the luminance L_S at the center of the sun's disk. After adjustment, $\phi_X = \phi_R$ and we have

$$\frac{L_X}{L_S} = KA \cos b \left(\frac{d_2}{d_1}\right)^2 \cdot \left(\frac{1}{f}\right)^2 \cdot \sin^2(\alpha - \alpha_0).$$

The Fabry lens N produces for both beams a single image of the objective at the pupil P in the plane of the modulator; this pupil is then imaged on both photocathodes. A Polaroid, completely polarizing the beam, can be introduced at Q. It enables the instrument to operate as a photometer for the total flux. A half-wave plate having its axis at 45° to the direction of polarization may be introduced at R; the azimuth of polarization is thereby turned by 90° in order to measure polarization when the predominant electric vector is radial to the sun. A neutral density plate ($D \approx 4$) is placed at S to estimate the instrumental polarization; measurement is made at the center of the solar disk using a flux comparable to that of the sky background. The major components in Fig. 7 may be identified on the picture of the instrument shown in Fig. 3.

POLARIZATION MODULATOR

The purpose of the modulator is to generate alternating signals from the small polarized component of the incident beam, by switching either the intensity or the azimuth of vibration. The nonpolarized component of the incident light, even though it may have an intensity of more than 10^5 times that of the polarized component, is in principle not modulated.

Sinusoidal Modulation

A *rotating half-wave plate modulator* in front of a fixed analyzer (Öhman 1949) gives a variation in polarized flux of the form $\sin 4\omega t$. The energy modulated is $4/\pi^2$, and the selective amplifier transmits a fraction $8/\pi^2$ of the noise energy in the total photocurrent so that the resulting signal/noise ratio of the modulator is $(4/\pi^2)/(8/\pi^2) = 0.5$. Because the photocurrent fluctuations vary with the square root of the light flux, this imperfect modulator needs a light level four times greater than an ideal one in order to obtain the same threshold sensitivity. For wavelengths λ (not equal to λ_0) for which the plate is half-wave, the modulation efficiency is reduced by a factor $\cos^2 \dfrac{\pi}{2}\left(\dfrac{\lambda_0}{\lambda} - 1\right)$. Similar comments, apart from the wavelength dependence, apply to the *rotating polarizer modulator* with variation of the form $\sin 2\omega t$. The *ADP* and *KDP electro-optic modulators* fed by sinusoidal supply behave in a slightly different mode.

Square-Wave Modulation

To improve modulation efficiency, one can replace the sinusoidal modulation by a device designed to generate a rectangular wave-form, the Fourier expression for which is

$$\frac{4}{\pi}\left[\sin X + \frac{\sin 3X}{3} + \cdots + \frac{\sin(2n-1)X}{2n-1} + \cdots\right].$$

The A.C. amplifier selects only the fundamental term, with a modulation amplitude of $(4/\pi)(2/\pi)$ and a modulated intensity of $(8/\pi^2)^2$. The signal/noise ratio is then $(64/\pi^4)/(8/\pi^2) = 8/\pi^2 = 0.81$. The efficiency is improved by a factor $(4/\pi)^2 = 1.62$ relative to sinusoidal modulation.

The Fan-Shaped Half-Wave Modulator

The square-wave modulator used with the solar photoelectric polarimeter between 1952 and 1957 was of the type described by Lyot (1948 a,b). However, this polarimeter used a rotating disk spinning at the modulation frequency and generated residual modulation of the unpolarized component at the rectification frequency of the demodulator. This residual arises from a small deflection of the beam produced by the prismatic effect of the rotating disk, internal reflection between its two faces, dust on the surface of the rotating disk, mechanical vibration, residual dichroism in the half-wave plates, irregularities in the form of these plates, etc. Under the best conditions, spurious modulation limits the polarization sensitivity to about 10^{-4}. If very great care is taken (Dollfus 1963b), laboratory measurements are feasible with the required threshold of 10^{-5} but not with the reliability demanded by telescopic observation.

After 1958 the newly designed so-called *fan-shaped half-wave modulator* was used. This device gives the required accuracy of 10^{-5} at the telescope, without special precautions. Description of the modulator is given by Dollfus (1963b), and computations by Marin (1965), and Tinbergen (1972).

The principle is based on the following experiment. Linearly polarized light passes through two successive quarter-wave plates with axes oriented at $+45°$ and $-45°$, respectively, to the azimuth of polarization. It is circularly polarized by the first plate and again becomes linearly polarized after the second plate. On introducing a half-wave plate between the quarter-wave plates, the sense of circular vibration is reversed; after the second quarter-wave plate, the emergent beam is once more linearly polarized, but in the direction perpendicular to that of the incident beam. This $\pi/2$ rotation is independent of the orientation of the half-wave plate axis.

One can easily understand the behavior of the light by using the geometrical representation of the Poincaré sphere, in which all the angles are doubled. In Fig. 8, OA is the direction of vibration of the linearly polarized incident light. The first quarter-wave plate has its axis at $45°$ to OA and is represented by twice this angle, or OM. The point A rotates

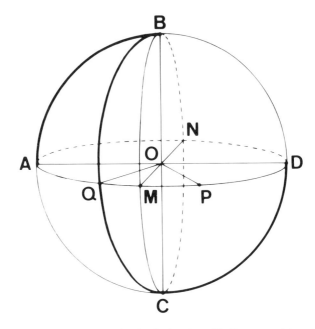

Fig. 8. Poincaré sphere representation for the fan-shaped half-wave modulator (see p. 707).

about OM by twice $\pi/4$ and reaches B (circular polarization). Without the half-wave plate, the light reaches the second quarter-wave plate with its axis in ON, and B returns to A. When the half-wave plate is introduced, with axis OP, B rotates about OP by twice $\pi/2$ and reaches C (inverse circular polarization). This is irrespective of the azimuth OP and of the axis of the half-wave plate. The second quarter-wave plate translates C to D, and OD corresponds to linearly polarized light with the same amplitude as the incident flux OA but with the azimuth rotated by $\pi/2$.

The modulator is made from a large half-wave sheet cut into a disk with n sectors and n equal intervals spaced alternately. This sheet is cemented between two plane optical blanks; it is rotated about its center, at a frequency $1/n$ that of the modulation frequency, by a small synchronous motor and positioned between the two fixed quarter-wave plates so as to intercept the incident beam near its periphery (see Fig. 6). The azimuth of the polarization in the emergent beam is therefore switched n times per cycle, producing square-wave modulation of the polarized component of the incident light at the modulation frequency. Of the residual defects listed above, some are modulated at $1/n$ times the amplification and demodulation frequency, and others at frequency $2/n$, but none are rectified. Usually, the fan-shaped disk has four half-wave sectors and four intervals ($n = 4$).

In the laboratory, this modulator has a threshold of detectable polarization of about 10^{-5}. Small amounts of polarization of the order of a few 10^{-6} can be detected, provided careful integration and a series of crossed measurements are made. At the telescope, a routine accuracy of 10^{-5} is achieved.

Extension to Large Wavelengths Ranges

The fan-shaped modulator gives its theoretical modulation efficiency at wavelength λ_0, for which the plates behave as perfect quarter- or half-wave plates; but this ratio falls off at wavelengths different from λ_0. For some applications, it is desirable to cover a large wavelength range with a single modulator.

First, the quarter-wave plates may easily be replaced either by Fresnel rhombs or by the so-called "achromatic" combination of a half-wave and a quarter-wave plate described by Destriau and Prouteau (1949). The spectral modulation efficiency is then limited by the rotating half-wave plate only. For $\lambda_0 = 0.5\ \mu\text{m}$, this efficiency falls to 0.70 at wavelengths .35 μm and 0.8 μm. The following improvement raises the modulation efficiency to more than 0.90 over the entire wavelength range. The fan-shaped half-wave plate is replaced by a split element comprising a quarter-wave plate followed by a half-wave plate whose axis is inclined at 45°, and by a second quarter-wave plate with its axis parallel to the first. All three sheets are cemented together between the glass plates of the rotating disk (Dollfus 1963b).

The Poincaré sphere representation (Fig. 9) is again useful for understanding the behavior of light in this modulator. As before, plane polarized light passes through a quarter-wave plate, and we assume it is completely circularly polarized. The problem now is to start from this circularly polarized light at B and, by means of the modulator, to move to C for a wide range of λ. The first quarter-wave plate, with axis OP, moves B along the great circle with pole P. For $\lambda = \lambda_0$, the plate is exactly quarter-wave and moves B by $\pi/2$ to Q. For other λ, B reaches R, where $BR = (\pi/2)(\lambda_0/\lambda)$. The light next passes through the half-wave plate whose axis is inclined by $\pi/4$; in the sphere the representation of this axis lies $2(\pi/4)$ from OP and coincides with OQ. R therefore moves about the axis OQ and reaches T, where the rotation angle $RT = 2BR = \pi(\lambda_0/\lambda)$. The second quarter-wave plate translates T to U, where $UC(=ST)$ is small (U does not lie exactly on the great circle $BACD$). The second fixed quarter-wave plate moves U to V, where $VD = UC = ST$. Resolving the spherical triangle TSQ, we have

$$\cos ST = \cos^2 \frac{\pi}{2}\left(\frac{\lambda_0}{\lambda} - 1\right) + \sin^2 \frac{\pi}{2}\left(\frac{\lambda_0}{\lambda} - 1\right) \cdot \cos \pi \left(\frac{\lambda_0}{\lambda} - 1\right).$$

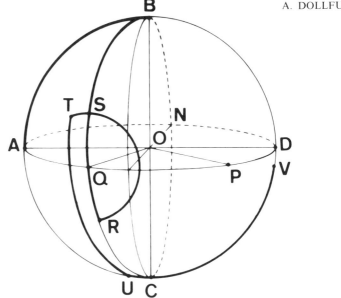

Fig. 9. Poincaré sphere representation for the achromatized modulator. (See p. 707.)

The modulation efficiency, $\cos^2(VD/2)$, is plotted as a function of λ, together with the case of simple modulation in Fig. 10.

ATMOSPHERIC SEEING COMPENSATOR

Light fluctuations result from atmospheric seeing. They are particularly enhanced when observing near the limb of the sun, at the edge of sunspots, or in other areas with brightness gradients. Fluctuations having a frequency similar to that of the demodulation frequency are partially rectified and produce spurious signals that reduce the sensitivity. To minimize these effects, incident light is split by the Wollaston prism into two components that are collected by two electrically opposed photocells, so that the nonmodulated part of the beam (along with its fluctuations) is exactly equalized, even though there are small sensitivity differences in the two cathodes. This is achieved by the use of a rotating dihedral polarization compensator introduced into the beam between the Wollaston and the photocells (Fig. 6). The dihedral plates give rise to a polarization of about $P_0 = 0.20$.

If α is the azimuth of the polarization introduced by the plates (reckoned from the bisector of the orthogonal polarizations produced by the Wollaston prism), the intensity ratio of the two beams after the compensator is $I_I/I_0 = P_0 \tan^2 \alpha$. This ratio is adjusted to minimize fluctuations resulting from atmospheric seeing.

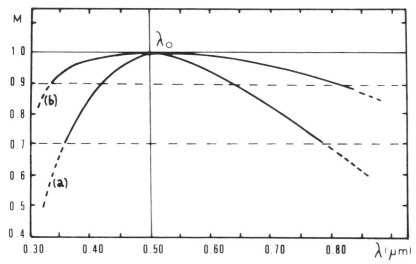

Fig. 10. Spectral variation of the modulator efficiency M for a modulator with achromatic quarter-wave plates and (a) a fan-shaped plate half-wave at $\lambda_0 = 0.5\,\mu\mathrm{m}$; (b) an achromatized half-wave plate.

PHOTOELECTRIC COMPONENTS

For solar work, during the past twenty years, we have considered it efficacious to use a pair of nonphotomultiplying photocells rather than a photomultiplier. The advantages are as follows:

1. Ordinary photocells are easily paired, each collecting one of the polarized components from the Wollaston prism; both the flux and the modulation amplitude are doubled, and the gain in the signal/noise ratio is $\sqrt{2} = 1.41$. The unpolarized part of the incident beam does not produce any A.C. photoelectric current, and atmospheric seeing fluctuations are reduced as already explained.
2. Simple photocells are easier to construct than photomultipliers, and their quantum efficiency is usually greater. Our cells have been specially made by Lallemand at Observatoire de Paris, and carefully selected by him from a series of several dozen tubes.
3. Photocells are free of the additional noise associated with photomultiplication. They work with a D.C. supply of only 4 volts, and current leakage noise, etc., is almost absent.
4. Photocells accommodate a large range of fluxes, up to 0.1 lumen, without requiring gain control.

However, the paired photocell device must be combined with a preamplifier of very high equivalent noise resistivity. The associated electronics have been described by Leroy (1962). The photocathodes and electrometer grid circuit are completely insulated and current free. The pentode electrometer 6J7 originally used was replaced by Leroy in 1960 with a ME.1400 tube. At input, the noise produced was 5×10^{-15} A/Hz.

New electronic systems could now replace those of the earlier period. Progress on photomultiplier design, stabilized supply, automatic gain control, etc., now make it easier to use paired photomultipliers adapted to cover a large range in light flux. By this means the domain AD in Fig. 1 would be more fully covered, but the domain DE (corresponding to solar work) cannot be improved upon in terms of quantum efficiency.

POLARIZATION-FREE TELESCOPE

Use of the polarimeter for studying the sun with a precision of 10^{-5} in polarization requires a telescope almost free from instrumental polarization. Residual polarization must be independent of azimuth and the angular distance from the center of the sun; it must be stable to better than 10^{-5} (see Dollfus 1958b). In view of the significant polarization engendered by coated mirrors, only refracting telescopes have been considered to meet these severe requirements.

The field-stop is located on the optic axis of the refractor. The telescope and polarimeter may be rotated as a single unit about the optic axis so as to select any desired azimuth. The objective, of 10-cm aperture, is an achromat with crown and flint elements cemented to reduce reflection between optical surfaces. A high focal length of 400 cm was chosen so that polarization due to light transmitted at the edge of the objective would be kept below about 4×10^{-5}; symmetry then reduces the residual polarization by refraction to less than 10^{-5}. When observing near the solar limb, the diffraction fringe produced by the edge of the objective is nonuniform in brightness and generates spurious polarization. The fringe is intercepted by a Lyot field-stop at the focus of a Fabry lens; Fig. 6 shows the stop with its adjusting screws. A small residual polarization is also apparent when observing strong brightness gradients. This arises from asymmetric diffraction at the edge of the field-stop. Its elimination can be assured by the use of a Lyot stop, but this procedure has not been found useful in practice.

To operate this type of polarization-free telescope, an area near the center of the solar disk, well away from any photospheric centers of activity, should first be observed; the residual signal is zeroed using the compensator. The polarimeter then registers zero polarization, apparently to within the threshold sensitivity of 10^{-5}, for all regions of the solar disk except those closer than about 15 arcsecs from the limb.

For observations of the solar corona, the achromatic objective is replaced by a coronograph lens of 15-cm aperture and similar focal length; the field-stop arrangement represented in Fig. 2 is replaced by that shown in Fig. 3, operating according to the principle indicated by Fig. 7. The entire instrument rotates about the line joining the center of the objective and the center of the metal plate (Figs. 7 and 4); the latter also corresponds to the center of the sun's image, so that the optic axis, along which observations are made, scans the sky around the sun at a chosen fixed distance from the limb. The coronal polarization is, therefore, always measured in the direction tangent to the sun's limb.

The polarization-free telescope associated with the polarimeter was supplied by Ets. R. Danger in 1963. A photograph of the instrument has already been published (Dollfus 1968). Figure 4 gives a detailed view of the rear part of this instrument, along with the polarimeter and scanning mechanism.

APPLICATIONS OF THE POLARIMETER IN SOLAR PHYSICS

Since the inception of the program of solar observation in 1956 (Dollfus 1958b, c), the polarimeter has been used to study the sun in three ways.

1. As a *polarimeter*, to measure polarization near the limb produced by the various scattering processes associated with the transfer of radiation through the upper layers of the photosphere.
2. As a *magnetometer*, to measure the azimuth and field-strength of the photospheric field component perpendicular to the line of sight, and to map the field in and around centers of activity.
3. As a *coronograph*, to detect, measure, and monitor coronal enhancements and streamers.

PHOTOSPHERIC POLARIZATION NEAR THE LIMB

A certain amount of polarization was expected near the sun's limb, arising from the combined effect of Thompson scattering by free electrons in the photosphere, Rayleigh scattering by photospheric molecules, resonance scattering by some Fraunhofer lines, and small-scale magnetic fields. Near the edge of the sun's disk, all these processes generate polarizations with predominant electric vectors tangential to the limb. An early attempt to detect such polarization effects was reported by Lyot (1948a).

Our first measurements were made at Meudon Observatory in March 1958 (Dollfus 1958b). A further 600 measurements were collected in 1959, and these have been reported by Dollfus and Leroy (1960) and by Leroy (1962). More recently, a series of about 2500 high-quality measurements were obtained by Leroy (1972) with the instrument operated at Pic-du-Midi.

In 1959, measurements were made using a field-stop of 15-arcsecs aperture, through six color filters covering the wavelength range from 0.41 μm to 0.60 μm; the lower wavelength limit was extended down to 0.38 μm for the Pic-du-Midi measurements. Radial scan sequences were made, from the center of the disk out to about 20 arcsecs from the limb. Some results from the earliest radial scans, previously published in 1960, are reproduced in Fig. 11.

In 1959, limb sequences were initiated; measurements were made at about 60 arcsecs above the limb, in steps of one degree in heliocentric position angle (about 15-arcsecs separation). A good example of such a scan has been published by Leroy (1962). Two additional examples, so far unpublished, are reproduced in Figs. 12 and 13. The west limb was observed between position angles 220° and 280° on 18 and 19 August 1959, at a constant height of 40 arcsecs above the limb. On 18 August (Fig. 12), for $\lambda = 0.53$ μm, the polarization was about 40×10^{-5} with noise fluctuations of about $\pm 2 \times 10^{-5}$. Several abrupt changes in polarization of amplitude 10×10^{-5} occurred at position angles 228° and 250°, and between 257° and 265°. This last corresponds to the location of a K_3 plage. Polarization can be produced by self-absorption in transverse Zeeman effects, as will be seen later. A radial magnetic field on the order of 100 gauss, distributed around the plage, may account for this departure.

On the following day (Fig. 13) measurements at the same wavelength showed the same departure at 260° (above the plage), along with a more widespread departure near 245° that may once more be connected with a bright filamentary K_3 feature. Measurements at $\lambda = 0.48$ μm gave an average polarization of 68×10^{-5} and showed the same feature at 260°, but this scan, probably made slightly closer to the limb, did not record the 245° departure.

In 1964, the polarimeter was equipped to study the effect of absorption lines on the signal measured. A tunable birefringent filter was added to select spectral regions only 20 Å wide. The field-stop, of 15-arcsecs aperture, was replaced by a slit of aperture 120×50 (arcsecs)^2, with its long axis parallel to the limb. The electronics supplied a flux integrator giving integration times of 60 secs. Several pairs of crossed integration sequences were needed to reach an accuracy of 10^{-5} with the low flux level available. The filter was first centered at 5834 Å, in a domain almost devoid of absorption lines, and then adjusted to transmit the spectral

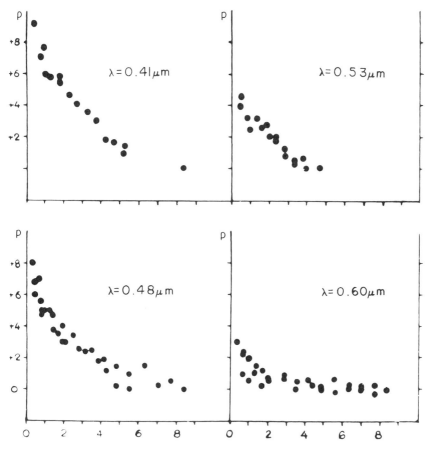

Fig. 11. Polarization p (in units of 10^{-4}) as a function of distance, d, inside the solar limb (in arcmins); four wavelengths (from Dollfus and Leroy 1960).

regions on either side, both rather rich in lines. Measurements 60 arcsecs from the solar limb, at five different position angles, gave an average polarization of 6×10^{-5} in the region of low-line abundance and 12×10^{-5} in the adjacent regions (Dollfus 1965). Polarization due to pure scattering processes in the photosphere was, therefore, about 6×10^{-5}. Spectral lines contributed a similar amount, the effect being due to resonance scattering (the D_2 line, partly measured, ought to give about 3.6×10^{-5} polarization) and possibly to localized magnetic fields with angular dimensions smaller than the scanning slit. Such fields would generate polarization by self-absorption in the transverse Zeeman effect (see below); near the center of the solar disk they should cancel by sym-

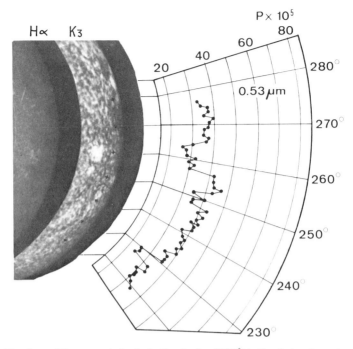

Fig. 12. Scan of the amount of polarization (units of 10^{-5}) in the photosphere at a constant distance of 40 arcsecs from the limb. Corresponding Hα and K_3 chromospheric images are shown. 18 August 1959, 07^h00^m–09^h00^m UT

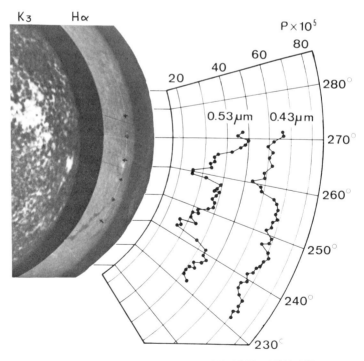

Fig. 13. As in Fig. 12. 19 August 1959, 07^h00^m–10^h00^m UT

metry but could produce residual polarization by foreshortening near the limb.

On the basis of these observations, theoretical work has been carried out under the direction of Pecker (1970), by Débarbat, Dumont, and Pecker (1970a, b), Dumont and Pecker (1971), and others. New measurements by Leroy (1972) have refined the observational contribution. Other observations have been obtained by Mickey and Orrall[1] using the polarimeter designed by the University of Hawaii. A balloon-borne polarimeter, designed by de Sainte Lorette of Service d'Aéronomie du CNRS (France), is scheduled to fly in 1973; it is hoped to extend ultraviolet measurements down to 0.2 μm.

APPLICATION TO THE STUDY OF PHOTOSPHERIC MAGNETIC FIELDS

An unexpected result of the first solar observations was the detection of localized polarizations, far from the sun's limb, of several tens of 10^{-5} (Dollfus 1958b). Polarizing regions show a complex distribution in azimuth and degree of polarization, and they are often associated with centers of activity.

Detailed analysis of these effects was undertaken at Meudon Observatory by Leroy. A preliminary account of his work describes the distribution of polarization in sunspots, based on 2500 measurements collected during the summer of 1959 (Leroy 1959). The polarimeter was later modified to give the determination of azimuth and degree of polarization on fields of 5-arcsecs diameter. A large number of observations were added in 1960 and 1961 (Leroy 1962).

A physical interpretation has been given by Leroy (1960a, b). The polarization results from self-absorption in Fraunhofer lines split by magnetic fields perpendicular to the line of sight. In the simple Zeeman effect, an absorption line is split into three components: one is not wavelength shifted and is totally polarized parallel to the field lines; the others, each of half intensity, are symmetrically shifted and polarized perpendicular to the field. In principle, no polarization is generated. But radiative transfer in the photosphere leads to a saturation effect that reduces the line widths of all three components, the strongest central component being affected more than the less-intense shifted components. Thus, a Zeeman-split line gives rise to a small amount of polarization perpendicular to the transverse component of the magnetic field. Every line in the observed spectral range contributes a certain polarization, the overall result being an average. For small fields

$$\mathbf{P} = K\lambda^2 \sum N_i p_i (\mathbf{H} \times \mathbf{e})^2,$$

---
[1] See p. 686.

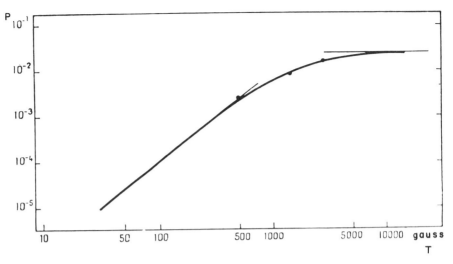

Fig. 14. Calibration curve relating the degree of polarization to the transverse component of the photospheric magnetic field, for an instrument working in the spectral range 0.45–0.60 μm (after Leroy 1960a).

where the sum is taken over all lines in the spectral range, p_i is the average polarization produced by lines of intensity N_i. **H** is the magnetic field strength, and **e** is a unit vector in the line of sight.

For fields exceeding 1000 gauss, a saturation effect occurs. A calibration curve, given by Leroy (1960a, b, 1962), is reproduced in Fig. 14. The result is that the polarimeter is used as a *solar magnetometer*. It does not, as is usual, measure the longitudinal magnetic field, but instead measures the transverse field, both in azimuth and intensity. This instrument, smaller and simpler than a large magnetographic spectrograph, has a sensitivity of about 40 gauss with a field diameter of 5 arcsecs. Calibration, referring to the averaged effects of a large number of lines, is completely independent of the instrumental adjustment and is stable with time; however, interpretation in terms of optical depth depends on the photospheric model used.

Figure 15, based on observations by Leroy (1962), shows examples of transverse magnetic field maps for single spots and their penumbrae, bipolar spots, and multiple spot fields. Figure 16 summarizes the average radial decrease in the horizontal component of the magnetic field in spot penumbrae and in the photosphere around spots. Figure 17 shows the average inclination of the field lines in penumbrae deduced from sunspot observations near the center and limb of the solar disk (all results adapted from Leroy 1962).

Recently, similar observations were obtained with the new University of Hawaii solar polarimeter, measuring the four Stokes parameters

THE FRENCH SOLAR POLARIMETER

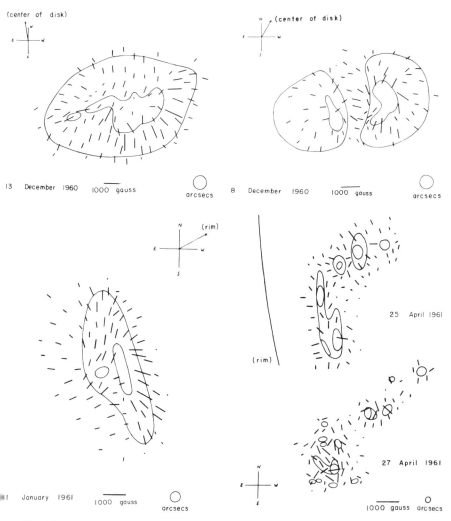

Fig. 15. Examples of photospheric transverse magnetic field maps. The lengths of segments are proportional to the components of the field lying perpendicular to the line of sight; azimuths are parallel to the field lines. Circles indicate diameter of field, namely 5 arcsecs (after Leroy 1962).

(Mickey and Orrall[2]). The Meudon Observatory polarimeter and its telescope have recently been improved by Koeckelenbergh; semi-automatic scanning has been provided with a view to extending the observational program.

---

[2] See p. 686.

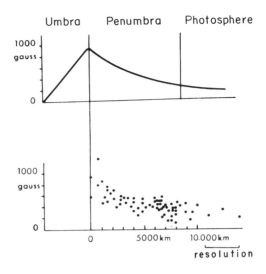
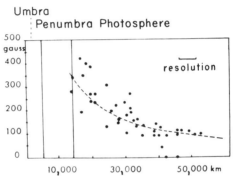

Fig. 16. Average magnetic fields in penumbrae and in the photosphere around sunspots. Intensities of the horizontal components are in gauss (after Leroy 1962).

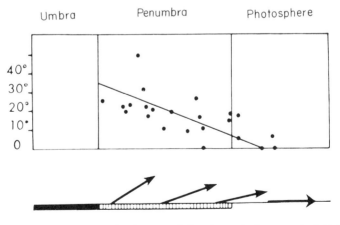

Fig. 17. Average inclinations, in degrees, of the photospheric magnetic field lines in sunspot penumbrae (after Leroy 1962).

MONITORING OF THE ELECTRON-SCATTERED CORONA

One of the major possibilities offered by the solar photoelectric polarimeter was the observation of coronal enhancements and streamers. Thompson scattering by the free electrons responsible for the solar corona generates large polarization, whereas sunlight scattered by the instrument, the terrestrial atmosphere, and interplanetary dust is relatively unpolarized, thus permitting discrimination of the true coronal component.

The observation of a bright north-east coronal streamer during the Khartoum eclipse of 25 February 1952 led to the idea that the photometric polarimeter, along with a coronograph, could be used outside of eclipse to detect and eventually monitor major coronal streamers up to three solar radii from the sun's center. It was subsequently realized, during initial design study for such a coronal streamer polarimeter, that the inner corona is accessible to observation at sea level without using a coronograph. Both these possibilities were indicated in 1953 (Dollfus 1953); the first observations were obtained in 1956, and the first results published in 1958 (Dollfus 1958b, c; 1959a, b).

Figure 18 shows the photoelectric polarimeter as it was in 1956, attached to Lyot's 20-cm aperture coronograph at Pic-du-Midi in order to detect and analyze coronal streamers around the sun, up to three radii from the sun's center. The polarimeter was rotated in steps about the mechanical axis in order to scan the sky background around the sun at radially increasing distances up to 35 arcmins from the limb. The prototype instrument for monitoring the inner corona without using a coronograph was set up at Meudon in 1958.

Early Photoelectric Observations of the Electron-Scattered Corona

The first complete series of coronal observations made at Pic-du-Midi with the instrument shown in Fig. 18 was obtained on 29 September 1956. Some results are shown in Fig. 19 (see caption). A coronal streamer at 110° position angle is easily detected at 8 arcmins from the limb. An example of the detection of coronal streamers 35 arcmins from the solar limb has been given by Dollfus (1958c).

At Meudon Observatory, working without coronograph and near sea level, the corona was first detected on 13 April 1958. Figure 20 refers to a complete scan around the sun at 2 arcmins from the limb on 19 August 1959. Coronal streamers were usually monitored at Meudon between 2 and 10 arcmins above the solar limb. Our first result was noticing that coronal enhancements and streamers persisted for days if not weeks. Later on, some streamers were recognized at the west limb 14 days after observation at the east limb (Dollfus 1959a, b). The axial rotation of the sun, therefore, provides a way of scanning in heliographic longitude, so

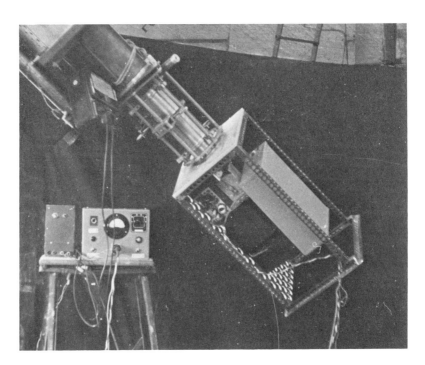

Fig. 18. The photoelectric polarimeter adapted to the Pic-du-Midi 20-cm aperture Lyot coronograph, in 1956, for the study of electron-scattered coronal enhancements and streamers.

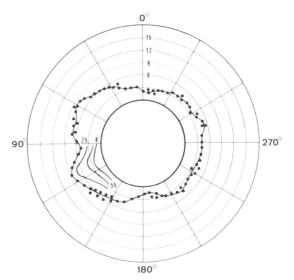

Fig. 19. First observation, 29 September 1956, of the electron-scattered corona with the photoelectric polarimeter attached to the Pic-du-Midi coronograph. The intensity of polarized light due to the corona is plotted radially. The complete scan is at 1 arcmin outside the solar limb. Additional partial scans of a streamer are 2.5, 5.5, and 8.0 arcmins.

THE FRENCH SOLAR POLARIMETER

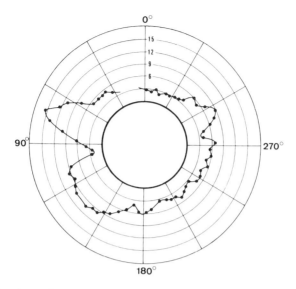

Fig. 20. First observation, at Meudon on 19 August 1959, of the electron-scattered corona made without using a coronograph (see Dollfus 1959a). Polarization is plotted radially. The scan was made 2 arcmins outside the solar limb.

that coronal enhancements may be mapped both in latitude and longitude.

Figure 21 shows coronal heliographic maps obtained at Meudon in 1959. The calibration refers to the degree of polarization in units of 10^{-4} (the absolute photometry of polarized light, referred to above, had not been established in Meudon in 1959). Observations made at the east and west limbs gave two distinct maps. Observation dates are indicated at the top of each map; gaps in data are due to adverse weather conditions. The evolution of a coronal enhancement may be seen at long. $0°$, lat. $15°$ between 12 August 1959 (west limb, upper map) and 28 August 1959 (east limb, lower map).

Combined observations between Pic-du-Midi and Meudon Observatories from 1959 to 1962 refined the results; some coronal streamers were followed for three consecutive months, giving pointers to their evolution and life-time (Leroy 1960a, b). On the occasion of the solar eclipse of 15 February 1961, almost daily combined observations during the preceding and following weeks provided complete mapping of the solar corona at 3 arcmins above the photosphere, photometrically calibrated by eclipse photographs (Dollfus, Marin, and Leroy 1961). Figure 22 reproduces the east limb coronal heliographic map together with an eclipse plate obtained by Laffineur. The eclipse date is represented by a solid line in the map; two of the photographic coronal streamers are recognized. The two maps for the east and west limbs help to locate the coordinates

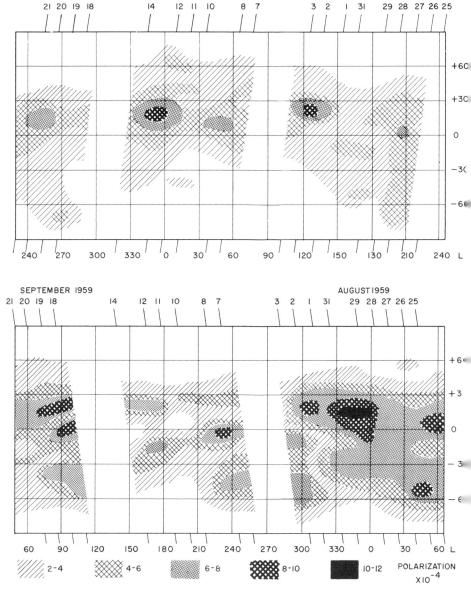

Fig. 21. The first heliographic maps of the electron-scattered coronal features, recorded at Meudon Observatory from 25 August to 21 September 1959. Scans, as in Fig. 20, are made 2 arcmins outside the limb on the dates indicated. Solar rotation provides the coverage in longitude, L. Top: west limb (unpublished). Bottom: east limb (from Dollfus 1959b).

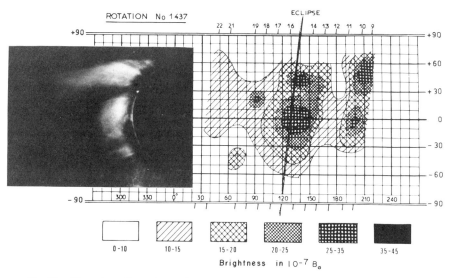

Fig. 22. The solar eclipse of 15 February 1961. Combined coronal observations at 3 arc-mins from the limb with the Pic-du-Midi and Meudon polarimeters. The coronal heliographic map (east limb), with eclipse date indicated, is compared with the eclipse photograph of the coronal streamers by Laffineur.

of the roots of coronal streamers; these data are needed in the computation of electron density models for coronal streamers. Hitherto unpublished results from this eclipse are given in Fig. 23 (curves 3, 4, and 5). Accounts of the work done in this period have also been given by Dollfus (1963a, b).

I.Q.S.Y. Survey of the Solar Electron Corona

In 1963, it was decided that the French contribution to the "International Quiet Sun Years" (I.Q.S.Y.) covering 1964–1967 would include a survey of solar coronal enhancements and streamers with the photoelectric polarimeters, to be combined with American and Japanese observations. The purpose was to establish experimentally a well-founded basis for astrophysical studies of electron density distribution in the solar corona and its variation with location and time. An attempt was made to sort out the complicated nature of enhancement and streamer fluctuations.

The solar photoelectric polarimeter, in its final form with calibration devices, was manufactured by Danger of Paris (Figs. 2, 3, 4), and two identical instruments were put into operation at Pic-du-Midi and Meudon Observatories. The survey, initially operated at Pic-du-Midi by Meudon Observatory, started in December 1963. In January 1968, after the I.Q.S.Y. period, the management of the survey was transferred to Pic-du-Midi under the responsibility of Leroy, with the aim of covering a full cycle

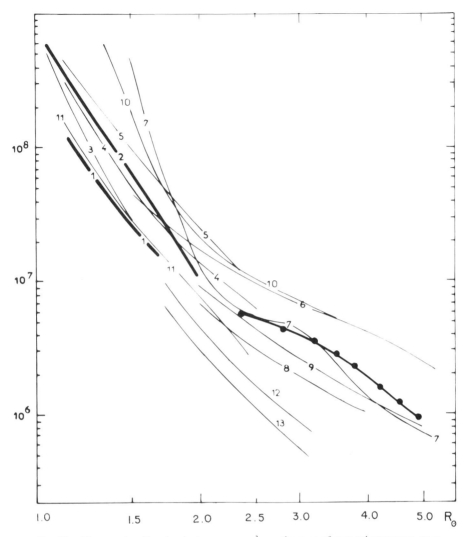

Fig. 23. Electron densities, in electrons per cm^3, on the axes of coronal streamers, as a function of distance from the center of the sun. *(1)* polarization observation at Pic du Midi on 10 March 1965; *(2)* same, 6 Sept. 1966; *(3)* similar measurements from eclipse plates, 15 Sept. 1961, at position angle PA = 218° by A. Dollfus; *(4)* as for *(3)*, but for PA = 95°; *(5)* as for *(3)*, but for PA = 55°; *(6)* as for *(3)*, by Michard, 25 Feb. 1952, at PA = 45°; *(7)* as for *(3)*, by Kouchmy, 7 March 1970, at PA = 45°; *(8)* as for *(3)*, by Kouchmy, 22 Sept. 1968, at PA = 110°; *(9)* as for *(3)*, by Saito, 20 June 1955, 4 streamers; *(10)* as for *(3)*, by Saito et al., 5 Feb. 1962, at PA = 260°; *(11)* as for *(3)*, by Hepburn, 25 Feb. 1952, at PA = 45°; *(12)* as for *(3)*, by Newkirk et al., 12 Nov. 1966, at PA = 135°; *(13)* as for *(12)*, at PA = 225°. The heavy line with dots indicates the measurements made with the French balloon-borne externally occulted coronograph flight of 13 Sept. 1971, at position angle PA = 50°, by A. Dollfus.

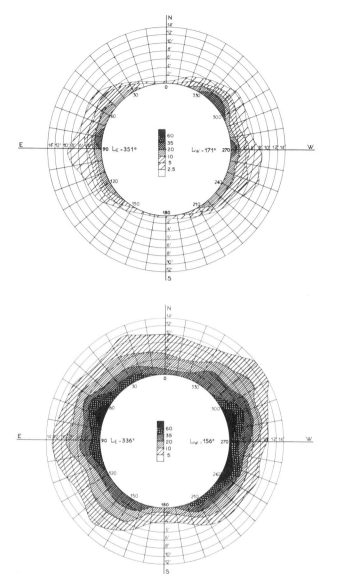

Fig. 24. Isophote maps of solar corona polarized light intensity around the disk (units of 10^{-8} times the luminance of the disk center). Top: 9 June 1964 at minimum solar activity (from scans at 1.5, 3, 5, and 8 arcmins from the limb). Bottom: 5 March 1968 for higher solar activity (from scans at 1.5, 5, and 10 arcmins from the limb). Pic-du-Midi solar polarimeter.

of solar activity. During the 1963–1967 I.Q.S.Y. period, about 500 observing days were accrued. A preliminary account of results was published by Dollfus (1968).

Figure 24 gives two examples of isophote maps of the distribution of polarized light around the sun, between 1.5 and 12 arcmins above the photosphere, obtained at times of minimum solar activity (9 June 1964) and higher activity (5 March 1964). The isophotes are given on an absolute scale in units of 10^{-8} of the luminance of the sun's center. Figure 25 gives a further example of the evolution of a coronal condensation at long. 90°, lat. +20°. Observations on 22, 23, and 24 July 1967 (east limb, upper map) and on 17 and 18 August 1967 (lower map), along with the corresponding isophote map of the streamers at the east limb on 22 July 1967, are given. The whole set of observations covers the evolution of this long-lasting active coronal feature.

In favorable circumstances, numerical electron-density distributions can be derived from observation. As an example, Fig. 23 (heavy curves 1 and 2) shows two determinations of electron densities on the axes of coronal streamers between 1.1 and 2.0 solar radii from the disk center. For comparison, the other published results are given: curves 6 to 13 are derived from photographic photometry of eclipse plates obtained by various authors; curves 3, 4, and 5 are derived from 1961 eclipse plates using supporting polarimetric observations to locate the streamers (see above). A result from the balloon-borne externally occulted coronograph, obtained by the author on 13 September 1971, is also included.

More results from the I.Q.S.Y. coronal survey period are in preparation. Since the transfer of the survey management to Pic-du-Midi in January 1968, refinements to the corrections for atmospheric polarization have been evaluated (Leroy 1972). Extension of the spectral range to the near infrared has once more improved the photometric accuracy (Leroy and Ratier[3]).

Acknowledgments. During the past two decades I have been indebted to a large number of people, a list of whom, even were it practical, would take up a sizeable portion of this paper. This work was instituted at Meudon Observatory after the death of Bernard Lyot in 1952, and its evolution is a natural consequence of his founding spirit.

REFERENCES

Debarbat, S.; Dumont, S.; and Pecker, J-C. 1970*a*. Interpretation of optical solar polarization. *Astrophys. Letters* 6: 251–256.
———. 1970*b*. La polarisation du spectre continu au bord du disque solaire. *Astron. Astrophys.* 8: 231–242.
Destriau, G., and Prouteau, J. 1949. Réalisation d'un quart d'onde quasi achromatique par juxtaposition de deux lames cristallines de même nature. *J. Phys. Radium* 8: 53–55.

---

[3] See p. 762.

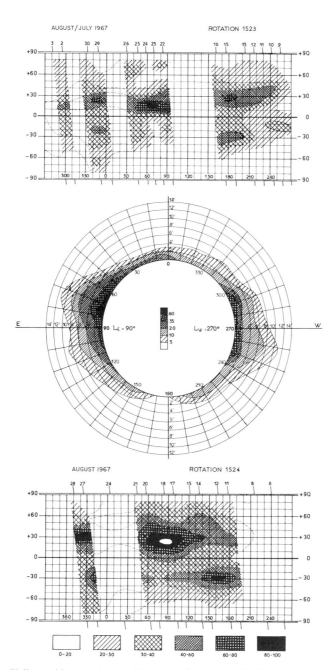

Fig. 25. Heliographic maps (top and bottom) of the coronal polarization at 1.5 arcmins above the solar disk (east limb), for rotation no. 1523 (July 1967) and 1524 (August 1967), showing the development of a coronal condensation with streamers at long. 90°; lat. +20°N. The isophote map of 22 July (center) shows the streamer at the east limb from scans made at 1.5, 3, 5, 8, and 14 arcmins from the limb. Intensities and the polarization gradations (below the bottom figure) are in units of 10^{-8} times the luminance of the disk center. Pic-du-Midi solar polarimeter.

Dollfus, A. 1953. 1) Nouveau procédé d'observation de la couronne solaire intérieure, en lumière blanche, par une journée ordinaire de beau temps. 2) Procédé permettant d'observer la couronne solaire blanche et ses jets jusqu'à une grande distance du bord solaire. Pli Cacheté n° 13056, Académie des Sciences Paris, deposited 2 Février 1953, opened 9 June 1958. Archives Acad. Sci. Paris.

———. 1955. Etude des planètes par la polarisation de leur lumière. Thèse Paris, *Ann. Astrophys. Suppl.* 4. English Transl. NASA-TT F-188.

———. 1958*a*. Un polarimètre photoélectrique très sensible pour l'étude du soleil. *C. R. Acad. Sci. Paris* 246: 2345–2347.

———. 1958*b*. Premières observations avec le polarimètre solaire. *C. R. Acad. Sci. Paris* 246: 3590–3593.

———. 1958*c*. Procédé permettant d'observer la couronne solaire et ses jets jusqu'à une grande hauteur. *C. R. Acad. Sci. Paris* 247: 42–45.

———. 1959*a*. Observation de la couronne solaire à l'Observatoire de Meudon en 1959. *C. R. Acad. Sci. Paris* 249: 2273–2275.

———. 1959*b*. Quelques propriétés des jets de la couronne solaire observées en lumière blanche. *C. R. Acad. Sci. Paris* 249: 2722–2724.

———. 1963*a*. Etude des grands jets de la couronne solaire observés en lumière blanche sans corongraphe. *The solar corona*. (J. Evans, ed.) pp. 243–246. New York: Academic Press.

———. 1963*b*. Un modulateur de lumière polarisée pour polarimètre photoélectrique. *C. R. Acad. Sci. Paris* 256: 1920–1922.

———. 1965. Recherches sur la structure fine du champ magnétique photosphérique. *Stellar and solar magnetic fields* (R. Lüst, ed.) Amsterdam: North Holland Publ. Co.

———. 1968. Observation des jets et concentrations de la couronne au-dessus des régions actives. *Structure and development of solar active regions* (K. O. Kiepenheuer, ed.) Dordrecht, Holland: D. Reidel Publ. Co. pp. 359–378.

Dollfus, A., and Leroy, J-L. 1960. Polarisation de la lumière au bord du disque solaire. *C. R. Acad. Sci. Paris* 250: 665–667.

Dollfus, A.; Marin, M.; and Leroy, J-L. 1961. Observation de la couronne solaire avant, pendant et après l'éclipse totale du 15 Février 1961. *C. R. Acad. Sci. Paris* 252: 3402–3404.

Dumont, S., and Pecker, J.-C. 1971. Influence de l'absorption dans les raies sur la polarisation du spectre continu. *Astron. Astrophys.* 10: 118–127.

Leroy, J.-L. 1959. Polarisation de la lumière des taches solaires. *C. R. Acad. Sci. Paris* 249: 2492–2494.

———. 1960*a*. Interprétation de la lumière des taches solaires. *C. R. Acad. Sci. Paris* 251: 1720–1722.

———. 1960*b*. Observation d'un jet coronal persistant. *Ann. Astrophys.* 23: 567–573.

———. 1962. Contribution à l'étude de la polarisation de la lumière solaire. Thèse, Paris; *Ann. Astrophys.* 25: 127–165.

———. 1972. Nouvelles mesures de la polarisation de la lumière au bord du disque solaire. *Astron. Astrophys.* 19: 287–292.

Lyot, B. 1929. Recherches sur la polarisation de la lumière des planètes et de quelques substances terrestres. Thèse, Paris, *Ann. Obs. Meudon* Tome VIII, Fasci. 1. English Transl. NASA-TT F-187.

———. 1948*a*. Un polarimètre photoélectrique. *C. R. Acad. Sci. Paris* 226: 25–28.

———. 1948*b*. Un polarimètre photoélectrique intégrateur utilisant des cellules à multiplicateurs d'électrons. *C. R. Acad. Sci. Paris* 226: 137–140.

Marin, M. 1965. Mesures photoélectrique de polarisation à l'aide d'un télescope coudé de 1 m. *Rev. Opt. Paris* 44: 115–144.

Öhman, Y. 1949. Photoelectric work by the flicker method. *Stockholm Obs. Ann.* 15: No. 8.
Pecker, J.-C. 1970. Some considerations from the direct comparison between the observations and the theory of solar disk polarization. *Solar Phys.* 15: 88–96.
Tinbergen, J. 1972. Development of a wide-band version of the Dollfus polarization modulator. Dissertation Leiden Univ., The Netherlands.

POLARIZATION OF SOLAR EMISSION LINES

E. TANDBERG-HANSSEN
High Altitude Observatory

The theory of polarization of solar emission lines produced by absorption and scattering is reviewed. Special emphasis is placed on the application of the theory for deducing the nature of magnetic fields in the solar atmosphere. In the general case, the interpretation of the state of polarization in an emission line is complicated by coherence effects due to overlapping of atomic levels. It is pointed out that for the forbidden coronal lines the coherence effect (Hanle effect) disappears, since we are in an effectively strong magnetic field case. The interpretation is then simplified considerably. In the case of prominences we are not in the effectively strong field case, and the interpretation of the fluorescence radiation is very complicated, due to the Hanle effect, collisions, and multiple scattering. The discussion is cast in the framework of the Stokes parameters which are often used in solar magnetograph work.

The sun's atmosphere produces a multitude of observable emission lines that may be catalogued into several groups. We are concerned here with two of them, viz. (1) lines due to permitted transitions originating in prominences seen above the solar limb, and (2) lines due to forbidden transitions emitted by the solar coronal plasma. In addition, at eclipses, emission lines from the chromosphere can be observed (the flash spectrum), and bright flares at times produce emission lines when observed on the disk. When flares are seen above the limb one observes emission lines similarly as for prominences, while the latter normally produce absorption lines on the disk. We shall not consider any of the additional groups of lines in the present paper.

In general, emission lines will be polarized when the incident flux that excites the upper level of the transition in question is polarized or anisotropic. In the solar atmosphere the exciting flux often is a radiative flux, yielding fluorescence polarization. Excitation produced by a flux of charged particles can lead to impact polarization.

HISTORY

More than 40 years ago Öhman (1929) suggested that, even though there was little hope of observing polarization of emission lines close to the sun's limb, i.e., in chromospheric lines, one should look for the effect in higher prominences. His assumption was that the expected fluorescence polarization would be due to the anisotropy of the incident photospheric radiation field. Öhman also mentioned the possibility that solar magnetic fields can affect the polarization in prominence lines. As we shall see later, much of the modern work on the theory of prominence-line polarization has been concerned with this influence of magnetic fields.

The first successful measurements of a small amount of linear polarization in prominence lines (Hα and He I D$_3$) were reported by Lyot (1934, 1937) and confirmed by Newkirk (1958). Zanstra (1950) was the first to give a quantitative treatment of the polarization of the Hα and D$_3$ prominence lines, combining the effects of limb darkening and an anisotropic radiation field. He considered resonance line scattering for an optically thin case with no magnetic field, while Thiessen (1951) included the effect of a magnetic field. Both Zanstra and Thiessen predicted amounts of polarization much larger than the observed values, and both suggested that collisional depolarization may be at work. The first detailed consideration of impact polarization in prominence lines and depolarizing effects due to collisions and to magnetic fields were given by Hyder (1964a).

Also the emission lines of the corona are polarized. As we shall see later, the interpretation of this polarization is simpler than in the case of prominence-line polarization, but the observations are difficult.

From Lyot's unpublished notes, it is clear that this pioneer of observational coronal physics observed a small polarization in the green coronal line from FeXIV at 5303 Å in 1936 (Charvin 1965), and his findings were confirmed by Karimov (1961) and Charvin (1963).

The first theoretical analysis of coronal emission-line polarization is due to Charvin (1964a, b). A more general treatment, including radiative and collisional effects and with magnetic fields present, has been given by Charvin (1965); Hyder (1965, 1966) has discussed magnetic effects in some detail.

Extensive theoretical treatments of emission-line polarization, including a number of depolarizing effects, have been published since the mid-1960s, and they will be considered below. As we shall see, the effects of magnetic fields are crucial for a complete understanding of the polarization phenomenon in the solar atmosphere. Conversely, given the state of polarization by observation, we may ask whether we can infer the nature of a magnetic field in the plasma producing the emission line. This is the approach taken both in prominence and coronal physics today, and several polarimeters have been or are being built for the express purpose of deducing the configuration of solar magnetic fields. Consequently, in this paper we emphasize the effect of magnetic fields on the state of polarization of emission lines.

DESCRIPTION OF POLARIZED LIGHT

The pure polarization of electromagnetic waves can be defined in terms of either the electric or the magnetic field vector in planes perpendicular to the direction of propagation. If the vector is not randomly oriented (and the radiation therefore is not unpolarized), the vector will either be oriented always in the same direction (linearly polarized waves) or the projection of the tip of the vector will describe an ellipse (elliptically polarized waves, of which circularly polarized waves form a subset). The percentage polarization can be defined as

$$P = 100 \frac{I_a - I_b}{I_a + I_b}, \tag{1}$$

where I_a is the intensity of the radiation with the electric vector along the semimajor axis, a, of the ellipse; and I_b is the intensity with the vector along the semiminor axis, b. The ellipticity of the polarization ellipse is $\sin \chi$, where $\tan \chi = \dfrac{b}{a}$ (Born and Wolf 1965).

Much of the modern work on polarization is cast in the framework of the Stokes parameters, which completely and elegantly describe polarized light.[1] Consider two orthogonal plane waves propagating in the z-direction and having amplitudes E_{ox} and E_{oy}, respectively:

$$E_x(t) = E_{ox} \exp\left[i(\omega t + \delta_x)\right] = \varepsilon_{ox} e^{i\omega t}$$

and

$$E_y(t) = E_{oy} \exp\left[i(\omega t + \delta_y)\right] = \varepsilon_{oy} e^{i\omega t},$$

where ε_{ox} and ε_{oy} are complex amplitudes containing the phase factor δ_x and δ_y. The Stokes parameters (Stokes 1852) can be written as a column matrix (Perrin 1942; see also Collett 1968),

$$S = \begin{bmatrix} I \\ Q \\ U \\ V \end{bmatrix} = \begin{bmatrix} \varepsilon_{ox}\varepsilon_{ox}^* + \varepsilon_{oy}\varepsilon_{oy}^* \\ \varepsilon_{ox}\varepsilon_{ox}^* - \varepsilon_{oy}\varepsilon_{oy}^* \\ \varepsilon_{ox}\varepsilon_{oy}^* + \varepsilon_{ox}^*\varepsilon_{oy} \\ \varepsilon_{ox}\varepsilon_{oy}^* - \varepsilon_{ox}^*\varepsilon_{oy} \end{bmatrix} = \begin{bmatrix} E_{ox}^2 + E_{oy}^2 \\ E_{ox}^2 - E_{oy}^2 \\ 2E_{ox}E_{oy}\cos\delta \\ 2E_{ox}E_{oy}\sin\delta \end{bmatrix}, \tag{2}$$

---

[1] Also see p. 45. Note that Clarke proposes to define ellipticity as $(a - b)/a$.

where $\delta = \delta_y - \delta_x$. The last part of Equation (2) shows how the Stokes four-vector representation is related to the description using the polarization ellipse, since the ellipticity can be written

$$\sin \chi = \frac{V}{(Q^2 + U^2 + V^2)^{1/2}}. \tag{3}$$

We have further (Chandrasekhar 1950)

$$I^2 \geq Q^2 + U^2 + V^2,$$

where the equality sign pertains to completely polarized light.

The Stokes representation of polarized light is related to the matrix representation (McMaster 1961), which is particularly useful in a quantum-mechanical treatment. Since we are concerned with the interaction of photons and electrons, we can express the expansion of the wave function, which describes a pure state of polarization, with only two orthonormal eigenfunctions

$$\psi = \varepsilon_{ox} \psi_x + \varepsilon_{oy} \psi_y.$$

For example, we could use two orthogonal linear polarization states or two orthogonal (left- and right-handed) states of circular polarization. For the wave functions for pure states, we can choose $\psi_x = \begin{pmatrix} 1 \\ 0 \end{pmatrix}$ and $\psi_y = \begin{pmatrix} 0 \\ 1 \end{pmatrix}$, and the wave function for the incoming light beam will be $\psi = \begin{pmatrix} \varepsilon_{ox} \\ \varepsilon_{oy} \end{pmatrix}$. This leads to the following expectation values for the unit matrix and for Pauli's spin matrices

$$\langle 1 \rangle = \varepsilon_{ox}^* \varepsilon_{oy}^* \begin{pmatrix} 1 & 0 \\ 0 & 1 \end{pmatrix} = \varepsilon_{ox} \varepsilon_{ox}^* + \varepsilon_{oy} \varepsilon_{oy}^*$$

$$\langle \sigma_z \rangle = \varepsilon_{ox}^* \varepsilon_{oy}^* \begin{pmatrix} 1 & 0 \\ 0 & -1 \end{pmatrix} = \varepsilon_{ox} \varepsilon_{ox}^* - \varepsilon_{oy} \varepsilon_{oy}^*$$

$$\langle \sigma_x \rangle = \varepsilon_{ox}^* \varepsilon_{oy}^* \begin{pmatrix} 0 & 1 \\ 1 & 0 \end{pmatrix} = \varepsilon_{ox} \varepsilon_{oy}^* + \varepsilon_{oy} \varepsilon_{ox}^*$$

$$\langle \sigma_y \rangle = \varepsilon_{ox}^* \varepsilon_{oy}^* \begin{pmatrix} 0 & -i \\ i & 0 \end{pmatrix} = i(\varepsilon_{ox} \varepsilon_{oy}^* - \varepsilon_{ox}^* \varepsilon_{oy}).$$

From Equations (2) and (3) we see that the Stokes parameters represent physically measurable quantities that completely describe the light beam.

If the photons are not completely in one or the other of the pure state, we have partially polarized light. We then need a statistical description, using an "ensemble" average in terms of a density matrix, ρ, (Fano 1957; ter Haar 1961):

$$\rho = \begin{pmatrix} \varepsilon_{ox}\varepsilon_{ox}^* & \varepsilon_{ox}\varepsilon_{oy}^* \\ \varepsilon_{oy}\varepsilon_{ox}^* & \varepsilon_{oy}\varepsilon_{oy}^* \end{pmatrix} = \begin{pmatrix} \rho_{11} & \rho_{12} \\ \rho_{21} & \rho_{22} \end{pmatrix}. \quad (4)$$

Equations (2) and (4) show the relationship between the Stokes vector and the density matrix.

When an incoming exciting beam of photons interact with the atom, the Stokes parameters describing the beam will be transformed. To express this mathematically one defines a column vector, D, which has as elements the components of the density matrix (Eq. [4]). From the vector D we can obtain the Stokes vector by the action of a transformation (or interaction) matrix T

$$S = TD \quad (5)$$

or

$$\begin{bmatrix} I \\ Q \\ U \\ V \end{bmatrix} = \begin{pmatrix} 1 & 0 & 0 & 1 \\ 1 & 0 & 0 & -1 \\ 0 & 1 & 1 & 0 \\ 0 & i & -i & 0 \end{pmatrix} \begin{bmatrix} \rho_{11} \\ \rho_{12} \\ \rho_{21} \\ \rho_{22} \end{bmatrix},$$

and the new transformed Stokes vector, due to the interaction beam-atom and characterized by S', is given by the Mueller matrix M (see, for instance, Schurcliff 1962):

$$S' = MS. \quad (6)$$

We shall return to the functional form of the Mueller matrix later.

BASIC CONCEPTS IN EMISSION-LINE POLARIZATION

Classical Treatment

It is exactly 50 years since Lord Rayleigh (Rayleigh 1922) published his discovery that if one excites gaseous mercury by polarized radiation, the emitted resonance line at 2537 Å will exhibit linear polarization. This resonance polarization effect, which is a special case of fluorescence polarization, has played a considerable role in the development of emission-line polarization research. Wood and Ellett (1923) and Hanle (1924, 1925) studied how a magnetic field affects the observed resonance polar-

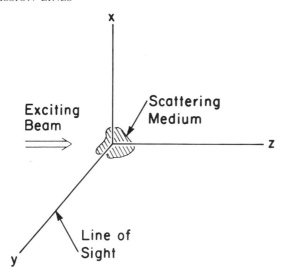

Fig. 1. The frame of reference used to discuss the geometry of observation of scattered radiation relative to incident, exciting beam.

ization, and we shall review briefly some results of this so-called Hanle effect.

Consider the scattering medium to be at the origin of a reference system whose z-axis is in the direction of the exciting beam of linearly polarized radiation (see Fig. 1). One observes the scattering mercury vapor along the y-axis and finds that the resonance emission is polarized in direction and amount according to the direction of polarization of the exciting radiation and to both the direction and strength of the magnetic field. Sometimes one observes strong polarization when an essentially unpolarized line would appear in the field-free case, while in other instances the field depolarizes the line. This Hanle effect in mercury can be explained qualitatively by classical theory of the atom (see, for instance, Mitchell and Zemansky 1934), but for other atoms a quantum-mechanical treatment is necessary.

We shall consider a few cases observed by Hanle and explained by classical theory. Let the direction of vibration of the electric vector in the exciting radiation be parallel to the x-axis in Fig. 1, and let there be a magnetic field (25–100 gauss) in the same direction. The polarization of the resonance-scattered light will be in the same direction as in the field-free case, i.e., we see that a high degree of incident polarization along the x-axis produces scattered light that has strongly enhanced vibrations in the x-direction. This follows immediately from classical theory of the atom in which the scatterer behaves like a classical oscillator.

Now, let a weak (less than a few gauss) magnetic field be parallel to the y-axis. We then observe a strong resonance-scattered line that has strongly enhanced vibrations in the x-direction. If the field strength increases, one observes that the percentage polarization decreases. As long as the field is weak, the direction of vibration only deviates slightly from that of the field-free case, but as the field increases—and the polarization decreases—the plane of polarization rotates through large angles.

We can explain the sensitivity of the observed polarization to the orientation of the magnetic field in the framework of the classical oscillator when we remember that the atom will precess about the magnetic vector, and that the atomic oscillator emits circularly polarized light along the magnetic field and linearly polarized light normal to the field.

Finally, if we assume that the atom is a damped oscillator, we explain the sensitivity of polarization to the strength of the field. Due to the precession, the direction of vibration will deviate from the original direction, and this causes the plane of vibration of the emitted light to rotate. The oscillator describes a rosette when viewed along the magnetic field, and the shape of the rosette—and therefore the nature of the polarization—will depend on the ratio between the angular velocity of precession, ω, and the radiative damping constant, Γ (or the reciprocal of the mean radiative lifetime, τ), of the oscillator. If $\omega \gg \Gamma$, or

$$\omega\tau \gg 1, \qquad (7)$$

the rosette will be axially symmetric, since the atom will have ample time to precess before it is damped out (see Fig. 2). Consequently, there is no polarization observed along the magnetic field if the field is strong enough to make the inequality (7) hold. If $\omega \approx \Gamma$, or $\omega\tau \approx 1$, the oscillator describes an asymmetrical rosette (see Fig. 3), which means that the degree of polarization is reduced relative to the value in the field-free case, and the direction of polarization has been rotated an angle Ω with respect to the direction of polarization of the exciting beam. Finally, if $\omega\tau \ll 1$, the oscillator hardly has time to precess before it is damped out, which explains the weak-field results.

Breit (1925) derived expressions for the percentage polarization, P, and the angle of rotation, Ω. For polarization observed along the magnetic field of strength B, the polarization is

$$P = \frac{P_0}{1 + (g\omega_e\tau)^2}, \qquad (8)$$

where P_0 is the percentage polarization for the field-free case, and $\omega_e =$

SOLAR EMISSION LINES

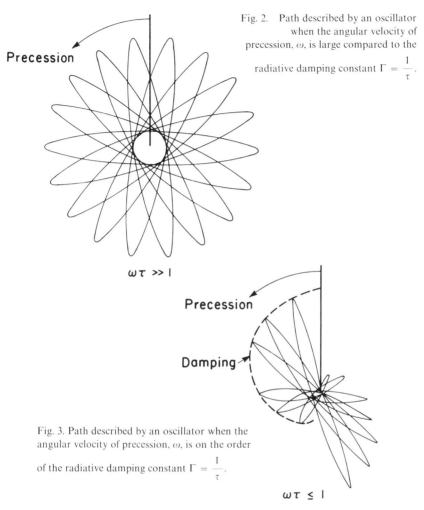

Fig. 2. Path described by an oscillator when the angular velocity of precession, ω, is large compared to the radiative damping constant $\Gamma = \dfrac{1}{\tau}$.

Fig. 3. Path described by an oscillator when the angular velocity of precession, ω, is on the order of the radiative damping constant $\Gamma = \dfrac{1}{\tau}$.

$\dfrac{eB}{2mc}$ is the Larmor precession velocity for an electron of mass m and charge e. The angle of rotation is given by

$$\tan 2\Omega = g\omega_e \tau. \tag{9}$$

In both Equations (8) and (9) g is a correction factor necessary to obtain the observed results from classical theory, and Breit (1933) showed that classical and quantum-mechanical results agree if this correction factor is Lande's g-factor for the upper level of the transition in question.

Quantum-Mechanical Treatment

Even though classical theory is capable of explaining most of the resonance polarization effects as exhibited by the mercury resonance line, it fails miserably for other resonance transitions, such as those involved in the Na I, D_1 and D_2 lines. Furthermore, for the more general fluorescence polarization a more sophisticated theory is necessary.

Hanle (1923) was the first to explain some of these phenomena using an elementary quantum-mechanical treatment, involving the Zeeman states of the levels for the transition in question. The behavior of the mercury resonance-line polarization can be understood from the appropriate energy-level diagram for the $^1S_0 - {}^3P_1$ transition responsible for the 2537 Å line (see Fig. 4). Using the frame of reference of Fig. 1, we let the $^3P_1$ level be excited by light polarized parallel to the x-axis. Then, if there is a strong magnetic field in the same direction, only the π-component (which, according to the classical Zeeman effect, is polarized parallel to the field) can be absorbed. Hence, we populate only the $m = 0$ Zeeman state of the $^3P_1$ level. Then, if there is no interlevel transition taking place, the emitted radiation will be a π-component whose electric vector is parallel to the field. On the other hand, if the magnetic field is parallel to the z-axis, the circularly polarized components will be absorbed and subsequently re-emitted. These will be circularly polarized, but along the line of sight (the y-axis), which is perpendicular to the field, the radiation appears linearly polarized parallel to the x-axis.

The difficulty that arises in predicting which components will be absorbed in the field-free case where the upper level is degenerate is resolved by Heisenberg's (1926) principle of spectroscopic stability, according to which the absorption of radiation is the same in the field-free case as when the magnetic field is parallel to the direction of vibration of the incident polarized beam. In other words, one may assume that absorption occurs in π-components.

The resonance polarization effect in sodium is of great importance in solar physics since the lines D_1 and D_2 are well observed in prominences. The appropriate energy-level diagram, with relative intensities of the transitions given below, is shown in Fig. 5. It can be used to explain why the D_1 line ($^2S_{1/2} - {}^2P_{1/2}$) is always unpolarized, while the D_2 line ($^2S_{1/2} - {}^2P_{3/2}$) shows polarization that depends on the magnetic field, which the classical theory cannot account for. Since both upper m-states for the D_1 line are connected with both lower m-states (see Fig. 5a), and since the intensities of the transitions are equal, we see that the resulting radiation must be linearly unpolarized and that the presence of a magnetic field will make no difference. However, the unequal population of the upper m-states and the difference in intensity of the emitted components will cause polarization effects in the D_2 line (see Fig. 5b).

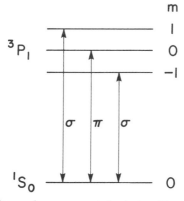

Fig. 4. The σ and π components due to transitions $^1S_0 - {}^3P_1$.

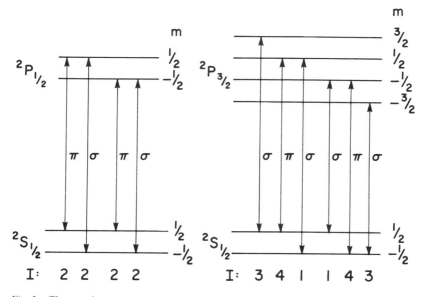

Fig. 5. The σ and π components and their relative intensities, I, due to transitions $^2S_{1/2} - {}^2P_{1/2}$ and $^2S_{1/2} - {}^2P_{3/2}$.

From the transition probabilities for the different transitions between Zeeman states, one can calculate the polarization of any resonance line (van Vleck 1925) for strong fields. Let A_π be the transition probability for the linearly polarized Zeeman component originating in the Zeeman state m, and let A_σ be the sum of the two circularly polarized σ components. The number of electrons excited to state m is proportional to $I(A_\pi^m \cos^2 \gamma + \frac{1}{2} A_\sigma^m \sin^2 \gamma)$, where I is the intensity of the exciting radiation and γ is the angle between the electric vector and the magnetic field direction. During the subsequent emission, a fraction $\dfrac{A_\pi^m}{A_\pi^m + A_\sigma^m}$ of the electrons will give rise to linearly polarized light, and a fraction $\dfrac{A_\sigma^m}{A_\pi^m + A_\sigma^m}$ to circularly polarized light. Van Vleck (1925) showed that if one observes at right angles to a strong magnetic field, the percentage polarization can be written as

$$P = 100 \frac{\left[3 \sum_m (A_\pi^m)^2 - \dfrac{2j+1}{3}(A^m)^2\right](3\cos^2\gamma - 1)}{(3\cos^2\gamma - 1)\sum_m (A_\pi^m)^2 + \dfrac{2j+1}{3}(A^m)^2(3 - \cos^2\gamma)}, \qquad (10)$$

where $A^m = A_\sigma^m + A_\pi^m$ and where we have used the related $\sum_m (A^m) = (2j+1)A^m$. Equation (10) yields the general result that the percentage polarization is zero for $\gamma = 54°45'$, and this holds for all resonance lines.

THE TRANSPORT OF POLARIZED RADIATION

The theory of radiative transfer of unpolarized light has been developed to a high degree of sophistication, and by using such theory one can, from an observed spectrum, deduce valuable information on the state of the emitting and absorbing gas. A similar goal is further from realization when it comes to polarized light. Ultimately we want to observe the detailed polarization in prominence emission lines (i.e., the value of the four Stokes parameters at several points in the line profile) and, using the appropriate radiative transfer equations, derive the vector magnetic field. However, the equations of transport of polarized radiation depend on the interaction coefficients between polarized radiation and matter in the presence of magnetic fields, and only in the last few years are we beginning to understand the nature of these coefficients. We shall discuss them below. Here we give a brief account of the transport equations.

Absorption

If one assumes that the lines are formed in pure absorption, the solution of the transfer equations is fairly straightforward under certain conditions. Unno (1956) gave the first solutions in terms of the Stokes parameters for a homogeneous magnetic field and a Milne-Eddington model atmosphere. The equation can be written in matrix formulation

$$\cos\gamma \frac{d}{dt}\mathbf{S} = (\mathbf{1} + \boldsymbol{\eta})(\mathbf{S} - \mathbf{B}), \qquad (11)$$

where γ is the heliocentric angle, τ the optical depth, **1** the unit matrix, **B** the matrix giving the Planck function $B(\tau)$ and η the absorption matrix, which contains the ratio between the coefficients of line absorption and continuum absorption for the Zeeman components. Independently Stepanov (1958a, b) developed a two-component theory for line formation giving two independent equations for two mutually orthogonal beams. Rachkovsky (1961a, b) showed that Stepanov's equations can be derived from Unno's as a subset. Different numerical solutions have been discussed by Mattig (1966), Kjeldseth Moe (1968), Kjeldseth Moe and Maltby (1968), Evans (1968, 1969a, b), Evans and Dreiling (1969), and Staude (1970a, b).

In a magnetic field, variations of the refractive index will be different for differently polarized Zeeman components of a line. Anomalous dispersion was first included by Rachkovsky (1962); see also Beckers (1969) and Staude (1970b), who included the effects of an inhomogeneous magnetic field. If we use Equation (11), the effect of dispersion can be accounted for by redefining the absorption matrix η.

Scattering

The use of Equation (11) leads to reasonably good approximations for the line wings when one observes near the central parts of the solar disk. However, for studies of emission-line polarization, radiative scattering must be considered in the line formation. This was first done by Stepanov (1958b, 1960, 1962) in a rough way by using his two-component approximations. Rachkovsky (1965) showed that the results obtained from Stepanov's simplified scattering theory and from his (Rachkovsky 1963) more rigorous model are essentially the same. However, both disagree significantly with pure absorption models. The inclusion of scattering in the transfer equations has been developed further by Obridko (1965), Rachkovsky (1967a, b), Domke (1971), and Rees (1969). For more details see Stenflo (1971) and Lamb (1972).

When we include both absorption and scattering, the transfer equation changes from its form given by Equation (11) to

$$\cos\gamma \frac{d}{d t} \mathbf{S} = (\mathbf{1} + \boldsymbol{\eta})\mathbf{S} - (1 - \varepsilon)\int_{4\pi} \mathbf{MS} \frac{d\omega}{4\pi} - (\mathbf{1} + \varepsilon\boldsymbol{\eta})\mathbf{B}, \quad (12)$$

where **M** is the Mueller matrix for scattering (to be discussed later), and $(1 - \varepsilon)$ gives the fraction of absorbed radiation that is being scattered.

In the case of true absorption, the transfer problem is simplified by diagonalizing the absorption matrix. One of the complications encountered when scattering plays a role is seen from the fact that the matrices η and **M** do not diagonalize simultaneously. House and Cohen (1969) and Rees (1969) have used numerical methods to obtain solutions to Equation (12). House and Cohen used the Monte Carlo technique to treat multiple scattering (of importance in dense prominences), while Rees included noncoherent scattering.

QUANTUM-MECHANICAL THEORY FOR THE HANLE EFFECT

When fluorescence polarization is analyzed using simple scattering theory in normal Zeeman pattern, we are not able to account for the observed phenomena, except when dealing with the simplest cases of resonance radiation. In more general cases, where scattering takes place between states only weakly removed from degeneracy by a magnetic field, the coherence effect (Hanle effect) must be treated with more sophisticated tools. This has been pointed out by Hyder (1964a, b) and by Warwick and Hyder (1965), and the first attempt to include the effect was made by Obridko (1968) in a phenomenological description. General discussions of the coherence problem has been given by House (1970a, b, 1971), Lamb (1970, 1971), Stenflo (1971), and Omont, Smith, and Cooper (1972).

Referring to Fig. 6, we consider both a simple and a more complex case of scattering. On the left in Fig. 6 the sublevels are far removed from degeneracy, so that each m state can be treated independently of the others. If we neglect collisions and interlocking with other levels, we see that if an excitation raises an electron from the m state of level a to the excited state m'' of level b, the subsequent emission of radiation will be due to a transition from this state m'' back to a state m' of level a. However, if we have a situation in which the states are not far removed from degeneracy, the different m'' states are more or less indistinguishable, and one must sum over those states that could be excited by the incident radiation. As pointed out by House, whose work we partly follow in this section, the summation over the excited states is carried out before one squares the amplitudes to obtain the scattered intensity. From a mathematical point of view, it is the cross-terms that may result from this procedure that produce the coherence—between the exciting radiation and the fluorescence emission—which gives the physics behind the Hanle effect.

Referring back to a scattering process of the type shown on the right in Fig. 6, we find that a time-dependent perturbation theory describes the behavior of the quantum system of interacting radiation and matter. Solution of the corresponding equations yields the scattering redistribution function, W (and the corresponding Mueller matrix for scattering, **M**), which gives the probability for the occupation of a final state of the system as a function of the initial state. This probability was given by

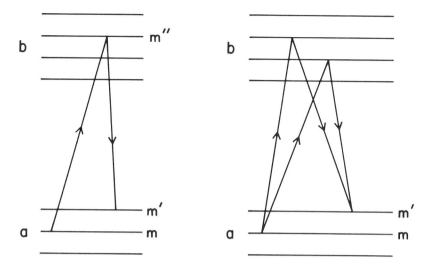

Fig. 6. Energy level diagram showing *on the left*: quantum numbers and transitions where coherency effects are not encountered, and *on the right*: where such effects are important.

Weisskopf (1931) in the dipole approximation when multiple scattering is neglected. It is expressed in terms of the matrix elements of interaction between matter and the fields of quantized radiation. Let $I(\alpha\nu\kappa)$ be the intensity of the exciting radiation of frequency ν and direction of propagation κ. The state of polarization is indicated symbolically by α, and ε^α denotes the polarization vector. As mentioned above, we can use two orthogonal linear vectors, and this will lead to a description of the scattering matrix for linear polarization, or we can use two orthogonal states of circular polarization to describe the scattering redistribution for circularly polarized light. Since the Stokes vector is often used for resonance polarization work on solar emission lines, House (1971) has cast the formulation in terms of the Stokes vector (Eq. [2]). Let unprimed and primed quantities refer to the exciting beam and the scattered radiation, respectively.

The expression for the scattering redistribution function is

$$W(\alpha\alpha'\nu\nu'\kappa\kappa') = \frac{4\pi^2 e^4}{h^4 c^2} \sum_m \sum_{m'} I(\alpha\nu) \left[I(\sigma'\nu') + \frac{h\nu'^3}{c^2} \right]$$

$$\times \frac{1}{\gamma_m^2 + (\nu' - \nu + \nu_m - \nu_{m'})^2}$$

$$\times \left| \sum_{m''} \frac{\langle n_b J''m'' | \mathbf{r} | n_a Jm \rangle \cdot \varepsilon^\alpha \langle n_a Jm' | \mathbf{r} | n_b J''m'' \rangle \cdot \varepsilon^{\alpha'}}{i\gamma_{m''} + \nu' - (\nu_{m'} - \nu_{m''})} \right|^2, \quad (13)$$

where, as in Fig. 6, m'' pertains to the intermediate state of atomic level b whose principal quantum number is n_b. The dipole matrix elements, $\langle|\mathbf{r}|\rangle$, are functions of the quantum numbers J, m, n_a, and n_b, and the damping constants, γ, depend on the radiation field, the dipole matrix elements, and the polarization vectors.

If the atom is excited by isotropic unpolarized light and if there is no splitting of the a and b levels, the damping constants reduce to

$$\gamma_m \to \gamma_a = \frac{B_{ab} I_{ab}}{4\pi} \quad \text{and} \quad \gamma_{m''} \to \gamma_b = \frac{B_{ba}(I_{ab} + A_{ba})}{4\pi},$$

where A and B are Einstein coefficients, and

$$I(\alpha, v_m - v_{m''}) = I(\alpha, v_{m''} - v_{m'}) = \tfrac{1}{2} I_{ab}.$$

The splitting of the levels can be given in terms of the frequencies of the lines between the different resulting states, i.e., in terms of expressions like $v_m - v_{m'}$, $v_{m'} - v_{m''}$ etc., which can be written, for example,

$$v_m - v_{m'} = \frac{\omega_e \Delta(mg)}{2\pi}. \tag{14}$$

We describe polarization in terms of Stokes parameters and transform the scattering matrix to the Mueller matrix, M, that describes the interaction of the beam with the scattering atom (Eq. [6]). The Mueller matrix is given by the operation of the transformation matrix T, in Equation (5), on the scattering probability W

$$\mathbf{S}' = \mathbf{MS} = \mathbf{T}(\mathbf{W} \otimes \mathbf{W}^*)\mathbf{T}^{-1}\mathbf{S}, \tag{15}$$

where $\otimes$ indicates the direct product of the matrices. The Mueller matrix will have components of the form $M(\alpha\alpha'\nu\nu'\kappa\kappa')$ that are built up from products of the form

$$w \otimes w^* = \begin{pmatrix} w_{11} & w_{12} \\ w_{21} & w_{22} \end{pmatrix} \otimes \begin{pmatrix} w_{11} & w_{12} \\ w_{21} & w_{22} \end{pmatrix}^*.$$

The components $W(\alpha\alpha'\nu\nu'\kappa\kappa')$ depend on initial and final levels of the atom and must be summed over the corresponding states m and m'.

House showed that one may write the Mueller matrix in the form

$$M(\alpha\alpha'\nu\nu'\kappa\kappa') = M^{(0)}(\alpha\alpha'\nu\nu'\kappa\kappa')$$
$$+ M^{(1)}(\alpha\alpha'\nu\nu'\kappa\kappa') + M^{(2)}(\alpha\alpha'\nu\nu'\kappa\kappa'), \tag{16a}$$

where $M^{(0)}$ corresponds to no quantum jump in m, i.e., $\Delta m = 0$, and

$M^{(1)}$ and $M^{(2)}$ correspond to the jumps $\Delta m = \pm 1$ and $\Delta m = \pm 2$, respectively. The term $M^{(0)}$ contains elements pertaining to nonoverlapping states and represents the scattering function for the simple Zeeman effect; in other words, M reduces to $M^{(0)}$ for strong magnetic fields. The terms $M^{(1)}$ and $M^{(2)}$ modify the scattering to include the effects of the partial degeneracy of the levels; i.e., the terms $M^{(1)}$ and $M^{(2)}$ are the expression for the Hanle effect.

We are often interested in the frequency-independent scattering, and to obtain the appropriate expression for M, we integrate over incoming, v, and outgoing, v', frequencies, which leads to the expression

$$M(\alpha\alpha'\kappa\kappa') = M^{(0)}(\alpha\alpha'\kappa\kappa') + M^{(1)}(\alpha\alpha'\kappa\kappa') + M^{(2)}(\alpha\alpha'\kappa\kappa'), \quad (16b)$$

and, again, the coherence effect due to level overlap is contained in $M^{(1)}$ and $M^{(2)}$.

A closer study of $M^{(1)}$ and $M^{(2)}$ shows that the coherence effect depends on the ratio of the frequency-differences, like Equation (14), and the damping constant for the level, γ_b, and both $M^{(1)}$ and $M^{(2)}$ vanish for large values of this ratio, i.e., for

$$\frac{\omega_e \Delta(mg)}{\gamma_b} \gg 1. \quad (17)$$

In the classical approximation, expression (17) reduces to the inequality (7). Under these strong-field conditions only $M^{(0)}$ remains, as noted above. With House we define the "effectively strong field case" by modifying inequality (17) to read

$$\frac{\omega_e}{\gamma} = \frac{eB}{m_e cA} \approx 2 \times 10^7 \frac{B}{A} \gg 1, \quad (18)$$

where B is given in gauss. The Einstein coefficient A is of the order $10^7 - 10^8 \text{ sec}^{-1}$ for many of the prominence lines used in polarization studies, which means that normally we are not in the effectively strong field case, and the full impact of the Hanle effect is felt in fluorescence polarization work. On the other hand, the A-values pertaining to the forbidden coronal lines are of the order $10 - 100 \text{ sec}^{-1}$, and we are always dealing with effectively strong magnetic fields, which greatly simplifies the interpretation of fluorescence polarization in coronal lines.

The application of the theory of fluorescence polarization to observations meets with several complicating factors. In the next section we mention the effects of line-of-sight integration and collisions. Also, the broadening of the emission lines due to Doppler motions must be included. This has been done by House (1973).

INTERPRETATIONS OF OBSERVATIONS

Coronal Lines

Because of observational difficulties, the variation of polarization over the line profile has not been studied for any coronal line. Hence, all measurements pertain to the integrated light and are analyzed in terms of the Stokes matrix Equation (16b), with only the $M^{(0)}$ term retained.

Hyder has predicted the maximum amount of polarization to be expected in a number of lines. In those lines that can be polarized, the percentage polarization will increase with distance from the solar limb as the photospheric radiation field becomes more and more directional. The asymptotic values reached at great distances and computed by Hyder (with some corrections) are given in Table I. The reader is referred to Firor and Zirin (1962) for details concerning these lines.

Two lines lend themselves well for observations, the green line at 5303 Å and, in particular, the infrared line at 10747 Å. Hyder, Mauter, and Shutt (1968) studied the 5303 Å line at eclipse and found polarizations reasonably consistent with field configurations inferred from coronal density structures; but eclipse data do not lend themselves too well for this purpose. Charvin (1971) observed the 5303 Å line outside of eclipse and made the first systematic inferences of magnetic fields from polarization measurements. The main results agree with the more accurate and simpler (simpler atomic structure) interpretations of the 10747 Å line observations.

Eddy and Malville (1967) first measured the polarization of the 10747 Å line, and the work was extended during the 12 November 1966 eclipse (Eddy, Lee and Emerson 1973). They found that the observed amount of polarization ranged from near zero at the limb to 80% at 1.6 R_0, and argued that depolarization effects are present in the corona. We shall return to this below. The other main conclusion regards the direction of the observed polarization. To discuss these results, we refer to the simple quantum-mechanical picture of the polarization effect discussed earlier. The incident radiation on coronal ions will have the electric field vector in a plane tangent to the sun's limb, and this directionality increases with distance from the limb. The ions see an unequal number of incoming σ and π photons, and the resultant emission possesses the sense of incident polarization. While electric dipole transitions are involved in the lines mentioned earlier, in the forbidden coronal lines we encounter magnetic dipole radiation. For these it is the magnetic field vector that conserves its direction. It is easily shown (Eddy 1967) that as a result of the behavior of the field vectors, the fluorescence radiation has its E-vector perpendicular to the limb.

As emphasized by Charvin (1965), Hyder (1965), Perche (1965), and House (1972), observed deviations of the E-vector from the radial direction can be interpreted as due to the coronal magnetic field, which is outlined by the observed field vector for the effectively strong field case. Hence, the observations of Eddy, Lee, and Emerson may indicate the magnetic field directions in the corona at the 12 November 1966 eclipse. They also showed that the percentage polarization tended toward zero in certain locations where the incident radiation field and the magnetic field intersected at the van Vleck angle (Eq. [10]).

The observed percentage polarization indicates that depolarization occurs in the corona. Some of this effect is due to the line-of-sight integration involved in any coronal observation. Of greater physical importance are collisional effects.

If collisions dominate the excitation of the lines, which may be the case in the lower corona, we expect the percentage polarization to be decreased relative to the values given by Equation (10). To indicate the nature of the effect, we write the depolarization factor as $R/(R + C)$, where R and C are the rates of radiative and collisional excitation, respectively. The problem of collisional depolarization is one of the most important features to be properly included in the general treatment of emission-line polarization (Lamb 1971; Omont, Smith, and Cooper 1972).

If there is a significant anisotropy in the velocity distribution of the exciting particles, we expect to observe impact polarization effects. The resultant polarization can be expressed as a vector sum of the weighted fluorescence polarization and impact polarization (Hyder 1965):

$$\mathbf{P} = \frac{R\,\mathbf{P}(R) + C\,\mathbf{P}(C)}{R + C},$$

where $\mathbf{P}(R)$ and $\mathbf{P}(C)$ are the percentage polarizations due to radiative and collisional processes, respectively (see also Percival and Seaton 1958). Eddy, Lee, and Emerson conclude that for coronal densities greater than $5 \times 10^7 \mathrm{cm}^{-3}$, collisional effects cannot be neglected.

Prominence Lines

The last statement in the preceding section tells us immediately that in prominences, whose derived densities always seem to exceed $10^9 \mathrm{cm}^{-3}$, collisional effects are of importance. In addition, inequality (18) tells us that any interpretation of observed polarization must be done with the full Mueller matrix, i.e., including the interference terms $M^{(1)}$ and $M^{(2)}$ (Eq. [16]). For the time being this is beyond our capabilities. Nevertheless, preliminary observations are available and some rough interpretations may be made. Brückner (1963) studied the polarization in the Ca II,

K-line and found a percentage polarization of about 1% to 2% in quiescent prominences, increasing with height. Since these values are less than the theoretical maximum values, depolarization—due to collisions, multiple scattering, and/or magnetic fields—is taking place. It is obvious that much more work is necessary, in particular on the effects of collisions and multiple scattering, before the wealth of information available in detailed observations of the polarization across prominence emission lines can be adequately interpreted.

Acknowledgments. I am indebted to J. A. Eddy and L. L. House for valuable discussions and for reading this manuscript.

TABLE I

Maximum Percentage Polarization, P, of Coronal Emission Lines

Line (Å)	Ion	Transition	P (percent)
3388	Fe XIII	$^1D_2 - {}^3P_2$	28
5303	Fe XIV	$^2P_{3/2} - {}^2P_{1/2}$	43
6374	Fe X	$^2P_{1/2} - {}^2P_{3/2}$	0
7892	Fe XI	$^3P_1 - {}^3P_2$	0.375
10747	Fe XIII	$^3P_1 - {}^3P_0$	100
10798	Fe XIII	$^3P_2 - {}^3P_1$	28

REFERENCES

Beckers, J. M. 1969. The profiles of Fraunhofer lines in the presence of Zeeman splitting. *Solar Phys.* 9: 372–386.

Born, M., and Wolf, E. 1965. *Principles of optics.* New York: Pergamon Press.

Breit, G. 1925. Polarization of resonance radiation in weak magnetic fields. *J. Opt. Soc. Am.* 10: 439–452.

———. 1933. Quantum theory of dispersion, parts VI and VII. *Rev. Mod. Phys.* 5: 91–140.

Brückner, G. 1963. Photoelektrische Polarisations-messungen am Resonanzlimen in Sonnenspektrum. *Z. Ap.* 58: 73–81.

Chandrasekhar, S. 1950. *Radiative transfer.* Oxford: p. 28. Oxford Univ. Press.

Charvin, P. 1963. Le nouveau coronomètre photoélectrique utilisé à l'observatoire de Paris. *C. R. Acad. Sci. Paris* 256: 368–370.

———. 1964a. Sur l'intensité et la polarisation des raies interdites de la couronne solaire. *C. R. Acad. Sci. Paris* 258: 1155–1158.

———. 1964b. Calcul de l'intensité et de la polarisation des raies interdites de la couronne en présence d'un champ magnétique non radial. *C. R. Acad. Sci. Paris* 259: 733–736.

———. 1965. Etude de la polarisation des raies interdites de la couronne solaire. *Ann. d'Astroph.* 28: 877–934.

———. 1971. Experimental study of the orientation of magnetic fields in the corona, IAU Symp. no. 43. *Solar magnetic fields*. (R. Howard, ed.), pp. 580–587. Dordrecht, Holland: D. Reidel Publ. Co.
Collett, E. 1968. The description of polarization in classical physics. *Am. J. Phys.* 36: 713–725.
Domke, H. 1971. Line formation in the presence of a magnetic field. *Astrofizika* 7: 39–56.
Eddy, J. A. 1967. Polarization of coronal emission lines. Lecture Notes on Summer Program on the Solar Corona, NCAR and Dept. of AG, Univ. of Colo., 1967.
Eddy, J. A.; Lee, R. H.; and Emerson, J. P. 1973. The 10747 coronal line at the 1966 eclipse I. *Solar Phys.* In press.
Eddy, J. A., and Malville, J. M. 1967. Observations of the emission lines of Fe XIII during the solar eclipse of May 30, 1965. *Astrophys. J.* 150: 289–297.
Evans, J. C. 1968. Report MASUA Theor. Phys. Conf. Univ. of Nebr., Nov. 2–3, 1968.
———. 1969a. Solar line formation in a magnetic field. *Bull. Am. Astron. Soc.* 1: 276.
———. 1969b. Rep. IAU Symp. on Lab. Astrophys. Toronto, November, 1969.
Evans, J. C., and Dreiling, L. A. 1969. Rep. 130th AAS Meeting. New York, August 1969.
Fano, U. 1957. Description of states in quantum mechanics by density matrix and operator techniques. *Rev. Mod. Phys.* 29: 74–93.
Firor, J., and Zirin, H. 1962. Observations of five ionization stages of iron in the solar corona. *Astrophys. J.* 135: 122–137.
Hanle, W. 1923. *Naturwiss* 11: 691.
———. 1924. Über Magnetische Beeinflussung der Polarisation der Resonanzfluoreszenz. *Z. Phys.* **29–30**: 93–105.
———. 1925. Ergebn. d Exakz. *Naturwiss* 4: 214.
Heisenberg, W. 1926. Über eine Anwendung des Korrespondenzprinzips auf die Frage nach der Polarisation des Fluoreszenzlichtes. *Z. Phys.* 31: 617.
House, L. L. 1970a. The resonance radiation of polarized radiation. I. *J. Quant. Spectros. Radiat. Transfer* 10: 909–928.
———. 1970b. The resonance radiation of polarized radiation. II. *J. Quant. Spectros. Radiat. Transfer* 10: 1171–1189.
———. 1971. The resonance radiation of polarized radiation. III. *J. Quant. Spectros. Radiat. Transfer* 11: 367–383.
———. 1972. Coronal emission line polarization. *Solar Phys.* 23: 103–119.
———. 1973. The resonance fluorescence of polarized radiation. IV. *J. Quant. Spectros. Radiat. Transfer*. In press.
House, L. L., and Cohen, L. C. 1969. The treatment of resonance scattering of polarized radiation in weak magnetic fields by the Monte Carlo technique. *Astrophys. J.* 157: 261–274.
Hyder, C. L. 1964a. The polarization of prominence helium I D_3 emission. Ph.D. Thesis, Univ. of Colorado, Boulder.
———. 1964b. Magnetic fields in the loop prominence of March 16, 1964. *Astrophys. J.* 140: 817–818.
———. 1965. The polarization of emission lines in astronomy. III. *Astrophys. J.* 141: 1382–1389.
———. 1966. Plane polarization and solar magnetic fields. *Atti del Convego Sui Campi Magnetici Solari*, pp. 110–119. Rome Obs., 1964.
Hyder, C. L.; Mauter, H. A.; and Shutt, R. L. 1968. Polarization of emission lines in astronomy. IV. *Astrophys. J.* 154: 1039–1058.
Karimov, M. G. 1961. An apparatus to observe coronal emission lines outside of eclipses to great distances. *Izv. Ap. Inst. Akad. Nauk. Kazan* 12: 65–76.

Kjeldseth Moe, O. 1968. A generalized theory for line formation in a homogeneous magnetic field. *Solar Phys.* 4: 267–285.

Kjeldseth Moe, O., and Maltby, P. 1968. A model for the penumbra of sunspots. *Solar Phys.* 8: 275–283.

Lamb, F. K. 1970. Line formation in magnetic fields. *Solar Phys.* 12: 186–201.

———. 1971. The effects of collisions on spectral line formation in solar magnetic regions. IAU Symp. No. 43. *Solar magnetic fields.* (R. Howard, ed.), pp. 149–161. Dordrecht, Holland: D. Reidel Publ. Co.

———. 1972. The theory of line formation in the presence of a magnetic field. *Rep. on Conf. on Line Formation*, Boulder, Colorado, 1971.

Lyot, B. 1934. Polarisation des protubérances solaires. *C. R. Acad. Sci. Paris* 198: 249–251.

———. 1937. Quelques observations de la couronne solaire et des protubérances en 1935. *L'Astronomie* 51: 203–218.

McMaster, W. H. 1961. Matrix representation of polarization. *Rev. Mod. Phys.* 33: 8–28.

Mattig, W. 1966. Line formation in sunspots. *Atti del Convegno sulle Macchie Solari*, pp. 194–208. Firenze 1964.

Mitchell, A. C. G., and Zemansky, M. W. 1934. *Resonance radiation and excited atoms.* London: Cambridge University Press.

Newkirk, G. A. Jr. 1958. Emission-line polarization in prominences. *Publ. Astron. Soc. Pac.* 70: 185–190.

Obridko, V. N. 1965. The matrix of radiative scattering in a magnetic field. *Astr. Zhur.* 42: 102–106.

———. 1968. Magnetic-field radiation-scattering matrix derived with allowance for the phase couplings of the upper level wave function. *Soln. Aktivnostj.* No. 3 (NASA Tech. Transl. TT-F-581).

Öhman, Y. 1929. Astronomical consequences of the polarization of fluorescence. *Mon. Not. R. Astr. Soc.* 89: 479–482.

Omont, A.; Smith, E. W.; and Cooper, J. 1972. Redistribution of resonance radiation. I. The effect of collisions. *Astrophys. J.* 175: 185–199.

Perche, J. C. 1965. Sur l'intensité et la polarisation des raies interdites de l'ion Fe XIII dans la couronne solaire. *C. R. Acad. Sci. Paris* 260: 6037–6040.

Percival, I. C., and Seaton, M. J. 1958. The polarization of atomic line radiation excited by electron impact. *Phil. Trans. Roy. Soc. London* A251: 139–160.

Perrin, F. 1942. Polarization of light scattered by isotropic opalescent media. *J. Chem. Phys.* 10: 415–427.

Rachkovsky, D. N. 1961a. On the formation of absorption lines in a magnetic field. *Izv. Krym. Ap. Obs.* 25: 277–280.

———. 1961b. A system of radiative transfer equations in the presence of a magnetic field. *Izv. Krym. Ap. Obs.* 26: 63–73.

———. 1962. Magnetic rotation effects in spectral lines. *Izv. Krym. Ap. Obs.* 28: 259–270.

———. 1963. The development of absorption lines in sunspots, taking into account scattering and absorption. *Izv. Krym. Ap. Obs.* 29: 97–117.

———. 1965. On the theory of formation of absorption lines in a magnetic field. *Izv. Krym. Ap. Obs.* 33: 111–117.

———. 1967a. The formation of absorption lines in a magnetic field. *Izv. Krym. Ap. Obs.* 36: 9–21.

———. 1967b. The reduction for anomalous dispersion in the theory of absorption line formation in a magnetic field. *Izv. Krym. Ap. Obs.* 37: 56–61.

Rayleigh, (Strutt, J. W., 3rd Baron Rayleigh) 1922. Polarization of the light scattered by mercury vapor near the resonance periodicity. *Proc. Roy. Soc.* 102: 190.

Rees, D. E. 1969. Line formation in a magnetic field. *Solar Phys.* 10: 268–282.

Schurcliff, W. A. 1962. *Polarized light.* Cambridge, Mass.: Harvard Univ. Press.
Staude, J. 1970*a*. Line formation in a magnetic field and the interpretation of magnetographic measurements. II. *Solar Phys.* 12: 84–94.
———. 1970*b*. Line formation in a magnetic field and the interpretation of magnetographic measurements. III. *Solar Phys.* 15: 102–112.
Stenflo, J. O. 1971. The interpretation of magnetograph results. *IAU Symp. no. 43. Solar magnetic fields.* (R. Howard, ed.), pp. 101–129. Dordrecht, Holland: D. Reidel Publ. Co.
Stepanov, V. E. 1958*a*. The absorption coefficient of atoms in the case of reverse Zeeman effect for an arbitrary directed magnetic field. *Izv. Krym. Ap. Obs.* 18: 136–150.
———. 1958*b*. On the theory of the formation of absorption lines in a magnetic field and the profile of Fe $\lambda 6173$ in the solar sunspot spectrum. *Izv. Krym. Ap. Obs.* 19: 20–45.
———. 1960. The absorption coefficient in the inverse Zeeman effect for arbitrary multiplet splitting and the transfer equation for light with mutually orthogonal polarization. *Astr. Zhur.* 37: 631–641.
———. 1962. Radiative equilibrium equations in atmospheres of magnetic stars. *Izv. Krym. Ap. Obs.* 27: 140–147.
Stokes, G. G. 1852. On the composition and resolution of streams of polarized light from different sources. *Trans. Cambr. Phil. Soc.* 9: 399.
ter Haar, D. 1961. Theory and application of the density matrix. *Rep. on Progress in Phys.* 24: 304–362.
Thiessen, G. 1951. Zur Theorie der Polarisation von Hα und D_3 in Protuberanzen. *Z. Ap.* 30: 8–16.
Unno, W. 1956. Line formation of a normal Zeeman triplet. *Publ. Astr. Soc. Jap.* 8: 108–125.
Vleck, J. H. van 1925. On the quantum theory of the polarization of resonance radiation in magnetic fields. *Proc. Nat. Acad. Sci.* 11: 612–618.
Warwick, J. W., and Hyder, C. L. 1965. The polarization of emission lines in astronomy. I. *Astrophys. J.* 141: 1362–1373.
Weisskopf, V. 1931. Zur Theorie der Resonanz-fluoreszenz. *Ann. Phys.* 9: 23–66.
Wood, R. W., and Ellett, A. 1923. On the influence of magnetic fields on the polarization of resonance radiation. *Proc. Roy. Soc.* 103: 396.
Zanstra, H. 1950. An attempt to explain the polarization in Hα and D_3 for prominences. *Mon. Not. R. Astr. Soc.* 110: 491–500.

POLARIZATION MEASUREMENTS WITHIN STELLAR LINE PROFILES

D. CLARKE AND I. S. McLEAN
The University, Glasgow

Polarizational effects occurring within stellar line profiles are discussed briefly. A photoelectric wavelength-scanning high-resolution spectropolarimeter, capable of investigating these effects, is described. The wavelength scan is achieved by tilting a narrow-band interference filter. Polarimetry is affected by rotation of a phase plate before a fixed polarizer. The data are accumulated by a chopping and integration technique employing photon counting. Preliminary results are reported, which show that photon shot noise sets the limit to the detectivity of polarization.

Besides the well-known polarization associated with Zeeman splitting, other polarizational effects have also been predicted to occur within stellar line profiles, and indeed, there is observational evidence to give qualitative support to such theories.

The first work in this field would seem to be that of Öhman (1934) who noted that one wing of the $H\gamma$ line of β Lyr exhibited some polarization that depended on the phase of this eclipsing binary. Other more recent observations are those of Tamburini and Thiessen (1961), who found that the equivalent widths of the Balmer lines of some early-type stars depend on the selected direction of vibration for observation. Clarke and Grainger (1966) have also reported a polarization effect across the $H\beta$ line in γ UMa. Effects in solar emission lines are described by Tandberg-Hanssen.[1]

On the theoretical side, following Chandrasekhar's (1946) famous paper predicting polarizational effects at the limbs of early-type stars,[2] Öhman (1946) suggested that the effects might be detected in such stars that also present Doppler broadened absorption features as a result of rapid rotation. Öhman's simple theory suggested that the degree of polarization might vary by about 0.8% across a broadened profile. In an

---

[1] See p. 730. [2] See, however, p. 663.

investigation of the effect of thermal Doppler broadening by an electron scattering atmosphere, Sen and Lee (1961) have shown that small amounts of polarization are likely to be present and to vary across both emission and absorption lines.

A recent observational UBV study by Serkowski (1970) has shown that early-type stars with extended atmospheres can have intrinsic polarization and that their wavelength dependences are perhaps related to the Balmer and Paschen series of hydrogen. One might surmise that some of this polarization results from effects within the absorption or emission features of these series and that higher spectral resolution studies of these stars are called for.

None of the line-profile observations described above have been completely satisfactory in that values of the Stokes parameters have either not been obtained uniquely or have not been determined with sufficient precision. However, from these observations and from the broader-band measurements by Serkowski, and from the theoretical prediction, there is good reason to pursue more definite measurements. Our approach to the observational problem is given below.

INSTRUMENTAL REQUIREMENTS

The anticipated polarization effects are small, and extremely accurate photometry is therefore required; the detector must be a photoelectric device. (The observations by Öhman [1934] and Tamburini and Thiessen [1961] were performed photographically.)

For any general spectropolarimetric study, one needs a system for achieving spectral isolation and a system for performing polarimetry. In making the combination, one has choice of the placement of the systems relative to each other.

When polarimetric observations are limited by noise generated by the earth's atmosphere, then the noise can be removed very effectively by using a double-beam polarizer and two detectors (e.g., see Hiltner 1951). With such a two-channel system, spectropolarimetry can be performed by using filters, having no polarizing effects, prior to the double-beam polarizer, so that each resolved component contains the identical spectral passband.

If spectropolarimetry is contemplated at medium-to-high spectral resolution or with a scanning monochromator, then the application of double-beam techniques with prior spectral isolator loses its attraction. Both prism and grating monochromators have polarizing properties and would introduce parasitic polarization that must be calibrated out. Grating monochromators are particularly troublesome in that their strong polarizances are very dependent on wavelength. Perhaps the only type of scanning monochromator that might be applied successfully prior to a double-beam polarimeter is the Fabry-Pérot interferometer. However, such a device is reserved for achieving very narrow spectral passbands, and the noise on the measurements is likely to result from photon shot noise rather than from effects of the atmosphere; in this case, provision of the second channel results only in a $\sqrt{2}$ advantage.

For a monochromator giving rise to polarizing effects, it is obviously better to place it after the polarimetric system so that it receives a constant direction of vibration. In such a position, its effects do not disturb the polarimetry, but they may control the signal level according to the wavelength being passed. However, it is now difficult to consider using double-beam polarimetry because of the awkwardness of directing the two resolved components into the same entrance aperture of the spectral isolator and recouping them for separate detection after the exit aperture. Except for simple devices such as filters, one would not normally think of employing two separate monochromators for each beam. Again, as mentioned above, at high spectral resolution, we are likely to be limited by photon shot noise, and the loss of the second channel is perhaps of marginal importance.

The considerations above suggest that medium-to-high resolution polarimetric studies might be performed simply by using a single-channel polarimetric system whose spectral passband is controlled by a following monochromator.

In order to perform the intended observations of stellar line profiles, a spectral resolving power of the order of 5×10^3 is required. This kind of value does not readily go hand in hand with photoelectric spectrometry in that for conventional medium-sized angular dispersive instruments, it is difficult to ensure acceptance of the complete seeing disk by the entrance aperture without losing the potential resolving power. According to the seeing conditions at Glasgow, it was calculated that the grating spectrometer accepting all the seeing disk and providing a spectral resolving power of 5×10^3 would be as large as the telescope itself and impossible to mount.

A system providing sufficient throughput and resolving power is the Fabry-Pérot interferometer or, indeed, a very narrow passband interference filter. Eather and Reasoner (1969) have shown that by tilting an interference filter to alter the position of its wavelength passband, it has application as a scanning monochromator. Because of the simplicity of its method of scanning, its stability, its small size and low weight, we have applied a tilting-filter instrument to line-profile work.

Before describing our tilting-filter spectropolarimeter, an additional comment should be made relating to angular-dispersive monochromators.

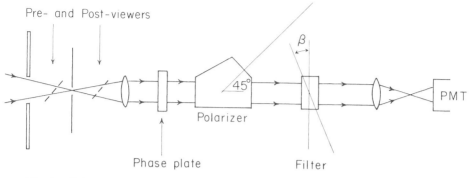

Fig. 1. Schematic layout of tilting-filter spectropolarimeter. An eyepiece or a second filter can be placed in the 45° beam. The angle of tilt (β) of the narrow-band interference filter is shown, and the photomultiplier tube is indicated by PMT.

If the polarimetric equipment has devices in it that rotate, it is very difficult to prevent small movements of a star's seeing disk on the monochromator's entrance aperture; such movements produce "spectral jitter" which in turn gives rise to spurious changes of signal level according to the detail in the spectrum, thus invalidating any polarimetry. A Fabry-Pérot or interference filter monochromator does not suffer from this defect to any extent.

THE TILTING-FILTER SPECTROPOLARIMETER

A schematic diagram of the spectropolarimeter is shown in Fig. 1. The polarization modulator consists of a rotatable phase plate (half-wave for linear polarization measurements) followed by a fixed double-beam polarizer which is a variant of the Foster (1938) design.

The first beam is used to record the polarimetric data, the spectral scan being achieved by controlled tilt of an interference filter in this beam. An eyepiece can be inserted in the second beam to facilitate telescope guidance. Alternatively, another filter, perhaps set to isolate a small part of the continuum, can be placed in the second beam, thus allowing two different spectral picture points to be recorded simultaneously.

Polarimetry is performed by a photon counting system; separate intensity measurements at three positions of the phase plate are used to determine the linear polarization parameters by Fessenkov's technique (see e.g., Clarke and Grainger 1971). With the rotatable half-wave plate, intensity measurements are required at angular separations of 30° or their equivalents. Our mechanical arrangement uses angular separations of 120°. A stepping motor is employed to rotate the half-wave plate to the click-stop positions, and an electronic control system (using integrated circuits) guides the output pulses from the photomultiplier and amplifiers to one of three digital counters for a preset integration time. After the third position (one mechanical rotation of the phase plate), the cycle is repeated a predetermined number of times before the data are printed out. This whole sequence followed by the print-out procedure may itself be repeated until the total counts accumulated from each channel are statistically significant, before a new wavelength position is selected.

By our method of chopping and integration (similar to that of Clarke and Ibbett 1968), we are able to record the polarimetric data to an accuracy limited by photon shot noise rather than by scintillation and variations in atmospheric transparency.

To prevent spread of bandwidth, the filter accepts a collimated beam. The position of the wavelength passband is altered by driving a micrometer, by means of a small stepping (or synchronous) motor, against a lever arm attached to the filter. In our exploratory observations, we have used a 2.5 Å half-width filter centered on a wavelength 10 Å greater than Hβ (4861.3 Å). Some points related to the tilting-filter technique are sketched below. (A fuller discussion of the technique can be found elsewhere; see Clarke and McLean, 1973.)

The advantage of the high throughput of an interference filter over prism and grating spectrometers has already been referred to. The shift to shorter wavelengths of the position of the filter passband with tilt is nonlinear, but calibration to a linear wavelength scale is easily effected. Although the filter half-width increases with tilt, i.e., the scan is recorded with a varying instrumental profile, the distortion by this effect on the profiles of broad features is small; the broadening of the passband with tilt is accompanied by a decrease of transmittance, thus offsetting serious signal level changes when scanning a continuum. With unpolarized incident light, the passband develops into a doublet as the filter is tilted, the component corresponding to the direction of vibration perpendicular to the plane of incidence transmitting the longer wavelength. Further, the half-width increases at a slightly different rate for each of the two directions of vibration, the perpendicular component having the slower broadening rate. Such effects are negligible over the small angles of tilt necessary to record the majority of line profiles, but in any case, the relative orientation of the polarizer and tilt axis has been chosen so that the most favorable component is selected.

OBSERVATIONS TO DATE

For comparison, Fig. 2 reproduces normalized records of Hβ line profiles of two stars. The broad absorption feature of η UMa and the emission profile of γ Cas were recorded by D.C. amplifier and pen-recorder with the spectropolarimeter (phase plate stationary) attached to the 50-cm telescope at Glasgow. To demonstrate the effectiveness of our tilting-filter method, the normalized record of the Hβ profile of η UMa obtained by a large photoelectric grating spectrometer (Grainger and Ring, 1963) attached to a telescope of larger aperture is also shown in Fig. 2. Our initial investigations consisted of recording Hβ profiles by D.C. amplifier and pen-recorder at each position of the half-wave plate. However, although reasonable wavelength scans were obtained, transparency changes required to record three scans occurring during the period limited the detectivity of polarization to about 2%–3%.

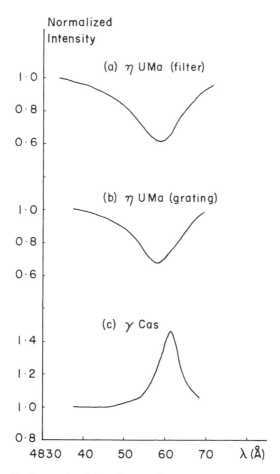

Fig. 2. Normalized records of Hβ line profiles of η UMa and γ Cas (emission feature). Curve b shows the Hβ line profile recorded by a large photoelectric grating spectrometer attached to a telescope of aperture 120 cm. The scans were obtained by D.C. amplifier and pen-recorder.

To investigate the source of noise occurring with the chopping technique already described, photometric data were collected into the three recording channels without rotation of the phase plate. From measurements of the Hβ line center of α Lyr (A0, $m_v = 0^m.03$), comparison of totals in the three channels revealed a divergence of less than 1%. Since the total per channel was $\sim 10^4$ counts (integration time, 8 seconds per channel), this kind of divergence is to be expected as a result of photon shot noise.

To observe the effect of photon shot noise on the polarimetric data, it is convenient to evaluate the normalized Stokes parameters Q and U defined by

$$Q = p \cos 2\zeta \text{ and } U = p \sin 2\zeta,$$

where p is the degree of linear polarization and ζ is the angle of the direction of vibration relative to the position of the first intensity measurement. When the light source is unpolarized and the instrumental polarization is zero, the average values of Q and U over a very large number of measurements should then converge to zero. If the light source is unpolarized and Q and U converge to constants, then these are the normalized Stokes parameters of the instrument (plus telescope) at that orientation and wavelength passband.

For small degrees of polarization, the three intensity measurements required by Fessenkov's technique are approximately equal. If $\triangle I/I$ represents the typical fractional uncertainty in any one of them, then uncertainty $\triangle p$ associated with p is given by

$$\triangle p = \sqrt{\frac{2}{3}} \frac{\triangle I}{I}.$$

Now, for the star ϵ Cas (B3, $m_v = 3^m.3$), which from Tamburini and Thiessen's (1961) list of stars, exhibits a 5% polarization effect in the equivalent width of Hβ, typical counts at the line center are 1.5×10^2 per second. In one second we would expect to obtain a value of $\triangle p = \pm 0.067$. With the aim of obtaining $\triangle p = \pm 0.001$, which should for example reveal the mechanism predicted by Öhman (1946), we would therefore require a total integration time of 75 minutes per spectral picture point. However, with the spectropolarimeter attached to a 250-cm telescope, the time required reduces to 3 minutes per spectral picture point.

The spectropolarimetric observations so far have only been made with relatively short integration times at two wavelength positions corresponding to the center of the line and a point in the continuum on the short-wavelength side. Our preliminary observations include the five stars listed in Table I.

TABLE I

Preliminary Observations of Q and U Parameters at the Hβ Line and in the Continuum (1972 Oct. 19)

Star	Wavelength Position	Mean Value of Q	Mean Value of U	Total Integration Time (Sec.)	Comments
α Lyr (A0, 0ᵐ03)	Hβ	+ 0.0041 ± 0.0020	+ 0.0082 ± 0.0023	264	Bright star, hopefully exhibiting no polarization effect and therefore useful as a standard
	Continuum	− 0.0005 ± 0.0017	+ 0.0044 ± 0.0034	120	
γ Cas (B1, 1ᵐ6–3ᵐ0 var.)	Hβ	− 0.0162 ± 0.0037	+ 0.0038 ± 0.0044	216	Well-known emission-line variable star
	Continuum	− 0.0054 ± 0.0046	+ 0.0031 ± 0.0057	120	
ε Cas (B3, 3ᵐ3)	Hβ	− 0.0077 ± 0.0189	− 0.0055 ± 0.0073	144	Listed by Tamburini and Thiessen (1961)
	Continuum	− 0.0071 ± 0.0089	+ 0.0167 ± 0.0072	120	
γ UMa (A0, 2ᵐ5)	Hβ	− 0.0049 ± 0.0071	− 0.0046 ± 0.0106	240	Listed by Clarke and Grainger (1966)
	Continuum	+ 0.0012 ± 0.0136	− 0.0043 ± 0.0044	96	
ζ Ori (O9.5, 2ᵐ0)	Hβ	− 0.0077 ± 0.0043	− 0.0005 ± 0.0056	288	Suspicious from our early measurements using continuous (D.C.) scanning
	Continuum	− 0.0050 ± 0.0027	+ 0.0059 ± 0.0038	288	

NOTE: The uncertainties listed have been obtained from repetitive measurements of Q and U taken over 24 second periods. The total integration time and values of Q and U are built up from measurements taken over this period.

No attempt has yet been made to use the long integration times necessary to achieve a detectivity of $p \sim 0.001$; the preliminary observations were made to explore the capabilities of the instrument and to check for the presence of large values of polarization, i.e., $p \sim 0.01$.

The observations clearly show that any instrumental polarization is less than 1%. The values of the Stokes parameters Q and U obtained for γ Cas and α Lyr appear to be statistically significant. For γ Cas, it is well known that in white light a significant polarization ($p \sim 0.008$) is observable; our values of Q and U are likely to be related to this polarization, at least to some extent. For α Lyr, again polarization has possibly been detected and perhaps varies between the absorption feature and the continuum; at this stage we cannot say whether this possible difference is intrinsic to the star or whether it is an instrumental effect.

For the other stars, the measurements are only meaningful in that they give an upper limit to the polarization and differential polarization that might be present between absorption feature and continuum. From our measurements, we are confident that $p \sim 0.03$ could have been detected if it had been present in any of the listed stars.

This work is continuing with the aim of extending the list of observed stars to a detectivity of $p \sim 0.01$. For some stars of special interest, we hope to apply long integration times with the aim of improving the detectivity to $p \sim 0.001$.

Acknowledgment. McLean has been supported by a Science Research Council grant over the period of this work.

REFERENCES

Chandrasekhar, S. 1946. On the radiative equilibrium of a stellar atmosphere. X. *Astrophys. J.* 103: 351–370.

Clarke, D., and Grainger, J. F. 1966. Polarization effects in stellar absorption lines. *Ann. d'Astroph.* 29: 355–359.

———. 1971. *Polarized light and optical measurement.* Oxford: Pergamon Press.

Clarke, D., and Ibbett, R. N. 1968. A three-channel astronomical photoelectric spectropolarimeter. *J. Sci. Inst.* (J. Phys. E) Ser. 2. vol. 1, 409–412.

Clarke, D., and McLean, I. S. 1973. A simple scanning monochromator for line profile studies. *Astron. Astrophys.* In press.

Eather, R. H., and Reasoner, D. L. 1969. Spectrophotometry of faint light sources with a tilting-filter photometer. *Applied Optics* 8: 227–242.

Foster, L. V. 1938. A polarizing vertical illuminator. *J. Opt. Soc. Am.* 28: 124–126.

Grainger, J. F., and Ring, J. 1963. A photoelectric grating spectrometer. *Mon. Not. R. Astr. Soc.* 125: 93–104.

Hiltner, W. A. 1951. Compensation of seeing in photoelectric photometry. *The Observatory* 71: 234–237.

Öhman, Y. 1934. Effects of polarization in the spectrum of β Lyrae. *Nature* 134: 534.
———. 1946. On the possibility of tracing polarization effects in the rotational profiles of early-type stars. *Astrophys. J.* 104: 460–462.
Sen, K. K., and Lee, W. M. 1961. The broadening and polarization of spectral lines due to the thermal Doppler effect in an electron scattering atmosphere. *Pubs. A. S. Japan* 13: 263–275.
Serkowski, K. 1970. Instrinsic polarization of early-type stars with extended atmospheres. *Astrophys. J.* 160: 1083.
Tamburini, T., and Thiessen, G. 1961. On the existence of a new polarization effect in stellar spectral lines. *Rendiconti delle Classe di Scienze Fisiche, Matematiche e Naturali,* Serie VIII, vol. XXX, 492–496.

DISCUSSION

WITT: Is there a possibility of obtaining false modulation when using a tilting filter in a polarized beam?

CLARKE: No, not to any serious extent. The polarimetric optics are placed prior to the filter, and so the latter does not introduce any parasitic polarization. The filter receives perfectly linearly polarized light, and the tilt axis of the filter has been chosen so that the filter receives the direction of vibration giving the slowest rate of bandpass increase with tilt. Tilting the filter may alter the position of the image of the aperture stop on the cathode of the photomultiplier, but as the polarimetric data are recorded at particular spectral points and, therefore, at fixed positions of the filter, any movement of the aperture-stop image should not affect the polarization data.

K-CORONA AND SKYLIGHT INFRARED POLARIMETRY

J.-L. LEROY and G. RATIER
Pic du Midi and Toulouse Observatories

K-corona polarimetric observations are made in the spectral range 0.8–1.0 μm at the Pic du Midi Observatory. Contributions of single and multiple scattering by molecules and aerosols in the skylight signal near the sun are considered, with reference to previous work where the same effects were studied for visible wavelengths.

The electron corona is observed at high-altitude observatories with the aid of sensitive photoelectric polarimeters which are able to separate the faint, strongly polarized, signal of the K corona from the much brighter, but faintly polarized, sky background. Thus, the knowledge of the intensity and polarization properties of the atmospheric "aureole" near the sun is very useful for applying appropriate corrections to rough observations and deriving true values of the coronal brightness.

In a previous paper (Leroy, Muller, and Poulain 1972) we have drawn some conclusions about the influence of skylight polarization on K-corona measurements; this study was limited to the visible spectral range since K-coronameters in operation were working near 0.5 μm. However, it was clear to us then that better results could be obtained at longer wavelengths. In fact, R. T. Hansen[1] and his colleagues in Hawaii have experienced a noticeable gain when observing near 0.62 μm. Our next effort, in order to reduce the atmospheric perturbations in K-corona measurements at the Pic du Midi, has been to perform measurements in the near-infrared.

Recent technological progress with silicon photodiodes has made it easier to design a photoelectric detector operating around 0.9 μm. We have independently found a solution similar to that described by Hamstra and Wendland (1972). In our case two photodiodes (UDT PIN 10), followed by Analog Devices operational amplifiers, make a low-noise,

---

[1] Personal communication, 1971.

small-sized detector that is very suitable for photoelectric polarimetry. An important feature of our instrument is that measurements are performed by a zero-method so that we have no trouble with the possible nonlinear response of the apparatus.

The polarimeter is associated with a coronagraph of 10-cm effective aperture; the scanning aperture is 1.3 arcmins; the useful spectral bandwidth is 0.8–1.0 μm. Under such circumstances, the detector noise alone — which is due to the photodiodes rather than to the amplifiers — would limit the accuracy of measurements towards faint values of coronal brightness, near the threshold of 1×10^{-8} times the sun's brightness.[2] In fact, the final accuracy of measurements depends mainly upon other errors due to the skylight polarization. This polarization is what had been found previously to be dominant in the case of observations in visible light. Thus, the next task was to see whether around 0.9 μm the skylight perturbations were noticeably smaller than those in the visible range. Reliable observations obtained during the summer of 1972 have led to the following conclusions:

1. As could easily be predicted, the effect of molecular multiple scattering becomes very small at 0.9 μm. The quantity of polarized light due to this mechanism is less than 1×10^{-8} for all solar elevations higher than 10°. This result agrees quite well with the value 20×10^{-8} we had found near 0.5 μm for a 10° solar elevation, owing to the λ^{-4} variation of molecular scattering.

   As a consequence, and since our observations are always performed for solar elevations higher than 10°, it is no longer necessary to perform additional observations at 20 arcmins from the solar limb in order to subtract the contribution of multiple scattering. Therefore, the observing time for the corona itself is increased and possible errors in the process of suppression of the multiple scattering contribution disappear.

2. Primary scattering at 20 arcmins from the solar limb now gives a signal that is equal to or less than 1×10^{-8} in most cases. Formerly we had found a value of 2×10^{-8} around 0.5 μm, and we suspected that this effect resulted from the primary scattering by aerosols in the whole atmosphere so that only a weak (λ^{-1}) wavelength dependence was to be expected. Our present result agrees quite well with this prediction. In any case, with such a small amount of less than

---

[2]Since this manuscript was written improvements of the detector give an accuracy of measurements near 0.5×10^{-8} times the sun's brightness (April 1973).

1×10^{-8}, the parasitic signal due to primary scattering at 20 arcmins becomes almost negligible in our application.

3. The polarization due to primary scattering by aerosols at 2 arcmins from the solar limb is harder to distinguish since its contribution is superimposed upon the K corona. Nevertheless, by studying series of observations of the sun's polar regions, we are able to get an evaluation of the skylight signal that has been found to reach 2 to 3×10^{-8}. Again we get a small gain with respect to former 0.5 μm measurements; this result is compatible with our previous explanation of the phenomenon (scattering by large-sized aerosols concentrated in a local cloud, around the Pic du Midi). On the other hand, even if the above values are not negligible, they show that direct observations still allow reasonably accurate measurements at 2 arcmins from the limb. We have, therefore, abandoned the complex and not very reliable reduction procedure (a more complete description is given in our previous paper) involving eclipse determinations of the polar coronal brightness.

4. Atmosphere-originated fluctuations, which give random errors (in contrast to the systematic effects considered above), had been found to amount to $\pm 1 \times 10^{-8}$ for 0.5 μm observations. For the infrared observations, their contribution is smaller than $\pm 1 \times 10^{-8}$, and in the present work random fluctuations from the instrument are dominant, at least in good sky conditions.

When the different sources of errors that have just been considered are combined, taking into account their character (random or systematic) and the time constant of every measurement (a few seconds), we obtain Table I. This table, a part of which is found in our previous paper (Leroy, Muller, and Poulain 1972) gives the final accuracy of measurements around 0.9 μm for average sky conditions.

In best atmospheric conditions, a decrease in uncertainty by a factor two is achieved both at 0.5 and 0.9 μm. In any case, a gain is noticeable

TABLE I

Intensity and Precision of K-Corona Measurements

Distance from the Limb (arcmins)	Active Corona Intensity	Former Uncertainty (0.5 μm)	Present Uncertainty (0.9 μm)
1.5 or 2	60×10^{-8}	$\pm 4 \times 10^{-8}$	$\pm 3 \times 10^{-8}$
4	30×10^{-8}	$\pm 3 \times 10^{-8}$	$\pm 2 \times 10^{-8}$
8	10×10^{-8}	$\pm 2 \times 10^{-8}$	$\pm 1 \times 10^{-8}$

at 0.9 μm, especially at 8 arcmins from the solar limb where it is most needed. We wish to point out again that the present measurements do not require additional observation for multiple-scattering elimination, and this allows more time to observe the K corona. Thus, our near-infrared polarimeter seems, in its present form, to be well suited to coronal observations and has, in fact, already given excellent results during the summer of 1972.

A further step toward longer wavelengths could be achieved with a PbS detector, which would allow us to work, for instance, through the 1.7 μm atmospheric window. But, in this range, the sun's brightness is four times less than at 0.9 μm so that measurements would require a sensitive detector. Present data concerning the skylight polarization for this wavelength (Bennett, Bennett, and Nagel 1960, 1961) do not allow a precise prediction of the possible gain if observations were performed in this spectral range.

REFERENCES

Bennett, H. E.; Bennett, J. M.; and Nagel, M. R. 1960. Distribution of infrared radiance over a clear sky. *J. Opt. Soc. Am.* 50: 100–106.
———. 1961. Question of the polarization of infrared radiation from the clear sky. *J. Opt. Soc. Am.* 51: 237.
Hamstra, R. H. Jr., and Wendland, P. 1972. Noise and frequency response of silicon photodiode operational amplifier combination. *Applied Optics* 11: 1539–1547.
Leroy, J.-L.; Muller, R.; and Poulain, P. 1972. The polarization of skylight near the sun and its influence on polarimetric K corona observations. *Astron. Astrophys.* 17: 301–308.

EXTRATERRESTRIAL POLARIZATION OF THE ZODIACAL LIGHT ROCKET MEASUREMENTS AND THE HELIOS PROJECT

C. LEINERT, H. LINK and E. PITZ
Landessternwarte, Heidelberg-Königstuhl

In the early stages of the *Helios* project, a rocket experiment was designed to close the observational gap between the outer corona and the inner zodiacal light. The experiment was launched on 2 July 1971, near noon, from the ESRO range in Sardinia and reached a peak altitude of 224 km. The experiment consisted of five photometers, two of them measuring the polarization of the zodiacal light in the blue spectral region at elongations of 15° and 21°, by means of three polarization filters changed by a filter wheel. The data proved to be nearly free of stray light and airglow components and showed a rather high degree of polarization at small elongations. The flat shape of the polarization curve at small elongations admits or even requires large particles, with respect to the wavelength of light, which is in accordance with color measurements made on the same payload and which gave a slightly reddish zodiacal light at small elongations. The results will be published in detail in *Astronomy and Astrophysics*.[1]

Helios A and B are two identical spacecraft that will be launched in the fall of 1974 and the fall of 1975 to penetrate the inner solar system to within 0.25 AU of the sun. The *Helios* zodiacal light experiment consists of three photometers (Fig. 1), mounted rigidly on the spacecraft at angles of 16°, 31°, and 90° with respect to the spacecraft equator. By the spin of the spacecraft, which is oriented with its axis perpendicular to the ecliptic plane, the first two photometers scan on circles of ecliptic latitude 16° and 31° respectively, while the third is observing the region of the south ecliptic pole. The fields of view are 1° × 1°, 2° × 2°,

---

[1] For review of other rocket programs that are studying the zodiacal light, see various papers in *The Zodiacal Light and the Interplanetary Medium* (ed. J. L. Weinberg) NASA SP–150, U.S. Government Printing Office, Washington, D.C.

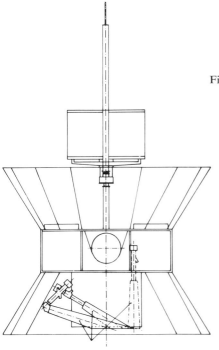

Fig. 1 Zodiacal light photometers in the *Helios* spacecraft.

and 3° diameter respectively. Angular resolution in the scanning photometers is given by 32 sectors, distributed symmetrically around the sun, the smallest of which measure 5°6 in longitude. Intensity and polarization of the zodiacal light are measured in the UBV system.[2] In the scanning photometers, the polarization is measured by means of three differently oriented polarization filters in a filter wheel. In each filter position a measurement for all 32 sectors is obtained. After the third measurement, the polarization is available for all 32 sectors at once. With the photometer looking parallel to the spin axis, the method of rotating Polaroid is used. The polarization filter is fixed with respect to the photometer, and the rotation is given by the spin of the spacecraft. The signal is integrated over intervals of 45°, which allows us to determine the Stokes parameters I, Q, U directly. Polarization measurements in different wavelength bands over an extended range of elongations should show the signature of the scattering dust particles, although it still may be difficult to identify the unique properties of the particle mixture. However, as *Helios* approaches the sun, the line of sight of our photometers will probe different regions of interplanetary space, and the polarization measurements should clearly show any change in the particle mixture with distance from the sun.[3]

---

[2] See the Glossary.

[3] For references to other zodiacal light observations from spacecraft, see p. 781 and p. 191.

MULTICOLOR POLARIMETRY OF THE NIGHT SKY

RAMON D. WOLSTENCROFT
Institute for Astronomy, University of Hawaii
and
JOHN C. BRANDT
Goddard Space Flight Center, Greenbelt, Maryland

Five-color observations of the linear polarization and single-color observations of the circular polarization are presented for the night sky. The pattern of orientations of the plane of vibration and the presence of the circular polarization in regions away from the galactic plane both imply that partially aligned nonspherical dust grains contribute substantially to the zodiacal light.

The light of the night sky comes from four main sources: airglow, zodiacal light, the Milky Way (integrated galactic light), and the light from these sources scattered by the troposphere. Three of these sources are polarized, the unpolarized source being the airglow. The astronomical objectives of polarimetric investigations of the night sky are to determine the state of polarization and its distribution over the sky separately for both the zodiacal light and the Milky Way and, hence, to increase our knowledge of the interplanetary and interstellar grains, particularly in regard to such properties as their shape and orientation.

Special problems, not experienced in point-source polarimetry, apply to the polarimetry of the night sky for the following reasons: (1) The unwanted sky background, which is principally unpolarized airglow emission (strictly foreground sky), is generally brighter than the source of interest by factors commonly in the range 1 to 10, depending on direction and wavelength; this causes a severe dilution of the degree of polarization. (2) The airglow has an irregular angular dependence, superposed on the idealized azimuth-independent increase of brightness toward the horizon (Van Rhijn law), and irregular time variations of appreciable amplitude occur; this makes it very difficult to determine the dilution factor with adequate precision. (3) The scattering of light from the extended sources by the troposphere introduces a spurious polarization whose polarized flux is, to the first order, proportional to the product of the zenith optical depth, air mass, and total night-sky brightness above the troposphere; to the second order, it depends on the details of the distribution of the Stokes parameters over the sky. The position angle of the scattered flux is also a sensitive function of this distribution.

The solution of these problems for ground-based observers is as follows:

1. Dilution Factor: correlations can be sought between line and continuum airglow emissions (Dumont 1965). Unfortunately these correlations are sometimes weak, and indeed, there is some dispute about the existence of certain of these covariance groups (Weinberg 1967); furthermore, correlations usually disappear when the airglow is "disturbed." Although this method has produced some useful results, it has done so at the expense of large amounts of observing time. The principal disadvantage is that a weak correlation leads to only a modest precision of the dilution factor. The alternative approach, and the one used in this paper, is to use the observing time not to measure the airglow correlations but rather to make high-precision measurements of the Stokes parameters Q and U of the astronomical sources, which are independent of the dilution factor; in our view, this is more useful than only moderately precise measurements of I, Q, and U.

2. Tropospheric Scattering: the method that has sometimes been used is to restrict observations to those air masses for which the polarized flux of the scattered light is less than some small value, as deduced from model calculations (Fesenkov 1963; Wolstencroft and van Breda 1967; Divari 1968). The main disadvantage of this approach is that brightness changes in the airglow can strongly influence the polarized flux; thus, it is desirable to determine the polarization of the scattered light empirically. This will be illustrated in the following section.

The best solution to the above problems, although the most expensive, is of course to take measurements from above the airglow layers, preferably using a satellite or space probe.[1] The remaining problem is then one of separating the galactic and zodiacal components of polarization. This can best be achieved by correlating changes in galactic and solar ecliptic coordinate systems as the two move relative to each other.

---

[1] See p. 781.

GROUND-BASED OBSERVATIONS OF LINEAR POLARIZATION

We have carried out an extensive polarimetric study of the night sky during July and August 1964 from a high-altitude site in the tropics, namely, Mount Chacaltaya in the Bolivian Andes. Because of the high altitude (5380 m), the problems of tropospheric scattering were minimized, and because of the tropical location, good coverage of both celestial hemispheres was possible. Observations were made with two scanning instruments. One instrument which made almucantar scans carried six parallel polarimeters with effective wavelengths, defined by narrow (100/400 Å) interference filters, at 3650 Å, 4510 Å, 5305 Å, 7075 Å, 9515 Å, and 5577 Å. The second instrument scanned through the zenith and carried four parallel polarimeters with effective wavelengths at 3650 Å, 5305 Å, 9515 Å, and 5577 Å. Further details of the instruments are given by Wolstencroft and Brandt (1967).

The values of the polarized flux to be discussed in this paper will be given[2] in the units $S_{10}(U, A0\ V)$, $S_{10}(B, A0\ V)$, $S_{10}(V, A0\ V)$, and $S_{10}(R, A0\ V)$ and have been deduced from the measurements at 3650 Å, 4510 Å, 5305 Å, and 7075 Å respectively. The calibration has been obtained from the measurement of over 100 unreddened standard stars and from the absolute spectrophotometric data of Code (1960) and Willstrop (1965). Measurements have also been made in $S_{10}(I, A0\ V)$ units but will not be presented here.

Tropospheric Scattering

In order to determine the influence of tropospheric scattering, we have studied the dependence of the apparent state of polarization on air mass for a representative set of points of constant astronomical coordinates. From this study we have deduced the polarized flux and position angle of the scattered component as a function of air mass for each region. Although the position angle of the scattered component depends quite sensitively on the particular astronomical direction, the polarized flux of the scattered component does not. For all these astronomical regions we have therefore determined the average dependence of polarized flux versus air mass for the scattered component. To correct the night-sky data for the scattered component, we have applied a strong weighting factor, which is the observed polarized flux divided by the scattered polarized flux appropriate to the direction of observation, when averaging the Stokes parameters Q and U obtained at various air masses in a given small area of sky. This very effectively eliminates the influence of scattering.

---

[2] Equivalent number of stars of spectral type A0 V of magnitude $U = 10.0$, etc., per square degree (Johnson's $UBVRI$ system).

The scattered polarized fluxes for 70° zenith distance (below the atmosphere) are as follows: 3.5 $S_{10}(U, \text{A0 V})$, 0.4 $S_{10}(B, \text{A0 V})$, 0.5 $S_{10}(V, \text{A0 V})$ and 1.1 $S_{10}(R, \text{A0 V})$.

The Position Angle of Polarization

The orientation of the plane of vibration for the hemisphere centered on the sun is shown in Fig. 1 for 4510 Å and 7075 Å. The average standard deviation of the position angle, θ, for both colors is in the range $10° \pm 2°$; this error includes an allowance for the weighting procedures used in correcting for tropospheric scattering. The clear circular symmetry on

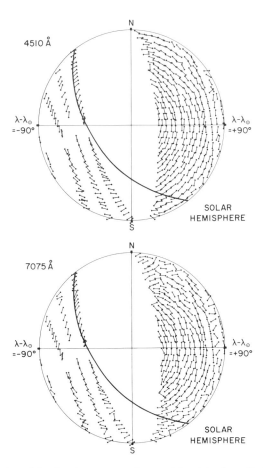

Fig. 1. Orientation of the electric-vector maximum of the night sky radiation in the hemisphere centered on the sun at 4510 Å and 7075 Å. The horizontal diameter is the ecliptic and the north ecliptic pole is at the top; the arc inclined to the ecliptic is the galactic equator.

the evening side of the flow pattern defined by the polarization (**E**) vectors confirms the widely held view that the zodiacal light is the principal source of polarization within 90° of the sun. On the morning side, the symmetry is apparently less pronounced; however, the Milky Way slightly influences θ on the morning side so that the apparent morning-to-evening difference may not be significant. There is a definite tendency for the vectors near the poles (yet removed from the galactic equator) to turn toward the sun, and this is also present at 5305 Å.

The orientation in the hemisphere centered on the antisolar point is shown in Fig. 2 for 4510 Å and 7075 Å. The average standard deviation of θ in this hemisphere is $15° \pm 2°$ for both colors. Although the patterns

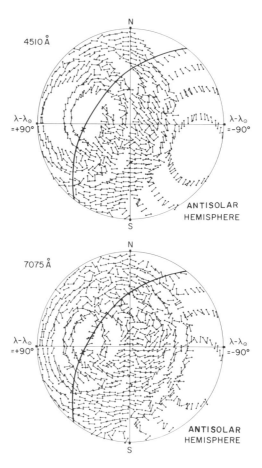

Fig. 2. The same arrangement as for Fig. 1 except for the hemisphere centered on the antisolar point. The galactic center is indicated by an X.

THE NIGHT SKY

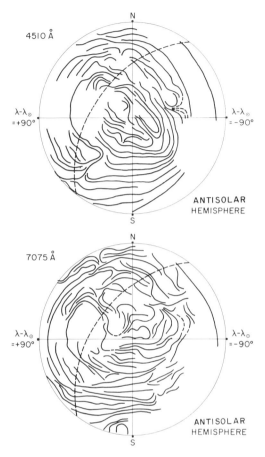

Fig. 3. The flow pattern defined by the polarization vectors shown in Fig. 2 with the vectors removed.

are complex, two features are easily discernible: (1) there is a circular flow pattern in the outer parts of the solar hemisphere; and (2) there is an ordered flow pattern approximately parallel to the ecliptic in a region centered on[3] about $\beta = -55°$, $\lambda - \lambda_\odot = +170°$. The overall flow pattern is more readily seen in Fig. 3. Additional features of the distribution are as follows: (3) The pattern is much more irregular within about 30° of the Milky Way than it is in the rest of the distribution; an appreciable polarized flux from the Milky Way, with an irregular distribution perhaps dominated by nearby dense dust clouds, is probably responsible. Evidence

---

[3]In this paper, λ is ecliptic longitude ($\lambda_\odot$ for the sun), and β is the ecliptic latitude.

for this galactic polarization has been presented by Schmidt and Leinert (1966), Sparrow and Ney (1972), Staude, Wolf, and Schmidt (1973), and Wolstencroft (1973). (4) The flow passes through the antisolar region, and although this flow is somewhat irregular, the flow direction is approximately parallel to the ecliptic at 7075 Å; this is also the case at 5305 Å. However, at 4510 Å the flow is inclined to the ecliptic; the pattern at 3650 Å is similar to that at 4510 Å.

Making the plausible assumption that the flow patterns away from the Milky Way (galactic latitudes greater than $|30°|$) are associated with the zodiacal light, we may draw an important conclusion concerning the interplanetary dust. If the interplanetary dust grains were all spherical, then the patterns would be combinations of circular arcs and radial lines. In fact, the observed orientations deviate by more than 2σ from either the radial or tangential directions for about 15% of the directions in the antisolar hemisphere and for about 4% of the directions in the solar hemisphere for 7075 Å. We must therefore conclude that the interplanetary dust grains that produce the zodiacal light are nonspherical and partially aligned. Note that the plane of polarization of the light scattered by an infinitely long cylinder is either parallel or perpendicular to the projection of the long axis on to the plane normal to the line of sight to the particle (see, e.g., Kerker et al. 1966), which is unrelated to the scattering plane. The greater number of deviations from the radial or tangential directions in the antisolar hemisphere perhaps implies that the degree of alignment increases with heliocentric distance. Alignment by either solar-wind-momentum transfer (the Gold mechanism [1952]) or by magnetic alignment of charged grains is possible in principle; however Harwit (1971) has argued that the Gold mechanism would dominate in the interplanetary situation, although the question of the alignment time cannot be considered settled. The apparent increase in the degree of alignment with heliocentric distance would in fact favor the magnetic alignment since the effectiveness of the Gold mechanism is proportional to r^{-2}, while that of the magnetic mechanism is proportional to the strength of the magnetic field and therefore proportional to r^{-1}.

The Linearly Polarized Flux

The distribution of the polarized flux over the night sky corrected for tropospheric scattering is shown in Fig. 4 for 7075 Å. It is immediately evident that the region of lowest polarized flux (or intensity) is centered not on the antisolar point but rather on about $\beta = 30°$, $\lambda - \lambda_\odot = 160°$. It is very probable that the polarized light from the Milky Way is strongly influencing the distribution in this area: in fact, these data (though with

THE NIGHT SKY

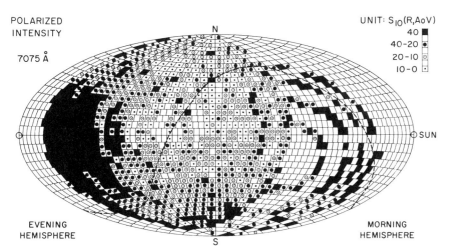

Fig. 4. The distribution of the linearly polarized flux at 7075 Å drawn on an equal-area solar-ecliptic coordinate projection with the antisolar point at the center. The ecliptic is the horizontal diameter and the north ecliptic pole is at the top. The galactic equator is indicated by the dashed line.

less angular resolution than is presented here) have been used to draw conclusions about the approximate magnitude of the polarization of the galactic component on the assumption that the zodiacal component has north-south symmetry with respect to the ecliptic (Wolstencroft 1973). However, it is evident from Fig. 4 that some intrinsic asymmetry of the zodiacal light may also be present. These two factors influencing the distribution of polarized flux can only be completely disentangled by a detailed polarimetric study of the night sky over at least six months as the zodiacal light moves relative to the Milky Way; a first attempt at this separation is discussed by Wolstencroft (1973).

CIRCULAR POLARIZATION OF THE NIGHT SKY

Measurements of a small fractional circular polarization of the night sky in the range $|q| = 0.1\%$ to 1% have been reported recently by Staude and Schmidt (1972) and by Wolstencroft and Kemp (1972). The values of q for zodiacal light close to the ecliptic, which have been corrected for dilution by the foreground airglow, are shown in Fig. 5a, which is taken from Wolstencroft and Kemp (1972). Although the standard error bars are in some cases comparable with the measured values, there is a clear trend with $\lambda - \lambda_\odot$ as indicated. There are changes of sign of q at $\lambda - \lambda_\odot = 75° \pm 5°$, $-175° \pm 5°$, and $-90° \pm 5°$, and in particular the data strongly

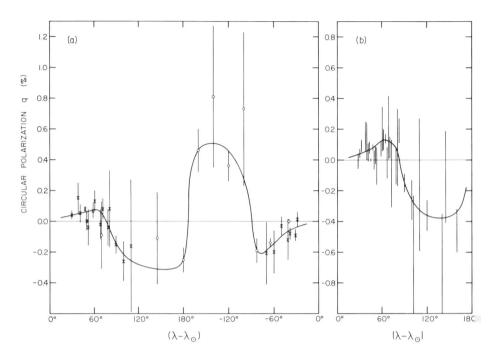

Fig. 5. (a) Fractional circular polarization of the zodiacal light close to the ecliptic (measured in a broadband in the visible) versus solar ecliptic longitude: the circles are the data of Wolstencroft and Kemp (1972) and the crosses are the data of Staude and Schmidt (1972). The convention for sign of q in this figure is that adopted in this book.
(b) $q(\lambda - \lambda_\odot)$ for $(\lambda - \lambda_\odot) > 0°$ and $-q(\lambda_\odot - \lambda)$ for $\lambda - \lambda_\odot \leq 0°$ versus $|\lambda - \lambda_\odot|$.

suggest an antisymmetry between morning and evening hemispheres of the form

$$q(\lambda - \lambda_\odot) = -q(\lambda_\odot - \lambda).$$

This conclusion is strengthened when $q(\lambda - \lambda_\odot)$ and $-q(\lambda_\odot - \lambda)$ are plotted against $|\lambda - \lambda_\odot|$ for $\lambda - \lambda_\odot = 0°$ to $180°$ and $\lambda_\odot - \lambda = 0°$ to $180°$ respectively (Fig. 5b).

The origin of the circular polarization can be understood in terms of the partial alignment of nonspherical particles. The theory for absorbing Rayleigh ellipsoids has been presented by Schmidt (1973) and for transparent needles of arbitrary size by Bandermann and Kemp (1973). The conclusion that circular polarization may result from the single scattering of unpolarized light is at first sight unexpected and therefore merits a simplified physical explanation as follows (see Fig. 6). An elongated grain is at the origin of a right-handed (x,y,z) coordinate system with its long axis in the (x,z) plane. Unpolarized radiation incident along the $-x$ direction is represented by two beams orthogonally polarized along the y and z direction with arbitrary phase difference. E_{oz} excites dipole moments in the z direction with components along both the long axis and the short axis, which lies in the (x,z) plane. These components will lead to scattered radiation along the y direction which has nonzero values for both the parallel and perpendicular components. E_{oy}, however, excites a dipole moment along y only, and this contributes no radiation along the y direction. Thus, the scattered light in this special case arises entirely from the E_{oz} component. If the grain is absorbing, the complex part of the polarizability will differ along the parallel and perpendicular directions in the grain and thus introduce a phase shift, $\delta_{par} - \delta_{per}$, between the parallel and perpendicular components (which would be zero if the grain did not absorb), and hence, a circularly polarized flux proportional to $\sin (\delta_{par} - \delta_{per})$. This simplified discussion describes the Schmidt mechanism for absorbing grains.

In a study in progress Bandermann and Wolstencroft are examining a variety of models in order to find a model for the grain alignment satisfying the constraints imposed both by the circular polarization and the linear polarization orientation. To explain the antisymmetry of the circular polarization, an alignment mechanism is required that has perfect azimuthal symmetry, i.e., it should produce an orientation relative to the outgoing radius vector from the sun that depends only on the heliocentric distance. A model has been examined in which the precessional axes of the grains are aligned along the magnetic field direction. This model explains most of the observed features but predicts an annual variation that at this stage of the study appears to be incompatible with the actual times of observation.

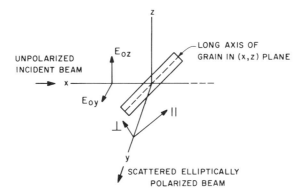

Fig. 6. The production of partially elliptically polarized radiation through the single scattering of unpolarized radiation by an obliquely oriented nonspherical grain.

A final remark is needed concerning the observed circular polarization in the Milky Way. Staude, Wolf, and Schmidt (1973) have found definite circular polarization in all eleven regions of the Milky Way studied by them. Wolstencroft and Kemp (1972) measured one of these regions independently and found agreement with their value. The origin of the polarization is discussed by Wolstencroft and Kemp (1972) who conclude that the diffuse galactic light rather than the integrated starlight is responsible.

Acknowledgments. This study was supported by the National Science Foundation under Grant GA-28201 and by the National Aeronautics and Space Administration under Contract NAS5-11303. The untiring efforts of Jerry D. Wilson who performed the computations is gratefully acknowledged.

REFERENCES

Bandermann, L. W., and Kemp, J. C. 1973. Circular polarization by single scattering from lossless non-spherical particles. *Mon. Not. R. Astr. Soc.* In press.

Code, A. D. 1960. Stellar energy distributions. *Stellar Atmospheres.* (J. L. Greenstein, ed.), pp. 50–87. Chicago: Univ. of Chicago Press.

Divari, N. B. 1968. Polarization of certain components of night-sky radiation scattered by the troposphere. *Sov. Astr. A. J.* 12: 503–511.

Dumont, R. 1965. Separation of the atmospheric, interplanetary and stellar components of the night sky radiation at 5000 Å. Application to the photometry of the zodiacal light and gegenschein. *Ann. d'Ap.* 28: 265–320.

Fesenkov, V. G. 1963. A table for the reduction of photometric observations of zodiacal light for the effect of tropospheric scattering. *Astron. Zhurnal* 40: 882–888.

Gold, T. 1952. The alignment of galactic dust. *Mon. Not. R. Astr. Soc.* 112: 215–218.

Harwit, M. 1971. Alignment of dust particles in comet tails. *Bull. Astr. Inst. Czech.* 22: 18–21.

Kerker, M.; Cooke, D.; Farone, W. A.; and Jacobsen, R. A. 1966. Electromagnetic scattering from an infinite circular cylinder at oblique incidence. I. Radiance functions for $m = 1.46$. *J. Opt. Soc. Am.* 56: 487–491.

Schmidt, T. 1973. Elliptical polarization by light scattering by submicron ellipsoids. IAU Symp. 52, *Interstellar dust and related topics.* (J. M. Greenberg and H. C. van de Hulst, eds.) Dordrecht, Holland: D. Reidel Publ. Co. In press.

Schmidt, T., and Leinert, C. 1966. A study concerning the surface polarimetry of the Milky Way. *Zeit. f. Astr.* 64: 110–115.

Sparrow, J. G., and Ney, E. P. 1972. Polarization of diffuse galactic light. *Astrophys. J.* 174: 717–720.

Staude, J., and Schmidt, T. 1972. Circular polarization measurements of the zodiacal light. *Astron. Astrophys.* 20: 163–164.

Staude, J.; Wolf, K.; and Schmidt, T. 1973. A surface polarization survey of the Milky Way and the zodiacal light. IAU Symp. 52, *Interstellar dust and related topics.* (J. M. Greenberg and H. C. van de Hulst, eds.) Dordrecht, Holland: D. Reidel Publ. Co. In press.

Weinberg, J. L. 1967. A program of ground-based studies of the zodiacal light. *The zodiacal light and the interplanetary medium.* (J. L. Weinberg, ed.), pp. 3–8, NASA SP-150. Washington, D.C.: U.S. Government Printing Office.

Willstrop, R. V. 1965. Absolute measures of stellar radiation. *Mem. Roy. Astron. Soc.* 69: 83–143.

Wolstencroft, R. D. 1973. Surface polarimetry of the Milky Way. IAU Symp. 52, *Interstellar dust and related topics*. (J. M. Greenberg and H. C. van de Hulst, eds.) Dordrecht, Holland: D. Reidel Publ. Co. In press.

Wolstencroft, R. D., and Brandt, J. C. 1967. A five-color photometry and polarimetry of the zodiacal light: a preliminary report. *The zodiacal light and the interplanetary medium* (J. L. Weinberg, ed.), pp. 57–62. NASA SP-150, Washington, D.C.: U.S. Government Printing Office.

Wolstencroft, R. D., and Kemp, J. C. 1972. Circular polarization of the night sky radiation. *Astrophys. J.* 177: L137–L140.

Wolstencroft, R. D., and van Breda, V. G. 1967. The determination of isophotes of extended sources in the night sky: Scattering in the earth's atmosphere. *Astrophys. J.* 147: 255–270.

DISCUSSION

SCHMIDT: I wish to make two remarks concerning the Heidelberg research mentioned by Wolstencroft:

1. By an elliptical *surface* polarization survey of the night sky we found the Milky Way to be linearly polarized up to about 2%, in general parallel to the galactic equator. Eleven different fields seem also to show a circular polarized light component between $q = 0.2\%$ and 0.9%, but the circular polarization of the zodiacal light was generally found not to exceed 0.1% to 0.2%.
2. Rayleigh scattering calculations show that it is possible to get *elliptical* polarized light up to $q/p = 0.15$ from unpolarized incident radiation by single scattering at dirty quartz like submicron spheroidal particles with a refractive index of $1.54-0.4i$.

For more details see the papers of Staude, Wolf, and Schmidt (1973) and of Schmidt (1973), both presented at the I.A.U. Symposium 52 at Albany, and of Staude and Schmidt (1972) (the convention in all these papers is: $q > 0$ means clockwise rotation looking against the light source).

WOLSTENCROFT: I am pleased to hear that you find that the orientation of the galactic polarization lies nearly parallel to the galactic equator. This is in good agreement with measurements we have made at Haleakala, some of which were repeated at the same symposium mentioned in your remark. This of course implies that the polarization is very probably due primarily to the integrated starlight and not to the diffuse galactic light, as many people had expected.

BAUM: When making your photoelectric sky scans, could you get enough signal (for example, with larger sky samples or with a larger telescope aperture) to use a very narrow-band (<5 Å) color filter centered on a strong solar absorption line? The violet K-line of CaII might be a good candidate. This should suppress the zodiacal contribution by almost an order of magnitude, because the latter includes a lot of violet light from early-type stars whose spectra are not significantly depressed in the K-line region. The purpose would be to identify the galactic contribution more accurately.

WOLSTENCROFT: The measurements presented here were obtained with 200 Å resolution, and therefore, to get the same signal-to-noise ratio would require an increase in integration time of $(40)^2$ from about 1 minute to many hours, at least with the same equipment. In any case, I do not believe it is necessary; observations of the polarized flux and orientation of the night-sky polarization, in regions away from the Milky Way which are properly corrected for tropospheric scattering, will tell you almost all you need to know about the polarization of the zodiacal light.

ROACH: What integration time per point did you use in your linear polarization measurements?

WOLSTENCROFT: For the linear polarization data there were typically about 50 observations each of about 2 secs duration, making a total of 100 secs per point. These observations were stretched over a period of about one week.

CLARKE: With respect to the position of zero linear polarization close to the antisolar point, does this position correspond to the counterglow?

WOLSTENCROFT: No. The region of minimum polarized flux in the zodiacal light is centered at about $\beta = 30°$, $\lambda - \lambda_\odot = 160°$, i.e., well removed from the counterglow. The position of the minimum is evidently influenced by the polarization of the Milky Way, but one certainly cannot rule out the possibility of an asymmetry of the polarization distribution in the zodiacal light as well. The question can be answered by studying the behavior of the polarization during a one-year period as the galactic and solar ecliptic coordinates move through one cycle of relative motion.

GEHRELS: Are you planning to make such measurements?

WOLSTENCROFT: Yes, they are currently being made.

GEHRELS: How do you know that the airglow emission is unpolarized?

WOLSTENCROFT: If the airglow were polarized, you would expect to find a correlation between changes of airglow brightness and changes of the polarized intensity of the total night-sky radiation. Such correlations are not seen; in fact, the polarized intensity is independent of the airglow brightness. The only exception to this is for measurements close to the horizon where tropospheric scattering produces a spurious airglow polarization. These remarks apply to the continuum airglow only.

POLARIZATION OF THE ZODIACAL LIGHT

J. L. WEINBERG
Dudley Observatory, Albany, New York

The zodiacal light is described briefly and is compared with other components of the light of the night sky. A chronology is given of observations of the polarization of the zodiacal light, followed by a comparison of results which is used to describe the principal polarization features. Brief mention is made of current satellite experiments to measure the polarization of the zodiacal light.

In the absence of moonlight, the light of the night sky (nightglow) at moderate to low latitudes is dominated by zodiacal light (Fig. 1), starlight (integrated starlight and diffuse galactic light), and airglow line and continuum emission. The zodiacal light is brightest toward the ecliptic and toward the sun, the starlight is brightest near the galactic equator, and the airglow is generally brightest toward the horizon.

The zodiacal light has a fairly high degree of polarization (Fig. 2), and in the ecliptic at 30° from the sun it is approximately three times as bright as the brightest regions of the Milky Way. The starlight has negligible polarization, except for low galactic latitude regions where the polarization may reach a few percent.[1] The principal airglow line emissions (5577 and 6300 Å) are very bright, each having a small degree of polarization arising from atmospheric scattering, with the 6300 Å emission also having a small intrinsic polarization. The airglow continuum is relatively faint and there is no evidence that it has any intrinsic polarization. All of these sources are absorbed and scattered in the atmosphere, this process itself giving rise to polarization which has a distribution over the sky that is different in direction and amount from that of the incident radiations. We shall be concerned here only with the zodiacal light and particularly with its polarization. Discussions of the general nightglow/zodiacal light problem have been given by Elsässer (1963), Roach (1964), Dumont (1965), Divari (1965), Blackwell, Dewhirst, and Ingham (1967), and others.

---
[1] See p. 888.

Fig. 1. Photograph of the morning zodiacal light by P. Hutchison, Mount Haleakala, Hawaii, January 1967.

HISTORICAL BACKGROUND

Although the Italian astronomer Cassini (1685) is credited with having made the first systematic observations of the zodiacal light and with first explaining it in terms of dust distributed about the sun, the literature (see, for example, Lynn 1896) is alive with paintings, picturesque descriptions, and incorrect explanations dating back some 2000 years.

There were relatively few observations of shape, position, color, and brightness of the zodiacal light until more than 100 years after Cassini's investigations. Searches for zodiacal light polarization did not commence until the mid-19th century and were not really successful until 1874 (Wright). Wright summarized the results of his investigation as follows:

1. "The zodiacal light is polarized in a plane passing through the Sun.
2. The amount of polarization is, with a high degree of probability, as much as 15 per cent, but can hardly be as much as 20 per cent.
3. The spectrum of the light is not perceptibly different from that of sunlight except in intensity.

Fig. 2. Photograph of the zodiacal light taken through Polaroid strips arranged with their axes alternately parallel and perpendicular to the direction to the sun (Blackwell and Ingham 1961, reproduced from *Monthly Notices of the Royal Astronomical Society*, Vol. 122, by permission).

4. The light is derived from the Sun, and is reflected from solid matter.
5. This solid matter consists of small bodies (meteoroids) revolving about the Sun in orbits crowded together toward the ecliptic."

Today, a full century later, Wright's conclusions again (still) adequately describe the zodiacal light, in spite of innumerable studies contending that the zodiacal light is wholly or partly associated with electrons in interplanetary space, with geocentric dust clouds, with gas and dust tails associated with the earth, etc.

Although Wright's visual observations gave the first positive, quantitative results on the polarization of zodiacal light in 1874, 50 years passed before Dufay (1925) obtained similar but more detailed results from photographic observations. Huruhata (1948) ushered in the era of photoelectric observations of nightglow polarization and was the first to use polarization measurements as a means of separating zodiacal light from airglow and starlight.

OBSERVATIONS OF POLARIZATION

Definitions and Methods

The nightglow can be characterized by the orientation of the plane of polarization (χ), the total degree of polarization (p_{tot}), the total brightness or radiance (B_{obs}), and the brightness of the polarized component (B_{pol}).

These quantities are related by:

$$p_{tot} = \frac{\sum_j B_{pol,j}}{\sum_j B_{obs,j}} \neq \sum_j p_j \qquad (1)$$

or

$$p_{tot} = \frac{(I_\perp - I_\|)_{ZL} + \sum_i (I_\perp - I_\|)_i}{(I_\perp + I_\|)_{ZL} + \sum_i (I_\perp + I_\|)_i}, \qquad (2)$$

where $I_\perp$ and $I_\|$ are orthogonal components of brightness having their electric vectors perpendicular and parallel, respectively, to a particular scattering plane (e.g., the plane containing the sun, the observer, and the observed direction). ZL and i refer to the zodiacal light and other brightness components, respectively, and $j = ZL + i$ when all components are referred to this single plane.

These parameters are illustrated in Fig. 3 for celestial pole observations obtained at the Haleakala Observatory, Hawaii, from evening until morning twilight in a narrow band at 5080 Å and are "total"; i.e., they are given as observed at the ground without correction for atmospheric extinction and scattering.[2] Of particular interest is the variation of χ. Since the solid line shown in the plot of χ is the orientation expected for zodiacal light (i.e., perpendicular to the direction to the sun), it is clear that the zodiacal light extends to high ecliptic latitudes (the ecliptic latitude of the north celestial pole is 66.°6). The difference in slope between the observed and predicted χ arises from polarization originating in the atmosphere. In these data the observed values of χ are used to determine the polarized brightness.

Most observers of zodiacal light from the ground have measured brightness but not polarization, and most observers of polarization have assumed but have not measured the direction of polarization. The assumption of direction introduces a large uncertainty, since it is equivalent to assuming that the zodiacal light is the only source of polarization. We now know, however, that there are several competing sources of polarization in the nightglow, especially in observations near the horizon and in wavelength bands containing both line and continuum emission (Weinberg, Mann, and Hutchison 1968).

A Chronology

Table I contains a chronology of published observations of the polarization of zodiacal light. References to polarization observations of other extended sources in the nightglow (e.g., airglow line and continuum emis-

---

[2]See also p. 768.

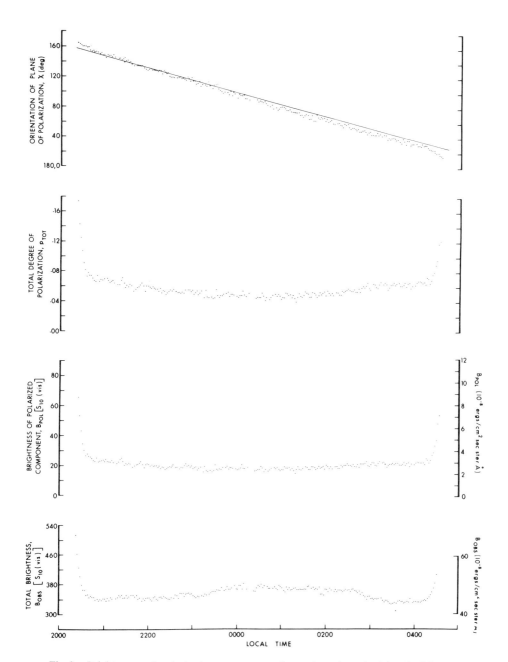

Fig. 3. Brightness and polarization parameters observed at the celestial pole, Mount Haleakala, 5–6 July 1967 (Weinberg, Mann, and Hutchison 1968). The orientation of the plane of polarization is measured in a clockwise direction from the vertical. Local time is Hawaii Standard Time.

TABLE I
A Chronology of Observations of the Polarization of the Zodiacal Light

Date of Observations	Observers	Place or Vehicle	Observation Type[a]	Color, System, or Wavelength Range (Å)
1870 Dec	Ranyard (and Burton) (1871)	Agosta, Sicily	Vis	
1874 Jan–Apr	Wright (1874)	New Haven, Connecticut	Vis	
1918 Apr, May	Strutt (Rayleigh) (1919)	Essex, England	Pg	
1919 Spring	Babcock (1919)	Mt. Wilson Observatory	Pg	
1924–25	Dufay (1925)	Montpellier, France	Pg	
?	Take-Uchi (Yamamoto 1939)	?	Pg	
1945 Jan, Feb	Huruhata (1948)	Azabu Observatory, Japan	Pe	visual (4000–7500) infrared (8500–11000) no filter (4000–11000)
1945–49	Huruhata (1951)	Tokyo University Observatory	Pe	blue (broadband) red (broadband)
1952 Feb (eclipse)	Rense, Jackson, and Todd (1953)	Aircraft, 9.8 km above Saudi Arabia	Pg	4460 (broadband)
1952 Feb, Mar	Behr and Siedentopf (1953)	Jungfraujoch, Switzerland	Pe	5430 (broadband)
1955 May	Blackwell (1956)	Aircraft, 2.7 km above South Pacific	Pg	red (broadband)
1955 Sept, Oct	Divari, Krylova, and Moroz (1964)	Kamenskoye Plato Observatory, USSR	Pe	4700 (broadband) 5200 (broadband)
1956 Aug, Sept	Elsässer (1958)	Boyden Station, South Africa	Pe	"visual spectral region"
1957 Sept, Oct	Nikolskii (1963)	Kazakh Steppe Station, USSR	Pe	yellow (broadband)
1957 Oct, Nov	Divari and Asaad (1960)	near Aswan, Egypt	Pe	5220 (85 half width) 5410 (840 half width)
1958 June–Aug	Blackwell and Ingham (1961)	Chacaltaya, Bolivia	Pg	4470 (broadband) 6200 (broadband)
1959 Nov	Peterson (1961)	Capillo Peak, New Mexico	Pe	4355 (broadband) 5425 (broadband)

Date	Reference	Location	Type	Wavelength
1961	Robley (1962)	Pic du Midi Observatory	Pe	4630 (120 half width)
				5280 (90 half width)
1961 June, July	Beggs, et al. (1964)	Chacaltaya, Bolivia	Pe	red (broadband)
1961–62	Weinberg (1964)	Mt. Haleakala, Hawaii	Pe	5300 (56 half width)
1962–65	Gillett (1966)	Balloons	Pe, Pg	B
1964 Feb	Dumont (1965)	Teide Observatory, Tenerife	Pe	5020 (240 half width)
1964 July, Aug	Wolstencroft and Brandt (1967)[c]	Chacaltaya, Bolivia	Pe	4510 (166 half width)
				5303 (128 half width)
				7073 (316 half width)
1964 Sept	Wolstencroft and Rose (1967)	Rocket, above New Mexico	Pe	7030 (180 half width)
1964–65	Dumont and Sanchez Martinez (1966, and Sanchez Martinez 1967)	Teide Observatory, Tenerife	Pe	5020 (240 half width)
1965	Gillett (1967)	OSO[b]-2	Pe	
?	Gillett (1967)	Balloon	Pe, Pg	
1965	Sparrow and Ney (1968)	OSO-B2	Pe	"astronomical visual"
1965 Nov, Dec	Regener and Vande Noord (1967)	Balloons, above New Mexico	Pe	B, V (UBV system)
1965 Nov, Dec	Vande Noord (1970)	Balloons, above New Mexico	Pe	B, V (UBV system)
1966 Mar				
1966 Oct	Ingham and Jameson (1968)	Aircraft, 5 km above North Atlantic	Pe	5100 (95 half width)
1967 May, Oct	Weinberg and Mann (1968)	Mt. Haleakala, Hawaii	Pe	multicolor (narrow-band)
1968 Mar	Jameson (1970)	Jungfraujoch, Switzerland	Pe	5100 (95 half width)
1969	Sparrow and Ney (1972)	OSO-5	Pe	4180 (broadband)
				6820 (broadband)
1971 July	Leinert, Link, and Pitz (1973)	Rocket, above Sardinia	Pe	4680 (500 half width)

[a] Vis = visual, Pg = photographic, Pe = photoelectric.
[b] (Earth) Orbiting Solar Observatory.
[c] See also p. 768.

sion and diffuse galactic light) are not included here. Also omitted are the recent studies concerning circular polarization of the zodiacal light (Wolstencroft and Rose 1967; Staude and Schmidt 1972; Staude, Wolf, and Schmidt, 1972; Wolstencroft and Kemp 1972).[3]

Difficulties attendant with observations from the ground, the decreasing number of dark-sky sites suitable for nightglow photometry, and the availability of vehicles for observation from above the earth's atmosphere have all contributed to the increasing tendency to observe from remote, high-altitude sites or from balloons, rockets, and satellites.

A Comparison of Results

In Fig. 4 we have collected results prior to 1967 on the degree of polarization of the zodiacal light in the ecliptic. Subsequent results by Gillett (1967), Wolstencroft and Rose (1967), and others agree with the results of Weinberg (1964) and of Dumont and Sanchez Martinez (1966).

The scatter in these data and in data on the brightness of zodiacal light does not result from differences in wavelength or from variations in the zodiacal light, although these may exist. As noted by Blackwell, Dewhirst, and Ingham (1967), there is a tendency for observers who find large polarizations to find low brightnesses and vice versa, suggesting errors of reduction more than of observation. The errors arise in corrections for atmospheric extinction and scattering and in the separation of zodiacal light from airglow and starlight (see, for example, Dumont and Sanchez Martinez 1973). These difficulties are increased for observations far from the sun and in off-ecliptic regions where the zodiacal light, airglow, and starlight all have approximately the same brightness. The result is that there are relatively few observations of the polarization of zodiacal light in these regions.

In spite of the differences among observers, a number of polarization features are apparent:

1. The electric vector is perpendicular to the scattering plane (the plane containing the observer, the observed direction, and the sun), except in those regions containing negative or circular polarization.
2. The degree of polarization has a relatively broad, flat maximum between $60°$ and $75°$ elongation,[4] and it has a maximum value between 0.17 and 0.23.
3. The polarization is not symmetric about this maximum; it falls off faster toward large elongations.

---

[3] See p. 779. [4] Angular distance from the sun.

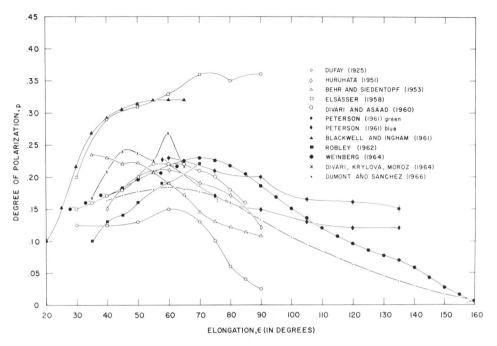

Fig. 4. The degree of polarization of the zodiacal light in the ecliptic.

4. Neutral points and negative polarization (electric vector parallel to the scattering plane) are found at large elongations.
5. The polarization tends to zero faster at longer wavelengths.

In a recent rocket experiment carefully designed to avoid stray light in measurements of the inner zodiacal light during daytime and outside of eclipse, Leinert, Link and Pitz (1973)[5] found polarizations of approximately 0.15 at 15° and 21° elongation. As noted by the authors, these high values for the inner zodiacal light have important implications for both the particle size distribution and chemical composition.

CONCLUDING REMARKS

There are essentially no observations of the polarization of zodiacal light in the ultraviolet or infrared, nor are there detailed results on the wavelength dependence of polarization even in the visible spectrum. Satellite-borne experiments during the 1970s (Table II), combined with

---

[5] See p. 766.

TABLE II
Current Satellite Experiments to Measure Polarization of the Zodiacal Light

Vehicle	Colors or Wavelength Range (Å)	Dates of Operation	Experimenters	Institutions
OSO-6	4000 5000 6100	Aug 1969–Jan 1972	Carroll and Roach	Rutgers University
Pioneer 10/11 Asteroid-Jupiter probes	blue and red	Pioneer 10: Mar 1972– Pioneer 11: Apr 1973–	Weinberg and Hanner[a]	Dudley Observatory
Apollo 17 (manned)	photographic	Dec 1972	Mercer and Dunkelman	Dudley Observatory, NASA Goddard Space Flight Center
Skylab (manned)	4000–8200, 10 colors	May–Dec 1973	Weinberg	Dudley Observatory
Helios close-in solar probes	3600 4200 5200	1974, 1975 launches	Leinert, Link, and Pitz[b]	Max Planck Institute for Astronomy, Heidelberg
MJS/77 (Mariner Jupiter-Saturn probes)	2200–7270, 7 colors	1977 launches	Lillie	University of Colorado

[a]See p. 781.
[b]See p. 766.

rocket observations and with coordinated ground-based observations should provide accurate results on the photometric properties of the zodiacal light and, in turn, more complete information on the optical properties and on the size and spatial distributions of the interplanetary grains.

Acknowledgments. This research has received support from the National Aeronautics and Space Administration (grant NGR 33-011-017) and from the Atmospheric Sciences Section, National Science Foundation (grant GA-27676).

REFERENCES

Babcock, H. D. 1919. Note on the polarization of the night sky. *Astrophys. J.* 50: 228–231.
Beggs, D. W.; Blackwell, D. E.; Dewhirst, D. W.; and Wolstencroft, R. D. 1964. Further observations of the zodiacal light from a high altitude station and investigation of the interplanetary plasma. III. Photoelectric measurements of polarization. *Mon. Not. R. Astr. Soc.* 128: 419–430.
Behr, A., and Siedentopf, H. 1953. Untersuchungen über Zodiakallicht und Gegenschein nach lichtelektrischen Messungen auf dem Jungfraujoch. *Zeits. f. Astrophys.* 32: 19–50.
Blackwell, D. E. 1956. Observations from an aircraft of the zodiacal light at small elongations. *Mon. Not. R. Astr. Soc.* 116: 365–379.
Blackwell, D. E.; Dewhirst, D. W.; and Ingham, M. F. 1967. The zodiacal light. *Advances in astronomy and astrophysics.* (Z. Kopal, ed.) Vol. 5, pp. 1–69, New York: Academic Press, Inc.
Blackwell, D. E., and Ingham, M. F. 1961. Observations of the zodiacal light from a very high altitude station: The average zodiacal light. *Mon. Not. R. Astr. Soc.* 122: 113–127.
Cassini, J. D. 1685. Découverte de la lumière céleste qui paroist dans le zodiaque. *Mem. Acad. Sci. Paris* 8: 121–209.
Divari, N. B. 1965. Zodiacal light. *Soviet Phys. Uspekhi* 7: 681–695.
Divari, N. B., and Asaad, A. S. 1960. Photoelectric observations of the zodiacal light in Egypt. *Sov. Astr. A. J.* 3: 832–838.
Divari, N. B.; Krylova, S. N.; and Moroz, V. I. 1964. Polarization measurements of the zodiacal light. *Geomagnetism and Aeronomy* 4: 684–687.
Dufay, J. 1925. La polarisation de la lumière zodiacale. *C. R. Acad. Sci. Paris* 181: 399–401.
Dumont, R. 1965. Séparation des composantes atmosphérique, interplanétaire et stellaire du ciel nocturne a 5000 Å. Application a la photométrie de la lumière zodiacale et du gegenschein. *Ann. d'Astrophys.* 28: 265–320.
Dumont, R., and Sanchez Martinez, F. 1966. Polarisation du ciel nocturne et polarisation de la lumière zodiacale vers 5000 Å, sur l'ensemble de la sphère céleste. *Ann. d'Astrophys.* 29: 113–118.
———. 1973. Photométrie de la lumière zodiacale hors de l'écliptique en quadrature et en opposition avec le soleil. *Astron. Astrophys.* 22: 321–328.
Elsässer, H. 1958. Neue Helligkeits- und Polarisationsmessungen am Zodiakallicht und ihre Interpretation. *Die Sterne* 34: 166–169.
———. 1963. The zodiacal light. *Planet. Space Sci.* 11: 1015–1033.

Gillett, F. C. 1966. Zodiacal light and interplanetary dust. Ph.D. Thesis, Univ. of Minnesota.

———. 1967. Measurement of the brightness and polarization of zodiacal light from balloons and satellites. *The zodiacal light and the interplanetary medium.* (J. L. Weinberg, ed.) pp. 9–15, NASA SP-150. Washington, D.C.: U.S. Government Printing Office.

Huruhata, M. 1948. Polarization of the night sky light and the zodiacal light. *Tokyo Astron. Bull.*, Second Series No. 10, 77–79.

———. 1951. Photoelectric study of the zodiacal light. *Publ. Astron. Soc. Japan* 2: 156–171.

Ingham, M. F., and Jameson, R. F. 1968. Observations of the polarization of the night sky and a model of the zodiacal cloud normal to the ecliptic plane. *Mon. Not. R. Astr. Soc.* 140: 473–482.

Jameson, R. F. 1970. Observations and a model of the zodiacal light. *Mon. Not. R. Astr. Soc.* 150: 207–213.

Leinert, C.; Link, H.; and Pitz, E. 1973. A rocket experiment to measure the zodiacal light at elongation angles between 15° and 30°. *Space Research XIII*, Akademie-Verlag, Berlin. In press.

Lynn, W. T. 1896. Early observations of the zodiacal light. *Observatory* 19: 274–275.

Nikolskii, G. M. 1963. Photoelectric observations of zodiacal light near Alma Ata. *Geomagnetism and Aeronomy* 1: 317–320.

Peterson, A. W. 1961. Three-color photometry of the zodiacal light. *Astrophys. J.* 133: 668–674.

Ranyard, A. C. 1871. On the zodiacal light. *Mon. Not. R. Astr. Soc.* 31: 171–172.

Regener, V. H., and Vande Noord, E. L. 1967. Observations of the zodiacal light by means of telemetry from balloons. *The zodiacal light and the interplanetary medium.* (J. L. Weinberg, ed.) pp. 45–47, NASA SP-150. Washington, D.C.: U.S. Government Printing Office.

Rense, W. A.; Jackson, J. M.; and Todd, B. 1953. Measurements of the inner zodiacal light during the total solar eclipse of February 25, 1952. *J. Geophys. Res.* 58: 369–376.

Roach, F. E. 1964. The light of the night sky: Astronomical, interplanetary and geophysical. *Space Sci. Rev.* 3: 512–540.

Robley, R. 1962. Photométrie des lumières zodiacale et anti-solaire. *Ann. de Geophys.* 18: 341–350.

Sanchez Martinez, F. 1967. Recent polarization measurements over the sky at Tenerife Island. *The zodiacal light and the interplanetary medium.* (J. L. Weinberg, ed.) pp. 71–73, NASA SP-150. Washington, D.C.: U.S. Government Printing Office.

Sparrow, J. G., and Ney, E. P. 1968. OSO-B2 satellite observations of zodiacal light. *Astrophys. J.* 154: 783–787.

———. 1972. Observations of the zodiacal light from the ecliptic to the poles. *Astrophys. J.* 174: 705–716.

Staude, J., and Schmidt, T. 1972. Circular polarization measurements of the zodiacal light. *Astron. Astrophys.* 20: 163–164.

Staude, J.; Wolf, K.; and Schmidt, T. 1973. A surface polarization survey of the Milky Way and the zodiacal light. *Interstellar dust and related topics.* (J. M. Greenberg and H. C. van de Hulst, eds.) Dordrecht, Holland: D. Reidel Publ. Co. In press.

Strutt, R. J. 1919. Polarization of the night sky. *Astrophys. J.* 50: 227–228.

Vande Noord, E. L. 1970. Observations of the zodiacal light with a balloon-borne telescope. *Astrophys. J.* 161: 309–316.

Weinberg, J. L. 1964. The zodiacal light at 5300 Å. *Ann. d'Astrophys.* 27: 718–737.

Weinberg, J. L., and Mann, H. M. 1968. Negative polarization in the zodiacal light. *Astrophys. J.* 152: 665–666.

Weinberg, J. L.; Mann, H. M.; and Hutchison, P. B. 1968. Polarization of the nightglow: Line versus continuum. *Planet. Space Sci.* 16: 1291–1296.

Wolstencroft, R. D., and Brandt, J. C. 1967. A five-color photometry and polarimetry of the zodiacal light: A preliminary report. *The zodiacal light and the interplanetary medium.* (J. L. Weinberg, ed.) pp. 57–62, NASA SP-150. Washington, D.C.: U.S. Government Printing Office.

Wolstencroft, R. D., and Kemp, J. C. 1972. Circular polarization of the night sky radiation. *Astrophys. J.* 177: L137–L140.

Wolstencroft, R. D., and Rose, L. J. 1967. Observations of the zodiacal light from a sounding rocket. *Astrophys. J.* 147: 271–292.

Wright, A. W. 1874. On the polarization of the zodiacal light. *Amer. J. Sci. and Arts.* Third Series 7: 451–459.

Yamamoto, I. 1939. Report of the sub-commission on zodiacal light. *Trans. Int. Astron. Union* (Stockholm, 1938), VI: 172–175.

DISCUSSION

WITT: Are the reported measurements at 0.52 μm disturbed by the air afterglow continuum?

WEINBERG: If you mean the reported measurements of zodiacal light, they are not "disturbed" but rather are based on a separation of the (starlight and) airglow continuum.

THE LINEAR POLARIZATION OF THE COUNTERGLOW REGION[1]

F. E. ROACH and B. CARROLL
Rutgers, The State University, Newark, New Jersey
L. H. ALLER
University of California, Los Angeles
J. R. ROACH
Ball Brothers Research Corporation, Boulder, Colorado

Measurements of the linear polarization of scattered sunlight in the antisun direction reveal patterns over a 13-day period centered on the 30 September 1970 new moon. Graphical representations of the patterns indicate that the orientation of the electric vector tends to favor approximate orthogonality to the ecliptic, but occasionally, systematic departures from orthogonality occur. The pattern effect is the consequence of changes in both the angle and the percentage of polarization. A comparison of the patterns for observations at six different solar elongations does not reveal any obvious interrelationships, suggesting that the physical source has lateral dimensions less than the 5° separation of the set of data. A speculative suggestion is made that the patterns are due to nonspherical particles that acquire a degree of preferred orientation in either the geomagnetic or the interplanetary magnetic field.

The Rutgers photopolarimeter on OSO-6 was launched in August of 1969 and continued in operation until it was shut off in January of 1972 after the completion of some 13,000 orbits. In each orbit over 5,000 individual sky readings were made through an optical system that permits the evaluation of the four Stokes parameters in three colors for a series of 38 solar elongations from the antisun region to within 10° of the sun (Rouy, Carroll, and Aller 1965; Ball Brothers Research Corp. 1970).

---

[1]*Editorial Note:* This paper has been kept in this book against the recommendations of referees from widely different backgrounds who have doubts on the instrumental performance, particularly on the effects of scattered light; the referees also doubt the results. The exception to editorial policy is made because this is an interesting paper, the results of which were obtained at great expense, and the merit should be discussed openly. The reader may be the judge. — T. Gehrels.

In this paper we have selected a sample of 200 contiguous orbits for a preliminary study of the linear polarization in the 0.5 μm region. The elongations selected were 195°, 190°, 185°, 180° (antisun), 175°, and 170°, giving a slice across the counterglow region as illustrated in Fig. 1. The selection of orbits was based on two criteria: (1) orbits during a time of year (September–October) when the antisun region was relatively far from the Milky Way; (2) orbits centered on the time of new moon when there was no possibility of any scattered moonlight entering the optical path. That we succeeded in getting a reasonably "clean" domain of the sky is illustrated in Fig. 2, which shows a comparison of the brightness for the several elongations from our present

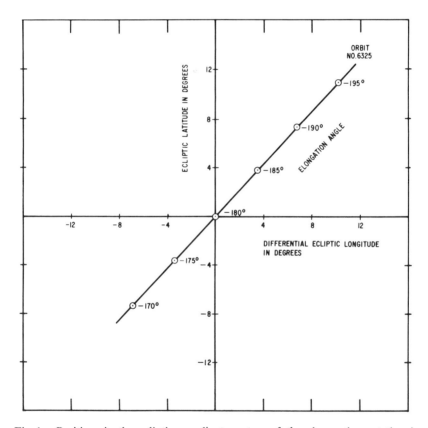

Fig. 1. Positions in the ecliptic coordinate system of the observations at the six solar elongations under study in this paper. The negative designation of the elongations is our convention to indicate that the satellite was in the "morning" part of its orbit.

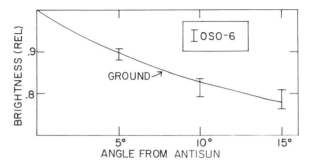

Fig. 2. Comparison of the relative brightness in the general antisun direction of the observations included in this paper with ground-based observations of the counterglow. The upper part of the OSO-6 error bars are the observed values and the lower are based on an assumed 20% contamination of the observations by integrated starlight.

space data and from ground observations recently summarized by Roach (1973). We have found that the absolute calibration of the instrument used as a photometer is possible by the use of star crossings. Also, we have reported preliminary calibration data in Roach et al. (1972). Ground-based studies of the Milky Way can be utilized for calibration purposes as we showed in a paper comparing photometric data from OSO-6 and OAO-2 in the region of the Milky Way near the direction to the galactic center (Roach and Lillie 1972).

POLARIMETRIC METHOD

A summary of the instrument characteristics is given in Table I and additional details are in the engineering report on the instrument (Ball Brothers Research Corp. 1970). The optical path of the photopolarimeter is shown schematically in Fig. 3. The polarimeter components include:

1. Quartz rotators in wheels 1 and 2, designed and sized as a function of wavelength to rotate the plane of polarization of the incoming light by combinations of $+45°$ and $-45°$.

2. A quarter-wave plate for 0.5 μm in wheel 2, with the fast axis kept always at $45°$ with respect to the pass axis of the analyzer (Rouy polarizing prism) when positioned into the optical path.

3. A polarizing prism consisting of a double cut of calcite designed by the late A. L. Rouy (Rouy et al. 1965).

TABLE I
Characteristics of the Rutgers Photopolarimeter

Instrument Characteristics	
Telescope:	f/1.6
Telescope aperture:	6.0 cm
Telescope field of view:	2°
Wavelength regions:	0.4, 0.5, and 0.61 μm
Wavelength bandpass:	0.04 μm full width at half maximum
Polarization:	1. Quartz rotators (multiples of 45°)
	2. Quartz quarter-wave retardation plate
Photomultiplier:	EMR-641E
Photometric dynamic range:	10^{-9} to 10^{-14} solar surface brightness
Operational Characteristics	
Observing sequence:	5° steps of elongation from sun
Dark current calibration each orbit	
Orbit inclination:	32°96
Orbit perigee altitude:	490 km
Orbit apogee altitude:	550 km
Orientation in celestial sphere varies with OSO-6 maneuvers	

4. Interface filters in wheel 3 to isolate nominal wavelengths of 0.4, 0.5, and 0.61 μm. Peak transmissions are greater than 60% and the width at half maximum is 400 ± 50 Å.

The instrument transmits less than 10^{-5} of the light intensity when a pair of the Rouy double-cut calcite prisms is crossed. At 0.5 μm a single prism has a transmission of 85%.

The rotators in conjunction with the Rouy prism pass polarized

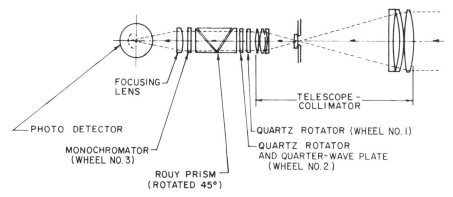

Fig. 3. Optical diagram of the zodiacal light photopolarimeter developed by Rutgers University for OSO-6.

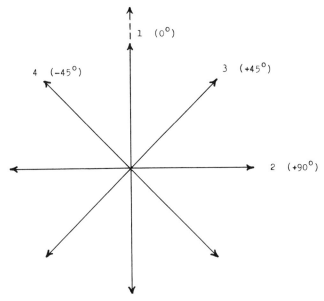

Fig. 4. Effective positions in the Rutgers polarimeter of the pass axis of the analyzing prism as modified by the insertion of quartz rotators into the optical path.

beams along the four axes illustrated in Fig. 4, yielding numerical values of the three Stokes parameters I, Q, and U. The fourth Stokes parameter, V, may be deduced from readings made with the positioning of the quarter-wave plate into the optical path along with an orthogonal set of rotators, as for example, the set corresponding to positions 1 and 2 (Fig. 4). The entire sequence of readings is repeated for each color (in quintuplicate), for each satellite orbit and for each solar elongation in the observing program. In this report we restrict ourselves to wavelength 0.5 μm; to linear polarizations from the parameters I, Q, and U; and to solar elongations 170°, 175°, 180°, 185°, 190°, and 195°. At these elongations the satellite shields direct sunlight from the instrument entrance aperture. Photometric studies of both the counterglow and the Milky Way, taken under these observing conditions, show that the instrumental scattered light is trivial (Roach et al. 1973a,b).

If one defines

$$p(1,2) = \frac{I(1) - I(2)}{I(1) + I(2)}$$

and

$$p(3,4) = \frac{I(3) - I(4)}{I(3) + I(4)},$$

where $I(i)$ is the observed intensity at position i, it can be shown that the polarization p can be written as

$$p = [p^2(1,2) + p^2(3,4)]^{1/2},$$

and the angle between the plane of polarization and the reference axis of the vehicle is

$$\psi = \tfrac{1}{2}\tan^{-1}\frac{p(3,4)}{p(1,2)}.$$

These are the relationships used to obtain the data presented in this report.

OBSERVATIONAL RESULTS

The data from the OSO-6 instrument were processed in several steps starting with the conversion of the "raw" readings which are on a quasi-logarithmic scale into linear readings. The individual readings are made as the satellite wheel rotates with a period of 2 secs. In order to obtain an acceptable spatial resolution, the duration of exposure to the sky is approximately 5 msecs. In each orbital pass five readings are made through each optical element, giving a total exposure time to the sky of about 25 msecs. At the low light level in the antisun direction, we have elected to further increase the effective exposure time by smoothing the linear readings by combining the results from 10 orbits. In order to estimate the "noise" introduced by the system using our procedure, it would have been desirable to make test runs on a light source of constant brightness and polarization. As a matter of fact, we had originally thought that the antisun observations might provide such a standard source, but its variability, which is the principal thrust of this paper, has precluded this possibility. Our judgment, based on working with the data, is that polarizations greater than 0.5% are physically real. In the case of the angle of orientation of the electric-vector maximum, we carried the calculations to $0°.1$, which is probably at least one power of ten better than justified by the quality of the observations.

In Figs. 5 and 6 we show the results in graphical form. We have ordered the data by orbit number and have made an indication of the percentage polarization and the orientation of the E vector for each orbit. The ecliptic plane is inclined some 12° with respect to the horizontal in Figs. 5 and 6; if each vector is rotated 12° clockwise, then the horizontal is the ecliptic. An inspection of the plots leads to some generalizations:

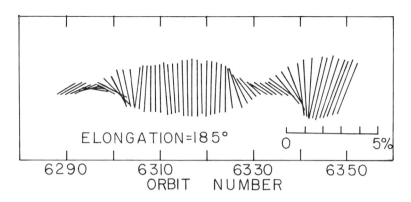

Fig. 5. Graphical representation of the percentage polarization and the orientation of the **E** vector for a sample of 60 orbits at solar elongation of 185°.

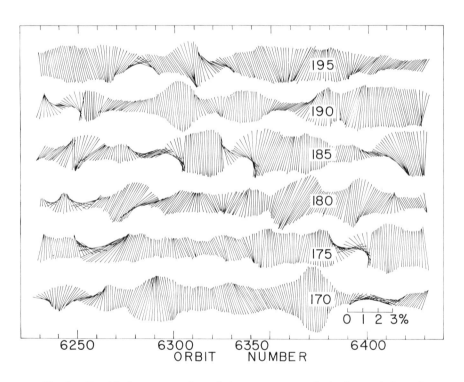

Fig. 6. Graphical representation of the percentage polarization and the orientation of the **E** vector for 200 contiguous orbits and for 6 solar elongations. Orbit 6330 occurred at the time of the new moon of 30 September 1970. See the text for the sense of the vector orientations.

1. There is a tendency for the orientation to be orthogonal to the plane of the ecliptic.
2. There are what we refer to as "patterns" evident in the data. The patterns seem to be the result of systematic changes in either or both the amount and orientation of polarization.
3. There is no obvious alignment or interrelationship of the patterns among the several elongations plotted.

COMMENTS

The significance of the observations of this study may be approached by reference to the meaning of "orbit number" as the abscissa in the plots. If the study were primarily of light from astronomical sources beyond the solar system, we would be dealing with a movement across the field of $1/15°$ per orbit. Thus, a star crossing the $2°$ field of the instrument requires about 30 orbits to traverse, and the 200 orbits cover a total astronomical field of some $13°$. We have chosen orbits when the antisun direction is well away from the Milky Way, and the evidence that we are dealing predominantly with the counterglow has already been stressed in discussion of Fig. 2.

If we are dealing with a cloud of scattering material at the lunar distance, then one orbit of the satellite in 90 mins corresponds to an orbital movement of the assumed cloud of about $0°.75$, assuming that the cloud is moving in a Keplerian orbit around the earth. At a lunar distance of 384,000 km, one satellite orbit corresponds to an extension along the lunar path of about 5,000 km. The full sweep of the 200 orbits corresponds to $150°$ or 1,000,000 km.

Another possibility is that the polarization patterns are due to a sweeping past the earth of clouds of dust moving more or less radially from the sun. If a cloud contains a significant fraction of nonspherical particles oriented by a magnetic field, we may get an approximate idea of the physical dimension in the sun-radial direction by the duration of a typical pattern. Taking 20 orbits as typical, we have a time duration pattern of 30 hours. If we assume a radial speed of 500 km/sec, the extent is some 54 million km (0.36 AU). An estimate of the transverse size comes from the fact that observations as close as $5°$ do not in general show similar polarization structure. The corresponding linear dimension is in the domain of tenths of an astronomical unit if the effective distance from the observer is greater than a tenth of an AU.

The polarization of the counterglow is surprising. Pure (i.e., $180°$) backscattering from randomly oriented irregular, amorphous or spherical particles should give zero polarization. The existence of a marked polarization effect in our observations implies scattering by elongated preferentially oriented particles. Needle-like particles are possible, but it is difficult to understand how they may be oriented systematically. One is tempted to think of a magnetic field as the arranging agency, but the conditions imposed on the field and particles are severe. The particles must be ferromagnetic or paramagnetic and respond rather quickly to the changing orientation of the field (i.e., the relaxation times must be less than a day). We are continuing the investigation of the linear polarization in the antisun direction from the large library of data available from OSO-6.

REFERENCES

Ball Brothers Research Corp. 1970. Rutgers zodiacal light analyzer for OSO-G. Final Report F 70-11.

Roach, F. E. 1972. A photometric model of the zodiacal light. *Astron. J.* 77: 887–891.

Roach, F. E.; Carroll, B.; Aller, L. H.; and Smith, L. 1972. Surface photometry of celestial sources from a space vehicle. *Proc. Nat. Acad. Sci.* 69 (No. 3): 694–697.

Roach, F. E.; Carroll, B.; Roach, J. R.; and Aller, L. H. 1973*a*. A photometric study of the counterglow from space. *Planetary and Space Science*. In press.

———. 1973*b*. A photometric perturbation of the counterglow. *Planetary and Space Science*. In press.

Roach, F. E., and Lillie, C. F. 1972. A comparison of surface brightness measurements from OAO-2 and OSO-6. Proc. of a NASA sponsored symposium, Amherst, August 23, 24, 1971. NASA-SP-310, 71–79. Washington, D.C.: U.S. Government Printing Office.

Rouy, A. L.; Carroll, B.; and Aller, L. H. 1965. Experiment for the determination of the brightness, polarization, and ellipticity of the zodiacal light from OSO-G. NSG 550. Rutgers, The State University.

DISCUSSION

WOLSTENCROFT: The irregular patterns and time variations you see in the linear polarization orientations close to the antisolar point are consistent with night-to-night variations in orientation that I have observed; the orientations shown in my paper[1] were averages over 7 nights.

GEHRELS: What is a quartz rotator and what is a Rouy prism?

J. R. ROACH: The circular retarder is a quartz disk cut perpendicular to the optical axis. To reduce rotary dispersion, the quartz rotators in this instrument have been ground to the proper thickness for each

---

[1] See p. 768.

color. The dextro and levo properties of the quartz rotators are due to the existence of quartz crystals that are mirror images of each other; this again due to atomic layers of SiO_2 built up in a clockwise direction in one case and counterclockwise in the other. Rotating the rotator rotates the plane of polarization of the incoming light by $+45°$ and $-45°$ in combinations as required by the system programing.

The prism is a Glan-Foucault type polarizer as modified by Archard and Taylor (1948). In place of two calcite elements the Rouy polarizer contains three in a rectangular array. This gives the prism the equivalent of two polarizers placed side by side and results in a polarizance exceeding 0.99999. The dimensions of the prism are 56 x 34 x 31 mm; the optical axis of each of the three triangular calcite elements is parallel to the entrance face and upper face of the prism. The air–calcite interfaces are held precisely by including a spacer of 0.2 mm. Along the spacer are provisions for outgassing. The elements were made of the best quality nonscattering calcite.

DISCUSSION REFERENCE

Archard, J. F., and Taylor, A. M. 1948. Improved Glan-Foucault prism. *J Sci. Instr.* 25:407–409.

ZODIACAL LIGHT MODELS BASED ON NONSPHERICAL PARTICLES

RICHARD H. GIESE and REINER ZERULL
Ruhr-Universität, Bochum

> The influence of different scattering functions on zodiacal light intensity and polarization is discussed. For simplified models of the interplanetary particle mixture, results based on Mie theory are compared with nonspherical particle models: microwave analogue measurements demonstrate that the contribution of zodiacal light by micron-size dielectric particles of spherical shape would be different from that of nonspherical particles. Absorbing particles of sizes much below and much above micron size, however, should give a contribution very similar to spheres.

Most model computations for zodiacal light analysis are based on the Mie theory of light scattering by homogeneous spherical particles. The results of such calculations are not realistic if the scattering pattern of interplanetary grains differs from the scattering functions of spherical particles. A complication is that the deviations are weighted in the procedure of integration over the line of sight to obtain the surface brightness of the zodiacal light.

To investigate the influence of these deviations, power laws were adopted for both the size distribution and the spatial-number distribution of interplanetary grains. To make the model even simpler, attention is limited to the ecliptic plane. If the number density $N(R)$ of dust particles varies with solar distance R according to $N \propto R^{-v}$, and if the scattering function $\sigma(\Theta)$, characterizing the average scattering behavior of the interplanetary grains as a function of the scattering angle Θ, is independent of R, one obtains the surface brightness I of zodiacal light (Giese and Siedentopf 1962) as a function of elongation ε in the ecliptic plane

$$I(\varepsilon) \propto (\sin \varepsilon)^{-(v+1)} \int_{\varepsilon}^{\pi} \sigma(\Theta)(\sin \Theta)^{v} d\Theta. \qquad (1)$$

This integration can be performed separately for the components of scattered light having the electrical vector perpendicular or parallel to the plane of scattering (sun, scattering-volume, observer). Thus, the polarized intensities $I_\perp$ and $I_\parallel$ are obtained from the corresponding scattering functions $\sigma_\perp$ and $\sigma_\parallel$, respectively. The degree of polarization can be defined for the scattering pattern as a function of Θ, i.e., $P(\Theta) = (\sigma_\perp - \sigma_\parallel)/(\sigma_\perp + \sigma_\parallel)$ and for the zodiacal light as a function of elongation

$$P(\varepsilon) = \frac{I_\perp(\varepsilon) - I_\parallel(\varepsilon)}{I_\perp(\varepsilon) + I_\parallel(\varepsilon)}. \qquad (2)$$

SCATTERING FUNCTIONS AND ZODIACAL LIGHT

As can be seen from Equation (1), the contribution to $I(\varepsilon)$ and $P(\varepsilon)$ of any portion of $\sigma(\Theta)$ is influenced by ε and ν. Regions of $\sigma(\Theta)$ at $\Theta < \varepsilon$ do not contribute at all, while, on the other hand, the portions for $\varepsilon \leq 180°$ are weighted by $(\sin \Theta)^\nu$. Figure 1a shows $\sigma(\Theta)$ for three extreme cases. "Small Spheres" (cf. Table I) corresponds to a polydispersion of dielectric ($m = 1.33$) mainly submicron-size spheres similar to that used by Weinberg (1964) in his zodiacal light model. "Large Spheres" represents a polydisperse mixture of spheres having the same refractive index but with sizes in the range of $\sim 10\mu$m. "Cubes" represents a mixture of randomly oriented, nearly dielectric cubes corresponding in volume to spheres in the size range $1.9 \leq x \leq 17.8$ (Zerull and Giese).[1] The percentage of $I(\varepsilon)$ contributed in these cases from the angular intervals $\Theta = 20°$ to $60°$, $60°$ to $100°$, $100°$ to $140°$, and $140°$ to $180°$ is shown in Table I for a very low ($\nu = 0.1$; see Fig. 1b) and a rather high ($\nu = 1.0$) decrease of particle numbers with solar distance.

In all cases Table I shows that the overwhelming contribution to $I(\varepsilon)$ at very low elongations ($\varepsilon = 20°$) stems from portions of $\sigma(\Theta)$ below $100°$. At $\varepsilon = 60°$ the main contribution to $I(\varepsilon)$ is scattered at $60° \leq \Theta \leq 140°$. Only in the case of extreme scattering efficiency for large ($\Theta > 140°$) scattering angles (the dielectric "Large Spheres" model) does most of the radiation come from $\Theta > 140°$. The effect is decreased by depletion of particles beyond 1 A.U. for $\nu = 1.0$. For $\varepsilon = 100°$ the shape of $\sigma(\Theta)$ governs whether the larger contribution to $I(\varepsilon)$ is scattered at $\Theta < 140°$ or at $\Theta > 140°$.

The differences in the contribution of angular scattering to the total intensity $I(\varepsilon)$ are found again in the run of $P(\varepsilon)$. Generally, of course, integration over the line of sight results in a smoother run[2] of $P(\varepsilon)$ compared to $P(\Theta)$. A tendency to have positive or negative polarization for

---

[1] See p. 901.   [2] These experiences are very similar to those encountered in the work on reflection nebulae (p. 867).—Ed.

TABLE I
Contribution of $\sigma(\Theta)$ to the Zodiacal Light Intensity $I(\varepsilon)$[a]

ε	Θ	20°				60°				100°			
		20°	60°	100°	140°	20°	60°	100°	140°	20°	60°	100°	140°
	Isotropic[b]	25	26	26	23	—	35	35	30	—	—	53	47
$v = 0.1$	Cubes[c]	53	21	15	11	—	46	31	23	—	—	57	43
	Small Spheres[d]	82	9	4	5	—	52	21	27	—	—	43	57
	Large Spheres[e]	83	5	3	9	—	28	19	53	—	—	27	73
	Isotropic	23	35	31	12	—	45	39	16	—	—	72	28
$v = 1.0$	Cubes	46	30	19	6	—	55	34	11	—	—	75	25
	Small Spheres	77	15	5	3	—	64	23	13	—	—	64	36
	Large Spheres	81	8	5	6	—	43	24	33	—	—	43	57

[a] Relative contribution (rounded, in percent) to total intensity at ε from four intervals $\Delta\Theta = 40°$ in Θ
[b] Isotropic scattering
[c] 11 cubes. $m = 1.574 - 0.006i$; $x = 1.91$ to (x 1.25) 17.79, $k = 2.5$ (cf. Zerull and Giese p. 901), x is size parameter (circumference/wavelength); size distribution: $dN \propto x^{-k} dx$
[d] dielectric spheres $m = 1.33$; $1 \leq x \leq 120$, $k = 4$ (Mie theory)
[e] dielectric spheres $m = 1.33$; $60 \leq x \leq 120$; $k = 2.5$ (Mie theory)

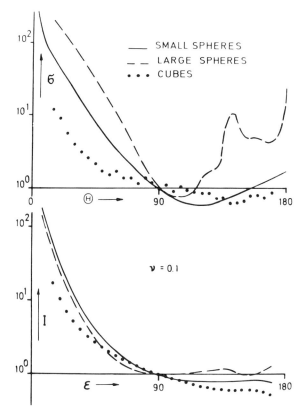

Fig. 1a. Average scattering function $\sigma(\Theta)$ and zodiacal light intensity $I(\varepsilon)$ derived from polydisperse mixtures of small and large spheres and the 11 cubes shown in Table I for $\nu = 0.1$ in spatial distribution $N \propto R^{-\nu}$, where R is the solar distance, σ and I are relative units of intensity, Θ is the scattering angle, and ε the elongation.

$P(\Theta)$ is found again in the $P(\varepsilon)$ diagram. But the capability of polarization features in the region of large Θ to influence $P(\varepsilon)$ below $\varepsilon = 90°$ is rather model dependent.

SPHERICAL VERSUS NONSPHERICAL PARTICLES

To obtain information about the influence of nonspherical particle shape on zodiacal light, it is useful to separate the possible constituents of the interplanetary dust mixture into a few types of components:

1. Submicron particles ($x \ll 1$): For submicron particles $\sigma(\Theta)$ produces a maximum of positive polarization at $\Theta \approx 90°$, even for randomly oriented cubes, fourlings (Donn and Powell 1963; Powell et al. 1967), graphite flakes, and cylinders (Hanner 1969, 1971). Therefore, it might

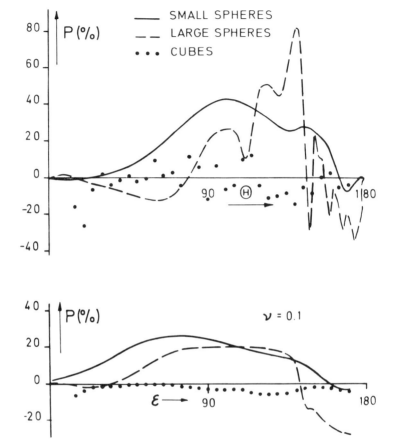

Fig. 1b. As in Fig. 1a for the percentage polarization (positive when electric-vector maximum is perpendicular to plane of scattering).

be concluded that such small particles in assemblies of random orientation will produce approximately the same run of $P(\varepsilon)$ as the spherical particles.

2. Large absorbing particles ($x \gg 1$): Big absorbing particles, having an imaginary part of the refractive index large enough to damp out refracted radiation entering the particle, will scatter outside the diffraction lobe by reflection only. Since for assemblies of randomly oriented convex particles there is no preference for any direction of the reflecting surfaces, scattering by reflection takes place as in the case of spheres (van de Hulst 1957). Therefore, no differences are expected in the prediction of $P(\varepsilon)$ for such particles (metallic, dirty ices) compared to spherical particle models.

3. Larger dielectric particles ($x > 1$): Contrary to the cases referred to above, deviations from spherical particle models are expected (Hodkinson 1966) for dielectric particles. They are confirmed in optical measurements (Holland and Gagne 1970) and by microwave analogue experiments (Zerull and Giese[3]). The forward region ($\Theta < 90°$) of the scattering diagram remains in reasonable agreement with predictions derived from Mie theory for spheres of equivalent size, but the pronounced enhancement of $\sigma(\Theta)$ towards backscattering in the case of dielectric spheres is gone. Furthermore there are differences in polarization.

Microwave experiments allow manufacturing of scattering bodies of definite shape, so it is possible to demonstrate how the scattering function and (after integration over the line of sight) $I(\varepsilon)$ and $P(\varepsilon)$ deviate from the predictions obtained by Mie theory as the scattering objects deviate from spheres. In Fig. 2, for example, $I(\varepsilon)$ and $P(\varepsilon)$ are shown as a mixture of spherical particles (five particle sizes, $6.33 \leq x \leq 14.02$) and as a corresponding assembly of spheres whose surfaces were roughened statistically. Typical deviations from a smooth surface are about 0.1 wavelength. For single rough spheres in a given orientation, deviations from Mie theory are observable. But a mixture of such objects in random orientation produces $I(\varepsilon)$ and $P(\varepsilon)$ indistinguishable from spherical particles.

An assembly of randomly oriented octahedrons (Fig. 3) gives an $I(\varepsilon)$ differing in backscattering from an assembly of exact spheres. There is good agreement for $20° \leq \varepsilon \leq 110°$. More striking is the decrease in magnitude of $P(\varepsilon)$. Negative polarization persists as in the corresponding example of spheres.

Finally, in the case of randomly oriented cubes (Fig. 4), the effects referred to above are much more pronounced. There is reasonable agreement of $I(\varepsilon)$ only for $\varepsilon \leq 90°$, no backward-scattering effects, and no considerable negative polarization. Admittedly, simulation of the interplanetary polydisperse particle mixture by a few examples of regular bodies is imperfect. Comparison with corresponding examples of spheres suggests, however, that the general features can be demonstrated in this range of x by use of a rather limited number of sizes.

CONSEQUENCES FOR ZODIACAL LIGHT MODELS

Zodiacal-light models using nonspherical particles were computed by Little, O'Mara, and Aller (1965) using the scattering functions obtained by Richter (1962) in optical laboratory experiments. The results, however, did not agree with the observed zodiacal light. Zodiacal light models approxi-

---

[3] See p. 901.

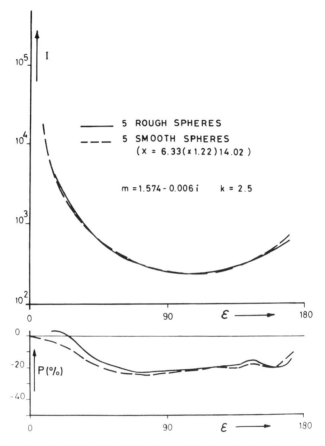

Fig. 2. As in Fig. 1, for a hypothetical assembly of randomly oriented particles.

mating observations will usually imply not only one single component of material but a mixture of materials of different refractive indices. For such models the following conclusions may be stated:

1. Components producing positive polarization having the maximum approximately at the correct elongation should not be affected by nonspherical shape. For spherical particles, such possible components are large absorbing particles ($x \geq 10$) or very small ($x \leq 1$) grains (Giese 1972), i.e., cases referred to above are rather invariant in scattering to shape, if extreme examples are excluded and if random orientation is adopted.
2. Since the scattering diagram of randomly oriented, nonspherical dielectric micron-size grains differs from the scattering properties of equivalent spheres, one has to be careful with interpretations of

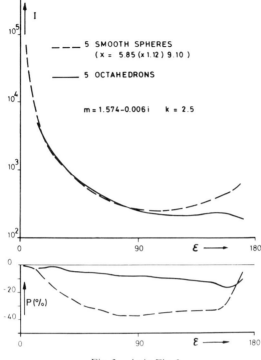

Fig. 3. As in Fig. 2.

zodiacal light models based on Mie theory. Differences are mainly expected at high elongations ($\varepsilon > 140°$) for the zodiacal light. The differences are primarily in backscattering and polarization phenomena. Therefore, any interpretation of the counterglow and of the wavelength dependence or even the existence of a neutral point in polarization (Greenberg 1970) should not be based on Mie theory. In the forward region ($\varepsilon \leq 40°$) of the zodiacal light, predictions by Mie theory should be rather reliable due to the strong increase of $\sigma(\Theta)$ for $\Theta < 90°$.

3. The results on scattering by nonspherical, micron-size dielectric grains encourage us to try zodiacal light models using μm particles as optically important components. Since polarization by absorbing particles of this size is usually too strong, some additional dielectric component has to be added to adjust $P(\varepsilon)$ to the observational curve. Large spherical particles, however, introduce some difficulties by their strong polarization and backward-scattering features. Since the scattering diagrams of equivalent assemblies of nonspherical particles are more isotropic and neutral in polarization, it will be much easier to understand the optical properties of the zodiacal light by assuming

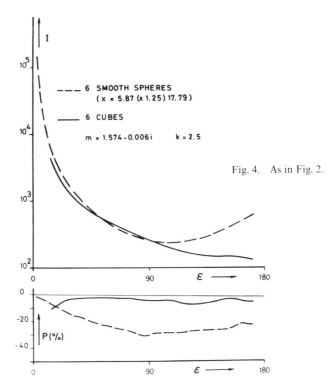

Fig. 4. As in Fig. 2.

nonspherical grains. This is satisfying since nonspherical particle models are a more realistic approach to actual interplanetary grains.

Acknowledgments. This is an interim report of results obtained during investigations supported by the *Deutsche Forschungsgemeinschaft* and by the *Bundesminister für Bildung und Wissenschaft*. We thank C. E. Ken-Knight for his suggestions for this manuscript.

REFERENCES

Donn, B., and Powell, R. S. 1963. Angular scattering from irregularly shaped particles with application to astronomy. *Electromagnetic scattering* (M. Kerker, ed.), pp. 151–158. New York: Pergamon Press.

Giese, R. H. 1972. Single component zodiacal light models. Dudley Observatory Report No. 7, Albany, N.Y.

Giese, R. H., and Siedentopf, H. 1962. Optische Eigenschaften von Modellen der Interplanetaren Materie. *Zeit. f. Astrophys.* 54: 200–216.

Greenberg, J. M. 1970. Models of the zodiacal light. *Space Research X*, pp. 225–231. Amsterdam: North-Holland Publ. Co.

Hanner, M. S. 1969. Light scattering in reflection nebulae. Ph.D. Thesis, Rensselaer Polytechnic Inst., Troy, New York.

———. 1971. Physical properties of the interplanetary dust. *Physical studies of minor planets* (T. Gehrels, ed.), pp. 377–386. NASA SP-267. Washington, D.C.: U.S. Government Printing Office.

Hodkinson, J. R. 1966. The optical measurements of aerosols. *Aerosol science* (C. N. Davies, ed.), pp. 287–357. New York: Academic Press.

Holland, A. C., and Gagne, G. 1970. The scattering of polarized light by polydisperse system of irregular particles. *Appl. Optics* 9: 1113–1121.

Hulst, H. C. van de 1957. *Light scattering by small particles.* New York: Wiley.

Little, S. J.; O'Mara, B. J.; and Aller, L. H. 1965. Light scattering by small particles in the zodiacal cloud. *Astron. J.* 70: 346–352.

Powell, R. S.; Woodson III, P. E.; Alexander, M. A.; Circle, R. R.; Konheim, A. G.; Vogel, D. C.; and McElfresh, T. W. 1967. Analysis of all available zodiacal-light observations. *The zodiacal light and the interplanetary medium* (J. L. Weinberg, ed.), pp. 225–241. NASA SP-150. Washington, D.C.: U.S. Government Printing Office.

Richter, N. B. 1962. The photometric properties of interplanetary matter. *J. Roy Astron. Soc.* 3: 179–186.

Weinberg, J. L. 1964. The zodiacal light at 5300 Å. *Ann. d'Astrophys.* 27: 718–738.

DISCUSSION

GIESE (in reply to a question regarding nephelometer measurements of scattering functions of nonspherical particles): Are you thinking about measurements as performed by Holland and Gagne? Such experiments are useful and they are faster than microwave analogue measurements. However, the advantages of microwave experiments are that one can systematically change and select the particle shapes and the orientation, and one can simulate by computer calculation any type of particle mixtures: polydisperse mixtures of random orientation; polydisperse mixtures of partially or completely aligned grains; monodisperse particles at any random or nonrandom orientation; mixtures of particles of different refractive index, etc.

DUBIN: Suppose nonspherical particles did not exist in significant concentrations in the submicron ($r \sim 0.1$ to 0.01 μm) sizes? In the case of sounding rocket collections of cosmic dust (by Hemenway et al.), nearly all the particles are almost spherical. Have you computed when effective nonsphericity would result in terms of polarization, if the ensemble of submicron particles were moving together at speeds of 10 to 100 km/sec? The effect on particle interaction or scattering with the photons may become polarizing for very small particles near 100 Å in size.

GIESE: We have not yet included any dynamic effects in our calculation. Even with models considering only particle number densities there remain some ambiguities about the main component producing the polarization of the zodiacal light.

DISCUSSION REFERENCE

Hemenway, C. L.; Hallgren, D. S.; and Schmalberger, D. C. 1972. Stardust. *Nature* 238: 256–260.

COMETARY POLARIZATIONS

D. CLARKE
The University, Glasgow

Comets were one of the first astronomical objects to have their polarized light observed. Arago (1855) found polarization in both Comets 1819 III and 1835 III (Halley).

Measurements made by Öhman (1941) using an objective prism followed by a photographic polarimeter clearly indicated for the first time that the polarization in the cometary light is produced by two mechanisms: scattering of sunlight by the cometary particles and fluorescence emission by the cometary molecules. Subsequent measurements confirm Öhman's observations, the relative contributions by the two mechanisms depending on the particular comet.

Because of the high degree of polarization of cometary light, it is possible to make spatial measurements of the polarization of the brighter comets by photography. Although the measurements cannot provide extreme precision, they show that the polarization varies significantly over the recorded images.

The comas around Comets Arend-Roland 1956h and Mrkos 1957d were studied by Elvius (1958), but the results were not presented in any detail. For Arend-Roland, however, it was noted that the amount of polarization was smaller on the sunward side of the head than in the tail, the polarization in both cases being perpendicular to the plane of scattering. This comet exhibited very little fluorescence emission, its light being mainly due to scattered sunlight. Elvius suggested that as the degree of polarization was dependent on the size of the scattering particles, the observed percentage would vary along the comet; the large particles expelled from the nucleus might later be split into smaller particles as they flowed away along the tail. A second explanation involved elongated dust particles that are preferentially oriented by weak magnetic fields.

More detailed measurements were made by Martel (1960a,b) of Comets Mrkos 1957d in various spectral domains and Giacobini-Zinner

1959b. (The presented diagrams displaying the direction of vibration over the images suggest that the depicted vectors should be rotated through 90° for correspondence with the direction of the electric vector.) For Mrkos, the directions of vibration (electric vector) generally lie within a few degrees of each other, close to being at right angles to the scattering plane, although small departures from this rule are apparent on the sunward side. For Giacobini-Zinner, one of the measured regions on the sunward side gives a value of the position angle significantly different from 90°. With similarity to Elvius' results for Arend-Roland, Martel notes that the percentage polarization is lower on the sunward side of the nucleus of Mrkos than in the tail.

More recently, Comets Honda 1968c (Osherov 1970) and Bennett 1969i (Clarke 1971; Osherov 1971) (see Fig. 1) have displayed strong

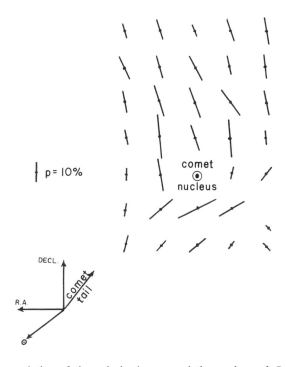

Fig. 1. The variation of the polarization around the nucleus of Comet Bennett, 01ʰ–10 UT of 27 April 1970 (Clarke 1971). The percentage polarization and the directions of electric vector maximum are marked by bars. Each measurement was made with a circular diaphragm of approximately 13 arcsecs diameter; the measured points are taken 25 arcsecs apart. The systematic uncertainty in relating directions of vibration to the coordinate system is approximately ± 5°. The nucleus itself was overexposed.

variations of the direction of vibration around the region of the coma, before becoming aligned at right angles to the scattering plane farther in the tail.

Harwit and Vanysek (1971) have shown that a significant degree of alignment of grains can be expected as a result of bombardment by solar wind protons. A high degree of alignment is also to be expected by the outward flow of molecules from the cometary nucleus. The rates at which alignment occurs by these mechanisms in relation to the velocities of the grains allow alignment to be first controlled by the gas flow from the nucleus and then to be modified by the solar wind as the grains flow into the tail.

Depending on the contribution to the scattered light by the aligned grains, the direction of vibration will be modified appropriately. For Comet Bennett, streamers and whirls of gas were seen to emanate from the nucleus (Roemer 1970). The observed values of the position angle may to some extent reflect first this flow from the nucleus and then the alignment caused by the solar wind.

In relation to alignment mechanisms, reference can be made to the measurements of Matyagin et al. (1968) in the tail of Comet Ikeya-Seki 1965f. It is interesting to note that the amount of polarization did not change smoothly along the tail but seemed to exhibit "waves." These waves may reflect positions in the tail where the degree of alignment is enhanced, perhaps as a result of bursts in the flow of energy of the solar wind. They may also correspond to pockets in which the dust particles are significantly different in size from the general grain size found in the tail.

In order to pursue these ideas, it is important to obtain further polarimetric observations with good spatial resolution. In particular, it would be of interest to measure the heads of all dusty comets to see if the polarization pattern is correlated with the whirls of material coming from their nuclei.

REFERENCES

Arago, F. 1855. *Popular astronomy* (W. H. Smyth and R. Grant, trans. and eds.) Vol. I. London.

Clarke, D. 1971. Polarization measurements of the head of Comet Bennett (1969i). *Astron. Astrophys.* 14: 90–94.

Elvius, A. 1958. Preliminary results of polarization measurements in comets. *Arkiv. Astr. Bd.* 2 (Nr. 19): 195–197.

Harwit, M., and Vanýsek, V. 1971. Alignment of dust particles in comet tails. *Bull. Ast. Inst. Czech.* 22: 18–21.

Martel, M.-T. 1960a. Variations de la polarisation observée du 18 au 29 Août dans la Comète 1957d pour des régions situées à moins de 4' du noyau. *Ann. Astrophys.* 23: 480–497.

———. 1960b. Observations polarimétriques de la Comète Giacobini-Zinner 1959b. *Ann. Astrophys.* 23: 498.

Matyagin, V. S.; Sabitov, S. N.; and Kharitonov, A. V. 1968. Polarimetry of the tail of Comet Ikeya-Seki. *Sov. Astron. A. J.* 11: 863–867.

Öhman, Y. 1941. Measurements of polarization in the spectra of Comet Cunningham (1940c) and Comet Paraskevopoulos (1941c). *Stock. Obs. Ann., Bd.* 13: No. 11.

Osherov, R. S. 1970. Polarization measurements of Comet Honda 1968c. *Dokl. AN. Tadzh SSR* 13 (No. 1): 15–18. (In Russian)

———. 1971. The distribution of polarization in the head and tail of Comet Bennett 1969i. *Dokl. AN. Tadzh SSR* 14 (No. 3): 17–20. (In Russian)

Roemer, E. 1970. Comet notes *Publ. Astron. Soc. Pac.* 82: 768–773.

DISCUSSION

WOLSTENCROFT: I also have unpublished observations of Comet Bennett which reveal that for a given wavelength, the degree of polarization changes along the tail.

COMET BENNETT 1970 II

L. R. DOOSE and D. L. COFFEEN
University of Arizona

Accurate photoelectric measurements were made of the linear polarization and color of scattered sunlight from the inner coma of Comet Bennett. Filters centered at 0.36 μm and 0.66 μm were specially designed to avoid emission bands in the comet's spectrum. Most observations were made with a 26-arcsec circular aperture. The amount of polarization was larger in the red than in the blue, reaching a maximum of 41% at 0.99 μm on 3 April 1970. Measurements taken with the aperture centered on the central condensation and 20 arcsecs both north and south of the central condensation showed the polarization to be strongest in the central region, nearly the same in the north region, but considerably less in the south region. The direction of vibration was always rigorously perpendicular to the scattering plane to within 1° for all 19 of our observations. The color index ($m_{0.36}-m_{0.66}$) showed the comet to be about 0.25 mag redder than the star λ Ser (representative of the sun). The red color is confirmed by other observers (e.g., Stokes 1972).

We have compared the observations with Mie calculations, which are strictly valid *only* for uncharged, homogeneous, isotropic spheres. Multiple scattering would cause dilution of the polarization of the central condensation with respect to surrounding regions. Since no dilution was observed, and since the surface brightness of the coma would have been much greater if the scattering optical thickness were not very small, only single scattering can be important.

Simple Rayleigh scattering is ruled out by the observed color. The trend of the observed variation of polarization with phase angle is similar to that of previously observed comets and resembles the curves for Fresnel reflection from surfaces. We were, therefore, encouraged to try Mie calculations for moderately large absorbing spheres.

To avoid the problem of the scatterers changing with time, we tried to fit the wavelength dependence of polarization at a single phase angle (90°). A narrow size distribution of spheres centered on 2.0 μm radius was assumed. The linear polarization at the observed phase angle was then calculated for all possible values of the refractive index, $m = n - in'$. At 0.36 μm wavelength a continuum of refractive indices was found which yielded exactly the measured polarization (25%), varying from $m = 0.8 - 1.75i$ at one extreme to $m = 5.0 - 2.0i$ at another. Similarly a continuum of complex refractive indices was found at 0.66 μm, where 34% polarization was observed. Any pair of refractive indices (one from each continuum) then gives a satisfactory fit to the polarization. Next the color was calculated for *all* such pairs of refractive indices and found to be the same for *all* combinations — equal to -0.26 mag. This is to be compared with the observed value of $+0.25$ mag. Thus we found a total contradiction. The comet has higher polarization in the red than in the blue and is brighter in the red (relative to the sun). But for these moderately large absorbing spheres, when the refractive index is made wavelength dependent in order to match the polarization, the scattering is brighter in the *blue* (analogous to the Umov effect typical of Fresnel reflection by absorbing materials).

As this analysis was limited by the choice of a specific mean particle size, we selected as a next step specific materials and, using their known refractive indices, varied the particle size and sought a fit between the Mie calculations and the observed polarizations and color. No match has yet been found. A number of possibilities are now excluded; e.g., there is no fit for iron spheres of any size.

Additional comparisons must be made, but it is doubtful that we will find agreement with any single type of spherical particles. Dielectric particles would not be "constrained" by the Umov effect, and it is conceivable that certain small dielectric spheres would work. Perhaps nonspherical particles are the answer; they tend to give positive polarization even for dielectrics.[1] Alternatively, a mixture of particles may be required, but we know that simple models are not sufficient.

REFERENCE

Stokes, G. M. 1972. The scattered-light continuum of Comet Bennett 1969*i*. *Astrophys. J.* 177: 829–834.

DISCUSSION

DUBIN: How is the gas contribution to the intensity of the coma, as a function of wavelength, subtracted from that by particle scattering?

---

[1] See p. 548 and p. 549.

COFFEEN: We have made no attempt to include a gas component in our rather simple models. Molecular scattering cannot be dominant, for it would make the comet brighter in the ultraviolet than in the red, contrary to the observations. Also the large polarizations characteristic of Rayleigh scattering would be enhanced in the ultraviolet compared to the red. We observe just the reverse for Comet Bennett.

GEHRELS: A. H. Delsemme and D. C. Miller (1971; see also Vanýsek 1972) believe that the particles of the coma are hydrate clathrates which are icy conglomerate particles. That is a convincing model as it is based on careful work in physics and chemistry. An attempt is therefore urgent to compare these particles to the photometry and polarimetry. It is not clear, however, how to make such a comparison, either in the laboratory or with some modification of the scattering theory.

During October and November, 1965, polarimetric observations of the sungrazing Comet Ikeya-Seki 1965VIII were obtained in several intermediate bandwidth filters both before and after perihelion. Because this comet was a morning twilight object, only one or two filters could be used each day.

As for Comet Bennett, the direction of electric vector maximum was always perpendicular to the scattering plane. On some occasions the phase angles of pre- and post-perihelion observations nearly coincided. In nearly all such cases the polarization was higher before perihelion. This lack of symmetry about perihelion is due to an actual physical change in the coma, because heliocentric and geocentric distances of the comet as well as aperture size used were very similar for the compared observations. Unfortunately, no filter used for this comet can be guaranteed to be free of emission-band effects. However, it is tempting to speculate that the close passage of the comet to the sun changed the nature of the scattering particles in the coma.

DISCUSSION REFERENCES

Delsemme, A. H., and Miller, D. C. 1971. Physico-chemical phenomena in comets. IV. The C_2 emission of Comet Burnham (1960 II). *Planet. Space Sci.* 19: 1259–1274.

Vanýsek, V. 1972. The structure and formation of comets. *From plasma to planet.* (A. Elvius, ed.) Stockholm: Almqvist and Wiksell.

POLARIMETRY OF LATE-TYPE STARS

STEPHEN J. SHAWL
University of Kansas

This paper presents a review of the research that has been carried out on late-type stars. Particular stellar types such as carbon stars and R CrB stars, T Tauri objects, RV Tauri stars, novae, and luminous oxygen-rich M-type variable stars are considered. The paper presents some previously unpublished work on particular stars such as o Ceti, X Her, V CVn, and g Her. The position angle variations for many stars (both time and wavelength variations) are often large.

The applications of theory to the interpretation of the observations are also reviewed. Some unpublished calculations of circumstellar scattering envelopes are presented. These results show the average particle size to be near 0.08 μm for silicate particles and slightly smaller for iron and graphite particles. It is shown that late-type stars may supply the assumed condensation nuclei for interstellar particle formation.

Many different types of stars have been found to be intrinsically polarized. Examples are the Be stars, RV Tauri and T Tauri stars, R CrB stars, novae, W-R stars, white dwarfs, and luminous red variable stars. In addition to the intrinsic linear polarization, all these types of objects (except the white dwarfs) have been found to show an excess of radiation at 10 μm when compared with similar unpolarized stars. The infrared excess in early-type stars is attributed to thermal free-free emission from ionized gas surrounding the star (Woolf, Stein, and Strittmatter 1970), while in late-type stars it is attributed to emission by silicate grains (Woolf and Ney 1969).

What is the best way to study the mechanism(s) producing polarization in these various types of objects? The dependence of polarization on wavelength could be expected to be important in determining the specific mechanism(s) producing the observed polarization. Since the polarization is known to change with time (indeed, it is these very time variations that are used to determine whether or not the polarization is intrinsic or interstellar), the specific form of the variation could be helpful in understanding what is happening with these objects.

Since these stellar types are mostly variable stars, any correlations between the polarization changes and changes of other observed parameters (such as light and spectral changes) will be of special importance.

The purpose of this paper is to review the observational and theoretical work on intrinsic polarization in late-type stars and to present some thoughts on additional work to be done [For other review articles on intrinsic polarization measurements, see Serkowski (1971b), and Zellner and Serkowski (1972)]. Since the greatest amount of work has been done on luminous red variable stars and on oxygen-rich (M-type) Mira variables, a large part of this paper is devoted to them; however, other types of stars will be considered first.

OBSERVATIONS OF VARIOUS STELLAR TYPES

Carbon Stars and R CrB Stars

Since carbon stars are faint, especially in ultraviolet light, not many of them have been studied polarimetrically. Kruszewski, Gehrels, and Serkowski (1968) found that the wavelength dependence of polarization for carbon stars does not seem to be as steep as it is for M-type stars; in particular, their observations for V CrB and R Lep show that the polarizations in the B and V spectral regions are equal. Observations of V CrB by Shawl (1972) and Dyck (unpublished) show that this is not always true. Thus, the wavelength dependence changes with time, indicating that if the polarization is produced by scattering from graphite, then the particle size varies from about 0.05 to 0.09 μm (Shawl 1972). Continued accurate observations are needed to detail the polarization-light correlations.

R CrB stars are hydrogen-deficient carbon stars (Warner 1967). The light variations consist of a long time of constant light and then a decrease of some 6 magnitudes to minimum; it then slowly returns to the constant maximum. For example, in March of 1972 (*Sky and Telescope* 1972) R CrB changed from 6th to 12th magnitude in one month; the magnitude had been constant since October 1969, which was the maximum following the June 1962 deep minimum. An explanation for the sudden light decrease was given by Loreta (1934) and O'Keefe (1939) as being due to a sudden condensation of graphite in the star's atmosphere. To quote Merrill (1960): "Their ideas should have received more attention." If they had, perhaps the idea of intrinsic stellar polarization would have been thought of and found earlier than it was.

Intrinsic polarization has been observed in the R CrB star RY Sgr (Serkowski and Kruszewski 1969) during the minimum of 1967–1968.

The polarization in the V filter varied from 0.3% to 1.3%, while the position angle varied from 10° to 150°. As is typical for late-type stars (Kruszewski, Gehrels, and Serkowski 1968), the polarization increases with decreasing wavelength. No other R CrB stars have been studied in detail, although polarization data from the spring 1972 minimum are presently being analyzed by Coyne and Shawl (1973).

T Tauri Objects

Since T Tauri stars are young objects still condensing out of interstellar gas and dust, one might expect the light to be polarized. The star T Tau shows chaotic polarization changes (Vardanian 1964; Serkowski 1969a, 1971a). Serkowski (1969a) found the position angle for AK Sco to change by more than 80° with time, while the wavelength variation between the blue and yellow spectral regions was less than 23°. The object R Mon, which consists of a star and a cometary nebula, was found by Hall (1965) and Zellner (1970a) to be polarized 10% independently of wavelength; this rules out Rayleigh scattering but may be compatible with scattering from large dust grains. More work is definitely needed.

RV Tauri Stars

Regular and striking changes in the polarization of the light from the RV Tauri stars U Mon and R Scu were reported by Serkowski (1970). For U Mon, which had previously been observed polarimetrically by Shakhovskoj (1964), the amount of polarization in the blue changes between nearly 0% and 1.5%; in addition, the position angle changes were found to be large and regular. In particular, the position angle changes from 0° to 105° between phases 0.1 and 0.22 and from about 90° to 0° between phases 0.62 and 0.73. This variation is shown in Fig. 1, taken from Serkowski (1970). U Mon has two light minima; the position angle is 0° near the deep minimum and 105° near the alternate shallow minimum. In addition, both position angle changes occur at the phase when the spectral lines are doubled. Since the amount of polarization changes considerably from cycle to cycle, while variations in the position angle repeat from cycle to cycle, the variability must be due to a change in the aspect with which the observer views the polarization-producing material.

Serkowski suggests that the observations may be explained by nonspherical oscillations of a star that is imbedded in a circumstellar cloud. [The large infrared excess observed for U Mon by Gehrz and Woolf (1970) is evidence of the presence of a cloud.] From von Zeipel's theorem, the surface brightness is greatest along the shortest axis of an ellipsoid of revolution; if we have nonspherical oscillations, the plane of greater surface brightness changes by some 90°. The illumination of the circumstellar material thus changes with time and explains the observations. A confirmation of these ideas was presented by Gehrz (1971) who combined visual and infrared observations for four RV Tauri stars and concluded that there must be nonradial pulsation occurring.

It is important to note that not all RV Tauri stars show the behavior exhibited by U Mon. Data presented by Serkowski (1970) for R Scu and by Peterson (1972) for T Cen show a different type of position-angle variation. Sufficient data are not available for any other RV Tauri stars.

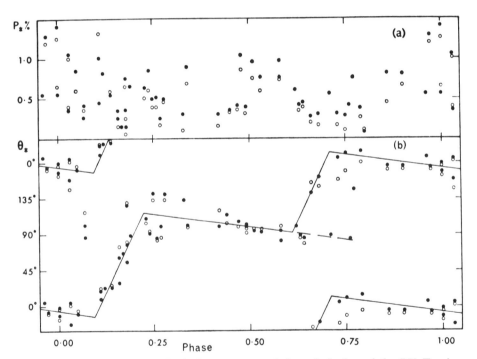

Fig. 1. Changes in the intrinsic component of the polarization of the RV Tauri star U Mon, observed by Serkowski (1970) in the yellow (open circles) and the blue (filled circles) spectral regions, with the phase of light variations. (a) Percentage polarization, (b) position angle. (Reprinted from *Astrophys. Journal*. Univ. of Chicago Press.)

Novae

Since novae are known to throw off matter in a relatively short time, one might expect that the radiation could be polarized and that the polarization would be time varying. A constant polarization was observed for Nova Herculis 1960 by Grigorian and Vardanian (1961), while Eggen, Mathewson, and Serkowski (1967) reported irregular variations in Nova T Pyxidis of several tenths of a percent in a time scale of hours and variations of 1.8% over a few weeks. Zellner and Morrison (1971) reported observations for Novae HR Delphini 1967 and Serpentis 1970. They found that the intrinsic polarization (i.e., the observed polarization corrected for an interstellar polarization) of HR Delphini was strongly peaked at 0.6% in the red during the first 120 days following eruption. During the later phases the intrinsic component was generally low and irregularly variable. They found variations of 1% for Nova Serpentis 1970, but no variations for Nova Vulpeculae No. 1 1968. In another paper, Zellner (1971) interpreted the observations in terms of scattering by solid particles surrounding the nova. This interpretation is reasonable in light of the strong infrared excess later observed by Geisel, Kleinmann, and Low (1970). Zellner interpreted the wavelength dependence as being caused by metallic (not silicate) particles with a radius near 0.1 μm in a 10^{-8} $M_\odot$ cloud of optical depth 0.25 in visible light. Geisel et al.'s infrared observations indicated a more massive cloud; Zellner explained this by suggesting that the 0.1 μm polarizing particles grew to become the larger infrared-excess producing particles. For future novae, coordinated polarization and infrared photometric observations should be made.

Luminous Red Variable Stars

Luminous red variable stars constitute the largest group of cool objects studied polarimetrically. Some of the more important papers include those by Capps and Dyck (1972), Coyne and Kruszewski (1968), Dombrovskij (1970), Dyck (1968), Dyck, Forbes, and Shawl (1971), Dyck et al. (1971), Dyck and Jennings (1971), Dyck and Sandford (1971), Kruszewski, Gehrels, and Serkowski (1968), Serkowski (1966a,b, 1969a,b,c, 1970, 1971 a,b), Shakhovskoj (1964), Shawl (1969, 1972), Shawl and Zellner (1970), Wolf (1970), Zappala (1967), and others. All the observations show that the polarization generally increases with decreasing wavelength, and both the polarization and the position angle vary with time, as do their wavelength dependencies. These variations are illustrated with data for X Her (Shawl 1972) in Fig. 2.

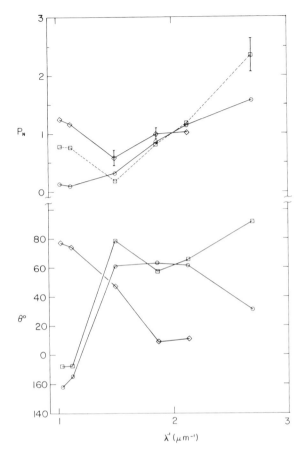

Fig. 2. X Her: Wavelength dependence of normalized polarization and position angle for selected times. O, JD 2440277; □, JD 2440339; ◊, JD 2440665. (From Shawl 1972.)

The position in the HR diagram of K and M giant and supergiant stars that are intrinsically polarized was studied by Dyck and Jennings (1971) and is shown in Fig. 3. They found that nearly all of the supergiants observed show intrinsic polarization, while no giant earlier than M2 was found to be polarized. In general, the hotter the star the greater the luminosity must be for intrinsic polarization to be present. They found a weak increase in the frequency of occurrence of intrinsic polarization with decreasing temperature; this was also noted by Serkowski (1971a).

The polarization for luminous red variables is usually found to increase with decreasing wavelength. Sometimes the increase is proportional to λ^{-4}, which is indicative of Rayleigh scattering. Often, however, a maximum is found in the ultraviolet. In addition, many cases exist of an increase toward 1 μm; in the case of VY CMa, L_2 Pup, and perhaps V CVn, a peak is present near 1.6 μm. An example is shown in Fig. 4 for Mira. Capps and Dyck (1972) searched for polarization in red variable stars at 10 μm and observed only the long-wavelength tail of the polarization curve. The interpretation of the wavelength-dependence observations will be considered later.

The existence of correlations between the polarization and other observed parameters in late-type stars should help to explain the nature of these objects. Zappala (1967) found no general correlation between the changes in the polarization and visual luminosity. Dyck (1968), on the other hand, reported observations for five Mira variables which all showed a decreasing polarization with increasing visual light. Kruszewski, Gehrels, and Serkowski (1968) found a maximum polarization at minimum light for V CVn. One can only conclude that no *general* correlation seems to exist between the polarization and the brightness changes.

The presence of intrinsic polarization is always associated with the emission lines in the stellar spectrum (Serkowski 1969a; see also

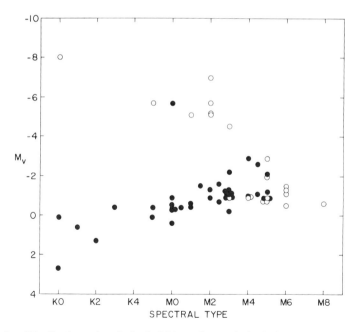

Fig. 3. Distribution of polarized (O) and unpolarized (●) stars in the HR diagram. (From Dyck and Jennings 1971.)

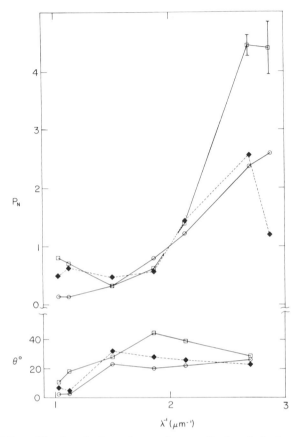

Fig. 4. Mira: Wavelength dependence of normalized polarization and position angle for selected times. ○, JD 2439115; □, JD 2440500; ◆, JD 2440523. (From Shawl 1972.)

Jennings and Dyck 1971, 1972). Dyck and Johnson (1969) found a correlation between the change in the polarization and the strength of the chromospheric H and K emission such that strong H and K emission implied small or no intrinsic polarization and vice versa. More recent work by Jennings and Dyck (1972) supports the earlier work but finds that the relation is more of a smoothed step function than a continuous function (Fig. 5). As to the hydrogen emission lines, Jennings and Dyck conclude that there is indeed a correspondence between intrinsic polarization and Balmer emission but that the two phenomena need not exist simultaneously.

There are some startling individual cases of strong correlations between the polarization, light, and spectrum changes. Mira has one of the most complete sets of polarization observations of any star. Shawl (1972) combined all the available data in the ultraviolet spectral region for six light cycles (including many unpublished data by Serkowski) into a plot of the polarization versus phase of visual light maximum; this plot is shown in Fig. 6. The polarization consistently decreases from phases 0.0 to 0.8, where it suddenly increases to a maximum at about phase 0.9; the decrease then sets in. This rapid increase in the polarization occurs at the same time as the eruption point on the light curve (Fischer 1968, 1969) defined as the beginning of the steep, nearly linear part of the ascending branch of the lightcurve. This is also the time of the minimum of the $U-B$ color (Serkowski 1971a,b) and the appearance of the hydrogen emission lines. Shawl (1972) argues that the observations are explained by there being a sudden production of small particles which then grow and dissipate; this is consistent with the observed changes of the wavelength dependence with phase. Detailed infrared observations are now needed.

A somewhat similar type of observation has been made for V CVn (Shawl 1972), where the polarization in the B filter has suddenly increased by more than a factor of 1.5 between phase 0.6 and 0.8. During one of these cycles, at phase 0.68, a spectrogram showed no emission (and

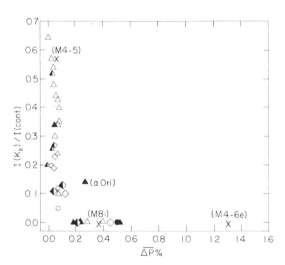

Fig. 5. Mean change in visual polarization versus K reversal intensity in units of continuum intensity. ○, K0-K5; △, M0-M4; ◊, M5-M9. Filled symbols are supergiants, half filled are bright giants, unfilled are giants. × indicates no luminosity classification. (From Jennings and Dyck 1972.)

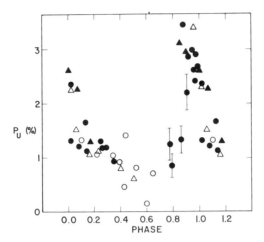

Fig. 6. Mira: Observations in the U filter versus phase of visual light maximum. ●, Serkowski (1966a,b, 1971a, unpublished); ▲, Kruszewski, Gehrels, and Serkowski (1968, unpublished); △, Shawl (1972); ○, Dyck et al. (1971, unpublished).

the polarization was low), while one month later at phase 0.84, another spectrogram showed very strong emission (and the polarization was high). Much better time resolution is needed to give a tight correlation.

An important correlation was found between the presence of intrinsic polarization and the presence of an excess of radiation at 11 μm by Dyck et al. (1971). They found that intrinsic polarization is always accompanied by an infrared excess, but an infrared excess is not necessarily accompanied by intrinsic polarization. Why were no stars reported with low polarization and high infrared excess? A large scatter might be expected because the geometrical aspects presented to an observer by a large sample of stars would be expected to differ from one star to another, thus producing high and low polarizations for a given infrared excess. Qualitatively, this implies that the general circumstellar cloud must have particles concentrated in a strongly nonuniform shell so that the cloud appears asymmetric when viewed from any angle. By assuming an optically thin cloud, they derived a linear relationship between the ratio of the observed flux at 11 μm to that at 3.5 μm and the average optical polarization. If the polarizations are plotted against $F(11\ \mu\mathrm{m})/F(3.5\ \mu\mathrm{m})$ (Fig. 7), the observed points do not deviate far from the theoretical linear relation (which, for this plot, assumed that all the stars have the same effective temperature). The correlation is thus clear.

The position angle is observed to vary with time; indeed, Zappala (1967) pointed out that the position angle variation is often better defined than the variation in the amount of polarization. For instance, Serkowski (1971a) noted that the position angle for Mira, R Hya and R Lep may show alternate high and low values of the position angle with each cycle in a manner not unlike that shown by U Mon.

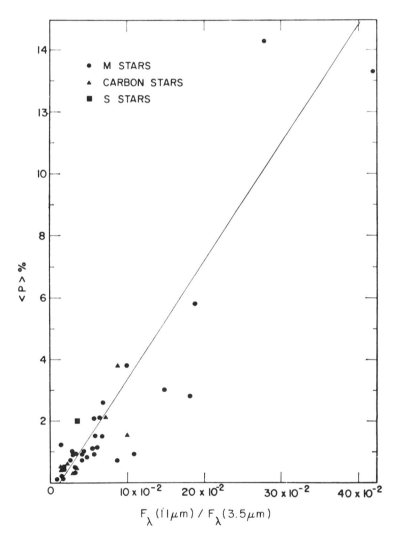

Fig. 7. Average polarization as a function of the flux ratio at 11 μm and 3.5 μm for comparison with the theoretical linear relation. (From Dyck et al. 1971.)

Shawl (1972) found that the position angle of Mira changes abruptly near minimum light (Fig. 8); this observation may be explained as the influence of the close companion VZ Cet. Thus, either VZ Cet is intrinsically polarized or it is inside Mira's circumstellar cloud. If the latter is true, we have a *lower limit* to the cloud radius of 50 A.U. Assuming a spherical cloud, the maximum density of scatterers is then 5×10^{-6} particles/cm^3 (Shawl 1972). The position angle of V CVn was found to vary cyclically (Kruszewski, Gehrels, and Serkowski 1968; Shawl 1972). Observations by Shawl (1972) showed the time variation of the position angle of X Her to be vastly different in the blue and red spectral regions (Fig. 2). A few other examples include g Her, VY CMa, L$_2$ Pup, and R Hya.

OBSERVATIONS OF SPECIFIC STARS

IRC+10216

Becklin et al. (1969) described IRC+10216 as an "unusual object." It is visually fainter than 18th magnitude but is very bright at 5 μm; it has an extended elliptical photographic image of 2×4 arcsecs on a 103a-E photographic plate (0.5–0.7 μm). Becklin et al. suggested that it is a Mira variable surrounded by an optically thick envelope; subsequently, Herbig and Zappala (1970) found it to be a carbon star

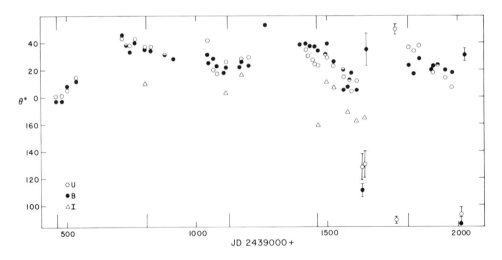

Fig. 8. Mira: Position angle versus time. The position angle data in the U, B, and I filters are from Serkowski (1966a, unpublished), Kruszewski (unpublished), Dyck and Sandford (1971), Dyck (unpublished) and Shawl (1972).

similar to the C6$_3$ star V CrB. Microwave emission by the CO molecule has been found (Solomon et al. 1971); neither H$_2$O nor OH have been observed.

Polarization observations were reported by Shawl and Zellner (1970) who found a polarization of 20.3% at 1 μm with a position angle of 123°, which is perpendicular to the long axis of the photographic image. They suggested that, if the polarization is due to only Rayleigh scatterers, the envelope will be transparent at 5 μm. The amount of polarization varies inversely with the light variation (Zellner and Serkowski 1972). Measurements have been extended into the infrared by Dyck, Forbes, and Shawl (1971) and Capps and Dyck (1972), who found that the polarization at 1 μm was variable and that the polarization decreases smoothly toward longer wavelengths so that it is absent at 10 μm. Zellner (Zellner and Serkowski 1972) observed with small diaphragms and found that most of the polarized intensity arises in the central 1 arcsec of the object. This difficult observation agrees exactly with the results of an occultation observed by Toombs et al. (1972); they observed at 2.2, 3.5, 4.8, and 10 μm and found that a composite cloud exists such that the excess 10 μm flux comes from an outer 2 arcsecs diameter shell, while the flux from the other wavelengths comes from an inner 0.4 arcsec diameter shell. The existence of two shells (composed of two different particle sizes) can be explained in two ways: firstly, the small particles in the smaller cloud acted as condensation nuclei to produce the larger particles, or secondly, both large and small particles were produced together and then separated by radiation pressure. Although we cannot determine which process is responsible, the data do show that at least one of these processes actually does occur.

VY CMa

VY CMa is one of the more unusual objects in our Galaxy. Spectroscopically, emission lines of Na, K, Ca, and probably Ba II have been identified, as well as emission of ScO, YO, and TiO (Wallerstein 1958; Herbig 1970); in the microwave region, OH (Eliasson and Bartlett 1969) and H$_2$O (Knowles et al. 1969) have been found. The polarization was discovered independently by Shawl (1969) and Serkowski (1969*b*); they found a polarization of 15%–19% in the blue spectral region with the usual decrease to the red and increase to the blue. A smooth rotation of position angle of about 60° was also found between the *U* and *I* spectral regions. Dyck, Forbes, and Shawl (1971) observed a polarization peak at 1.6 μm and a reversal in the sign of the polarization (change of the position angle by 90°) near 1 μm

(Fig. 9). Serkowski (1969c) found the reflection nebula about the star to be polarized 40%; Herbig (1972) subsequently found the nebula to be polarized up to 70%.

An explanation of the observations was offered by Dyck et al. (1971). Using the computer programs developed by Shawl (1972), they found that the wavelength dependence of polarization could be explained by a bimodal distribution of silicate scattering particles, the sizes being near 1.0 and 0.1 μm and the ratio of large to small particles of 8×10^{-4}. In addition, after removing an assumed interstellar polarization component, the models explained the change of position angle. It is important to note that calculations for iron and graphite will *not* reproduce the observations. Also, this model explains the wavelength dependence and not the actual amount of polarization observed. A different explanation has been offered by Herbig (1972); he suggested that the position angle rotation with wavelength can be explained by a wavelength variation of the relative brightness of the star and various parts of the nebula. This model is consistent with the color changes between the star and the nebula reported by Serkowski (1973). However, it would be difficult for Herbig's mechanism to reproduce the observed *smoothness* of the

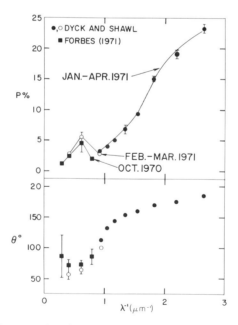

Fig. 9. VY CMa: Wavelength dependence of polarization and position angle. (From Dyck, Forbes, and Shawl 1971.)

position angle variation, since this requires the geometry to change smoothly with wavelength. Serkowski (1973) explained the 90° position angle rotation by requiring the polarization in the blue to be produced by light scattered around the polar axis of a disk-shaped nebula; the polarization cannot come from the disk itself because of the high ultraviolet optical depth. In the red, the polarization is produced by the disk itself, since at the poles the optical depth in the red is small. However, this model also does not explain the observed smoothness of the rotation. Additional observations are needed to sort out the true explanation.

A circular polarization component has been reported by Gehrels (1972); he observed 0.4 ±0.05% at 1 μm and 0.1% at 0.53 μm. Serkowski (1973) observed the circular component further in the infrared, finding 0.1% at 2.2 μm and 0.2% at 1.7 μm; thus the circular component has a maximum near 1 μm. Serkowski (1973) also reported a circular component to the polarization of NML Cyg of 0.6% at 1.7 μm. The maximum circular polarization occurs at the wavelength of minimum linear polarization, as expected from the Mie theory. The circular polarization will be a maximum if the light incident on a spherical dielectric is linearly polarized at 45° to the scattering plane (van de Hulst 1957; Martin 1972). This will occur mostly when multiple scattering is important and when the position angle of linear polarization varies strongly with wavelength, as is the case for VY CMa and NML Cyg. Observations of circular polarization are given by Gehrels (1972), Kemp, Wolstencroft, and Swedlund (1972), Serkowski (1973), and Wolf (1972).

INTERPRETATION OF OBSERVATIONS

The interpretation of the observations falls into four classes: synchrotron emission, scattering by gas in a stellar photosphere, scattering by gas in a circumstellar envelope, and scattering by solid particles in circumstellar clouds. The synchrotron mechanism as the cause of the polarization in late-type stars was first mentioned by Shakhovskoj (1964); since neither the expected radio emission nor a changed continuum energy distribution was known for these stars, he considered the mechanism unlikely. Kruszewski, Gehrels, and Serkowski (1968) argued that synchrotron emission was unlikely because if the spectral index of the emission were similar to that observed in objects known to have synchrotron emission, the wavelength dependence would be steeper than observed.

Scattering from neutral hydrogen in an asymmetric cloud was suggested by Shakhovskoj (1964) as an explanation for the observed polarization. However, models were not available until Kruszewski, Gehrels, and Serkowski (1968) made calculations to find the maximum polarization possible from a circumstellar shell; they found a maximum polarization of 5.5% for Rayleigh scattering from H_2, with lower values for dielectric and metallic particles. Small dielectric particles were found to give a polarization approaching the 5.5% maximum. However, they reported some stars with polarization greater than 5.5%. Therefore, Harrington (1969) proposed that a polarization higher than 5.5% could be obtained if the polarization were caused by scattering in a photosphere that has a source function that varies strongly with optical depth, has appreciable absorption, and has a temperature variation of several hundred degrees between the pole and the equator. Harrington pointed out that his model predicts a variation of the polarization across spectral absorption features. Although this prediction has not been truly tested, Dyck and Sandford (1971) and Dyck and Jennings (1971) ruled out Harrington's photospheric model on the grounds that the TiO bands in late-type stars should cause a change in the wavelength dependence of polarization as one considers stars of later spectral types; no such change was found by them. Serkowski (1971b), on the other hand, believes that the UBV spectral regions are too wide to detect the effects of TiO bands on the wavelength dependence even if the effects are large. However, since TiO has a large effect on UBV photometry (Smak 1964), it should likewise affect the polarization. Also, because of the infrared excess, solid circumstellar material is known to exist around these stars. Therefore, it is likely that the polarization is primarily circumstellar rather than photospheric in origin.

Is the polarization caused by scattering from gas or from solid particles? As evidence against molecular scattering, Dyck and Jennings (1971) pointed out that some stars (η Gem, π Aur, and α Her) show evidence of gas in a circumstellar cloud but no polarization or infrared excess, and that it is unlikely that particles that absorb (to produce the 11 μm excess) would not also scatter.

What kinds of solid particles might explain the observed polarization? The possibility that aligned graphite platelets might explain the observations of Serkowski (1966a,b) was suggested by Donn et al. (1966, 1968); they found that a magnetic field of 100 gauss acting on submicron graphite disks was sufficient to produce a few percent visual polarization. The difficulties of obtaining large amounts of graphite grains in an oxygen-rich star were pointed out by Kruszewski, Gehrels, and Serkowski (1968). If the oxygen-to-carbon ratio, O/C, is greater than unity, almost all of the carbon will be tied-up in CO or CO_2. However, if $O/C \simeq 1$, some of the oxygen could be tied-up with elements other than carbon, e.g. silicon, thus freeing carbon to condense into graphite (Wickramasinghe 1968). However, C_2 and perhaps CN should form before graphite platelets and deplete the carbon vapor. Therefore, it is not clear that graphite can form in oxygen-rich stars.

Kruszewski, Gehrels, and Serkowski (1968) pointed out that Kamijo (1963) had suggested that SiO_2 molecules could produce solid particles in the atmospheres of M-type variables. Although the optical depth of the particles would be low ($\simeq 10^{-5}$), they suggested that SiO_2 particles should be given further study. Observationally, the existence of silicate particles in circumstellar shells was suggested by Woolf and Ney (1969) on the basis of a comparison between the observed excess of radiation at 10 μm and the emissivity of silicate mineral grains. Silicate grains are known to be formed easily in oxygen-rich atmospheres (Gilman 1969). Low and Krishna Swamy (1970) found the infrared excesses at 10 and 20 μm observed in α Ori to be consistent with the silicate hypothesis.

Since the above interpretations are mostly qualitative, Shawl (1972) developed a computer code to calculate the polarization produced by scattering from grains in a circumstellar envelope. His method is a generalization of the one developed by Zellner (1971) for novae. Thus, the polarization can be calculated from an optically thin cloud of any specified geometry composed of spherical Mie particles illuminated by a point source; both particle size and spatial distributions may be specified. He found that the wavelength dependence of polarization is essentially independent of geometry (so long as there are an equal number of scatterers toward and away from the observer). The results for enstatite (a silicate) and graphite are shown in Figs. 10 and 11 for a delta-function size distribution. The features of the observed wavelength dependence are well reproduced. The models are able to fit the observations of 27 out of 30 stars. The results show that the particles are small, the average being 0.08 μm. Since the particles are small, one cannot distinguish between particle species. Cloud masses of $\simeq 10^{-10}$ $M_\odot$ are sufficient to explain many of the observations.

These calculations do a good job of reproducing the shape of the polarization curve with wavelength (Fig. 12) but are not sufficient for explaining high polarizations unless a special geometry is allowed. For example, the polarization of scattered light from a disk seen edge-on (that is, there is no contribution from direct stellar light) is 33% for a silicate. Since there are not very many stars with spectacular polarizations, perhaps a special geometry is not impossible; however, other

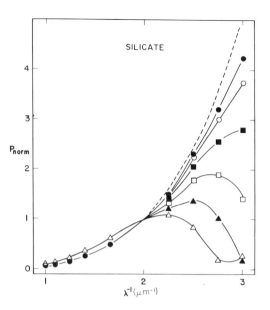

Fig. 10. Calculated normalized polarization for enstatite (a silicate). The optical depth at 0.4 μm is 0.2. Following the symbol is the particle radius in microns and the polarization at $\lambda^{-1}=2$. ●, 0.05, 1.08; ○, 0.06, 1.10; ■, 0.07, 1.15; □, 0.08, 1.19; ▲, 0.09, 1.17; △, 0.10, 1.02. (From Shawl 1972.)

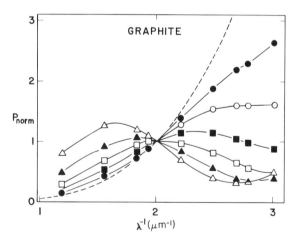

Fig. 11. Calculated normalized polarization for graphite. The optical depth at 0.641 μm is 0.25. Following the symbol is the particle radius in microns and the polarization at $\lambda^{-1}=2.0$. ●, 0.05, 1.43; ○, 0.06, 1.62; ■, 0.07, 1.73; □, 0.08, 1.58; ▲, 0.09, 1.30; △, 0.10, 0.98. (From Shawl 1972.)

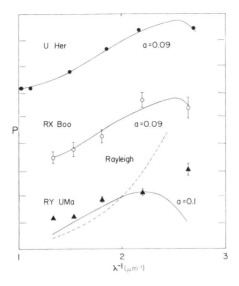

Fig. 12. Enstatite models fitted to the observations of U Her, RX Boo, and RY UMa. Each curve is normalized to unity at 2 μm^{-1}; they have been shifted for clarity. Each tick mark corresponds to 0.5% in the normalized polarization. The polarization at 2 μm^{-1} is 3.25%, 0.96%, and 1.30% for U Her, RX Boo, and RY UMa, respectively.

methods of obtaining high polarizations need to be considered quantitatively. In addition, calculations are needed for high optical depths.

Capps and Dyck (1972) found that the lack of polarization at 10 μm does not admit either significant polarized emission from aligned grains or significant polarization from scattering by high-albedo grains.

It is important to note that magnetic fields are not required to explain the observations. The presence and strength of magnetic fields in late-type stars are generally unknown quantities. Babcock (1958, 1960) has observed magnetic fields in the M stars HD 4174 (1200 gauss) and WY Gem (500 gauss); probable but not firmly established fields were found for R Leo and α Ori. Shawl (1972) has observed HD 4174 and WY Gem polarimetrically; he found a polarization of 0.44 $\pm$ 0.08% for HD 4174 in the visual and 1.36% and 1.38% in the blue and visual for WY Gem. The results for WY Gem indicate an interstellar origin, since the galactic latitude is $+2°$ and $P_B \cong P_V$. A few percent polarization is often observed for R Leo; α Ori shows up to 1% polarization. These stars, however, can have their polarizations explained by simple scattering models. Thus, a mechanism requiring a magnetic field to produce polarization is improbable, although not impossible.

A spin-off of the calculations of the late-type-star circumstellar scattering envelopes presented by Shawl (1972) is the possibility of explaining the origin of condensation nuclei for the interstellar medium. The interstellar medium has a dust component of $10^{-4}M_{gal}$ or 2×10^{40} gms. If the average interstellar particle has a density of 1 gm/cm^3 and a radius of 0.1 μm, there are then 5×10^{54} dust particles in the galaxy. Assuming each particle required a single condensation nucleus, that the column density of the particles for each late-type star is 2×10^9 cm^{-2} (from Shawl's calculations), a cloud radius of 50 A.U. (similar to Mira's), and 2×10^6 Mira variables projected onto the galactic plane (Gehrz and Woolf 1971), then 5×10^{55} particles are produced in 5×10^9 years by Mira variables alone. Thus, late-type stars may indeed supply the interstellar medium with condensation nuclei.

FUTURE WORK

The observations presented in this paper for Mira indicate what is needed observationally for the entire problem of intrinsic polarization. For Mira, data covering six years were necessary to clearly reveal certain events. In addition, many types of observing techniques have been combined. Simultaneous observations, using all the techniques known, should be applied to selected objects. Only then can the problems be unraveled. The causes for both time and wavelength variations in the position angles are not clear; this explanation is of fundamental importance. More detailed observations of the circular component to the polarization of VY CMa and NML Cyg are needed.

Theoretical problems are many. What is needed is a unified model explaining the formation of particles, polarization, and infrared excess; their spatial and temporal relations must be explained more fully. How can the observed large polarizations be explained without the necessity of a special geometry? As to circular polarization, why is it so rare in late-type stars? The search for the answers to these questions will provide years of engaging research.

REFERENCES

Babcock, H. W. 1958. A catalog of magnetic stars. *Astrophys. J. Supp.* 3:141–210.
———. 1960. Stellar magnetic fields. *Stellar atmospheres* (J. L. Greenstein, ed.), pp. 282–320. Chicago: Univ. of Chicago Press.
Becklin, E. E.; Frogel, J. A.; Hyland, A. R.; Kristian, J., and Neugebauer, G. 1969. The unusual object IRC+10216. *Astrophys. J.* 158:L133–L137.
Capps, R. W., and Dyck, H. M. 1972. The measurement of polarized 10-micron radiation from cool stars with circumstellar shells. *Astrophys. J.* 175:693–697.

Coyne, G. V., and Kruszewski, A. 1968. Wavelength dependence of polarization. XI. Mu Cephei. *Astron. J.* 73:20–25.
Coyne, G. V., and Shawl, S. J. 1973. Polarimetry of R Coronae Borealis at visual light minimum. *Astrophys. J.* In press.
Dombrovskij, V. A. 1970. Polarization of light of red variable stars of high luminosity. *Astrophys. J.* 6:207–224.
Donn, B.; Stecher, T. P.; Wickramasinghe, N. C.; and Williams, D. A. 1966. Comments on the intrinsic polarization of late-type stars. *Astrophys. J.* 45:949–951.
———. 1968. Erratum for "Comments on the intrinsic polarization of late-type stars." *Astrophys. J.* 153:L143.
Dyck, H. M. 1968. Photometric polarimetry of late-type stars. *Astron. J.* 73:688–696.
Dyck, H. M.; Forbes, F. F.; and Shawl, S. J. 1971. Polarimetry of red and infrared stars at 1 to 4 microns. *Astron. J.* 76:901–915.
Dyck, H. M.; Gillett, F. C.; Forrest, W.; Stein, W. A.; Gehrz, R. D.; Woolf, N. J.; and Shawl, S. J. 1971. Polarization and infrared excess of cool stars. *Astrophys. J.* 165:57–66.
Dyck, H. M., and Jennings, M. C. 1971. Intrinsic polarization in K and M giants and supergiants. *Astron. J.* 76:431–444.
Dyck, H. M., and Johnson, H. R. 1969. Chromospheres and polarization in late-type stars. *Astrophys. J.* 156:389–392.
Dyck, H. M., and Sandford, M. T. 1971. Multicolor polarimetry of some Mira variables. *Astron. J.* 76:43–49.
Eggen, O. J.; Mathewson, D. S.; and Serkowski, K. 1967. Changes in polarization of light and colour during the outburst stage of the recurrent Nova T Pyxidis. *Nature* 213:1216–1217.
Eliasson, B., and Bartlett, J. R. 1969. Discovery of an intense OH emission source. *Astrophys. J.* 155:L79–L81.
Fischer, P. L. 1968. The irregularities in the light-changes of Mira Ceti. Non-periodic phenomena in variable stars, I.A.U. Colloquium, Budapest. (L. Detre, ed.) pp. 331–338. London: Academic Press.
———. 1969. Der Lichtwechsel von Mira Ceti. *Annalen der Universitäts-Sternwarte Wien* 28:139–209.
Gehrels, T. 1972. On the circular polarization of HD 226868, NGC 1068, NGC 4151, 3C273, and VY Canis Majoris. *Astrophys. J.* 173:L23–L25.
Gehrz, R. D. 1971. Infrared radiation from RV Tauri stars. *Bull. Am. Astron. Soc.* 3:454.
Gehrz, R. D., and Woolf, N. J. 1970. RV Tauri stars: a new class of infrared objects. *Astrophys. J.* 161:L213–L217.
———. 1971. Mass loss from M stars. *Astrophys. J.* 165:285–294.
Geisel, S. L.; Kleinman, D. E.; and Low, F. J. 1970. Infrared emission of novae. *Astrophys. J.* 161:L101–L104.
Gilman, R. C. 1969. On the composition of circumstellar grains. *Astrophys. J.* 155:L185–L187.
Grigorian, K. A., and Vardanian, R. A. 1961. Photometric, colorimetric, and polarimetric study of Nova Herculis 1960. *Soob. Byurakan Obs.* 29:39–48. In Russian.
Hall, R. C. 1965. Polarization and color measures of NGC 2261. *Publ. Astron. Soc. Pac.* 77:158–163.

Harrington, J. P. 1969. The intrinsic polarization of Mira variables. *Astrophys. Letters* 3:165–168.

Herbig, G. H. 1970. Introductory remarks. *Mém. Soc. Roy. Sci. Liège* 19:13–26.

———. 1972. VY Canis Majoris. III. Polarization and structure of the nebulosity. *Astrophys. J.* 172:375–381.

Herbig, G. H., and Zappala, R. R. 1970. Near infrared spectra of NML Cygni and IRC+10216. *Astrophys. J.* 162:L15–L18.

Hulst, C. H. van de 1957. *Light scattering by small particles.* New York: Wiley.

Jennings, M. C., and Dyck, H. M. 1971. Polarization and emission lines in late-type giants and supergiants. *Kitt Peak Contrib. No. 554,* pp. 203–212.

———. 1972. The consequences of grains in the atmospheres of late-type stars. I. Intrinsic polarization, infrared excess, and emission lines. In press.

Kamijo, F. 1963. A theoretical study on the long period variable stars. III. Formation of solid or liquid particles in the circumstellar envelope. *Publ. Astron. Soc. Japan* 15:440–448.

Kemp, J. C.; Wolstencroft, R. D.; and Swedlund, J. B. 1972. Circular polarimetry of fifteen interesting objects. *Astrophys. J.* 177:177–189.

Knowles, S. J.; Mayer, C. H.; Cheung, A. C.; Rank, D. M.; and Townes, C. H. 1969. Spectra, variability, size, and polarization of H_2O microwave emission sources in the galaxy. *Science* 163:1055–1057.

Kruszewski, A.; Gehrels, T.; and Serkowski, K. 1968. Wavelength dependence of polarization. XII. Red variables. *Astron. J.* 73:677–787.

Loreta, E. 1934. Nota sulle stella variabili R Coronidi. *Astron. Nachr.* 254:151.

Low, F. J., and Krishna Swamy, K. S. 1970. Narrow-band infrared photometry of α Ori. *Nature* 227:1333–1334.

Martin, P. G. 1972. Interstellar circular polarization. *Mon. Not. R. Astr. Soc.* 159:179–201.

Merrill, P. W. 1960. Spectra of long-period variables. Stellar atmospheres (J. L. Greenstein, ed.), pp. 509–529. Chicago: Univ. of Chicago Press.

O'Keefe, J. A. 1939. Remarks on Loreta's hypothesis concerning R Coronae Borealis. *Astrophys. J.* 90:294–300.

Peterson, A. V. 1972. The variable polarization of T Cen. *Proc. Astr. Soc. Australia* 2:108–110.

Serkowski, K. 1966a. Intrinsic polarization of the Mira variables. *Astrophys. J.* 144:857–859.

———. 1966b. Long period variable stars with large intrinsic polarization. *I.A.U. Info. Bull. on Variable Stars No. 141.*

———. 1969a. Changes in polarization of T Tauri. *Astrophys. J.* 156:L55–L57.

———. 1969b. Large optical polarization of the OH emission source VY Canis Majoris. *Astrophys. J.* 156:L139–L140.

———. 1969c. Polarization of reflection nebulae associated with VY Canis Majoris and R Coronae Austrinae. *Astrophys. J.* 158:L107–L110.

———. 1970. Polarimetric observations of the RV Tauri stars U Monocerotis and R Scuti. *Astrophys. J.* 160:1107–1116.

———. 1971a. Polarimetric observations of red variable stars. *Kitt Peak Contrib. No. 554,* pp. 107–126.

———. 1971b. Polarization of variable stars. *Proc. I.A.U. Colloquium No. 15, Rameis Sternw. Bamberg* 9:11–31.

———. 1973. Infrared circular polarization of NML Cygnus and VY Canis Majoris. *Astrophys. J.* 179:L101–L106.

Serkowski, K., and Kruszewski, A. 1969. Change in polarization of the R Coronae Borealis star RY Sagittarii. *Astrophys. J.* 155:L15–L16.

Shakhovskoj, N. M. 1964. Investigation of the polarization of radiation of variable stars. I. *Sov. Astr. A. J.* 7:806–812.

Shawl, S. J. 1969. Strong optical polarization observed in VY Canis Majoris. *Astrophys. J.* 157:L57–L58.

———. 1972. Observations and models of polarization in late-type stars. Ph.D. dissertation, Univ. of Texas; *Astron. J.* In preparation.

Shawl. S. J., and Zellner, B. H. 1970. Polarization of IRC+10216. *Astrophys. J.* 162:L19–L20.

Smak, J. 1964. Photometry and spectrophotometry of long-period variables. *Astrophys. J. Supp.* 9:141–184.

Solomon, P.; Jefferts, K. B.; Penzias, A. A.; and Wilson, R. W. 1971. Observation of CO emission at 2.6 millimeters from IRC+10216. *Astrophys. J.* 163:L53–L56.

Toombs, R. I.; Becklin, E. E.; Frogel, J. A.; Law, S. K.; Porter, F. C.; and Westphal, J. A. 1972. Infrared diameter of IRC+10216 determined from lunar occultations. *Astrophys. J.* 173:L71–L74.

Vardanian, R. A. 1964. Polarization of T and RY Tauri. *Soob. Byurakan Obs.* 35:3–23.

Wallerstein, G. 1958. The spectrum of the irregular variable VY Canis Majoris. *Publ. Astron. Soc. Pac.* 70:479–484.

Warner, B. 1967. The hydrogen deficient carbon stars. *Mon. Not. R. Astr. Soc.* 137:119–139.

Wickramasinghe, N. C. 1968. On the formation of graphite particles in the atmospheres of Mira variables. *Mon. Not. R. Astr. Soc.* 140:273–280.

Wolf, G. W. 1972. A search for elliptical polarization in starlight. *Astron. J.* 71:576–583.

Woolf, N. J., and Ney, E. P. 1969. Circumstellar infrared emission from cool stars. *Astrophys. J.* 155:L181–L184.

Woolf, N. J.; Stein. W. A.; and Strittmatter, P. A. 1970. Infrared emission from Be stars. *Astron. Astrophys.* 9:252–258.

Zappala, R. R. 1967. Fluctuating polarization in long-period variable stars. *Astrophys. J.* 148:L81–L85.

Zellner, B. H. 1970a. Wavelength dependence of polarization. XXI. R Monocerotis. *Astron. J.* 75:182–185.

———. 1970b. Polarization in reflection nebulae. Ph.D. Dissertation, Univ. of Arizona; *Astron. J.* In preparation.

———. 1971. Wavelength dependence of polarization. XXIII. Dust grains in novae. *Astron. J.* 76:651–654.

Zellner, B., and Morrison, N. D. 1971. Wavelength dependence of polarization. XXII. Observations of novae. *Astron. J.* 76:645–650.

Zellner, B. H., and Serkowski, K. 1972. Polarization of light by circumstellar material. *Pub. Astron. Soc. Pac.* 84:619–626.

DISCUSSION

SERKOWSKI: The observed wavelength dependence of degree of linear polarization for red stars can be legitimately compared with theoretical models only at those times when the observed position angle is independent of wavelength. Usually the position angle depends strongly on wavelength which means that the observed wavelength dependence of the degree of polarization results from superposition of polarization produced in various regions of the circumstellar envelope. In each of such regions the grain size distribution and position angle of the plane of scattering would be different.

COLLINS: In calculating the production of grains from your model, you have completely ignored the destruction rate. Therefore, the result is completely fortuitous.

SHAWL: The calculation of the grain production depends strongly on the assumed cloud radius and can therefore be adjusted by changing the radius. If cloud-cloud collisions are the main cause of grain destruction, the average lifetime of a grain is $\approx 10^8$ years or about $1/100$ the lifetime of the galaxy (Greenberg 1968). The stars still produce nearly the required number in 10^8 years.

DISCUSSION REFERENCE

Greenberg, J. M. 1968. Interstellar grains. *Nebulae and interstellar matter.* (B. M. Middlehurst and L. H. Aller, eds.), pp. 201–402. Chicago: Univ. of Chicago Press.

CIRCUMSTELLAR POLARIZATIONS OF EARLY-TYPE STARS AND ECLIPSING BINARY SYSTEMS

ANDRZEJ KRUSZEWSKI
Polish Academy of Sciences, Warsaw

The linear polarization produced by circumstellar envelopes of early-type stars and eclipsing binaries is discussed. The observational data are surveyed and the theoretical interpretations are reviewed using mostly β Lyr as an example. New observations for W Ser are presented. Possible implications of intrinsic polarization are discussed.

The objects reviewed in this paper have in common a tenuous cloud of detached matter in the form of nonspherical shell, ring, or disk. In addition, the temperature in that cloud is sufficiently high so that the electron scattering could be an important contribution to the opacity.

Many early attempts to detect intrinsic polarization, mostly among eclipsing binaries, were unsuccessful because of the instrumental limitations. We may consider γ Cas as the first early-type star with detected intrinsic polarization after Behr (1959) reported the variability of its linear polarization. However, this finding escaped the attention of astronomers, and it took Shakhovskoj's discovery of the variability of linear polarization in β Lyr to open a new field of investigation (Shakhovskoj 1962a, 1964). The existence of intrinsic polarization in β Lyr has been confirmed by Appenzeller (1965) and Serkowski (1965). Further work (Ruciński 1966; Appenzeller and Hiltner 1967; Shulov 1967a; Coyne 1970a,b) provided ample observational data, so that β Lyr is by far the object best investigated polarimetrically among the eclipsing binaries. Many other eclipsing variables have been found to have intrinsic linear polarization: RY Per (Shakhovskoj 1964; Shulov and Gudkova 1969); V444 Cyg (Shakhovskoj 1964; Hiltner and Mook 1966; Shulov 1966); RY Sct, RZ Sct (Shakhovskoj 1964); U Sge (Shulov 1966, 1967b); Z Vul, GG Cas (Shulov 1967b); DQ Her (Dibay and Shakhovskoj 1966); RW Tri (Efimov 1967); TT Hya, V453 Sco (Serkowski 1970); U Oph (Coyne 1970c); and W Ser (Kruszewski 1972).

It was relatively more difficult to find intrinsic polarization in early-type stars that are not eclipsing binaries because of the lack of periodicity which would have helped in discriminating the variations of intrinsic polarization from the observational and instrumental errors. Thus, the first reports of polarization variability in γ Cas (Behr 1959), χ Oph (Shakhovskoj 1962b), and χ^2 Ori (Vitriczenko and Efimov 1965) met with little interest. Coyne and Gehrels (1967) have found several new early-type objects with variable polarization. A new tool in identifying early-type stars with intrinsic polarization was introduced by Serkowski (1968), who found that the emission-line stars have abnormally low polarization in the ultraviolet. This feature had already been observed in β Lyr by Appenzeller and Hiltner (1967) who suggested that it is caused by self-absorption in circumstellar hydrogen gas. Coyne and Kruszewski (1969) have shown that the wavelength dependence of polarization in Be-type stars as observed in the region 0.3–1.0 μm, which covers the Balmer and Paschen series limits, can be explained by electron scattering attenuated by self-absorption in a hydrogen shell. The values of electron density and electron temperature consistent with the polarimetry coincide with commonly accepted values based on spectroscopy. Further work by Serkowski (1970) and by Strittmatter and Serkowski (1972) made it clear that the intrinsic polarization in early-type stars with emission lines is a rule rather than an exception.

We should say here a few words about the properties of eclipsing binaries and of early-type stars with emission lines and the problem of their intrinsic polarizations. Both observational (Underhill 1960) and theoretical (Marlborough 1969) arguments lead to a picture of a Be-type star in which circumstellar matter is concentrated in the equatorial plane. The densest region is situated, in the equatorial plane, close to the star's surface. More diluted parts of an envelope extend several stellar radii from the star, and the whole structure gets progressively more extensive in the z-dimension away from the star. The electron density is estimated as 10^{11}–10^{13} cm^{-3}, and this combined with a dimension of 10^{12}–10^{13} cm results in the conclusion that the envelope can be optically thick due to electron scattering. The electron temperature in the envelope is determined by the radiation of the central star, and in case of Be-type stars it exceeds 10^4 °K.

If we imagine a Be-type star with a flat equatorial envelope, and another larger star, which is a binary component, moving in an orbit whose plane coincides with the equatorial plane of our Be star, then we get a good picture of an eclipsing binary that is typical for those exhibiting intrinsic polarization. Of course, the size and importance of a flat disk around the smaller binary component varies from object to object. The disk may be small with little matter in it and be detectable only when the smaller brighter binary component is totally eclipsed; or, on the other hand, it may be like that in β Lyr where it is a dominant feature of the system. The fact that there is an anisotropic geometry together with sufficiently large optical thickness for electron scattering ascertains the presence of intrinsic polarization in these objects.

TIME VARIABILITY

The periodicity of the variations of the polarization in eclipsing binaries is very useful because an accurate polarization curve can be constructed by reducing all observations to one cycle. Up to now, this advantage was fully exploited only in the case of β Lyr.

Figure 1 shows the observational data obtained for this object with the V filter of the UBV system. The uppermost curve depicts the variability of brightness according to Wood and Walker (1960). Individual observations are not shown because they fall closely along the curve. The degrees of polarization and the position angles are plotted utilizing observations of Appenzeller and Hiltner (1967), Serkowski (1965), and Ruciński (1966). At the bottom the polarized flux in intensity units on an arbitrary scale is presented. No correction for the interstellar contribution to the polarization has been applied. While this correction is relatively well known from the measurements of the optical components of β Lyr, it is still insufficient to avoid spurious variations in position angles. On the other hand, the general features in the degree of polarization and the polarized flux curves are not changed when the influence of the interstellar polarization is taken into account. The similar curves constructed with the help of observations obtained with other filters show larger scatter but are generally similar to those obtained with the V filter.

The intrinsic linear polarization in β Lyr was explained by scattering of stellar light on electrons in an optically thin disk situated around the fainter, smaller component (Shakhovskoj 1962a, 1964). This model was modified in some details by other astronomers (Ruciński 1966; Appenzeller and Hiltner 1967; Shulov 1967a; Coyne 1970b). We shall discuss how the polarized flux curve should behave in such models. If the disk is optically thin and only the light of the secondary component is scattered in it, the polarized flux should stay constant but for a sharp drop during the secondary minimum when the disk is eclipsed by the primary component. This does not match the observations because the polarized flux

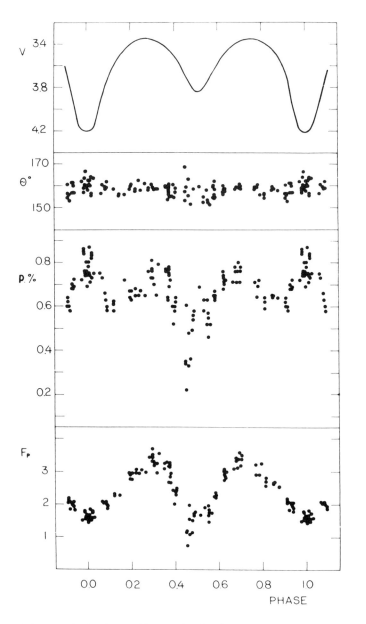

Fig. 1. The magnitude, the position angle of polarization θ, the degree of polarization $p\%$, and the polarized flux Fp, all with a yellow V filter as functions of the phase for eclipsing binary β Lyr. The polarized flux was calculated on the basis of the above lightcurve and polarimetric observations and is expressed in arbitrary units.

changes during the whole cycle and the polarized flux eclipse is wider than expected. The radiation from the primary scattered in the disk must also contribute to the observed polarization. In case of an optically thin disk this contribution has maxima at phases 0.25 and 0.75 and is zero at phases 0.0 and 0.5. The observed maxima fall at 0.3 and 0.7 or even closer to the secondary minimum. This indicates that the light from the primary is scattered in an optically thick medium with a strong phase effect. This scattering medium may be the secondary star or an optically thick disk about it. The latter possibility is more likely because in that case the parts of the disk closest to the primary are the most effective scatterers, and this explains the strong phase effect and the unusually wide eclipse in the polarized flux curve. The disk, besides being optically thick, must extend in the equatorial plane almost to the limits of the Roche lobe. And exactly such a disk is called for in the Huang-Woolf model for β Lyr (Huang 1963; Woolf 1965). The light scattering on the secondary star surface is unlikely to be an important factor because this does not explain the width of the polarized flux eclipse; the phase effect is smaller and in addition there is no indication that we directly observe the secondary star at all. Thus the light of the primary scattered in nearby parts of an optically thick disk seems to be the dominant factor. The scattered light of the secondary star may still play some role. Even when the secondary is small and completely hidden by the disk, the regions above and below the equatorial plane may be optically thin, and light from the secondary star scattered there might be highly polarized.

None of the remaining eclipsing binaries is as well observed as β Lyr. However, a few have polarization curves that show some features besides the mere fact of the variability. RY Per is a typical Algol system with a relatively inconspicuous ring structure around the smaller and brighter component. The polarized flux curve (Shulov and Gudkova 1969) is consistent with the polarization produced by scattering in the ring by the light of the brighter component lying inside the ring. Another eclipsing binary, V444 Cyg which has a Wolf-Rayet component, shows polarization minima at times of both light minima and polarization maxima at quadratures (Hiltner and Mook 1966). This would be consistent with the scattering agents situated between the components.

Figure 2 gives the light, polarization, and polarized flux curves obtained with a green filter for W Ser (Kruszewski 1972). This binary system was known to be very peculiar (Sahade and Struve 1957) and the polarization properties seem to be peculiar too. From the polarized flux curve one can conclude that the scattering matter is concentrated outside the axis joining the centers of both components.

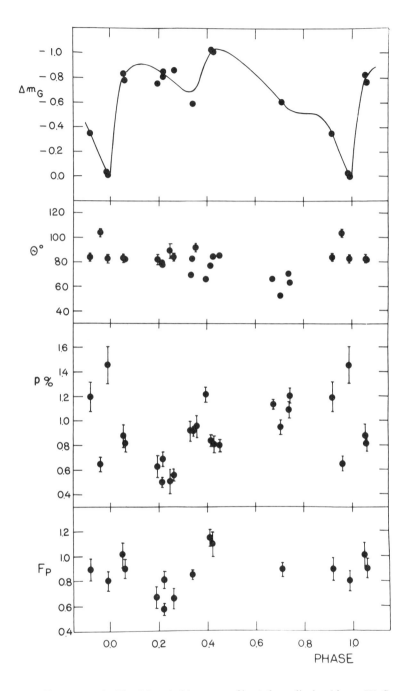

Fig. 2. The same as in Fig. 1 but (with a green filter) for eclipsing binary W Ser.

We see that the polarization behaves differently in different eclipsing variables. We may expect that when observing other binaries we will encounter still other types of polarization variability.

Be-type stars that are not close binaries show only irregular variations in polarization. Only a few of them have been sufficiently well observed to follow the time variability continuously (Strittmatter and Serkowski 1972). For many others the variability is deduced from the scatter in observations which is much larger than the observational errors. This variability is evidently caused by changes with time in the amount of material in the envelope, as indicated most clearly by correlation between intrinsic polarization and visual magnitude observed for an A-type shell star HR 6000 (Bessell and Eggen 1972).

WAVELENGTH DEPENDENCE

Figure 3 gives the observed wavelength dependence of linear polarization for a sample of early-type stars that have emission lines. For a majority of these objects, there is clearly a decrease in the polarization in the ultraviolet spectral region beyond the Balmer jump. A similar, though smaller, decrease is also visible beyond the Paschen series limit. These features are evidently caused by the absorption in the hydrogen gas. Such absorption may occur on the way from the star before and after scattering. The observed wavelength dependence imposes some limitations on the physical conditions in the envelope. If the electron scattering is the dominant source of opacity, then the intrinsic polarization should not depend strongly on the wavelength. If the hydrogen absorption is dominating, then the intrinsic polarization will be very small. We adopt the following values for the physical electron parameters: $T_e = 10^4$ °K, $N_e = 10^{12}$ cm^{-3}, which seem to be reasonable estimates.

Figure 4 gives the opacity for hydrogen bound-free and free-free absorption and for electron scattering as functions of inverse wavelength together with the average normalized polarization for several Be-type stars. We can see that for these physical parameters both sources of opacity play comparable roles and that the polarization varies approximately inversely with the hydrogen absorption. Coyne (1972) has pointed out that the unpolarized radiation emitted by the envelope modifies the wavelength dependence of the polarization. Capps, Coyne, and Dyck (1972) have constructed a model for the scattering envelope in ζ Tau, taking into account the unpolarized envelope emission. The unpolarized radiation of the hydrogen gas helps to explain the decrease of the polarization at wavelengths longer than 1 μm observed by them in ζ Tau.

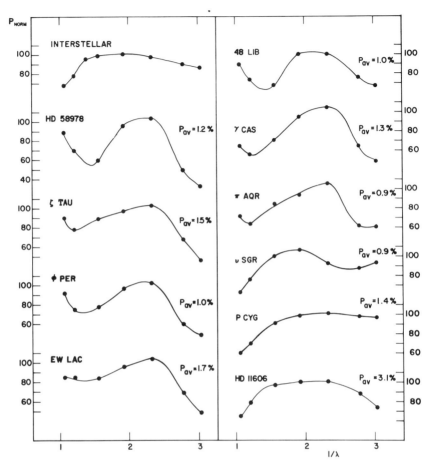

Fig. 3. The wavelength dependence of polarization for a sample of early-type stars with emission lines. The characteristic interstellar curve is shown for comparison.

Figure 5 gives the wavelength dependence of the degree of polarization for two eclipsing binaries, β Lyr and W Ser. In the case of β Lyr this dependence is very similar to that in Be-type stars, as should be expected. However, W Ser is peculiar also in that respect. Its wavelength dependence is similar to that of red variables, such as V CVn, rather than to that of early-type Be stars.

CONCLUSIONS

The intrinsic polarization among Be-type stars and eclipsing binaries is well documented observationally, and its origin seems to be satisfactorily explained by electron scattering of starlight in shells, disks, or

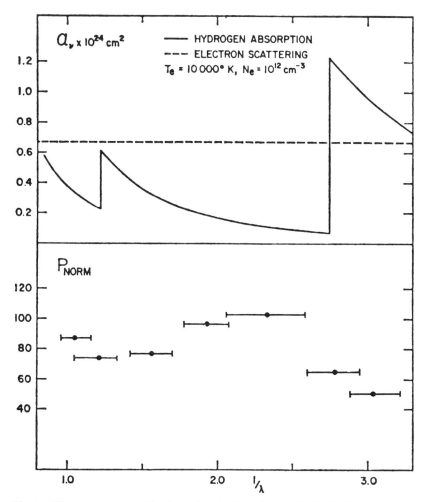

Fig. 4. The average normalized wavelength dependence of polarization for several Be-type stars plotted together with the wavelength dependence of opacities due to electron scattering and hydrogen gas absorption. The horizontal bars denote the half-power widths of the spectral bands used.

rings composed of partly ionized gas. In fact, with our present knowledge of these objects, we are in a position to predict the existence of intrinsic polarization with properties not very different from the observed ones. We might now ask what new information about these objects we can get from the observed polarization.

A most straightforward application consists in the possibility of determining the orientation of the orbital plane of close binaries and the equatorial plane of shell stars. Such a possibility existed previously

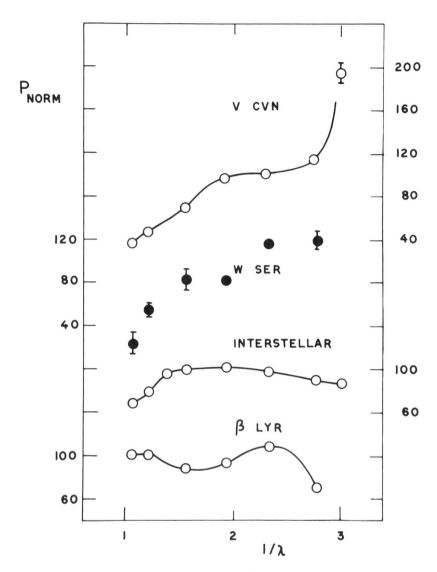

Fig. 5. The wavelength dependence of polarization for β Lyr (Coyne 1970b) and W Ser (Kruszewski 1972). The interstellar curve and the wavelength dependence for the red variable V CVn are shown for comparison.

only for visual double stars. The first attempt to use polarimetry in this manner was made by Shakhovskoj (1969) who, considering data for eclipsing binaries, arrived at the conclusion of random orientation of the orbital planes.

Another interesting possibility is to obtain information about the distribution of detached matter in close binary systems. This information would be complementary to information that can be obtained with spectroscopy. In principle, it is possible from the observed wavelength dependence of the polarization to reach some conclusions about the physical conditions in the scattering envelopes; again, this information is supplementary to spectroscopic results.

There is one problem in which the polarimetric measurements could be very helpful: when a close binary star is not affected by the interstellar polarization, it is possible to determine polarimetrically the inclination of the orbit. The addition of interstellar polarization makes the problem more difficult. At least in one case, that of Cygnus X-1, the information about the orbital inclination is badly needed. However, the intrinsic polarization even if present at all is dominated by a much larger interstellar contribution (Gehrels 1972); hence this opportunity most likely cannot be exploited in this case.

The characteristic shape of the wavelength dependence of polarization associated with the emission-line stars can serve as a means of detecting stars with electron scattering disks (Breger and Dyck 1972).

Finally, there is the possibility of using polarimetry for estimating the mass of gas in the circumstellar envelope. The analysis, however, is based on an assumption that the envelope is optically thin, and therefore, the resulting masses are only lower limits.

REFERENCES

Appenzeller, I. 1965. The polarization of Beta Lyrae. *Astrophys. J.* 141:1390–1392.
Appenzeller, I., and Hiltner, W. A. 1967. True polarization curves for Beta Lyrae, *Astrophys. J.* 149:353–362.
Behr, A. 1959. Die Interstellare Polarization des Sternlichts in Sonnenumgebung, *Nachr. Akad. Wiss. Göttingen II Math-Phys. Kl.* No. 7, 185–219. *Veröffentlichungen Göttingen* No. 126.
Bessell, M. S., and Eggen, O. J. 1972. Rapid changes in the new shell star HR 6000. *Astrophys. J.* 177:209–217.
Breger, M., and Dyck, H. M. 1972. Pre-main-sequence stars. II. Stellar polarization in NGC 2264 and the nature of circumstellar shells. *Astrophys. J.* 175:127–134.
Capps, R. W.; Coyne, G. V.; and Dyck, H. M. 1972. A model for the observed polarization flux for ζ Tauri. Submitted to *Astrophys. J.*
Coyne, G. V. 1970a. Multi-color polarimetry of Beta Lyrae. *Ric. Astr. Spec. Vat.* 8:85–102.
———. 1970b. Mass exchange for Beta Lyrae. *Astrophys. J.* 161:1011–1014.
———. 1970c. Variable polarization in U Ophiuchi. *Ric. Astr. Spec. Vat.* 8:105–116.
———. 1972. Balmer continuous emission and polarization in Be stars. *Scientific results from the orbiting astronomical obs.* NASA SP-310, 495–504.
Coyne, G. V., and Gehrels, T. 1967. Wavelength dependence of polarization. X. Interstellar polarization. *Astron. J.* 72:887–898.

Coyne, G. V., and Kruszewski, A. 1969. Wavelength dependence of polarization. XVII. Be-type stars. *Astron. J.* 74:528–532.

Dibay, E. A., and Shakhovskoj, N. M. 1966. Polarization observations of DQ Herculis (N Her 1934). *Astron. Zhurn.* 43:1319 (Transl. *Soviet Astr.* 10: 1059–1060.

Efimov, J. S. 1967. Polarization observations of variable stars. *Izv. Crimean Astrophys. Obs.* 37:251–261.

Gehrels, T. 1972. On the circular polarization of HD226868, NGC 1068, NGC 4151, 3C273 and VY Canis Majoris. *Astrophys. J.* 173:L23–L25.

Hiltner, W. A., Mook, D. E. 1966. Variable polarization in V444 Cyg. *Astrophys. J.* 143:1008–1009.

Huang, S. S. 1963. An interpretation of Beta Lyrae. *Astrophys. J.* 138:342–349.

Kruszewski, A. 1972. Intrinsic polarization of the eclipsing binary W Serpentis. *Acta Astronomica* 22:405–410.

Marlborough, J. M. 1969. Models for the envelopes of Be stars. *Astrophys. J.* 156:135–155.

Ruciński, S. M. 1966. Changes in polarization of Beta Lyrae. *Acta Astronomica.* 16:127–136.

Sahade, J., and Struve, O. 1957. The spectrum of W Serpentis. *Astrophys. J.* 126:87–98.

Serkowski, K. 1965. Changes in polarization of Beta Lyrae. *Astrophys. J.* 142: 793–795.

———. 1968. Correlation between the regional variations in wavelength dependence of interstellar extinction and polarization. *Astrophys. J.* 154:115–134.

———. 1970. Intrinsic polarization of early-type stars with extended atmospheres. *Astrophys. J.* 160:1083–1105.

Shakhovskoj, N. M. 1962a. Polarization observations of Beta Lyrae. *Astron. Zhurn.* 39:755–758. (Transl. *Soviet Astron.* 6:587–589.)

———. 1962b. Polarimetric observations of Chi Ophiuchi. *Astr. Circular* (USSR), No. 228, pp. 16–17.

———. 1964. Polarization in variable stars. Part II. Eclipsing variables. *Astron. Zhurn.* 41:1042–1055. (Transl. *Soviet Astron.* 8:833–842.)

———. 1969. The spatial orientation of the orbit planes of close binary stars. *Astron. Zhurn.* 46:386–388. (Transl. *Soviet Astron.* 13:303–305.)

Shulov, O. S. 1966. Variability of polarization in eclipsing binaries V444 Cyg and U Sge. *Astr. Circular* (USSR), No. 385, pp. 5–7.

———. 1967a. On the interpretation of polarization variability of Beta Lyrae. *Astrofizika* 3:233–244.

———. 1967b. Polarimetric observations of close binary systems. *Trudy Leningrad Univ. Observ.* 24:38–53.

Shulov, O. S., and Gudkova, G. A. 1969. On the intrinsic polarization of RY Persei. *Astrofizika* 5:477–485.

Strittmatter, P., and Serkowski, K. 1972. Intrinsic polarization of early-type stars. In preparation.

Underhill, A. B. 1960. Early type stars with extended atmospheres. *Stellar atmospheres.* (J. L. Greenstein, ed.), pp. 411–435. Chicago: Univ. of Chicago Press.

Vitriczenko, E. A., and Efimov, Y. S. 1965. Electropolarimetric observations of Chi^2 Ori. *Izv. Crimean Astrophys. Obs.* 34:114–117.

Wood, D. B., and Walker, M. F. 1960. Photoelectric observations of Beta Lyrae. *Astrophys. J.* 131:363–384.

Woolf, N. J. 1965. The problem of Beta Lyrae. II. The masses and the shapes. *Astrophys. J.* 141:155–160.

DISCUSSION

GEHRELS: I once observed YY Gem and the negative results may be of interest. On 24 January 1971 YY Gem was observed during about 1½ hours centered at the time of primary maximum, and no intrinsic polarization effects were found to a confidence level of ±0.05% probable error. The day before, at the time of YY Gem being out of eclipse entirely, the linear polarization was found to be 0.38 ± 0.15% (p.e.) which again is considered a negative result.

ASTRONOMICAL POLARIMETRY FROM 1 TO 10 µm

H. M. DYCK
Kitt Peak National Observatory

Polarimetric observations in the infrared are briefly reviewed. The value of making such observations is discussed for some specific examples.

With the recent advances in photometric techniques in the infrared, it is natural to expect astronomical interest to turn toward the more difficult and higher precision measurement of the state of polarization of the infrared radiation. The acquisition of such data represents a logical and fundamental extension of our knowledge of the character of polarized light measured in the visible part of the spectrum. In addition, there is the anticipation of unexpected discovery that exists when any new field is explored.

As far as instrumentation is concerned, astronomical polarimetry in the infrared can be neatly segregated into groups of measurements at wavelengths either side of $\lambda = 3.5$ µm. For observations in the range $1 \, \mu m \leq \lambda \leq 3.5 \, \mu m$, the techniques are straightforward; stable detectors of high quantum efficiency are available as are good polarizing materials. Also, background radiation problems are not so severe at these wavelengths. Polarization measurements with precisions of $\pm 0.1\%$ or better now exist in the literature. To date, only relatively simple, single-channel devices have been built, but it is possible to construct multichannel devices because of the ready availability of natural birefringent materials. Descriptions of operating polarimeters using sheet polarizers can be found in Dyck, Forbes, and Shawl (1971), and one using an MgF_2 Wollaston prism in Serkowski (1973). At wavelengths longer than 3.5 µm, the problems become more severe because of the increased thermal background from the telescope and photometer. Furthermore, the simplest polarizers that are available commercially (wire grids) are poor optical elements even though their polarizance is large. They are extremely fragile, exhibit cold flow, and are mostly wedge-shaped, producing severe beam wander at the focal plane. Nonetheless, it seems probable that measurements having precisions of a few tenths of a percent will soon be possible for this spectral region.

The purpose of this short review, however, is not to describe the problems[1] that one may encounter with the techniques, but to concentrate upon what astronomical problems have been investigated. One can summarize the results at the outset: Because one can more easily measure those stars that are intrinsically very red or are heavily reddened, most of the observational effort has been spent in that direction. Only recently has any attempt been made to measure, accurately, the polarization of hot, relatively unreddened stars.

THE OBSERVATIONS
Interstellar Polarization

No published data yet exist at wavelengths longer than 1 μm for stars showing typical interstellar polarization. I have listed in Table I some of the near-infrared data obtained at Kitt Peak during the past two years for VI Cygni No. 12 and for HD 183143; also listed are some previously unpublished data taken in the visible with polarimetric equipment described by Dyck and Sandford (1971). These data are plotted in Fig. 1 along with other published data for the stars. One sees immediately that both stars show a smoothly decreasing amount of polarization toward the infrared. This result is not unexpected since the short-wavelength data indicate particle sizes on the order of a few tenths of a micron, for which the polarizing efficiency decreases toward longer wavelengths. Thus, one should observe those stars that show evidence

TABLE I

Linear Polarization Data for HD 183143 and VI Cygni No. 12

$1/\lambda$ (μm)$^{-1}$	HD 183143 $P \pm \epsilon$ (percent)	θ (degree)	VI Cygni No. 12 $P \pm \epsilon$ (percent)	θ (degree)
2.65	5.02±.06	179	—	—
2.20	5.81±.05	179	9.2±.6	120
1.82	6.08±.07	180	9.1±.1	118
1.56	5.73±.09	180	8.7±.2	118
1.33	5.17±.11	179	7.0±.2	118
0.95	—	—	4.8±.1	118
0.62	1.2 ±.3	13	2.5±.2	115
0.45	0.6 ±.2	171	1.1±.2	122

[1] See p. 318.

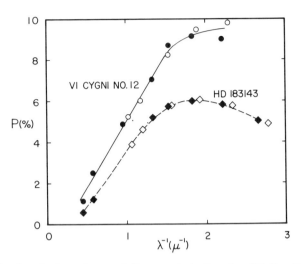

Fig. 1. Wavelength dependence of linear polarization for VI Cyg No. 12 and HD 183143. The open circles are data from Kruszewski (1971) and the open diamonds are from Coyne and Wickramasinghe (1969).

of larger particle sizes,[2] such as HD 147889 (Serkowski, Gehrels, and Wiśniewski 1969) and some of the other stars in the ρ Oph dark cloud (Carrasco, Strom, and Strom 1973). On the other hand, it seems improbable that near-infrared observations of interstellar polarization will reveal any striking departures from the characteristics inferred from visual measurements.

The Be Stars

It has recently been demonstrated that many emission-line Be stars show intrinsic polarization (Serkowski 1968; Coyne and Kruszewski 1969; Coyne 1971) that is exhibited as a peculiar wavelength dependence of the amount of polarization and as time-variations.[3] For stars that are relatively uncontaminated by interstellar polarization, the polarization-wavelength dependence shows a maximum in the blue and a rapid drop toward the ultraviolet. In general, a second minimum is seen in the region of the Paschen limit with a slight rise toward longer wavelengths. Coyne and Kruszewski (1969) proposed that the intrinsic polarization mechanism is electron scattering in a circumstellar disk modified as a function of wavelength by the continuous emission of atomic hydrogen (free-bound emission) to levels $n = 2$ and $n = 3$. The effect ought to be observable into the infrared because the levels become more closely

---

[2] See p. 888.   [3] See p. 845.

spaced with wavelength. Furthermore, there is evidence that most Be stars show infrared excesses arising from free-free emission in the circumstellar disk (Woolf, Stein, and Strittmatter 1970) which should be unpolarized and also affect the observed wavelength dependence of polarization, especially in the infrared. For one particular star, ζ Tau, Capps, Coyne, and Dyck (1973) have obtained observations from 0.33 to 2.2 μm and find that the data can be fitted by a simple model consisting of electron scattering in a circumstellar disk with the wavelength dependence modified by hydrogen free-bound and free-free emission (presumed to be unpolarized) arising within the cloud. The observed data and one of the models are shown in Fig. 2; also shown, with the solid line, is the mean relation for high-latitude Be stars from Coyne (1971). It is interesting to note that the observed rapid drop of the amount of polarization between about 0.8 and 2.2 μm cannot be repro-

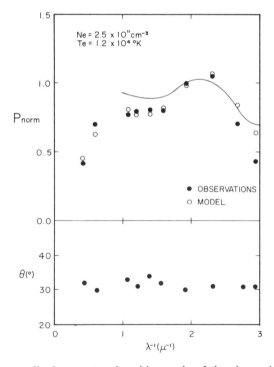

Fig. 2. The normalized amount and position angle of the observed linear polarization (filled circles) and one model (open circles) for ζ Tau. The model corresponds to a circumstellar disk of ionized hydrogen with a radius of 3 stellar radii and a thickness of 2 stellar radii, $N_e = 2.5 \times 10^{11}$ cm^{-3}, and $T_e = 1.2 \times 10^4$ °K. The solid line is the mean relation for high-latitude Be stars from Coyne (1971).

duced by free-bound emission alone but *requires* the addition of free-free emission as well. A useful future investigation would be the observation of additional Be stars to see what differences exist in the polarization in the near-infrared because, as one can see from Fig. 2, many Be stars appear to be remarkably different from ζ Tau in the 0.8 to 1 μm region.

The Cool Stars

The first linear-polarization observations of cool stars in the infrared were made by Forbes (1967) who reported large linear polarization of NML Cyg in the 1 to 4 μm region. This M6 giant (Herbig and Zappala 1970) shows strong position angle rotation as a function of wavelength and a rapid rise in polarization toward shorter wavelengths (Kruszewski 1971; Dyck, Forbes, and Shawl 1971). More recently, Serkowski (1973) has observed circular polarization that peaks between 1 and 2 μm and decreases on either side.

One of the most interesting stars, polarimetrically, is VY CMa (Shawl 1969; Serkowski 1969) which shows a large amount of linear polarization in the blue and ultraviolet and a strong wavelength dependence of both the percentage and position angle. Hashimoto et al. (1970) first observed it at 1.6 and 2.2 μm and noted that their data did not agree with the smoothly extrapolated visible data that had been published. Dombrovskij (1970) also observed VY CMa in the infrared, but only at 2.2 μm, and consequently failed to notice the discrepancy. Forbes (1971*b*) then showed that the polarization-wavelength dependence for this star decreases smoothly from short wavelengths to about 1.2 μm, where it reaches a minimum, and then rises to a maximum at about 1.6 μm and declines further toward longer wavelengths. The details of the variation of the position angle with wavelength, which is also very peculiar in the 1 μm region, has been discussed by Dyck, Forbes, and Shawl (1971). Linear polarization observations have been extended to wavelengths as long as 10 μm by Capps and Dyck (1972); these observations show a continued decline from the 1.6 μm peak in the amount of polarization and continued rotation of the position angle. Circular polarization has also been detected for this star, measured near 1 μm by Gehrels (1972) and extended to 2.6 μm by Serkowski (1973). I have shown the available data for VY CMa in Fig. 3, plotted in the form of normalized Stokes parameters as a function of wavelength. The linear polarization data are from Dyck, Forbes, and Shawl (1971) and Capps and Dyck (1972). The important features of this diagram have been interpreted as the result of scattering from dielectric particles in the circumstellar shell. Dyck, Forbes, and Shawl (1971) argued that the point where Q/I crosses U/I is displaced from 0.0 by interstellar

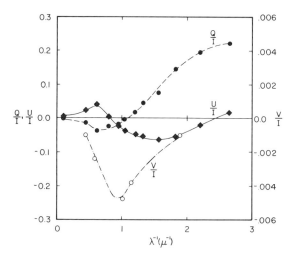

Fig. 3. Observed wavelength dependence of linear and circular polarization parameters for VY CMa.

polarization and that the crossing represents a true sign reversal of the polarization characteristic for dielectrics. Serkowski (1973) argued that the circular polarization arises from multiple scattering on dielectric particles and ought to peak where the linear polarization is a minimum. Indeed, Martin's (1972)[4] calculations for metals and dielectrics show no such sharply peaked circular polarization curves for metals but only for dielectrics.

Linear polarization data for about 60 other stars in the 1 to 4 μm region have been discussed by Dyck, Forbes, and Shawl (1971) and for eight other stars through the 8–14 μm atmospheric window by Capps and Dyck (1972).

Other Objects

There has been a smattering of observations of other kinds of objects longward of 1 μm. Forbes (1971a) has observed Venus to wavelengths as long as 3.5 μm and has found that the linear polarization rises toward longer wavelengths.[5] Low et al. (1969) have reported an attempt to measure the polarization of the galactic center at 10 μm, which yielded a null result. Kemp and Swedlund (1970) observed the circular polarization of the white dwarf Grw+70°8247 with a broad passband covering 0.9–1.6 μm.

---

[4] See p. 926.   [5] See p. 637.

FUTURE OBSERVATIONS

Two important areas of theoretical work have indicated that polarimetric observations in the infrared may be extremely valuable. One is the prediction that emission or absorption by lattice vibration bands in aligned particles will be strongly linearly polarized (Martin 1971; Gilra 1971, 1972). It has been argued that lattice bands of various silicates are primarily responsible for the observed 8–14 μm flux excesses observed in oxygen-rich stars (Woolf and Ney 1969; Gammon, Gaustad, and Treffers 1972) and that SiC produces the observed excesses in carbon stars (Gilra 1971). Thus, if there is any preferential alignment of the particles in the circumstellar shell, one ought to see strongly polarized emission. Preliminary attempts to observe the phenomenon at very low spectral (broadband) resolution (Capps and Dyck 1972) have produced no positive results, but higher resolution must be used to be certain the phenomenon is not 'washed out" in the broadband observations. If the effect can be observed, it may offer a useful constraint for the identification of the emitting materials. Similarly, those objects that show the same feature in absorption ought to be observed with high spectral resolution.

The other interesting prediction is that dielectric particles of the sizes indicated by circumstellar and interstellar (visual) polarization observations ought to produce maxima of circular polarization in the infrared (Martin 1972). The stars VY CMa and NML Cyg may demonstrate the relatively frequent occurrence of the phenomenon.

CONCLUSION

It would seem that the field of infrared polarimetry, now in an embryonic stage, will provide useful astronomical data during the next few years. As an example of what to expect, one may recall that observations of ζ Tau basically confirmed the suggestions of Coyne and Kruszewski (1969) regarding the wavelength dependence in Be stars but provided new information about what dominates the wavelength dependence in the near-infrared. The near-infrared observations may also be taken as independent confirming evidence for the existence of free-free emission from Be stars as suggested by Woolf, Stein, and Strittmatter (1970). The peculiar nature of the polarization of VY CMa in the infrared, on the other hand, must be regarded as a complete surprise. There is already a hint (Shawl 1972) that a few other late-type stars show somewhat the same sort of behavior, and the possibility that a whole class of such objects exists is an exciting one indeed. One may hope, by comparing these kinds of stars with the more normal ones, to find out something about the evolution of circumstellar envelopes. Herbig's (1970) suggestion that VY CMa is very young ($\sim 10^6$ years) prompts one to inquire whether the observed phenomenon in that star is somehow a consequence of the early stages of star formation. Finally, the prospects for measuring circular polarization in this spectral region are good, and the observations ought to yield much useful information.

REFERENCES

Capps, R. W.; Coyne, G. V.; and Dyck, H. M. 1973. A model for the observed polarized flux for ζ Tauri. To be submitted to *Astrophys. J.*

Capps, R. W., and Dyck, H. M. 1972. The measurement of polarized 10-micron radiation from cool stars with circumstellar shells, *Astrophys. J.* 175:693–697.

Carrasco, L.; Strom, S. E.; and Strom, K. M. 1973. Interstellar dust in the ρ Ophiuchus dark cloud. *Astrophys. J.* In press.

Coyne, G. V. 1971. Polarization of Be stars at high galactic latitudes. *Ric. Astr.* 8:201–210.

Coyne, G. V., and Kruszewski, A. 1969. Wavelength dependence of polarization. XVII. Be-type stars. *Astron. J.* 74:528–532.

Coyne, G. V., and Wickramasinghe, N. C. 1969. Wavelength dependence of polarization. XVIII. Interstellar polarization and composite interstellar particles. *Astron. J.* 74:1179–1190.

Dombrovskij, V. A. 1970 Polarization of the emission of red variable stars with high luminosity. *Astrophysika* 6:207.

Dyck, H. M.; Forbes, F. F.; and Shawl, S. J. 1971. Polarimetry of red and infrared stars at 1 to 4 microns. *Astron. J.* 76:901–915.

Dyck, H. M., and Sandford, M. T. 1971. Multicolor polarimetry of some Mira variables. *Astron. J.* 76:43–49.

Forbes, F. F. 1967. The infrared polarization of the infrared star in Cygnus. *Astron. J.* 147:1226–1229.

———. 1971*a*. Infrared polarization of Venus. *Astrophys. J.* 165:L21–L25.

———. 1971*b*. The infrared polarization of Eta Carinae and VY Canis Majoris. *Astrophys. J.* 165:L83–L86.

Gammon, R. H.; Gaustad, J. E.; and Treffers, R. R. 1972. Ten-micron spectroscopy of circumstellar shells. *Astrophys. J.* 175:687–691.

Gehrels, T. 1972. On the circular polarization of HD 226868, NGC 1068, NGC 4151, 3C273, and VY Canis Majoris. *Astrophys. J.* 173:L23–L-25.

Gilra, D. P. 1971. The violet opacity in S, C-S and N stars and circumstellar silicon carbide grains. *Bull. Am. Astron. Soc.* 3:379.

———. 1972. *Collective excitations and dust particles in space. The scientific results from the Orbiting Astronomical Observatory (OAO-2).* (A. D. Code, ed.), pp. 295–319. NASA SP-310, U.S. Government Printing Office, Washington, D.C.

Hashimoto, J.; Maihara, T.; Okuda, H.; and Sato, S. 1970. Infrared polarization of the peculiar M-type variable VY Canis Majoris. *Publ. Astr. Soc. Japan* 22: 335–340.

Herbig, G. H. 1970. Introductory remarks. *Mem. [8°] Soc. Roy. Sci. Liege* 19:13–26.

Herbig, G. H., and Zappala, R. R. 1970. Near-infrared spectra of NML Cygni and IRC+10216. *Astrophys. J.* 162:L15–L18.

Kemp, J. C., and Swedlund, J. B. 1970. Large infrared polarization of Grw+70° 8247. *Astrophys. J.* 162:L67–L68.

Kruszewski, A. 1971. Wavelength dependence of polarization. XXIV. Infrared objects. *Astron. J.* 76:576–580.

Low, F. J.; Kleinmann, D. E.; Forbes, F. F.; and Aumann, H. H. 1969. The infrared spectrum, diameter and polarization of the galactic nucleus. *Astrophys. J.* 157: L97–L101

Martin, P. G. 1971. On the infrared spectra of interstellar grains. *Astrophys. Letters* 7:193–198.

———. 1972. Interstellar circular polarization. *Mon. Not. R. Astr. Soc.* 159: 179–190.

Serkowski, K. 1968. Correlation between the regional variations in wavelength dependence of interstellar extinction and polarization. *Astrophys. J.* 154:115–134.

———. 1969. Large optical polarization of the OH emission source VY Canis Majoris. *Astrophys. J.* 156:L139–L140.

———. 1973. Infrared circular polarization of NML Cygnus and VY Canis Majoris. *Astrophys. J.* 179:L101–L106.

Serkowski, K.; Gehrels, T.; and Wiśniewski, W. 1969. Wavelength dependence of polarization. XIII. Interstellar extinction and polarization correlations. *Astron. J.* 74:85–90.

Shawl, S. J. 1969. Strong optical polarization observed in VY Canis Majoris. *Astrophys. J.* 157:L57–L58.

———. 1972. Ph.D. dissertation, Univ. of Texas, Austin, Tex.

Woolf, N. J., and Ney, E. P. 1969. Circumstellar infrared emission from cool stars. *Astrophys. J.* 155:L181–L184.

Woolf, N. J.; Stein, W. A.; and Strittmatter, P. A. 1970. Infrared emission from Be stars. *Astron. Astrophys.* 9:252–258.

POLARIZATION STUDIES OF REFLECTION NEBULAE

BEN ZELLNER
University of Arizona

Photoelectric studies of color and polarization in reflection nebulae are briefly reviewed. In most cases the observations can be explained by scattering from dielectric dust grains similar to those in the general interstellar medium. More realistic models that take into account the effects of high optical depth are needed.

Angular scattering by dust grains in reflection nebulae produces some of the strongest polarizations in astronomy. The amount and wavelength dependence of this polarization, together with color and brightness data, provide the only available means of sampling the optical properties of interstellar grains in a relatively small volume of space.

Results from the principal observational programs to date are reviewed in the first section. In the second section, progress on the microscopic (particle scattering) and macroscopic (radiative transfer) problems necessary to understand colors and polarizations in reflection nebulae is discussed. The deficiencies of existing theoretical techniques and some possibilities for future improvement are pointed out in the final section.

OBSERVATIONS

Reflection nebulae generally have surface brightnesses in the magnitude range 20 to 22 per square arcsecond, comparable to the brightness of the night sky. Thus, early photographic studies of color and polarization were notoriously subject to systematic errors. The first photoelectric data became available with photometry and polarimetry in several nebulae by Martel (1958) and Johnson (1960), and polarimetry in NGC 7023 by Gehrels (1960). Vanýsek and Svatoš (1964) reported colorimetry in NGC 7023 from photometric scans through the nebula. O'Dell (1965) measured brightnesses and colors in the Merope nebula in four narrow-band filters between 0.34 and 0.56 μm. Rather complete areal coverage of polarization and color in three filters was provided

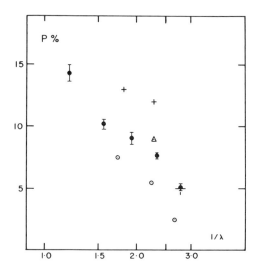

Fig. 1. Polarization versus wave number (waves per micrometer) in a region of NGC 2068. Crosses are from Johnson (1960), the triangle from Glushkov (1965), open circles from Elvius and Hall (1966), and filled circles with error bars from Zellner (1970).

by R. C. Hall (1965) for NGC 2261. In these studies it was established that reflection nebulae are usually bluer than their illuminating stars by up to a magnitude in $B-V$, that the polarization tends to increase with offset distance from the star, and that with rare exceptions, the polarization increases monotonically toward longer wavelengths.

Additional measures in the Pleiades nebulosities have been provided by Roark (1966), Dahn (1967), Artamonov and Efimov (1968), and O'Dell (1969). Zellner (1970) obtained color and polarization data in a few selected regions of NGC 2068, IC 5076, and NGC 7023, with emphasis upon high photometric precision over a wide spectral range (0.33 to 0.83 μm). Racine (1971) reported UBV photometry in 34 regions of the nebula Cederblad[1] 201. For sheer volume of homogeneous data, no study has surpassed the three-color photometry and polarimetry of numerous regions in the Merope nebula, NGC 7023, and NGC 2068 carried out at Lowell Observatory by A. Elvius and J. S. Hall (1966, 1967).

The observational problem of reflection nebulae is essentially one of collecting a sufficient number of photons with proper correction for the

---

[1] The principal catalogs of reflection nebulae are those of Cederblad (1946), Dorschner and Gürtler (1964), van den Bergh (1966), and Rozhkovski and Kurchakov (1968).

sky noise. In my work, for example, approximately 50 hours of integration time at a 1.5-meter telescope were necessary to obtain complete color and polarization data in a single nebular region. Work on reflection nebula should never be begun until a high-resolution, low-density photograph is available to aid the choice of homogeneous and star-free regions. Dark-sky fields may be located from high-density prints, such as those of the Palomar Observatory Sky Survey atlas; if possible, they should be chosen in the same dust-cloud complex as the reflection nebula in order to minimize the effects of faint background stars and diffuse galactic light. Variability of night sky emissions, which become especially troublesome at wavelengths longer than about 0.77 μm, necessitate switching between nebular and sky regions with a frequency on the order of five to ten times per hour; this procedure becomes awkward for large nebulae like the Pleiades complex. Halation from bright field stars has received considerable attention in the past, but in most cases this problem is easily handled by appropriate calibrations at the telescope.

Examples of the observed spectral dependence of brightness and polarization are given in Figs. 1 and 2. The systematically low polarizations

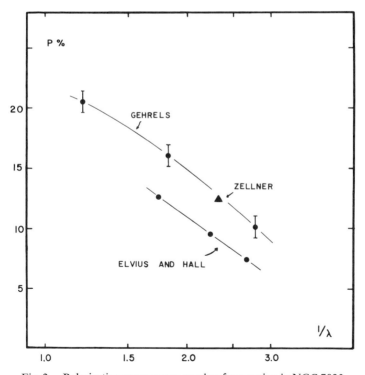

Fig. 2. Polarization versus wave number for a region in NGC 7023.

obtained by Elvius and Hall are puzzling; on several occasions I have tried to resolve this discrepancy by new observations of the same nebular regions using Lowell Observatory equipment, with inconclusive results. A similar discrepancy exists for photometric observations in the Merope nebula: the UBV colors found by Elvius and Hall (1966) are in good agreement with those from O'Dell (1965) but differ systematically from results obtained by Dahn and partially confirmed by Zellner (unpublished). Clearly, existing color and polarization data cannot be used uncritically even for the brightest and best-observed reflection nebulae.

When spherical, isotropic grains or grains of arbitrary shape but random orientation are illuminated by unpolarized light, symmetry requires that the plane of polarization (strongest electric vector) of the scattered light either coincides with the plane of scattering or lies perpendicular to it. Projected against the plane of the sky, all reflection nebula polarizations should then have position angles that show a radial symmetry about the illuminating star, either along lines drawn from the star to the nebular regions (negative) or perpendicular to such lines (positive polarization). While a well-established case of negative polarization has never been found in ordinary reflection nebulae, substantial deviations from perfect positive symmetry occur. For the outer regions of the Merope nebula, Elvius and Hall (1967) found that the position angle of polarization "does not depend so strongly on the direction to the illuminating star, but is more closely related to the direction of nebulous filaments in the area. This indicates that the light is scattered by non-spherical particles oriented in a magnetic field associated with the filaments." The same effect was noted independently by Artamonov and Efimov (1968). Less dramatic cases of nonradial polarization are often seen and can probably be explained by multiple scattering from adjoining dense regions in the nebula. Multiple scattering should also give rise to weak circular polarizations, as yet detected only in the nebulosity around VY CMa (Serkowski 1973).

INTERPRETATIONS

The rather strong spectral dependence of polarization in reflection nebulae immediately rules out scattering by free electrons, neutral particles of atomic dimensions, or large solid bodies. Thus we are dealing with the Mie-scattering domain. Suppose that a plane wave of intensity $I_0(\lambda)$ is incident upon a spherical particle of radius a. Then the scattered wave at distance R from the grain has intensity components, polarized perpendicular and parallel to the plane of scattering, given by

$$I_1(\lambda) = I_0(\lambda) \frac{\lambda^2}{8\pi^2 R^2} i_1(m,x,\Theta),$$

$$I_2(\lambda) = I_0(\lambda) \frac{\lambda^2}{8\pi^2 R^2} i_2(m,x,\Theta)$$

(van de Hulst 1957, p. 36). Computation of the intensity functions i_1 and i_2 in terms of the complex refractive index $m = n - in'$, the dimensionless size parameter $x = 2\pi r/\lambda$, and the scattering angle Θ has been reduced to a trivial matter by high-speed digital computers; a widely used program is that of Shah (1967). A distribution of particle sizes is easily incorporated and, in fact, is necessary to smooth over optical resonances that are present only for particles in the form of perfect spheres. The distribution

$$N(a) = N_0 \exp[-5(r/r_0)^3] \tag{1}$$

introduced by Greenberg (1968) provides a close analytic description of the size distribution originally tabulated by Oort and van de Hulst (1946). By the transformation $x/x_0 = r/r_0$, where $x_0 = 2\pi r_0/\lambda$, it is possible to leave the scale radius r_0 as a free parameter to be determined by observation.

The radiative transfer problem for reflection nebulae, however, is not so easily handled. The plane of scattering for observations in a single nebular region is sketched in Fig. 3. (The cloud geometry is intended

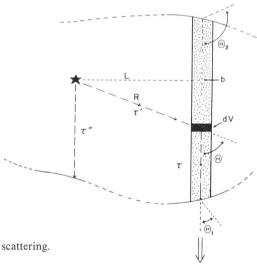

Fig. 3. The plane of scattering.

to be completely general. In particular, the illuminating star is not constrained to lie inside the dust cloud.) A priori, one has no knowledge of the range of scattering angles Θ_1 to Θ_2 or the internal attenuations τ, τ', and τ'' involved, nor of the volume distribution of particle density. Also, multiple-scattering effects are likely to be important in bright nebulae.

These geometrical problems have been attacked in several ways. Most early investigators assumed a simple idealized shape for the dust cloud, usually a sphere with the illuminating star at its center (Sobolev 1960; Minin 1962, 1965; Vanýsek and Svatoš 1964). Predating the work of Elvius and Hall, these studies depended upon inadequate observational data and were often restricted to the distribution of surface brightness as a function of offset distance from the illuminating star; the elegant analytic models of Sobolev and Minin, for example, yielded little in the way of practical results. Krishna Swamy and O'Dell (1967) suggested that the problems of volume integration could be sidestepped, for nebulae believed to lie wholly behind the illuminating star, by extrapolation of the intrinsic colors to zero offset angle and, hence, scattering angle 180°. The deficiencies of this "radar" method were pointed out by Hanner and Greenberg (1970). The color at zero offset was shown to be model-dependent and impossible to ascertain observationally with any confidence; also, because of the strongly forward-throwing phase function of Mie particles, a foreground dust layer too thin to cause noticeable reddening of the illuminating star can make a substantial contribution to the total nebular luminosity.

Considerable success for the Merope nebula has been achieved by modeling the dust cloud as a homogeneous plane slab whose thickness, tilt, and position with respect to the illuminating star can be adjusted, along with the grain size and refractive index, to best match the observed *offset dependence* of color and polarization (Dahn 1967; Greenberg and Roark 1967; Greenberg and Hanner 1970; Hanner 1971). Hanner (1969) also explored the effects of a second order of scattering and of grains in the form of randomly oriented infinite cylinders. A glance at Fig. 4 should suffice to illustrate the inapplicability of an idealized cloud geometry for objects that are less homogeneous than the Merope nebula. In NGC 2068 one has the additional problem of a large and very irregular foreground extinction. Thus, Gehrels (1967) and Zellner (1970) followed a rather different theoretical approach. In each separate nebular region, the effective range of scattering angles Θ_1 to Θ_2 was freely adjusted to best match the observed *wavelength dependence* of polarization, a parameter that is relatively free of parasitic effects; color data provide a check on the resulting model. While this second approach is free from geometrical constraints, it is clearly ill-suited to the exact

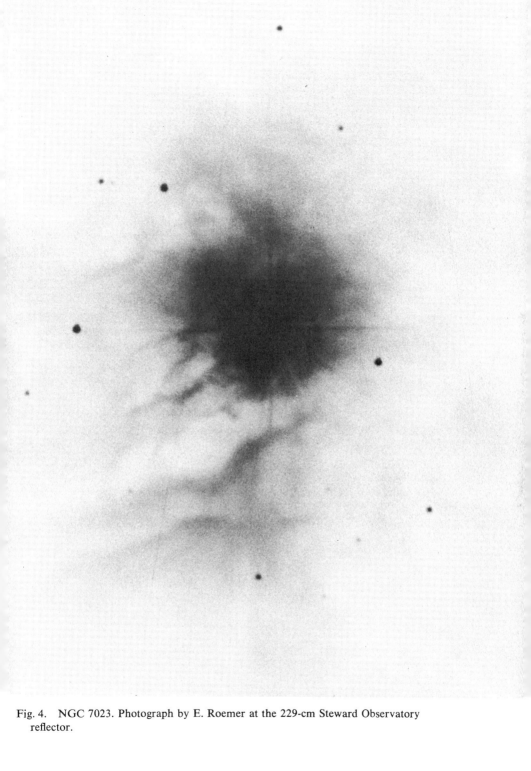

Fig. 4. NGC 7023. Photograph by E. Roemer at the 229-cm Steward Observatory reflector.

treatment of internal attenuations or the inclusion of multiple scattering.

Highly absorbent materials ($n' \sim n$) can roughly reproduce the interstellar extinction law only with an effective grain size so small that their angular scattering properties do not depart markedly from the Rayleigh case. Strong blue colors ($B\text{-}V \approx 1.0$ at any scattering angle) and nearly wavelength-independent polarizations would be expected from such particles. The commonly observed linear drop in nebular polarization toward larger wave number, on the other hand, is a signature of forward scattering dielectric spheres of diameter comparable to the wavelength. Thus, recent studies have uniformly rejected graphite or metallic grains as the principal scattering agents in reflection nebulae. Gehrels (1967) demonstrated that a moderate imaginary component, or more probably composite particles, would be required to explain both the (classical) interstellar extinction curve and his polarization observations in NGC 7023. Figure 5 reproduces the fit to color and polarization data in the Merope nebula obtained by Greenberg and Hanner for pure ice grains with scale radius near 0.5 μm in the above size distribution. Hanner (1971) demonstrated that silicate grains with $r_0 = 0.25$ μm give negative polarizations, contrary to all observations. However, it should be noted that dielectric spheres with refractive indices between 1.1 and 2.0 have essentially identical scattering properties except for adjustments in $2\pi r/\lambda$; the difficulties of negative polarization can be avoided by taking an appropriately smaller scale radius. Zellner (1970) found good agreement with his polarization data in NGC 7023, and both colors and polarizations in NGC 2068 and IC 5076, for nearly-pure dielectric grains of any sort, but only if the scale radius were a factor of two or three smaller than that necessary to match the classical extinction curve. Examples from Zellner's models are given in Figs. 6 and 7.

FUTURE WORK

Good observations now exist at least for the larger and brighter reflection nebulae. First-order theoretical investigations have been fairly successful in explaining the observed colors and polarizations, though, as noted above, with mixed results in terms of the physical properties

Fig. 5. Observed intrinsic colors and polarizations in the Merope nebula as a function of angular offset distance from the illuminating star. Open circles are from Greenberg and Roark (1967), dots from Elvius and Hall (1967). The theoretical line is computed for a dust cloud in the form of a homogeneous plane slab with ice particles of scale radius $r_0 = 0.5$ μm in the size distribution $N(r) = N_0 \exp[-5(r/r_0)^3]$. Adapted from Greenberg and Hanner (1970)

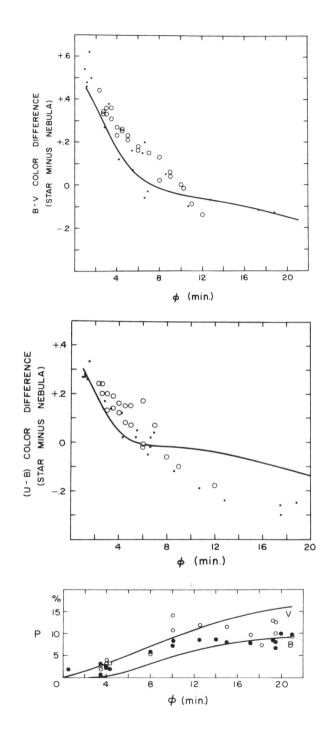

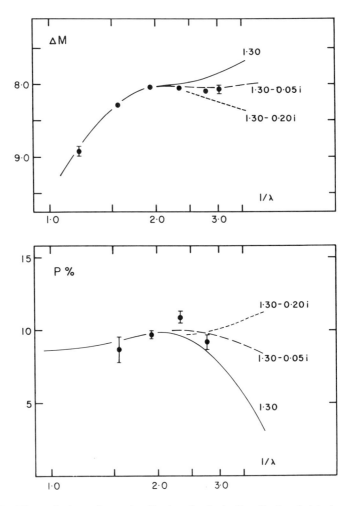

Fig. 6. Theoretical match to color (top) and polarization (bottom) data in a region of NGC 2068. ΔM is the magnitude difference between the nebular region and the illuminating star. Theoretical curves are for ice grains in the size distribution of Equation (1), with range of scattering angles 10° to 50°. Solid curve: $m = 1.30$ and $r_0 = 0.26$ μm. Dashed curve: $m = 1.30 - 0.10i$, $r_0 = 0.29$ μm. Adapted from Zellner (1970).

of the grains. However, no modern study has properly taken into account the easiest observational quantity to measure, namely the absolute surface brightness. Simple brightness calculations show that, for most well-observed nebulae, the internal optical depths are greater than unity. Hence, multiple scattering should be quite important, and it is rather

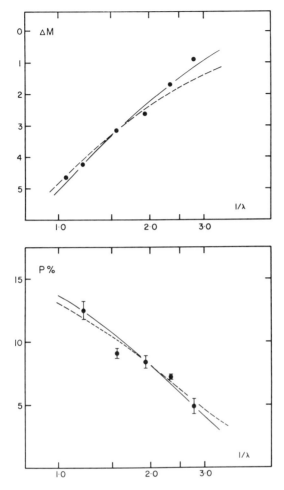

Fig. 7. Theoretical match to color (top) and polarization (bottom) data in a region of IC 5076. The nebula is behind the illuminating star, with range of scattering angles 150° to 165°. The scale radius is near 0.16 μm for all three refractive indices. Adapted from Zellner (1970).

puzzling that the optically thin approximation works as well as it does. Exact integration over two orders of scattering, as done by Hanner (1969), helps to indicate the potential effects of multiple scattering, but third and higher orders become important soon after second order. The transfer problem for polarized light in dust clouds of arbitrary optical depth can be solved by the straightforward but inefficient Monte Carlo method; to my knowledge, such computations for reflection nebulae are being carried out by T. Roark at Ohio State, by Mattila at Helsinki Observatory, and by Vanýsek at the Charles University in Prague. However, we must remember that space-density inhomogeneities, which have only minor effects in first-order models, will play a dominant role for higher orders of scattering. For thick, chaotic objects like NGC 7023, the prospects for realistic models seem dim. Cometary nebulae like NGC 2261, NGC 2245, and Cederblad 201 probably differ cosmogonically from ordinary reflection nebulae but tend to have smooth isophotes and should be adequately modeled by simple homogeneous cones or paraboloids. Perhaps the best solution would be for observers to confine their attentions to optically thin, and hence rather faint, reflection nebulae.

Acknowledgments. I am indebted to Martha Hanner for helpful discussions and the detection of errors in the manuscript. This work was supported by the National Science Foundation.

REFERENCES

Artamonov, B. P., and Efimov, Yu. S. 1968. Photoelectric observations of the polarization and color of the reflection nebula near Merope. *Sov. Astron. A. J.* 11:607–608. (1967: *Astron. Zhurnal* 44:755–756.)

Bergh, S. van den 1966. A study of reflection nebulae. *Astron. J.* 71:990–998.

Cederblad, S. 1946. Studies of bright diffuse nebulae. *Medd. Lund Obs. Ser II*, Vol. 12, No. 119.

Dahn, C. 1967. Ph.D. Thesis, Case Institute of Technology.

Dorschner, J., and Gürtler, J. 1964. Untersuchungen über Reflexions-nebel am Palomar Sky Survey. I. Verzeichnis von Reflexionsnebeln. *Astron. Nachr.* 287:257–260.

Elvius, A., and Hall, J. S. 1966. Observations of the color and polarization of the reflection nebulae NGC 2068, NGC 7023, and the Merope Nebula obtained in three spectral regions. *Lowell Obs. Bull.* 6 (No. 135):257–269.

———. 1967. Observations of polarization and color in the nebulosity associated with the Pleiades cluster. *Lowell Obs. Bull.* 7 (No. 139): 17–29.

Gehrels, T. 1960. Measurements of the wavelength dependence of polarization. *Lowell Obs. Bull.* 4:300.

———. 1967. Wavelength dependence of polarization. IX. Interstellar particles. *Astron. J.* 72:631–641.

Glushkov, Yu. I. 1965. On the polarization and continuous spectrum of NGC 2068. *Trud. Astrophiz. Inst., Akad. Nauk Kazakh. S. S. R.* 5: 277–284.

Greenberg, J. M. 1968. Interstellar grains. *Nebulae and interstellar matter.* (B. M. Middlehurst and L. H. Aller, eds.), pp. 221–364. Chicago: Univ. of Chicago Press.

Greenberg, J. M., and Hanner, M. S. 1970. Light from reflection nebulae. II. Colors and polarization. *Astrophys. J.* 161:947–959.

Greenberg, J. M., and Roark, T. P. 1967. Light from reflection nebulae. I. Dielectric versus metallic particles. *Astrophys. J.* 147:917–936.

Hall, R. C. 1965. Polarization and color measures of NGC 2261. *Publ. Astron. Soc. Pac.* 77:158–163.

Hanner, M. S. 1969. Light scattering in reflection nebulae. Ph.D. Thesis, Rensselaer Polytechnic Institute.

———. 1971. Light from reflection nebulae. IV. Scattering by silicate grains. *Astrophys. J.* 164:425–435.

Hanner, M. S., and Greenberg, J. M. 1970. Light from reflection nebulae. III. Color at 0° offset. *Astrophys. J.* 161:961–963.

Hulst, H. C. van de 1957. *Light scattering by small particles.* New York: Wiley.

Johnson, H. M. 1960. Photoelectric photometry of diffuse galactic nebulae and comet Arend-Roland. *Publ. Astron. Soc. Pac.* 72:10–23.

Krishna Swamy, K. S., and O'Dell, C. R. 1967. Distinction between models of interstellar grains. *Astrophys. J.* 147:937–942.

Martel, M.-T. 1958. Polarisation des nébuleuses diffusantes. *Publs. de l'Obs. de Haute-Provence* 4 (No. 20):39–80.

Minin, I. N. 1962. On the optical properties of dust nebulae. *Sov. Astron. A. J.* 5:487–490. (1961: *Astron. Zhurnal* 38:641–646.)

———. 1965. Light scattering in dust nebulae. *Sov. Astron. A. J.* 8:528–532. (1964: *Astron. Zhurnal* 41:662–668.)

O'Dell, C. R. 1965. Photoelectric spectrophotometry of gaseous nebulae. II. The reflection nebulae around Merope. *Astrophys. J.* 142:604.

———. 1969. Optical backscattering functions in the Pleiades. *Astrophys. J.* 156:381–384.

Oort, J. H., and Hulst, H. C. van de 1946. Gas and smoke in interstellar space. *Bull. Astron. Inst. Netherlands.* 10:187–204.

Racine, R. 1971. UBV photometry of the reflection nebula Ced 201. *Astron. J.* 76:321–323.

Roark, T. P. 1966. Ph.D. Thesis, Rensselaer Polytechnic Institute.

Rozhkovski, D. A., and Kurchakov, A. V. 1968. A catalog of reflection nebulae. *Trud. Astrophiz. Inst., Akad. Nauk Kazakh. S. S. R.* 11:3–42.

Serkowski, K. 1973. Infrared circular polarization of NML Cygnus and VY Canis Majoris. *Astrophys. J.* 179:L101–L106.

Shah, G. 1967. Ph.D. Thesis, Rensselaer Polytechnic Institute.

Sobolev, V. V. 1960. The brightness of a spherical nebula. *Sov. Astron. A. J.* 4:1–6. (1960: *Astron. Zhurnal* 37:3–8.)

Vanýsek, V., and Svatoš, J. 1964. Solid particles in reflection nebulae. *Acta Univ. Carolinae — Math. et Phys.* No. 1:1–18.

Zellner, B. 1970. Polarization in reflection nebulae. Ph.D. Thesis, Univ. of Arizona; *Astron J.* In preparation.

DISCUSSION

HANNER: I have done some calculations of the second-order scattering for a plane-parallel nebular model, with $\tau \cong 1$. The second-order scattering makes a significant contribution to the emergent intensity, comparable in some cases to the first-order scattering. The general effect is to decrease the intensity gradient with distance from the illuminating star, compared to that based on first-order scattering alone, but the decrease in intensity gradient is not nearly enough to fit the Merope nebula.

COLLINS: In some Monte Carlo models made by T. and B. Roark and myself, we have found that it is possible to match the surface brightness for Merope with reasonable accuracy.

SCHMIDT: A Heidelberg elliptical polarization survey of the 30 Doradus nebula in the Large Magellanic Cloud gave for one region $V/I = 0.40 \pm 0.13\%$ or $V/p = 0.15 \pm 0.05$ (see Schmidt 1971).

SANDFORD: Your data are plotted as essentially linear functions in semilog axes, and it does not surprise me that the single scattering models that fit particle size, r, refractive index, m, and integrated exponential geometrical path ($e^{-\pi r^2 Q} \text{ext}^{\Delta x}$) can reproduce the data. Since the polarization is relatively large, are photographic measures such as those made by Beebe of Venus, Herbig of VY CMa, and Keller of the solar corona not easier and more valuable? These measures could be made with narrow-band filters to yield the spatial- as well as the wavelength-dependence of polarization.

ZELLNER: The semilogarithmic coordinates were chosen so that plots of observed polarization versus $\log 1/\lambda$ and theoretical curves of P versus $\log 2\pi r/\lambda$ may be matched by a simple shift of $\log 2\pi r$ in the abscissa. It is a useful technique for any Mie-scattering problem. Linearity or the lack of it in the functions is of no consequence.

Photographic polarimetry is useful for small, bright, highly polarized objects such as VY CMa. But for ordinary reflection nebulae, with surface brightness comparable to that of the night sky, an absolute precision better than ± 0.01 would be very difficult by photographic techniques. We generally need ± 0.005 to adequately delineate wavelength dependencies. Also, some of the most useful data are obtained at wavelengths of 0.8 μm and longer, outside the range of sensitive photographic emulsions.

DISCUSSION REFERENCE

Schmidt, T. 1971. Elliptical polarization measurements near 30 Doradus and η Carinae. *Astron. Astrophys.* 12:456–463.

ORION NEBULA POLARIZATION

RICHARD HALL
Northern Arizona University, Flagstaff

Polarization measures of NGC 1976 have been made in the visible part of the spectrum at four emission lines and at four regions of the continuous spectrum. The central regions of this nebula are found to have low emission line polarization (1%) but moderate continuum polarization (4%). Continuum polarization of the central regions appears higher to the west and northwest of θ_c Orionis. Polarization at 5007 Å [O III] is equal to polarization at 3727 Å [O II] close to the Trapezium but is markedly higher in the outer regions of the nebula. A general increase in polarization at all wavelengths with increasing distance from θ_c Orionis is found.

Since the Fall of 1966, I have made measurements of the linear polarization of the Orion Nebula, NGC 1976, with the 61-cm telescope on the campus of Northern Arizona University in Flagstaff, Arizona. Although others have measured this nebula in white light, particularly for the inner and brightest regions (Hall 1951), to my knowledge there have been no measures that attempt to separate emission-line from continuum polarization. References concerning Orion nebula polarization have been given by Johnson (1968).

The present observations were made using a Glan-Foucault analyzer in a polarimeter utilizing interference filters with the following peak wavelengths and half-intensity widths in angstroms: 3727, 100; 4210, 80; 4655, 110; 4860, 60; 5007, 30; 5300, 130; 5445, 200; 6563, 6; 6900, 100.

A nearly identical polarimeter has been described elsewhere (Hall 1965). In the present work, EMI 6256SA and EMI 9558A "end-on" window phototubes were employed. No depolarizer was used anywhere in the light path. Checks on instrumental polarization were obtained from nightly observations of unpolarized stars. A "single" measurement consisted of eight deflections on the recorder chart: a double measure of one Stokes parameter, a centering check, and a double measure of the

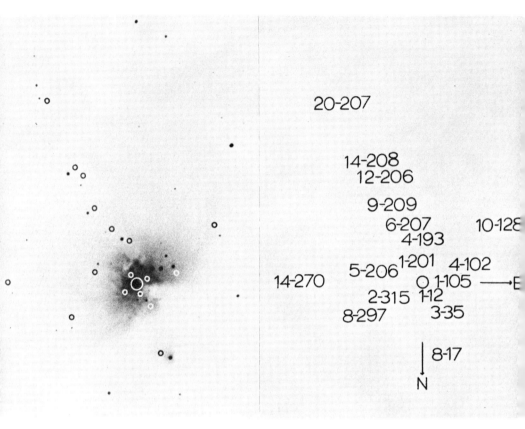

Fig. 1. Areas observed on the Orion Nebula, NGC 1976. The circles represent measurements with a 55 arcsec aperture. Measures with a 27 arcsec aperture were made along the circumference of the 100 arcsec diameter circle centered on θ_c Orionis. Areas are identified in the right-hand diagram by the number of minutes of arc and the position angle from θ_c Orionis.

other parameter. All measures were made within two hours of the meridian. All blue measures were made under moonless conditions.

There have been four sets of observations: 1) repeated measurements of the five areas measured for intensity by O'Dell and Hubbard (1965). O'Dell's areas I–V are renamed here according to the number of minutes of arc and the position angle in degrees from θ_c Orionis as: 2-315, 1-105, 4-193, 14-208, 20-207, respectively; 2) single measurements of regions not sampled by the measurements of O'Dell and Hubbard; 3) single measurements along a 100 arcsec diameter circle surrounding θ_c Orionis; 4) single measurements of a few bright areas at red wavelengths.

Figure 1 shows the areas measured and the positional notation for each. In Fig. 2 can be seen the low values of emission-line polarization (1%) found for the central regions of the nebula. The straight reductions of the recorder charts are plotted with only sky polarization removed. These results should be considered in terms of the estimated contamination of the line filters by the continuum. In the 3727 filter about 30% of the measured intensity was due to continuum light in all regions of the nebula. In the 4860 filter, about 10% of the light was continuum light for the inner regions and 16% for the outer regions; in the 5007 filter, about 2% was the case for the inner regions and 8% for the outer regions. In a most severe case, for 3727 Å, a 5% continuum polarization could appear as a 1.5% "line" polarization even if the line were unpolarized. A portion of the 3727 Å polarization shown can probably be explained away on this basis. Such corrections involve interpolation of the intensity and Stokes parameters for the continuum on either side of the line. There was uncertainty in a correction made on this basis and in the need to extrapolate in the case of 3727 Å. I decided not to include such corrections. Similarly, the continuum filters record numerous faint emission lines, an atomic continuum, and a scattered light continuum of

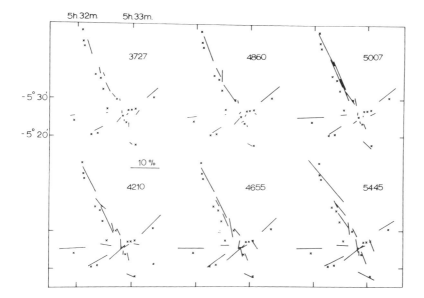

Fig. 2. Polarization of the areas shown in Fig. 1. The area measured on NGC 1976 is at line center; line length is proportional to the percentage polarization. Line orientation is that of the electric vector plus 90°. The coordinates are 1950 right ascension and declination.

stellar and atomic light. Although it might be expected that the scattered light continuum would dominate; especially in the outer parts of the nebula, no attempt is made here to distribute the observed continuum polarization among these possible sources. The continuum filters were chosen to exclude strong emission lines. There may be as much as 10% H gamma intensity contamination in the 4210 filter. The continuum is polarized about 4% in the central regions and shows an asymmetry to the west and northwest with higher values of polarizations for these directions. These findings are seen, as well, in Fig. 3 which shows separate measurements along the 100 arcsec circle surrounding θ_c Orionis. Unusual orientations are noted in both sets of observations for measurements in the "dark bay" at position angles of about 70°. Whether these indicate a separate source for light in this region of the nebula or whether there is another interpretation is not certain. The scattered light of θ_c Orionis from sky and telescope is similar in intensity to that of the dark bay at this angular distance: about 10^{-4} of the stellar intensity.

Figure 4 shows the wavelength dependence of polarization. One should regard the labeled points as the most reliable; excepting the single red wavelength measures and area 10-128. The continuum wavelength dependence can be characterized as small with some notable exceptions. The "blue" (as in color photographs) region 10-128 shows a decrease

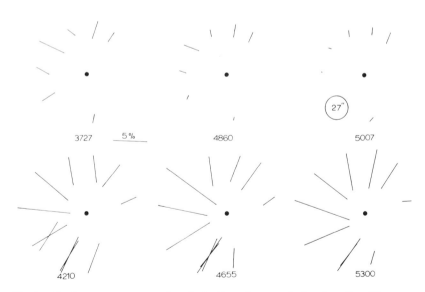

Fig. 3. Single measurements of polarization along a circle of radius 50 arcsec centered on θ_c Orionis. The 27 arcsec diameter aperture used is shown. Position angle is oriented with north down. The dark bay is at position angle 70°.

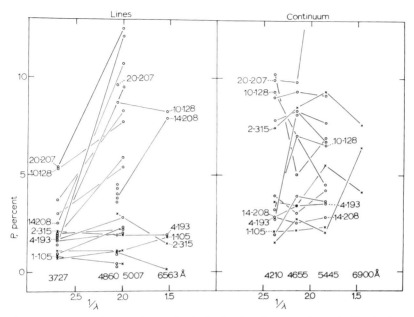

Fig. 4. Wavelength dependence of polarization for the areas shown in Fig. 1. Areas represented by an "x" are those within two minutes of arc of θ_c Orionis. Labeled areas are those with four or more measurements. Area 10-128, however, is a single measurement of a "blue" area.

in continuum polarization with increasing wavelength, and it has been noted (Hall 1967) that for the outermost region, 20-207, there is an increase in polarization from short to long wavelengths — a finding possibly characteristic of reflection nebulae. The line polarization 5007/ 3727 Å changes character from the inner regions where these polarizations are similar in value to the outer regions where 5007 polarization is much larger than 3727 polarization. Zellner has pointed out the difficulty of interpretation of reflection nebula polarization.[1] There is the suggestion from these data, however, that 5007 [0 III] radiation originates closer to θ_c Orionis than the less polarized and possibly less severely scattered 3727 [0 II] radiation.

In Fig. 5 is plotted the increase in polarization with increasing distance from θ_c Orionis. Data from all regions, except the 100 arcsec circle, are included here. An especially uniform increase in polarization with distance is noted for the 5007 Å wavelength. The 5007 Å polarization appears to be equal to the continuum polarization at adjacent wavelengths.

---

[1] See p. 867.

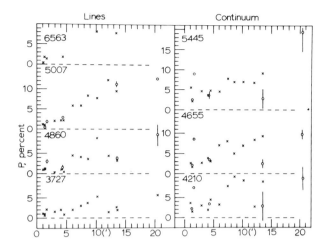

Fig. 5. Variation in percentage polarization for NGC 1976 with increasing distance from θ_c Orionis. All areas on the nebula shown in Fig. 1 are included here. Error bars are shown for areas with four or more measurements; such areas are represented by circles.

Acknowledgments. Thanks are due Jane Hall and Karl Karlstrom for their help in reducing the majority of the recorder charts.

REFERENCES

Hall, J. S. 1951. Polarization measurements in astronomy. *J. Opt. Soc. Am.* 41: 963–966.

Hall, R. C. 1965. Polarization and color measures of NGC 2261. *Pub. Astron. Soc. Pac.* 77: 158–163.

———. 1967. Orion Nebula polarization of O'Dell's and Hubbard's five areas at 4210, 4655, 4860, 5007, 5445 Å. *Astron. J.* 72: 802 (Abstract only).

Johnson, H. M. 1968. Diffuse nebulae. *Nebulae and interstellar matter,* Vol. VII of *Stars and stellar systems* (B. M. Middlehurst and Lawrence H. Aller, eds.), pp. 65–118. Chicago: Univ. of Chicago Press.

O'Dell, C. R., and Hubbard, W. B. 1965. Photoelectric spectrophotometry of gaseous nebulae. I. The Orion Nebula, *Astrophys. J.* 142: 591–603.

DISCUSSION

WOLSTENCROFT: You mentioned the observation of anomalous polarization orientations in the dark bay of the Orion Nebula. Would not the strong linearly polarized flux from the foreground sky dominate in such a faint part of the nebula?

R. HALL: I certainly rejected more measures in studying this region than for any other; measures here were especially affected by poor seeing. The large surface brightness gradient at this location on the nebula, combined with seeing and telescope drive rate effects, conspires to produce spurious polarization. Subsequently, I have measured the polarization in the dark bay with a larger aperture (55 arcsecs) at a larger distance from θ_c Orionis. At this location, 4.3 arcmins away at position angle 70°, these effects are much less pronounced. Averages of six measures show 1.0% polarization at 4655 Å with a 28° position angle orientation ($\pm$ 35° probable error) and 1.1 % polarization at 5007 Å with an orientation of 58° ($\pm$ 40° probable error). Although largely nebular, a portion of this polarization could be interstellar. The polarization of θ_c Orionis, HD 37022, has been reported as 0.3% at 69° (Hall 1958).

DISCUSSION REFERENCE

Hall, J. S. 1958. Polarization of starlight in the galaxy. *Publ. U.S. Naval Observatory*, Vol. XVII-Part VI. p. 299.

POLARIZATION BY INTERSTELLAR GRAINS

GEORGE V. COYNE, S.J.
University of Arizona
and
Vatican Observatory

In this review emphasis is placed upon observations concerning the nature of birefringence in the interstellar medium and upon the wavelength, λ_{max}, at which the maximum linear polarization occurs as a characteristic parameter of the interstellar grains. The rotation with wavelength of the plane of vibration of the interstellar linear polarization may reveal a peculiar twisted grain alignment in opposite senses to either side of about 140° galactic longitude, out to a distance of approximately one kiloparsec from the sun. Measures of circular polarization in this region are needed. Coincident measures of λ_{max} and circular polarization will allow us to study the distribution in the galaxy of both particle size and composition.

Among the most important observational developments during the past two decades in our knowledge of polarization by interstellar grains are: (1) the discovery in 1948 of interstellar polarization by Hall (1949) and Hiltner (1949); (2) the subsequent study by Gehrels (1960) of the wavelength dependence of the interstellar polarization; and (3) the recent discovery of interstellar circular polarization (Kemp 1972[1]; Martin, Illing, and Angel 1972[2]).

Concomitant theoretical studies on the composition, size, and shape of interstellar grains and on alignment mechanisms have been the subject of several recent reviews (Wickramasinghe and Nandy 1972; Greenberg 1968; Greenberg and van de Hulst 1973). The studies of van de Hulst (1957) on particle scattering properties, of Greenberg (1968) and his colleagues and of Wickramasinghe (1967) on the fitting of grain models to the observed interstellar polarization and extinction curves, and the discussion of alignment mechanisms for the interstellar grains by Davis and Greenstein (1951), by Purcell and Spitzer (1971), and by Cugnon (1971) have been significant stages in the development of the theory of polarization by interstellar grains. Serkowski (1973)

---

[1] See also p. 607.   [2] See also p. 939.

has recently reviewed the more important observational facts concerning the interstellar polarization. The purpose of the current review is principally to present observations concerning the birefringence of the interstellar medium and to discuss the extent to which λ_{max}, the wavelength at which the maximum polarization occurs, yields information about the size and composition of interstellar particles. All of the important observational facts cited by Serkowski (1973) are presented here under the discussion of either birefringence or the wavelength of maximum polarization.

The presence of grains in the atmospheres or stellar environments of both early- and late-type stars, as shown by circumstellar polarization and infrared excesses in these stars, contributes significantly to our knowledge of the origin and nature of the interstellar grains. Reviews of circumstellar polarization (Zellner and Serkowski 1972) and of infrared excesses in stars (Herbig 1972; Woolf 1973) have been published recently. (See also reviews by Kruszewski[3] and by Shawl[4] in this book.)

DIRECTION OF THE PLANE OF VIBRATION

Practically all of the available data on the linear polarization of the interstellar medium[5] have been assembled by Mathewson and Ford (1970) in a combined plot of the direction and magnitude of the linear polarization for more than 7000 stars as a function of galactic latitude and longitude. This is shown in Fig. 2 of Verschuur.[6] One sees the well-known general preference of the plane of polarization to lie in the galactic plane; the degree of alignment varies, however, with galactic longitude. The high degree of alignment parallel to the galactic plane at about longitude 130° is interpreted as due to looking across a spiral arm; likewise, the patterns of alignment near 50° longitude indicate a line of sight along a spiral arm. In this interpretation, the magnetic field, which is along the spiral arm, aligns the short (spin) axis of elongated interstellar particles in this direction. Thus, the linear dichroism of the interstellar medium is such that the electric vector maximum is in the direction of the magnetic field and the alignment and linear polarization are maximum in directions of propagation transverse to the magnetic field. There are many details in the polarization "flow patterns" of Fig. 2 of Verschuur.[7] The random alignment seen near 80° longitude has been, at times, interpreted as due to viewing

---

[3] See p. 845.
[4] See p. 821.
[5] A major survey has been published recently by Klare, Neckel, and Schnur (1972)
[6] See p. 964.
[7] See p. 960.

along the magnetic lines of force in a spiral arm. Verschuur[8] shows that this is not the case, but that what we see near 80° longitude is rather a local disturbance associated with the Cygnus X-1 source. Verschuur[9] offers convincing arguments against Mathewson's (1968) proposal of a helical magnetic field, directed along longitudes 90° and 270°, to explain the overall "flow patterns" of the polarization. Currently there appears to be no overall magnetic field structure that consistently explains all of the systematic trends in the directions of polarization.

Any general structure of the magnetic field must suffer local small-scale disturbances. Circular polarization, for instance, has been observed to have opposite senses for stars in the Scorpio-Centaurus cluster which lie only one degree apart (Kemp and Wolstencroft 1972). From variations in the directions of the plane of vibration for the linear polarization in the cluster Stock 2, Krzeminski and Serkowski (1967) deduce a microscale of 0.3 parsec for fluctuations in the direction of the magnetic field.

For the vast majority of early-type stars for which multicolor polarimetry is available, the direction of polarization does not vary with wavelength in the spectral range, 0.3 to 1.0 μm. Gehrels and Silvester (1965) found, however, a number of stars whose polarization appeared to be interstellar and for which the plane of polarization varied with wavelength. The combined effect of any two or more sources of linear polarization that have different dependencies of the polarization on wavelength and different directions of the plane of vibration can cause a rotation of the direction of the plane of vibration for the resultant polarization. One can think, for instance, of the linear polarization from a Be star modified by the interstellar polarization, or of a line of sight through several interstellar clouds, each of which has a different mean particle size and particle alignment. On the other hand, the rotation of the plane of vibration with wavelength may be due to birefringence of the interstellar medium — i.e., the projection of the grain alignment vector on the plane transverse to the propagation direction varies systematically about this direction as a function of distance.[10] Since such a twisted grain-alignment model is adduced to explain the recently observed interstellar circular polarization (Kemp and Wolstencroft 1972), an analysis of the observations showing the rotation of the plane of linear polarization with wavelength is important. The analysis is complicated, of course, by the knowledge we now have of the incidence of intrinsic polarization in early-type stars (Serkowski 1968, 1970; Coyne and Kruszewski 1969; Coyne 1971a,b).[11]

---

[8] See p. 960.   [9] See page 960.   [10] See p. 926.   [11] See p. 845.

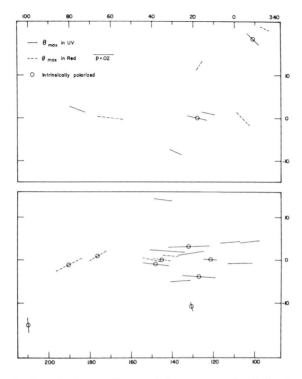

Fig. 1. A plot in galactic coordinates of the amount of polarization (length of line) and direction of the plane of vibration (direction of line) for all of those stars observed in Arizona that show a rotation of the plane of vibration with wavelength. Mean values over seven spectral bands (from 0.3 to 1.0 μm) are plotted. The direction defined by $\theta_{\max}$, the maximum value for the position angle of the plane of vibration, is counter-clockwise from those plotted. Stars that are intrinsically polarized, but that also have a significant interstellar polarization are indicated. The data for this figure are published by Coyne (1974).

In order to establish a sound observational basis for the discussion of the rotation with wavelength of the plane of vibration in terms of birefringence in the interstellar medium, the stars for which Gehrels and Silvester (1965) had reported a rotation with wavelength of the plane of polarization were reobserved, and other stars, especially in the galactic longitude range 135° to 145°, were observed. Coyne (1974) has analyzed these data, and a summary of these results is given in Figs. 1 and 2. No late-type stars are included since it is clear that any such effect in these stars takes place mainly in an extended atmosphere or shell about the star.[12]

---

[12] See p. 821.

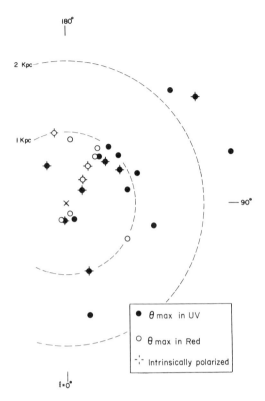

Fig. 2. The same data as for Fig. 1 but plotted on the galactic plane. The sun is at the center and the photometric distances are determined using a value of 3 for total-to-selective absorption. Galactic longitudes are indicated with: 0°, 90°, and 180°. The distribution of points on either side of approximately 140° longitude is of peculiar interest (see text).

Of the stars we have observed in Arizona, 15% show this effect. The total amount of rotation of the plane of vibration from the ultraviolet to the red spectral regions for the various stars ranges from about 5° to 15°. The distribution of the sense of rotation at about 140° galactic longitude and 0° latitude is of particular interest since it coincides with other well-known discontinuities at this galactic longitude (Reddish 1967; Seaquist 1967). Because of the coincidence of these various discontinuities, it appears that we are seeing a structural effect in the galaxy rather than simply an aspect effect. Circular polarization measurements will be made of these stars in order to see whether a twisted grain alignment along the line of sight explains the observations.

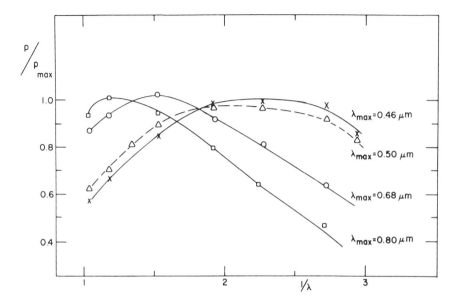

Fig. 3. A plot of the mean polarization at various wavelengths normalized to the maximum polarization for the following groups of stars: for $\lambda_{max} = 0.46$ μm (HD 159975, 161961, 204827, and VI Cyg No. 12); for $\lambda_{max} = 0.50$ μm (HD 25291, 25443, 207538, 209481, 213470, 237211); for $\lambda_{max} = 0.68$ μm (HD 147932, 147933); for $\lambda_{max} = 0.82$ μm (HD 147889). The values of p_{max} and λ_{max}, the maximum polarization and the wavelength at which this occurs, are published by Coyne, Gehrels, and Serkowski (1974).

WAVELENGTH DEPENDENCE OF THE POLARIZATION AND GRAIN PARAMETERS

The wavelength dependence of interstellar polarization is different for different stars. This is seen in Fig. 3 where the mean polarization (normalized to the maximum polarization) is plotted for various groups of stars that have similar values of λ_{max}, the wavelength at which the maximum polarization occurs. The details of these observations are published by Coyne, Gehrels, and Serkowski (1974). It has been shown, however, by Serkowski (1973) that all of the multicolor measurements of the interstellar polarization can be represented by an empirical formula:

$$p/p_{max} = \exp[-1.15 \ln^2(\lambda_{max}/\lambda)]. \qquad (1)$$

Figure 4 shows a plot for a sample of the stars measured at the University of Arizona. Stars with larger than average and smaller

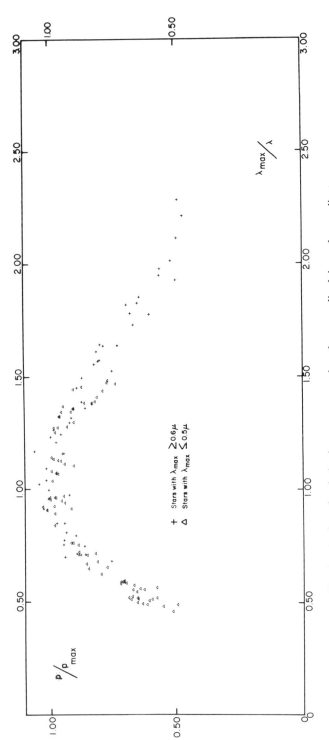

Fig. 4. A plot of polarization versus wavelength, normalized in each coordinate by p_{max}, the maximum polarization, and λ_{max}, the wavelength at which the maximum polarization occurs, respectively. Only observations for stars with larger and smaller than average values of λ_{max} are plotted. Similar curves representing the wavelength dependence of the interstellar polarization have been given for all stars observed in Arizona (Coyne, Gehrels, and Serkowski 1974) and in Australia (Serkowski, Mathewson, and Ford 1974) for which multicolor polarimetry is available.

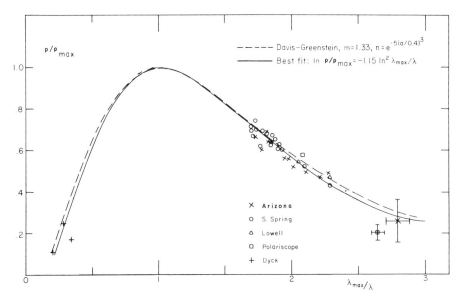

Fig. 5. The best fit (solid curve) to virtually all multicolor polarimetric observations of the interstellar polarization (Serkowski 1973) is compared to model calculations (dashed curve) for spinning dielectric cylinders whose radii follow an Oort–van de Hulst size distribution. These grains are aligned by the Davis-Greenstein (1951) mechanism, and curves for different values of ξ, the orientation parameter, coincide. Some observations at the extreme value of λ_{max}/λ are plotted to show the scatter among various stars.

than average values of λ_{max} are plotted. Attempts have been made to fit this curve with various theoretical curves given by Greenberg (1968). The best fit was found for a model of dielectric cylinders with an Oort–van de Hulst size distribution of radii, aligned by the Davis-Greenstein mechanism. This is shown in Fig. 5 where the solid curve is the best fit to the observations given by Equation (1). Individual observations are plotted for points at the extreme ends of this curve in order to show the observational scatter about the mean curve. These individual observations will be discussed in detail elsewhere (Coyne, Gehrels, and Serkowski 1974; Serkowski, Mathewson, and Ford 1974; Gehrels 1974). The dashed curves are for the model described above with an index of refraction, $m = 1.33$, a mean particle radius of 0.4 μm. Curves for different values of the orientation parameter, $\xi = 0.1$ to 0.5 (see Greenberg 1968) coincide. The smaller values of ξ represent a greater degree of orientation; a value of $\xi = 0.2$ represents a magnetic field of about 5×10^{-5} gauss.

Martin (1972) has suggested that fits to the observed linear polarization curve are not very sensitive to the index of refraction used in the model calculations. In Fig. 6 we compare curves computed by Martin for various refractive indices with a representative sample of the polarimetric observations. A curve for silicates from Wickramasinghe and Nandy (1972) is also shown. All of these models are for the "picket fence" static alignment model.

From Figs. 5 and 6 it is clear that there is no unique model that best fits the observations of the wavelength dependence of the interstellar polarization. On the other hand, highly absorbing particles are excluded since the curves for these drop too steeply into the ultraviolet as compared to the mean observed curve of Fig. 4. The circular polarization by interstellar particles may be a more discriminating technique for determining the index of refraction, once a particle size parameter is determined from the linear polarization (Martin 1972[13]).

There appear to be regions of characteristically high and low values of the wavelength of maximum polarization in the galaxy (Serkowski, Mathewson, and Ford 1974), and Serkowski (1973) has shown that there is a correlation between variations in λ_{max} and variations in the extinction law for various regions of the galaxy. In fact, if the wavelength scale of the extinction curve is normalized by using the respective mean values of λ_{max} for the Perseus-Cepheus and Scorpius regions of the galaxy, then the extinction curves coincide. This is shown in Fig. 7, adapted from Serkowski (1973). The probable interpretation of these curves is that both variations in λ_{max} and in the extinction curves reflect variations in the mean size of the interstellar particles for different regions of the galaxy.

It is expected that λ_{max} is proportional to the mean value of $(m-1)r$, where m is the index of refraction and r is the mean particle radius. The normalization of the extinction curves which yields Fig. 7 assumes that m is approximately constant with wavelength. The normalization will not, therefore, work in spectral regions where the extinction is dominated by discrete absorptions. On the other hand if, as Martin (1972)[14] has indicated, the circular polarization of the interstellar medium is more discriminatory of m, then concomitant studies of circular polarization and λ_{max} will complement one another in determinations of m and r and the distribution of these parameters in the galaxy. It is, of course, extremely important for studies of the nature and evolution of the interstellar medium to know what values of the index of refraction and mean particle size are characteristic of various regions of the galaxy.

---

[13] See also p. 926.   [14] See also p. 926.

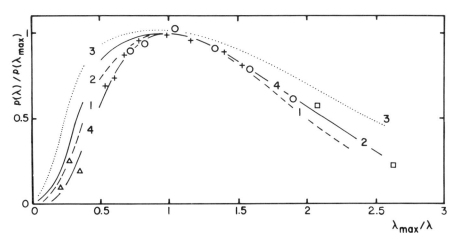

Fig. 6. Curves computed for grains with various refractive indices in a "picket fence" alignment are compared to observations (□, Gehrels, 1974; 0 and +, Coyne, Gehrels, and Serkowski, 1974; Serkowski, Mathewson, and Ford 1974; △, Dyck [unpublished]). Curves 1, 2 and 3 are from Martin (1972) for the following refractive indices, respectively: 1.5, 1.1–0.1i, 1.1; curve 4 is from Wickramasinghe and Nandy (1972) for $m = 1.7$–$0.05i$.

REFERENCES

Coyne, G. 1971a. Intrinsic polarization in the atmospheres of supergiant stars. *Colloquium on supergiant stars*. (M. Hack, ed.). Osservatorio Astronomico di Trieste, Italy.
———. 1971b. Polarization of Be stars at high galactic latitudes. *Ricerche Astronomiche* 8:201.
———. 1973. Wavelength dependence of polarization. Rotation of the direction of vibration of interstellar polarization by the interstellar medium. *Astron. J.* In draft.
Coyne, G.; Gehrels, T.; and Serkowski, K. 1974. Wavelength dependence of polarization. The wavelength of maximum polarization as a characteristic parameter of the interstellar grains. *Astron. J.* In draft.
Coyne, G., and Kruszewski, A. 1969. Wavelength dependence of polarization. XVII. Be-type stars. *Astron. J.* 74:528–532.
Cugnon, P. 1971. On the orientation of dust grains in the interstellar space. *Astron. Astrophys.* 12:398–420.
Davis, L., and Greenstein, J. L. 1951. The polarization of starlight by aligned dust grains. *Astrophys. J.* 114:206–240.
Gehrels, T. 1960. The wavelength dependence of polarization. II. Interstellar polarization. *Astron. J.* 65:470–473.
———. 1974. Wavelength dependence of polarization. XXVIII. Interstellar polarization from 0.22 to 2.2 μm. *Astron. J.* In draft.
Gehrels, T., and Silvester, A. B. 1965. Wavelength dependence of polarization. V. Position angles of interstellar polarization. *Astron. J.* 70:579–580.

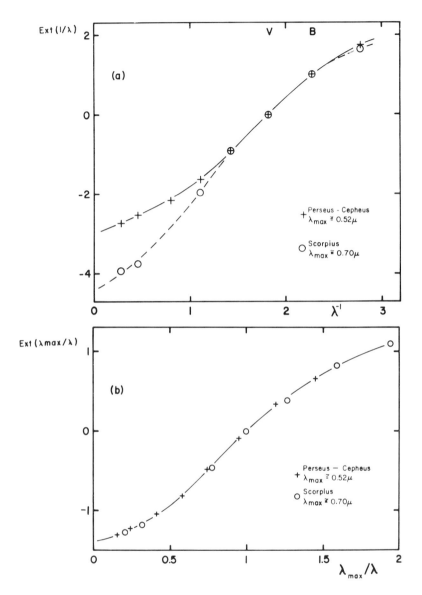

Fig. 7. These curves are adapted from Serkowski (1973). (a) The traditional extinction curves for the Perseus-Cepheus and Scorpius regions of the galaxy. The zero point and scale are set so that the function Ext $(1/\lambda)$ is zero in the visual spectral region and one in the blue. (b) The extinction curves normalized to the respective values of λ_{max}, the wavelength at which the maximum interstellar polarization occurs. The zero point and scale are set so that the function Ext(λ/λ_{max}) is zero for $\lambda = \lambda_{max}$ and about one for $\lambda = 0.5\,\lambda_{max}$.

Greenberg, J. M. 1968. Interstellar grains. *Stars and stellar systems. Vol. VIII: Nebulae and interstellar matter.* (B. M. Middlehurst and L. W. Aller, eds.), Ch. 6, pp. 211–364. Chicago: University of Chicago Press.
Greenberg J. M., and Hulst, H. C. van de (eds.) 1973. *Interstellar dust and related topics.* Dordrecht, Holland: D. Reidel Publ. Co. In preparation.
Hall, J. S. 1949. Observations of the polarized light from stars. *Science.* 109: 166–167.
Herbig, G. 1972. Symposium: Infrared excesses in stars. *Publ. Astron. Soc. Pacific* 84. In press.
Hiltner, W. A. 1949. Polarization of light from distant stars by interstellar medium. *Science* 109:165.
Hulst, H. C. van de 1957. *Light scattering by small particles.* New York: John Wiley and Sons.
Kemp, J. C. 1972. Circular polarization of Omicron Scorpii: Possible interstellar origin. *Astrophys. J.* 175:L35–L37.
Kemp, J. C., and Wolstencroft, R. A. 1972. Interstellar circular polarization: Data for six stars and the wavelength dependence. *Astrophys. J.* 176:L115–L118.
Klare, G.; Neckel, T.; and Schnur, G. 1972. Polarization measurements of 1660 southern OB-stars. *Astron. and Astrophys. Suppl.* 5:239–262.
Krzeminski, W., and Serkowski, K. 1967. Photometric and polarimetric observations of the nearby strongly reddened open cluster Stock 2. *Astrophys. J.* 147: 988–1002.
Martin, P. G. 1972. Interstellar circular polarization. *Mon. Not. R. Astr. Soc.* 159:179–190.
Martin, P. G.; Illing, R.; and Angel, J. R. P. 1972. Discovery of interstellar circular polarization in the direction of the Crab Nebula. *Mon. Not. R. Astr. Soc.* 159: 191–201.
Mathewson, D. S. 1968. The local galactic magnetic field and the nature of radio spurs. *Astrophys. J.* 153:L47–L53.
Mathewson, D. S., and Ford, V. L. 1970. Polarization observations of 1800 stars. *Mem. R. Astr. Soc.* 74:139–182.
Purcell, E. M., and Spitzer, L. 1971. Orientation of rotating grains. *Astrophys. J.* 167:31–62.
Reddish, V. C. 1967. A galactic discontinuity at $l^{II} = 140°$. *Nature* 213:1107–1108.
Seaquist, E. R. 1967. Linear polarization at 42 cm of the galactic background in the vicinity of $l^{II} = 140$, $b^{II} = 6°$. *Astron. J.* 72:1359–1365.
Serkowski, K. 1968. Correlation between the regional variations in wavelength dependence of interstellar extinction and polarization. *Astrophys. J.* 154: 115–134.
———. 1970. Intrinsic polarization of early type stars with extended atmospheres. *Astrophys. J.* 160:1083–1105.
———. 1973. Interstellar polarization. *Interstellar dust and related topics.* I.A.U. Symposium No. 52. (J. M. Greenberg and H. C. van de Hulst, eds.). Dordrecht, Holland: D. Reidel Publ. Co. In preparation.
Serkowski, K.; Mathewson, D.; and Ford, V. 1974. Observations of the wavelength dependence of interstellar polarization with a multi-channel polarimeter. *Astrophys. J.* In press.
Wickramasinghe, N. C. 1967. *Interstellar grains.* London: Chapman and Hall.
Wickramasinghe, N. C., and Nandy, K. 1972. Recent work on interstellar grains. *Reports on Progress in Physics.* 35:157.

Woolf, N. J. 1973. Dust around stars. *Interstellar dust and related topics.* I.A.U. Symposium No. 52, (J. M. Greenberg and H. C. van de Hulst, eds.). Dordrecht, Holland: D. Reidel Publ. Co. In preparation.

Zellner, B., and Serkowski, K. 1972. Polarization of light by circumstellar material. *Publ. Astron. Soc. Pacific* 84:619–626.

DISCUSSION

DICKEL: Is there any correlation of the amount of rotation of the linear polarized vector with distance, extinction, or any depth quantity?

COYNE: No, there is no clear correlation of the position-angle rotation with distance, extinction, or 21-cm hydrogen density studies; nor should we expect such correlations. The rotation is apparently due to the simultaneous existence of both a twisted grain alignment and a change in particle size along the line of sight to the respective stars. There is a good correlation of this effect with circular polarization (Coyne 1974). Not all stars that are circularly polarized show the effect. This means that there was no differentiation in grain size along the line of sight. The model predicts that all stars that show the effect should be circularly polarized, but only a few stars have been measured for circular polarization (see Kemp[1]).

DISCUSSION REFERENCE

Coyne, G. V. 1974. Wavelength dependence of polarization. XXV. Rotation of the direction of vibration of the linear polarization by the interstellar medium. *Astron. J.* 78. In press.

---

[1] See p. 607.

MICROWAVE ANALOGUE STUDIES

REINER ZERULL and RICHARD H. GIESE
Ruhr-Universität, Bochum

Equipment for microwave analogue studies is described, and recent results concerning scattering properties of nonspherical particles are presented. Differences of the averaged scattering functions with respect to spheres (Mie theory) are pointed out for rough spheres, octahedrons, and cubes. Averaged scattering functions are presented for single particles and for mixtures.

Spherical particles are adopted for most calculations concerned with the optical properties of interstellar and interplanetary dust. Micrometeorites, collected by rocket experiments, and interstellar polarization effects show, however, that this assumption is not quite correct. In order to obtain results on the scattering properties of nonspherical particles for astronomical problems, microwave analogue experiments were performed by Greenberg, Pedersen, and Pedersen (1961), Lind, Wang, and Greenberg (1965), and by Giese and Siedentopf (1962). In such experiments, the sizes of the wavelength and of the scattering particle are both multiplied by the same factor, say 10^4, in order to transform the optical problem into the microwave region, where the same classical electromagnetic laws are valid. In that way scattering bodies attain a size that can be easily handled in the workshop and during the measuring procedure. Of course, the materials selected for the microwave measurements have to simulate the optical refractive indices of interest for the astronomical problem.

As an example, we will describe microwave equipment used for such analogue studies at the University of Bochum (Bereich Extraterrestrische Physik). We will also present recent results for scatterers that deviate from spherical geometry in order to demonstrate the fundamental differences of the scattering properties with respect to smooth spheres (Mie-theory). Figure 1 shows some scattering bodies (rough spheres, octahedrons, cubes) that were used in the analogue experiments.

EQUIPMENT

As in the optical case, the scatterer has to be illuminated by a plane wave. Therefore, the far-field condition (Rhodes 1954)

$$R_{min} \geq \frac{2(D + d)^2}{\lambda}$$

has to be satisfied. Here R_{min} is the minimum distance from the antenna to the scatterer, D the maximum dimension of the antenna, d the maximum dimension of the scatterer, and λ is the wavelength. In a laboratory of 6 m × 6 m size, a wavelength of 8 mm (Ka-band, 35 GHz) turned out to be an economical compromise between the far-field condition and the particle sizes.

Figure 2 shows the equipment in the laboratory which has absorbing walls. Both the transmitting (right) and the receiving antenna are rectangu-

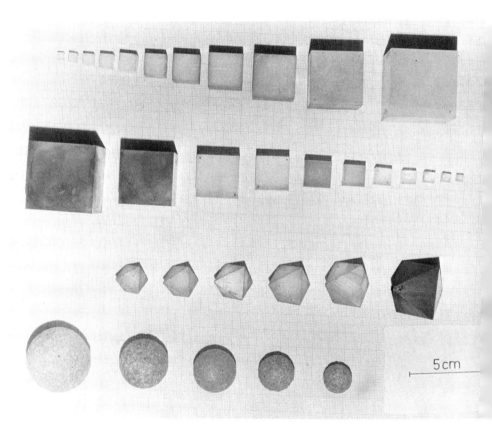

Fig. 1. Scattering bodies used for microwave analogue measurements.

Fig. 2. Scattering laboratory.

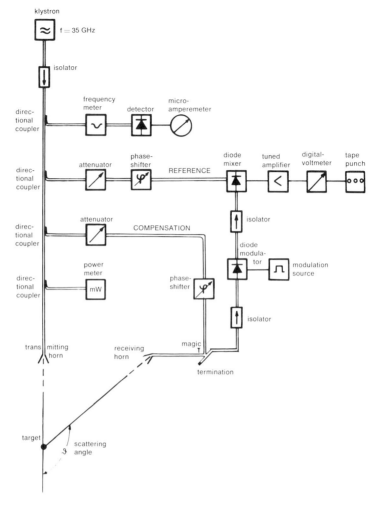

Fig. 3. Block diagram of the apparatus for transmitter and receiver.

lar horns. The receiving antenna can be moved around the target. Thus, scattering angles Θ between 15° and 170° can be obtained in both directions. The scatterer is suspended by four thin nylon threads. By means of these threads and the two stepper motors, to which they are connected, any desired spatial orientation of the scatterer with respect to the antennas can be obtained. The direction of linear polarization of the antennas is variable by means of rotary joints.

Figure 3 is a schematic diagram of the components of the transmitter and receiver. The microwave energy is produced by a reflex klystron (Varian VA 322). The scattered energy is intercepted by the receiving horn and modulated (2 KHz). After passing the diode mixer and a selective amplifier (tuned to 2 KHz), the signal is recorded on punch tape.

Unfortunately, the true scattered signal is blended with secondary signals either reflected from the walls and parts of the construction (not quite avoidable in spite of using absorbing materials) or directly transmitted via side lobes of the antenna pattern. Such unwanted signals are still present when the scattering body is removed. This permitted us to eliminate them by adding a signal of the same amplitude but of opposite phase. Therefore, a part of the klystron energy is branched off into a compensation line and superposed by means of a "magic T." This principle (Lind, Wang, and Greenberg 1965) guarantees that only the unbalanced signal of the scattering body is recorded.

Most parts of the measuring and recording procedure are performed automatically as directed by a preset program. The time for a single measurement is about 2 secs. The size range of scattering bodies measurable with sufficient accuracy is between 0.5 and 10 wavelengths. The lower limit is due to the sensitivity of the receiver and the error signals produced by the threads; the upper limit is determined by the far-field condition. The microwave refractive index of the scattering bodies was evaluated approximately by the reflection coefficient in a waveguide and, after this, more exactly by comparing the scattering functions of a calibration sphere with the Mie theory.

DEFINITIONS

In the figures of the following section, the scattering properties are described by plotting the first Stokes parameter I and the degree of polarization P, both derived from the measurements of $I_{\|}$ and $I_{\perp}$ (van de Hulst 1957):

$$I = I_{\perp} + I_{\|}$$

$$P = \frac{I_{\perp} - I_{\|}}{I_{\perp} + I_{\|}}.$$

As the scattering function I^{ν} of a target in one single spatial attitude is of minor interest, scattering data of targets in n positions can be worked up by computer to obtain for each nonspherical particle a mean scattering function $\bar{I}$ representing an average over these positions, as follows:

$$\bar{I} = \frac{\sum_{\nu=1}^{n}(I_{\perp}^{\nu} + I_{\|}^{\nu})}{n} = \bar{I}_{\perp} + \bar{I}_{\|}$$

$$\bar{P} = \frac{\sum_{\nu=1}^{n}(I_{\perp}^{\nu} - I_{\|}^{\nu})}{\sum_{\nu=1}^{n}(I_{\perp}^{\nu} + I_{\|}^{\nu})} = \frac{\bar{I}_{\perp} - \bar{I}_{\|}}{\bar{I}_{\perp} + \bar{I}_{\|}}$$

For astronomical problems, a knowledge of the scattering properties of polydisperse mixtures is usually necessary. The size distribution of the particle sizes is assumed in this paper to be a power law $dn = x^{-\kappa} dx$, where x is the circumference $2\pi r$ divided by the wavelength. The average function of a mixture of many particles is then

$$\sigma = \frac{\sum_{\mu=1}^{N} I_\mu \cdot x_\mu^{-\kappa}}{\sum_{\mu=1}^{N} x_\mu^{-\kappa}}.$$

In the same manner, average scattering functions $\sigma_\perp$ and $\sigma_\parallel$ can be defined for the two directions of polarization separately:

$$\sigma = \sigma_\perp + \sigma_\parallel.$$

So the degree of linear polarization can be defined as

$$P_\sigma = \frac{\sigma_\perp - \sigma_\parallel}{\sigma_\perp + \sigma_\parallel}.$$

For practical reasons, the sizes of the scattering bodies were chosen in this work according to a geometric series. This choice involves that the weighting exponent κ' is increased by 1 compared to the exponent κ valid for the usual case of equidistant sizes. In all figures presented here this effective value of κ is given.

RESULTS

In this section $I(\Theta)$ and $P(\Theta)$ are presented for rough spheres, octahedrons, and cubes. The results are compared either with spheres of the same volume or with a polydisperse mixture of spheres in the same size interval, calculated with a Mie program (Giese 1971). The choice of the volume as the basis of comparison seems to be reasonable since it is simple, independent of orientation, and rather generally applicable. In our examples, this choice is further justified empirically by the good agreement seen in the details of the corresponding scattering diagrams, such as the shape of the diffraction lobe or the position of maxima and minima. In all measurements presented in this section, the transmitting and the receiving horn had the same direction of polarization, i.e., cross-polarization was not included in the routine measurements. Preliminary measurements of the cross-polarization, however, suggest that its influence is negligible for cubes and rather insignificant in the case of octahedrons. To give a realistic idea of the averaged scattering function, the results given in this section are based on at least 36 different positions. For reason of measuring economy, random orientating could only be approximated (for example cubes and octahedrons were turned around the vertical axis of the equipment in 18 steps, starting from two markedly different inclinations of the body to the vertical axis, thus receiving 36 positions for averaging). For economic reasons, too, polydisperse mixtures of non-spherical particles were simulated by a rather limited number (5 to 11) of scattering bodies.

Rough Spheres

To produce spheres of statistical roughness, smooth spheres were bombarded by metal particles, using the installation of a foundry. Because of this method of production, size and roughness could not be predicted exactly. In particular, the sizes are no longer exactly a geometric series. The roughness is defined by $r_{max} - r_{min} \simeq 2(r_{min} - r)$, where r is the radius of the equivalent smooth sphere (Fig. 4 and Table I). The roughness, obtained by the bombardment, is about 1.5 mm.

TABLE I
List of Rough Spheres Nos. 1 to 5
$m = 1.57 - 0.006\,i$

No.	1	2	3	4	5
r (mm)	8.6	10.9	12.9	14.7	19.1
$x = 2\pi r/\lambda$	6.3	8.0	9.5	10.8	14.0
$(r_{max} - r_{min})/\lambda$	0.13	0.20	0.06	0.17	0.18

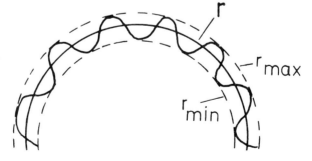

Fig. 4. Definition of roughness: $r_{max} - r_{min}$.

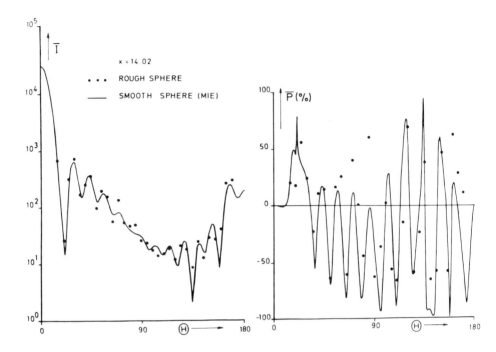

Fig. 5. Averaged scattering function $\bar{I}$ and degree of polarization $\bar{P}$ of a rough sphere, compared to an equivalent smooth sphere; $m = 1.57 - 0.006i$.

In Fig. 5, the averaged scattering properties of the rough sphere No. 5 are compared to an equivalent smooth sphere. There is a good agreement especially in intensity. To find out whether the deviations in polarization are typical, Fig. 6 compares $I(\Theta)$ and $P(\Theta)$ for five rough spheres (Nos. 1–5) first to the results obtained from five equivalent smooth spheres and second to a corresponding polydisperse mixture of smooth spheres. No typical differences are seen.

Comparison of the Mie curves in Figs. 6, 7, and 8 shows that mixtures of only a few particles, each representing a size interval, can give a surprisingly good example of the scattering properties of a corresponding continuous polydisperse mixture. The grade of agreement, of course, increases with the number of particles. In Fig. 8, for example, the correspondence is excellent.

Octahedrons

Because of their plane surfaces, specific differences in the scattering properties of octahedrons (Table II) and spheres might be expected.

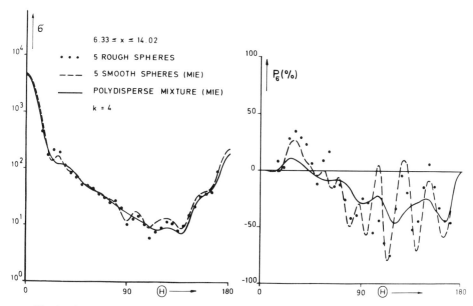

Fig. 6. Averaged scattering function σ and degree of polarization P_σ of a mixture of rough spheres compared to equivalent smooth spheres and to a polydisperse mixture of smooth spheres; $m = 1.57 - 0.006i$.

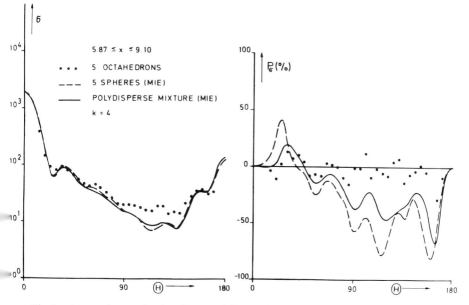

Fig. 7. Averaged scattering function σ and degree of polarization P_σ of a mixture of octahedrons compared to equivalent smooth spheres and to a polydisperse mixture of smooth spheres; $m = 1.57 - 0.006i$.

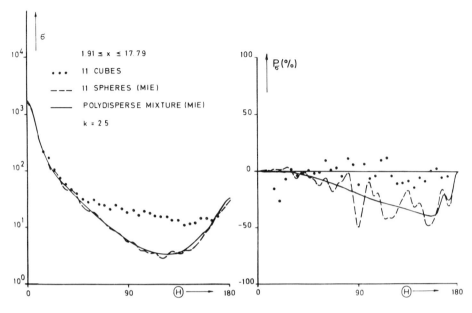

Fig. 8. Averaged scattering function σ and degree of polarization P_σ of a mixture of cubes compared to equivalent smooth spheres and to a polydisperse mixture of smooth spheres; $m = 1.57 - 0.006i$.

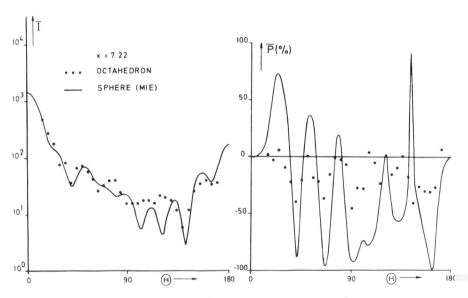

Fig. 9. Averaged scattering function $\bar{I}$ and degree of polarization $\bar{P}$ of an octahedron, compared to an equivalent smooth sphere; $m = 1.57 - 0.006i$.

TABLE II
List of Octahedrons Nos. 1 to 9

a	16.6	18.5	20.6	23.0	25.7	32.1
x	5.9	6.5	7.2	8.1	9.1	11.4
$m = 1.57 - 0.006i$	No. 1	2	3	4	5	—
$m = 1.7 - 0.015i$	No. 6	—	7	—	8	9

But in Fig. 9 the average scattering of octahedron No. 3, for example, is compared to an equivalent sphere. Besides the good agreement in the diffraction pattern, correspondence in singularities, i.e., positions of several maxima and minima, are evident. The same idea is given by Fig. 7, which shows the scattering properties of octahedron Nos. 1–5 compared with five equivalent spheres. Here the specific differences are more definite: less polarization, increased intensity for $50° \lesssim \Theta \lesssim 150°$, and decreased intensity in the backscattering direction.

Cubes

The particles most different from spheres were the cubes, as can be seen in Table III. In Fig. 10, the averaged scattering function of the rather small cube No. 4 is compared to an equivalent sphere. The agreement is very good in intensity and also in polarization, but a tendency to be more neutral in polarization may already be seen. Figure 11 shows the scattering functions of six "small" cubes ($1.9 \lesssim x \lesssim 5.9$) compared to an equivalent polydisperse mixture of spheres. Correspondence in intensity is quite

TABLE III
List of Cubes Nos. 1 to 22

a	4.2	5.3	6.6	8.3	10.3	12.9	16	20	25	31.3	39.1
x	1.9	2.4	3.0	3.6	4.7	5.9	7.3	9.1	11.4	14.2	17.8
$m = 1.57 - 0.006i$	No. 1	2	3	4	5	6	7	8	9	10[a]	11[a]
$m = 1.7 - 0.015i$	No. 12	13	14	15	16	17	18	19	20	21	22

[a] Refractive index cubes 10 and 11 were $1.5 - 0.005i$

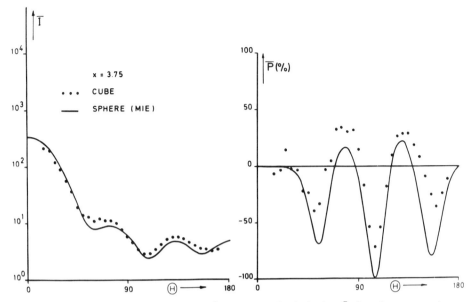

Fig. 10. Averaged scattering function $\bar{I}$ and degree of polarization $\bar{P}$ of a cube, compared to an equivalent smooth sphere; $m = 1.57 - 0.006i$.

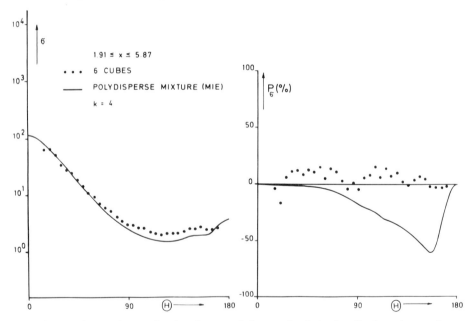

Fig. 11. Averaged scattering function σ and degree of polarization P_σ of a mixture of cubes compared to an equivalent polydisperse mixture of smooth spheres; $m = 1.57 - 0.006i$.

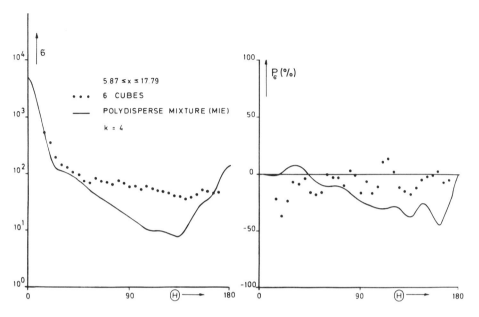

Fig. 12. Averaged scattering function σ and degree of polarization P_σ of a mixture of cubes compared to an equivalent polydisperse mixture of smooth spheres; $m = 1.57 - 0.006i$.

good, apart from a slight increase at medium scattering angles. Polarization is already completely different. Figure 12 shows the averaged scattering functions of six "big" cubes, Nos. 6–11 ($5.9 \lesssim x \lesssim 17.8$). Here the differences with respect to spheres become very evident. Figure 8 shows the scattering properties of 11 cubes, Nos. 1–11. Contrary to spheres, there is a tendency to practically neutral polarization. The total scattered intensity agrees with the mixture of spheres in the forward pattern but is nearly isotropic at larger scattering angles ($\Theta \lesssim 50°$). No increase in the backscattering regions occurs as in the case of dielectric spheres.

CONCLUSIONS

In the literature, the examples of scattering by assemblies of nonspherical particles do not correspond exactly to the refractive indices, shapes, and sizes of the samples used in this investigation. Some qualitative comparisons are, however, possible. The results of experimental observations on silver bromide sols of octahedral and cubic particles (Napper and Ottewill 1963), for example, are not contradictory to those presented here.[1] The maxima and minima for spheres have the same location for octahedrons and cubes, but there are quantitative deviations from the theoretical functions of spheres, especially by the cubes. Furthermore, the fact that the greatest deviation occurs in the backward hemisphere agrees with our results. Another paper (Holland and Gagne, 1969) presents results for irregular flat particles. Their properties are similar to those given here for cubes: good agreement with spheres concerning the diffraction pattern, higher intensity for medium scattering angles, and no increase toward backscattering. In another part of this book, the scattering measurements presented here are applied to zodiacal light model calculations.[2]

While predictions for assemblies of extremely small ($x \ll 1$) particles and for larger ($x \gg 1$) absorbing particles in random orientation are possible from theoretical arguments, it is advisable to extend the experimental investigations also to absorbing particles, especially in the size region $1 \lesssim x \lesssim 20$.

Acknowledgment. This work was supported by the Deutsche Forschungsgemeinschaft (Contract Gi 52).

REFERENCES

Giese, R. H., and Siedentopf, H. 1962. Ein Modellversuch zur Bestimmung der Streufunktionen nicht kugelförmiger Teilchen mit 3 cm-Wellen. *Zeitschr. Naturforschung* 17a: 817–819.

———. 1971. *Tabellen von MIE-Streufunktionen.* Forschungsbericht W 71–23, Bundesminister für Bildung and Wissenschaft, Bonn.

Greenberg, J. W.; Pedersen, N. E.; and Pedersen, J. C. 1961. Microwave analog to the scattering of light by non-spherical particles. *J. of Appl. Physics,* 32: 233–242.

Hulst, H. C. van de. 1957. *Light scattering by small particles.* New York: Wiley.

Lind, A. C.; Wang, R. T.; and Greenberg, J. M. 1965. Microwave scattering by nonspherical particles. *Applied Optics* 4: 1555–1561.

Napper, D. H., and Ottewill, R. H. 1963. Light scattering studies on monodisperse silver bromide sols. *Electromagnetic waves* (M. Kerker, ed.), pp. 377–386. New York: Pergamon Press.

Rhodes, D. R. 1954. On minimum range of radiation patterns. *Proceedings of the I.R.E.,* pp. 1408–1410.

---

[1] Such optical scattering measurements have the advantage to give an average over all random orientations in much less measuring time. On the other hand microwave analogue measurements provide a chance to simulate assemblies of partially oriented particles, processing data once recorded by simple computation. Furthermore, shapes and refractive indices of the scatterers can be varied independently and systematically.

[2] See p. 804.

DISCUSSION

GEHRELS: Can circular polarization be studied with your microwave analogue techniques? Do you plan such measurements, which seem important now that circular polarization is being observed on interstellar and interplanetary particles?

GIESE and ZERULL: We are quite aware that the measurement of circular polarization is highly desirable. We plan to extend investigations in that direction in the very near future but are still working on the problem, to keep alterations of the equipment and measuring procedure within economic limits.

EFFECTS OF PARTICLE SHAPE ON THE SHAPE OF EXTINCTION AND POLARIZATION BANDS IN GRAINS

J. MAYO GREENBERG and SEUNG SOO HONG
State University of New York at Albany

Exact and approximate calculations are made for extinction by spheres, spheroids, and cylinders within absorption bands. Calculations are also made for the wavelength dependence of polarization by variously oriented cylinders. It is found that the oscillation of the band extinction by spheroids and cylinders varies with particle size essentially like that for spheres. The oscillations of the polarization by cylinders have a significantly different size dependence.

It is well known that the shape of absorption bands as produced by spherical particles whose size is comparable to the wavelength is distinctly different from the characteristic bulk absorption. For nonspherical particles whose size to wavelength ratio $x = \frac{2\pi a}{\lambda} \ll 1$, there exist nontrivial absorption modifications (Greenberg 1972). We consider here to what extent the particle shape influences the absorption when the particles are such that $x \gtrsim 1$. In particular, we inquire how the degree of orientation of the particles as indicated by the interstellar polarization may introduce differences in the shape of the 4430 Å diffuse band if this band originates in an imbedded constituent of the grains.

The calculations for spheres are performed exactly. Spheroidal calculations are made using a ray (Wentzel-Kramers-Brillouin; W.K.B.) approximate method (Greenberg 1960, 1968). In order to justify the approximation, we first show that the exact Mie theory results are, in terms of band shape, very close to the ray approximation results. The ray calculation for spheres was first performed by van de Hulst (1946).

The bulk optical properties are those for "dirty ice" with the complex index of refraction around the absorption bands as given in Table I.

TABLE I
Complex Index of Refraction $m = m' - im''$ for a 4430 Å Absorption Band in "Dirty Ice" as Used in Greenberg and Stoeckly (1971)

$\lambda(\mu m)$	m'	m''
0.4380	1.324	0.0219
.4340	1.316	.0254
.4375	1.306	.0333
.4400	1.296	.0525
.4415	1.299	.0792
.4430	1.334	.1034
.4445	1.372	.0837
.4460	1.378	.0562
.4485	1.369	.0354
.4520	1.359	.0266
0.4580	1.351	0.0226

EXACT AND APPROXIMATE BAND SHAPES FOR SPHERES

Figure 1 contains the band shapes for various values of the size parameter x. In these and in all other cases, we define x by $x = 2\pi a/4430$ Å. The calculations use (1) Mie theory, (2) the ray approximation, (3) the Rayleigh approximation, and (4) an expansion in x as given by van de Hulst (1957). The shape of the latter differs negligibly from that for the Mie theory at $x = 1$. It should be noted that both the Rayleigh and modified Rayleigh approximations greatly overestimate the extinction for $x > 1$ even when the band shape is reproduced reasonably well. The ray results are seen to reproduce the Mie theory band shapes quite faithfully. The largest difference in absolute value is at $x = 1$. Otherwise the differences are not very large.

As plotted, all the extinctions are normalized by dividing by

$$\frac{Q(\lambda_1) + Q(\lambda_2)}{2}$$ where $\lambda_1 = 4280$, $\lambda_2 = 4580$. The quantity

$$\bar{Q} = \frac{Q(\lambda_1) + Q(\lambda_2)}{2}$$ is close to the average extinction in the absence of an absorption band. The ratios of the approximate to the exact (Mie) values of $\bar{Q}$ are given in Table II.

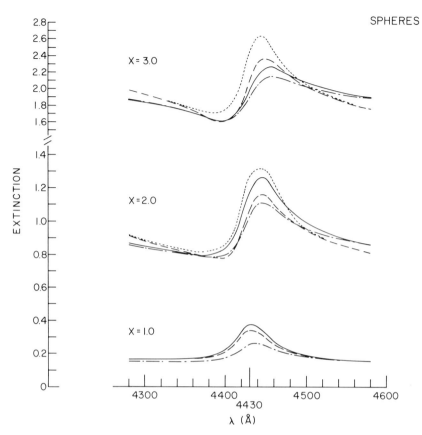

Fig. 1a. Comparison of the ray approximation (—·—·—·), Rayleigh approximation (———), and an extended Rayleigh approximation (·····) with the Mie theory for spheres S. For $x = 1$, the extended Rayleigh approximation gives the same result as the Mie theory. The approximate extinction efficiencies are given as relative to the Mie result by $Q_{rel}(\lambda) = Q(\lambda) \times (\bar{Q}_{Mie}/\bar{Q})$, where $\bar{Q}$ is as defined in the text $\bar{Q} = [Q(\lambda_1) + Q(\lambda_2)]/2$.

BAND SHAPES FOR SPHEROIDS

Let us consider prolate spheroidal grains of semimajor axis b and semiminor axis a such that $b/a = 2$. Let $Q(\chi)$ be the efficiency when the spheroid symmetry axis (b) makes an angle χ with respect to the direction of the incident radiation.

Perfect spinning (Davis-Greenstein) orientation giving maximum polarization requires the particles to be spinning about their short axes with the plane of the particle symmetry axes containing the propagation vector. Thus, the extinction produced when there is maximum polarization is given by

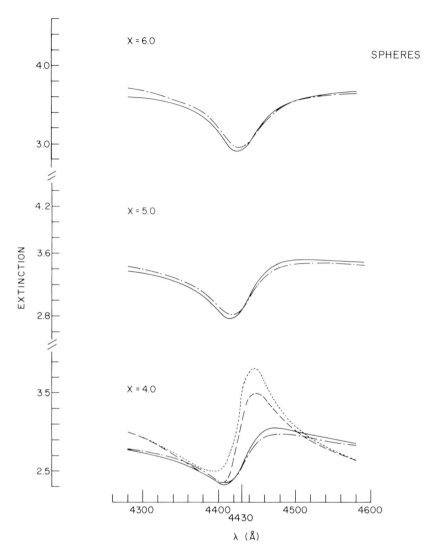

Fig. 1b. Same as for Fig. 1a.

$$Q_{DG,\perp} = \frac{2}{\pi} \int_0^{\pi/2} Q(\chi) d\chi. \quad (1)$$

If the magnetic field is along the line of sight, then the spin plane is normal to the line of sight and

$$Q_{DG,\parallel} = Q(\chi = 90°). \quad (2)$$

TABLE II
Extinction Efficiencies $\bar{Q}$ (See Text) for Spheres Calculated by Mie Theory, Ray Approximation, Rayleigh Approximation, and An Extended Rayleigh Approximation. (The latter three are given as relative values to the Mie theory results.)

		Relative value		
x	Q_{Mie}	ray	Rayleigh	Mod. Rayleigh
1	0.160	1.70	1.04	1.16
2	0.860	1.03	2.28	2.46
3	1.88	0.87	5.10	5.38
4	2.82	0.83	10.7	11.1
5	3.43	0.81	21.3	22.0
6	3.63	0.81	41.7	42.9

This situation produces zero polarization. Zero polarization is also produced by random orientation, for which we have

$$Q_R = \frac{2}{\pi} \int_0^{\pi/2} Q(\chi) \sin \chi \, d\chi. \tag{3}$$

The values of $Q_{DG,\perp}$, $Q_{DG,\|}$, and Q_R are in Fig. 2. The curves are again normalized by division by $\bar{Q}$, and the relative values to $Q_{DG,\|}$ are given in Table III. We have defined the value of x by the smaller dimension of the spheroid, i.e., $x = 2\pi a/\lambda$ where again $\lambda = 4430$ Å.

It appears that there are no *major* differences in band shapes between the three types of orientation. It should be noted that the nature of the ray approximation implies that the $Q_{DG,\|}$ shape is exactly that for a sphere of radius a. In general, the modification of the band shape seems to be somewhat depressed for the spinning spheroid as compared with the result for spheres.

BAND SHAPES FOR INFINITE CYLINDERS

In Fig. 3 are shown the normalized extinction band shapes for picket fence (perpendicular incidence) alignment and for perfect Davis-Greenstein alignment. The normalization is made by division by $\bar{Q}$, and the values are given in Table IV. We see that increasing the randomness of orientation leads, as in the case of the spheroids, to a reduction in

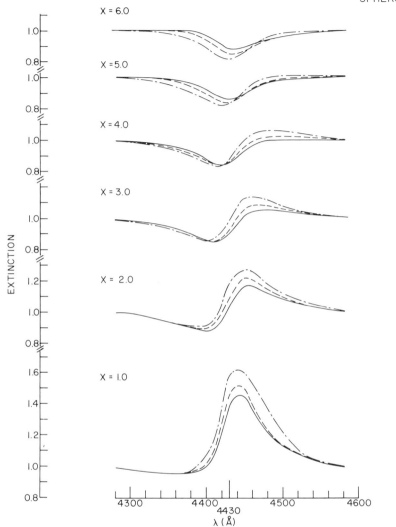

Fig. 2. Normalized extinction efficiencies ($Q/\bar{Q}$) for variously oriented spheroids. The solid line represents the case of perfect Davis-Greenstein spinning alignment when the magnetic field is perpendicular to the line of sight ($Q_{DG,\perp}$). The dot-dash line represents the case $Q_{DG,\parallel}$. The dashed line is for Q_R. See text.

the amplitude of the band oscillation. We also see that the "center" of the band is shifted toward longer wavelengths. This again was predicted by the spheroid results in Fig. 2.

TABLE III
Extinction Efficiencies $\bar{Q}$ for Variously Oriented Spheroids

x	ABSOLUTE			RELATIVE		
	$\bar{Q}_{DG,\parallel}$	$\bar{Q}_{DG,\perp}$	$\bar{Q}_R$	$\bar{Q}_{DG,\parallel}$	$\bar{Q}_{DG,\perp}$	$\bar{Q}_R$
1	0.679	0.546	0.622	1.00	0.805	0.917
2	2.04	1.77	1.94	1.00	0.866	0.952
3	3.31	3.27	3.36	1.00	0.988	1.01
4	4.02	4.65	4.38	1.00	1.16	1.09
5	4.17	5.57	4.79	1.00	1.34	1.15
6	3.94	5.87	4.67	1.00	1.49	1.19

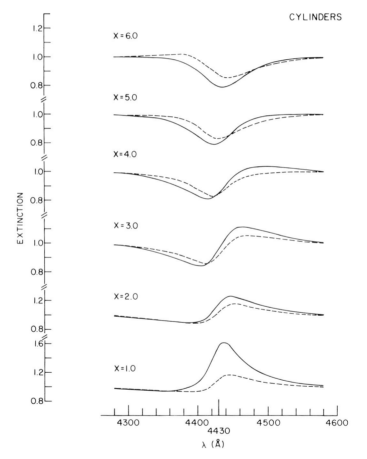

Fig. 3. Normalized infinite cylinder extinction efficiencies ($Q^{ext}/\bar{Q}^{ext}$) for perfect alignment (———) and perfect Davis-Greenstein alignment (— — —).

TABLE IV
Extinction Efficiencies, $\overline{Q}$, for Picket Fence and Davis-Greenstein Oriented Infinite Cylinders

x	PF	DG	DG/PF
1	0.359	0.824	2.29
2	1.22	1.32	1.08
3	2.25	1.90	0.84
4	3.09	2.18	0.70
5	3.50	2.13	0.60
6	3.41	1.88	0.54

Perhaps the most interesting and unexpected results in this paper are contained in Fig. 4. First of all it is clear that the shape of the polarization oscillation is more *strongly* dependent on size than is the shape of the extinction oscillation. Note also that for increasing size, the degree of randomness of alignment produces much more significant differences in the polarization wiggle than in the extinction. However, for the size range of interest in interstellar polarization ($x \cong 1.5$), the earlier theoretical prediction on the polarization wiggle band based on picket fence orientation (Greenberg and Stoeckly 1971) is essentially unaffected.

The principal difference between the extinction and polarization band shapes for spinning cylinders is that, whereas the extinction goes from a positive increment (absorption) to a dispersion curve and finally to a negative increment (apparent emission) as the size increases, the polarization goes from a dispersion *through* a depression to a kind of reversed dispersion for increasing x. A reversed dispersion is defined here as a positive increment at shorter wavelengths and a negative increment at longer wavelengths about the "center" of the band.

The relative values of the polarization efficiencies for picket fence and Davis-Greenstein orientation are given in Table V.

SUMMARY

In general, there are no qualitative differences in extinction band shapes for smooth spherical and nonspherical particles. The degree of oscillation within a band is, as expected, less for randomly oriented nonspherical particles than for equivalent-sized spheres. On the other hand, the shape of the polarization band produced by grains having a realistic amount of interstellar alignment depends *differently* on the particle size than does the extinction.

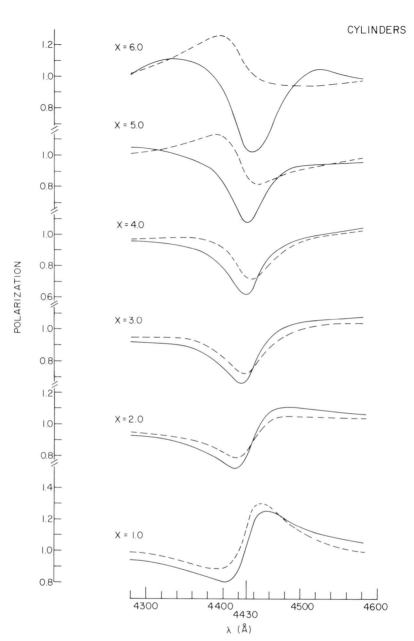

Fig. 4. Normalized infinite cylinder polarization efficiencies ($Q^{pol}/\bar{Q}^{pol}$) for perfect alignment (———) and perfect Davis-Greenstein alignment (— — —).

Acknowledgment. This work was supported in part by the National Aeronautics and Space Administration, Grant NGR 33-011-043.

TABLE V
Polarization Efficiencies, $\bar{Q}$, for Picket Fence and Davis-Greenstein Oriented Infinite Cylinders

x	PF	DG	DG/PF
1	0.321	0.209	0.651
2	0.395	0.415	1.05
3	0.342	0.370	1.08
4	0.250	0.235	0.943
5	0.167	0.124	0.743
6	0.101	0.077	0.761

REFERENCES

Greenberg, J. M. 1960. Scattering by nonspherical particles. *J. Applied Phys.* 31: 82–84.
———. 1968. Interstellar grains. *Stars and stellar systems.* (B. M. Middlehurst and L. M. Aller, eds.) Vol. VII, Ch. 7. Chicago: The Univ. of Chicago Press.
———. 1972. Absorption and emission of radiation by nonspherical particles. *J. Colloid and Interface Sci.* 39: 5–13, 513–519.
Greenberg, J. M., and Stoeckly, R. 1971. Shape of diffuse interstellar bands. *Nature Phys. Sci.* 230: 15–16.
Hulst, H. C. van de 1946. Optics of spherical particles. *Rech. Astr. Obs. Utrecht* 11 (Part 1): 1–87.
———. 1957. *Light scattering by small particles.* New York: Wiley.

DISCUSSION

SERKOWSKI: I would like to call attention to a paper by Rudkjøbing who predicts strong polarization in some interstellar absorption features which he attributes to forbidden lines of the negative hydrogen ion enforced by the Paschen-Back effect in the galactic magnetic field.

DISCUSSION REFERENCE

Rudkjøbing, M. 1969. Diffuse interstellar absorption bands as due to quadruple transitions enforced by a magnetic field. *Astrophys. Space Sci.* 5: 68–70.

INTERSTELLAR CIRCULAR POLARIZATION: A NEW APPROACH TO THE STUDY OF INTERSTELLAR GRAINS

P. G. MARTIN
University of Toronto

Interstellar circular polarization has been discovered recently in the directions of the Crab Nebula and certain stars. As a basis for analyzing these observations, we present details of how aligned elongated interstellar grains govern the optical properties of the interstellar medium and cause linear dichroism and birefringence. It is emphasized that the wavelength dependences of both the circular and linear polarization are needed for a complete analysis, leading to a knowledge of the composition of the grains and their size. Present observations of circular polarization clearly show that the grains are dielectric.

Circular polarization of the light from stars arises when the direction of grain alignment changes along the line of sight. In such a twisting medium, if the grain size also changes along the line of sight, the position angle of linear polarization can be wavelength dependent. Strong evidence is obtained when both phenomena are observed in the same star. One known example is analyzed to illustrate how the amount of twist and the variation in grain size can be obtained.

That the interstellar medium exhibits linear dichroism because of the presence of aligned anisotropic (elongated) grains is well known from studies of interstellar linear polarization. A large amount of data is now available on this aspect of the medium.[1] The same alignment causes linear birefringence as well. Thus, the medium may be thought of as a wave plate, albeit a very inefficient one, that can convert suitably oriented linearly polarized light into elliptically polarized light. The resulting component of circular polarization is so small ($V/I \lesssim 10^{-3}$) that until recently it has been largely ignored; however, interstellar circular polarization has now been detected, at first in the direction of the Crab Nebula (Martin, Illing, and Angel 1972), and more recently in several stars (Kemp 1972).[2]

In this paper we present some theoretical considerations of interstellar circular polarization as a basis for discussing the existing observations.

---
[1]See p. 888.   [2]See also p. 939.

It is shown how such observations, taken in conjunction with observations of interstellar linear polarization and extinction, provide important information about the chemical composition of the interstellar grains, their size, and their alignment.

OPTICAL PROPERTIES OF THE INTERSTELLAR MEDIUM

The interstellar medium containing grains is equivalent to a medium with complex refractive index m; m can be calculated from the grain properties (van de Hulst 1957). When the grains are aligned there are two refractive indices, m_1 and m_2, corresponding to orthogonal orientations of the incident electric vector, along and perpendicular respectively to the long axis of the grain profile.

In terms of $S_j(O)$, the complex amplitude functions for radiation scattered in the forward direction by a grain are

$$m_j = 1 + \frac{\lambda}{4\pi} NC_{p,j} - i\frac{\lambda}{4\pi} NC_{e,j}; j = 1, 2, \qquad (1)$$

and

$$C_j = C_{e,j} + iC_{p,j} = \frac{\lambda^2}{\pi} S_j(O), \qquad (2)$$

where N is the number density of grains, C_e is the extinction cross section per grain, and C_p is the analogous quantity describing the phase shift (Serkowski 1962; Martin 1972). It follows that the relative phase shift and the differential extinction per unit pathlength are determined by

$$\Delta\varepsilon = \tfrac{1}{2}N(C_{p,1} - C_{p,2}) \qquad (3)$$

and

$$\Delta\sigma = \tfrac{1}{2}N(C_{e,1} - C_{e,2}) \qquad (4)$$

respectively. Calculations of both the birefringence and the dichroism can be carried out using the Mie theory for infinite circular cylinders (Martin 1972); size and grain orientation distributions can be included when desirable.

Serkowski (1962) has given the four differential equations describing the transformation of the Stokes parameters in such a medium. When the degree of polarization produced is small ($p^2 \sim V/I \ll p$), as is usually the case for interstellar polarization, then Martin (1972) gives for pathlength s:

$$I^{-1}\frac{dI}{ds} = -\frac{1}{2}N(C_{e,1} + C_{e,2}) \tag{5}$$

$$\frac{d(Q/I)}{ds} = \Delta\sigma \tag{6}$$

$$\frac{d(U/I)}{ds} = 0 \tag{7}$$

$$\frac{d(V/I)}{ds} = \Delta\varepsilon U/I = \left(\frac{\Delta\varepsilon}{\Delta\sigma}\right)\frac{d(Q/I)}{ds}U/I \tag{8}$$

when the position angle is measured from axis two. The right hand side of Equation (8) is convenient for estimating $d(V/I)/ds$ since $\Delta\varepsilon/\Delta\sigma \sim 1$.

Note that, as for any wave plate, only the U component of linear polarization is relevant for linear-to-circular conversion. Thus, in evaluating observational possibilities, we require radiation linearly polarized at an inclination to the principal axes 1 and 2. In the case of the Crab Nebula, which is intrinsically strongly polarized, many individual small regions are found with different position angles relative to the axes of the foreground medium. For stars that have no intrinsic polarization, it is still possible to obtain circular polarization if the grain alignment "twists" along the line of sight (Serkowski 1962). One model of this type of medium is discussed below. The observed effect for the stars is about an order of magnitude lower than for the Crab Nebula because the degree of linear polarization "incident" on the medium is smaller by that amount.

RESULTS OF MODEL CALCULATIONS

Circular polarization observations are of particular importance for determining the grain composition, since the wavelength dependence predicted for different materials varies markedly (Martin 1972). This is made clear by Fig. 1 which shows several curves of the relative phase shift versus wavelength for different grain materials. Before we discuss the differences in the birefringence, it is important to note that each grain model was constrained (i.e., the size of the grain was chosen) so that the predicted wavelength dependence of linear polarization would match the mean curve observed by Coyne and Gehrels (1967); also each model was scaled (i.e., the column density of grains was chosen) so that the same degree of linear polarization was produced, namely 2% at maximum. Therefore Fig. 1 displays the real differences in birefringence between grain models, all of which will nevertheless produce the same linear polarization.

CIRCULAR POLARIZATION: A NEW APPROACH

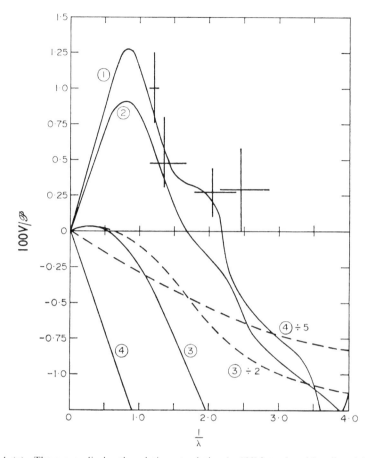

Fig. 1. (a) The curves display the relative retardation ($\times 100$) introduced by aligned interstellar grains with different refractive indices. Dielectric grains: ①$m = 1.5$, ②$m = 1.5 - 0.1i$; Metallic grains: ③$m = 1.5 - 0.5i$, ④$m = 1.5 - 1.5i$. The ordinate also gives the circular polarization produced when linearly polarized light, degree $\mathscr{P}$, is incident on the medium at position angle $45°$ relative to the wave plate axis.

(b) The four points represent observed values of the relative retardation in front of the Crab Nebula.

Although the models (picket-fence alignment, single size, constant refractive index) are not the most accurate approximations to the actual aligned grains, they are sufficient to demonstrate the large composition-dependent variations in the relative phase shift. The shape of the curve is most sensitive to the imaginary part of the complex refractive index of the grain material. The most important feature to note is the sign change, characteristic of dielectric materials but absent for highly absorbing (metallic) grains. The predicted change in $|V|$ from 1 μm toward shorter wavelengths is a decrease for dielectrics but an increase for metallic grains. Thus, the wavelength dependence of circular polarization should be measured. Also if the sign of $\Delta\varepsilon$ can be found, as in the special case of the Crab Nebula, then another choice between the types of grain material is possible. Observations of circular polarization can therefore be used to discriminate between proposed grain compositions.

OBSERVATIONS OF INTERSTELLAR CIRCULAR POLARIZATION

The few existing observations of interstellar polarization are discussed here to show what has already been learned about the grain composition.

The Crab Nebula

Interstellar circular polarization was first detected in the direction of the Crab Nebula; these observations are discussed in detail by Martin, Illing, and Angel (1972). There are several reasons why the Crab Nebula was chosen to study this phenomenon. First, from observations of neighboring stars, the interstellar linear polarization in this direction is known[3] to be about 2% at position angle 145°. Thus, the orientation θ_0 and the retardation (Fig. 1) of the wave plate are known. Second, the intrinsic (synchrotron) linear polarization of small regions of the Nebula is large, so that the circular polarization is expected to be relatively large $[V/I \sim 0.1\%$; Equation (8)]. Finally, the position angle of the intrinsic polarization $\mathscr{P}$ of the Crab Nebula relative to the wave plate axis (i.e., $\theta - \theta_0$) varies from point to point in the Nebula[4] so that, as described below, the circular polarization can be shown to be of interstellar origin.

Circular polarization from interstellar grains is expected to change for different regions of the Nebula according to

$$V/\mathscr{P} = A(\lambda)\sin 2(\theta - \theta_0), \qquad (9)$$

where $A(\lambda) = \Delta\varepsilon s$ is the total retardation of the medium. Many different regions of the Nebula were observed to ensure a variety of position angles. The curve $A(\lambda)\sin 2(\theta - \theta_0)$ fitted to plots of $V/\mathscr{P}$ versus θ for each passband gives the magnitude and sign of the retardation $A(\lambda)$ and the orientation of the wave plate θ_0. Results from a preliminary analysis of data obtained in November 1972 by Angel and myself are shown in Fig. 2. It can be seen that the fit to a sine curve is good, with the exception of one highly polarized point. The value of θ_0 from this curve is near

---
[3] See p. 997.   [4] See p. 1014.

CIRCULAR POLARIZATION: A NEW APPROACH

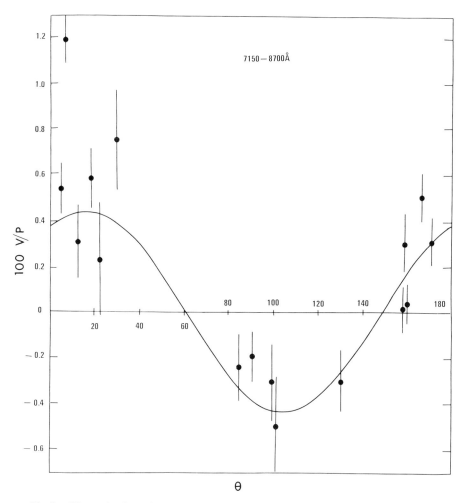

Fig. 2. Observed values of the circular polarization V in small regions of the Crab Nebula, normalized by the degree of intrinsic linear polarization $\mathscr{P}$, and plotted against θ, the position angle of the intrinsic linear polarization. A sine curve is fitted to these data. The intercept gives the orientation of the wave plate, near 145°, the value obtained independently from observations of interstellar linear polarization. The amplitude can be compared to the predictions of the grain models in Fig. 1. (This figure is only a preliminary analysis of recent observations.)

145°, in agreement with the value of θ_0 determined from the interstellar linear polarization observations.

The characteristic sine wave dependence on θ, based on points scattered indiscriminantly over the Nebula, and the agreement of θ_0 require that the circular polarization be of interstellar origin and not intrinsic to the

Nebula. The intrinsic circular polarization, which is predicted by the "synchro-Compton" theory for the Nebula emission, would be of opposite sign on opposite sides of the pulsar (Rees 1971; Blandford 1972; Arons 1972). Furthermore, this theory predicts more than 1% polarization (an order of magnitude larger than found here) and requires that it be independent of wavelength, in contradiction to the observations discussed below.

From sine curves fitted to previous data (Martin, Illing, and Angel 1972), the retardation $A(\lambda)$ for four different passbands was determined (θ_0 was also determined but with less accuracy than in Fig. 2). These values are plotted in Fig. 1 which, it will be recalled, shows curves of the relative phase shift predicted for a medium producing 2% linear polarization. All three factors—the observed sign, the magnitude, and the wavelength dependence of the circular polarization—indicate a dielectric composition for the grains; each independently rules out highly absorbing grains.

A more detailed study of the wavelength dependence of the circular polarization is being made by Angel and myself. Also the interstellar linear polarization will be investigated using several emission lines from the filaments of the Crab Nebula. These combined data are to be compared with the predictions of more detailed models of the aligned interstellar grains to determine both the grain size and the complex refractive index of the grain material. Already, at least along this 2 kpc path towards the galactic anticenter, the grains are restricted to being dielectric, with a low imaginary part of the refractive index.

Stars

It is important to realize that stars can also show interstellar circular polarization, since then many different directions in our galaxy can be examined. For this to happen, however, the grain alignment must twist along the line of sight (Serkowski 1962). In that case, the alignment of the grains can be studied, as well as their composition and size.

Because this effect is so small ($V/I \sim 0.01\%$) it has been detected only recently (Kemp 1972; Kemp and Wolstencroft 1972),[5] although several attempts had been made previously. The degree of circular polarization actually detected agrees with that expected by scaling down the values measured in the Crab Nebula. Other indirect arguments that it is of interstellar origin are presented in the next section.

The circular polarization for each of the six stars in which it has been found has opposite signs in the red and blue spectral regions. It is important to recognize that these observations can be used to identify the

---

[5] See also p. 939.

grain composition, if interpreted by means of Fig. 1 or similar calculations. For instance as remarked earlier, the observed sign reversal is a characteristic of dielectric grains only. Thus, even though the sign of $\Delta\varepsilon$ is unknown, we can conclude that the grain material in front of these stars is dielectric, as it is in front of the Crab Nebula.

More detailed observations of the wavelength dependence of circular polarization would be important since the position of the crossover is dependent on both the grain size and the composition. If, in addition, observations of the wavelength dependence of linear polarization were available, both of these grain parameters could be determined with some uniqueness using model computations. From a detailed study of the interstellar polarization of a wide distribution of stars, much could be learned about the variation of these grain parameters with position in our galaxy.

Several other kinds of investigation support the conclusion that the grains are dielectric. Serkowski (1972) has found an empirical formula [see Eq. (14)] that fits detailed observations of the wavelength dependence of the interstellar linear polarization of many stars. As he has pointed out, grain models using dielectric compositions are required to reproduce this curve. Discrimination based on colors of reflection nebulae is possible, although some ambiguity may be introduced by the (unknown) geometry of the dust cloud. In some cases a clear distinction in favor of dielectric grains has been possible (Greenberg and Roark 1967; Zellner[6]). Analysis of observations of diffuse galactic light gives a locus of possible values of the albedo and the asymmetry parameter for scattering (van de Hulst and de Jong 1969; Mattila 1971). Nevertheless, highly absorbing materials like iron and graphite are ruled out. On the other hand, dielectric grains are acceptable if they are slightly absorbing (albedo not unity). The surface brightness of dark nebulae has been interpreted in the same way (Mattila 1970).

POLARIZATION FROM A MEDIUM WITH TWISTING ALIGNMENT

Because interstellar circular polarization can be produced when the direction of grain alignment changes along the line of sight, it can be a probe of the source of grain alignment, usually assumed to be the galactic magnetic field. Several simple models of such a medium are discussed elsewhere (Martin 1973). The only one that will be described here is where there is a regular twist in the alignment direction in an otherwise uniform medium. Kemp (1972) has presented this model in the approximation of small twist. Actually, the amount of twist is apparently large for the stars

---

[6]See p. 867.

observed, so that the general solution is more appropriate. From Equations (6)–(9) it can be shown that for total twist ϕ in a medium that would produce polarization $p_0 = (Q/I)_0 = \Delta\sigma s$, $V_0 = 0$ for $\phi = 0$, the polarization is given by

$$Q/Ip_0 = \sin 2\phi/2\phi, \ U/Ip_0 = \sin^2\phi/\phi; \quad (10)$$

$$p/p_0 = \sin\phi/\phi \equiv D(\phi), \ \theta = \phi/2; \quad (11)$$

$$V/Ip_0\Delta\varepsilon s = -\frac{2\phi - \sin 2\phi}{(2\phi)^2} \equiv -v(\phi); \quad (12)$$

$$V/Ip^2 = -\frac{\Delta\varepsilon}{\Delta\sigma}\frac{2\phi - \sin 2\phi}{4\sin^2\phi} \equiv -\frac{\Delta\varepsilon}{\Delta\sigma}F(\phi). \quad (13)$$

The functions D, F, and v are plotted in Fig. 3. Notice the broad range in ϕ for which there is significant circular polarization.

Application to σ Sco. As an illustration of how these equations might be used, we analyze the polarization data for σ Sco. The linear and circular polarization have both been measured as a function of wavelength by Coyne and Gehrels (1966) and Kemp and Wolstencroft (1972) respectively. The wavelength of the maximum of linear polarization, λ_{max}, is 0.555 μm, indicating that the grains are of "normal" size. As discussed above, the sign change in V shows that the grains are dielectric. Clearly a more detailed analysis of the size and composition could be carried out.

Since the measurements in the red and the blue passbands give $|V/Ip^2| \sim 0.5$ and for similar passbands and dielectric materials, $|\Delta\varepsilon/\Delta\sigma| \sim 0.3$ (from Martin 1972, Figs. 1 and 2), we can calculate that $|F(\phi)| \sim 1.7$. Thus, the amount of twist in the medium is found (Fig. 3) to be very large, $\phi \sim 115°$, and nearly the optimum amount (90°). Again a more detailed analysis is possible, using a better estimate of $\Delta\varepsilon/\Delta\sigma$, which would be available from the determination of the composition and size.

An independent check on the amount of twist is obtained by observing p_V/E_{B-V}. This ratio is usually observed to be less than a "limiting" value 0.19 (Hiltner 1956; p_V is in magnitudes and the V is the visual wavelength of the UBV system). One of the several reasons this ratio might be lower than 0.19 is the depolarization resulting from a change in the direction of grain alignment. The ratio of p_V/E_{B-V} to 0.19 gives a lower limit to D and, hence, an upper limit to ϕ. For σ Sco, $E_{B-V} = 0.39$ and $p_V = 0.0347$ so that $D \geq 0.47$ and $\phi \leq 115°$. Here the close agreement in ϕ must be considered fortuitous since the analysis is only approximate; nevertheless it is indicative of the validity of the model. Note that this upper limit on ϕ could also be used to give an upper limit to F (i.e., 1.7) and hence a lower limit to $\Delta\varepsilon/\Delta\sigma$.

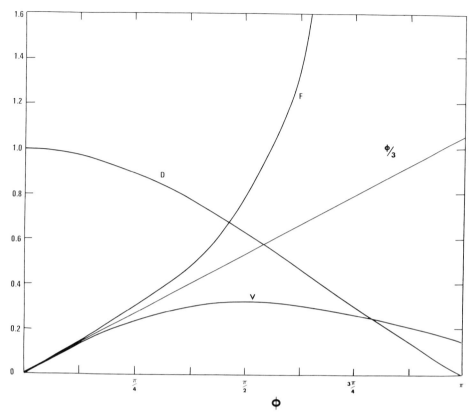

Fig. 3. Three polarization parameters [Equations (11)–(13)] for a medium in which the direction of grain alignment twists by an amount ϕ along the line of sight. Depolarization of the linear component p is described by D. The normalized circular polarization is v, and F is proportional to the ratio V/Ip^2.

Wavelength Dependence of the Position Angle of Linear Polarization

In a medium with twisting alignment there is the interesting possibility that the position angle of the linear polarization will change with wavelength.[7] In an otherwise uniform medium there is no effect if the degree of polarization is not large [cf. the solution $\theta = \phi/2$ in Equation (11)], although possibly small amounts of rotation could occur if the medium is highly polarizing. However, even in the usual limit of small polarization, rotation can occur if the wavelength dependence of linear polarization changes along the line of sight (Treanor 1963; Gehrels and Silvester 1965).

---

[7] See p. 888.

Such a change is not improbable since from star to star the observed wavelength dependence is not constant. In fact, the single-parameter empirical formula

$$p/p_{max} = \exp(-1.15 \ln^2 \lambda_{max}/\lambda) \tag{14}$$

has been found to fit a large number of interstellar linear polarization curves (Serkowski 1972); the differences in λ_{max} are attributable to different sizes of the polarizing grains.

Thus measurements of $\theta(\lambda)$, together with an estimate of ϕ [from $D(\phi)$ and/or $F(\phi)$], show how much the grain size varies along the line of sight. This in turn can be compared to the variation in size estimated by comparing λ_{max} values of neighboring stars.

For example, σ Sco shows $\theta_U - \theta_V = 4°.3$ (D. S. Mathewson[8]). Recall that σ Sco is also circularly polarized. These two observations together are most significant, since they provide the first clear evidence that the correct explanation for both effects is indeed a twisting of the grain alignment along the line of sight.

A simple analysis on the basis of a two-slab model limits the size change along the line of sight to less than a factor of 1.1 (Martin 1973). For comparison, the maximum change in λ_{max} for bright stars near σ Sco in the ρ Oph region is a factor of 1.5. These various lines of evidence support the interpretation that both the direction of alignment and the grain size are changing along the line of sight to σ Sco.

CONCLUSIONS

Interstellar circular polarization has been presented as a new means of investigating several properties of the interstellar medium. For example, observations of the wavelength dependence of circular polarization in the directions of the Crab Nebula and certain stars show that the grain material is dielectric. Further, more detailed observations, together with the wavelength dependence of extinction (or the ratio of total-to-selective extinction) and of linear polarization (or λ_{max}) will be important in determining both the grain size and the grain composition.

In a twisting medium, the amount of twist is measured by the degree of circular polarization relative to linear polarization, and the depolarization of the linear component. When the twist is known, changes in the grain size along the line of sight are detectable in the wavelength dependence of the position angle of linear polarization.

---

[8] Personal communication; U and V indicate the wavelengths in the UBV system.

What is needed is as complete a set of observations as possible for a large number of stars at different galactic longitudes. In particular, interstellar circular polarization measurements of stars must be developed. One way to ensure a high rate of success in these observations would be to look only at stars for which θ changes with wavelength; a list of many such stars will be available soon (G. V. Coyne[9]). However, these stars form only a subset of the possible candidates. Stars in regions showing a high disorder of the electric vectors should be examined as well. One obvious region is near 0° galactic latitude and 80° longitude; others can be chosen from the maps of electric vectors[10] published by Mathewson and Ford (1970). Finally, it would be interesting to look in small selected regions such as clusters where the dispersion in polarization is large, or in the ρ Oph complex, for which the size of the grains has already been investigated (Carrasco, Strom, and Strom 1973).[11]

Acknowledgments. I thank D. A. MacRae for his comments on the manuscript. Support from the National Research Council of Canada is acknowledged.

REFERENCES

Arons, J. 1972. Nonlinear inverse Compton radiation and the circular polarization of diffuse radiation from the Crab Nebula. *Astrophys. J.* 177: 395–410.

Blandford, R. D. 1972. The polarisation of synchro-Compton radiation. *Astron. Astrophys.* 20: 135–144.

Carrasco, L.; Strom, S. E.; and Strom, K. M. 1973. Interstellar dust in the ρ Oph dark cloud. *Astrophys. J.* 182: 95–109.

Coyne, G. V., and Gehrels, T. 1966. Wavelength dependence of polarization. VIII. Interstellar polarization. *Astron. J.* 71: 355–363.

———. 1967. Wavelength dependence of polarization. X. Interstellar polarization. *Astron. J.* 72: 887–898.

Gehrels, T., and Silvester, A. B. 1965. Wavelength dependence of polarization. V. Position angles of interstellar polarization. *Astron. J.* 70: 579–580.

Greenberg, J. M., and Roark, T. 1967. Light from reflection nebulae. I. Dielectric versus metallic grains. *Astrophys. J.* 147: 917–936.

Hiltner, W. A. 1956. Photometric, polarization and spectroscopic observations of O and B stars. *Astrophys. J. Supp.* 2: 389–462.

Hulst, H. C. van de 1957. *Light scattering by small particles.* New York: Wiley.

Hulst, H. C. van de, and Jong, T. de 1969. The interpretation of the diffuse galactic radiation. *Physica.* 41: 151–162.

Kemp, J. C. 1972. Circular polarization of o Scorpii: Possible interstellar origin. *Astrophys. J.* 175: L35–L37.

Kemp, J. C., and Wolstencroft, R. D. 1972. Interstellar circular polarization: Data for six stars and the wavelength dependence. *Astrophys. J.* 176: L115–L118.

---

[9] Personal communication. [10] See p. 964. [11] See also p. 954.

Martin, P. G. 1972. Interstellar circular polarization. *Mon. Not. R. Astr. Soc.* 159: 179–190.
———. 1973. Interstellar polarization from a twisting medium. In draft.
Martin, P. G.; Illing, R.; and Angel, J. R. P. 1972. Discovery of interstellar circular polarization in the direction of the Crab Nebula. *Mon. Not. R. Astr. Soc.* 159: 191–201.
Mathewson, D. S., and Ford, V. L. 1970. Polarization observations of 1800 stars. *Mem. R. Astr. Soc.* 74: 139–182.
Mattila, K. 1970. Interpretation of the surface brightness of dark nebulae. *Astron. Astrophys.* 9: 53–63.
———. 1971. The interpretation of the absolute intensity of the diffuse galactic light. *Astron. Astrophys.* 15: 292–298.
Rees, M. J. 1971. The non-thermal continuum from the Crab Nebula: The "synchro-Compton" interpretation, *IAU Symp. 46*, pp. 407–413, (R. D. Davies and F. G. Smith, eds.) Dordrecht, Holland: D. Reidel Publ. Co.
Serkowski, K. 1962. Polarization of starlight. *Adv. Astron. and Ap.* 1: 289–352.
———. 1972. Interstellar polarization. *Interstellar dust and related topics.* H. C. van de Hulst and J. M. Greenberg, eds.) Dordrecht, Holland: D. Reidel Publ. Co.
Treanor, P. J. 1963. Wavelength dependence of interstellar polarization. *Astron. J.* 68: 185–189.

INTERSTELLAR CIRCULAR POLARIZATION OF EARLY-TYPE STARS

JAMES C. KEMP and RAMON D. WOLSTENCROFT
Institute for Astronomy, University of Hawaii

Circular polarization believed to be caused by varying grain-alignment orientation along the line of sight has recently been discovered (Kemp 1972, 1973; Kemp and Wolstencroft 1972). These authors obtained data for six early-type stars, generally in the Scorpio-Centaurus association. These stars are all heavily reddened ($E_{B-V} \gtrsim 0.3$) and all have substantial interstellar linear polarization ($p \gtrsim 2\%$). The circular polarization values are in the range 0.01–0.04%. The data were interpreted in terms of a model (Kemp 1972b) for propagation through a medium in which the eigendirections of the effective refractive-index tensor are twisted along the z-axis (line of sight) in accordance with $\phi = \phi(z)$, where ϕ is an angle in the transverse xy plane that gives the twist relative to an initial orientation $\phi = z = 0$. The starlight is taken as initially unpolarized (i.e., at $z = 0$). The propagation equations for a plane wave with complex-amplitude components $E_x(z)$, $E_y(z)$ are:

$$\frac{dE_x}{dz} = ik(-\sin^2\phi \cdot E_x + \sin\phi \cos\phi \, E_y)$$

$$\frac{dE_y}{dz} = ik(\sin\phi \cos\phi \cdot E_x + \cos^2\phi \, E_y), \quad (1)$$

where $k = (2\pi/\lambda)(n_l - n_r)$. Here $n_{l,r} = n'_{l,r} + in''_{l,r}$ are complex refractive indices; l, r are effective, projected eigendirections in the transverse x, y plane. It is useful to think of l, r as the xy-plane projections of the long and short axes respectively of aligned, rod-shaped particles. The latter interpretation is not essential to the model, which is phenomenological and requires only an interstellar medium composed of a continuous sequence of two-dimensionally anisotropic slabs normal to the line of sight, the eigenaxes rotating as a function of z. The medium is "optically active" on a galactic scale; in contrast to the usual chemical optical activity, however, in our case the optical wavelength is very small compared to the distance over which the medium rotates by 360° (pitch interval). Equations analogous to Equations (1) but written in Stokes-vector components, were discussed by Serkowski (1962). The reader should note that the independent variables $E_x(z)$, $E_y(z)$ in Equations (1) contains a suppressed z-dependence $\exp(ik_r z)$ common to both these variables.

For a small total twist angle $\phi_0 \ll 1$, and with the convenient choice of a linear twist $\phi(z) = \alpha(z)$, Equations (1) are easily solved by power series in z, $E_i(z) = \sum_n a_{in} z^n$. Initial conditions are included by setting ($a_{x0} = 1$, $a_{y0} = 0$) and ($a_{x0} = 0$, $a_{y0} = 1$) respectively for statistically independent components $E_x(0)$, $E_y(0)$, of the incident, unpolarized light. The lowest-order result for the transmitted circular polarization, and its relation to the first-order linear polarization, is:

$$q_0 \cong \frac{1}{3} \left(\frac{2\pi}{\lambda}\right)^2 (n_l'' - n_r'')(n_l' - n_r') \phi_0 z_0^2 \qquad (2)$$
$$\cong \frac{1}{3} \frac{(n_l' - n_r')}{(n_l'' - n_r'')} p_0^2 \phi_0.$$

We can visualize this result also in terms of a "two-cloud" picture. Light passes through the first interstellar cloud, in which particles are all aligned in some direction, becoming linearly polarized by virtue of the dichroism $(n_l'' - n_r'')$; it then traverses a second cloud in which the alignment is rotated about z by an amount ϕ_0, and the wave-plate action proportional to the birefringence $(n_l' - n_r')$ in the second cloud produces ellipticity.

The total twist angles ϕ_0 in the first six stars studied are not small compared to unity, but they are small enough ($\lesssim 180°$) so that the estimate in Equation (2) is meaningful. Exact solutions of Equations (1) would seem to involve Mathieu-type or similar higher functions, owing to the trigonometric coefficients (for the case $\phi = \alpha z$). Martin[1] gives closed formulae for the polarization in the twisting medium, as solutions of Serkowski's (1962) transfer equations. Whether Equations (1) are in fact precisely equivalent to Serkowski's, under the transformation $(E_x, E_y) \to (I, Q, U, V)$, has not been checked. Derivation of our equations is quite simple and unambiguous (based merely on the forward propagation of plane waves through an infinitesimal slab). The equivalence and the correctness of Martin's solution (which he does not prove) need to be verified.

---

[1] See p. 926.

If the usual peak in the interstellar linear polarization in the mid-visible (sometimes toward the red) is attributed to a strong spectral peak in the dichroism ($n_l'' - n_r''$), we can infer the structure of the spectral birefringence ($n_l' - n_r'$) by pure Kronig-Kramer inversion, entirely independent of the actual nature of the interstellar grains, i.e., whether they are dielectric or metallic. Using such an assumption, the birefringence should be (at least crudely) antisymmetric in wavelength dependence about a wavelength not far from the linear-polarization peak, and from Equations (1), $q(1/\lambda)$ should tend to resemble the derivative of $p(1/\lambda)$. This is in fact the kind of thing we observed (Kemp and Wolstencroft 1972).

Because of the dispersion relation connecting q and p, it appears that the circular polarization gives us only a limited amount of new information for determining the nature of the interstellar grains. This view is different from that advanced by Martin, Illing, and Angel (1972)[2] in interpreting their data on the mixed intrinsic-interstellar elliptical polarization of the Crab Nebula. Of course, very accurate Kronig-Kramer inversion of the dichroism would require knowledge of $p(1/\lambda)$ far out in the ultraviolet and infrared, but such data are not very extensive in the literature. More careful use of the $q(1/\lambda)$ data can therefore help to deduce the behavior of the linear polarization in the less accessible regions.

For an extended survey of the interstellar circular polarization in reddened early-type stars, we advocate use of a two-color system (blue and far-red), as suggested by the wavelength dependences of σ Sco-A and o Sco; we used this simple scheme for the other four of the first stars studied. The intent is to discriminate against intrinsic effects in some stars. We found that the q values at $\lambda \sim 4000$ and 7000 Å respectively had comparable magnitudes but opposite signs; the far-red values were usually smaller. We take the sign change (for the present at least) to be a signature of this type of polarization. Contributions to such a survey by various observers are invited. A question of special interest to us is whether there is a preferred "handedness" in the galactic medium. This will only be answered statistically, because there are obviously effects of random handedness due to local clouds.

Finally, an aspect of this discovery is that a note of caution is now injected into studies of circular polarization in miscellaneous objects, particularly those of unknown structure and distance (quasars, etc.). Discrimination against an interstellar contribution will have to rely on the wavelength dependence and on the rough estimate $q \sim p^2$ connecting the interstellar circular and linear effects.

---

[2] See also p. 926.

Acknowledgments. Our work is partially supported by the National Science Foundation (GA-28201) and indirectly by the National Aeronautics and Space Administration through its support of the Mauna Kea Observatory.

REFERENCES

Kemp, J. C. 1972. Circular polarization of Omicron Scorpii: Possible interstellar origin. *Astrophys. J.* 175: L35–L37.

———. 1973. Interstellar circular polarization of upper Scorpius stars. I.A.U. Symp. No. 52. *Interstellar dust and related topics* (J. M. Greenberg and H. C. van de Hulst, eds.) Dordrecht, Holland: D. Reidel Publ. Co. In press.

Kemp, J. C., and Wolstencroft, R. D. 1972. Interstellar circular polarization: Data for six stars and the wavelength dependence. *Astrophys. J.* 176: L115–L118.

Martin, P. G.; Illing, R.; and Angel, J. R. P. 1972. Discovery of interstellar circular polarization in the direction of the Crab Nebula. *Mon. Not. R. Astr. Soc.* 159: 191–201.

Serkowski, K. 1962. Polarization of starlight. *Adv. Astron. Astrophys.* 1: 289–301.

DISCUSSION

WOLSTENCROFT: I wish to point out that there may be a short method of determining whether systematic screw alignments of interstellar grains occur in many directions. Instead of studying a large number of individual stars, it should be possible to see the circular polarization in the integrated light of the Milky Way in various directions, provided that we look toward directions of small optical depth, e.g., toward the anticenter and perhaps at latitudes greater than 10°. The fractional circular polarization so far seen in the Milky Way, which is large (0.5% to 1.0%), is seen in directions of large optical depth and is due undoubtedly to multiple scattering by grains. If the optical depth is sufficiently small (say <0.5), the integrated starlight component may prevail.

NOVICK: Is there any hope of directly detecting the spin motion of the interstellar grains? This might show up as a broad peak in the photon noise power spectrum in the frequency region of the spin rate. Perhaps this could be detected by measuring either the time-interval distribution or the Fourier power spectrum of the photoelectrons produced by a bright star that is subject to large interstellar polarization. It would be helpful if the star had large intrinsic polarization. In any case, a large telescope with a linear polarization analyzer and fast photon-counting electronics will be required to detect this effect. It is important to note that the various grains will be spinning with different phases so that the modulation will result only from the *fluctuations* in the grain spin phase distribution. Clearly, it will be necessary to collect a very large number of photoelectrons to detect this effect, if it can be detected at all.

KEMP: This is an extremely interesting suggestion and might relate to questions of "star noise," namely noise in excess of normal photon statistics. Offhand, I would guess (as you do also) that such an effect would be too delicate to detect.

MARTIN: I have four separate points I should like to make.

First, there is a fundamental point on which your conclusions are in disagreement with those expressed in my paper,[1] namely, whether observations of interstellar circular polarization add anything to our knowledge about the composition of interstellar grains. I have said they do.

The linear birefringence is related to the linear dichroism through the Kronig-Kramer relations, as you point out. However, the required Kronig-Kramer inversion can be performed properly only if the linear dichroism is known at all frequencies. Equivalently, it is not possible to carry out the inversion merely on the basis of the observed interstellar linear polarization in the visible window of the spectrum. Your conclusion based on the latter approach—the wavelength dependence of the birefringence is independent of the grain material—demonstrates this impossibility. (In fact, even your model of the production of the dichroism is not accurate, as can be seen by comparing the upper part of Fig. 2 in Kemp [1972] with, for example, Fig. 38 in Greenberg [1968].)

In pursuing this problem of inversion further, let me first emphasize that the linear dichroism predicted by models of interstellar grains is dependent on the grain composition. It is possible to construct models of differing composition that predict roughly the same dichroism over a limited frequency range, for example the visible range. However, when this match in the visible is obtained, there remain marked differences in the infrared and the ultraviolet. In fact, the basis for hoping to distinguish between metallic and dielectric grains by extending the observations of linear polarization into the ultraviolet lies in this different behavior for different materials. (Recent results using this approach are reported by Gehrels [1974]; he finds the grains are not metallic, in agreement with my interpretation of the interstellar circular polarization observations.) Also, in the infrared the increase in the dichroism with frequency is much less steep for metals than for dielectrics.

Since metallic and dielectric grains do not produce the same wavelength dependence of linear dichroism throughout the spectrum, one should not be surprised to find (using the Kronig-Kramer relations) that the wavelength dependences of the birefringence are also different, even in the visible region of the spectrum. On the basis of these differences, I have proposed that observations of interstellar circular polarization in

---

[1] See p. 926.

the visible range provide an additional method of distinguishing between dielectric and metallic particles.

In an earlier paper (Martin 1972) I have shown one way in which the different behaviors of the birefringence can be calculated, using models similar to those required to predict the dichroism. This method is reliable and bypasses the problem of Kronig-Kramer inversion. Since these models make quantitative predictions of both the linear dichroism and birefringence, they seem to be a suitable theoretical basis for the study of interstellar polarization.

The second point on which I wish to comment is that the closed form solutions I have given for the polarization from the twisting medium were derived assuming small amounts of polarization, which usually is a good approximation in the interstellar medium. Details are to be published separately. In the limit $\phi \to 0$, my solution (Equation [13]) becomes identical to the approximate solution you have given in Equation (2).

Then I have a minor point: In the "two-cloud" interpretation of Equation (2), the rotation of alignment from one cloud to the next is $2/3\phi_0$ rather than ϕ_0.

As a final comment I would like to question the appropriateness of using the term "optically active" to describe this medium. My understanding was that an optically active crystal or solution rotates the linear polarization vector of the incident radiation (without altering the degree of polarization), because the medium is circularly birefringent. This model of the interstellar medium does not exhibit circular birefringence. It is true that the position angle of the linear polarization increases with pathlength in the medium. However, this only arises from the addition of the linear polarization vectors produced by successive linearly dichroic slabs of the medium that are oriented so that the axes of the refractive-index tensor twist along the line of sight.

KEMP: Martin has gone deeply into the question of interpretation of the interstellar circular polarization in terms of explicit models, indeed rather more deeply than I have. Our apparent disagreement as to the value of the circular polarization data for establishing the nature of the grains is really a matter of emphasis. We understand that the circular and linear polarization are related, at least for the case of small twist angles, by Kronig-Kramer inversion. Your point is well taken, however (and probably not well enough appreciated by us), that the range of the known data for the wavelength dependence of the dichroism (i.e., the linear polarization) is small enough so that the $q(\lambda)$ data may indeed be critical for deciding, e.g., between metallic versus dielectric grains. We have not gone into this aspect as you have, and we leave the matter for future developments by you and others.

Our work on the interstellar circular polarization of reddened early stars was, first of all, strictly observational. We were, however, pleased to find that the basic appearance of the $q(\lambda)$ curves—with peaks of opposite signs in the blue and far red—does follow from Kronig-Kramer inversion of a grossly idealized spectral dichroism, consisting of a simple broad peak in the visible. This qualitative fact, we feel, is the basis for believing that we are observing the twisted grain-alignment effect rather than some totally different source of circular polarization, e.g., magnetic circular dichroism or intrinsic circular dichroism. Whether or not the latter possibilities could be ruled out by other considerations, the $q(\lambda)$ for such mechanisms would most likely bear quite different relationships, or even no typical relationship at all, to the spectral linear polarization. Our appeal to the Kronig-Kramer relations here was, therefore, directed toward identifying the physical mechanism for the circular polarization, rather than the nature of the particles.

Our use of the term "optically active" purposely included quotation marks. One may consider optical activity in the following generalized sense, for which the technically accepted term might be "gyrotropic." A medium with such a property is defined by the fact that infinitesimal slabs of the medium, cut normal to some axis, are optically identical apart from a screw rotation along the axis. If the pitch interval d is $\ll \lambda$, we have the situation in an optically active single crystal; if it is $\gg \lambda$, we have the situation in the twisted interstellar medium. The intermediate case $d \sim \lambda$ is perhaps unimportant optically but can be met with in microwave electronics (certain slow-wave structures in a circular waveguide, etc.).

DISCUSSION REFERENCES

Gehrels, T. 1974. Wavelength dependence of polarization. XXVIII. Interstellar polarization from 0.22 to 2.2 μm. *Astron. J.* In draft.

Greenberg, J. M. 1968. Interstellar grains. *Nebulae and interstellar matter* (B. M. Middlehurst and G. P. Kuiper, eds.) Chicago: The Univ. of Chicago Press.

Kemp, J. C. 1972. Circular polarization of o Scorpii: Possible interstellar origin. *Ap. J.* 175: L35–L37.

Martin, P. G. 1972. Interstellar circular polarization. *Mon. Not. R. Astr. Soc.* 159: 179–190.

POLARIZATION OF STARS IN ORION AND OTHER YOUNG REGIONS

MICHEL BREGER
University of Texas, Austin

More than 20% of the 120 very young stars near the Orion Nebula measured to date show linear polarization significantly above the average interstellar value. The existence of polarization is well correlated with the presence of stellar infrared excesses. These stars, as well as pre-main-sequence objects in other young regions, show a wide variety in the wavelength dependence of their polarization.

The position angles of polarization in Orion are not randomly distributed, although a wide range is present. This presents a powerful argument for the concept that most of the polarization in the Orion cluster originates in intracluster dust rather than in extended envelopes, circumstellar shells, or reflection nebulae.

The new polarization data for pre-main-sequence stars in many regions suggest that in general the polarization originates from at least three causes: intracluster dust, often of unusually large grain sizes; circumstellar shells, presumably by electron scattering; and reflection nebulae contaminating the observed starlight.

Pre-main-sequence and other young stars are often situated in heavily obscured regions. This interstellar — actually intracluster — absorption, as well as the presence of thick circumstellar shells, enormously affects observed magnitudes and colors and often leads to high linear polarization. Polarization measurements provide important clues to the nature of this absorption (or even emission). In particular, we expect that polarization measurements of starlight will allow us to differentiate between several different possible origins of observed infrared excesses and optical extinction, such as circumstellar gas and dust shells (Strom, Strom, and Yost 1971; Breger and Dyck 1972), intracluster dust, or even scattering in "cometary" nebulae (Zellner 1970).

In the case of interstellar or intracluster extinction, the wavelength of maximum polarization, λ_{max}, is related to the ratio, R, of total to selective absorption (Serkowski 1968). Serkowski (1973) has also found that the curve of interstellar polarization against wavelength has a constant shape.[1] This should provide an excellent check on whether any high polarization (and by inference, optical extinction) is of interstellar origin. One should also be able to study grain size in some thick clouds as a function of optical depth. Finally, if the polarization is of "local" origin, one would expect the position angles of polarization to be randomly distributed within any young region. This, of course, requires statistical analyses of a large group of highly polarized stars in a few populous regions, such as the Orion Nebula region.

We are presently engaged in a large program of polarization and light variability measurements of very young stars in selected regions. The polarization measurements are made with the Dyck polarimeter at the Kitt Peak National Observatory, while the light variability measurements are carried out at the McDonald Observatory. The observed light variability has already provided important evidence for extremely high ratios of total to selective absorption (Breger 1972), assuming of course that the light variability of some young shell stars of spectral types B and A originates in the heavy obscuring material. We have already measured 120 stars in and near the Orion Nebula so far for polarization, and several conclusions have been drawn.

PROPERTIES OF POLARIZATION

Only about 20% of the 120 stars observed in the Orion region, mostly of spectral type B to F, show significantly large linear polarization (in excess of 0.6%). We would like to use the data in four groups: (1) polarization and infrared excesses; (2) polarization and reddening; (3) the position angles of polarization; and (4) the wavelength dependence of polarization. We have previously found in NGC 2264 that the presence of high polarization was also accompanied by a large infrared excess (though the converse is not necessarily true). Comparison with infrared flux measurements of Orion stars made by Penston (1972) gives the good relationship shown in Fig. 1. We have matched $E(B-V)$ values, the color excess on the UBV photometric system, computed from spectral types with the K (2.2 μm) and L (3.5 μm) excesses. The stars with significant excesses that could not be explained by normal interstellar reddening are called IR excess stars in the figure. Two explanations come to mind: first, interstellar absorption with unusual reddening laws ($R > 3$) and second, absorption by circumstellar shells. Both processes may be present, since some values of R are greater than 10.

---

[1] See p. 888.

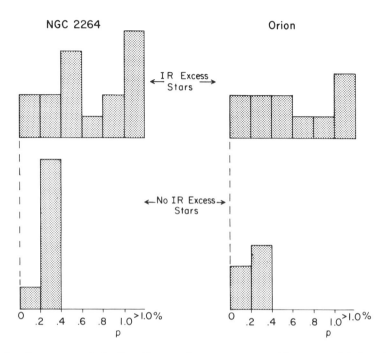

Fig. 1. Frequency-polarization diagram for young stars with measured infrared colors, in NGC 2264 and generally in Orion. An average between the measured polarization in the V and B filters (p_V, p_B) was taken. High polarization is accompanied by a significant infrared excess.

If polarization and extinction are of interstellar origin due to grains aligned by a magnetic field, a correlation between the amount of polarization, p, and reddening may be expected. A diagram of these quantities is shown in Fig. 2 and is, unfortunately, not very instructive. We note, however, that several shell stars (e.g., W 90 in NGC 2264) lie above the "maximum" line which represents the previous observations of *interstellar* reddening and polarization (Serkowski 1965). The position of these stars is not surprising if one considers the very large value of R seen in them and the fact that the polarization of these stars is presumably of circumstellar origin. Several stars show considerable reddening but no polarization. For the Trapezium stars this has also been noted by Hall (1958).

Although polarization is accompanied by an infrared excess, the size of p is not well correlated with the size of $E(V-K)$, the infrared excess at 2.2 μm. This is in contrast to results found by Carrasco, Strom, and

Strom (1973)[2] for the stars imbedded in the ρ Oph cloud. In particular, we find several stars with very low $p/E(V-K)$ ratios. It is attractive to speculate that this is caused by circumstellar shells around some Orion stars.

The position angles of the stars with high polarization in Orion show a wide range of values but are not distributed at random. This is shown in Fig. 3. The strong peak near 85° definitely suggests a nonlocal origin of most of the polarization. Should the polarization originate in a stellar envelope, circumstellar shell, or in a reflection nebula around the star, there should exist no preferred position angle. The nonrandom position angles therefore suggest an interstellar or intracluster origin.

It therefore seems plausible that most of the polarization and, consequently, some of the extinction and infrared excesses are caused by

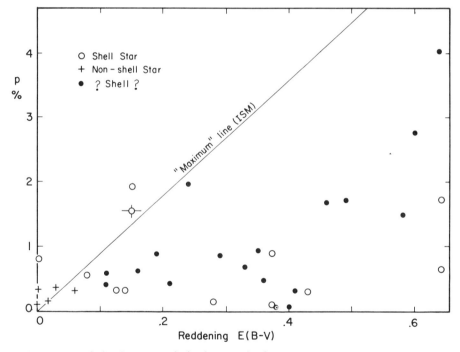

Fig. 2. Correlation between polarization p and color excess $E(B-V)$. The "shell" stars plotted have circumstellar shells or show unusual reddening curves. The line drawn refers to the observed maximum line for the *interstellar* medium (ISM).

---

[2] See p. 954.

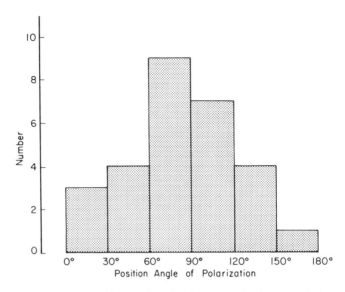

Fig. 3. Distribution of position angles of polarization for highly polarized stars in or near the Orion Nebula. The nonrandom distribution suggests at least partial intracluster dust origin.

interstellar dust. The large values of R observed near the Trapezium region (Lee 1968) predict grain sizes considerably larger than the average interstellar value. This in turn predicts that the wavelength of maximum polarization, λ_{max}, is shifted to larger wavelengths. Evidence for this may be seen in the present observations of polarization as a function of wavelength in Fig. 4, which shows λ_{max} greater than 0.7 μm for several of these stars. While these results give us confidence that the interstellar dust interpretation may be correct, a final conclusion has to await further observations, hopefully soon to be obtained. Especially the wavelength dependence of polarization and the determination of λ_{max} should be instructive.

Other polarization mechanisms must also be operative, since interstellar dust cannot explain all the observed polarization of pre-main-sequence stars. The shape of the wavelength dependence of polarization of the highly polarized stars in NGC 2264 differs from the shape of the interstellar curve. Furthermore, λ_{max} peaks in the blue or visual region of the spectrum. In the interstellar dust absorption model, this is incompatible with the extremely high value of total-to-selective absorption also found for these stars. Analogous to Be stars, the polarization of W 90 in NGC 2264 has recently been interpreted as arising in a variable circumstellar shell with free-free emission providing the observed infrared excess (Breger and Dyck 1972). Reflecting nebulae around some pre-main-sequence objects also give rise to high polarization. An excellent example is the cometary nebula R Mon (Hall 1965; Zellner 1970). This combination of reflection nebulae and circumstellar shells with the proper position angles of both the star and the nebula has also been observed by us for several other field stars that show emission, the Ae/Be pre-main-sequence stars.

CONCLUSIONS

The preliminary results of this investigation can be summarized as follows:

1. Only about 20% of the stars in or near the Orion Nebula show significantly high linear polarization.

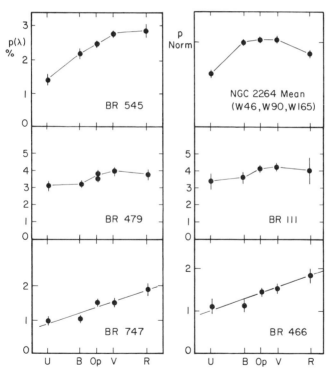

Fig. 4. Observed wavelength dependence of polarization for some Orion stars. Brun (1935) numbers are used. A mean curve of normalized polarization for the three pre-main-sequence stars in NGC 2264 is also shown. The filters are labeled U (effective wavelength 0.36 μm), B (0.42 μm), Op (Open, no filter, $\sim$0.48 μm), V (0.56 μm), and R (0.7 μm).

2. The occurrence of high polarization is accompanied by an infrared excess.

3. The nonrandom distribution of position angles of polarization in Orion suggest an intracluster origin, presumably due to aligned interstellar dust. Both the measured ratio of total-to-selective absorption and the wavelength of maximum polarization argue in favor of large dust grain sizes.

4. The radiation of pre-main-sequence stars is often also polarized by circumstellar shells and reflection nebulae.

Acknowledgments. It is a pleasure to thank H. M. Dyck and S. E. Strom for helpful discussions. Part of this investigation was made while the author was at the State University of New York at Stony Brook and supported by NSF Grant GP-31027.

REFERENCES

Breger, M. 1972. Pre-main-sequence stars. I. Light variability, shells and pulsation in NGC 2264. *Astrophys. J.* 17:539–548.

Breger, M., and Dyck, H. M. 1972. Pre-main-sequence stars. II. Stellar polarization in NGC 2264 and the nature of circumstellar shells. *Astrophys. J.* 175:127–134.

Brun, A. 1935. La nebuleuse d'Orion et ses étoiles variables. *Pub. Obs. Lyon* 1: No. 12.

Carrasco, L.; Strom, S. E.; and Strom, K. M. 1973. Interstellar dust in the ρ Ophiuchus dark cloud. *Astrophys. J.* 182:95–109.

Hall, J. S. 1958. Polarization of starlight in the galaxy. *Pub. U. S. Naval Obs.* 17:275–342.

Hall, R. C. 1965. Polarization and color measures of NGC 2261. *Publ. Astron. Soc. Pac.* 77:158–163.

Lee, T. A. 1968. Interstellar extinction in the Orion association. *Astrophys. J.* 152:913–941.

Penston, M. V. 1972. Multicolour observations of stars in the vicinity of the Orion Nebula. In preparation.

Serkowski, K. 1965. Polarization of galactic clusters M25, NGC 869, 884, 1893, 2422, 6823, 6871, and association VI Cygni. *Astrophys. J.* 141:1340–1361.

———. 1968. Correlation between the regional variations in wavelength dependence of interstellar extinction and polarization. *Astrophys. J.* 154:115–134.

———. 1973. Interstellar polarization. *Interstellar dust and related topics.* (J. M. Greenberg and C. H. van de Hulst, eds.) Dordrecht, Holland: D. Reidel Publ. Co. In press.

Strom, K. M.; Strom, S. E.; and Yost, J. 1971. Circumstellar shells in the young cluster NGC 2264. *Astrophys. J.* 165:479–488.

Zellner, B. 1970. Wavelength dependence of polarization. XXI. R Monocerotis. *Astron. J.* 75:182–185.

DISCUSSION

BREGER (in reply to a question on possible contamination of starlight by nebulosity): Circumstellar shells are generally too close to the star (closer than one arcsec; distance of 500 pc) to be resolvable. One therefore measures star plus shell. In Orion, however, one has to consider the effect of the polarized nebulosity. In the present investigation, nebulosity contamination has a negligible effect as was found by nebula counts. I have also measured the polarization of several stars with a range of diaphragm sizes, typically from 1.7 to 25 arcsecs, and no change in polarization was found.

BREGER: (in reply to a comment that infrared shell of NU Ori [BR 747] has a diameter of many arcsecs): It is fortunate that BR 747 was one of the test stars. However, no variation of polarization with diaphragm size was detected, which implies that the infrared shell does not contribute much light relative to the star (at least in the optical region).

ANONYMOUS: Is any of the polarization variable?

BREGER: Quite a few stars were measured more than once during 1971 and 1972, and little variability was found. One star, however, BR 545, showed a remarkable change within 60 days, during which the polarization changed from 2.74% to 2.44%, while the position angle changed by 80°.

INTERSTELLAR DUST IN DARK CLOUDS

L. CARRASCO
University of California, Berkeley
K. M. STROM and S. E. STROM
Kitt Peak National Observatory

The dust component in several dark clouds has been studied. It is shown that important density-dependent characteristics of the grains can be detected by combining infrared and optical photometry with measurements of the wavelength dependence of linear polarization. The observations suggest that the grains are larger in the higher-density regions of the clouds. It is argued that the size increase results from depletion of heavy elements from the gas onto the grains. The observed polarization is highest for the most heavily reddened stars; this suggests that the grains are well aligned even in the densest observable regions of the clouds. A magnetic field appears to be the most likely source of alignment, and at the densities prevailing near the cloud centers, fields as high as several times 10^{-4} gauss may be required.

A study of several dark-cloud complexes has been undertaken in order to determine the properties of interstellar dust in these high-density regions (Carrasco, Strom, and Strom 1973). We are particularly interested in determining whether the expected depletion of heavy elements from the interstellar gas onto grains could be demonstrated observationally. If this process is efficient, it could have a significant effect on the thermal balance of the interstellar medium by removing the major coolants from the gas. Because depletion would be expected to be more efficient under high density conditions, the choice of the dark-cloud regions seems the most appropriate "laboratory" for such an observational test.

Heavy element depletion onto the grains would result in shifting the maximum in the grain size distribution toward larger sizes. This, in turn, would have the following observational consequences:

1. the ratio of total to selective extinction, A_V/E_{BV}, would be expected to rise, primarily because the larger grains act to decrease the selective extinction in the blue and visual regions;

2. the measured polarization in the optical region should peak toward longer wavelengths.

With these expectations in mind, we have chosen to study the dust properties in several dark-cloud complexes: the region surrounding ρ Ophiuchi (R.A. $= 16^h24^m$; Decl. $= -24°00'$), a region near the center of the II Per association, IC 348 (3^h43^m; $+32°00'$), where the extinction in this association reaches a maximum, and the Taurus clouds (R.A. $\sim 4^h30^m$; Decl. $\sim 30°$). A study of star counts in these regions and the assumption of spherical symmetry provide us with a very crude estimate of the density variations in each of the clouds (implicitly we assume that the gas and dust are always well mixed). The density appears to vary by over a factor of 100 from the peripheral regions of each of these clouds to the heavily obscured centers. In absolute terms, densities appear to range from about 10 atoms/cm³ to over 1000 per cm³. These estimated variations seem confirmed by the identification at radio wavelengths of density-sensitive molecular transitions near the cloud centers (for example, the center of the ρ Ophiuchi complex is located near Heiles OH-cloud 4 while the bright OH-source, Heiles 2, is located in the Taurus aggregate). Hence, we are observing clouds in which the density is considerably above typical interstellar conditions and where heavy element depletion onto the grains should be very important if it occurs at all.

In order to determine the shape of the interstellar extinction curve in these regions, we have carried out a program of broadband infrared and optical photometry and spectral classification for a sample of stars in each of these regions. Furthermore, the Dyck polarimeter at Kitt Peak National Observatory has been used to measure the wavelength dependence of polarization for many of these stars. Our own data were significantly supplemented by polarization measurements kindly provided to us by Serkowski.

The results of our investigation can be summarized as follows:

1. The ratio of total-to-selective extinction has been found to increase in the more heavily obscured, denser regions of the clouds. As an example, we plot in Fig. 1 the color excess ratio $R_K \equiv E_{VK}/E_{BV}$ against E_{VK} for stars located in the II Per complex near IC 348; there is a clear tendency for R_K to increase for higher values of E_{VK}. Note that under typical interstellar conditions, R_K is observed to be approximately 2.80. Similar results have been found for the Ophiuchus regions with values of R_K reaching as high as 4.30. The Taurus region shows no such obvious trend, although one star, near Heiles Cloud 2, has a value of R_K near 3.50.

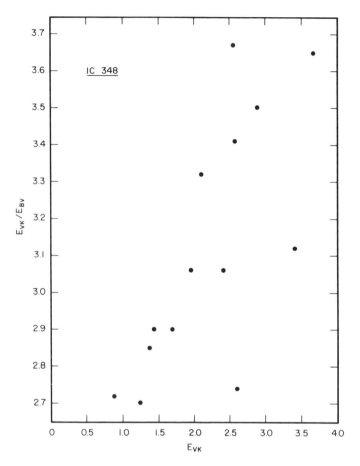

Fig. 1. The observed color excess ratio $R_K = E_{VK}/E_{BV}$ plotted against E_{VK} for the IC 348 region. Under normal interstellar conditions, 1.1 R_K is equal to the ratio of total-to-selective extinction, and the color excess E_{VK} is approximately equal to the optical depth of the cloud at visual wavelengths.

2. The increase of R_K is accompanied by an increase in the wavelength of maximum polarization. This is illustrated in Fig. 2, where the relation between these properties is illustrated for stars in the Ophiuchus region. Similar results are found for the two other regions, although because the stars were fainter, far fewer could be sampled.
3. For a few stars in the Ophiuchus region, observations are available from the Wisconsin Orbiting Astronomical Observatory experiment.

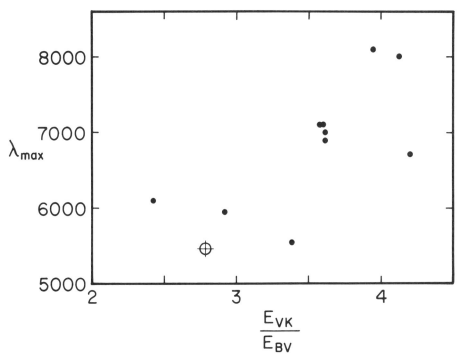

Fig. 2. The wavelength dependence of polarization (λ_{max}) plotted against color-excess ratio R_K for the region of the ρ Oph cloud. The symbol ⊕ indicates the location in this diagram of a star viewed through "typical" interstellar material.

As the wavelength of maximum polarization and R_K increase, the slope of the ultraviolet extinction curve decreases; that is, the absorption becomes significantly grayer.

The above results are best interpreted by assuming that the grains are larger in the regions of high density and high extinction. Further interpretation of the Ophiuchus cloud observations are possible in terms of a very simple grain-depletion model as outlined by Spitzer (1968). It is found that the increase in particle size with increasing extinction (density) within the cloud can be explained if the cloud age is less than 10^7 years and if the probability of a heavy atom "sticking" to the grain is greater than 0.1 per collision of a heavy atom with a grain. The cloud age is a conservative upper limit based both on kinematical studies of the stars in the region and on a nuclear age deduced for the stars in the nearby Upper Scorpius region. Hence, depletion appears to be a very efficient mechanism under the range of conditions that prevail in the dark-cloud regions included in our study. It is very likely an important factor in determining the heavy element content of the gas under more normal interstellar cloud conditions.

Some more direct supporting evidence in favor of the heavy element depletion picture is provided by some recent spectroscopic studies by Chaffee et al. (1972) and by Cohen (1973). They find a general tendency for calcium to be depleted in dark-cloud regions to a greater degree than in lower-density regions of the interstellar medium. Moreover, Cohen finds that sodium also appears to be depleted in dark-cloud regions. Although neither of these surveys is as yet complete, the results to date offer further evidence in favor of greater depletion of heavy elements onto the grains in dense, dark-cloud regions.

Our investigations have provided, as a by-product, information on two points that bear on the problem of grain alignment. In the first place, from the measured orientations of electric vectors for stars measured in the Ophiuchus complex, it appears possible to rule out alignment both by exchange of photon angular momentum with the grains and by the effects of radiation pressure. Secondly, the largest polarization values have been found for stars observed through the densest regions of the clouds. This observation is not unique to the regions surveyed in this study. In particular, the stars NGC 2024-1 and Herschel 36 (in M8) both show very high values ($\gtrsim 8\%$) of polarization and a large value of R_K; and both objects are located in regions in which the local gas density probably exceeds 10^4 particles/cm^3. Even more dramatic is the high value of polarization ($\sim 30\%$) recently measured at infrared wavelengths for the Becklin-Neugebauer object. This object is believed to be obscured by a cloud in which the gas density exceeds 10^5 particles/cm^3.

It would appear from these observations that alignment of particles persists into regions of very high density within interstellar clouds. If alignment results from a paramagnetic relaxation mechanism, then the values of the magnetic field required must be almost two orders of magnitude higher than the typical interstellar value of about several microgauss. If the higher-density regions represent regions in which collapse has taken place, then the field would appear to increase with increasing density, at least during the early collapse phases. This result is consistent with the conclusion of Goldreich, Keeley, and Kwan (1973) that the field must be "frozen" into the gas (to density values as high as 10^8 particles/cm^3) in order to explain the observed intensity and polarization characteristics of OH and H$_2$O masers.

For a more detailed description of our study of dust properties in the Ophiuchus complex see Carrasco et al. (1973). A discussion of the results for the IC 348 region is currently under preparation.

Acknowledgments. We wish to thank the following colleagues for several helpful discussions: K. Serkowski, H. M. Dyck, G. Grasdalen, B. Lynds, C. Heiles, P. Solomon, F. Shu, B. Turner, and B. Zuckerman. W. Piegorsch and M. Bailey provided considerable assistance at the Kitt Peak National Observatory 213-cm telescope.

REFERENCES

Carrasco, L.; Strom, S. E.; and Strom, K. M. 1973. Interstellar dust in the Ophiuchus dark cloud. *Astrophys. J.* In press.

Chaffee, F. H.; Strom, S. E.; Lutz, B.; and Cohen, J. G. 1972. A spectroscopic study of the molecular constituents of dark clouds. *Publ. Astron. Soc. Pac.* 84: 639–640.

Cohen, J. G. 1973. Optical interstellar lines in dark clouds. In draft.

Goldreich, P.; Keeley, D. A.; and Kwan, J. Y. 1973. Astrophysical masers. II. Polarization properties. *Astrophys. J.* 179: 111–134

Spitzer, L. Jr. 1968. *Diffuse matter in space.* Interscience Publishers, New York: Wiley.

DISCUSSION

GEHRELS: Is the polarization of the Becklin-Neugebauer object not an intrinsic effect, perhaps due to symmetry in the object rather than to interstellar polarization as you seem to imply?

CARRASCO: It is an effect of interstellar origin, as proven by the Dyck et al. (1973) paper. The wavelength dependence of linear polarization, for the 10 μm silicate feature, is positively correlated with the wavelength dependence of the extinction coefficient in the Becklin-Neugebauer object, which is just the opposite to what one would expect for a shell phenomenon.

DISCUSSION REFERENCE

Dyck, H. M.; Capps, R. W.; Forrest, W. J.; and Gillett, F. C. 1973. Large 10 μ linear polarization of the Becklin-Neugebauer source in the Orion Nebula. *Astrophys. J.* (letters). In press.

INTERSTELLAR POLARIZATION AND THE GALACTIC MAGNETIC FIELD

G. L. VERSCHUUR
National Radio Astronomy Observatory

The direction of the local magnetic field implied by observations of starlight polarization and radio-continuum polarization suggest a longitudinal field toward $\ell = 50°$. This is compared with the field direction derived from an examination of the Faraday rotation measures of extragalactic radio sources, which indicate a field toward $\ell = 90°$. It is pointed out that the assumption that rotation measures are produced in the solar neighborhood, as would be the case if our galaxy were a flat disk, is incorrect because there is considerable evidence for the existence of distant spiral arms extending to very high latitudes. The field direction derived from radio-source-rotation measures should, therefore, not be compared with measurements of the more local field. The possible existence of a helical field configuration is also briefly discussed.

This brief review will be concerned with the structure of the galactic magnetic field but will not discuss its strength.

OPTICAL POLARIZATION DATA AND SUBSEQUENT FIELD MODELS

There are basically two schools of thought concerning the possible configuration of the local galactic magnetic field. There are the longitudinal field supporters and the helical field proponents. I shall examine both models in the light of the optical polarization data available and then see to what extent other polarization data, such as that of the radio background and the rotation measures of extragalactic sources and pulsars, lend support to either of these models. The one common denominator in the papers discussing the magnetic field is that it is the local field that is usually discussed rather than a more general large-scale field in our galaxy. Local is taken to mean fields within distances on the order of several hundred parsecs.

Several papers, such as that by Jokipii, Lerche, and Schommer (1969), have shown that there is a characteristic scale length of several hundred parsecs in the structure of the magnetic field "irregularities." I have found support for their conclusions from an examination of the dependence of the mean polarizations of stars in clusters and associations on distance. Furthermore, Krzeminski and Serkowski (1967) found that the polarization of the cluster Stock 2, lying only 300 pc from the sun, is almost as great as the mean polarization of the stars in h and χ Per lying several kpc from the sun beyond the cluster Stock 2. This strongly suggests that in this direction the polarization is predominantly produced within several hundred parsecs of the sun. This is also borne out in the plots made by Mathewson and Ford (1970) which show E vectors of the optical polarization of all stars, observed up to that time, grouped in various distance intervals. The lengths of the vectors on the diagrams showing stars in the 200 to 400 pc distance interval are not very different from those in the 2000 to 4000 pc interval. Figure 1 shows that the polarization is predominantly produced in the solar neighborhood. In addition, the orientation of the polarization vectors at distances of a few hundred parsecs already illustrates the basic pattern in the polarization which Mathewson (1968) felt was indicative of a local helical magnetic field. These general remarks tie in well with the conclusion of Hiltner (1956) who found no distance dependence of optical polarization for stars beyond 1 kpc.

We, therefore, take it as adequately illustrated that most of the optical polarization is introduced relatively locally and that the scale length of the structures in the polarizing medium is probably on the order of 150 pc. We should bear in mind that these structures must be predominantly in the orientation of the field direction rather than simply in the amount of polarizing material, for if the field were everywhere uniform we would still expect polarization to be a linear function of distance.

Let us examine just where in the sky we can be certain we are seeing a well-ordered field that might be construed as being parallel to the plane and longitudinal. When one examines the plots of Mathewson and Ford, there does not seem to be any overwhelming pattern suggesting field lines parallel to the plane for stars in the distance intervals up to 600 pc. We will confine our attention to the data for stars between 2 and 4 kpc away (Fig. 1). Here the well-ordered patterns noted by Hiltner (1956) and Hall (1958) are obvious between galactic longitude $\ell = 100°$ and $\ell = 150°$. It is in the direction $\ell = 140°$ that the radio-continuum polarization data also indicate that we are looking normal to the local field. Around $\ell = 80°$ one sees the confusion of vectors often associated with our line of sight being along a spiral arm. This statement is ques-

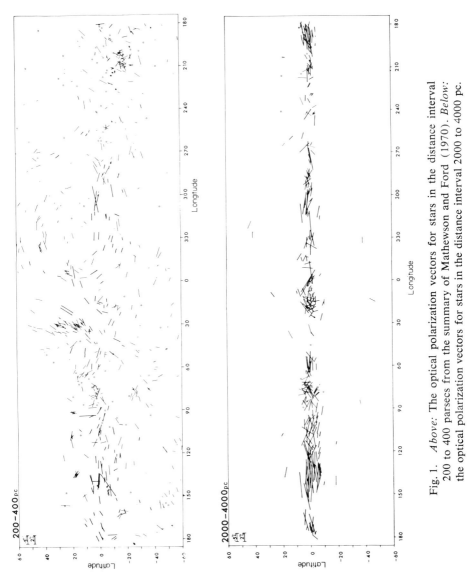

Fig. 1. *Above:* The optical polarization vectors for stars in the distance interval 200 to 400 parsecs from the summary of Mathewson and Ford (1970). *Below:* the optical polarization vectors for stars in the distance interval 2000 to 4000 pc.

tionable because it is not clear which spiral arm this is supposed to be. Let us digress and recall the pictures showing the distribution of young stars, clusters, etc., thought to be spiral arm tracers (e.g., Becker 1963). The local feature runs toward $\ell = 50°$ and not to $\ell = 80°.$ Indeed, if one looks at the polarization of stars within 1 kpc, then the effect of looking down the local spiral feature toward $\ell = 50°$ is also very obvious. Mathewson simulated the polarization vector pattern here using a sheared helical field model, but this is the direction along which a longitudinal field in the local spiral feature would also be expected to show the patterns seen in Fig. 2. But what of the $\ell = 80°$ region? It is the Cygnus X complex. It is possible that this irregularity has nothing to do with looking down an arm at all but is merely the Cygnus X region, which is complex for reasons not yet understood. One need only look at any of the maps of spiral structure based on 21-cm data to see that the Cygnus X direction is not along the axis of any arm at all.

Continuing to examine the map of Mathewson and Ford, we find the chaotic pattern around $\ell = 50°$ even in the starlight polarization vectors for stars up to 4 kpc distant, and it is difficult to recognize any longitudinal field effects again until one reaches $\ell = 330°.$ Then the vectors show some semblance of order down to $\ell = 270°$, below which they again are smaller and more disordered to as far as $\ell = 210°.$ Since this region of disorder is about 180° away from the region at $\ell = 50°$, it again fits the longitudinal field rather well. Around $\ell = 180°$ the vectors appear ordered but not as impressively so as around $\ell = 140°.$

Mathewson distinguishes between the helical and longitudinal field on the basis of the fact that the polarization vectors of many stars are nearly perpendicular to the plane at low latitudes in directions well away from the axis of any possible local longitudinal field. In other words, a longitudinal field would show perspective effects in the projected field lines, directed toward $\ell = 50°$ and 230°, but nowhere except at these longitudes would the field ever appear perpendicular to the plane. However, although the helical field model accounts for those vectors rather well, it does not in my opinion account too well for the highly ordered field parallel to the plane at $\ell = 140°.$ Both models, therefore, have their shortcomings.

If one recognizes the Cygnus X region as an anomaly and removes those data from consideration, then there seems to me to be less reason to favor the helical model over the longitudinal. Also, Appenzeller (1968) claims that the observed vectors near the poles are the critical test in that they show a disagreement with the helical model in the south polar cap. But then the inferred longitudinal field axis, as derived from the data at the two poles, is different for the two poles. However, these

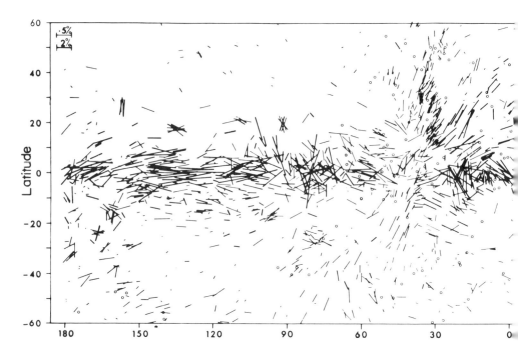

Fig. 2. The polarization vectors of all the stars with known polarizations collected by Mathewson and Ford.

differences are perhaps not quite as much as between the helical field model and the data for the south pole. The data for the north pole fit both models extremely well.

However, it was not Mathewson's intention to fit perfectly the observations in every region of the sky with his helical model. Indeed, the original reason for his model was to explain the field seen in the galactic spurs of continuum emission and to explain the spurs themselves.

MAGNETIC FIELDS AND THE GALACTIC SPURS

The problem we have to consider is whether the helical field model is better for explaining the spurs and the polarization of starlight (as well as the polarization of the radio continuum, incidently), or whether the longitudinal field plus a series of supernovae in the solar neighborhood can do the job better.

Let us consider the longitudinal field plus supernovae. They have been frequent enough that about 4 to 6 remnants are found within

THE GALACTIC MAGNETIC FIELD

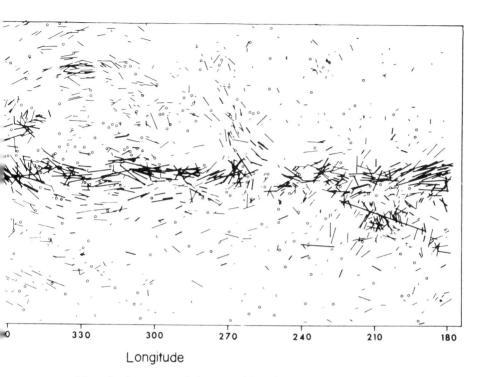

several hundred parsecs of the sun. Also they have, in fact, not seriously affected the ambient magnetic field because the obvious spurs have axes that follow the field lines in either of our two models fairly well. Consider that the field lines in either the helical or longitudinal field in the region $\ell = 30°$ have a curvature projected on the sky very similar to that of the north polar spur (NPS) axis. A longitudinal field line has to be distorted a little to fit this spur, while the spur is a natural part of the helical field. We cannot distinguish between the two models based on a close examination of the shape of this spur, or any of the other two major spurs, which also follow the ambient magnetic field lines rather well. This is why Mathewson fitted the spurs so readily into his model. One recent bit of information that affects this discussion is that Spoelstra (1971) finds that the magnetic field lines in the NPS are normal to the continuum ridge at $b < 40°$, which destroys the good fit with the helical or longitudinal field altogether, suggesting that the NPS is an irregularity in any general field configuration.

The study of the map of the optical polarization vectors of all stars (Fig. 2) does not of itself give the impression that we are seeing a longitudinal field plus supernova remnants. It does look more like either a helical field with small irregularities or a longitudinal field with small irregularities. I wish to emphasize that the way the field lines cross $\ell = 45°$ is very striking and not simply an observational selection effect in the location of the stars studied. Based on the optical polarization studies alone we would join the NPS loop I to loop II south of the plane at $\ell = 45°$, which is contrary to the way Berkhuijsen and others would have us draw loops through these two spurs independent of one another.

A strong argument for the supernova model for the spurs is that Spoelstra (1972) appears to have simulated some of the loops rather well using the van der Laan model for shells expanding into a surrounding medium. However, his model applies to what might be called the lesser loops and the examples that he shows are not very striking features in the continuum emission.

FARADAY ROTATION DATA

To resolve the dilemma we can always turn to another source of information in our attempts concerning the local magnetic field: that is, the radio source Faraday rotation data. I need not review the whole controversy about this; a partial discussion has been given by Verschuur (1969). Instead we may examine the latest data and in particular the summary map made by Mitton (1972), showing the rotation measures for 192 radio sources, and the pulsar rotation data given by Manchester (1972).

In Fig. 3 we show Mitton's map, and at last one thing clearly emerges. The sources between $\ell = 0°$ and $180°$ below the plane show predominantly negative rotation measures, and those between $\ell = 180°$ and $\ell = 360°$ show positive rotation measures (apart from the exceptions, which are mostly sources with small rotation). Clearly this indicates a longitudinal field toward $\ell = 90°$. Unfortunately, the data above the plane destroy this conclusion because there we see little order apart from some groups of positive and some groups of negative rotation measures. If the northern sources had followed the pattern, we would again have been left with the problem of what is so important about $\ell = 90°$, because that is neither the direction of any spiral arm feature nor the field inferred from optical polarization data. It is impossible to find any indication of a simple longitudinal field toward $\ell = 50°$ in Mitton's map. (The helical field, too, is in question when one considers that the axis of the helix is supposed to be toward $\ell = 90°$, which is not readily related to the direction of the local spiral feature.)

Manchester claimed that the rotation measures of 21 pulsars indicated a longitudinal field toward $\ell = 90°$ and that his data were inconsistent with a helical field. He notes that if one ignores the rotation measures

THE GALACTIC MAGNETIC FIELD

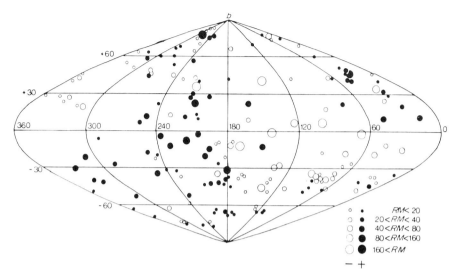

Fig. 3. The distribution of sources with known rotation measurements of Faraday, in galactic coordinates. A negative sign (open circles) indicates a field away from the observer. (From Mitton 1972.)

(two) of the pulsars in the direction of the NPS, then his conclusion is not unreasonable. Reinhardt (1972) concluded that the field direction inferred from pulsar rotation measures was toward $\ell = 120°$, $b = -30°$, while the quasar rotation data indicated a field toward $\ell = 110°$, $b = 10°$ to $20°$.

Superimposing Manchester's data on Mitton's map shows that there are no obvious inconsistencies, in that the pulsar rotation measure signs agree very well with those of radio sources in nearby directions. Many of the new pulsar data (12 of 21 pulsars) refer to directions within 15° of the plane, while many of the remaining ones have small rotation measures which do not really affect the overall impression one gains from the data shown in Fig. 3.

If we wish to look for simple patterns, then I should point out that of those sources lying inside the best-fit circles to the three major spurs, twice as many have negative as positive rotation measures, while those outside the loops show the opposite distribution. Perhaps the model that appears to best fit the rotation data alone is that proposed by Gardner, Morris, and Whiteoak (1969). They suggested a field toward $\ell = 80°$ with a loop distorting it in the line of sight toward the galactic-center direction (above the plane) so that the rotation measures from $\ell = 60°$ to $\ell = 300°$ wouldn't follow the longitudinal field pattern but be opposite to it. Their model seems fair, even now, although one would revise the basic field direction to be toward $\ell = 90°$ according to the data shown in Fig. 3. This again lands us back with the same dilemma concerning the uniqueness of $\ell = 80°$ or $90°$.

SOME PROBLEMS

There are still many unanswered questions concerning the local field, and one of the most neatly defined problems concerns the field in the direction of $\ell = 140°$. It is undoubtedly true that in that direction in the plane we are looking normal to local field lines, that not only the data based on optical polarization but also the radio continuum polarization data show there is a highly ordered field here, and that the rotation measure is zero at $\ell = 140°$. In this direction the field is truly normal to the line of sight within several hundred parsecs of the sun. On the other hand, the pulsar PSR 0329+54 is located at $\ell = 145°$, $b = -1°$ and shows a large rotation measure of -63 rad m^{-2} and a mean field of 2.9 μ gauss away from the sun in a region in which the field is supposed to be normal to the line of sight (bear in mind that Manchester inferred a mean field of 3.5 μ gauss toward $\ell = 90°$ from his data). This pulsar, therefore, reveals that the local field has considerable irregularities in it, which makes it surprising that we are able to recognize any order in it at all.

If the model for the local field is one of a longitudinal field along the arm, then we would expect something similar in the next spiral arm with the field parallel to the local field. But this is not true. In the Perseus arm, the field according to the Zeeman effect data is toward the observer; in other words, the field is directed outward along the spiral arm, whereas locally it goes toward lower longitudes or is directed inward along a spiral arm. Also in the direction of the Crab Nebula, the mean field inferred from pulsar data and source rotation data is toward the nebula, whereas the Zeeman effect data show a field toward the observer. Clearly the field has completely reversed at some point along that line of sight.

FARADAY ROTATION IN DISTANT ARMS IN THE GALAXY

It has been customary not to be concerned over rotation effects in distant arms for sources at latitudes above about $10°$ because the line of sight there was expected to pass only through the local spiral arm and through no other arms. However, that assumption is based on a model in which our galaxy is flat with a half-density thickness of only a few hundred parsecs. Recently a considerable amount of evidence has been building up to show that distant spiral arms are far from planar and indeed increase in their half-width as one moves out from the galactic center. Verschuur (1973) has suggested that the high-velocity clouds are in these distant arms. Examination of neutral hydrogen data shows that the center of gravity of distant arms moves up to many kpc above the plane beyond about 14 kpc from the galactic center and the arms themselves can have half-density widths of many kpc, as has indeed been predicted by Kellman (1972) in a recent paper. Giovanelli, Verschuur, and Cram (1973) show that the average density in the H I clouds in these distant arms is on the order of 0.2 cm^{-3} and that the temperatures may be as high as 10,000° K in these clouds. If the clouds are to be in pressure equilibrium with their surroundings, say in a medium of 50,000° K (Giovanelli and Brown 1973), then the intercloud density will be $\sim$0.04 cm^{-3} and largely ionized. Since we might expect lines of sight at latitudes up to 50° to pass through the high-latitude extensions in several arms (this is certainly true at latitudes below 40°), a path length through these arms of 1 kpc is not an unreasonable estimate (see Verschuur 1973). For a field of 1 μ gauss, one then can derive a rotation measure of 30 rad m^{-2}. Obviously this introduces a considerable contribution to the rotation measures of extragalactic sources in high-latitude regions in distant, nonplanar arms; this, in turn, will cause us to find patterns in the rotation measure data that are not consistent with the more local field structures which we inferred by examining local phenomena.

It is interesting to note that if the radio-source-rotation measures do give information about the field in distant arms, then we would indeed expect the inferred field to be directed toward $\ell = 90°$; this derived direction is dependent on where the rotation measures change sign, and for a longitudinal field in distant arms, this should occur around $\ell = 180°$. Lines of sight toward the galactic center would not suffer contributions to the rotation measures from distant arms since the galactic plane is only 200 pc thick inside the solar circle. The rotation measures above and below the plane are opposite to one another in the longitude range from $\ell = 300°$ through the center to $\ell = 60°$. It seems to me that the rotation data for extragalactic radio sources should not be used to derive too much about the local field because the lines of sight to these sources cut too many arms (especially in the longitude range 60 to 300) that reach up to latitude 50° around $\ell = 160°$. The sources outside these boundaries show very low rotation measures in general, and no obvious pattern remains. This point of view should be examined much more closely when a more detailed picture of the distant spiral arms emerges, especially after southern hemisphere neutral hydrogen data are added. Only then can the rotation-measure distribution be more carefully studied. Of course, more pulsar-rotation measures will be essential if we are to relate them to the optical polarization data.

CONCLUSIONS

Let us sum up the information about the local field that we now have. The optical polarization data near the plane are consistent with a field directed along the local spiral feature toward $\ell = 50°$, with the Cygnus X region an anomaly. The optical polarization data away from the plane indicate that the local field may be longitudinal with irregularities associated in part with the spurs, or it may be helical with the spurs part of the fields. However, the axis of the helix is toward $\ell = 90°$ which is not the local spiral feature axis and, therefore, is somewhat artificial. Radio-continuum polarization data suggest a field toward $\ell = 50°$ in the plane or at least normal to the line of sight at $\ell = 140°$ (and $320°$ from the data of Mathewson and Milne [1965]), but data away from the plane show the effect of the spurs, distorting either a longitudinal field or being part of and also distorting a possible helical field. Spoelstra (1972) has managed to show that the field lines associated with the spurs are consistent with their being supernovae expanding into a longitudinal field initially directed toward $\ell = 40°$, which is in good agreement with the optical and radio data in the plane.

Does any of this tie in with Gould's belt? In this context one would merely point out that the Gould's belt phenomenon is a local one in the distribution of young stars and that the magnetic fields we are discussing are also local, but aside from that similarity it is not clear whether the field configuration is closely tied to the distribution of stars or why it should be.

My conclusion, therefore, is that the local field does appear to be longitudinal since it leads to an axis down the local spiral feature, and that this is to be favored over a helical field with an axis toward $\ell = 90°$. It fits well with optical and radio-continuum polarization data but not too well with pulsar rotation data. Extragalactic radio source data should not be assumed to give information only about the local field and, therefore, should not be directly used for comparison with the other polarization data.

REFERENCES

Appenzeller, I. 1968. Polarimetric observations of nearby stars in the directions of the galactic poles and the galactic plane. *Astrophys. J.* 151:907–918.

Becker, W. 1963. Die räumliche Verteilung von 156 galaktischen Sternhaufen in Abhängigkeit von ihrem Alter. *Z. fur Astrophys.* 57:117–134.

Gardner, F. F.; Morris, D.; and Whiteoak, J. B. 1969. The linear polarization of radio sources between 11 and 20 cm wavelength. *Aust. J. Phys.* 22:813–819.

Giovanelli, R., and Brown, R. L. 1973. *Astrophys J.* In press.

Giovanelli, R.; Verschuur, G. L.; and Cram, T. 1973. High resolution studies of high velocity clouds of neutral hydrogen. *Astron. Astrophys. Suppl.* In press.

Hall, J. S. 1958. Polarization of starlight in the galaxy. Publ. U.S. Naval Observ., 2nd Series, 17:275–342.
Hiltner, W. A. 1956. Photometric polarization and spectrographic observations of O and B stars. *Astrophys. J. Suppl.* 2: 389–462.
Jokipii, J. R.; Lerche, L.; and Schommer, R. A. 1969. Interstellar polarization of starlight and the turbulent structure of the galaxy. *Astrophys. J.* 157: L119–L124.
Kellman, S. A. 1972. The equilibrium configuration of the gaseous component of the galaxy. *Astrophys. J.* 175: 353–362.
Krzeminski, W., and Serkowski, K. 1967. Photometric and polarimetric observations of the nearby strongly reddened open cluster Stock 2. *Astrophys. J.* 147: 988–1002.
Manchester, R. M. 1972. Pulsar rotation and dispersion measures and the galactic magnetic field. *Astrophys. J.* 172: 43–52.
Mathewson, D. S. 1968. The local galactic magnetic field and the nature of the radio spurs. *Astrophys. J.* 153: L47–L53.
Mathewson, D. S., and Ford, V. L. 1970. Polarization observation of 1800 stars. *Mem. Roy Astr. Soc.* 74: 139–182.
Mathewson, D. S., and Milne, D. K. 1965. A linear polarization survey of the southern sky at 408 Mc/s. *Aust. J. Phys.* 18: 635–653.
Mitton, S. 1972. The polarization properties of 65 extra-galactic sources in the 3C catalogue. *Mon. Not. Roy. Astr. Soc.* 155: 373–381.
Reinhardt, M. 1972. Interpretation of rotation measures of radio sources. *Astron. Astrophys.* 19: 104–108.
Spoelstra, T. 1971. A survey of linear polarization at 1415 MHz. II. Discussion of results for the north polar spur. *Astron. Astrophys.* 13: 237–248.
———. 1972. A survey of linear polarization at 1415 MHz. IV Discussion of the results for the galactic spurs. *Astron. Astrophys.* 21: 61–84.
Verschuur, G. L. 1969. Observational aspects of galactic magnetic fields. IAU Symp. No. 39, *Cosmical gas dynamics* (H. J. Habing, ed.), pp. 150–167. Dordrecht, Holland: D. Reidel Publ. Co.
———. 1973. High velocity clouds and "normal" galactic structure. *Astron. Astrophys* 22: 139.

DISCUSSION

DICKELL: With regard to the lack of a 21-cm arm in the direction of Cygnus, I should point out that there is an H II-region spiral arm feature in that direction extending out to about 4 kpc from the sun centered on galactic longitude $77°$ with a width of about 0.4 kpc (see Dickell, Wendker and Bieritz 1970).

DISCUSSION REFERENCE

Dickell, H. R.; Wendker, H. J.; and Bieritz, J. 1970. The distribution of H II regions in the local spiral arm in the direction of Cygnus. *Proceedings of the IAU Symposium #38, The spiral structure of our galaxy.* (W. Becker and G. Contopoulos, eds.), pp. 213–218. Dordrecht, Holland: D. Reidel Publ. Co.

THE ORIENTATION OF THE LOCAL INTERSTELLAR DARK CLOUDS WITH RESPECT TO THE GALACTIC MAGNETIC FIELD

W. SCHLOSSER and TH. SCHMIDT-KALER
Ruhr-Universität, Bochum

A comparison of the distribution pattern and filamentary structure of the local dark clouds, in the longitude range from 315° to 45°, with the pattern of the optical polarization of starlight shows a high degree of correlation. The highest degree of correlation is obtained by using stars in the range 200–600 parsecs. The average distance of the dark clouds is 200 pc.

Photographs of the Milky Way with a spherical mirror wide-angle camera (135° field diameter) reveal low-contrast structures that extend far beyond the fields of classical astronomical instruments. The spectral region chosen determines within certain limits the range of distances reached. In the ultraviolet (0.32–0.38 μm) most of the light from beyond 2 kpc is absorbed, and thus structures within this distance dominate the photographs. Therefore, in the galactic center region the dark clouds between the sun and the next inner spiral arm are most clearly exhibited.

Local dark clouds are recognized by (a) heavy absorption very well visible against the rich background of stars in the center region with virtually no foreground stars; (b) large extension in galactic latitude; and (c) their small distances from the sun.

A visual inspection of the ultraviolet photographs (Fig. 1) shows the following general features in the distribution and orientation of the local dark clouds:

1. A pronounced north-south asymmetry in the general distribution of dust;
2. A complex of clouds from galactic longitudes $\ell = 350°$ to $45°$, and latitudes from $b = -20°$ to $+40°$, including the well-known Ophiuchus dark clouds;
3. A cloud complex of roughly annular shape linked to the Norma dark cloud ($\ell = 315°–345°$, $b = 0–30°$);

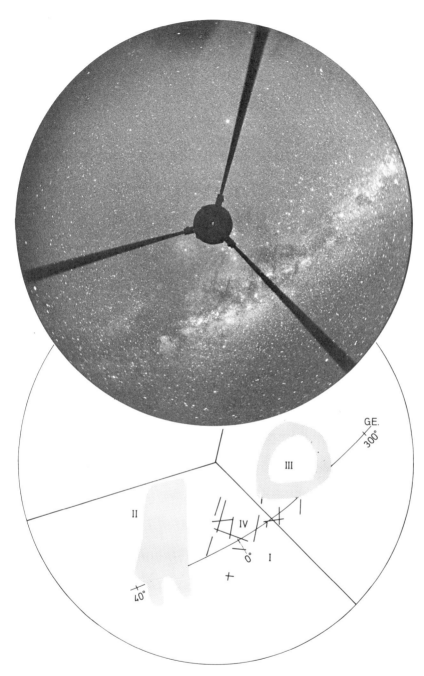

Fig. 1. Photograph of the southern Milky Way in the ultraviolet. The identification chart below gives the galactic coordinates and refers to the various phenomena discussed, e.g., north-south asymmetry (I), Ophiuchus-complex (II), Norma-complex (III) and filamentary network (IV).

4. A network of at least 16 conspicuous filamentary dark clouds between $\ell = 330°$ and $30°$ intersecting the galactic equator preferentially at two distinct angles, e.g., $35 \pm 5°$ and $127 \pm 14°$ rms deviation. Most of the filaments are very sharp, with a mean length of $10°$.

A comparison of these features with the basic pattern of the vectors of optical polarization (Mathewson and Ford 1970) shows a high degree of correlation:

1. The north-south asymmetry of the dark clouds is seen again in the smooth distribution and alignment of the vectors south of the galactic equator as compared to their complex behavior north of it.[1]
2. The general run of the Ophiuchus complex (inclined $65°$ to the galactic equator) is exactly followed by the polarization vectors.
3. The annular form of the Norma cloud is displayed again by the apparent rotation of the polarization vectors.
4. The delicate filamentary network covers only a small part of the sky as compared to the aforementioned structures; stars with known polarization in these filaments are scarce. Nevertheless, a general agreement may be noted.

The highest degree of correlation between the general orientation of the dark clouds and that of the polarization vectors is obtained by using stars in the range 200–600 pc. This, therefore, represents an upper limit to the distances of the clouds studied. Indeed, those clouds whose distances have been determined (Bok and Cordwell 1971) are at about 200 pc.

We conclude that the interstellar dust particles do not only act as indicators of the magnetic field but that the mechanism, such as streaming, producing the structural features of the local dark clouds is connected to the local magnetic field. Further analysis is in progress.

REFERENCES

Bok, B. J., and Cordwell, C. S. 1971. A study of dark nebulae. Presented at Symposium on *Interstellar Molecules,* Charlottesville, Virginia, Oct. 1971. In press.
Mathewson, D. S., and Ford, V. L. 1970. Polarization observations of 1800 stars. *Mem. R. Astron. Soc.* 74: 139–182.

DISCUSSION

VERSCHUUR: The elongation of dust clouds in directions parallel to those indicated by optical polarization vectors of nearby stars was first noted by Shajn (1955a), and Verschuur (1970) found confirmation for this effect.

---
[1] See p. 964.

SCHLOSSER AND SCHMIDT-KALER: First indications of alignment of bright and dark filaments parallel to the interstellar magnetic field have been given by Behr (1955) and by Behr and Tripp (1955) and Shajn (1955b). In most cases these refer to small features. Verschuur's (1970) comparison of the directions of some clouds with those of the magnetic fields is based on a highly idealized picture of the magnetic flow pattern by Mathewson (1968); it is a theoretical helix model. For example, the whirl corresponding to feature III (Fig. 1) is centered on $\ell = 326°$, $b = +18°$ as derived from Mathewson's (1970) observations while his idealized picture puts it at $\ell = 308°$, $b = +4°$. The theoretical model deviates even more from the observations. Small wonder, therefore, that Verschuur notes differences of the directions up to $70°$.

The center of the feature III is at $\ell = 327°$, $b = +15°$, i.e. almost exactly coinciding with the center of the whirl. The same applies to a feature associated to the whirl around $\ell = 160°$. It is obvious that the flow pattern is probably not due to a general helix field but to a number of individual features.

DISCUSSION REFERENCES

Behr, A. 1955. The polarization of stellar light in selected area 40. Symposium Mém. Soc. Roy. Liège XV, pp. 547–550.

Behr, A., and Tripp, W. 1955. Polarisation des Sternlichts in einem Gebiet interstellarer Wolken filamentartiger Struktur. *Naturwiss.* 42: 9.

Mathewson, D. S. 1968. The local galactic magnetic field and the nature of the radio spurs. *Astrophys. J.* 153: L47–L53.

Shajn, G. A. 1955a. On the magnetic fields in the interstellar space and in the nebulae. *Astron. Zh.* 32: 381–394.

———. 1955b. Diffuse nebulae and interstellar magnetic field. *Astr. Zh. SSR* 32: 110–117.

Verschuur, G. L. 1970. Observational aspects of galactic magnetic fields. *Interstellar Gas Dynamics* (H. J. Habing, ed.), pp. 150–167. Dordrecht-Holland: D. Reidel Publ. Co.

STARLIGHT POLARIZATION BETWEEN BOTH MAGELLANIC CLOUDS

THOMAS SCHMIDT
Max-Planck-Institut für Astronomie, West Germany

Within three regions between the main bodies of the Large Magellanic Cloud (LMC) and the Small Magellanic Cloud (SMC), the polarization angles have been found nearly similar for galactic foreground and Magellanic member stars. The LMC-SMC magnetic field connection derived by Mathewson and Ford and by Schmidt must still be confirmed by more extensive polarimetric measurements. The polarization angles of the galactic foreground stars are in contradiction to Mathewson's (1968, 1969) model of a galactic magnetic helix, the most probable interpretation of the data being a longitudinal field directed toward longitude $60° \pm 15°$, superimposed by considerable irregularities.

From the Magellanic Cloud optical polarization surveys of Mathewson and Ford (1970) and Schmidt (1970), a large-scale magnetic field connecting both Magellanic Clouds seems to be indicated. To clarify the situation, further starlight polarization measurements have been undertaken within three fields between the Magellanic Clouds; two in the SMC far-wing region and one in the LMC southwestern outskirts, using the 1-m photometric telescope and polarimeter (Behr 1968) of the European Southern Observatory at La Silla, Chile.

MEASUREMENTS

Special weight was given to measurements of galactic foreground stars for a more accurate separation of galactic effects from intrinsic Magellanic Cloud polarizations. Altogether 66 cloud members and 47 galactic stars have been observed, all except five on two or more different nights. Together with the work of Mathewson and Ford (1970), 160 stars were available for the analysis of these three fields. The results are summarized in Figs. 1, 2, and 3. Assuming the mechanism of Davis and Greenstein (1951) for interstellar dust alignment, the magnetic field direction can be traced from angles of polarization.

THE MAGELLANIC MAGNETIC FIELD

The separation of the intrinsic Magellanic magnetic field from galactic foreground effects is complicated by the very similar alignment of galactic and Magellanic polarizations. Only in the case of the LMC field does there exist a significant difference between Magellanic and galactic amounts and angles of polarization. For both SMC regions a significant difference of the amounts of polarization is present, while the angles are identical within the *rms* error of measurement.

For the LMC field a Magellanic magnetic field parallel to the LMC-SMC connection seems, therefore, to be confirmed (see Fig. 2). For the SMC regions (Fig. 3) a similar Magellanic field could be deduced with certainty only if the difference between the amounts of polarization could be attributed positively to the SMC itself and not to our galaxy. The averaged distance of all observed galactic foreground stars from the galactic plane is greater than 300 pc and should, therefore, apparently be located outside the general galactic dust layer; however, this still needs to be studied more extensively, especially as absorbing matter at distances

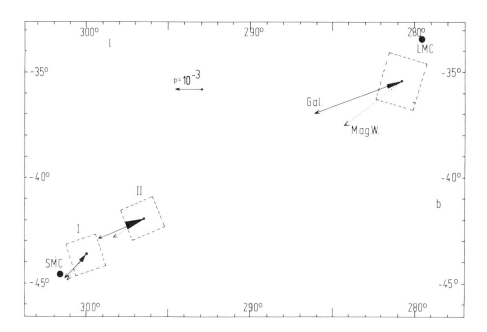

Fig. 1. The three fields covered by starlight polarization measurements in the galactic coordinate system. *Drawn Arrows:* averaged galactic foreground polarizations with *rms* errors of the polarization amount and angle. *Dotted Arrows:* intrinsic polarization of LMC and SMC.

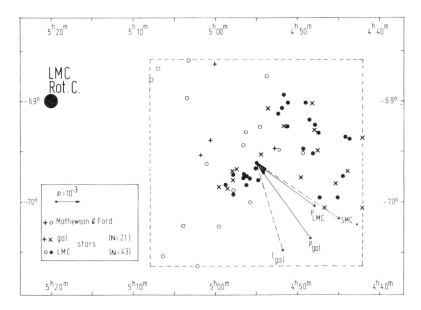

Fig. 2. Stars with known polarization in the LMC field; right ascension in the abscissa and declination in the ordinates. *Dashed Arrows:* direction parallel to the galactic equator (l_{gal}) and to the LMC-SMC connection. *Drawn Arrows:* averaged intrinsic LMC starlight polarization (p_{LMC}) and the galactic foreground contribution (p_{gal}), with the respective *rms* errors of polarization amounts and angles (indicated by the black sector connected with the polarization arrows). *Rot. C* stands for Rotation Center.

greater than 700 pc from the galactic plane seems to be indicated by recent analyses (see, for instance, Pfau 1972).

Although within the main bodies of both Magellanic clouds net intrinsic amounts of polarization still seem to be well confirmed, more accurate galactic foreground corrections are necessary in these regions to decide whether the polarization angles are really intrinsically aligned parallel to the LMC-SMC connections; this is important for understanding the cosmogonical history of the Magellanic system.

THE GALACTIC MAGNETIC FIELD

From the polarization angles of the galactic foreground stars, some conclusions can be drawn concerning the galactic magnetic field structure. The inclinations of the galactic polarization angles of 20° to 45° against the galactic plane is not in accordance with the helical magnetic field model of Mathewson (1968, 1969) which requires angles parallel (or even with an opposite inclination) to the galactic equator for the sky regions under consideration.

Assuming a uniformly aligned magnetic field parallel to the galactic plane, the galactic longitude ℓ of the field direction can be deduced separately from polarization angles of each single region. As shown in the first row of Table I, ℓ varies between $43°$ and $85°$.

For the data of the lower part of Table I the hypothesis of a helical magnetic field has been assumed, taking the axis of symmetry parallel to the galactic plane but pointing at different galactic longitudes ℓ. From the polarization angles of the three regions, the respective pitch angles $\mathcal{P}$ of the helix have been calculated. The resulting differences between the three fields seem to indicate that the general alignment of the galactic magnetic field is superimposed by local irregularities rather than by a large-scale helical structure. This result is in accordance with recent starlight polarization data near the south galactic pole obtained by Behr and Schröder (1973).

The observational part of this survey has just been completed with polarization measurements of an additional 193 stars of the Magellanic

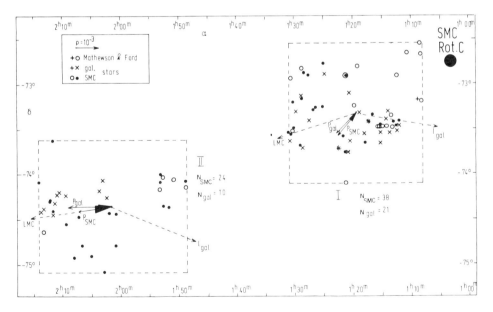

Fig. 3. Stars with known polarization in the SMC fields. Notations as in previous figs. The asterisks indicate stars observed by both Mathewson and Ford (1970) and by Schmidt (1970).

TABLE I

The Galactic Magnetic Field Structure Derived From the Polarization Data

		SMC		LMC
		Field I	Field II	
$\mathcal{P} = 0°$	$\ell =$	$85° \pm 4°$	$55° \pm 11°$	$43° \pm 4°$
$\ell = 90°$	$\mathcal{P} =$	$-4° \pm 2°$	$-17° \pm 5°$	$-14° \pm 1°$
$\ell = 60°$	$\mathcal{P} =$	$+3° \pm 2°$	$-8° \pm 4°$	$-4° \pm 1°$

$\mathcal{P} =$ pitch angle of a fitting magnetic helix. ($\mathcal{P} = 0°$: pure longitudinal field.)
$\ell =$ galactic longitude of the axis of field symmetry.

Cloud regions (the main bodies included), most of them being galactic foreground stars preferably with known spectral types and luminosity classes. The purpose was to deduce the dependence of the galactic polarization on the distance from the galactic plane with more detail.

REFERENCES

Behr, A. 1968. The polarimeter of the 1 m photometric telescope. *European Southern Obs. Bull.* No. 5: 9–13.

Behr, A., and Schröder, R. 1973. Polarization measurements in the south galactic pole and the structure of the magnetic field in the local spiral arm. *Interstellar dust and related topics* (H. C. van de Hulst, J. M. Greenberg, eds.), Dordrecht, Holland: D. Reidel Publ. Co. In press.

Davis, L. Jr., and Greenstein, J. L. 1951. The polarization of starlight by aligned dust grains. *Astrophys. J.* 114: 206–240.

Mathewson, D. S. 1968. The local galactic magnetic field and the nature of the radio spurs. *Astrophys. J.* 153: L47–L53.

———. 1969. The galactic magnetic field. *Proc. Astron. Soc. Australia* 1: 209–210.

Mathewson, D. S., and Ford, V. L. 1970. Polarization measurements of stars in the Magellanic Clouds. *Astron. J.* 75: 778–784.

Pfau, W. 1972. Interstellar reddening in the north galactic polar cap. II. A cloud model. *Astron. Nachr.* 293: 275–280.

Schmidt, T. 1970. Polarization measurements and magnetic field structure within the Magellanic Clouds. *Astron. Astrophys.* 6: 294–308.

DISCUSSION

STROM: Do you know the p/A_V ratios for the foreground stars in the field of the Magellanic Clouds and for the south galactic pole region?

T. SCHMIDT: There is not enough photometry available in the literature, but some information could be extracted from Mathewson (1970) and from Behr and Schröder (1973).

OBSERVATIONS OF MAGNETIC CIRCULAR POLARIZATION OUTSIDE THE SOLAR SYSTEM

J. D. LANDSTREET
University of Western Ontario

A review is presented of recent searches for and observations of circular polarization arising from magnetic fields outside the solar system. The fields discussed include the circular polarization observed in the continua and spectral features of several white dwarfs, photoelectric Zeeman observations of bright stars, and searches for circular polarization in extragalactic objects, particularly quasi-stellar objects and the nuclei of Seyfert galaxies.

Although linear polarization of the light from stars, planets, and nebulae has been known for some years and has been extensively studied, until recently little has been known observationally about the occurrence of circular polarization in astronomical objects except for that due to the longitudinal Zeeman effect in the magnetic Ap stars and the sun. Serkowski (1965) and Serkowski and Chojnacki (1969) report negative results for circular polarization observations of 15 bright stars. Gehrels[1] and Wolf (1972) have both carried out surveys for appreciable amounts of circular polarization (more than $\sim 0.3\%$) covering many kinds of objects, again with negative results. Only in specific classes of objects has circular polarization been detected: in a few white dwarfs, presumably due to the presence of a strong magnetic field; in the Crab Nebula and in a few stars with strong reddening, due to birefringence in the interstellar medium; in infrared objects, probably caused by multiple scattering in the inhomogeneous nebulae surrounding these stars; and in spectral lines of magnetic stars, where it is produced by the longitudinal Zeeman effect. An interesting case of circular polarization not found, is in the optical radiation of quasi-stellar objects that show weak circular polarization of their radio emission. The observations of circular polarization arising from magnetic fields will be surveyed below.

---

[1] Personal communication.

WHITE DWARFS

Techniques and General Searches

It has been thought for some years that magnetic fields of $\sim 10^6$ gauss might be produced in white dwarfs by magnetic flux conservation during the collapse of a normal star to the white-dwarf state; and the discovery of apparently large magnetic fields in pulsars, perhaps produced in this way, has stimulated several observational searches for strong magnetic fields in white dwarfs. Three approaches have been tried: (1) Angel and Landstreet (1970a) searched for circular polarization due to the linear Zeeman effect in the wings of Hγ of nine DA-type white dwarfs but found only upper limits of $1–5 \times 10^4$ gauss; (2) Preston (1970) observed that the quadratic Zeeman effect would systematically displace higher members of the Balmer series toward shorter wavelengths, and he used the absence of such an effect to place limits of $\sim 5 \times 10^5$ gauss on the fields of DA stars generally; recent work by Trimble and Greenstein (1972) has reduced this limit to $1–2 \times 10^5$ gauss; and (3) Kemp (1970) pointed out that continuum radiation emitted in the presence of a longitudinal magnetic field should have a circular polarization on the order of Ω_L/ω, where $\Omega_L = eH/2mc$ is the Larmor frequency and ω is the angular frequency of observation, which leads to a polarization of 0.15% for $H \sim 10^6$ gauss at $\lambda \sim 0.5$ μm. A search for broadband circular polarization in white dwarfs was accordingly launched by Kemp which almost immediately led to the discovery by Kemp and Swedlund of circular polarization of $\sim 3\%$ in the white dwarf Grw+70°8247 (Kemp et al. 1970). This was interpreted as evidence for a magnetic field of $\sim 10^7$ gauss. Further searches have led to the discovery of three other circularly polarized white dwarfs, G195-19, G99-37, and G99-47.

Grw+70°8247 = EG 129

This star shows a unique spectrum especially near 4135 Å (Greenstein and Matthews 1957; Greenstein 1970) which may be due to a combination of atomic and molecular helium features (Angel 1972). The circular polarization has been measured in several broad bands in visible light by Angel and Landstreet (1970b) and Gehrels (1973) and at $\lambda > 1$ μm by Kemp and Swedlund (1970). The polarization spectrum has been obtained with a resolution of 0.008–0.036 μm over the range 0.32 μm–1.1 μm by Angel, Landstreet, and Oke (1972). The circular polarization reaches a maximum of 5.5% at 0.45 μm and may become even larger beyond 1.1 μm. Sharp steps in the polarization spectrum are present at the λ4135 and λ4470 features. This star is also linearly polarized with a maximum of $P = 3.7\%$ at 0.38 μm (Angel and Landstreet 1970b; Gehrels 1973).

$G195$-$19 = EG\ 250$

This DC star has no readily apparent spectral features (Greenstein, Gunn, and Kristian 1971). Circular polarization was discovered by Angel and Landstreet (1971a) and confirmed by Kemp, Swedlund, and Wolstencroft (1971). The polarization was found to be periodically variable with a period of 1.331 days (Angel and Landstreet 1971b; Angel, Illing, and Landstreet 1972), which is assumed to be the rotation period of the star. At 0.46 μm the circular polarization varies nearly sinusoidally between 0.0% and 0.5%; at 0.74 μm the variation is between 0.6% and 1.3%, and is about a quarter cycle out of phase with the variation at 0.46 μm. The magnetic field estimated using Kemp's (1970) theory is about 10^6 gauss at maximum polarization.

$G99$-$37 = EG\ 248$

G99-37, like Grw+70°8247, has a unique spectrum, showing only absorption bands of C_2 and CH (Greenstein 1970; Greenstein, Gunn, and Kristian 1971). The spectrum of circular polarization in broad bands was reported by Landstreet and Angel (1971). More recently Angel and Landstreet (unpublished) have obtained the polarization spectrum with 0.016–0.036 μm resolution. The continuum polarization goes through a minimum of about 0.5% near 0.4 μm, rising to near 1% in the ultraviolet and infrared. A large Zeeman effect is found across the G-band of CH at 0.43 μm, with the polarization departing by $\pm 3\%$ from the continuum value in the two 160 Å channels adjacent to the band center. Fitting the absorption and polarization profiles leads to a field strength of $(3\pm 1) \times 10^6$ gauss.

$G99$-$47 = GR\ 289$

Circular polarization is reported in this very cool DC white dwarf (Greenstein, Gunn, and Kristian 1971) by Angel and Landstreet (1972). The polarization in broad bands is nearly constant from 0.35 μm to 0.85 μm at about 0.4%.

Other White Dwarfs

In addition to the stars discussed above, more than fifty other white dwarfs have been observed by Angel and Landstreet for circular polarization, mostly with a standard error of about 0.1% (Angel and Landstreet 1970a,b; Landstreet and Angel 1971). The list of white dwarfs with circular polarizations $\lesssim 0.2\%$ (including unpublished data) now contains 1 DO-B, 6 DB's, 14 DA's, 2 DF's, 1 DG, 1 DMp, 4 λ4670 stars, 22 DC's, 2 SS Cyg stars, and 1 old nova. In particular, the photometrically variable white dwarfs G44-32, HL Tau 76, Ross 548, HZ 29, DQ Her, and Z Cam have no detectable circular polarization at the 0.1% level (for a discussion of these stars see Warner and Robinson 1972).

Theoretical Work

Kemp's (1970) original study of the circular polarization of thermal radiation of various idealized radiators ("grey-body magnetoemission"), which led to the discovery of the circularly polarized white dwarfs, gives a circular polarization of $q \sim \Omega_L/\omega$ and a linear polarization of $p \sim (\Omega_L/\omega)^2 \sim q^2$, where $\Omega_L = eH/2mc$ is the electron Larmor frequency and ω is the angular frequency of light. This order of magnitude estimate should probably be regarded as a lower limit for the polarization to be expected from a radiator in the presence of a magnetic field H, as specific mechanisms may be more efficient at polarizing the emitted light. Thus, the fields given by this estimate are probably upper limits.

The wavelength dependence of the polarization ($q \sim \lambda$, $p \sim \lambda^2$) given by this model is not in good agreement with that observed in any of the polarized white dwarfs. The possibility that this may be due to the effects of radiative transfer has been considered by Shipman (1971). He finds that the grey-body magnetoemission model together with a strongly wavelength-dependent source of continuous opacity such as hydrogen may introduce structure (somewhat similar to that observed in Grw+70°8247) into the calculated polarization spectrum. However, the sources of the continuous opacity in the polarized white dwarfs are still quite uncertain, making realistic calculations difficult. (It is unlikely that neutral hydrogen is an important opacity source).

The possibility that other mechanisms than Kemp's are at work is suggested by both the strong wavelength dependence of polarization and the large linear polarization which is on the order of q (not q^2 as expected) in Grw+70°8247. Continuum polarization due to bound-free transitions and cyclotron absorption is being investigated by Lamb and Sutherland (1972); bound-free and free-free transitions are also being studied by the group at Louisiana State University (O'Connell 1972*a*; Chanmugan, O'Connell, and Rajagopal 1972). Cyclotron absorption has been considered by Landstreet (1965). In hydrogenic atomic processes it is found that away from the absorption edges the behavior is similar to that given by Kemp's model, but near an edge sharp polarization changes occur. Cyclotron absorption is only likely to be important in the optical range for $B \gtrsim 10^8$ gauss, but it can lead to large circular (and comparable linear) polarization.

It has been suggested by Angel (1972) that bands in a complex molecule such as He_2 could be an important source of continuous opacity, and that the asymmetry of the molecular Zeeman patterns could give rise to continuum circular polarization. This process is substantially more efficient than grey-body magnetoemission at polarizing emitted radiation and might lead to the observed complex wavelength dependence with smaller fields than that needed by the Kemp model. Lack of information about atmospheric compositions and about the relevant molecules (especially He_2) make detailed computations difficult.

PHOTOELECTRIC STELLAR ZEEMAN OBSERVATIONS

The circular polarization produced in stellar absorption line wings by the longitudinal Zeeman effect has long been recognized as offering the possibility of detecting a magnetic field weak enough ($H \lesssim 10^3$ gauss on the main sequence) that the separation of the σ components is small compared to the line width, in which case detection from the absorption profile alone is impossible. This effect has been extensively used for observations of the magnetic fields of the sun and of the magnetic Ap stars. Stellar observations have usually been done photographically, and due to the limited accuracy of photographic intensity measurements and the small effects involved ($q \sim 1\%$ per 100 gauss in favorable cases), standard errors of 100–200 gauss are the best that can be achieved (Preston 1969). Much lower errors (1–10 gauss) have been routinely obtained in solar observations through the use of electro-optic (Pockels cell) analyzers of the type perfected by Babcock (1953). Presently, efforts are being made by several groups to use Coudé photoelectric Pockels cell polarimeters of the Babcock type to search for magnetic fields weaker than those detectable photographically. Severny (1970) reports a survey of eight stars with standard errors between 12 and 50 gauss and claims positive detection of magnetic fields of ~ 40 gauss in Sirius and of ~ 200 gauss in γ Cyg and β Ori. Borra and Landstreet (1973) and Borra, Landstreet, and Vaughan (1973) have surveyed 23 stars, mostly later than A0, with standard errors ranging from 4 to 200 gauss; they detected a field in γ Cyg on several occasions but none in Sirius, and found no other magnetic fields in their sample. Walker has employed a television camera with a solar spectrograph to search for magnetic fields.[2]

Angel and Landstreet (1970a) and Kemp and Wolstencroft (1973)[3] have measured magnetic fields in magnetic Ap stars by observing the wings of Balmer lines through narrow (5–10 Å) interference filters. This eliminates the necessity of having a high-resolution spectrograph, thus increasing the speed of the system and eliminating the spurious polarization due to oblique reflections in the Coudé train, but the effect

[2] See p. 237.   [3] See p. 988.

is much smaller in the broad hydrogen lines ($q \sim 1\%$ per 10^4 gauss) than in the sharp absorption lines observed with Coudé instruments so that the efficiency is poor.

EXTRAGALACTIC OBSERVATIONS

It has recently been found that the radio emission from several quasars is circularly polarized by a few tenths of a percent (references are given in Landstreet and Angel 1972). If the nonthermal optical radiation of quasars and Seyfert galaxies is due to Compton scattering from energetic electrons, then optical circular polarization might be detectable in some of these objects. Optical circular polarization at the 1% level in 3C273 and some Seyfert nuclei has been reported by Nikulin, Kuvshinov, and Severny (1971), but these observations seem to be in error. No effect is found by Gehrels (1972) or Kemp and Wolstencroft (1973). A search of seven quasars, three Seyfert nuclei, and BL Lac (several of which seem to show radio circular polarization) by Landstreet and Angel (1972) revealed no effect at the 0.1% level.

REFERENCES

Angel, J. R. P. 1972. Interpretation of the Minkowski bands in Grw+70°8247. *Astrophys. J.* 171: L17–L21.

Angel, J. R. P.; Illing, R. M. E.; and Landstreet, J. D. 1972. New measurements of circular polarization and an ephemeris for the variable white dwarf G195–19. *Astrophys. J.* 175: L85–L87.

Angel, J. R. P., and Landstreet, J. D. 1970a. Magnetic observations of white dwarfs. *Astrophys. J.* 160: L147–L152.

———. 1970b. Further polarization studies of Grw+70°8247 and other white dwarfs. *Astrophys. J.* 162: L61–L66.

———. 1971a. Detection of circular polarization in a second white dwarf. *Astrophys J.* 164: L15–L16.

———. 1971b. Discovery of periodic variations in the circular polarization of the white dwarf G195–19. *Astrophys. J.* 162: L71–L75.

———. 1972. Discovery of circular polarization in the red degenerate star G99–47. *Astrophys. J.* 178: L21–L22.

Angel, J. R. P.; Landstreet, J. D.; and Oke, J. B. 1972. The spectral dependence of circular polarization in Grw+70°8247. *Astrophys. J.* 171: L11–L15.

Babcock, H. W. 1953. The solar magnetograph. *Astrophys. J.* 118: 387–396.

Borra, E. F., and Landstreet. J. D. 1973. A search for weak stellar magnetic fields. *Astrophys. J.* In press.

Borra, E. F.; Landstreet, J. D.; and Vaughan, A. H. 1973. High-resolution Zeeman polarimetry. *Astrophys. J.* In press.

Chanmugam, G.; O'Connell, R. F.; and Rajagopal, A. K. 1972. Polarized radiation from magnetic white dwarfs: Exact solution of Kemp's model. *Astrophys. J.* 175: 157–159.

Gehrels, T. 1972. On the circular polarization of HD 226868, NGC 1068, NGC 4151, 3C273, and VY Canis Majoris. *Astrophys. J.* 173: L23–L25.

———. 1973. Photopolarimetry of planets and stars. *Vistas in astronomy*. Vol. 15 (A. Beer, ed.), London and New York: Pergamon Press.
Greenstein, J. L. 1970. Some new white dwarfs with peculiar spectra. VI. *Astrophys. J.* 162: L55–L59.
Greenstein, J. L.; Gunn, J. E.; and Kristian, J. 1971. Spectra of white dwarfs with circular polarization. *Astrophys. J.* 169: L63–L69.
Greenstein, J. L., and Matthews, M. S. 1957. Studies of the white dwarfs. I. Broad features in white dwarf spectra. *Astrophys. J.* 126: 14–18.
Kemp, J. C. 1970. Circular polarization of thermal radiation in a magnetic field. *Astrophys. J.* 162: 169–179.
Kemp, J. C., and Swedlund, J. B. 1970. Large infrared circular polarization of Grw+70°8247. *Astrophys. J.* 162: L67–L68.
Kemp, J. C.; Swedlund, J. B.; Landstreet, J. D.; and Angel, J. R. P. 1970. Discovery of circularly polarized light from a white dwarf. *Astrophys. J.* 161: L77–L79.
Kemp, J. C.; Swedlund, J. B.; and Wolstencroft, R. D. 1971. Confirmation of the magnetic white dwarf G195-19. *Astrophys. J.* 164: L17–L18.
Kemp, J. C., and Wolstencroft, R. D. 1973. HD 215441 and 53 Cam: Intrinsic polarization of Hβ and the continuum. *Astrophys. J.* 179: L33–L37.
Lamb, F. K., and Sutherland, P. G. 1972. Continuum polarization in magnetic white dwarfs. Paper presented at IAU Symp. No. 53, *Physics of dense matter*, Aug. 1972, Boulder, Colorado (Proceedings edited by W. Brittin).
Landstreet, J. D. 1965. Effects of a large magnetic field on energy transport in white dwarf stars. Ph.D. thesis (unpublished), Columbia University.
Landstreet, J. D., and Angel, J. R. P. 1971. Discovery of circular polarization in the white dwarf G99-37. *Astrophys. J.* 165: L71–L75.
———. 1972. Search for optical circular polarization in quasars and Seyfert nuclei. *Astrophys. J.* 174: L127–L129.
Nikulin, N. S.; Kuvshinov, V. M.; and Severny, A. B. 1971. On the circular polarization of some peculiar objects. *Astrophys. J.* 170: L53–L58.
O'Connell, R. F. 1972. Polarized radiation from magnetic white dwarfs. Paper presented at IAU Symp. No. 53, *Physics of dense matter*, Aug. 1972, Boulder, Colorado.
Preston, G. W. 1969. The periodic variability of 78 Virginis. *Astrophys J.* 158: 243–249.
———. 1970. The quadratic Zeeman effect and large magnetic fields in white dwarfs. *Astrophys. J.* 160: L143–L145.
Serkowski, K. 1965. Polarization of galactic clusters M 25, NGC 869, 884, 1893, 2422, 6823, 6871, and Association VI Cygni. *Astrophys. J.* 141: 1340–1361.
Serkowski, K., and Chojnacki, W. 1969. Polarimetric observations of magnetic stars with two-channel polarimeter. *Astron. Astrophys.* 1: 442–448.
Severny, A. B. 1970. The weak magnetic fields of some bright stars. *Astrophys. J.* 159: L73–L76.
Shipman, H. L. 1971. Polarization of Grw+70°8247: The transfer problem. *Astrophys. J.* 167: 165–168.
Trimble, V., and Greenstein, J. L. 1972. The Einstein redshift in white dwarfs. III. *Astrophys. J.* 177: 441–452.
Warner, B., and Robinson, E. L. 1972. Non-radial pulsations in white dwarfs. *Nature Phys. Sci.* 239: 2–7.
Wolf, G. W. 1972. A search for elliptical polarization in starlight. *Astron. J.* 77: 576–583.

INTRINSIC POLARIZATION AND THE TRANSVERSE ZEEMAN EFFECT IN MAGNETIC Ap STARS

JAMES C. KEMP and RAMON D. WOLSTENCROFT
Institute for Astronomy
University of Hawaii

We have recently found the first clear evidence for intrinsic linear polarization in magnetic Ap stars. The preliminary report (Kemp and Wolstencroft 1972) dealt with HD 215441 and 53 Cam. Later work in progress has confirmed and extended the findings on the latter star, and analogous effects have been found in α^2CVn; widespread discoveries are anticipated in other magnetic main-sequence stars.

Evidence for this type of polarization has been sought for many years; see, for example, Hiltner and Mook (1967), and Serkowski and Chojnacki (1969). The problem was the smallness of the intrinsic effects and the possible masking by large interstellar polarization in some stars. Virtually all the early searches involved broad or unfiltered passbands. We reasoned that the place to look for intrinsic linear polarization in magnetic Ap stars was in the spectral lines. Two mechanisms were envisioned: (1) the obvious one of the transverse Zeeman effect (routinely seen in sunspots by the magnetograph method); and (2) asymmetrical scattering caused by so-called hot spots (a source of the magnitude variations), which would not generally be centered on the visible disk. Mechanism (1) would surely reflect the line structure. Mechanism (2) could under certain conditions (resonant scattering) also produce a sharp peaking in the lines; alternatively, there could be a strong suppression of the polarization in the lines due to absorption. In either event, we would certainly expect a *difference* in polarization between the lines as compared with broadbands or the continuum. Detection of such differences would immediately reveal the intrinsic effects, regardless of masking by either interstellar or instrumental contributions, since both the latter would vary slowly with wavelength. We were prepared, as well, to look for correlation of time variations with the phase curves, which was the classical test applied by others in earlier work.

We selected the Balmer lines, specifically Hβ, for our investigations. The hydrogen lines have many times the equivalent width of other individual lines in all B and A stars and would, we felt, give the best signal-to-noise ratio for the measurements. The Hβ filter had a width of 15 Å. For comparison with Hβ we chose also a broadband, 3800–5400 Å, roughly centered on Hβ and taken to represent the average continuum. This band is blanketed by many more-or-less shallow lines in the Ap stars, besides the Balmer lines. It is not clear at this time whether what we have called the continuum polarization is really an unresolved but nonvanishing effect of the many lines, or whether it is a true continuum effect, i.e., due to free-free or free-bound transitions. In either case, a large polarization difference was observed.

To summarize our above-mentioned paper, striking results were obtained, and quite different behavior was seen in the first two stars. In HD 215441, the broadband polarization is on the order of 2% and is at least approximately constant in time, as measured by us and previous observers, but we found the Hβ polarization to vary wildly, over the approximate limits 0%–3%, on time scales as short as a few hours. The changes were so large that there was absolutely no doubt as to their reality. Curiously, only small changes in the *angle* of the Hβ polarization were seen; the direction remained almost the same as that of the broadband polarization. We could discern no pattern in the changes in the Hβ amount of polarization, nor did we establish a correlation with the magnetic phases. The behavior seems reminiscent of polarization variations in Be-type shell stars, although, to our knowledge, there is no evidence for a shell nor for emission in HD 215441. For the moment we can only guess that the Hβ effects in this star, which is atypical among the Ap stars, are due to some kind of turbulent process and not to the ordered magnetic field that causes the Zeeman splittings.

In the more typical Ap star, 53 Cam (preliminarily also in α^2CVn), we find a regular pattern. In 53 Cam the Hβ and broadband (blue-green) polarizations have mean amplitudes 0.15% and 0.05% respectively; both rotate through large angles, on the order of 180°, over the course of the magnetic phase curve, and are generally at a large angle relative to each other except for a coincidence at two phases. The broadband points have been reproduced with good agreement through three cycles, and complete phase curves will be reported in the near future. A calculation using the theory of Unno (1956) for the transverse Zeeman effect, and involving an integration over the disk, is in progress to attempt a quantitative comparison. The first data for α^2CVn show a varying Hβ polarization on the order of 0.10% but very small broadband polarization, $\leq 0.01\%$.

In order to carry the search to a wide range of other stars, it is important to reduce the instrumental polarization as far as possible. We have recently made many tests on relatively unpolarized stars, including ι Peg, Sirius, α Psa, β Cas, and Procyon. The results were consistent with a net instrumental linear polarization of not more than 0.008%, including effects in our photoelastic polarimeter and either of two telescopes we used at Mauna Kea.

An interesting use of the transverse Zeeman effect will be the detection of new magnetic stars. The effect has a tendency to be larger than the better known longitudinal effect for two reasons: (1) it appears to be enormously "amplified" by the line-saturation or growth-curve mechanism (assuming the present interpretation of our findings for 53 Cam and α^2CVn to be correct), and (2) there is, roughly speaking, a two-to-one chance that a given magnetic field is transverse to the line of sight rather than parallel to it. It will be worthwhile, for example, to search for linear polarization in the Balmer lines of DA-type white dwarfs, whereas a previous search was concerned with only the circular polarization (Angel and Landstreet 1970).

Acknowledgments. Our research is partially supported by the National Science Foundation (GA-28201) and indirectly by the National Aeronautics and Space Administration through its support of the Mauna Kea Observatory.

Note added in proof: Later analysis indicates that the linear polarization we see in 53 Cam (and perhaps in α^2CVn as well) is fairly certainly *not* due to the transverse Zeeman effect, at least not in the sense of Unno's model, i.e., magneto-absorption under conditions of local thermodynamic equilibrium (LTE). Explicit calculations for 53 Cam using that model have been carried out by E. F. Borra (*Astrophys. J.*, in press); they predict polarization values about two orders of magnitude smaller than the observed values, and the phase dependence is wrongly predicted. A promising alternate model involves non-LTE magneto-scattering, and is presently under consideration. The principal problem is to account for the curious fact that the linear polarization in 53 Cam, both in Hβ and in the broad band, is a *minimum* (or zero) when the magnetic field is transverse, not a *maximum* as Unno's model or a simple absorbing-layer model would predict.

REFERENCES

Angel, J. R. P., and Landstreet, J. D. 1970. Magnetic observations of white dwarfs. *Astrophys. J.* 160: L147–L152.

Hiltner, W. A., and Mook, D. E. 1967. Plane polarization in magnetic variables. *The magnetic and related stars.* (R. C. Cameron, ed.), pp. 123–129. Baltimore: Mono Book Corp.

Kemp, J. C., and Wolstencroft, R. D. 1972. HD 215441 and 53 Camelopardalis: Intrinsic polarization of Hβ and the continuum. *Astrophys. J.* 179: L33–L37.

Serkowski, K., and Chojnacki, W. 1969. Polarimetric observations of magnetic stars with two-channel polarimeter. *Astron. Astrophys.* 1: 442–448.

Unno, W. 1956. Line formation in a normal Zeeman triplet. *Publ. Astron. Soc. Japan* 8: 108–125.

COMPUTATION OF STRONG MAGNETIC FIELDS IN WHITE DWARFS

R. F. O'CONNELL
Louisiana State University

Our work on the properties of atoms, ions, and electrons in the presence of a strong magnetic field is briefly surveyed.

The discovery of polarized radiation from certain white dwarfs has brought the study of white dwarfs into prominence once again. So far, a total of four white dwarfs has been found to exhibit a fractional polarization[1]

$$q \equiv [P_+(\omega) - P_-(\omega)]/[P_+(\omega) + P_-(\omega)], \tag{1}$$

where $P_\pm(\omega)$ are the intensities of right and left circularly polarized light of angular frequency ω. As far as I am aware, no attempt has been made to explain the observations other than by invoking the presence of a magnetic field in the white dwarf.[2]

In his initial pioneering work, Kemp (1970) predicted that

$$q \simeq -(\Omega/\omega), \text{ if } \Omega \ll \omega \tag{2}$$

where $\Omega = (eB/2\mu c)$ is the Larmor frequency, and where B and μ denote the magnetic field and electron mass, respectively. From the observed q values, it was deduced that B in all cases is in the range 10^6–10^8 gauss. However, it soon became apparent that the spectral dependence of q is rather complicated and not in conformity with the predictions of Kemp's model.[3] Attempts made to improve the theoretical predictions by taking into account radiative transfer and by a more exact treatment of Kemp's model have met with only limited success. A detailed report of this work has recently been given (O'Connell 1972) and thus will not be repeated here. As a result, it became clear that the way to proceed was by a detailed examination of the properties of atoms, ions, and electrons

---

[1] See p. 981.   [2] Also see p. 54.   [3] See p. 982.

in the presence of a strong magnetic field. The progress already made in this direction is also summarized by O'Connell (1972). In particular, the energy spectrum of the hydrogen atom in a strong magnetic field has been obtained (Smith et al. 1972). In addition, the theory of transition probabilities in a strong magnetic field has been formulated and applied to bound-bound transitions in the hydrogen atom (Smith et al. 1973). Similar calculations have been carried out for He II (Surmelian and O'Connell 1973).

The thrust of our present efforts is mainly two-pronged: (a) calculation of photoionization in a strong magnetic field, and (b) calculation of the quadratic Zeeman effect (QZE) to an accuracy greater than that given by perturbation theory, when the magnetic field is large.

The photoionization calculation is being carried out by Henry, Roussel, and O'Connell. We believe that this process will be the basic ingredient of a realistic physical model that will explain the many intriguing facets of the polarized radiation from magnetic white dwarfs. Introducing several simplifying assumptions, we have obtained analytic results that give a q value with the same spectral dependence as predicted by Kemp's model. We expect to obtain a more complicated—and hence, in the present context, more interesting—q dependence on wavelength when we include B^2 terms and Coulomb field effects in our calculation.

An accurate calculation of the QZE is being carried out by Surmelian and O'Connell. Displacements of *line* spectra due to this effect may be used, as another method, to detect B fields (Preston 1970). The existing expression used for the QZE is obtained by treating the B^2 term in the Hamiltonian as a first-order perturbation. It follows that, for a hydrogen atom in a magnetic field, the energy of the electron may be written (in units $\hbar = c = \mu = 1$)

$$E \equiv E_0 + E_1 + E_2$$
$$= \frac{\alpha^2}{2n^2}\left[-1 + mn^2\left(\frac{B}{B_0}\right) + F_{nlm}n^6\left(\frac{B}{B_0}\right)^2\right], \quad (3)$$

where

$$B_0 \equiv \frac{\mu^2 c e^3}{\hbar^3} \equiv \alpha^2 B_c = 2.350 \times 10^9 \text{ gauss}, \quad (4)$$

α is the fine-structure constant, *nlm* are the usual hydrogen-atom-quantum numbers, and

$$F_{nlm} = \frac{5\left\{1 + \frac{1}{5n^2}[1 - 3l(l+1)]\right\}\{l(l+1) + m^2 - 1\}}{4(2l-1)(2l+3)}. \quad (5)$$

For large n and $l = 1$, we have $F = (1 + m^2)/4$, so that (in Rydbergs)

$$E_2 = \frac{1}{4} n^4 (1 + m^2) \left(\frac{B}{B_0}\right)^2, \qquad (6)$$

which is the formula used by Preston (1970).

Now, it is generally stated that, since

$$(E_2/E_1) \sim [(B/B_0)n^4], \qquad (7)$$

perturbation theory breaks down at a critical field, B_H say (H for hydrogen), given by (B_0/n^4). However, this is not quite correct, as the value of the coefficients of the B terms also play an important role in determining the exact value of B_H. It is important to determine B_H exactly so that we know precisely when Equation (7) is no longer trustworthy. A knowledge of B_H is also very useful in considering other problems such as photoionization, for example, because much work may be saved if perturbation-theory results are known to be reliable. On the other hand, this knowledge will also prevent us from using perturbation theory blindly in situations where it is not justified. Thus, we proceed to an accurate evaluation of B_H.

We define B_H as the magnetic field for which

$$|E_2| = |E_0 + E_1|. \qquad (8)$$

This is certainly an upper limit, and we emphasize that perturbation-theory results will probably not be very accurate at values of B say five times smaller than B_H. It is advantageous to consider separately the cases $m = 0$ and $m \neq 0$, in the derivation of B_H from Equations (3) and (8). It is also convenient to write

$$B_H \equiv (B_0/y). \qquad (9)$$

1. The case where $m = 0$. Thus,

$$y = n^3 |F_{nl0}^{1/2}|. \qquad (10)$$

   For large n and $l = 1$, it follows that $y \approx n^3$—not n^4.

2. The case where $m \neq 0$. For many cases of interest $|E_0| \gg |E_1|$. Hence,

$$y = n^3 |F_{nlm}^{1/2}|. \qquad (11)$$

   For large n and $l = 1$ (hence, $|m| = 1$), it follows that $y \approx n^3/\sqrt{2}$.

For use in interpreting the observational data, a knowledge of the QZE is useful for n values up to 10. Thus we have calculated the corresponding values of B_H over this range. In Tables I and II, we present, for selected values of n, the values of B_H corresponding to both $l = n - 1$, $m = 1$ and $l = |m| = 1$.

TABLE I
Values of B_H *From Equations (9) and (10)*

n	l	m	B_H (gauss)
2	1	0	6.8×10^8
2	0	0	4.4×10^8
5	4	0	4.7×10^7
5	0	0	2.9×10^7
6	5	0	2.7×10^7
6	0	0	1.7×10^7
9	8	0	8.4×10^6
9	0	0	5.0×10^6
10	9	0	6.2×10^6
10	0	0	3.6×10^6

TABLE II
Values of B_H *From Equations (3) and (8)*

n	l	m	B_H (gauss)
2	1	1	3.2×10^8
5	4	4	1.7×10^7
5	1	1	2.3×10^7
6	5	5	9.9×10^6
6	1	1	1.4×10^7
9	8	8	2.8×10^6
9	1	1	4.2×10^6
10	9	9	2.1×10^6
10	1	1	3.1×10^6

Acknowledgments. The author would like to thank G. Surmelian for computing the numbers appearing in the tables.

REFERENCES

Kemp, J. C. 1970. Circular polarization of thermal radiation in a magnetic field. *Astrophys. J.* 162: 169–179.

O'Connell, R. F. 1972. Polarized radiation from magnetic white dwarfs and atoms in strong magnetic fields. *Proceedings of the IAU symposium on the physics of dense matter.* (W. Brittin and C. Hansen, eds.) Dordrecht, Holland: D. Reidel Publ. Co.

Preston, G. W. 1970. The quadratic Zeeman effect and large magnetic fields in white dwarfs. *Astrophys. J.* 160: L143–L145.

Smith, E. R.; Henry, R. J. W.; Surmelian, G. L.; O'Connell, R. F.; and Rajagopal, A. K. 1972. Energy spectrum of the hydrogen atom in a strong magnetic field. *Phys. Rev. D* 6: 3700–3701.

Smith, E. R.; Henry, R. J. W.; Surmelian, G. L.; and O'Connell, R. F. 1973. Hydrogen atom in a strong magnetic field: bound-bound transitions. *Astrophys. J.* 179: 659–663.

Surmelian, G. L., and O'Connell, R. F. 1973. Energy spectrum of He II in a strong magnetic field and bound-bound transition probabilities. *Astrophys. Space Sci.* 20: 85–91.

POLARIZATION OF PULSAR RADIATION

W. J. COCKE
University of Arizona

A review of observations and theories concerning the polarization of pulsar radiation is given, with emphasis on the Crab Nebula pulsar. Both physical emission mechanisms and geometrical rotating vector models are considered.

It is now generally believed that pulsars are rotating neutron stars, with the observed pulsation period being equal to the rotational period. Further, it is thought that there exists a beaming mechanism of some sort, so that the sharp pulses picked up in our radio telescopes are caused by the beams sweeping across our line of sight. The beaming mechanism would then arise in the magnetosphere of the neutron star and would be associated with the magnetic poles (or perhaps the magnetic equator) of an off-axis field frozen into the neutron star.

We shall not go into the genesis of these basic ideas but suggest that the reader consult earlier review papers (Hewish 1970; "Proceedings of the Flagstaff Crab Nebula Symposium," 1970). Interesting summaries of observed pulsar properties and theoretical requirements have also been given by Smith (1970a,b).

In this paper we review specifically the diverse observations and theories relating to the polarization of pulsar radiation, with particular emphasis on the Crab Nebula pulsar, since it is unique in many respects and is the pulsar about which we know the most.

The acceleration of high-energy particles by pulsars should be mentioned since they may be the particles that produce the pulsed radiation. Since relativistic particles radiate predominantly in their direction of motion, a beam of energetic particles traveling nearly parallel to a magnetic field line would be an excellent mechanism for generating the observed pulses. The production of high-energy particles is particularly important for understanding the energetics of the Crab Nebula, where the total amount of nebular synchrotron radiation produced by high-energy particles has been shown (Finzi and Wolf 1969) to be about equal to the rate at which the pulsar is losing rotational energy (about 2×10^{38} erg/sec).

There are two important "pulsar" mechanisms of high-energy particle acceleration. The first is basically an electrostatic mechanism and would operate even if the magnetic field were symmetric about the rotational axis. It is called the "unipolar inductor" and relies on the presence of strong electric fields generated in the pulsar magnetosphere by the rotating magnetized neutron star (Gold 1969; Goldreich and Julian 1969). Particle energies up to the maximum observed in cosmic rays could theoretically be generated by this mechanism. The second mechanism relies on the fact that a spinning off-axis magnetic dipole radiates dipole radiation at the pulsar rotational frequency (Pacini 1967, 1968; Gunn and Ostriker 1969). This radiation is inherently capable of accelerating charged particles to very high energies, since the electromagnetic fields of this radiation are very strong and since the frequencies are so low. However, these frequencies are certainly much less than the formal plasma frequencies in the neighborhood of pulsars, and thus this radiation may not be able to propagate. It is possible to produce very high energy cosmic ray particles in this way, but not necessarily those of more moderate energy (Gunn and Ostriker 1971).

The magnetic fields near neutron stars are certainly very strong. Estimates of 10^{11}–10^{13} gauss have been obtained by various arguments (Pacini 1968; Gold 1969; Gunn and Ostriker 1969; Sturrock 1971), and we assume these to be typical values.

OBSERVED PULSAR POLARIZATION PROPERTIES

Pulsar radio observations are generally made from around 20 MHz to 10 GHz, where most pulsars become almost completely unobservable. Throughout this spectral region, nearly all pulsars show large amounts of linear polarization, and a great many show strong circular polarization as well. The reader is referred to the article by Lyne, Smith, and Graham (1971) for collections of pulse shapes and polarization properties. The intensities and pulse shapes of many pulsars tend to fluctuate rather erratically, and hence, the polarization patterns also tend to be variable. However, some pulsars show strikingly regular pulse shapes and intensities, as well as regular sweeps in the position angle of the linear component. PSR 0833-45 (the "Vela" pulsar) is notorious in this regard (Radhakrishnan et al. 1969; Ekers and Moffet 1969), as illustrated in Fig. 1. Figure 2 shows the position-angle sweep of nearly 180° through the pulse. Other examples of such regular sweeping may be found in the article by Lyne, Smith, and Graham (1971).

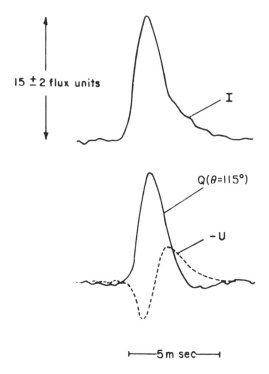

Fig. 1. The Stokes parameters *I, Q, U*, of the Vela pulsar PSR 0833-45 that has a period of 0.083 sec; 1720 MHz, $\triangle f$ is 7 MHz, TC is 1 msec. (Radhakrishnan et al. 1969.)

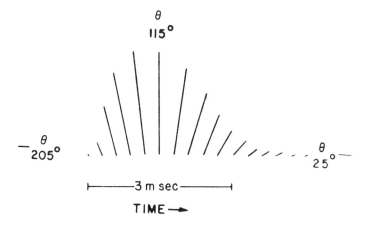

Fig. 2. As for Fig. 1, but giving position angle of the linear component.

Manchester (1971c) has measured the polarization of various pulsars at 410 and 1665 MHz. His results are consistent with the position angle sweeps being independent of frequency, and he mentions that this implies that the position angle sweeps are not propagation effects. He also finds that the percentage linear polarizations are generally lower at 1665 MHz than at 410 MHz. He interprets this as being due to a plasma depolarization effect similar to Faraday rotation, but with the cyclotron frequency large compared to the wave frequency.

Circular polarization generally tends to be weaker than linear, especially if an average profile over many pulses is taken. Graham (1971) finds strong circular polarization in many pulsars (V/I on the order of 10%) and notes a tendency for leading and trailing edges to differ in sign of the circular component. Particularly striking is PSR 1508, which shows at 408 MHz an average linear polarization of 15% and a maximum circular of 30%. The position angle of its linear polarization also shows a regular sweep of about 180° through the pulse.

The Crab Nebula pulsar (NP 0532) is especially interesting and is the fastest of all known pulsars with a period of 0.033 sec. It is unfortunately difficult to observe because of the strong nebular background radiation and because of the narrow width of the primary and secondary pulses ($\approx$ 300 μsec). Thus observers often have obtained contradictory results. A good summary of earlier observations of its radio properties has been given by Drake (1971). The radio pulse shape is quite complex and has three components (see Fig. 3): a broad, highly linearly polarized "precursor"; a narrow, intense primary pulse; and a less intense secondary pulse (also called the "interpulse"). Manchester, Hugenin, and Taylor (1972) show the presence of possible additional components at low frequencies (110–160 MHz). The precursor is about 80% linearly polarized, and the primary and secondary less so, and Manchester et al. show that this is true even at low frequencies, in contradiction to the results of Heiles, Rankin, and Campbell (1970b) at 111.5 MHz.

The position angle of the precursor seems to be constant, but Schönhardt (1971) has observed a strong sweep in position angle through the primary after averaging over many pulses. This sweep was not seen by Campbell, Heiles, and Rankin (1970), who also used a time-averaging technique, but their time resolution was not quite as good as Schönhardt's. Campbell et al. also do not observe any significant circular polarization.

A particularly interesting property of the Crab pulsar is the occurrence of occasional "giant" radio pulses, which are roughly 2000 times more intense than the average pulse and which occur only about once every

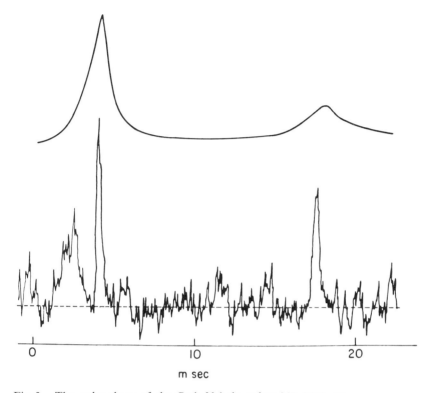

Fig. 3. The pulse shape of the Crab Nebula pulsar NP 0532. The upper curve is the optical light curve (the secondary peak is actually sharper than pictured). The lower curve is the radio pulse shape at 430 MHz. (From Drake 1971.)

10,000 times. Heiles, Campbell, and Rankin (1970a) found that the primary and possibly the secondary are involved in this peculiar phenomenon, but Rankin and Heiles (1970) show that it is only the primary. These giant pulses are strongly polarized, with large amounts of circular present (Heiles, Campbell, and Rankin 1970a). Graham, Lyne, and Smith (1970) did not observe any circular polarization in these strong pulses, but their time resolution was not as good as that of Heiles et al.

Giant pulses arriving earlier than the average show right-handed circular polarization, and those arriving later are left-handed. The linear polarization position angle does not seem to show any regular sweep (Graham, Lyne, and Smith 1970). Other pulsars also show occasional strong pulses, but not in nearly as striking a way as the Crab (Manchester 1971a).

The Crab pulsar apparently is the only radio pulsar emitting detectable amounts of energy in the optical band. It also emits substantial amounts in the UV and X-ray regions,[1] and in fact most of its pulsed energy is output in extreme UV and soft X rays, at a time-averaged rate of 2×10^{37} erg/sec (Rappaport 1971). The time-averaged radio output for this pulsar is about 10^{32} erg/sec, and the total of all the emissions from the nebula itself is about 2×10^{38} erg/sec. See Rappaport (1971) and Bradt (1972) for reviews of the X-ray properties. The X-ray polarization has not yet been measured, and we discuss only the optical results from here on.

The most striking difference between the radio and optical properties of the Crab pulsar is that the optical pulses are much more stable than the radio pulses. No variations of the pulse shape with time have been reliably detected (Horowitz, Papaliolios, and Carleton 1972), and Kristian (in "Proceedings of the Flagstaff Crab Nebula Symposium," 1970) reports that no short-term intensity variations have been observed that could not be attributed to photon-counting statistics. The lightcurve is also considerably different from the radio pulse shape (Fig. 3).

The polarization in the optical is also different. There is a large, well-defined sweep in the position angle across both primary and secondary pulses (Wampler, Scargle, and Miller 1969; Cocke et al. 1970, 1972; Kristian et al. 1970). The rate of sweeping in the radio is about $100°$/msec (Schönhardt 1971), whereas in the optical it is about $30°-40°$/msec. Figure 4 shows the data corrected for interstellar polarization. The secondary intensity peak occurs 13.4 msec after the primary, so that the placement of the two peaks is not symmetric. The polarization minima occur about 1 msec after the primary peak and 2 msec after the secondary peak. Possible secular variations in the optical polarization have been reported by Cocke, Ferguson, and Muncaster (1973), but these variations may have been instrumental in origin. Circular polarization has not as yet been detected (Cocke, Muncaster, and Gehrels 1971).

The striking position-angle sweeps of many pulsars are of great interest and are highly suggestive of the rotational nature of these objects. This is discussed in more detail below.

THEORIES OF PULSAR-EMISSION MECHANISMS

Such complex and striking observations have naturally provoked a voluminous flood of theoretical work. Indeed, the theoretical picture is even more untidy than the observational one, and I shall attempt to

---

[1] See p. 262.

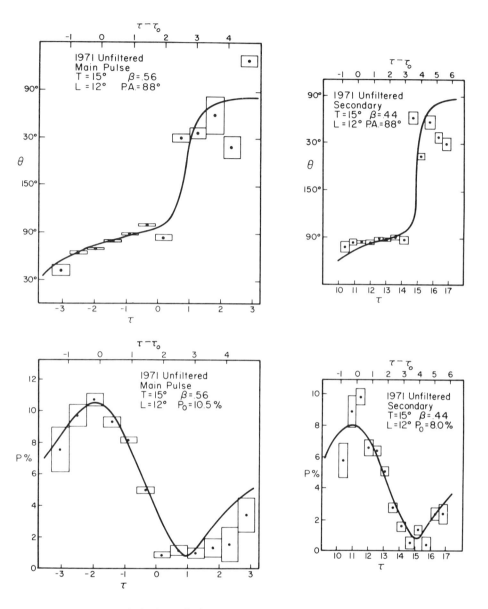

Fig. 4. Linear polarization of the primary and secondary pulses of the Crab Nebula pulsar in the optical, corrected for interstellar polarization. The abscissal parameter τ is time, in msec, taken relative to the main pulse peak. τ_0 is the time when the emission region is moving most nearly toward the observer. θ is the position of the electric vector maximum of the observed polarization corrected for interstellar polarization. P.A. is the position angle of the rotation axis. For definitions of other parameters, see Fig. 5 and the text.

cover only those theories that seem to me to be the more plausible ones and that in addition have something interesting to say about the polarization.

It is generally agreed that the very high brightness temperatures of the radio pulses (10^{24}–10^{30} °K) require the use of coherent and/or stimulated processes. Accordingly, many authors have discussed such mechanisms.

Bertotti, Cavalieri, and Pacini (1969), Pacini and Rees (1970), and Komesaroff (1970) discuss the "curvature radiation" from coherently moving bunches of particles. Curvature radiation is radiation emitted by particles moving in a curved path. It resembles synchrotron radiation, except that the curved path is defined geometrically, the radius of curvature being the same for all particles regardless of energy. This mechanism is capable of giving linear and circular polarization; however, the mechanism for producing the bunching remains obscure. Sturrock (1971) discusses coherent radiation from moving sheets of charge, but the details of the "sheeting" mechanism have not been worked out.

Chiu and Canuto (1971) have proposed a maser mechanism involving quantized synchrotron radiation from the neutron star surface, which is capable of emitting linearly polarized radio waves, but perhaps not circular. However, it is possible that linear polarization can be partially converted into circular in the surrounding plasma (Ginzburg, Zheleznyakov, and Zaitsev 1969, Pacholczyk and Swihart 1970).

Ginzburg, Zheleznyakov, and Zaitsev (1969) discuss coherent plasma oscillations in a general way, as well as their conversion to vacuum waves by induced scattering. Eastlund (1970, 1971) discusses coherent synchrotron radiation and gets both single- and double-humped pulse shapes, as observed in some pulsars, as well as very suggestive circular and linear polarization patterns. His process suggests a continual heating of the magnetosphere by incident radio energy.

Cocke (1973) has investigated stimulated linear acceleration radiation as a possible pulsar-emission mechanism. Linear acceleration in an electric field is ordinarily not an effective way of producing radiation, but one can show that stimulated effects in an ensemble of particles in a linear accelerator give rise to a negative absorption coefficient. This mechanism produces linearly polarized radiation, but if the particle trajectories are curved, the radiation should have a circular component as well. This mechanism seems suited to produce pulses in conjunction with the unipolar inductor mechanism of Goldreich and Julian (1969).

Staelin (1972) has considered Cherenkov radiation by relativistic particles moving into a dense plasma. He gets elliptical polarization with double or single pulses. However, details about the energetics seem to be missing. Smith (1970a) proposes a mildly relativistic cyclotron mechanism but runs into difficulty with the fact that such radiation is almost entirely monochromatic. However, his polarization patterns and pulse shapes are similar to many that have been observed.

Incoherent synchrotron radiation would not seem appropriate to pulsar radio emissions because the high fields involved (10^{12} gauss near the neutron star to 10^6 in the outer reaches of the magnetosphere) would give radiation in the X-ray region. However, Pacini and Rees (1970) have proposed proton synchrotron radiation as a possible explanation for the radio pulses. Pacini (1971) has stated that proton synchrotron would be particularly appropriate for the precursor of the Crab pulsar and predicted low polarization at low frequencies. This is confounded by the observations of Manchester, Hugenin, and Taylor (1972) showing that the precursor remains highly polarized down to 110 MHz.

Optical and X-ray frequencies are probably too high to involve stimulated, coherent, or plasma effects, and so the theoretical picture of this spectral region of the Crab pulsar emission is much simpler. The fact that the amount of energy received in the optical and X-ray regions is much greater than what would be expected by simply extrapolating the radio is a good argument for thinking that the emission mechanisms here are quite different.

The most likely candidate for the optical and X-ray emission seems to be synchrotron emission of some type. Shklovskij (1970) suggested that the spectral index of the Crab pulsar in the X-ray range might really be the same as that observed in the nebular radio observations (-0.28) and, hence, would identify the emitting particles as being the same. He invoked interstellar scattering of the X rays (Slysh 1969) as a means of steepening the pulsar X-ray spectrum to the observed value current at that time. Unfortunately, interstellar scattering does not seem to be effective in this way (e.g., Bowyer, Mack, and Lampton 1970). Also, Shklovskij wrongly applied conventional synchrotron theory to the case of very small pitch angles.

The chief difficulty with conventional synchrotron theory in this case is the interpretation of the infrared cutoff in the observed spectrum, which Shklovskij assumed to be due to synchrotron self-absorption. O'Dell and Sartori (1970a,b) maintain that for the Crab pulsar a more likely explanation for the cutoff is that it is due to cyclotron turnover, which appears only when very small pitch angles are considered. Melrose (1971) has shown, however, that the radiation in the neighborhood of the turnover would be strongly circularly polarized, contrary to observation (Cocke, Muncaster, and Gehrels 1971), but Tademaru

(1972) points out that special particle distributions with respect to pitch angle can result in rather peculiar effects. Also, synchrotron self-absorption in the small pitch angle case does not yet seem to have been considered.

One should emphasize that the cutoff could be a result of conventional synchrotron self-absorption if pitch angles on the order of 0.1 rad are allowed (O'Dell and Sartori 1970b). However, it is difficult to see how the sharp beaming of the optical pulses could be produced by such large pitch angles. The peak of the main pulse is still "sharp" down to a time resolution of 32 μsec (Horowitz, Papaliolios, and Carleton 1972). It is possible that relativistic compression and beaming effects (see above) would remedy this situation, but then the selfabsorption cutoff frequency would be Doppler-shifted up.

Gunn and Ostriker (1971) treat the radiation from particles accelerating in the pulsar-frequency dipole radiation, which we have discussed above. This mechanism seems theoretically capable of providing optical and X-ray emissions, which would be elliptically polarized, with the relative amounts of circular and linear depending on the polarization of the accelerating dipole radiation.

Sturrock (1971) proposes that γ-rays are produced by the curvature radiation process close to the pulsar, where the radius of curvature of the field would be about 10 km. These γ-rays would then interact with the 10^{12} gauss magnetic field, producing electron-positron pairs in cascade. These secondary particles would then radiate by the synchrotron process. However, this model is unable to produce the observed amount of optical radiation.

Komesaroff (1970) has noted that if the optical is synchrotron radiation and if the radio is coherent curvature, then their linear polarization ought to be perpendicular to each other. Manchester (1971b) remarks that this seems to be true, but his radio data show none of the sweeping reported by Schönhardt (1971).

GEOMETRIC MODELS OF POLARIZATION

The regular sweep of the position angle of the linear polarization of the Vela pulsar led Radhakrishnan and Cooke (1969) to develop a rotating magnetic dipole vector model. They proposed that the position angle, as a function of time, be that of the apparent projection on the celestial sphere of a radius vector that is carried around rigidly with the pulsar as it rotates. This is an attractive proposition and is based on general properties of sources of linearly polarized radiation. For curvature radiation, for example, the position angle is parallel to the projection of the path along which the particle moves. For synchrotron radiation, it is perpendicular to the magnetic field, a fact that has led other authors to take the position angles to be perpendicular to the rotating vector, with the result that computed orientations of the pulsar rotation axes differ by 90° from the parallel models. All other parameters, however, would be computed to be the same.

Let T be the tilt of the pulsar rotation axis out of the plane of the sky, and let L be the angle between the rotating vector **B** and the equatorial plane of the pulsar (if the vector is a radius vector, L is the "latitude" of the point of emission). The reader may refer to Fig. 5, which has been drawn to show aberration effects for relativistic motion. Then for the Vela pulsar, Radhakrishnan and Cooke (1969) find $L > 10°$ and $T < L < T + 10°$. Thus, the magnetic dipole vector passes between the line of sight and the rotational axis.

Wampler, Scargle, and Miller (1969) apply the same model to the optical radiation of the Crab pulsar, getting $L = 0°$ and $T = 13°$. They took **B** to be perpendicular to the polarization position angles and obtained a position angle of the rotation axis of 30°. Kristian et al. (1970) generalize the model to allow the rotating vectors (there are two, one for the primary and one for the secondary) to make non-zero angles with the local meridian plane. Then they obtain $L = 0°$ for the primary pulse, $L = 6°$ for the secondary, and $T = 7°$. They find that the polarization position angle at the "moment of symmetry" (i.e., when **B** points most nearly along the line of sight) is $160° \pm 7°$. If Wampler et al. had used the parallel model they would have found 120° for this position angle, and so we see that there are substantial differences between the two sets of data. The data by Wampler et al. are generally thought to be the less reliable because of unsteady sky conditions and because of certain assumptions made in data reduction.

Kristian et al. note the coincidence of their observed "point of symmetry" position angle with that of the nebular synchrotron radiation in the neighborhood of the pulsar and, hence, infer a causal connection between them. Wampler et al. also mention this coincidence but express it as being a coincidence between the position angle of the rotation axis and the position angle of the nebular wisps and magnetic field near the pulsar. Unfortunately, neither of these groups corrected their data for interstellar polarization, and so their computed values may be off by substantial amounts.

Gehrels (see Cocke et al. 1970) has measured the polarization of ten faint stars within a few minutes of arc of the pulsar, getting $p = 2.0 \pm 0.2\%$, $\theta = 147° \pm 3°$ for the interstellar polarization. Disney (1971) has used the corrected data of Cocke et al. (1970)

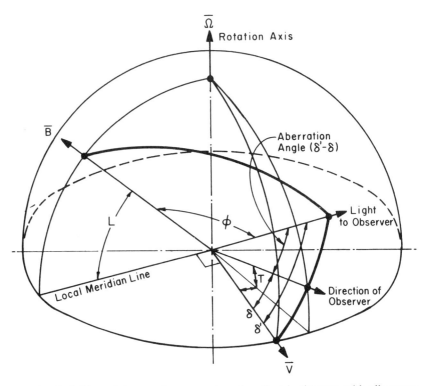

Fig. 5. Definition of the angular parameters described in the text, with allowance made for aberration effects. (From Ferguson 1973.)

to fit the rotating perpendicular vector model, obtaining a value of 50° for the position angle of the rotation axis. This disagrees by 20° with the value obtained by Kristian et al. The raw data of Cocke et al. and of Kristian et al. are in agreement, and the difference is caused by the interstellar polarization corrections and by the fact that Kristian et al. used an additional free parameter. Disney took an average between primary and secondary pulses and claims an agreement with the nebular magnetic field direction obtained by Woltjer (1957), which disagrees with the measurements of this parameter by Kristian et al. For further discussion of this question see Forman and Visvanathan (1971).

Lyne, Smith, and Graham (1971) apply the rotating vector model to several pulsars showing position-angle sweeps. They conclude that some of the position angle curves are so complicated that it is necessary to construct composite models in which there are several vectors oriented at different angles to the rotation axis.

The rotating vector model has been extended to the relativistic case, to include the possibility that the emission region is in the outer reaches of the magnetosphere and, yet, still co-rotating with the neutron star. The co-rotation limit is reached at the "velocity of light cylinder," defined as the cylindrical surface parallel to the rotation axis on which any object rotating rigidly with the neutron star would be traveling at the velocity of light. The position angle of the polarization is then parallel (or perpendicular) to the apparent orientation of the rigidly rotating vector as seen by a stationary observer, due allowance being made for aberration effects.

If the motion of the source is relativistic, then its emission pattern will be compressed and its emissions intensified in the direction of motion. Smith (1970a) demonstrates these effects for cyclotron radiation and shows good correspondence with observed pulse shapes. Other aspects of his theory are discussed above.

Ferguson (1971) shows that relativistic beaming effects can displace a polarization minimum later in time with respect to a pulse intensity maximum, as is observed in the Crab pulsar. The extreme "cusps" at the primary and secondary maxima suggest that these optical intensity peaks may arise from locations very near the velocity of light cylinder, where relativistic compression and beaming are strongest. Manchester (1971c) states that it would be difficult for the various pulsar parameters to be as stable as they are if the emission were to take place near the velocity of light cylinder, but little is known about the stability of pulsar magnetospheres.

Smith (1971) suggests that the finite velocity of light is responsible for the fact that the primary and secondary peaks are not symmetrically placed with respect to each other.

Ferguson (1973) has extended the relativistic vector model to try to fit the curves of the optical percentage polarization of the Crab Nebula pulsar, as well as the position-angle curves, by assuming that the percentage polarizations through the primary and secondary pulses are proportional to the apparent projected lengths of the two rotating vectors. This method was applied by Cocke, Ferguson, and Muncaster (1973) to fit three separate years of observations. The result for data taken in 1971 is shown in Fig. 4. The values of $\beta = v/c$ are almost exactly the same over the three years, with $\beta = 0.56 \pm 0.01$ for the primary pulse, and 0.43 ± 0.02 for the secondary.

CONCLUSIONS

We have seen that the radio polarization properties of the pulsars are very diverse and also that the theory of coherent plasma processes at radio frequencies is frustratingly complex. Therefore, it would be premature to try to draw any firm conclusions about the radio-emission mechanism, and we must wait for further observational and theoretical developments.

Indeed, there may be several mechanisms involved, since some pulsars show regular pulses and a regular sweep of polarization position angle through the pulse, and others do not. Perhaps this reveals a fundamental difference between two distinct emission mechanisms. The Crab pulsar at radio frequencies seems to exhibit these two different mechanisms, one for producing the ordinary pulses, which show a regular position-angle sweep (Schönhardt 1971), and another for producing the anomalous strong pulses. Perhaps the precursor, which has very high linear polarization but shows no position-angle sweep, is produced by yet a third mechanism.

On the other hand, the optical and X-ray pulses of the Crab pulsar present a far tidier picture, and we can probably say that coherent and stimulated mechanisms are not apt to be at work. It appears to me that the synchrotron mechanism is the best guess for explaining this radiation. Perhaps it is small pitch-angle synchrotron, with the infrared cutoff being due to self-absorption rather than cyclotron turnover. As we have mentioned, self-absorption for the case of small pitch angles does not yet seem to have been thoroughly investigated.

It might be well to make a few remarks about the rotating vector models, which have been applied to explain the position-angle sweeps of several pulsars. The general applicability of the rotating vector model depends, unfortunately, on the premise that the emission region has an axis of symmetry. If, however, this is not the case, the polarization pattern will depend on the detailed structure of the magnetosphere. Thus, the "regularity" of a pulsar may depend on the symmetry properties of the magnetosphere and not on any fundamental differences in emission mechanisms.

Acknowledgments. The author would like to thank T. Gehrels and G. W. Muncaster for their help in obtaining the observations shown in Fig. 4. He would also like to thank D. C. Ferguson for fitting the rotating relativistic vector model to these data, and F. G. Smith for helpful criticism.

REFERENCES

Bertotti, B.; Cavalieri, A.; and Pacini, F. 1969. Electromagnetic spectrum of NP 0532. *Nature* 223:1351–1352.

Bowyer, C. S.; Mack, J.; and Lampton, M. 1970. X-ray scattering by grains in the direction of the Crab pulsar and Sco XR-1. *Nature* 225:1125–1127.

Bradt, H. 1972. X-ray observations of pulsars. *The physics of pulsars*. pp. 33–44. New York: Gordon and Breach, Inc.

Campbell, D. B.; Heiles, C.; and Rankin, J. M. 1970. Pulsar NP 0532: average polarization and daily variability at 430 MHz. *Nature* 225:527–528.

Chiu, H.-Y., and Canuto, V. 1971. Theory of radiation mechanisms of pulsars. *Astrophys. J.* 163:577–594.

Cocke, W. J. 1973. Stimulated linear acceleration radiation: A possible pulsar emission mechanism. *Astrophys. J.* 184:291–300.

Cocke, W. J.; Disney, M. J.; Muncaster, G. W.; and Gehrels, T. 1970. Optical polarization of the Crab Nebula pulsar. *Nature* 227:1327–1329.

Cocke, W. J.; Ferguson, D. C.; and Muncaster, G. W. 1973. Optical polarization of the Crab pulsar. II. Observational results and fits by the relativistic vector model. *Astrophys. J.* 183:987–996.

Cocke, W. J.; Muncaster, G. W.; and Gehrels, T. 1971. Upper limit to circular polarization of optical pulsar NP 0532. *Astrophys. J.* 169:L119–L121.

Disney, M. J. 1971. The optical emission of the Crab Nebula pulsar. *Astrophys. Letters* 9:9–12.

Drake, F. D. 1971. Radio observations of the Crab Nebula pulsar. *The Crab Nebula,* pp. 73–83. Dordrecht, Holland: D. Reidel Publ. Co.

Eastlund, B. J. 1970. Low-mode coherent synchrotron radiation and pulsar phenomena. *Nature* 225:430–434.

———. 1971. Low-mode coherent synchrotron radiation and pulsar models. *The Crab Nebula,* pp. 443–448. Dordrecht, Holland: D. Reidel Publ. Co.

Ekers, R. D., and Moffat, A. T. 1969. Polarization of pulsating radio sources. *Astrophys. J.* 158:L1–L8.

Ferguson, D. C. 1971. Mechanism for the delay of polarization minima in the optical pulsar NP 0532. *Nature Phys. Sci.* 234:86–87.

———. 1973. Optical polarization of the Crab pulsar. I. A relativistic vector model. *Astrophys. J.* 183:977–986.

Finzi, A., and Wolf, R. A. 1969. Possible conversion of rotational energy of the neutron star in the Crab Nebula into energy of relativistic electrons. *Astrophys. J.* 155:L107–L114.

Forman, W., and Visvanathan, N. 1971. Magnetic field structure around the Crab pulsar. *Nature* 229:39–40.

Ginzburg, V. L.; Zheleznyakov, V. V.; and Zaitsev, V. V. 1969. Coherent mechanisms of radio emission and magnetic models of pulsars. *Astrophys. Space Sci.* 4:464–504.

Gold, T. 1969. Rotating neutron stars and the nature of pulsars. *Nature* 221:25–27.

Goldreich, P., and Julian, W. H. 1969. Pulsar electrodynamics. *Astrophys. J.* 157:869–880.

Graham, D. A. 1971. Circular components of polarization in pulsar radiation. *Nature* 229:326–327.

Graham, D. A.; Lyne, A. G.; and Smith, F. G. 1970. Polarization of the radio pulses from the Crab Nebula pulsar. *Nature* 225:526.

Gunn, J. E., and Ostriker, J. P. 1969. Magnetic dipole radiation from pulsars. *Nature* 221:454–456.

———. 1971. On the motion and radiation of charged particles in strong electromagnetic waves. *Astrophys. J.* 165:523–542.

Heiles, C.; Campbell, D. B.; and Rankin, J. M. 1970a. Pulsar NP 0532: properties and systematic polarization of individual strong pulses at 430 MHz. *Nature.* 226:529–531.

Heiles, C.; Rankin, J. M.; and Campbell, D. B. 1970b. Absence of average polarization at 111.5 MHz in pulsar NP 0532. *Nature* 228:1074.

Hewish, A. 1970. Pulsars. *Ann. Rev. Astron. Astrophys.* 8:265–296.

Horowitz, P.; Papaliolios, C.; and Carleton, N. P. 1972. Stability of the Crab pulsar. *Astrophys. J.* 172:L51–L54.

Komesaroff, M. M. 1970. Possible mechanism for pulsar radio emission. *Nature.* 225:612–614.

Kristian, J.; Visanathan, N.; Westphal, J. A.; and Snellen, G. H. 1970. Optical polarization and intensity of the pulsar in the Crab Nebula. *Astrophys. J.* 162:475–483.

Lyne, A. G.; Smith, F. G.; and Graham, D. A. 1971. Characteristics of the radio pulses from the pulsars. *Mon. Not. R. Astr. Soc.* 153:337–382.

Manchester, R. N. 1971a. Crab pulsar radiation characteristics. *The Crab Nebula,* pp. 209–210. Dordrecht, Holland: D. Reidel Publ. Co.

———. 1971b. Rotation measure and intrinsic angle of the Crab pulsar radio emission. *Nature Phys. Sci.* 231:189–191.

———. 1971c. Observations of pulsar polarizations at 410 and 1665 MHz. *Astrophys. J. Suppl.* 23:283–322.

Manchester, R. N.; Hugenin, G. R.; and Taylor, J. H. 1972. Polarization of the Crab pulsar at low radio frequencies. *Astrophys. J.* 174:L19–L23.

Melrose, D. B. 1971. Synchrotron radiation from particles with small pitch angles. *Astrophys. Letters* 8:35–37.

O'Dell, S. L., and Sartori, L. 1970a. Limitation on synchrotron models with small pitch angles. *Astrophys. J.* 161:L63–L64.

———. 1970b. Low-frequency cut-offs in synchrotron spectra. *Astrophys. J.* 162:L37–L42.

Pacholczyk, A. G., and Swihart, T. L. 1970. Polarization of radio sources. II. Faraday effect in the case of quasi-transverse propagation. *Astrophys. J.* 161:415–418.

Pacini, F. 1967. Energy emission from a neutron star. *Nature.* 216:567–568.

———. 1968. Rotating neutron stars, pulsars, and supernova remnants. *Nature.* 219:145–146.

———. 1971. A possible interpretation of the precursor pulse in NP 0532. *Astrophys. J.* 169:L11–L12.

Pacini, F., and Rees, M. 1970. The nature of pulsar radiation. *Nature* 226:622–624.

"Proceedings of the Flagstaff Crab Nebula Symposium." 1970. *Publ. Astron. Soc. Pac.* Vol. 82.

Radhakrishnan, V., and Cooke, D. J. 1969. Magnetic poles and the polarization structure of pulsar radiation. *Astrophys. Letters* 3:225–229.

Radhakrishnan, V.; Cooke, D. J.; Komesaroff, M. M.; and Morris, D. 1969. Evidence in support of a rotational model for the pulsar PSR 0833–45. *Nature* 221:443–446.

Rankin, J. M., and Heiles, C. 1970. Pulsar NP 0532: polarization of strong pulses at 430 MHz as seen with 300 k Hz bandwidth. *Nature* 227:1330–1331.

Rappaport, S. 1971. X-ray observations of NP 0532. *The Crab Nebula,* pp. 84–86. Dordrecht, Holland: D. Reidel Publ. Co.

Schönhardt, R. E. 1971. Radio observations of the Crab pulsar at 408, 240, and 151 MHz. *The Crab Nebula,* pp. 110–113. Dordrecht, Holland: D. Reidel Publ. Co.

Shklovskij, I. S. 1970. Pulsar NP 0532 and the injection of relativistic particles into the Crab Nebula. *Astrophys. J.* 159:L77–L80.

Slysh, V. I. 1969. Scattering of pulsar X-ray radiation by interstellar dust particles. *Nature* 224:159–160.

Smith, F. G. 1970a. The beaming of radio waves from pulsars. *Mon. Not. R. Astr. Soc.* 149:1–15.

———. 1970b. Generation of radio waves in pulsars. *Nature* 228:913–916.
———. 1971. Rotation axis and magnetic field axis of the Crab Nebula pulsar PSR 0532 + 22. *Nature Phys. Sci.* 231:191–193.
Staelin, D. H. 1972. Searching for pulsars — results and techniques. *The Physics of Pulsars,* pp. 57–68. New York: Gordon and Breach, Inc.
Sturrock, P. A. 1971. A model of pulsars. *Astrophys. J.* 164:529–556.
Tademaru, E. 1972. Cyclosynchrotron radiation at small angles. *Astrophys. J.* 172:327–330.
Wampler, E. J.; Scargle, J. D.; and Miller, J. S. 1969. Optical observations of the Crab Nebula pulsar. *Astrophys. J.* 157:L1–L10.
Woltjer, L. 1957. Polarization and intensity distribution in the Crab Nebula. *B.A.N.* 13:301–311.

POLARIZATION AND STRUCTURE OF THE CRAB NEBULA

JAMES E. FELTEN
University of Arizona

Present knowledge of the optical, radio, and X-ray polarization of the Crab Nebula is reviewed and discussed as it bears on the structure of the magnetic field, time scales in the nebula, and relations between nebula and pulsar. Not as much high-resolution polarimetry has been done on the Crab Nebula as might have been expected. Loops of field and a large-scale structure can be recognized, but it is not known whether the fields generally are smooth or chaotic on a small scale. Field lines tend to curl around the filaments. The large angular size of the X-ray source poses a difficulty to conventional theory. The form of the nebula does not single out the pulsar as its source, and the exact relation between pulsar and nebula is uncertain. The wave-field or "synchro-Compton" interpretation of the continuum emission is erroneous but has led to interesting observations of circular polarization. Circular polarization of the ordinary synchrotron radiation might be observable in the radio band. Magnetic flux may have been generated in the nebula by winding of lines around the rotating pulsar. Polaroid photographs then suggest that the pulsar rotation axis is roughly NW-SE, but confirmation is lacking.

The prediction by Soviet theorists that the optical continuum of the Crab supernova remnant should be strongly polarized and the detection of this polarization by Soviet observers in 1953[1] can be said to mark the birth of "high-energy astrophysics." It is the most dramatic and one of the most important contributions yet made by polarimetry to astronomy and deserves commemoration in any book on polarimetry.

POLARIZATION OF THE NEBULA

In this age of pulsarology[2] we tend to imagine that the form and polarization of the Crab Nebula itself are well understood, but this is far from the truth. In the sixteen years since the high-resolution

---

[1] The history of these developments is discussed by Shklovskij (1960) and by Oort and Walraven (1956).
[2] See p. 997.

polarimetric map by Woltjer (1957), based on Baade's photographs, and the photoelectric map by Hiltner (1957), remarkably little observational or theoretical advance has been made. Woltjer mapped the linear polarization of the optical continuum over most of the nebula with a diaphragm 5.25 arcsecs in diameter and a grid spacing of 5.59 arcsecs. Figure 1 (Woltjer 1958) shows this map superimposed on a photograph taken in a band including several nebular emission lines so that the network of gaseous "filaments" (Trimble 1968) is visible. Within this network (in projection) lies the roughly elliptical (4 × 2 arcmins) mass of continuum emission, seen most clearly when the lines are suppressed by an appropriate filter. Figures 2 and 3 show two photographs of this continuum taken through a Polaroid filter by Baade (1956); the arrows indicate the E-vector transmission directions of the filter. These are two of Baade's first series of eight, only five of which have been published (Baade 1956; Scargle 1969a). Additional photos through Polaroid by Münch, Baade, and Arp, showing time changes, have been presented by Scargle (1969a).[3] A positive print of the continuum without Polaroid is shown in Fig. 4, again with Woltjer's map superimposed. In general, these photos and map strongly support the nearly universal belief that the continuum is synchrotron radiation (Ginzburg and Syrovatski 1965) from a more-or-less tangled magnetic field that fills the nebula.

The net optical polarization of the entire Crab Nebula lies roughly in a NW-SE direction. It is about 9.2% in position angle 159.6°, according to Oort and Walraven (1956). The polarization of the continuum alone must be somewhat larger because their band included some filamentary radiation. A better figure could be derived by a summation over Woltjer's map, but apparently this has not been done. Burn (1966) quotes figures of 12% and 14%, but the source of these is not made clear. Woltjer first pointed out that the interstellar polarization toward the Crab Nebula is fortuitously in nearly the same position angle, so that about 2% must be subtracted from the observed polarization to get the intrinsic synchrotron polarization (Trimble 1971; cf. Cocke et al. 1970).[4] The net intrinsic optical polarization of the

---

[3] Unfortunately the captions of these photographs are scrambled. The correct caption for the figure on page 198 is found on page 206; that for 199 is on 199; 202 is on 212; 203 on 215; 204 on 204; 205 on 210; 206 on 198; 207 on 208; 208 on 213; 209 on 203; 210 on 207; 211 on 211; 212 on 205; 213 on 214; 214 on 202; and 215 on 209 (J. D. Scargle, personal communication). The figures on pp. 198 and 199 are printed upside down with respect to the others. Position angle 45° should read 135°, and 67°.5 should be 157°.5.

[4] See p. 997.

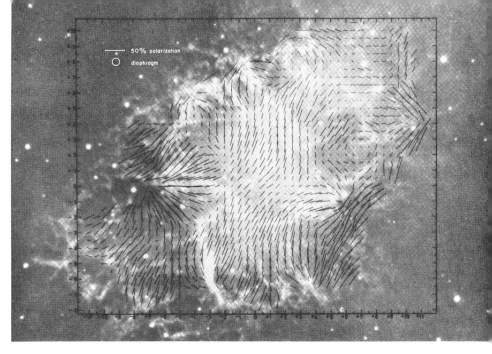

Fig. 1. Optical polarizations in the Crab Nebula measured photographically and projected against a photograph of the nebula taken with the 5-m Hale Telescope by Baade in the light of the [SII] lines (Woltjer 1958). North is at top, east at left. The length of each line segment shown is proportional to the percentage polarization at the point centered on the segment; the segment shown at upper left to establish the scale represents 50% polarization. The small circle shows the diaphragm size.

Fig. 2. A photograph of the Crab Nebula taken in the continuum (5400 to 6400 Å) through a Polaroid filter (Baade 1956). Printed to the same scale as Fig. 1. The arrow indicates the electric-vector transmission direction.

Fig. 3. Same as Fig. 2, but with the Polaroid rotated through 90°. Printed to the same scale as Figs. 1 and 2.

Fig. 4. The polarizations of Fig. 1 projected against a continuum photograph of the nebula without Polaroid (Woltjer 1957). Printed to the same scale as Figs. 1, 2, and 3. The white arrow indicates the 33 msec optical pulsar NP 0532.

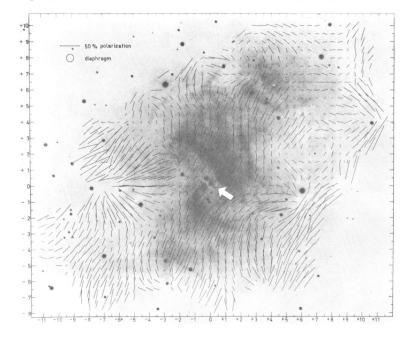

synchrotron continuum from the whole nebula is thus probably $p \approx 7\%$ to 10%. At short centimeter wavelengths the corresponding quantity is $p \approx 8\%$ (Wilson 1972a), though this refers to a different and larger spatial distribution of emission. At longer wavelengths p is lower; this is probably due mainly to Faraday depolarization, especially in the filaments (Burn 1966; Wilson 1972a). There is also X-ray polarization, to which I shall return shortly.

STRUCTURE OF THE MAGNETIC FIELD

Despite many discussions, the details that these photos and map show have probably not been exploited fully by theorists. Thread-like features in the continuum have long been reported by optical observers (Scargle 1969b), and the Polaroid photos seem to show fibrous or ropy structure elongated perpendicular to the electric vector transmission direction (i.e., the axis of the Polaroid filter). Note, for example, the bright isolated loops at the northeast margin of the nebula in Fig. 3. Since the Polaroid at a given orientation tends to pick out regions where the local projected magnetic field is perpendicular to the transmission direction, it appears that these fibers trace magnetic field lines. This impression is strengthened by the fact that, as the Polaroid is rotated through successive photos, some of the loops can be traced through large angles (Baade 1956). A sketch of some of the field lines is given by Oort and Walraven (1956).

How chaotic is the nebular field? In addition to these sizable and well-defined loops visible on the Polaroid plates, we have, as the Woltjer map shows, a rather uniform polarization pattern over the central part of the nebula; the magnitude and direction of polarization change, but not on a scale much smaller than the nebular diameter. Of course, we see a projected summation: along any one line of sight we see from the front of the nebula to the back. Nevertheless, the polarization orientation pattern suggests that the field is rather smooth and simple — comparable, perhaps, to a dipole with a few big twists and loops added. The *percentage* polarizations revealed by Woltjer's 5.25 arcsecs diaphragm can be examined further for a rough check on this. Typically the values in the central region of the nebula are $p \approx 20\%$. On the standard theory (Ginzburg and Syrovatski 1965), the value expected for a *uniform* field is

$$p = \frac{\gamma + 1}{\gamma + \frac{7}{3}}, \qquad (1)$$

where γ is the usual spectral index of the electrons. Baldwin's (1971) review suggests that the effective γ in the optical band may be about 2.8, implying $p \approx 74\%$; recent work on the reddening by Miller (1973) indicates $\gamma \approx 2.1$ and $p \approx 70\%$. We naturally interpret the difference between 70% and the observed 20% as due to projection averaging (plus a smaller contribution from finite resolution). From this we can understand why values $p \geq 70\%$ are in fact measured at a few places on the periphery of the nebula where we perhaps see only the last outward-looping line of force, while p is much lower near the center where several loops are seen crossing in projection along the line of sight.

Let us examine this idea semiquantitatively. If there are n independent, randomly oriented regions of uniform field along a line of sight, each gives rise to a linearly polarized beam with $p \approx 70\%$. By adding Stokes parameters and considering the statistics, we readily see that if n such beams of roughly equal strength are superimposed, the expected linear polarization is $p_n \approx p/\sqrt{2n}$. (We can see a posteriori and perhaps even a priori that this is just an example of a random walk.) Setting $p_n \approx 20\%$ as the observations suggest, we find $n \approx 6$ for the number of independent elements in a column along a typical line of sight. On the other hand, looking at the polarization patterns *transverse* to the line of sight, as revealed by Fig. 4, we might judge the correlation length to be such that $n \approx 3$, certainly no larger than 4, across the bright part of the nebula. This small discrepancy in n is probably not significant, and I believe no studies of this sort have been made; the model of superimposed independent regions is probably a bad one. But we certainly cannot exclude the possibility that there is small-scale structure in the field, causing some depolarization, even within regions where the mean field on a scale of 5 arcsecs appears to have a constant direction. It is not difficult to imagine substantial polarization of synchrotron radiation even from a region where the magnetic lines execute random walks, provided there is a preferred axis (cf. Burn 1966). For example, a gas cloud containing a tangled, chaotic, initially isotropic magnetic field, which is then allowed to expand anisotropically (say, along one axis only) with the fields frozen in (not at all implausible in an astrophysical context), can generate a substantial amount of polarization even though the field remains quite tangled and the propagation of cosmic rays through the field is quite slow.

It would be interesting to know how smooth the field is in the bright northeasterly loops mentioned earlier. Is the polarization in these loops anywhere near the theoretical 70% for a uniform field? Of course projection effects must be allowed for (Scargle 1971), but in isolated features like these they should not be large. Woltjer's map suggests that the polarization is much smaller, 10%–30%. Maximum-resolution studies of a few selected regions of the nebula such as this would be of interest. Although a lot of work is involved, it is surprising that such studies have not been undertaken.

Another interesting point about these bright loops is this: the reason for their brightness is probably not solely that the field lines are properly aligned. Continuity would suggest that neighboring lines that do not show up brightly are nevertheless similarly aligned. Probably the visible loops represent regions of enhanced field or relativistic-electron density, or both. Possibly these particular lines are connected to sources of the particles, while neighboring lines are not. In this case the bright loops would illustrate that cosmic rays diffuse (or stream) faster along the field than across it.

The general polarization pattern in Fig. 1 is obviously related to the positions and orientations of the gaseous filaments. Woltjer (1958) suggested the most successful interpretation, namely that the magnetic lines tend to circle around the filaments. This view is supported to some extent by recent radio polarization maps (Wilson 1972a,c; cf. Conway 1971). It is not obvious, however, that the great radial "fans" of polarization vectors at the margins of the nebula can be explained in this way. To my eye, the fan at the eastern edge seems to be centered on the dark bay in the continuum rather than on the nearby filament. Certain features like this relate more closely to the continuum than to the filamentary network.

ANGULAR SIZES AND TIME SCALES

Recent Cherenkov-light observations of gamma rays (Fazio et al. 1972) have pinned down the mean magnetic field in the bright parts of the Crab Nebula fairly well as $B \approx 1 \times 10^{-3}$ gauss. In such a field, an electron whose critical synchrotron frequency is optical ($\nu_c \approx 10^{15}$ Hz) will lose a large fraction of its energy in a characteristic time $t_s \approx 40$ years. Thus, the optical electrons we see in the Crab Nebula cannot be much older than this; a more elaborate treatment (Oort and Walraven 1956) does not change the order of magnitude of the result. It follows that the optical electrons cannot be remnants of the supernova explosion 919 years ago. They must be generated continuously or at least recently, probably at or near the pulsar. At the distance of the Crab Nebula (1.5 to 2 kpc) 1 arcmin is about 2 light years; then if the optical electrons are to fill the 4×2 arcmins continuum region in 40 years, they must stream outward fairly efficiently along field lines. There is not much time for slow diffusive motion in a chaotic field.

This argument becomes progressively more restrictive as higher frequencies are considered, since $t_s \propto \nu_c^{-1/2}$. Indeed, there is evidence that the size of the continuum region *is* smaller in the blue band than in the red (Scargle 1969b).

In view of this it is surprising that the Crab Nebula is also a synchrotron *X-ray* source of *large size*.[5] It is at least 1.1 arcmin in diameter (Kellogg 1971). The measured polarization of 15.4 ± 5.2% at position angle 156° ± 10° (Novick et al. 1972) is consistent with that of the optical continuum radiation from a central circular region of the same size. This is a strong argument for the synchrotron character of the radiation. While granting this, we should note that the location and spatial extent of the X radiation are not well known either along or across the line of sight. A line source along the major axis of the nebula would be compatible with the data. I believe the data also still allow the hypothesis that the spatial distribution of X-ray emission is the *same* as that of the optical synchrotron radiation.

The X-ray source is large despite the fact that t_s is only about 1 year at $\nu_c = 10^{18}$ Hz. There is always the possibility that particle acceleration is widespread in the nebula. It is difficult, however, to construct a theory of particle circulation that will produce a size dependence on color in the optical and yet permit a large size in the X-ray band. In a recent attempt, Wilson (1972b) is very disturbed by the X-ray size, and his summary of the X-ray results appears to squeeze the data uncomfortably into his mold. Improved knowledge of the X-ray structure, especially as a function of frequency, is essential. The "knee" near the infrared band in the power-law spectrum of the Crab Nebula (Baldwin 1971) is obviously related to these lifetime considerations, but a discussion of this would carry us beyond the scope of this paper. The recent reddening observations by Miller (1973) pose difficulties with joining the optical and X-ray spectra smoothly and suggest that there may be additional structure in the spectrum that will force us to construct more complicated models.

PULSAR AND NEBULA

Most people believe that the cosmic rays are produced at or near the pulsar; nearly everyone believes that the present *field* in the nebula was generated somehow by the pulsar or by its supernova predecessor. The details of this are a mystery. Wilson (1972a) asserts that the optical polarization pattern over the face of the nebula is symmetric about

---

[5] See p. 262.

the pulsar (see Fig. 4), but I do not find this symmetry very striking. There is little in the gross features of the nebula to single out this star as its prime mover. The star is reasonably bright ($V \approx 16.5$; Kristian et al. 1970) and does lie near the center of gravity of the continuum emission and, perhaps more important, near the center of expansion of the filaments (Trimble 1968). The kinematical and spectroscopic evidence was, however, inconclusive (Minkowski 1968; Trimble 1968) until after the discovery of the optical pulses. Piddington (1957), who was quite prescient, nevertheless did not identify this star (nor any other) as the supernova remnant. Even the "wisps" very near the pulsar, which show changes correlated with pulsar activity (Scargle and Harlan 1970), are not strictly symmetric with respect to the pulsar position. The usual theories of relativistic-particle production by the pulsar are in conflict with the observed accelerations of the filaments, as shown by Trimble and Rees (1970). Altogether it is safe to say that if theorists had had to *predict* the optical appearance of the nebula, given only what is known about the pulsar and supernova, they would have been quite inaccurate.

It has long been realized that if the usual synchrotron-radiation theory of the nebular continuum is correct, the magnetic energy of the nebula is so large at present that the nebula cannot have expanded isomorphically, with fields frozen in, from an initial star or other state stabilized by gravitation. The ratio of magnetic to gravitational energy is preserved in such an expansion, and since it is presently $\sim 10^6$, the magnetic flux could not have been contained in a stable object (Woltjer 1971). This difficulty created a favorable climate for an alternative suggestion (Gunn and Ostriker 1971) that the "classical" theory of the radiation, imputing it to fast electrons in a static magnetic field, was completely erroneous, and that the nebular continuum was, instead, "synchro-Compton" radiation from fast electrons in a huge cavity filled with 30-Hz electromagnetic waves generated by the rotating pulsar. Indeed a favorable climate was essential for this idea, because it receives almost no support from Figs. 1–4. But since no one had understood the form of the nebula in any detail anyway, the theoreticians were not deterred. Rees (1971*a,b*) carried the suggestion further and, laudably, offered a prediction: that the "synchro-Compton" optical continuum from the nebula should show a few percent *circular* polarization at source points near the rotation axis of the pulsar — which, in this theory, lies (at least in projection) parallel to the general linear polarization in the central region of the nebula, i.e., roughly along the NW-SE (major) axis. The theory of the radiation has been developed quantitatively by Arons (1972) and Blandford (1972), who confirm Rees's approximate results.

This prediction led to the most interesting polarimetry that has recently been done on the Crab Nebula. Landstreet and Angel (1971) looked for optical circular polarization and found none at the level predicted. Martin, Illing, and Angel (1972), in a definitive study, have detected circular polarization q of order 0.2% — too small to fit the "synchro-Compton" theory.[6] All are now agreed that this theory of the nebula must be abandoned. The variation of q with wavelength and with position in the nebula indicates that it is due to birefringence of the interstellar grains, investigated theoretically by Martin (1972).[7] Thus circular polarization becomes a powerful new tool for study of the interstellar grains. Rees (1971b) remarked that "If it were found that no parts of the Crab Nebula displayed even $\sim 1\%$ circular polarization ... it would ... suggest that the popular 'oblique rotator' magnetic dipole model for NP 0532 would require some reappraisal." This reappraisal by Gunn and Rees apparently is now in progress.

It is possible that there is a *small* cavity around the pulsar that does fit the wave-field description. Scargle (1971) observed that there is an elliptical region of reduced surface brightness near the center of the nebula which might be identified with such a cavity. Perhaps circular polarization of about 1% could indeed be manifested here. The measurements by Martin, Illing, and Angel (1972) do not encourage this belief, but perhaps they do not yet eliminate the possibility.

To explain the gross features of the nebula, we are forced back onto the conventional synchrotron theory. It is worth recalling that synchrotron radiation too produces circular polarization (Sciama and Rees 1967; Legg and Westfold 1968; Pacholczyk and Swihart 1971). The expected q is proportional to $\nu^{-1/2}$, so that it is very small in the optical band, but at radio frequencies, where the Crab Nebula has $\gamma \approx 1.5$ (Baldwin 1971), it would be about

$$q \sim 2 \left(\frac{B}{\nu}\right)^{1/2} \qquad (2)$$

if the field were uniform. Here B is in gauss and ν in MHz. With B as large as 10^{-3} gauss (Fazio et al. 1972), we would have $q \sim 1\%$ at 40 MHz and 0.25% at 610 MHz. This should be observable (cf. Andrew, Purton, and Terzian 1967; Gilbert and Conway 1970; Conway et al. 1971; Berge and Seielstad 1972).

Depolarization due to field nonuniformity could be large; recall that

---
[6] See also p. 54.   [7] See p. 926.

the net *linear* polarization of the whole Crab Nebula, even in the optical band where Faraday depolarization can be neglected, is only $\sim 1/8$ of the theoretical value for a uniform field. But the circular depolarization can be smaller, and in any case the remaining polarization might be detectable. Parts of the nebula with high linear polarization should particularly be examined. Observation of the pattern of q over the face of the nebula would give valuable new information on the field structure, since its sense and degree depend on the magnitude and orientation of B (Legg and Westfold 1968).

In a remarkable paper, Piddington (1957), aware of the difficulty in accounting for the large magnetic flux of the nebula, suggested that what was needed, in addition to general expansion of the nebula, was amplification of the field by stretching of the lines due to winding around a rotating supernova remnant. He estimated the rotation period of the star to be as short as 5 mins — in 1957! Other authors have advanced the same idea, especially since the discovery of the 30-Hz pulsar (Kardashev 1965, 1970; Sturrock 1970). It may well be that theorists will soon settle upon a model of the nebula not very different from Piddington's. If the lines of force, continually being stretched by wrapping, are also carried outward by a kind of stellar wind, the wind will wrap them around the windward sides of the denser and more resistive filaments — in general agreement with Woltjer's (1958) observation (M. J. Rees[8]; cf. Wilson 1972c). We may hope to learn more about this by detailed polarimetry.

Can we tell the orientation of the pulsar rotation axis by looking at pictures of the nebula? After looking at Figs. 2 and 3, I feel that the lines in the nebula tend to wind more around the NW-SE (major) axis, and it seems that the rotation axis should then lie in that general direction. Kardashev (1970), in an argument that I do not understand, concludes that the lines should and do wind around the NW-SE axis, even though he apparently believes the pulsar rotation axis to be NE-SW. This paper possibly involves a misinterpretation of the conclusions of Wampler, Scargle, and Miller (1969).

There have been attempts to deduce the orientation of the rotation axis from "rotating-vector" models of the pulsar polarization; they are reviewed by Cocke.[9] The results are model-dependent. In particular there is a model-determined ambiguity of 90°, plus additional ambiguity in fitting the data. In general the analyses seem to indicate that the pulsar rotation axis is nearly in the plane of the sky and roughly parallel

---

[8] Personal communication.
[9] See p. 997.

or perpendicular to the major axis of the nebula. But there is an additional disagreement of 40° among three groups, so this does not mean much, and we must conclude for now that the orientation of the rotation axis is unknown. There is a surprising disagreement between Cocke et al. and Kristian et al.[10] about a simpler matter, namely a disagreement of 20° in the position angle of the nebular polarization in the central region, in the sense that Cocke et al. agree with Woltjer's original result while Kristian et al. disagree. More data are obviously needed.

In closing this review I should mention two recent high-resolution polarimetric studies on the Crab Nebula. Forman and Visvanathan (1971) made observations in the central region and found an "island" in the linear polarization pattern near "Wisp 2," which they associate with a hydromagnetic wave in the wisp. Scargle (1971) emphasized that such work is difficult because suitable corrections must be made for projection effects. Attempting such corrections himself, he measured linear polarizations in the wisps and concluded that they are high enough to be consistent with a uniform field. The nature of these wisps, and their relation to the pulsar and to the nebula at large, is not understood. We hope that high-resolution polarimetry, applied more extensively, will resolve some of these questions.

Acknowledgments. This research received essential support from the State of Arizona. I thank W. J. Cocke, P. Goldreich, M. J. Rees, J. D. Scargle, G. A. Seielstad, T. L. Swihart, V. Trimble, and R. V. Wagoner for advice and for confirming some of my prejudices about the Crab Nebula.

REFERENCES

Andrew, B. H.; Purton, C. R.; and Terzian, Y. 1967. An upper limit to circularly polarized radiation from the Crab Nebula. *Nature* 215:493–494.
Arons, J. 1972. Nonlinear inverse Compton radiation and the circular polarization of diffuse radiation from the Crab Nebula. *Astrophys. J.* 177:395–410.
Baade, W. 1956. The polarization of the Crab Nebula on plates taken with the 200-inch telescope. *Bull. Astron. Insts. Neth.* 12:312.
Baldwin, J. E. 1971. The electromagnetic spectrum of the Crab Nebula. *The Crab Nebula* (IAU Symposium No. 46) (R. D. Davies and F. G. Smith, eds.) Dordrecht, Holland: D. Reidel Publ. Co., pp. 22–31.
Berge, G. L., and Seielstad, G. A. 1972. Measurements of the integrated Stokes parameters of compact radio sources. *Astron. J.* 77:810–818.

---

[10] References are given by Cocke, see p. 1010.

Blandford, R. D. 1972. The polarization of synchro-Compton radiation. *Astron. Astrophys.* 20:135–144.

Burn, B. J. 1966. On the depolarization of discrete radio sources by Faraday dispersion. *Mon. Not. R. Astr. Soc.* 133:67–83.

Cocke, W. J.; Disney, M. J.; Muncaster, G. W.; and Gehrels, T. 1970. Optical polarization of the Crab Nebula pulsar. *Nature* 227:1327–1329.

Conway, R. G. 1971. The radio polarization of the Crab Nebula. *The Crab Nebula* (IAU Symposium No. 46) (R. D. Davies and F. G. Smith, eds.) Dordrecht, Holland: D. Reidel Publ. Co., pp. 292–295.

Conway, R. G.; Gilbert, J. A.; Raimond, E.; and Weiler, K. W. 1971. Circular polarization of quasars at λ21 cm. *Mon. Not. R. Astr. Soc.* 152:1P–4P.

Fazio, G. G.; Helmken, H. F.; O'Mongain, E.; and Weekes, T. C. 1972. Detection of high-energy gamma rays from the Crab Nebula. *Astrophys. J.* 175:L117–L122.

Forman, W., and Visvanathan, N. 1971. Magnetic field structure around the Crab pulsar. *Nature* 229:39–40.

Gilbert, J. A., and Conway, R. G. 1970. Circular polarization of quasars at λ49 cm. *Nature* 227:585–586.

Ginzburg, V. L., and Syrovatski, S. I. 1965. Cosmic magnetobremsstrahlung (synchrotron radiation). *Ann. Rev. Astron. Astrophys.* 3:297–350.

Gunn, J. E., and Ostriker, J. P. 1971. On the motion and radiation of charged particles in strong electromagnetic waves. I. Motion in plane and spherical waves. *Astrophys. J.* 165:523–541.

Hiltner, W. A. 1957. Polarization of the Crab Nebula. *Astrophys. J.* 125:300–305.

Kardashev, N. S. 1965. Magnetic collapse and the nature of intense sources of cosmic radio-frequency emission. *Sov. Astr. A. J.* 8:643–648 (*Astron. Zhur.* 41:807–813, 1964).

———. 1970. Pulsars and nonthermal radio sources. *Sov. Astr. A. J.* 14:375–384 (*Astron. Zhur.* 47:465–478).

Kellogg, E. M. 1971. X-ray observations of the Crab Nebula. *The Crab Nebula* (IAU Symposium No. 46) (R. D. Davies and F. G. Smith, eds.) Dordrecht, Holland: D. Reidel Publ. Co., pp. 42–53.

Kristian, J.; Visvanathan, N.; Westphal, J. A.; and Snellen, G. H. 1970. Optical polarization and intensity of the pulsar in the Crab Nebula. *Astrophys. J.* 162:475–483.

Landstreet, J. D., and Angel, J. R. P. 1971. Search for optical circular polarization in the Crab Nebula. *Nature* 230:103.

Legg, M P. C., and Westfold, K. C. 1968. Elliptic polarization of synchrotron radiation. *Astrophys. J.* 154:499–514.

Martin, P. G. 1972. Interstellar circular polarization. *Mon. Not. R. Astr. Soc.* 159:179–190.

Martin, P. G.; Illing, R.; and Angel, J. R. P. 1972. Discovery of interstellar circular polarization in the direction of the Crab Nebula. *Mon. Not. R. Astr. Soc.* 159:191–201.

Miller, J. S. 1973. Reddening of the Crab Nebula from observations of [SII] lines. *Astrophys. J.* 180:L83–L87.

Minkowski, R. 1968. Nonthermal galactic radio sources. *Nebulae and interstellar matter*. (B. M. Middlehurst and L. H. Aller, eds.) Chicago: Univ. of Chicago Press, pp. 623–666.

Novick, R.; Weisskopf, M. C.; Berthelsdorf, R.; Linke, R.; and Wolff, R. S. 1972. Detection of X-ray polarization of the Crab Nebula. *Astrophys. J.* 174:L1–L8.

Oort, J. H., and Walraven, T. 1956. Polarization and composition of the Crab Nebula. *Bull. Astron. Insts. Neth.* 12:285–308.

Pacholczyk, A. G., and Swihart, T. L. 1971. Polarization of radio sources. III. Absorption effects on circular polarization in a synchrotron source. *Astrophys. J.* 170:405–408.

Piddington, J. H. 1957. The Crab Nebula and the origin of interstellar magnetic fields. *Austral. J. Phys.* 10:530–546.

Rees, M. J. 1971a. The non-thermal continuum from the Crab Nebula: The "synchro-Compton" interpretation. *The Crab Nebula* (IAU Symposium No. 46) (R. D. Davies and F. G. Smith, eds.) Dordrecht, Holland: D. Reidel Publ. Co., pp. 407–413.

———. 1971b. Implications of the "wave field" theory of the continuum from the Crab Nebula. *Nature Phys. Sci.* 230:55–57.

Scargle, J. D. 1969a. Evidence for continued activity in the Crab Nebula. *Supernovae and their remnants.* (P. J. Brancazio and A. G. W. Cameron, eds.) New York: Gordon and Breach, pp. 197–225.

———. 1969b. Activity in the Crab Nebula. *Astrophys. J.* 156:401–426.

———. 1971. Magnetic field structure around the Crab Nebula pulsar. *Nature Phys. Sci.* 230:37–38.

Scargle, J. D., and Harlan, E. A. 1970. Activity in the Crab Nebula following the pulsar spin-up of 1969 September. *Astrophys. J.* 159:L143–L146.

Sciama, D. W., and Rees, M. J. 1967. Possible circular polarization of compact quasars. *Nature* 216:147.

Shklovskij, I. S. 1960. *Cosmic radio waves.* (R. B. Rodman and C. M. Varsavsky, trans.) Cambridge, Mass.: Harvard Univ. Press, pp. 292–316 et passim.

Sturrock, P. A. 1970. Discussion remark, Crab Nebula Symposium, June 1969, Flagstaff. *Pub. Astron. Soc. Pac.* 82:557.

Trimble, V. 1968. Motions and structure of the filamentary envelope of the Crab Nebula. *Astron. J.* 73:535–547.

———. 1971. Optical observations of the Crab Nebula. *The Crab Nebula* (IAU Symposium No. 46) (R. D. Davies and F. G. Smith, eds.) Dordrecht, Holland: D. Reidel Publ. Co., pp. 3–11.

Trimble, V., and Rees, M. 1970. The expansion energy of the Crab Nebula. *Astrophys. Lett.* 5:93–97.

Wampler, E. J.; Scargle, J. D.; and Miller, J. S. 1969. Optical observations of the Crab Nebula pulsar. *Astrophys. J.* 157:L1–L10.

Wilson, A. S. 1972a. The structure of the Crab Nebula at 2.7 and 5 GHz - I. The observations. *Mon. Not. R. Astr. Soc.* 157:229–253.

———. 1972b. The structure of the Crab Nebula - II. The spatial distribution of the relativistic electrons. *Mon. Not. R. Astr. Soc.* 160:355–371.

———. 1972c. The structure of the Crab Nebula - III. The radio filamentary radiation. *Mon. Not. R. Astr. Soc.* 160:373–379.

Woltjer, L. 1957. The polarization and intensity distribution in the Crab Nebula derived from plates taken with the 200-inch telescope by Dr. W. Baade. *Bull. Astron. Insts. Neth.* 13:301–311.

———. 1958. The Crab Nebula. *Bull. Astron. Insts. Neth.* 14:39–80.

———. 1971. Relationship between pulsar and nebula. *The Crab Nebula* (IAU Symposium No. 46) (R. D. Davies and F. G. Smith, eds.) Dordrecht, Holland: D. Reidel Publ. Co., pp. 389–391.

DISCUSSION

GEHRELS: I have an isolated observation that has never as yet been published but that may be of interest here. The observation was made on 8 December 1969, on a region of 30 arcsecs diameter, with the diaphragm centered 20 arcsecs in the direction of position angle 315° from the pulsar. The region was chosen as it appeared to be free of emission effects so that the results should pertain to the continuum. The observations have not been corrected for interstellar polarization. Probable errors are listed.

Wavelength (μm)	Polarization (%)	Position Angle (degrees)
0.36	21.5 ± 0.3	161.8 ± 0.4
0.52	21.9 ± 0.3	161.8 ± 0.3
0.83	22.1 ± 0.5	160.0 ± 0.7

FELTEN: Your results are consistent with $p(\lambda) = $ constant. Few or no other observations of $p(\lambda)$ in the optical band have been made. One expects the linear polarization of synchrotron radiation emitted in a uniform field to be independent of wavelength if Faraday depolarization can be neglected (which is certainly true of the Crab Nebula in the optical band) and if the particle spectral index γ is independent of particle energy (which is less well established but may hold for the optical electrons; cf. Baldwin 1971, Miller 1973). The Crab field, however, is nonuniform, and the state of polarization is determined by superposition of radiation from regions of variously oriented fields along a line of sight. This could produce a dependence of p on λ if the electrons radiating at short λ occupy a smaller volume than those radiating at longer λ, as suggested by Scargle's (1969b) observations. The effect would probably be rather small.

RADIO POLARIMETRIC OBSERVATIONS OF SUPERNOVA REMNANTS

D. K. MILNE
Division of Radiophysics, CSIRO
and
JOHN R. DICKEL
University of Illinois Observatory

Polarimetric maps of thirty supernova remnants (SNR) have been made using the 64-m radio telescope at Parkes at a wavelength of 6 cm; over half of these sources have data at other wavelengths obtained by us and/or others. Combinations of the data at two or more frequencies allow a determination of the Faraday rotation and thus the intrinsic position angles of the magnetic field distribution across the SNR. In general the fields show a very irregular or tangled structure but with small net alignment. The most common structure shows a more or less radial orientation of the field lines when projected onto the plane of the sky. Details of this paper have been submitted to the Australian Journal of Physics.

POLARIZATION OF EXTRAGALACTIC RADIO SOURCES

A. G. PACHOLCZYK
University of Arizona

This paper reviews the theory of polarization of radiation of synchrotron sources, problems of transfer of polarized radiation in a magnetoplasma, and radio observations and interpretation of the linear and circular polarization of extragalactic sources.

Results of investigation of linear and circular polarization of extragalactic radio sources are reviewed together with their interpretation in terms of synchrotron radiation theory and theory of its propagation in a magnetoplasma. Early data and a bibliography on radio polarization are presented in a review by Gardner and Whiteoak (1966); the bibliography following the present paper contains mainly items published since 1966. In the first part of this paper, the propagation of polarized radiation in a magnetoactive plasma is discussed. The second part is dedicated to a discussion of the linear polarization of extragalactic sources, while at the end of the paper the few cases of circular polarization of radio sources that are presently known are mentioned.[1]

PROPAGATION OF POLARIZED RADIATION IN A MAGNETOACTIVE PLASMA

The Transfer Problem

We will consider the propagation of electromagnetic radiation in a plasma medium composed of low energy "thermal" electrons with number density N_c (cold plasma), relativistic electrons with number density N_r (relativistic plasma), a magnetic field **H**, and an appropriate number of heavy positive charges neutralizing the total charge. When referring to quantities characterizing (or due to) cold and relativistic plasma com-

---

[1] *Editorial note:* We have tried to unify symbols throughout this book but in the case of this paper we let the symbols stand because they are consistent with the usage in the field of radio astronomy.—T. Gehrels.

ponents, we will use respectively an index "c" and "r"; we will omit indices when referring to quantities related to the plasma as a whole. The passage of radiation through such a plasma medium can be described by the transfer equation (Sazonov and Tsitovich 1968)

$$I'_{\alpha\beta} = e_{\alpha\beta} + i(\tau_{\alpha\sigma}\delta_{\beta\tau} - \delta_{\alpha\sigma}\tau^*_{\beta\tau})I_{\sigma\tau} \tag{1}$$

written in a right-handed Cartesian system of coordinates with the direction of wave vector **k** along the 3 axis and with the projection of the magnetic field **H** onto the plane (1,2) parallel to the 2 axis (the Greek indices denote[2] the axes 1 and 2). In the equation of transfer, prime denotes the derivative $\frac{1}{c}\frac{\partial}{\partial t} + \frac{\mathbf{k}}{k}\frac{\partial}{\partial \mathbf{r}}$ and δ_{kl} is the Kronecker symbol. $I_{\alpha\beta}$ stands for the polarization tensor of the wave,

$$\begin{aligned} I_{\alpha\beta} &= a_{\alpha i}a_{\beta j}I_{ij}(k,\omega), \\ I_{ij} &\propto \langle E_i(k,\omega)E_j^*(k,\omega) \rangle, \end{aligned} \tag{2}$$

the brackets denote averaging over the statistical ensemble (wave phases), and an asterisk denotes a complex conjugate; $a_{\alpha i}$ and $a_{\beta j}(\alpha,\beta = 1,2,3)$ are the transformation coefficients from the 123 system to 1'2'3' system in which the magnetic field is directed along the 3' axis and the wave vector **k** lies in the (1', 3') plane. The latin indices range through numerals 1', 2', 3'. $E_s(k,\omega)$ is a Fourier transform of the electric field $E_s(\mathbf{r},t)$, which fulfills the Maxwell equations

$$\varepsilon_{pqi}\varepsilon_{ijs}E_{s,jq} + (\epsilon_{ps}E_s)^{\cdot\cdot} = -4\pi\dot{j}_p, \tag{3}$$

in which ϵ_{ps} is the dielectric tensor, j_p is the external current density, ε_{pqi} is the antisymmetric permutation tensor; comma and a dot represent respectively a time derivative and a derivative with respect to the coordinate following the comma. The dielectric tensor ϵ_{ps} is

$$\epsilon_{ps} = \delta_{ps} + 4\pi\kappa_{ps}, \tag{4}$$

where κ_{ps} is the dielectric susceptibility tensor; we will consider here the propagation of high frequency waves and assume that the wave frequency $\omega = 2\pi\nu$ is much larger than the plasma frequency ω_0 and the electron gyrofrequency ω_G, so that

$$4\pi|\kappa_{ps}| \ll 1 \tag{5}$$

and the wave can be considered as transverse. The quantity $\tau_{\alpha\beta}$ is the transfer tensor, related to the dielectric susceptibility tensor κ_{ij}, trans-

---

[2] Since $I_{\alpha 3} = I_{3\beta} = I_{33} = e_{\alpha 3} = e_{3\beta} = e_{33} = \kappa_{\alpha 3} = \kappa_{3\beta} = \kappa_{33} = 0$ in this system of coordinates.

formed to the system 1, 2, 3, in which it is two dimensional; the relation is

$$\tau_{\alpha\beta} = \frac{4\pi\omega}{c}\kappa_{\alpha\beta}. \tag{6}$$

The transfer tensor can be written in the form

$$\tau_{\alpha\beta} = h_{\alpha\beta} + a_{\alpha\beta} = \begin{Vmatrix} \Delta k & -2i\beta \\ 2i\beta & -\Delta k \end{Vmatrix} + i\begin{Vmatrix} \kappa + q & iv \\ -iv & \kappa - q \end{Vmatrix}, \tag{7}$$

in which $h_{\alpha\beta}$ is Hermitian and $a_{\alpha\beta}$ is anti-Hermitian. For a cold plasma (Sazonov 1969a),

$$\Delta k_c = -\frac{1}{4\pi}\frac{\omega_0^2}{\omega}\left(\frac{\omega_G \sin\vartheta}{\omega}\right)^2 = -6 \times 10^{10}\frac{N_c H_\perp^2}{v^3}, \tag{8}$$

$$\beta_c = -\frac{1}{4\pi}\frac{\omega_0^2}{\omega}\frac{\omega_G \cos\vartheta}{\omega} = -2.5 \times 10^4\frac{N_c H_\parallel}{v^2}, \tag{9}$$

$$\kappa_c = \frac{1}{4\pi^2}\frac{\omega_0^2}{\omega}\frac{\omega_{\text{coll}}}{\omega} \cong 2 \times 10^{-1}\frac{N_c^2}{v^2 T^{3/2}}, \tag{10}$$

$$q_c = \frac{3}{8\pi^2}\frac{\omega_0^2}{\omega}\frac{\omega_{\text{coll}}}{\omega}\left(\frac{\omega_G \sin\vartheta}{\omega}\right)^2 \cong 2 \times 10^{12}\frac{N_c^2 H_\perp^2}{v^4 T^{3/2}}, \tag{11}$$

$$v_c = -\frac{1}{2\pi^2}\frac{\omega_0^2}{\omega}\frac{\omega_{\text{coll}}}{\omega}\frac{\omega_G \cos\vartheta}{\omega} \cong -9 \times 10^5\frac{N_c^2 H_\parallel}{v^3 T^{3/2}}, \tag{12}$$

where

$$\omega_0 = \sqrt{\frac{4\pi e^2 N_c}{mc}},$$

$$\omega_G = \frac{eH}{mc}, \tag{13}$$

$$\omega_{\text{coll}} \cong 2\pi \cdot 3.6\frac{N_c}{T^{3/2}}\left[18 + \ln\frac{T^{3/2}}{v}\right],$$

are plasma frequency, electron gyrofrequency, and effective collision frequency, respectively; and T is the effective temperature. For a relativistic plasma

$$\Delta k_r = -8.5 \times 10^{-3}\frac{2}{\gamma-2}(mc^2)^{1-\gamma}\left[\left(\frac{v}{v_L}\right)^{(\gamma-2)/2} - 1\right]$$
$$[4\pi N_0 \phi(\vartheta)](v_G \sin\vartheta)^{(\gamma+2)/2}v^{-(\gamma+4)/2}, \tag{14}$$

$$\beta_r = -8.5 \times 10^{-3}\frac{\ln\mathscr{E}_L}{(\gamma+1)\mathscr{E}_L^{\gamma+1}}(mc^2)^{1-\gamma}$$

$$[4\pi N_0 \phi(\vartheta)] \left(\cot \vartheta + \frac{\phi'}{\phi} \right) v_G \sin \vartheta v^{-2}, \quad (15)$$

$$\kappa_r = c_6 (H \sin \vartheta)^{(\gamma+2)/2} [4\pi N_0 \phi(\vartheta)] \left(\frac{v}{2c_1} \right)^{-(\gamma+4)/2}, \quad (16)$$

$$q_r = \frac{\gamma+2}{\gamma+10/3} \kappa_r, \quad (17)$$

$$v_r = -c_{16}(H \sin \vartheta)^{(\gamma+3)/2} [4\pi N_0 \phi(\vartheta)] \left[(\gamma+2) \cot \vartheta + \frac{\phi'}{\phi} \right] \left(\frac{v}{2c_1} \right)^{-(\gamma+5)/2}$$

$$\phi' = \partial \phi / \partial \vartheta, \; \mathscr{E} = E/mc^2, \quad (18)$$

where

$$c_6 = \frac{3^{1/2}\pi}{72} em^5 c^{10} (\gamma + \tfrac{10}{3}) \Gamma\left(\frac{3\gamma+2}{12}\right) \Gamma\left(\frac{3\gamma+10}{12}\right),$$

and

$$c_{16} = \frac{3^{1/2}\pi}{54} em^6 c^{12} \frac{\gamma+3}{\gamma+1} \Gamma\left(\frac{3\gamma+11}{12}\right) \Gamma\left(\frac{3\gamma+7}{12}\right). \quad (19)$$

The functions are tabulated in Table I. The above coefficients are valid for a power-law distribution of electrons in which the number density of relativistic electrons with energies between E and $E + dE$ and directed within the solid angle Ω is

$$N_r(\theta, E) dE \, d\Omega = \phi(\theta) N_0 E^{-\gamma} dE \, d\Omega. \quad (20)$$

The quantity θ is the angle between the magnetic field and the direction of the electron velocity; for an isotropic distribution of electrons $\phi = 1/4\pi$. ϑ is the angle between the magnetic field and the line of sight; c_1 is a constant ($= 6.27 \times 10^{18}$ cgs).

TABLE I
Special Functions of the Electron Energy Exponent γ

			γ		
Function	1	2	3	4	5
$10^{23} c_5$	4.88	1.37	0.752	0.556	0.498
$10^{40} c_6$	1.18	0.861	0.797	0.855	1.03
$10^{30} c_{15}$	8.21	2.66	1.44	1.02	0.88
$10^{47} c_{16}$	3.05	2.07	1.76	1.76	1.99
c_{17}	0.646	0.328	0.231	0.183	0.153

(From Pacholczyk 1970 and Pacholczyk and Swihart 1971a)

The quantity $e_{\alpha\beta}$ is the emission tensor e_{ij} transformed to the 1, 2, 3 system, and defined as

$$e_{ij} \propto \langle j_i^k j_j^{k*} \rangle, \tag{21}$$

where j_i^k is the Fourier component of the current density. The tensor $e_{\alpha\beta}$ is related to the emission coefficients expressed as Stokes parameters ε_I, ε_Q, ε_U, ε_V by

$$e_{\alpha\beta} = \tfrac{1}{2} \begin{Vmatrix} \varepsilon_I + \varepsilon_Q & \varepsilon_U + i\varepsilon_V \\ \varepsilon_U - i\varepsilon_V & \varepsilon_I - \varepsilon_Q \end{Vmatrix}. \tag{22}$$

For a cold plasma (thermal bremsstrahlung)

$$\varepsilon_I^c = 3.0 \times 10^{-39} \frac{N_c^2}{T^{1/2}} \left(17.7 + \ln \frac{T^{3/2}}{\nu} \right), \tag{23}$$

$$\varepsilon_Q^c = \varepsilon_U^c = \varepsilon_V^c = 0. \tag{24}$$

For a relativistic synchrotron plasma with a power-law distribution of electron energies,

$$\varepsilon_I = c_5 (H \sin \vartheta)^{(\gamma+1)/2} [4\pi N_0 \phi(\vartheta)] \left(\frac{\nu}{2c_1} \right)^{-(\gamma-1)/2}, \tag{25}$$

$$\varepsilon_Q = \frac{\gamma + 1}{\gamma + 7/3} \varepsilon_I, \tag{26}$$

$$\varepsilon_U = 0 \tag{27}$$

$$\varepsilon_V = -c_{15} (H \sin \vartheta)^{(\gamma+2)/2} [4\pi N_0 \phi(\vartheta)] \left[(\gamma + 2) \cot \vartheta + \frac{\phi'}{\phi} \right] \left(\frac{\nu}{2c_1} \right)^{-\gamma/2}, \tag{28}$$

where

$$c_5 = \frac{3^{1/2}}{16\pi} \frac{e^3}{mc^2} \frac{\gamma + 7/3}{\gamma + 1} \Gamma\left(\frac{3\gamma - 1}{12}\right) \Gamma\left(\frac{3\gamma + 7}{12}\right), \tag{29}$$

$$c_{15} = \frac{3^{1/2}}{12\pi} e^3 \frac{1}{\gamma} \Gamma\left(\frac{3\gamma + 8}{12}\right) \Gamma\left(\frac{3\gamma + 4}{12}\right),$$

are functions of the exponent γ, tabulated in Table I. The polarization tensor $I_{\alpha\beta}$ is related to the Stokes parameters I, Q, U, V through the equation

$$I_{\alpha\beta} = \tfrac{1}{2} \begin{Vmatrix} I + Q & U - iV \\ U + iV & I - Q \end{Vmatrix}. \tag{30}$$

The total intensity of radiation is

$$I = \operatorname{Sp} I_{\alpha\beta} = I_{11} + I_{22}. \tag{31}$$

the degree of linear polarization is

$$\Pi_L = \frac{\sqrt{Q^2 + U^2}}{I^2} = \frac{\sqrt{(I_{11} - I_{22})^2 + (I_{12} + I_{21})^2}}{I_{11} + I_{22}}, \qquad (32)$$

and the angle between the electric vector and 3 axis is

$$\chi = \frac{1}{2} \arctan \frac{U}{Q} = \frac{1}{2} \arctan \frac{I_{12} + I_{21}}{I_{11} - I_{22}}, \qquad (33)$$

while the degree of circular polarization is

$$\Pi_V = \frac{V}{I} = -i \frac{I_{12} - I_{21}}{I_{11} + I_{22}}. \qquad (34)$$

If $\Pi_V > 0$, the electric vector rotates clockwise as seen by the observer. The total degree of polarization is

$$\Pi = \frac{\sqrt{Q^2 + U^2 + V^2}}{I} = \sqrt{1 - \frac{\mathrm{Det}\, I_{\alpha\beta}}{\mathrm{Sp}\, I_{\alpha\beta}}}. \qquad (35)$$

In terms of the Stokes parameters, the transfer Equation (1) can be rewritten (Sazonov 1969a) as

$$\begin{aligned}
I' &= \varepsilon_I - \kappa I - qQ \qquad\qquad - vV, \\
Q' &= \varepsilon_Q - qI - \kappa Q + 2\beta U, \\
U' &= \varepsilon_U \qquad\quad - 2\beta Q - \kappa U - \Delta k V, \\
V' &= \varepsilon_V - vI \qquad\qquad + \Delta k U - \kappa V.
\end{aligned} \qquad (36)$$

Polarization of a Synchrotron Source

Let us consider now a homogeneous, stationary synchrotron source with a uniform magnetic field, disregarding all effects due to cold plasma (all quantities superscripted or subscripted with "c" are assumed zero) and propagation effects due to relativistic plasma ($\Delta k_r = \beta_r = 0$). Since $\varepsilon_U = 0$ and the incident $U_0 = 0$, $U' = 0$ and the equations of transfer reduce to

$$\begin{aligned}
\frac{dI}{ds} &= \varepsilon_I - \kappa I - qQ - vV, \\
\frac{dQ}{ds} &= \varepsilon_Q - qI - \kappa Q, \\
\frac{dV}{ds} &= \varepsilon_V - vI - \kappa V.
\end{aligned} \qquad (37)$$

The solutions of these equations (Pacholczyk and Swihart 1971a) are

$$I = I_p + Ge^{-(\kappa - g)s} + He^{-(\kappa + g)s},$$

$$Q = Q_p - \frac{q}{g}Ge^{-(\kappa - g)s} + \frac{q}{g}He^{-(\kappa + g)s} + De^{-\kappa s}, \quad (38)$$

$$V = V_p - \frac{v}{g}Ge^{-(\kappa - g)s} + \frac{v}{g}He^{-(\kappa + g)s} - \frac{q}{v}De^{-\kappa s},$$

where

$$g = (q^2 + v^2)^{1/2},$$

$$G = \frac{1}{2g}[g(I_0 - I_p) - q(Q_0 - Q_p) - v(V_0 - V_p)],$$

$$H = \frac{1}{2g}[g(I_0 - I_p) + q(Q_0 - Q_p) + v(V_0 - V_p)], \quad (39)$$

$$D = \frac{v}{g^2}[v(Q_0 - Q_p) - q(V_0 - V_p)],$$

and

$$I_p = \frac{\kappa \varepsilon_I - q\varepsilon_Q - v\varepsilon_V}{\kappa^2 - q^2 - v^2}, \quad Q_p = \frac{(\kappa^2 - v^2)\varepsilon_Q - \kappa q \varepsilon_I + qv\varepsilon_V}{\kappa(\kappa^2 - q^2 - v^2)}$$

$$V_p = \frac{(\kappa^2 - q^2)\varepsilon_V - \kappa v \varepsilon_I + qv\varepsilon_Q}{\kappa(\kappa^2 - q^2 - v^2)}. \quad (40)$$

Let us discuss the degree of linear polarization. If the source is optically thin, the degree of linear polarization (Westfold 1959) is

TABLE II
Polarization Parameters of Synchrotron Sources for Different Electron Energy Exponents γ

	γ				
Function	1	2	3	4	5
Π_L (thin)	0.69	0.75	0.79	0.82	0.84
Π_L (thick)	0.12	0.10	0.08	0.07	0.06
τ_Q	4.2	6.2	8.2	10.4	12.7
τ_V	2.3	4.1	6.2	7.9	9.5
τ_m	—	0.35	0.65	0.88	1.08
v_Q/v_m	—	0.38	0.49	0.54	0.58
v_V/v_m	—	0.44	0.53	0.58	0.62

(From Pacholczyk 1970 and Pacholczyk and Swihart 1971a)

$$\Pi_L \text{ (thin)} \cong \frac{qI_p + \kappa Q_p}{\kappa I_p + q Q_p} = \frac{\varepsilon_Q}{\varepsilon_I} = \frac{\gamma + 1}{\gamma + 7/3}, \quad (41)$$

(neglecting v and ε_V as compared with κ, q, and ε_I, ε_Q, respectively) and is given in Table II.

If the source is optically thick (Pacholczyk and Swihart 1967),

$$\Pi_L \text{ (thick)} = \frac{|Q_P|}{I_P} = \frac{3}{6\gamma + 13} \quad (42)$$

and is given in Table II. Note that the Stokes parameter Q is negative in the optically thick case; the polarization position angle in the optically thick regime differs by 90° from that for a thin regime; in the former case the electric vector is parallel to the projection of the magnetic field onto the plane perpendicular to the direction of wave propagation.

The degree of linear polarization Π_L decreases to zero at the frequency v_Q in the neighborhood of the spectral turnover (Fig. 1). The optical depth τ_Q at which $\Pi_L = 0$ is, approximately,

$$e^{-\tau_Q} \cong \gamma + \tfrac{8}{3} \quad (43)$$

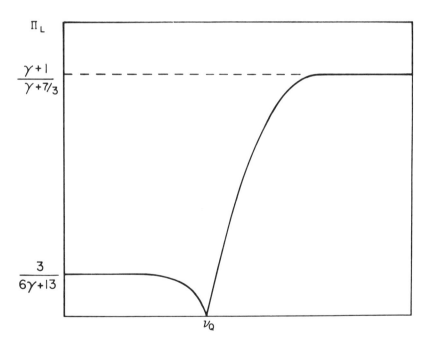

Fig. 1. The degree of linear polarization for a homogeneous synchrotron source without Faraday rotation.

and is given in Table II. The ratio v_Q/v_m, where v_m is the frequency of the maximum in the Stokes parameter I is given by

$$\frac{v_Q}{v_m} = \left(\frac{\tau_m}{\tau_Q}\right)^{2/(\gamma+4)}, \qquad (44)$$

where τ_m is the optical depth at v_m (see e.g., Pacholczyk 1970, p. 98). This ratio is also given in Table II.

The degree of circular polarization has a $v^{-1/2}$ frequency dependence in the optically thin source (Roberts and Komesaroff 1965; Legg and Westfold 1968),

$$\Pi_V(\text{thin}) = \frac{\varepsilon_V}{\varepsilon_I} = -\frac{c_{15}}{c_5}(H\sin\vartheta)^{1/2}\left[(\gamma+2)\cot\vartheta + \frac{\phi'}{\phi}\right]\left(\frac{v}{2c_1}\right)^{-1/2}. \qquad (45)$$

For an optically thick medium (Pacholczyk and Swihart 1971a),

$$\Pi_V(\text{thick}) = \frac{(q^2-\kappa^2)\varepsilon_V + \kappa v\varepsilon_I - qv\varepsilon_Q}{\kappa(v\varepsilon_V - \kappa\varepsilon_I + q\varepsilon_Q)}. \qquad (46)$$

For relativistic electrons, $v \ll \kappa, q$ and $\varepsilon_V \ll \varepsilon_I, \varepsilon_Q$ and

$$\Pi_V(\text{thick}) = -c_{17}(\gamma)\Pi_V(\text{thin}), \qquad (47)$$

where c_{17} is a positive function of γ (in the relevant range of γ) given by

$$c_{17} = \frac{c_{16}c_5}{c_6c_{15}} - \frac{(\gamma+7/3)[(\gamma+2)^2 - (\gamma+10/3)^2]}{(\gamma+10/3)[(\gamma+1)(\gamma+2) - (\gamma+7/3)(\gamma+10/3)]}. \qquad (48)$$

The numerical values of c_{17} are given in Table I. The degree of circular polarization in an optically thick source has the same $v^{-1/2}$ frequency dependence and the same dependence on magnetic field as in an optically thin region. There is, however, a sign reversal of Π_v (i.e., of V) between the transparent and the thick regimes (Pacholczyk and Swihart 1971a; Ramaty and Holt 1970; Melrose 1971); the polarization degree goes through zero at a frequency v_V, corresponding to an intermediate optical thickness (cf. Fig. 2). This optical thickness, τ_V, is a function of the index γ as given in Table II. The ratio v_V/v_m is given by

$$\frac{v_V}{v_m} = \left(\frac{\tau_m}{\tau_V}\right)^{2/(\gamma+4)}, \qquad (49)$$

where τ_m is the optical depth at v_m. This ratio is also represented in Table II.

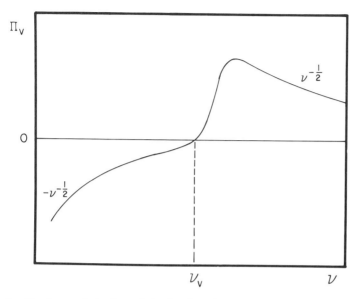

Fig. 2. The degree of circular polarization for a homogeneous synchrotron source.

Faraday Rotation and Depolarization

Consider thermal plasma ($\beta_c \gg \beta_r$, $\Delta k_c \gg \Delta k_r$) in which a high frequency wave ($\omega \gg \omega_G$) is propagating. From Equations (8) and (9) it follows that $\beta^2 \gg \Delta k^2$ gives

$$\frac{\sin^4 \varphi}{4 \cos^2 \varphi} \frac{\omega_G^2}{\omega^2} \ll 1, \tag{50}$$

when $\omega \gg \omega_0$. With $\omega \gg \omega_G$, the inequality holds for a large range of angles φ and determines a condition of quasi-longitudinal propagation in plasma. Under this condition we have (neglecting emission and absorption of elliptical polarization, $\varepsilon_V \sim V \sim 0$) the transfer equation in the form:

$$\begin{aligned} I' &= \varepsilon_I - \kappa I, \\ Q' &= \varepsilon_Q - \kappa Q + 2\beta U, \\ U' &= - \kappa U - 2\beta Q. \end{aligned} \tag{51}$$

If I_0, Q_0 and U_0 are the initial values of Stokes parameters, the solution for an optically thin, uniform, homogeneous and stationary medium is

$$I = \varepsilon_I s + I_0,$$

$$Q = \left(U_0 + \frac{\varepsilon_Q}{2\beta}\right) \sin 2\eta + Q_0 \cos 2\eta, \quad (52)$$

$$U = \left(U_0 + \frac{\varepsilon_Q}{2\beta}\right) \cos 2\eta - Q_0 \sin 2\eta - \frac{\varepsilon_Q}{2\beta},$$

where $\eta = \beta s$.

Assume no emission within the medium ($\varepsilon_I = \varepsilon_Q = 0$). We have

$$\Pi = \frac{\sqrt{Q^2 + U^2}}{I} = \frac{\sqrt{Q_0^2 + U_0^2}}{I_0} = \Pi_0 \quad (53)$$

$$\tan 2\chi = \frac{U}{Q} = \tan(2\eta + \text{const.}) \quad (54)$$

The polarization degree remains unchanged in the process of transfer through the medium, but the polarization position angle rotates by an angle proportional to

$$N_c H_\parallel s v^{-2} \quad (55)$$

This propagation effect is called Faraday rotation (of the polarization position angle).

Faraday rotation can cause change of the polarization degree if radiation emitted at various depths within the source undergoes rotation in different amounts. Indeed, from the solutions of Equation (52) (with no incident radiation, $I_0 = Q_0 = U_0$) we have

$$\Pi = \frac{\varepsilon_Q}{\varepsilon_I} \left| \frac{\sin \eta}{\eta} \right|. \quad (56)$$

The behavior of the degree of polarization as a function of the Faraday rotation depth $\eta = \beta s$ is illustrated in Fig. 3.

Faraday Pulsation

In this section we will describe a propagation effect capable of converting linear polarization of a wave in a thermal plasma into circular polarization. The effect is referred to as Faraday pulsation (Pacholczyk and Swihart 1970), because of its analogy to Faraday rotation discussed in the preceding section and because of the changes in ellipticity of the polarization ellipse as the wave propagates (the ellipse periodically "pulsates"). We assume $\beta_r = \Delta k_r = 0$, and $\varepsilon_I^c = \kappa_c = q_c = v_c = 0$ and omit the subscripts or superscripts in what follows. The effect is characteristic for the case of quasi-transverse propagation, in which $(2\beta)^2 \ll (\Delta k)^2$. Indeed, this inequality is equivalent to [cf. Equations (8) and (9), $\omega_0 \ll \omega$]

$$\frac{\sin^4 \varphi}{4 \cos^2 \varphi} \frac{\omega_G^2}{\omega^2} \gg 1. \tag{57}$$

Neglecting $(2\beta^2)$ as compared with $(\Delta k)^2$ and also ε_v and v as compared with ε_I, ε_Q and κ, q in order to simplify the calculation (contribution to circular polarization due to ε_v and v are negligible as compared with the amount of circular polarization generated through the effect to be described), we see that pairs of equations of transfer are uncoupled:

$$\begin{aligned} I' &= \varepsilon_I - \kappa I - qQ, \\ Q' &= \varepsilon_Q - qI - \kappa Q, \\ U' &= -\kappa U - \Delta k V, \\ V' &= \Delta k U - \kappa V. \end{aligned} \tag{58}$$

The latter pair can be easily integrated to give

$$\begin{aligned} U &= (U_0 \cos \Delta k s - V_0 \sin \Delta k s) e^{-\kappa s}, \\ V &= (U_0 \sin \Delta k s + V_0 \cos \Delta k s) e^{-\kappa s}, \end{aligned} \tag{59}$$

where the subscript zero indicates incident values.

The solutions of Equations (58) (for simplicity we assume that there is no absorption) indicate that the degree of polarization, Π, changes from its initial value, Π_0, for $s = 0$ to the value $\varepsilon_Q/\varepsilon_I$ for $s \to \infty$, and is not

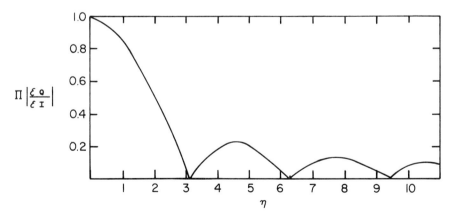

Fig. 3. The degree of polarization Π for an optically thin medium with Faraday rotation.

influenced by the Faraday effect. The position angle χ (counted from the projection of the magnetic field),

$$\tan 2\chi = \frac{U}{Q} = \frac{U_0 \cos \Delta ks - V_0 \sin \Delta ks}{Q_0 + \varepsilon_Q s}, \tag{60}$$

and the ellipticity δ

$$\sin 2\delta = \frac{V}{I} = \frac{U_0 \sin \Delta ks + V_0 \cos \Delta ks}{I_0 + \varepsilon_I s}, \tag{61}$$

are very much influenced by the Faraday effect. The position angle of initially linearly polarized radiation will oscillate periodically between χ_0 and $-\chi_0$ if there is no polarized emission within the plasma itself. If there is polarized emission within the plasma, the amplitude of oscillation will gradually decrease from χ_0 to zero with increasing s. An originally linearly polarized beam will acquire certain ellipticity through the Faraday effect, and the maximum ellipticity δ will be equal to χ_0 if there is no emission within the plasma. If there is emission, the periodically reached maxima of ellipticity will be decreasing with increasing s. It is seen from the above that Faraday pulsation does not occur for plane-polarized radiation if the plane of polarization is either parallel to or normal to the magnetic field.

For initially circularly polarized radiation ($I_0 = V$, $U_0 = 0$) we have

$$\tan 2\chi = \frac{I_0}{\varepsilon_Q} \sin \Delta ks, \tag{62}$$

and

$$\sin 2\delta = \frac{I_0}{I_0 + \varepsilon_Q} \cos \Delta ks. \tag{63}$$

When there is no emission within the plasma, the radiation varies between the two extremes of linear and circular polarization periodically, and the position angle stays at 45° to the field.

The equations of this section with $\varepsilon_I = \varepsilon_Q = 0$ can describe changes in the polarization characteristics of synchrotron radiation emitted in a region of lower magnetic field and passing through a uniform plasma with strong magnetic field under the condition of quasi-transverse propagation. In this case, however, one has to use the expression for Δk

$$\Delta k = \frac{\omega_0^2 \sin^2 \theta}{2\omega c} = \frac{2\pi e^2 N_e}{mc\omega} \sin^2 \theta \tag{64}$$

valid for $\omega \ll \omega_H$, which is different than that given by Equation (8). If the radiation originates from high-energy electrons (and is therefore initially linearly polarized), the Faraday pulsation will introduce circular polarization, which will be large for appropriate path lengths within the plasma, if the angle between the original plane of polarization and the magnetic field is close to $\frac{1}{4}\pi$. Synchrotron radiation emitted by high-energy electrons in a plasma with a uniform magnetic field will not undergo Faraday pulsation from that plasma, since its plane of polarization is normal to the magnetic field and since the radiation is emitted mainly at higher harmonics of electron gyrofrequency.

Rotation and Pulsation in Relativistic Plasma

In the two preceding sections we discussed the principal polarization effects accompanying propagation of waves in a cold (thermal plasma): Faraday rotation and Faraday pulsation. The processes can also be described in a different way, by considering the normal modes of propagation of electromagnetic radiation in plasma. It can be shown that under the conditions of Equation (50) the normal modes of propagation are circularly polarized (right- and left-handed), while under the conditions of Equation (57) they are linearly polarized (polarization angles are mutually perpendicular). The presence of relativistic plasma, in general, changes the character of propagation (Sazonov 1969b). If, however, $|2\beta_r| \gg |\Delta k_r|$, which for a power-law distribution of relativistic electrons implies

$$\left| \frac{\ln \mathscr{E}_L}{\mathscr{E}_L} \cot \varphi + \frac{\phi'}{\phi} \right| \gg \frac{2}{\gamma - 2} \left[\left(\frac{v}{v_L} \right)^{(\gamma - 2)/2} - 1 \right] \left(\frac{v}{v_L} \right)^{-(\gamma/2)}, \quad (65)$$

where $\mathscr{E}_L$ is the lower cutoff in the energy distribution and v_L the corresponding critical frequency, the Hermitian part of the propagation tensor will be antidiagonal and the modes of propagation will be circular, regardless of the concentration of cold plasma. The dominant propagation effect will be that of Faraday rotation. In this case, the principal contribution to the Hermitian propagation tensor is made by the low-energy relativistic electrons, with energies near the lower cutoff of the spectrum $\mathscr{E}_L$; those electrons differ little from cold electrons in the sense that the critical frequency of their synchrotron radiation is much lower than the frequency of the wave the propagation of which is being considered.

If, however, $|2\beta| \ll |\Delta k|$ and $|v| \ll |q|$, where v and q are coefficients that appear in Equation (36), which in a relativistic plasma may be fulfilled at frequencies of propagation close to the critical frequency of electrons with energies close to the lower cutoff in the absence of cold plasma, the modes of propagation will be linearly polarized and Faraday pulsation will take place.

In a more general case, the modes of propagation are elliptical, and with a large contribution of cold plasma to a radio source, the modes are almost circular. This small deviation from circularity might, however, be responsible for contribution to observed elliptical polarization of compact sources through Faraday pulsation.

LINEAR POLARIZATION OF RADIO SOURCES

We discuss first results and interpretations of measurements of integrated polarization of radio sources. Integrated polarization at 49-cm wavelength ranges from 0% to 6% with the median value of 1%. Polarization generally decreases with increasing wavelength; median value at 73 cm is $\sim 0.7\%$ (Conway et al. 1972). In the last section we will discuss measurements of the distribution of linear polarization of extended sources. In well-resolved sources, linear polarization amounts locally to tens of percent and may be as high as 70% (Gardner and Whiteoak 1971).

Distribution of Orientation Differences and Correlation with Brightness

Early polarization studies (Gardner and Whiteoak 1966; Macdonald, Kenderdine, and Neville 1968; Gardner and Whiteoak 1963; Davies 1966; Gardner and Whiteoak 1969; Seielstad and Weiler 1971) indicated a specific distribution of the difference between the polarization position angle and the position angle of the major axis of double radio sources $|\Delta\chi|$, with concentrations around $0°$ and $90°$. Recent data with improved resolution and more accurately determined rotation measures do not confirm the trend (Mitton 1972); there is a gradual increase of number of sources as $|\Delta\chi|$ increases (Fig. 4). In a number of cases $|\Delta\chi|$ does not have a clear physical significance as there are substantial differences in intrinsic polarization angles and rotations between components of double sources. The reported (Gardner and Davies 1966) correlation of the orientation difference with brightness temperature is not confirmed (Mitton 1972; Seielstad and Weiler 1971) (Fig. 5).

Polarization and Magnetic Field

Early polarization data indicated correlation between the polarization degree and magnetic field derived from the assumption of equipartition between magnetic and particle energies within the source, which correlation was thought to be due to Faraday depolarization processes within sources (Seielstad, Morris, and Radhakrishnan 1964). It was shown later (Pacholczyk 1964) that this was a reflection of apparent correlation between polarization and separation of components of double sources to which partially resolved sources contributed substantially (equipartition magnetic field is strongly dependent on apparent size, which is correlated

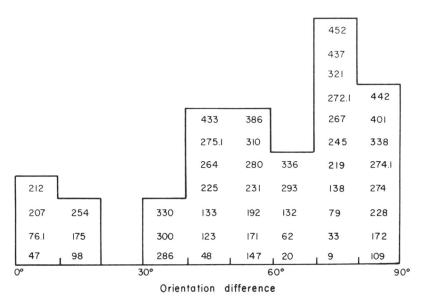

Fig. 4. The histogram of the distribution of the orientation difference of double radio sources (numbers refer to 3C catalogue) (Mitton 1972).

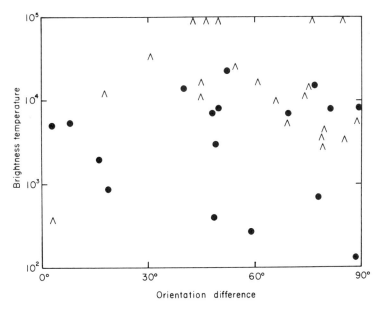

Fig. 5. Brightness temperature (at 1.4 GHz) versus orientation difference (arrow heads indicate a lower limit on brightness temperature) (Mitton 1972).

with apparent separation of double sources); unresolved sources as well as completely resolved components did not show statistically significant correlation between equipartition magnetic field and polarization.

Depolarization in Radio Sources

The average polarization properties of extragalactic sources yield the product $\langle N_e H_\| \rangle$ typically less than 2.5×10^{-9} cm^{-3} G (Mitton 1972), which does not put special constraints on models of extended sources. For very compact sources, however, the situation is different (Pacholczyk and Swihart 1973a). The flux emitted in an optically thin region can be written as [see Eqs. (3.50) and (3.28) in Pacholczyk (1970)]

$$F_\nu \gtrsim (\gamma - 1)c_5 H_\perp N s \Omega \left(\frac{2\nu_m}{\nu}\right)^{(\gamma - 1)/2}, \qquad (66)$$

where $(\gamma - 1)c_5$ is almost a constant equal to 1.4×10^{-23} in the relevant range of γ; s and Ω are the size and angular extent of the emitting region, and N is the number density of relativistic electrons assumed to be not larger than the same order of magnitude as the number of thermal electrons. This is a reasonable assumption if one considers energy losses of relativistic electrons and their thermalization; also some of the difficulties in understanding why these sources are not expanding with the speed of light are alleviated by assuming that a considerable amount of non-relativistic material is associated with the relativistic particles. ν_m is the frequency of maximum in the spectrum; ν_m is not less than the critical frequency corresponding to the lower cutoff energy in the power-law electron energy distribution. If there are any electrons with energies lower than that corresponding to the frequency ν_m, they will contribute to N but not to F_ν. The Faraday rotation β is proportional to $NH_\|$. If there are no special orientations of the magnetic field, $H_\| \approx H_\perp$ and $H_\perp N$ can be eliminated from the inequality of Expression (66), since for optically thin synchrotron emission

$$\beta s = \frac{3\gamma + 3}{3\gamma + 7} \frac{\sin \beta s}{\Pi_L} \gtrsim \frac{0.7}{\Pi_L}. \qquad (67)$$

With these substitutions, the flux inequality becomes

$$F_\nu \gtrsim 4 \times 10^{-28} \frac{\Omega \nu_L^2}{\Pi_L} \left(\frac{2\nu_m}{\nu}\right)^{(\gamma - 1)/2}, \qquad (68)$$

where ν is the frequency at which the flux is F_ν, and ν_L is the frequency for which the linear polarization is Π_L (cgs units are used in the above). Observational data indicate that the inequality in Equation (68) is violated by source PKS 1148-00 by a factor of 2 and by PKS 0237-23 by a factor of about 4. Measurements of the linear polarization of PKS 2134+004 are not available at frequencies higher than the spectral maximum at 6.5 GHz, but the indications are that Equation (68) is not satisfied by this source either. While all of the data that are used here are not of high numerical accuracy, the upper limit for the flux given by Equation (68) is a very generous one, and it is significant that the inequality is not satisfied by any of the three sources. This points toward the conclusion that even apparently simple compact sources have an appreciably nonuniform structure, i.e. that linear polarization must be produced essentially in a different region of a compact source (shell) than most of synchrotron flux (core).

The relationship between the average $N_c H_\parallel$ and size $\langle s \rangle$ of radio sources is presented in Fig. 6 (Strom 1972), on which lines of constant mass of depolarizing ionized gas M times the parallel component of the field $H_\parallel$ are also plotted. The values of $MH_\parallel$ are in the range $10^{2\pm0.5} M_\odot G$. The figure indicates that no systematic change in the mean $MH_\parallel$ occurs over the interval $4 \text{ kpc} \leq \langle s \rangle \leq 40 \text{ kpc}$ and that the quasars do not differ

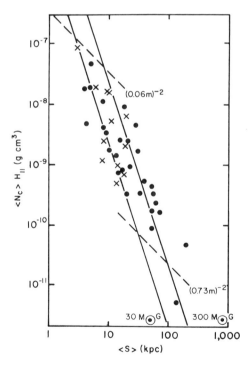

Fig. 6. The average $N_c H_\parallel$ versus the mean size S of the component for radio galaxies (dots) and quasars (crosses) (the solid lines are lines of constant $MH_\parallel$) (Strom 1972).

considerably from radio galaxies in the value and dispersion of $MH_\parallel$ if assumed to be cosmological, thus arguing in favor of such an assumption (Strom 1972). There is, however, evidence described earlier that compact sources are not homogeneous as far as emission of polarized radiation is concerned, and that most of the polarized radiation may be produced in a thin shell surrounding the source. In this case, the discussed values of $MH_\parallel$ would refer to the shell rather than to the entire source. The figure indicates the trend that depolarization decreases with increasing size of source.

Polarization Versus Spectral Index and Depolarization in Flat Spectra Objects

It was pointed out by Gilbert et al. (1969) and by Conway et al. (1972) that high values of polarization at 49 cm are found mainly among flat spectrum quasars, while high values at 73 cm (being less common) are restricted to variable sources, which are themselves restricted to objects with flat or peculiar spectra (Fig. 7). A suggested interpretation required that little Faraday depolarization occurred in flat spectra sources as compared with steep spectra sources. Recent work (Pacholczyk and Gregory 1973), however, seems to indicate that depolarization in individual com-

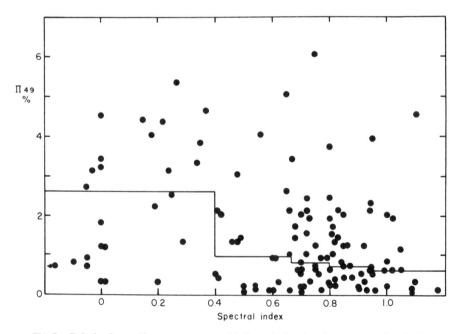

Fig. 7. Polarization at 49 cm versus spectral index calculated at the same wavelength (line indicates the trend in the median value of polarization) (Conway et al. 1972).

ponents of flat spectra objects is similar to that in steep spectra objects, and that the typical trend of polarization of flat spectra objects (gentle peak with drop-offs that approach zero, with possible rises on the extreme sides of the drop-offs) can be interpreted as a superposition of spectral components with a polarization trend typical for steep spectra sources (monotonic decrease of polarization with frequency) (Fig. 8).

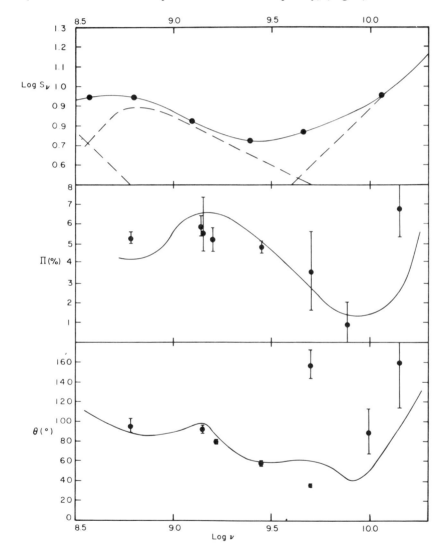

Fig. 8. Polarization spectrum of 3C 345 (solid line) interpreted as a superposition (broken line) of three components with polarization spectra of the 3C 48 type with orientations differing by 30° (Pacholczyk and Gregory 1973).

Correlation with Red Shift

Sources with red shift $z < 1$ have lower frequency $v^R_{1/2}$ (in the rest frame) at which Π_L falls to half of its maximum value, reached at a somewhat higher frequency (the median $v^R_{1/2}$ is 1400 MHz, the median dispersion is 140 rad/m^2) than sources with $z > 1$ (the median frequency 3000 MHz, dispersion 45 rad/m^2), implying that quasars with $z > 1$ have a tendency to depolarize more strongly than sources with lower red shift (Fig. 9) (Kronberg, Conway, and Gilbert 1972). This tendency seems to be intrinsic to the source, since in order to have at least one galaxy in the line of sight to a quasar with $z = 2$ (such a galaxy would contribute about 30 rad/m^2 to the Faraday dispersion), one would need the number density of galaxies to be greater by an order of magnitude than that generally accepted. Therefore, quasars at distant cosmological epoch would contain or be surrounded by a larger mass of plasma than the nearer ones. The correlation becomes less pronounced when one takes into account the superposition of polarizations of spectral components in complex sources (Pacholczyk and Gregory 1973).

Polarization and Luminosity

Although quasars with higher z have also high luminosity L, there is no correlation between $v^R_{1/2}$ and L for sources with low red shift (Kronberg, Conway, and Gilbert 1972).

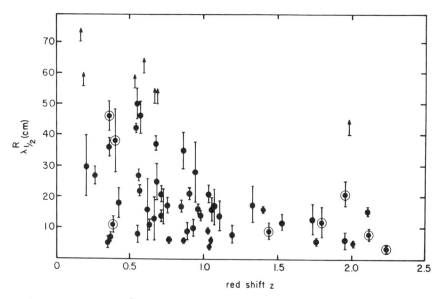

Fig. 9. The values of $\lambda^R_{1/2}$ (in the rest frame) versus red shift for quasars (circles indicate presence of absorption lines) (Kronberg, Conway, and Gilbert 1972).

Polarization and Absorption Lines

Earlier data (Gardner and Whiteoak 1970; Conway and Gilbert 1970) suggested that high red shift sources with absorption lines have lower polarization at all wavelengths than those with no absorption lines, but more recent data (Conway and Gilbert 1970) indicate that this tendency is not statistically significant.

Variability of Polarization

Measurements of polarization of compact sources indicate variability of linear polarization (both degree and position angle) in some sources that also have variable flux (Aller 1970a, b, c; 1971). Interpretations of variability of polarization were made in terms of the Shklovski–Van der Laan–Kellermann model of expanding source (Aller 1970b). It should be pointed out that interpretation of data even for simple compact sources is not entirely straightforward when data concerning all Stokes parameters are taken into account (Pacholczyk and Swihart 1973a), and the situation is far more complex in multicomponent variable sources. It will therefore require considerably more polarization studies with simultaneous observations of all four Stokes parameters before a coherent picture of polarization structure of a variable source emerges.

Faraday Rotation in Extragalactic Sources

Faraday rotation and galactic magnetic fields are reviewed elsewhere in this volume.[3] We will limit ourselves here to a few remarks pertaining mainly to internal Faraday rotation within the sources. Earlier analyses of the rotation data (Kawabata et al. 1969; Reinhardt and Thiel 1970) indicated some correlation between the rotation measure and the red shift of a source, which were discussed in terms of a possible intergalactic magnetic field. Subsequent data (Brecher and Blumenthal 1970; Reinhardt 1972; Mitton and Reinhardt 1972; Mitton 1972) confirmed the dependence of the rotation measures of quasars on their position in the sky, implying a uniform component of the magnetic field directed toward $110°$ longitude and latitude between $-10°$ and $-20°$, close to the direction $120°$ longitude and $-30°$ latitude indicated by analysis of pulsars that are galactic sources. Rotation measures for radio galaxies do not indicate dependence on position, but this may be due to the fact that their intrinsic rotations are larger than those of quasars and, therefore, dominate the contributions from outside the sources (Reinhardt 1972). The smaller rotation measures of quasars can be interpreted through the dependence of the intrinsic rotations on red shift because of the wavelength dependence of the rotation measure. After a correction for this red-shift de-

---
[3]See p. 960.

pendence of the rotation measures of quasars, the data seem to be compatible with the assumption that intrinsic rotations of quasars are similar to those of radio galaxies and not dependent further on red shift. The question of intergalactic magnetic field as derived from rotation data needs further study and much more observational material.

Distribution of Polarization in Extended Sources

Since the last review (Gardner and Whiteoak 1966), many investigations of distribution of polarization over extended sources have been reported (Morris and Whiteoak 1968; Mayer and Hollinger 1968; Conway and Kronberg 1969; Fomalont 1970, 1972; Gardner and Whiteoak 1971; Wardle 1971a, b; Conway and Stannard 1972; Mitton 1971). For a bibliography concerning Cyg A, see Mitton (1971). An interesting feature observed in some sources is high polarization in low brightness areas. In 3C 120 intense components are only slightly polarized, while the central regions have polarization as large as 20% (Fomalont 1970). In Cen A (at 6 cm), there are regions in the northern area where polarization exceeds 50%, reaching a maximum close to 70%, while the northeastern component exhibits polarization of only 14.2% (Gardner and Whiteoak 1971).

Another interesting feature observed in several sources (3C 353, 3C 219, and Fornax A) is the presence of a component with distribution of polarization indicating a circumferential magnetic field (Fomalont 1972) (Fig. 10). The component with such a field appears to be more extended in a

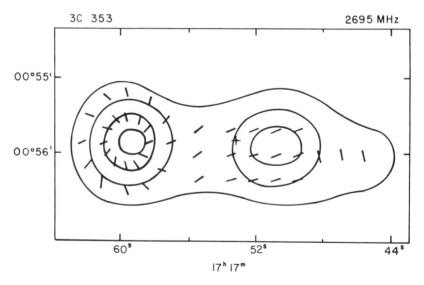

Fig. 10. Distribution of linearly polarized radiation at 2695 MHz across 3C 353 illustrating circumferential magnetic field in the eastern component (Fomalont 1972).

direction perpendicular to the source major axis than the other normal component. Moreover, components with circumferential field do not show any significant fine structure. The integrated polarization of the peculiar component is low since a perfect circumferential geometry yields no net polarization. A possible formation of circumferential field may be due to differential rotation of the component with an initially uniform field winding the field into helices to appear circumferential when viewed along the direction of rotation (Fomalont 1972; Piddington 1966).

A relatively well studied (at many frequencies) source is Cyg A. The distribution of polarization at 6 cm is shown in Fig. 11. The observations show that the magnetic field is coherent and ordered over several kiloparsecs and that there are large differences in the rotation measures of each component, as well as a marked change in the rotation across the axis of each component. A model describing the rotation and depolarization in each component in Cyg A, in terms of two vectors that rotate in opposite senses, is represented in Fig. 12 (Mitton 1971).

The currently available polarization data are rather sketchy and incomplete. It seems to be premature to attempt a discussion of the polarization distribution in terms of theoretical models. At the present moment it would perhaps be enough just to point out that among the ram pressure-confined models of extended sources (De Young and Axford 1967; Mills and Sturrock 1970), the Mills-Sturrock model has characteristic magnetic lines of force extending from the radio component toward the parent galaxy; the Rees model (Rees 1971a, b) involving synchro-Compton radiation implies lack of depolarizing thermal gas within radio components. Detailed comparisons of the ram pressure-confined models and synchro-Compton models of sources with observations require both

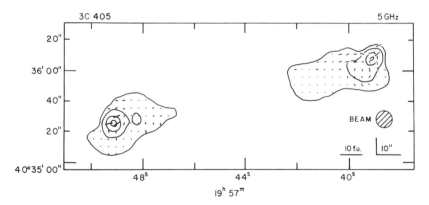

Fig. 11. Distribution of linearly polarized radiation at 5 GHz across Cygnus A (Mitton 1971).

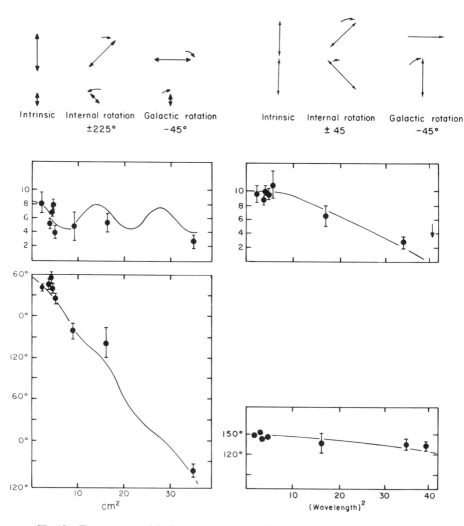

Fig. 12. Two-vector model of each component of Cygnus A (solid line) as compared with observations (Mitton 1971).

better understanding of the polarized brightness distribution predicted by various models and more high resolution observations.

CIRCULAR POLARIZATION OF RADIO SOURCES

Observations for circular polarization of radio sources were made as long as a decade ago, and the results were to set an upper limit of a few percent. It was not until recently that significantly nonzero values for the circular polarization have been measured (Gilbert and Conway 1970; Conway et al. 1971; Roberts et al. 1972; Berge and Seielstad 1973; Seaquist 1969, 1971, 1973; Seaquist et al. 1973). At the present moment about a dozen sources have nonzero circular polarization measured; the highest degree of circular polarization is 0.4%; in the variable source BL Lacertae it reaches at times as much as 2%. The results of these measurements indicate that linear and circular polarizations are not correlated; flat-spectrum (compact) objects exhibit relatively high circular polarization. This last result can be understood in terms of circular synchrotron emission that depends on the magnetic field; compact sources are thought to have higher field than extended objects. Polarization generally increases with decreasing frequency, but errors at the present time are too large to discuss fitting of the data to the $v^{-1/2}$ dependence required by synchrotron theory of an optically thin source. For a thick source, the theory predicts a change in the sense of rotation (see section on Polarization of Synchrotron Source), but at least one source (PKS 2134 + 004) with a simple turned-over intensity spectrum shows polarization rising with decreasing frequency below turnover. PKS 1510-08 has different senses of circularity at different frequencies, but its intensity spectrum is rather complex. Some sources with variable intensity show possible variability of circular polarization (PKS 1127-14, VRO 422201, CTA 102; and possibly 3C 84, 3C 120 and 3C 345).

The observed circular polarization is too small to suggest explanation through Faraday pulsation in cold plasma (where Π_V would be statistically of the same order as Π_L); it also seems to be too small for models involving small pitch angles (Berge and Seielstad 1973) or for those with cyclotron turnover (Pacholczyk and Swihart 1971b). The fields computed from the degree of circular polarization are generally much higher than fields implied by the self-absorption turnover, at least by an order of magnitude (Pacholczyk and Swihart 1973a; Berge and Seielstad 1973). This may indicate considerable nonuniformity in structure of such sources, polarization being produced in different regions than most of the flux (Pacholczyk and Swihart 1973a). High fields required by the interpretation through synchrotron theory together with the lack of change of sense of the rotation may suggest that all circular polarization is not due solely to synchrotron emission and absorption. Part of the observed polarization may be due to the Faraday pulsation effect (Pacholczyk and Swihart 1973a) from relativistic electrons within the source. In a relativistic plasma with no low-energy electrons for frequencies close to that of the spectral turnover, the effect is similar to that in a cold plasma under the quasi-transverse approximation, since in both those cases the normal modes of propagation of radiation are linearly polarized (Sazonov 1969b). It is however not realistic to require the absence of cold plasma in the source;

moreover, in the purely relativistic plasma case, the circular component of polarization due to the effect discussed would be of the same order of magnitude as linear polarization. If thermal electrons and low-energy relativistic electrons are predominant in the source, the modes of propagation are elliptically polarized, but close to circularly polarized. The Faraday pulsation then takes place to the extent permitted by the deviation from circularity of the propagation modes, which is a function of relativistic and cold electron densities. Thus, measurements of circular polarization of radio sources, when interpreted along those lines are a valuable source of information about the relationship between the thermal electron density and that of relativistic electrons (Pacholczyk 1973, Pacholczyk and Swihart 1973b).

Acknowledgment. The author thanks T. L. Swihart for his help and comments.

REFERENCES

Aller, H. D. 1970a. The polarization of variable radio sources at 8 GHz. I. Observations. *Astrophys. J.* 161: 1–19.

———. 1970b. The polarization of variable radio sources at 8 GHz. II. Interpretation. *Astrophys. J.* 161: 19–30.

———. 1970c. Variations of linear polarization in extragalactic radio sources. *Nature* 225: 440.

———. 1971. The evolution of 8 GHz of the linear polarization of 3C 279. *Astron. J.* 671–676.

Berge, C. L., and Seielstad, C. A. 1973. Measurements of the integrated Stokes parameters of compact radio sources. *Astron. J.* 77: 810–818.

Brecher, K., and Blumenthal, G. R. 1970. On the origin of cosmic magnetic fields. *Astrophys. Letters* 6: 169–173.

Conway, R. G., and Gilbert, J. A. 1970. Linear polarization of distant quasars. *Nature* 226: 332–333.

Conway, R. G.; Gilbert, T. A.; Kronberg, P. P.; and Strom, R. G. 1972. Polarization of radio sources at $\lambda 49$ cm and $\lambda 73$ cm. *Mon. Not. R. Astr. Soc.* 157: 443–459.

Conway, R. G.; Gilbert, J. A.; Raimond, E.; and Weiler, K. W. 1971. Circular polarization of quasars at $\lambda 21$ cm. *Mon. Not. R. Astr. Soc.* 152: 1–4.

Conway, R. G., and Kronberg, P. P. 1969. Interferometric measurement of polarization distributions in radio sources. *Mon. Not. R. Astr. Soc.* 142: 11–32.

Conway, R. G., and Stannard, D. 1972. Brightness distribution and polarization of 3C273. *Nature Phys. Sci.* 239: 22–23.

Davies, R. D. 1966. Magnetic fields in galaxies. *Proc. IAU Symp. No. 29*, pp. 178–184.

De Young, D. S., and Axford, W. I. 1967. Inertial confinement of extended radio sources. *Nature* 216: 129–131.

Fomalont, E. B. 1970. Polarization-brightness distribution for 3C20. *Astrophys. J.* 160: L73–L77.

———. 1972. Radio components with a circumferential magnetic field configuration in 3C 219 and 3C 353. *Astrophys. Letters* 12: 187–192.

Gardner, F. F., and Davies, R. D. 1966. The polarization of radio sources. I. Observations of small diameter sources, *Austr. J. Phys.* 19: 441–459.

Gardner, F. F., and Whiteoak, J. B. 1963. Polarization of radio sources and Faraday rotation effects in the galaxy. *Nature* 197: 1162–1164.
———. 1966. The polarization of cosmic radio waves. *Am. Rev. Astr. Aph.* 4: 245–292.
———. 1969. The linear polarization of radio sources between 11 and 20 cm wavelength. II. Polarization and related properties of extragalactic sources. *Astr. S. Phys.* 22: 107–119.
———. 1970. Quasar absorption lines and radio polarization. *Nature* 227: 585.
———. 1971. The polarization of extended radio sources at 6 cm wavelength. I. Extragalactic sources. *Astr. J. Phys.* 24: 899–911.
Gilbert, J. A., and Conway, R. G. 1970. Circular polarization of quasars at $\lambda 49$ cm. *Nature* 227: 585–586.
Kawabata, K.; Fujimoto, M.; Sofue, Y.; and Fukin, M. 1969. A large-scale metagalactic magnetic field and Faraday rotation for extragalactic radio sources, *Publ. Astron. Soc. Japan* 21: 293–305.
Kronberg, P. P.; Conway, R. G.; and Gilbert, J. A. 1972. The polarization of radio sources with appreciable redshift. *Mon. Not. R. Astr. Soc.* 156: 275–282.
Legg, M. P. C., and Westfold, K. C. 1968. Elliptic polarization on synchrotron radiation. *Astrophys. J.* 154: 499–513.
Macdonald, G. H.; Kenderdine, S.; and Neville, A. C. 1968. Observations of the structure of radio sources in the 3C catalogue—I. *Mon. Not. R. Astr. Soc.* 138: 259–311.
Mayer, C. H., and Hollinger, J. P. 1968. Polarization brightness distribution over Cassiopeia A, the Crab Nebula, and Cyg A at 1.55-cm. *Astrophys. J.* 151: 53–65.
Melrose, D. B. 1971. On the degree of circular polarization of synchrotron radiation. *Astrophys. Space Sci.* 12: 172–192.
Mills, D. M., and Sturrock, P. A. 1970. A model of extragalactic radio sources. *Astrophys. Letters* 5: 105–110.
Mitton, S. 1971. Observations of the distribution of polarized emission of Cygnus A at 6-cm wavelength. *Mon. Not. R. Astr. Soc.* 153: 133–143.
———. 1972. The polarization properties of 65 extragalactic sources in the 3C catalogue. *Mon. Not. R. Astr. Soc.* 155: 373–381.
Mitton, S., and Reinhardt, M. 1972. Interpretation of rotation measures of radio sources. II. *Astron. Astrophys.* 20: 337–340.
Morris, D., and Whiteoak, J. B. 1968. The distribution of linear polarization over 13 extended sources at 21.2 cm wavelength. *Austr. J. Phys.* 21: 475–492.
Pacholczyk, A. G. 1964. Polarization and physical properties of radio sources, JILA Report (Boulder, Colo.) No. 23.
———. 1970. *Radio astrophysics: Nonthermal processes in galactic and extragalactic sources.* San Francisco: W. H. Freeman and Co.
———. 1973. Circular repolarization in compact radio sources. *M.N.R.A.S.* In press.
Pacholczyk, A. G., and Gregory, S. A. 1973. Depolarization in flat spectra quasars. *Mon. Not. R. Astr. Soc.* 161: 31p–34p.
Pacholczyk, A. G., and Swihart, T. L. 1967. Polarization of radio sources. I. Homogeneous source of arbitrary optical thickness: *Astrophys. J.* 150: 647–650.
———. 1970. Polarization of radio sources. II. Faraday effect in the case of quasi-transverse propagation. *Astrophys. J.* 161: 415–418.
———. 1971*a*. Polarization of radio sources. III. Absorption effects on circular polarization in a synchrotron source. *Astrophys. J.* 170: 405–408.
———. 1971*b*. Circular polarization of quasars. *Mon. Not. R. Astr. Soc.* 153: 3–5.
———. 1973*a*. Polarization of radio sources. IV. The compact source PKS 2134+004. *Astrophys. J.* 179: 21–28.
———. 1973*b*. Polarization of radio sources. V. Absorption effects on circular repolarization. In draft.

Piddington, J. H. 1966. Galactic explosions, radio galaxies and quasi-stellar sources. *Mon. Not. Astr. Soc.* 133: 163–180.
Ramaty, R., and Holt, S. S. 1970. Polarization reversal of solar microwave bursts. *Nature* 226: 68–69.
Rees, M. J. 1971a. New interpretation of extragalactic radio sources. *Nature* 229: 312–317.
———. 1971b. Erratum to above. *Nature* 229: 510.
Reinhardt, M. 1972. Interpretation of rotation measures of radio sources. *Astron. Astrophys.* 19: 104–108.
Reinhardt, M., and Thiel, M. A. F. 1970. Does a primaeval magnetic field exist? *Astrophys. Letters* 7: 101–106.
Roberts, J. A., and Komesaroff, M. M. 1965. Observations of Jupiter's radio spectrum and polarization in the range from 6 cm to 100 cm. *Icarus* 4: 127–155.
Roberts, J. A.; Ribes, J. C.; Murray, J. D.; and Cooke, D. J. 1972. Circular polarization at 1.4 and 5 GHz. *Nature Phys. Sci.* 236: 3–4.
Sazonov, V. N. 1969a. The generation and transfer of polarized synchrotron emission. *Astron. Zhurnal* 46: 502–511.
———. 1969b. Polarization of normal waves and synchrotron radiation transfer in a relativistic plasma. *Zh. Exp. Theor. Fiz.* 56: 1075–1086.
Sazonov, V. N., and Tsitovich, V. N. 1968. Polarization effects during generation and radiation transfer of relativistic electrons in a magnetoactive plasma. *Radiofizika* 11: 1287–1299.
Seaquist, E. R. 1969. Circular polarization of Jupiter at 9.26 cm. *Nature* 224: 1011–1012.
———. 1971. PKS 1127-14-circular polarization at 3,240 MHz. *Nature Phys. Sci.* 231: 93–97.
———. 1973. Circular polarization studies of selected compact sources at 3240 MHz. *Astron. Astrophys.* 22: 299–308.
Seaquist, E. R.; Gregory, P. C.; Biraud, F.; and Clarke, T. R. 1973. Frequency dependence of circular polarization in three compact radio sources. *Nature Phys. Sci.* 242: 20–23.
Seielstad, G. A.; Morris, D.; and Radhakrishnan, V. 1964. On the linear polarization of the microwave radiation from radio sources. *Astrophys. J.* 140: 53.
Seielstad, G. A., and Weiler, K. W. 1971. Dual-frequency orthogonal strip distributions of linearly polarized and total radiation in eight extragalactic radio sources. *Astron. J.* 76: 211–221.
Strom, R. G. 1972. Faraday depolarization of extragalactic radio sources. *Nature Phys. Sci.* 239: 19–21.
Wardle, J. F. C. 1971a. The polarization of strong radio sources at 9.5 mm wavelength. *Astrophys. Letters* 8: 183–186.
———. 1971b. The structure and polarization of 3C459 at 610 MHz. *Astrophys. Letters* 8: 53–55.
Westfold, K. C. 1959. The polarization of synchrotron radiation. *Astrophys. J.* 130: 241–258.

EXTRAGALACTIC OPTICAL POLARIMETRY

NATARAJAN VISVANATHAN
Hale Observatories

The polarization produced by the scattering of dust particles in spiral nebulae, interstellar polarization in the Large and Small Magellanic Clouds, origin of linear polarization in the continuum and hydrogen filaments in the peculiar galaxy M 82, and the polarization in the synchrotron continua of Seyfert N-type, QSS, and BL Lacertae-type objects are discussed.

At any one moment light is considered unpolarized when the electric field changes and oscillates in all directions in the plane perpendicular to the line of sight. If the electric field oscillates in only one direction, the light is considered polarized, and we are concerned about the direction of the oscillation of the electric vector in the plane perpendicular to the line of sight. If both x and y components oscillate in phase, then the resultant vibration is called linear polarization. In this paper we will be mainly concerned with the measurement of this important property of light, namely the state of polarization in extragalactic objects.

IMPORTANT PROCESSES RESPONSIBLE FOR PRODUCING POLARIZATION IN EXTRAGALACTIC OBJECTS

Some of the important processes that are responsible for the observed polarization in celestial objects are as follows:[1]

1. *Scattered Radiation:* When an unpolarized source lights a particle anisotropically, then the scattered radiation emerging from the particle is linearly polarized. The scattered radiation is composed of two components whose electric vectors are perpendicular and parallel to the scattering plane (Chandrasekhar 1950). (This plane is defined as that which contains both the direction of incident and scattered light.) For both Rayleigh and Thomson scattering, the intensity ratio between the parallel and perpendicular components is $\cos^2\alpha$ to 1, if

---

[1] Also see p. 54.

the angle between the incoming unpolarized light and the scattered polarized beam is α. The scattered light then is partially plane polarized and the percentage polarization is equal to $(1 - \cos^2 \alpha)/(1 + \cos^2 \alpha)$. For $\alpha = 90°$ the intensity of the parallel component is 0, and hence the light is 100% polarized with the electric vector perpendicular to the scattering plane.

2. *Interstellar Polarization:* The discovery of interstellar polarization by Hall (1949) and Hiltner (1949) and its subsequent theoretical interpretation by Davis and Greenstein (1951) (the most widely accepted one and that which best explains the observations in our Galaxy) has opened a way of mapping the magnetic field in the galaxies. According to Davis and Greenstein, the elongated interstellar particles assumed to be paramagnetic are in a state of rapid rotation resulting from collisions of interstellar gas atoms. In the presence of a magnetic field, part of the rotational energy is dissipated by paramagnetic absorption in such a way as to align the short axis with the direction of the magnetic field. When we observe the star straight, the light is partially absorbed along the line of sight by the elongated dust particles between the star and the observer; we will receive the strongest radiation in the plane containing the short axis and the line of sight. This direction of the electric vector indicates the direction of the projection of the magnetic field on the plane of the sky.

3. *Synchrotron Radiation:* The third process arises from the fact that many of the extragalactic objects exhibit nonthermal continuum radiation. This has been interpreted as the synchrotron radiation since the continuum of these objects can be represented by a power law of the form $F_\nu \alpha \nu^{-\alpha}$, where α is the spectral index. In this case the continuum will be linearly polarized and the direction of the electric vector is perpendicular to the magnetic field.

SPIRAL NEBULAE AND POLARIZATION

The spiral nebula whose plane makes a large angle with the line of sight was considered a favorable object for polarization observations as the occulting matter in the surrounding spiral arm produces polarization by scattering the light from the bright nucleus. Thus, the early work in extragalactic objects has been done in the spiral nebulae NGC 224, 1068, 3628, and 4736 by Reynolds (1912) and Mayer (1920) using photographic techniques. No clear evidence of polarization was obtained. Smith (1935) put a limit of 2% of polarization for the elliptical galaxy M 32. Öhman (1942) was the first who successfully established the presence of 6.2% polarization with the electric vector parallel to the minor axis of the galaxy in the dust-obscured northwest side of M 31. He used an ingenious device for his observation — scanning and acquiring simultaneous photographs in perpendicular and parallel planes of vibration of one or more strips across the nebula. He interpreted his results of polarization with the aid of a scattering model and decided that M 31 is titled to the line of sight such that the obscured northwest side of the galaxy is farther from us than the brighter side of the galaxy. Elvius (1951, 1956) extended this type of observation to more galaxies with large diaphragms and found 7.4% polarization in the dust clouds in NGC 5055 and 5.5% in NGC 7331.

The interpretation of her results became complicated as both the processes of polarization due to scattered radiation as well as the interstellar polarization contributed to the observed result. More polarization observations of many spirals were done by Elvius and Hall (1964) between 1961 and 1964 by photoelectric techniques using small apertures. NGC 185, 205, 221, 253, 891, 2841, 3031, 3077, 4258, 4565, 4631, 5195, 7331 showed no polarization within the uncertainty of observations. In NGC 5128, which is a strong radio source, up to 6% polarization with the electric vector parallel to the dark band was observed (Fig. 1). They interpreted this polarization as due to absorption of dust particles, as they found $A_v = 3$ mag absorption in this part of the dark band. It should be mentioned, however, that in this strong radio source the bright spot at the northern edge of the dark lane, in which Burbidge and Burbidge (1959) found a strong blue continuum, showed very little polarization.

More recent observations of polarization by dust on spiral galaxies in NGC 3067, 3718, 4216, 4438, 4565 have been reported by Elvius (1972b). In NGC 3718 the light from the dark bar is about 4% polarized with the electric vector along the bar, indicating that large-scale magnetic fields are associated with bar-like structures.

INTERSTELLAR POLARIZATION IN M 31 AND LARGE AND SMALL MAGELLANIC CLOUDS

Hiltner (1958) observed 21 discrete objects (globular clusters) in M 31 and proved that polarization due to absorption by dust particles is present in this galaxy. The direction of the electric vector has a tendency to be preferentially aligned parallel to the major axis of M 31, indicating the presence of a magnetic field along the spiral arms. The maximum polarization observed is about 3%, and the average ratio of polarization to absorption for the whole galaxy is 0.03.

Fig. 1. A reproduction of Fig. 7 from Elvius and Hall (1964). Polarization measures of NGC 5128 have been plotted after correction for galactic polarization of 1.4% at 70°. Note that the vectors in the dark lane just to the east of the center are nearly parallel to this lane. The vectors shown elsewhere have a more nearly random distribution, and the polarization in the northeast region is zero. (From Elvius and Hall 1964.)

In 1963 a double-channel photoelectric polarimeter was constructed at Mount Stromlo Observatory, and 30 individual stars distributed uniformly in and around the large Magellanic Cloud were observed for polarization (Visvanathan 1966). It should be mentioned that it was possible to observe individual stars in the large Magellanic Cloud (LMC), as LMC is less than one tenth of the distance of M 31 and the line of sight makes a large angle with the plane of the cloud. The results show that the polarization varies from 0.5% to 3%. Interstellar polarization in the direction of LMC due to dust in our Galaxy is very small because of the high galactic latitude of LMC. Hence, most of the polarization observed in the stars in LMC does belong to LMC itself. The average value of p/A_V derived is 0.05, which is close to the maximum value of p/A_V in our Galaxy; this can be taken to support the premise that the cloud is seen nearly face on. Another important conclusion was that the alignment of electric vectors seems to be uniform over large distances on the order of kiloparsecs, indicating the uniformity of a magnetic field over large areas in the cloud. With the same polarimeter, Mathewson and Ford (1970) observed 180 stars both in LMC and the Small Magellanic Cloud (SMC). In SMC, apart from some electric vectors at the southern end of the bar, the remainder of the electric vectors have position angles close to 115°, which is the direction of LMC. In LMC, outside the emission nebular regions, a fair proportion of the electric vectors lies fairly parallel to the position angle of the line directed toward the SMC (Fig. 2).

Mathewson and Ford interpreted these results to mean that both the clouds are enveloped in a large-scale regular magnetic field with the lines of force parallel to the line joining the two galaxies. However, they were careful to point out that the above direction is also nearly the same direction of the electric vectors as derived from sixty galactic foreground stars from 100 pc to 2 kpc in the direction of LMC and SMC. Strongest polarization, up to 3%, is found in 30 Doradus Nebula, and the electric vectors show a remarkable focusing of the magnetic lines of force of this giant emission nebula.

POLARIZATION IN THE JET OF M 87

The discovery of a large amount of polarization in the Crab Nebula, along with the fact that the polarization increased with smaller apertures (Oort and Walraven 1956), has demonstrated that the Crab Nebula in our Galaxy is the first source in which the optical continuum radiation is of synchrotron origin. It has been suggested (Shklovskij 1955) that the synchrotron mechanism may be operative in the curious jet of M 87. Hence, polarization observations were made photographically by Baade

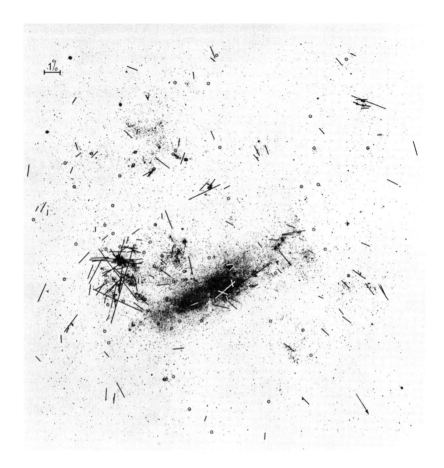

Fig. 2. A reproduction of Fig. 2 from Mathewson and Ford (1970). The percentage polarization and the position angle of the electric vector θ are marked on a red-light photograph taken with a 20-cm Schmidt telescope. Arrow points toward the direction of the Small Magellanic Cloud. (From Mathewson and Ford 1970.)

(1956). Later Hiltner (1959) made more accurate photoelectric observations of the jet with small diaphragms (Fig. 3). The results are as follows:

1. The large amount of polarization (about 20%) is found in the central condensation of the jet with position angle perpendicular to the direction of the jet.
2. The other two condensations in the jet show high polarization, but angles are different from that for the central condensation.

The fact that the large amount of polarization is found in the jet and no significant polarization is seen in other parts of M 87 is a confirmation

EXTRAGALACTIC OPTICAL POLARIMETRY

Fig. 3. A reproduction of Fig. 2 from Hiltner (1959). Mean polarization observations of the Jet M 87. The upper part of the figure is in the same scale as the photograph of M 87 below, obtained by Baade with the Hale 5.08-m telescope. The open circles refer to observations of areas other than the jet where no significant polarization is detected. (From Hiltner 1959.)

that the continuum of the jet of M 87 is of synchrotron origin, as predicted by Shklovskij.

ORIGIN OF POLARIZATION IN THE FILAMENTARY STRUCTURE OF M 82

The discovery of an extensive faint filamentary system in the peculiar galaxy M 82 (Johnson 1963; Sandage and Miller 1964) emitting continuum radiation extending 4000 parsecs above and below the fundamental plane, with a delicate looped pattern that is assumed to outline magnetic field lines, and the similarity of this with the Crab Nebula prompted polarization observations of these filaments to prove that these continuum filaments in M 82 are of synchrotron origin and that this continuum is the energy source needed to ionize the extensive hydrogen emission filaments (Lynds and Sandage 1963) that are found to coexist with the continuum filaments. The initial photographic polarization observations in all these faint outer filaments by Sandage and Miller (1964) showed that the brighter part of the filaments is very highly polarized, in some regions nearly 100% with the electric vector perpendicular to the predominant direction of the filament. They considered these results to support the theory that the radiation of continuum filaments is of optical synchrotron emission. Previously Lynds (1961) had suggested that synchrotron radiation was present on the basis of his identification of the galaxy as 3C 231; the synchrotron hypothesis was further developed by Lynds and Sandage (1963).

Elvius (1962) made a photoelectric polarimetric study in 1961–62 and found little polarization in the main body of the galaxy and more polarization as she went from the nucleus both below and above the fundamental plane of the galaxy. The polarization increased from 1% to 16%, mainly in the direction of the minor axis. Her observations were confined only to the inner part of the filamentary structure. She interpreted her results to mean that the polarization is produced by scattering in bright regions of the galaxy by the dust clouds in M 82. Her further observations (Elvius 1969) in the lower filamentary structure have been put together by Solinger (1969), who showed that the data of Elvius are consistent with a model in which light from a small nucleus in the center of the galaxy is scattered by free electrons in a heated gas behind a shock wave that is postulated to be propagating through the halo of M 82. In the spring of 1968, 11 patches with surface brightness $V = 22.4$ to 25.0 mag per square arcsec in the outer filamentary structure of M 82 at distances varying as far as 68 to 193 arcsecs from the nucleus of the galaxy were observed with a single-channel polarimeter with apertures 12.1 sec to 30.6 sec, and the results showed the polarization varying from 12% to 32% (Sandage and Visvanathan 1969). Although these values of polarization are high, it should be noted that they are much lower than those suggested by Sandage and Miller (1964). The techniques as well as the faintness of the filaments made their determination of the amount of polarization unreliable. The position angle of the electric vector is, in general, perpendicular to the filamentary structure. The data were consistent both with the synchrotron hypothesis as well as with the scattering hypothesis because all the measurements were near the minor axis where the angles are generally perpendicular not only to the filamentary structure but also to the radius vector from the nucleus.

In the spring of 1970, to provide tests of a more positive nature to decide between a synchrotron and a scattering model, five new regions in the outer filamentary structure were chosen so that the perpendicular to the filaments on these regions makes an appreciable angle to the perpendicular to the radius vector (Visvanathan and Sandage 1972). These five regions are exceedingly faint and for the faintest patch the signal was less than 1% of the night-sky radiation. The measurements of θ agree to within 2 σ with the scattering model but differ by 4 to 9 σ from the prediction for the synchrotron hypothesis.

Figure 4 summarizes the data for these five patches along with eleven patches measured in 1968. In Fig. 5 the perpendiculars to the direction of the electric vector have been drawn for all the patches in the outer filaments. It is found that except for the patches C and J the difference between the radius vector and the observed angle of the electric vector is not greater than 2 σ. The data suggest that the intermediate and outer continuum filaments cannot be synchrotron radiators and that some scattering process is involved and the source is confined to the central region of M 82. The patch RD (taken near the patch D) has been observed for polarization in the wavelength range of 0.34–0.8 μm with the multichannel scanner and polarimeter at the Cassegrain focus of the 5.08-m telescope. In one of the channels the Hα emission line has been centered and the 0.004 μm bandwidth used. The results showed that the Hα emission line is polarized 27% at $\theta = 54°$.

Further, from the same patch the continuum radiation was found to be as polarized as the Hα emission at the same angle. These polarization data suggested some form of scattering of both the continuum and Hα line emission from a source in the main body of the galaxy. Because no difference between the line and continuum polarization was detected, a limit to any recombination radiation in the halo of M 82 can be assumed to be less than 20% of the observed Hα intensity. The lack of recombination radiation in the halo proves that at least parts of the synchrotron model by Lynds and Sandage (1963) and the scattering model by Solinger (1969) do not fit the observations. Van Blerkom, Castor, and Auer (1973) proposed to explain the Hα emission in M 82 by intrinsically polarized fluorescence. We feel, however, that the same mechanism should explain polarization both in the continuum and in the lines.

Further, the equivalent width of Hα in the filaments (patches RD and RJ) and the nucleus M 82A (O'Connell 1970) is found to be similar. If more accurate measurements support these preliminary results, the central region of M 82A may then be the source of the scattered

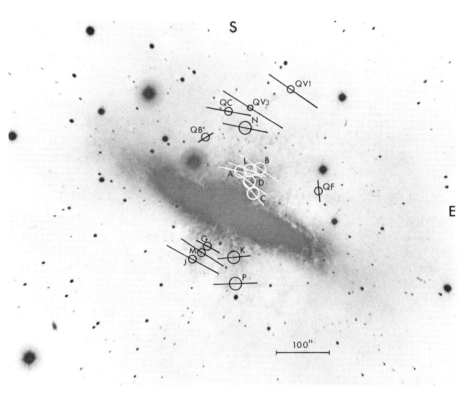

Fig. 4. A reproduction of Fig. 1 from Visvanathan and Sandage (1972). Composite negative print from three superposed plates of M 82 taken with the Hale reflector in the wavelength interval 0.38–0.50 μm. Polarization of 16 regions is shown by the length of the lines oriented to the angle of the electric vector. Three aperture diameters of 30″.6, 18″.8 and 12″.1 are shown. (From Visvanathan and Sandage 1972.)

light in the halo. Further comparison of the filamentary structure, on photographs in the continuum and Hα light, are quite similar (they may be identical). This again supports the theory that both the Hα and the continuum are scattered from a central source. The polarization of the continuum over the spectral range 0.36–0.80 μm in patch RD is found to be constant within the ± 3% accuracy of data. This requires either that scattering particles be electrons or that they be grains with much larger dimensions than the wavelength of the measured light, unlike those in our Galaxy. If the scattering is by electrons and if the central region of the galaxy M 82A is the source of scattered light, the electron density in the patch RD is 3400/cm^3. The filamentary surface brightness decreases with distance from the center as $d^{-2.65}$, which

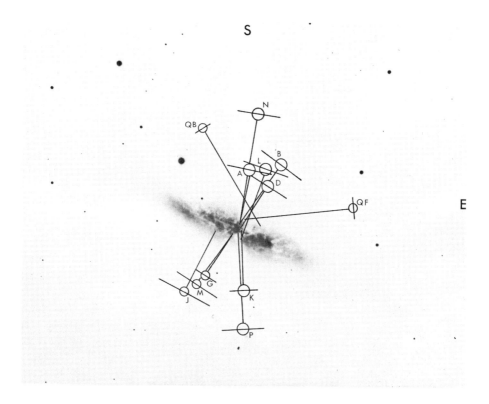

Fig. 5. A reproduction of Fig. 2 from Visvanathan and Sandage (1972), a print from a short exposure photograph of M 82 taken with the 5.08-m Hale reflector on an Eastman 103aO plate. Lines that are perpendicular to the direction of the electric vector are shown for 12 of the 16 regions. (From Visvanathan and Sandage 1972.)

requires that the density of scattering particles in the halo declines at a rate between $d^{-0.65}$ and $d^{-1.65}$, depending on the geometry. The total electron mass in the halo would be 2×10^4 $\odot$, if we adopt a volume-filling factor of 38^3. Otherwise, it is 10^{10} $\odot$, but in this case the halo would be optically thick and we could not explain the high polarization at large distances from the center. If the scattering is by dust, 4×10^4 $\odot$ will be the mass of dust required in the halo when we assume the total volume of the filaments is equal to 7×10^{62} cm^3 (Sanders and Balamore 1971).

It should be mentioned that the complex velocity field of M 82 which led to the original explosion hypothesis is not understood in the presence of scattering. According to Lynds and Sandage's (1963) explosion model, if all the clouds of scattering particles in the filaments are moving away from the light source in the nucleus, the velocity in the filaments should be positive with respect to the nucleus both above and below the galactic plane. The observed radial velocity in such a case represents the actual expulsion velocity and this cannot be negative in any part of the filamentary structure. However, actual observations (Lynds and Sandage 1963; Heckathorn 1972) show that the light in the southern region is highly blue-shifted relative to the light from the center part of the region. A scattering model with motions of the expelling matter along the curved orbits has been suggested by Sanders and Balamore (1971). Elvius (1972a) has suggested that a model based on the assumption of the galaxy moving through a cloud of scattering particles can explain the observed velocity field.

POLARIZATION IN SEYFERT GALAXIES AND N GALAXIES CONTINUA

It has been noted that certain physical properties relate Seyfert galaxies, N galaxies, and quasi-stellar sources (QSS), the important factor being the degree of quasar activity in the galactic nucleus. All these objects do show a stellar-like nucleus superposed over the galaxy. In the case of Seyfert galaxies and N-type galaxies, we see a mixture of galaxy radiation and the radiation from the bright nucleus, while in QSS almost all the radiation seems to come from the central nucleus. The stellar-like nucleus emits continuous radiation, and in most cases this radiation is found to be nonthermal. Polarization observations of these types of galaxies have been made with the aim to prove whether this nonthermal radiation is of synchrotron origin.

Seyfert galaxies NGC 1068, 1275, 4151 (Walker 1968; Visvanathan and Oke 1968; Kruszewski 1971) 3C 120, II ZW 2130 + 09, Markarian galaxies 9 and 10, N-type galaxies 3C 371, 3C 390.3 (Visvanathan 1967; Kinman et al. 1968; Visvanathan unpublished) have been observed for polarization. Except for NGC 1068, 1275, 3C 371, 3C 390.3, all the other galaxies are polarized by about 1% or less. Polarization in NGC 1068 and 1275 varies from 3% to 5%. In general, polarization in ultraviolet light is more than that in blue and visual light, and also, the smaller the aperture used, the higher is the polarization measured.

Detailed observations in U, B, V, R colors with varying apertures have been made for NGC 1068, as seen in Fig. 6 (Visvanathan and Oke 1968; Kruszewski 1971). In U, polarization varied from 5% to 2.5% when the aperture is changed from 12.5 to 28 arcsecs. Further, the polarization varied from 5% to 0.3% with an aperture of 12.5 arcsecs when the wavelength is varied from U to R. The position angle

EXTRAGALACTIC OPTICAL POLARIMETRY

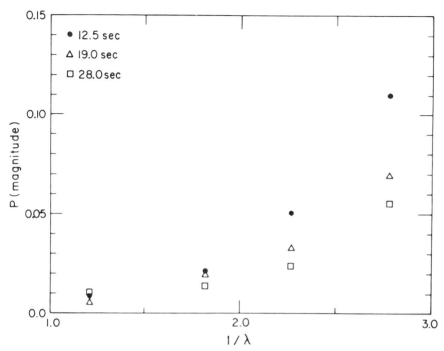

Fig. 6. A reproduction of Fig. 1 from Visvanathan and Oke (1968). Polarization measured in NGC 1068, with 12."5, 19."0, and 28."0 apertures, in U, B, V, and R colors have been plotted. (From Visvanathan and Oke 1968.)

of the electric vector remained constant in all wavelengths and in all apertures. The higher polarization, found in the U wavelength and smaller apertures, has been interpreted as due to the fact that the contribution from polarized nonthermal continuum radiation from the nucleus is more than that from the underlying galaxy radiation in U wavelength as well as in smaller apertures.

The above facts have been used to separate the nonthermal component in NGC 1068, by assuming the underlying galaxy radiation is unpolarized. A first trial contribution of nonthermal radiation was removed from the observed flux at 0.36 μm, and the scale factor between this component and the observed flux of underlying galaxy radiation (we have assumed that this is similar to M 31) was computed. By applying the same scale factor to all other wavelengths, the entire scan of NGC 1068 was separated into a nonthermal component and a galaxy component. This procedure was repeated until the polarization of the nonthermal component became constant at all wavelengths. The results indicate that the derived nonthermal component is smooth and similar to the energy distribution of QSS 3C 273.

It is interesting to note that Oke, Neugebauer, and Searle (1972) have concluded from a detailed study of 18 representative Markarian galaxies, which includes several classical Seyfert galaxies, that the integrated spectrum of a Seyfert galaxy contains a component closely resembling the spectrum of QSS 3C 273.

However, more recent observations of the SII line ratio $I(\lambda\ 10320/\lambda\ 4072)$ by Wampler (1971) has shown that interstellar reddening occurs in or near the nuclei of several Seyfert galaxies, including NGC 1068 where A_V derived is 1.3 mag. Kaneko (1972) corrected the nonthermal continuum of the stellar nucleus (Visvanathan and Oke 1968) for the reddening by dust and found the corrected continuum agrees with the energy distribution of a hot blackbody and estimated its temperature to be $2 \times 10^{5\circ}$K. If the above results are true, the polarization measurements of NGC 1068 (Visvanathan and Oke 1968; Kruszewski 1971) should be interpreted to show the presence of a dense shell of dust grains surrounding the nucleus of NGC 1068 (Nandy and Wolstencroft 1970). The observed polarization may be explained in terms of scattering due to dust particles. The fact that polarization of the nonthermal component is 7.5% indicates that the dust shell is asymmetrical in shape.

NGC 4151 is found to have variable polarization (0.4% to 1.95%) positively correlated with brightness variation (Kruszewski 1971). No variation in polarization is found in NGC 1068 (Visvanathan, unpublished; Nordsieck 1972).

POLARIZATION IN QSS AND VIOLENT VARIABLE QSS

As we have already noted, most QSS show a power law spectrum (Oke, Neugebauer, and Becklin 1970) indicating that most of the radiation is nonthermal and probably of synchrotron or inverse Compton origin. Twenty-six QSS have been observed for polarization (Appenzeller and Hiltner 1967; Visvanathan 1968). The polarization of these QSS varies from 1% to 3%, which is more than interstellar polarization due to dust in our Galaxy in the direction of QSS, and hence the polarization measured in these QSS can be taken to be inherent to the objects themselves. The fact that most of the QSS show measurable intrinsic polarization lends support to the idea that the nonthermal continuum may be of synchrotron origin. Further, it should be noted that the integrated polarization is not high in most of the sources and probably means that the distribution of the magnetic field is not uniform throughout the region emitting the radiation.

Even though most QSS are variable, a few constitute a special class because of their large amplitude variations of more than 1 mag. The objects PKS 0403-13, 3C 279, PKS 1510-08, 3C 345, PKS 2135-14, 3C 446, and 3C 454.3 belong to this class. Continua of most of these QSS show linear polarization that varies with time. In 1966–67 Kinman and collaborators (1966, 1968; Kinman 1967) discovered the polarization as well as the electric vector θ change in large amounts in the variables QSS 3C 279, 345, 446. Further, in 1967 when 3C 345 was bright with frequent bursts of 10-to-20 days duration, Kinman et al. (1968) made observations of the polarization about twice a month from March to September. They explained their observations with a model for 3C 345 continuum consisting of the three following components: (1) a short-burst component of 10-to-20 days duration with high polarization; (2) a slowly varying component polarized about 17% extending over several hundred days; and (3) a third faint component that is unpolarized. Further, they found that whenever a burst occurred, the polarization increased with change in the **E** vector θ (Fig. 7).

Regular study of the interrelation among polarized flux, electric vector θ, total flux, the wavelength dependence of polarization, line intensity, and line polarization during different phases of variations in 3C 279, 345, 446, 454.3 was undertaken from 1967 to 1970 (Visvanathan 1973). The observations were obtained mainly at the Mount Wilson 2.5-meter telescope with the single-channel polarimeter and at the Palomar 5-meter telescope with the multichannel scanner and polarimeter.

Observations of the polarization and B magnitude of 3C 279, 3C 345, 3C 454.3, 3C 446 were obtained on several occasions (Fig. 8), and thirteen of our observations were made during burst periods. The ratio of the total flux in the burst period to that in the nonburst period varies from 1.2 to 7.1. On all these occasions the continuum was linearly polarized, and whenever the total flux increased, an increase in polarized flux was observed. The ratio of polarized flux between the burst and nonburst periods varied from 1.1 to 7.6. Also the position angle of the electric vector θ changed during the burst, as compared with the period before or after the burst period. Thus, the change in the magnetic-field configuration of the regions emitting the continuum is apparently a characteristic feature of the burst phase of these objects.

The objects 3C 345 and 3C 454.3 reached minimum brightness during our observations. It should be noted that the continuum was linearly polarized on these occasions, just as in other nonburst periods. In the

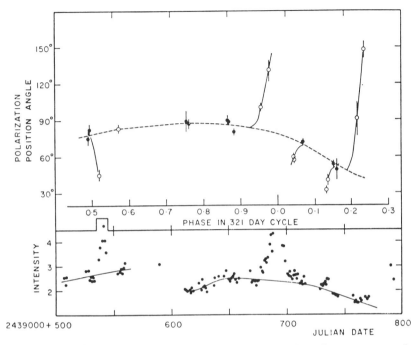

Fig. 7. A reproduction of Fig. 8 from Kinman et al. (1968). The upper part of the diagram shows the variation of the position angle electric vector θ with Julian date of the QSS 3C 345. On the three-component model, the dashed curve shows the trend of θ for the slowly varying component. The lower diagram shows the observed intensities of 3C 345 in units of the constant component of the three-component model. The solid curve is an estimate of the contribution of the slowly varying component. (From Kinman et al. 1968.)

case of 3C 345, there is possible evidence for the presence of a constant polarized component, as indicated by the nearly constant total flux, polarized flux, and θ of the 1968 observations. In the case of 3C 454.3, a slowly varying component extending from October 1968 to December 1969 is evident. During this period θ changed smoothly from 90° to 30°.

All our polarization observations of the objects, along with available observations of the same objects by other observers (Fig. 9), show that the ratio between the polarized flux and the total flux does not exceed a value of about 0.16. If in these objects the continuum is emitted by several active spots embedded in an extensive region, we can interpret the low ratio as being due to the small resultant polarization caused by these active spots, each exhibiting high polarization but with different directions of the magnetic field.

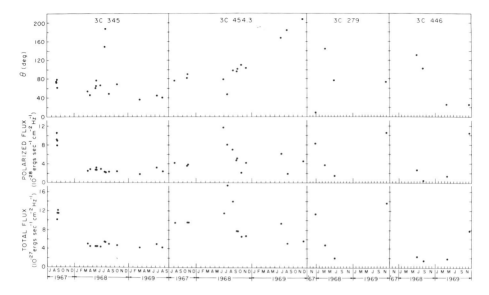

Fig. 8. Total flux, polarized flux, and position angle of the electric vector θ plotted against date of observation for QSS 3C 345, 3C 454.3, 3C 279, and 3C 446.

The wavelength dependence of polarization observations from 0.35–0.80 μm of 3C 345, 3C 454.3, and 3C 446 in 1968 and 1969 show that p and θ remained constant with wavelength within the errors of observations. This result strongly suggests that the mechanism responsible for continuum radiation in these objects is the same mechanism that operates in all parts of the spectrum from the ultraviolet to the infrared.

The absolute energy distribution of the continuum of these four objects was observed several times in the period 1968 to 1970 and shows that a straight line fitting for the continuum in the plot of log f_ν versus log ν is valid, so that the continuum can be represented by a power law of the form $f_\nu \alpha \nu^{-\alpha}$. This can be taken as evidence that the radiation in these objects in the observed frequency range is of nonthermal origin in both active and quiet periods. The fact that this radiation is linearly polarized and that the polarization is independent of wavelength supports the view that this nonthermal radiation is optical synchrotron radiation.

In 3C 345, the emission line Mg II was observed twice in 1969 for polarization and the results indicate that the line radiation is either unpolarized or only slightly polarized. This suggests that the regions from which the emission line originates are randomly arranged so that

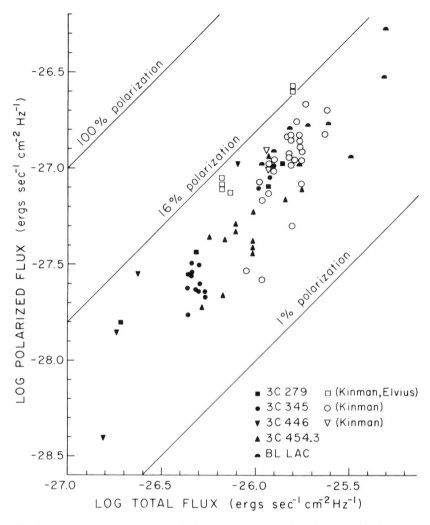

Fig. 9. Total flux and polarized flux for objects 3C 279, 3C 345, 3C 446, 3C 454.3, and BL Lac observed in B wavelengths are plotted against each other. Lines represent limits of 1%, 16%, and 100% polarization levels.

the resultant polarization of the scattered radiation from all these regions is small.

POLARIZATION IN BL LAC TYPE OBJECTS

The radio position of the source VRO 42.22.01 is of interest because the inverted radio spectrum and variability has been determined accurately by MacLeod and Andrew (1968), and Schmitt (1968) identified this radio source with the star BL Lac in the variable star catalogue. The continuous spectrum from ultraviolet to infrared was featureless without any emission or absorption lines and was found to be highly linearly polarized. The polarization is constant with wavelength (Visvanathan 1969). The continuum from ultraviolet to infrared can be represented by a power law (Visvanathan 1969; Oke, Neugebauer, and Becklin 1969). From the above facts it can be easily seen that BL Lac is very similar to the variable QSS that we discussed above, except in that they show no spectral lines so that no red shift can be measured. It has been conjectured, as in QSS, that the emission lines in BL Lac are washed out by a strong nonthermal continuum but should become visible when the continuum intensity becomes minimum. But in the past, spectral observations in faint phases of this object have not revealed any lines. However an upper limit on the distance based on the importance of Compton scattering in compact, highly luminous nonthermal sources indicates that this object is extragalactic (Stein, Gillet, and Knacke 1971). Hence, it seems that BL Lac can be taken to represent a separate class in itself, different from the variable QSS (Strittmatter et al. 1972).

During 1968, 1969, 1970 the polarization of this object was observed six times. From Table I it can be seen that the object has varied between 14.8 to 16.5 in B magnitude, while the polarization has varied from 3.6% to 11.1%. The position angle of the electric vector also shows a change from 96° to 165°. It is important to note that the continuum is polarized in the faint as well as in the bright phase of the object. In the same period, the wavelength dependence of polarization from 0.35–0.8 μm shows constancy, and in all these occasions the continuum can be represented by a power law. The spectral index varied from 2.85 to 3.17 during this period. These facts show the continuum from this object is of synchrotron origin, both in active and quiet periods.

Recent observations have brought to light five more objects similar to BL Lac. They are OJ 287 (Blake 1970); ON 231, visual object W Com; ON 325 (Browne 1971); PKS 1514 − 24, visual object AP Lib (Bolton, Clarke, and Ekers 1965; Westerlund and Wall 1969); and possibly compact galaxy I ZW 1727+50 (Le Squéren, Biraud, and Lauqué 1972). Of these, more detailed observations have been made of OJ 287. Initial study by Kinman and Conklin (1971) revealed the optical continuum to have a variable polarization of 9% to 15% during April and May 1971. Nordsieck (1972) observed in September 1971, November 1971, and the beginning of 1972 and found a polarization varying from 2.3% to 13.5%. Moreover, on September 20–21 his observations show large variable polarization from 8.4% to 2.3% and the position angle changed from 116° to 167°. It is interesting to note

TABLE I

Polarization Observations of BL Lac Type Objects

Objects	B (mag)	p %	θ (deg)	Date (UT)	Reference
BL Lac	16.5	9.7	165	1968 Oct 20	Visvanathan (1973)
	16.1	10.8	162	1968 Nov 23	
	15.9	9.0	96	1969 May 16	
	14.8	11.1	107	1969 July 16	
	14.8	6.1	132	1969 Aug 19	
	15.3	3.6	109	1969 Dec 5	
OJ 287	13.2	15.1	26	1971 April 23	Kinman and Conklin (1971)
	13.2	9.1	24	1971 April 24	
	13.0	13.3	39	1971 May 27	
	12.67	8.4	116	1971 Sep 20	Dyck et al. (1971)
	12.67	2.3	167	1971 Sep 21	
	12.62	3.0	0	1971 Sep 22	
	13.2	2.8	59	1971 Nov 23	Nordsieck (1972)
	13.2	4.3	28		
	13.2	8.2	108	1972 Jan 15	
		7.3	102		
	13.2	13.5	96	1972 Jan 16	
		11.9	97		
	13.3	6.5	84	1972 Jan 18	Visvanathan (unpublished)
	12.8	4.6	10	1972 Feb 19	Nordsieck (1972)
	12.8	2.9	39	1972 Feb 18	Strittmatter et al. (1972)
	13.2	5.0	92	1972 Mar 17	
	13.9	10.3	87	1972 April 16	
		10.8	87	1972 April 17	
ON 325 (B2 1215+30)	16.2	7.2	140	1972 Feb 20	Strittmatter et al. (1972)
	16.1	4.4	146	1972 March 17	
	15.9	4.6	119	1972 Apr 16	
PKS 1514−24 (AP LIB)	16.1	5.6	173	1972 Mar 17	
	16.1	5.8	139	1972 Apr 16	
I ZW 1727+50	15.81	5.0	175	1967 Nov 3	Visvanathan (unpublished)
	∼16.18	3.4	91	1968 Apr 21	
		1.0	142	1968 May 26	
		1.3	61	1968 May 28	
		2.2	48	1968 May 29	
		2.1	52	1968 Sep 26	

that this variation on the time scale of one day occurred when the object was in maximum brightness of about 48 flux units (Visvanathan and Elliot 1973). The polarization and position angle of the electric vector remain constant with wavelength on the few occasions when the polarization was measured in many wavelengths (Strittmatter et al. 1972). The multichannel scanner observations from 0.33–1.0 μm with the Palomar 5-meter telescope during February 1972 showed that the continuum can be represented by a straight line in a log $F\nu$ versus log ν

plot with a slope of 1.25 and that the polarization remained constant with wavelength from 0.34 to 0.8 µm. The four days of continuous observations did not show any change in the spectral index (Visvanathan unpublished). Wide-band photometry observations by Dyck et al. (1971) in the range 0.36 to 3.4 µm show a spectral index of about 1.0. It is interesting that in the wavelength region between 3.5 to 10 µm, the energy distribution of this source is very similar to that of BL Lac, while in the optical region BL Lac is much redder (spectral index 2.8) than OJ 287 (spectral index 1.25). Other objects show featureless continuum and variable polarization with time, as is shown in Table I.

CONCLUSION

The past 15 years of the extragalactic polarization observations have witnessed a growth in the development of highly accurate polarimeters using photoelectric techniques that have led to the measurement of polarization with an accuracy on the order of 0.01%. Hence, the search for the circular polarization in linearly polarized galactic nuclei that lie behind the thick interstellar clouds can now be realized and will provide an opportunity to investigate whether the grains in other galaxies are characteristically dielectric or metallic (Martin 1972).[2]

More observations of linear polarization of Seyfert galaxies and N-type galaxies with different apertures in different colors similar to the NGC 1068 observations reported in this paper will be made in an attempt to separate thermal and nonthermal components in these galaxies.

Sciama and Rees (1967) pointed out that a significant circularly polarized component in the optical radiation could arise if Compton scattering is present in QSS. Magnetic fields on the order of 200 gauss would be required to produce even 0.1% circular polarization in visible wavelengths by this mechanism. An attempt to detect circular polarization in six QSS, BL Lac, and OJ 287 met with negative results at the 0.1% level (Landstreet and Angel 1972; Gehrels 1972).

It has been suggested that QSS are magnetic, rotating supermassive objects (spinars) with an emission mechanism similar to that of pulsars (Morrison 1969). In the Crab pulsar the pulses are linearly polarized and the position angle of the electric vector θ rotates 150° from the beginning to the end of the pulse (Visvanathan 1971). This has been interpreted by assuming that the observed pulses are produced by a polarized radiation pattern fixed in the pulsar, which is scanned as the pulsar rotates. Even though our results with QSS and BL Lac type

---

[2] See p. 926.

objects indicate that the polarized flux increases and θ changes during the burst in these objects, we do not have sufficient data to see how they behave with the different phases of the burst. Hence, it is very important to make continuous observations of p and θ in different phases of many bursts to see whether any periodicity is present in the variation of p and θ in these objects.

Acknowledgments. I thank the Observatories for the support of this work. The Hale Observatories are operated jointly by the Carnegie Institution of Washington and the California Institute of Technology.

REFERENCES

Appenzeller, I., and Hiltner, W. A. 1967. Polarimetric observations of 14 quasi-stellar objects. *Astrophys. J.* 149:L17–L18.

Baade, W., 1956. Polarization in the jet of M 87. *Astrophys. J.* 123:550–551.

Blake, G. M. 1970. Observations of extra-galactic radio sources having unusual spectra. *Ap. Letters* 6:201–205.

Bolton, J. G.; Clarke, M. E.; and Ekers, R. D. 1965. Identification of extragalactic radio sources between declinations $-20°$ and $-44°$. *Austr. J. Phys.* 18:627–633.

Browne, I. W. A. 1971. Two bright new quasi-stellar radio sources. *Nature* 231: 515–516.

Burbidge, E. M., and Burbidge, G. R. 1959. Rotation and internal motions in NGC 5128. *Astrophys. J.* 129:271–281.

Chandrasekar, S. 1950. *Radiative transfer,* pp. 35–37. Oxford: Clarendon Press.

Davis, L., Jr., and Greenstein, J. L. 1951. The polarization of starlight by aligned dust grains, *Astrophys. J.* 114:206–240.

Dyck, H. M.; Kinman, T. D.; Lockwood, G. W.; and Landolt, A. U. 1971. Observations of OJ 287 between .36 and 3.4 μm. *Nature Phys. Sci.* 234:71–72.

Elvius, A. 1951. A polarigraphic study of the spiral nebula NGC 5055 (M 63). *Stockholm Ann.* 17, No. 4:1–17.

———. 1956. Polarization of light in the spiral galaxy NGC 7331 and the interpretation of interstellar polarization. *Stockholm Ann.* 19, No. 1:1–34.

———. 1962. A polarimetric study of the galaxy M 82. *Lowell Obs. Bull.* V, No. 119:281–294.

———. 1969. Photoelectric observations of polarization and color in the galaxy M 82. *Lowell Obs. Bull.* VII, No. 149:116–128.

———. 1972a. Polarization and velocity field in galaxy M 82. *Astron. Astrophys.* 19:193–196.

———. 1972b. Polarization of light by dust in galaxies. *Interstellar dust and related topics.* (J. M. Greenberg and H. C. van de Hulst, eds.) Dordrecht, Holland: D. Reidel Publ. Co.

Elvius, A., and Hall, J. S. 1964. Polarimetric observations of NGC 5128 (Cen A) and other extra-galactic objects. *Lowell Obs. Bull.* VI, No. 123:123–134.

Gehrels, T. 1972. On the circular polarization of HD 226868, NGC 1068, NGC 4151, 3C 273 and VY Canis Majoris. *Astrophys. J.* 173:L23–L25.

Hall, J. S. 1949. Observations of the polarized light from stars. *Science* 109: 166–167.

Heckathorn, H. M. 1972. The emission line velocity field in M 82, *Astrophys. J.* 173:501–527.
Hiltner, W. A. 1949. Polarization of light from distant stars by interstellar medium. *Science* 109:165.
———. 1958. Interstellar polarization in M 31, *Astrophys. J.* 128:9–13.
———. 1959. Photoelectric polarization observations of the jet in M 87. *Astrophys. J.* 130:340–343.
Johnson, H. M. 1963. The structure of M 82. *Publ. Nat. Radio Astr. Obs.* 1, No. 17:283–293.
Kaneko, N. 1972. The nonstellar continuum of the Seyfert galaxy NGC 1068. *Pub. Astr. Soc. Japan* 24:145–148.
Kinman, T. D. 1967. Optical polarization of 5 radio sources. *Astrophys. J.* 148:L53–L56.
Kinman, T. D., and Conklin, F. K. 1971. Observations of OJ 287 at optical and millimeter wavelengths, *Ap. Letters* 9:147–149.
Kinman, T. D.; Lamla, E.; Ciurla, T.; Harlan, E.; and Wirtanen, C. A. 1968. The variability of the optical brightness and polarization of the quasi-stellar radio source 3C 345. *Astrophys. J.* 152:357–374.
Kinman. T. D.; Lamla, E.; and Wirtanen, C. A. 1966. The optical brightness variations and polarization of quasi-stellar radio source 3C 446. *Astrophys. J.* 146:964–969.
Kruszewski, A. 1971. Optical polarization of two Seyfert galaxies, *Acta Astron.* 21:311–328.
Landstreet, J. D., and Angel, J. R. P. 1973. Search for optical circular polarization in QSS and Seyfert nuclei. *Astrophys. J.* 174:L127–L129.
Le Squéren, A. M.; Biraud, F.; and Lauqué, R. 1972. A new Lacertid? *Nature Phys. Sci.* 240:75–76.
Lynds, C. R. 1961. Radio observations of the peculiar galaxy M 82, *Astrophys. J.* 134:659–661.
Lynds, C. R., and Sandage, A. 1963. Evidence for an explosion in the center of the galaxy M 82. *Astrophys. J.* 137:1005–1021.
MacLeod, J. M., and Andrew, B. H. 1968. The radio source VRO 42.22.01. *Ap. Letters* 1:243–246.
Martin, P. G. 1972. Interstellar circular polarization. *Mon. Not. R. Astr. Soc.* 159:179–190.
Mathewson, D. S., and Ford, V. L. 1970. The magnetic field structure of the Magellanic Clouds, *Astrophys. J.* 160:L43–L46.
Mayer, W. F. 1920. A study of certain nebulae for evidences of polarization effects. *Lick Bull.* 10:68–78.
Morrison, P. 1969. Are quasi-stellar radio sources giant pulsars? *Astrophys. J.* 157:L73–L76.
Nandy, K., and Wolstencroft, R. D. 1970. Origin of the optical polarization in the nucleus of NGC 1068. *Nature* 225:621–622.
Nordsieck, K. H. 1972. A search for circular polarization in extra-galactic objects. *Astrophys. Letters* 12:69–74.
O'Connell, R. W. 1970. Stellar contents of M 82. Thesis, California Institute of Technology.
Öhman, Y. 1942. A polarigraphic study of obscuring clouds in the great Andromeda Nebula M 31. *Stockholm Ann.* 14, No. 4:1–29.
Oke, J. B.; Neugebauer, G.; and Becklin, E. E. 1969. Spectrophotometry and infrared photometry of BL Lacertae. *Astrophys. J.* 156:L41–L43.

———. 1970. Absolute spectral energy distribution of quasi-stellar objects from 0.3 to 2.2 microns, *Astrophys. J.* 159:341–355.

Oke, J. B.; Neugebauer, G., and Searle, L. 1972. Annual Report of the Director, Hale Observatories, p. 680.

Oort, J., and Walraven, T. 1956. Polarization and composition of the Crab Nebula. *Bull. Astron. Netherlands* 12:285–308.

Reynolds, J. H. 1912. Preliminary observations of spiral nebulae in polarized light. *Mon. Not. R. Astr. Soc.* 72:553–555.

Sandage, A., and Miller, W. C. 1964. The exploding galaxy M 82: Evidence for the existence of large scale magnetic field. *Science* 144:405–409.

Sandage, A., and Visvanathan, N. 1969. Colors, linear polarization and preliminary mapping of the magnetic field for the outer filaments in the exploding galaxy M 82. *Astrophys. J.* 157:1065–1074.

Sanders, R. H., and Balamore, D. S. 1971. Dust scattering in the filaments of M 82. *Astrophys. J.* 166:7–12.

Schmitt, J. L. 1968. BL Lac identified as a radio source. *Nature* 218:663.

Sciama, D. W., and Rees, M. J. 1967. Possible circular polarization of compact quasars. *Nature* 216:147.

Shklovskij, I. S. 1955. Radio galaxies, *Soviet Phys. A. J.* 4:885–896.

Smith, S. 1935. Some notes on the structure of elliptical nebulae. *Astrophys. J.* 82:192–201.

Solinger, A. B. 1969. The galaxy M 82; I. Analysis of the observations. *Astrophys. J.* 155:403–415.

Stein, W. A.; Gillet, F. C.; and Knacke, R. F. 1971. Possible upper limit to the distance of BL Lacertae. *Nature* 231:254–255.

Strittmatter, P. A.; Serkowski, K.; Carswell, R.; Stein, W. A.; Merrill, K. M.; and Burbidge, E. M. 1972. Compact extragalactic nonthermal sources. *Astrophys. J.* 175:L7–L13.

Van Blerkom, D.; Castor, J. I.; and Auer, L. H. 1973. Hα fluorescence in the filaments of M 82. *Astrophys. J.* 179:85–88.

Visvanathan, N. 1966. Interstellar polarization in the Large Magellanic Cloud. *Mon. Not. R. Astr. Soc.* 132:423–432.

———. 1967. Polarization observations in the radio galaxy 3C 371 and the x-ray source Cyg X-2. *Astrophys. J.* 150:L149–L151.

———. 1968. Optical polarization in quasi-stellar sources. *Astrophys. J.* 153:L19–L22.

———. 1969. The continuum of BL Lacerta. *Astrophys. J.* 155:L133–L137.

———. 1971. Photometry and polarimetry of the Crab pulsar. *The Crab Nebula*. I.A.U. Symp. No. 46 (R. D. Davis and F. G. Smith, eds.), pp. 152–156. Dordrecht, Holland: D Reidel Publ. Co.

———. 1973. Long term variations of total and polarized fluxes, absolute energy distribution and line strength of BL Lacerta and four quasi-stellar sources. *Astrophys. J.* 179:1–20.

Visvanathan, N., and Elliot, J. L. 1973. Variations of the radio source OJ 287 at optical wavelengths. *Astrophys. J.* 179:721–730.

Visvanathan, N., and Oke, J. B. 1968. Nonthermal component in the continuum of NGC 1068. *Astrophys. J.* 152:L165–L168.

Visvanathan, N., and Sandage, A. 1972. Linear polarization of the Hα emission line in the halo of M 82 and the radiation mechanism of the filaments. *Astrophys. J.* 176:57–74.

Walker, M. F. 1968. Studies of extragalactic nebulae. V. Motions in the Seyfert galaxy NGC 1068. *Astrophys. J.* 151:71–97.

Wampler, E. J. 1971. Photoelectric spectrophotometry of Seyfert galaxies. *Astrophys. J.* 164:1–9.

Westerlund, B. E., and Wall, J. V. 1969. Three color photometry of southern QSO's, radio galaxies and normal galaxies. *Astron. J.* 74:335–351.

Glossary

List of Contributors

Index

GLOSSARY[1]

Å 10^{-10} meter.

α orbital phase angle, which is the angle (negative before opposition as seen from the earth) at the object between the radius vectors to the sun and the observer.

A.U.   astronomical unit, the semi-major axis of the earth's orbit, 1.5×10^8 km.

i $\sqrt{-1}$.

I intensity, the first Stokes parameter; $I_\perp$ stands for the intensity in the perpendicular direction and $I_\parallel$ in the parallel direction to the plane of scattering (see θ_r, below).

IAU   International Astronomical Union.

λ wavelength of the light.

m refractive index, $n - in'$.

mag   astronomical magnitude, which is $-2.5 \log_{10} I/I_0$.

μm   1 μm = 1 micrometer = 1 micron.

nm   nanometer, 10^{-9} meter.

p the degree of linear polarization, $p = \dfrac{(Q^2 + U^2)^{1/2}}{I}$.

---

[1] The symbols of this Glossary are commonly used, unless defined differently by individual authors. The definitions of positive circular polarization, and a few other definitions, are described in the discussion following the paper by Clarke (see p. 52).

GLOSSARY

P	the degree of polarization when referred to the plane of scattering (see θ_r, below): $P = \dfrac{I_\perp - I_\parallel}{I_\perp + I_\parallel}$.
$\varphi - \varphi_0$	azimuth difference.
P_{norm}	normalized polarization.
q	degree of circular polarization: $q = V/I$.
Q	the second Stokes parameter.
r	radius of particle.
rms	root mean square.
σ	standard deviation.
τ	optical depth.
θ	position angle, of the direction of vibration as defined by the electric-vector maximum, measured from north through east, etc.
θ_r	position angle for reflecting objects with respect to a normal to the plane of scattering (which is the plane through source, object, and observer); the sense is the same as for θ, counterclockwise as seen by the observer.
ϑ	angle of emergence, with respect to the local vertical.
ϑ_0	angle of incidence, with respect to the local vertical.
Θ	scattering angle; $\Theta = 180° - \alpha$.
U	the third Stokes parameter.
$UBVRI$	the photometric system described by H. L. Johnson (see Iriarte et al. *Sky and Tel.* 30: 21–31, 1965).
v	linear velocity.
V	the fourth Stokes parameter.
ω	angular frequency.
x	$2\pi r/\lambda$.

LIST OF CONTRIBUTORS

S. Aiello, Univ. di Firenze, Italy
L. H. Aller, Univ. of Calif., Los Angeles, Calif.
J. R. P. Angel, Columbia Univ., New York, N.Y.
J. R. Auman, Univ. of British Columbia, Vancouver, Canada
S. Bashkin, Univ. of Ariz., Tucson, Ariz.
T. Baur, High Altitude Obs., Boulder, Colo.
R. Beebe, New Mexico State Univ., Las Cruces, N. Mex.
B. J. Bok, Univ. of Ariz., Tucson, Ariz.
E. Bowell, Obs. de Paris, Meudon, France
J. C. Brandt, Goddard Space Flight Ctr., Greenbelt, Md.
J. B. Breckinridge, Kitt Peak Nat. Obs., Tucson, Ariz.
M. Breger, State Univ. of New York, Stonybrook, N.Y.
P. F. Buerger, Ohio State Univ., Columbus, Ohio
O. I. Bugaenko, Main Astron. Obs., Kiev, U.S.S.R.
D. J. Caldwell, Univ. of Utah, Salt Lake City, Utah
R. W. Capps, Univ. of Ariz., Tucson, Ariz.
L. Carrasco, Univ. of Calif., Berkeley, Calif.
B. Carroll, Rutgers State Univ., Newark, N.J.
C. R. Chapman, Planetary Sci. Inst., Tucson, Ariz.
H. S. Chen, Univ. of Calif., Los Angeles, Calif.
D. Clarke, The University, Glasgow, U.K.
W. J. Cocke, Univ. of Ariz., Tucson, Ariz.
D. L. Coffeen, Univ. of Ariz., Tucson, Ariz.
D. S. Colburn, NASA Ames Res. Center, Moffett Field, Calif.
G. W. Collins, II, Ohio State Univ., Columbus, Ohio
R. G. Conway, Nuffield Radio Astron. Lab, Jodrell Bank, U.K.
K. L. Coulson, Univ. of Calif., Davis, Calif.
G. V. Coyne, S. J., Vatican Obs., Roma, Italy
D. P. Cruikshank, Univ. of Hawaii, Honolulu.
G. W. Curtis, High Altitude Obs., Boulder, Colo.
G. Dall'Oglio, Instituti di Fisica, Firenze, Italy
J. R. Dickell, Univ. of Illinois, Urbana, Ill.
A. Dollfus, Obs. de Paris, Meudon, France
L. R. Doose, Univ. of Ariz., Tucson, Ariz.
M. Dubin, NASA Headquarters, Washington, D.C.

H. M. Dyck, Kitt Peak National Obs., Tucson, Ariz.
H. Eyring, Univ. of Utah, Salt Lake City, Utah
G. G. Fahlman, Univ. of British Columbia, Vancouver, Canada
G. Feix, Ruhr Universität, Bochum, W. Germany
J. E. Felten, Univ. of Ariz., Tucson, Ariz.
U. Fink, Univ. of Ariz., Tucson, Ariz.
F. F. Forbes, Univ. of Ariz., Tucson, Ariz.
J. W. Fountain, Univ. of Ariz., Tucson, Ariz.
A. L. Fymat, Jet Propulsion Lab., Pasadena, Calif.
T. Gehrels, Univ. of Ariz., Tucson, Ariz.
R. H. Giese, Ruhr Universität, Bochum, W. Germany
J. W. Glaspey, Univ. of British Columbia, Vancouver, Canada
A. F. H. Goetz, Jet Propulsion Lab., Pasadena, Calif.
J. M. Greenberg, Dudley Obs., Albany, N.Y.
J. S. Hall, Lowell Obs., Flagstaff, Ariz.
R. Hall, Northern Ariz. Univ., Flagstaff, Ariz.
D. Hannemann, Ruhr Universität, Bochum, W. Germany
M. S. Hanner, Dudley Obs., Albany, N.Y.
J. E. Hansen, NASA Inst. for Space Studies, New York, N.Y.
B. W. Hapke, Univ. of Pittsburgh, Pittsburgh, Pa.
M. Harwit, Cornell Univ., Ithaca, N.Y.
S. S. Hong, State Univ. of New York, Albany, New York
H. Hull, High Altitude Obs., Boulder, Colo.
O. Jensen, Univ. of British Columbia, Vancouver, Canada
A. R. Kassander, Univ. of Ariz., Tucson, Ariz.
J. C. Kemp, Univ. of Hawaii, Honolulu.
C. E. KenKnight, Univ. of Ariz., Tucson, Ariz.
K. E. Kissell, Wright-Patterson AFB, Ohio
A. Kruszewski, Polish Academy of Sci., Warsaw, Poland
R. Landau, Univ. of Calif., Berkeley, Calif.
J. D. Landstreet, Univ. of Western Ontario, London, Canada
H. H. Lane, National Science Foundation, Washington, D.C.
C. Leinert, Landessternwarte, Heidelberg, W. Germany.
J.-L. Leroy, Pic du Midi Obs., Bagnères-de-Bigorre, France
H. Link, Landessternwarte, Heidelberg, W. Germany
C. H. Liu, Univ. of Ariz., Tucson, Ariz.
P. G. Martin, Univ. of Toronto, Toronto, Canada
M. S. Matthews, Univ. of Ariz., Tucson, Ariz.
I. S. McLean, The University, Glasgow, Scotland
J. Meeus, Heuvestraat 31, Erps-Kwerps, Belgium
B. Melchiorri, Instituti di Fisica, Firenze, Italy
F. Melchiorri, Instituti di Fisica, Firenze, Italy
F. Mencaraglia, Univ. di Firenze, Firenze, Italy
D. L. Mickey, Univ. of Hawaii, Honolulu.
D. K. Milne, Univ. of Illinois, Urbana, Ill.
A. V. Morozhenko, Main Astron. Obs., Kiev, U.S.S.R.
S. Mukai, Kyoto Univ., Kyoto, Japan

LIST OF CONTRIBUTORS

V. Natale, Instituti di Fisica, Firenze, Italy
R. Novick, Columbia Univ., New York, N.Y.
R. F. O'Connell, Louisiana State Univ., Baton Rouge, La.
F. Q. Orrall, Univ. of Hawaii, Honolulu.
A. G. Pacholczyk, Univ. of Ariz., Tucson, Ariz.
C. E. Pieters, Mass. Inst. of Tech., Cambridge, Mass.
E. Pitz, Landessternwarte, Heidelberg, W. Germany
C. W. Querfeld, High Altitude Obs., Boulder, Colo.
C. R. N. Rao, Univ. of Calif., Los Angeles, Calif.
E. Raschke, Ruhr Universität, Bochum, W. Germany
G. Ratier, Pic du Midi Obs., Bagnères-de-Bigorre, France
D. Rea, Jet Propulsion Lab., Pasadena, Calif.
L. A. Riley, Lowell Obs., Flagstaff, Ariz.
F. E. Roach, Rutgers State Univ., Newark, N.J.
J. R. Roach, Ball Brothers Res. Corp., Boulder, Colo.
J. Rush, High Altitude Obs., Boulder, Colo.
M. T. Sandford, Los Alamos Sci. Lab., Los Alamos, N. Mex.
W. Schlosser, Ruhr Universität, Bochum, W. Germany
T. Schmidt, Max Planck Institut, Heidelberg, W. Germany
Th. Schmidt-Kaler, Ruhr Universität, Bochum, W. Germany
K. Serkowski, Univ. of Ariz., Tucson, Ariz.
S. J. Shawl, Univ. of Kansas, Lawrence, Kan.
C. P. Sonett, Univ. of Ariz., Tucson, Ariz.
K. M. Strom, Kitt Peak Nat. Obs., Tucson, Ariz.
S. E. Strom, Kitt Peak Nat. Obs., Tucson, Ariz.
T. L. Swihart, Univ. of Ariz., Tucson, Ariz.
W. Swindell, Univ. of Ariz., Tucson, Ariz.
T. Takashima, Univ. of Calif., Los Angeles, Calif.
E. Tandberg-Hanssen, High Altitude Obs., Boulder, Colo.
J. Tinbergen, Sterrewacht te Leiden, Netherlands
M. G. Tomasko, Univ. of Ariz., Tucson, Ariz.
G. L. Tyler, Stanford Univ., Palo Alto, Calif.
J. E. Tyler, Scripps Inst. of Oceanography, La Jolla, Calif.
S. Ueno, Univ. of Southern Calif., Los Angeles, Calif.
G. L. Verschuur, Nat. Radio Astron. Obs., Green Bank, W. Va.
N. Visvanathan, Hale Observatories, Pasadena, Calif.
G. A. H. Walker, Univ. of British Columbia, Vancouver, Canada
A. P. Wang, Ariz. State Univ., Tempe, Ariz.
T. H. Waterman, Yale Univ., New Haven, Conn.
A. B. Weaver, Univ. of Ariz., Tucson, Ariz.
J. L. Weinberg, Dudley Obs., Albany, N.Y.
R. Weymann, Univ. of Ariz., Tucson, Ariz.
G. Witt, Meteorologiska Institut., Stockholm, Sweden
R. D. Wolstencroft, Univ. of Hawaii, Honolulu.
E. G. Yanovitskii, Main Astron. Obs., Kiev, U.S.S.R.
B. Zellner, Univ. of Ariz., Tucson, Ariz.
R. Zerull, Ruhr Universität, Bochum, W. Germany

INDEX

The names of authors in this volume are listed in *italic* type in the index, and the page numbers of their contributions are in **boldface** type.

Abhyankar, K. D., 524, 541, 574, 576, 583, 589, 618, 627, 634, 635, 638, 643, 647
Absorption
 bands, 37, 69, 76, 77, 82, 84, 235, 331, 337, 348, 405–18, 525, 540, 558, 573, 638–40, 645, 655, 836, 916–25, 983
 coefficients, 346, 497
 function, 68
 general, 64, 66, 67, 76, 237, 238, 327, 328, 332–34, 372, 637, 741
 line spectrum, 521, 524, 558, 559, 617, 649, 741
 Lorentz, 651
 modes, 523
 parameters, 72, 73
 telluric, 652
 unsaturated, 647
 wings (edges), 984, 985
Abu-Shumays, I. K., 583, 589
Accuracy versus precision, 13, 15, 135, 137, 138, 142, 160, 164, 165, 167, 168, 204, 209–11, 224, 230, 233–35, 237, 239, 245, 249–52, 267, 268, 288, 303, 306, 307, 310, 321, 406, 474, 515, 519, 543, 598, 695, 699, 700, 705, 706, 712, 726, 753, 756, 769, 868, 1079
Adams, C. N., 541, 554, 559, 578, 603, 605
Adams, J. B., 386, 402, 405, 406, 410, 417, 418
Adams, T. F., 265, 267, 314
Adler, H. E., 475, 492
Adolph, K. W., 337, 349
Aeropol, 196, 204, 206, 211, 549, 565.
 See also Airborne instruments

Aerosols
 absorption and emission, 618
 concentrations, 562, 563, 566, 567, 763
 effects, 457, 462, 618
 general, 11, 12, 29, 30, 445, 500, 502–04, 525, 562–64
 layer, 500, 502–08, 614
 particles, 462, 474, 599, 608, 764
 properties, 461, 462, 506, 541, 608, 611, 613, 616–18, 645, 762, 764
 scattering. *See under* Scattering
 size, shape, 518
 type, 506
Aiello, S., 11, 22, 35, 38, 193, **322–51**, 1087
Airborne instruments and participating aircraft lists, 22, 189, 194–97, 204, 206, 211, 213, 549, 564, 565, 653, 656
Airglow
 afterglow continuum, 793
 continuum, 780, 781, 784, 793
 emission, 199, 515, 769, 780, 781, 784
 general, 197, 199, 213, 768, 775, 788
Air-water interface, 434
Albedo
 Bond, 383, 392, 556
 bulk, 381, 399, 403
 diffuse, 372, 375
 general, 227, 229, 330, 384–87, 391, 392, 396–99, 452, 504, 523, 556, 585, 587
 geometric, 381–83, 393–95, 398, 399, 702
 -polarization relationship, 224, 227, 229, 384–88, 392–98, 403
 single scattering, 542, 543, 637, 651, 652
 -slope law, 390, 391, 395, 396, 400, 403, 404

Albedo, *cont.*
  surface, 457, 505–08
Alexander, M. A., 807, 813
Alfvén, H., 33
Alfvén fields, waves, 423, 425, 426, 430, 432
Algol system, 849
Alguard, M. J., 103, 106
Alignment
  Davis-Greenstein mechanism, 895
  Gold mechanism, 774
  grains, 33, 59, 60, 334–36, 774–77, 813, 816, 836, 839, 864, 888–96, 900, 926–39, 943, 948, 954, 958, 976
  magnetic, 774, 954, 1020
  mechanisms, 8, 36, 38, 88, 91, 97, 98, 105, 888
  patterns, 889, 890
  "picket fence" model, 36, 896, 897, 923, 929
  random orientation, 917, 920, 927
  twisted grain, 890, 892, 900
Allen, D. A., 393, 400
Aller, H. D., 266, 314, 1051, 1056
Aller, L. H., 21, 34, **794–803**, 809, 813, 1087
Almucanter, 456
Altschuler, M. D., 254, 260
Ambarzumian, V. A., 582, 589
Amplifier, 700, 705, 756, 757, 762, 763
Analyzers
  channels, 479, 485–91
  dichroic, 472, 473, 476–81, 489–91
  electro-optic, 985
  fixed orientation, 138, 139, 141
  general, 8, 20, 135, 138, 142, 147, 150, 157, 190, 192, 194, 196, 198, 200, 202, 701
  Glen-Foucault, 881
  HR-sheet, 649
  perfect, 10, 144
  polarization, 476, 477, 479
  wire grid, 19, 207
Andrä, H. J., 88, 93, 101, 103, 105
Andrew, B. H., 1023, 1025, 1076, 1081
Angel, J. R. P., xv, 11, 15, 33, 35, 37, 39, 40, 48, 51, **54–63**, 157, 171, 175, 186, 256, 266–68, 281, 294, 314–17, 334, 349, 445, 463, 684, 926, 930, 932, 938, 941, 942, 982–87, 990, 991, 1023, 1026, 1079, 1081, 1087
Animal polarimeters. *See under* Polarimeters
Anisotropic
  oscillator molecules, 83, 84
  parameters, 627
  particles. *See under* Particles
  velocity distribution, 747
Antennas
  general, 431, 676, 679, 681, 684
  helical, 354
  orthogonal, 360, 363–65
Anti-matter, 40, 86
Antisolar point, antisun, 5, 444, 445, 454, 455, 460, 461, 473, 772, 774, 780, 794–96, 799, 801, 802
Apodization, 640
Appenzeller, I., 155, 169, 171, 845–47, 855, 963, 970, 1072, 1080
Arago, D. F. J., 5, 445, 450, 463, 814, 816
Arago point (also double Arago point), 444, 450, 453–55, 457, 458, 460, 461
Archard, J. F., 803
Archimedean spiral, 420
Arctic "white-out," 438
Arking, A., 554, 555, 557, 559, 560, 574, 577, 603, 605, 631, 636
Arnett, D. W., 490, 493
Arnquist, W. N., 201, 217
Arons, J., 55, 61, 932, 937, 1023, 1025
Aronson, R., 541, 574
Arp, H. C., 1015
Arsenijevic, J., 157, 173
Artamonov, B. P., 868, 870, 878
Arthropods, 472, 473, 477, 490
Arvesen, J. C., 197, 212, 215
Asaad, A. S., 786, 791
Ashburn, E. V., 504, 509
Asteroid
  diameters, 26
  Mars-crossing, 392, 399
  polarimetry, polarization, 24, 26, 30, 34, 318, 388–99, 612
Astigmatism, 357
Asymmetric molecules. *See under* Molecules
Asymmetric radiance, 267, 421, 422, 424, 438, 439
Atmosphere
  aerosol. *See* Aerosols
  CO_2, 525, 526, 558
  earth. *See* Terrestrial atmosphere
  extinction, 784, 788
  fluctuations, perturbations, 762–64
  gray (also nongray), 663–75
  illuminated, nonilluminated, 663, 669–73, 675
  inhomogeneous, 541, 587
  ionized, 672, 673
  layers, 518, 520, 647
  Mie, 630
  models, 502, 505, 508, 554–56, 622
  molecules, 446–62, 564
  optical thick, 382, 447, 450, 455, 504, 505, 518, 574
  optical thin, 382
  particulate, 444, 460, 526, 564
  planetary (thick). *See under* various planet

INDEX 1093

listings, and *under* Planetary
planetary turbidity, 453, 457, 500–13
polarization, 472–75
pure gas, 613
Rayleigh, 448–62, 522, 553, 583, 633
scattering, nonscattering, 622, 781, 784.
    See also Scattering
seeing, scintillation, 137–40, 152, 160, 701, 708–10
semi-finite, 523
stellar, 510, 642, 666–68, 670–74, 889
structure, 550, 667–69, 672
studies, 546
top, 544
transparent, 457, 519, 756, 757
Atmosphereless bodies, 381, 382, 398
Atomic
    energy levels, 405, 409, 730, 738, 744
    oscillator, 735–37
    resonance lines, 59
    transitions (permitted, forbidden), 730, 738, 739, 742, 746
Atoms
    decay, 59
    excited state, 59
    general, 65, 66, 88, 93, 96, 103, 105, 993
Auer, L. H., 1067, 1082
Auman, J. R., 16, **237–45**, 1087
Aumann, H. H., 330, 342, 349, 570, 574, 863, 866
"Aureole," 762
Aurière, M., 552
Auton, J. P., 338, 348
Axford, W. I., 1053, 1056

Baade, W., 1015, 1016, 1018, 1025, 1063, 1065, 1080
Babcock, H. D., 786, 791
Babcock, H. W., 6, 156, 171, 237, 245, 839, 840, 985, 986
Babinet, J., 445
Babinet point, 29, 444, 445, 453–55, 457, 458, 460, 461
Backscatter radiation. *See under* Radiation
Bahcall, J. N., 264, 314
Bahcall, N. A., 264, 314
Baker, A. L., 552, 553, 605
Balamore, D. S., 1069, 1082
Baldwin, J. E., 1019, 1021, 1023, 1025
Balick, B., 265, 315
Ballard, S. S., 4, 27, 42
Balloon observations, 22, 189–91, 205, 210, 213, 307, 322, 342–44, 348, 563, 564, 641. *See also* Polariscope balloon system
Balmer
    absorption edges, 668

emission, 828
    jump, 672, 673, 851
    lines, 752, 985, 989, 990
    series, 753, 846, 982
324 Bamberga, 390, 393, 397
Band (rotational) structures of CO_2, 637
Bandermann, L. W., 498, 499, 609, 613, 614, 776–78
Bangs, L., 126, 133
Barbier, D., 29, 40
Barkla, C. G., 291, 314
Barlett, J. R., 833, 841
Barnard, E., 392, 393, 400
Barrett, J. W., 230
Barrick, D. E., 363, 366
Barth, C. A., 599, 605, 618, 635
Bartholinus, E., 5, 43
Bary, E. de, 448, 468, 500, 504, 508, 523, 574
Bashkin, S., 11, **88–106**, 1087
Basu, S., 526, 577
Batterman, B. W., 278, 314
Baum, W. A., 779
Baur, T., 15, 32, **246–53**, 1087
Bazilevskii, K. K., 197, 201, 213, 564, 574
Beam-foil spectroscopy, 88–106
Beaming mechanism (relativistic), 997, 1009
Beamsplitter transmission, 151, 238, 257, 319, 320, 640, 651, 686
Beardsley, G. F., 442
Beauchemin, G., 103, 105
Beaver, E. A., 136, 164, 171
Becker, W., 963, 970
Beckers, J. M., 148, 159, 160, 165, 171, 179, 186, 247, 253, 254, 260
Becklin, E. E., 832, 833, 840, 843, 1072, 1077, 1081
Beckmann, P., 446, 468, 549, 574
Beebe, R., 15, **218–22**, 880, 1087
Beer, R., 642, 659
Bees, 474
Beeson, D., 199, 216
Beetles, 473, 495, 496
Beggs, D. W., 787, 791
Behr, A., 9, 17, 18, 33, 40, 155, 171, 786, 791, 845, 846, 855, 975, 976, 980
Belcher, J. W., 430, 432
Belleville, R. E., 475, 492
Bellman, R., 582, 583, 585, 589
Belton, M. J. S., 619, 620, 635
Benedict, W. S., 642, 659
Bennett, H. E., 446, 468, 765
Bennett, J. M., 179, 186, 765
Berge, G. L., 1023, 1025, 1055, 1056
Bergh, S. van den, 868, 878
Berkeley infrared polarization survey, 318–21
Berkhuijsen, E. M., 966

Bernard, G. D., 475, 491, 493
Berry, H. G., 88, 92, 101, 103, 105, 106
Berthelsdorf, R., 55, 62, 267, 280, 294, 303, 316, 317, 1021, 1026
Bertotti, B., 1004, 1011
Bessel functions (spherical), 115, 161, 248, 423
Bessell, M.S., 851, 855
Beychok, S., 473, 492
Biaud, F., 326, 348
Bickel, W. S., 93, 101, 103, 105, 106
Biggerstaff, J. A., 93, 106
Bijvoet, J. M., 65, 87
Billings, B. H., 141, 152, 160, 171, 183, 184, 186
Binary systems. *See under* Stars
Biot, J. B., 5
Biraud, F., 1055, 1058, 1077, 1081
Bird, G. R., 337, 338, 348
Birefringence (electrical, magnetic), 6, 79, 81, 204, 888–91, 926–28, 940, 941, 943, 944, 981
Birefringent
  component, 435
  filter, 712
  material, 138, 141, 147–49, 160, 187, 858
Blackbody (emission, radiation), 323, 332, 335, 585, 1071
Black hole, 265
Blackwell, D. E., 195, 213, 781, 783, 786, 788, 791
Blake, G. M., 1077, 1080
Blake, R. P., 337, 348
Blandford, R. D., 55, 61, 932, 937, 1023, 1025
Blank, J. L., 421, 432
Blättner, W. G., 541, 575
Blau, H. H., 549, 574
Bless, R. C., 37, 41
Blumenthal, G. R., 546, 577, 1056
Bohn, H., 475, 492
Bohr magneton, 98
Bok, B. J., 974, 1087
Boldt, E. A., 300, 314
Bolometer, 319, 340, 342
Bolton, C. T., 264, 314
Bolton, J. G., 1077, 1080
Boltzmann constant, 624
Bond, G. R., 524, 574
Bondasenko, L. N., 562, 574
Borghesi, A., 332, 348
Borgogno, V., 260
Born, G. H., 391, 402
Born, M., 48, 49, 51, 360, 366, 391, 402, 510, 513, 639, 659, 676, 684, 732, 748
Borra, E. F., 985, 986, 990
Borrmann effect, 276–80
Bouligand, Y., 496, 499

Bouricius, G. M., 464, 468
Bousquet, J., 232, 236
Bowell, E., 11, 13, 25, 28, **381–404**, 464, 1087
Bow shock wave, 420, 432
Bowyer, C. S., 263, 314, 1005, 1010
Bracewell, R., 679, 684
Bradt, H., 1002, 1010
Bragg
  angle, 277, 281, 284
  crystal, 262, 272, 276, 277, 281
  energy, 278
  equation, 277
  polarimeters, 280–91
Brandt, J. C., 33, **768–80**, 787, 793, 1087
Brault, J. W. 242, 245
Braun, B., 504, 508
Brecher, K., 264, 314, 1051, 1056
Breckinridge, J. B., 136, 171, 185, 186, **232–36**, 1087
Breda, V. G. van, 769, 779
Breger, M., 37, 855, **946–53**, 1087
Breit, G., 90, 105, 736, 737, 748
Bremsstrahlung
  linear, 262, 270, 271, 287
  thermal, 57, 58, 324, 1034
Brewster
  angle, 24, 337, 384, 434, 444
  law, 5
  point, 5, 444, 445, 453–55, 458
Brewster, D., 5, 445, 456, 463, 468
Bridges, T. S., 337, 348
Brinkman, A. C., 265, 316
Brockelman, R. A., 363, 366
Broglia, P., 140, 173
Bromander, J., 88, 92, 101, 103, 106
Brouw, W. N., 354, 358
Brown, G. H., 496, 499
Brown, R. L., 969, 970
Browne, I. W. A., 1077, 1080
Browning, S. R., 500, 509, 541, 577, 618, 636
Brückner, G., 747, 748
Brun, A., 951, 952
Bryant, C. H., 533, 539, 574, 576
Buchholz, V., 239, 245
Buckingham, A. D., 77, 87
Buell, J., 583, 589
Buerger, P. F., 31, **663–75**, 1087
Bugaenko, O. I., 30, 157, 171, 573, 574, **599–606**, 1087
Bullrich, K., 448, 457, 468, 500, 504, 508, 523, 564, 574, 575
Burbidge, E. M., 136, 164, 171, 1061, 1077, 1080, 1082
Burbidge, G. R., 323, 328, 348, 1061, 1080
Burke, B. F., 55, 61
Burn, B. J., 1015, 1018, 1019, 1026

Burnett, G. B., 201, 213
Burnight, T. R., 263, 314
Burns, R. G., 406, 418
Burton, C. E., 786
Busbridge, I. W., 582, 583, 589
Byram, E. T., 263, 295, 315

Caldwell, D. J., 12, 21, 39, 40, **64–87**, 1087
Calibration
  in-flight, 207, 209-11
  instrument, 18–20, 135, 209, 355, 357, 406, 410, 418, 512, 699–702, 716, 723, 753, 756, 770, 796, 869
  polarization-albedo relationship, 24, 393–97
  spectra, 652
Callisto, 391, 392, 397
Campbell, C., 230
Campbell, D. B., 1000, 1001, 1011
Campbell, M. J., 363, 367
Cann, M. W. P., 568, 575
Canuto, V., 1004, 1011
Capps, R. W., 58, 61, 342, 348, 825, 827, 833, 839, 840, 851, 855, 859, 861, 862, 864, 865, 1087
Carleton, N. P., 1002, 1006, 1012
Carli, B., 341, 348
Carpenter, M. S., 195, 215
Carrasco, L., 37, 39, 860, 865, 937, 948, 952, **954–59,** 1087
Carre, M., 101, 106
Carroll, B., 21, 34, 790, **794–803,** 1087
Carswell, R., 1077, 1078, 1082
Cassinelli, J. P., 668, 674
Cassini, J. D., 782, 791
Casti, J., 583, 589
Castor, J. I., 1067, 1082
Catalogues and lists. *See also* Star lists
  Behr, 18
  Hiltner, 16
  polarized starlight, 7
  standard stars, 20
Catchings, F. E., 541, 578, 580
Cauchy system, 582–86, 588
Cavalieri, A., 1004, 1010
Cederblad, S., 868, 878
Cephalopods, 472, 473, 477, 489, 490
1 Ceres, 26, 390, 393, 395, 397, 400
Chaffee, F. H., 958, 959
Chaldu, R. S., 219, 222
Chamberlain, J. W., 568, 575, 619, 620, 635
Chambers, F. W., 278, 279, 314
Chandrasekhar, S., 7, 8, 30, 31, 45, 51, 58, 59, 61, 179, 186, 445, 447, 448, 451, 454, 468, 481, 492, 496, 499, 503, 504, 508, 520, 522, 523, 541, 562, 575, 582–84, 589, 618, 635, 637, 659, 663, 668, 669, 674, 733, 748, 752, 760, 1059, 1080
Chandrasekhar's planetary problem, 29, 30, 582–92
Chandrasekhar's theory, 562
Chapman, C. R., 403, 1087
Chapurskii, L. I., 195, 214
Charette, M. P., 405, 417, 418
Charvin, P., 255, 259, 260, 731, 746–48
Chaumugam, G., 984, 986
Chemical composition. *See under* Particles
Chen, H. S., 29, 446, 466, 468, 470, **500–09,** 1087
Cherenkov radiation, 204, 1004, 1020
Cheung, A. C., 833, 842
Chevalier, R. A., 260
Chirality
  biological systems, 85, 86
  nuclear, 86, 87
Chiu, G., 526, 577
Chiu, H.-Y., 1004, 1011
Chlorophyll *a* and *b,* 495, 497
Choisser, J. P., 136, 164, 172
Chojnacki, W., 60, 63, 138, 155, 173, 981, 987, 988, 991
Cholesteric liquid crystals, 495, 496
Christie, A. D., 517
Chromophores, 473, 477
Chromosome structure, 496
Chubb, T. A., 295, 315
Church, D. A., 98, 101, 105, 106
Circle, R. R., 807, 813
Circular dichroism. *See under* Dichroism
Circular polarization
  animals, insects, 473, 481
  antisymmetry, 777, 780
  general, 4, 5, 7, 9, 11, 17, 21, 27, 31, 33–35, 37, 40, 48, 49, 51–54, 64, 65, 67, 69, 71–73, 99, 147, 174, 176, 189, 237, 238, 250, 264, 266, 268, 270, 326, 334, 352, 356, 358, 362, 363, 433, 495–99, 510, 521, 522, 530, 563, 568, 607–16, 681, 686, 688, 690, 691, 693, 705–07, 738, 739, 768, 775–79, 835, 840, 862–65, 888, 890, 892, 896, 900, 915, 926–45, 990, 998, 1000–02, 1004, 1006, 1014, 1023, 1030, 1035, 1037–40, 1042, 1043, 1054–56, 1079, 1084, 1085
  handedness, clockwise and counterclockwise. *See* Handedness
  magnetic, 981–87
  mechanism, 54–63, 615, 616
  night sky, 788
  partial, 9
  percentage amount, 56, 57, 552, 1084. *See also* amount *under* Polarization
  radio, 59

Circular polarization, *cont.*
  reflected, 498
  sign reversal, 41, 776, 933, 1085. *See also* Sign conventions
  spectral dependence, 993
  transmitted, 498
  wavelength dependence. *See under* Polarization
Circumstellar
  particles, 33–36
  polarization, 35, 889
  shells, extended envelopes, 31, 33, 35, 37, 453, 946–52, 959
Ciurla, T., 1073, 1074, 1081
Clapham, P. B., 179, 186
Clarke, D., 4, 7, 10, 12, 13, 15, 23, 32, 34, 41, 43, 44, **45–53,** 135, 136, 155, 171, 174, 179, 186, 353, 481, 492, 638, **752–61,** 780, **814–17,** 1085, 1087
Clarke, M. E., 1080
Clarke, T. R., 1055, 1058
Clouds (terrestrial and planetary)
  chamber, 548
  cirrus, 548, 562, 569
  convective, 558
  cumulus, 537, 543, 562
  dense, 526, 543, 560
  general, 191, 195, 197, 201
  H I, 969
  high-velocity, 969
  ice, 510, 548
  layer, 518, 616
  particle type, 526, 533, 540, 553, 557–60, 562, 564, 573
  stratus, 535, 537
  studies, 564
  thin, 138, 160
  top, 526, 544, 552, 553, 556, 558, 560
  water. *See under* Water
Clusters
  dust. *See* intracluster *under* Dust
  Scorpio-Centaurus, 890
  Stock 2, 890, 961
  young, 963, 970
Cocke, W. J., 39, **997–1013,** 1015, 1024–26, 1087
Code, A. D., 663, 668, 674, 770, 778
Coffeen, D. L., xv, 10, 12–14, 16, 21, 22, 27, 30, 31, 35, 40, 154, 187, **189–217,** 218, 221, 222, 227, 230, 380, 383, 386, 387, 389, 393, 401, 463, 468, **518–81,** 605, 615, 616, **818–20,** 1087
Cohen, J. G., 958, 959
Cohen, L. C., 741, 749
Cohen, M. H., 41, 51, 327, 348
Coherence
  area, 676, 677, 683
  degree, 677, 679
  effect, 730, 742, 743, 745
  emission, 677, 678
  general, 11, 23, 88–104, 742, 1009
  matrix, 360, 364, 366, 677, 679
  radio, 676–85
Coherent excitation, 88, 90, 91, 742
Colburn, D. S., **419–33,** 1087
Cole, H., 278, 279, 314
Collett, E., 749
Collins, D. G., 541, 575
Collins, G. W., 31, 513, **663–75,** 844, 879, 1087
Collisional depolarization, 730, 731, 742, 745
Collisional effects, 731, 747, 844
Collisions (beam-foil), 88–90, 97, 105
Color (intrinsic), 868, 869, 872, 874
Colorimetry, 867
Comet
  Bennett, 815, 816, 818–20
  color index, 818–20
  coma, 35, 818–20
  dust, 816
  emission bands, 818
  general, 5, 24, 33–35, 518, 814–17
  heads, 816
  molecules, 814, 816
  nuclei, 400, 814–16
  particle composition, 814
  polarization mechanism, 34, 814–18
  spectrum, 818
  tail, 814–17
Cometary nebula, 823, 878, 946, 951
Computer
  analyses, calculations, xv, 15, 16, 21, 121, 259, 318, 406, 407, 445, 813, 834, 837, 840, 871, 916
  controlled instruments, 251, 252, 256–58, 318, 320
  data processing, 297, 298, 360
  types, 164, 204, 206, 230, 232, 252, 254, 258–60, 319, 320, 412, 504, 515, 520, 547, 582, 686
Condensation
  atmospheric origin, 516
  extraterrestrial, 516
  nuclei, 516
Conklin, F. K., 1077, 1078, 1081
Conn, G. K. T., 340, 349
Connes, J., 638, 642, 659, 660
Connes, P., 638, 642, 659, 660
Connes' resolving power advantage, 641
Conrath, B. J., 618, 635, 642, 659, 660
Conti, P. S., 265, 316
Continuum
  emission, 1014, 1015, 1018, 1020, 1022
  optical, 1014

INDEX 1097

polarization. *See under* Polarization
Contributors (list), 1087–89
Converters (polarization), 179–83
Conway, R. G., 23, 51, 326, 327, 349, 350, **352–58,** 1020, 1023, 1026, 1044, 1048, 1050–52, 1055–57, 1087
Cooke, D. J., 55, 63, 759, 761, 998, 1012, 1055, 1058
Cooper, J., 742, 747, 750
Cordwell, C. S., 974
Corliss, W. R., 201, 217
Cornu, A., 456, 460, 468
Corona. *See under* Solar
Coronagraph
  Mount Haleakala, 686, 687, 690, 692, 693
  Sacramento Peak, 246, 251, 252, 254–56, 259, 260
  various, 695, 710, 719–21, 762, 763
Cosmic gamma rays, 302, 306, 307
Cosmic ray induced background, 300, 302, 303, 306, 307
Cosmic rays, 204, 998, 1019–21
Coulomb field effects, 993
Coulson, K. L., 7, 12, 27, 28, 30, 41, **444–71,** 473, 504, 505, 508, 509, 520, 523, 562, 567, 575, 1087
Counterglow, 34, 780, 794–803, 811
Cox, A. J., 539, 574
Coyne, G. V., xv, 23, 26, 29, 36, 37, 41, 164, 172, 334, 349, 386, 387, 389, 401, 828, 829, 841, 845–47, 851, 854, 855, 860, 861, 864, 865, **888–900,** 928, 934, 937, 1087
Crab Nebula
  continuum polarization, 1018, 1020–22, 1028
  dark bay, 1020, 1061, 1062
  emission, 267, 268, 932, 1018, 1021, 1022
  filaments, 1015, 1018, 1020, 1024, 1034
  general, 7, 37, 39, 40, 926, 928–33, 936, 941, 981, 1063
  loops, 1014, 1018, 1020, 1066
  magnetic field, 268, 1014, 1018–22, 1028
  optical polarization, 317, 1014–16, 1021–23, 1028
  particle acceleration, 1021
  radio polarization, 7, 317, 1014, 1020
  structure, 1014–28
  supernova explosion, 1020–22, 1024
  time scales, 1020, 1021
  X-ray polarization, 317, 1014, 1018, 1021
Crab pulsar
  general, 33, 39, 40, 317, 932, 997–1002, 1005–07, 1009, 1010, 1014, 1017, 1021–28, 1079
  great pulse, 1001
  properties (optical, radio), 1000–02, 1010

rotation axis orientation, 1007, 1010, 1024
time changes, 1015
X-ray, 268, 288, 289, 310, 312, 1002, 1005, 1021
Cram, T., 969, 970
Crayfish, 477, 478, 480
Crossover effect, 240, 933
Crosstalk terms, 249, 250, 353, 357
Cruikshank, D. P., 391, 402, 570, 579, 598, 1087
Cruise, A. M., 275, 316
Crustaceans, 473, 474, 479, 489
Crystals, 281–84, 288, 312
Cudaback, D. D., 331, 349
Cugnon, P., 888, 897
Culhane, J. L., 263, 275, 314, 316
Curran, R. J., 195, 214, 263, 468, 500, 509, 541, 577
Curtis, G. W., 15, 32, **246–53,** 1087
Cygnus A, 356, 1052, 1053

Dahn, C., 868, 872, 878
Dalgarno, A., 324, 325, 349
Dall'Oglio, G., 11, 22, 35, 38, 193, **322–51,** 1087
Damany, H., 179, 186
Damping constant (radiation), 736, 737, 744, 745
Dandekar, B. S., 562, 575
Danforth, H. H., 363, 366
Danger, R. (Etablissements), 697, 710, 723
Dark clouds
  age, 957
  density variation, 955
  distance, 972
  distribution, 972–75
  general, 332–35, 954–59, 972, 975
  Heiles II, 955
  high-density regions, 954, 957, 958
  low-density regions, 958
  Norma, 972, 974
  ρ Oph, 860, 949, 955, 957, 972, 974
  II Per (IC 348), 955
  Scorpius, 957
  structure, 972–75
  Taurus, 955
  See also Nebulae
Dave, J. V., 7, 41, 451, 452, 454, 457, 463, 468, 504, 505, 509, 520, 523, 533, 541, 542, 575, 583, 589
Davidson, A., 264, 314
Davidson, K., 264, 314
Davies, R. D., 357, 358, 1044, 1056
Davies, W. O., 568, 575
Davis, L., 38, 41, 334, 349, 430, 432, 888, 897, 976, 980, 1060, 1080
Davis, M. A., xvi
Davis-Greenstein alignment mechanism, 38, 918, 920–25, 970, 1060

Débarbat, S., 687, 693, 715, 726
Debye, P., 6
Definitions (polarimetric), 12–14, 43–54, 174, 353, 362, 608, 677, 732, 773, 779, 1030, 1085
Deimos, 391, 395, 396
Deirmendjian, D., 457, 460, 470, 504, 509, 521, 575, 631, 635, 652, 660
Delbouille, L., 642, 660
Delouis, H., 642, 659
Delsemme, A. H., 820
De Marrais, G. A., 505, 509
Depolarization. *See* Collisional depolarization *and* depolarization *under* Polarization
Depolarizers
 linear, circular, 26, 183–86, 210, 392
 Lyot, 135, 147, 148, 152–54, 162, 187
 wave, 363
Desai, U. D., 300, 314
Destriau, G., 707, 726
Detectors
 anticoincidence, 302, 305, 308
 cross-correlation, 352, 354
 crossed linear, 353, 354
 general, 284
 germanium photodiodes, 257–59, 762, 763
 multiple, multichannel, 190, 192, 194, 196, 198, 200, 202, 207, 223
 photoelectric, 753, 762
 photoemissive, 520
 total power, 352
 windows, 291
de Vries, H., 495, 496, 499
de Vries theory, 495, 496
Dewar, 165, 257
Dewhirst, D. W., 781, 788, 791
Dextrogyrate, 237, 239, 240, 244
De Young, D. S., 1053, 1056
Diameters
 infrared, 381, 391, 392, 396, 397, 399, 400
 measurements, 391–93, 395–97, 400
 polarimetric, 393, 397, 403
Diaphragms
 dielectric versus metallic, 135, 141–43
 general, 19, 162, 164, 166, 174
Dibay, E. A., 845, 856
Dichroism
 circular, 5, 64, 67, 69, 76, 77, 84, 473, 497, 498
 curve, 64, 80
 general, 5, 152, 182, 251, 480, 940, 941, 943–45
 linear, 147, 889, 926, 927
 magnetic circular, 67, 82–84, 945
 molecular, 472
Dickel, J. R., 39, 317, 900, 917, **1029,** 1087

Dielectric
 interfaces, 362, 363
 materials, 5, 19, 36, 38
 properties, 70, 73, 407
 surface, 47, 371, 372, 377, 378, 384, 612, 616
 tensor. *See under* Tensor
Diem, M., 532, 576
Diffraction
 fringe, 710
 grating polarization, 232–36
 lobe, 808
Digicon image tubes, 16, 136, 142, 164–66
Dilution factor, 775
Dilution of polarization, 768, 769, 818
Dione, 391
Dipoles
 electric, 76, 92, 96, 98
 general, 353
 Hertzian, 281
 magnetic, 76, 96
 oscillating, 10, 73–76, 96, 97, 277, 278, 522, 532
 radiation, radiators, 277, 278, 539
 scattering, 530
 transitions (electric), 746
Disney, M. J., 1002, 1007, 1008, 1011, 1015, 1025, 1026
Dispersion
 behavior, 69, 80
 general, 64, 66–69
 magnetic optical rotary (MORD), 67, 82
 optical rotary (ORD), 67, 69
 parameters, 68, 72, 73, 80
Dissymmetric media. *See* asymmetric *under* Molecules
Divari, N. B., 564, 576, 769, 778, 781, 786, 791
Djerassi, C., 77, 87
Dolan, J. F., 276, 279, 313, 314
Dollfus, A., xv, 6, 9, 14, 15, 17, 22, 26, 32, 156, 158, 161, 172, 191, 204, 205, 214, 218, 221, 250, 253, 382–88, 391–95, 400, 401, 463, 469, 470, 519, 543, 550–52, 555, 557, 564, 566–68, 570, 573, 576, 577, 593, 594, 598, 605, 649, 687–89, 693, **695–729,** 1087
Dombrovskij, V. A., 7, 155, 172, 825, 841, 862, 865
Domke, H., 583, 589, 741, 749
Donahue, T., 514, 516
Donn, B., 807, 812, 836, 841
Doose, L. R., 34, 35, 568, 576, 598, **818–20,** 1087
Doppler
 broadening, 90, 100, 745, 752, 753
 motions, 359, 745
Dorno, C., 456, 469

Dorschner, J., 868, 878
Doschek, G. A., 270
"Doubling" method, 520, 541, 550, 584
Downes, D., 329, 349
Downs, M. J., 179, 186
Drake, C. W., 103, 106
Drake, F. D., 1000, 1001, 1011
Dreiling, L. A., 741, 749
Drinkwine, M., 546, 577
Druetta, M., 100, 104–06
Dubin, M., 813, 819, 1087
Dufay, J., 783, 786, 791
Dufay, M., 101, 104, 106
Duhamel principle, 583
Dumont, R., 769, 778, 781, 787, 788, 791
Dumont, S., 687, 694, 715, 726, 728
Dunkelman, L., 790
Dunlap, J. L., 26, 41, 390
Dunn, R., 260
Dust
  cloud complex, 869, 872, 874, 877, 933, 972, 974, 975
  clouds (geocentric), 783, 801
  cosmic, 813
  distribution, 782, 791, 807
  -gas interaction, 334
  interplanetary, 32, 791, 804, 812, 901, 915
  interstellar, 11, 18, 33, 34, 36–39, 262, 266, 330–37, 397, 398, 719, 767, 768, 774, 791, 804, 807, 840, 867, 888–901, 926–38, 949, 952–59, 974, 976, 1023, 1060, 1079
  intracluster, 949, 950, 952, 953
  particle absorption, 1061–63
  particles, 504, 506, 517, 564, 565, 782, 791, 807, 816, 1061, 1063
  shell, 1071
  size, 36, 791. *See also* Size distribution
  See also Grains, Particles
Duverney, R., 337, 351
Duxbury, T. C., 391, 402
Dyal, P., 419, 433
Dyck, H. M., xv, 22, 23, 35, 37, 58, 61, 157, 172, 318, 321, 342, 348, 825–34, 836, 839, 851, 855, **858–66**, 897, 946, 947, 952, 955, 959, 1078–80, 1087
Dye, J. E., 199, 216

Earth
  atmosphere. *See* Terrestrial atmosphere
  -based observation. *See* ground-based *under* Observations
  gas and dust tail, 34, 783
  global scattering properties, 565
  shine polarization, 26, 566
  surface studies, 3, 6, 191, 195, 213
  twilight, 562

Eastlund, B. J., 1004, 1011
Eather, R. H., 754, 760
Eaton, G. K., 340, 349
Echelle, 165, 167
Echoes
  depolarization, 362–64
  polarized, 362–64, 432
  radar. *See* Radar
  unpolarized, 364
Eddy, J. A., 146–49, 195, 214, 254, 260
Edelson, S., 57, 61
Edwards, H. D., 191, 214
Effantin, C., 642, 660
Efimov, Yu. S., 157, 173, 845, 846, 856, 868, 878
Eggen, O. J., 825, 841, 851, 855
Eguchi, E., 479, 480, 492
Eiden, R., 457, 463, 468, 469, 563, 576
Eigendirections, 939
Eigenfunctions, 91
Einstein coefficients, 744, 745
Ekers, R. D., 998, 1011, 1077, 1080
Elbert, D. D., 448, 451, 454, 468, 504, 508, 523, 575
Electric field, 25, 46, 70, 91–95, 103, 105, 110, 114, 118–20, 277, 278, 360, 423, 523, 998, 1004, 1031, 1059
Electric vector, 10, 13, 24, 25, 29, 32, 33, 43, 46–50, 53, 58, 61, 73, 84, 118, 122, 129, 145, 227, 234, 268, 270, 273, 275, 280, 281, 291, 294, 337–39, 343, 382, 423, 434, 439, 472–77, 479, 480, 482, 485–91, 510, 593, 594, 651, 677, 687, 690, 693, 698, 701, 704, 711, 732, 735, 738, 739, 746, 747, 771, 772, 788, 789, 794, 799, 800, 805, 808, 815, 820, 889, 929, 959, 961, 1003, 1035, 1037, 1059–67, 1069, 1071, 1073, 1077–79, 1085
Electromagnetic
  dissymmetry, 64–87
  field, 10, 77, 78, 332, 522
  induction fields, 419
  radiation, 64–87, 435, 472
  theory, 64, 108, 419, 421, 530
  wave scattering, 58, 108, 110–13, 115–21, 125, 277, 291, 360, 361, 445, 539, 732
Electron
  density, density distribution, 260, 723, 724, 726, 846, 1020
  energetic, 986, 993, 1042
  energy distribution function, 268, 270, 271, 292, 1020
  excitation, 742
  free (around the sun), 695, 719–26
  general, 733, 737, 739, 783, 845, 1020
  mass, 992
  polarized, 280

Electron, *cont.*
  properties, 992
  relativistic, 54, 55, 58, 59, 692, 1020, 1030, 1033, 1038, 1046, 1055, 1056
  temperature, 846, 847
  thermal, 1030, 1046
  scattering, 946
  sub-relativistic, 692
Element depletion, 954, 957, 958
Elements (abundances), 89
Eliasson, B., 833, 841
Elipticity, 13, 47, 52, 60, 72, 73, 476, 511, 513, 733, 1040, 1042
Ellett, A., 734, 751
Elliot, J. L., 1078, 1082
Elliott, W. G., 549, 580
Elliptical polarization (also partial), xv, 13, 14, 19, 29, 46, 47, 49, 50, 60, 61, 72, 144, 151, 372, 432, 433, 441, 462, 510–13, 520, 607, 613, 732, 777, 779, 926, 941, 1004, 1006, 1039, 1044, 1056
Ellipticity, 13, 47, 52, 60, 72, 73, 476, 511, 513, 733, 1040, 1042
Elsässer, H., 781, 791
Elterman, L., 504, 509
Elvius, A., xv, 155, 172, 814–16, 868, 870, 872, 878, 1061, 1062, 1066, 1070
Elwert, G., 270, 314
Emerson, J. P., 195, 214, 255, 260, 746, 747, 749
Emission
  atomic, 642
  Auger, 275
  bands, 638, 818, 820
  fluorescence, 814
  general, 617, 730, 746, 833
  grain (interstellar), 323, 324, 331–36, 348,
  infrared (thermal), 58, 266, 331–36, 348, 392, 545
  mechanism, 262, 266, 267, 322–31, 345, 684, 997, 1004
  nonthermal, 54–56, 262, 265, 268, 352, 692
  optical, 88, 91, 92, 96, 97
  polarized, 23, 54, 254–62, 267, 680, 681, 683, 684, 730–52, 828, 833, 839, 881, 883, 884
  radio. *See under* Radio
  synchrotron, 262, 265–70, 328, 331, 332, 348, 352, 835
  thermal. *See under* Thermal
  variable, 829, 830
  X-ray. *See under* X ray
Emission-line
  broadening, 746
  identification, 833
  polarization, 639, 640, 645
  solar, 352, 354, 730
Emmons, G. H., 337, 349
Energy levels. *See under* Atomic
Engberg, M., 155, 172
Epps, H. W., 136, 164, 171
Epstein, G., 280, 317
433 Eros, 390
Errors
  instrumental, 16, 18, 25, 135, 136, 139, 147, 152, 170, 174, 209, 289, 291, 513, 846
  observational, 18, 25, 137, 140, 160, 168, 170, 395, 403, 406, 441, 442, 513, 519, 560, 605, 788, 846, 851, 867, 985, 986
Etalon (passband), 257–59
15 Eunomia, 390, 397
Europa, 391, 395, 396
Evans, J. C., 741, 749
Evans, J. V., 359, 362, 366
Ewald, P. P., 277, 314
Excitation
  collisional, 747
  ions, 88–91, 93, 96–98, 101, 103–05
  lines, 747
  partial waves, 426, 427
  steady state, 421, 431, 432
  transverse magnetic, 419, 421, 423, 426
Extended sources. *See under* Sources
Extinction
  anisotropic, 546, 926, 927, 936
  band shapes, 920
  calculations, 916–20
  coefficient, 9, 407, 952
  cross section, 927
  curves, 331, 874, 888, 896, 898, 900, 916–25, 955
  efficiencies, 121, 920, 922, 923
  elimination, 135, 138
  foreground, 872
  grain size relation, 107, 957
  interstellar, 330, 874, 946, 949, 950, 955
  intracluster, 946, 949
  optical, 927
  properties, 124
Extragalactic objects
  30 Doradus, 1063
  general, 3, 322, 323, 325, 327, 329, 334, 335, 347, 348, 355, 960, 969, 970, 981, 986, 1030, 1059–83
  BL Lac type objects, 1076–79
  list, 1060, 1061, 1070, 1077
  M 31, 1061, 1063
  M 32, 1060
  M 82 (filamentary structure, loops, halo), 1066–72
  M 87 Jet, 1064, 1065
  Markarian galaxies, 1071, 1072

N galaxies, 1070–72, 1079
Seyfert galaxies, 59, 263, 981, 986, 1070–72, 1079
Eye
  animal (insect), 14
  human, 14
Eyring, H., 11, 21, 39, 40, **64–87,** 1087

Fabelinskii, I. I., 526, 576
Fabian, A. C., 263, 314
Fabry lens, 138, 143, 162
Fabry-Pérot (étalon, also filter), 257, 341, 640, 641, 755
Fahlen, T. S., 533, 576
Fahlman, G. G., 16, **237–45,** 582–92, 1088
Fairchild, C. E., 103, 106
Fano, U., 734, 749
Faraday
  cup, 89
  depolarization, 1018, 1019, 1024, 1028, 1044, 1047, 1048, 1050, 1052
  dispersion, 327, 328, 330, 1050
  effect, 65, 67, 77–84, 352, 1042
  pulsation, 326, 1040–44, 1055
  rotation, 39, 80, 327, 362, 960, 966–69, 1018, 1019, 1023, 1028, 1029, 1039, 1040, 1046
Faraday, M., 5
Faraggio, N., 496, 499
Farlow, N. H., 517
Farone, W. A., 774, 778
Fastrup, B., 100, 106
Fazio, G. G., 1020, 1023, 1026
Fechtig, H., 517
Fedulin, A. V., 457, 469
Feeds
  crossed linear (also circular), 354–58
  linear polarized, 356
  opposite circular, 355, 356
  orthogonal (also circular), 356, 357
  single linear, 354, 355, 357
Feix, G., 11, 23, 32, **676–85,** 1088
Fellgett, P. B., 30, 638, 642, 660
Fellgett's multiplex advantage, 640, 641
Felten, J. E., 39, 265, 267, 314, **1014–28,** 1088
Feofilov, P. P., 4, 12, 41
Fergason, J. L., 496, 499
Ferguson, D. C., 1002, 1008–11
Fernald, F. G., 195, 214, 463, 469
Fernández, H. R., 477, 478, 480, 481, 494
Ferry, G. V., 517
Fesenkov, V. G., 564, 576, 755, 758, 769, 778
Field
  far, 122, 423, 430, 527, 902, 905
  near, 421, 423, 427, 431, 433
  photon, 433
Field, G. B., 263, 314

Filice, A. L., 410, 418
Films
  mylar, 337, 338
  polyethylene, 337
Filters
  analyzing, 278
  color, 341, 342, 347, 456, 519, 562, 701, 712, 779, 830, 832, 847, 848
  dichroic, 136, 138, 162, 207
  general, 141, 219, 223, 224, 234, 242, 243, 258, 340–42, 348, 456, 515, 609–11, 645, 712, 752–57, 766, 767, 818
  interface, 797
  interference, 162, 250, 257, 501, 610, 615, 752, 754–56, 770, 867, 985
  narrow-band (also intermediate band), 820, 867, 880
  neutral-density, 250, 257, 319
Fink, D., 101, 103, 105
Fink, U., 403, 1088
Finzi, A., 997, 1011
Firor, J. W., 254, 260, 746, 749
Fischer, P. L., 829, 841
Fix, J. D., 573, 578
Flies (dipterau), 489, 490
8 Flora, 390, 397
Focas, J., 382, 401, 567, 576
Fog (also dry fog), 474, 548
Foitzik, L., 508, 509
Fomalont, E. B., 1052, 1056
Forbes, F. F., xv, 23, 30, 157, 172, 321, 330, 342, 349, 551, 554, 558, 576, 617, 618, **637–60,** 825, 833, 834, 841, 845, 863, 865, 866, 1088
Ford, V. L., 136, 155, 162, 169, 173, 895–97, 889, 899, 937, 938, 961–63, 971, 976, 979, 980, 1063, 1064, 1081
Forman, W., 1008, 1011, 1025, 1026
Forrest, W. J., 825, 830, 834, 841, 859
Forward, R. B., 490–92, 494
Foster, L. V., 755, 760
Fountain, J. W., 15, 25, 30, **223–31,** 570, 576, 1088
Fourier
  analysis, 11, 30, 174, 242, 483, 638–42
  coefficients, components, 365, 424, 428, 679, 686, 704, 1034
  interferometer, 23, 637, 638, 640, 650, 658
  power spectrum, 419, 942
  spectropolarimetry, 23, 637–60
  transform spectroscopy, 638–42, 647, 653, 1031
Fowles, G. R., 407, 418
Franceschini, N., 489, 492
Franklin, K. L., 55, 61
Franz, O. G., 598

Fraser, R. S., 450, 455, 457, 460, 463, 469, 470
Fraunhofer lines, 687, 689, 711, 715
Frecker, J. E., 161
Fredholm type equation, 582
Fresnel
  equations, 24, 234, 372, 385, 407, 476, 527, 533
  laws, 5, 47, 384, 406, 466, 700, 701
  peak, 385, 404
  reflection coefficient, 528, 818
  reflectivity, 372, 406, 416, 446, 451, 540
  rhomb, 179, 707
  type reflector, 463, 466
Fresnel, A., 5
Friedman, H., 263, 295, 315
Friend, J. P., 558, 576
Frisch, K. von, 474, 483, 492
Fritz, G., 263, 295, 315
Frogel, J. A., 832, 833, 840, 843
Frost, K. J., 269, 314
Fujimoto, M., 1051, 1057
Fujita, S., 337, 342, 350
Fukin, M., 1051, 1057
Furukawa, P. M., 3, 14, 448, 469
Fymat, A. L., 23, 30, 31, 524, 525, 541, 558, 574, 576, 583, 584, 586, 589, 590, **617–36, 637–60**, 1088

Gabsdil, W., 193, 214
Gagen-Torn, V. A., 155, 172
Gagne, G., 549, 577, 809, 813, 916
Gaillard, M. L., 101, 106
Galactic
  background, 22
  center, 40, 325, 328, 343, 345, 863, 932, 969, 972, 1079
  diffuse light, 38, 61, 778, 779, 781, 788, 869, 933, 942
  dust layer, 977
  foreground stars, 976–80
  loops I and II, 966, 967
  magnetic field, 933, 960–76, 978–80
  polarization, 774, 775, 778–80, 863
  spurs (north polar spur), 964–67, 970
  young regions, 946–53
Galatry pressure profile, 624
Galaxies
  cluster, 263
  general, xiii, 352, 354, 355, 892
  radio, 263, 265, 266, 960, 969, 970, 1047, 1048, 1051, 1052
  See also Extragalactic objects
Galkin, L. S., 157, 171, 573, 574, 604, 605
Gallaway, W. S., 232, 236
Galler, S. R., 475, 492

Gamma rays, 21, 223, 306, 520, 1020
Gammon, R. H., 864, 865
Gandolfi, E., 342, 343, 349
Ganymede, 391, 395, 396,
Garbury, M., 407, 418
Gardiner, R. B., 101, 105, 106
Gardner, F. F., 51, 53, 327, 349, 357, 358, 967, 970, 1030, 1044, 1051, 1052, 1056, 1057
Garlick, G., 388, 394, 401
Garmire, G. P., 274, 316
Gas
  circumstellar, 846, 851
  cloud, 1019
  comet, 819
  ionized, 853
Gauss-Seidel iteration, 541
Gaustad, J. E., 331, 349, 864, 865
Gay, M., 260
Geake, J. E., 385, 388, 395, 400, 401
Gegenschein. *See* Counterglow
Gehrels, T., xv–xvi, **3–44**, 48, 50–53, 61, 62, 150, 155, 164, 172, 191, 214, 218, 222, 227, 230, 334, 349, 353, 383, 384, 386, 387, 389, 390, 392, 393, 397, 400, 401, 457, 469, 520, 551–53, 555, 557, 560–62, 567, 568, 571, 575, 577, 593, 598, 615, 649, 694, 780, 802, 820, 822, 825, 827, 830, 832, 835–37, 841, 842, 846, 855–57, 860, 862, 866, 867, 872, 878, 888, 890, 891, 893–95, 897, 915, 928, 934, 935, 937, 943, 945, 959, 981, 982, 986, 987, 1002, 1005, 1007, 1010, 1011, 1015, 1024–26, 1028, 1079, 1080, 1088
Gehrz, R. D., 823–25, 830, 831, 834, 840, 841
Geise, R. H., 12, 23, 30, 33, 36, 549, **804–13, 901–15**, 1088
Geisel, S. L., 825, 841
Gelbstoff, 435
Gemperlein, R., 490, 493
Geneux, F., 93, 106
1620 Geographos, 390, 397
Geomagnetic field, 794
Geometric optics, 526–28, 531, 533, 536, 540, 546
Gex, F., 160, 172
Giacconi, R., 263, 265, 289, 310, 314, 316
Gilbert, J. A., 326, 327, 349, 350, 356, 358, 1023, 1026, 1044, 1048, 1050, 1051, 1055–57
Gillespie, C. M., 570–74
Gillet, F. C., 191, 195, 201, 214, 215, 324, 349, 787, 788, 790, 792, 825, 830, 841, 959, 1077, 1082
Gilman, R. C., 837, 841
Gilra, D. P., 331, 349, 864, 865
Ginzberg, D. M., 337, 350

INDEX

Ginzburg, V. L., 324, 349, 1004, 1011, 1015 1018, 1026
Ginzerl, M., 341, 349
Giovanelli, R., 969, 970
Glare, 27, 474
Glaspey, J. W., 16, **237–45,** 1088
Glennon, B. M., 104, 106
Gliese, W., 168, 169, 172
Glints (optical), 371, 372, 375
Globular clusters, 1061
"Glory," 533, 534, 539, 556, 558, 568
Glossary, 1084, 1085
Glushkov, Y. I., 878
Gnedin, Y. N., 59, 62, 264, 315
Gockel, A., 456, 469
Goetz, A. F. H., xv, 568, 577, 1088
Gold, T., 226, 315, 334, 349, 774, 778, 998, 1011
Goldberg, B., 240, 245
Goldberg, L. G., 637, 638
Goldhaber, M., 280, 315
Gold mechanism. *See under* Alignment
Goldreich, P., 958, 959, 998, 1004, 1011, 1025
Goldsmith, T. H., 477, 478, 480, 481, 492, 494
Goldstein, R. M., 24, 41, 359, 366, 397, 401
Goody, R. M., 620, 635
Gorbalskii, V. G., 582, 591
Gorchakov, G. I., 462, 469, 470, 563, 580
Gorenstein, P., 263, 302, 315
Gould's belt, 970
Govi, G., 6
Graham, D. A., 355, 358, 998, 1000, 1001, 1008, 1011, 1012
Graham, H. A., 337, 351, 650, 660
Grainger, J.F., 4, 41, 45, 51, 135, 136, 172, 481, 492, 752, 755, 757, 759, 760
Grains
  absorbing, 335, 929, 930, 932, 941, 943 944, 946
  alignment. *See* Alignment
  composition, chemical properties, 331–33, 927–30, 932–34, 936, 941, 943
  density, 331
  dielectric, 60, 335, 337, 926, 930, 941, 943
  elongated, 926, 939, 940
  emission. *See under* Emission
  graphite (also iron), 933
  interplanetary. *See under* Dust
  interstellar. *See under* Dust
  large, 946, 950, 953, 954
  magnetic, 54, 58, 322, 324, 334
  nonspherical, 60
  shape, 331, 332, 334, 916, 925
  size, 36, 926, 927, 928, 932–34, 936, 937, 947, 950, 954, 957
  spin, 942

surface structure, 331
temperature, 331–35
velocity, 816
See also Particles, Dust
Grant, I. P., 541, 577, 584, 590
Grasdalen, G., 959
Grating
  angle, 138, 232
  instruments, 19, 232, 337, 338, 340
  spectrograph, 138, 232, 233, 250
Gray, E. L., 464, 468
Gray, L. D., 618, 635
Gray, P. R., 191, 209, 215
Green, P. E., 359, 366
Greenberg, J. M., xv, 12, 14, 23, 36, 37, 60, 62, **107–34,** 331, 332, 349, 547, 549, 577, 811, 812, 871, 872, 874, 878, 879, 888, 895, 899, 901, 905, 914, **916–25,** 933, 937, 943, 945, 1088
Greene, T. F., 568, 577
Greenler, R. G., 337, 349, 546, 577
Green plants, 495–98
Green's function, 111, 113
Greenspan, J. A., 568, 575
Greenstein, J. L., 38, 41, 334, 349, 888, 897, 976, 980, 982, 983, 987, 1060, 1080
Green's theorem, 111, 121
Gregory, P.C., 265, 315, 1055, 1058
Gregory, S. A., 1048, 1050, 1057
Griffin, P. M., 93, 106
Grigorian, K. A., 825, 841
Grodzins, L., 280, 315
Grossjean, M., 77, 87
Grossman, K., 584, 590, 618, 636
Gucker, F. T., 526, 577
Guelachvili, G., 642, 659
Guenther, B. T., 514, 516
Gugkova, G. A., 845, 849, 856
Gunn, J. E., 267, 315, 983, 987, 998, 1006, 1011, 1022, 1023, 1026
Gursky, H., 263, 265, 289, 310, 315, 316
Gürtler, J., 868, 878
Gutkevich, S. M., 155, 172

Hadeishi, T., 92, 93, 98, 101, 106
Hadni, A., 340, 349
Hagfors, T., 359, 361, 363, 366
Hagins, W. A., 480, 481, 493
Haidinger, W. K., 5
Haidinger's brush, 5, 14, 27
Halation, 869
Hale, G. E., 6
Half-wave plate, 701, 704–07, 709
Hall, J., 886
Hall, J. S., **xiii,** 4, 7, 15, 18, 30–32, 41, 155, 172, 175, 186, 187, 227, 230, 388, 402, 552,

Hall, J. S., *cont.*
  554, 570, 573, 577, **593–98**, 865, 870, 872, 878, 881, 886–88, 899, 961, 971, 1060–62, 1080, 1088
Hall, R. C., 23, 823, 842, 879, **881–87**, 948, 950, 951, 1088
Hallgren, D. S., 29, 42, 517, 813
"Halo," 546, 548, 559
Hämeen-Anttila, J., 157, 161, 172, 197, 214
Hamilton, D. R., 624, 635
Hamilton's coefficient, 624, 993
Hamstra, R. H., 762, 765
Handedness (clockwise, counterclockwise), 9, 12, 21, 45–51, 53, 54, 56, 67, 71–73, 210, 237, 253, 354, 363, 495–97, 511, 512, 522, 570, 573, 607, 733, 779, 941, 992, 1001, 1035, 1043. *See also* Sign conventions
Hanel, R. A., 618, 635, 659, 660, 642
Hankel function, 115
Hanle effect, 59, 101, 730, 735, 742–45
Hanle, W., 734, 735, 738, 749
Hannemann, D., 8, 29, **510–13**, 563, 1088
Hanner, M. S., 29, 34, 41, 42, 516, 790, 807, 812, 872, 874, 877–79, 1088
Hansen, J. E., 10, 12, 16, 30, 61, 62, **518–81**, 584, 590, 603, 605, 613, 614, 616, 618, 619, 631, 636, 762, 1088
Hansen, R. T., 762
Hanson, L. B., 363, 366
Hapke, B. W., 24, 42, 1088
Harlan, E. A., 1022, 1027, 1073, 1074, 1081
Harper, E., 253, 260
Harrick, N. J., 337, 349
Harrington, J. P., 663, 667, 668, 674, 836, 842
Harris, D. L., 386, 391, 392, 401, 621, 636
Harris, O., 649, 650, 660
Harris, S. E., 179, 182, 186
Hart, R. W., 121, 126
Hartmann, W. K., 391, 402
Harvey, J. W., 157, 172, 233, 235
Harwit, M., 43, 44, 334, 349, 774, 778, 816, 1088
Hashimoto, H., 490
Hashimoto, J., 157, 172, 862, 866
Hatanaka, T., 354, 358
Hayakawa, S., 199
Hays, D., 492
Haze, 525, 526, 540, 553, 558, 560, 567.
  See also under Water
Heath, D. F., 197, 214
6 Hebe, 390
Heckathorn, H. M., 1070, 1081
Heeschen, D. S., 40, 42
Heiles, C., 959, 1000, 1001, 1011, 1012
Heisenberg, W., 738, 749
Heisenberg's uncertainty principle, 640, 738
Heitler, W., 273, 294, 295, 315

Helical magnetic field, 39, 48–50, 495, 496, 960, 961, 963, 965, 966, 970, 975, 976, 978–80, 1053
Heliograph, 692, 727
Helios project, 766, 767
Helmholtz equation, 108
Helmken, H. F., 1020, 1023, 1026
Hemenway, C. L., 29, 42, 516, 517, 813
Henderson, M. E., 217
Henry, J., 334, 349
Henry, R. C., 263, 264, 295, 314, 315
Henry, R. J., 993, 996
Hepburn, N., 724
Herapath, W. B., 6
Herbig, G. H., 38, 42, 832, 834, 842, 862, 865, 866, 880, 889, 899
Herman, B. M., 195, 214, 463, 469, 500, 509, 524, 541, 570, 571, 576, 577, 598, 618
Herman, M., 541, 577
Hermitian, 1032, 1043
Herzog, A., 222
Hesstvedt, E., 516
Hetherington, N., 364, 367
Hewish, A., 997, 1011
High altitude sites, 33, 246–53, 254–62, 770, 782, 784, 786, 787
Hiltner, W. A., 7, 16–18, 38, 42, 140, 154, 172, 266, 315, 753, 760, 845–47, 849, 856, 888, 899, 934, 937, 961, 971, 988, 991, 1015, 1026, 1060, 1061, 1064, 1065, 1072, 1080, 1081
Hjellming, R. M., 265, 315
Hodgdon, E. B., 191, 214
Hodkinson, J. R., 548, 577, 809, 813
Holback, B., 199, 208, 216
Holdridge, D., 24, 41
Holland, A. C., 549, 577, 809, 813, 916
Hollinger, J. P., 1052, 1057
Holmes, D. A., 147, 172, 187, 188
Holmes effect, 187
Holt, H. E., 201, 214, 565, 578
Holt, S. S., 300, 314, 1038, 1058
Holzworth, G. C., 457, 460, 461, 469, 505, 509
Honeycutt, R. K., 219, 222
Hong, S. S., 12, **916–25**, 1088
Horak, H. G., 199, 216, 541, 575, 622, 627, 632
Horch, K. W., 479, 490–92, 494
Hord, C. W., 599, 605, 618, 635
Horn (feed), 353, 904, 906
Horowitz, P., 1002, 1006, 1012
Horridge, G. A., 489, 492
Hosler, R. C., 505, 509
Houck, J. R., 334, 349
Houghton, J. T., 656, 660
House, L. L., 193, 253, 254, 259, 261, 742–45, 747–49

Houssier, C., 497, 499
Hovenier, J. W., 520, 544, 550, 554, 557, 561, 578, 583, 590, 603, 605, 618, 636
Hovis, W. A., 618, 635, 642, 660
Howard, H. T., 363, 366
Howard, R., 32, 42
Howell, H. B., 520, 541, 581, 584, 591
Howes, M. L., 26, 41
Hoyle, F., 330, 349
HR diagram, 826, 827
Huang, S. S., 849, 856
Hubbard, W. B., 331, 350, 882, 886
Hudson, H. S., 692, 694
Huffman, P., 548, 578
Hugenin, G. R., 1000, 1005, 1011
Hughes, N. D. P., 656, 660
Hughes, V. A., 265, 315
Hulbert, E. O., 195, 216
Hull, H., 15, 32, **246–53**, 260, 1088
Hulst, H. C. van de, 7, 9, 10, 12, 36, 37, 41, 42, 58, 60, 62, 109, 117, 179, 186, 266, 315, 520, 521, 526, 530, 533, 539, 540, 550, 578, 583, 584, 590, 618, 636, 808, 813, 835, 842, 871, 879, 888, 899, 905, 914, 916, 917, 925, 927, 933, 937
Hummer, D. G., 582, 590, 668, 674
Hunt, G. E., 521, 540, 541, 577, 578, 584, 590, 618, 619, 636
Hunten, D. M., 568, 620, 635
Huriet, J. R., 201, 214
Huruhata, M., 783, 786, 792
Hutchison, P. B., 782, 784, 793
Hutchison, R. B., 659
Hutley, M. C., 338, 348
Huygens, C., 5
Hvelplund, P., 100, 106
Hyde, G. M., 363, 366
Hyder, C. L., 59, 62, 63, 195, 214, 254, 261, 731, 742, 746, 747, 749
Hydrosphere, 472, 473, 475
Hyland, A. R., 832, 840

Iapetus, 391, 392, 397
Ibbett, R. N., 756, 760
1566 Icarus, 359, 390, 397
Ice clouds (Earth), 510, 559, 565, 569
Ice spheres, 516, 547, 548
Ikhsanov, R. N., 159, 160, 172
Illing, R., 48, 51, 55, 60–62, 266, 316, 334, 350, 888, 899, 926, 930, 932, 938, 941, 942, 983, 986, 1023, 1026
Image
  devices, 522
  Isocon, 237–40, 244, 245
  registration, 220–22
  selection, 224, 227, 228
  slicer, 238
Imbedding (invariant), 582–92
Incan, J. d', 642, 660
Index of refraction. *See under* Refractive Index
Infrared
  diameters. *See under* Diameters
  excess, 344, 348, 821, 823, 825, 830, 836, 840, 861, 889, 946–52
  frequencies, 65
  objects, 191, 199, 981, 982
  observations. *See under* Observations
  photometric measurements, 318, 393, 954
  polarimetry, 318–21, 762–65, 858–67, 941
  polarizers, 336–38, 342
  spectroscopy, 235, 259, 638
Ingersoll, A. P., 402, 567, 578
Ingham, M. F., 217, 781, 787, 788, 790–92
Insects, 4, 28, 473, 489, 490. *See also*
  Beetles, Flies
Instrument
  calibration. *See under* Calibration
  high-resolution, 23
  linear-to-circular conversion, 140, 210
  noise level, 629. *See also under* Noise
  polarization (also spurious polarization), 17–20, 25, 135, 138–41, 168, 169, 232, 234, 235, 249, 250, 306, 318, 320, 321, 330, 340–42, 345, 353, 357, 358, 376, 515, 650, 651, 658, 686, 691, 694, 703, 710, 760, 769, 881, 988, 990
  sensitivity (also sensitivity diagram), 67, 73, 207, 262, 266, 272, 291, 293, 300, 302, 303, 306–08, 310–13, 343, 647, 656, 695, 696, 699, 700, 704, 708, 716, 752, 755, 757, 760, 794, 845
  spatial resolution, 270, 312, 799
Instrumental
  errors. *See under* Errors
  scattered light, 798
Interference
  fringes, 677
  optical, 536, 539, 543, 644
  patterns, 435, 531, 533
Interferogram, 644
Interferometers
  Block Engineering Model 1500, 649–51
  Fabry-Pérot, 754
  Fourier. *See under* Fourier
  general, 23, 126, 340, 352, 354, 355–57, 400, 648, 649
  Michelson, 638, 643
  polarimeter, 638, 643, 644
Interplanetary
  dust. *See under* Dust
  electric field, 423
  grains. *See under* Dust

Interplanetary, *cont.*
  magnetic field, 419–33, 794–803
  particles. *See under* Dust
  space, 25, 767, 774, 783
Interstellar
  absorption features, 925, 946
  cloud, 890, 958
  dust alignment. *See* Alignment
  gas, 954–59
  grains, dust. *See under* Dust
  magnetic field, 38, 39, 266
  medium, 867
  particle formation. *See under* Particle
  polarization, 3, 4, 7, 9, 17, 18, 31, 32, 35–37, 39, 40, 170, 266, 834, 839, 847, 855, 859, 860, 862–64, 887–91, 894–96, 900, 915, 916, 923, 926–39, 941, 943–46, 954–71, 988, 1002, 1003, 1005, 1007, 1008, 1015, 1028, 1060, 1063
  reddening, 36, 947, 948
Intracluster dust, 946, 949, 952
Intrinsic polarization, 7, 31, 33, 35, 127, 781, 821, 824–26, 828, 830, 832, 840, 845–47, 851–53, 855, 857, 860, 890, 891, 928, 930, 931, 941, 942, 945, 959, 988–91, 1015, 1044, 1072
Invariant imbedding technique, 582–92
Inversion angle, points, 28. *See also* Neutral point
Io, 391
Ion
  beam (collisions), 88–90, 93, 97, 103
  metal, 405
  properties, 88–93, 99–101, 103, 992
  -water clustering, 516
Iriarte, B., 1085
Iron spheres. *See* metallic *under* Particles
Irtran 2 and 4, 337
Irtrons, 329
Irvine, W. M., 24, 42, 386, 402, 583, 592, 619, 636
Isaev, G. S., 457, 469
Isherwood, B., 239, 245
Israel, H. I., 305, 316
Ivanoff, A. I., 437, 440–42, 457, 469, 473, 474, 492, 500, 509
Ivanov, V. D., 58, 63, 270, 294, 316
Ivanov, V. V., 582, 591

Jackson, J. D., 55, 62
Jackson, J. M., 195, 216, 786, 792
Jacobowitz, H. J., 520, 541, 546, 578, 581, 584, 591
Jacobs, G. J., 475, 492
Jacobsen, R. A., 774, 778
Jacquinot, P., 638, 660

Jacquinot's luminosity advantage, 641
Jaenicke, R., 457, 468
Jäger, F. W., 340, 349, 512, 513
Jameson, R. F., 787, 792
Jander, R., 143, 146, 152, 492
Jefferts, K. B., 833, 843
Jenkins, E. B., 559, 578
Jennings, M. C., 825, 826–28, 836, 841, 842
Jensen, C., 455, 456, 469
Jensen, O., 16, **237–45,** 1088
Jeppesen, M. A., 149, 172
Jerlov, N. G., 437, 442, 443
Johnson, F. H., 85, 87
Johnson, H. L. (Johnson's UBVRI system), 642, 647, 649, 652, 653, 660, 770, 1085
Johnson, H. M., 867, 879, 881, 886, 1081
Johnson, H. R., 828, 841
Johnson, T. V., 405, 417, 418
Jokipii, J. R., 960, 971
Jones calculus, 4, 179, 186
Jones, R. V., 235, 334, 349
Jong, T. de, 933, 937
Jovicevic, S., 232, 235
89 Julia, 390, 397
Julian, W. H., 998, 1004, 1011
Junge power law, 505
3 Juno, 390, 393, 395–97
Jupiter
  circular polarization, 570, 607, 608, 610, 611, 613, 616
  decametric bursts, 55
  flyby mission, 189, 204, 205
  Great Red Spot, 31, 593–98
  ground-based observations, 593
  infrared observations, 593
  latitudinal variations of polarization, 518, 519, 570–73, 594, 597
  polarization measures, 223, 224, 227, 359, 518, 519, 552, 569, 593–98, 601, 605
  poles, 29
  satellites, 24, 381, 382, 399
  spectra, 642
  studies, 359, 568–71
  temperature, 570
  variation (polarization), 595–97, 606
Jupiter's atmosphere and clouds
  aerosol scattering, 593, 601, 608, 613
  ammonia layer, 604
  composition, 569, 570, 608, 611, 613
  general, 11, 28, 30, 203, 205, 569, 570, 593, 602, 604, 618
  molecular scattering, 593
  particles, 569, 599, 601, 606, 618
  single scattering, 519
  top, 570, 598

INDEX 1107

Kagiwada, H., 583, 585, 589, 590
Kahle, A. B., 523, 580, 583, 590, 591
Kalaba, R. E., 582, 583, 585, 589, 590, 627, 635
Kalhor, H. A., 232, 235
Kalitin, N. N., 456, 470
Kalle, K., 435, 443
Kamijo, F., 837, 842
Kaneko, N., 1071, 1081
Kano, M., 500, 508, 509, 618, 636
Kanter, H., 274, 275, 315
Kaplan, L. D., 642, 659, 660
Kardashev, N. S., 1024, 1026
Karimov, M. G., 731, 749
Karlstrom, K., 886
Karp, A. H., 667, 674
Kassander, A. R., xv, 1088
Katmai eruption, 460
Kattawar, G. W., 491, 541, 554, 559, 578, 580, 603, 605
Kawabata, K., 1051, 1057
Kawata, Y., 583, 592
Keeley, D. A., 958, 959
Keller, C. F., 195, 214, 215, 880
Kellman, S. A., 969, 971
Kellogg, E. M., 263, 265, 289, 310, 315, 316, 1021, 1026
Kelsey, L. A., 573, 578
Kelvin (W. Thomson, Lord Kelvin), 43
Kemp, J. C., 7, 13–15, 31, 33, 37, 56, 57, 62, 157, 160, 172, 175, 186, 498, 499, 520, 552, 568, 570, 573, 578, 581, **607–16**, 775, 776, 778, 779, 788, 793, 835, 842, 863, 866, 888, 890, 899, 900, 926, 932–34, 937, **939–45**, 982, 984, 985, 987, **988–91**, 992, 993, 995, 1088
Kenderdine, S., 1044, 1057
KenKnight, C. E., 126, 127, 203, 215, 379, 383, 384, 388, 394, 396, 402, 812, 1088
Kerker, M., 521, 539, 546, 547, 579, 774, 778
Kernahan, J. A., 104, 106
Kerr, J., 6
Kestenbaum, H., 265, 289, 312, 313, 315
Khangil'din, U. V., 57, 62
Kharitonov, A. V., 817
Kida, K., 175, 187
Kifune, T., 275, 276, 313, 315
King, R. J., 179, 186
Kinman, T. D., 40, 42, 1070, 1073, 1074, 1077, 1080, 1081
Kirschfeld, K., 480, 485, 486, 489, 490, 492, 493
Kissell, K., 24, **371–80,** 1088
Kjeldseth Moe, D., 741, 750
Klare, G., 889, 899
Kleinmann, D. E., 330, 342, 350, 825, 841, 863, 866
Klüver, J. W., 337, 348

Knacke, R. F., 331, 349, 1077, 1082
Kneubuhl, F. K., 344, 350
Knowles, S. J., 833, 842
Koeckelenberg, A., 717
Komesaroff, M. M., 326, 350, 999, 1004, 1006, 1012, 1038, 1058
Konheim, A. G., 807, 813
Korchak, A. A., 58, 62, 270, 271, 315
Kouchmy, S., 724
Kourganoff, V., 582, 590
Krakatoa eruption, 460
Kramer, H. A., 67, 87
Kraus, J. D., 47, 50, 51
Krishna Swamy, K. S., 332, 334, 350, 837, 842, 872, 879
Kristian, J., 47, 51, 266, 268, 310, 315, 832, 840, 983, 987, 1002, 1007, 1008, 1012, 1022, 1025, 1026
Kronberg, P. P., 265, 315, 327, 350, 356, 358, 1044, 1048, 1050, 1052, 1056, 1057
Kronecker delta function, 586
Kronecker symbol, 1031
Kronig, R. de L., 67, 87
Kronig-Kramer theorem, 67, 941, 943–45
Kruszewski, A., xv, 35, 822, 825, 827, 830, 832, 835–37, 841–43, **845–57,** 860, 862, 864–66, 889, 890, 897, 1070, 1071, 1081, 1088
Krylova, S. N., 786, 791
Krzeminski, W., 890, 899, 961, 971
Kubicela, A., 157, 173
Kuhn, W., 75, 87
Kuiper, G. P., 230, 392, 402, 559, 579
Kunde, V. G., 618, 635, 642, 659, 660
Kundu, M. R., 57, 62
Kurchakov, A. V., 868, 879
Kuroshio (submersible vessel), 440
Kurucz, R. L., 665, 670, 674
Kuvshinov, V. M., 986, 987
Kwan, J. Y., 959

Laboratory measurements, experiments, 56, 85, 126, 274, 291, 381–85, 388, 393, 394, 396–99, 403, 405, 406, 408–10, 412–15, 464–66, 495, 519, 521, 522, 545, 548, 553, 566, 569, 573, 809
Lacey, L., 253, 260
Laffineur, M., 721, 723
Lal, D., 25, 42
Lallemand, A., 695
Lamb, D. Q., 264, 315
Lamb, F. K., 57, 62, 264, 315, 741, 742, 747, 750, 984, 987
Lamb, W., 388, 394, 401
Lambert
  ground, surface, screen, 382, 452–55, 505, 507, 522, 525, 568

Lambert, *cont.*
  law, 503, 507, 585
  reflection (also non-Lambert), 446, 455, 462, 503
Lambert, D. L., 260, 642, 659
Lamla, E., 1070, 1073, 1074, 1081
Lampton, M., 1005, 1010
Lancoz spectral window, 242
Land, E. H., 6
Landau, R., 19, 26, 35, **318–21,** 551, 675, 1088
Landecker, P. B., 294, 298, 303, 313, 315
Landé factors, 88, 96–99, 101, 103, 105, 689, 737
Landerer, J.-J., 550, 579
Landolt, A. U., 1078–80
Landstreet, J. D., 33, 55–57, 59, 61, 62, 157, 172, 175, 186, 238, 266, 268, 315, 316, **981–87,** 990, 1023, 1026, 1079, 1081, 1088
Lane, A. L., 618
Lane, A. P., 24, 42
Lane, H. H., 1088
Lang, K. R., 329, 350
Langer, H., 475, 480, 489, 493
Laramore, L., 260
Larmor frequency, 97, 98, 982, 984, 992
Larmor precession, 96, 737
Larson, S. M., 230, 570, 576
Laser illumination, source, 291, 338, 549
Lauque, R., 1077, 1081
Law, S. K., 833, 843
Leavitt, J. A., 90, 106
Lebofsky, L. A., 405, 417, 418
Lee, H. J., 421, 422, 432
Lee, I., 253
Lee, R. H., 195, 214, 254, 260, 746, 747, 749
Lee T. A., 950, 952
Lee, W. M., 753, 761
Legg, M. P. C., 55, 62, 1024, 1026, 1038, 1057
Le Grand, M., 77, 87
Le Grand, Y., 435, 443
Leinert, C., 21, 34, 199, 203, **766–67,** 774, 778, 787, 789, 790, 792, 1088
Lemke, D., 193
Lenoble, J., 525, 579, 618, 630, 635, 636, 638, 660
Lequex, J., 329, 350
Lerche, L., 960, 971
Leroy, J.-L., 15, 32, 687, 688, 694, 710, 712, 713, 715–18, 721, 723, 728, **762–65,** 1088
Leslie, J. D., 337, 350
Le Squéren, A. M., 1077, 1081
Levin, G. V., 618, 635, 642, 660
Levinson, R., 263, 316
Levogyrate, 237, 239, 240, 244
Lewis, J. S., 559, 560, 579
Liais, E., 6

Lieske, J. H., 24, 41
Ligands, 504
Light
  curve, 375, 376
  diffracted, 277, 281, 526, 527, 531, 540, 557
  diffuse, 405–07, 409–12, 416, 417, 447, 448, 453
  environment, 444–71
  incident, 58, 60, 61, 84, 89, 233, 234, 495, 503, 526, 530, 531
  monochromatic, 68
  polarized. *See* Linear, Circular, and Elliptical polarization. *See also under* Polarization
  pulse, 90, 352
  reflected. *See* Reflection
  scattering. *See under* Scattering
  specular, 405–07, 412, 416, 417
  transmitted, 84, 496, 497, 502, 503
  unpolarized (natural), 8, 9, 47, 58, 60, 140, 146, 152, 169, 176, 211, 222, 233–35, 250, 372, 377, 407, 418, 434–36, 438, 441, 453, 464, 481, 495–98, 502, 512, 523, 527, 531, 533, 538, 544, 583, 616, 651, 670, 732, 735, 740, 744, 756, 758, 769, 851, 861, 870, 939, 1059, 1060, 1071, 1073, 1075
  See also Radiation
Liller, W., 264, 316
Lillie, C. F., 201, 216, 790, 796, 802
Lin, C. C., 104, 106
Lin, S. H., 77, 87
Lind, A. C., 119, 121, 126, 901, 905
Lindblad, B. A., 517
Linear polarization
  curve, 896
  direction, 889
  partial, 9
  percentage, amount, degree, 37, 46, 50, 52, 54–66, 71, 72, 80, 97–99, 127–30, 132, 134–36, 139, 141, 147, 174, 176, 177, 184, 185, 187, 189, 209, 232–34, 238, 239, 247, 250, 254, 264–66, 268, 270, 278, 294, 320, 321, 326, 334–36, 342, 352, 354–56, 372, 432, 436, 437, 439–41, 462, 472, 473, 481–83, 501, 502, 510, 512, 518, 521, 522, 530, 552, 563, 564, 571, 573, 607, 612, 614, 616, 628, 645, 681, 686–91, 693, 705, 707, 731, 732, 734, 740, 743, 758, 784, 794–803, 845, 859, 860, 863, 888–90, 926–31, 934–36, 939, 941, 943, 944, 946, 947, 952, 981, 982, 984, 990, 1015, 1019, 1024, 1025, 1028, 1030, 1034–38, 1040–53, 1055, 1059, 1060, 1073, 1077, 1079, 1084
  observation of night sky. *See* observations *under* Night sky
  rotation of plane, 889–91, 900
Line saturation mechanism, 990

INDEX								1109

Link, F., 517
Link, H., 21, 34, 199, 203, **766–67,** 787, 789, 790, 792, 1088
Linke, R. A., 55, 62, 267, 280, 294, 303, 313, 316, 317, 1021, 1026
Liou, K. N., 532, 533, 546, 547, 579, 618, 636
Lipovka, N. M., 329, 350
Lipskij, Yu. N., 607, 614
Little, S. J., 809, 813
Liu, C. H., 11, **88–106,** 1088
Livingston, W., 156, 157, 172, 233, 235
Livshitz, G. Sh., 457, 469, 500, 509
Livshitz, M. A., 270, 317
Llewellyn, E., 199, 208, 216
Lock, C., 195, 215
Lockwood, G. W., 1078, 1079, 1080
Loewenstein, E. V., 638, 660
Lomonosov (Mikhail Lomonosov, research vessel), 440
Lonke, A., 340, 350
Lorentz profile, 524, 624
Loreta, E., 822, 842
Loskutov, V. M., 554, 569, 579
Low, F. J., 329, 330, 340, 350, 570, 574, 825, 837, 841, 842, 863, 866
Lowery, T. M., 64, 87
Lowman, P. D., 618, 635, 642, 660
LTE and non-LTE calculations, 260
Luminescence, 12, 25
Lunar. See Moon
Lundgren, B., 439, 443
Lutz, B., 958, 959
Lyman absorption, 668, 672
Lynch, D. J., 103, 106
Lynds, B. T., 346, 350
Lynds, C. R., 1066, 1067, 1069, 1081
Lyne, A. G., 355, 358, 998, 1001, 1008, 1012
Lynn, W. T., 782, 792
Lyot
  coronagraph, 686, 720
  depolarizer, 15, 16, 18
  diaphragm, 256
  field-stop, 701, 710
Lyot, B., 4, 6, 14, 15, 25, 26, 42, 156, 161, 172, 184, 186, 381, 382, 386–89, 392, 394, 402, 403, 463, 470, 519–21, 543, 548, 549, 550, 552, 553, 556, 557, 566–68, 570–73, 579, 593, 601, 605, 686, 695, 701, 705, 710, 711, 726, 728, 731, 750

Macdonald, G. H., 1044, 1057
Macek, J., 88, 106
Mack, J. E., 263, 314, 1005, 1010
Magellanic Clouds
  Large Magellanic Cloud (LMC), 39, 288, 880, 976–80, 1061, 1063, 1064
  region between Clouds, 976–80
  Small Magellanic Cloud (SMC), 976–80, 1061, 1063, 1064
Magnetic birefringence. See Birefringent
Magnetic circular polarization. See under Circular polarization
Magnetic field
  degeneracy, 745
  detection, 985
  equations, 421–23
  frozen, 958, 997, 1019, 1022
  galactic, 32, 37, 38, 925, 960-71, 1030, 1033, 1035, 1037, 1038, 1042–44, 1046, 1051–53, 1055, 1060–62, 1066, 1072, 1074, 1079
  general, 6, 25, 33, 54, 57, 59, 65, 70, 77, 79, 81, 82, 85, 88, 93, 96–103, 105, 237–45, 254, 259, 260, 266, 268, 270, 326, 327, 334, 336, 352, 674, 686, 691, 693, 715, 716, 740, 741, 777, 801, 802, 814, 836, 839, 870, 889, 890, 895, 919, 925, 948, 954, 958, 976–80, 985, 989, 990, 1014, 1015, 1022, 1024
  helical, 890, 960, 961, 1053
  inhomogeneous, 741
  interplanetary, 419–33, 794
  local, 960, 961, 963, 968, 970, 974
  lunar, 419, 423, 424
  maps, 33, 716, 966, 967, 974, 975, 1029, 1060
  measurements, 237–45, 419
  outside solar system, 981–87
  patterns (interstellar), 266
  radial, 712
  solar. See under Solar
  stellar. See under Stellar
  strength, 716, 774
  strong, 735, 736, 739, 745, 747, 992–96, 998
  structure, 960, 1029
  variations, 419, 420, 426
Magnetic tail (earth), 420
Magnetographs, 237, 238, 254, 255, 716, 730, 988
Magnetoionic theory, process, 56, 684
Magnetometer experiments, 419, 424, 430, 431, 433, 695, 710, 716
Magnetoplasma, 1030, 1031
Maihara, T., 157, 172, 862, 866
Maillard, J.-P., 642, 660
Makas, A. S., 337, 348, 350
Malcolm, B., 499
Malitson, I. H., 149
Mallmann, A. J., 546, 577
Maltby, P., 741, 750
Malus, E.-L., 5
Malus' law, 5, 10, 20
Malville, J. M., 254, 261, 746, 749
Manchester, R. M., 966–68, 971, 1000, 1001, 1005, 1006, 1012

Mandel'stam, S. L., 58, 63, 270, 294, 316
Mann, H. M., 784, 787, 790, 792
Mantis, H. T., 195, 215
Mapping, maps
  corona, 32
  magnetic field. *See under* Magnetic field
  photoelectric, 1015
  polarization, 32, 218, 230, 255, 259, 266, 562, 570, 695, 710, 716, 721, 723, 725, 727, 967, 1015, 1018, 1020, 1029
  radio polarization, 1020
  range-Doppler, 359–64
Marcoux, J. E., 570, 579
Margon, B., 265, 316
Marin, M., 157, 159, 172, 700, 705, 721, 728
Mariner Jupiter-Saturn mission. *See under* Space missions
Mariner radio-occultation data, 567
Mariner Venus-Mercury vidicon system, 202, 562
Markovian process, 583
Marlborough, J. M., 846, 856
Mars
  aerosol amount, 566, 567, 599, 600
  atmosphere, 560, 565, 599, 600, 642
  circular polarization, 568, 608, 610, 611
  disk variable polarization, 599, 600, 605
  dust storm, 27, 519, 568, 599, 610, 611, 618
  general, 17, 18, 21, 27, 382, 397, 499, 519, 567, 599, 605
  satellites, 381, 382, 391, 463, 464
  spectra, 617, 618
  surface, 566, 567
Marsh, J. C. D., 324, 350
Martel, M.-T., 814–17, 868, 879
Martin, D. H., 338, 350
Martin, P. G., 23, 34, 35, 37, 40, 48, 51, 53, 55, 60–62, 266, 316, 334, 350, 835, 842, 864, 866, 888, 896, 897, 899, **926–38**, 940–42, 944, 945, 1023, 1026, 1081, 1088
Martinson, I., 88, 92, 101, 103, 106
Masers, 352, 958
Mason, S. F., 499
20 Massalia, 390
Masursky, H., 201, 215, 363, 367
Mathewson, D. S., 136, 155, 162, 169, 173, 825, 841, 889, 895–97, 899, 936–38, 961–65, 970–80, 1063, 1064, 1081
Mathieu, J., 496, 499
Mathieu function, 940
Matilsky, T., 265, 316
Matrix
  absorption, 741
  density, 734
  dipole, 744
  interaction, 734, 744

Jones, 651
Mie, 541, 557, 623, 631, 634, 652
Mueller, 6, 8, 175, 182, 247, 361, 734, 741, 742, 744, 747
order of scattering, 586
Pauli's spin, 733
phase, 519, 520, 531, 541, 544, 550, 557, 623, 631, 652
Rayleigh-Cabannes, 582, 587, 623, 625, 632, 634, 637
reflection, 446, 448, 503, 504
scattering, 361, 447, 463, 549, 582, 584, 586–88
source, 588
Stokes, 446, 746
transfer, 541
transform, 144, 146
transmission, 448, 503, 504, 587, 625
Matson, D. L., 393, 402
Matsumoto, M., 583, 590
Matthews, M. S., xvi, 40, 982, 987, 1088
Mattig, W., 741, 750
Mattila, K., 878, 933, 938
Matyagin, V. S., 816, 817
Mauter, H. A., 195, 214, 254, 261, 746, 749
Maxwell, A., 349, 350
Maxwell equations, 70, 110, 1031
Mayer, C. H., 7, 354, 358, 833, 842, 1052, 1057
Mayer, W. F., 1060, 1081
Mayfield, E. B., 57, 61
Mc Cann, G. D., 490, 493
Mc Cleese, D. J., 549, 574
Mc Cord, T. B., 386, 402, 405, 406, 417, 418
Mc Cullough, T. P., 7, 57, 62, 354, 358
Mc Elfresh, I. W., 807, 813
Mc Elroy, M. B., 570, 579
Mc Ilwain, C. E., 136, 164, 171
Mc Intyre, C. M., 179, 182, 186
Mc Kenzie, D. L., 268, 316
Mc Knight, R. V., 314, 350
Mc Lean, I. S., 15, 23, 32, 157, **752–61**, 1088
Mac Leod, J. M., 1076, 1081
Mc Mahon, D. H., 153, 174
Mc Master, W. H., 730, 750
Mc Phedran, M. C., 232, 236
Mac Rae, D. A., 937
Mead, J., 605
Mean free path, 407, 410, 412, 414, 526, 557
Meekin, J. F., 263, 295, 315
Meeus, J., 13, 53, 1088
Melchiorri, B., 11, 22, 35, 38, 193, **322–51**, 1088
Melchiorri, F., 11, 22, 35, 38, 193, **322–51**, 1088
Melrose, D. B., 1005, 1012, 1038, 1057

INDEX

Mencaraglia, F., 11, 22, 35, 38, 193, **322–51**, 1088
Mendez, M. E., 642, 652, 653, 660
Menzel, D. H., 386, 402
Mercer, R. D., 790
Mercury
  polarization measurements, 318, 321, 381, 382, 386–88, 392, 395–97, 610
  studies, 23, 26, 203, 417, 610
Merrill, K. M., 1077, 1078, 1082
Merrill, P. W., 822, 842
Mertz, L., 642, 660
Mesopause, 514–17
Meteors, xiii
9 Metis, 390, 397
MgO surface, 410, 414
Michael, M., 93, 106
Michard, R., 724
Michel, G., 642, 659
Michelson, A. A., 495, 496, 499
Michelson's visibility technique, 639
Mickey, D. L., 15, 32, **686–94**, 715, 717, 1088
Mickiewicz, S., 302, 315
Micrometeorites, 901
Micron, 13, 1084
Microvilli, 472, 477–81, 490, 491
Microwave
  analogue technique, measurements, experiments, 12, 23, 126, 127, 549, 804, 809, 813, 901–15
  bursts, 268
  electronics, 945
  emissions, 268, 833
  regions, 692, 833
Middelditch, J., 264, 314
Middleton, W. E. K., 505, 509
Mie
  calculations, 126, 528, 532, 548, 569, 572, 599, 818, 819, 906, 917, 918
  curves, 908
  infinite cylinders, 927
  particles (spherical), 837, 872
  scattering. *See under* Scattering
  theory, 6, 10–12, 17, 26, 28, 30, 35, 36, 125–27, 397, 419–33, 502, 528, 532, 548, 569, 572, 614, 804, 806, 809, 818, 819, 835, 901, 905, 906, 916, 918, 920, 927
Mie, G., 6
Mifflin, R., 364, 367
Mihalas, D., 670, 674
Mihalov, J. D., 419, 433
Mikesell, A. H., 156, 173, 175, 186
Milky Way
  photographs, 972–75
  structure, 972–75
  studies, 34, 39, 768, 770, 772–75, 778–81, 795, 796, 798, 801, 942
  See also diffuse light *under* Galactic
Millard, J. P., 197, 212, 215
Miller, D. C., 820
Miller, J. S., 266, 310, 317, 1002, 1007, 1013, 1019, 1021
Miller, W. C., 1065, 1066, 1082
Mills, D. M., 1053, 1057
Milne, D. K., 39, 970, 971, **1029**, 1088
Milton, D. J., 391, 402
Mineral
  absorbing, 405, 406
  absorption band polarization, 23, 405–18
  opaque, 405, 412
  samples, 406, 408–10, 412, 416
  transparent, 405, 406, 412
Mingle, J. O., 584, 590
Minin, I. N., 567, 579, 872, 879
Minkowski, R., 1022, 1026
Minnaert, M. G. J., 14, 28, 42
Minton, R. B., 230
Mirrors
  aluminizing, 140
  flat, 340, 341
  spherical, 340–42
  tilted plane, 135, 140, 141
Mitchell, A. C. G., 59, 62, 750
Mitchell, D., 280, 317
Mitsuishi, A., 337, 342, 350
Mitton, S., 356, 358, 966, 967, 971, 1045, 1046, 1051–54, 1057
Moak, C. D., 93, 106
Model
  absorbing layer, 990
  absorption (pure), 741
  acretion, 262, 264–67
  aligned grains, 59, 60, 777, 890–96, 900, 928–30, 932–36, 939, 940, 943. *See also* Alignment
  atmosphere (general), 447, 521–23, 541, 573, 574, 618, 622–24, 627
  atmosphere (non-LTE), 990
  atmosphere (planetary), 445, 447, 518, 541, 601–04
  atmosphere (solar), 680
  atmosphere (stellar), 663, 666–68, 670, 672, 836, 839, 847, 849, 851, 855, 861, 862, 864
  binary system, 267
  biological, 485–87, 490
  cloud, two-cloud, 514, 516, 517, 552, 553, 556, 559, 560, 570, 623, 940, 944
  cloud (2-layer), 525, 541, 559
  complex oscillator, 64
  conductivity, 419, 425, 428
  dielectric (large sphere, small sphere, cubes), 805–11

Model, *cont.*
  electron density, 723
  explosion, 1069
  grain, 888, 895, 896, 928, 931–33
  grain deflection, 957
  interstellar dust absorption, 950
  locally plane-parallel, 541
  lunar, 423, 424, 426, 428, 432, 500, 502
  magnetic field models (helical, longitudinal), 960–70, 975, 976
  magneto-absorption (LTE) scattering, 990
  Mills-Sturrock, 1053
  Milne-Eddington, 740, 741
  nebular (dark), 332
  noise, 241, 242
  non-LTE magneto-scattering, 672, 674, 990
  nonspherical particles, 804
  nonthermal, 266
  oblique rotator, 1023
  photospheric, 716
  Puka, 614
  pulsar, 266
  radiator, 984
  reflection, 407, 408
  rotating vector, 997, 1006–10
  scattering (polar), 607–09, 612–14
  scattering distribution (stacked sphere), 126, 364, 426, 502, 505, 520, 540, 556, 630, 631, 667, 834–40, 844, 1061, 1066, 1067, 1070
  Shklovski-Van der Laan-Kellermann, 1051
  solid, 397
  spectral, 627
  sphere-sphere, 614
  spherical, 541
  structure (planetary), 608, 609, 612–14
  structure Crab Nebula, 1024, 1025
  surface (rough), 36, 362, 612
  synchrotron, 1066, 1067
  thermal, 266
  turbid, 500
  Unno's, 990
  Van der Laan, 966
  zodiacal, 34, 804–13
Modulators, modulation
  achromatic, 175–79
  circular, 176, 178, 182
  efficiency, 138, 139, 142, 152, 153, 156, 158, 160, 161, 164, 168, 175, 176, 179, 704–09
  fan-shaped, halfwave, 701, 705–07
  general, 175, 176, 246–51, 253, 256, 258, 704–06, 709, 755
  linear, 176, 177
  rotary, 204
  square-wave, 160, 176, 704–08
  transfer function, 238
Moffat, A. T., 998, 1011

Molecular
  anisotropy, 59, 445, 457, 480, 515, 525, 526, 562. *See also* parameters *under* Anisotropic
  atmosphere, 398, 524, 564
  bands, 56
  composition, 526, 556
  isotropic, 525
  layers, thickness, 500, 502–08, 570
  parameters, 69
  scattering, scattering law, 12, 387, 444, 522, 567, 570
Molecules
  asymmetric, dissymmetric, 64, 65
  CO, 833
  gaseous, 446, 526, 552, 553, 560, 563, 568, 616
  general, 76, 82, 85, 86, 93, 496, 525, 762
  organic, 473
  SiO_2, 837
  symmetric, 496. *See also* Symmetric media
Molenkamp, C. R., 524, 579
Möller, K. D., 341, 350
Mommaerts, W. F. H. M., 473, 493
Monochromatic light. *See under* Light
Monochromator
  grating, 753
  prism, 753
  scanning, 753–55
Monomers, 76, 559
Monte Carlo method, 305, 541, 742, 877, 880
Montroll, E. W., 121, 126
Moody, M. F., 485, 487, 489, 493
Mook, D. E., 845, 849, 856, 988, 991
Moon (lunar)
  circular polarization, 607, 614, 615
  conductivity, 426, 431
  interior, 419, 422, 423, 428, 431, 432
  magnetic induction, 421, 423, 424, 427
  mirror, 565, 566
  particle size, 388, 412
  photometry, 25, 31, 34, 35
  polarimetry, 223, 224, 228–30, 381, 382, 386–88, 430–32, 463, 464, 566
  radius, 419, 423, 430
  samples, 381, 385, 388, 394–96
  studies, xiii, 5, 12, 18, 21, 23–26, 47, 191, 201, 211, 213, 359, 361, 381, 384, 385, 387, 394, 395, 398, 403, 406, 653
  surface, 362–64, 388, 405, 406, 417, 419–33
  whole, 24, 26, 395, 419
Moore, J. G., 195, 215, 216, 562, 580
Moore, W. E., 274, 316
Moreno, G., 341, 343, 349
Moroz, V. I., 570, 579, 786, 791
Morozhenko, A. V., 30, 157, 171, 554, 567–69, 573, 574, 579, **599–606**, 1088
Morris, D., 356, 358, 967, 970, 998, 999, 1012,

INDEX 1113

1044, 1052, 1057, 1058
Morrison, D., 391–93, 400, 402
Morrison, N. D., 825, 843
Morrison, P., 267, 316, 1079, 1081
Morton, D. C., 559, 578
Moss, T. S., 656, 660
Mossop, S. C., 558, 579
Mounting
　alt-azimuth, 358
　Ebert, 165
　equatorial, 358
Mueller
　calculus, 4, 175–88
　matrices. See under Matrix
Mueller, H., 6, 8
Mukai, S., 12, 30, 522, **582–92**, 1088
Müller, E. A., 350
Muller, R., 762, 764, 765
Multipole expansion, 539
Muncaster, G. W., 1002, 1005, 1007, 1009–11, 1015, 1025, 1026
Münch, G., 1015
Murphy, R. E., 391, 392, 402, 570, 578, 607, 614
Murray, B., 203, 215
Murray, J. D., 55, 63, 1055, 1058
Murray, S., 265, 314, 316
Murrell, S., 222

Nagel, M. R., 765
Nagirner, D. I., 668, 674
Najita, K., 692, 694
Nandy, K., 330, 332, 351, 888, 896, 897, 900, 1071, 1081
Napper, D. H., 913, 914
Natale, V., 11, 22, 35, 38, 193, **322–51**, 1089
Nebulae, nebular objects
　Cederblad, 201, 868, 878
　VY CMa, 38, 870, 880
　cometary, 951
　dark, 331, 332, 334–36, 344, 346, 348
　dark cloud. See Dark cloud
　extragalactic, 3, 40. See also Extragalactic objects
　flare events, 262, 267
　general, 38–40
　IC 5076, 868, 874, 877
　Merope, ii, 867, 868, 870, 872, 874, 879
　NGC 1976, 881–87
　NGC 2068, 868, 872, 874
　NGC 2261, 868, 878
　NGC 2264, 947, 948, 950, 951
　NGC 7023, 867–69, 874, 878
　reflection. See Reflection nebulae
　spiral, 1059–61
　See also Galaxies
Neckel, T., 889, 899

Nephelometer, 540, 813
Neptune, 518, 573
Neugebaur, G., 346, 453, 457, 470, 832, 840, 1072, 1077, 1081
Neupert, W. M., 268, 270, 294, 316
Neureuther, R. A., 232, 235
Neutral points, 5, 28, 29, 191, 789, 811.
　See also Inversion angle, points
Neutron stars, 55, 57–59, 264, 266, 997, 998, 1004, 1005, 1009
Neven, L., 642, 660
Neville, A. C., 1044, 1057
Newkirk, G., 195, 215, 254, 260, 261, 724, 731, 750
Newton, I., 44
Ney, E. P., 34, 38, 42, 201, 213, 216, 692, 694, 774, 778, 787, 792, 821, 837, 843, 864, 866
Night sky
　-glow, 781, 783, 784, 788
　light (emission, radiation), 33, 768–93, 867, 869, 1067
　photometry, 768, 770, 771, 777, 779, 780
　polarimetry, 768–80
Nikolskii, G. M., 786, 792
Nikulin, N. S., 986, 987
Nimbus 3, 642
Noctilucent clouds, 29, 34, 199, 208, 471, 514–17
Noise
　level, 209, 699, 700, 704, 752, 754, 755, 758, 763
　photon shot, 754, 756, 758
　signal-to-noise ratio. See under Signal
Noland, M., 391, 396, 402, 403
Nordsieck, K. H., 159, 173, 1071, 1077, 1078
Norton, R. H., 642, 659
Nova remnants (also supernova remnant), 3, 39, 825, 1029
Novick, R., 4, 21, 32, 39, 40, 55, 63, **262–317**, 694, 942, 1021, 1026, 1088
Nowak, W., 457, 468
Nucleation process, 516
Nuclei condensation, 840
Number density (dust particle), 804, 813

Oblateness, 404
Obridko, V. N., 741, 742, 750
Observations
　aircraft, airborne, 22, 549, 562, 564, 565, 641
　balloon. See Balloon
　color range, 786, 787
　errors. See under Errors
　ground-based, 22, 23, 26, 34, 35, 333, 345, 416, 417, 500, 510, 561–64, 573, 641, 769–75, 791, 796
　infrared (near, far), 22, 23, 562, 786, 789, 824, 825, 829, 833, 858–67, 941

Observations, *cont.*
  photoelectric. *See under* Photoelectric
  photographic. *See under* Photographic
  radio, 39, 1029, 1030
  rocket. *See under* Rocket
  satellite, 641, 642
  solar. *See under* Solar
  space. *See under* Space
  spectroscopic, *See under* Spectroscopy
  ultraviolet, 789
  visual, 783, 786, 789, 824, 858, 859, 862, 864
  wavelength range, 786, 787
  X-ray, 32, 39
Observing techniques, 135, 189–217
Oceanography (polarimetry), 28, 434, 435, 437, 439, 441
O'Connell, R. F., 33, 984, 986, 987, **992–96**, 1089
O'Connell, R. W., 1038, 1081
O'Dell, C. R., 331, 350, 867, 868, 870, 872, 879, 882, 886
O'Dell, S. L., 1005, 1006, 1012
Oetken, L., 340, 349, 512, 513
Öhman, Y., 4, 6, 156, 173, 519, 695, 704, 729, 730, 750, 752, 758, 761, 814, 817, 1060, 1081
Okami, N., 440, 443
Okazaki, N., 175, 187
Oke, J. B., 982, 986, 1070–72, 1077, 1081, 1082
O'Keefe, J. A., 822, 842
Okuda, H., 157, 172, 862, 866
O'Leary, B. T., 363, 367, 559, 579
O'Mara, B. J., 809, 813
O'Mongain, E., 1020, 1023, 1026
Omont, A., 742, 747, 750
Oort, J. H., 871, 879, 895, 1014, 1015, 1018, 1020, 1026, 1063, 1082
Oort-van de Hulst size distribution, 895
Opacity, 57, 381, 382, 384, 387, 399, 406, 668, 845, 984
Opposition effect, surge, 24, 383, 386
Optical
  active medium, 6, 939, 940
  active surface, 495–99
  activity (also magnetic), 12, 64, 65, 67, 69–79, 81, 83, 84, 608, 611–13
  cascades, 88, 100–04
  decay, lifetimes, mean life, 88–90, 93, 97, 100–04
  depth, thickness, 8, 27, 28, 38, 57, 447, 448, 505, 518, 519, 522–24, 526, 542–44, 550, 556, 558, 565, 570, 616, 623, 626, 877, 947, 1037, 1085
  rotation dispersion, 66, 67, 69
  rotation dispersion, magnetic, 67, 82
  thick medium, 55, 57, 61, 584, 587, 599, 847, 849, 1038, 1055
  thin medium, 55, 57, 58, 847, 849, 855, 878, 1036, 1041, 1046, 1055
Orbit inclination (close binaries), 853–55
Orientation (also random), 59, 527, 528, 546, 547, 550, 809–13, 917, 920, 927
Orion Nebula
  dark bay, 884, 887
  general, 324, 881–87, 946, 948, 950, 952
  region stars, 946–53
Orrall, F. Q., 15, 32, 158, 159, 173, **686–94**, 715, 717, 1089
Oscillation
  band extinction, 916, 921, 923
  nonspherical, 824
  polarization, 510, 916, 923
Oscillator. *See under* Atomic
Osherov, R. S., 815, 817
Oshiba, G., 440, 443
Oskanjan, V., 157, 173
Ostriker, J. P., 264, 268, 314, 315, 998, 1006, 1011, 1022, 1026
Otsuka, Y., 342, 350
Ottewill, R. H., 913, 914
Owen, T. C., 568, 570, 571, 575, 576, 580, 593, 598
Owings, D., 383, 386, 387, 393, 401

Pacholczyk, A. G., 11, 14, 24, 40, 55, 327, 350, 1004, 1012, 1023, 1027, **1030–58**, 1089
Pacini, F., 998, 1004, 1005, 1010, 1012
Packer, D. M., 195, 215
Pakhomov, L. A., 197, 213, 564, 574
Palik, E. D., 149, 173
2 Pallas, 390, 393, 395–97
Pancharatnam, S., 149, 173, 182, 186
Papaliolios, C., 1002, 1006, 1012
Parker, E. N., 420, 433
Parker field, 420
Parkin, C. W., 419, 433
Parrish, M., 337, 338, 348
Parriss, J. R., 485, 487, 489, 493
Parsignault, D. R., 265–67
Partial-wave analysis, 539
Particles
  acceleration, 692, 693
  aerosol, atmospheric, 30, 444, 445, 518
  aligned. *See* grains *under* Alignment
  anisotropic, 81–84
  carbon, 385
  charged, 25, 691, 693, 730, 736, 747
  chemical composition, 10, 30, 36–38, 399, 521, 616, 767, 789, 888
  cloud, 514–16, 519, 520, 560, 816
  collision, 957
  counting technique, 582

INDEX 1115

crystal, 386, 546, 550
cubes, 901, 906, 907, 911, 913
cylinders, 107, 117–24, 546, 774, 872, 916, 920–23, 925
destruction, 844
dielectric, 36, 37, 804–06, 809, 811, 819, 836, 862–64, 874, 895, 1079
dust, 27, 33, 60, 558, 564, 767, 768
elliptical, 10
formation, 37, 514, 516, 821
graphite, 36, 337, 821, 822, 834, 837, 838, 874
haze. *See under* Water
hexagonal, 546, 548, 559
high-albedo, 839
high-energy, 997, 998
homogeneous, 115–17, 815
ice, 36, 548, 559, 874, 876, 916
identification, 543
interplanetary. *See under* Dust
interstellar. *See under* Dust
irregular, 36, 521, 546, 916
isotropic, 114–17, 818, 870
large polydispersed, 521, 522
large solid, 526–28, 870
liquid droplet, 547, 550, 553, 554, 558, 566
metallic, iron, 38, 819, 825, 836, 863, 874, 1079
microstructure, 568
Mie, 627, 633
mineral, 405–18
needle-like, 802
nonabsorbing, 37, 381, 524, 527, 550, 562, 777, 808, 810, 811, 818, 819, 896, 914
nonspherical, 107, 521, 545–50, 623, 627–30, 768, 774, 776, 777, 794, 801, 804–13, 819, 916, 923
octahedrons, 901, 906–13
orientation. *See* Orientation
origin, nature, 889, 896
packed, 26
parameters, 893, 895, 896
production, 829, 833, 836, 837, 840, 844
radius, 525, 528–30, 532, 534–36, 547, 548, 553, 555, 557, 558, 565, 569, 572, 1084
Rayleigh-Cabannes, Rayleigh, 776
relativistic, 997, 1004, 1022
rough, 521, 546, 901, 906–09
scattering. *See under* Scattering
shape, 8, 30, 547, 556, 559, 570, 573, 888, 913, 916–25
silicate, 837, 838, 874, 896
size, 6, 10, 34, 36, 37, 61, 399, 405, 406, 409–11, 417, 444, 518, 522, 545, 549, 550, 554–59, 561, 564, 565, 569, 570, 573, 819, 821, 822, 837, 844, 859, 860, 872, 888–90, 895, 896, 900, 902, 908, 913, 916
size distribution. *See* Size distribution
small, nonspherical, 804–09, 813
smoke, 6, 564
smooth, 37, 809, 908, 909, 916
spheres, spherical, 10, 107, 124–26, 447, 462, 518, 540, 545, 550, 569, 623, 630–32, 779, 801, 804, 805, 807–11, 813, 818, 819, 870, 916–20, 922, 923
surface, 526
transparent spheres, needles, 528
type, 547, 550, 559, 560
water haze. *See* Haze; *see also under* Water
water-suspended, 434, 435
See also Grains, Dust
Partington, J. R., 526, 580
Paschen back effect, 925
Paschen series, 753, 846, 851, 860
Pascu, D., 391, 402
Pasteur, L., 6, 64, 87
Paterson, E. W., 337, 351
Pavageau, J., 232, 236
Pavlov, V. Ye., 500, 509
Peake, W. H., 362, 366
Pearl, J. C., 618, 635, 642, 660
Pecker, J.-C., 323, 324, 350, 687, 694, 715, 726, 729
Pederson, A., 199, 216
Pederson, J. C., 547, 549, 577, 901, 914
Pederson, N. E., 547, 549, 577, 901, 914
Peebles, G. H., 520, 580
Pellicori, S. F., 29, 41, 191, 203, 209, 215, 230, 386–89, 401, 402, 463, 470
Penn, R. D., 480, 481, 493
Penston, M. V., 947, 952
Penzias, A. A., 833, 843
Pepin, T. J., 195, 215
Perche, J. C., 747, 750
Percival, I. C., 747, 750
Perkins, B., 16
Perrin, F., 750
Perry, R. M., 255, 260
Peterson, A. V., 824, 842
Peterson, A. W., 786, 792
Peterson, E. W., 650, 660
Pethick, C. J., 57, 62, 264, 315
Pettengill, G. H., 359, 362, 366
Pfau, W., 978, 980
Pfund condensing optics, 340
Phase angle
  asteroid, 388, 392
  brightness-phase relation, 24
  general, 8, 9, 205, 374, 376, 378, 382–85, 388–90, 404–08, 412, 414, 416, 518, 524, 525, 528, 531, 566, 569, 599, 604, 608, 613, 614, 617, 632, 928, 1084
  Jupiter, 568–71

Phase angle, *cont.*
  Mercury, 387, 417
  polarization-phase relation. *See under*
    Polarization
  Saturn, 573
  Uranus and Neptune, 573, 574
  variation, 621, 990
  Venus, 218, 223, 551–53, 555
Phase curve, 25, 387, 525, 533, 549, 551, 552,
    559, 566, 568, 608, 616, 988, 990
Phase difference, 50
Phase effect, 849
Phase function, 24, 383, 527, 529, 531–33, 535,
    537, 543, 663, 668, 672, 872
Phase integral, 383
Photocathode, 137, 142, 148, 153, 160, 164
Photodiodes. *See under* Detectors
Photoelectric
  components, 709, 710
  observations, scanning, xiii, 167, 224, 232,
    233, 520, 695–729, 779, 783, 786, 787, 818,
    867, 985, 986, 1064, 1066
  photometers, 374, 500, 501, 515
  polarimeters (also French solar), xvi, 14, 15, 36,
    246–62, 392, 520, 686, 695–729, 762,
    1063, 1066
  scanners, 14, 15, 238
  techniques, 695
Photographic
  emulsions, 224, 880
  measurements, observations, 38, 786, 787,
    814, 985
  plates, 38, 218, 234, 1014
  studies, 15, 25, 31, 561, 570, 598, 1014, 1060
  subtraction, 223, 224, 227
Photoionization, 275
Photometer
  filter, 406, 637
  general, xiii, xv, 179, 208, 704, 721, 766, 796
  imaging, 21
  scanning, 30, 767, 868
  under-water, 435
Photometry
  comet, 820
  general, 3, 7, 8, 16, 21, 647
  infrared, 954
  lunar. *See under* Moon
  night-sky. *See under* Night sky
  observations, measures, data, 265, 318, 371,
    374, 376, 392, 640, 702, 825, 868, 870
  UBV, 836
  visual, 954
  wide band, 234, 1079
Photomultiplier, phototubes, 22, 141, 143, 160,
    162, 164, 183, 204, 207, 210, 234, 238, 246,
    249, 300, 374, 412, 413, 686, 695, 709,
    710, 755
Photon
  absorption, 479–81
  angular momentum, 39, 958
  collecting, 209, 868, 942
  collisions, 38
  counting (electronic), 942
  counting (process), 93, 136, 151, 160, 164, 165,
    167–69, 272–75, 298, 414, 541, 544, 752,
    755, 1002
  counting image tubes and television, 230
  detectors, 284. *See also* Detectors
  energy, 270, 271, 273, 274, 279, 292–98
  field, 433
  general, 273, 274, 291, 295–97, 308, 433, 449,
    477, 544, 699, 700, 733, 734, 746
  noise, 136–38, 140
  oscillating, 474
  polarized, 274, 275, 282, 294, 308, 334
  power spectrum, 942
  scattering. *See under* Scattering
Photopolarimeters
  general, 213, 388, 638, 645, 648
  imaging, 204, 205, 207
  Rutgers, 794–98
  two-color spin-scan, 205
Photopolarimetric
  accuracy, 230, 650, 651
  data, 7, 8, 14, 28, 43, 371, 376, 645, 647, 792,
    796–98
  high-resolution, 647, 648, 658
  quantities, 382, 383, 403
Photoresist, 338, 881
Piddington, J. H., 1022, 1024, 1027, 1053, 1058
Pieters, C., 11, 21, 23, 26, **404–18,** 1089
Pikoos, C., 386, 402
Pinard, J., 641, 660
Pines, D., 57, 60, 264, 315
Pinnington, E. H., 104, 106
Pioneer 10 and 11. *See under* Space missions
Pipher, J. L., 334, 349
Pitz, E., 21, 34, 199, 203, **766–67,** 787, 789, 790,
    792, 1089
Planck function, 664, 741
Planck's constant, 98, 281
Planetary
  albedos, reflectivities, 405, 406, 417, 621, 622,
    625, 627, 628
  atmospheres, 3, 7, 8, 28–31, 85, 223, 224, 227,
    228, 381, 382, 445, 518–81, 608, 615–19,
    637, 599–606
  circular polarization, 61, 607–11
  disks, 628, 630, 632, 634
  light scattering, 607, 608
  linear polarization, 608
  polar effect, 610–12, 617

INDEX
1117

polarization spectra, 621
problem. *See* Chandrasekhar's planetary problem
satellites, 381–404
subsurface studies, 362, 363
surface, 359, 361, 362, 364, 381, 397, 405, 519, 608–10
studies, 4, 6, 12, 47, 191, 213, 362, 417, 599–616, 619, 695, 700

Plasma
  cold, 1030, 1032, 1034, 1035, 1043, 1055
  dense, 1004
  frequencies, 998
  general, 11, 85, 267, 420, 421, 426, 433, 692
  hot, 57
  magnetic, 1030
  noise, 424, 425, 430
  optical thin, 57
  oscillation, 324, 1004
  relativistic, 1030, 1032, 1033, 1035, 1043, 1050, 1055
  solar, 730, 731
  thermal, 1039, 1040, 1043, 1056

Plass, G. N., 541, 554, 559, 578, 580, 603, 605

Plates
  achromatic, 150, 158, 161, 162, 164, 246, 247, 250, 251
  halfwave encoder, 258
  halfwave (general), 138, 140, 146, 150, 160, 175, 184, 185, 207
  halfwave Pancharatnam achromatic, 135, 150
  halfwave quartz, 148, 256
  halfwave rotating, 139, 141, 150, 154, 156, 158, 164, 165, 174, 254, 256, 755, 756
  halfwave superachromatic, 150, 161, 162, 164
  MgO-smoked, 383
  phase, 174, 752, 755, 757, 758
  quarter-wave, 138, 143, 146, 150, 151, 154, 156, 158, 161, 165, 238, 239, 246, 247, 353, 701, 705, 707, 709, 796, 798
  wave, 204, 207

Platonov, Y. P., 159, 160, 173
Plechaty, E. F., 274, 316
Plesset, M. S., 520, 580
Pluto, 26, 573
Pockels cell, 159, 160, 404, 985
Poincaré sphere, 9, 705–07
Poisson counting statistics, 272
Polar effect (phase dependence, mechanisms), 610–13

Polarimeters
  Aeropol, 196, 204, 206, 549, 565
  animal, 27, 472–94
  Borrmann effect, 272, 276–80
  Bragg-crystal, 262, 272, 277, 280–91
  calibration, 183, 248–52, 256, 343. *See also* Calibration
  circular polarization, 141, 154, 156, 161–63, 280
  comparison, 151
  coronal emission-line, 254–62
  Coudé, 985
  counters, 291
  design, 135–75, 179, 186, 255–57, 260, 319, 342, 345
  errors and remedies, 137, 140
  French solar, 695–729
  gamma-ray, 262, 680
  Hall's scanning, 552, 554, 593
  High Altitude Observatory, 246–61
  high-energy X-ray, 262
  imaging, 262, 312
  infrared, 318, 322, 340, 342, 344
  interferometer, 637, 638, 640, 650, 658
  linear, 154, 159, 161
  linear-plus-circular, 161, 162, 164, 174
  list, 189, 190, 192, 194, 196, 198, 200, 202
  Lyot's, 15
  multichannel, 136, 175, 1073, 1078
  multilayer reflection, 280
  OSO-1, 266–68, 280, 282, 284, 288, 310
  photoelectric photographic, 15, 246–53, 272–75, 392, 453, 456, 457, 460, 510–13, 695–729, 814, 1079
  principles, 696–99, 710
  rapid-signal modulation, 154–61
  scanning, 31, 135, 161–67, 561
  secondary fluorescence, 272, 275
  signals, 45, 255, 258, 319, 344, 345, 347, 352, 354, 358. *See also under* Polarization
  single, 1073
  solar, 246–62, 695–729, 731
  space, 21, 189–217
  Thomson-scattering (incoherent). *See under* Thomson-scattering polarimeters
  ultraviolet, 210
  University of Hawaii, 716
  Visual, visual hand-held, 189, 204, 205, 456, 457, 463, 519, 548
  Wollaston, 16, 153, 154, 160
  X-ray (also circular), 189, 262, 270, 272–80, 312

Polarimetry (polarimetric studies)
  animal, 472–94
  area scanning, 164
  basic concepts, 3, 4, 7–10
  broadband, wide-band, 161, 162, 164, 232
  comet, 820
  discoveries, 135
  efficiencies, 151–53, 160, 174
  extragalactic, 1059–83
  general, xv, xvi, 3–44, 232, 519, 562, 568, 867, 1014, 1023–25

Polarimetry (polarimetric studies), *cont.*
  high-resolution, 3, 39, 647, 648, 1025
  infrared, 318–51, 762–65, 858–66
  late-type stars, 821–44
  lunar, 381
  measurements, observations, 54, 167, 382–85, 387, 388, 392, 395, 398, 405, 498, 510–13, 518, 753, 762, 768, 816, 825, 839, 854, 855, 1025
  method, 796–98
  multichannel, 175, 768–80, 890, 894, 895
  night sky. *See* Night sky
  planets, 27, 599
  photographic, 880
  radar, 359–67
  radio, 24, 352–58
  solar, 730–32, 734–41
  spectral, 405, 406, 417, 752–61
  standard stars, 168, 169
  stellar, 262–67, 318
  techniques, 608
  visual, 15, 456, 457, 463, 519
  X-ray, 262–317
Polariscope balloon system, 22, 210, 211, 567
Polarization
  -aerosol contribution, 563
  amount, degree, percentage, 13, 46, 47, 52, 168–70, 209, 211, 227–29, 232, 236, 320–22, 326–30, 334–37, 341–43, 348, 352, 361, 371–74, 376, 378, 379, 381, 382, 384, 404, 406–16, 434, 436–41, 444, 449, 451–54, 456, 458, 459, 463–67, 471, 474, 476, 485, 496, 501, 502, 505–07, 524, 525, 529, 531, 532, 535, 537, 538, 542–45, 547–49, 565–67, 571, 603, 606, 671–79, 684, 686–94, 700–04, 714, 716, 732, 736, 740, 747, 748, 794, 799, 800, 805, 808, 814, 816–19, 823, 824, 833, 847, 848, 851, 859, 891, 894, 905, 906, 909, 910, 912, 913, 1016, 1018, 1060, 1064, 1084
  analyzers. *See* Analyzers
  artificial satellites, 371–80
  biological materials (animals, plants), 495–99
  broadband, 686–93
  circular. *See* Circular polarization
  -coherence relationship, 677, 678
  -color relationship, 229, 874, 876, 877
  compensators, 700–04, 708
  composition determinator, 388, 398, 399
  continuum, 687, 881–85, 960, 961, 964, 983, 984, 988, 989, 1059
  cross, 906
  curve shape, slope, 382, 398, 464–66, 544, 766
  depolarization effects, 211, 326, 327, 340, 341, 363, 377, 384, 387, 437, 447, 449, 457, 474, 505, 513, 651, 731, 735, 746–48, 881, 934–36, 1023. *See also* Collisional depolarization
  diluted, 32, 498, 768
  earth's environment, 229, 444–71, 500–13, 515, 976
  effects. *See* depolarization effects *under* Polarization
  elliptical. *See* Elliptical polarization
  extraterrestrial, 766, 767
  field, 450–53, 464–66
  fluorescent, 730, 734, 738, 742, 745, 747, 1067
  flux (also flux curve), 271, 283, 288, 289, 298, 299, 615, 616, 702, 769–75, 777, 778, 847–49, 886, 1073, 1075, 1080
  impact, 730, 731, 747
  infrared, 548, 947–49
  instrumental. *See under* Instrument
  intensity, state, 552, 554, 637
  interstellar. *See under* Interstellar
  light, 483, 698, 703, 731–34
  light unpolarized. *See under* Light
  line, 687, 731
  linear. *See* Linear polarization
  mechanism, 54–63, 434, 435, 444, 463, 814, 816, 821, 950
  modulators. *See* Modulators
  negative, 13, 24–26, 47, 50, 52, 53, 268, 270, 382, 384, 449, 450, 453, 458, 529, 533, 548, 565, 568, 571, 788, 805, 809, 870, 874
  negative branch (phase curve), 381, 385–87, 390, 392, 394, 395, 397, 398, 400, 537, 539, 548, 565, 614, 653, 657, 658
  nonradial, 870
  observations, measurements, 44, 209, 218–32, 235, 246–61, 288, 294, 295, 302, 303, 306–10, 313, 318, 320, 321, 381, 382, 386, 388, 521, 534
  optical data, 960–67, 970, 972, 974
  partial, 435, 437
  patterns, distribution, 434–43, 446, 450, 453, 454, 466, 475, 593, 595, 598, 715, 725, 768, 772–74, 794, 801, 802, 889, 890, 961, 963–66, 989, 998, 1018, 1020, 1021, 1024, 1025, 1052, 1053
  peak, 59
  -phase angle relationship, 519, 651
  phase curve structure (Venus, Saturn), 381–85, 387, 392, 398, 519, 573
  plants. *See* biological materials *under* Polarization
  -position angle variation, 262, 267, 284, 303, 519–21, 552, 712, 713, 719, 771–74
  positive, 24, 25, 47, 50, 52, 53, 268, 270, 382, 384, 387, 449, 458, 512, 529, 533, 548, 549, 554, 571, 606, 805, 807, 808, 810, 819, 870, 874

INDEX 1119

positive branch (phase curve), 386, 390, 392, 465, 537–39, 548, 653, 657, 658
properties of materials, 463, 566, 593, 598
radial, 227–29, 570, 593, 870, 1029
reflectivity, 25, 26
resonance, 734, 738–40, 743, 747
reversal, 617, 632, 634
signals, 432, 433
spatial distribution, 318, 320, 321, 382, 386, 679, 684, 686, 881, 883–85, 887
spectral, 617–36, 645, 647, 649–51, 653, 657, 658, 684, 746, 752–61, 827, 828, 982–84
spurious, 174, 272–74, 289, 291, 303, 306, 686–88, 780, 887, 985. *See also* polarization *under* Instrument
sunlight, 371, 518
theory, 730, 731, 734–41, 1030
tilt, 434, 437
time variability, 32, 35
underwater measurements, 28, 439–42
uses, 88, 324, 359, 371, 805
variations, variability (also differences), 25, 28, 161, 211, 212, 223, 227–30, 250, 255, 266, 270, 327, 387, 388, 392, 400, 404–07, 412, 414–17, 436, 444, 450, 513, 523–25, 540, 551–53, 557, 563, 569–71, 573, 593, 595–97, 629–32, 634, 681, 684, 746, 753, 794, 809, 815–18, 821–23, 825, 829, 830, 833, 839, 845–47, 849, 851, 884, 885, 896, 953, 983, 989, 1051–53, 1070, 1071, 1077
versus brightness distribution, 552, 679, 804, 827, 869, 870
versus grain characteristics, 34
versus line strength, 617
versus magnetic field strength, 735, 736
versus reflection properties, 444
versus source size, particle size, 388, 405, 406, 417, 533–35, 557, 685, 814, 907, 908, 911, 913
versus spectral variations, 322, 323, 348, 414, 416, 869, 870
wavelength dependence, 7, 8, 16–18, 32–37, 40, 60, 163, 167, 170, 326, 348, 388, 392, 405, 408, 414, 418, 434, 440, 441, 444, 450, 453, 454, 466, 496, 514, 515, 518–20, 522, 526, 551, 553, 555, 556, 559, 561, 563, 567, 570, 571, 573, 609, 611, 637, 649, 663, 671, 687, 753, 754, 763, 789, 811, 818, 819, 821–23, 825–29, 833–40, 846, 851, 852, 855, 861–65, 867, 872, 874, 880, 884, 885, 888, 890, 891, 916, 926, 928–36, 946–52, 954, 957, 959, 984, 985, 993, 1075
zodiacal light, 788–90, 804
Polarizers
birefringent, 336, 337
Brewster angle, 337
dichroic, 6, 336, 337
far infrared, 204, 206, 210, 322
fixed, 174, 562
general, 211, 224, 232, 234, 246, 248, 250, 319, 320, 336–38, 376, 404, 798, 803
grid, wire grid, 337–39, 858
linear, 98, 99, 101, 176, 352
Polaroid. *See* Polaroid
rotating, 704
Rouy, 796, 797, 802, 803
sheet, 6, 858
Polaroid (plates, filters), 6, 16, 20, 21, 27, 152, 156, 165, 187, 207, 209, 219, 222–24, 250, 337, 374, 412, 413, 565, 702, 704, 767, 790, 1014, 1015, 1017, 1018
Polarotaxis, 483, 485–87
Polar scattering effect, 608–12
Poljakova, T. A., 155, 172
Pollack, J. B., 363, 367, 556, 580
Polymer, 76, 77, 559
Poon, P. T. Y., 584, 591
Porter, F. C., 833, 843
Porteus, J. O., 446, 470
Position angle
calibration, 20, 135, 153
general, 9, 19–21, 26, 35, 52, 53, 141, 144, 146, 165, 167–69, 189, 210, 320, 321, 352–54, 385, 473, 476, 483, 489, 769–74, 815, 847, 884, 887, 926, 928, 930, 931, 998, 1000–02, 1015, 1021, 1025, 1028, 1037, 1040, 1042, 1044, 1063, 1064, 1066, 1070, 1073, 1074, 1077, 1079, 1084
-polarization relationship. *See under* Polarization
sweeps, variations, variability, 35, 821, 823–26, 828, 831–33, 835, 840, 844, 847, 848, 861, 862, 891, 900, 928, 930, 935, 936, 944, 946–53, 998, 1000, 1002, 1006, 1008, 1010, 1044
variation for thick atmospheres, 519
-wavelength relationship, 35, 37, 150, 935, 936
Positive polarization. *See under* Polarization
Pospergelis, M. M., 50, 51, 141, 173, 175, 179, 183, 186, 499, 607, 609, 614
Potter, J. F., 619, 636
Poulain, P., 762, 764, 765
Poulizac, M. C., 104, 106
Poulsen, O., 136, 173
Powell, R. S., 807, 813
Poynting flux, vectors, 110, 120, 143
Prabhakara, C., 618, 635, 642, 659, 660
Prasad, J. S., 496, 499
Preisendorfer, R. W., 438, 443
Preisendorfer's linear interaction principle, 625, 626

Pressure
  cloud top, 526, 552, 553, 556, 558
  Doppler, 624
  Galetry, 624
  gaseous atmosphere, 521, 524, 552, 553, 556, 558, 624
  Lorentz, 624
  Voigt, 624
Preston, G. W., 238, 245, 982, 985, 987, 993, 994, 996
Prestrud, M. C., 582, 589
Priebe, J. R., 179, 186
Pringle, J. E., 264, 316
Prism
  biprism, 256
  dichroic, 162
  general, 16, 207
  Glan-Thomson, 143, 154, 162, 501
  Rouy, 796, 797, 802, 803
  Wollaston, 16, 138, 141, 142, 146, 151, 153, 161, 162, 164, 174, 210, 256, 686, 696, 701, 708, 858
Pritchard, B. S., 549, 580
Prouteau, J., 707, 726
Provin, S., 392, 402
Puiseun-Das, S., 496, 499
Pulido, A. A., 526, 577
Pulsars
  Cen X-3, 262, 308
  Crab Nebula, 262, 264
  Cyg X-1, 264
  electrostatic mechanism, 998
  emission mechanism, 1002, 1004–06, 1010
  general, 3, 33, 57, 262–64, 267, 308, 310, 352, 355, 357, 982, 997–1013, 1005, 1051, 1079
  Her X-1, 262, 264
  magnetic dipole radiation mechanism, 998, 1006
  magnetosphere, 998, 1004, 1005, 1009, 1010
  PSR 1508, 998
  PSR 0327+54, 968
  radio observations, 998
  rotation axis data, 966, 968–70, 1007, 1014
  theories, 267
  Vela, 998, 999, 1007
  X-ray, 57, 267, 310, 1005
Pulsation nonradial (stellar), 824
Pulses overlapping, 160, 162
Puplett, E., 338, 350
Purcell, E. M., 60, 62, 335, 350, 888, 890
Purton, C. R., 1023, 1025
Pyashovskaya-Fesenkova, E. V., 456, 470

Quantum
  beats, 88, 90–94, 96–105
  efficiency, 710
  jump, 744, 745
  theory, mechanical treatment, 735, 737, 738, 742–46
Quarter-wave plate. *See under* Plates
Quartz rotator, 798, 802, 803
Quasars, 3, 55, 59, 352, 358, 941, 967, 986, 1047, 1048, 1050, 1051
Quasi stellar object (QSO), 263, 266, 981
Quasi stellar sources (QSS), 1070–75
Querfeld, C. W., 15, 32, 157, **254–61**, 1089

Rachkovsky, D. N., 741, 750
Racine, R., 868, 879
Radar
  frequencies, 65
  polarimetry, 23, 24, 359–65
  pulse, 84
  range, 359
  signal, 359–64, 366, 779
Radhakrishnan, V., 356, 358, 998, 999, 1006, 1012, 1058
Radial velocity measurements, 165, 166
Radiation
  absorption, 738
  background, 355, 858, 1000
  backscatter, 361, 671, 801, 809, 811, 911, 913
  Cherenkov. *See* Cherenkov radiation
  coherent, 1004
  continuum, 1063–68, 1070, 1072–75, 1077
  curvature, 1004, 1006
  cyclotron, 55, 57, 1005, 1009
  diffuse, 619, 622, 623, 625, 628, 670–72
  dipole, 746, 998, 1006
  electric, 419, 431
  emission, 1004
  fields, 430, 448, 584, 585, 617, 618, 637, 665, 667, 668, 672–74
  fluorescence, 730, 742, 746
  incoherent, 1005
  magnetic, 430
  maser, 1004
  monochromatic, 447
  nonthermal, 1070–72, 1075, 1077, 1079
  pressure, 39, 958
  resonance, 742
  synchrotron, 7, 55, 59, 270, 666, 997, 1004–07, 1010, 1015, 1019–23, 1028, 1030, 1042, 1043, 1055, 1060, 1063, 1065–67, 1070, 1072, 1075, 1077
  thermal. *See under* Thermal
Radiative transfer, 7, 54, 57, 58, 60, 445, 447, 500, 502–04, 508, 553, 664, 715, 740, 741 867, 871, 877, 984, 992
Radiators (graybody magnetic) (magneto-emission), 984, 985

INDEX

Radio
  background data, 960
  bursts, 268, 269
  circular polarization, 986, 1030
  21-cm data, 963, 969, 971
  emission, 55, 265, 268, 981, 986
  extragalactic sources. *See under* Sources
  galaxies. *See under* Galaxies
  polarization, 23, 24, 40, 262, 352–58, 397, 676–85, 1030, 1056
  -source-rotation measurements, 960, 966–70
  techniques, 40
  waves, wavelengths, 55, 56, 329, 676, 680
Radiometry, 213, 233, 397, 400
Radziemski, L. J., 100, 106
Raimond, E., 1023, 1026, 1055, 1056
"Rainbows"
  angle, 562
  peak, 534
  primary, 528, 533, 534, 537, 540, 546, 548, 549, 556, 558, 562
  secondary, 528, 534, 537
  size dependence, 537
  supernumerary, 533, 534
Rajagopal, A. K., 986, 993, 996
Ramachandran, G. N., 146, 173
Ramaseshan, S., 146, 173
Ramaty, R., 1038, 1058
Rank, D. M., 833, 842
Rankin, J. M., 1000, 1001, 1011, 1012
Ranyard, A. C., 786, 792
Rao, C. R. N., 29, 191, 195, 215, 216, 445, 446, 457, 460, 461, 466–70, **500–09**, 559, 562, 574, 580, 1089
Rappaport, S., 1002, 1012
Raschke, E., 8, 29, 445, **510–13**, 563, 1089
Raster scans, 686, 690
Ratcliffe, J. A., 56, 62
Ratier, G., 15, 32, 726, **762–65**, 1089
Rauser, P., 517
Ray calculations, 916
Ray optics, 531–33
Ray (WKB) approximation, 916–18, 920
Rayleigh (J. W. Strutt, Lord Rayleigh), 6, 30, 232, 236, 445, 470, 734, 750, 780, 792
Rayleigh
  atmosphere, 446–55, 457
  -Chandrasekhar theory, 28–30, 398
  color, 516
  elementary theory, 445, 503, 917, 918, 920
  ellipsoid, 776
  limit, 426, 430
  phase function, 618, 663, 668, 672
  points, 232
  -type scattering. *See under* Scattering
Rea, D., xv, 230, 364, 367, 1089
Reasoner, D. L., 754, 760

Rebillot, R. J., 379
Reddish, V. C., 892, 899
Redheffer, R., 520, 541, 580, 584, 591
Red shift, 1050, 1051, 1077
Rees, D. E., 741, 750
Rees, M. J., 55, 57, 59, 63, 264, 265, 267, 314, 316, 932, 938, 1004, 1005, 1012, 1022–25, 1027, 1053, 1058, 1079, 1082
Reflection, reflectivity
  coefficient, 905
  diffuse, 582, 584, 588, 617, 626
  Fresnel. *See under* Fresnel
  functions, 541
  irradiance, 474, 477
  isotropic, 582, 584
  light, 47, 65, 234, 371, 373, 382, 405, 495–97, 503, 527, 528, 531, 534, 562
  magneto, 608
  measurements, 47, 340, 405–07, 409, 410, 412, 414, 416–18
  nebula. *See* Reflection nebula
  polarized, 562
  properties, 24, 150–52, 164–66, 444–46, 463, 473
  specular, 455, 582–84, 587, 588, 612, 614
  surfaces, 12, 24, 150–52, 164–66, 444–46, 463, 473, 1085
  theory, 371–74, 378, 445
Reflection nebulae
  Becklin Neugebauer object, 958, 959
  brightness, 867, 870, 874, 876, 880
  catalogues, 867, 868
  color, 867, 868, 870
  30 Doradus, 880
  filaments, 870
  general, 3, 36–38, 834, 867–80, 885, 933, 946, 949, 951
  illuminating star, 870, 872, 877, 879
  polarization, 867–70
Reflectors
  diffuse, 583
  double, 613
  specular, 583, 585
Refracted light, 378, 379, 382, 384, 528, 529
Refractive index
  aerosol particles, 518, 520
  double, 5, 43, 528
  general, 5, 11, 30, 60, 66, 67, 80, 107, 127, 145–47, 187, 372, 378, 400, 407, 409, 417, 466, 504, 527–30, 536, 540–42, 546, 553–60, 569, 570, 602, 604, 631, 701, 741, 779, 805, 808, 810, 813, 819, 871, 872, 874, 877, 895–97, 901, 905, 911, 913, 916, 917, 927, 929, 932, 939, 944, 1084
  imaginary part, 556
  real part, 406, 407

Refractive index, *cont.*
  variable polarization, 146, 147, 151, 540
Regas, J., 619, 636
Regener, V. H., 191, 216, 787, 792
Reichardt, W., 489, 493
Reinhardt, M., 327, 350, 967, 971, 1051, 1058
Remote sensing
  general, 521
  optically active materials, 213, 498
  scattered sunlight, 518
Rennilson, J. J., 201, 214
Rense, W. A., 195, 216, 786, 792
Resonance, 12, 334
Retallack, D., 265, 315
Retardation frequencies, 174, 248, 251, 510–12
Retarders
  achromatic, 148–50, 160
  general, 166
  halfwave plates. *See under* Plates
  materials, 148, 149, 165
  Pancharatnam, 149, 150
  perfect, 144, 145
  Polaroid plastic, 187
  quarter-wave plate. *See under* Plates
  quartz, 216
  rotating, 141, 142, 146–48, 184, 185
  single, 184, 185
  theory, 187
  total-internal-reflection, 179
  variable, 160, 161
  wedge-shaped, 404
Retina (retinular cells), 472, 473, 476–80, 482, 489, 490
Reynolds, J. H., 1060, 1082
Rhabdoms, 472, 473, 477, 478, 480, 489–91
Rhea, 391
Rhodes, D. R., 902, 914
Rhomb, 247
Ribes, J. C., 55, 63, 1055, 1058
Riccati type, 582
Rice, L., 195, 215
Richards, J. C. S., 235
Richardson, J. H., 20
Richter, H. L., 201, 217
Richter, N. B., 809, 813
Riddle, A. C., 329, 350
Riegler, G. R., 274, 316
Riley, L. A., 30, 31, 41, 227, 230, 388, 402, 570, 573, 577, **593–98,** 1089
Ring, J., 757, 761
Roach, F. E., 21, 34, 42, 201, 216, 780, 781, 790, 792, **794–803,** 1089
Roach, J. R., 21, 34, **794–803,** 1089
Roark, B., 880
Roark, T. P., 868, 872, 874, 878–80, 933, 937
Roberts, J. A., 55, 63, 326, 350, 1035,
  1055, 1058
Robinson, C., 496, 499
Robinson, E. L., 984, 987
Robley, R., 787, 792
Roche lobe, 849
Rocket
  -borne polarimeters, 196–99
  observations, 22, 189, 207, 208, 213, 471, 515, 517, 766, 767, 787–89, 813, 901
  programs, 766, 767
Rodgers, S., 253
Roemer, E., 392, 401, 816, 817, 873
Roland, E. H., 161
Roland, G., 642, 660
Ron, A., 340, 350
Roosen, R. G., 34, 42
Rose, L. J., 197, 216, 787, 788, 793
Rosenberg, D. L., 383, 384, 388, 394, 396, 402
Rosenberg, H., 326, 350
Rosse, W. P., (Lord Rosse), 6, 550, 580
Rotating neutron star. *See* Neutron stars
Rotational variation of polarization, 392, 400
Rotation measurements, 988, 1044, 1051, 1052
Roussel, K. M., 993
Roux, F., 642, 660
Rouy, A. L., 796, 802
Rowell, R. L., 521, 580
Rozenberg, G. V., 462, 470, 500, 507, 509, 562–64, 580
Rozhkovski, D. A., 868, 879
Rucinski, S. M., 668, 674, 845, 847, 856
Ruderman, M. A., 268
Rudkjøbing, M., 925
Rupprecht, G., 337, 350
Rush, J., 15, 32, **246–53,** 260, 1089
Russell, E. E., 203, 215
Rybicki, G., 582, 590

Sabitov, S. N., 816, 817
Sagan, C., 391, 402, 556, 580, 619, 636
Sahade, J., 849, 856
Sakai, H., 638, 660
Salpeter, E. E., 332, 351
Sanchez Martinez, F., 787, 788, 791, 792
Sandage, A., 1065–69, 1081–83
Sandeman, D. C., 489, 492
Sanders, R. H., 1069, 1082
Sandford, M. T., 157, 172, 199, 216, 825, 832, 836, 841, 859, 1089
Sanford, P. W., 275, 316
Sartori, L., 267, 316, 1005, 1006, 1012
Sasaki, T., 440, 443
Satellite (artificial) polarization effect, 371–80
Satellites (planetary). *See under* Planetary
Sato, S., 157, 172, 862, 866

Saturation
  bands, 410
  effect, 689, 693, 715, 716
  lines, 617, 628, 629, 632, 634, 645
Saturn
  atmospheric studies, 30, 571–73, 599, 608
  circular polarization, 573, 607, 612, 613, 615, 616
  disk, 604, 605
  latitudinal variation of polarization, 518, 573
  phase angle, 388, 392
  polarization measurements, 223, 228, 388, 392, 518, 519, 572, 607, 610, 611, 613, 615
  rings, 24, 26, 518, 604, 612, 613
  satellites, 381, 382, 391, 392, 399
  second scattering in a gas, 519, 613
  single scattering clouds, 519, 613
  spectra, 613, 642
Sauer, K., 497, 499
Sautter, C. A., 100, 106
Savage, B. D., 37, 41
Sayer, M.-O., 496, 499
Sazonov, V. N., 612, 615, 1031, 1032, 1035, 1043, 1055, 1058
Scanning techniques, 642, 710, 717, 721
Scargle, J. D., 266, 310, 317, 1002, 1007, 1013, 1015, 1018, 1019, 1021–25, 1027, 1028
Scattering
  aerosol, 541, 613, 617, 618
  albedo, 933
  anisotropic, 306, 663, 667–72
  antisymmetric, 776, 777
  asymmetric, 307, 610, 611, 933
  atomic, 59
  block, 297, 299–301, 303–05, 307, 308, 310
  calculations, solutions, 107–34, 570
  circumstellar envelope, 835, 836
  Compton, 59, 280, 986, 1077, 1079
  diagrams, 906
  dielectric, 862, 867
  distributions (angular scattering), 124–26
  double, 608, 612, 613
  dust particle, 38, 59–61, 767, 823
  efficiency, 117
  electron, 31, 32, 58, 245, 246, 267, 667, 668, 673, 687, 753, 846, 851–53, 855, 860, 861, 870
  enhanced, 533
  formulation, 108–14
  forward, 112, 114, 361, 533, 539, 652, 874, 927
  incoherent, 9, 742
  isotropic, 29, 30, 623, 667–69, 671–73, 806, 913
  large particle, 520, 823
  light, 47, 58, 107, 526, 548, 562
  materials, 10, 295–97, 307
  matrix, 361
  Mie, 419–33, 502, 531–34, 541, 613, 616, 617, 631, 652, 870, 880
  model. *See under* Model
  molecular, 9, 11, 29, 30, 37, 38, 59, 387, 444, 445, 502, 503, 505, 515, 522, 563, 564, 763, 820, 836
  multiple, 9, 11, 29, 30, 37, 38, 307, 308, 400, 445, 447, 449, 450, 454, 461, 510, 513, 520–22, 525, 534, 540–45, 557, 562–64, 568–72, 599, 616, 617, 730, 742, 743, 748, 762, 763, 818, 835, 863, 874, 942
  non-Rayleigh, 455, 462, 524
  nonspherical, 262, 522, 777, 901, 905
  2nd and 3rd order scattering, 877, 879
  particle, 107–34, 442, 518, 521, 526, 819, 867, 870, 871, 874, 876, 877
  photon, 324, 540, 543, 553
  plane, 9, 127–30, 132–34, 227, 374, 382, 385, 412, 447, 450, 530
  planetary, 519, 637
  primary, 447, 450, 763, 764
  processes, 54, 58–61, 362–64, 372, 423
  properties, 359, 362–64, 373, 379, 382, 437–39, 888
  radiative, 741
  Rayleigh, 10, 30, 59, 451, 461, 472, 473, 475, 520–23, 525, 526, 528, 539, 541, 552, 553, 555–57, 560, 562, 570, 599, 613, 615, 624, 631, 687, 693, 818, 823, 827, 833, 836, 1059
  Rayleigh-Cabannes, 617, 624, 631
  Rayleigh-Chandrasekhar, 12, 16, 17, 28
  resonance, 693, 711, 713, 731, 734–36, 738
  single-particles, 9, 11, 436, 437, 508, 519, 520, 522, 524, 529–31, 533, 534, 537, 538, 553, 572, 587, 612, 616, 762, 777, 818
  small-particle, 107–134
  solid-particle, 835–37
  spherical-particle, 421, 430, 520, 530–33, 537, 538, 540–45, 547, 553, 568–70
  surfaces, 371, 372, 382, 384
  theory, 12, 28, 108, 112, 446, 584, 585, 730, 741, 742, 820
  Thomson, 279, 294, 295, 687, 693, 719, 1059
  underwater, 434–36, 439, 442, 473, 474
Schiffer, R. A., 525, 580
Schild, R., 140, 172
Schlachman, B., 618, 635, 642, 660
Schlosser, W., 39, **972–75,** 1089
Schmalberger, D. C., 29, 42, 517, 813
Schmidt, T., 39, 60, 63, 174, 774–79, 788, 792, **976–80,** 1089
Schmidt-Kaler, Th., 39, **972–75,** 1089
Schmidt-Koenig, K., 475, 492
Schmidt mechanism for absorbing grains, 770
Schmitt, J.-L., 1076, 1082

Schneider, L., 474, 489, 493
Schnur, G., 889, 899
Schommer, R. A., 960, 971
Schönhardt, R. E., 1000, 1002, 1006, 1010, 1012
Schreier, E., 263, 265, 289, 310, 314, 316
Schröder, R., 979, 980
Schrödinger wave equation, 109, 113
Schubart, J., 26, 42
Schubert, G., 419, 421, 422, 432, 433
Schwartz, K., 419, 421, 422, 432, 433
Schwarzschild-Milne integral equations, 663, 664
Sciama, D. W., 59, 63, 1023, 1027, 1079, 1082
Seaman, C., 457, 460, 470
Seaquist, E. R., 265, 266, 892, 899, 1055, 1058
Searle, L., 1071, 1082
Seaton, M. J., 747, 750
Seeger, C. L., 354, 358
Seeley, J. S., 656, 660
Seielstad, G. A., 356, 358, 1023, 1025, 1044, 1055, 1056, 1058
Sekera, Z., 7, 41, 191, 216, 448, 451, 452, 454, 457, 460, 463, 468, 470, 473, 474, 493, 500, 503–05, 508, 509, 520, 523, 541, 564, 575, 580, 583, 585, 591, 618, 626, 636, 649
Sellin, I. A., 93, 106
Sen, K. K., 753, 761
Serkowski, K., xv, 7, 9, 15, 16, 18–20, 35, 36, 40, 42, 60, 61, 63, 126, 127, **135–74,** 179, 182, 186, 187, 216, 224, 230, 404, 433, 675, 753, 761, 822–25, 827, 829–37, 841–47, 851, 856, 858, 860, 862, 866, 870, 879, 889, 890, 893–900, 927, 932, 933, 936, 938, 940, 942, 946, 948, 952, 955, 959, 961, 971, 981, 987, 988, 991, 1077, 1078, 1082, 1089
Sesnic, S., 232, 235
Severny, A. B., 238, 245, 985–87
Seyfert galaxies. See under Extragalactic objects
Shadow pattern, 139
Shah, G., 871, 879
Shajn, G. A., 975
Shakhovskoj, N. M., xv, 7, 135, 157, 173, 823, 825, 835, 843, 845–47, 854, 856
Shaw, S. R., 480, 493
Shawl, S. J., 36, 157, 172, 321, **821–44,** 858, 862, 864–66, 1089
Shieh, D. J., 77, 87
Shifrin, K. S., 521, 580
Shimabukuro, F. I., 57, 61
Shipman, H. L., 984, 987
Shklovskij, I. S., 7, 1005, 1012, 1014, 1027, 1063, 1082
Shoemaker, E. M., 201, 216
Shorthill, R. W., 363, 367
Shu, F., 959
Shulov, O. S., 155, 172, 845, 847, 849, 856

Shurcliff, W. A., 4, 9, 14, 27, 42, 45, 52, 175, 187, 247, 253, 337, 350, 649, 660, 734, 751
Shuryghin, A. I., 58, 63, 270, 294, 316
Shutt, R. L., 195, 214, 254, 261, 746, 749
Siedentopf, H., 786, 791, 804, 812, 901, 914
Siegel, S., 499
Siewert, C. E., 524, 574
Signal
 analogue, 258
 general, 139, 143, 151, 153, 160, 209, 359, 779, 905
 radar. See under Radar
 -to-noise ratio, -to-background ratio, 237–44, 258, 274, 275, 282, 285, 288, 291, 295, 300, 302, 303, 306, 307, 313, 342, 362, 364, 638, 640, 641, 645, 653, 656–58, 699, 700, 704, 709, 780, 989
 spurious. See signals and spurious under Polarization
Sign conventions, 12, 45, 51, 53, 82, 353, 691. See also Handedness, and sign reversal under Circular polarization
Sill, W. R., 421, 432
Silvester, A. B., 890, 891, 899, 935, 937
Simon, T., 386, 402
Simpson, R. A., 363, 367
Sinton, W. M., 559, 580
Sirovantzkii, S. I., 324, 349
Sites (dark), 788. See also High altitude sites
Size. See under Particles, Grains, Dust
Size determination of asteroids and satellites, 25, 26, 465, 504
Size distribution
 bimodal, 536, 537
 effect, 533–40
 general, 8, 11, 30, 31, 34, 107, 504, 515, 516, 518, 520, 528–30, 533–40, 543, 544, 546, 548, 555–58, 572, 789, 804, 806, 807, 819, 871, 874, 895, 906, 954
 Oort-van de Hulst, 895
Skliarevskii, V. G., 197, 213, 564, 574
Sky, skylight
 background brightness, 134, 138, 142, 143, 162, 300, 374, 434, 435, 437, 439, 440, 444–46, 451, 500, 541, 562, 564, 615, 704, 719, 762, 768
 dark regions, 869
 foreground, 886
 infrared polarimetry, measurements, 562, 762–65
 multicolor polarimetry, 768–80
 neutral points, 444–46, 450, 453, 455, 457–66, 473, 474
 night. See Night sky
 noise, 869
 overcast, cloudy, 444

INDEX

polarization, 59, 344, 346, 447, 449, 451, 453, 456, 459, 461, 463, 474, 475, 500, 502, 505, 507, 508, 763–65, 883
subtraction (polarization removed), 318, 319
sunlit blue, 3, 5, 6, 17, 18, 22, 28, 29
transparency, 687
unpolarized, 143
unsteady, variable, 137, 142, 143, 1007
Sloanaker, R. M., 7, 354, 358
Slysh, V. I., 1005, 1012
Smak, J., 836, 843
Smith, B. A., 391, 402
Smith, B. F., 419, 433
Smith, E. R., 993, 996
Smith, E. W., 742, 747, 750
Smith, F. G., 327, 358, 997, 998, 1001, 1005, 1008, 1009, 1011, 1012
Smith, H. E., 264, 265, 314, 316
Smith, L., 802
Smith, M. W., 104, 106
Smith, R. C., 441, 443
Smith S., 1060, 1082
Smith, S. M., 217
Smog, 505, 513, 563. *See also* Aerosol
Smola, U., 490, 493
Snellen, G. H., 47, 51, 266, 268, 310, 315, 1002, 1007, 1012, 1022, 1025, 1026
Snell's law, 439, 528, 533, 540
Snyder, A. W., 480, 493
Soberman, R. K., 516
Sobolev, V. V., 445, 470, 554, 556, 580, 582, 583, 591, 601, 605, 872
Sofue, Y., 1051, 1057
Soifer, B. T., 334, 349
Solar
  absorption, 730
  active regions, 676, 680
  almucanter. *See* Almucanter
  area sources. *See under* Sources
  atmosphere, 687
  beams, 447
  cell arrays, 371
  center, 681
  chromosphere, chromospheric flash spectrum, 692
  corona, electron corona, 6, 32, 58, 59, 195, 197, 199, 211, 692, 695, 702, 710, 719, 721, 762, 746, 762, 763, 766
  coronal lines, 730, 731, 745–48
  coronal streamers, 695, 710, 719, 721, 723, 724, 726
  D1 and D2 lines, 738
  disturbance, 420, 428
  eclipse, 6, 29, 32, 562, 719, 721, 723, 724, 726, 730, 746, 747, 764
  emission, 692, 730–51
  flares, 32, 262, 268, 270, 680, 686, 687, 691–94
  Fraunhofer spectrum, 32
  K-corona measurements, 762–65
  limb, limb darkening, 32, 513, 686–88, 693, 695, 702, 708, 710–12, 719, 763, 765
  magnetic field, 695, 710, 712, 713, 715–18, 730, 731, 735, 736, 738, 746–48, 985
  magnetograph. *See* Magnetographs
  microwave source, 680
  nonthermal radiation, 687
  observations during eclipse, 719, 721, 723, 724, 726, 730, 746, 747
  penumbrae multiple spot fields, 716
  penumbrae spots (bipolar), 716
  photosphere, photospheric radiation field, 32, 446, 692, 695, 731, 746, 748
  photospheric polarization, 513, 711–16, 721, 725
  pillars (sun), 546
  plages, 680–83, 712
  polar regions, 764
  prominences, prominence lines, 32, 59, 730, 731, 747, 748
  quiet, 420, 723, 725, 726
  radius, 420
  rotation, 426
  spectroscopy. *See under* Spectroscopy
  spectrum, 244, 656
  studies, 3, 4, 6, 15, 23, 29, 33, 47, 160, 211, 246–317, 343, 510–13, 695–751
  sunspots (also magnetic field), 6, 57, 686, 688–90, 693, 708, 715, 716, 718, 719, 988
  surface, 420
  -system objects, 211, 359, 381, 382, 394–400, 766, 767
  wind. *See* Solar wind
  X-ray flares, 58
Solar wind
  experiments, 419, 432
  magnetic field, 420, 422–26, 432, 433
  momentum transfer (Gold mechanism), 774
  pressure, 420, 421, 423
  protons, 816
  vector, 424
  velocity, speed, 420, 422–25, 428–30, 433
Solc filters, 182
Solinger, A. B., 1066, 1067, 1082
Solomon, P., 833, 843, 959
Sonett, C. P., xv, **419-33**, 1089
Soref, R. A., 153, 174
Soret, J. L., 455, 456, 470
Sorvari, J. M., 264, 315
Sources
  area, 676–81
  compact, dense, 265, 313, 1046–48, 1055

Sources, *cont.*
  Cyg A. *See* Cygnus A
  double radio, 1043–45
  extended, 22, 1043, 1044, 1046, 1052–55
  extragalactic, 7, 40, 1046, 1051, 1059–83
  far-infrared, 328–30
  microwave, 680
  observations, 50, 57, 264, 265, 288, 293, 306, 307, 310, 313, 675
  optical (thick, thin), 266, 326–28
  radio, 7, 40, 265, 266, 328, 352, 1044–56
  synchrotron, 1035–37, 1039
  thermal, 262
  X-ray, 262–67, 287, 288, 291–93, 296, 298, 299
  young, 963, 970
South Atlantic anomaly, 287, 306
Spacecraft, satellites
  Ablestar, 375–77
  Apollos, 204, 394, 396, 424, 430
  Explorer-35, 419, 424, 428, 432
  general, 8, 189, 207, 213, 263, 270, 280, 284, 289, 291, 303, 371–80, 392, 790
  HEAO, 303, 310
  Helios, 202
  list of craft carrying polarimeters, 200–03
  lunar orbiting, 363
  Mariner-5 (-6, -7, -9), 202, 391, 599, 618, 642
  Meteor-8, 564
  OAO-2, 796
  observations, 22, 34, 303, 306, 562, 569, 766, 767, 788–90, 794
  OSO-1 (-B2, -5, -6, -7), 34, 189, 200, 207, 213, 280, 284, 286, 288, 290, 291, 303, 307, 310, 787, 794, 796, 797, 799, 802
  Pioneer 10 and 11, 202, 205, 565, 568
  Skylab, 202, 204
  Surveyor, 200, 565
Space environment effects, 371, 381
Space missions, probes
  Apollo, 417, 419
  general, 27, 28, 37, 97, 189–217, 294, 312, 400, 417, 517, 518, 520, 562, 564, 565, 568, 573, 574, 769, 791
  Mariner Jupiter-Saturn, 202, 565, 568, 569, 574, 790
  Pioneer, 34
  research, 22, 766, 767
  schedules, 262
  Soviet, 28, 270, 294
Spankuch, D., 508, 509
Sparrow, E. M., 446, 471
Sparrow, J. G., 34, 38, 42, 201, 213, 216, 774, 778, 787, 792
Spectral lines
  profile, shape, 620, 624, 628, 634, 637, 640, 649, 654, 656, 657, 740, 746, 752–61
  saturation. *See under* Saturation
  strength, variable strength, 634, 639, 653, 657, 658, 686
  weak, 642, 651
  wings, 741
Spectra, spectral type
  absorption line. *See under* Absorption
  atom, 88, 89, 93, 101, 105
  bands, 654, 686, 836
  broadband, wide-band, 638, 639
  circular, circular dichroic, 69, 421
  continuum, 232, 618, 625, 627
  curve-of-growth, 620
  difference, 237–43
  Doppler, 624, 634
  emission. *See* Emission
  equivalent widths, 640
  Fourier, 640, 642, 643
  Galatry profile, 624, 634
  infrared, 640, 641
  intensity, 643, 649
  jitter, 174, 755
  laws of darkening, 620
  line. *See* Spectral lines
  Lorentz profile, 624, 627, 634
  lunar, 244
  narrow-band, 639
  polarization, 617–37, 644, 645, 656, 657
  resolution studies, 637, 638, 640, 647–49
  resonance-fluorescence line, 617, 624, 625, 632
  solar. *See under* Solar
  stellar, 234, 238, 239, 241, 244, 672, 673
  thermal, 545
  ultraviolet, 641
  variations center-to-limb, 620
  Voigt, 624, 634
  weak line. *See* Spectral lines
Spectrograph (narrow-slit), 233, 234, 238, 239, 246, 249, 250, 340, 570, 716
Spectrometers
  Cary, 409, 410, 415
  double beam, single beam, 246, 251, 252
  general, 90, 136, 179, 185, 213, 406
  grating, 165, 167, 174, 450, 641, 647, 754, 756, 757
  Littrow mount, 251–53
  prism, 756
  scanning, 165–68
Spectropolarimetry, 23, 31, 35, 166, 168, 174, 213, 405, 477, 479, 519, 525, 558, 617, 619, 627–32, 634, 637–60, 752–57
Spectroscopy
  beam-foil. *See* Beam-foil spectroscopy
  Fourier. *See under* Fourier
  observations, 21, 247, 598, 611, 852–55, 958

INDEX 1127

photographic, 237
solar, 238
Zeeman, 100
Spiller, E., 280, 316
Spin-orbit coupling, 237
Spiral arm
  distant spiral arms, 968, 969
  galactic, 889, 890, 960, 961, 963, 966, 968, 972
  H II-region, 971
  Perseus arm, 968
Spitzer, L., 60, 62, 334, 349, 888, 899, 957, 959
Spizzichino, A., 446, 468
Spoelstra, T., 970
Spurious polarization. *See under* Polarization
Staelin, D. H., 1004, 1013
Stairway rule, 22
Stamov, D. G., 457, 470, 500, 509
Stannard, D., 1052, 1056
Star lists
  nearby stars (spectral type G), 168, 169
  standard stars, 135, 140, 144, 168–70
  unpolarized nearby stars, 168, 169, 211
Star observations
  53 Cam, 988–90
  γ Cas, 756, 757, 759, 760, 845
  ε Cas, 758, 759
  R CrB, 821–23
  V CrB, 822, 833, 862–65
  a² CVn, 988–90
  V CVn, 35, 821, 827, 829, 832, 852, 854
  Cyg X-1, 855, 963, 970, 971
  NML Cyg, 835, 840, 862, 864
  HD 215441, 988, 989
  X Her, 821, 825, 826, 832
  R Leo, 839
  R Lep, 822, 831
  α Lyr, 758–60
  β Lyr, 752, 845, 847–49, 852
  R Mon, 36, 823
  U Mon, 823, 824, 831
  ρ Oph complex, 936, 937
  α Ori, 642, 837, 839
  θi Ori, 881, 882, 884, 885, 887
  σ Sco, 60, 934, 936, 941
  W Ser, 845, 849, 850, 852
  ζ Tau, 851, 861, 862, 864
  γ UMa, 752, 759
  η UMa, 756, 757
  VY UMa, 827, 832–35, 840, 862–65
Stars. *See also* Stellar
  A and B shell, 947
  Ap, 237, 318, 320, 981, 982, 985, 988–91
  B dwarfs, 668, 670, 672, 673
  Be, 36, 58, 318, 674, 821, 846, 851–53, 861, 862, 864, 890, 951, 989
  binary systems, 31, 313
  binary systems (close), 404, 668, 670, 672, 673, 752
  binary systems (Orientation orbital plane), 854
  binary systems (X-ray), 262
  BL Lac objects, 36
  carbon, 821, 822, 864
  catalogues. *See* Catalogues
  circumstellar shells (envelopes, clouds), 126, 127, 674, 821, 823, 825, 830, 832, 833, 835–37, 839, 840, 844–47, 849, 851, 852, 855, 860, 862, 864, 865, 891
  cool, 342, 862, 863
  early-type, 7, 31, 36, 667, 668, 673, 675, 752, 753, 821, 845–57, 889, 890, 939–45
  eclipsing binary, xiii, 58, 845–57
  emission-line, 23, 36, 846, 851, 852, 855, 860
  flare, 262
  formation, 865
  giants (M giants), 36, 826
  hot, 31, 36, 323
  late-type, 674, 821–44, 864, 889–91
  -light contaminated, 946, 953
  -light integrated, 779, 781, 788, 793, 796
  -light polarized (also unpolarized), 126, 127, 330, 332, 335, 546, 827, 881, 932, 933, 960, 961, 963, 964, 972, 976–80, 989
  magnetic, 33, 240, 243, 245, 981, 982, 985, 988–90
  Mira-type, 7, 821, 822, 827, 828, 830–32, 839, 840
  neutron. *See* Neutron star
  noise, 943
  novae, 36, 821, 825, 837, 1029. *See also* Nova remnants
  nuclei condensation, 821
  spectral type, 320
  supergiant (F, G, M), 318, 334, 668, 672, 674, 826, 829
  supernova remnants. *See* Nova remnants
  supernovae. *See* Supernovae
  RV Tauri objects, 821, 823, 824
  T Tauri objects, 821, 823
  variable (also luminous red), 35, 167, 821–32, 839, 840
  white dwarfs. *See* White dwarfs
  Wolf-Rayet objects, 821, 849
  X-ray, 59
  young, 946–53
Staude, J., 61, 63, 741, 751, 774–76, 778, 779, 788, 792
Stead, R. P., 379
Stecher, T. P., 199, 216, 836, 841
Stegman, J., 199, 208, 216

Steigmann, G., 388, 394, 401
Steinmetz, D., 649, 650, 660
Stein, R. S., 521, 580
Stein, W. A., 323, 324, 328, 332, 334, 348–50, 821, 825, 830, 834, 841, 861, 864, 866, 1077–79, 1082
Stellar
  effects, 31–33
  line profiles. *See under* Spectral lines
  magnetic fields, 33, 165
  magnitudes, 238, 374–76
  masses, 264, 265
  oscillations, 823, 824
  photosphere, 9, 835, 836
  polarization, 330, 332, 335, 668, 670–75
  shells. *See* circumstellar *under* Stars
  spectra. *See under* Spectra
  X-ray polarimetry. *See* X-ray, *and* stellar *under* Polarimetry
  See also Stars
Stellar atmospheres
  chemical composition, 670
  color temperature, 27, 672
  general, 642, 666–68, 670–74
  nongray, 663–75
  polarization, 168, 169, 663–75
  surface gravity, 667, 668, 672
  temperature, 663, 668. *See also* Temperature
  theory, 663
  See also under Atmosphere
Stenflo, J. O., 741, 742, 751
Stepanov, V. E., 741, 751
Stephens, P. J., 77, 87
Sternglass, E. J., 274, 275, 315
Stettler, P., 344, 350
Stevens, J. C., 274, 316
Stewart, A. I., 618, 635
Stewart, J. E., 232, 236
Stiles, G. J., 372, 379
Stoeckly, R., 923, 925
Stokes, G. G., 6, 8, 43, 45, 52, 663, 675, 732, 751
Stokes, G. M., 818, 819
Stokesmeter (polarimeter), 246–53
Stokes parameters, 35, 36, 43, 45–47, 50–52, 55, 136, 141, 145–47, 149, 157, 159, 161, 174, 175, 183, 184, 189, 214, 246, 247, 249–52, 315, 355, 442, 446, 472, 476, 481, 485, 487, 501, 510, 521, 530, 584, 608, 618, 626, 637, 643, 644, 652, 654, 656, 658, 663–65, 670, 677, 682, 688, 690, 717, 730, 732, 734, 740, 744, 753, 760, 767, 769, 770, 794, 798, 862, 881, 883, 905, 927, 1019, 1034, 1035, 1037, 1039, 1051, 1084, 1085
Stokes vector
  emission-line, 259, 359, 361, 363, 584, 619–21, 686, 692, 734

flux, 619–21, 940
Stone, S. N., 217
Stoner, J. O., 90, 100, 106
Storm, E., 305, 316
Strait, B., 195, 215
Stratton, J. A., 423, 433
Streeter, J. R., v, xvi
Strittmatter, P. A., 136, 164, 171, 821, 843, 846, 851, 856, 861, 864, 866, 1077, 1078, 1082
Strom, K. M., 37, 39, 846, 848, 860, 865, 937, 952, **954–59,** 1089
Strom, R. G., 356, 358, 1044, 1047, 1048, 1056, 1058
Strom, S. E., xv, 37, 39, 846, 848, 860, 865, 937, 952, 953, **954–59,** 980, 1089
Strömgren, B., 17, 42
Struve, O., 849, 856
Sturrock, P. A., 998, 1004, 1006, 1013, 1024, 1027, 1053, 1057
Subtil, J. L., 103, 105
Sun. *See* Solar
Sun compass, 472, 475
Sunspot. *See under* Solar
Sunyaev, R. A., 59, 62, 264, 315
Sunyar, A. W., 280, 315
Supernovae, 263, 964–66, 970. *See also* Nova remnants
Surface
  atmospheric, 610
  density, 364
  dielectric. *See under* Dielectric
  diffuse, 446
  general, 396, 405–07, 417, 444, 446, 503, 521
  isotropic, 361
  low-albedo, 417
  optically active. *See under* Optical
  planetary, 553
  polarization, 463–67
  polished, 446
  properties, 371, 463
  reflectivity, 364, 444–46, 453
  rough, 361–63, 371, 378, 381, 384, 385, 392, 399, 446
  smooth, 362, 372, 384, 385, 398, 403, 466, 610
  solid, 298, 519, 610
  specular, 371–74
  structure, 382, 384, 398, 403
  subsurface, 362, 363
Surmelian, G. L., 993, 995, 996
Surveyor 7. *See under* Spacecraft
Sussman, M., 372, 379
Sutherland, P. G., 57, 62, 984, 987
Suzuki, S., 354, 358
Svatoš, J., 867, 872, 879
Svechnikov, M. A., 155, 172
Svestka, Z., 692, 694

INDEX

1129

Swedlund, J. B., 61, 62, 157, 173, 498, 499, 570, 573, 578, 581, 607, 610, 611, 613–15, 835, 842, 866, 983, 987
Sweet-type feedback power supply, 374
Sweigart, A. V., 559, 578, 583, 591
Swihart, T. L., 327, 350, 1004, 1012, 1023, 1025, 1027, 1033, 1035–38, 1040, 1046, 1051, 1055, 1056, 1089
Swindell, W., 21, 40, 43, 1089
Symmetric media, symmetry, 70, 77, 79, 80, 83, 84, 86, 211, 267, 421, 422, 438, 550, 608–10, 612, 614, 777
Synchro-Compton theory, radiation, models, 932, 1014, 1022, 1023, 1053–55, 1072
Syrovatski, S. I., 1015, 1018, 1026

Tademaru, E., 1005, 1013
Takakura, T., 268, 316
Takashima, T., 29, 195, 216, **500–09**, 562, 580, 583, 584, 591, 618, 636, 1089
Takazaki, H., 175, 187
Tamburini, T., 752, 753, 758, 759, 761
Tanabe, H., 193, 199
Tanaka, M., 618, 636
Tananbaum, H., 263, 265, 289, 310, 314, 316
Tandberg-Hanssen, E., 15, 23, 32, 52, 253, 260, 638, **730–51**, 1089
Tashenov, B. T., 457, 469
Tassart, J., 160, 172
Täuber, U., 474, 489, 492
Taylor, A. M., 809
Taylor, J. H., 329, 350, 1000, 1005, 1012
Taylor, R. C., 392, 401
Teifel, V. G., 570, 581
Telescope
 aluminizing, 19
 conversion of linear-to-circular polarization, 19, 140, 141
 cosmic ray, 302
 inaccurate guiding, 137, 160
 lens objective, 141
 metallic reflections (also unnecessary reflections), 19, 137, 141, 142
 mirrors, 19, 53
 polarization-free, 698, 710
 refracting, 710
 rotatable tubes, 18, 140, 168, 169
 suitable for polarimetry, 18, 19
 X-ray incidence grazing, 262, 274, 312
Television camera, 16, 230, 237–45, 319
Temperature
 correction (Avrett-Krook), 666
 gaseous atmosphere, 521–24, 563
 general, 514, 515, 585, 624
 gradient, 515, 667
 variations, 622

Tensor
 dielectric, 70, 79, 1031
 emission, 1034
 general, 78
 polarization, 78, 79, 1034
 propagation, 1043
 transfer, 1032
Ter Haar, D., 734, 751
Terizan, Y., 1023, 1025
Terrall, J. R., 274, 316
Terrestrial atmosphere
 clear, 413, 457–59, 461, 462, 510, 513, 520, 574
 cloudy, 510, 513, 520, 570, 571, 630, 634. *See also under* Clouds
 Davis, 457, 458, 460, 461
 general, 28, 29, 191, 195, 213
 Los Angeles, 457–61, 505–09, 562, 563
 polluted, 413, 457–59, 461, 562, 564. *See also* Aerosols
 Ruhr, 510, 513
 scattering, 208, 213
 stratosphere, 556, 558–60, 564
 studies, 191, 195, 213, 307, 562–66, 615, 617, 618, 652, 657, 719, 753
 tropospheric haze, 558
 See also Clouds
Terrestrial samples, 381, 385, 388, 394–96
Teska, T. M., 18, 41, 155, 172, 457, 469, 520
Teyfel, Ya. A., 500, 509
Theobald, J. K., 199, 216
Thermal
 bremsstrahlung, 57
 emission (mechanism), 55–57, 59, 267, 268, 321, 322, 324, 340, 585, 692, 693, 821
 processes, 55, 56
 radiation, 392
Thiel, M. A., 1051, 1058
Thiessen, G., 139, 173, 731, 751–53, 758, 759, 761
Thompson, F. I., 649, 650, 660
Thompson, R. A., 329, 350
Thompson, T. W., 363, 367
Thomson scattering. *See under* Scattering
Thomson-scattering polarimeters
 general, 291–310
 incoherent, 165, 270, 272, 291–310
 rocket, 294–303
 satellite, 303–10
Thorell, B., 480, 493
Tichanowski, J. J., 456, 470
Timofeeva, V. A., 440, 441, 443
Tinbergen, J., 16, 150, 157, 159, 173, **175–88**, 354, 358, 705, 729, 1089
Tindo, I. P., 58, 63, 270, 294, 316, 317
Titan, 382

Titulaer, C., 384, 385, 388, 394, 401
Todd, B., 195, 216, 786, 792
Tomasko, M. G., 1089
Toolin, R. B., 508, 509
Toombs, R. I., 833, 843
1685 Toro, 24, 26, 359, 390, 397
Torrance, K. E., 446, 471
Torrey, R. A., 217
Toubhans, R. H., 161, 197, 214
Tousey, R., 195, 197, 216
Townes, C. H., 833, 842
Tozer, W., 199, 216
Transfer (radiative). *See* Radiative transfer
Transfer equations, functions, 423–25, 429, 541, 582, 592, 1030–35, 1039, 1041
Transformation equations, 144–46
Transmission-line theory, 520
Transport theory, 663
Transverse electric mode, 421, 423
Transverse magnetic mode, 421, 423, 431, 432
Transverse magnetic response peak, 432
Trask, N. J., 387, 394, 403
Treanor, P. J., 935, 938
Treffers, R. R., 864, 865
Tricker, R. A. R., 547, 581
Trimble, V., 982, 987, 1015, 1022, 1025, 1027
Tripp, W., 975
Troposphere scattering light, 769–71, 774, 780
Trujillo-Cenóz, O., 455, 490 493
Tsitovich, T. A., 197, 213, 564, 574
Tsitovich, V. N., 1031, 1058
Tsuchiya, A., 354, 358
Tucker, W. H., 263, 315
Turner, B., 959
Turtle, J. P., 562, 575
Twomey, S., 471, 520, 581, 584, 591
Tyler, G. L., 24, **359–67**, 1089
Tyler, J. E., 12, 28, **434–43**, 473, 474, 1089
Tyndall, D., 6

UBV photometric system, 767, 787, 847, 934, 947, 1085
UBVRI colors, xiii, 1071, 1072
Ueno, S., 12, 30, 522, **582–92**, 618, 635, 636, 1089
Uesugi, A., 583, 592, 619, 636
2 Uhuru catalogue, 263
Ukrainian Academy of Sciences, 599–605
Ultraviolet excess, 334
Umov effect, law, 12, 25, 223, 421, 465, 466, 819
Underhill, A., 670, 675, 846, 856
Unno, W., 740, 741, 751, 989–91
Unpolarized. *See under* Light
Unz, F., 191, 216, 563, 564, 581
Uranus, 518, 573

Valnicek, B., 270, 317
Vanasse, G. A., 638, 660
Van Blerkom, D., 1067, 1082
van Cittert, P. H., 676, 684
van Cittert-Zernicke theorem, 679
Vanden Bout, P., 263, 267, 294, 314, 317
Vandenburgh, R. C., 380
Vande Noord, E. L., 191, 215, 216, 787, 792
Van Rhijn law, 768
Vanýsek, V., 816, 820, 867, 872, 878, 879
Vardanian, R. A., 823, 825, 843
Vasudevan, R., 583, 589
Vaughan, A. H., 985, 986
Vectors
 diagrams, 596
 displacement, 79
 electric. *See* Electric vector
 equation, 78, 110
 field, 420, 421, 582, 619, 627, 746
 force, 73–75
 gyration, 71, 73
 intensity, 447, 584–86, 588
 maximum radiance, 438
 polarization, 30, 46, 70, 83, 272, 275, 284, 483, 743, 744
 Poynting. *See* Poynting flux, vectors
 Stokes. *See under* Stokes
 wave equation, 107, 108
Velluz, L., 77, 87
Venera 4, 642
Venus
 albedo, 555, 556, 631
 atmosphere, 11, 28, 321, 519, 520, 553, 559, 560, 561, 618, 627, 642, 653, 655
 circular polarization, 552, 610, 613
 clouds. *See* Venus clouds
 CO_2, 655, 657, 658
 color, 558
 diffraction features, 557
 general, 6, 18, 21, 30
 global polarization, 551, 560
 infrared observations, 863
 linear polarization, 553
 photographic measurements, 880
 polarization-phase curve structure, 519, 551, 568
 refractive index, 558–60
 spectra, 619, 637, 638, 642, 653–59
 spectral reflectivity, 558
 variable polarization, 552, 556
Venus clouds
 composition, nature, 30, 648, 651
 general, 203, 223, 224, 226, 227, 520, 540, 541, 556, 558, 559, 562
 layer, 604
 markings, 223, 224, 226, 227
 particles, 518, 555, 558, 568, 569, 606, 618

INDEX 1131

polarization measurements, 6, 18, 21, 30, 218–24, 226, 227, 318, 321, 518, 519, 539, 550–62, 610, 613, 616, 631, 653, 655, 658
pressures, 518, 553, 556, 558, 631
temperature, 560
top, 560
Vergnoux, A. M., 337, 351
Verschuur, G. L., 39, 56, 63, 889, **960–71,** 974, 975, 1089
4 Vesta, 26, 390, 393, 395, 397, 400
Veverka, J., 390–92, 396, 400, 402, 403, 559, 581
Vibration direction, 47, 53, 64
Vibration plane tilt, 434–37, 439, 440
Vidicons, 213
Viezee, W., 523, 580
Vincent, R. K., 340, 351
Viner, M. R., 265, 315
Vinnichenko, N. K., 201, 213, 564, 574
Visual measurements of polarization, 15, 456, 457, 463, 519
Visvanathan, N., 36, 38, 40, 47, 51, 157, 173, 266, 268, 310, 315, 1002, 1007, 1008, 1011, 1012, 1022, 1025, 1026, **1059–83,** 1089
Vitriczenko, E. A., 846, 856
Vleck, J. H. van, 739, 747, 751
Vogel, D. C., 807, 813
Voigt profiles. *See under* Spectra
Volcanic ash, 460, 519
Volz, F., 521, 581
von Zeipel's theorem, 823

Waddell, J. H., 217
Wagner, W. J., 255, 260
Wagoner, R. V., 1025
Walborn, N. R., 169, 173
Walker, G. A. H., 16, **237–45,** 388, 985, 1089
Walker, M. F., 847, 857, 1070, 1083
Wall, J. V., 1077, 1083
Wallerstein, G., 833, 843
Walraven, R. L., 446, 463, 469
Walraven, T., 1018, 1026, 1063, 1082
Walsh, T. E., 149, 174
Wampler, E. J., 310, 317, 1002, 1007, 1013, 1024, 1027, 1071, 1083
Wang, A. P., 10, 30, 522, **582–92,** 1089
Wang, R. T., 126, 133, 901, 905, 914
Wangsness, R. K., 101, 103, 105
Wardie, J. F. C., 1052, 1058
Warner, B., 822, 843, 984, 987
Warwick, J. W., 59, 63, 742, 751
Watanabe, S., 440, 443
Water
 cloud, 510, 518, 535, 536, 543, 548, 553, 554, 558, 559, 564, 565
 cloud particle, 518, 566

droplets, 548
haze, 504, 506, 514, 515, 540, 553. *See also* Haze
underwater polarization, 434–43
Waterman, T. H., 28, 435, 439–41, 443, **472–94,** 1089
Waterworth, M. D., 232, 236
Watson, D., 549, 574
Watts, L. A., 203, 215
Watts, R. N., 216
Wave
 circularly polarized, 79
 cylindrical, 116, 122
 equation, 421, 427
 field, 426, 433, 1023
 front, 430, 431
 function, 76, 112, 116, 733
 -guide rectangular, circular, 353, 905
 interaction, 232
 plane, 108, 111, 114, 115, 614
 plate, 928, 930, 931
 polarization, 71, 72, 79, 340–42
 scalar equation, 108, 110, 112, 113, 115–21, 124, 125
 spherical, 122
 standing, 277
Wavelength
 absorption peak, 498
 dependence, composition and symmetry of mineral, 406, 407, 416
 dependence, grain size, 11, 27, 890, 941
 dependence, orientation of optical axis, 135
 dependence, polarization. *See under* Polarization
 dependence, position angle. *See under* Position angle
 effective, 143
 microwave region, 549
 range, xiii, 8
 sensitivity, 396
 spectrum, 419, 427, 430
Weaver, A. B., xv, 1089
Weekes, T. C., 1020, 1023, 1026
Wehner, G. K., 383, 384, 388, 394, 396, 402
Weiler, K. W., 357, 1023, 1026, 1044, 1056, 1058
Weinberg, J. L., 21, 23, 33, 34, 42, 203, 216, 769, 778, **781–93,** 805, 813, 1089
Weisskopf, M. C., 55, 62, 267, 280, 281, 294, 303, 313, 314, 316, 317, 743, 751, 1021, 1026
Wells, M. B., 541, 575
Wendland, P., 762, 765
Wendler, R., 260
Werner, M. W., 332, 351
West, C. D., 337, 348
Westell, W. E., 443, 445, 449, 475, 494

Westerhout, G., 354, 358
Westerlund, B. E., 1077, 1083
Westfold, K. C., 55, 62, 1023, 1026, 1036, 1038, 1057, 1058
Westphal, J. A., 47, 51, 266, 268, 310, 315, 833, 843, 1002, 1007, 1008, 1012, 1022, 1025, 1026
Weymann, R., xv, 1089
Whitaker, E. A., 224, 230
White, O. R., 242, 245
White dwarfs
 circular polarization, 981–87
 continua, spectral features, 981, 982
 G 99-37, 982, 983
 G 99-47, 982, 983
 G 195-19, 982, 983
 general, 7, 33, 56, 57, 59, 266, 821, 863, 981–85, 990, 992–96
 GrW$^+$ 70 8247, 983, 984
 magnetic fields (also strong magnetic fields), 982, 992–96
Whitehill, L., 363, 367, 560, 561, 581
Whiteoak, J. B., 51, 53, 327, 349, 967, 970, 1030, 1044, 1051, 1052, 1057
Whitney, L. V., 437, 443
Wickramasinghe, N. C., 330, 332, 334, 336, 349, 351, 837, 841, 843, 860, 865, 888, 896, 897, 899, 900
Widorn, T., 382, 384, 403
Wiese, W. L., 104, 106
Wigand, A., 22, 191, 216
Wilhelm, N., 199, 208, 216
Wilhelms, D. E., 387, 394, 403
Williams, A. P., 199, 208, 216
Williams, D. A., 836, 841
Willstrop, R. V., 770, 779
Wilson, A. S., 1018, 1020, 1021, 1024, 1027
Wilson, J. D., 778
Wilson, L., 388, 403
Wilson, R. W., 833, 843
Wing, M. G., 582, 592
Wing, T. E., 300, 307, 317
Winslow, O., 195, 215
Wirick, M. P., 149, 174
Wirtanen, C. A., 1070, 1073, 1074, 1081
Wiśniewski, W., 860, 866
Witt, G., 29, 199, 208, 216, 471, **514–17**, 761 793, 1089
Wolf, E., 48, 49, 51, 360, 366, 510, 513, 639, 659, 676, 684, 732, 748
Wolf, G. W., 825, 835, 843, 981, 987
Wolf, K., 61, 63, 774, 778, 779, 788, 792
Wolf, R. A., 997, 1011
Wolff, M., 400
Wolff, R. S., 55, 62, 267, 268, 280, 294, 295, 298, 300, 303, 313, 314, 316, 317

Wollaston prism. *See under* Prism
Wolstencroft, R. D., 27, 33, 37, 61, 62, 157, 172, 174, 187, 197, 216, **495–99,** 570, 573, 578, 581, 607–11, 613–15, **768–80,** 787, 788, 793, 802, 817, 835, 842, 886, 890, 899, 934, 937, **939–45,** 983, 985–87, **988–91,** 1072, 1081, 1089
Woltjer, L., 266, 351, 1015, 1017, 1020, 1022, 1024, 1025
Wood, C. G., 278, 279, 314
Wood, D. B., 847, 857
Wood, R. W., 232, 236, 734, 751
Wood's anomalies, 232, 235
Woodson, P. E., 807, 813
Woodworth, A., 265, 315
Woody, R. W., 499
Woolf, N. J., 821, 823, 825, 830, 834, 837, 840, 841, 843, 849, 857, 861, 864, 866, 889, 900
Worthing, A. G., 340, 351
Wright, A. W., 782, 783, 786, 793
Wright, E. L., 324, 325, 349
Wronskian, 124
Wysoczanski, W., 136, 164, 172

X-ray
 diffuse background, 263, 300, 303
 emission, emission mechanism, nature, 264–68, 270, 310, 312, 694
 flux, 288, 294, 295, 297, 300, 310
 hard, 269, 692
 high-energy events, 268, 287
 imaging technique, 65, 289
 polarization, polarimetry, 21, 57, 262–68, 270, 317, 675
 pulsars. *See under* Pulsars
 soft, 334, 1002
 sources. *See under* Sources
 stars. *See under* Stars

Yamada, Y., 337, 350
Yamamoto, I., 786, 793
Yanovitskii, E. G., 30, 554, 568, 570, 579, **599–606,** 1089
Yarger, D. N., 524, 577
Yarmush, D. L., 541, 574
Yeh, C., 547, 581
Yellin, J., 93, 106
Yentis, D. J., 263, 317
Yoshinaga, H., 337, 341, 342, 350
Yost, J., 946, 952
Young, A. T., 139, 173, 386, 402, 560, 568, 577, 581
Young, J. B., 337, 351, 650, 660
Young, T., 5
Young clusters. *See under* Stars
Young stars. *See under* Stars

Zaitsev, V. V., 1004, 1011
Zane, R., 686, 694
Zanstra, H., 731, 751
Zappala, R. R., 825, 827, 832, 842, 843, 862, 866
Zeeman
 components, 56, 237, 739, 741
 effect, 6, 33, 54, 56, 105, 239, 244, 686, 715, 738, 745, 900, 943, 968, 981–83, 985
 magnetic sublevels, 96
 molecular, 56
 patterns, 100, 237, 742, 984
 quadratic effect, 993
 shift, 238, 240, 244
 spectroscopy. *See under* Spectroscopy
 splitting, 752
 states, 738, 739
 transverse, 513, 688, 693, 712, 713, 988–91
 triplet, 688, 715
Zeeman, P., 6, 33
Zellner, B. H., 11, 13, 25, 28, 36, 38, 40, 43, **381–404,** 464, 822, 823, 825, 833, 837, 843, **867–80,** 885, 889, 900, 933, 946, 951, 952, 1089
Zemansky, M. W., 59, 62, 735, 750
Zerin, H., 746, 749

Zernike, F., 676, 684
Zerull, R., 12, 23, 30, 33, 36, 549, **804–13, 901–15,** 1089
Zheleznyakov, V. V., 1004, 1011
Zisk, S. H., 363, 367
Zodiacal light
 asymmetry polarization, 780
 brightness, color, 782–84, 788, 790
 circular polarization, 779
 cloud, 3
 color, 782, 783
 dust, grain contribution, 768
 experiments, 790
 general, 21, 22, 29, 30, 33, 34, 191, 193, 195, 197, 199, 213, 518, 766, 767, 769, 772, 774–76, 779–93
 intensity, 804, 805, 807, 813
 linear polarization, 791
 models, 804–13
 photometric properties, 791
 photopolarimeter, 797
 polarization. *See under* Polarization
 shape, position, 782, 783
Zuckerman, B., 959
Zwicky, F., 224, 230